Advances in Intelligent Systems and Computing

Volume 1073

AF881011

The series "Advances in Intelligent Systems and Computing" contains publications on theory, applications, and design methods of Intelligent Systems and Intelligent Computing. Virtually all disciplines such as engineering, natural sciences, computer and information science, ICT, economics, business, e-commerce, environment, healthcare, life science are covered. The list of topics spans all the areas of modern intelligent systems and computing such as: computational intelligence, soft computing including neural networks, fuzzy systems, evolutionary computing and the fusion of these paradigms, social intelligence, ambient intelligence, computational neuroscience, artificial life, virtual worlds and society, cognitive science and systems, Perception and Vision, DNA and immune based systems, self-organizing and adaptive systems, e-Learning and teaching, human-centered and human-centric computing, recommender systems, intelligent control, robotics and mechatronics including human-machine teaming, knowledge-based paradigms, learning paradigms, machine ethics, intelligent data analysis, knowledge management, intelligent agents, intelligent decision making and support, intelligent network security, trust management, interactive entertainment, Web intelligence and multimedia.

The publications within "Advances in Intelligent Systems and Computing" are primarily proceedings of important conferences, symposia and congresses. They cover significant recent developments in the field, both of a foundational and applicable character. An important characteristic feature of the series is the short publication time and world-wide distribution. This permits a rapid and broad dissemination of research results.

**** Indexing: The books of this series are submitted to ISI Proceedings, EI-Compendex, DBLP, SCOPUS, Google Scholar and Springerlink ****

More information about this series at http://www.springer.com/series/11156

Faisal Saeed · Fathey Mohammed ·
Nadhmi Gazem
Editors

Emerging Trends in Intelligent Computing and Informatics

Data Science, Intelligent Information Systems and Smart Computing

Springer

Editors
Faisal Saeed
College of Computer Science
and Engineering
Taibah University
Medina, Saudi Arabia

Fathey Mohammed
School of Computing
Universiti Utara Malaysia (UUM)
Sintok, Kedah Darul Aman, Malaysia

Nadhmi Gazem
Management of Information
Systems Department
College of Business Administration
Taibah University
Yanbu, Saudi Arabia

ISSN 2194-5357 ISSN 2194-5365 (electronic)
Advances in Intelligent Systems and Computing
ISBN 978-3-030-33581-6 ISBN 978-3-030-33582-3 (eBook)
https://doi.org/10.1007/978-3-030-33582-3

This Springer imprint is published by the registered company Springer Nature Switzerland AG
The registered company address is: Gewerbestrasse 11, 6330 Cham, Switzerland

Preface

We are honored to welcome you to the Fourth International Conference of Reliable Information and Communication Technology 2019 (IRICT 2019) held at Pulai Springs Resort, Johor, Malaysia, on September 22–23, 2019, and organized by the Yemeni Scientists Research Group (YSRG), Information Engineering Behavioral Informatics Research Group (INFOBEE) in Universiti Teknologi Malaysia (Malaysia), Data Science Research Group in College of Computer Science and Engineering at Taibah University (Kingdom of Saudi Arabia), School of Science and Technology in Nottingham Trent (UK), College of Engineering, IT and Environment at Charles Darwin University (Australia), and Association for Information Systems – Malaysia Chapter (MyAIS).

IRICT 2019 is a forum for the presentation of technological advances in the field of information and communication technology. The main theme of the conference is "Toward Reliable Intelligent Computing and Informatics."

Many researchers have been attracted to submit 175 papers to IRICT 2019 from 29 countries including Algeria, Australia, China, Egypt, Fiji, Germany, India, Indonesia, Iraq, Iran, Jordon, Malaysia, Morocco, Myanmar, Nigeria, Oman, Pakistan, Saudi Arabia, Singapore, Somalia, South Africa, Sri Lanka, Sudan, Sweden, Taiwan, Tunisia, UK, USA, and Yemen. Of those 175 submissions, 109 submissions (62%) have been selected to be included in this book.

The book presents several research topics which include artificial intelligence, machine learning, data science, big data analytics, business intelligence, Internet of things (IoT), information security, intelligent communication systems, health informatics, information systems theories and applications, computational vision and robotics technology, software engineering and multimedia applications and services.

We would like to express our appreciation to all authors and the keynote speakers for sharing their expertise with us. And we would like to thank the organizing committee for their great efforts in managing the conference. In addition, we would like to thank the technical committee for reviewing all the submitted papers; Prof. Dr. Janusz Kacprzyk, AISC series editor; and Dr. Thomas Ditzinger, Arumugam Deivasigamani from Springer.

Finally, we thank all the participants of IRICT 2019 and hope to see you all again in the next conference.

IRICT 2019 Organizing Committee

Honorary Co-chairs

Rose Alinda Alias (President)	Association for Information Systems – Malaysia Chapter, Head of the Information Service Systems and Innovation Research Group (ISSIRG) in Universiti Teknologi Malaysia, Malaysia
Ahmad Hawalah (Dean)	Deanship of Information Technology, College of Computer Science and Engineering, Taibah University, Kingdom of Saudi Arabia

International Advisory Board

Abdul Samad Haji Ismail	Universiti Teknologi Malaysia, Malaysia
Ahmed Yassin Al-Dubai	Edinburgh Napier University, UK
Ali Bastawissy	Cairo University, Egypt
Ali Selamat	Universiti Teknologi Malaysia, Malaysia
Ayoub AL-Hamadi	Otto von Guericke University, Germany
Eldon Y. Li	National Chengchi University, Taiwan
Kamal Zuhairi Zamil	Universiti Malaysia Pahang, Malaysia
Kamalrulnizam Abu Bakar	Universiti Teknologi Malaysia, Malaysia
Mohamed M. S. Nasser	Qatar University, Qatar
Srikanta Patnaik	SOA University, Bhubaneswar, India

Conference General Chair

Faisal Saeed (President)	Yemeni Scientists Research Group (YSRG), Head of Data Science Research Group in Taibah University, Kingdom of Saudi Arabia

Program Committee Co-chairs

Fathey Mohammed	Universiti Utara Malaysia (UUM), Malaysia
Nadhmi Gazem	Taibah University, Kingdom of Saudi Arabia

Technical Committee Chair

Tawfik Al-Hadhrami	Nottingham Trent University, UK

Publications Committee

Abdulaziz Al-Nahari (Chair)	Universiti Teknologi Malaysia, Malaysia
Abdullah Abdurahman Mohamed Ahmed	Universiti Utara Malaysia, Malaysia

Publicity Committee

Abdullah Aysh Dahawi (Chair)	Universiti Teknologi Malaysia, Malaysia
Abdulalem Ali	Universiti Teknologi Malaysia, Malaysia
Mohammed Sultan Ahmed Mohammed	Universiti Teknologi Malaysia, Malaysia
Maged Naeser	Universiti Teknologi Malaysia, Malaysia
Mohammed Abdulrahman Ebrahim Abdo Nashtan	University of Malaya, Malaysia
Ali Saleh Amer Maaodhah	Universiti Teknologi Malaysia - KL, Malaysia
Hamzah Amin Ahmed Alhamidi	University International Islamic Malaysia, Malaysia
Taha Hussein Qasem Dahawi	Multimedia University, Malaysia
Mohammed Omar Awadh Al-Shatari	Universiti Teknologi PETRONAS, Malaysia, Malaysia
Ahmed Tawfik Alqadami	Universiti Teknologi PETRONAS, Malaysia
Abdullah Faisal Abdulaziz Al-shalif	Universiti Tun Hussein Onn Malaysia, Malaysia
Ali Ahmed Ali Salem	Universiti Tun Hussein Onn Malaysia, Malaysia
Bassam Mstafa Alhammadi	Universiti Teknikal Malaysia Melaka, Malaysia
Salman Ameen Ali Abdullah Alabd	Universiti Malaysia Pahang, Malaysia
Yahya Ayesh Qasem Dahawi	Universiti Malaysia Pahang (UMP), Malaysia
Ahmed Abdullah Ahmed Alhurabi	UniMAP, Malaysia
Hammam Abdullah Abdurabu Thabit	Universiti Sains Malaysia, Malaysia

IT and Multimedia Committee

Yunes Abdulwahab Lutf Al-Dailami (Chair)	Universiti Teknologi Malaysia, Malaysia
Fuad Abdeljalil Al-shamiri	Universiti Teknologi Malaysia, Malaysia
Bander Ali Saleh Al-rimy	Universiti Teknologi Malaysia, Malaysia
Amer Alsaket	Universiti Putra Malaysia, Malaysia
Abdulrahman Ali Mohammed Bin-Break	Universiti Teknologi Malaysia, Malaysia
Manea Mohammed Ahmed Musleh Al-Asaadi	Universiti Teknologi Malaysia, Malaysia
Mehedi Hasan	Universiti Teknologi Malaysia, Malaysia
Sulaiman Mohammed Abdulrahman	Taibah University, KSA

Treasure Committee

Hamzah Gamal Abdo Allozy (Chair)	Universiti Teknologi Malaysia, Malaysia
Md Hafiz bin Selamat	Universiti Teknologi Malaysia, Malaysia
Abdullah Aysh Dahawi	Universiti Teknologi Malaysia, Malaysia

Logistic Committee Chair

Wahid Al-Twaiti	Universiti Teknologi Malaysia (UTM), Malaysia

Registration Committee Chair

Sameer Hasan Albakri	Universiti Teknologi Malaysia (UTM), Malaysia

International Technical Committee

Abdelhamid Emara	Taibah University, Kingdom of Saudi Arabia
Abdelrahman Elsharif	Taibah University, Kingdom of Saudi Arabia
Abdelsalam Busalim	Universiti Teknologi Malaysia (UTM), Malaysia
Abdulbasit Darem	Northern Border University, Kingdom of Saudi Arabia
Abdullah Ali	Universiti Teknologi Malaysia (UTM), Malaysia
Abdullah Gharib	University of Sheba Region, Yemen
Abdulrahman Alsewari	Universiti Malaysia Pahang (UMP), Malaysia
Abdulrahman Elsharif	Taibah University, Kingdom of Saudi Arabia
Abdulrazak Alhababi	UNIMAS, Malaysia
Abrar Mohammed	Bacha Khan University, Charsadda, KPK, Pakistan

Ahmed Al-Samman	Universiti Teknologi Malaysia (UTM), Malaysia
Ahmed Sayegh	Universiti Tun Hussein Onn Malaysia (UTHM), Malaysia
Adel Alshabi	Universiti Kebangsaan Malaysia (UKM), Malaysia
Aladdein Amro	Taibah University, Kingdom of Saudi Arabia
Ashraf Osman	Alzaiem Alazhari University, Sudan
Asma Alhashmi	Northern Border University, Kingdom of Saudi Arabia
Bander Al-Rimy	Universiti Teknologi Malaysia (UTM), Malaysia
Errais Mohammed	Hassan II University of Casablanca, Morocco
Essa Hezzam	Taibah University, Kingdom of Saudi Arabia
Fatimah Albalushi	Universiti Teknologi Malaysia (UTM), Malaysia
Fuad Ghaleb	Universiti Teknologi Malaysia (UTM), Malaysia
Hael Al-bashiri	University Malaysia Pahang (UMP), Malaysia
Hussein Hussein Abualrejal	Universiti Utara Malaysia (UUM), Malaysia
Mohammed Alsarem	Taibah University, Kingdom of Saudi Arabia
Mohammed Al-Sharafi	Universiti Malaysia Pahang (UMP), Malaysia
Mohammed Mumtaz Aldabbagh	Tishk International University, Iraq
Muaadh Mukred	Universiti Kebangsaan Malaysia (UKM), Malaysia
Murad Rassam	Taiz University, Yemen
Nasrin Makbol	Universiti Malaysia Pahang (UMP), Malaysia
Nibras Abdullah	Universiti Sains Malaysia (USM), Malaysia
Ola Al-wasabi	Hodeidah University, Yemen
Osama Sayaydeh	University Malaysia Pahang (UMP), Malaysia
Qais Al-Nuzaili	Al-Nasser University, Yemen
Sabri Hanish	Universiti Sains Malaysia (USM), Malaysia
Samir Algailani	Universiti Sains Malaysia (USM), Malaysia
Shaima Mustafa	University of Mosul, Iraq
Taha Hussain	Universiti Malaysia Pahang (UMP), Malaysia
Waseem Alromimah	Taibah University, Kingdom of Saudi Arabia
Yahya Al-Dheleai	Universiti Sains Malaysia (USM), Malaysia
Yazan Al-Khassawneh	Universiti Kebangsaan Malaysia (UKM), Malaysia

Contents

Data Science and Big Data Analytics

Arabic Text Stemming Using Query Expansion Method

Nuhu Yusuf[1,2], Mohd Amin Mohd Yunus[1(✉)], and Norfaradilla Wahid[1]

[1] Faculty of Computer Science and Information Technology, Universiti Tun Hussein Onn, Malaysia (UTHM), Parit Raja, Malaysia
{aminy, faradila}@uthm.edu.my
[2] Management and Information Technology Department, Abubakar Tafawa Balewa University (ATBU), Bauchi, Nigeria
ynuhu@atbu.edu.ng

Abstract. Information retrieval is an activity of finding needed information that satisfies the information need of the users. Stemming is one of the technique uses to improve information retrieval search performance for different languages. The Arabic language is one of these languages that utilize stemming to reduce various words to their stems for both the query and corpus documents. However, most of the approaches that were based on stemming do not provide better results due to poor stemming algorithms. Specifically, root-based stemming produced low precision results as compared to higher recall. This paper presents a new query expansion approach to improve relevant search results. The new approach utilizes word synonyms and stemming. For word synonyms; a set of Arabic word synonyms were used to expands the query. For stemming, both the expanded query and the original corpus were then stemmed using Arabic stemming algorithm. The experiment on Quran datasets for classical and modern standard Arabic shows that the proposed approach achieved almost 15% MAP improvement results on modern standard Arabic dataset.

Keywords: Query expansion · Quran search · Information retrieval · Synonyms · Stemming

1 Introduction

Information retrieval exploits the use of resource to obtain relevant information. The aim is to retrieve information relevant to a particular user query. Information can be in the form of text, image, audio, video or their combinations. Many organizations now provide an information retrieval system in the form of a search engine. This attracts many researchers and practitioners to conduct research in order to improve relevant search results.

Generally, there are many methods used to improve the relevant search results. Query expansion is widely used in information retrieval. In query expansion information retrieval, stemming's are used to reduce words into their stems or roots.

F. Saeed et al. (Eds.): IRICT 2019, AISC 1073, pp. 3–11, 2020.
https://doi.org/10.1007/978-3-030-33582-3_1

Stemming in query expansion information retrieval has been categorized into document-based and query based.

Commonly, information can be retrieved when stemming both the document (document-based) and query (query-based) through mapping various words to stems. Although word synonym is frequently used in a query-based to improve search, it has not been relating to stemming both document and query-based. Also, the root-based stemming produced low precision results as compared to higher recall values. This study tried to expand the query using synonyms, stemming this expanded query and then combined with the document-based stemming to improve relevant search results. In our experiment, we considered Quran text information retrieval in both modern and classical Arabic. In comparisons to other query expansion approach, the new proposed approach provides better performance in retrieving Quran text in classical Arabic.

Other parts of this paper are organized as follows: Sect. 2 provides related work; Sect. 3 presents the proposed approach; Sect. 4 describes the experiments conducted. Finally, the conclusion is presented in Sect. 5.

2 Related Work

Query expansion, as a method of improving search result, has been extensively used in many application areas [1–5]. However, query expansion has faced challenges of term selection and lack of precision results. This paper tried to improve precision results. Query expansion adds additional information to the query for better relevant results. Consequently, various methods are now available for selecting and expanding query words. However, the results of these methods still required improvement. This motivates researchers to concentrate on proposing better methods for improvements.

In term of improving search results through relevant feedback, Ben Guirat et al. [6] suggest pseudo-relevant feedback with result aggregation as an alternative approach. This work, present how the results would be re-sorted with regard to words stems. In addition to that, El Mahdaouy et al. [7] propose how term dependencies for proximity and multi-word would help to improve the results. Their work focused on retrieving Arabic text collected from TREC. Stemming does not consider any term selection for expanding the query. Alhaj et al. [8] examine how different stemming technique in retrieving Arabic characters. Their results indicate better performance on proposed ARLstem algorithm.

Beirade et al. [9] suggest how Quran search engine can be designed to accommodate semantic ontology. This will ensure that meaning of each in relation to user query can produce better performance. However, Chandra and Dwivedi [10] examine how a query can be reformulated to search and retrieve information using Hindu query in English search engine. Ali et al. [11] to proposed Urdu stemmer for generating Urdu text instead of considering only English language text. Yusuf et al. [12] investigate the query types that would provide relevant search results. Therefore, Azad et al. [13] suggest individuals and Wikipedia term expansion for query in order to better search results. Moreover, Yusuf et al. [14] present a query expansion for improving Quran English translation results.

3 Proposed Approach

Arabic information retrieval helps in obtaining relevant Arabic search results. Document and query based with synonyms stemming for query expansion (DQSSQE) is our new approach to query expansion designed to improve the Arabic text search results, specifically Quran text.

DQSSQE combines document and query based on synonyms query expansion. It does not only consider the original query. Instead, DQSSQE considers the stemming process in order to reduce words to their stems in two ways: one containing document and original user query and then combine with new query with synonyms.

We used Arabic light stemmer algorithms obtained from Tashaphyne [15] which is the most popular stemming algorithm for Arabic character. Although this stemming algorithm required Pyarabic [16] to detect Arabic text. All the prefixes and suffixes were also removed using predefined Arabic word affixes list provided by the Tahspyne configuration of Tashaphyne [15]. Once the stemming has been done, the stem root and un-stem word would be available for term weighting.

Lucene search based on TF-IDF for weighting and vector space model was used for indexing and searching. Equation (1) was used to compute the weighting. In this weighting, the $xy(a, b, C) =$ represent the term frequency tf for inverse document frequency (a, C) is estimated using the below equation:

$$xy(a, b, C) = tf(a, b) \cdot idf(a, C) \tag{1}$$

To determine the similarity between two documents obtained, we used the cosine similarity. Specifically, the similarity between document x and y based on the general vector space model was computed using Eq. (2):

$$Sim(X, Y) = \sum\nolimits_v \sum\nolimits_w X_v \cdot Y_w \cdot S_{v,w} \tag{2}$$

Where $S_{v,w}$ represent the similarity of document x and y. While the tfidf weights of v in X, w in Y.

4 Experiments

4.1 Datasets

This study considers two Quran datasets collected from Tanzil [17] for comparing our proposed approach with other similar ones. Quran simple clean is one of the datasets which is recognized as a copy of the holy Quran verses without marks such as fat-ha, dhamma, kasra, sukoon, pause marks, sajdah signs as well as rub-el-hizb signs. The Quran simple clean is the Quran written in classical Arabic language. Tafsir Al-muyassar is another dataset used which is the copy of the Quran translated and printed by the King Fahad Quran complex, Kingdom of Saudi Arabia.

Datasets come in a single file; we, therefore, split the datasets where each verse forms a document. There are 6236 verses in each of the two datasets.

4.2 Evaluation Metrics

This paper use means reciprocal rank, average precision, average recall and mean average precision [18] measure to evaluate the performance of the proposed approach. The mean reciprocal rank was also computed using Eq. (3). It computes the reciprocal rank average above queries sets.

$$MRR = \frac{1}{N}\sum\nolimits_{q=1}^{N} \frac{1}{rank_q} \tag{3}$$

Mean average precision computed using Eq. (4) based on the rankings from different users' queries and then average these rankings to obtain average precision.

$$MAP = \frac{1}{Q}\sum\nolimits_{Z=1}^{Q} \frac{1}{D_Z}\sum\nolimits_{i=1}^{D_Z} P(d_i) \tag{4}$$

Average recall utilizes the position at which each ranked document provides the retrieved document and was computed using Eq. (5):

$$AR = \frac{1}{n}\sum\nolimits_{i-1}^{n} recall(R_i) \tag{5}$$

The position at which each ranked document provide the relevant document determine average precision and also computed using Eq. (6):

$$AP = \frac{1}{n}\sum\nolimits_{i-1}^{n} precision(P_i) \tag{6}$$

4.3 Benchmark

In this paper, the Quran relevant judgement provided by Moawad et al. [18] will be used as our benchmark. This document was provided by the Quran translation experts in relations to 36 queries. Table 1 gives a summary relevant judgement used.

4.4 Performance

In this experiment, two different methods; namely original Lucene and proposed QSEES have been considered to measure average recall, average precision, mean average precision and reciprocal ranks on Quran clean and Tafsir datasets.

Figure 1 shows the average recall performance on Quran clean dataset with two different methods. It was noticed that the proposed DQSSQE outperform the original Lucene method in term of average recall. This performance was specifically obtained from query 18, 27 and 32. However, the proposed DQSSQE produce worse results on other queries.

Table 1. Summary of the relevant judgement used as benchmark.

Query no	Total documents	Query no	Total documents	Query no	Total documents
1	19	14	3	27	225
2	74	15	51	28	550
3	26	16	8	29	114
4	3	17	7	30	90
5	43	18	18	31	17
6	10	19	43	32	294
7	13	20	14	33	8
8	8	21	102	34	77
9	10	22	11	35	36
10	6	23	77	36	10
11	39	24	31		
12	1252	25	1		
13	8	26	395		

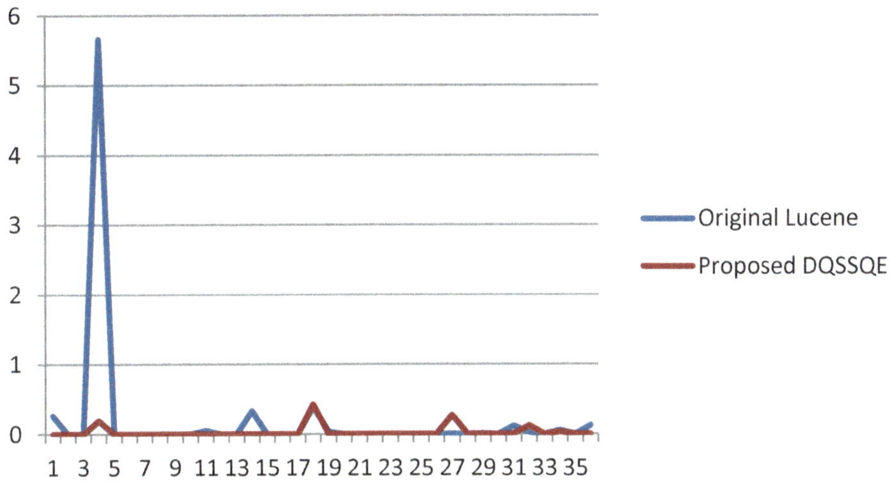

Fig. 1. Average recall for Quran clean dataset

Figure 2 shows the performance of average precision measures on Quran clean dataset using two different methods. It can be seen from Fig. 2 that the original Lucene method performs best on Quran clean dataset for almost all the 36 queries except query 27. This indicates that the Quran in classical Arabic doesn't require additional query word synonyms to be able to retrieve information.

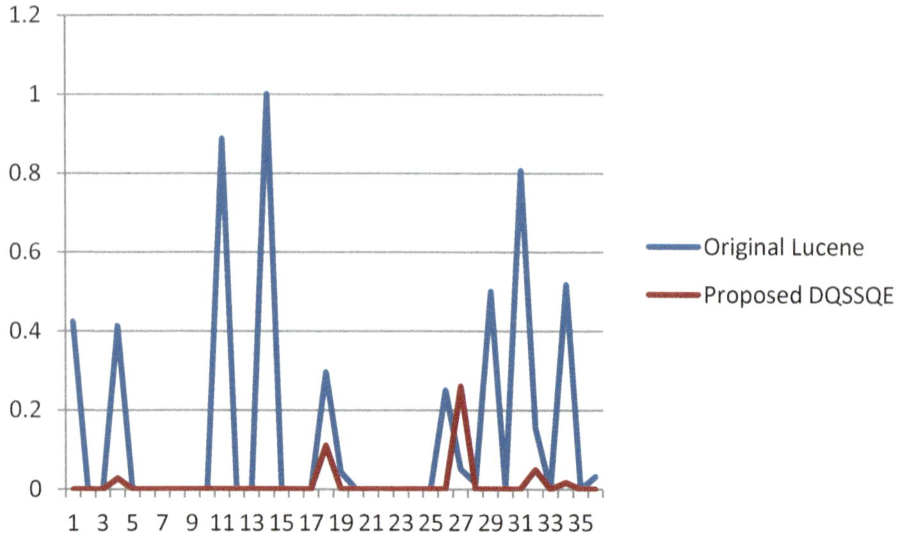

Fig. 2. Average precision performance for Quran clean dataset

Figure 3 gives the average recall performance on Tafsir dataset. It can also be noticed that the proposed DQSSQE provide better performance compared to the original Lucene method. This shows that more results can be retrieved from Tafsir dataset if additional information were added to queries. However, the original Lucene method also obtains the best result in query 31.

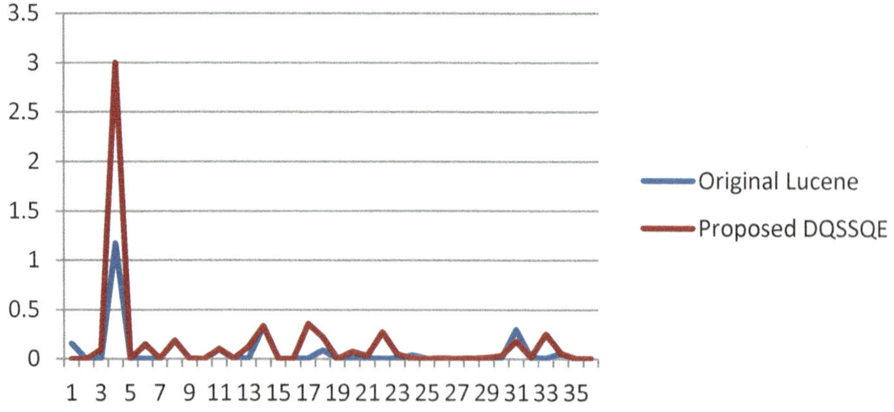

Fig. 3. Average recall for Tafsir dataset

Figure 4 has similarity with Fig. 4-13. In Fig. 4-15, better results have been shown for the proposed DQSSQE on Tafsir dataset using average precision. However, the original Lucene method only gives better results on query 4, 24, 31 and 34.

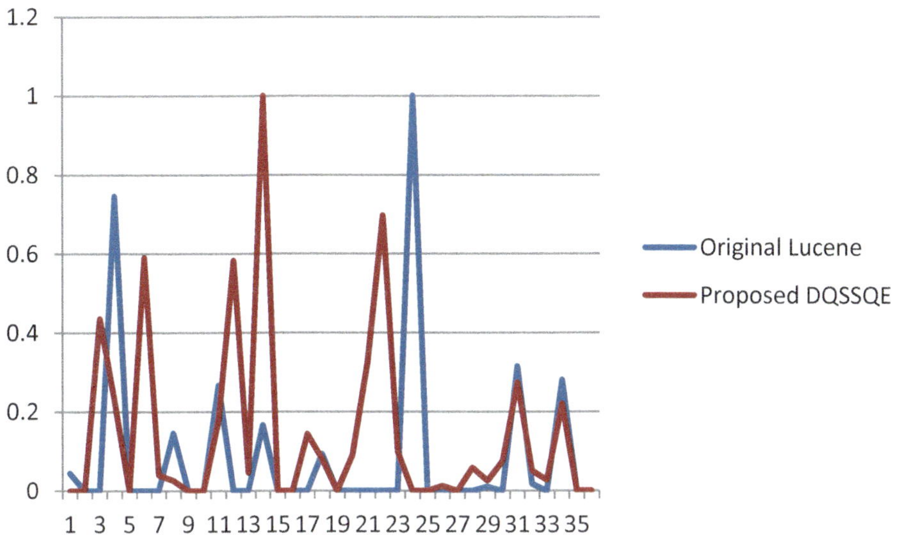

Fig. 4. Average precision for Tafsir dataset

Figures 5 and 6 show the MRR and MAP performance with two different methods on Quran clean and Tafsir datasets. In Fig. 5, it was noticed that the Tafsir dataset gives more results. However, Quran clean dataset also shows an improvement for original Lucene but below Tafsir dataset.

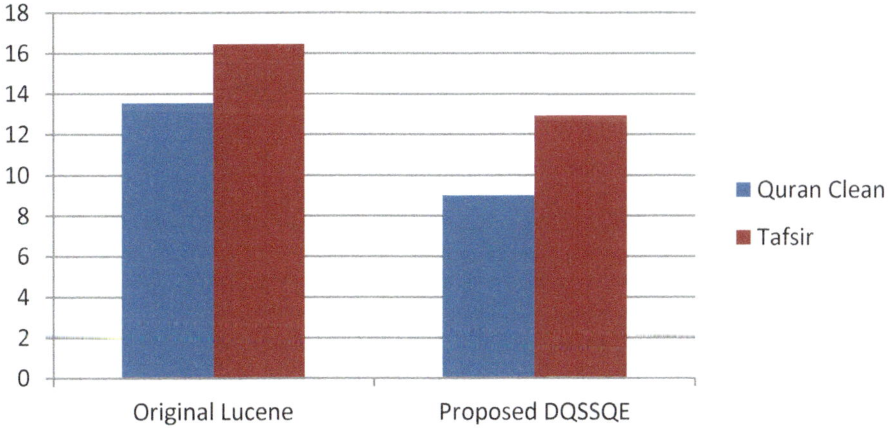

Fig. 5. MRR results for Quran clean and Tafsir datasets

In Fig. 6, the MAP performance on the proposed DQSSQE shows that better results have been achieved on Tafsir dataset. However, the original Lucene gives better results on Quran clean dataset.

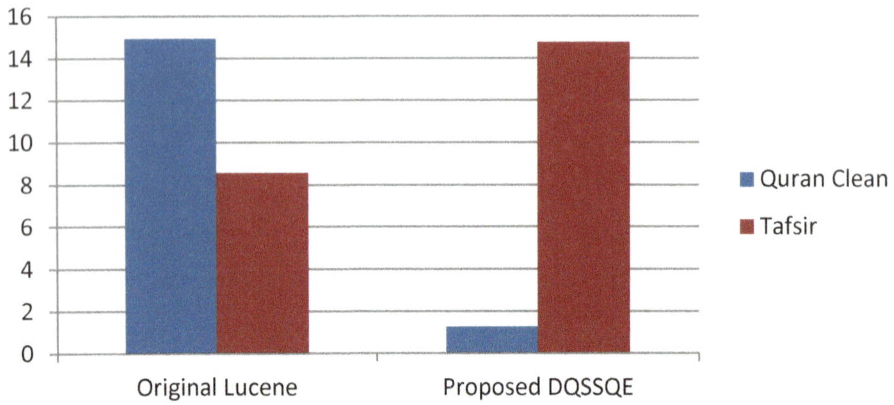

Fig. 6. MAP results for original Lucene and proposed DQSSQE methods

5 Conclusion

This paper present a new query expansion method called Document and query-based with synonyms and stemming query expansion (DQSSQE). The DQSSQE combines query expanded with synonyms and document stemming's to retrieve Arabic text. The paper shows that the performance of proposed DQSSQE provides better performance on Tafsir dataset. However, the original Lucene method gives better results on Quran clean dataset.

Acknowledgment. This research project has been sponsored by Universiti Tun Hussein Onn Malaysia (UTHM) for financially supporting this Research under Tier 1 vote no. U898, Enhancing Quran Translation in Multilanguage using Indexed References with Fuzzy Logic.

References

1. Alqahtani, M., Atwell, E.: Evaluation criteria for computational Quran search. Int. J. Islam. Appl. Comput. Sci. Technol. **5**(1), 12–22 (2017)
2. Elayeb, B.: Arabic word sense disambiguation: a review. Artif. Intell. Rev. 1–58 (2018)
3. Mohd Yunus, M.A., Mustapha, A., Samsudin, N.A.: Analysis of translated query in Quranic Malay and English translation documents with stemmer. In: MATEC Web of Conferences, vol. 135, p. 00069, November 2017
4. Jabbar, A., Iqbal, S., Khan, M.U.G., Hussain, S.: A survey on Urdu and Urdu like language stemmers and stemming techniques. Artif. Intell. Rev. **49**(3), 339–373 (2018)

5. Khalifi, H., Elqadi, A., Ghanou, Y.: Support vector machines for a new hybrid information retrieval system. Procedia Comput. Sci. **127**, 139–145 (2018)
6. Ben Guirat, S., Bounhas, I., Slimani, Y.: Enhancing hybrid indexing for Arabic information retrieval. In: 32nd International Symposium on Computer and Information Science, vol. 935, no. 1, pp. 247–254 (2018)
7. El Mahdaouy, A., Gaussier, E., El Alaouib, S.O.: Should one use term proximity or multi word terms for Arabic information retrieval? Comput. Speech Lang. **58**(1), 76–97 (2019)
8. Alhaj, Y.A., Xiang, J., Zhao, D., Al-Qaness, M.A.A., Elaziz, M.A., Dahou, A.: A study of the effects of stemming strategies on Arabic document classification. IEEE Access **7**, 1 (2019)
9. Beirade, F., Azzoune, H., Zegour, D.E.: Semantic query for Quranic ontology. J. King Saud Univ. Comput. Inf. Sci. (2019)
10. Chandra, G., Dwivedi, S.K.: Query expansion for effective retrieval results of Hindi-English cross-lingual IR. Appl. Artif. Intell. **33**(7), 1–27 (2019)
11. Ali, M., Khalid, S., Saleemi, M.: Comprehensive stemmer for morphologically rich Urdu language. Int. Arab J. Inf. Technol. **16**(1), 138–147 (2019)
12. Yusuf, N., Yunus, M.A.M., Wahid, N.: A comparative analysis of web search query: informational vs navigational queries. Int. J. Adv. Sci. Eng. Inf. Technol. **9**(1), 136–141 (2019)
13. Azad, H.K., Deepak, A.: A new approach for query expansion using Wikipedia and WordNet (2019)
14. Yusuf, N., Amin, M., Yunus, M., Wahid, N.: Query expansion based on explicit-relevant feedback and synonyms for English Quran translation information retrieval. Int. J. Adv. Comput. Sci. Appl. **10**(5), 227–234 (2019)
15. Zerrouki, T.: Tashaphyne, Arabic light stemmer2018))
16. Zerrouki, T.: Pyarabic, an Arabic language library for Python (2010)
17. Hamid, Z.-Z.: Quran translations. Tanzil (2007)
18. Moawad, I., Alromima, W., Elgohary, R.: Bi-gram term collocations-based query expansion approach for improving Arabic information retrieval. Arab. J. Sci. Eng. **43**(12), 7705–7718 (2018)

The Adoption of Business Intelligence Systems in Textile and Apparel Industry: Case Studies

Sumera Ahmad[1(✉)] and Suraya Miskon[2]

[1] School of Computing, Universiti Teknologi Malaysia, 81310 Johor Bahru, Johor, Malaysia
sumera_vjm@yahoo.com

[2] Information Systems, Azman Hashim International Business School, Universiti Teknologi Malaysia, 81310 Johor Bahru, Johor, Malaysia
suraya@utm.my

Abstract. The textile and apparel industry are characterized by highly labor-intensive operations, short production lead times, huge capital investment, seasonal demand and frequent style changes. In the recent decade, BI systems have been broadly adopted and implemented to achieve the true effectiveness of various systems and emerging technologies that are integrated to enhance the strategic, management and operational efficiency of textile and apparel industry to cope with the rapid growing challenges of globalization and expanding international competitive business environment. This research is first attempt to investigate the adoption of BI systems and discussing the real textile and apparel industry cases based on qualitative research method and also highlight the improved processes with some leading BI solutions. In addition, some major barriers and critical success factors for BI systems adoption are identified. The study limitation is also discussed with conclusion.

Keywords: Textile industry · Apparel industry · BI systems adoption

1 Introduction

Dynamic global trade drives the industries to adopt modern and emerging technologies to become more agile and innovative. Currently, the textile and apparel industry have also been influenced by critical auxiliary changes in the world of manufacturing and trade trends [1]. Thus, international trade is becoming more complex, with dispersed functional industry units that require internal and external data analysis of textiles and apparels for efficient and timely distribution of products and services. A substantial integration of systems and technologies is vital not only for effective communication among different information systems and emerging technologies but also significant for improving the inventory management, products delivery, forecasting, planning production and manufacturing of textile and apparel industry. It is worthwhile to identify the significance of modern technologies in industry that leads to generate structured and unstructured data exponentially. Companies need to access and analyze the huge data sets to make wise business decisions for understanding the competitive business markets [2]. In contemporary trade, most of textile and apparel companies are facing the strategic,

F. Saeed et al. (Eds.): IRICT 2019, AISC 1073, pp. 12–23, 2020.
https://doi.org/10.1007/978-3-030-33582-3_2

management and operational challenges in textile quota free system. To deal with these challenges, BI system is one of the proved proposed solutions, and it has been adopted broadly, particularly in the large scale of textile and apparel industry. The BI system is used as an umbrella term that incorporates models, architectures, databases, tools, techniques and applications with the objective of aggregating and analyzing the information regarding market trends to support the decisions and functions of organizations [3]. BI systems improve not only the organizational functionality but also escalate its revenue and competitiveness [4]. With these features of BI systems, it is confirmed that it has great potential to be applied in the textile and apparel industry. On the other hand, the praxis demonstrated that the accomplishment from BI systems is still a big question mark. The failure ratio of BI systems is still high. The businesses don't harness the expected advantages from the usage of BI systems [3]. In addition, the adoption of BI systems in the textile and apparel industry is not greatly investigated yet [5]. As a result, it is pertinent to explore the reasons, improved processes, enabling technologies, best BI solutions, barriers and critical success factors for the adoption of BI systems in the textile and apparel industry. The pattern of this paper is designed as follows: Literature review of BI systems adoption is provided in following Sect. 2. Section 3 explains the research problem. Section 4 describes the research methodology. Some specific real-world cases of textile and apparel industry about the successful adoption of BI systems are discussed in Sect. 5. In addition, some leading BI solutions, some major barriers and critical success factors of the BI systems adoption are also explored in this section. Last Sect. 6 presents concluding remarks with limitation of this study.

2 The Literature Review

The textile and apparel industry is one of the mature and oldest industries in the world [1]. The rapid changing market is the characteristic of textile and apparel industry. Most of textile and apparel products are seasonal in nature and consumers' tastes are changing at frequent bases [6, 7]. Customers are demanded apparel with a customized style, design, fit, color and print. Along these lines, the textile and fashion companies can lose a considerable amount of cash because of excessive stock, which ends up with outdated apparels because of dynamic changes [8]. In competitive world, every business enterprise, big or small needs authentic and precise information for decision making. Today's world of social media and multimedia, data is generated exponentially by everyday business activities. This great growth of data creates the challenges and opportunities for individuals and as well as for organizations. This data generation is referred as big data [9]. The usage of big data for decision making attracted the attention of industry practitioners to adopt BI systems.

BI systems benefit the decision-making concepts in various sectors of textile and apparel industry that is the sole reason to gain the attention of industry experts to adopt and implement this innovation in textile and apparel companies. Some prominent benefits of BI systems are following.

- ***Manufacturing and Production Management:*** BI systems enable the Executives and managers to identify economic and technical information on different levels of manufacturing and production BI systems reduce the production lead time, improve product quality by analysis of material consumption [10].

- ***Production Planning and Control:*** BI systems produce key performance indicators (KPI) and metrics to analyze and monitor production plans and control factors through automated alerts, whenever a critical situation arises. Quick decisions are taken with full confidence and critical issues are solved before they aggravated.
- ***Finance:*** The textile and apparel industry can improve the financial reporting and decision making by analyzing financial data in the Data Warehouse (DW). BI systems support budgeting through well-established cost analysis methods and analytical models that compare to provisions, actual cost of promotional campaigns and risk evaluation expenses.
- ***Sales and Distributions:*** BI tools can analyze the history of processes and sales results of salesmen and distribution channels of textile and apparel industry. BI tools enable the executives of industry to compare the results of different channels' activities and perform analysis of products and salesmen. As a result, activities can be tracked easily in order to make quick and precise decisions.
- ***Inventory Management:*** BI systems can improve the inventory management. BI systems also help the merchandisers and inventory managers to decrease the obsolete and unsold stocks and supply in demand products of textiles and apparels. Inventory managers can be aware of critical operations like returns, back orders and stock shortage due to optimization analysis of physical warehouse and location selection [11].

2.1 Enabling Technologies for BI Systems Adoption

The introduction of smart textiles and integration of emerging technologies in textile and apparel industry generate huge amount of data on daily bases that is called big data It implies an extensive gathering of structured (numbers, text, records, documents, financial data, personal data, etc.) or unstructured data (text messages, audio and videos, geographical location, photos, 3D models, simulations, and so on) that is generated and transmitted over the internet as exemplified by data gathering from social media, gigantic range of sensors from the smaller scale (atomic) to the larger scale (worldwide). It is believed by researchers that the integration of big data and BI systems is the way from insights to value [12] and put pressure on companies to adopt BI systems. The application of Radio-Frequency Identification (RFID), Internet of Things (IOT), mobile technology, and cloud technology become the essential parts of today's the world of big data. The adoption of these emerging technologies in textile and apparel companies give rise to process the data for decision making and supports the adoption BI systems.

- **RFID (Radio-Frequency Identification):** RFID technology is utilized in textile and apparel industry for manufacturing, warehousing, distribution, inventory control, logistics, automatic item tracking and supply chain management [13]. For analyzing large amount of data that is generated by the adoption of RFID systems with other applications, the enterprises need to adopt BI systems techniques to grasp the true economical value of RFID systems.
- **Cloud Technology:** The emerging cloud technology is one of the most significant innovative technology that has strong potential to enhance the strategic, management and operational efficiency of manufacturing industries [14]. As textile and

apparel industry are prone to adopt different emerging technologies by PaaS cloud service. Cloud based BI systems remove the technology obsolescence issues in the adoption of costly BI in term of maintenance and updating systems [15].

- **Mobile Technology:** A mobile application has become a popular trend and has taken the apparel world by a storm of emerging technologies. Tracking and monitoring of the processes of textiles and apparels, including manufacturing, become easy with the application of NFC (mobile technology) and RFID technology. Mobile utilization has finally achieved the goal where they can meet the different needs of BI systems users in terms of collaboration and the convenience.
- **Internet of Things (IOT):** Smart cities, smart factories, smart devices and smart clothing are buzz words in these days. IOT infrastructure consists of numerous technologies. This interconnection of technologies make it possible the modern monitoring, traceability, coordination and collaborative decision making between business partners [16]. Huge amount of data is collected on daily bases by smart, connected devices in store or outside geographically dispersed equipment. Thus, it is very important to convert raw data into knowledgeable information As a result, companies are leading to adopt BI systems for better predictive analytics and decisions making [17].

3 Research Problem

There is always a requirement of data accuracy and visibility for real-time information at each phase of business network of textile and apparel industry. In 21st century, if industries are not able to harness the chances to develop and discover appropriate solutions along with expansion of opportunities and threats, they will be out of the competition [18]. According to the Gartner and Forrester surveys, enterprises including textile and apparel companies are adopting and integrating various Information Systems (IS) and emerging technologies to maintain their sustainability in global competitive market. The emerging technologies generate huge amount of data at daily bases. The trues effectiveness of the emerging technologies is just possible by the adoption of BI systems but the adoption of BI systems in the textile and apparel industry is not greatly investigated yet [5]. In addition, the integration of BI systems in textile and apparel industry is rare. Some textile companies had integrated BI systems, but researchers did not give much attention to this area [19]. The main gap observed from previous research just focused on adoption of components of BI systems, such as OLAP and data mining however the research about the adoption of BI systems is still scarce [20]. Although the primary objective of the study is to investigate the adoption of BI systems in textile and apparel industry however, the sub-objectives of this study are as follow;

- To investigate the reasons to adopt BI systems in textile and apparel industry.
- To examine the improved processes by the adoption of BI systems in textile and apparel industry.
- To explore the major barriers and critical success factors for the adoption of BI systems in industry.
- To find out the best BI solutions with vendors for textile and apparel industry.

4 Research Methodology

The subject of case studies is textile and apparel industry. Theses case studies are based on interviews. The interviews are considered as the most common way of exploring the required information in IS field [21]. Interviews are comprised of open-ended questions as further elaboration is possible by the open nature of questions [21]. The potential of this research method enables the participant to design their answers by sharing, what is important to them. The qualitative research method empowers informants to think about the core content and topics in a new way and to reflect upon their perception and experience [22]. It is confirmed by the studies that the adoption of BI systems in textile and apparel industry is rare [19]. As a result, the textile and apparel companies were chosen only with successful adopted BI systems or ready to adopt BI systems. The requests were sent to thirty-five industries for interviews and received responses from eleven industries only. The interviewees were selected based on their designation and positions such as owner/managers and Chief IT officers in the company. (Table 1 shows the detail information about the informants). Finally, Out of eleven, seven interviewees were selected by using snowballing method [12]. All of them have awareness about BI systems and most of them were using BI systems in their everyday work. The objective of the case studies is to analyze and understand the adoption of BI systems in textile and apparel industry. The seven semi-structured interviews were conducted for the adoption of BI systems in textile and apparel industry. The main structure of the interviews was as follows: Reasons to adopt BI systems, processes improved by BI systems, adopted BI solutions with vendors, barriers and critical success factors according to the interviewees' point of view. The questions were asked and answered in open-ended manner.

Table 1. Informants' data.

Company	Interviewee position	Country	Company	Mode
Kohinoor Textile Mills	IT Executive	Pakistan	Textile & Apparel	On site
Kayseria	Retail Manager	Pakistan	Clothing brand	On site
Chaudhary Textile Mill	Owner & IT Manager	Pakistan	Embroidery Textile	On site
Adidas	IT manager	Spain	Apparel	On skype
Nishat Mills Limited	CEO (IT)	Pakistan	Textile & Apparel	On site
Zara	Regional Manager	Saudi Arabia	Clothing brand	Onsite
Abercrombie & Fitch	Senior Manager	America	Apparel	On skype

Every informant was also asked for other details that might be related to the topic except specific questions. Although, some interviewees forbade recording their interviews as a result, detailed topic notes had been taken during the interviews and completed with complementary notes immediately. A systematic analysis of notes was completed after each interview. The content analysis approach was used for the analysis of interviews.

5 Findings

5.1 BI Systems in Textile and Apparel Industry

To investigate the recent industrial scenario about the adoption of BI systems in textile and apparel industry, the results of some interviews are following.

Case 1: Kohinoor Textile Mills: Kohinoor textile mill is one of the largest textiles and apparel industry of Pakistan that is famous with the name of Saigol group of industries including, Kohinoor textile Mills Rawalpindi. The informant was IT executive who is working on the implementation of BI systems project in the company. He explained that company exports apparels to USA and European countries etc. Recently, company has been started working on feasibility for the BI system project and ready to invest in BI system project with the goal of enhancing the efficiency of company's supply chain efficiency such as improving sourcing abilities, reducing lead time and to support the growing retail business of the company. The ultimate objective is to implement BI systems successfully that can be used for decision making by collecting internal (ERP, SCM, HRM, CRM) and as well as external data like web and market. The increasing demand of market trend and high intensity of cooperation with suppliers and customers are also desirable.

Case 2: Kayseria (Apparel Company): Informant was retail manager who told about the company that Kayseria is one of the market famous extended brand of Bareeze in Pakistani industry, that is high-end fashion retailer of Pakistan. Bareeze established its outlets almost in every city of Pakistan and as well in four other countries that are India, Malaysia, United Kingdom and United Arab Emirates. The informant also explained that market environment created lot of challenges for the company in term of inventory management such as uncertain production yield rate of fabrics and apparels that leads to the uncertainty of production output. On the other hand, company customers are constantly demanding for a shorter lead time. Highly volatile demands headed to a lot operational problems and delivery scheduling. To cope with these challenges, company has implemented versatile Business Objective (BOBJ) running on SAP for retail operations that is provided by leading software vendor. This BOBJ solution is helpful in improving company 'financial and retailing operations'. These technology solutions are also enhancing the production forecasting and partnership with customers and suppliers. With BOBJ running on SAP solutions, company is growing rapidly and expanding its business all over the world.

Case 3: Chaudhary Textile Embroidery: Chaudhary Textile Embroidery falls in medium category of industry with 200 employees. The interviewee was IT manager and owner of the company. He described that the impact of market pressure has led to the shortening delivery lead time, inventory management and supply chain management for the company. The ever-changing market, raw material price fluctuation and quality of product inconsistency are challenges faced by this company before adopted Oracle solutions. The owner of the company realized that better collaboration with retailers, logistics partners, customers and vendors, are necessary for the company. Thus, the company adopted the Oracle free solutions to solve the issues like delays in production, delivery orders, inaccurate demand forecast, sales management, supply chain management and a productivity to compete the market. The Oracle solutions also allow the company to synchronize with its customers. The Oracle solutions have significantly improved the sales and inventory of the company. As motivated by the technological solutions, the company aims to install more cloud-based technology solutions such as BI systems for decision making because of cost affordability in term of IT infrastructure, maintenance and software updating facilities.

Case 4: Adidas Group (Adidas, Rockport, Reebok Apparel Brands): The informant was IT manager at Adidas Group in Spain, he explained! That Adidas (world well renowned apparel sportswear brand that manufactures clothing, shoes and accessories) has adopted BI systems to harness a large range of data by using Micro-strategy analytics as the front end of BI systems. Company integrated the Micro-strategy tools with the BI systems running on SAP HANNA to help staff by providing fast delivery of customer insights within business group. Four independent data ware houses were consolidated into one single platform based on the HANA n- memory database. It was part of big data analytics strategy that develops and implements advance methodologies to optimize customer experience, loyalty, acquisition and engagement. The company used this data to improve the understanding of the customer behavior. The company analyzes the financial reports that help to predict about market place and to know about the consumers experiences what they like and what influence on their buying decisions. The company empowers its staff to access valuable data on mobile devices by self-service. Staff members allowed to create dashboards for quick response according to changes occurred in the market. The ultimate aim is to make sure that customers get a great brand experience. Traditional systems and databases are also connected to new platform for getting customer data from CRM and information from its Hadoop systems.

Case 5: Nishat Mills Limited: The informant was Chief Executive Officer (CEO) of IT at Nishat Mills Limited, he stated that company consists of spinning, weaving, dying and stitching units with remarkable production capabilities. The company has renowned top brand of Pakistan with the name of Nishat Linen (NL). NL opened 72 outlets in Pakistan and also has international retail stores in Canada, Saudi Arabia, Abu Dhabi and Dubai with online stores as well. The company is a long-term supplier of world's top brands like John Lewis, Hugo Boss, Gap, Crate and Barrel, Ocean garments, Next, Levis and some others. Company has expanded its presence in big markets and enlarged its production base to more countries upon the adoption of BI systems providing by leading Software Company. BI systems have been integrated with multiple legacy systems in the company to approach its target markets. As a result, customer services are improved; overhead costs reduced and achieved better control on international trade. BI systems also provide decision support ability to the company with financial analysis, horizontal and vertical analysis for profit and loss, Ratio analysis, market analysis, competitor analysis and environmental analysis. These technological solutions enabled company to enhance its strategic vision.

Case 6: Zara (Clothing Brand): The informant designation is regional manager of Zara clothing brand. The manager revealed that Zara brand has 1770 outlets in all over the world. The Zara is considered as most innovative retailer in the world. The Zara has implemented advanced IS including BI systems that include technologies, collection, analysis and present information for business decision making in multiple areas of the company. The company has well-adopted BI systems majorly for distribution management, replenishment control, supply chain management size/color management and inventory management. With the BI solutions in inventory management, the company has reduced delivery lead time successfully across its various outlets in all over the world. Consequently, reduced the logistics cost of supply chain. This resulted as great success for the company. The main objective of BI systems adoption is to support better business decision making.

Case 7: Abercrombie & Fitch (Apparel Brand): Abercrombie & Fitch is renowned international American apparel and accessories retailer for kids, women and men with 865 worldwide outlets and stores. The fast merchandise planning is critical for the company. To cope with this challenge, the company selected Tableau BI solutions. Senior Manager of Product Facing Solutions also explained that Abercrombie & Fitch improved their merchandising analysis by using Tableau BI solutions with quick identification of customer purchasing decisions at locations with visibility and clarity. Merchandise teams see true picture of demands sales and ships that were almost impossible previously to visualize over time. The company can identify key metrics and make sure the availability of right product before its demand and ship. The company improved the seamless experience between shop and online to deliver best customer services.

5.2 BI Systems in Textile and Apparel Industry

A survey on BI Data Analytics revealed that market is overwhelmed by proprietary BI solutions and BI Data analytics [23]. Gartner and others, based on the research by Dresner Advisory Services, confirmed that selection of appropriate vendor can drive towards the greater success of BI systems adoption. As a result, some adopted and implemented BI systems with functionalities and vendor companies in textile and apparel industry are summarized in following Table 2.

Table 2. BI solutions for textile and apparel companies

Names	Description
Oracle www.oracle.com	Oracle BI systems are prebuilt complete solutions that support role-based intuitive intelligence throughout a company. Its central architecture is delivering and generating information to business partners, employees and customers. Chaudhary embroidery textile and Kohinoor textile mills are using oracle solutions
Trade gecko www.tradegecko.com	TradeGecko BI solutions are cloud-based inventory management platform for wholesalers and retailers. It improves inventory management, retail operations and online business operations. Zara, H&M and Pink Boutique are using TradeGecko solutions
Dematic www.dematic.com	Dematic solutions improve the company's performance in term of productivity, be responsive to the market demands, assets, space utilizations and labor challenges. Next, Gap Inc and Adidas are using the Dematic BI solutions for their supply chain management
SAP BW/4HANA www.altivate.com	SAP HANA in-memory technology depends on single data warehouse for business analytics tools. SAP solutions can integrate the Hadoop, sensor, unstructured and third-party databases. Cloud based SAP provides real-time high-volume data processing at high speed and simplified administration and data modeling. Adidas and Kayseria are using SAP solutions
Micro strategy www.microstrategy.com	MicroStrategy analytics are used as front end of BI systems. Companies integrated the Micro-strategy tools with the BI systems running on SAP HANNA to help staff by providing fast delivery of customer insights within business group. Multiple independent dataware houses can be consolidated into one single platform based on the HANA n- memory database. Adidas is benefiting by MicroStrategy
TIBCO spotfire www.tibco.com	Mark & Spencer benefited by TIBCO Spotfire. The TIBCO solutions enabled the company analysts to mashup data sources, data warehouses, spreadsheets, Hadoop databases, without needing IT assistance
Tableau www.tableau.com	Tableau BI solutions empower the company to take true picture of customer demand, sales and ship. The company can achieve faster insights of customers and products to respond quickly for demand and ship. Abercrombie and fitch are using Tableau solutions for improving merchandising

5.3 Some Major Barriers for BI Systems Adoption

After reviewing the interviews and full discussion with seven industrialists (including one textile embroidery manufacturer, three fabric manufacturers and three apparel producers) it appears to us that the textile and apparel industry is striving hard to maintain its sustainability by improving the analytical strategies but still facing some barriers in the way of BI systems adoption.

- Four interviewees revealed that cost and complexity are considered the major barriers by industry experts in term of BI systems adoption. Mostly vendors are unable to demonstrate the exact budget and real benefits of using BI systems.
- Three interviewees said that leaders, who mostly rely solely on experience and prone to instinct-based-decision-making rather than analytic insights, are directly affected to the poor adoption of BI systems in companies. Mostly old age leaders are unaware that how much necessary to modernize the organizational strategy.
- The manager at Adidas said, change management is another significant barrier. Employee resist to adopt new technology, fear of misinterpretation of data resulted in using old spreadsheet technology.
- All interviewees agreed that major barrier is often hard to demonstrate the success of BI adoption in term of return on investment (ROI) in advance and resulted in underinvestment for BI project and lack of the funding by company.
- It is revealed by the five interviewees that greatest technological barrier is lack of integration of BI tools with 3rd-party databases and legacy systems. The integration of gadgets from various hardware suppliers, diversity of communication protocols, diverse level of equipment openness, and distinctive intelligent standards are major barriers. BI systems running on single database deliver limited benefits to organizations. It is leading to poor adoption of BI systems.
- It is confirmed by all informants that sometime companies have a sound BI strategy and appropriate BI systems with company requirements but legging in technical training and skills such as supporting, designing, building and maintaining BI solutions.

5.4 Critical Success Factors for the Adoption BI Systems

Some critical success factors are presented here after deep analysis of interviews. These critical success factors may be helpful for the adoption and implementation of BI systems successfully.

- Data quality can make or break any BI systems success. Data quality should be top priority of the company to achieve the maximum potential results from BI systems adoption.
- Selection of BI systems, agile and flexible architecture including supporting embedded application that can accommodate to changing requirements of the user/company that delivers best user experience in term of price, value, performance, approaches, preferences and policies are essential.
- Organizations approach the higher level of strategic BI systems adoption if they integrate already existed systems and technologies together. It results in attaining real-time analysis and reporting of company.
- Interactive tools and self-service BI systems in the hands of non-technical users lead towards the BI systems adoption.

6 Discussion and Conclusion

The case studies of textile and apparel companies can be beneficial not only for BI vendors but also for companies to enhance the process of BI adoption to be on-budget, on-time and most important according to the business requirements. In future the barrier and critical success factors of the BI adoption can be considered seriously that can lead to the successful ROI for BI systems adoption. Real textile and apparel company cases based on interviews that explored the reasons to adopt BI systems and improved process such as supply and retail chain efficiency and reduced lead time and solved the issues like delay in production, late delivery orders, inaccurate demand forecast, sales management, supply chain management and optimize the productivity and customer experience with more fast and informed decisions to compete the market. All case studies revealed that major benefits of BI systems adoption were seen in retail chain processes. The well-established and large textile and apparel companies installed advance analytical systems and preferred to adopt proprietary BI systems. This study is limited to seven textile and apparel companies. The situation and infrastructure are different of each company and case studies cover only a portion of textile and apparel industry with specific scenario. Nonetheless, it is believed that these case studies can be directly applied to other textile and apparel companies within the same sector for BI systems adoption. Therefore, the textile and apparel companies can achieve desired efficiency in various sectors without trial and error. This study can also helpful to other industries for the successful adoption of BI systems with their existing infrastructure without additional investments. Future research can be extended in research area with practical knowledge and experience by sharing similar case studies. Researchers also benefit from the study findings in term of direction to identify their present research position and focusing the future research directions that need attention.

References

1. Balogun, A.L., Tyler, D., Han, S., Moora, H., Paco, A., Boiten, V.J., Ellams, D., Leal Filho, W.: A review of the socio-economic advantages of textile recycling. J. Clean. Prod. **218**, 10–20 (2019)
2. Bordeleau, F.È., Mosconi, E., Santa-Eulalia, L.A.: Business intelligence in industry 4.0: state of the art and research opportunities. In: Proceedings of the 51st Hawaii International Conference on System Sciences, pp. 3944–3953 (2018)
3. Gaardboe, R., Svarre, T.: Business intelligence success factors: a literature review. J. Inf. Technol. Manag. **29**(1), 1 (2018)
4. Moreno Saavedra, M.S., Bach, C.: Factors to determine business intelligence implementation in organizations. Eur. J. Eng. Res. Sci. **2**(12), 1 (2018)
5. Rocha, Á., Correia, A.M., Costanzo, S., Reis, L.P.: New Contributions in Information Systems and Technologies. Advances in Intelligent Systems and Computing, vol. 353, pp. III–IV. Springer, Switzerland (2015)
6. Choi, T.M.: Launching the right new product among multiple product candidates in fashion: optimal choice and coordination with risk consideration. Int. J. Prod. Econ. **202**, 162–171 (2018)

7. Safra, I., Jebali, A., Jemai, Z., Bouchriha, H., Ghaffari, A.: Capacity planning in textile and apparel supply chains. IMA J. Manag. Math. **30**(2), 209–233 (2018)
8. Jain, S., Bruniaux, J., Zeng, X., Bruniaux, P.: Big data in fashion industry. In: IOP Conference Series: Materials Science and Engineering, vol. 254, no. 15, p. 152005 (2017)
9. Iqbal, M., Soomrani, A.R., Butt, S.H.: Opportunities & Challenges (2018)
10. Hänel, T., Felden, C.: Operational business intelligence meets manufacturing. In: Americas Conference on Information Systems, pp. 1–9 (2013)
11. Rostek, K.: Business intelligence for insurance companies. Found. Manag. **1**(1), 65–82 (2009)
12. Božič, K., Dimovski, V.: Business intelligence and analytics for value creation: the role of absorptive capacity. Int. J. Inf. Manag. **46**, 93–103 (2019)
13. Nayak, R., Singh, A., Padhye, R., Wang, L.: RFID in textile and clothing manufacturing: technology and challenges. Fash. Text. **2**(1), 1–16 (2015)
14. Ooi, K.B., Lee, V.H., Tan, G.W.H., Hew, T.S., Hew, J.J.: Cloud computing in manufacturing: the next industrial revolution in Malaysia? Expert Syst. Appl. **93**, 376–394 (2018)
15. Tole, A.A.: Cloud computing and business intelligence. Database Syst. J. **4**, 49–57 (2014)
16. Molano, J.I.R., Lovelle, J.M.C., Montenegro, C.E., Granados, J.J.R., Crespo, R.G.: Metamodel for integration of internet of things, social networks, the cloud and industry 4.0. J. Ambient Intell. Humaniz. Comput. **9**(3), 709–723 (2018)
17. Jouriles, N.: IoT and the fashion industry. Personal Interview, 10 May 2016
18. Pool, J.K., Jamkhaneh, H.B., Tabaeeian, R.A., Tavakoli, H., Shahin, A.: The effect of business intelligence adoption on agile supply chain performance. Int. J. Prod. Qual. Manag. **23**(3), 289–306 (2018)
19. Istrat, V., Lalić, N.: Association rules as a decision making model in the textile industry. Fibres Text. East. Eur. **25**(4), 8–14 (2017)
20. Bach, M.P., Čeljo, A., Zoroja, J.: Technology acceptance model for business intelligence systems: preliminary research. Procedia Comput. Sci. **100**, 995–1001 (2016)
21. Yusof, A.F., Miskon, S., Ahmad, N., Alias, R.A., Hashim, H., Abdullah, N.S., Ali, N.M., Maarof, M.A.: Implementation issues affecting the business intelligence adoption in public university. ARPN J. Eng. Appl. Sci. (2015)
22. Webster, L.: Using narrative inquiry as a research method (2014)
23. Poba-Nzaou, P., Uwizeyemungu, S., Saada, M.: Critical barriers to business intelligence open source software adoption. Int. J. Bus. Intell. Res. **10**(1), 59–79 (2018)

Review on Feature Selection Methods for Gene Expression Data Classification

Talal Almutiri(✉) and Faisal Saeed

College of Computer Science and Engineering, Taibah University, Medina, Saudi Arabia
Almutiri.Talal@hotmail.com, fsaeed@taibahu.edu.sa

Abstract. Microarray technology makes it easier for scientists to rapidly measure thousands of gene's expression levels. By analyzing these data, we can find the altered genes, thereby facilitating easy diagnosis and classification of the genetic-related diseases. However, predicting and identifying cancer types is a great challenge in the medical field. Gene expression microarray contains information that can help in this regard, but microarray data have high dimensionality problem which means a large number of genes or features and a small number of samples, also there are redundant and irrelevant features that increase the challenge of microarray analysis. This study reviewed recent studies about methods, algorithms, and limitations of feature selection for microarray gene expression classification. This study compared and focused on four aspects for each related study: datasets, feature selection methods, classifiers, and accuracy results. Feature selection methods are considered as a pre-processing step which plays a vital role in the effectiveness of a classification. This paper showed that applying filter methods such as t-Test, Pearson's Correlation Coefficient (PCC), and Bhattacharyya distance eliminate irrelevant features that help to increase classification performance and accuracy. Therefore, applying wrapper or embedded methods such as Genetic Algorithm (GA) without applying filter methods in advance could affect the effectiveness of a classification negatively.

Keywords: Cancer classification · Gene expression · Feature selection · Microarray data

1 Introduction

Organisms are classified according to their cell components into two large families: Prokaryote; organisms that are located within the single cell and are simple in composition. And eukaryote; organisms with more complex cells. DNA is the most important molecule in the cell. Deoxyribonucleic acid (DNA) is a combined molecule that holds all the necessary information required in building and maintaining an organism. All the livings have DNA in their cells. In fact, almost all the cells in a multicellular organism have the whole set of DNA needed for that same organism [1]. We can find DNA inside a special place in the cell which is known as the nucleus. As the cell is extremely small, and as the organisms have many DNA molecules inside each cell, all the DNA molecules have to be well packaged. This packaged form of

F. Saeed et al. (Eds.): IRICT 2019, AISC 1073, pp. 24–34, 2020.
https://doi.org/10.1007/978-3-030-33582-3_3

DNA is known as the chromosome [2]. A human being's normal cell contains 46 chromosomes, which are corresponded accordingly into 23 pairs. These chromosomes consist of some smaller genetic material units known as DNA [3]. These units can be defined as a sequence of letters that spell out the genetic code. DNA is arranged into words and sentences known as genes. Human beings have an approximate of 20,000 to 25,000 of genes, and each gene influences a specific part of development. Genes are the parts responsible for determining how the cells are going to live and function, and in deciding how proteins are going to carry on the process of building and reproducing in our body [4]. All the livings depend their genes to determine how they are going to develop in their lives and how to pass on their genetic traits to their breed [3]. We can call the change in the sequence or gene of the DNA spelling a mutation. All human's DNA contains some harmless mutations. However, some mutations may be responsible for rising specific diseases [5].

The "Curse of Dimensionality" is one of the key issues facing using Gene Microarray Data. Gene Microarray Data may consist of fewer samples figures, but on the other hand, it has higher numbers of genes. It is also spotted that so many genes may be either redundant or irrelevant, and either way, they are not helpful in classifying cancer. But a few genes may be considered related. This leads us to know the importance of features selection. It is still challenging for the gene expression microarray to have a high dimensionality [6]. The process of choosing the informative genes among thousands of genes is usually rendered to as gene selection, which can also be called as feature selection. Feature selection techniques work on saving the original semantics of features. It is exceedingly believed, as indicated by literature, that in the majority of microarray gene expression data, a small number of genes play an important role in the classification, while the rest are irrelevant to it. Thus, to know most of the important genes of all the others that were measured, feature selection techniques are needed. Selecting important genes for the classification of a variety of phenotypes with microarray data, such as the types of cancer, aims to provide better recognition of the underlying biological system and to improve the prediction of the classifiers [7].

Section 2 introduced knowledge and background about DNA, gene expression, and transcription and translation steps to generate proteins. Section 3 presented feature selection methods such as filter, wrapper and embedded. Section 4 elaborated discussion on feature Selection methods for gene expression data classification which focused on recent studies. The conclusion and the main review findings were presented in Sect. 5.

2 Gene Expression

Genes are made up of DNA segments which carry the genetic information required to encode a particular ribonucleic acid cellular (RNA) along with proteins. The molecular biology central dogma says that proteins can be generated out of DNA by using a two-stage process. First, transcription which describes the fact that a DNA gene can be exploited by moving its coded information or data into messenger ribonucleic acid (mRNA). The second step is the translation which characterizes that Ribosomes generate proteins after the information was decoded from the mRNA. This two-stage procedure, usually known as gene expression, permits expressing genes as proteins [6]. All the organism's body cells have the exact DNA print, while the level of mRNA and the types of cells differ over time and under different circumstances. The amount of the expressed mRNA has a significant part because the process of producing proteins depends on the process of producing mRNA, while the gene expression can be measured by using the level of mRNA. So, we can indicate the amount of mRNA that a cell produces through protein synthesis by using the gene expression level. A mutation of specific genes may cause tumors. This unusual alteration inside the cells is a result of the alteration or change in the expression level of the same mutated genes that are expressed to be unusual in some cells. A gene is said to be upregulated whenever it has a high level of expression, while it is said to be down-regulated whenever it is expressed at a lower level. A gene also may not be expressed at all. However, developing a test to detect these mutations or the difference between the gene expressions levels can be very challenging and can produce or generate a particular profile for each gene, because mutations can occur in many regions of most of the large genes. For example, a belief among the researchers indicates that in the genes BRCA1 and BRCA2, mutations can cause about 60% of all the cases of ovarian and hereditary breast cancers. However, there is no particular mutation or change accountable for all of these cases. Researchers have discovered more than 800 different mutations only in BRCA1. DNA microarray is a technology used in determining whether a specific individual's DNA holds a mutation in genes like BRCA1 and BRCA2 [8].

Gene expression analysis is concerned with the way used in transcribing genes in order to synthesize the functional gene products - functional RNA species or protein products. Some insights into the regular cellular processes may be provided while studying gene regulation. Researchers are capable of performing gene expression analysis at any of the levels where the gene expression is at: "transcriptional, post-transcriptional, translational, and post-translational protein modification".

Transcription is the procedure of making up a complementary RNA copy of a DNA sequence that is capable of being regulated in several ways. In the experiments related to the regular gene expression analysis, transcriptional regulation procedures are usually studied and manipulated ones.

Furthermore, the binding of regulatory proteins to DNA binding sites is the most direct method by which transcription is modulated in a natural way. Instead, regulatory processes can also make interaction with the cell's transcriptional machinery. Lately, researchers have found out that epigenetic regulation's impact, such as how the variable DNA methylation affects gene expression, is a strong tool for gene expression profiling. Different methylation degrees are known for their impact on chromatin folding and of their strong effect on genes capability of accessing to active transcription. After making the transcription, eukaryotic RNA is usually spliced in order to get rid of the non-coding intron sequences and capped with a poly (A) tail. At this stage, RNA constancy has a huge impact on the functional gene expression, which is, the production of functional protein. "Small interfering RNA (siRNA)" has double-stranded nucleic acid molecules that participate in the RNA interference pathway, in which the expression of specific genes is modulated (usually by decreasing activity). It is still not completely understood how this alteration is exactly accomplished. There is a growing field of gene expression analysis which is located in the microRNAs (miRNAs) area. Short RNA molecules that can also work as eukaryotic post-transcriptional regulators and gene silencing agents [9–11].

3 Feature Selection

Feature selection - variable selection or attribute selection – is a technique or process of choosing a subset of main features, so based on a specific evaluation criterion, the feature will be reduced [12]. Feature selection was proved to be an efficient way in eliminating all the irrelevant and redundant attributes or features, raising the efficiency in the tasks related to learning, enhancing the performance of learning as in predictive thoroughness, and improving the clearness of the learned results. Feature selecting is a necessary process for applying a classification algorithm for high dimensional data set. Feature selection methods are categorized into three modes or types, "filter, wrapper and embedded" methods, based on how they combine or integrate the feature selection with the building and training of classification learning model.

3.1 Filter Methods

Filter methods evaluate the relevance or relationship of features by choosing only the valuable properties or characteristics of the data. In general, they apply the process of feature selection in two stages. In the first one, depending on calculating a score for features relies on scoring criteria, a degree or index measuring the relevance or dependence of feature will be calculated. After that in the second stage, features that have a low score are excluded or features not reach certain criterion will be eliminated, thereafter, the subset of residual features will be used as the input to the classification model [12].

Filter methods can be classified into univariate and multivariate. Univariate filter methods ignore feature dependencies which can lead to the selection of redundant features and worst classification performance compared to other feature selection techniques. On the other hand, multivariate filter methods model feature dependencies independent of the classifier. In addition to evaluating class relevance like univariate, they also calculate the dependency between each feature pair [13].

3.2 Wrapper Methods

While filter methods select the best features without relying on a classification model that means separate from the learning model, wrapper methods are using a subset of features and train a learning model using them. Depending on the inferences from the learning model, therefore, a model determines which feature will be added or removed from the subset [12].

3.3 Embedded Methods

Embedded methods react with the learning algorithms which are of a lower computational cost than the wrapper approach. It captures feature dependencies and in addition to considering the relations between an input feature and the output feature, it also searches for the features locally in order to allow better local discrimination. To decide the optimal subset for a known cardinality, it uses some independent criteria. Using the learning algorithm, the final optimal subset is selected from the optimal subsets across different cardinality. This approach has the advantage of including the interaction with the classification model while being much less computationally intensive than wrapper methods [13].

4 Discussion to Feature Selection Methods for Gene Expression Data Classification

Mahmoud and Maher [14] integrated (T-test & K-Nearest Neighbor KNN) for leukemia which recorded higher classification accuracy of 97.06% using 10 genes. For the Small Round Blue-Cell Tumor (SRBCT) dataset, they integrated (T-test & Support Vector Machine SVM) which reached 100% accuracy with 3 genes. Similarly, for Lymphoma dataset, they integrated (T-test & SVM) and reached 100% accuracy with only 2 genes. Finally, Lung dataset reached 98.65% accuracy using (T-test & SVM). Therefore, T-test and Class Separability (CS) ranking techniques played a vital role in selecting important genes and enhancing classification accuracy.

Lu et al. [15] applied "mutual information maximization (MIM) and adaptive genetic algorithm (AGA)" called MIMAGA-Selection algorithm which showed effective results to reduce the dimensionality of the original gene expression datasets and removes the redundancies of the data. The ReliefF and Extreme Learning Machine (ELM) showed the lowest accuracy results for all the datasets. In addition, Sequential Forward Selection (SFS) and ELM recorded low results but still higher than ReliefF and ELM.

Zhong et al. [16] investigated a distance-based feature selection method for two-group classification problem. In order to select the genetic markers, the Bhattacharyya distance is implemented to gauge the contrast in the levels of gene expression between groups. Also they compared the average number of the genes selected, the average number of misclassifications in testing set and average misclassification rate for the three methods Bhattacharyya distance with SVM (B/SVM), Supervised Weighted Kernel Clustering/SVM (SWKC/SVM), and SVM with Recursive Feature Elimination (SVM-RFE). For the Colon dataset the B/SVM recorded 90.5% classification accuracy with only the top 6 ranked genes out of the total 2000. Also, for the leukemia dataset, the B/SVM reached 96.9% classification accuracy with only the top 9 or 10 ranked genes out of the total 3571.

Hameed et al. [17] applied a combination of filter and wrapper that showed better results, when compared with applying classifiers directly on the original and filtered data. Least Absolute Shrinkage and Selection Operator (LASSO) showed better performance and results when comparing with the combination of filter and wrapper for high dimensional data. However, for low dimensional data, applying the wrapper method on the original datasets showed better performance when compared with other methods. However, the classification effectiveness of the low dimensional dataset, Yeast and E. coli still need to be enhanced.

Hameed et al. [18] presented a three-phase hybrid approach especially proposed for the features selection and classification of microarray data with high dimensionality problem. The method uses Pearson's Correlation Coefficient (PCC) in combination with Binary Particle Swarm Optimization (BPSO) or Genetic Algorithm (GA), along with the various classifiers, forming a "PCC-BPSO/GA-multi classifiers" approach. The PCC filter showed a noticeable enhancement in the accuracy and effectiveness of classification when it was combined with BPSO or GA. BPSO is working faster than GA, but BPSO has better performance than GA when it is combined with PCC feature selection.

Uma [19] applied SVM and KNN classifiers with/without feature selection method, also, the proposed Hybrid Genetic-Firefly Algorithm improved accuracy from 86.49% (without F.S methods) to 91.67%, but still need to be enhanced and tested on different cancer type's datasets.

Liu et al. [20] introduced an effective feature selection method, based on double Radial Basis Function (RBF-kernels) with weighted analysis which modified the kernel-based clustering method for gene selection (KBCGS) called DKBCGS (Double-kernel KBCGS). They applied their method on different datasets such as Diffuse Large B-cell Lymphoma (DLBCL), Colon, Lymphoma, Gastric cancer, and mixed tumors. To validate the effectiveness of DKBCGS method, they compared it with some commonly used filter-based feature ranking methods namely $\chi 2$ -Statistic, Maximum relevance and minimum redundancy (MRMR), Relief-F, Information Gain and Fisher Score. Their method showed superior performance in accuracy and run-time for both two-class datasets and multiclass datasets.

Khan et al. [21] proposed a method called "GClust" to select the most important genes on gene expression microarray. They applied their method in two stages: In the first stage, a minimum set of genes using a greedy approach was identified. The second stage includes dividing the remaining genes that are not selected in the minimum subset of genes in the first stage into a specific number of clusters. The k-means algorithm was used based on the Euclidean distance to find dissimilarity between genes. Random forest and SVM were applied on 7 gene expression datasets. Five feature selection methods were used: Proportional Overlapping Score analysis (POS), Masked Painter approach (MP), Wilcoxon Rank Sum technique (Wil-RS), Relative Simplicity (RS) method, Double Kernel-Based Clustering for Gene Selection (DKBCGS), which compared with the minimum subset of genes selected via the greedy approach.

Table 1 shows a comparison between the results of most recent studies in Microarray Gene expression data classification:

Table 1. Summary of the recent studies in Microarray Gene expression data classification

Author	Dataset	F.S method	Classifiers	Accuracy	Limitations
Mahmoud, and Maher [14]	Lymphoma Leukemia SRBCT Lung	t-Test and Class Separability (CS)	SVM KNN	97.06% for leukemia, 100% for Lymphoma, 100% for SRBCT, and 98.65 for lung	F.S methods showed good results to reduce dataset dimensionality, but the datasets which have the largest number of features such as Breast and Ovarian were not tested in this research
Lu et al. [15]	Leukemia Colon Prostate Lung Breast SRBCT	Relieff, SFS, MIM, and Integrate AGA and MIM: MIMAGA-Selection algorithm	BP SVM ELM RELM	Higher than 80% for all datasets.	The best accuracy results were obtained by MIMAGA-Selection and ELM, but Colon and Breast need to be enhanced. The accuracy rate for the colon 80%–89%
Zhong et al. [16]	Leukemia Colon	Bhattacharyya distance and SVM (B/SVM), SWKC/SVM and SVM-RFE	SVM	Colon: 90.5% Leukemia: 96.9%	The proposed method reduced dataset dimensionality, but Colon still needs to be improved. The datasets which have the largest number of features such as Breast and Ovarian were not tested in this research

(*continued*)

Table 1. (*continued*)

Author	Dataset	F.S method	Classifiers	Accuracy	Limitations
Hameed et al. [17]	Leukemia Colon Prostate Mice Protein Yeast E. coli	ReliefF, Wrapper subsetEval, and LASSO	Bayes Net, Naive Bayes, KNN SVM	57.9%–100%	This research achieved high accuracy rate with high dimensional datasets but the other datasets were not included. Also, low dimensional datasets such as Yeast and E.coli have low accuracy rate (52% to 57.9 for Yeast and 80% to 85.4 for E.coli)
Hameed et al. [18]	Brain Breast CNS Colon Leukemia Lung Lymphoma MLL Ovarian Prostate SRBCT	PCC BPSO GA Hybrid PCC-BPSO/GA-multi classifiers	Bayes Net KNN Naïve Bayes Random Forest SVM	90.72% to 100%	The research included most of the datasets, and five classifiers were applied which means having different views and comparisons. But they did not recommend feature selection or classification methods of each dataset. The results could be improved if they use other classification methods
Uma [19]	Leukemia (AML-ALL)	Firefly Algorithm, Genetic Algorithm, Hybrid Genetic-Firefly Algorithm	SVM KNN	91.67%	91.67% still need to be enhanced. Also, the other datasets were not tested

(*continued*)

Table 1. (*continued*)

Author	Dataset	F.S method	Classifiers	Accuracy	Limitations
Liu et al. [20]	DLBCL, Colon, Lymphoma, Gastric cancer, and mixed tumors	Double RBF-kernels with weighted analysis (DKBCGS)	SVM KNN	DKBCGS showed high accuracy results which the lowest result was 91.51%	This research has shown satisfactory results and applied different performance measures, such as accuracy, the true positive rate (TPR) and the true negative rate (TNR). But it will be useful to apply this method on the other datasets such as Breast, CNS, and Ovarian
Khan et al. [21]	NKI, Colon, Breast, Leukemia, OVA Ovary, GSE4045, and AP Breast Colon	GClust, POS, MP, Wilcoxon (Wil-RS), RS, DKBCGS, and Min	RF SVM	GClust showed high accuracy results but with NKI dataset showed 72.50% using SVM	The GClust has achieved the highest values of classification accuracy most of the times for the datasets. But this method was applied on only binary classification datasets, also some results such as NIK dataset need to be enhanced

A summary of reviewing of the previous studies in Microarray Gene expression data classification:

- There is no common scheme or a procedure to evaluate which method or algorithm is best for this type of cancer dataset.
- A Microarray is high dimensional data, therefore selecting a feature selection method or algorithm is playing a vital role in prediction accuracy. That means the power of classifiers algorithm is not enough to get high prediction results.

- The perfection of SVMs method outperforms other several classifiers. Most studies use SVM as classifier method with different feature selection methods.
- Most studies applied their proposed methods on a few numbers of datasets, therefore, it is needed for applying the proposed methods on different cancer datasets to test and evaluate methods.

5 Conclusion

Implementing DNA microarray technology in the studies has shown huge success in both diagnosing and explaining the pathological mechanism. Cancer diseases are caused by some changes in the genes controlling the way our cells function, especially the way they grow and divide. Thereby, if the diagnosis of cancer is determined in a late stage of the disease, then it is difficult to control and eventually causes death in most cases. Gene expression microarray holds data that could help to get an accurate and fast diagnose, however, because of a high dimensionality problem – a large number of genes with few numbers of instances - also irrelevant and redundant features the challenges of analyzing microarray will be increased. Applying feature selection methods such as filter, wrapper, and embedded are a necessary step that affects the effectiveness of classification. Also, applying filter methods such as PCC or Bhattacharyya distance help to reduce features before applying wrapper or embedded methods such as GA. Therefore applying wrapper or embedded methods without applying filter methods in advance could affect negatively by decreasing classification performance and accuracy. Also, there are different datasets of cancer types with no common scheme to test and evaluate which method or algorithm is best for this type of cancer dataset, consequently, there is no a recommended method or combination of methods for each dataset. Future research can study the application of combined or hybrid methods of feature selection for gene expression data classification, which could obtain notable results compared to the existing methods.

References

1. Miko, I., LeJeune, L.: Essentials of Genetics. NPG Education, Cambridge (2009)
2. Khurana, S., Singh, M.: Biotechnology: Principles and Process, 12th edn. Studium Press LLC, Houston (2015)
3. Difference between DNA and Genes. http://www.differencebetween.net/science/difference-between-dna-and-genes. Accessed 20 May 2019
4. Matilainen, M.: Identification and characterization of target genes of the nuclear receptors VDR and PPARs. Doctoral dissertation, University of Kuopio, Finland (2007)
5. Gene editing: a molecular miracle. https://www.thedailystar.net/news/health/gene-editing-molecular-miracle-1627630. Accessed 20 May 2019
6. Babu, M., Sarkar, K.: A comparative study of gene selection methods for cancer classification using microarray data. In: Second International Conference on Research in Computational Intelligence and Communication Networks (ICRCICN), Kolkata, pp. 204–211. IEEE (2016). https://doi.org/10.1109/ICRCICN.2016.7813657

7. Srivastava, S., Joshi, N., Gaur, M.: A review paper on feature selection methodologies and their applications. Int. J. Comput. Sci. Netw. Secur. (IJCSNS) **14**(5), 78–81 (2014)
8. Plunkett, J.: Plunkett's Biotech and Genetics Industry Almanac. Plunkett Research Ltd., Houston (2006)
9. Bustin, S., Benes, V., Garson, J., Hellemans, J., Huggett, J., Kubista, M., Vandesompele, J.: The MIQE guidelines: minimum information for publication of quantitative real-time PCR experiments. Clin. Chem. **55**(4), 611–622 (2009)
10. Lefever, S., Hellemans, J., Pattyn, F., Przybylski, D., Taylor, C., Geurts, R.: RDML: structured language and reporting guidelines for real-time quantitative PCR data. Nucleic Acids Res. **37**(7), 2065–2069 (2009)
11. Vandesompele, J., De Preter, K., Pattyn, F., Poppe, B., Van Roy, N., De Paepe, A., Speleman, F.: Accurate normalization of real-time quantitative RT-PCR data by geometric averaging of multiple internal control genes. Genome Biol. **3**(7), 1–12 (2002)
12. Zhong, W.: Feature selection for cancer classification using microarray gene expression data. Doctoral dissertation, University of Calgary, Canada (2014)
13. Mwadulo, M.: A review on feature selection methods for classification tasks. Int. J. Comput. Appl. Technol. Res. **5**(6), 395–402 (2016)
14. Mahmoud, A., Maher, B.: A hybrid reduction approach for enhancing cancer classification of microarray data. Int. J. Adv. Res. Artif. Intell. (IJARAI) **3**(10), 1–10 (2014)
15. Lu, H., Chen, J., Yan, K., Jin, Q., Xue, Y., Gao, Z.: A hybrid feature selection algorithm for gene expression data classification. Neurocomputing **256**(2017), 56–62 (2017)
16. Zhong, W., Lu, X., Wu, J.: Feature selection for cancer classification using microarray gene expression data. Biostat. 03 Biom. Open Acc. J. **1**(2), 1–7 (2017)
17. Hameed, S., Petinrina, O., Hashi, A., Saeed, F.: Filter-wrapper combination and embedded feature selection for gene expression data. Int. J. Adv. Soft Comput. Appl. **10**(1), 91–105 (2018)
18. Hameed, S., Muhammad, F., Hassan, R., Saeed, F.: Gene selection and classification in microarray datasets using a hybrid approach of PCC-BPSO/GA with multi classifiers. J. Comput. Sci. (JCS) **14**(6), 868–880 (2018)
19. Uma, S.: A hybridization of genetic – firefly algorithm technique for feature selection in micro array gene data. Int. J. Innov. Adv. Comput. Sci. IJIACS **7**(4), 70–81 (2018)
20. Liu, S., Xu, C., Zhang, Y., Liu, J., Yu, B., Liu, X., Dehmer, M.: Feature selection of gene expression data for cancer classification using double RBF-kernels. BMC Bioinform. **19**(1), 396 (2018)
21. Khan, Z., Naeem, M., Khalil, U., Khan, D., Aldahmani, S., Hamraz, M.: Feature selection for binary classification within functional genomics experiments via interquartile range and clustering. IEEE Access **7**(1), 78159–78169 (2019)

Data Governance Support for Business Intelligence in Higher Education: A Systematic Literature Review

Soliudeen Muhammed Jamiu[1](✉), Norris Syed Abdullah[2](✉), Suraya Miskon[2](✉), and Nazmona Mat Ali[2](✉)

[1] School of Computing, Universiti Teknologi Malaysia, 81310 Johor Bahru, Johor, Malaysia
jamiumuhammed@yahoo.com

[2] Azman Hashim International Business School, Universiti Teknologi Malaysia, 81310 Johor Bahru, Johor, Malaysia
{norris, suraya, nazmona}@utm.my

Abstract. Business Intelligence (BI) is important for achieving effective decision-making in higher education. This study, however, advocated the need to support BI with data governance in higher education. The systematic literature review was conducted using a qualitative approach. The study cover 2005–2019. A total of 483 papers were retrieved and after exclusion and inclusion criteria, two hundred and three were removed due to lack of relevance. Some of the removed papers were those written in other languages other than the English language. Finally, one hundred and eighty were analyzed for this study. Some of those sources used for the study include Scopus, Springer, science direct, IEEE explore, Web of science. The results were arranged under word cloud, word frequency, Year-source by attribute, matrix coding by methodology, business intelligence, and its benefits, critical success factor, data governance, and its benefits, an overview of higher education and need to support business intelligence with data governance. The study provides information to higher education business intelligence experts on the need to support their BI with data governance.

Keywords: Data governance · Data quality · Business Intelligence · Higher education

1 Introduction

Higher education plays a vital role in the creation of various innovations and human capital development in any given country. In Malaysia, higher education is referred to as institutions that award diploma, degrees, and certificates [1]. In this study, we refer to universities that award degrees only. Universities generate data in their daily routines. Those data are stored using information and communication technology. The existence of ICT leads to the storage of data in multiple media, which made the

F. Saeed et al. (Eds.): IRICT 2019, AISC 1073, pp. 35–44, 2020.
https://doi.org/10.1007/978-3-030-33582-3_4

retrieval of such data difficult at the time of need. This portrayed scenario portends serious problems for higher education since there was the need to use the existing data to take vital decisions but this was made impossible due to lack of data quality. Data quality is the ability of information systems to produce correct, complete, dependable and consistent data, in conformity with the already laid down rules and regulation, based on the business insight [2]. The foregoing need a data warehouse that could store and produce data with the aforementioned qualities and the tendencies to provide access to those data without wasting time [3]. Scholttz et al. [4] Identified business intelligence as that tool with the inherent tendency and the capacity to provide the higher education with the desired outcome of solving the problem of lack of data quality needed for taking reliable decisions, which could help in promoting higher education business [5]. Business intelligence is a data visualization tool, which gathers, process and produces data that could be used to evaluate the performance of higher education. It plays vital roles in the management of higher education data. Despite the huge purpose business intelligence can serve, there have been reports of its failure. Management of higher education still finds it difficult to get dependable data for their decision-making process. In view, this, Combita et al. [6] observe that higher education is still confronted with the problem of lack of access to accurate data. Similarly, [7] stated that many institutions of higher learning has implemented business intelligence but did not achieve the expected result. In addition to these, [5] observes that though business intelligence was implemented with the aim of achieving a single source of truth it was discovered that low-level staff can change the truth. Ong et al. [8] opine that data governance can serve as the solution to effective and efficient business intelligence. This is because it gives room for the management of data as a vital asset of the organization. In order word, data governance can be regarded as a critical success factor for business intelligence. It defines the data, who will use it and for what purpose. It defines, data related rules, the users of the data, the steward, etc. Data governance will open so many benefits for higher education. Those benefits include: saving cost, assigning responsibility and making people accountable to their duties, laid down a written method that can be repeatedly followed over time. It establishes a good structure line of communication, commitment among staff both at a managerial and strategic level as well as overall management of data in higher education. This study review related study that revolves around the above scenario. Some of the objectives of these studies include (a) what are the frequently used words in the studies. (b) What the prominent text used in the existing study. (c) What is the percentage of publications published during the years under this review. (d) What are those methodologies used in those studies. (e) What are the benefits of Business intelligence in higher education. (f) What are the critical success factors. (g) What is data Governance and how can it be used to support business intelligence?

Next section presents related works in Sect. 2, Sect. 3 methodology used for the review and review of related literature on data governance for business intelligence in higher education. Section 3.1 analysis, Sect. 4 conclusions.

2 Related Works

Like every other commercial organization, higher education generates data on daily bases but could not make use of the data because they are scattered in multiple storage facilities [4]. The overlapping benefits of those data cannot be easily seen. Some of the data were redundant because they are forgotten [5]. Technology such as business intelligence was deemed necessary because of its potential of being able to extract data from the large data generated in higher education [6]. It has the tendencies to gather data, analyses, transform and generate reports useful for the decision making the process. Those reports reflect the objectives, opportunities, and the comparative advantages [12], Anjariny [7] stated that Business intelligence is a mechanism for intelligence exploration of data, aggregation, and integration of data from multiple sources. It treats data as a valuable asset and transforms them from quantity to quality [8]. Business intelligence is a tool that makes higher education business run smoothly, successfully and profitably [9]. It creates more access and understanding of data. Guesswork is eliminated, it improves communication, and allow quick response to challenges emanating from the financial situation, customers interest, supply chain and improve the performance of higher education through the usage of reliable information to make decisions. Business intelligence speed up the rate at which decisions are taken [10]. The use of this technology can help in detecting the comparative advantage of higher education and know loyal customers. It can also help in alerting the fraudulent activities of staff. Mohamadina et al. [10] observes that some of the public organizations who used business intelligence still faced data quality problem due to the fact that attention was not given to those who will ensure the organizations have data quality and in some cases some do not know what the data quality was meant for. Based on this, Jusoh et al. [16], identified the importance of data governance as the solution to the problem of data quality. Data governance could help in ensuring the security of data; accountability among staff, reliable planning, preventing higher education from running against the law, generating more money and prevention of loses. Data governance aids in good and reliable decisions and taking into consideration the information needs of the stakeholders, remove frictions, provision of standard, transparency and common approach to be followed by all staff. According to [17] a good data governance program should draw a line of communication among the stakeholders, provision of metadata, standard, procedures, and infrastructure for the program. Tutunea and Rus [18] are of the opinion that commitment from senior management is very crucial to the success of data governance. Educating the stakeholders on the importance of data warehouse, needed resources and the period is very germane. Hočevar and Jaklič [19] noted the critical success factors for data governance as standard, accountability, managing data complexity, cross-level participation, metric explanation and definition, control and monitoring of all stakeholders, aligning technology, process, and business objectives.

3 Methodology

This study uses a qualitative approach to conduct a systematic literature review. In order to conduct the review, the first step was taken by formulating the keywords. The identified keywords are "Business intelligence AND higher education, Business intelligence AND university, business intelligence AND college, Business intelligent AND higher education, Business intelligent and university, Business intelligent and polytechnic, Business intelligent AND college, Business analytic AND higher education, business analytic AND university, Business analytic AND polytechnic, Business analytic AND college, data governance OR information governance, data governance for business intelligence in higher education and information governance for business intelligence in higher education". The search was not limited to Information systems. This is because data governance is also known as information governance by some people. Some of the Sources identified and used for the gathering of data include Scopus, Science Direct, IEEE explore, Web of science and springer. This study covers between 2005–2019. The result of these decisions yielded 483 dataset. 119 were removed based on lack of relevancy. Consequently, 364 datasets were left behind. Exclusion of 84 papers was also made based on papers written in other languages different from English and those that have abstract only. Only 180 were left.

3.1 Analysis

Analysis of the data was done using Nvivo 11. All the datasets were imported into Nvivo. Headings were made under nodes. Both parent and child nodes were created. The parent headings include data problem in higher education, the definition of business intelligence and definition data governance. The child nodes consist of benefits of business intelligence, challenges of business intelligence, critical success factors, benefits of data governance and implications of not having data governance in higher education. Consequently, nodes were compared against nodes; analyses of the data set based on year of publications, the methodology used in the data set were also compared and represented graphically. The results of the analysis are presented below: The results were arranged under word cloud, word frequency, Year-source by attribute, matrix coding by methodology business intelligence, and its benefits, critical success factor, data governance, and its benefits, overview of higher education and need to support business intelligence with data governance. The arrangement was made under the aforementioned factors because they form the bases of what this study intends to show in the findings.

4 Results and Findings

Word Frequency. Table 1 shows word frequency found from the searching. They are represented by both numbers and percentages. Data appeared (23331) with 2.04%, information (11376) with 0.99%, business (10855) with 0.95%, management (5810) with 0.51%, intelligence (5112) with 0.45 and governance (4007) with 0.35%. The outcome of these, however, is that the keywords selected for the study are valid.

Word Cloud. The second analysis was made under text word search Fig. 1. This was meant to authenticate the search terms used. The text search revealed most frequently used text with a minimum of four-letter words. The most prominent text includes data, information, business, management, governance, intelligence. Based on this a word cloud was generated through query wizard.

Table 1. Word frequency

Word	Count	Weighted percentage
Data	23331	2.04
Information	11376	0.99
Business	10855	0.95
Management	5810	0.51
Intelligence	5112	0.45
Governance	4007	0.35

Fig. 1. Word cloud

Year-Source by Attributes Value. Figure 2 shows the distributions of publications by year. 2005 has the highest number of publication with 19.2%. This is followed by 2012, which has 18.1%. 2014 has 15%, in 2011 is 9.6, 2010 and 2015 has the same number of publications with 7.5% respectively. Similarly, 2007 and 2013 recorded 4.3% each. 2008, 2009 and 2018 have 3.2% in common. 2006 and 2017-recorded 2.1% and 2019 has just 1.1% of publications.

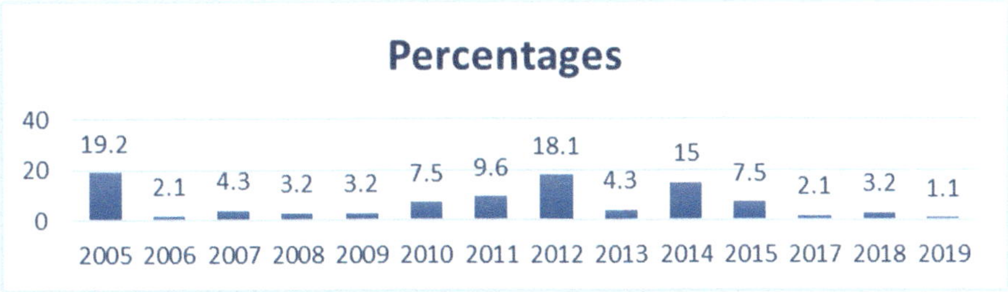

Fig. 2. Distribution of relevant publications by year

Matrix Coding by Methodology. Table 2 depicts the compared nodes by methodology. Four types of methodologies were considered. They are qualitative, quantitative, mix-method and design science. The table shows that a benefit of data governance node has 15 qualitative methods and definition of data governance used 12 qualitative methods and 22 quantitative. A benefit of data governance node has 15 qualitative and 10 quantitative. Data problem in higher education node has 2 mix-methods and benefits of business intelligence node have 2 design science methods. The summary is that most of the paper used both qualitative and quantitative methodology. Table 2 below shows the summaryoud. The second analysis was made under text word search Fig. 2. This was meant to authenticate the search terms used.

Table 2. Matrix coding by methodology

	Qualitative	Quantitative	Mixed	Design science
Data problem in higher education	0	0	2	0
Benefits of business intelligence	0	0	0	2
Benefits of data governance	15	10	0	0
Definition of data governance	12	22	0	0

Overview of Higher Education Context. Higher education employed the use of various multiple media ushered in by the emergence of ICT to store their data. Cheong and Chang [24] observed that this opportunity afforded higher education to store their data in multiple medium but accessing this data at the time of planning become a serious problem. The more the data in those sources, the more the problem of retrieval for planning. Sometimes such data are as good as not being in existence. This is because the frequent data are sighted, the more they are remembered and used. Devlin [25] identified that the aftermath of this ICT is what could be tag information explosion. Higher education has many data in their disposal but using such data became another problem. Cardoso [15] stated that because of the medium of storage for the data, data seem to be too much, but instead of using the available data for decision making, higher education is still information starved [19]. Hawking and Sellitto [20] stated that because of this multiple media of storage some problems such as data validity, data duplication, lack of control over data and difficulty in making such data available at the time of need, therefore, higher education need a better way of managing these data. A medium that could provide intelligent data, which can also be, used for decision making and planning hence, the need for business intelligence. Business intelligence has that capability to integrate data from various sources and make them available at the point of need. It is synonymous with having data, which will be useful for planning and not the one that can be used for record only.

Business Intelligence. Business intelligence is that technology with a great potential of extracting valuable information from mass data that exist within an organization [6]. It is an information system that gathered, process and transforms data from various sources into useful information for the decision-making process and for the purpose of efficiency [15]. Business and intelligent system is a tool for improving the decision process with credible based information. Business intelligence can also be described as a decision support system while some saw it as Executive information systems [13]. Business intelligence has the capacity to gather, analyze, and transform and generate reports that could aid the strategic and managerial decision-making process. The reports generated are based on the objectives, opportunities, and the position of an organization. However, business intelligence is a collection of tools, technologies, applications, and processes used to gather data, process it and generate meaningful reports [7, 11]. Based on this, Business intelligence can be vary based on its purpose, complexity, and functionality. Hawkings and Sellitto [20] itemize some of the business intelligence tools to include enterprise reporting tools, ad-hoc query tool, statistical analytical tool, OLAP tool, spatial-OLAP analysis tool, data mining tool, text mining tool, dashboard, scorecard, predictive analytic/advance analytic. The business intelligence architecture also includes people, metadata and master data, data source, data integration layer, and data access layer. Morien, Mungree and Rudra [21] assert that business intelligence architecture includes data source, data warehouse, and tools for data access and analyses. Business intelligence has many benefits, which are discussed below.

Benefits of Business Intelligence. Some of the benefits of business intelligence include the presentation of business report in a fast, simple and efficient manner. It also provides flexibility, accurate reporting, improved decision-making process. It provides customer satisfaction, improves revenue generation and safe cost [15]. The implementation of business intelligence would create autonomy and flexibilities for higher education. After the implementations of the project, higher education will be able to rely on their data. They would be able to get reports without wasting time. It provides a varying form of data that could give them insight into the hidden truth about the situation at hand. One of the major reasons for the implementation of business intelligence is the cost benefits it has. Business intelligence saves the time of the higher education through the integration of all needed data and after processing them produce the required reports in a varying format which can later be used for planning and decision-making process. The prompt reports, which has the decision-making and planning value, which will lead to operational efficiency and give competitive advantages to higher education. The so call reports can be used to analyze the customers' needs. The proper understanding of the market and lead to an increase in revenue generation. Business intelligence, if used correctly, can help in the detection of fraud. Lastly, because, business intelligence pulls data from various sources. Based on all the integrated data from various sources, intelligence reports are generated for the purpose of decision-making and planning. The success of BI is dependent on some critical success factors, which are, identify below.

Critical Success Factors for Business Intelligence. The success of business intelligence depends on some major factors. Studies have revealed that if higher education wants to be successful with their business intelligence project, there must be absolute support from the management of higher education. The support can come in the form of training, provision of funds; a clear channel of communication must be in place. However, the will must be apparent instead of paying leaps service to the project. Other factors such as a show of commitment of the management of higher education, technical skill required for the immediate take-off and the maintenance of the project as stated by [22]. Norfal and Yusof [17] Identified other factors in addition to those already stated such as alignment of business intelligence strategy with business objectives, a good data governance program. Norfal and Yusof [17] Itemized more factors, which include user's participation, change management, data quality, and methodology, the scope of the project and sources systems.

Data Governance. Data governance has the potential to increase organizational efficiency and effectiveness through the elimination of duplicate data and provisions of quality control. It also provides the organization with the opportunity to respond to the market. The comparative advantage is made possible through the appropriate creation and use of data. The decision-makers, on the other hand, are more confident with the data they are using. It also enables members of the same organization to collaborate [16]. Establishing data governance is a means of providing risk management. This can be achieved in three ways: (a) provision of privacy and adherence to security standard (b) operation based on best practices and (c) safeguarding the customers' interest [23].

Data Governance Support for Business Intelligence. Business intelligence exists in order to meet business needs. However, to succeed in this task, the knowledge of information needs of the stakeholders is paramount. The required information needs must be prioritized and if there are any conflicts of interest, there must be an established way to resolve them. The authorized persons who will speak for each of the stakeholders must be well spelled out so that when resolving the conflicts and providing resolutions, those to meet will not be a problem. In doing these, also, there must be a good spell out policies, standard and rules to follow when building, maintaining and enhancing business intelligence [21]. In order to achieve this, the data governance process becomes handy because it helps in defining rules and regulations, policies and procedures to be used for the effective data management for business intelligence. Data governance will provide a module Operandi for higher education collaborations at all levels when managing their data and provides the opportunities to align data problem with the institutional objectives [22]. Lately, the relevance of data governance to business intelligence has increased due to a large percentage of business intelligence implementation's failure arising from poor data quality. Business value from big data and organizational standing in the business environment can only be analyzed using business intelligence technology with the support of a good data governance program [23].

The importance of data governance is the ability for driving business growth initiatives, it also associates data with applications and business line. It ensures data are easier to find and build trust in the available data by ensuring that they are of good value for decision-making. Other factors that make data governance to be more relevant is the need for a regulatory and legal framework and the corporate governance initiative. However, data governance can help higher education in defining their business problem, identify executive sponsors, classify business terms, determine critical elements of business, establish policies, rules and give room to allowable values. Integration of glossary and metadata as well as management of stewardship process and workflow are facilitated by data governance [24]. Appropriate data governance guarantees data quality. It ensures data integrity, usability, and security in higher education. The stakeholders are made accountable in a DG program. It also provides audit tray of those that have access to the data [25]. Data governance provides access to data based on job descriptions [27]. Rule and regulations that state in a clear term the appropriate structure for data access and control [26]. Similarly, rules relating to data quality, data stewards and data profiling are also essential. The rules on both profiling and standardization are synchronized and enforced to ensure data conform with the laid down rules at the point of entry [19, 27].

5 Conclusion

This study was able to investigate the importance of business intelligence and the needs to support it with data governance. We, however, advocated the need for higher education to support their business intelligence with data governance. We present that this will help higher education to have reliable information, which they can be turned to during decision-making. Although, business intelligence has the capacity to provide a varying form of intelligent information which can provide business insight of the higher education there have been various arguments that the case of business intelligence is garbage in garbage out. Therefore, higher education must make a vital effort at ensuring that they have clean data imputed into the BI so that they can have reliable information as output. Based on this, therefore, this study is a review paper; it is assumed that it has focused the attention on the need for the higher education to put data governance as a priority during business intelligence implementations. Future research should focus more on the empirical investigation on the importance of data governance to business intelligence in higher education.

References

1. Yilmaz, Y.: Higher Education Institutions in Thailand and Malaysia - Can They Deliver? pp. 1–54 (2010)
2. Muntean, M., Bologa, A.R., Bologa, R., Florea, A.: Business intelligence systems in support of university strategy. In: Proceedings of WSEAS/IASME International Conference Education Technology, pp. 118–123 (2011)
3. Kabakchieva, D.: Business intelligence systems for analyzing university students data. Cybern. Inf. Technol. **15**, 104–115 (2015)
4. Scholtz, B., Calitz, A., Haupt, R.: A business intelligence framework for sustainability information management in higher education. Int. J. Sustain. High. Educ. **19**, 266–290 (2018)
5. Riggins, F.J., Klamm, B.K.: Data governance case at KrauseMcMahon LLP in an era of self-service BI and Big Data. J. Account. Educ. **38**, 23–36 (2017)
6. Combita Niño, H.A., Cómbita Niño, J.P., Morales Ortega, R.: Business intelligence governance framework in a university: Universidad de la Costa case study. Int. J. Inf. Manag. 1 (2018). https://doi.org/10.1016/j.ijinfomgt.2018.11.012
7. Anjariny, A.H., Zeki, A.M., Hussin, H.: Assessing organizations readiness toward business intelligence systems: a proposed hypothesized model. In: Proceedings of 2012 International Conference on Advanced Computer Science Applications and Technologies (ACSAT), pp. 213–218 (2013). https://doi.org/10.1109/acsat.2012.57
8. Ong, I., Siew, P., Wong, S.: A five-layered business intelligence architecture. Commun. IBIMA **2011**, 1–11 (2011)
9. Clavier, P.R., Lotriet, H.H., Van Loggerenberg, J.J.: Towards a 'BI value coin': applying service research to address business intelligence challenges. In: Proceedings of Annual Hawaii International Conference on System Science, pp. 1324–1333 (2014). https://doi.org/10.1109/hicss.2014.170
10. Mohamadina, A.A., Ghazali, M.R.B., Ibrahim, M.R.B., Harbawi, M.A.: Business intelligence: concepts, issues and current systems. In: 2012 International Conference on Advanced Computer Science Application Technology, pp. 234–237 (2012). https://doi.org/10.1109/acsat.2012.94

11. Wieder, B., Ossimitz, M.: The impact of business intelligence on the quality of decision making – a mediation model. Procedia Comput. Sci. **64**, 1163–1171 (2015)
12. Yeoh, W., Koronios, A.: Critical success factors for business intelligence systems. J. Comput. Inf. Syst. **50**, 23–32 (2010)
13. Guster, D., Brown, C.G.: The application of business intelligence to higher education: technical and managerial perspectives. J. Inf. Technol. Manag. **XXIII**, 42–62 (2012)
14. Otto, B., Weber, K.: Data Governance. Daten- und Informationsqualität 269–286 (2015). https://doi.org/10.1007/978-3-658-09214-6_16
15. Cardoso, E.: The current status of business intelligence initiatives in European Higher Education Institutions, pp. 1–2 (2014)
16. Jusoh, J.A., Endot, N., Hamid, N.A., Bongsu, R.H.R., Muda, R.: Conceptual framework of business intelligence analysis in academic environment using BIRT. In: International Conference on Informatics Application (ICIA 2012), Social Digital Information Wireless Communication, pp. 390–396 (2012)
17. Nofal, M.I., Yusof, Z.M.: Integration of business intelligence and enterprise resource planning within organizations. Procedia Technol. **11**, 658–665 (2013)
18. Tutunea, M.F., Rus, R.V.: Business intelligence solutions for SME's. Procedia Econ. Financ. **3**, 865–870 (2012)
19. Hočevar, B., Jaklič, J.: Assessing benefits of business intelligence systems – a case study. Management **15**, 87–119 (2010)
20. Hawking, P., Sellitto, C.: Critical success factors of business intelligence (BI) in an ERP systems environment. In: Proceedings of Conference on Research Practise Issues Enterprise Information System (2010)
21. Morien, D., Mungree, D., Rudra, A.: A framework for understanding the critical success factors of enterprise business intelligence implementation. In: Proceedings of Nineteenth American Conference on Information Systems, pp. 1–9 (2013)
22. Winterberry Group: The New Rules of the Road: Marketing Data Governance in the Era of "Big Data", p. 27 (2013)
23. Thomas, G.: How to use the DGI data governance framework to configure your program. Data Governance Institute, p. 17 (2009)
24. Cheong, L.K., Chang, V.: The need for data governance: a case study. In: 18th Australasian Conference on Information Systems, pp. 999–1008 (2007)
25. Devlin, B.: BI data governance the secret of successful business decision making, pp. 1–11 (2017)
26. Oracle Data Quality Solutions: Data Governance with Oracle (2015)
27. Guster, D., Brown, C.: The application of business intelligence to higher education: technical and managerial perspectives. J. Inf. Technol. Manag. **23**, 42–62 (2012)

Big Data Analytics Adoption Model for Malaysian SMEs

Eu Lay Tien, Nazmona Mat Ali(✉), Suraya Miskon(✉), Norasnita Ahmad(✉), and Norris Syed Abdullah(✉)

Universiti Teknologi Malaysia, Johor Bahru, Malaysia
laytien27@gmail.com, {nazmona, suraya, norasnita, norris}@utm.my

Abstract. Big Data Analytics (BDA) was utilized to analyze and examine big data sets. This system is able to analyze enormous data sets which exist in different formats and to extract useful information within the data which may be used to improve business decision-making, predict sales, enhance customer relationships, and ultimately lead to generating increased revenues and profits. Multinational and large companies are starting to adopt BDA to acquire the benefits and advantages from this technology. However, the rate of adoption of BDA by Small and Medium Enterprises (SMEs) is low. There is a need and desire that SMEs should start to adopt BDA in order to stay one step ahead of their rivals, and at the same time, to remain competitive in the market. Hence, this study aims to identify the factors influencing the adoption of BDA in Malaysian SMEs and propose a BDA adoption model for Malaysian SMEs.

Keywords: Big Data Analytics · Adoption model · Malaysian SMEs

1 Introduction

Along with globalization and the digital revolution, the lives of people all over the world have been changed dramatically. Now, the term "data" is vital in people's daily lives. Data is generated everywhere and at any time with any device or even when simply browsing through a social application. Every day, this enormous volume of valuable data grows exponentially and is known as Big Data (BD) [1]. According to Coleman et al. [2], approximately 2.5×10^6 terabytes of data are generated in every corner of the world on a daily basis and it is thought that the data volume is doubling every 3.5 years. These valued assets are very much useful if people can analyze the data set and gain insight from the data. Big Data Analytics (BDA) was introduced to analyze these complex data sets. In general, BDA is the utilization of superior analytic techniques to examine enormous, diverse data sets in the format of structured, semi-structured and unstructured data, from different sources, and in different sizes from terabytes to zettabytes [3]. BDA's worth is realized only when leveraged to drive decision making. Consequently, firms and companies demand efficient as well as pragmatic processes to extract information from this diverse data and enable such evidence-based decision making [4].

F. Saeed et al. (Eds.): IRICT 2019, AISC 1073, pp. 45–53, 2020.
https://doi.org/10.1007/978-3-030-33582-3_5

To adapt to the change of the business market nowadays, international organizations from different sectors such as pharmaceutical, clothing, automotive, retail, healthcare, financial services are adopting BDA system to increase their performance, productivity and efficiency [5]. This is also supported by Columbus [6] who states that BDA plays a vital role in technology firms with data warehouse optimization in sectors such as financial services, healthcare, and customer or social analysis. Approximately 53% of companies adopted BDA in 2017 which is a great increase compared to 17% of companies in 2015 [6]. Two of the earliest sectors and industries to implement and adopt this beneficial system were hypermarkets and banks [2]. Supermarkets used the data to determine customer preferences and to offer personalized sales or to promote products based on different seasons and festivals [2]. The banking sector has been utilizing the information to measure the trustworthiness and predict the financial condition of clients [2]. There are some basic features of the BDA system which help to gather and analyze large and different types of data sets in order to discover schematic plan such as data processing, predictive applications, analytics, reporting features, security features and technologies support [7].

There are many large international companies using BDA, for example Amazon, Netflix, American Express and Starbucks [8], but the adoption of BDA in SMEs is comparatively low in all over the world [2]. Prior study also found out that there is a need for that SMEs to consider adopting BDA in their organization in order to help them predict their target customer's preferences and requirements [1]. The poor adoption of BDA in SMEs will cause these companies to fall behind compared to the developments in large organizations [2].

Implementing BDA would benefit an organization from all perspectives and aspects, but it is a quite massive and extensive challenge to organizations especially SMEs [2, 9]. Some of the problems and challenges were identified from prior study when SMEs intend to adopt and use BDA. The first problem encountered was the lack of IT infrastructure [10, 11]. According to Wang et al. [11], a well-developed and advanced IT infrastructure is essential in order to utilize BDA. However, most of the IT infrastructure of SMEs consists of basic and fundamental functions which are not capable of processing BDA [10]. Moreover, SMEs are concerned about the data security issues of BDA [2, 12]. This is due to the fact that there are no standard policies and regulations monitoring the usage and ethics of BDA [12]. Data security is more serious for SMEs compare to large enterprises as SMEs usually use outdated and unsupported security software and Database Management Systems which make them are highly exposed to security breaches and they become more vulnerable [2].

In addition, financial status or budgets are the main challenge and obstacle when SMEs adopt and implement BDA [2, 9, 10, 12, 13]. In fact, in general, SMEs will be very cautious and careful about new investment because they have limited financial resources [2]. The sophisticated and advanced functions of BDA make implementing BDA costly, especially for those companies located in developing countries with a limited IT budget or financial support [10]. Additionally, unskilled and inexperienced IT employees or a shortage of permanent BDA specialists are barriers against take up of BDA in SMEs [2, 11, 12]. According to Coleman et al. [2], most SMEs have few or no specialists to approach the taking up of BDA and professionals such as data scientists, programmers or analysts play an important role when implementing BDA in an organization [12]. Due to the shortages and inexperience of employees, it is difficult for SMEs to deal with technical issues when using BDA [11]. Besides that, competition is one of the concern of SMEs as competition is cited as the reason to adopt BDA [12]. The strong competition in the market will force SMEs to adopt BDA in order to enhance their organizational operation and performance [12]. These pitfalls and challenges may lead to the failure of implementing BDA in organization.

There are some research and papers which have studied the adoption model of BDA [11, 14–16], but these studies were focused on larger organizations and firms in other countries such as India, South Africa and Portugal. For instances, Bhattacharyya's [14] studies focused on Indian firms which applied a qualitative approach and suggested that future research could examine BDA adoption in different sectors as well as different countries. Also, a solid and appropriate model of BDA adoption for SMEs is a trend yet to be investigated and studied [14]. Cruz-Jesus et al. [15] addressed the factors that affect the adoption of business analytics and intelligence (BAI) among firms in Portugal. Raguseo [16] mainly focused on probing the extent of implementation of BD technologies in French companies, and the BD software employed by them. Raguseo [16] proposed that future research studies the adoption of BD technologies and software in other contexts, in order to examine the similar and dissimilar factors between different countries. Generally, in Malaysia, there are few papers concerning applied theory or framework regarding BDA, and they are focused on the public sector and its readiness for BD [17, 18]. It can be concluded that none of the papers have studied a BDA adoption model for SMEs in the current Malaysian. Thus, this paper will study and examine all the available research and established adoption models then design and validate the determinants for BDA adoption in Malaysian SMEs to fulfil this research gap.

This paper is organized into 4 sections. Section 1 is the introduction of this paper, followed by Sect. 2 which explains the adoption model widely used by scholars in adopting BDA. Section 3 discusses the proposed BDA adoption model for Malaysian SMEs in detail and Sect. 4 contains the conclusion and future work is explained.

2 Theoretical Background

The primary goal of BDA is to elicit the useful information inside huge and complex data which can support proper decision making as well as uncover hidden patterns. The use of BDA can help an organization to gain competitive advantages as it would benefit the organization in facilitating their business operation regardless of whether it is a large enterprise or small company. It is generally believed that BDA is the new trend for transformation, competitiveness, and efficiency [19]. Subsequently, the utilization of BDA to achieve the organizational goals has grasped the attention of researchers as well as practitioners [20]. The rise of the BDA era was caused by the surge of data sets with a high degree of complexity generated by utilization of smart gadgets such as mobile phones, smart pads, and Internet of Things (IoT) devices by people all over the world [9]. For example, E-mails; every story uploaded on social media such as a document or paragraph, photo, video, purchase of products using a credit card, are pieces of BD, and this information is collected by digital machines or devices [21].

It is crucial and critical to understand the factors or constructs that influence the BDA adoption model. Generally, most of the researchers applied prominent frameworks or models such as the Technological, Organizational and Environmental (TOE) framework, Technology Acceptance Model (TAM), Diffusion of Innovation (DOI) model, or Motivation, Opportunity and Ability (MOA) model. Each framework and model will be discussed in the following section.

Technological, Organizational, Environmental (TOE) Framework
The TOE framework was designed and developed in 1990 by Tornatzky and Fleischer to investigate the adoption and utilization of technological products and services [13]. The framework encompasses 3 components of an organization's context that affect the process of adopting technological innovation namely technological, organizational, and environmental contexts [15]. The technology domain refers to the technologies obtainable in organizations for feasible adoption of a new system. The organization domain deals with the attributes and properties of the company, such as size, skills and experience. The environment domain points to the market components, competitors, and the regulatory environment [22]. The TOE framework has solid and significant empirical support in the literature as well as study and research. Therefore, it provides a guideline and references to analyze appropriate and suitable constructs to achieve a better understanding of an innovation-adoption decision [23].

Technology Acceptance Model (TAM)
The TAM is used to examine the behavioral intentions and the utilization of IT technology. It is widely applied by scholars in conducting studies related to adoption of new technology in the IT field. In 1967, Fishbein and Ajzen developed the TAM based on reasoned action model theory. The main purpose of this model is to investigate the user's willingness to accept the information systems or their willingness to use the system [9]. The TAM has been continuously studied and expanded by Davis in 1989. In fact, scholars have studied this model based on previously used constructs and they have extended it. In 2004, Amoako-Gyampah and Salam stated that to increase user acceptance and user willingness to use the system, the researcher must clearly comprehend the belief constructs, and this could assist in developing a more effective model that truly can be used by organizations. Only by using this method, would the degree of user acceptance to adopt the system be increased and result in real and actual utilization [24]. In addition, prior studies have stated that the degree of user belief that the system would be actually useful to them affect their intention to adopt the system. Lastly, as the user's attitude is an important determinant in deciding their intention to adopt a system, the users' belief in its usefulness is affected by the perceived ease of use. Put simply, the perspective and viewpoint of the user is a crucial element between perception and behavioral intention. Venkatesh, Morris, Davis, and Davis [25] claimed that attitude is important only when the determinants relevant to capability and effort anticipation are not included in the model. Due to this, there are many variations and extension models from TAM, causing uncertainty about the catalyzing purpose of attitude in TAM and diluting the actual attitude towards intention of adopting the system.

Diffusion of Innovation (DOI)

DOI was developed by Rogers in 2005 with the aim of offering a complete analysis of innovation diffusion drivers and issues along with insights into the process of adopting, or not, an innovation, either technological or non-technological, both for individual and organizations. Innovation diffusion is a complex model or framework, because it relies on multiple determinants, consisting of relative advantage, compatibility, complexity, trialability and observability [15]. Relative advantage refers to the degree to which the user believes that the innovation is better and would benefit the user or firm more than the idea it replaced. Compatibility refers to the degree to which an innovation can integrate with existing and legacy systems. Complexity is the degree that an innovation is hard to use or not user-friendly and causing the entity difficulty in adopting the new technology. Trialability means the extent to which an innovation can be tested, investigated and assessed. Lastly, observability refers to the extent to which the outcome and achievement of an innovation, once implemented, are visible to entities. According to Agrawal [23], the relative advantage, compatibility and complexity are the factors most used by scholars to examine and explain innovation diffusion in organizations.

Motivation, Opportunity and Ability (MOA) Model

The MOA model is a popular theoretical framework in the IT field to examine the adoption of technology by organizations or individuals. In 1995, Ölander and Thøgerson analyzed and described the three main constructs of the MOA model and the connection between them. Motivation, opportunity and ability are the main constructs in the MOA framework in order to bring about the desired result. The MOA framework justifies individual and organizational behaviors from subjective and objective viewpoint by providing a comprehensive analysis. Thus, the framework is widely applied by scholars in many studies area such as adoption of innovation technology, planning business strategy and marketing as the framework is highly stable and has good predictability [11].

3 Research Model

In Sect. 2, adoption models and frameworks such as TOE, TAM, DOI, and MOA model have been discussed and explained. These are the popular models or frameworks applied by scholars to study the adoption by an organization of new technologies. However, there are many determinants used and applied by researchers to construct their model. Some researchers applied the same theory such as TOE framework and TAM model, but the determinants selected were not the same. This may be due to the different context they examined, for example, studied were carried out in various countries and the scope of the study focused on dissimilar group and organizations. Therefore, this study identifies all the factors and determinants discussed from the previous papers and articles collected. Then, the proposed factors have been chosen according to the problems encountered by Malaysian SMEs as stated in Sect. 1.

Generally, there are 4 major models used by scholars to develop a BDA adoption model namely TOE, TAM, DOI and MOA. Based on the problem background addressed in the previous section, the author concluded that TOE was the most relevant and appropriate model to use in this study. The TOE is a useful theoretical framework for understanding the technology system adoption in an organization [14]. In addition, Bhattacharyya [14] proposed that the TOE framework is useful to explain and justify intra-firm innovation adoption as it includes technological, organizational and environmental contexts. As the TOE framework has the solid theoretical basis and the reliable empirical support, it has been selected as the base model in this study. Based on the proposed determinants, an initial model for BDA adoption in Malaysian SMEs is illustrated in Fig. 1.

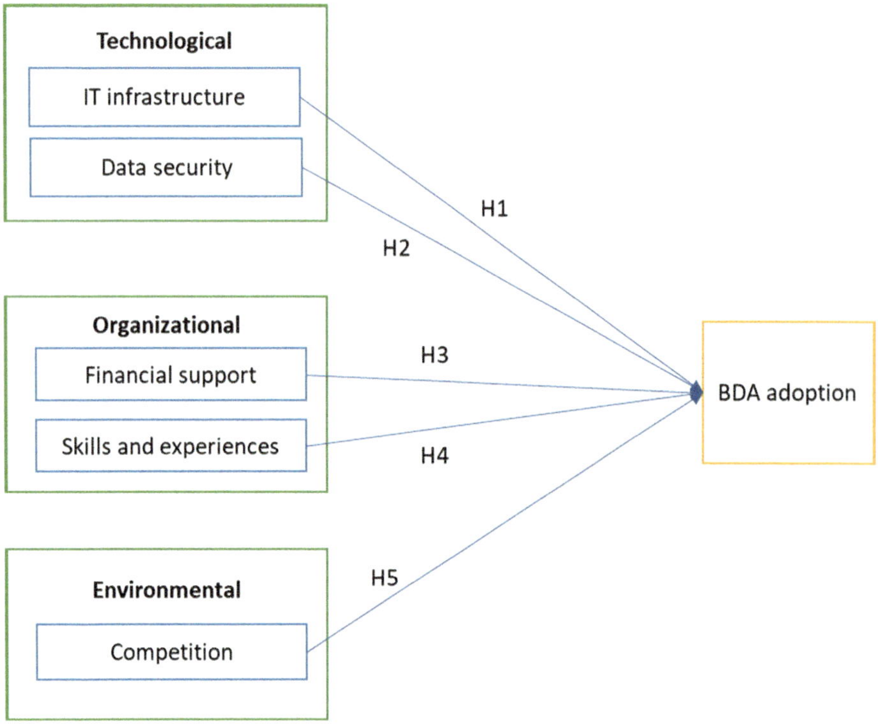

Fig. 1. Proposed BDA adoption model for Malaysian SMEs

Technological Context

Technological context describes the technology used by an organization in the internal and external environment. The intention to utilize and implement a technology is influenced by what material and sources are obtainable in the business, as well as the way the technologies are able to integrate with the legacy or existing system that the organization already owns and has implemented [14, 15, 28].

IT Infrastructure

The foundation or assets for an enterprise's technologies, applications, hardware, software and services that is comprised of data, network, and processing architectures [10, 11, 14, 26, 27].

H1: The IT infrastructure is essential and positively influences adoption of BDA [10, 22, 26].

Data Security

Defined as how the enterprises handled the challenge and issues of offering effective methods for secure management of distributed data and data sharing such as personally identifiable information as well as customer data [10, 13, 26, 27].

H2: Data security is an essential antecedent and positively affects the adoption of BDA [10, 12, 13, 22, 26].

Organizational Context

The organizational context describes the internal characteristics and elements of a firm which affects the adoption and implementation of innovations [14, 22, 28].

Financial Support

This is defined as degree to which the capability or ability organizations can invest in a project such as implementing the system or software [11, 13, 14, 22, 26].

H3: Availability of financial resources and support positively influences the adoption of BDA [10, 12, 13, 22, 26].

Skills and Experiences

The availability of employees with the essential skillset, competencies, knowledge, engagement, experience and capability [10, 12–14, 22, 28].

H4: Employees' skills and experience have a significant positive effect on the adoption of BDA [10, 12, 22, 23, 28].

Environmental Context

The environmental context describes the exterior condition namely the environmental requirements or prerequisite in which the enterprise conducts their business and services [14, 22].

Competition

This refers to the degree and intensity with which firm is affected or influenced by competitors in the market [13, 22, 23, 28].

H5: Competitive intensity has a significant positive effect on BDA adoption [13, 22, 23, 26, 28].

4 Conclusion and Future Work

In conclusion, the factors influencing the adoption of BDA were identified and a BDA adoption model for Malaysian SMEs was developed. The model was developed based on the understanding of the problem background as well as studies related to adoption models applied by various scholars. The proposed model comprises five factors: IT infrastructure, data security, financial support, skills and experience, and competition. Later, the proposed model will be empirically tested and analyzed using SmartPLS 3.

Acknowledgement. The authors would like to thank the Ministry of Higher Education (MOHE) and the Universiti Teknologi Malaysia (UTM) for the UTM Transdisciplinary Research Grant (vote number: 07G33) that had supported this research.

References

1. Iqbal, M., Soomrani, A.R., Butt, S.H.: A study of big data for business growth in SMEs: opportunities and challenges. In: International Conference on Computing, Mathematics and Engineering Technologies (iCoMET), pp. 1–7. IEEE (2018). https://doi.org/10.1109/ICOMET.2018.8346368
2. Coleman, S., Göb, R., Manco, G., Pievatolo, A., Tort-martorell, X., Seabra, M.: How can SMEs benefit from big data? Challenges and a path forward. Qual. Reliab. Eng. **32**(6), 2151–2164 (2016). https://doi.org/10.1002/qre.2008
3. IBM: Big Data Analytics (2018). https://www.ibm.com/analytics/hadoop/big-data-analytics
4. Gandomi, A., Haider, M.: Beyond the hype: big data concepts, methods, and analytics. Int. J. Inf. Manag. **35**(2), 137–144 (2015). https://doi.org/10.1016/j.ijinfomgt.2014.10.007
5. Moktadir, M.A., Ali, S.M., Paul, S.K., Shukla, N.: Barriers to big data analytics in manufacturing supply chains: a case study from Bangladesh. Comput. Ind. Eng. (2018). https://doi.org/10.1016/j.cie.2018.04.013
6. Columbus, L.: 53% of companies are adopting big data analytics (2017). https://www.forbes.com/sites/louiscolumbus/2017/12/24/53-of-companies-are-adopting-big-data-analytics/#4863223539a1
7. Adair, B.: Features of big data analytics and requirements (2019). https://selecthub.com/big-data-analytics/big-data-analytics-requirements/
8. O'Neill, E.: 10 companies that are using big data (2016). https://www.icas.com/ca-today-news/10-companies-using-big-data
9. Verma, S., Sekhar, S., Kumar, S.: An extension of the technology acceptance model in the big data analytics system implementation environment. Inf. Process. Manag. **54**(5), 1 (2018). https://doi.org/10.1016/j.ipm.2018.01.004
10. Kalema, B.M.: Developing countries organizations' readiness for big data analytics **15**(1), 260–270 (2017). https://doi.org/10.21511/ppm.15(1-1).2017.13
11. Wang, L., Yang, M., Pathan, Z.H., Salam, S., Shahzad, K.: Analysis of influencing factors of big data adoption in Chinese enterprises using DANP technique. Sustainability **10**(11), 3956 (2018). https://doi.org/10.3390/su10113956
12. Malaka, I., Brown, I.: Challenges to the organisational adoption of big data analytics: a case study in the South African telecommunications industry. In: Proceedings of the 2015 Annual Research Conference on South African Institute of Computer Scientists and Information Technologists, p. 27 (2015). https://doi.org/10.1145/2815782.2815793

13. Park, J.H., Kim, M.K., Paik, J.H.: The factors of technology, organization and environment influencing the adoption and usage of big data in Korean firms. In: 26th European Regional Conference of the International Telecommunications Society (ITS), Spain (2015). http://hdl.handle.net/10419/127173
14. Bhattacharyya, S., Verma, S.S.: Perceived strategic value based adoption of big data analytics in emerging economy: a qualitative approach for Indian firms. J. Enterp. Inf. Manag. **30**(3), 354–382 (2017)
15. Jesus, F.C., Oliveira, T., Naranjo, M.: Understanding the adoption of business analytics and intelligence. In: World Conference on Information Systems and Technologies, pp. 1094–1103. Springer, Cham (2018). https://doi.org/10.1007/978-3-319-77703-0
16. Raguseo, E.: An empirical investigation on their adoption, benefits and risks for companies. Int. J. Inf. Manag. Big Data Technol. **38**(1), 187–195 (2018). https://doi.org/10.1016/j.ijinfomgt.2017.07.008
17. Luen, W.K., Hooi, C.M., Fook, O.S.: Are Malaysian companies ready for the big data economy? A business intelligence model approach, August 2015
18. Haslinda, R., Mohd, R., Mohamad, R., Sudin, S.: A proposed framework of big data readiness in public sectors. In: AIP Conference Proceedings (2016). https://doi.org/10.1063/1.4960929
19. Manyika, J., Chui, M., Brown, B., Bughin, J., Dobbs, R., Roxburgh, C., Byers, A.H.: Big Data: The Next Frontier for Innovation, Competition, and Productivity (2011). https://www.mckinsey.com/~/media/McKinsey/BusinessFunctions/McKinseyDigital/OurInsights/BigdataThenextfrontierforinnovation/MGI_big_data_full_report.ashx
20. Mikalef, P., Boura, M., Lekakos, G., Krogstie, J.: Big data analytics and firm performance: findings from a mixed-method approach. J. Bus. Res. **98**(January), 261–276 (2019). https://doi.org/10.1016/j.jbusres.2019.01.044
21. Sen, D., Ozturk, M., Vayvay, O.: An overview of big data for growth in SMEs. Procedia Soc. Behav. Sci. **235**, 159–167 (2016). https://doi.org/10.1016/j.sbspro.2016.11.011
22. Sam, K.M., Chatwin, C.R.: Understanding adoption of big data analytics in China: from organizational users perspective. In: IEEM (2018)
23. Agrawal, K.: Investigating the determinants of big data analytics (BDA) adoption in Asian emerging economies. In: AMCIS 2015, pp. 1–18 (2015)
24. Liao, C., Tsou, C.: Expert systems with applications user acceptance of computer-mediated communication: the SkypeOut case. Expert Syst. Appl. **36**(3), 4595–4603 (2009). https://doi.org/10.1016/j.eswa.2008.05.015
25. Venkatesh, V., Morris, M.G., Davis, G.B., Davis, F.D.: User acceptance of information technology: toward a unified view. MIS Q. **27**(3), 425–478 (2003). https://doi.org/10.2307/30036540
26. Motau, M., Kalema, B.M.: Big data analytics readiness: a South African public sector perspective. In: 2016 IEEE International Conference on Emerging Technologies and Innovative Business Practices for the Transformation of Societies (EmergiTech), pp. 265–271 (2016)
27. Surbakti, F.P.S., Wang, W., Indulska, M., Sadiq, S.: Factors influencing effective use of big data: a research framework. Inf. Manag. (2019). https://doi.org/10.1016/j.im.2019.02.001
28. Schüll, A., Maslan, N.: On the adoption of big data analytics: interdependencies of contextual factors, vol. 1, pp. 425–431 (2018). https://doi.org/10.5220/0006759904250431

Aedes Entomological Prediction Analytical Dashboard Application for Dengue Outbreak Surveillance

Yong Keong Tan[1], Noraini Ibrahim[1,2(✉)], and Shahliza Abd Halim[1]

[1] School of Computing, Faculty of Engineering, Universiti Teknologi Malaysia (UTM), Skudai, Malaysia
yong.keong@live.com, {noraini_ib, shahliza}@utm.my

[2] Centre for Engineering Education, Universiti Teknologi Malaysia (UTM), Skudai, Malaysia

Abstract. Entomological surveillance is used in majority of the countries around the world. The main objective of this process is to monitor, control and prevent the occurrence of a dengue outbreak. In Malaysia, Entomology and Pest Unit (EPU) from the State Health Office is a local division that is responsible for performing the dengue surveillance operations. There exists human-related risks due to the procedures of predicting cryptic breeding sites and upcoming dengue outbreak locations are done manually. This paper discusses the implementation and results of Aedes Entomological Predictive Analytic Dashboard (AePAD) application. The AePAD multi-platform application provides several features especially historical ovitrap data retrieval and visualization to predict the possible cryptic breeding sites of Aedes mosquitoes and upcoming dengue outbreak locations prediction using Deep Neural Network (DNN). It is hoped that the application will be able to help the EPU team in performing better prevention and strategic control operations, thus enhancing the existing dengue surveillance systems and minimizing future dengue outbreaks in Malaysia.

Keywords: Dengue · Entomological surveillance · Dengue prediction

1 Introduction

Dengue is known as a dangerous disease especially in countries that experience tropical climate like Brazil, Central Africa, Malaysia and so on [1]. Dengue outbreaks do not only lead to death cases among the community, it also affect the local tourism industry [2] and poses a threat to socio-economy aspect [3]. In Malaysia, 68,950 dengue cases were reported with 102 deaths for the first half of 2019 which is more than double compared to the same period in 2018 [4].

It is essential to perform preventive and control actions towards the dengue population to prevent the occurrence of future dengue outbreaks [6]. In the National Dengue Strategic Plan (2015–2020) of Malaysia, dengue surveillance and prevention has been declared as part of the seven methods required to improve the existing detection and response interventions at the state or national level [5]. There is a local division in Malaysia which is known as EPU from the State Health Office that is

F. Saeed et al. (Eds.): IRICT 2019, AISC 1073, pp. 54–65, 2020.
https://doi.org/10.1007/978-3-030-33582-3_6

accountable to perform the dengue monitoring and prevention operations. It is imperative to gather significant information on dengue and recognize the potential cryptic breeding sites of Aedes mosquitoes in order to conduct suitable preventive and control actions [3]. This information includes entomological data such as indices of density, previous dengue cases, ovitrap data and so on. Public health experts have the chance to implement machine learning or even deep learning mechanism in predicting dengue outbreak patterns [7] due to the headway in artificial intelligence and big data technologies [8]. Telenor research highlighted the potential of using big data to prevent the spread of dengue disease by accessing the city authority in monitoring the population of mosquitoes [9].

The main objective of this paper is to present the implementation process and testing results of Aedes Entomological Predictive Analytical Dashboard (AePAD) application, in assisting EPU team to detect future dengue outbreaks earlier and helping them to perform better strategic control actions. AePAD application aims to improve and extend the functionalities of existing Dengue Entomological Surveillance (DES) system [10] by providing prediction on cryptic breeding sites of Aedes mosquitoes and forecasting upcoming dengue outbreak locations entomologically. Detailed requirements used in this study are gathered with the help of the EPU team of Johor Bahru District Health Office.

2 Methods

2.1 Requirement Phase

Firstly, in order to document all the requirements needed to be fulfilled by the developed AePAD application, two requirements elicitation techniques were used namely: survey (using questionnaire) and closed-ended interview. In addition to that, detailed information on the workflow representing the outcome of this requirement phase can be referred to as in the previous paper [11].

2.2 Analysis Phase

Use case is drawn during the requirement analysis phase to document all the elicited requirements in order to understand user requirements.

Figure 1 shows the revised use case diagram of AePAD application that document all the fundamental requirements in increasing the productivity of users to perform their daily tasks. Compared to the previous paper [11], the revised use case diagram for this study documents several extra actors including Spark, Weather API, Map API and so on.

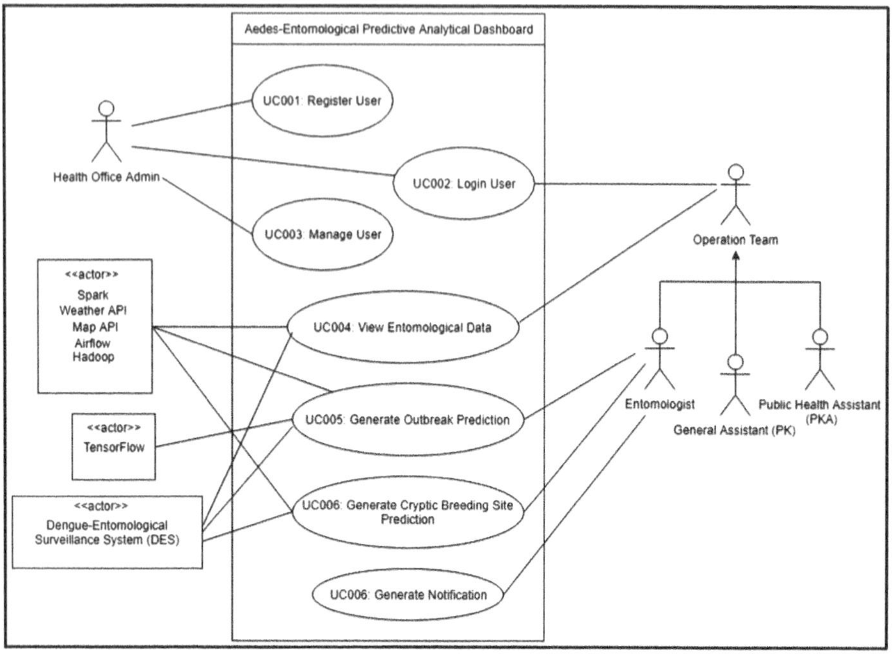

Fig. 1. Use case diagram of AePAD application

There are seven use cases included in the AePAD application which are:

1. Register User - AePAD allows Admin to register new users into the system;
2. Login User - AePAD allows users to login into the system;
3. Manage User - AePAD allows Admin to edit or delete users in the system;
4. View Entomological Data - AePAD allows users to view the historical ovitrap data;
5. Generate Outbreak Prediction - AePAD allows users to view the prediction on upcoming dengue outbreak locations;
6. Generate Cryptic Breeding Site Prediction - AePAD allows users to view mosquitoes cryptic breeding; and
7. Generate Notification - AePAD allows users to receive notification once new prediction has been performed.

2.3 Design Phase

The architecture style of AePAD is Model-View-Controller (MVC) in the system development phase. As shown in Fig. 2, the AePAD is divided into two sides which are server side and client side. This is an improvement of the previous version of AePAD system architecture as in the previous paper [11]. The overall architecture of

AePAD application is deployed in Kubernetes Cluster with the use of Docker container for visualizing the micro-services concept for easy scaling and easy maintainability purposes.

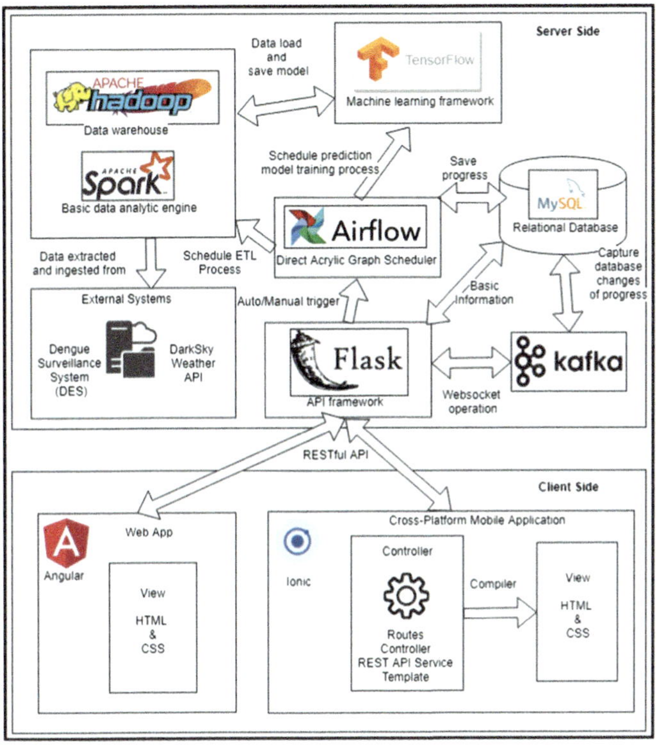

Fig. 2. System architecture of AePAD application

The server side is deployed and hosted on the cloud while the mobile application will be installed in physical devices owned by users. The backend algorithm and prediction mechanism is hosted at the server side to reduce the burden of the mobile devices to run AePAD application. Flask will act as the main gateway for the front-end AePAD application in interacting with the backend services. Flask also acts as controller in the server side to handle requests directed to the server, for example user authentication, map display, data extraction and so on.

The detailed workflow of the machine learning pipeline starts with Apache Spark cluster, an open-sourced unified analytic engine that extracts entomological and weather data from external sources using SparkSQL mechanism with suitable configuration and SQL query. Then, the extracted data will be stored in the Hadoop File System (HDFS) in Apache Hadoop cluster. The purpose of using HDFS instead of Hive is to store both structured and unstructured data. The most important aspect of HDFS is it supports saving the prediction model and progress of TensorFlow model training process. Next, data cleaning and data featuring will be done using Apache Spark together with the Python SKLearn library and packages.

Since AePAD requires machine learning and deep learning, Google's TensorFlow framework is used in order to ease the process of prediction pipeline development. Apache Airflow, an open source pipeline scheduler or directed acrylic graph scheduler, is the main scheduler that helps AePAD application to smoothly schedule the machine learning pipeline. Apache Kafka is used to detect database changes and inform users whether the machine learning data pipeline has been successfully run. Websocket is used in this situation to display real time data in front-end pages. Besides, third party cloud components such as Cloud Messaging in Firebase are also used to generate real time notification and display pop-up in AePAD mobile application.

For the client side, AePAD provides two options for the users, namely web application and mobile application. Web application is developed using Angular 7 framework whereas the mobile application is developed using Ionic 4 application. Both of them use REST API which is responsible to handle the request and response operation of the data from the server. This process will return the results in JSON format for easy data query and validation. After the controller received the data, the related services will pass the required information and data to the display pages on the screen. REST API will be predefined in the web application's service file and is ready to be called by the controllers.

After the server returns the data and file to the client, the data cannot be updated in real-time using traditional web technology. However, the use of Angular based framework will provide two ways data binding which will allow the data to be automatically updated in real-time.

3 System Development and Implementation

3.1 Development of System Functionalities

The AePAD web application is developed using Angular 7 framework while the backend API is developed using Flask framework. The prediction algorithm used in AePAD is Deep Neural Network (DNN). On the other hand, the AePAD mobile application is developed using Ionic 4 framework in conjunction with Firebase Cloud Messaging features.

Figure 3 shows the dashboard view displayed to the users once they have login into the system. The dashboard view shows the basic historical information in graph or chart. Overall data is also shown in the dashboard view such as total ovitrap collected, total egg collected and so on.

Fig. 3. AePAD web – dashboard view

Figure 4 shows the historical ovitrap and cryptic breeding sites displayed in the AePAD web application. The historical map view shows the cryptic breeding sites in 50 m radius. User is required to select the location and date of prediction, and AePAD will generate results based on these requests. Date and the ovitrap index of the area will be displayed to indicate the density of Aedes mosquitoes at the particular day. A map is shown to indicate the installed ovitrap devices. The overlapping areas are the locations with the highest probability of detecting cryptic breeding sites of Aedes mosquitoes. The pop-up shows detailed information about the prediction result for the installed ovitrap locations.

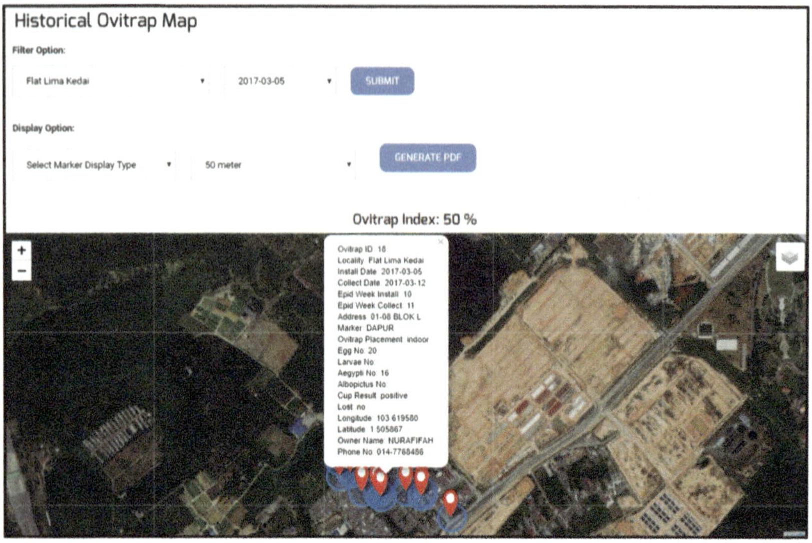

Fig. 4. AePAD web – historial ovitrap and cryptic breeding sites view

Figure 5 shows the prediction map for the upcoming dengue outbreak location. User is required to select the location and date of prediction. Then, a map is shown based on the user requests. User can choose the display radii such as 200 meters or 400 meters as shown in red circle and blue circle respectively. These radii are essential for entomologists to plan for future surveillance operations. The pop-up shows detailed prediction result of the upcoming dengue outbreak location. The pop-up shows the average predicted number of eggs for that location in the following week. The average predicted number of eggs for that location in three months can also be calculated based on the formula provided by the stakeholder.

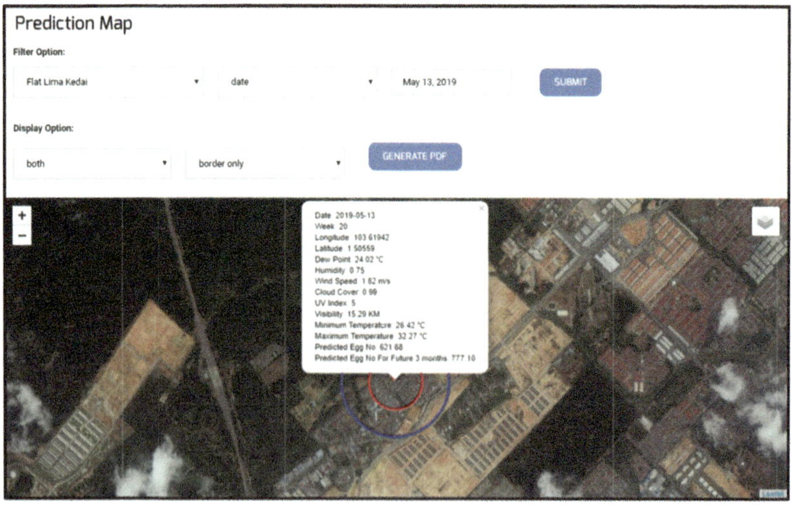

Fig. 5. AePAD web – prediction map for the upcoming dengue outbreak location

Figure 6 shows some of the interfaces of the AePAD mobile application. The functionalities in the AePAD mobile application are only limited to user login, view historical ovitrap data, view dengue outbreak prediction, view cryptic breeding sites prediction and generate notification. Report generation is prohibited in AePAD mobile.

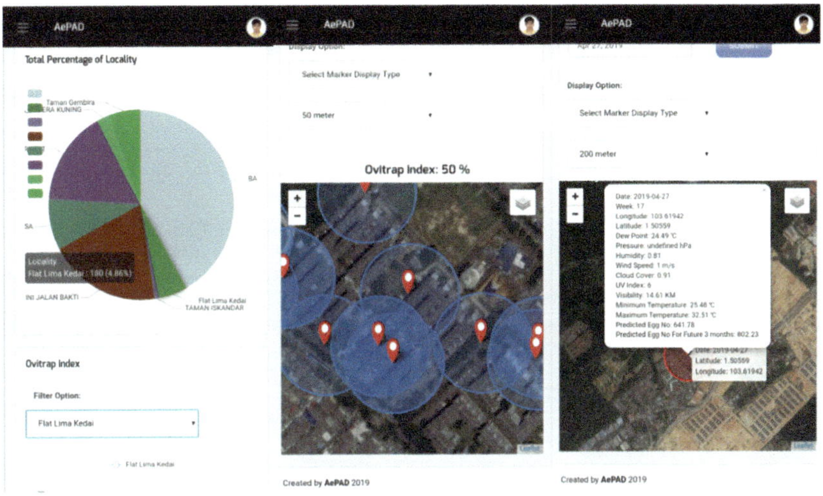

Fig. 6. Interfaces of AePAD mobile

3.2 Prediction Algorithm Using Google's TensorFlow

Figure 7 shows the pseudocode of the prediction algorithm used by AePAD application to predict the upcoming dengue outbreak locations. The algorithm used in the backend of AePAD is Deep Neural Network (DNN) regressor using Google's TensorFlow framework.

It is observed that the DNN supports the predictive analysis of cryptic breeding site and dengue outbreak location where the output is the ovitrap egg number and predicted number of ovitrap egg in specific locality in the next three months. This in turn results to higher possibility of dengue cases in that locality. There are a lot of machine learning and deep learning models that have been applied in this field of study. However, most of these studies only use dengue cases to predict dengue outbreaks unlike the focus of this system which included entomological data in its prediction. DNN algorithm is very flexible and useful to deal with larger dataset while maintaining the main features and performance of normal neural network algorithm. Specifically, DNN regressor is used instead of DNN classifier because the expected prediction result is in number format instead of a category.

Specific dataset which contains entomological data such as locality name, epidemic week of ovitrap installation and collection and so on is requested from the State Health Office before conducting any predictive model construction for the AePAD prediction engine. Weather data which include dew point, humidity, temperature, cloud cover, pressure and so on are collected from DarkSky API and then combined with the ovitrap dataset. The dataset acts as the raw dataset for the model training and testing process. The dataset is divided into two groups of 80% and 20% where the former is used for model training purpose and the latter is used for model validation purpose. Before the model training, some required parameters are initialized such as dataset features and model parameters.

```
INPUT: Entomological dataset, weather dataset
OUTPUT: Forecasting cryptic breeding sites and dengue outbreak lo-
cations

START
1.  Read data from the dataset
2.  Randomly partition 80% of the dataset for training and 20% for
    testing
3.  Clean the null or missing values from the dataset
4.  Normalize the dataset
5.  Select the features or columns for model training
6.  Determine the targeted column
7.  Prepare the pattern matrix
8.  Initialize all required parameters and features for the Tensor-
    Flow Deep Neural Network Regressor
9.  WHILE the termination condition is not achieved
        a.  Continue the training process for model prediction using
            Deep Neural Network Regressor
10. END WHILE
11. Validate the model using the testing dataset
12. Check the model's performance and accuracy
13. Fine-tune the model with feedbacks obtained from the validation
    process
```

Fig. 7. Pesudocode of the prediction algorithm

A predictive model is produced with the aid of Python 3 programming language machine and deep learning libraries. The validation dataset is fed into the predictive model after the model training process is done in order to validate the performance and accuracy of the model outcome. Further fine-tuning is done reduce time required and improve the accuracy of AePAD application. The predictive model is integrated and deployed to AePAD application server. The process of model training and validation is done every day at 12.00 am together with new prediction results.

4 System Testing

System testing which includes black box testing and user acceptance testing (UAT) are conducted for AePAD application. The main objective of system testing is to perform system validation and verification with the stakeholders. Stakeholders' opinions are gathered in order to further improve the system.

Black box testing is a testing method that only involves the input and the output of the system, and this kind of testing method does not involve the internal algorithm or structure of the system. Normally, a black box testing is conducted by system developers in order to identify the errors or bugs that may affect the system from producing the desired output.

Table 1 shows the black box testing result of historical ovitrap map information under certain conditions. The success combination from Table 1 shown for displaying the historical ovitrap data on the map is HOR002 while the HOR004 shows the additional radius around the marker.

Table 1. Black box testing of historical ovitrap map information

Test Id	HOR001	HOR002	HOR003	HOR004
Inputs and actions				
Select locality	–	Flat Lima Kedai	Flat Lima Kedai	Flat Lima Kedai
Select date	–	2017-03-05	2017-03-05	2017-03-05
History ovitrap radius checkbox	–	–	checked	checked
Internet connection	Online	Online	Offline	online
Result	*Inputs and actions*			
1. Error: required input not provided	/			
2. Error: no internet connection			/	
3. No changes in the map display	/		/	
4. Display marker information successfully		/		/
5. Display radius and information successful				/
Expected result	1,2	4	2,3	3,4

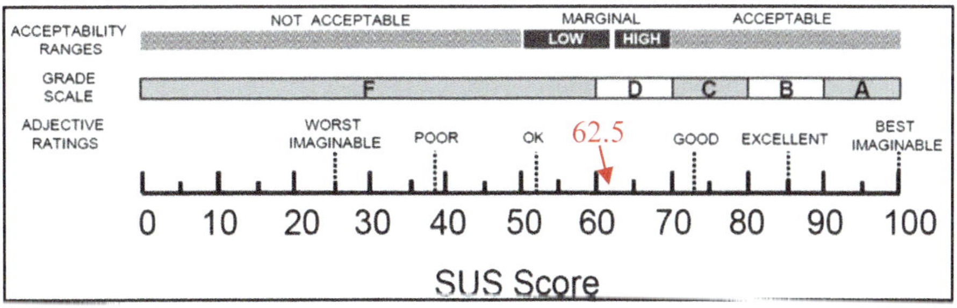

Fig. 8. System Usability Scale (SUS) score

Figure 8 shows the SUS score scale for the AePAD application. AePAD application scores 62.5% which is classified as "GOOD". This low score may be due to human norm of not giving full mark in assessments and not understanding the questionnaire. In addition, this is the first time the participants interact with AePAD application and they may not be familiar with the flow of the system.

5 Conclusion

In conclusion, the detailed development and testing of AePAD application has been documented in this paper. AePAD application is designed and developed based on its specific user's requests and requirements; namely the EPU team of State Health Office. The AePAD application offers two important functionalities namely (i) providing historical cryptic breeding sites of Aedes mosquitoes prediction, and (ii) predicting upcoming dengue outbreak locations using Deep Neural Network (DNN). These novel functionalities and justification of DNN as prediction algorithm are described in details in Sect. 3. With the developed AePAD application, it is hoped that there will be improvements in the efficiency and productivity of the EPU teams in performing their daily tasks of predicting the cryptic breeding sites and upcoming dengue outbreak locations. This in turn will lead to earlier detection and better strategic planning in eradicating or minimizing the occurrence of future dengue outbreaks in Malaysia.

Acknowledgement. This study is partially funded by the Ministry of Education Malaysia's Research University Grant (RUG) and High Impact Grant (HIG) of Universiti Teknologi Malaysia (UTM) under Cost Centre No. R.J130000.7728.4J237 & Q.J130000.2451.04G70. In particular, the authors wish to thank especially the entomologists from Johor Bahru District Health Office for their involvement during requirements elicitation activities and survey for the development of AePAD application. The historical entomological information used in this study is approved by the National Medical Research Register (NMRR ID: NMRR-16-2837-31417), Ministry of Health Malaysia.

References

1. Dom, N.C., Ahmad, A.H., Ishak, A.R., Ismail, R.: Assessing the risk of dengue fever based on the epidemiological, environmental and entomological variables. Procedia Soc. Behav. Sci. **105**, 183–194 (2013)
2. Malaysia Ministry of Health (MoH): Strategic Plan for Dengue Prevention and Control 2009–2013 (2009). http://www.moh.gov.my/english.php/pages/view/118. Accessed 20 Apr 2018
3. Lee, H., et al.: Dengue vector control in Malaysia - challenges and recent advances. Int. Med. J. Malaysia **14**(1), 11–16 (2015)
4. World Health Organization (WHO): Dengue Situation Updates Number 573 (2019). https://www.who.int/docs/default-source/wpro—documents/emergency/surveillance/dengue/dengue-20190718.pdf?sfvrsn=5160e027_12. Accessed 26 July 2018
5. Suli, Z.: Dengue prevention and control in Malaysia. In: International Conference on Dengue Prevention and Control and International Dengue Expert Consultation Meeting (2015)
6. GSMA: Tainan: Dengue Fever (2016). https://www.gsma.com/iot/tainan-dengue-fever/. Accessed 04 Apr 2018

7. Munasinghe, A., Premaratne, H., Fernando, M.: Towards an early warning system to combat dengue. Int. J. Comput. Sci. Electron. Eng. **1**(2), 252–256 (2013)
8. Zainudin, Z., Shamsuddin, S.M.: Predictive analytics in Malaysian dengue data from 2010 until 2015 using BigML. Int. J. Adv. Soft Comput. Appl. **8**(3), 18–30 (2016)
9. Telenor: Telenor research deploys big data against dengue (2015). http://www.telenor.com/media/press-releases/2015/%0Dtelenor-research-deploys-big-data-against-dengue. Accessed: 28 Apr 2018
10. Ibrahim, N., Quan, T.W.: The development of multi-platforms application for dengue-entomological surveillance system. In: 6th ICT International Student Project Conference. Elevating Community Through ICT, ICT-ISPC 2017, January 2017–April 2017, pp. 1–4 (2017)
11. Ibrahim, N., Yong Keong, T.: Development of aedes-entomological predictive analytical dashboard application. In: Proceeding 2018 7th ICT International Student Project Conference, ICT-ISPC 2018 (2018)
12. Salehzadeh, R., Shahin, A., Kazemi, A., Shaemi Barzoki, A.: Is organizational citizenship behavior an attractive behavior for managers? A Kano model approach. J. Manag. Dev. **34**(5), 601–620 (2015)
13. Kumar, C.V.S., Routroy, S.: Demystifying manufacturer satisfaction through kano model. Mater. Today Proc. **2**(4–5), 1585–1594 (2015)

A Study on the Impact of Crowd-Sourced Rating on Tweets for the Credibility of Information Spreading

Nur Liyana Mohd Ramlan, Nor Athiyah Abdullah(✉), Kamal Karkonasasi(✉), and Seyed Aliakbar Mousavi

School of Computer Sciences, Universiti Sains Malaysia, USM, 11800 Pulau, Pinang, Malaysia
athiyah@usm.my, asasi.kamal@gmail.com

Abstract. Social media has been used extensively for information spreading during disasters. The problem occurs nowadays is overloaded of information from social media that is spread by people and confused the other people. Current function in Twitter allows people to 'favorite' tweets, perform re-tweet instantly, or added their own opinion on the tweet by quote and re-tweet. It is allowed the user to report tweets, sensitive or harmful tweets, and non-interested tweet, but there are no features to identify the credibility and accuracy of those tweets. Therefore, we proposed a technical solution where a prototype was developed in a Twitter-like environment by adding crowdsource rating features. The purpose of this research is to investigate the influence of crowd-sourced rating on tweets for decision making for the credibility of information spreading. A pilot study was conducted to a small real sample group of 31 respondents to know the respondent's feedback on the prototype. The pilot study shows that most of the respondents were agree with the capabilities of crowdsourcing rating feature in identifying the accuracy and credibility of that information. The prototype design and the questionnaires have to be modified again, but the rating features was still the same. This prototype then redistributed to 139 respondents, and the respondents need to answer the control scenario part, prototype part, and the questionnaires. Control scenario design was similar like a prototype but without adding the crowdsource rating features. The questionnaires have divided into four parts, which is demographic information, social media usage, evaluation of the prototype, and evaluation on the usability of the prototype. Overall, the result from the questionnaires answer was a positive result where most of the respondents were agreed that crowdsourced rating feature was useful and helped them to identify and determine the accuracy and credibility of the information and help to prevent from spreading the misinformation in social media.

Keywords: Crowdsourcing · Information overload · Information spreading · Twitter · Social media

1 Introduction

Situation happens in that disaster area, and one of the communication ways for them to get to know the information is through social media since many people used it such as Facebook and Twitter. Social media is a powerful tool for spreading information

F. Saeed et al. (Eds.): IRICT 2019, AISC 1073, pp. 66–78, 2020.
https://doi.org/10.1007/978-3-030-33582-3_7

because it reaches faster and farther than any communication method to date. However, people sometimes wrote some issue in social media then share and make it viral. In this study, we would help people to avoid the spread of rumors and false information from social media by using crowdsourced rating in a Twitter-like environment to examine the credibility of information spreading. We want to investigate the influence of crowdsourced rating for decision making of information spreading. One of the benefits from this study is to assist people to make better decision-making regarding information spreading where they can rate that info either it is true or false so that the spread of inaccurate information in social media could be minimized.

In this research, we want to see the influence of crowdsourcing in an individual's decision making on disaster information spreading on Twitter. Our passion in this study is to prepare a prototype which includes crowdsourced rating feature in a Twitter-like environment. Besides that, we also want to know whether crowdsourced rating features proposed will affect people's decision making on tweets for information spreading.

The problem that we highlight in this study is how we would help people to prevent from spreading rumors and misinformation social media. One possible solution that can be used to overcome this problem is by using crowdsourced rating to examine the credibility of information for information spreading.

The motivation of this research is that it can contribute to the knowledge in human behavior of information spreading. People need to know and be responsible for what information they spread since there is much information (information overload) spread by a person or anonymous on social media. Other than that, this will benefit the individuals in terms of deciding the truthful information since there are much information overloads in social media such as Twitter, Facebook, and Instagram.

Besides that, it is crucial for a designer who has design apps or another website related to disaster information. This could help the designer to design suitable and useful apps based on the result that we get. For the example eBanjir Kelantan's Portal. That portal updated various of flood's information such as level of water in Sungai Kelantan, rescue efforts, government response, assistance and aid, and reuniting missing individuals. This portal is beneficial to people and the victims. Since our passion is to explore about crowdsourcing, possibly it is helpful to be included into catastrophe site/application later on.

Finally, this study will contribute to the knowledge of social media study because people still rely on social media as one of their essential communication tools.

This research will undergo six procedure before it concludes in order to complete the objective of this research were to investigate the influence of crowd-sourced rating for decision making of information spreading and the most rating (true, false & doubtful) that influence the most sharing. A prototype was created by adding crowd-source rating features by displaying it together with what other people have a rate for that information in a Twitter-like environment. This prototype has used sample disaster tweet happen in Malaysia, such as during the Sabah earthquake and fire eruption at Hospital Sultanah Aminah Johor Bharu. This research was focused more on disaster compared to the other issues.

The primary user survey has been conducted to 139 respondents where they need to use and answer the proposed prototype and some questionnaire. The questionnaire answered are strictly confidential and will only be used for this research purpose. To retrieve a meaningful conclusion from it, we aim to analyze the gathered data. Some analysis will be performing to find out either proposed rating features can minimize the spreading of rumors and inaccurate information about disaster or not.

1.1 Background Study

Social media have been used extensively for information spreading during disasters. People around the world commonly used Facebook and Twitter as their social media. Kaplan has defined social media as a "group of internet-based application that allows individuals to create and exchange information in a network" (Kaplan and Haenlein 2010). Ideas, opinion, and experience are being shared faster in an online platform as a fast communication (Eksil et al. 2014). Social media also was a powerful tool for spreading information because it reaches faster and farther than any communication channel such as mobile messaging, radio, and newspaper.

In Japan, Twitter is one of the social media that commonly used by people to share or spread something related with the current issue for the example during the great Japan earthquake and Fukushima Daiichi plant accident (Thomson et al. 2010). Previous research by Huaye Li said by displaying both retweets counts, and collective truthfulness rating could reduce the spread of inaccurate information based on health-related statement since their research is more to health-related. This research gets a positive result where when people know that others think the information is false and know what number of individuals have shared that information, they may refuse to forward it since they would prefer not to spread false information (Li and Sakamato 2015).

Besides that, information overload also leads people either to trust or ignore that information. Kellingley said the causes of information overload today are because of the "huge amount of new information that being constantly created and the simplicity of creating, duplicating and sharing of information online without know where the source come from" (Kellingley 2016). People usually give a lack of attention to the information provided by anonymous since they have a lack of trust in them. This could prevent the transformation of information provided into usable knowledge (Fisher 2013; Jaeger et al. 2007). Hence, it could be concluded that trustworthiness of information is required in achieving a rapid decision making during the critical condition (Murayama et al. 2013).

In this study, our focus is to investigate the impact of individual crowd-sourced on decision making on tweets for the credibility of information spreading and the most rating (true, false and doubtful) that influence the most sharing. Crowdsource is "the practice of getting collective intelligence from the online community for problem-solving to achieve specific organizational goals" (Brabham 2009). One possible solution that can use to overcome this problem is by using crowdsourced rating. A prototype similar to Twitter will be developed by adding new features which are crowdsourced rating features. This feature enables the user to rate for any information that appears on their timeline during disaster happen. This rating features will enable more comfortable people to decide since there are information overloads, and they do not know the credibility and the accuracy of that information to be trusted.

2 Research Method

Information overload's issue on social media has been raised especially during disaster happen where people are not knowing whom and which information to trust. The cause of overloading of information today is because "the huge amount of new information that being constantly created and the simplicity of creating, duplicating and sharing of information online without knowing where the source comes from" (Kellingley 2016). This might be the problem for people on spreading the right and correct information on social media because it may lead to the spreading of rumors, false information, and misinformation on social media. Figure 1 shows the procedure and steps in the current research methodology.

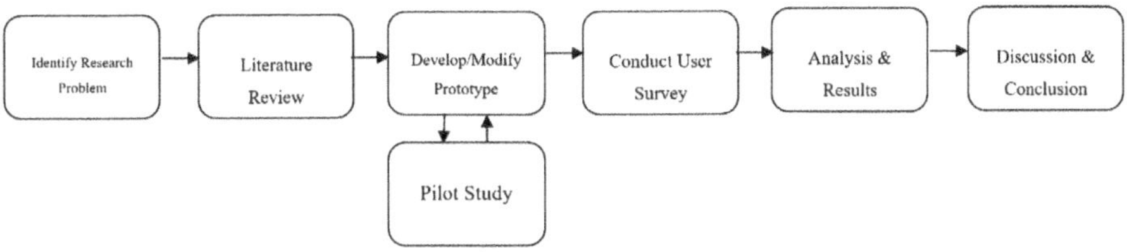

Fig. 1. Research procedure

The first step is identifying the research problem and issue, which is how we would help people to avoid spreading of false, fake, and suspicious information from social media. Therefore, we seek to answer the research question:

1. Does crowdsource rating features proposed affect people's decision making to spread disaster information?
2. Which rating (e.g., true, false, and doubtful) will influence the most sharing?

The second step is the literature review. The purpose of this step is to find out and know whether there are any method or solution to overcome this problem and to make sure the proposed method is never attempted before by other researchers. Previous research has proposed few methods on how to reduce the spread of information in order to improve the reliability of information especially during a disaster such as by using tweet categorization framework, by using critical thinking experiment and by using crowdsource credibility evaluation by displaying it together with re-tweet counts. The difference in our study is that we propose a technical solution where a prototype with rating features are developed. This technical solution was different from the previous study by Li because we are not displaying together re-tweet counts in the prototype. The solution from Tanaka also different where they are focused on experimental design with critical thinking is crowdsourcing, where people could benefit from other's criticism.

The next step is developing a prototype. In this study, we have developed a prototype in a Twitter-like environment by adding the rating feature where we displayed what other respondents have a rate for that tweet. The reason for adding the new feature is because we want to test how rating features will influence people to make a better decision regarding information spreading to minimize the spreading of rumors, inaccurate and false information on social media.

Then is conducted a pilot study to a small real sample group. Thirty-one respondents have taken part in this survey where there is 18 female, and the rest are male. This pilot study was needed because we want to know how the user will respond to the prototype and how the results we will get either it will be a positive or negative result. 14 sample of tweets were collected from each of case during Sabah earthquake and fire eruption at Hospital Sultanah Aminah Johor Bharu, which have caused few deaths to be inserted into the prototype. However, before respondents answer the prototype, they need to answer control scenario part first where respondents need to rate for a likelihood of sharing of that information such as "rare," "unlikely," "possible," "likely" and "almost certain." There will only have one tweet given for this control scenario, and it is not like a random tweet that appears on the prototype part. The tweet selected came from actual incident based on case from Sabah earthquake in 2015 where it has caused 19 death, and the reason why we selected the actual tweet is, we want to see whether the respondents will likely to share that info or not even though they are not shown with what others respondents' rate for that tweet. The figure below shows the interface of the control scenario.

After finished answer the control scenario part, they need to answer the proposed prototype where they will have a new feature, which is the rating feature that showed what the other respondents have a rate for those tweets. There are three types of rating which are "true," "doubtful," and "false." Once the rate for that tweet, what other respondents have rate will be shown in the interface. Four sample tweets need to be the rate, which is two from HSA fire eruption case and two from the earthquake in Sabah. Lastly, the respondents need to answer a questionnaire that consists of three-part, which are demographic information, social media usage, and the usability of the prototype.

Next, the control scenario design and the questionnaire have been modified after we have conducted the pilot study. The reason why we have modified it because the design is not similar to the proposed prototype. It should be similar like the prototype but without adding the rating features that showed what other respondents have rate about that tweet after they rate it. The previous design is to know the likelihood of respondents to share that tweets, so that, we do a few changes and redesign it similar to the proposed prototype. The figure below shows the interface of the control scenario after being modified. Besides that, some of the questionnaires also have been changes to make it more suitable for the scope of this research study (Fig. 2).

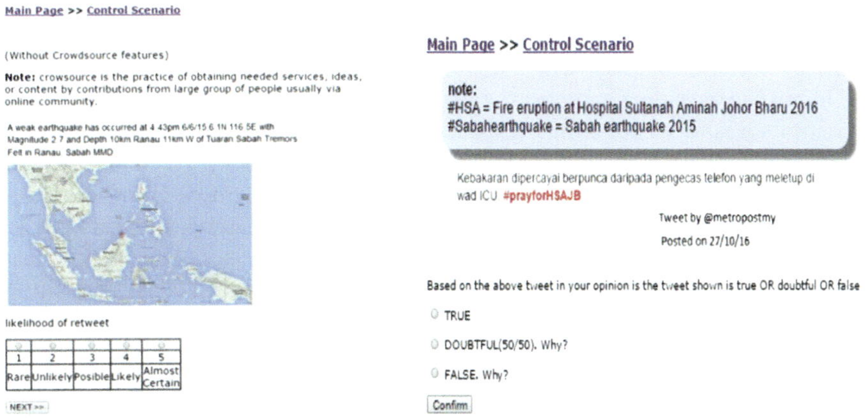

Fig. 2. On the left is the control scenario interface of old and on the right is the interface of a modified version of the control scenario

3 Experimental Design

In this research, the methodology used to investigate the impact of crowdsourced rating features on tweets for the credibility of information spreading is by using the proposed prototype. Hence, 1. Answer Control Scenario, 2. Answer Prototype, 3. Questionnaire part 1 (demographic information), 4. Questionnaire part 2 (social media usage), 4. Questionnaire part 3 (prototype evaluation), 6. Questionnaire part 4 (usability testing).

Firstly, they need to rate two sample tweets came from Sabah earthquake and fire eruption at HSAJB case for control scenario and prototype part. Then they need to answer questionnaires that have been divide into four-part, which are demographic information, social media usage, prototype evaluation, and usability testing. The prototype design was developed in a Twitter-like environment.

The main page of the experimental design where there are few explanations about what this research was about and the flowchart to guide the respondents during answer the prototype to make sure the respondents understand and know what they need to do after they click the consent button. As the figure above shows, there are six steps need to do to complete the experimental design where the first one is answering the control scenario.

Control scenario was created to make a differentiation of sharing the result with a proposed prototype where it was created similar to a prototype but without adding the rating features. This method is to find out whether how the user will make their decision on those tweets for information spreading without seeing what others rate about that tweet was. In this step, the respondents were asked to rate the tweet either it is true, doubtful, or false, and then they are given a choice either they want to share this tweet or not. There will have twenty-eight of sample tweets where it has been categories into three, which are accurate tweets, doubtful tweets, and false tweets. The tweet selected came from actual incident based on a case from Sabah earthquake on 2015 where it has caused nineteen death and from fire eruption at Hospital Sultanah

Aminah Johor Bharu on 2016 where it has caused six patients' death. The reason why we choose that incident because during the incidents happen, few rumors are being spread by social media user where this might confuse the others user where this can lead to misunderstanding and misinformation to happen. The figure below shows how to control scenario works on the respondents. Each respondent was shown with different sample tweets because the tweet was organized randomly to minimize the bias among respondents where each of them will answer different sample tweets for each case. For control scenario and prototype, there is one tweet that was selected from each case, Sabah earthquake and fire eruption at Hospital Sultanah Aminah Johor Bharu. There are three types of rating which are true, doubtful, and false. The reason why only three types of rating that has been given in this control scenario and prototype are because as a user, we usually think is the information is true or false, and sometimes we might confuse whether about half agreeing and half disagree about that information. That is how doubtful present as the type of rating where the user might feel unsure, unconfirmed, and undecided about that information. After the respondent rate the tweet, they need to choose whether they want to share that tweet or not. The figure below shows the respondents were asked whether they want to share that tweet or not without they know what other respondents' rate about that tweet was.

Next is the second step; respondents need to answer the prototype. The prototype design was similar like control scenario where it still has the rating features but was added new thing which is it shows the result of rating by the other respondents (show in percentage) once the respondents click the confirm button.

The result might influence the respondent's decision making on re-tweet that tweet or not. If the respondent rate as doubtful or false, they will need to write the reason why it could be false or doubtful so that other respondents would know the reason. Once they rate it, there will have a horizontal bar graph showed the percentage of respondent's rating on that tweet and the reason for false and doubtful choices. It can be if that respondent was unsure with his rating on that tweet and he rates for doubtful, once he clicks the submit button, the results for other's rating show there are higher percentage said exactly for that tweet, so this maybe could influence him to re-tweet that since there are many people rate for real. This features also might help the respondent to determine the accuracy and the credibility of the tweets. This situation might help the community on social media from the spread or re-tweet the rumors or misinformation. Thus, this can increase the credibility of information spreading on social media. Few samples of tweets selected from two cases which is Sabah earthquake on 2015 and fire eruption at Hospital Sultanah Aminah Johor Bharu on 2016. Each sample that has been categorizing into three types, which are true, false, and doubtful was selected based on the date of that tweets been posted on Twitter. For real tweets, it has been selected from a tweet posted by people that can trust, such as Ministry of health or from an organization such as "Jabatan Bomba dan Penyelamat." The table below shows the sample of real tweets.

For selecting the doubtful tweet, we select the tweet that posted on the day that the incident case happens. This is because the social media user has made their assumption about the incident happen before hearing what is happen and caused the incident from police or policeman that do the rescue process with their team. For the example, during the HSA case, Twitter was flooded with rumors that said the fire eruption at Hospital

Sultanah Aminah was caused by the explosion of phone charger in ICU ward. After the investigation is done, the cause of the fire eruption was coming from the burning of capacitor lamp in that building, which then spread to flammable material and then burning happen and cannot be controlled. The table below shows a few samples of doubtful tweets.

For selecting the false tweets, we select the tweet that posted after the media or rescue department such as police, "bomba" and the medic team from that hospital gave their statement about the incident such as the total death and others. For example, few false tweets posted about the total of the patient died during the fire eruption. Only six people dead were four of them are women and others are men, but the tweet posted said seven patients and one doctor was dead. This false information was spread by others on Twitter by re-tweeting, and this is not a suitable environment for social media user when the false information is spread and share everywhere, and this might confuse the other people on social media. The table below shows the sample of false tweets.

3.1 Expected Deliverable

The main objective of this research is to develop a prototype which includes rating features in a Twitter-like environment and to investigate the influence of crowdsourcing rating on tweets for the credibility of information spreading. At the end of this study, we would like to deliver a prototype in a Twitter-like environment.

In this prototype, there will have a new feature that current Twitter does not have, which is the rating feature. Rating feature will be including to investigate either it will influence people to make better decision making for information spreading or not since there are many information overloads in social media and some info's credibility is hesitant to be trusted. Our motivation to develop this prototype is because current features do not have this but have re-tweet or quote tweet which bring new people into a thread, inviting them to engage without directly addressing them. This feature (rating feature) enables the user to rate tweet about a disaster such as a tweet during the Sabah earthquake. It also might influence individual to be responsible to each tweet they post because people (their follower) will rate and it will show either most people trust or not with that tweet. If they are uncertain with that tweet, they will ask for evidence to prove that information is credible. Hence, this could prevent misunderstanding of information and the spreading of rumors and misinformation and increase the proposition of useful information on social media during disaster happen.

4 Results and Analysis

4.1 Questionnaires Results

The results from Questionnaires are as follow:

1. *Demographic info:* This survey has stated publish online for nine days from 20th April until 28th April 2017. Total respondents for this survey was 141. Unfortunately, only 139 respondent's data were collected because of irrelevant and illogic data from 2 of the respondents. 2 of them was said they are not the social media

social, but they have ticked for social media they used such as Instagram and Facebook. From the pie chart above, most of the respondents was a female where the percentage of female respondents was 69.1% while the percentage for male was only 30.9%. Respondents who have bachelor's degree was the highest among the level of the other where there is 73 participation, while for those who have Ph.D. was the lowest since there are only two respondents who participate in this research questionnaires and prototype. For Diploma/Matriculation/STPM and STAM level, there are 39 respondents where this is the second-highest frequency of respondents. Next, are Master and SPM or equivalent level where 16 and nine respondents have participated in answering the prototype and questionnaires.

2. *Usage of Social media:* Most of the respondents, which are 64% can differentiate whether it is rumors or right information in social media. Unfortunately, the rest of the respondents was unable to differentiate both of that information, hence the proposed prototype which came with new features which is the rating features by displayed it together with what other people were rate might be useful for them to know and differentiate rumors and correct information before they want to spread that information.
3. *Evaluation of the prototype:* The respondents were questioned either they will share or re-tweet that information or not if most of the crowd was rate it as accurate even though they have rated it as doubtful. There is an equal frequency for respondents that agree or strongly agree and disagree or strongly disagree. Forty respondents were agreed, and 40 respondents disagreed to retweet the information if most of the crowd was rated it as real even though they have rated it as doubtful. The highest percentage select by the respondents was neutral where there is 42.4% which is equal with 59 respondents were select neutral as their choice for this question
4. *Evaluation of the usability of prototype design:* Most of the respondents agreed that this prototype design had satisfied them. There are about 73 respondents were agreed and strongly agree they were satisfied with this prototype, while 47 respondents were neutral and only 19 respondents disagreed because this prototype had not satisfied them.

4.2 Analysis of the Frequency of the Respondent's Rating for Sample Tweets

In this section, the analysis of the occurrence of respondent ratings for a tweet is given.

1. Control scenario: Bar chart below shows the frequency of respondents who have a rate for sample tweets for Sabah earthquake case. For real tweets, most of the respondents were rate it as true, and only 12 and one respondents were rated is as doubtful and false. Then, for the doubtful tweets, respondents who have rated it as doubtful was the highest compared with respondents who have rated it as true. Next, for false tweets, most of the respondents were rated it as doubtful while the frequency of respondents who have rated as false has the approximate frequency like doubtful where there are 17 respondents was rate as false.

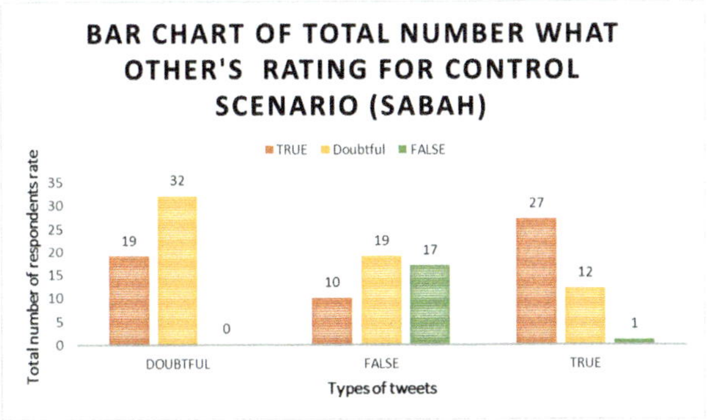

Fig. 3. Bar chart of the total number of what other's rating for control scenario (Sabah)

2. Proposed Prototype: Bar chart below shows the difference of the type of sample tweets that have been rated by the respondents either it is faithful, doubtful of false for HSA case. For real tweets, the majority of the respondents know it is actual tweets, and most of them were rated it as real. Just a few of 7 respondents rate the original tweets as doubtful, and only one respondent rate it as false. For doubtful tweets, as shown in Fig. 3 above, most of them were rated as doubtful, and only 9 of the respondents rated it as accurate and false. The last on is for false tweets, even though it is false tweets, majority of the respondents has rated it as true, but the frequency of respondents who have rated as doubtful and false was approximately similar like who have rated as real. Next, we moved to analysis for the frequency of respondent's rating on Sabah earthquake tweets (Fig. 4).

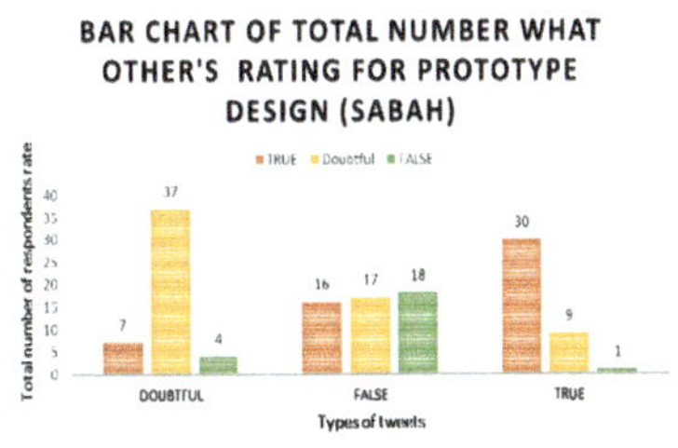

Fig. 4. Bar chart of the total number of what other's rating for the prototype (Sabah)

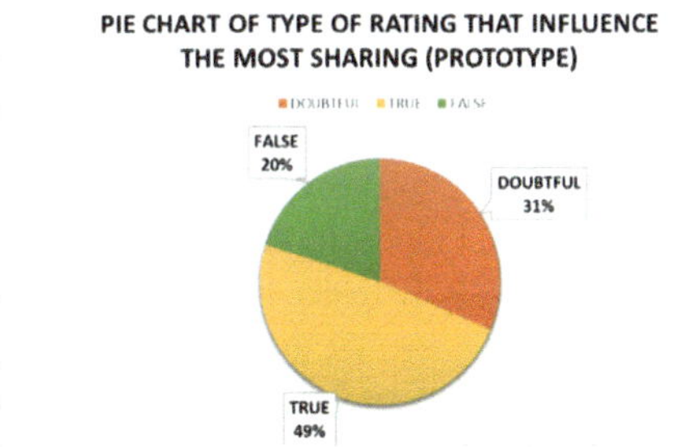

Fig. 5. Pie chart of rating that influences the most sharing (prototype)

4.3 Analysis of the Type of Tweets that Influence the Most Sharing

In this part, we would like to investigate which type of tweets that influence the most sharing among the respondents. Below we would like to show the differences in the type of tweets that have the most sharing between the control scenario and prototype part. This is because the only prototype has the crowdsource rating where the respondents were able to see what others have rated for those tweets before they decide to share it. Figure 5 shows the pie chart of type of rating that influences the most sharing among the respondents for prototype part where this part was included the crowdsourced rating feature where the respondents were shown what other people have rated for that tweets before they decide to share it. As the pie chart shows, the respondents who decide to share tweets that rated as real is higher, which is about 49% compared to those who share the tweets that rated as false and doubtful. However, the percentage of respondents who shared the tweets rated as doubtful was still higher than that who have share tweets rated as false.

5 Discussion and Conclusion

The focus of this research is how to crowdsource rating features will influence people's decision making on spreading of credible information on social media. The evolution of social networking sites such as Facebook and Twitter have allowed people around the world to engage and being connected. For example, during the Great East Japan Earthquake, people have used Twitter to post, share, and get updated information (Tanaka et al. 2013). This show social media have acted as an essential communication channel, especially during a disaster because it reached faster and farther than another communication channel such as newspaper. Information overloaded also cause rumors and misinformation to be spread; this happens because too many amounts of new information that being create, duplicating and sharing without know where the source come from (Kellingley 2016). Therefore, a technical solution was proposed to overcome the credibility of information spreading in social media.

We will give more detail instruction for them and add some popup message if they are making any mistake so that they can recover from that mistake by themselves. Lastly, we have found out that 52.52% of respondents were agreed that this prototype was satisfied with their needs on how its work. This has shown that the respondent's impression of the overall prototype design was good. In this study, we are concerned on finding out a way to help social media user from spreading rumors and false information on Twitter by proposed a technical solution where crowdsourced rating works on tweets to examine either it influences people's decision making for the credibility of information spreading or not. People need to know and be responsible for what information they spread since there is much information (information overload) spread by a person or anonymous on social media. This will benefit the individuals in terms of deciding the truthful information since there are many information overloads in social media such as Twitter. In real life, Twitter does not have this feature. Thus we developed a prototype and trying it to 139 respondents. The outcome of the study is, this feature gives a positive result among the respondents where it helps them to determine and identify the credibility, accuracy and prevent them from spreading the rumors and misinformation in social media. The usability evaluation is used to evaluate the prototype on respondent's impression about the proposed prototype. As for the sharing information, we can conclude that most of the respondents were decided to share the tweets that have been rated as real, so that it can minimize from spreading the rumors and inaccurate information in social media.

References

Abdullah, N.A., Nishioka, D., Murayama, Y.: Questionnaire testing: identifying Twitter user's information sharing behaviour during disaster. J. Inf. Process. **24**(1), 20–28 (2016)

Acar, A., Muraki, Y.: Twitter for crisis communication: lesson learn from Japan's tsunami disaster. Int. J. Web Based Communities **7**(3), 392–402 (2011)

Boyd, D., Golder, S., Lotan, G.: Tweet, tweet, retweet: conversational aspects of retweeting on Twitter, pp. 1–10 (2010)

Brabham, D.C.: Crowdsourcing the public participation process for planning projects. Plann. Theory **8**(3), 242–262 (2009)

Chakraborty, R., Agrawal, M., Rao, H.R.: An exploratory study of information processing under stress: the case of Mumbai Police Control and first responders during terrorist attacks of November 26, 2008. In: Pre-ICIS ISGSEC Workshop on Information Security and Privacy, St. Louis, Missouri, 12 December 2010 (2010)

DiFonzo, N., Bordia, P.: Rumor Psychology Social and Organizational Approaches, pp. 13–34. American Psychological Association, Washington, D.C. (2007)

Dijkc, J.V., Poell, T.: Understanding social media logic. Media Commun. **1**, 2–14 (2013)

Ekşil, A., Celikli, S., Kiyan, G.S.: The effects of social networking on disaster communication used by the emergency medical and rescue staff. J. Acad. Emerg. Med. **13**, 58–61 (2014)

Ennis, R.H.: Critical thinking dispositions: their nature and assessability. Informal Log. **18**(2), 165–182 (1996)

Fisher, R.: A gentleman's handshake: the role of social capital and trust in transforming information into usable knowledge. J. Rural Stud. **31**, 13–22 (2013)

Jaeger, P., Shneiderman, B., Fleischmann, K., Preece, J., Qu, Y., Fei, W.: Community response grids: e-government, social networks, and effective emergency management. Telecommun. Policy **31**(10–11), 592–604 (2007)

Kaplan, A.M., Haenlein, M.: Users of the world, unite! The challenges and opportunities of social media. Bus. Horiz. **53**(1), 59–68 (2010)

Kelllingley, N.: Information overloads, why it matters and how to combat it. https://design.org/literature/article/information-overloadwhy-it-matters-and-how-to-combat-it. Accessed 28 Sept 2016

Li, H., Sakamato, Y.: Computing the veracity of information through crowds: a method for reducing the spread of false messages on social media, pp. 2003–2011 (2015)

Lister, M.: 40 Essential Social Media Marketing Statistics for 2017, 20 January 2017. http://www.wordstream.com/blog/ws/2017/01/05/social-media-marketing-statistics. Accessed 24 Apr 2017

Mendoza, M., Poblete, B., Castillo, C.: Twitter under crisis: can we trust what we RT?, pp. 71–79 (2010)

Murayama, Y., Saito, Y., Nishioka, D.: Trust issues in disaster communications. In: Proceeding of the 46th Hawaii International Conference on System Science, pp. 335–342 (2013)

Oh, O., Agrawal, M., Rao, H.R.: Community intelligence and social media services: a rumor theoretic analysis of tweets during social crisis. MIS Q. **37**(2), 407–426 (2013)

Perrin, A.: Social media usage: 2005–2015, 8 October 2015. http://www.pewinternet.org/2015/10/08/social-networking-usage-2005-2015/. Accessed 13 Nov 2016

S. V. President: Asia Pacific: daily social media usage 2015 | statistic (2016). https://www.statista.com/statistics/412462/apac-daily-socialmedia-usage/. Accessed 15 Oct 2016

Spence, P.R., Lachlan, K.A., Lin, X., Greco, M.D.: Variability in Twitter content across the stages of a natural disaster: implications for crisis communication. Commun. Q. **63**(2), 171–186 (2015)

Sreedhar, G. (ed.): Design Solutions for Improving Website Quality and Effectiveness. IGI Global, Hershey (2016)

Tanaka, Y., Sakamato, Y., Matsuka, T.: Toward a social-technological system that inactivates false rumors through the critical thinking of crowds, pp. 649–658 (2013)

Thomson, R., Ito, N., Suda, H., Lin, F., Liu, Y., Hayasaka, R., Isochi, R., Wang, Z.: Trusting tweet: the Fukushima disaster and information source credibility on Twitter, pp. 1–10 (2010)

Uslaner, E.: Trust but verify: social capital and moral behavior. Soc. Sci. Inf. (Information sur les Sciences Sociales) **38**(1), 29–55 (1999)

A Study of Deterioration in Classification Models in Real-Time Big Data Environment

Vali Uddin[1], Syed Sajjad Hussain Rizvi[1], Manzoor Ahmed Hashmani[2], Syed Muslim Jameel[2(✉)], and Tayyab Ansari[1]

[1] Faculty of Engineering Sciences and Technology, Hamdard University, Karachi, Pakistan
[2] Department of Computer and Information Sciences, University Technology PETRONAS, Seri Iskandar, Malaysia
muslim_16000370@utp.edu.my

Abstract. Big Data (BD) is participating in the current computing revolutions. Industries and organizations are utilizing their insights for Business Intelligence (BI). BD and Artificial Intelligence are one of the fundamental pillars of Industrial Revolution (IR) 4.0. IR 4.0 demands real time BD analytic for prediction and classification. Due to complex characteristics of BD (5 V's), BD analytics is considered a difficult task in offline mood. However, in real time or online mood, BD analytic become more challenging and requires Online Classification Models. In real time mood, the nature of input streams (input data) and target classes (output class) are dependent and non-identically distributed, which cause deterioration in OCM. Therefore, it is necessary to identify and mitigate the causes of this deterioration in OCM and improve OCM performance in RTBDE. This study investigates some fundamental causes of deterioration of Online Classification Models and discusses some possible mitigation approaches. This study also presents some experimental results to show the deterioration in OCM due to real time big data environment. In the future, this study will propose a method to mitigate deterioration in Online Classification Models.

Keywords: Big Data · Real-time machine learning · Stream analytics · Online Classification

1 Introduction

The usage of data for business process improvement has been adapted for four decades. Through Data mining and Data Analytics, organizations use to enhance their business performance or estimate their potential risks. Fundamentally, this historical data was well structured (i.e. relational) and normalized in nature. Even though the data preparation process was costly enough (in terms of computational processing), the prepared data was limited in size too [1]. Due to the limitation of data size, the traditional data mining algorithms/models were bound to predict in certain data space, and therefore, these models were less accurate as compared to the current models. From

F. Saeed et al. (Eds.): IRICT 2019, AISC 1073, pp. 79–87, 2020.
https://doi.org/10.1007/978-3-030-33582-3_8

the time of traditional Data Mining, the industries also use to generate a massive amount of Big Data. However, due to the unawareness towards the potential and capabilities of Big Data, the management and storage of Big Data use to considered as burden for Organizations. Gradually, researchers investigated further in the area of Data Mining and Data Analytics and gradually found the significance of Big Data for industries and organizations, which can help them to minimize their risk and cost or predict things more accurately as compared to the traditional Data Mining approaches. Along with the benefits associated with utilization of Big Data for Business Intelligence, researchers highlighted some critical issues, which arise due to complex and versatile characteristics of Big Data initially 3 V's (Velocity, Variety, Volume). To cope up this issue, new Classification approaches introduced, for example, Deep Learning based Classification is well stable to predict/classify from Big Data [2]. Additionally, the Real-Time Big Data Environment (in which Big Data is being analyzed in real-time) increased the complexity of Big Data until 5 V's (Variability and Veracity). The complexities due to Real-Time Big Data Environment (RTBDE), urged researchers to improve the classical Classification Models towards Online Classification Models (OCM). This OCM supposed to be robust enough to work efficiently in RTBDE. Certainly, still, the unpredicted behavior of Variability and Veracity characteristics, cause deterioration in OCM [3]. Currently, several studies highlight the issues related to this problem [4, 5] and demand the more robust OCM to avoid this deterioration. The robustness is defined as a key attribute of OCM models, whose performance does not degrade substantially even after changes in feature wise distribution of data due to Variability and Veracity. The principal object to handle these complex characteristics is to improve accuracy performance in OCM caused by Variability and Veracity in input date in these classification models.

1.1 Motivation

As per the report published by Tractica research agency, that till 2025, the world revenue generated by AI will reach $36.8 billion. Currently, industries and organizations, including automotive, education healthcare, are utilizing the cutting-edge technologies Machine Learning and Big Data to compete in the race of digitization, which in result make their systems fast, reliable, cost and time-efficient [6]. Interestingly, after the Industrial Revolution 4.0, more than 19000 companies are currently harnessing the power of Big Data through Classification Models to advance their legacy systems. However, the current processes (in IR4.0) demands Classification in Real-Time Big Data Environment (RTBDE). RTBDE is dynamically changing and non-stationary environment and requires Online approach in Classification. In RTBDE the input data may come across from many data sources. Therefore, it is more likely feature-wise or class-wise distribution may change over time, which adversely affect the performance Online Classification models in such a way that these models may face deterioration with time [7]. The issue of deterioration in OCM in RTBDE is radical in nature and affects several critical applications, including Health, Education and Business, etc. Hence, the complex characteristics of Big Data upfront more challenging situation for researchers to explore new ways to mitigate the deterioration of OCM in RTBDE.

1.2 Dynamic Behavior of Big Data and Improvement in Classification Models

The Big Data in nature is very complex and versatile. Whereas, the Variability and Veracity characteristics of Big Data are unpredicted. Due to which the Intelligent Systems (for prediction, classification, etc.) unable to adjust these dynamic behaviours of Big Data. The concept of Dynamic Classification model is essential to handle the issues cause by Big Data complex and versatile characteristics. These dynamic approaches can be categories into partial and fully dynamic. The concept of dynamic nature of Big Data Classification (BDC) models is to make classification models self-regulate, which will adjust the BDC models dynamically if the data trends change (data trends change due to Variability and Veracity). Moreover, the improvement of this improvement in BDC models results in better robustness. The robustness is defined as a key attribute of BDC models whose performance does not degrade substantially even after changes in feature wise distribution of data due to Variability and Veracity.

1.3 Classification Using Big Data

In Real-Time Big Data Environments, the input data may come from different sources on a temporary basis. Therefore, the Online Classification Models may utilize the limited amount of memory and inspect/process each instance individually. OCM must ensure the quick response with acceptable accuracy, and able to adapt temporal changes (which are supposed to be encounter due to the non-stationary environment).

2 Related Work

Generally, the classification models are divided into two significant types;

1. Offline or Batch Learning
2. Online or Real-Time Learning

In offline learning, the classification algorithm learns from the available dataset (testing data) and then classify the unseen dataset (testing data). Typically, during the algorithm development process, the available data set is divided into two sets (testing and training). These sets are shuffled using several cross-validation techniques to verify the accuracy of build models and later utilize for classification using the input (unseen) data. In offline mode, the behavior input data for classification is deterministic, and feature of class distribution of input data will always be the subset of the training dataset. However, in the online setting, particularly real-time stream classification using the significant data streams are not deterministic. In bid data streams classification, the uncertainty in input streams changes the input feature behavior and class distribution, which in return causes a massive deterioration in the classification algorithm, which significantly affects the performance. For example, an online shop may require classifying product purchasing behavior (whether a customer buys of not a product) for every customer. However, it is evident, the customer behavior is not a static attribute and changes with time, there are several variables which contribute for a final decision, for example, weather, season, saving and mood, etc. Therefore, due to un deterministic behavior of customer classification requires the online mode of learning. The system ought to get the "label" indicating whether indeed the customer bought the product or not, and this feedback can be used to tune the classification algorithm to avoid any deterioration (Table 1).

Table 1. Major techniques & significance of classifier evaluation in data streams

Techniques	Significance	Ref.
Majority class classifier	The most straightforward mechanism of majority classifier stores a count for each class labels and classify based on most frequent class labels	[8]
No-change classifier	Predict the last class in the data stream. It exploits autocorrelation in the label assignments, which is very common	[9]
Lazy classifier	Some of the instances are seen in this classifier and classify based on closest class boundary	[10]
Naive Bayes	This approach works on probability-based and attribute values. It computes the probability for every available class and predicts based on the high probability	[11]
Decision trees	Decision tree learners build a tree structure from training examples to predict class labels of unseen examples	[12]
Ensembles	Ensemble approach aggregates the power of all the individual instances and classifies based on several criteria	[13]

3 Research Methodology

This research study has gone through comprehensive literature studies to identify the most significant Machine Learning classification models for streaming datasets. This research concern was to see the performance of existing models during the uncertain behavior of Big Data Stream. Therefore SEA data stream is selected (benchmark datastream) to make input stream much challenging. This study followed below objectives;

1. A comprehensive literature review of existing Big Data Stream classification approaches.
2. Perform the experiments to investigate the performance of existing Big Data classification models
3. Acquired results deduction and analysis.

3.1 Dataset

The issue of concept drift is mostly evaluated using the synthetic data stream. Synthetic data streams are a good source to measure the performance of our machine learning algorithm given different types of drifts. This experiment uses the SEA [14] data stream (generator). SEA dataset is considered as one of the benchmark dataset for determining the deterioration in the classification models in real time big data environment [15].

SEA data stream contains 60,000 examples, three attributes, and three classes. Attributes are numeric between 0 and 10; only two are relevant. There are four concepts, 15,000 examples each, with different thresholds for the concept function, which is if relevant_feature1 + relevant_feature2 > Threshold then class = 0. Threshold values are 8, 9, 7, and 9.5. Dataset has about 10% of the noise.

3.2 Models and Tools

In this experiment, 22 different classification models are evaluated. These classifications are belonging to below seven categories; (1) Tree, (2) Discriminant Analysis, (3) Support Vector Machine, (4) Subspace Discriminant, (5) Subspace KNN, (6) Rusboosted Trees, (7) KNN. All the experiments in this study are carried out using the MATLAB: especially its, statistical and machine learning toolbox and deep learning toolbox.

3.3 Performance Parameters

This study selected six different parameters to evaluate the performance of the 22 selected machine learning algorithms. These parameters include True Positive Rate, False Negative Rate, Training Accuracy, Testing Accuracy, Prediction Speed, Train Time. These selected parameters are appropriate to determine the deterioration in classification models after occurrence of uncertain data in BD stream. Moreover, this experiment carried out into two moods of Principal Component Analysis (PCA), for example, PCA Disable and PCA Enable.

4 Results

To validate the performance of existing models using SEA data stream, this study undergoes through a systematic investigation. This experiment mainly finds the Training and Testing accuracy along with the model's prediction speed and Train Time. Mainly, the nature of experiments is categorized into two-mode PCA enabled and PCA Disabled, as mentioned below. Through the critical analysis of the results obtained after various experiments, we deduct some significant findings for PCA disable mode. Different models of a machine learning app are applying on SEA training data set to discover the best model. The method of detecting the best model by calculating the training, testing accuracy, prediction speed and confusion matrix in both conditions disabled PCA and enabled PCA. In disabled PCA, cubic support vector machine has minimum training accuracy 75.7, and maximum training accuracy has complex tree, linear, quadratic, median and coarse Gaussian support vector machine was 87.4%. Also, RUS Boosted maximum testing accuracy was 84.9433 while minimum testing accuracy was 37.3067 of the elaborate tree, linear SVM and median SVM. Furthermore, the peak prediction speed was detected in quadratic discriminant while lowest prediction speed in cosine KNN. To sum up, the RUS Boosted tree model was found the best model, as shown in Table 2.

Table 2. Performance of SEA datastream on existing models using PCA disable

	Confusion matrix			True class		Tue positive rate	False negative rate	Training accuracy (%)	Testing accuracy (%)	Prediction speed (Obs/sec)	Train time (sec)
				0	1						
Tree	Fine tree	Predict class	0	97	3	97	3	86.2	39.3117	~290000	5.3569
			1	86	14	14	86				
	Medium tree	Predict class	0	99	1	99	1	87.2	37.4033	~230000	1.2703
			1	96	4	4	96				
	Complex tree	Predict class	0	100	0	100	0	87.4	37.3067	~270000	0.88521
			1	100	0	0	100				
Discriminant analysis	Linear discriminant	Predict Class	0	95	5	95	5	86.3	53.7033	~430000	1.2803
			1	72	28	28	72				
	Quadratic discriminant	Predict class	0	95	5	95	5	86.3	54.1000	~1100000	1.002
			1	72	28	28	72				
Support vector machine	Linear SVM	Predict class	0	100	0	100	0	87.4	37.3067	~43000	18.159
			1	100	0	0	100				
	Quadratic SVM	Predict class	0	100	0	100	0	87.4	54.2067	~27000	1110.5
			1	100	0	0	100				
	Cubic SVM	Predict class	0	79	21	79	21	75.7	58.9183	~23000	1475.5
			1	47	53	53	47				
	Fine Gaussian SVM	Predict class	0	98	2	98	2	86.9	42.5583	~11000	72.023
			1	88	12	12	88				
	Medium Gaussian SVM	Predict class	0	100	0	100	0	87.4	37.3067	~11000	74.932
			1	100	0	0	100				
	Coarse Gaussian SVM	Predict class	0	100	0	100	0	87.4	62.9650	~15000	87.884
			1	100	0	0	100				
KNN	Fine KNN	Predict class	0	91	9	91	9	83.6	62.9650	~100000	7.053
			1	65	35	35	65				
	Medium KNN	Predict class	0	97	3	97	3	86.6	46.9583	~54000	7.2397
			1	85	15	15	85				
	Coarse KNN	Predict class	0	>99	<1	>99	<1	87.3	38.0083	~21000	9.477
			1	99	1	1	99				
	Cosine KNN	Predict class	0	97	3	97	3	86.8	45.3233	~7000	19.488
			1	87	13	13	87				
	Cubic KNN	Predict class	0	97	3	97	3	86.6	46.7800	~26000	10.372
			1	85	15	15	85				
	Weighted KNN	Predict class	0	94	6	94	6	85.5	58.4833	~79000	8.3205
			1	73	27	27	73				
Boosted tree	Boosted tree	Predict class	0	98	2	98	2	86.8	39.6550	~68000	9.8194
			1	90	10	10	90				
Bagged tree	Bagged tree	Predict class	0	95	5	95	5	86.0	56.2817	~55000	12.634
			1	75	25	25	75				
Subspace discriminant	Subspace discriminant	Predict class	0	98	2	98	2	87.0	42.4250	~58000	6.7217
			1	91	9	9	91				
Subspace KNN	Subspace KNN	Predict class	0	96	4	96	4	85.8	52.7650	~14000	10.06
			1	82	18	18	82				
RUSBoosted tree	RUSBoosted tree	Predict class	0	79	21	79	21	79.9	84.9433	~110000	10.931
			1	13	87	87	13				

Table 3. Performance of SEA datastream on existing models using PCA enable

	Confusion matrix			True class 0	True class 1	Tue positive rate	False negative rate	Training accuracy (%)	Testing accuracy (%)	Prediction speed (Obs/sec)	Train time (sec)
Tree	Fine tree	Predict class	0	98	2	98	2		40.8800	~170000	5.1464
			1	89	11	11	89				
	Medium tree	Predict class	0	99	1	99	1	87.2	40.7067	~910000	1.1431
			1	98	2	2	98				
	Coarse tree	Predict class	0	100	0	100	0	87.4	37.3067	~950000	0.71467
			1	100	0	0	100				
Discriminant analysis	Linear discriminant	Predict class	0	95	5	95	5	86.3	53.7033	~450000	1.1494
			1	72	28	28	72				
	Quadratic discriminant	Predict class	0	95	5	95	5	86.3	54.1000	~700000	1.1466
			1	72	28	28	72				
Support vector machine	Linear SVM	Predict class	0	100	0	100	0	87.4	37.3067	**~12000**	30.123
			1	100	0	0	100				
	Quadratic SVM	Predict class	0	100	0	100	0	87.4	44.6200	~31000	759.53
			1	100	0	0	100				
	Cubic SVM	Predict class	0	71	29	71	29	68.4	55.0133	~29000	1033
			1	51	49	49	51				
	Fine Gaussian SVM	Predict class	0	98	2	98	2	86.9	42.5417	~13000	61.18
			1	88	12	12	88				
	Medium Gaussian SVM	Predict class	0	100	0	100	0	87.4	37.3067	~17000	86.907
			1	100	0	0	100				
	Coarse Gaussian SVM	Predict class	0	100	0	100	0	87.4	37.3067	~17000	75.887
			1	100	0	0	100				
KNN	Fine KNN	Predict class	0	91	9	91	9	83.6	62.9600	~150000	2.3245
			1	65	35	35	65				
	Medium KNN	Predict class	0	97	3	97	3	86.6	46.9733	~72000	2.7134
			1	85	15	15	85				
	Coarse KNN	Predict class	0	>99	<1	>99	<1	87.3	38	~26000	5.4074
			1	99	1	1	99				
	Cosine KNN	Predict class	0	97	3	97	3	86.8	45.3183	**~6700**	16.757
			1	87	13	13	87				
	Cubic KNN	Predict class	0	97	3	97	3	86.5	46.7650	~26000	7.732
			1	85	15	15	85				
	Weighted KNN	Predict class	0	94	6	94	6	85.5	58.4817	~97000	8.6364
			1	73	27	27	73				
Boosted tree	Boosted tree	Predict class	0	98	2	98	2	86.9	41.0667	~68000	8.262
			1	90	10	10	90				
Bagged tree	Bagged tree	Predict class	0	95	5	95	5	86.3	56.4383	**~71000**	10.114
			1	76	24	24	76				
Subspace discriminant	Subspace discriminant	Predict class	0	98	2	98	2	86.9	44.4983	**~70000**	5.5106
			1	89	11	11	89				
Subspace KNN	Subspace KNN	Predict class	0	95	5	95	5	85.5	53.64	~21000	7.6529
			1	82	18	18	82				
RUSBoosted tree	RUSBoosted tree	Predict class	0	79	21	79	21	79.9	85.0250	~130000	8.984
			1	13	87	87	13				

While in PCA enabled, training accuracy of all models are more than 80% except RUS Boosted tree was 79.9. Also, the testing accuracy result got slightly increase the accuracy of each model with the accuracies of each model when PCA disabled. Furthermore, training speed drops by 30 to 50% in the fine tree, quadratic discriminant, linear SVM, bagged tree and subspace discriminant while the minor drop in cosine KNN by 5%. Also, medium tree prediction speed drastically increases by 400%. In conclusion, RUS Boosted tree model training, and testing accuracy remains the same in both conditions, as shown in Table 3.

5 Conclusion

Through the experimental results presented in this paper, an attempt has been made to highlight the performance degradation issue and their origins in the real-time big data environment. However, the deterioration of the classification model is directly relevant to the severity and frequency of change occurrence (Concept Drift) in the input stream, which happens due to variability and veracity characteristics of Big Data environment. When ever a model is realized a sudden or gradual change in real time streams, it doesn't sustain its performance accuracy, which also impacts on overall performance accuracy, as shown in Tables 2 and 3. Furthermore, we also can conclude that it is necessary to identify and mitigate the causes of this deterioration in OCM and improve OCM performance in real time bid data environment. In addition, sudden, gradual or abrupt changes in stream needs to be investigated further to avoid the deterioration of Online Classification Models. In the future, this study will enhance towards to improvement of classification accuracy during Concept Drift.

Acknowledgment. This research study is a part of the funded project under a matching grant scheme supported by University Technology Petronas (UTP), Malaysia and Hamdard University, Pakistan. This is the second phase, which focuses on problem formulation, preliminary simulations, and identification of possible problem mitigation approaches.

References

1. Jin, X., et al.: Significance and challenges of big data research. Big Data Res. **2**(2), 59–64 (2015)
2. LeCun, Y., Bengio, Y., Hinton, G.: Deep learning. Nature **521**(7553), 436 (2015)
3. L'Heureux, A., Grolinger, K., Elyamany, H.F., Capretz, M.A.: Machine learning with big data: challenges and approaches. IEEE Access **5**, 7776–7797 (2017)
4. Zhang, Q., et al.: An adaptive droupout deep computation model for industrial IoT big data learning with crowdsourcing to cloud computing. IEEE Trans. Ind. Inf. **15**(4), 2330–2337 (2018)
5. Jameel, S.M., et al.: A fully adaptive image classification approach for industrial revolution 4.0. In: International Conference of Reliable Information and Communication Technology, vol. 843, pp. 311–321. Springer, Cham (2018)
6. Sun, Y., Tang, K., Zhu, Z., Yao, X.: Concept drift adaptation by exploiting historical knowledge. IEEE Trans. Neural Netw. Learn. Syst. **29**(10), 4822–4832 (2017)

7. Gama, J., Žliobaitė, I., Bifet, A., Pechenizkiy, M., Bouchachia, A.: A survey on concept drift adaptation. ACM Comput. Surv. **46**(4), 1–37 (2014)
8. Kononenko, I.: Semi-naive Bayesian classifier. In: European Working Session on Learning, vol. 482, pp. 206–219. Springer, Heidelberg (1991)
9. Ng, W.L.: A simple classifier for multiple criteria ABC analysis. Eur. J. Oper. Res. **177**(1), 344–353 (2007)
10. Veloso, A., Meira Jr., W., Zaki, M.J.: Lazy associative classification. In: Sixth International Conference on Data Mining (ICDM 2006), pp. 645–654. IEEE (2006)
11. Wang, Q., et al.: Naive Bayesian classifier for rapid assignment of rRNA sequences into the new bacterial taxonomy. Appl. Environ. Microbiol. **73**(16), 5261–5267 (2007)
12. Jain, V., Phophalia, A., Bhatt, J.S.: Investigation of a joint splitting criteria for decision tree classifier use of information gain and gini index. In: 2018 IEEE Region 10 Conference (TENCON 2018). IEEE (2018)
13. Bertini Junior, J.R., do Carmo Nicoletti, M.: An iterative boosting-based ensemble for streaming data classification. Inf. Fusion **45**, 66–78 (2019)
14. Sayed-Mouchaweh, M. (ed.): Learning from Data Streams in Evolving Environments: Methods and Applications, vol. 41. Springer, Heidelberg (2018)
15. Elwell, R., Polikar, R.: Incremental learning of concept drift in nonstationary environments. IEEE Trans. Neural Netw. **22**(10), 1517–1531 (2011)

Missing Data Characteristics and the Choice of Imputation Technique: An Empirical Study

Oyekale Abel Alade[1,2(✉)], Roselina Sallehuddin[1], Nor Haizan Mohamed Radzi[1], and Ali Selamat[1]

[1] Faculty of Computing, Universiti Teknologi Malaysia, Skudai, Johor, Malaysia
kaleabel@gmail.com, {roselina, haizan, aselamat}@utm.my
[2] Computer Science Department, Federal Polytechnic, Bida, Nigeria

Abstract. One important characteristic of good data is completeness. Missing data is a major problem in the classification of medical datasets. It leads to incorrect classification of patients, which is dangerous to health management of patients. Many imputation techniques have been employed to solve this problem, but these techniques are without recourse to the characteristics that cause the missingness. In this paper, we investigated the causes of missing data in a medical dataset and proposed multiple imputation technique to solving the problem of missing data. A 5-fold-iteration multiple imputation was employed. The whole missing values in the dataset was regenerated 100%. The imputed datasets were validated using extreme learning machine (ELM) classifier. The results show improvement on the accuracy of the imputed datasets. The work can, however, be extended to compare the accuracy of the imputed datasets with different classifiers.

Keywords: Imputation techniques · Mechanism of missingness · Missing data · Multiple imputations

1 Introduction

Missing data describes the absence of important datapoints in datasets. Data are collected at various points for medical investigation. The pervasive nature of missing data has been described as one of the most difficult tasks in data science [1].

The accuracy of classification from incomplete dataset is unreliable, because some vital information relevant to the analysis might be lost. Besides, some classifiers find it difficult to run in the face of missing datapoints [2]. Therefore, there is a necessity for missing value imputation. Imputation estimates the possible value of the missing datapoints from observed data and fixed the estimated values as a replacement of the lost values.

Some authors used the percentage of missingness in a dataset as a measure for the choice of imputation technique [3]. Some implementations treat missing data implicitly, this brings about different results when such treatments are replicated using different applications. This work examined the characteristics and treatment of missing

F. Saeed et al. (Eds.): IRICT 2019, AISC 1073, pp. 88–97, 2020.
https://doi.org/10.1007/978-3-030-33582-3_9

datapoints to choose an imputation technique; ELM classifier was used to validate the imputed datasets because of its speed and generalization of performance [4].

The rest of this paper is organised as follows: Sect. 2 reviews relevant work on imputation of medical datasets; characteristics and treatments of missing data are discussed in Sect. 3; in Sect. 4, ELM classifier was discussed, Sect. 5 proposed multiple imputation; Sects. 6 and 7 are based on experimental setup and discussion of results respectively, while Sect. 8 concluded the study.

2 Review of Literature

Medical datasets have been classified by many researchers. Some datasets are complete while some are incomplete. Several studies had been carried out on missing data and imputation techniques from different perspectives. Some authors explicitly treat missing data using different imputation techniques, while some are passive about it leading to assignment of zeros (0), deletions of cases/features or completely ignore missing values.

Armina *et al.* in [5] analysed various imputation techniques. They grouped imputation techniques into four (4) broad categories: global, local, hybrid and knowledge assisted approaches, but there was no experimental proof of the imputation techniques in their study. In [6], Tsai and Chang investigated the effects of filtering outliers from datasets on imputation tasks using instance selection on categorical, numerical and mixed type attributes. The effectiveness of the method was tested with k-NN and SVM classifiers. To compare the performance of three Bayesian imputation techniques, [7] placed prior distribution on attributes with missing values. Monte Carlo simulation model was used to examine the performance of the sibling models. The result showed that mean and mean square error of logistic and Bayesian models depend on risk factors examined, and the mechanism of missing data been used. This gives an insight to the necessity for consideration of types of missingness for a right choice of imputation technique [8] which is discussed in Sect. 3. Gaussian mixture model and extreme learning machines (GMM-ELM) was proposed as a reliable approximation technique for multiple imputation of data by Sovilj *et al.* in [9]. The result of their work improved imputation of missing datapoints over mean imputation technique, however, the causes of missingness was not considered, and the execution time was longer. Multiple imputations with Pohar-Perme method was used in [10] to estimate the net survival for stage-specific colorectal cancer. However, the mechanism of missingness of data in the dataset was not taken into consideration.

Although, most of the imputation techniques mentioned above attempted to fill the missing values by approximately conforming to the distribution of the datasets. However, the types of missing values were not explicitly modelled, these may lead to bias result. Therefore, this study examines the nature of missingness in a dataset for the right choice of imputation technique. This is considered in the next section.

3 Characteristics and Treatments of Missing Data

The theory of missing characteristics is examined in this section. The models determine the causes of the missingness in the dataset. The knowledge enhance the choice of appropriate imputation technique [3].

3.1 Missing Data Characteristics

The characteristics of missingness is refers to as mechanism of missing data [8]. It explains the reason for the missingness of datapoints in a dataset. This comes in three categories: Missing at random (MAR), missing completely at random (MCAR), and missing not at random (MNAR). With MAR, the probability of a missing datapoint in an instance is a function of other observed values, and not on the missing datapoint itself. In MCAR, the probability that an instance has a missing datapoint does not depend on the observed datapoints or itself. For $\boldsymbol{X}$ missing data indicator matrix and $\boldsymbol{Z}$ complete data matrix, MAR and MACR are given in Eqs. 1 and 2 respectively:

$$P(\boldsymbol{X}|Z) = P(\boldsymbol{X}|\boldsymbol{Z}_{obs})\,for\ all\ Z \tag{1}$$

$$P(\boldsymbol{X}|Z) = P(X)\,for\ all Z \tag{2}$$

According to [6], any imputation technique can be applied to MCAR. For MNAR, the probability of missingness may depend on the value of a missing datapoint and some/all other observed datapoints. Tsai in [6] observed that MNAR mechanism will be difficult to judge because the missing data is unknown.

The treatment of missing data should be based on the mechanism of missingness as explained in the next section.

3.2 Treatments of Missingness

Two broad approaches common in literature for the treatment of missing data are omission, and imputation of missing data [9]. We considered the approaches for the choice of suitable treatment of missing values (MV).

Omission scheme deletes instances with missing data [9]. This approach reduces the sample size and limits the study power. Therefore, omission of missing values would be inadequate in this work.

Imputation replaces the missing datapoints with analytical plausible values. Method of imputation can be classified as single or multiple [2]. Single imputation regenerate the missing datapoint only ones using mean, median, mode, last observation carried forward or conditional mean [11]. It is simple, but it underestimates the variance, ignores relationships among variables and complicates statistical inferences in the dataset [1]. Multiple imputation scheme replaces missing data with scientifically plausible values, such as Multiple Imputation by Chain Equation and Regression Tree (MICE-CART) [2]. Different models are used to predict the replacement values [12].

It is suitable for all mechanism of missingness, because it accounts for uncertainty behind missing values [2]. Therefore, we proposed multiple imputation technique in Sect. 5 of our study.

4 Extreme Learning Machine Classifier

ELM classifier is a single layer feed-forward neural network (SLFN). The learning parameters (input weights W and hidden neuron biases b) of the network are randomly initialised and are never adjusted. The hidden neuron output weights (β) are analytically determined using least square solution of a linear system. This makes the implementation very simple. Since the random initialized input weights W are fixed, the output connection weight β does not depend on input weights and biases. It is generally accepted for its high generalization of performance because different activation function can be used in the hidden layer of the networks [13].

The training of the SLFN is equivalent to the least square solution $\hat{\beta}$ of the linear system $\boldsymbol{H}\beta = \boldsymbol{T}$ for the fixed input weights $\boldsymbol{w}_i$ and hidden neuron biases $\boldsymbol{b}_i$ as in Eq. 3

$$\begin{aligned} &\| H(w_1, \ldots, w_L, b_1, \ldots, b_L)\hat{\beta} - Y \| \\ &= \min_{\beta} \| \boldsymbol{H}(w_1, \ldots, w_L, b_1, \ldots, b_L)\beta - \boldsymbol{Y} \| \end{aligned} \tag{3}$$

ELM theory proposed the smallest norm least square solution in Eq. 4

$$\hat{\beta} = \boldsymbol{H}^{\dagger} Y \tag{4}$$

$$\boldsymbol{H}^{\dagger} = (\boldsymbol{H}^T \mathbf{H})^{-1} \mathbf{H}^T \tag{5}$$

ELM analytically solved the least square solution linearly as in Eq. 6

$$\hat{\beta} = (\boldsymbol{H}^T \mathbf{H})^{-1} \mathbf{H}^T Y \tag{6}$$

5 Proposed Multiple Imputation Model

In this section, we proposed regression model for multiple imputation of missing data, because it draws values randomly from donor instances to predict values that are closed to the missing values to be predicted; also, regression creates inter-data variability using stochastic elements. The results of this data pooling produce correct standard error estimates.

Given an incomplete dataset $\boldsymbol{X}$ with m variables (X_1, …, X_m) and complete dataset Y, an incomplete random sample $\boldsymbol{x}$ can be estimated with regression model in Eq. 7:

$$E(Y|X_1,\ldots,X_m) = \beta_0 + \sum\nolimits_{m=1}^{M} \beta_m X_m \quad (7)$$

where $X = (X_1,\ldots,X_m)$ is the correlation coefficient, $\beta = (\beta_1,\ldots,\beta_m)$ the relative size of coefficient of regression to one another, and β_0 is the intercept.

To make an inference of the regression coefficient, β_m of the missing value and the standard errors $e_1 \ldots e_M$ are obtained for each dataset in M. The mean dataset estimates $(\hat{\beta})$ is given in Eq. 4:

$$\hat{\beta} = \frac{1}{m} \sum\nolimits_{m=1}^{M} \beta_m \quad (8)$$

The variation within (V) and between (W) the imputation is shown in Eq. 5 below:

$$V_\beta = W + \left(1 + \frac{1}{m}\right) E \quad (9)$$

where

$$W = \frac{1}{m} \sum\nolimits_{m=1}^{M} e^2, E = \frac{1}{m-1} \sum\nolimits_{m=1}^{M} (\beta_m - \hat{\beta}) \quad (10)$$

6 Experimental Setup

A. Dataset Description

Pima Indians diabetes dataset from UCI repository [14] was used in this study. The dataset consists of 768 instances with 8 features expressed as V1 to V8, and a target variable V9, which is binary (1 and 0). The description of the dataset is given in Table 1. The distribution of completeness/missingness of datapoints by instances, features and values in the dataset is shown in Fig 1.

Table 1. Dataset description showing number and percentage of instances with missing datapoints in each features of Pima Indians diabetes dataset.

Feature	Description	No. missing	% missing
V1	Number of times pregnant	0	0
V2	Plasma glucose concentration a 2 h in an oral glucose tolerant test	5	0.7
V3	Diastolic blood pressure (mmHg)	35	4.6
V4	Triceps skin fold thickness (mm)	227	29.6
V5	2-hours serum insulin (mm U/ml)	374	48.7
V6	Body mass index (weight in Kg/weight in m)^2	11	1.4
V7	Diabetes pedigree function	0	0
V8	Age in years	0	0
V9	Class (binary: 1-diabetes, 0-not diabetes)	0	0

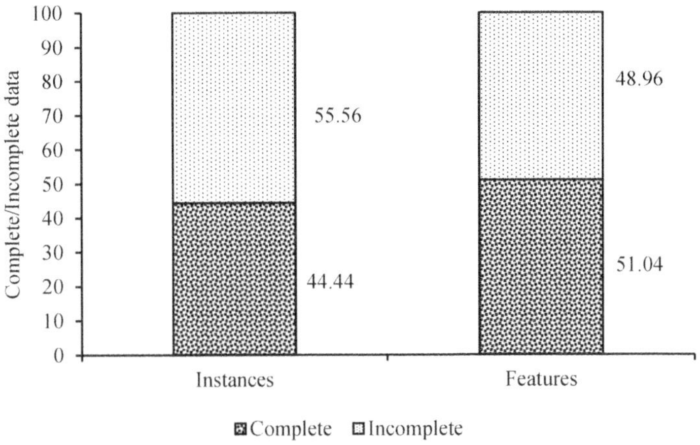

Fig. 1. Percentage of complete/incomplete data in instances, features and datapoints of Pima Indians diabetes dataset

B. Missing Value Characteristic

The pattern of missingness of Pima India Diabetes was investigated to know whether the missingness is monotonous or non-monotonous. Expectation maximum (EM) and the statistical properties were used to test the characteristics of its random missingness i.e. MAR, MCAR or MNAR. From the EM test, we have Little's MCAR test result (the significance of the missingness) [15].

C. Imputation Parameter Setting

The number of imputation was set to five (5). Linear regression was used to impute the features in the original dataset as proposed in Sect. 5 because they are all continuous variables, and the features with missing datapoints (V2, V6, V3, V4, and V5) were imputed accordingly in an increasing order. The maximum percentage of missing data was 100%. The analyses of the multiple imputations were generated in tables – a table for a feature with missing data, i.e. Tables 2, 3, 4, 5 and 6.

D. Classification of Imputed Datasets

In this study, we used extreme learning machine (ELM) classier to validate the result of imputation on the original data and five sets of imputed datasets generated. The pooled output of multiple imputation for each dataset was divided into training and testing sets in 70:30 ratio. The ELM classifier was implemented in Matlab 2018a on a HP computer with Core i3-4030U CPU @ 1.9 Hz X64-based processor, RAM 6.00 GB, TOSHIBA MQ01AB075 HDD and 64-bit Windows OS. The number of ELM hidden neurons was varied from 10 to 200, but the optimal results were obtained at 180; and sigmoid activation function was used.

In the validation, P0 represents the result classification of the original dataset, P1–P5 represent pooled output of the imputed datasets. The results were discussed in the next section.

7 Discussion of Results

Figure 1 shows that 44.44% of the features in the data set is complete, while 55.56% has missing data. Based on the instances, 51.04% is complete, while 48.96% have incomplete data. From EM test, Little's MCAR significance is less than 0.05, and the pattern is non-monotonous, this show that the mechanism of missingness is MCAR [15]. These inform the reason for multiple imputation.

Table 2 shows multiple imputations for feature V2. The five (5) instances with missing values were imputed in 5 iterations (totalled 25 runs). The mean, standard deviation, minimum and maximum values for the original data set are shown in the tabulations. The same treatment is done for V3, V4, V5, and V6 in Tables 3, 4, 5 and 6 respectively. The only variation is the number of missing values in each variable, which brings about different numbers of imputed values: V3 with 35 missing values has 175 imputed values, V4 with 227 has 1135 imputed values, V5 with 374 missing values has 1870 imputations, and V6 with 11 missing values has 55 imputations.

Table 2. Multiple imputation for V2

Imputation		N	Mean	Dev.	Min.	Max.
Original data		763	121.69	30.536	44.00	199.00
Imputed values	1	5	125.52	38.578	88.84	170.01
	2	5	119.22	10.672	111.21	137.24
	3	5	145.21	35.676	104.01	191.63
	4	5	111.99	27.614	76.61	140.40
	5	5	135.09	27.640	95.38	172.70
Complete data after imputation	1	768	121.71	30.565	44.00	199.00
	2	768	121.67	30.446	44.00	199.00
	3	768	121.84	30.603	44.00	199.00
	4	768	121.62	30.511	44.00	199.00
	5	768	121.77	30.520	44.00	199.00

Table 3. Multiple imputation for V3

Imputation		N	Mean	Dev.	Min.	Max.
Original data		733	72.41	12.382	24.00	122.00
Imputed values	1	35	72.55	14.902	43.49	100.49
	2	35	72.25	13.165	44.49	97.34
	3	35	74.56	12.447	43.87	101.37
	4	35	70.85	11.039	45.90	95.96
	5	35	69.06	11.674	50.13	96.20
Complete data after imputation	1	768	72.41	12.497	24.00	122.00
	2	768	72.40	12.410	24.00	122.00
	3	768	72.50	12.385	24.00	122.00
	4	768	72.33	12.322	24.00	122.00
	5	768	72.25	12.363	24.00	122.00

Table 4. Multiple imputation for V4

Imputation		N	Mean	Dev.	Min.	Max.
Original data		541	29.15	10.477	7.00	99.00
Imputed values	1	227	27.84	10.680	−3.03	63.88
	2	227	28.47	10.895	−1.92	55.99
	3	227	27.45	10.373	−.62	56.51
	4	227	27.85	10.190	−6.94	58.87
	5	227	27.74	11.160	.75	57.31
Complete data after imputation	1	768	28.77	10.547	−3.03	99.00
	2	768	28.95	10.600	−1.92	99.00
	3	768	28.65	10.469	−.62	99.00
	4	768	28.77	10.403	−6.94	99.00
	5	768	28.74	10.696	.75	99.00

Table 5. Multiple imputation for V5

Data	Imputation	N	Mean	Dev.	Min.	Max.
Original data		394	155.55	118.776	14.00	846.00
Imputed values	1	374	146.07	111.231	−113.11	489.68
	2	374	153.85	122.138	−227.40	488.04
	3	374	154.91	120.382	−257.32	488.35
	4	374	144.23	126.086	−182.66	540.58
	5	374	134.68	119.561	−180.91	521.16
Complete data after imputation	1	768	150.93	115.186	−113.11	846.00
	2	768	154.72	120.349	−227.40	846.00
	3	768	155.24	119.483	−257.32	846.00
	4	768	150.03	122.441	−182.66	846.00
	5	768	145.39	119.538	−180.91	846.00

Table 6. Multiple imputation for V6

Data	Imputation	N	Mean	Dev.	Min.	Max.
Original data		757	32.457	6.9250	18.200	67.100
Imputed values	1	11	32.271	7.4383	18.047	46.451
	2	11	34.026	7.7246	22.682	45.062
	3	11	29.469	7.2017	18.687	43.716
	4	11	32.797	6.2029	22.493	42.757
	5	11	28.430	6.8445	13.355	38.423
Complete data after imputation	1	768	32.455	6.9274	18.047	67.100
	2	768	32.480	6.9340	18.200	67.100
	3	768	32.415	6.9333	18.200	67.100
	4	768	32.462	6.9117	18.200	67.100
	5	768	32.400	6.9360	13.355	67.100

Each of the imputed datasets was pooled with the original dataset to form the complete datasets. Comparing the results of the complete datasets after the imputation with the original dataset, it can be observed throughout Tables 2, 3, 4, 5 and 6 that the statistical mean, standard deviation, and maximum values are close to the corresponding original values. This shows that multiple imputation approximation is reliable.

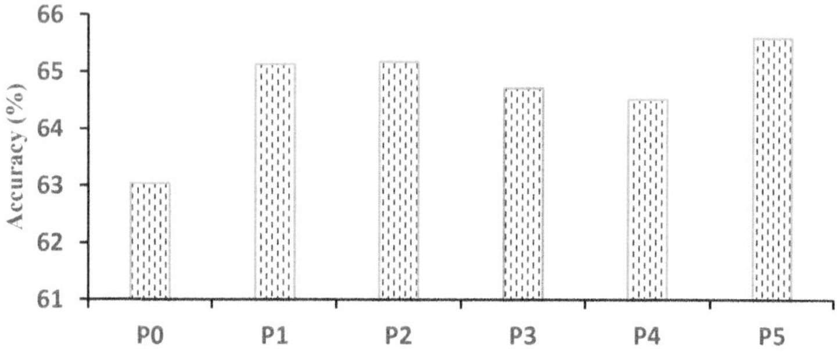

Fig. 2. Extreme learning machine (ELM) validation of multiple imputation on Pima Indian Diabetes showing that the results of classification of imputed datasets (P1–P5) are more accurate than the uncompleted dataset (P0)

ELM classification in Fig. 2 shows that the original data P0 has the least accuracy of 63.0431, while the imputed datasets P1–P5 have higher accuracies 65.1303, 65.1738, 64.7129, 64.5210, 65.5912 percent respectively. Therefore, the results of ELM classification show that P5 is the best imputed dataset to represent the original complete dataset since it has the highest accuracy.

8 Conclusion

This paper examined the characteristics of missing data on the choice of imputation technique. The mechanism of missingness was examined as the bases for a choice of imputation technique for filling missing data in a medical dataset. In this study, we considered the various mechanism for missing values - MAR, MCAR and MNAR. Different treatments of the missing data based on the mechanisms of missingness were discussed. Regression model was used as multiple imputation technique to handle the missing values. Our further research work will be extended to compare the accuracy of the imputed datasets and the original dataset with different classifiers like support vector machines (SVM), random forest (RF), radial basis function (RBF) and ELM.

References

1. Simpson, J.A., Moreno-Betancur, M., Lee, K.J., et al.: Multiple imputation methods for handling missing values in a longitudinal categorical variable with restrictions on transitions over time: a simulation study. BMC Med. Res. Methodol. **19**, 1–14 (2019). https://doi.org/10.1186/s12874-018-0653-0
2. Zhang, Z.: Multiple imputation with multivariate imputation by chained equation (MICE) package. Ann. Transl. Med. **4**, 30 (2016). https://doi.org/10.3978/j.issn.2305-5839.2015.12.63
3. Choi, J., Dekkers, O.M., le Cessie, S.: A comparison of different methods to handle missing data in the context of propensity score analysis. Eur. J. Epidemiol. **34**, 23–36 (2019). https://doi.org/10.1007/s10654-018-0447-z
4. Zhang, F., Tian, S., Chen, S., et al.: Voxel-based morphometry: improving the diagnosis of Alzheimer's disease based on an extreme learning machine method from the ADNI cohort. Neuroscience **414**, 273–279 (2019). https://doi.org/10.1016/j.neuroscience.2019.05.014
5. Armina, R., Mohd Zain, A., Ali, N.A., Sallehuddin, R.: A review on missing value estimation using imputation algorithm. J. Phys. Conf. Ser. **892**(1), 012004 (2017). https://doi.org/10.1088/1742-6596/892/1/012004
6. Tsai, C.F., Chang, F.Y.: Combining instance selection for better missing value imputation. J. Syst. Softw. **122**, 63–71 (2016). https://doi.org/10.1016/j.jss.2016.08.093
7. Austin, P.C., Escobar, M.D.: Bayesian modeling of missing data in clinical research. Comput. Stat. Data Anal. **49**, 821–836 (2005). https://doi.org/10.1016/j.csda.2004.06.006
8. Jakobsen, J.C., Gluud, C., Wetterslev, J., Winkel, P.: When and how should multiple imputation be used for handling missing data in randomised clinical trials - a practical guide with flowcharts. BMC Med. Res. Methodol. **17**, 1–10 (2017). https://doi.org/10.1186/s12874-017-0442-1
9. Sovilj, D., Eirola, E., Miche, Y., et al.: Extreme learning machine for missing data using multiple imputations. Neurocomputing **174**, 220–231 (2015). https://doi.org/10.1016/j.neucom.2015.03.108
10. Falcaro, M., Carpenter, J.R.: Correcting bias due to missing stage data in the non-parametric estimation of stage-specific net survival for colorectal cancer using multiple imputation. Int. J. Cancer Epidemiol. Detect. Prev. **48**, 16–21 (2017). https://doi.org/10.1016/j.canep.2017.02.005
11. Tran, C.T., Zhang, M., Andreae, P., et al.: An ensemble of rule-based classifiers for incomplete data. In: Proceedings 21st Asia Pacific Symposium on Intelligent and Evolutionary Systems IES, pp. 7–12. IEEE (2017)
12. Rodwell, L., Lee, K.J., Romaniuk, H., Carlin, J.B.: Comparison of methods for imputing limited-range variables: a simulation study. BMC Med. Res. Methodol. **14**, 1–11 (2014). https://doi.org/10.1186/1471-2288-14-57
13. Yin, Y., Zhao, Y., Zhang, B., et al.: Enhancing ELM by Markov boundary based feature selection. Neurocomputing **261**, 57–69 (2017). https://doi.org/10.1016/j.neucom.2016.09.119
14. Strack, B., Deshazo, J.P., Gennings, C., et al.: Impact of HbA1c measurement on hospital readmission rates: analysis of 70,000 clinical database patient records. Biomed. Res. Int. **2014**, 11 (2014)
15. Little, R.J.A.: A test of missing completely at random. J. Am. Stat. Assoc. **83**, 1198–1202 (1988)

Semantic Annotation of Scientific Publications Based on Integration of Concept Knowledge

Shwe Sin Phyo(✉) and Nyein Nyein Myo(✉)

University of Computer Studies, Mandalay, Myanmar
shwesinphyo.060@gmail.com, vicky.mdy@gmail.com

Abstract. Discovery of knowledge plays a crucial role in large volumes of data for extracting the valuable knowledge units. The indexing activity with the meaning of contents instead of character strings has become the motivation of searching documents in the information retrieval field. The process of finding and selecting the relevant concepts are the main objectives for the semantic indexing activity. This paper proposes a semantic annotation strategy to support the semantic indexing activity of academic community. The proposed activity extracts the corresponding concepts of a specific document from the semantic network. The annotation activity is based on the semantic degree value of each concept. The knowledge-based approach is used to calculate the degree value of concepts and this approach only rely on the concepts structure of knowledge graph. The proposed annotation activity can be applied as part of the semantic web application and semantic search engines for analyzing and characterizing the meanings of contents.

Keywords: Semantic annotation · WordNet ontology · Semantic similarity · Taxonomic knowledge

1 Introduction

The integration of semantic information attracts a great deal of attention for large volumes of the research area in the information retrieval field. The indexing of documents with their semantic meaning is the fundamental process in a great variety of semantic web applications. The current indexing activities modeled as a bag of words design for cataloging the contents [1]. Their limitation is tokens and simple words. In general, the results may be high level of noise and low recall. The actual meaning of specific document may lead to ambiguous problems.

For this reason, various knowledge structures have been developed for different domains and applications such as Wikipedia, Dbpedia, and other knowledge resources [2]. Moreover, some domain limited ontologies have been developed for various domain areas like the medical field, agricultural field and so no [3]. These ontologies are applied to different domains of knowledge representation mechanisms. For example, it utilized the notation of biomedical application such as genes, proteins, and their relationships.

Ontology is the main element of semantic relation, which is the formal representation of domain knowledge by linking words or phrases in plain text. The semantic

F. Saeed et al. (Eds.): IRICT 2019, AISC 1073, pp. 98–109, 2020.
https://doi.org/10.1007/978-3-030-33582-3_10

annotation activity finds the link between the text fragment of the document and the instances (individuals) of the ontology. Many researchers utilized the domain-specific ontologies for the extraction of semantic knowledge for different domains [3–6].

But the previous annotation activities are designed without considering the semantic meaning of the context [6, 7]. The authors assumed that text corpora are sufficient to express the meaning of the document terms. Several researchers have analyzed the content of the document for the improvement annotation process [8, 9]. The problems associated with semantic annotation remain a research topic.

The annotation process for the academic landscape area is also analyze based on the concepts of each publication. The consideration of the concepts is based on the bibliographic information of each publication [10, 11]. This paper contributes to supporting the indexing activity by adding the semantic concept to the specific document. The content of the document is used for finding the concept of knowledge. The concept of knowledge consideration is not limited to the specific domain area and it depends on the document content. Any domain based document can be annotated with the conceptual knowledge of document content.

To get the domains unlimited concept knowledge, the system uses the popular semantic network, WordNet. WordNet database groups the terms with the same meaning or similar meaning of terms into a set of synonyms (synsets), records the semantic relations between them [12]. Among many different research areas, the proposed system focus on the exploration of the academic landscape. Firstly, the system connects textual expressions (terms) and semantic information (semantic concepts). The semantic meaning of textual expressions is identified by using the direct knowledge of the semantic network.

The document terms are corresponded to more than one categories in WordNet database. The synsets collection of all these categories are not correspond to the actual content of the research articles. Based on the POS category of each terms, the system collect the suitable synsets and finds their semantic similarity value and then selects the most important one. By using the first one, the system selects the suitable synsets of the remaining terms of the article. Then filtering process is performed for improving the quality of annotation process. Finally, the collected synsets are used to annotate the scientific article. The proposed system can be applied not only semantic indexing of academic landscape area and also other different areas of web search applications.

Section 2 introduces the several research background. Section 3 expresses the basic approaches of semantic similarity measures. Section 4 presents the semantic similarity methods based on WordNet semantic Knowledge Graph. Section 5 explains the overview of the proposed annotation activity. Section 6 presents the architecture design of the proposed system. Then, the system discusses the evaluation of the proposed system. The final section describes the conclusion and recommendations for future work.

2 Related Work

The indexing activity with the concepts of the documents rather than character strings of documents is necessary to improve the performance of information retrieval services. The researchers introduced the various strategies with the use of different knowledge representation structures for the semantic annotation of documents.

Castells, Fernandez, and Vallet [16] presented the model for providing search capabilities over large document repositories. The indexing of their research work is based on specific domain ontology knowledge and weighted procedure. Ruiz, Valencia, Fernández [17] introduced the semantic annotation approach based on the UMLS ontology concepts and provided the concept retrieval framework. The annotation process identifies the parts of the text related with each concept of UMLS ontology and expressed as XML tags over the document.

Nebot and Berlanga [4] proposed the semantic annotation activity for identifying and characterizing the binary relations of unstructured texts in the biomedical domain. They used the probabilistic approach for measuring the relationships of synonyms and searching for abstract semantic relations.

Rahman and Siddiqi [6] proposed the semantic annotation activity for multimedia data based on the lightweight ontology of acoustic metadata. Albukhitan, Alnazer, and Hermy [5] proposed web documents annotation activity with the knowledge of food nutrition and health ontologies for the biomedical field. Teixeira, Belloze, and Cavalcant [18] illustrated the annotation activity of scientific articles. The scientific articles are annotated with the distinct ontologies and presented with the design methodology of the data mart, i.e., smaller data ware-house.

All of the above activities attempted to avoid the ambiguous problems of semantic annotation activity. But, the integration of semantic information attracts a great deal of attention to the scientific community. Most of the literature discovery tools, (Google Scholar, Cite seer, ACM Digital Library), stand search-engine-like interface and permit the student to search the articles with the keyword or faceted search on characteristics such as author name, year of publication and so on [1]. Previous research works attempt to analysis the scientific publications based on the various field of knowledge such as author, knowledge, subject, time, place and other bibliographic information [10, 11]. For this reason, the proposed annotation system attempts to support the academic area of scientific publications with their semantic concept knowledge. The concepts are defined by the content of each publication. Among many resources, the proposed system uses the popular large-scale effort, WordNet [12].

There are several computational measures for quantifying the similarity of the WordNet concepts. The path-based measure considered the semantic distance of concepts in the WordNet hierarchy. The information content-based approach is based on the definitions of two concepts or terms. The feature-based approach is characterized by the total number of common features between the concepts. The system uses concepts similarity of path with depth relative semantic similarity measure. The strength of this measure is the independent of corpora availability and human preprocessing of data [13–15]. The process connects textual expressions (terms) and semantic information (semantic concepts) to provides the meaning of each document terms. Finally, the system creates semantic annotation of each research publication.

3 Semantic Similarity

In general, the degree of semantic similarity measures is defined between concepts, words, and text segments. There are two main approaches for measuring the words similarity: (1) corpus-based techniques (2) knowledge-based techniques. The former approach define the similarity values based on the models of distributional similarity learned from large text corpora. Knowledge-based approaches rely primarily on the conceptual structure of the knowledge-graph [19].

3.1 Corpus-Based Approaches

The similarity value of two words has been defined by the knowledge derived from large text collection such as Wikipedia. This approach is based on contextual information of words. They consider the concepts in the form of words and without taking into account the different meaning of words (word senses). The strength of corpus-based approaches is the better coverage of vocabulary since the computational model can be totally applied to several updated corpora. But the researchers need to consider the complexity of creating own corpus and the weakness of existing large text corpora.

3.2 Knowledge-Based Approaches

The degree of similarity value between concepts is quantified by the semantic structure of knowledge graphs. All these measures take two concepts as input and compute the similarity value of two concepts based on the hierarchy of knowledge graph [19]. The proposed system uses the WordNet knowledge graph for getting the concept knowledge of each document. The following section presents the basic knowledge-based methods.

4 Semantic Similarity Measures: WordNet

WordNet is a database of words information. It categorized the words as the sets of cognitive synonyms (synsets), each synsets has a specific concept. There are four categories in the WordNet lexical network (Noun, Verb, Adjective and Adverb). This categories organized the synsets into a hierarchy of 'is-a relation' or 'whole-path relation'. The similarity measures only considered for the categories of nouns and verbs categories in WordNet [12]. The categories of adjective and adverb concepts do not exist in the semantic similarity measures. The basic similarity measures of WordNet are Path Length based measure, Information Content-based measure, Feature-based measure, Hybrid measure [13, 14].

4.1 Path Length-Based Measure

The semantic degree value is specified by the path length that links to two concepts and their positions in the taxonomic structure. The shorter the path, the more similar they are. Thus, the degree value of path length based measure is mainly depended on the minimum number of their path length.

And then, the researcher presented the similarity approach that is also depended on the shortest path but it also considered the depth level of the edges that connect to two concepts in the semantic structure of the knowledge graph [15]. The depth level is calculated by the level of root entity to the target concepts. The measure considered by the position of two concepts and the position of their lowest level of the common ancestor in the WordNet hierarchy.

4.2 Information Content-Based Measure

This measure based on the integration of ontology and corpus and referred to as the corpus based or information theoretic based approach [14]. The similarity of two concepts is defined by the brief definition of their lowest common sub-summer. This measure quantified the similarity degree by "the number of common terms".

4.3 Feature-Based Measure

The two concepts of feature-based measure is quantified by the number common characteristics of the two concepts. WordNet concepts are defined by a set of terms indicating their properties or characteristics, such as their 'glosses', relating with the common concepts and so on. The more common properties they have, the more similar they are [13].

4.4 Hybrid Measure

Hybrid measure based on the combinations of knowledge from characteristics of above measures, such as path length, depth of concept and so on. This measure can be used when the knowledge of an information source is incomplete then it may be combined with the alternative information sources.

The above measures are the basic theory of similarity measures between the concepts. The first two measure has the advantageous of independent on corpora availability and human pre-processing of data, the system uses the path with depth-based measure (Wu and Palmer) for measuring the similarity of all of the synsets [13–15]. The evaluation of semantic annotation activity is based on these two measures.

5 Overview of the Approach

The creation of semantic annotation process requires some preliminary activities which are the basic execution processes of the system.

Initially, the collection of published articles is input to the system. The text-based parsing of the articles creates a meaningful set of key-terms associated with each article. In general, NLP analyzes performed for the preprocessing steps such as tokenization, stop-words remove and duplicate-word removes, lemmatization and define the part-of-speech (POS). Based on the POS category, the semantic features of key terms are collected from the semantic lexical database. And then, these features are used for finding the most suitable concept by using the computational model of semantic similarity value. The semantic annotation process mainly depends on the similarity value of the retrieve concepts for getting the accurate and appropriate concept knowledge for each document. This process continues through the translation of the semantic concepts generation from the lexical database to accurate and meaningful semantic annotation process for each research document. Following sections are devoted to describing the detail representation of the phases.

6 Semantic Annotation Architecture

There are five phases in semantic annotation architecture. (1) Data Preparation (2) Features Analysis (3) Computation Model of Semantic Similarity (4) Concepts Integration and (5) Document Annotation.

6.1 Data Preparation

The system uses the two datasets of the UCI Machine Learning Repository. This repository includes the feature values of research articles, such as title, author, abstract and keywords, with the CSV format. The first one has about 548 of research articles with abstract and the second has over 100000 of research articles. The process of finding the key features is the basic fundamental step in the phases of semantic annotation. Generally, preprocessing steps are applied to the abstract of research articles. Natural Language Processing is used for the analysis of research articles and then extract term features of each abstract content.

Firstly, the system subdivides the abstract text of the articles into separate sentences. The next step is to divide the sentence into separate words or tokens. This is called tokenization. Knowing the role of each word is one of the major role in semantic annotation activity. The system examines each token and try to guess the part of speech of each token, whether it is a noun, a verb, an adjective and so on.

The two general approaches of POS tagging process are Rule-based and Learning-based approaches. In general, Learning-based approaches are more effective, because it take into the consideration of the total amount of human expertise and effort involved. Thus POS tagging of the system is based on HMM model, one of the learning-based categories [20]. And then, the system lemmatizes the above result of POS tagging. Finally, stop words and repeated words are removed for each document. Among the four categories of the WordNet lexical database, the similarity measures of WordNet only considered for the categories of nouns and verbs [13]. Therefore, the system collects the noun and verb words of each research article.

Let $P_i = \{p_1, p_2 \ldots p_i\}$ be the abstracts set of original research articles collection. After NLP analyzes and the part-of-speech tagging of each article, the system selects only the abstract of noun and verb words for each article. The content of each research article is prepared by the following equation:

$$Pi \leftarrow \{w_{ni}, w_{vi}\}, w_{ni}, w_{vi} \in w_i, w_i \in P_i \quad (1)$$

6.2 Features Analysis

The collection of noun and verb words are the basic lexical units and also called the keyword features of each publication. The system used the semantic lexical database to annotate the keyword features with their semantic meaning.

Table 1. Synsets retrieved for the word 'Document'.

Synsets	Synonyms	Definitions
document.n.01	document	Writing that provides information (especially information of an official nature
document.n.02	document	Anything serving as a representation of a person's thinking by means of symbolic marks
document.n.03	document	A written account of ownership or obligation
text_file.n.01	document	A computer file that contains text (and possibly formatting instructions) using seven-bit ASCII characters
document.v.01	document	Record in detail
document.v.02	document	Support or supply with references

Example:

1. We checked the **document** again.

('we', 'PRP'), ('checked', 'VBD'), ('the', 'DT'), ('document', 'NN'), ('again', 'RB'), ('.', '.')

2. Can you **document** your claims?

('can', 'MD'), ('you', 'PRP'), ('document', 'VB'), ('your', 'PRP$'), ('claims', 'NNS'), ('?', '.')

This process attaches the keyword features with the semantic concept of the WordNet database. The semantic concepts are used in computational model of semantic similarity value. Nevertheless the selected noun and verb words are related with both noun and verb categories in the WordNet database. The above two sentences include the 'document' term. The word 'document' visualized as noun and verb concept in WordNet database. Table 1 illustrates all of retrieving synsets for the word 'document'. These retrieving synsets are also called the semantic features of the word.

The system isn't considered all of these synsets as the semantic features of the word 'document'. The feature analysis phase extracts the suitable synsets based on the POS tagging category. The first sentence of the word attaches with 'noun' category, thus the system considers the four noun concepts synsets of the word 'document'. The word 'document' of the second sentence attaches with the 'verb' category. Therefore the system retrieves two synsets of verb concept from the semantic lexical database. The selected synsets are used to quantify the degree of semantic similarity value.

The root concept of noun and verb concept are difference in WordNet hierarchy. Therefore these separation will supports the calculating of semantic similarity values and the calculation can get the accurate and meaningful concepts for each word. Thus the analyzing of the corresponding semantic features of each word is one of main important step in the semantic annotation activity. From these steps, the system gets the right semantic features and these features are applied for searching the most suitable synsets for each noun and verb word. The next sections clearly explained the selection of valuable concepts.

6.3 Computational Model of Semantic Similarity

The quantification of similarity value between words is defining a method that can simulate the process of human thought. The similarity measures are used to examine the concepts, whether they are enclosed or infinite values, whether they are symmetric and whether they give different perspectives.

The computational model of the system attempts to quantify the similarity value of each synset based on WordNet knowledge-based similarity measure. Among the basic categories of WordNet similarity measures, path length based measure has the advantageous of the lack of dependence on the availability of corpora and human preprocessing of data. The computation model of the system considers these two similarity measures [15]. Path length similarity measure is only rely on number of path length between two concepts. Path with depth relative measures are more specific than path only based measures because it combined the shortest path length of two concepts with depth information of common parents of the two concepts [16]. Thus the similarity measure of the system is rely on path length with depth relative measure. The similarity measure of Concepts 'c_1' and Concept 'c_2' was considered by the position of c_1 and c_2 in the knowledge graph, and the position of the closest common ancestor 'c' of c_1 and c_2. The closest common ancestor or least common subsumer 'c' is the common parent of c_1 and c_2 and related with the minimum number of 'is-a' links with concepts c_1 and c_2. The following equation is the estimation of semantic similarity value between two concepts.

$$sim(a,b) = \frac{(2 \times depth(LCS(a,b)))}{(depth(a) + depth(b))} \tag{2}$$

For above equation, depth of 'a' and 'b' is the distance of concept 'a' and 'b' from the root node or depth (a), depth (b) are shortest 'is-a' path (root, x). LCS (a, b) is the distance of most specific common subsumer of concept 'a' and 'b' from the root node.

6.4 Concepts Integration

The aim of the proposed system is to annotate each research article with the most relevant concepts. The procedure of semantic annotation activity is shown in below:

1: organize the noun and verb words of each article

$$P \leftarrow \{w_n, w_v\} \tag{3}$$

2: for every noun words in the word collection
  2.1: collect the noun synset of each word

$$w_n \leftarrow \{w_{n1}, w_{n2}, \ldots, w_{nj}\} \tag{4}$$

  2.2: find the first general synset

$$w_n(ss) \leftarrow \{ss_1, \ldots, ss_k\} \tag{5}$$

  2.3: find the synset for remaining noun words;

$$w_n(ss_k) \leftarrow max\left(avg\left(\sum wp\left(w_{nj}(ss_{total}), w_n(ss)\right)\right)\right) \tag{6}$$

3: annotate each noun word with the above synsets

$$w_v \leftarrow \{w_{v1}, w_{v2} \ldots, w_{vi}\} \tag{7}$$

4: for every verb words in the collection of word
  4.1: collect the verb synset of each word

$$w_v(ss) \leftarrow \{ss_1, \ldots, ss_l\} \tag{8}$$

  4.2: find the first general synset;

$$w_v(ss_1) \leftarrow max\left(avg\left(\sum wp(w_{vi}(ss_{total}), w_v(ss))\right)\right) \tag{9}$$

  4.3: find the synset for remaining verb words;
5: annotate each noun word with the above synsets;

$$w_v(ss_i) \leftarrow max\left(avg\left(\sum wp(w_v(ss_1), w_{vi}(ss_{total}))\right)\right) \tag{10}$$

Firstly, the system selects the research paper ‘p’ of noun and verb words from the set of p_i. Feature analysis phase extracts the set of synsets that correspond to the noun and verb words of p_1. By using the basic theory of computational model phase, the system finds the degree value of each synset. This is done separately for noun and verb words. Firstly, the system find the most important synset that has the greatest similarity value. These two synsets are used to collect the suitable synsets (semantic meaning) of remaining noun and verb words. Based on the most important synsets of each noun and verb synset, the suitable synsets are defined for each of remaining noun and verb words. Concepts knowledge integration for each document is based on the above instructions.

6.5 Document Annotation

From previous section, the system collects the related concepts knowledge for each of research publication. The range of similarity value between two synsets is between ‘0’ and ‘1’. If the concept of ss_1 is totally related to the concept of ss_2, then the value is ‘1’. If ss_1 and ss_2 are not related to each other, then the value becomes ‘0’ [14]. To annotate each document, the system selects the synsets with the similarity of nearly similar to ‘1’. Therefore, the system defines the threshold value ‘0.7’ and gets the synsets with the degree value of ‘0.7’ and above ‘0.7’. Thus, the concepts of selected synsets are closely related to each other for each research article. The selected synsets are used to annotate each research article. Finally, each article is annotated by the following equation.

$$P \leftarrow \{w_{ni}(ss_j), w_{vi}(ss_i)\} \tag{11}$$

7 Result and Discussion

The evaluation is based on the most related instances over the number of collected instances and the portion of most related instances that have been retrieved over the total number of related instances. The similarity measure of the system relies on the WordNet knowledge graph structure. Because of the system uses the advantage of the lack of dependence on the availability of the corpus and the pre-processing of human data, the evaluation compare the path only similarity measure and the path length with depth relative based measure for the evaluation of the system. These two measures have the same advantage. The standard measures Precision, Recall, and F-measure are used for evaluating the annotation activity.

Table 2. Summary of semantic annotation with WordNet synsets

#Doc	Noun word	Verb word	Synsets (above 0.7)			
			Path with depth based measure		Path only measure	
			Noun	Verb	Noun	Verb
1.	50	12	45	11	43	11
2.	29	7	25	7	25	6
3.	46	13	40	12	38	12
4.	42	10	38	9	38	9
5.	47	12	41	12	39	12
6.	38	14	32	13	30	13
7.	37	12	33	12	30	12
8.	54	13	50	13	48	13
9.	35	8	31	7	28	7
10.	62	16	56	16	52	16

Table 2 shows the result of the first 10 annotated research articles. Column 2 and 3 are the total number of the noun and verb words by Natural Language Processing. The remaining columns are the synsets collection that have the similarity value of ‘0.7’ and above ‘0.7’ with path-only measure and path with depth relative measure.

Table 3. Precision, recall and F-measure of semantic annotation.

Standard measures	Path with depth relative measure	Path only measure
Precision	0.9578	0.822
Recall	0.8959	0.739
F-measure	0.9273	0.772

Table 3 illustrates the evaluation result of the semantic annotation activity with first 10 annotated documents. Column 1, 2 and 3 are the evaluation results of the path with depth based measure and column 4, 5 and 6 are the results of the path only measure.

Precision is the positive predictive rate between the annotated synsets and the total number of detected synsets. Recall is the ratio of correctly predicted positive observations to all observations in the actual class. Precision and recall are reciprocal measures and generally the increase of one effect the decrease to another.

Based on the result of Table 2, the system quantify the performance of two similarity measures. TP (True Positives) is set of retrieved synsets that have the similarity value of ‘0.7’ and above ‘0.7’ and FP (False Positives) are below ‘0.7’. FN (False Negative) is the set of synsets that are wrongly defined as the non-relevant.

Instead of the keyword-based annotation process, the system considers the semantic meaning of each term in the document. Based on the result of Table 3, the proposed semantic annotation activity can retrieve the valuable concepts for each document.

8 Conclusion and Future Work

This system presents the annotation technique for improving the indexing activity of specific documents. To annotate the research articles, the system used the WordNet domain knowledge, which has the taxonomic hierarchy of the synonym sets with the related semantic relation. The system uses semantic similarity value for selecting a suitable concept for each document. Finally, the system annotates each research article with the corresponding concept of knowledge.

Furthermore, the overall framework will be applied with desire domains and can also be applied for the semantic tagging of other documents in many different landscape areas. Based on the nature of different domain areas, the more suitable similarity methods can be elected in the next future of semantic annotation activity.

References

1. Lu, M., Bangalore, S., Cormode, G., Hadjieleftheriou, M., Srivastava, D.: A dataset search engine for the research document corpus. In: IEEE 28th International Conference on Data Engineering. IEEE (2012)
2. Abramowicz, W. (ed.): Knowledge-Based Information Retrieval and Filtering from the Web. Springer, Heidelberg (2013)
3. Nebot, V., Berlanga, R.: Exploiting semantic annotations for open information extraction: an experience in the biomedical domain. Knowl. Inf. Syst. **38**(2), 365–389 (2014)
4. Albukhitan, S., Alnazer, A., Helmy, T.: Semantic annotation of arabic web documents using deep learning. Procedia Comput. Sci. **130**, 589–596 (2018)
5. Rahman, F., Siddiqi, J.: Semantic annotation of digital music. J. Comput. Syst. Sci. **78**(4), 1219–1231 (2012)
6. Aronson, A.R., Lang, F.M.: An overview of MetaMap: historical perspective and recent advances. J. Am. Med. Inform. Assoc. **17**(3), 229–236 (2010)
7. Dai, M., Shah, N.H., Xuan, W.: An efficient solution for mapping free text to ontology terms. AMIA Summit on Translational Bioinformatics, San Francisco, CA (2008)
8. Agirre, E., López de Lacalle, O., Soroa, A.: Random walks for knowledge-based word sense disambiguation. Computational Linguistics, 40(1), 57–84 (2014)
9. Navigli, R., Lapata, M.: An experimental study of graph connectivity for unsupervised word sense disambiguation. IEEE Trans. Pattern Anal. Mach. Intell. **32**(4), 678–692 (2009)
10. Pavlovskiy, I.S.: Using concepts of scientific activity for semantic integration of publications. Procedia Comput. Sci. **103**, 370–377 (2017)
11. Hood, Z., Sahari, N.: Researchers annotation collections and practices. Procedia Technol. **11**, 354–358 (2013)
12. Miller, G.A., Beckwith, R., Fellbaum, C., Gross, D., Miller, K.J.: Introduction to WordNet: an on-line lexical database. Int. J. Lexicogr. **3**(4), 235–244 (1990)
13. Elavarasi, S.A., Akilandeswari, J., Menaga, K.: A survey on semantic similarity measure. Int. J. Res. Advent Technol. **2**(3), 389–398 (2014)
14. Poorna, B., Ramkumar, A.S.: Semantic similarity measures: an overview and comparison. Int. J. Adv. Res. Comput. Sci. **9**(1), 100 (2018)
15. Wu, Z., Palmer, M.: Verbs semantics and lexical selection. In: Proceedings of the 32nd annual meeting on Association for Computational Linguistics, pp. 133–138. Association for Computational Linguistics (1994)
16. Castells, P., Fernandez, M., Vallet, D.: An adaptation of the vector-space model for ontology-based information retrieval. IEEE Trans. Knowl. Data Eng. **19**(2), 261–272 (2006)
17. Ruiz-Martínez, J.M., Valencia-García, R., Fernández-Breis, J.T., García-Sánchez, F., Martínez-Béjar, R.: Ontology learning from biomedical natural language documents using UMLS. Expert Syst. Appl. **38**(10), 12365–12378 (2011)
18. Teixeira, M.A.C., Belloze, K.T., Cavalcanti, M.C., Silva-Junior, F.P.: Data mart construction based on semantic annotation scientific articles, a case study for the prioritization of drug targets. Comput. Methods Programs Biomed. **157**, 225–235 (2018)
19. Mihalcea, R., Corley, C., Strapparava, C.: Corpus-based and knowledge-based measures of text semantic similarity. In: AAAI. Vol. 6. (2006)
20. Martin, J.H., Jurafsky, D.: Speech and Language Processing: An Introduction to Natural Language Processing, Computational Linguistics, And Speech Recognition. Pearson/Prentice Hall, Upper Saddle River (2009)

Genetic Algorithm Based Feature Selection for Predicting Student's Academic Performance

Al Farissi[1(✉)], Halina Mohamed Dahlan[2], and Samsuryadi[1]

[1] Fakultas Ilmu Komputer, Universitas Sriwijaya, Palembang, Indonesia
{alfarissi,samsuryadi}@unsri.ac.id
[2] Information Systems Department, Azman Hashim International Business School, Universiti Teknologi Malaysia, Skudai, Johor, Malaysia
halina@utm.my

Abstract. Recently, student's academic performance prediction has become an increasingly prominent research topic in the field of Educational Data Mining (EDM). The prediction of student's academic performance aims to explore information that is beneficial to the learning process of student. Therefore, accurate prediction of student's academic performance provide benefits for education institutions to improve the quality of their institutions by improving the learning process of students. In predicting the student's academic performance, the problem of high dimensional dataset is often faced in the datasets which significantly impacts the accuracy of student academic performance prediction. This paper proposed Genetic Algorithm based Feature Selection (GAFS) along with selected single classifier for classification in order to improve the accuracy in predicting student academic performance. Kaggle dataset is used in this paper and two phase of experiment have been conducted, single classifier without GAFS, and single classifier with GAFS. Results from the experiments show that, the accuracy of the proposed GAFS for classification makes an impressive performance in predicting student academic performance in terms of accuracy compare to existing techniques.

Keywords: Student academic performance · Feature selection · Genetic Algorithm · Classification · Prediction

1 Introduction

The challenge in predicting student academic performance is increasing due to the greater data in the education database. Prediction and analysis of students academic performance has an important role for students academic development. Identifying factors that influence students academic performance is a complex research task [1]. Predicting student academic performance with high accuracy values will be useful for educational institutions to be able to distinguish the academic performance of the students they have [2]. Educational data mining is able to produce information used by

F. Saeed et al. (Eds.): IRICT 2019, AISC 1073, pp. 110–117, 2020.
https://doi.org/10.1007/978-3-030-33582-3_11

educational institutions in developing educational strategies to improve the quality of education.

Problems that are often encountered in dataset are: high dimensional dataset and noisy attributes that can influence the predictive results. Feature selection is commonly been employed in solving the problem involving high dimensional dataset and noisy attributes [3].

In this paper, four classifiers have been applied, which are Decision Tree (DT), NaïveBayesian (NB), k-NearestNeighbor (k-NN) and Random Forest (RF) with the proposed GAFS. In order to measure the accuracy of the combination, four evaluation measurements are used which are: Accuracy, Precision, Recall and F-Measure.

The rest of this paper is organized into several sections, Sect. 2 presents related work in the field of student academic performance prediction, the student academic performance prediction framework in Sect. 3. Experimental design are presented in Sect. 4 and results and discussions are detailed out in Sect. 5. Final section of this paper, is conclusion, in Sect. 6.

2 Related Work

2.1 Feature Selection

Feature selection (FS) has been widelybeen applied in various applications, except that it has not been so widely applied in predicting student academic performance applications. The main purpose of using feature selection techniques is to minimize redundancy and maximize the subset of relevant features while maintaining high accuracy without losing important information.

Amrieh et al. [4] applied Filter Based Selection (FBS) as a feature selection in their study and combined with the esemble method of the classification: Boosting & Bagging. They concluded that the accuracy produced using FS & ensemble methods increased by 25.8% where the Visited Resource feature was the most effective feature on they student academic performance.

Punlumjeak, W & Rachburee, N [3], in their study applied two feature selection approaches: filter & wrapper approach. They stated that the results of the experiments showed the choice of the most relevant features from the list of features used in the student dataset affecting the prediction model of student academic performance.

Luthfia et al. [5] applying Wrapper and Information Gain approaches as feature selection techniques with a single classification algorithm: NB (Naive Bayes), DT (Decision Tree) and NN (Neural Network). They get better accuracy values after applying feature selection techniques to the models they developed. The combination of Information gain and ANN resulted in an accuracy value of 79.375%.

Zaffar et al. [6] used six feature selection techniques and fifteen single classfiers. From the results of their study, the Principal Components Analysis feature selection technique with the Random Forest classifier gets the best results. Their experimental results illustrate that, a way is needed to set parameters in the feature selection method, in order to produce better performance.

Calderon et al. [7], in their experiments, Genetic Algorithm (GA) feature selection was applied with the ANN algorithm for 1271 data sets with 39 attributes, the data set is the property of the university in southern Peru. The parameters of the Genetic Algorithm adjustment: population size is 25, the crossover level is 0.6, the mutation level is 0.3. For the ANN algorithm parameters are applied three layers (1 input layer, 1 hidden layer, and 1 output layer). The experimental results selected 14 attributes used for predictions. Accuracy value result 88.50% increased by 8% compared to predictions using only ANN classifier.

There have been many study in data mining that have been produced very well by using feature selection techniques. In data mining, feature selection at the pre-processing data has an important role. The feature selection technique is to get a feature subset in order to reduce the occurrence of prediction errors. In addition, the selection of features aims to produce attributes that have strong relevance in order to produce high accuracy predictive values.

However, the majority of feature selection algorithms work in getting feature selection solutions that range between sub-optimal and almost optimal regions, these searches are carried out locally throughout the search process, rather than searching globally. Therefore, it is difficult to get an almost optimal solution to the optimal solution if using this algorithm [8].

The purpose of this paper is to improve the accuracy of prediction in the context of students' academic performance by using GAFS as well as to overcome the problems facing by the current feature selection. GA capable to obtain solutions throughout the search space by using global search capabilities, therefore feature selection using genetic algorithm can produce good solutions in a reasonable period of time [9].

3 Method

3.1 Framework

In this section, the proposed framework as shown in Fig. 1. This framework divided into two steps, which are feature selection and modeling. In the first step which is feature selection, dataset is separated into training data and testing data. Next, GAFS is employed in order to get features subset to improvement accuracy prediction.

In the second step which is modeling, data training with feature subset selected is trained with the selected classifier. Classification accuracy is measured by testing set with selected feature subset. Then a fitness function is constructed using classification accuracy of classifier, the number of selected features and the feature cost.

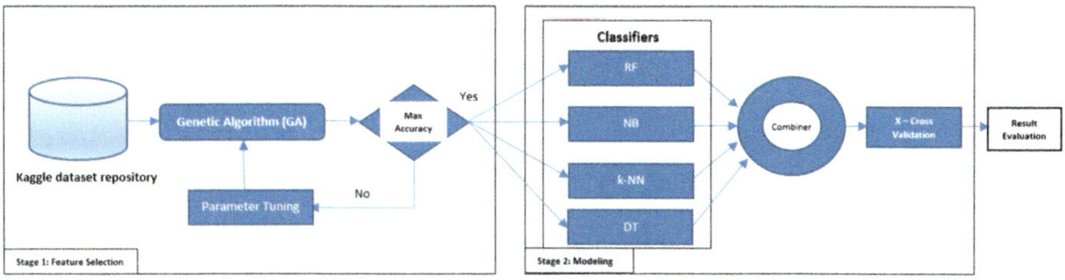

Fig. 1. Framework for student academic performance prediction

Parameters tuning for the Genetic Algorithm Feature Selection (GAFS) and classifiers is shown in the Table 1. These parameters are as follows; initial population values (30), maximum number of generations (30), mutations (0.01), probability of crossover (0.9), k (10), and random number of seeds (1). When the final condition reached, the operation will be stopped.

However, if the conditions are not met, it will go forward with the following generation of operations. The framework searches for better solutions by genetic operations, selection and and mutation, including crossover.

For the value of k in the k-NN classifiers, are k = 3, 5, 7, and 9. The best accuracy results obtained using the k-NN algorithm are with the value k = 5. The performance of base classifiers as single classifier is validated using X-fold cross validation.

Table 1. Parameters tuning for GAFS and selected classifiers

GAFS	RF	NB	k-NN	DT
Initial population: 30 Number of generations: 30 Mutations: 0.01	Number of trees: 1000 Criterion: gain_ratio Maximal depth: 30 Pruning number: 3	Laplace correction: true	Value k training: 5 Weighted voted: true Measure type: Mixed Measure Euclidean Distance	Criteria: gain ratio. Maximal depth: 10. Confidence level: 0,1. Minimal gain:0,01.

4 Experimental Design

4.1 Experimental Setup

Two phase experiments have been done. Phase one is classification without GAFS. In this phase, all classifier which are DT, NB, k-NN and RF are employed without GAFS. In Phase two, GAFS is employed to each classifier.

This experiment conduct used Intel Processor i7 fifth generation with 2.4 GHz CPU, 12 GB Memory RAM and Microsoft Win 10 Edition 64Bit Operating System.

4.2 Dataset Description

This study uses student academic performance dataset from the Kaggle repository. This data collection contains 480 student academic performance data and consists of 16 attributes with one class label that has three intervals: low, medium and high [10]. The features and descriptions of the data set are shown in Table 2.

Table 2. Features and descriptions of student dataset

No	Feature name	Description
1	Nationality	Nationality
2	Gender	Male or female
3	POB	Birthplace
4	Parent responsible for student	Status parent
5	Educational levels	Levels of school
6	Grade levels	Student class group
7	Section ID	Register classroom
8	Semester	Academic year
9	Topic	Subject course
10	Student absence days	Attendance
11	Parent answering survey	Parent participation on survey
12	Parent school satisfaction	Level satisfaction of parent
13	Discussion activity	Student interaction with e-learning system
14	Active visiting resources	
15	Raised hand on class	
16	Seeing announcements	

4.3 Model Validation

This paper employed x-fold cross validation to validate training and testing data. Usually, x-fold cross validation is used because it can reduce computing time while maintaining the accuracy of estimates. This method distributes training data into 10 equal parts. The next step, the learning process is carried out ten times as shown in Table 3.

Table 3. X-fold cross validation

n-validation	Dataset Partition									
1	■									
2		■								
3			■							
4				■						
5					■					
6						■				
7							■			
8								■		
9									■	
10										■

4.4 Evaluation Measures

Four general measurements to evaluate quality of the classification are employed: Accuracy, Precision, Recall and F-Measure. Measures calculated using confusion matrix classification based on the Eqs. 1, 2, 3 and 4, respectively.

Accuracy values refer to how accurate the classification can classify data correctly. Equation 1 is used to obtain the accuracy value. In other words, the value of accuracy is a comparison between data that is correctly classified with the whole data. Precision values refer to the number of positive category data that are classified correctly divided by the total data classified as positive. Equation 2 is used to obtain precision. Meanwhile, recall shows how many percent of the positive category data is correctly classified by the classification. Equation 3 is used to obtain the recall value.

$$\text{Accuracy} = \frac{TP + TN}{TP + FN + FP + TN} \tag{1}$$

$$Precision = \frac{TP}{TP + FP} \tag{2}$$

$$\text{Recall} = \frac{TP}{TP + FN} \tag{3}$$

$$F - Measure = \frac{2TP}{TP + FP + FN} \tag{4}$$

5 Result and Discussion

The evaluation result using classifier with all feature is shown in Table 4, and the result of classifier using with GAFS is shown in Table 5. By using GAFS shows a higher increase in accuracy than classifiers without using feature selection. Therefore, the use

Table 4. Classification method result with all attributes

Evaluation measure	Decision tree (DT)	k-nearest neighbour (k-NN)	Naive Bayes (NB)	Random forest (RF)
Accuracy	62.71	61.04	57.50	**79.79**
Recall	62.70	62.04	59.96	**80.34**
Precision	64.73	61.61	58.14	**80.42**
F-measure	63.70	61.79	58.76	**80.28**

of GAFS in classifying techniques is a solution to improve accuracy in predicting student academic performance.

Table 5 shows the classification results using GAFS with classifiers: DT, k-NN, NB and RF. From the results obtained, RF classifier outperform compared to other classifiers. Overall, the most superior performance was shown when GA feature selection was combined with RF classifier. The results of the accuracy obtained were 82.29%, RF successfully classifies student data correctly with 395 data from a total of 480 student data.

Table 5. Classification method result with GA feature selection

Evaluation measure	Genetic algorithm feature selection (GAFS)			
	Decision tree (DT)	k-nearest neighbour (k-NN)	Naive bayes (NB)	Random forest (RF)
Accuracy	74.58	68.54	75.42	**82.29**
Recall	75.15	68.84	76.25	**82.81**
Precision	75.39	69.75	76.26	**82.70**
F-measure	75.26	69.29	76.25	**82.75**

6 Conclusion

In this paper, DT, k-NN, NB and RF classifier have been used with GAFS techniques with the aim of increasing the performance accuracy for student academic prediction. The dataset used containing the performance of students from Kaggle repository. This study succeeded in applying feature selection techniques to reduce the dimensions of data set with the intention of to improve predictive accuracy. From the experimental results show impressive improvements to predict student academic performance. Future research will measure the comparison of several proposed optimization techniques using other feature selection techniques with other ensemble methods.

References

1. Mueen, A., Zafar, B., Manzoor, U.: Modeling and predicting students' academic performance using data mining techniques. Int. J. Mod. Educ. Comput. Sci. **8**, 36–42 (2016)
2. Zimmermann, J., Brodersen, K.H., Heinimann, H.R., Buhmann, J.M.: A model-based approach to predicting graduate-level performance using indicators of undergraduate-level performance. JEDM-J. Educ. Data Min. **7**, 151–176 (2015)
3. Punlumjeak, W., Rachburee, N.: A comparative study of feature selection techniques for classify student performance. In: 2015 7th International Conference on Information Technology and Electrical Engineering (ICITEE), pp. 425–429. IEEE (2015)
4. Amrieh, E.A., Hamtini, T., Aljarah, I.: Mining educational data to predict student's academic performance using ensemble methods. Int. J. Database Theory Appl. **9**, 119–136 (2016)
5. Rahman, L., Setiawan, A.N., Permanasari, E.A.: Feature selection methods in improving accuracy of classifying students' academic performance. In: 2nd International Conferences on Information Technology, Information Systems and Electrical Engineering (ICITISEE), pp. 267–271. IEEE, Yogyakarta (2017)
6. Zaffar, M., Savita, K.S., Hashmani, M.A., Rizvi, S.S.H.: A study of feature selection algorithms for predicting students academic performance. Int. J. Adv. Comput. Sci. Appl. **9**, 541–549 (2018)
7. Echegaray-Calderon, O.A., Barrios-Aranibar, D.: Optimal selection of factors using genetic algorithms and neural networks for the prediction of students' academic performance. In: 2015 Latin America Congress on Computational Intelligence (LA-CCI). IEEE, Curitiba (2015)
8. Kabir, M.M., Shahjahan, M., Murase, K.: A new hybrid ant colony optimization algorithm for feature selection. Expert Syst. Appl. **39**, 3747–3763 (2012)
9. Yusta, S.C.: Different metaheuristic strategies to solve the feature selection problem. Pattern Recognit. Lett. **30**, 525–534 (2009)
10. Amrieh, E.A., Hamtini, T., Aljarah, I.: Preprocessing and analyzing educational data set using X-API for improving student's performance. In: IEEE Jordan Conference on Applied Electrical Engineering and Computing Technologies (AEECT). IEEE, Amman (2015)

Morphosyntactic Preprocessing Impact on Document Embedding: An Empirical Study on Semantic Similarity

Nourelhouda Yahi(✉) and Hacene Belhadef

NTIC Faculty, MISC Laboratory, University of Constantine 2-Abdelhamid Mehri, Constantine, Algeria
{nourelhouda.yahi, hacene.belhadef}@univ-constantine2.dz

Abstract. Word embedding technique is among the most widely known and used representations of text documents vocabulary. It serves to capture word context in a document, but in many applications the need is to understand the content of text, which is longer than just a single word, that's what we call "Document Embedding". This paper presents an empirical study that evaluates the morphosyntactic data preprocessing impact on document embedding techniques over textual semantic similarity evaluation task, and that by comparing the impact of the most widely known text preprocessing techniques, such as: (1) Cleaning technique containing stop-words removal, lowercase conversion, punctuation and number elimination, (2) Stemming technique using the most known algorithms in the literature: Porter, Snowball and Lancaster stemmer and (3) Lemmatization technique using Wordnet Lemmatizer. Experimental analysis on MSRP (Microsoft Research Paraphrase) dataset reveals that preprocessing techniques improve classifier accuracy, where Stemming methods outperforms other techniques.

Keywords: Word embedding · Document embedding · Doc2vec · Text preprocessing · Semantic similarity

1 Introduction

Digital representation of text documents is an important and difficult task in machine learning. Such a representation can be used for many purposes like: document retrieval [4], web search [10], topic detection [18], etc. For this purpose, many approaches have been proposed, but the method that have recently become very popular is based on recent developments in distributional semantics, particularly neural network-based approaches, that requires a large amount of untagged data. This data is used to create a semantic space. In this last, terms are represented as vectors that are called word embeddings [8]. Word embedding has opened up new perspectives through a reduction in dimensionality and a new semantic vision of words compared to traditional approaches, so that a certain number of applications were thus able to know an important improvement. These applications include: sentiment analysis, translation, speech recognition, etc.

F. Saeed et al. (Eds.): IRICT 2019, AISC 1073, pp. 118–126, 2020.
https://doi.org/10.1007/978-3-030-33582-3_12

Since preprocessing tasks have been found of positive impact on different data mining tasks in general - despite taking longer time, and may even having more challenges than data mining -, we realized this work which focuses on the role of data preprocessing on document embeddings.

The main goal of this paper is to answer the questions: Is preprocessing step needed before applying document embedding methods? and what is the most appropriate preprocessing technique that gives more accuracy in NLP tasks? To get an answer we adopted semantic similarity evaluation task between text fragments, this task was chosen because it is considered as a counter stone of different natural language processing tasks as: information extraction, machine translation, information retrieval, plagiarism detection, etc.

This paper is organized as follows: Previous related work is given in Sect. 2. Section 3 describes the notion of 'word and document embedding'. Section 4 provides some background on text preprocessing. The results of numerical experiments are presented in Sect. 5. Finally, Sect. 6 presents the conclusions of this work.

2 Related Work

Preprocessing is one of the key components in data mining tasks in general, and the most important is to know the right preprocessing techniques that improves every task separately. For this purpose, preprocessing impact has already been studied in text classification [19], sentiment analysis [6], and author identification literature [15]. The results show that preprocessing step is useful for getting better performance, taking into account the peculiarities of each task to choose the combination of techniques that suits it.

Concerning word and document embedding methods, the impact of text preprocessing on their performance has not been well studied, and we do not find in the literature a concentration on it. Among the few works that shed light on that we have [3] which evaluated the role of text preprocessing on word embeddings over text Categorization and sentiment Analysis tasks, and found that - in general - a simple tokenization gives equally or better results than more complex preprocessing techniques such as lemmatization or multiword grouping. However, [9] has found that Stemming yields a better overall performance than not reducing the feature granularity at all, and the increasing of the feature granularity of contexts by including Combinatory Categorical Grammar categories or Part Of Speech tags has no little improvement.

So, in our experiments we aim to evaluate the role of text preprocessing on word and document embeddings over another important task, which is semantic similarity.

3 Word Embedding

When speaking about digital representation of data in general, we find that it is a difficult task in machine learning, especially in semantic tasks where we find that the word is the most basic unit. In order to represent a word as a vector there are different techniques in literature for doing this. These techniques include: 'one-hot vectors' in which every string (word) is converted to a number. Although 'one-hot vectors' can be stored efficiently, their main problem is that they do not capture information about the similarity between strings (words). We also cite another technique: 'Distributional vectors' which is based on searching for all the occurrences of a certain word and its neighbors - selected via a predefined window-, then the word's vector is updated by increasing the number of occurrences in its neighbors indices. Although 'Distributional vectors' technique capture information about the similarity between strings (words), storing vectors requires a very massive memory space, and performing operations on all words is computationally expansive. From 2003 Bengio et al. [2] in order to reduce the high dimensionality of words representations in contexts, he provided in a series of papers a new representation method, that is based on saving the same contextual information in a low-dimensional vector by "learning a distributed representation for word". In 2013, a team at Google led by Tomas Mikolov created word2vec [14], a word embedding toolkit which gained extreme popularity.

3.1 Word2vec

Usually, when we build a model using words, we just code them. However, when you use such an encoding, the words lose their meaning. For example, if we code king as (2), Queen as (4) and Food as (6), Queen will have the same relationship to Food as King. We would prefer a representation in which Queen and King will be closer than Queen and Food. The word2vec, aims to give a digital representation for each word, which will be able to capture these relationships as above (such as synonyms, antonyms, analogies, etc.) (Fig. 1).

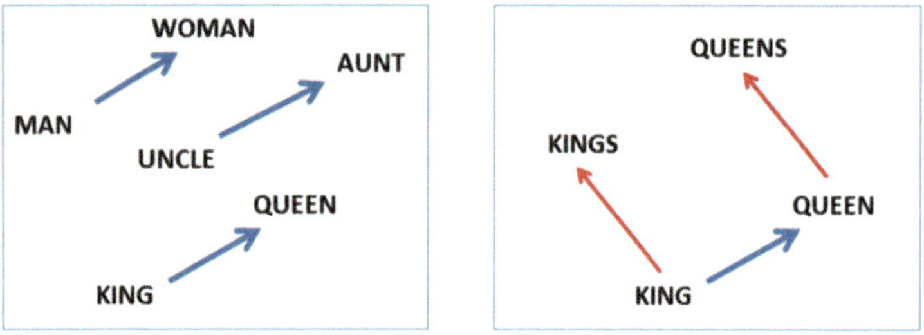

Fig. 1. In the left: vector offsets for three word pairs illustrating the gender relation. The right panel shows a singular/plural relation for two words [13].

The word2vec representation is created using two algorithms: one is the Continuous Bag Of Words model (CBOW) and the second is the Skip-Gram model. The first model predicts the current word from a sliding window of surrounding words -the context-. However, the second algorithm is the opposite of CBOW algorithm, it uses the current word to predict the surrounding window of context words (Fig. 2).

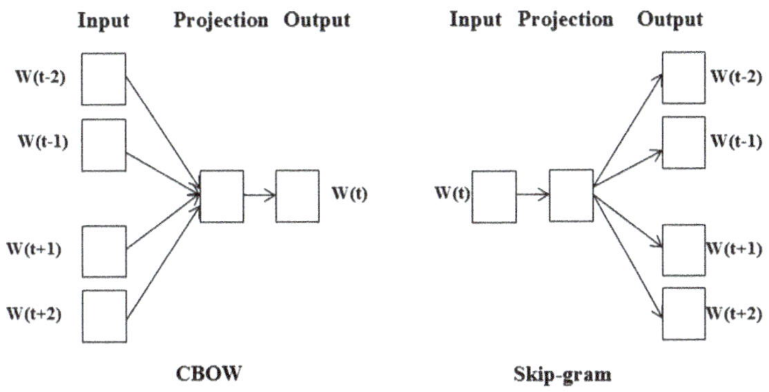

Fig. 2. CBOW and skip-gram architectures.

3.2 Doc2vec

For tasks like finding similarity between sentences, paragraphs or documents, Mikolov and Le [11] have proposed an extension of word2vec which allows to generate a numeric representation of a document, whatever its length. The doc2vec representation is created using two Paragraph Vector versions: Paragraph Vector with Distributed Memory Model (PV-DM), and Paragraph Vector with Distributed Bag of Words (PV-DBOW). In the first algorithm, every paragraph is mapped to a singular vector, represented by a column in matrix D and each word is additionally mapped to a singular vector, represented by a column in matrix W. Then a process of averaging and concatenation are performed in order to predict the next word in a context. The second method performs a paragraph vector concatenation with the word vectors to predict the following word in a text window. Another way is to ignore the context words in the input, but force the model to predict words randomly sampled from the paragraph in the output (Fig. 3).

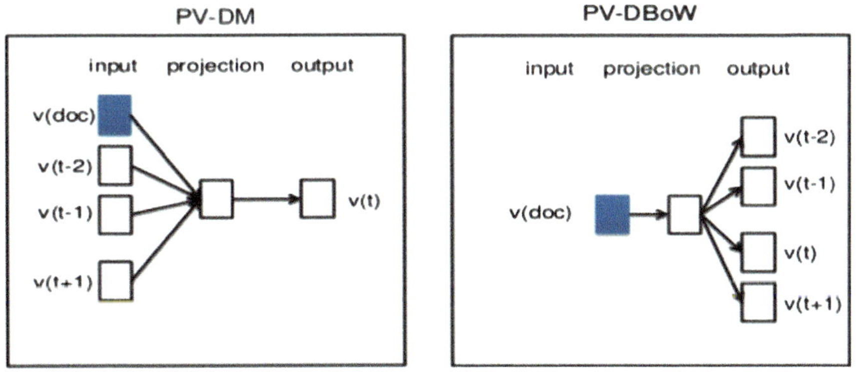

Fig. 3. PV-DM and PV-DBOW architectures.

4 Text Preprocessing

Given a textual information, tokenization is the task of splitting it up into pieces, called tokens, perhaps at the same time eliminating certain characters, such as punctuation, numbers, any markup (html) or tags, special characters, and extra white-space characters as a kind of text cleanup.

Another widely used preprocessing technique in text mining is lowercase conversion, which serves to convert all uppercase characters to their lowercase forms, in order to avoid considering the same words written in different forms as different words.

Among the most well-known preprocessing techniques is Stop-words removal which is based on removing words that are so common in a language that their information value is almost negligible like 'on', 'the', 'it'… etc. Therefore it is usual to remove them before further analysis.

Stemming technique is the process of reducing a word to its stem, generally by cutting off the end or the beginning of the word, based on a list of prefixes and suffixes commonly found in an inflected word. A stemming algorithm reduces the words "retrieval", "retrieved", "retrieves" to the stem "retriev". Among the stemming algorithms known in the literature, Martin Porter's Porter stemming algorithm [17] is the most popular one (i.e., the mostly used one), it's defined by the process for removing the commoner morphological and inflexional endings from words in English. It is worth noting that Martin Porter updated the original Porter stemmer. Nowadays, the Porter2 stemmer is called Snowball stemmer which is almost universally regarded as an improvement over porter. Slightly less computation time than porter, with a fully large community around it. Another known algorithm is Lancaster stemmer, which was developed at Lancaster University in the late 1980s, this simple and iterative stemmer removes the endings from a word in an indefinite number of steps. It is a very aggressive stemming algorithm. The word stems become so short that they are hardly readable by a human anymore. It is among the fastest algorithms, and will reduce the working set of words hugely, but it does not serve data distinction.

Lemmatization represents the process of reducing the different forms of a word to one single base form (lemma), based on vocabulary and morphological analysis of words, it means that lemmatization depends on correctly identifying the intended part of speech and meaning of a word in a sentence, for example, reducing "builds", "building", or "built" to the lemma "build".

5 Experiments

In general, semantic similarity aims at evaluating a score that represents the resemblance between the meanings of the compared entities. For that, similarity measures exploit one or several information or knowledge sources and rely on different theoretical principles [1]. Due to its importance, semantic similarity of texts has become a long time ago - And so far - a central task in Natural Language Processing (NLP) ranging from web page retrieval [16], question answering [7], text classification [12], etc. In this study we have completed several experiments on each Doc2vec parameter in order to find the optimal values. Note that we have applied the both training algorithms *PV-DM* and *PV-DBOW*, and the following values are used for all experiments: α = *0.025; min_*α = *0.025; vector_size = 20; window_size = 300; min_count = 0; epoch = 10.*

Each hyper-parameter is described in Table 1.

Table 1. A description of Doc2vec hyper-parameters

dm (int1,0)	Defines the training algorithm. if *dm = 1*, memory Distributed (PV-DM) is used. Otherwise, a distributed bag of words (PV- DBOW) is used
size (int)	Dimensionality of feature vectors
window (int)	The maximum distance between the current word and the expected word in a sentence
α *(float)*	The initial learning rate
*min_*α *(float)*	The learning rate will drop linearly to *min_*α as the training progresses
min_count (int)	Ignore all words whose total frequency is less than this
epoch (int)	Number of iterations (epochs) on the corpus

5.1 Corpora

The preprocessing techniques described in Sect. 4 were evaluated against the Microsoft Research Paraphrase Corpus[1] [5] which contains 5800 pairs of sentences taken from news sources on the Web (4076 for train and 1724 for test), The sentence pairs have been marked by human judges with a binary classification determining if each pair represents a paraphrase or not (Table 2).

Table 2. Sample pair of sentences of MSRParaphrase Corpus.

Pair of sentences	Semantic equivalence
PCCW's chief operating officer, Mike Butcher, and Alex Arena, the chief financial officer, will report directly to Mr So Current Chief Operating Officer Mike Butcher and Group Chief Financial Officer Alex Arena will report to So	1
A tropical storm rapidly developed in the Gulf of Mexico Sunday and was expected to hit somewhere along the Texas or Louisiana coasts by Monday night A tropical storm rapidly developed in the Gulf of Mexico on Sunday and could have hurricane-force winds when it hits land somewhere along the Louisiana coast Monday night	0

[1] http://research.microsoft.com/nlp/msrparaphrase.htm.

As the validation technique that we will apply is the cross validation, then this dataset is partitioned into three subsets. Knowing that cross-validation is a process for creating a distribution of pairs of learning and test sets from a single dataset. In cross-validation, the dataset is partitioned into k subsets, S_1 ... S_k, each partition called a fold. The folds are approximately the same size. The learning algorithm is then applied k times, for $i = 1$ to k, each time using the union of all non S_i subsets as the training set and using S_i as the test set.

5.2 Evaluation Metrics

In our experiments, k-nearest neighbors classifier is adopted as classification model, and standard metrics were evaluated for each preprocessing technique on the MSRPC: Precision, Recall and F-1 score. These are defined as follows:

$$Precision = \frac{TP}{TP + FP} \tag{1}$$

$$Recall = \frac{TP}{TP + FN} \tag{2}$$

$$F1 - score = \frac{2 * (Recall * Precision)}{Recall + Precision} \tag{3}$$

where TP, FP and FN represent numbers of true positives, false positives, and false negatives, respectively.

5.3 Results

During the experiments, all mentioned preprocessing tasks were considered where the Cleaning method gathers techniques that eliminates noise, such as lowercasing, numbers, punctuation and stop words removal. Concerning Stemming method, the three mentioned techniques are applied, namely: Snowball, Lancaster and Porter stemmer. Regarding Lemmatization, Wordnet lemmatizer was used with and without a POS tagger.

Table 3. Experimental results for preprocessing impact on 'Doc2vec' model.

		Without preprocessing	With preprocessing					
			Cleaning	Stemming			Lemmatization	
				Snowball	Porter	Lancaster	Wordnet	Wordnet with POS-tagger
PV-DM	Rec.	0.766	0.781	0.804	0.790	0.784	0.787	0.769
	Prec.	0.689	0.698	0.699	0.701	0.762	0.691	0.690
	F1	0.725	0.737	0.743	0.748	0.737	0.737	0.729
PV-DBOW	Rec.	0.742	0.746	0.764	0.752	0.758	0.740	0.729
	Prec.	0.685	0.675	0.679	0.676	0.676	0.673	0.675
	F1	0.712	0.709	0.719	0.712	0.715	0.705	0.701

The results of the experimental analysis on MSRParaphrase Corpus are shown in Table 3, this table includes Precision, Recall and F1-scores for each preprocessing technique applied on Doc2vec, using PV-DM and PV-DBOW algorithms, where values in bold are the best, and the underlined ones are the worst.

What is noted first, is that PV-DM algorithm performs better than PV-DBOW algorithm at all levels. For PV-DM it is worth noting that the preprocessed text fragments gave better results compared to the raw text, which means that the cleaning, the stemming and the lemmatization have improved the accuracy of the classifier by bringing more semantics to the data. Concerning the preprocessing techniques, the stemming outperforms the cleaning and lemmatization, where the best F1-score value is obtained by Porter stemmer, the best Precision was that of Lancaster and the best Recall is that of Snowball stemmer. In general, we can say that Stemming techniques bring a good compromise between classifier performance and precision, which is considered as a good impact on Doc2vec in its PV-DM form. For PV-DBOW, although the best Precision is given by the raw text, the best Recall and F1-score are obtained by a Stemming algorithm, which is Snowball stemmer. Even for PV-DBOW, Stemming technique gave the best compromise between performance and precision.

According to these experiments we can answer the first question "Is preprocessing step needed before applying document embedding methods?" by "YES, the preprocessing step is needed to improve document embedding performance", and "What is the most appropriate preprocessing technique that gives more accuracy in NLP tasks?" by "The most appropriate technique is Stemming".

6 Conclusion and Future Work

This paper presented an empirical study on the morphosyntactic data preprocessing impact on document embedding technique 'Doc2vec' over textual semantic similarity evaluation task. In our experiments we have applied the mostly used preprocessing techniques, namely: data cleaning, stemming and lemmatization. Experimental analysis on MSR Paraphrase dataset reveals that preprocessing techniques improve classifier accuracy, especially Stemming technique. The interesting line of future research would be to verify the impact of combination of the preprocessing methods mentioned in the paper, and check the consistency of the obtained findings on other corpora.

References

1. Batet, M., Sanchez, D.: A review on semantic similarity. In: Encyclopedia of Information Science and Technology, Third Edition, pp. 7575–7583. IGI Global (2015)
2. Bengio, Y., et al.: A neural probabilistic language model. J. Mach. Learn. Res. **3**, 1137–1155 (2003)
3. Camacho-Collados, J., Pilehvar, M.T.: On the role of text preprocessing in neural network architectures: an evaluation study on text categorization and sentiment analysis. arXiv preprint arXiv:1707-1780 (2017)

4. Chu, H.: Information Representation and Retrieval in the Digital Age. Information Today, Inc., Medford (2003)
5. Dolan, B., Quirk, C., Brockett, C.: Unsupervised construction of large paraphrase corpora: exploiting massively parallel news sources. In: Proceedings of the 20th International Conference on Computational Linguistics, p. 350. Association for Computational Linguistics (2004)
6. Duwairi, R., El-Orfali, M.: A study of the effects of preprocessing strategies on sentiment analysis for Arabic text. J. Inform. Sci. **40**(4), 501–513 (2014)
7. Ibrahim, A., Katz, B., Lin, J.: Extracting structural paraphrases from aligned monolingual corpora. In: Proceedings of the Second International Workshop on Paraphrasing, vol. 16, pp. 57–64. Association for Computational Linguistics (2003)
8. Kenter, T., De Rijke, M.: Short text similarity with word embeddings. In: Proceedings of the 24th ACM International on Conference on Information and Knowledge Management, pp. 1411–1420. ACM (2015)
9. Kiela, D., Clark, S.: A systematic study of semantic vector space model parameters. In: Proceedings of the 2nd Workshop on Continuous Vector Space Models and their Compositionality (CVSC), pp. 21–30 (2014)
10. Kosala, R., Blockeel, H.: Web mining research: a survey. ACM Sigkdd Explor. Newsl. **2**(1), 1–15 (2000)
11. Le, Q., Mikolov, T.: Distributed representations of sentences and douments. In: International Conference on Machine Learning, pp. 1188–1196 (2014)
12. Lodhi, H., et al.: Text classification using string kernels. J. Mach. Learn. Res. **2**(Feb), 419–444 (2002)
13. Mikolov, T., Yih, W., Zweig, G.: Linguistic regularities in continuous space word representations. In: Proceedings of the 2013 Conference of the North American Chapter of the Association for Computational Linguistics: Human Language Technologies, pp. 746–751 (2013)
14. Mikolov, T., et al.: Efficient estimation of word representations in vector space. arXiv preprint arXiv:1301.3781 (2013)
15. Pak, M.Y., Gunal, S.: The impact of text representation and preprocessing on author identification. Anadolu Üniversitesi Bilim Ve Teknoloji Dergisi A-Uygulamalı Bilimler ve Mühendislik **18**(1), 218–224 (2017)
16. Park, E.-K., Ra, D.-Y., Jang, M.-G.: Techniques for improving web retrieval effectiveness. Inf. Process. Manage. **41**(5), 1207–1223 (2005)
17. Porter, M.F.: An algorithm for suffix stripping. Program **14**(3), 130–137 (1980)
18. Sergienko, R., Shan, M., Minker, W.: A comparative study of text preprocessing approaches for topic detection of user utterances. In: Proceedings of the Tenth International Conference on Language Resources and Evaluation (LREC 2016), pp. 1826–1831 (2016)
19. Uysal, A.K., Gunal, S.: The impact of preprocessing on text classification. Inform. Process. Manage. **50**(1), 104–112 (2014)

Text Steganography with High Embedding Capacity Using Arabic Calligraphy

Ali A. Hamzah(✉) and Hanaa Bayomi

Faculty of Computers and Artificial Intelligence, Cairo University, Giza, Egypt
ali.a.hamzah@grad.fci-cu.edu.eg, h.mobarz@fci-cu.edu.eg

Abstract. Text is a challenging media for steganography due to the limited amount of redundancy bits in text and the ease of detecting minor changes in it. In linguistic steganography, the features of a language are used to hide data. The more features we have in a language, the more capacity of hiding data we can get. The Arabic language is rich in its features, many of which are still not used in text steganography. For example, Arabic calligraphy includes a variety of fonts and shapes for the Arabic alphabet. In this paper, by modifying a string matching algorithm and using suffix tree, the generation of Arabic stego text according to a secret message is implemented. The generation process is performed using a linguistic framework that utilizes Arabic calligraphy (AC-STEGA). In the presented work, multiple shapes of letters in Arabic font (Naskh) are used to increase the capacity of hiding data. This work comes as an extension of previous work to apply one shape of letters in (AC-STEGA). The evaluation showed an improvement in capacity of 65.4% compared to 25.4% presented in previous work. To satisfy security requirement, we did not use cover, the cover represents the secret message itself.

Keywords: Steganography · Information hidden · Linguistic steganography · Arabic text Steganography · Arabic Calligraphy

1 Introduction

Information security is becoming increasingly important nowadays. Steganography is a data hidden technique that is used in the information security domain. Steganography is the art of hiding secret messages in the cover medium. A cover medium can be text, image, audio, or video. The output of hiding a secret message in a cover medium is called the stego-message. To evaluate any steganography system, three criteria are used: capacity, security, and robustness. Capacity is the amount of hidden information in the cover. Security refers to the ability of detecting the secret information by eavesdroppers. Lastly, robustness is the resistance against the modification or destruction of hidden data in the stego-message [1–5].

Text is considered the most widely used way of communication among people [6]. Text steganography contains little redundancy and unused bits. Hence, it is more complicated than other steganography techniques. Moreover, changes in the text can be detected quite easily. In other words, when we embed secret data in text file, the text structure must remain the same. [7]. Multiple languages, such as English, Arabic, Hindi

F. Saeed et al. (Eds.): IRICT 2019, AISC 1073, pp. 127–138, 2020.
https://doi.org/10.1007/978-3-030-33582-3_13

and Persian are using the text in steganography as a cover. Each language has its own characteristics which are exploiting to hide data such as white spaces, word shifting and line shifting[2, 8–10]. Arabic language has many characteristics such as: (1) the upper letters are the same as the lower letters, (2) there are 15 letters of its alphabet containing dots, (3) different kinds of diacritics can be used, and (3) most of the letters can be connected together.

In literature, most of the Arabic-based text steganography methods exploits few of features to increase the capacity. For example, the dot steganography proposed in [11]. Furthermore, the authors in [12] have improved the capacity based on pointed Arabic letters. However, their improvement was not robust enough. The la steganography was proposed in [13], whereas the pointed letters with extension kashida were proposed in [14–16]. Although different kashida approaches have been used to improve the steganography parameters, they still have limitations. Besides, kashida is not normally used much in this era. Arabic text Steganography using diacritics was proposed in [17–19]. However, the increasing use of diacritics raises the level of suspicion for the embedded secrets because diacritics are not commonly used.

1.1 Arabic Calligraphy

Arabic calligraphy is defined as the art of drawing the letters that expresses the word in multiple structures. It is considered one of the components of the Arabic language. Arabic calligraphy forms a tool to demonstrate the linguistic content[20]. Arabic Calligraphy fonts have multiple uses in the language, for example in congratulations, blessings, Quran, poetry, and proverbs. Nowadays, Arabic calligraphy applications are increasing dramatically in different areas especially in social media to exchange daily messages among Arabic users. Figure 1 depicts the possible different shapes of the same Arabic letters in Arabic calligraphy.

Fig. 1. Example of Arabic calligraphy letters in multiple writing shapes of the same word and font (Naskh)

In this paper, we followed up on our previous work [21] which introduced the first linguistic steganography framework based on Arabic calligraphy (AC-STEGA). The first implementation of the framework has used only one shape of letters. In this study, we are using multiple shapes of letters to improve the capacity. The rest of the paper is organized as follows. Section 2 presents the related work of some Arabic text steganography methods and other similar approaches. In Sect. 3, the phases of AC-STEGA system using multiple shapes of letters are presented. Section 4 presents the experimental results and the evaluation of the AC-STEGA system using multiple letters shapes. Finally, Sect. 5 summarizes the paper and presents future work.

2 Related Work

In literature, several approaches of Arabic text steganography and other similar approaches were proposed. These approaches can be divided into the following types:

Using the Arabic Extension Character (Kashida) and Sharp-Edges: In [14–16, 22], kashida is added to increase the capacity and improve the security. In [11], secret bits in the sharp edges of Arabic letters were embedded. The "*kashida*" stego-cover method is modified by adding white spaces to embedded secret data in [23]. A small space character with kashida is proposed in [24] where a kashida is used to hide one bit of a secret message, while small spaces between words will hide three bits.

Adding kashida with four scenarios is presented in [25] where the text is divided into blocks, each block picks up one of the scenarios (e.g. different embedding methods). Kashida is also used in merge secret sharing technique in [26]. In this work, the kashida is added to hide data when the secret shares produced. In addition, Arabic (prophet's hadiths text) was used as cover text. In general, adding the kashida character in Arabic scripts leads to suspicions by eavesdroppers because the kashida is not commonly used nowadays.

Shifting Letter or Dots of Letters: In [12], pointed letters are used to hide secret bits. If the secret bit is zero, the letter remains unchanged. If it is one, the point of the letter is shifted upwards. Multipoint letters are used in [27] to hide secret data. Each letter can hide two bits of secret information. The drawback of these methods is low robustness.

Diacritics–Based Methods: The aim is to utilize the diacritics used on Arabic vowels. Multiple methods uses diacritics for steganography to hide secret bits are proposed in [17, 19]. 'Fathah' and 'Kasrah' diacritics are used to hide information In [18]. The increasing use of diacritics raises the level of suspicion for the embedded secrets because it is not commonly used nowadays also.

Linguistic Steganography Framework Based on Arabic Calligraphy (AC-STEGA): Our recent work represents the first linguistic steganography framework based on Arabic calligraphy using one shape of letters [21]. It includes three phases: preparation, embedding and extraction phases. The first implementation of AC-STEGA uses only one shape for Arabic letters. Figure 2 shows the main components of AC-STEGA.

Steganography Based Cover Generation: In [28], automatic text steganography is presented. Two steps are applied to generate steganography text. *The first step* is to generate text based on a Markov chain model. *The second step* is to use Huffman code technique to represent the generated text in binary form. This method suffers from low capacity. In [29], a paraphrasing technique is used to generate multiple tweets based on an original tweet. To encode the generated paraphrasing based on transition probability, the hash code is employed to encode each paraphrase. A manual filtering step is then applied where a human is presented several options to select the desired tweet.

3 Material and Methods

This section describes the phases of AC-STEGA where multiple shapes of letters are used instead of only one shape.

3.1 Data Set

By using multiple shapes of letters that are supported in Arabic calligraphy, the collected dataset (about 378 valid Arabic sentences (segments) as data set in [21]) can reach up to 196913730 segments due to the variety of shapes of letters and the combination between them where each segment can be written in different forms with different shapes of letters. In this work, we used three forms for each segment compared to one form in the previous work. Figure 3 depicts an example of writing the same segment in different forms.

3.2 AC-STEGA Using One Shape of Letters

AC-STEGA based on one shape of Arabic letters was proposed in previous work [21]. It consists of three main phases: preparation, embedding and extraction phases. In the preparation phase, a corpus is created from selected Arabic segments (Poems, Proverbs) that have valid syntax and semantics. Besides, the Aho-Corasick (AC) string matching algorithm is altered to match entire segments and was called AC*. The suffix tree is also used to represent the segments by a set of symbols.

In the embedding phase, a string-matching process is performed on the constructed AC* model to locate one or more binary encoding(s) that match the secret message. After that, the symbol(s) corresponding to these encoding(s) are used in searching the suffix tree to find how the corresponding segment(s) can be combined to form the entire secret message. The extraction phase reproduces the original secret message by replacing each letter shape in the stego-message with its encoding.

3.3 Preparation Phase

Figure 2 illustrates the preparation phase of AC-STEGA which consists of three steps. Firstly, is to collect Arabic sentences, or simply segments. Secondly, the system maps each letter in a given segment to its corresponding encoding and symbols. Thirdly, the AC* model and the suffix tree are built.

Modified Aho-Corasick Algorithm (AC*). The original algorithm of AC is proposed for multiple patterns within text. So, AC algorithm is not appropriate for selecting a complete encoding that corresponds to a complete segment in Arabic with valid syntax and semantic. Therefore, working with the original AC algorithm will pick parts of the encoding of the segment, which produces a bad segment containing Arabic letters without syntax and semantic. To solve this problem, an algorithm is adapted by Aho-Corasick (AC*) where the failure function is changed to match a complete encoding that matches a valid Arabic segment.

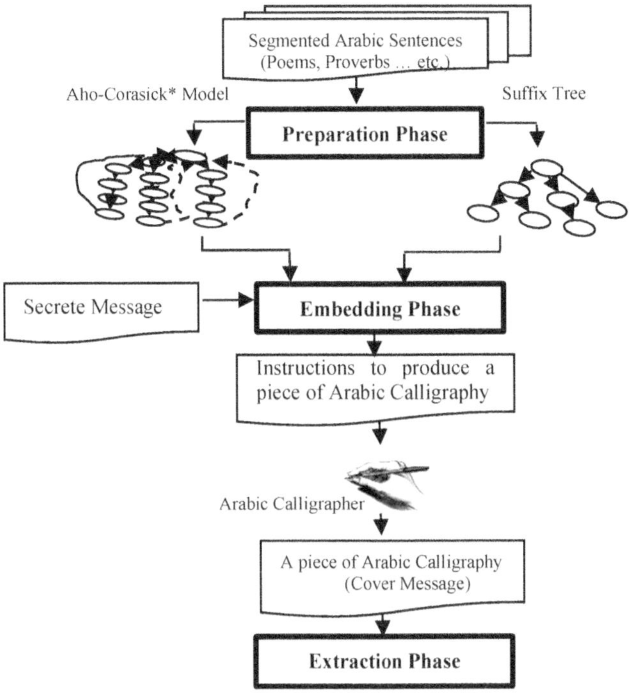

Fig. 2. Flowchart depicting the main phases of the (AC-STEGA)

3.4 Embedding Phase for Multiple Shapes of Letters

According to the entire secret message, in the embedding phase, the matching process is performed to generate a matching segment(s). The matched segments are presented to an Arabic calligrapher to handwrite them. The Arabic calligrapher handwrites the shape of each letter based on the shape encoding that matches the secret message. Finally, handwritten segments that represent the stego-message corresponding to the original secret message are created.

Variable Length Encoding. The shapes of letters in Arabic calligraphy are still without encoding in well-known encoding systems. Therefore, in the preparation phase among different encoding schemes, the variable length encoding has been used to represent different letter shapes in binary form. So, the problem that arises in the embedding phase is how to distinguish letter shapes of the matched binary segment(s) because there is a similarity in encoding different shapes of letters (e.g., **(00) ـسـ(00 ـحـ)**). It also has different lengths of various letter shapes (e.g. 1, 2, 3, 4 bits). So, the selection of the letter shapes of the matched segment (stego message) must be correctly written by the calligrapher. Figure 3 depicts an example.

Figure 3, shows multiple encodings for different shapes of the same segment. Note the encodings similarity for some shapes of letters. To solve this problem, the next subsection explains the process of matching the correct shape of the letter from the related segment encoding.

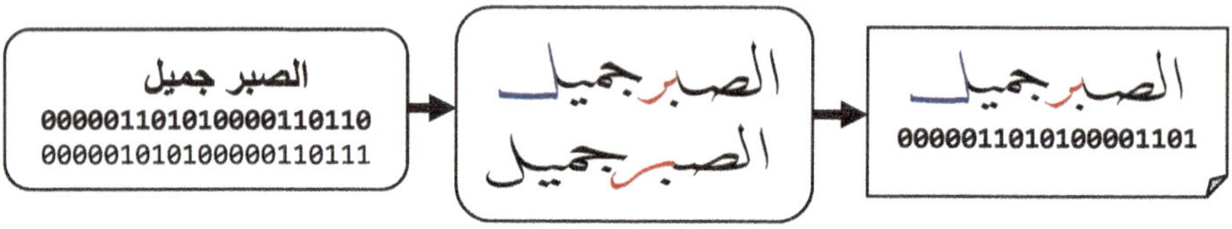

Fig. 3. Example of stego-message (matched segment(s)) before and after represent by shapes of letters

Generating Letters Shapes of the Matched Segment. To achieve the correct matching, an algorithm is created to select or generate the correct letters shapes of the stego-message as follows: Firstly, all the characters shapes of the Naskh font are collected. Secondly, a database that contains all characters with its shapes and encoding is created. Thirdly, a software is developed to check each character of the matched segment with its encoding and generate the character shape. The software is developed using Visual basic.net. Figures 4 depicts a snapshot of this process.

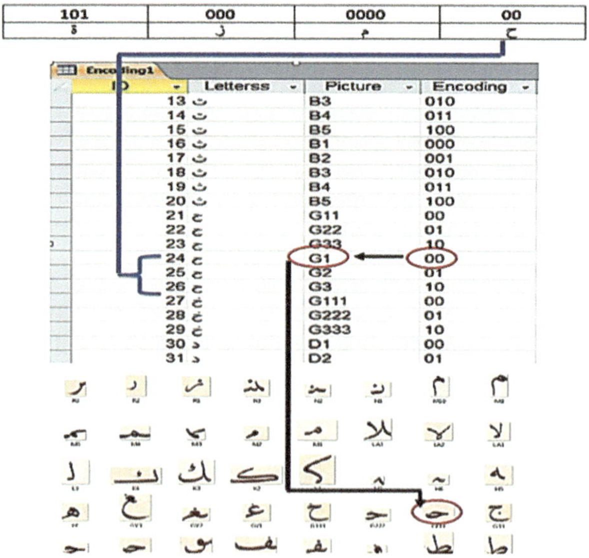

Fig. 4. Example of generating the letters shapes corresponding to encoding of matched segment

Human Filtering. In this work, the human filtering method is used to select desired Arabic segments. In the AC-STEGA system, some secret messages match more than one segment. This is due to the variable length encoding that represents the multiple shapes of letters (see Fig. 3). So, according to the aesthetic nature that characterizes the Arabic calligraphy, the human will select the segment(s) that he wishes to transmit. Figures 5 and 6 show human filtering process.

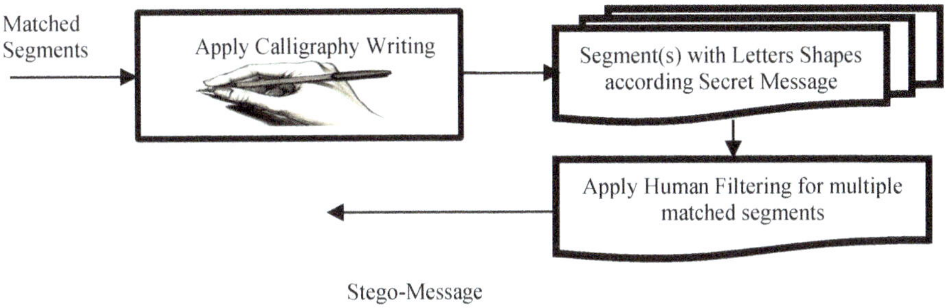

Fig. 5. The steps of human filtering when selecting stego-message

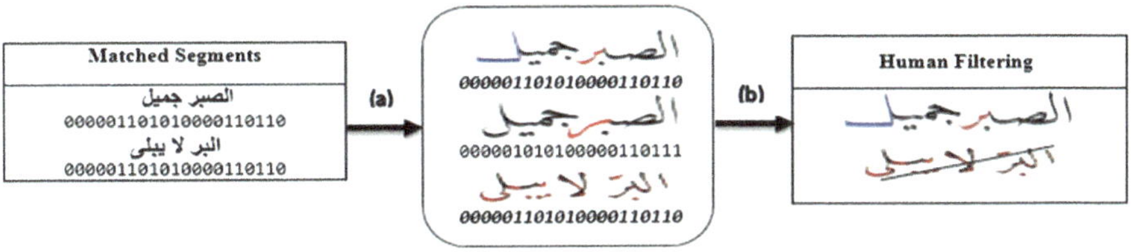

Fig. 6. Example of stego-message. The database contains all possible stego objects for a given secret message (a). These are then filtered according to the matched process (b). The final list is filtered by a human if multiple segments matched secret message, who goes on to choose the stego-message to complete writing and transmission process.

3.5 Extraction Phase for Multiple Shapes of Letters

In this phase, the extraction process of the secret message can be summarized in the following steps: (1) Get the piece of Arabic Calligraphy (i.e. segment) that represents the stego-message (2) For each shape of letter in the stego-message, extract its encoding from the encoding letters shapes (3) The binary encoding that is produced from step *(2)* is represented the original secret message. The snapshot of steps of the extraction phase is illustrated in Fig. 7.

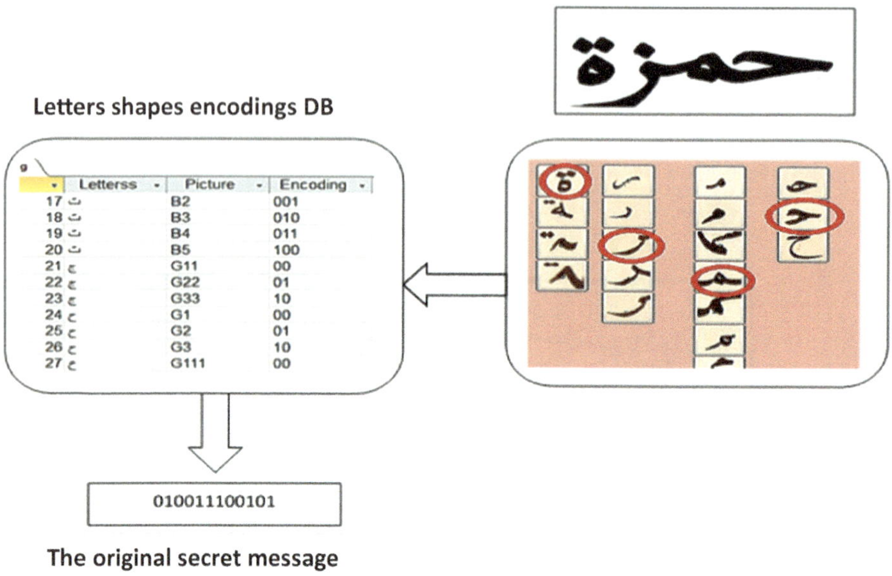

Fig. 7. An example of the extraction phase steps and showing multiple shapes for each letter

4 Experimental Results and Discussion

This section explains the experimental environment and the dataset that is used in testing. It also discusses the coverage capacity and the security of this work.

4.1 The Experimental Environment

The proposed system (AC-STEGA) which is based on multiple shapes of letters is implemented using Java (JDK 1.8) and the Eclipse IDE installed on a machine with Dell Core™ I7-3517U CPU @ 1.90 GHz 2.40 GHz Processor, 8 GB Memory, and Microsoft Windows 7 operating system. The collected dataset (provided in previous work entitled "ARS_1.pdf") is extended after representing each segment in multiple forms with multiple shapes of letters. The embedding and extraction phases are implemented using visual basic.net. The modified AC-STEGA system is evaluated using two parameters: the coverage capacity and security.

4.2 Coverage Capacity

In the methodology of the AC-STEGA system, a stego-message can be generated according to a given binary secret message. In this work, instead of one shape for each letter to each segment, multiple shapes for each letter are implemented. This implies that each segment will be represented more than once. In addition, to evaluate the capacity of the proposed methodology, the formula to calculate the number of secret messages was presented in previous work [21] and the scenarios to cover the secret bits based on the length were proposed in previous work as well. Figures 8 and 9 show the coverage capacity (CC) values calculated using Eq. (1). CC values of the proposed system were calculated over the collected dataset sample compared to previous work and other similar work.

$$\boldsymbol{CoverageCapacity}(\boldsymbol{M}) = \frac{1}{\boldsymbol{M}} \sum_{i=1}^{M} \frac{(\boldsymbol{Li}).\boldsymbol{k}}{\boldsymbol{D}(\boldsymbol{Si})} \times 100 \quad (1)$$

Where, M represents the covered secret messages (e.g. the number of matched segments) at maximum length N bits. Li refers to the length of matched segment (e.g. the number of letters in each matched segment). The average number of bits embedded in each letter of the segment is represented by K and the number of bits in a segment is represented by D (Si). The Arabic letter takes 8 bits according to the standard encoding system therefore, D (Si) = 8 Li. After applying this equation, the coverage capacity of our proposed system using multiple shapes of letters reached up to 65.4% with average 5.2 bits/letter.

4.3 Evaluation of the Results

The capacity rate of the AC-STEGA system using multiple shapes of letters is much higher than the capacity rate using one shape of letters. However, both AC-STEGA systems still produce better results than other similar linguistic steganography methods in other languages. Figure 8 shows the coverage capacity between AC-STEGA using one shape of letters in our previous work [21] compared to AC-STEGA using multiple shapes of letters in this work. Figure 9 shows the coverage capacity between AC-STEGA and other methods where the capacity rates of these methods [28, 30, 31] reaches up to 7.34% while the capacity rate of AC-STEGA using one letter shape reaches up to 25.7%, and using multiple shapes of letters in this paper is 65.4%.

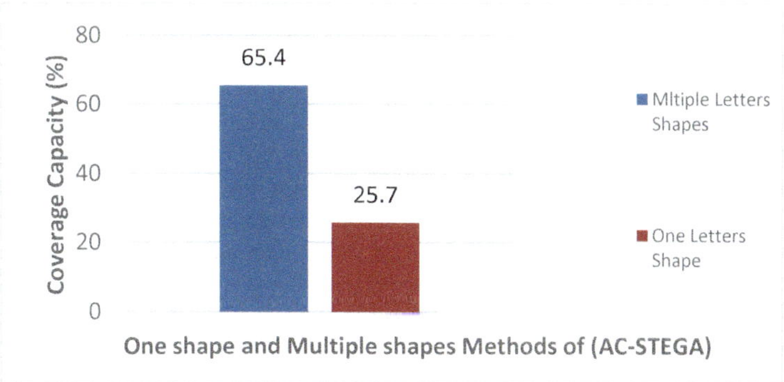

Fig. 8. Capacity rate of AC-STEGA with multiple shapes of letters compared to one shape of letters

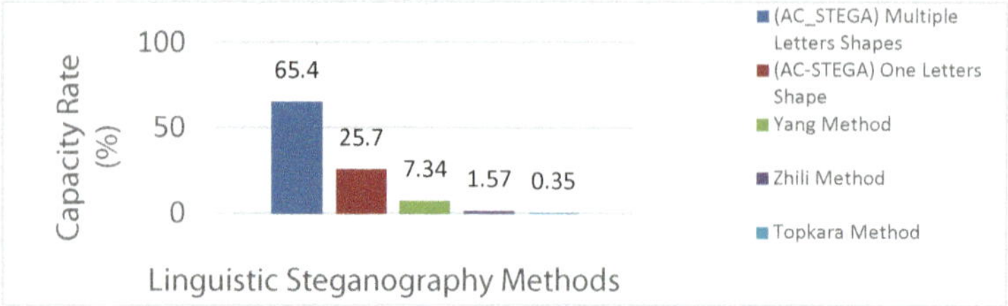

Fig. 9. Capacity rate of AC-STEGA with multiple letters shapes compared to one letter shape and other similar approaches

4.4 Security Evaluation

In linguistic steganography, a security measurement is performed between the cover and stego-cover. A high similarity between them refers to good security thus avoiding suspicions by eavesdroppers and vice versa. We consider the security of our system is satisfied because the cover represents the secret message itself and there is no cover used.

5 Conclusions and Future Work

In this paper, we implemented the Arabic text steganography using multiple shapes of letters. Multiple shapes of letters increased the capacity significantly as opposed to using one shape of letters in the previous and other related works. The evaluation with respect to capacity and security shows the highly satisfying results for steganography parameters. In future work, security will be enhanced using machine learning techniques to select meaningful statements. Moreover, deep learning techniques will be employed to implement the embedding and extraction phases' automatically.

References

1. Yang, Y.: Information analysis for steganography and steganalysis in 3D polygonal meshes. Durham University (2013)
2. Hassanein, M.S.: Secure digital documents using Steganography and QR Code. Brunel University (2014)
3. Por, L.Y., Delina, B.: Information hiding: a new approach in text steganography. In: WSEAS International Conference, Proceedings. Mathematics and Computers in Science and Engineering. World Scientific and Engineering Academy and Society (2008)
4. Hakak, S., Kamsin, A., Tayan, O., Idris, M.Y.I., Gilkar, G.A.: Approaches for preserving content integrity of sensitive online Arabic content: a survey and research challenges. Inf. Process. Manag. 1–14 (2017)
5. Gono, R.: Steganography in computer graphics. In: Faculty of Informatics. Masaryk University (2017)
6. Tatwadarshi, P., Nagarhalli, J.W.B., Jain, N.: A survey of Hindi text steganography. Int. J. Sci. Eng. Res. **7**(3), 55–61 (2016)

7. Chang, C.-Y., Clark, S.: Linguistic steganography using automatically generated paraphrases. In: Human Language Technologies: The 2010 Annual Conference of the North American Chapter of the Association for Computational Linguistics. Association for Computational Linguistics (2010)
8. Sharma, S., et al.: Analysis of different text steganography techniques: a survey. In: 2016 Second International Conference on Computational Intelligence & Communication Technology (CICT). IEEE (2016)
9. Bhattacharyya, S., Banerjee, I., Sanyal, G.: A survey of steganography and steganalysis technique in image, text, audio and video as cover carrier. J. Glob. Res. Comput. Sci. **2**(4), 1–16 (2011)
10. Ramakrishnan, B.K., Thandra, P.K., Srinivasula, A.S.M.: Text steganography: a novel character-level embedding algorithm using font attribute. Secur. Commun. Netw. **9**(18), 6066–6079 (2016)
11. Roslan, N.A., Mahmod, R., Udzir, N.I.: Sharp-edges method in Arabic text steganography. J. Theor. Appl. Inf. Technol. **33** (2011)
12. Shirali-Shahreza, M.H.: A new approach to Persian/Arabic text steganography. In: 5th IEEE/ACIS International Conference. IEEE (2006)
13. Shirali-Shahreza, M.: A new persian/arabic text steganography using "La" word. In: Sobh, T. (ed.) Advances in Computer and Information Sciences and Engineering, pp. 339–342. Springer, Dordrecht (2008)
14. Gutub, A., Fattani, M.: A novel Arabic text steganography method using letter points and extensions. In: WASET International Conference on Computer, Information and Systems Science and Engineering (ICCISSE) (2007)
15. Gutub, A., Al-Nazer, A.A.: High capacity steganography tool for Arabic text using 'Kashida'. ISC Int. J. Inf. Secur. **2**(2), 107–118 (2010)
16. Alotaibi, N., Gutub, A., Khan, E.: Stego-system for hiding text in images of personal computers. In: The 12th Learning and Technology Conference: Wearable Tech/Wearable Learning (2015)
17. Aabed, M.A., et al.: Arabic diacritics based steganography. In: IEEE International Conference on Signal Processing and Communications, ICSPC 2007. IEEE (2007)
18. Gutub, A.A., Elarian, Y.S., Awaideh, S., Alvi, A.: Arabic text steganography using multiple diacritics. In: Proceedings of the 5th IEEE International Workshop on Signal Processing and its Applications (WoSPA 2008), Sharjah, UAE (2008)
19. Ahmadoh, E., Gutub, A.: Utilization of two diacritics for Arabic text steganography to enhance performance. Lect. Notes Inf. Theory **3**(1), 42–47 (2015)
20. Uday Nazim Farman, K.A.D.: Aesthetic theory of Arabic calligraphy in the Islamic Art. J. Nabo **11–12**, 143–173 (2015)
21. Hamzah, A.A., Khattab, S., Bayomi, H.: A linguistic steganography framework using Arabic calligraphy. J. King Saud Univ.-Comput. Inf. Sci. (2019)
22. Gutub, A., et al.: Utilizing extension character 'Kashida' with pointed letters for Arabic text digital watermarking. In: International Conference on Security and Cryptography (2007)
23. Al-Nofaie, S., Fattani, M., Gutub, A.: Capacity improved Arabic text steganography technique utilizing 'Kashida' with whitespaces. In: The 3rd International Conference on Mathematical Sciences and Computer Engineering (ICMSCE2016) (2016)
24. Taha, A., Hammad, A.S., Selim, M.M.: A high capacity algorithm for information hiding in Arabic text. J. King Saud Univ.-Comput. Inf. Sci. (2018)
25. Odeh, A., Elleithy, K., Faezipour, M.: Steganography in Arabic text using Kashida variation algorithm (KVA). In: Proceedings of the 2013 IEEE Long Island Systems, Applications and Technology Conference (LISAT), May 2013

26. Alaseri, K., Adnan, A.: Merging secret sharing within Arabic text steganography for practical retrieval. IJRDO J. Comput. Sci. Eng. **4**, 1–17 (2018)
27. Odeh, A., et al.: Steganography by multipoint Arabic letters. In: Systems, Applications and Technology Conference (LISAT). IEEE, Long Island (2012)
28. Yang, Z., et al.: Automatically generate steganographic text based on Markov Model and Huffman Coding. arXiv preprint arXiv:1811.04720 (2018)
29. Wilson, A., Ker, A.D.: Avoiding detection on twitter: embedding strategies for linguistic steganography. Electron. Imaging **8**, 1–9 (2016)
30. Zhou, Z., et al.: Coverless information hiding method based on multi-keywords. In: International Conference on Cloud Computing and Security. Springer (2016)
31. Topkara, M., Topkara, U., Atallah, M.J.: Information hiding through errors: a confusing approach. In: Security, Steganography, and Watermarking of Multimedia Contents IX 2007. International Society for Optics and Photonics (2007)

A Genetic Algorithm-Based Grey Model Combined with Fourier Series for Forecasting Tourism Arrivals in Langkawi Island Malaysia

Abdulsamad E. Yahya[1], Ruhaidah Samsudin[2(✉)], and Ani Shabri Ilman[3(✉)]

[1] Northern Border University, Kingdom of Saudi Arabia, Arar, Saudi Arabia
Abdulsamad.qasem@nbu.edu.sa
[2] School of Computing, Faculty of Engineering, Universiti Teknologi Malaysia, Skudai, Malaysia
ruhaidah@utm.my
[3] Department of Mathematics, Faculty of Science, Universiti Teknologi Malaysia, Skudai, Malaysia
ani@utm.my

Abstract. Accurate prediction of tourism demand is an important issue for the tourism industry because it can efficiently provide basic information for better tourism planning, along with the desire to reduce risk and uncertainty. To successfully achieve an accurate prediction of tourism demand, this study develops a new forecasting model by combining the Fourier residual modification with an optimized Grey model GM (1, 1). Genetic algorithms (GA) were optimally used simultaneously to select the parameter of the GM (1, 1). For illustration of the proposed model, the monthly international tourist arrivals to Langkawi Island Malaysia has been used as a sample study. Empirical results indicate that the proposed forecasting model demonstrates a superior performance to other methods in terms of forecasting accuracy.

Keywords: ARIMA · Grey · Genetic algorithms · Artificial neural network · Fourier series

1 Introduction

In the past few decades, the tourism industry has emerged as the fastest growing sector, and has spread widely around the world. Tourism has become one of the major contributing sectors for Malaysia's economic growth, contributing to a large proportion of the national income and generating vast employment opportunities. Therefore, more accurate and reliable forecasting of tourism demand is essential for government and business sectors to help formulate responsive policies on tourism swiftly.

Numerous past studies concluded that most of the tourism demand series often exhibit patterns in terms of trend, cyclic and seasonal components. Therefore, the selection of a forecasting model is the important criteria that will influence the forecasting accuracy. Some traditional forecasting models, such as autoregressive integrated moving average (ARIMA) [1–6] model and artificial neural network

F. Saeed et al. (Eds.): IRICT 2019, AISC 1073, pp. 139–151, 2020.
https://doi.org/10.1007/978-3-030-33582-3_14

(ANN) [7–13] are the mostly widely used in tourism demand. However, their performance in a number of specific situations is inconsistent and suffers from some weakness, such as difficulty in parameter selection, requires complex calculations and involves selecting a large number of parameters experimentally, through trial and error, in order to derive good forecasting accuracy [30, 31].

Grey forecasting models are among the recent forecasting methods. Generally, the grey model is expressed as GM (u, v), which is the uth order partial differential equation of v variables. The simplest and widely used grey model is the GM (1, 1) model that denotes a first-order linear differential equation of a single variable. The GM (1, 1) was introduced by Deng [14] with no strict hypothesis for the distribution of the original data series, and has been successfully applied to various fields, such as tourism [15, 16], electricity demand [17], high-tech industry [18], stock market [19] and so on. Although GM (1, 1) is a powerful series forecasting technique, the GM (1, 1) in some cases cannot provide satisfactory performance for the data is highly nonlinear, irregular, and has seasonality [20].

In order to improve forecasting accuracy of the GM (1, 1) model, several improved versions based on residual modification have been noticeable and played an important role in grey forecasting. For example, Hsu and Chen [21] proposed the grey forecasting model that combines residual modification with ANN residual sign estimation to forecast power demand. Hsu et al. [22] proposed the Markov-Fourier grey model approach to forecasting the turning time of Taiwan weighted stock index for increasing the forecasting accuracy. Lee and Tong [23] proposed an approach that combines residual modification and residual genetic programming sign estimation to improve the precision of the residual sign estimator. Many researchers have proposed combined GM (1, 1) with Fourier residual modification, and gained more accuracy than the original [21, 24–28].

As described above, the traditional GM (1, 1) model is widely used for combination with modified residual for forecasting application. However, the estimated parameters of the GM (1, 1) may not be optimized to best fit the data. Therefore, this article by optimum theory of genetic algorithm as the optimal parameter estimation method is suggested to improve forecasting performance of the GM (1, 1) model. This paper proposes a tourist arrivals forecasting with a combination of genetic algorithm based GM (1, 1) model and Fourier series to modify the residuals of GM (1, 1) to derive an improved grey forecasting model. The genetic algorithm considered in this article is one of the most used algorithms for solving numerous optimization problems and the results are more robust and accurate [29].

2 Materials and Methods

2.1 Traditional GM (1, 1) Model

Grey systems theory is a generic theory that deals with problems with small samples or poor data [14]. The solution of differential equation is the form of exponential function that could fit observed data with more flexibility. The GM (1, 1) modeling algorithm can be summarized in the following steps:

Step 1: Assume that the original series of data with n entries is:

$$x^{(0)} = \left\{x_1^{(0)}, x_2^{(0)}, \ldots, x_k^{(0)}, \ldots, x_n^{(0)}\right\}, \tag{1}$$

where $x_k^{(0)}$ stands for the original historical time series data at time k.

Step 2: Construct $x^{(1)}$ by one time accumulated generating operation (1-AGO), which is

$$x^{(1)} = \left\{x_1^{(1)}, x_2^{(1)}, \ldots, x_k^{(1)}, \ldots, x_n^{(1)}\right\}, \tag{2}$$

where $x_k^{(1)} = \sum_{i=1}^{k} x_i^{(0)}$, $k = 1, 2, \ldots, n$

Step 3: The result of 1-AGO is monotonic increasing sequence, which is similar to the solution curve of first order linear ordinary differential equation. Therefore, the solution curve of following differential equation represents the approximation of 1-AGO data. A first-order differential equation is then fitted to the grey series, $x_t^{(1)}$, and is written as

$$\frac{dx^{(1)}}{dt} + ax^{(1)} = b \tag{3}$$

Step 4: The values of a and b are model parameters. According to the grey system theory, the whitening of grey derivatives for discrete data with unit time interval is given by

$$\left.\frac{dX^{(1)}(k)}{dk}\right|_{t=k} = x_k^{(1)} - x_{k-1}^{(1)} = x_k^{(0)} \tag{4}$$

Defining a new variable $z_k^{(1)}$, which is the whitening value of $\left.x_k^{(1)}\right|_{t=k}$, the Eq. (3) can be written in a discrete differential form as

$$x_k^{(0)} + az_k^{(1)} = b, \forall k = 2, 3, \ldots, n \tag{5}$$

where $z_k^{(1)}$ is defined as $z_k^{(1)} = \alpha x_k^{(1)} + (1-\alpha)x_{k-1}^{(1)}, \ \forall k = 2, 3, \ldots, n$.
The α values is within the range of 0 to 1, for the traditional GM (1, 1), α equals to 0.5.

Step 5: The Eq. (5) is called the basic form of GM (1, 1) model, and can be written as $Y = BW$, where

$$W = [a, b]^T, Y = \left[x_2^{(0)}, x_3^{(0)}, \ldots, x_n^{(0)}\right]^T \text{ and } B = \begin{bmatrix} -z_2^{(1)} & -z_3^{(1)} & \cdots & -z_n^{(1)} \\ 1 & 1 & \cdots & 1 \end{bmatrix}^T$$

The values of a and b can be estimated by least-squares method as

$$\hat{W} = \begin{bmatrix} a \\ b \end{bmatrix} = \left(B^T B\right)^{-1} B^T Y$$

After substituting values of a and b into Eq. (3) and set initial condition $\hat{x}^{(1)} = \hat{x}^{(0)}$, then the specified solution of Eq. (3) can be obtained as:

$$\hat{x}_k^{(1)} = \left(x_1^{(0)} - \frac{b}{a}\right) e^{-a(k-1)} + \frac{b}{a}$$

Hence, the forecasting series $\hat{x}^{(0)}$ can be obtained by inverse accumulated generating operation (IAGO), which is defined as $\hat{x}^{(0)} = \left\{\hat{x}_1^{(0)}, \hat{x}_2^{(0)}, \ldots, \hat{x}_k^{(0)}, \ldots, \hat{x}_n^{(0)}\right\}$, where

$$\hat{x}^{(0)} = \begin{cases} \hat{x}_1^{(0)} = \hat{x}_1^{(1)} \\ \hat{x}_k^{(0)} = \hat{x}_k^{(1)} - \hat{x}_{k-1}^{(1)} \end{cases}, k = 1, 2, \ldots, n$$

2.2 Fourier Series Based GM (1, 1) Model

The Fourier series has been widely used in modifying the residuals of GM (1, 1) in order to improve the accuracy of forecasting models [28]. The computational steps of constructing modified forecasting values by combining GM (1, 1) and residual GM (1, 1) are as follows:

Step 1: Construct a traditional GM (1, 1) model.

Step 2: Based on the forecasted series $\hat{x}^{(0)}$ obtained from the GM (1 ,1) model, generate the

sequence of a residual series $\varepsilon^{(0)} = \left\{\varepsilon_2^{(0)}, \varepsilon_3^{(0)}, \ldots, \varepsilon_k^{(0)}, \ldots, \varepsilon_n^{(0)}\right\}$, where

$$\varepsilon_k^{(0)} = x_k^{(0)} - \hat{x}_k^{(0)}.$$

Step 3: Establish a residual, $\varepsilon_k^{(0)}$ model using Fourier series as

$$\varepsilon_k^{(0)} = \frac{a_0}{2} + \sum_{t=1}^{M} a_t cos\left(\frac{2tk\pi}{n-1}\right) + b_t sin\left(\frac{2tk\pi}{n-1}\right), k = 2, 3, \ldots, n$$

where $M = \frac{n-1}{2} - 1$ only take integer number. The residual series can be written as

$$\varepsilon^{(0)} = P.C$$

where $P = \left(0.5_{(n-1)\times 1} P_1 \ldots P_k \ldots P_F\right)$,

$$P_k = \begin{pmatrix} cos\left(\frac{2\times 2\pi\times k}{n-1}\right) & sin\left(\frac{2\times 2\pi\times k}{n-1}\right) \\ cos\left(\frac{3\times 2\pi\times k}{n-1}\right) & sin\left(\frac{3\times 2\pi\times k}{n-1}\right) \\ \vdots & \vdots \\ cos\left(\frac{n\times 2\pi\times k}{n-1}\right) & sin\left(\frac{n\times 2\pi\times k}{n-1}\right) \end{pmatrix}$$

$$C = [a_0, a_1, b_1, a_2, b_2, \cdots, a_F, b_F]^T$$

The parameters $C = [a_0, a_1, b_1, a_2, b_2, \cdots, a_F, b_F]^T$ are obtained by using the ordinary least squares method (OLS), which results in the equation of:

$$\hat{C} = \left(P^T P\right)^{-1} P^T \left[\varepsilon^{(0)}\right]^T$$

Once the parameters are calculated, the forecasted series residual $\hat{\varepsilon}^{(0)}$ is then easily achieved based on the following expression:

$$\hat{\varepsilon}_k^{(0)} = \frac{a_0}{2} + \sum_{t=1}^{M} a_t cos\left(\frac{2tk\pi}{n-1}\right) + b_t sin\left(\frac{2tk\pi}{n-1}\right)$$

Step 4: Perform residual modification.

A value of forecasting FGM (1, 1) model can be obtained by subtracting the forecasting $\hat{x}^{(0)}$ obtained from GM (1, 1) model with $\hat{\varepsilon}^{(0)}$ obtained from Fourier series.

$$\tilde{x}_k^{(0)} = \hat{x}_k^{(0)} + \hat{x}_k^{(0)} \qquad k = 2, 3, \ldots, n$$

2.3 Genetic Algorithm Based Grey Model – GAGM (1, 1)

In the traditional GM (1, 1) model, the value of α is set 0.5 for each $z_k^{(1)} = \alpha x_k^{(1)} + (1-\alpha)x_{k-1}^{(1)}$ in the generated sequence $z^{(1)} = \left\{z_1^{(1)}, z_2^{(1)}, \ldots, z_k^{(1)}, \ldots, z_n^{(1)}\right\}$. It means that each data point has the equal weight. However, this value is not the optimal selection for all data sets in reality and often causes this model to perform poorly [29]. In order to provide an optimal selection for parameters α, in this paper, genetic algorithm was used as a parameter search technique to find near optimal solutions to estimate the parameter of the GM (1, 1) model. To find optimal solutions, the MAPE was used as a fitness function of the GA. The objective of our problem is to minimize the MAPE of the training data sets:

$$\min(MAPE) = \frac{1}{N}\sum\nolimits_{i=1}^{N}\left|\frac{x_i - \hat{x}_i}{x_i}\right| \times 100\%$$

A method based on the GA is developed to automatically determine the parameter for the improved traditional GM (1, 1) model. A genetic algorithm is the heuristic method, which is used to find the global minimum in the search domain. The mathematical form of GAs was first developed by John Holland [32]. Basically, GA is based on random processing. The searching process of the GA depends on the initialization, selection, and reproduction (crossover and mutation), all of which involve random factors. The strength of global searching within the searching space makes it a unique tool for any optimization problem, especially for those areas where exact mathematical solution is not possible.

The same Fourier Series based methodology steps are also fed into the GAGM (1, 1) model in order to build the combined tourism forecasting model, namely, the FGAGM (1, 1) model.

2.4 ARIMA Model

The ARIMA model was first developed by Box and Jenkins [33], and dominated many areas of time series forecasting. For modeling, the time series data contains the seasonal series, ARIMA $(p, d, q)\,(P, D, Q)_S$ model, known as seasonal ARIMA (SARIMA) model that is considered and represented as follows

$$\varphi_p(B)\emptyset_P\left(B^S\right)(1-B)^d\left(1-B^S\right)^D x_t = \theta_q(B)\vartheta_Q\left(B^S\right)a_t$$

where B is the backshift operator and is defined as $B^S x_t = x_{t-S}$; p, q are the number of parameters in the non-seasonal ARIMA model; P, Q are the number of parameters in the seasonal ARIMA model; S is the number of periods per season; d is the number of non-seasonal difference; D is the number of seasonal differences; and a_t is white noise. $\varphi_p(B) = 1 - \varphi_1 B - \varphi_2 B^2 - \ldots - \varphi_p B^p$ is the AR operator; $\theta_q(B) = 1 - \theta_1 B - \theta_2 B^2 - \ldots - \theta_p B^p$ is the MA operator; $\emptyset_P(B^S) = 1 - \emptyset_1 B^S - \emptyset_2 B^{2S} - \ldots - \emptyset_P B^{PS}$ is the seasonal AR operator; and $\vartheta_Q(B^S) = 1 - \vartheta_1 B^S - \vartheta_2 B^{2S} - \ldots - \vartheta_Q B^{QS}$ is the seasonal MA operator. The Box–Jenkins methodology used to formulate the ARIMA model consists of four steps: identification of the model, estimation of the parameters model, diagnostic checking and forecasting [33]. The best fitted ARIMA model is selected using minimum Akaike Information Criterion (AIC).

2.5 Evaluate Precision of Forecasting Models

In order to evaluate the forecasting accuracies of the FGAGM (1, 1) method compared with other methods on the tourism arrivals data series, the two criteria are computed to compare the models. The first one is the root mean squared percentage error (RMSPE), which is written as

$$RMSPE = \sqrt{\frac{1}{N-1}\sum_{i=1}^{N}\left(\frac{x_i - \hat{x}_i}{x_i}\right)^2}$$

where x_i is the source point, $\hat{x}_i$ is the forecasted point and N is the number of forecasted points. The alternative criterion is the mean absolute percentage error (MAPE). The MAPE is regarded as one of the standard statistical performance measures and takes the following form

$$MAPE = \frac{1}{N}\sum_{i=1}^{N}\left|\frac{x_i - \hat{x}_i}{x_i}\right| \times 100\%$$

The smaller the value of RMSPE and MAPE, the better the forecast is. Table 1 shows the summarized criteria of RMSPE and MAPE for model evaluation [34, 35].

Table 1. Typical MAPE Values for Model Evaluation

RMSPE and MAPE (%)	Forecasting power
<10	Excellent
10–20	Good
20–50	Reasonable
>50	Inaccurate

Source: Lewis [34], DeLurgio [35]

3 Results and Discussion

To demonstrate the effectiveness of the proposed grey forecasting model, the real case of the monthly arrivals of international tourists to the Langkawi Island is considered as an example. The Langkawi Island is one of the famous tourist destinations located in the northern part of Peninsular Malaysia. The data used in this study is the monthly arrivals of international tourists to the Langkawi Island from January 2002 to December 2016 (180 observations in total). The data is obtained from the monthly statistical data published on the website of the Langkawi Development Authority (LADA). The historical monthly data from January 2002 to December 2015, which contains 168 observations, was considered for model building phase, whereas the remaining data from January 2016 to December 2016 is used to assess their out-of-sample forecasting abilities. Figure 1 shows the time series plot of the monthly arrivals of international tourists to the Langkawi Island from January 2002 to December 2016. The time series data clearly shows a seasonal pattern.

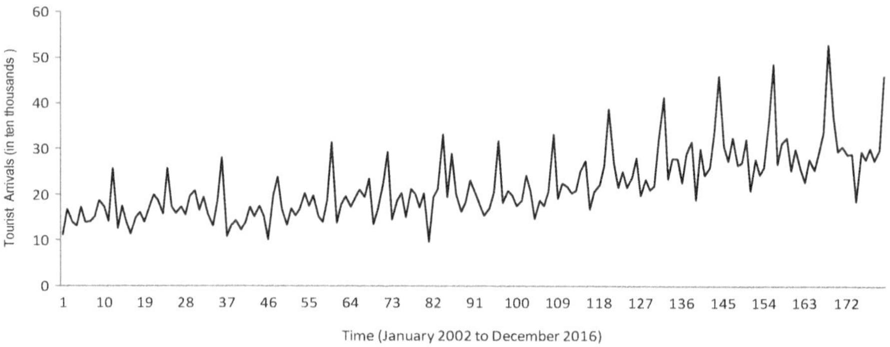

Fig. 1. The monthly tourist arrivals to Langkawi Island in Malaysia from 2002 to 2016

In order to compare the above models, the most widely used seasonal ARIMA model has been presented in this study. Figure 1 describes the curve of monthly arrivals of international tourists to the Langkawi Island from January 2002 to December 2016. The data shows trend, non-stationary and seasonal characteristics. The time series data was preprocessed by taking first-order regular differencing (d = 1) and first-order seasonal differencing (D = 1), with S =12 as the period to remove the growth trend and seasonality. The sample autocorrelation function (ACF) and sample partial autocorrelation function (PACF) after differencing is plotted in Fig. 2. ACF showed that q = 1 and Q = 3 (ACF are significant spikes at lags 1, 12 and 36). PACF revealed that p = 2 and P = 3 (PACF are significant spikes at lags 1, 2, 12 and 36). All combinations of ARIMA models are evaluated using the AIC and Ljung-Box tests. AIC is used for identification of the best model whereas the Ljung-Box test is required to verify that the residuals of the model is an adequate one for the series. Table 1 shows the value of AIC and the p-values of Ljung-Box test for several possible ARIMA models. Table 2 shows that the best model for the monthly international tourist arrivals to Langkawi Island time series was found to be ARIMA (0, 1, 1) × $(0, 1, 3)_{12}$ (i.e., it has the smallest AIC value). The model can be expressed as

$$(1 - B)(1 - B^{12})x_t = (1 - 0.840B)(1 - 0.405B^{12} + 0.208B^{24} - 0.252B^{36})a_t$$

The p-values of Ljung-Box of ARIMA (0, 1, 3) × $(0, 1, 1)_{12}$ model suggests the residuals are white noise and ready for forecasting.

The traditional GM (1, 1) model equation calculated using Matlab software for the monthly tourist arrivals to Langkawi Island from January 2002 to December 2015 are as follows:

$$\hat{x}_k^{(1)} = 28882177e^{-0.0048(k-1)} + 28770444$$

For the GAGM (1, 1) model, the parameters of GA are as follows: the population size is 50, the crossover rate is set to be 0.8. The default values of the MATLAB optimization toolbox are used for the other parameters such as mutation rate, stopping

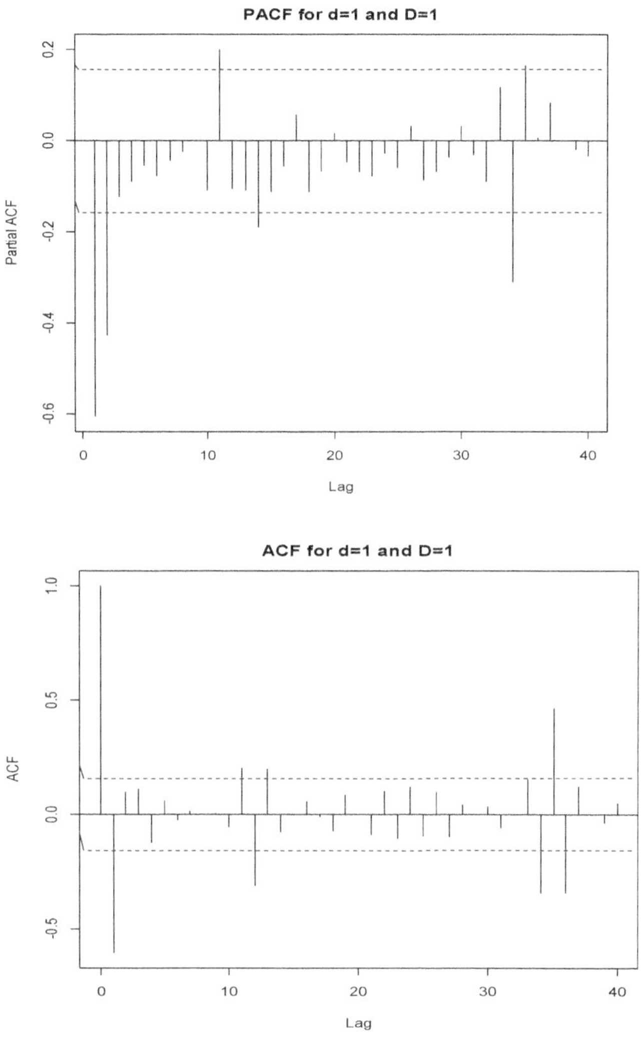

Fig. 2. ACF and PACF plots for $d = 1$ and $D = 1$ of tourist arrivals to Langkawi Island Malaysia

Table 2. Comparison of AIC and p-value of Ljung Box test for selected candidate models

No	Model		p-value of Q(r) test		
	ARIMA	AIC	Lag = 12	Lag = 24	Lag = 36
1	$(1,1,1) \times (3,1,0)_{12}$	3639.30	0.8559	0.8985	0.0516
2	$(1,1,0) \times (3,1,0)_{12}$	3672.77	0.0005	0.0035	0.00001
3	$(0,1,1) \times (3,1,0)_{12}$	3639.22	0.6210	0.7056	0.0108
4	$\mathbf{(0,1,1) \times (0,1,3)_{12}}$	**3636.55**	**0.7101**	**0.8188**	**0.0190**
5	$(1,1,1) \times (0,1,3)_{12}$	3637.29	0.8294	0.9025	0.0499

criteria, elite count and selection function. The GAGM (1, 1) model equation calculated using Matlab software is

$$\hat{x}_k^{(1)} = 28880411e^{-0.00483(k-1)} + 28768688$$

From models GM (1, 1) and GAGM (1, 1), its residual series were then modified with Fourier series to become new models named FGM (1, 1) and FGAGM (1, 1) respectively. The forecast results of GM (1, 1), FGM (1, 1), FGAGM (1, 1) and ARIMA models are shown in Table 3.

In Table 3, the RMSPE of the GM (1, 1), FGM (1, 1), FGAGM (1, 1) and ARIMA models applied to the forecasting data are 25.411%, 12.041%, 12.015% and 20.042%, respectively. The MAPE of the GM (1, 1), FGM (1, 1), FGAGM (1, 1) and ARIMA models are 18.021%, 10.915%, 10.596%, and 13.915%, respectively; these values indicated that the four models yielded good forecast accuracy. The above results indicate that the proposed FGAGM (1, 1) model has superior fitting and has a higher precision of forecasting than the other forecasting methods. The FGAGM (1, 1) model exhibits high forecasting ability according to the RMSPE and MAPE criteria. Thus, the FGAGM (1,1) model can enhance the model forecasting precision effectively.

Table 3. Performance of the four forecasting methods for monthly tourism arrivals data

Model	RMSPE (%)	MAPE (%)
GM(1,1)	25.411	18.021
FGM(1,1)	12.041	10.915
FGAGM(1,1)	**12.015**	**10.596**
ARIMA	20.042	13.915

The forecasts obtained from the GM (1, 1), FGM (1, 1), FGAGM (1, 1) and ARIMA models are plotted in Fig. 3, which demonstrates the superiority of the

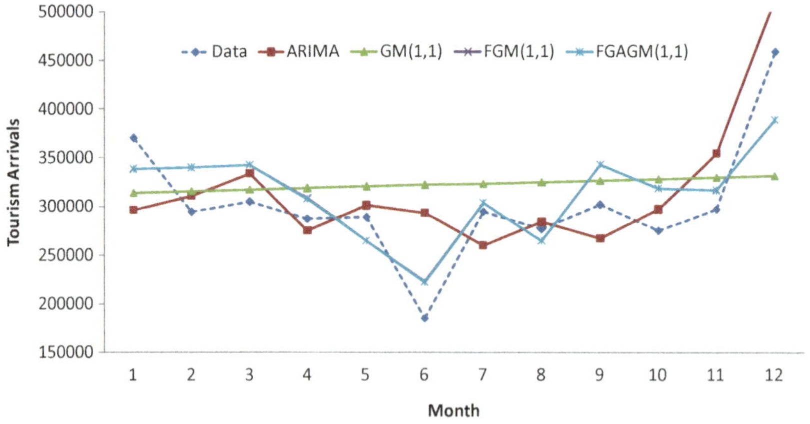

Fig. 3. The trend of actual vs. forecasted output for the four models.

FGAGM (1, 1), since the forecasts obtained from FGAGM (1, 1) very closely fit the actual data compared to the other three methods. According to the results described above, the improved FGAGM (1, 1) model on optimal parameter estimation by GA obtains the lowest forecasting errors among these models.

4 Conclusion

In this study, the GA-based GM (1, 1) and Fourier series (FGAGM (1, 1)) have been proposed for improving the forecasting of the tourist arrivals series. The proposed forecasting model approach is a three-stage procedure: the first stage involves using genetic algorithm (GA) to optimize the parameters of GM (1, 1), then using the GAGM (1, 1) model to derive the forecasted value, and finally using the Fourier series to modify the residual errors of the GAGM (1, 1) model. This approach can filter out high frequency terms, which are supposed to be noisy, resulting in better performance. A comparison of the forecast accuracy of the three grey models and ARIMA model according to the RMSPE and MAPE indicators showed that the FGAGM (1, 1) model had the highest forecast accuracy, followed by the FGM (1, 1) model, ARIMA model and GM (1, 1). This study confirmed that when developing tourist arrivals forecasting models with trend, cyclic and seasonal components in time series, the combining of modified residual by Fourier series with GAGM (1, 1) or with GM (1, 1) had higher forecast accuracy compared with the individual GM (1, 1) model. The result shows that the traditional GM (1, 1) model is unable to provide satisfactory performance when forecasting the time series with trend, cyclic and seasonal components. The forecasting accuracy of the FGAGM (1, 1) is the best and, therefore, the proposed optimized method by GA is effective for use in the tourist arrivals forecasting.

References

1. Aslanargun, A., Mammadov, M., Yazici, B., Yolacan, S.: Comparison of ARIMA, neural networks and hybrid models in time series: tourist arrivals fore-casting. J. Stat. Comput. Simul. **77**(1), 29–53 (2007)
2. Chen, C.F., Chang, Y.H., Chang, Y.W.: Seasonal ARIMA fore-casting of inbound air travel arrivals to Taiwan. Transportmetrica **5**(2), 125–140 (2009)
3. Ko, K.M., Chen, J.F., Nguyen, T.L., Hsu, B.M., Shu, M.H.: Forecasting inbound tourism demand in thailand with grey model. Wseas Trans. Math. **13**, 95–104 (2012)
4. Nguyen, T.L., Shu, M.H., Hsu, B.M.: Forecasting international tourism demand- an empirical case in Taiwan. Asian J. Empirical Res. **3**(6), 711–724 (2013)
5. Baldigara, T., Mamula, M.: modelling international tourism demand using seasonal arima models. Tour. Hosp. Manag. **21**(1), 19–31 (2015)
6. Zhang, G., Wu, J., Pan, B.: Improving daily occupancy forecasting ac-curacy for hotels based on EEMD-ARIMA model. Tour. Econ. **23**(7), 1496–1514 (2017)
7. Burger, C., Dohnal, M., Kathrada, M., Law, R.: A practitioners guide to time-series methods for tourism demand forecasting: a case study of Durban, South Africa. Tour. Manag. **22**, 403–409 (2001)

8. Law, R.: The impact of the Asian financial crisis on Japanese demand for travel to Hong Kong: a study of various forecasting techniques. J. Trav. Tour. Mark. **10**, 47–66 (2001)
9. Cho, V.: A comparison of three different approaches to tourist arrival fore-casting. Tour. Manag. **24**, 323–330 (2003)
10. Kon, S.C., Turner, W.L.: Neural network forecasting of tourism demand. Tour. Econ. **11**, 301–328 (2005)
11. Chen, C.F., Lai, M.C., Yeh, C.C.: Forecasting tourism demand based on empirical mode decomposition and neural network. Knowl.-Based Syst. **26**, 281–287 (2012)
12. Oscar, C., Salvador, T.: Forecasting tourism demand to Catalonia: Neural networks vs. time series models. Econ. Model. **36**, 220–228 (2014)
13. Cankurt, S., Subasi, A.: Developing tourism demand forecasting models using machine learning techniques with trend, seasonal, and cyclic compo-nents. Balkan J. Electr. Comput. Eng. **3**(1), 42–49 (2015)
14. Deng, J.L.: Control problems of grey systems. Syst. Control Lett. **1**, 288–294 (1982)
15. Wang, C.H.: Predicting tourism demand using fuzzy time series and hybrid grey theory. Tour. Manag. **25**, 367–374 (2004)
16. Yu, G., Schwartz, Z.: Forecasting short time-series tourism demand with artificial intelligence models. J. Travel Res. **45**(2), 194–203 (2006)
17. Akay, D., Atak, M.: Grey prediction with rolling mechanism for electricity demand forecasting of Turkey. Energy **32**, 1670–1675 (2007)
18. Hsu, L.C., Wang, C.H.: Forecast the output of integrated circuit industry using a grey model improved by the Bayesian analysis. Technol. Forecast. Soc. Ch. **74**, 843–853 (2007)
19. Wang, Y.F.: Predicting stock price using fuzzy grey prediction system. Expert Syst. Appl. **22**, 33–39 (2002)
20. Tseng, F.M., Yu, H.C., Tzeng, G.H.: Applied hybrid grey model to forecast seasonal time series. Technol. Forecast. Soc. Chang. **67**, 291–302 (2001)
21. Hsu, C.C., Chen, C.Y.: Applications of improved grey prediction model for power demand forecasting. Energy Convers. Manag. **44**(14), 2241–2249 (2003)
22. Hsu, Y.T., Liu, M.C., Yeh, J., Hung, H.F.: Forecasting the turning time of stock market based on Markov-Fourier grey model. Expert Syst. Appl. **36**, 8597–8603 (2009)
23. Lee, Y.S., Tong, L.I.: Forecasting energy consumption using a grey model improved by incorporating genetic programming. Energy Convers. Manag. **52**, 147–152 (2011)
24. Kan, M.L., Lee, Y.B., Chen, W.C.: Apply grey f prediction in the number of Tourist. In: The fourth international conference on Genetic and Evolutionary computing, Shenzhen, China, pp. 481–484 (2010)
25. Thanh, P.V.: Enhanced prediction accuracy of grey forecasting model: a case by tourism industry in Vietnam. Hue Univ. J. Sci. **113**(14), 157–167 (2015)
26. Huang, Y.L., Lee, Y.H.: Accurately forecasting model for the Stochastic volatility data in tourism demand. Mod. Econ. **2**, 823–829 (2011)
27. Askari, M., Fetanat, A.: Long-term load forecasting in power system. Grey system prediction -based models. J. Appl. Sci. **11**, 3034–3038 (2011)
28. Wang, C.N., Phan, V.T.: An improved nonlinear grey bernoulli model combined with fourier series. Math. Prob. Eng. **2015**, 1–7 (2015)
29. Wang, C.H., Hsu, L.C.: Using genetic algorithms grey theory to forecast high technology industrial output. Appl. Math. Comput. **195**(1), 256–263 (2008)
30. Zhang, G., Putuwo, B.E., Hu, M.Y.: Forecasting with artificial neural networks: the state of the art. Int. J. Forecast. **14**, 35–62 (1998)
31. Box, G.E.P., Jenkins, G.: Time Series Analysis: Forecasting and Control. Wiley, San Francisco (1970). Holden-Day

32. Holland, J.H.: Adaptation in Natural and Artificial Systems. The University of Michigan Press, Ann Arbor (1975)
33. Box, G.E.P., Jenkins, G.: Time Series Analysis: Forecasting and Control. Wiley, San Francisco (1976). Holden-Day
34. DeLurgio, S.A.: Forecasting principles and applications. McGraw-Hill, New York (1998)
35. Lewis, C.: Industrial and Business Forecasting Methods: A Practical Guito Exponential Smoothing and Curve Fitting. Butterworth Scientific, London (1982)

Drought Forecasting Using Gaussian Process Regression (GPR) and Empirical Wavelet Transform (EWT)-GPR in Gua Musang

Muhammad Akram Shaari[1], Ruhaidah Samsudin[1](✉), Ani Shabri Ilman[2], and Abdulsamad E. Yahya[3]

[1] School of Computing, Faculty of Engineering, Universiti Teknologi Malaysia, Skudai, Malaysia
arekram@live.com, ruhaidah@utm.my
[2] Department of Mathematics, Faculty of Science, Universiti Teknologi Malaysia, Skudai, Malaysia
ani@utm.my
[3] Northern Border University, Arar, Kingdom of Saudi Arabia
abdulsamad.qasem@nbu.edu.sa

Abstract. Drought forecasting is important in preparing for drought and its mitigation plan. This study focuses on the investigating the performance of Gaussian Process Regression (GPR) and Empirical Wavelet Transform-Gaussian Process Regression (EWT-GPR) in forecasting drought using Standard Precipitation Index (SPI). Daily rainfall data from Gua Musang, Kelantan from 1975 to 2008 was used in this study. SPI data of 3, 6, 9, and 12 months were then calculated using the rainfall data. EWT was employed to decompose the time series into several finite modes. EWT was used to create Intrinsic Mode Functions (IMF) which are used to create ARIMA models. The objective of this study is to compare the effectiveness of the proposed method against GPR model in the accuracy of forecasting drought in Arau, Malaysia. It was found that the proposed model performed better compared to GPR model in all SPI studied.

Keywords: Forecasting · Drought · Gaussian Process Regression (GPR) · Empirical Wavelet Transform (EWT)

1 Introduction

Drought forecasting is one of the important process that helps with the preparation of an incoming drought. There are two ways to forecast drought, either by using physical model or data driven model. The most used method for drought forecasting is by using data driven model since they require less information and less development effort [1].

Gaussian Process Regression (GPR), a kernel-based machine learning has been successfully used in forecasting in several fields. These fields includes forecasting streamflow [2], wind speed [3] and monthly solar radiation [4]. However, its use in drought forecasting is quite limited. GPR are able to model the nonlinear relationship

F. Saeed et al. (Eds.): IRICT 2019, AISC 1073, pp. 152–161, 2020.
https://doi.org/10.1007/978-3-030-33582-3_15

between predictand and predictor [2]. The main advantage of employing GPR to forecast drought is the ability to perform point forecast and also its uncertainty [5].

To improve forecasting performance where the data consists of linear and nonlinear components, data decomposition method is used. Empirical Mode Decomposition (EMD) is one of the signal decomposition method that is frequently explored in forecasting. Several studies in wind speed forecasting, annual precipitation forecasting and fishery landing forecasting [6–9] have shown that forecasting the Intrinsic Mode Functions (IMFs) decomposed using EMD creates a forecast that are more accurate when compared to directly forecasting on the time series data. This technique of forecasting is known as ensemble forecasting.

Jerome Gilles [10] proposed empirical wavelet transform (EWT) that identifies the many different intrinsic modes of a time series and extracts them. Compared to EMD, EWT provides a more consistent decomposition [10]. Conceptually, EWT is similar to DWT, which have been proven to increase the forecast accuracy for drought forecasting [11] and streamflow forecasting [12]. EWT has been used in wind speed forecasting along with Gaussian Process Regression (GPR), and it was found out that the models that utilized EWT have higher accuracy compared to the models that does not implement EWT [3, 13]. However, there are a lack of studies that utilize EWT to forecast drought.

The purpose of this study is to determine whether the utilization of EWT in EWT-GRP improves forecasting accuracy compared to GPR model in forecasting drought for Gua Musang, Kelantan.

2 Materials and Methods

2.1 EWT-GPR Model

The proposed model consists of using EWT and GPR to forecast SPI data. Figure 1 shows EWT-GPR development process. Firstly, SPI are calculated from the daily rainfall data. EWT is used to decompose the SPI timeseries into several IMFs. GPR models are trained for each of the IMFs. Then, the forecast is made for each IMFs independently using GPR. The result of the forecast is reconstructed into the original series by summing the forecasting result of each IMFs to obtain the final forecast.

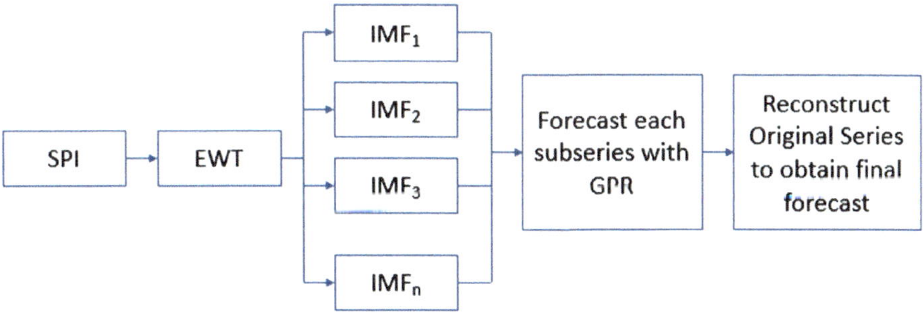

Fig. 1. EWT-GPR development process

2.2 Empirical Wavelet Transform

The empirical wavelets can be defined as bandpass filters on each Λ_n, where Λ_n denotes each segment $\Lambda_n = [w_{n-1}, w_n]$ and $\cap_{n=1}^{N} \Lambda_n = [0, \pi].\forall_n > 0$ The Eqs. (1) and (2) follows defines the empirical scaling functions and empirical wavelets [3, 10]

$$\hat{\phi}_n(\omega) = \begin{cases} 1 & \text{if} |\omega| \leq (1-\gamma)\omega_n \\ \cos\left[\frac{\pi}{2}\beta\left(\frac{1}{2\tau_n}(|\omega| - \omega_n + \tau_n)\right)\right] & \text{if}(1-\gamma)\omega_n \leq |\omega| \leq (1+\gamma)\omega_n \\ 0 & \text{otherwise} \end{cases} \tag{1}$$

$$\hat{\phi}_n(\omega) = \begin{cases} 1 & if(1+\gamma)\omega_n \leq |\omega| \leq (1+\gamma)\omega_{n+1} \\ \cos\left[\frac{\pi}{2}\beta\left(\frac{1}{2\tau_{n+1}}(|\omega| - \omega_{n+1} + \tau_{n+1})\right)\right] & if(1-\gamma)\omega_{n+1} \leq |\omega| \leq (1-\gamma)\omega_{n+1} \\ \sin\left[\frac{\pi}{2}\beta\left(\frac{1}{2\tau_n}(|\omega| - \omega_n + \tau_n)\right)\right] & if(1+\gamma)\omega_n \leq |\omega| \leq (1-\gamma)\omega_n \\ 0 & otherwise \end{cases} \tag{2}$$

The function $\beta(x)$. is an arbitrary $C^k([0, 1])$ function which satisfies the following properties $\beta(x) = \begin{cases} 0 \; if \; x \leq 0 \\ 1 \; if \; x \geq 0 \end{cases}$ and $\beta(x) + \beta(1-x) = 1, \forall\, x \in [0, 1]$.

The set $\{\phi_1(t), \{\varphi_n(t)\}_{n=1}^{N}\}$ is a tight frame of $L^2(R)$ when $\gamma < \min_n\left(\frac{\omega_{n+1} - \omega_n}{\omega_{n+1} + \omega_n}\right)$. The inner product of the signal and empirical wavelets yields the detail coefficients:

$$w_f^{\epsilon}(n, t) = \langle f, \varphi_n \rangle = \int f(\tau)\overline{\varphi_n(\tau - t)}d\tau = f(\tau))\left(\overline{\hat{\varphi}_n(\tau - t)}\right)^{\vee} \tag{3}$$

The approximation coefficients are then calculated using the inner products and the scaling function

$$\begin{aligned} w_f^{\epsilon}(0, t) &= \langle f, \phi_1 \rangle = \int f(\tau)\overline{\phi_1(\tau - t)}d\tau \\ &= f(\tau))\left(\overline{\hat{\phi}_1(\tau - t)}\right)^{\vee} \end{aligned} \tag{4}$$

The reconstruction can be obtained:

$$\begin{aligned} f(t) &= w_f^{\epsilon}(0, t) * \phi_1(t) + \sum_{n=1}^{N} w_f^{\epsilon}(n, t) * \varphi_n(t) \\ &= \left(\widehat{w_f^{\epsilon}}(0, \omega) * \phi_1(\omega) + \sum_{n=1}^{N} \widehat{w_f^{\epsilon}}(n, \omega) * \varphi_n(\omega)\right)^{\vee} \end{aligned} \tag{5}$$

Considering this formalism, the empirical mode is given by

$$\begin{aligned} f_0(t) &= w_f^{\epsilon}(0,t) * \varphi_1(t) \\ f_k(t) &= w_f^{\epsilon}(k,t) * \varphi_k(t) \end{aligned} \tag{6}$$

2.3 Gaussian Process Regression

GPR provides a principled, practical, and probabilistic approach to learning in kernel machines. The primary idea behind GPR is to use a Gaussian process (GP) to represent f, referred to as latent variables. The following shows how a GPR model can be constructed

$$y_i = f(x_i) + \varepsilon_i, i = 1, \ldots, n \tag{7}$$

Where $\varepsilon_i \sim N(0, \sigma^2), i = 1, \ldots, n, f(x)$ is given a GP prior. Assuming the predicted values $\tilde{f}$ with new input $\tilde{x}$ to the latent function, the joint prior for latent variables f and $\tilde{f}$ is

$$\begin{bmatrix} y \\ \tilde{f} \end{bmatrix} x, \tilde{x}, \theta \sim N\left(\begin{bmatrix} K_{f,f} + \sigma_n^2 I & K_{f,\tilde{f}} \\ K_{\tilde{f},f} & K_{\tilde{f},\tilde{f}} \end{bmatrix}\right) \tag{8}$$

Where $K_{f,\tilde{f}} = k(x, \tilde{x}|\theta)$. By defining the marginal distribution of $\tilde{f}$ as $p(\tilde{f}|\tilde{x}, \theta) = N\left(\tilde{f}|0, K_{\tilde{f},\tilde{f}}\right)$, the conditional distribution of $\tilde{f}$ given f is

$$\tilde{f}\Big|y, x, \tilde{x}, \theta \sim N\left(m(\tilde{x}|\theta), k\left(x, \tilde{x'}|\theta\right)\right)\Big| \tag{9}$$

Where the mean function $m(x|\theta) = k(x, x|\theta)\left(K_{f,f} + \sigma_n^2 I\right)^{-1} y$ and covariance function $k\left(\tilde{x}, \tilde{x'}\Big|\theta\right) = k\left(\tilde{x}, \tilde{x'}\Big|\theta\right) - k(\tilde{x}, x\theta)\left(K_{f,f} + \sigma_n^2 I\right)^{-1} k\left(x, \tilde{x'}|\theta\right)$, where the conditional distribution of the latent function is defined.

2.4 Performance Comparison

The performance of each method is evaluated by comparing the error statistics. The error statistics used were MAE and RMSE. Lower MAE and RMSE implies that the forecasting performance is better. These error statistics are given by:

$$\text{MAE} = \frac{1}{N}\sum_{i=1}^{N}\left|\text{EST}_i - \text{OBS}_i\right| \tag{10}$$

$$\mathrm{RMSE} = \sqrt{\frac{1}{N}\sum_{i=1}^{N}(\mathrm{EST}_i - \mathrm{OBS}_i)^2} \tag{11}$$

2.5 Study Area

The physical area covered in this study is the area around Gua Musang, Kelantan. The data consists of daily rainfall values from 1 January 1975 until 31 December 2008, for a total of 34 years.

3 Results and Discussion

Firstly, SPI 3, 6, 9, and 12 are calculated using the daily rainfall data. These SPIs are chosen since they are able to represent the short-term drought which are common in Malaysia [14]. Figure 2 below shows the SPI 3, 6, 9, and 12 data respectively vs time in months. The data is split into 2 parts, one part which consist of 80% of the data for training and another part consists of the rest of the data for testing the model.

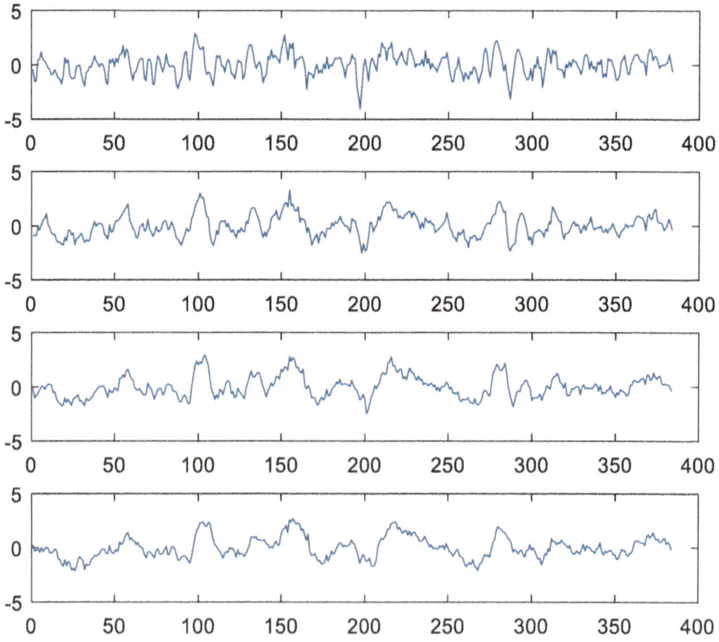

Fig. 2. SPI 3, 6, 9, and 12 for Gua Musang Region from 1975 to 2008

3.1 Gaussian Process Regression

In order to obtain the best input for GPR, partial autocorrelation function (PACF) was used to select the input to be used with GPR models [3]. Figure 3 below shows the PACF obtained for SPI3. From the PACF obtained, the inputs suitable for GPR model was determined to be lag 1, 2, 3, 4 and 6. The process is repeated for SPI 6, 9 and 12. The suitable inputs for GPR model for each SPI is as in Table 1 below. GPR model is then produced using the inputs found. Next, the forecast is performed using the model obtained earlier.

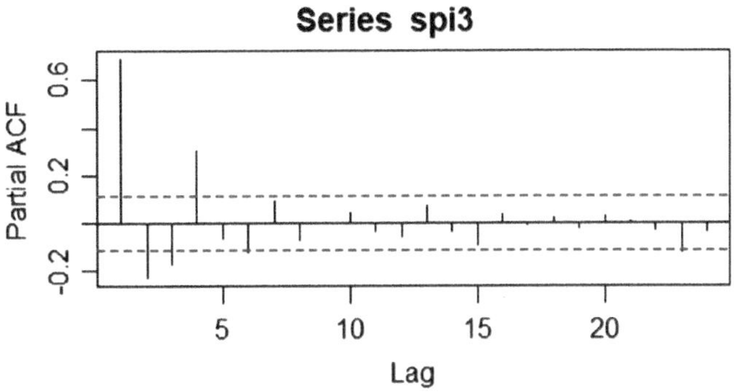

Fig. 3. PACF for SPI 3

Table 1. Number of IMFs produced by EWT

SPI	Input variables
SPI 3	$\{x_{t-1}, x_{t-2}, x_{t-3}, x_{t-4}, x_{t-6}\}$
SPI 6	$\{x_{t-1}, x_{t-3}, x_{t-4}, x_{t-6}\}$
SPI 9	$\{x_{t-1}, x_{t-2}, x_{t-6}, x_{t-7}\}$
SPI 12	$\{x_{t-1}, x_{t-3}, x_{t-6}, x_{t-7}\}$

3.2 EWT-GPR

EWT is used to decompose the SPI series into several subseries. The number of IMF depends on the data, where it is determined using Otsu's method [15]. Table 2 shows the number of IMFs generated by EWT process. Figure 4 shows SPI 9 decomposed to 5 IMFs using EWT. For each of the IMFs, a separate GPR model is developed which are then used to perform forecasts. Then, the results obtained for each IMFs are summed to obtain the final forecast. This process is repeated for each SPI studied.

Table 2. Number of IMFs produced by EWT

Area	SPI 3	SPI 6	SPI 9	SPI 12
Gua Musang	7	6	5	7

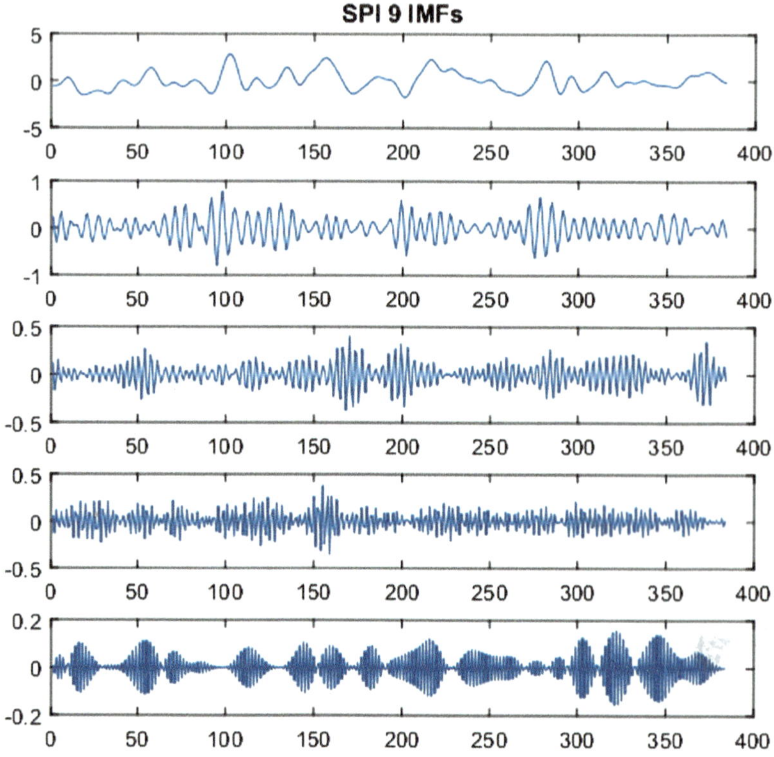

Fig. 4. EWT decomposition on SPI 9

3.3 Performance Evaluation

MAE and RMSE are calculated from the forecasting results of each models and are compared. Table 3 below shows the MAE and RMSE for testing data of the models on each SPI. Lower values of MAE and RMSE indicates a higher forecast accuracy.

EWT-GPR performed better compared to GPR on every SPI. GPR on the other hand, are unable to produce accurate results when compared to EWT-GPR. The biggest improvement is on SPI 12, where the MAE decreased from 0.2630 to 0.0397 and RMSE decreased from 0.3411 to 0.0516, which are 84.90% and 84.87% respectively. The least improvement is SPI 9, where the MAE decreased from 0.3036 to 0.0674 and RMSE decreased from 0.3993 to 0.0843, which are 77.80% and 78.89% respectively.

Table 3. MAE and RMSE of the models

Models	GPR		EWT-GPR	
SPI	MAE	RMSE	MAE	RMSE
SPI 3	0.4932	0.6372	**0.0873**	**0.1089**
SPI 6	0.3725	0.4728	**0.0672**	**0.0852**
SPI 9	0.3036	0.3993	**0.0674**	**0.0843**
SPI 12	0.2630	0.3411	**0.0397**	**0.0516**

The nature of GPR models which are able to provide good probabilistic predictive distribution. However, as GPR is better with forecasting persistent events rather than erratic events, GPR is unable to produce a good forecast in this study [2]. EWT separates the linear and nonlinear component of the data, thus GPR is able to forecast them with greater accuracy. Figure 5 below shows that EWT-GPR is able to forecast more accurate and more closely to the actual value, while GPR have some inconsistencies. The result agrees with prior research that using EWT improves the forecast accuracy [3, 13].

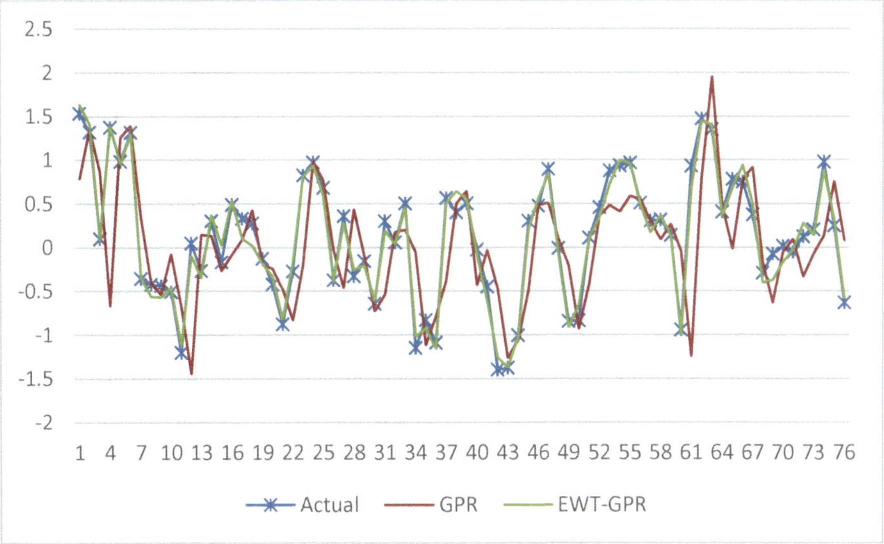

Fig. 5. EWT-GPR and GPR forecast value compared to actual value

4 Conclusion

This study compares the accuracy of drought forecasting using GPR and EWT-GPR for the region Gua Musang, Kelantan. SPI is used as the drought index and SPI 3, 6, 9, and 12 were calculated using daily rainfall data from Gua Musang, Kelantan. MAE and RMSE were used as the error statistic to measure the effectiveness of the forecast produced from the model studied. It was found out that EWT-GPR performed better compared to GPR model on all SPI tested. Therefore, the proposed model is suitable to be used to perform drought forecast.

Acknowledgement. The authors would like to express their deepest gratitude to Research Management Center (RMC), Universiti Teknologi Malaysia (UTM), Ministry of Higher Education (MOHE) and Ministry of Science, Technology and Innovation (MOSTI) for their financial support under Grant Vot 4F875.

References

1. Belayneh, A., Adamowski, J., Khalil, B., Ozga-Zielinski, B.: Long-term SPI drought forecasting in the Awash River Basin in Ethiopia using wavelet neural network and wavelet support vector regression models. J. Hydrol. **508**, 418–429 (2014). https://doi.org/10.1016/j.jhydrol.2013.10.052
2. Sun, A.Y., Wang, D., Xu, X.: Monthly streamflow forecasting using Gaussian Process Regression. J. Hydrol. **511**, 72–81 (2014). https://doi.org/10.1016/j.jhydrol.2014.01.023
3. Hu, J., Wang, J.: Short-term wind speed prediction using empirical wavelet transform and Gaussian process regression. Energy **93**, 1456–1466 (2015). https://doi.org/10.1016/j.energy.2015.10.041
4. Rohani, A., Taki, M., Abdollahpour, M.: A novel soft computing model (Gaussian process regression with K-fold cross validation) for daily and monthly solar radiation forecasting (Part: I). Renew. Energy **115**, 411–422 (2018). https://doi.org/10.1016/j.renene.2017.08.061
5. Rasmussen, C.E.: Gaussian processes for machine learning. Presented at the (2006)
6. Guang, Y.Y.U., Hu, Z.Y.U., Liu, X.S.I.: A novel strategy for wind speed prediction in wind farm. TELKOMNIKA Indones. J. Electr. Eng. **11**, 7007–7013 (2013)
7. Guo, Z., Zhao, W., Lu, H., Wang, J.: Multi-step forecasting for wind speed using a modified EMD-based artificial neural network model. Renew. Energy **37**, 241–249 (2012). https://doi.org/10.1016/j.renene.2011.06.023
8. Liu, Y.P., Wang, Y., Wang, Z.: RBF prediction model based on EMD for forecasting GPS precipitable water vapor and annual precipitation. Adv. Mater. Res. **765–767**, 2830–2834 (2013). https://doi.org/10.4028/www.scientific.net/AMR.765-767.2830
9. Shabri, A.: A modified EMD-ARIMA based on clustering analysis for fishery landing forecasting **10**, 1719–1729 (2016). https://doi.org/10.12988/ams.2016.6389
10. Gilles, J.: Empirical wavelet transform. IEEE Trans. Signal Process. **61**, 3999–4010 (2013). https://doi.org/10.1109/TSP.2013.2265222
11. Djerbouai, S., Souag-Gamane, D.: Drought forecasting using neural networks, wavelet neural networks, and stochastic models: case of the Algerois Basin in North Algeria. Water Resour. Manag. **30**, 2445–2464 (2016)

12. Pandhiani, S.M., Shabri, A.B.: Time series forecasting using wavelet-least squares support vector machines and wavelet regression models for monthly stream flow data. Open J. Stat. **3**, 183 (2013)
13. Hu, J., Wang, J., Ma, K.: A hybrid technique for short-term wind speed prediction. Energy **81**, 563–574 (2015). https://doi.org/10.1016/j.energy.2014.12.074
14. Shaaban, A.J., Low, K.S.: Droughts in Malaysia: a look at its characteristics, impacts, related policies and management strategies. Presented at the Water and Drainage 2003 Conference (2003)
15. Gilles, J., Heal, K.: A parameterless scale-space approach to find meaningful modes in histograms—application to image and spectrum segmentation. Int. J. Wavelets Multiresolution Inf. Process. **12**, 1450044 (2014). https://doi.org/10.1142/S0219691314500441

Xword: A Multi-lingual Framework for Expanding Words

Faisal Alshargi[1(✉)], Saeedeh Shekarpour[2], and Waseem Alromema[3]

[1] Universität Leipzig, Augustusplatz 10, 04109 Leipzig, Germany
alshargi@informatik.uni-leipzig.de
[2] University of Dayton, Dayton, USA
sshekarpour1@udayton.edu
[3] Taibah University, Madinah, Kingdom of Saudi Arabia
wromema@taibahu.edu.sa

Abstract. The word expansion task has applicability in information retrieval and question answering systems. It relieves the vocabulary mismatch problem leading to a higher recall. The recent word embedding models demonstrated merit for the word expansion task in comparison to the traditional n-gram models. However, to acquire quality embeddings in each language, the processes of corpus compilation, normalization and parameter tuning are time-consuming and challenging especially for poor resources languages such as Arabic. In this paper, we introduce Xword as an online multi-lingual framework for automatic word expansion. Xword relies on both pre-trained ad hoc word embedding models and n-gram models for the expansion task. Xword currently includes the two languages Arabic, and German. Xword represents the results of each model both individually and collectively. Additionally, Xword can filter out the result set based on sentiment and part of speech (POS) tag of every single word. Xword is available as a Web API along with the downloadable models and sufficient documentation on our public GitHub.

Keywords: X-word · Word expansion · Embedding · German language · Arabic language · Quality

1 Introduction

The word expansion is about taking into account words, which share similar semantics or context with respect to a given word. This task is highly essential for information retrieval or question answering tasks such as query expansion or query rewriting [14, 26, 27, 30]. The word expansion task resolves the vocabulary mismatch problem yielding in a higher recall. It commonly relies on deriving words from a huge corpus where the derived words expose a shared semantics or shared context. The words with shared semantics typically refer to a similar meaning, e.g., the verbs "*obtain*" and "*purchase*" have shared semantics in a particular sense. Whereas the words with shared context likely co-occur, e.g., concerning the sentence "*Alice drinks Apple juice*", the word "*apple juice*" likely co-occur with the verb "*drink*" while it unlikely appears with

F. Saeed et al. (Eds.): IRICT 2019, AISC 1073, pp. 162–175, 2020.
https://doi.org/10.1007/978-3-030-33582-3_16

the verb "*eat*". The existing approaches for the word expansion task lie in two directions. The first one, which is traditional relies on an external thesaurus such as WordNet [9] or Wikipedia[1] (a general encyclopedia) to derive words. Sample Heading (Third Level). Only two levels of headings should be numbered. Lower level headings remain unnumbered; they are formatted as run-in headings.

This approach is effective for finding words with synonymous and generalization (or specification) relationships. However, it has a trivial contribution for including contextual words. The other approach is based on the language model (n-grams model) which plays a better role in capturing the contextual words. Continuous space dense embeddings transform discrete representations of words into low dimensional numerical representations. The recent generation of embedding models on linguistic entities demonstrated higher quality regarding the proper encoding of the context as well as semantic patterns. For example, [19] indicated that the vector which separates the embeddings of Man and Woman is very similar to the vector which separates the embeddings of King and Queen (analogy task); this geometry disposition is consistent with the semantic relationship. Matrix factorization methods [16, 21] and neural network models [18] are two common approaches for learning dense embeddings for words. Thus, adjacent words acquire similar embedding. The pre-trained quality embedding models are typically in English. To acquire the quality embedding in other languages, it is required to compile a large corpus, normalize text, train models with various parameters for finding the optimum setting. This process is time-consuming and even challenging for languages such as Arabic which has various dialectal forms [10]. Having a comprehensible corpus including all dialects is essential. Moreover, due to limited NLP resources for Arabic dialectal text, text normalization is cumbersome.

This paper is organized as follows: Sect. 2 presents the related work. Section 3 presents the architecture of Xword. Section 4 lists the NLP resources employed by Xword. Section 5 discusses the details of the corpus compilation and prepossessing strategies. The details of training embeddings are provided in Sect. 6. We close with the concluding remarks and plan.

2 Related Work

In this section, we separately review the related literature regarding the word expansion task and the state-of-the-art embedding models for German and Arabic language.

Word Expansion. The task of word expansion is a long-standing task in Information Retrieval (IR) and Natural Language Processing (NLP). This task is mainly utilized for query expansion and query rewriting tasks in IR to solve the vocabulary mismatch problem. Some related work has been focused on the use of WordNet; however its use may raise several issues; e.g., lack of proper nouns. [20] Argued that extracting

[1] https://www.wikipedia.org/.

concepts belonging to the same semantic domain of the query is a better approach than the simple use of synonyms. Although the classical approach of using lexicon showed great achievement e.g., [6], it might yield in loading numerous irrelevant words. Thus, the heuristic approaches were employed to filter out the less likely words [27].

Training Embedding Models for Arabic and German Languages. The other common approach for word expansion is using word embeddings. The two recent approaches of this type i.e., (i) matrix factorization [16, 21] and (ii) neural network [19] demonstrated a great performance for finding semantically similar or related words. These two approaches learn dense embedding for words. The quality of the pre-trained embedding models for non-English languages is not clear. The work presented in [28] provides an unsupervised approach for learning morphological-based embedding for words. This work is language-agnostic and provides empirical study for six languages including German and Arabic. Similar morphological-based embedding approaches presented in [5] and [4]. In [1] the word embedding models are trained for 117 different languages including Arabic. Typically the work about embedding in the Arabic language is limited to MSA, and the size of the corpus is not very significant.

3 Xword Architecture

In this paper, we introduce Xword which integrates n-gram model and embedding models for expanding words. Furthermore, Xword is a configurable framework, thus it can be restricted to specific models or word characteristics such as sentiment and POS, or the type of gram (unigram, bigram).

So far, Xword is extended for Arabic and German language, however, we plan to extend it to more number of languages. Xword contributes the state-of-the-art in the following directions:

(1) It provides quality pre-trained ad hoc embedding models for German and Arabic (both modern standard Arabic and dialectal Arabic) language.
(2) It is empowered by a configuration approach which filters out the words and models.
(3) The embedding models are trained on large corpora. The parameters are carefully tuned. The output embeddings encode semantic relationships.
(4) It is an open Web API available at[2,3] and public GitHub[4]. To provide quick access, Xword relies on a SQL database in which the whole of the vocabulary, word annotations and models reside. Thus, it reduces a significant burden from researchers and developers for expanding words in various languages.

[2] http://alshargi.us/de.aspx.
[3] http://alshargi.us/ara.aspx.
[4] https://github.com/alshargi/xword.

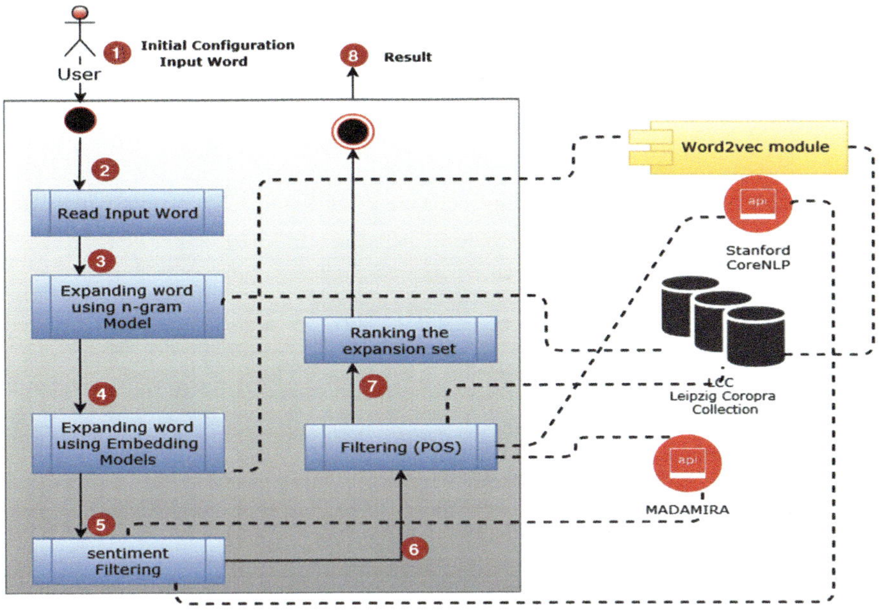

Fig. 1. The architecture of Xword: given the word expanded, and then various filtering strategies based on the initial configuration is applied to the result set

Xword is a web service which can be integrated into any other external tool. The pipeline of the Xword represented in Fig. 1 and it contains the following steps:

- **Input:** the input of Xword provided by the user is a single word along with settings such as appropriate filtering, e.g., sentiment or POS tags or a specific linguistic tool for expansion.
- **Expanding words:** the expansion list X which is empty in the beginning is initiated in this step. This component drives all the words from the specific linguistic tool configured by the user (i.e., n-gram and embedding models). Xword includes words having the cosine or dice similarity more than the given threshold η (the threshold comes from the configuration). Then the list of the derived words is added to the expansion list X.
- **Filtering words:** To provide high-quality results, the results go through filtering steps to automatically removing the undesired words. We filter out the words based on the sentiment i.e. (positive, negative and neutral) of words. Furthermore, the results also can be filtered out using the POS tags. Another type of filtering is based on uni-grams or bi-grams.
- **Representation:** the final shortlisted set of expanded words X is representing to the end user along with the similarity score. This representation is divided into several delivery modes (i) a single table per each linguistic tool used for the expansion, (ii) a combined table ranking the words using their similarly score.

Example**:** Figures 2 and 3 shows the expansion process respectively in German and Arabic words. It receives the given word coward (جبان in Arabic and **Feigling** in German). Then, it expands the input word using all the existing models integrated into Xword (currently n-gram model, and two embedding models). Regarding the input setting top-n, it obtains the top-n similar words from each model. The models might share some words. The common words are represented in the blue box. These words acquire higher priority in the ranking mechanism of Xword. The words which are in

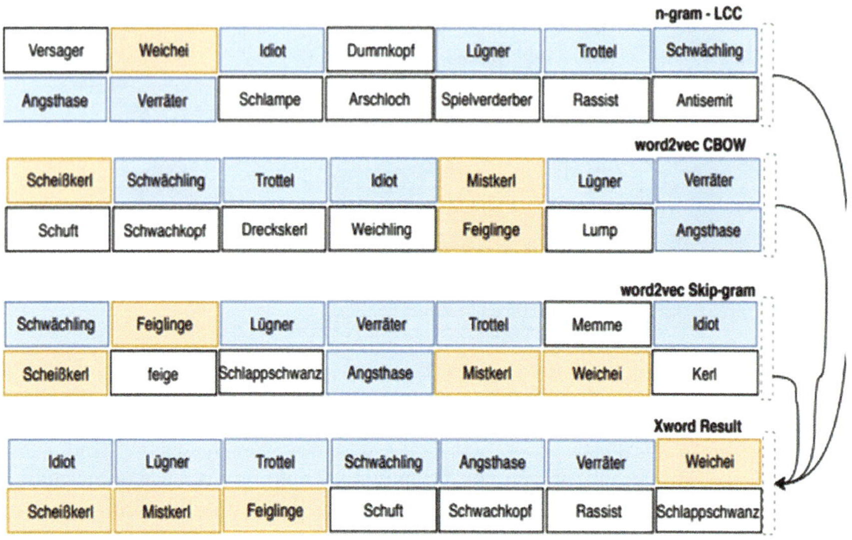

Fig. 2. Xword result for the German word Feigling 'Coward'

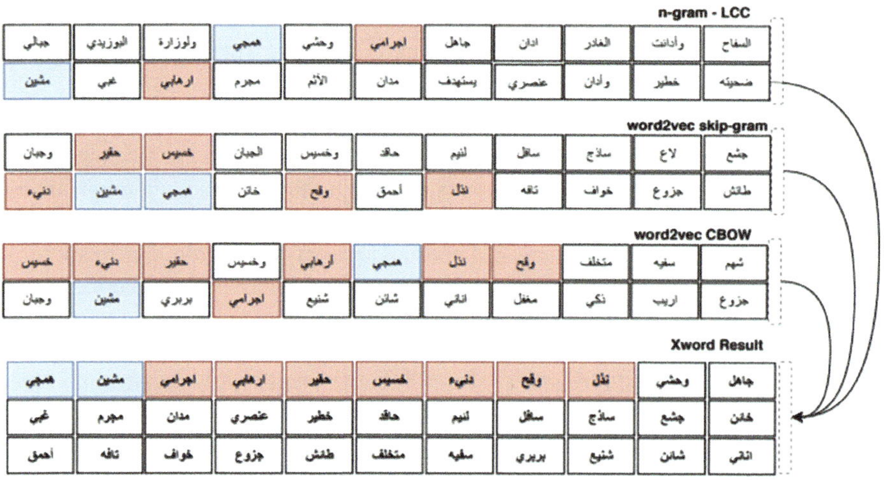

Fig. 3. Xword result for the Arabic word جبان 'jbAn', "coward"

common between two models have the orange (pink) box and they have secondary priority in the Xword ranking. Xword also performs multiple filtering approaches, for example, if the input sentiment is set up to negative thus the words ذكي 'smart' and شهم 'chivalrous' are removed (since they have positive sentiment). The other criterion for ranking is the similarity score provided by the models.

4 NLP Resources Used in Xword

In the following, we shortly list all the linguistics resources employed in Xword. These resources spans from dictionaries and NLP tools integrated in Xword.

Stanford POS API. Stanford Part-Of-Speech tagger (POS Tagger)[5] is a library, which annotates the input text with respect to the POS role of words in the given sentence such as noun, verb, adjective, etc. This software is an open source Java implementation of the log-linear part-of-speech taggers described in [29].

MADAMIRA POS API. MADAMIRA[6] is a POS tagger for Arabic language and it integrates the salient features of the two previously commonly used systems i.e., [3], and [12] for Arabic processing. MADAMIRA brings improvements over the preceding systems and in addition, it produces a streamlined Java implementation that is more robust, portable, and extensible is faster than its ancestors [3]. Thus, we use Stanford POS for English language and MADAMIRA for the Arabic Language.

Stanford CoreNLP. Stanford CoreNLP[7] [17] is a library which integrates several tools annotating text with respect to (i) POS Tagging, (ii) Named Entity Recognition (NER), (iii) syntactic dependencies and (iv) coreference. We employed Stanford CoreNLP for initial pre-processing and annotating of our underlying corpus.

Columbia SLSA. Large-scale Standard Arabic Sentiment Lexicon SLSA is a lexicon for the sentiment of words in the Arabic Language, which developed in Columbia University [8]. The existing sentiment lexicons in the Arabic language have major deficiencies whereas SLSA addresses these deficiencies. We employ SLSA for recognizing sentiment of words in the Arabic language. Please be noted that for the sentiment of words in English, we employ Stanford CoreNLP.

SentiWS. Sentiment Wortschatz[8] (abbreviated SentiWS) [22] is a publicly available German-language resource. It can be utilize for sentiment analysis, and opinion mining tasks. This resource contains around 1,650 positive words and 1,800 negative words, which is around 16,000. It represents polarity of words using the range [−1, 1]. Additionally, it also provides POS tag and inflection form.

[5] http://nlp.stanford.edu/software/tagger.shtml.

[6] https://camel.abudhabi.nyu.edu/madamira/.

[7] https://stanfordnlp.github.io/CoreNLP.

[8] http://wortschatz.uni-leipzig.de/en/download/.

Word2vec. The word2vec model [18, 19] feature two models for generating embedding: (i) a skip-gram model and (ii) a continuous bag of words (CBOW) model. The skip-gram model learns two separate embedding for each target word *wi*, (i) the word embedding and (ii) the context embedding. These embeddings are used to compute the probability of the word w_k (i.e. context word) appearing in the neighborhood of word *wi* (i.e. target word), P $(wk|wi)$. The skip-gram algorithm (with negative sampling) starts traversing the corpus for any given target word *wi*. For any occurrence of the target word, it collects the neighboring words as positive samples and chooses *n* noise samples as negative sampling (i.e., non-neighbor words). Eventually, the objective of the shallow neural network of the skip-gram model is to learn a word embedding maximizing its dot product with context words and minimizing its dot products with non-context words. The CBOW model is roughly similar to the skip-gram model as it is also a predictive model and learns two embedding for each word (a word embedding and a context embedding). The difference is that CBOW predicts the target word *wi* from the context words as $P(wi|w_k, wj)$. Thus, the input of the neural network is composed by the context words (e.g.w_{i-1},w_{i+1}] for the context with length 1); then, the algorithm learns the probability of *wi* appearing in the given context. Although the difference between these two algorithms is slight, they showed different performance in various tasks. State-of-the-art evaluations suggest that these algorithms are individually suited to particular tasks.

5 Corpus

Corpus Compilation. *Leipzig Collection Corpora (LCC)* [11, 23, 24] compiled from textual part of various sources such as worldwide Web, Wikipedia, and online newspapers. After compilation, the texts are token under several pre-processing steps. First, the language is automatically recognized. Next, the text segmented in both a sentence-wise and word-wise manners. This corpus currently contains around more than 360 languages. Furthermore, the necessary statistics of the corpus was computed e.g., the frequency of words. The German corpus relies on German data of LCC. Arabic corpus also includes the Arabic data of LCC [7]. However, this data is mainly the Modern Standard Arabic (MSA), which is a formal Arabic language and the Arabic classic data. The classic Arabic Data refers to the previous version of Arabic like Al-Quran Al-Kareem, old literature, old books. We enriched the Arabic corpus including 21 regional Arabic dialects [2] (more details hidden for blind review). The dialectal Arabic data compiled from various sources such as oral interviews, social media, wisdom, politic, humor, poems, etc. Table 1 represents the statistics of both Arabic and German corpora concerning the number of included documents, sentences and tokens. The German corpus contains more than six billion tokens and the Arabic corpus contains more than three billion tokens.

Table 1. Statistics of our underlying corpora.

Language	#Document	#Token	#Unique Words
German	401,026,032	6,478,352,662	636,569
Arabic	134,206,993	3,037,609,519	681,946

Pre-processing Strategies. To prepare our corpora for Xword, we initially preprocessed each corpus based on the specific consideration of each language. The generic tasks are typical text normalization tasks such as removing stopwords and punctuation. While the tasks such as removing stop words and punctuation in English or German is straightforward, in Arabic they are problematic (because some stop words are sometimes part of words). After that, we annotated our corpora by adding annotations such as (i) POS tags for each word for both Arabic and German languages using respectively MADAMIRA and Stanford CoreNLP, (ii) sentiment which is acquired from Columbia SLSA for Arabic language and SentiWs for the German language.

6 Training Embedding Models

We train both Skip-gram model and CBOW model of Word2Vec on the German and Arabic corpora. We rely on intrinsic task for evaluating the quality of embeddings. We employ SimLex-999 [13][9], which is a gold standard resource for the evaluation of models that learn the meaning of words and concepts. It provides 999 pairs of words with three types of POS (Noun, Adjective and Verb) with a similarity score. This gold standard also published in several other languages including German [15][10].

However, it does not provide an Arabic version. Thus, we translated the English version into Arabic and asked two trained annotators who are native in Arabic to validate that. This gold standard in Arabic is available for the research community. We compare the scores of the gold standard and cosine similarity of the embedding model using Spearman's correlation coefficient denoted by ρ which is a standard metric to compare ranking between variables in word similarity task. Furthermore, we scaled the scores of SimLex-999 to the range of 0-1 and computed the absolute error between the score of each pair s and the cosine similarly s' provided by the model as $e = |s - s'|$.We computed these scores for (i) whole of the gold standard, (ii) pairs having Adjective POS, (ii) pairs having Noun POS, (ii) pairs having Verb POS. Our key hyper-parameters are as follows:

Vector Dimension: the vector dimension represents the size of the learned embeddings. The higher dimension likely provides quality encoding while demands more computational resources.

Context Window Size: this size shows how far the context is spanned. For example, the window size of two means two words before and two words after the current word. The other hyper-parameters are set up as follows; the minimum count is 100 for Arabic corpus and 150 for German corpus. We did not apply negative sampling. Also the learning rate α is 0.1.

Evaluation on Intrinsic Similarity Task. Tables 2 and 3 show the parameter tuning for similarity (relatedness) task on SimLex-999 gold standard respectively for Arabic corpus and German corpus. The Spearman correlation ρ for the pairs with the type

[9] https://fh295.github.io/simlex.html.

[10] http://www.leviants.com/ira.leviant/MultilingualVSMdata.html.

Table 2. Arabic Corpus: intrinsic evaluation results of CBOW and Skip-gram models for vectors with different window size and dimension: Spearman's correlation coefficient (ρ) and absolute error (e). General column is over all pairs of SimLex-999 gold standard, respectively Adjective, Noun, Verb columns are over pairs with A, N, V POS tags.

Model	Parameters	General		Adjective		Noun		Verb	
	Window, dimension	ρ	$\|e\|$	ρ	$\|e\|$	ρ	$\|e\|$	ρ	$\|e\|$
CBOW	7, 100	0.19	**0.26**	0.26	0.28	0.28	0.25	0.12	0.30
	7, 200	**0.21**	0.28	**0.28**	**0.28**	0.30	**0.25**	**0.13**	0.31
	3, 50	0.17	0.26	0.22	0.30	**0.73**	0.27	0.11	**0.29**
Skip-gram	**7, 100**	**0.20**	**0.22**	**0.31**	**0.27**	**0.29**	**0.21**	**0.10**	**0.24**
	7, 200	−0.12	0.28	0.11	0.34	−0.23	0.27	−0.25	0.25
	3, 50	−0.13	0.27	−0.19	0.32	−0.28	0.25	−0.32	0.32
	5, 50	0.18	0.23	0.27	0.29	0.28	0.23	0.11	0.24

adjective is higher, then for pairs with Noun POS tag and then for pairs with Verb POS tag. Regarding Arabic models, the setting (window = 7, dimension = 100) for Skip-gram and the setting (window = 7, dimension = 200) for CBOW outperforms. Regarding German models, the setting (window = 7, dimension = 300) for both Skip-gram and CBOW outperforms which hits the state-of-art (best reported is 0.354).

Evaluation on Relationships. We ranked the word pairs of our gold standard using the best performing model. We looked at the pairs with the cosine similarity of more than 0.8. They represent a meaningful relationship. Table 4 shows a few samples for each POS type in Arabic, and Table 5 shows a few samples for each POS type in German. We labeled the relationship, they are typically related words, e.g., geo-spatial, having the same morpheme or share same semantics.

Table 3. German Corpus: intrinsic evaluation results of CBOW and Skip-gram models for vectors with different window size and dimension: Spearman's correlation coefficient (ρ) and absolute error (e). General column is over all pairs of SimLex-999 gold standard, respectively Adjective, Noun, Verb columns are over pairs with A, N, V POS tags.

Model	Parameters	General		Adjective		Noun		Verb	
	Window, dimension	ρ	$\|e\|$	ρ	$\|e\|$	ρ	$\|e\|$	ρ	$\|e\|$
CBOW	7, 300	**0.37**	**0.18**	**0.39**	**0.25**	**0.34**	**0.18**	**0.41**	**0.14**
	7, 50	0.27	0.22	0.33	0.26	0.26	0.22	0.26	0.20
	3, 100	0.32	0.20	0.38	0.25	0.20	0.20	0.35	0.16
Skip-gram	7, 300	**0.37**	**0.18**	0.43	**0.24**	**0.34**	**0.17**	**0.37**	**0.14**
	7, 50	0.26	0.25	0.30	0.28	0.26	0.25	0.22	0.24
	5, 200	0.36	**0.18**	**0.44**	**0.24**	0.33	0.18	0.36	0.15
	3, 100	0.31	0.20	0.37	0.25	0.30	0.20	0.28	0.19

Evaluation on Coherence. While the similarly or relatedness task is between pairs of words, coherence is about a group of words. In the coherence task, it is evaluated whether groups of words in a dense neighborhood of the embedding space are similar (related) or not [25]. To measure coherence, we rely on visualization. Figures 4 and 5 represent the neighborhood of the word coward (جبان) in Arabic, respectively for CBOW model and Skip-gram model. Similarly, Figs. 6 and 7 represent the neighborhood of the word coward (Feigling) in German, respectively for CBOW model and Skip-gram model. Interestingly, in all of the models the surrounding words hold the adjective POS and negative sentiment similar to the POS and sentiment of the input word. Moreover, skip-gram generally shows higher coherence with the surrounding words (Table 5).

Table 4. Arabic Corpus - Example of relationships.

POS	Relationship	Arabic	Translation
A	Antonym	طويل - قصير	Long-short
	synonym	مهم - ضروري	Important-essential
N	Geo-spatial	شمال - جنوب	South-North
	related	كوب - ملعقة	Cup-spoon
V	Similar base form	أكتشف-يكتشف	Discover-discovers
	Similar semantics	تحليل- تقييم	Analyze-evaluate

Table 5. German Corpus - Example of relationships.

POS	Relationship	German	Translation
A	Antonym	klug-blöd	Long-short
	synonym	selten-rar	Important-essential
N	Geo-spatial	Süden-Norden	South-North
	related	Tasse-Krug	Cup-spoon
V	Similar base form	prüfen-überprüfen	Discover-discovers
	Similar semantics	beschützen verteidigen	Analyze-evaluate

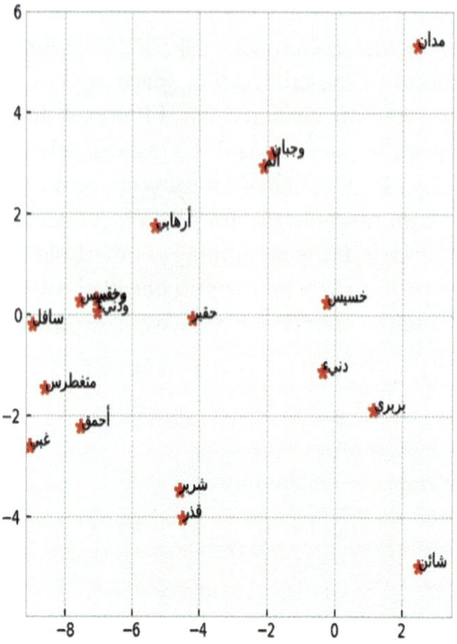

Fig. 4. CBOW: word coward 'جبان'

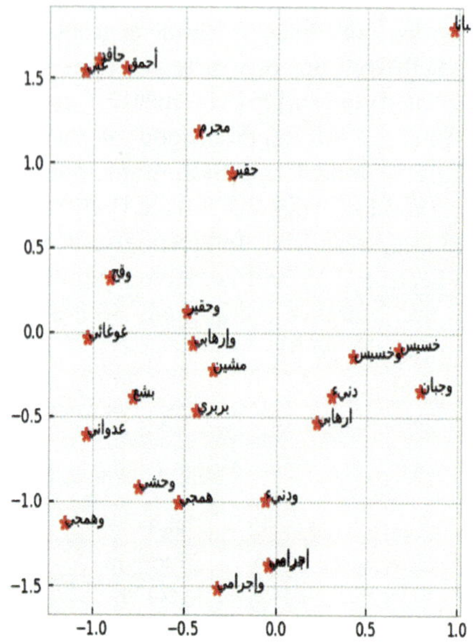

Fig. 5. Skip-gram: word coward 'جبان'

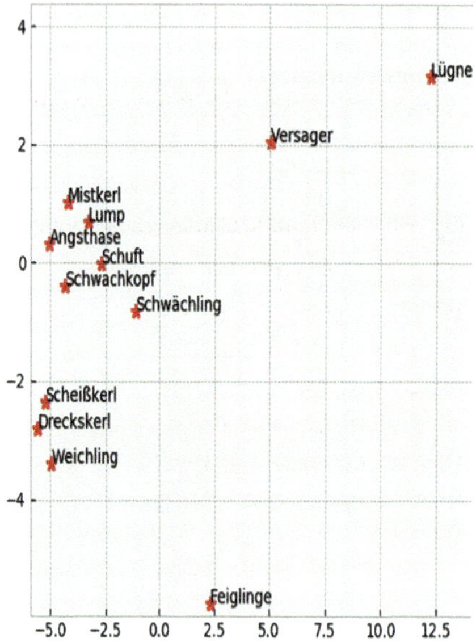

Fig. 6. CBOW: word coward 'Feigling'

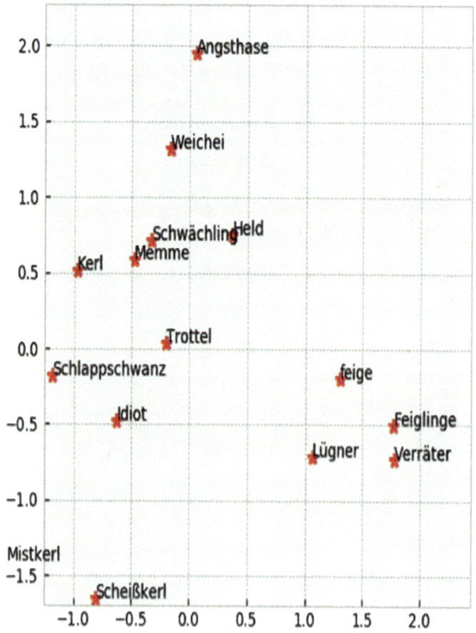

Fig. 7. Skip-gram: word coward 'Feigling'

7 Conclusion and Future Plan

In this paper, we presented the first version of Xword, which is a framework for expanding words in multiple languages. The current version adopted the Arabic and German language. However, we plan to extend Xword toward more number of languages. Xword is run on both n-gram and embedding models. We are aware of publishing newer embedding models. We plan to integrate the newer models into Xword soon. Xword is available as a Web API, thus easily can be integrate with various cross-lingual text-processing languages.

References

1. Al-Rfou, R., Perozzi, B., Skiena, S.: Polyglot: Distributed word representation for multilingual NLP. In: Proceedings of the Seventeenth Conference on Computational Natural Language Learning, pp. 183–192. Association for Computational Linguistics (2013)
2. Al-Shargi, F., Kaplan, A., Eskander, R., Habash, N., Rambow, O.: Morphologically annotated corpora and morphological analyzers for Moroccan and Sanaani Yemeni Arabic. In: Proceedings of the Tenth International Conference on Language Resources and Evaluation (LREC 2016) (2016)
3. Diab, M., El Kholy, A., Eskander, R., Habash, N., Pooleery, M., Rambow, O., Pasha, A., Al-Badrashiny, M., Roth, R.M.: Madamira: a fast, comprehensive tool for morphological analysis and disambiguation of Arabic. In: LREC (2014)
4. Bojanowski, P., Grave, E., Joulin, A., Mikolov, T.: Enriching word vectors with subword information. Trans. Assoc. Comput. Linguist. **5**, 135–146 (2017)
5. Cotterell, R., Schütze, H.: Morphological word-embeddings. In: Proceedings of the 2015 Conference of the North American Chapter of the Association for Computational Linguistics: Human Language Technologies, pp. 1287–1292 (2015)
6. Ding, X., Liu, B., Yu, P.S.: A holistic lexicon-based approach to opinion mining. In: Proceedings of the 2008 International Conference on Web Search and Data Mining, pp. 231–240. ACM (2008)
7. Eckart, T., Alshargi, F., Quasthoff, U., Goldhahn, D.: Large Arabic web corpora of high quality: the dimensions time and origin. In: Workshop on Free/Open-Source Arabic Corpora and Corpora Processing Tools, LREC, Reykjavík (2014)
8. Eskander, R., Rambow, O.: SLSA: a sentiment lexicon for standard arabic. In: Proceedings of the 2015 Conference on Empirical Methods in Natural Language Processing, Lisbon, Portugal, 17–21 September 2015, pp. 2545–2550 (2015)
9. Fellbaum, C. (ed.): WordNet: An Electronic Lexical Database. MIT Press, Cambridge (1998)
10. Ferguson, C.A.: Diglossia. Word: Journal of the International Linguistic Association (1959)
11. Eckart, T., Quasthoff, U., Goldhahn, D.: Large monolingual dictionaries at the Leipzig corpora collection: from 100 to 200 languages. In: Proceedings of LREC 2012, pp. 759–765 (2012)
12. Habash, N., Rambow, O.: Arabic tokenization, part-of-speech tagging and morphological disambiguation in one fell swoop. In: Proceedings of the 2015 Conference on Empirical Methods in Natural Language Processing, ACL 2005, Lisbon, Arbor, MI, USA, pp. 2545–2550 (2005)

13. Hill, F., Reichart, R., Korhonen, A.: Simlex-999: Evaluating semantic models with (genuine) similarity estimation. Computational Linguistics (2015)
14. Kuzi, S., Shtok, A., Kurland, O.: Query expansion using word embeddings. In: Proceedings of the 25th ACM International on Conference on Information and Knowledge Management, CIKM 2016, pp. 1929–1932. ACM (2016)
15. Leviant, I., Reichart, R.: Separated by an un-common language: towards judgment language informed vector space modeling (2015)
16. Levy, O., Goldberg, Y.: Neural word embedding as implicit matrix factorization. In: Advances in Neural Information Processing Systems, pp. 2177–2185 (2014)
17. Manning, C.D., Surdeanu, M., Bauer, J., Finkel, J., Bethard, S.J., Mc- Closky, D.: The Stanford CoreNLP natural language processing toolkit. In: Association for Computational Linguistics (ACL) System Demonstrations (2014)
18. Mikolov, T., Chen, K., Corrado, G., Dean, J.: Efficient estimation of word representations in vector space. CoRR, abs/1301.3781 (2013a)
19. Mikolov, T., Sutskever, I., Chen, K., Corrado, G.S., Dean, J.: Distributed representations of words and phrases and their compositionality. In: Advances in Neural Information Processing Systems 26: 27th Annual Conference on Neural Information Processing Systems 2013. Proceedings of a Meeting Held, 5–8 December 2013, Lake Tahoe, Nevada, United States, pp. 3111–3119 (2013b)
20. Navigli, R., Velardi, P.: An analysis of ontology-based query expansion strategies. In: Proceedings of the 14th European Conference on Machine Learning, Workshop on Adaptive Text Extraction and Mining, Cavtat-Dubrovnik, Croatia (2003)
21. Pennington, J., Socher, R., Manning, C.D.: Glove: Global vectors for word representation. In: Proceedings of the 2014 Conference on Empirical Methods in Natural Language Processing, EMNLP 2014, 25–29 October 2014, Doha, Qatar, A meeting of SIGDAT, a Special Interest Group of the ACL, pp. 1532–1543 (2014)
22. Heyer, G., Remus, R., Quasthoff, U.: SentiWS - a publicly available German-language resource for sentiment analysis. In: Proceedings of the 7th International Language Ressources and Evaluation (LREC 2010), pp. 1168–1171 (2010)
23. Hallsteinsdóttir, E., Biemann, C., Richter, M., Quasthoff, U.: Exploiting the Leipzig corpora collection. In: Proceedings of the IS-LTC, Ljubljana, Slovenia (2006a)
24. Hallsteinsdóttir, E., Biemann, C., Richter, M., Quasthoff, U.: Exploiting the Leipzig corpora collection. In: Proceedings of the IS-LTC. Ljubljana, Slovenia (2006b)
25. Schnabel, T., Labutov, I., Mimno, D.M., Joachims, T.: Evaluation methods for unsupervised word embeddings. In: Proceedings of the 2015 Conference on Empirical Methods in Natural Language Processing, EMNLP 2015, Lisbon, Portugal, 17–21 September (2015)
26. Shekarpour, S., Höffner, K., Lehmann, J., Auer, S.: Keyword query expansion on linked data using linguistic and semantic features. In: 2013 IEEE Seventh International Conference on Semantic Computing, Irvine, CA, USA, 16–18 September 2013 (2013)
27. Shekarpour, S., Marx, E., Auer, S., Sheth, A.P.: RQUERY: rewriting natural language queries on knowledge graphs to alleviate the vocabulary mismatch problem. In: Proceedings of the Thirty-First AAAI Conference on Artificial Intelligence, 4–9 February 2017, San Francisco, California, USA, pp. 3936–3943 (2017)
28. Soricut, R., Och, F.: Unsupervised morphology induction using word embeddings. In: Proceedings of the 2015 Conference of the North American Chapter of the Association for Computational Linguistics: Human Language Technologies. Association for Computational Linguistics (2015)

29. Toutanova, K., Manning, C.D.: Enriching the knowledge sources used in a maximum entropy part-of-speech tagger. In: Proceedings of the Joint SIGDAT Conference on Empirical Methods in Natural Language Processing and Very Large Corpora (EMNLP/VLC-2000), pp. 63–70 (2003)
30. Zamani, H., Croft, W.B.: Embedding-based query language models. In: Proceedings of the 2016 ACM International Conference on the Theory of Information Retrieval, ICTIR 2016. ACM (2016)

Artificial Intelligence and Soft Computing

Context-Aware Ontology for Dengue Surveillance

Siti Zulaikha Mohd Zuki, Radziah Mohamad(✉), and Nor Azizah Sa'adon

Universiti Teknologi Malaysia, 81310 Skudai, Johor, Malaysia
xiearf@gmail.com, {radziahm,azizahsaadon}@utm.my

Abstract. Ovitrap survey is one of dengue surveillance techniques employed by Ministry of Health Malaysia to identify current rate of vector population in a selected area. Result obtained from this activity is used to determine suitable control method to eradicate vector in the said area. This research study proposes a context-aware ontology that provides foundation of dengue surveillance system based on Ovitrap survey procedures exercised in Malaysia to assist entomologies on data analysis and decision-making. Ontology concepts are derived following guidelines provided by METHONTOLOGY methodology. The sub-activities of Ovitrap survey are categorized into two and formed two main concepts, which are field activities and lab activities. The other two main concepts represent instruments required to carry out Ovitrap survey and context information for the ontology. Context-awareness in ontology allows information flow and sharing between these main concepts. Result obtained could be used as support when requesting deployment of vector control activities. An entomologist expert validated the information flow of the proposed ontology while the ontology quality is measured following ontology quality metric suite. Accurate and correct representation of dengue surveillance activity via ontology could assist entomologist to carry out Ovitrap survey faster and easier.

Keywords: Ontology · Context-aware · Dengue surveillance · Ovitrap survey

1 Introduction

Vector-borne disease is a type of infectious disease transmitted by cellular contact with carriers from arthropod species [1]. The carriers, referred to as vectors, are living organisms that transmit disease-producing microorganisms or parasites during blood meal. Vectors, such as mosquitoes, inject the disease-producing microorganism into a new host that could be either human or another animal during their subsequent blood meal, thus infecting the host with viruses. Some examples of mosquito vector-borne disease (MVBD) are dengue fever, Malaria, and chikungunya. Dengue fever has caused significant morbidity and mortality with high number of confirmed cases and deaths every year around the world [2].

Advances in technologies lead to the development of information technology-based application for healthcare surveillance systems including dengue fever in MVBD domain. The system supports data collection, management and analysis [3]. Result

F. Saeed et al. (Eds.): IRICT 2019, AISC 1073, pp. 179–188, 2020.
https://doi.org/10.1007/978-3-030-33582-3_17

acquired from the data analysis then used to determine areas that would require vector control and other prevention measures to prevent or reduce the risk of another dengue outbreak. One of dengue surveillance techniques employed in Malaysia is Ovitrap survey, carried out by Entomology and Pest Unit from local District Health Office. Ovitrap survey is an activity comprises of planting Ovitrap cups within a specified area where dengue cases were reported to measure current mosquito population [4]. Result obtained from this activity is known as Ovitrap index. Based on Ovitrap index acquired, the Entomology and Pest Unit advises which vector control activity the Vector Unit should dispatch onto the said canvassed area. Ovitrap survey is done manually by hand. Thus, it takes time to complete data collecting, recording and analysis. This proposed context-aware ontology aims to assist entomologist with data collection and analysis activities where it could suggest vector control activities based on Ovitrap index obtained.

1.1 Dengue Fever

Dengue fever is an acute systematic arthropod-borne, which also known as vector-borne [5]. It is a type of viral infection in humans. This arboviral infection transmitted by Aedes genus of mosquito family, primarily Aedes aegypti, which commonly found in tropical and subtropical regions. Dengue fever is endemic in more than 100 countries around the world, predominantly the South East Asia region such as Indonesia, Thailand, Manila, Philippines and Malaysia. However, only Aedes aegypti and Aedes albopictus are viable in Malaysia [6]. Early symptoms of its infection are fever, headache, joint pain and rash. However, the key diagnostic features of a more severe infection case are thrombocytopenia, hemorrhage and plasma leakage [7].

Dengue virus transmitted by direct contact with infected mosquito such as Aedes aegypti. This species thrives in urban environment and breeds in small bodies of fresh water [2]. Common breeding sites for them are water containers found around residential areas, storage drums, water tanks, discarded waste items, polystyrenes, and buckets. Since the breeding sites of choice by this species normally found near residential area, the risk of dengue virus infection is high compared to other vector-borne diseases. Across the world, it is estimated that 50 to 100 million dengue cases reported each year, approximately half of the world's population [8]. This makes dengue fever as the most widespread mosquito vector-borne disease. In addition to that, study estimated that dengue infection cases, with clinically manifested at any level of severity, increased more than three times per year [9]. Since there is no specific treatment for dengue and severe dengue yet, surveillance on areas with case history is important for early vector control operations to prevent another infection breakout [1].

1.2 Dengue Surveillance in Malaysia

Dengue surveillance refers to health data collection that related to dengue fever and its carrier vector. An example of surveillance techniques employed in Malaysia is Ovitrap survey. This activity is usually carried out by Entomology and Pest Unit from local District Health Office. Ovitrap survey is done on areas that have previous reported dengue fever cases in span of five years. The purpose of this activity is to measure current mosquito population in the specified area. Result from this activity is known as Ovitrap index. Entomology and Pest Unit uses Ovitrap index to determine which vector control method is suitable for the canvasses area and uses the data to advice Vector Unit to dispatch vector control activities.

One Ovitrap survey lasts for seven days. Minimum of 30 Ovitrap cups are placed around the canvassed area. Each of the cups is labelled with identification number. Entomologist officer records location of the placed cup by taking note of the latitude and longitude. The type of spot cup placed at is also logged. For example, the cup is placed indoor, outdoor or semi-door. After seven days, entomologist officers will return to collect the cups and bring them back to lab for analysis. The species of larvae collected in cups are identified and the number of larvae is counted in order to determine Ovitrap index [4].

2 Methodology

Ontology development follows guideline as provided by METHONTOLOGY methodology [10]. **It allows construction of ontology at knowledge level and provides development guideline through five activities, which are specification, conceptualization, formalization, implementation and maintenance. Activity descriptions are illustrated in Table 1. Each activity acts as a development stage where it produces particular output to be used during the subsequent activity.**

Table 1. METHONTOLOGY activity description

Name	Description
Specification	**Identifies the purpose of ontology development, intended achievement and contribution. Activity result is basic set of concepts**
Conceptualization	**Organisation of knowledge acquired, information illustrated in intermediate representations such as tabular and graph notations for better delivery. Result from this activity is ontology conceptual model**
Formalization	**Conceptual model produced from previous activity is converted into formal model**
Implementation	**Ontology modelled into selected ontology language. Set of concepts organised, grouped and relationships established to formulate context and information flow**
Maintenance	Ontology updates, edits and corrections

3 Result and Discussion

3.1 Ontology Main Classes and Concepts

In general, Ovitrap survey activity is divided into two parts, which are field activity and lab activity. An entomologist officer explained the procedures of dengue vector surveillance technique via Ovitrap survey, as illustrated in Fig. 1, starting from placing cups in throughout a designated area until the analysis of data collected. Result obtained from this activity, known as Ovitrap index, is then used to determine suitable control method to be deployed on the said area.

Taking Ovitrap procedures in Malaysia into consideration, field activity and lab activity are made into main classes in the proposed ontology, as *Ovitrap survey* class and *Analysis activity* class respectively. Besides that, there are *Tools* class added into the ontology that signifies tools used during an Ovitrap activity and *Context* class that is responsible for the context-awareness of the ontology. Table 2 describes the main classes determined for this ontology. Figure 2 shows the overview of relations between main classes meanwhile Fig. 3 illustrates the general view of process and information flow through the ontology.

Table 2. Description of ontology main classes and concepts

Name	Description
Ovitrap survey	Ovitrap survey class in this ontology refers to survey activity in the field. It describes an exhibit's data such as Ovitrap cup placement and collection date
Analysis activity	Analysis activity class describes the analysis done on the collected Ovitrap cups in order to determine Ovitrap index of a canvassed area
Tools	Tools class signifies the hardware and software tools used to obtain and analyse data during an Ovitrap survey activity. It also describes exhibit items entomologist officers use during survey activity
Context	Context class refers to information that can be used to characterise or affect the situation of an entity or analysis result. In this ontology, context refers to user input, condition of surrounding environment, location and storage foundation

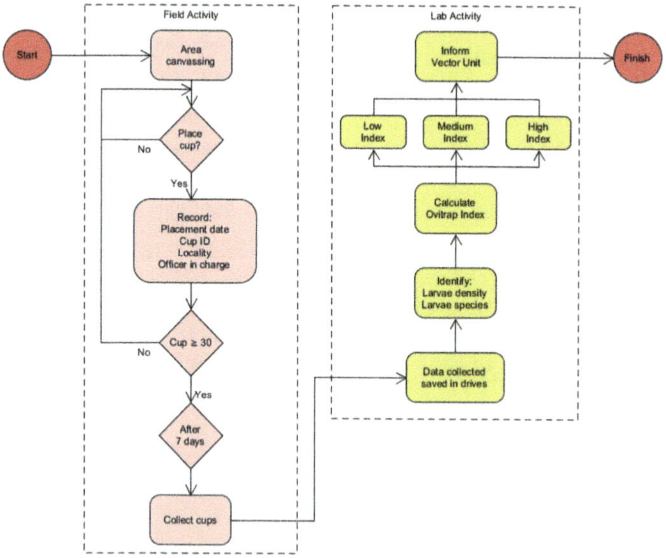

Fig. 1. Overview of vector surveillance procedures via Ovitrap survey in Malaysia

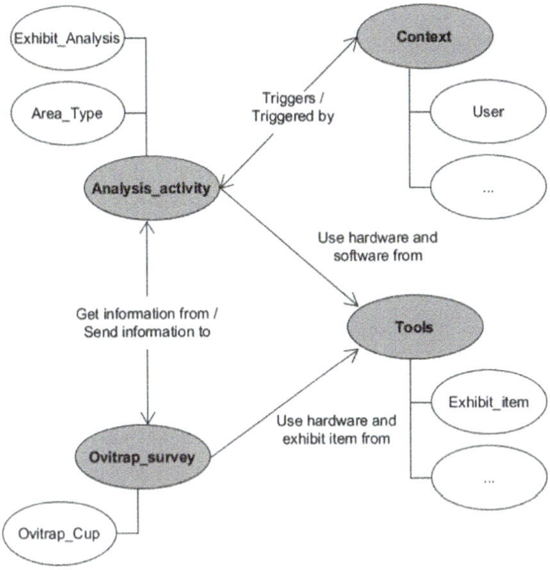

Fig. 2. Overview of relations between the main classes

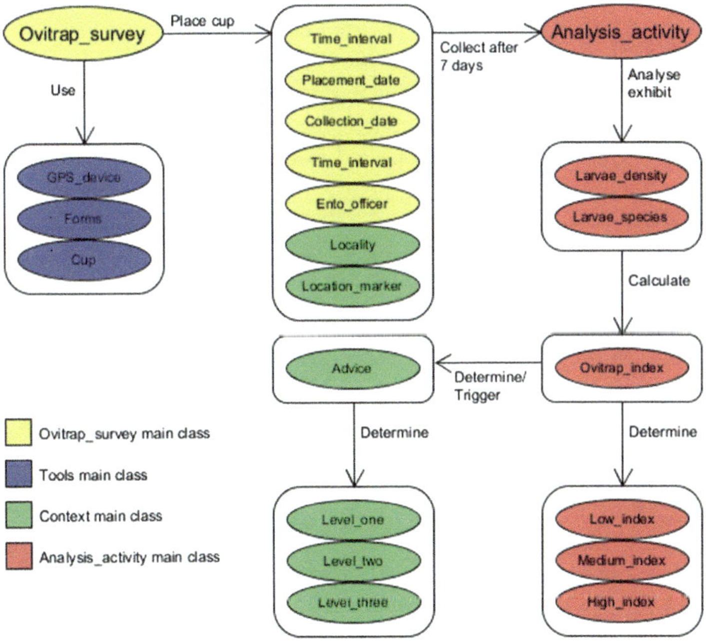

Fig. 3. Process and information flow through the ontology

***Ovitrap Survey* Class. *Ovitrap Survey class* shown in Fig. 4 refers to field activity, comprises of Ovitrap cup placement and collection.** Three sub-classes under Cup ID class are cup's attributes that are recorded for every cup placed in field. These attributes keep track of cup placement and collection date. Since Ovitrap survey could only determine current vector population in an area, date recording is important to avoid ambiguous information. Information collected is then passed to *Analysis activity* class.

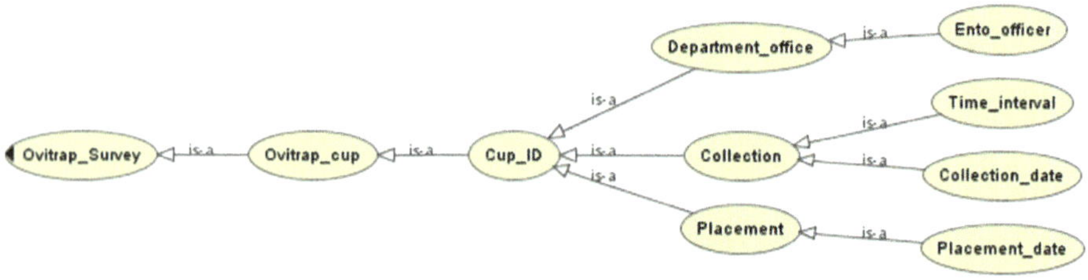

Fig. 4. *Ovitrap survey* class

***Analysis Activity* Class.** Figure 5 illustrates *Analysis activity* class. It branches out into two sub-classes, which are *Exhibit Analysis* and *Area Type. Area Type* is categorized into three; priority one, priority two and priority three, determined based on past dengue case reports. *Priority one* refers to area with fatal case between one to three years. *Priority two* is the area with infrequent reported dengue case without death while *Priority three* refers to area with no reported case in span of five years.

On the other hand, *Exhibit Analysis* uses collected data to calculate Ovitrap index. Sub-classes of *Ovitrap index* refer to the threshold of Ovitrap index result. Result from these classes will determine which advice type the Entomology and Pest Unit should deliver to Vector Unit. The result is passed to *Context* class.

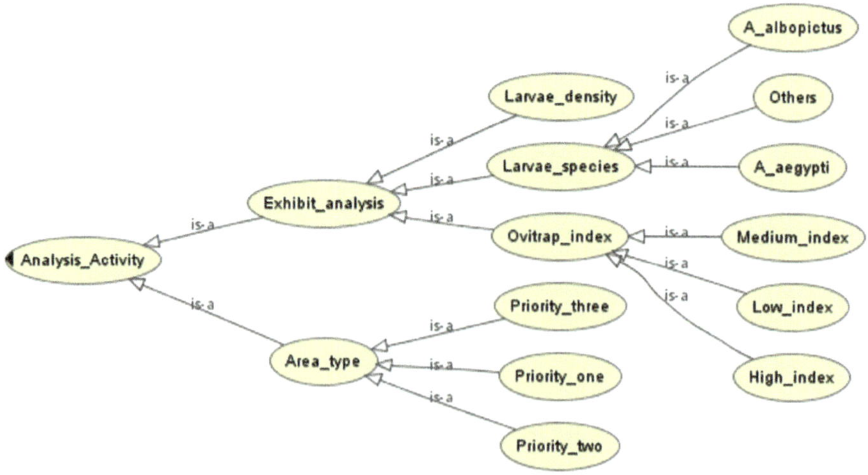

Fig. 5. *Analysis activity* class

***Context* class.** *Context* class is illustrated in Fig. 6. It is derived into five sub-classes, which are *Foundation*, *Location*, *Environment*, *User* and *Advice*. The *Foundation* class defines data storage for data collected and uploaded into the system. *Location* class refers to the locality and location marker of the canvassed area and the longitude and longitude value of where cups are placed. *Environment* class defines the environmental variables that effect life cycle and biology of vector and *User* class refers to the way data in inputted to the system.

Advice class refers to the type of decision to be made based on Ovitrap index result. *Low index*, *Medium index* and *High index* classes from *Analysis activity* class are linked to *Level one*, *Level two* and *Level three* classes in *Advice* class respectively. Level one refers to low intensity of control methods such as holding campaign for dengue prevention in residential areas and encouraging residents to monitor vector's breeding sites around their houses. Level two is medium intensity control method. It includes holding cleaning activities around the entire residential area to remove vector's breeding sites and several fogging activities. Level three refers to high intensity vector control method that such as more frequent fogging and stronger eradication of vector's breeding sites movements.

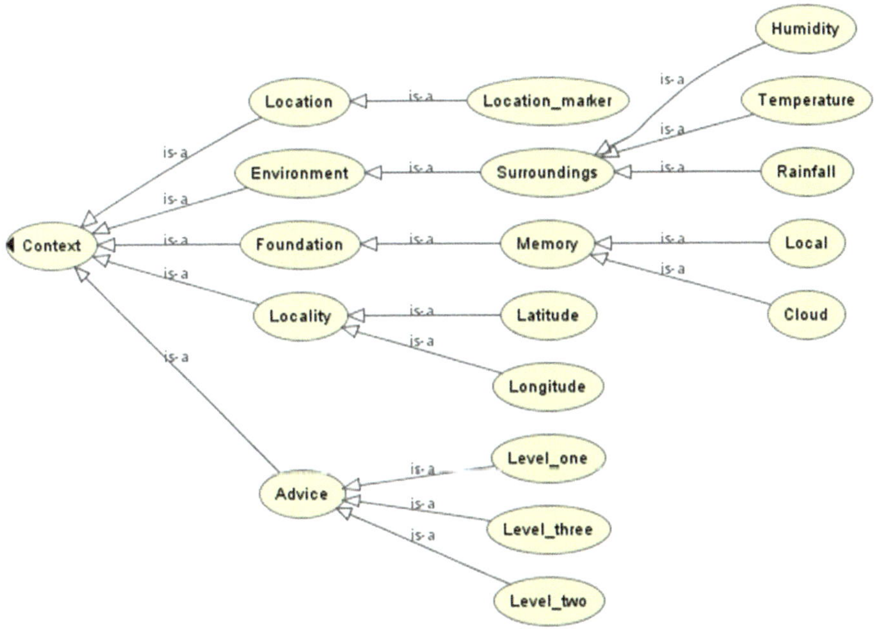

Fig. 6. *Context* class

***Tools* Class.** Figure 7 illustrates *Tools* class. It refers to hardware, software and exhibit items that are used during data collecting and data analysis. Hardware involved in Ovitrap survey activity are computer, WiFi and GPS device to record coordinates of cups placed. System platform is the software identified for this ontology while Ovitrap cup and official forms are the exhibit items involved for survey activity.

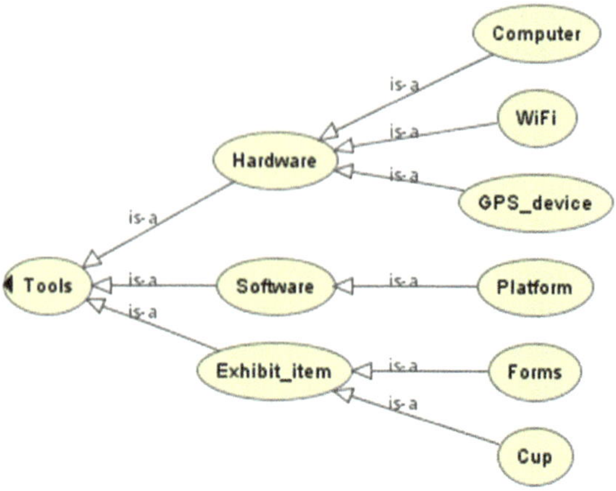

Fig. 7. *Tools* class

3.2 Ontology Validation

Validation by Entomologist Officer. After a meeting with Entomology and Pest Unit officer from local Health District Office, there are several changes made in the first version of proposed context-aware ontology. The removal of some instances that are not relevant with the domain is suggested. Table 3 below lists the instances removed from the proposed context-aware ontology and the reason of its removal.

Table 3. Instances removed from the proposed ontology after validation by expert

Instance Name	Removal Reason
Placement_Time	Time is not recorded when placing Ovitrap cup
Collection_Time	Time is not recorded when collecting Ovitrap cup
A_scutellaris	This Aedes species does not populate Malaysia
A_polynesiensis	This Aedes species does not populate Malaysia
Lighting	Lighting does not affect mosquito biology cycle

Evaluation of Ontology Quality. The evaluation of the proposed context-aware ontology quality is done by referring to a metric suite proposed by Burton-Jones et al. [11]. The metric suite contains ten attributes that could measure the syntactic, semantic, pragmatic and social quality of ontology. Table 4 lists the ten attributes as well as their corresponding description.

Table 4. Metric suite for ontology quality auditing

Metric suite	Attributes	Description
Syntactic quality	Lawfulness	Syntax correctness
	Richness	Breadth of syntax used
Semantic quality	Interpretability	Terms meaningfulness
	Consistency	Terms meaning consistency
	Clarity	Average number of word senses
Pragmatic quality	Comprehensiveness	Number of classes and properties
	Accuracy	Information accuracy
	Relevance	Relevance of information for a task
Social quality	Authority	Reliability of other ontologies on it
	History	Number of times ontology being used

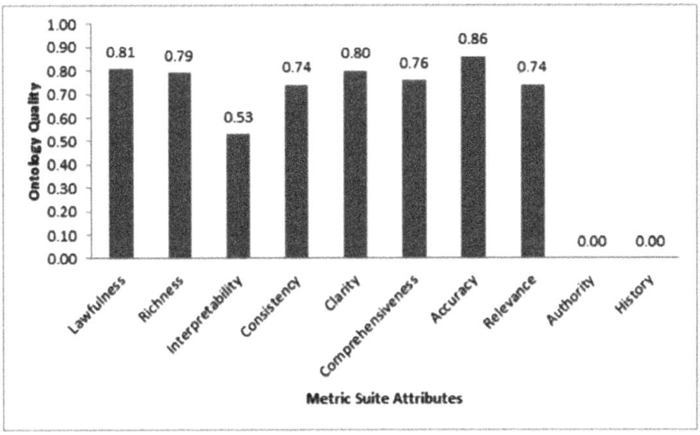

Fig. 8. Ontology quality evaluation result

Based on the criteria and determination value for the metric suites, the proposed context-aware ontology quality is calculated. Calculation result is scaled from 0.00 to 1.00 where the worst value is 0.00 and the best value is 1.00. As shown in Fig. 8, seven criteria achieved high scores, which are lawfulness, richness, consistency, clarity, accuracy, comprehensiveness, and relevance. On the other hand, interpretability criteria attained moderate score of 0.53 while both authority and history criteria scored 0.00, which is the worst score.

4 Conclusion

Dengue surveillance activity such as Ovitrap survey is important in order to determine the current mosquito population in an area that had experienced dengue outbreak in the past. However, both of data collection and analysis processes of Ovitrap survey are done manually by entomology officers. The proposed context-aware ontology provides assistance in both data collection and data analysis. It also determines the type of advice should Entomology and Pest Unit deliver to Vector Unit so a suitable vector control operation could be dispatched to the said area.

The proposed ontology only modelled a backbone of a context-aware system. Future work can include a development of system application that endorses this ontology as its decision support system. Besides that, this ontology could be expanded by adding other similar vector-borne related diseases such as Malaria.

Acknowledgements. We would like to thank the Ministry of Education (MOE) Malaysia for sponsoring the research through the Fundamental Research Grant Scheme (FRGS) with vote number 5F080 and Universiti Teknologi Malaysia for providing the facilities and supporting the research. In addition, we would like to extend our gratitude to the lab members in the EReTSEL Lab, School of Computing, Faculty of Engineering, Universiti Teknologi Malaysia for their invaluable ideas and support throughout this study.

References

1. World Health Organization. Dengue and Severe Dengue: Fact Sheet (2017). http://www.who.int/mediacentre/factsheets/fs117/en/. Accessed 3 Apr 2018
2. Bowman, L., Donegan, S., McCall, P.: Is dengue vector control deficient in effectiveness or evidence? Systematic review and meta-analysis. PLOS Negl. Trop. Dis. **10**(3), e0004551 (2016)
3. Blouin Genest, G.: World health organization and disease surveillance: jeopardizing global public health? Health: Interdiscip. J. Soc. Study Health Illn. Med. **19**(6), 595–614 (2014)
4. Ministry of Health Malaysia. GARIS PANDUAN: Penguatkuasaan Akta Pemusnahan Serangga Pembawa Penyakit 1975 untuk Pegawai-Pegawai Diberi Kuasa Dibawah APSPP 2975. Kuala Lumpur (2004)
5. Roth, A., Mercier, A., Lepers, C., Hoy, D., Duituturaga, S., Benyon, E., et al.: Concurrent outbreaks of dengue, Chikungunya and Zika virus infections – an unprecedented epidemic wave of mosquito-borne viruses in the Pacific 2012–2014. Euro Surveill. **19**(41), 20929 (2014)
6. Packierisamy, P., Ng, C.-W., Dahlui, M., Venugopalan, B., Halasa, Y., Shepard, D.: The cost of dengue vector control activities in Malaysia by different service providers. Asia Pac. J. Public Health **27**(8), 735–785 (2015)
7. Kularatne, S.: Dengue fever. BMJ: Br. Med. J. **351**, h4661 (2015)
8. Cucunawangsih, Lugito, N.P.H.: Trends of dengue disease epidemiology. Virol.: Res. Treat. **8**, 1178122X17695836 (2017)
9. Bhatt, S., Gething, P., Brady, O., Messina, J., Farlow, A., Moyes, C., et al.: The global distribution and burden of dengue. Nature **496**, 504–507 (2013)
10. Fernández-López, M., Gómez-Pérez, A., Juristo, N.: METHONTOLOGY: from ontological art towards ontological engineering. In: Proceedings of the Ontological Engineering AAAI 1997 Spring Symposium Series (1997)
11. Burton-Jones, A., Storey, V.C., Sugumaran, V., Ahluwalia, P.: Assessing the effectiveness of the DAML ontologies for the semantic web. In: Eighth International Conference on Applications of Natural Language to Information Systems, pp. 56–69 (2003)

A Proposed Gradient Tree Boosting with Different Loss Function in Crime Forecasting and Analysis

Alif Ridzuan Khairuddin(✉), Razana Alwee, and Habibollah Haron

Applied Industrial Analytics Research Group (ALIAS), School of Computing, Faculty of Engineering, Universiti Teknologi Malaysia, 81310 Johor Bahru, Malaysia
alifjam1991@gmail.com, {razana, habib}@utm.my

Abstract. Gradient tree boosting (GTB) is a newly emerging artificial intelligence technique in crime forecasting. GTB is a stage-wise additive framework that adopts numerical optimisation methods to minimise the loss function of the predictive model which later enhances it predictive capabilities. The applied loss function plays critical roles that determine GTB predictive capabilities and performance. GTB uses the least square function as its standard loss function. Motivated by this limitation, the study is conducted to observe and identify a potential replacement for the current loss function in GTB by applying a different existing standard mathematical function. In this study, the crime models are developed based on GTB with a different loss function to compare its forecasting performance. From this case study, it is found that among the tested loss functions, the least absolute deviation function outperforms other loss functions including the GTB standard least square loss function in all developed crime models.

Keywords: Loss function · Gradient tree boosting · Artificial intelligence · Crime forecasting and multivariate crime analysis

1 Introduction

Crime forecasting is an area of research that assists authorities in enforcing early crime prevention measures [1]. Forecasting in crime is an analysis technique used to predict and forecast crime patterns as accurate as possible so that it forms significant insights into possible future crime trends based on past crime data. It is very helpful in analysing and understanding the behaviour of crime trends that potentially occurs in the future. The advantages of crime forecasting is that it can prevent recurring crimes in specific areas or regions by analysing the pattern of past crime occurrences, help in allocating an appropriate resource management within a community for better police coverage, and provide useful information to the authorities when planning an efficient solution in crime prevention measures.

In the last decades, applications of artificial intelligence techniques in crime forecasting have been extensively studied by researchers due to their capability to produce a high forecasting performance accuracy. This is because artificial intelligence

F. Saeed et al. (Eds.): IRICT 2019, AISC 1073, pp. 189–198, 2020.
https://doi.org/10.1007/978-3-030-33582-3_18

techniques possess non-linear functions which can detect nonlinear patterns in data [2]. Hence, they are able to discover a new crime pattern that never occurred previously [3]. Among the known artificial intelligence techniques, gradient tree boosting (GTB) introduced by [4] is a newly emerging technique in crime forecasting. A few studies have found that there is limited implementation of GTB in crime analysis [5, 6].

In GTB, the loss function plays critical roles that determine its predictive capabilities and performance. The loss function consecutively minimises the GTB 'pseudo responses' value (error-fitting) over the response variable. The standard applied loss function in GTB is the least square function. Generally, the distributions of response variables (crime data) are varied and not constant [7]. Since the loss function is reflected in it, thus it is recommended to consider other potential mathematical functions to be used as the loss function in GTB instead of only depending on the least square function.

The appropriate application of a loss function is beneficial as it can provide flexibility in a model design that fits different application needs [8]. Hence, such an approach provides robustness to GTB that is suitable and fit for the proposed crime forecasting model. Motivated by this, this study is conducted to observe and identify a potential replacement for the current loss function in GTB by applying a different existing standard mathematical function.

The rest of this paper is organised into the following sections. Section 2 briefly explains GTB. Section 3 discusses the study on different loss functions in GTB in crime forecasting. Section 4 provides a detailed explanation of the experimental setup conducted in this study. Section 5 presents an analysis of the results and a discussion of this study. Section 6 provides the conclusion of the study.

2 Gradient Tree Boosting (GTB)

Gradient tree boosting (GTB) is an artificial intelligence technique that develops a prediction model based on boosting and decision tree learning techniques. It is inspired from another statistical framework introduced by [9] called the adaptive reweighting and combining (ARC) algorithm. GTB is a stage-wise additive framework that adopts numerical optimisation methods to minimise the loss function of the predictive model which later enhances it predictive capabilities. The advantage of GTB is that it is capable to produce highly competitive, robust and interpretable solutions for both regression and classification problems [4]. In addition, the application of boosting technique in GTB can avoid overfitting problems when new independent data is added [4].

3 Study on Loss Function in GTB for Crime Forecasting

GTB's main principle is to construct a new base or weak learner to be highly correlated with the gradient of loss function associated with the whole ensemble (boosting and decision tree). The applied loss function in GTB is to consecutively fit a new model in order to provide a more accurate prediction [10]. Hence, the loss function plays critical roles that determine GTB's predictive capabilities and performance.

As mentioned before, the GTB introduced by [4] applied the least square function as a standard loss function to consecutively minimise its 'pseudoresponses' value (error-fitting) over the response variable (crime data). Thus, motivated by this limitation, this study is conducted to observe and identify a potential replacement for the current least square loss function in GTB by applying a different existing standard mathematical function. Based on the literature study, 3 mathematical functions were found to be suitable as potential replacements for the GTB loss function in this case study. The identified mathematical functions are the least absolute deviation, huber and quantile function. Table 1 lists the identified loss function used in this study.

Table 1. List of identified GTB loss function

Mathematical function	Equation	Description
Least square	$\sum_{i=1}^{m}(k_i - F(j_i))^2$	• Applied standard loss function in GTB by [4] • Minimise the error loss between actual and predicted values by fitting the error differences
Least absolute deviation	$\sum_{i=1}^{m}\lvert k_i - F(j_i)\rvert$	• Based on implementation of least square function but it attempts to identify the best solution that approximate the target data
Huber	$\begin{cases} \frac{1}{2}(a)^2, & \text{for } \lvert a\rvert^2 \leq \delta \\ \delta\lvert a\rvert - \frac{1}{2}\delta^2, & otherwise \end{cases}$ Where, $a = k_i - F(j_i)$	• Specify the robust effect of the function by specifying the maximum value of error in the least square function • Less sensitive to the outliers compared to least square error function
Quantile	$\begin{cases} \tau k, & if\ k > 0 \\ (\tau - 1)k, & if\ k \leq 0 \end{cases}$ $\lvert k(\tau - I(k < 0))\rvert$	Predict the conditional median or other quantiles of the response variable

The aim of this study is to observe and identify which of these new potential loss functions are capable of improving the crime model performance in terms of quantitative measurement errors. The hypothesis made is that the lower the error made by the crime model when producing a forecast result, the better and more suitable the respective loss function selected for that respective crime model. This study focuses on solving the regression problem since this study is about forecasting crime where the main objective is to forecast or predict crime rate values for different crime types. Our proposed enhanced GTB crime model with a different loss function is based on multivariate analysis where several factors significantly influence crime rates. In addition, it is focussed on solving the regression problem where it is used to forecast or predict crime rate values for different crime types. Figure 1 shows the proposed crime forecasting model development framework of this study.

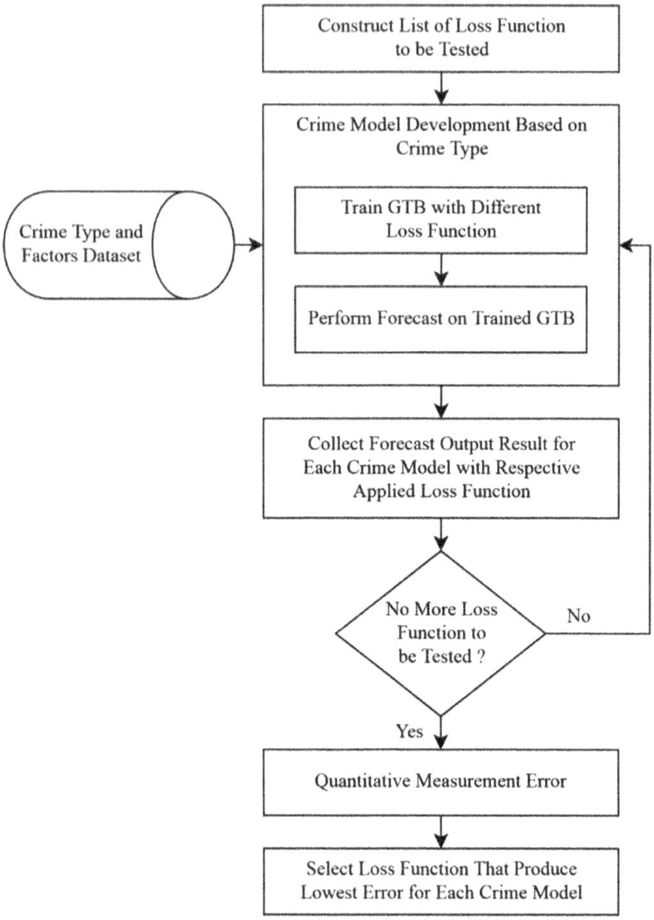

Fig. 1. Proposed crime forecasting model development framework

First, the loss functions to be tested were defined. Then, each crime type was modelled using GTB. In this process, GTB with different loss functions were trained using training data. Once trained, the model was tested by forecasting the crime rate value using the testing data that later produced the crime forecast output. These outputs were used in calculating the quantitative measurement error. The results were then used to analyse and compare the performance of the proposed crime model. Finally, the crime model with a loss function that produced the best measurement error was determined for each crime type.

4 Experimental Setup

The experiment was primarily conducted in a Python environment and the Scikit-learn tools were used in building the proposed crime model. Scikit-learn was developed by [11] and is a Python module package that implements a variety of state-of-the-art machine learning algorithms for various problem-solving solutions. It offers a good flexibility in configuring the parameters and produces consistent results. The Matlab platform is also used in calculating the quantitative measurement error for each crime model with a different loss function.

4.1 Data Definition

Two datasets were collected in this study; crime type and factors. The crime type dataset is the main dataset used since the crime model is developed based on this dataset. The 8 crime types dataset used in this study are murder and non-negligent manslaughter, forcible rape, aggravated assaults, robbery, burglary, larceny theft, motor vehicle theft, and total crime rate for all type of crime. The crime datasets were obtained from the United States Uniform Crime Reporting Statistics website. The factors dataset is used in developing the proposed model for multivariate analysis. The factors dataset is obtained from numerous United States government agencies and other related data repository websites. The selection of factors for each crime type was based on [12]. Both datasets consist of annual time series data from year 1960 to 2015 where each data series has 56 samples. Table 2 shows the selected factors dataset used in developing each crime model.

4.2 Data Preparation

The data preparation is conducted in two cases. The first case takes place before the crime model training while the second case is after crime forecasting. For the first case, this study implements a data normalisation technique using the feature scaling method with a scale range between 0 and 1 to pre-process the obtained raw datasets of crime rate and factors. The data normalisation used is defined in Eq. (1).

$$x' = \frac{x_i - min_x}{max_x - min_x} \tag{1}$$

From Eq. (1), x_i is an actual value of the selected sample in the respective data series x, max_x is the maximum actual value in the respective data series x, min_x is the minimum actual value in the respective data series x and x' is the normalised value for x_i. For the second case, after the forecasting process, the normalised forecast values are transformed back into actual raw data values through denormalisation. The denormalised forecast output values are then used to calculate the quantitative measurement error. The denormalisation of the data is based on the mathematical transformation in Eq. (1) and expressed in Eq. (2).

$$x_i = x'(max_x - min_x) + min_x \tag{2}$$

Table 2. Selected factor dataset for each crime model

Crime model	Factor dataset
Murder and non-negligent manslaughter	• US population, male (% of total)
Forcible rape	• US inflation rate • US gross domestic product, Net export services • US immigration statistic, total aliens apprehended
Robbery	• US population, female (% of total) • US immigration statistic, lawful permanent residence status
Aggravated assaults	• US consumer price index, durables in city average • US immigration statistic, total aliens apprehended • US poverty rate, children under 18, female householder no husband present • US population, ages 0–14 (% of total) • US tax revenue, income tax, business • US unemployment rate, women, 16 years old and over • US inflation rate • US consumer sentiment index
Burglary	• US consumer price index, apparel in city average • US immigration statistic, lawful permanent residence status • US poverty rate, under 18, related children in families • US population, male (% of total) • US tax revenue, excise taxes • US unemployment rate, women, 20 years old and over • US inflation rate • US consumer sentiment index
Larceny theft	• US consumer price index, purchasing power in city average • US immigration statistic, total aliens apprehended • US poverty rate, with children under 18, female householder no husband present • US unemployment rate, married women, spouse present, 16 years old and over • US consumer sentiment index
Motor vehicle theft	• US consumer price index, energy in city average • US gross domestic product, net exports of goods and services • US poverty rate, people between 1.00–1.25% of poverty level • US population, male (% of total) • US tax revenue, income tax, individual • US unemployment rate, women, 16 years old and over • US inflation rate • US consumer sentiment index
Total crime rate for all types of crime	• US consumer price index, purchasing power in city average • US gross domestic product, gross private domestic investment, fixed investment, residential • US immigration statistic, total aliens apprehended • US tax revenue, excise taxes

During the experiment, the obtained datasets of crime rate and selected factor are divided into two groups of training (in-sample) and testing (out-sample) data. Training data is used to train each crime model while testing data is used to test and forecast the crime rate values based on the trained crime model. In this study, 90% of the obtained dataset is used for training purpose (from year 1960 to 2009) while the remaining 10% is used for testing purpose (from year 2010 to 2015).

4.3 Parameter Configuration

Before the development and forecasting of the proposed crime model is conducted, several input parameters are required to be configured in GTB. In this study, number of trees, learning rate and size of individual tree in GTB is set to 100, 0.1 and 3 respectively.

4.4 Evaluation Analysis

In this study, 3 types of quantitative measurement error analysis are applied to measure and compare the performance of each crime model with a different loss function. The quantitative error measurements used are root mean square error (RMSE), mean absolute deviation (MAD) and mean absolute percentage error (MAPE). The formula to calculate the RMSE, MAD and MAPE are defined in Eqs. (3), (4) and (5) respectively.

$$RMSE = \sqrt{\frac{1}{n}\sum\nolimits_{t=1}^{n} (z_t - \widehat{z}_t)^2} \tag{3}$$

$$MAD = \sum\nolimits_{t=1}^{n} \frac{|z_t - \widehat{z}_t|}{n} \tag{4}$$

$$MAPE = \sum\nolimits_{t=1}^{n} \left| \frac{z_t - \widehat{z}_t}{z_t} \right| \times \frac{100}{n} \tag{5}$$

From Eqs. (3), (4) and (5), n is the total number of test data used during the testing process, z_t is the actual value of the selected element in the test data and $\widehat{z}_t$ is the denormalised forecast value of the selected element in the output test data.

5 Result Analysis and Discussion

The calculated forecast or predicted results for each developed crime model based on quantitative measurement error are used to evaluate and compare its performance. The results are shown in Table 3 and based on the observed results, several important findings were found and analysed for each developed crime model. In all developed models, the least absolute deviation function was proven to be more suitable and fitted the case study as it successfully outperformed other loss functions including the standard GTB least square function. This is attributed to its lowest quantitative measurement error (RMSE, MAD and MAPE) of all the all developed crime models.

Table 3. Quantitative measurement error for each crime model with different loss function

Crime model	Loss function	Quantitative measurement error		
		RMSE	MAD	MAPE
Murder and non-negligent manslaughter	Least square	0.2160	0.1667	3.6765
	Least absolute deviation	**0.1974**	**0.1638**	**3.5737**
	Huber	0.2246	0.1700	3.7147
	Quantile	1.0478	1.0339	22.3206
Forcible rape	Least square	2.4847	2.3480	8.7582
	Least absolute deviation	**1.1218**	**1.0038**	**3.7250**
	Huber	1.5987	1.1275	4.0861
	Quantile	8.7125	8.6833	32.1734
Robbery	Least square	22.2902	21.2509	19.7793
	Least absolute deviation	**21.1347**	**20.3792**	**18.8461**
	Huber	26.1019	25.3536	23.5125
	Quantile	39.2434	38.6939	35.7375
Aggravated assaults	Least square	30.0766	27.9605	11.7951
	Least absolute deviation	**21.3244**	**20.4725**	**8.6541**
	Huber	34.3026	31.5301	13.3976
	Quantile	97.0287	96.0332	40.3953
Burglary	Least square	92.9717	79.8025	12.5637
	Least absolute deviation	**83.7707**	**58.3076**	**10.9684**
	Huber	121.2909	92.7360	17.0614
	Quantile	324.503	309.7683	53.1959
Larceny theft	Least square	165.0034	148.3594	7.9073
	Least absolute deviation	**143.7844**	**118.2869**	**6.2926**
	Huber	199.2879	182.9202	9.7571
	Quantile	858.8883	840.2371	43.6885
Motor vehicle theft	Least square	55.5383	53.5530	23.8739
	Least absolute deviation	**46.4784**	**45.8346**	**20.3380**
	Huber	51.1281	49.5632	21.9751
	Quantile	183.5211	182.9248	81.0416
Total crime rate for all types of crime	Least square	232.3952	186.9429	6.2063
	Least absolute deviation	**229.6815**	**160.6459**	**5.4065**
	Huber	513.2459	406.3552	13.5706
	Quantile	1189.9246	1177.6505	37.9663

These findings show that the least absolute deviation function's efficiency to approximate the crime data is high and thus the forecast error can be further reduced. This indicates that the least absolute deviation is able to produce multiple solutions in approximating the crime data compared to the least square function that only approximates one solution. Overall, the GTB performance can be improved by identifying the appropriate mathematical functions in handling variations of data structure and representation. In this case study, the least absolute deviation is the most appropriate loss function replacement in enhancing the GTB forecasting performance. Hence, the hypothesis defined in this study was successfully achieved.

6 Conclusion

Forecasting in crime is very helpful in understanding the behaviour of crime trends that potentially occur in the future. In the last decades, it has been found that researchers have shifted their interest towards the application of artificial intelligence techniques in crime forecasting due to their capacity for producing high forecast accuracy. Among them, gradient tree boosting (GTB) is a newly emerging technique in crime forecasting.

GTB is a stage-wise additive framework that adopts numerical optimisation methods to minimise the loss function of the predictive model which later enhances it predictive capabilities. In GTB, the loss function plays critical roles that determine its predictive capabilities. In standard GTB, the least square function is used as the loss function. Thus, motivated by this limitation, this study was conducted to observe and identify a potential replacement for the current loss function in GTB by applying different existing standard mathematical functions.

The result shows that the least absolute deviation loss function outperformed other loss functions including the GTB least square loss function in all developed crime models. In this case study, it became the most suitable replacement loss function in enhancing the GTB crime forecasting performance when using the obtained crime rate data. Also, it was found that the least absolute deviation function was able to improve GTB performance when using a small dataset (small sample size with only 56 samples).

Overall, the proposed crime model with the least absolute deviation function as the new loss function in GTB was indeed able to improve its performance in forecasting crime. In conclusion, the application of different loss functions in enhancing GTB in forecasting crime is important as the loss function plays very critical roles that heavily influence GTB's overall performance. Hence, it is beneficial to select an appropriate GTB loss function as their performance are varied on different crime datasets.

Acknowledgement. This work was supported by FRGS research grant granted by Malaysian Ministry of Education with grant number R.J130000.7828.4F825 for School of Computing, Universiti Teknologi Malaysia (UTM).

References

1. Ismail, S., Ramli, N.: Short-term crime forecasting in Kedah. Procedia - Soc. Behav. Sci. **91**, 654–660 (2013)
2. Rather, A.M., Sastry, V., Agarwal, A.: Stock market prediction and Portfolio selection models: a survey. OPSEARCH **54**, 1–22 (2017)
3. Alwee, R.: Swarm optimized support vector regression with autoregressive integrated moving average for modeling of crime rate. Unpublished dissertation in partial fulfillment of the requirements for the degree of Doctor of Philosophy, Universiti Teknologi Malaysia, Johor Bahru, Malaysia (2014)
4. Friedman, J.H.: Greedy function approximation: a gradient boosting machine. Ann. Stat. **29**, 1189–1232 (2001)
5. Chandrasekar, A., Raj, A.S., Kumar, P.: Crime prediction and classification in San Francisco City. CS229 Technical report: Machine Learning. Stanford Computer Science Department: Stanford University (2015)
6. Nguyen, T.T., Hatua, A., Sung, A.H.: Building a learning machine classifier with inadequate data for crime prediction. J. Adv. Inf. Technol., 8 (2017)
7. Natekin, A., Knoll, A.: Gradient boosting machines, a tutorial. Front. Neurorobotics **7**, 21 (2013)
8. Guelman, L.: Gradient boosting trees for auto insurance loss cost modeling and prediction. Expert Syst. Appl. **39**, 3659–3667 (2012)
9. Breiman, L.: Arcing the edge. Technical report 486, Statistics Department, University of California, Berkeley (1997)
10. Freeman, E.A., Moisen, G.G., Coulston, J.W., Wilson, B.T.: Random forests and stochastic gradient boosting for predicting tree canopy cover: comparing tuning processes and model performance. Can. J. For. Res. **46**, 323–339 (2015)
11. Pedregosa, G.V.F., Gramfort, A., Michel, V., Thirion, B., Grisel, O., Blondel, M., Prettenhofer, P., Weiss, R., Dubourg, V., Vanderplas, J., Passos, A., Cournapeau, D., Brucher, M., Perrot, M., Duchesnay, E.: Scikit-learn: machine learning in Python. J. Mach. Learn. Res. **12**, 2825–2830 (2011)
12. Yang, W., Wang, K., Zuo, W.: Neighborhood component feature selection for high-dimensional data. J. Comput. **7**, 161–168 (2012)
13. Castelli, M., Sormani, R., Trujillo, L., Popovič, A.: Predicting per capita violent crimes in urban areas: an artificial intelligence approach. J. Ambient. Intell. Hum. Comput. **8**, 29–36 (2017)
14. Elith, J., Leathwick, J.R., Hastie, T.: A working guide to boosted regression trees. J. Anim. Ecol. **77**, 802–813 (2008)
15. Chen, T., Guestrin, C.: XGBoost: a scalable tree boosting system. In: Proceedings of the 22nd ACM International Conference on Knowledge Discovery and Data Mining (SIGKDD), pp. 785–794 (2016)
16. Chen, T., He, T.: Higgs boson discovery with boosted trees. In: NIPS Workshop on High-energy Physics and Machine Learning, pp. 69–80 (2015)
17. Chen, Y., Jia, Z., Mercola, D., Xie, X.: A gradient boosting algorithm for survival analysis via direct optimization of concordance index. Comput. Math. Methods Med. **2013**, 8 (2013)
18. Ding, F., Ge, Q., Jiang, D., Fu, J., Hao, M.: Understanding the dynamics of terrorism events with multiple-discipline datasets and machine learning approach. PloS One **12** (2017). Article no. e0179057
19. Dubey, N., Chaturvedi, S.K.: A survey paper on crime prediction technique using data mining. Int. J. Eng. Res. Appl. **4**(3), 396–400 (2014)

Derivation of Test Cases for Model-based Testing of Software Product Line with Hybrid Heuristic Approach

R. Aduni Sulaiman[1], D. N. A. Jawawi[2](✉), and Shahliza Abd Halim[1,2]

[1] Software Engineering Department, Faculty of Computing, University Technology Malaysia, 81300 Skudai, Johor, Malaysia
raduni2@live.utm.my

[2] Faculty of Computer Science and Information System, University Tun Hussein Onn Malaysia, 86400 Parit Raja, Johor, Malaysia
dayang@utm.my

Abstract. In Model-based testing (MBT) for Software Product Lines (SPLs), many algorithms have been proposed for test case generation. The test case is generated based on a test model which aims to achieve optimization. The heuristic search algorithm is one of the techniques that can be used to traverse the test model with a good quality of solutions. This paper describes our experience in using three types of search algorithm, which are Floyd's Warshall, Branch and Bound algorithm and Best First Search (FWA-BBA-BFS) which were integrated and hybridized in order to fully explore the test model. In this paper, this algorithm is validated based on test case results measured according to coverage criteria, generation time and size of test suite. Based on the experimental results, it is established that our proposed algorithm can generate test cases with reasonable cover-age, minimal execution time and appropriate size of test suite.

Keywords: Software Product Line · Model-based testing · Software testing · Branch and bound algorithm

1 Introduction

A Software Product Line (SPL) is defined as a group of products that share core features of functionality but differ slightly in specific requirements [1]. The main goal of SPLs is to provide reuse of software artefacts in terms of commonality and variability [2]. A variety of testing techniques have been applied to guarantee the quality, since testing an SPL is a very costly and time consuming task [3]. Hence, automated and systematic techniques are required to validate the quality of the software in an efficient manner. However, the current SPL testing techniques still suffer in handling large numbers of products, which has resulted in the common techniques becoming unacceptable for this purpose.

This issue relates to the scalability of products, that brings new demands for techniques that offer functionalities to handle large numbers of products, and also offer

F. Saeed et al. (Eds.): IRICT 2019, AISC 1073, pp. 199–208, 2020.
https://doi.org/10.1007/978-3-030-33582-3_19

reusability and cost reduction [4–6]. Model-based testing (MBT) is an approach that offers automation and reusability based on test cases [7]. The core of MBT is to have a test suite from the test model, based on the requirements. The test model defined can be used to derive a set of test cases based on the strategy chosen. The proposed approach is illustrated in Fig. 1. It starts with our previous work [8] and continues this work by considering test cases generation. We reuse the model constructed in mapping for test case generation since we are concerned about the cost of testing in terms of a reusable model. The idea is to have a good quality of test cases based on defined coverage criteria. In this paper, three types of coverage criteria are used, which are all-states, all-transition and transition-pair coverage. We explore the hybridization of three heuristic algorithms, utilizing the test case generation approach. Although some works that have proposed a hybrid search heuristic combining Floyd's Warshall Algorithm (FWA) and Branch and Bound Algorithm (BBA), to our best knowledge, there are no studies in the SPL domain that have focused on hybridization of three types of algorithm, which are FWA, BBA and Best First Search (BFS). The contribution of this paper can be summarized as follows:

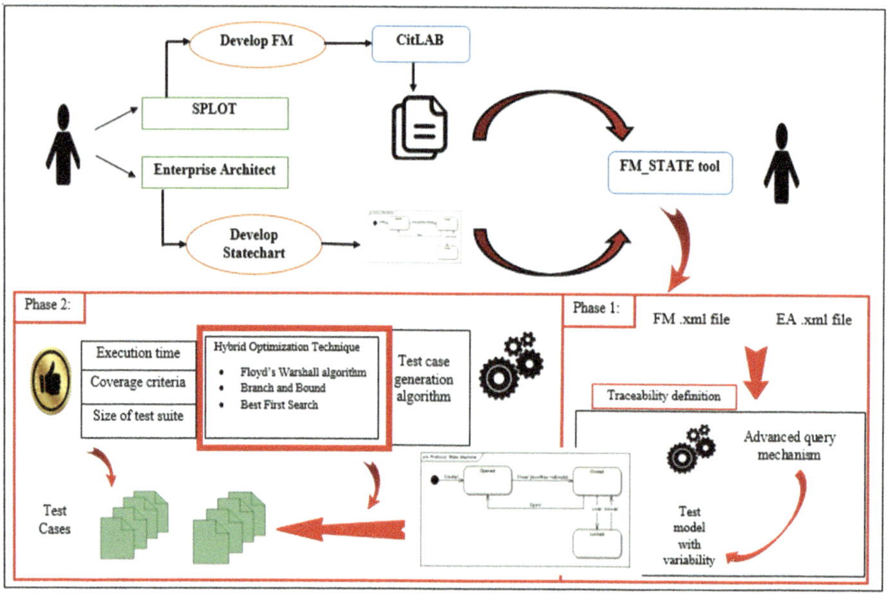

Fig. 1. Overview of proposed approach

- The implementation of a hybrid heuristic search in the SPL domain for MBT case studies that use statechart as a test model. Based on the result, we were able to analyze the performance of the proposed algorithm.

The rest of the paper is organized as follows. Section 2 reviews related works and a basic definition and concept is developed in Sect. 3. Section 4 presents the proposed framework and the results are described in Sect. 5. Section 6 draws conclusions from this study and suggests future work.

2 Related Works

Many studies have proposed frameworks for MBT for SPL. For example, Devroey et al. [9] proposed an MBT framework using a Feature Transition System (FTS). Three coverage criteria were defined, which comprised structural, dissimilarity and statistical coverage. This technique works by performing test case selection and prioritization based on a behavioral model in the SPL. This work has been further extended by considering test selection in terms of coverage and size of test cases. However, this method only supports random generation of test cases. In contrast to this approach, the current work does not seek t-wise generation, but conducts test case generation based on the adaptive Floyd Warshall algorithm, branch and bound algorithm and best first search (FWA-BBA-BFS).

Another technique that has been used to explore the test case in software testing area is the heuristic search. In this area, BBA and BFS have been previously implemented for test case generation. Xing et al., [10] tackled pairwise test case generation by proposing dynamic variable ordering to handle branching conditions. Wang et al., [11] also proposed BBA and BFS and focused on solving the constraint satisfaction problem. The current work is similar concept to [10], since it implements dynamic variable ordering in the algorithm. However, an additional algorithm, Hill Climbing, is used to improve the previous work. In MBT for SPL, the implementation of heuristic search is still in its early phase. In the context of the current study, there are studies that also used FWA and BBA as the technique for test case generation.

Devroey et al. [9] reported the used of these two algorithms for test case generation, based on the mathematical model known as Feature Transition System (FTS). FWA and BBA were used to fulfil the all-state coverage by using a score computed from BBA. Their study shows the potential of FWA and BBA as a technique that can traverse the nodes of graph. However, they faced scalability issues that caused the test case generation process to take a longer time. Although using a heuristic search, the present study requires only a test model that consists of a weighted graph, due to the concept of BBA, which will estimate the bounds. Specifically, our work deals with a heuristic search using FWA, BBA with BFS in order to contribute towards removing unrelated looping of similar nodes and help to improve the test generation. Thus, the process of generation only focuses on single looping and can minimize the execution time.

3 Basic Concept and Definition

In this section, we describe the basic concepts of SPL and MBT for test case generation based on a large scale of product variants.

3.1 Floyd-Warshall Algorithm in Model-Based Testing

The Floyd-Warshall Algorithm (FWA) is the simplest algorithm used to compute all shortest paths in a test model [12]. It covers the paths between graph pairs' vertices based on a directed weighted graph. It is the one of the famous algorithms that is capable of finding the shortest path. According to Aini et al. [13], the core of FWA starts with four main procedural steps. It starts with graphic data converted to a matrix form, followed by calculating the routes of the nodes. The process will continue by calculating the of matrices route of t-1. The accessibility matrix M consist of information between two states (s1, s2). Based on the path generated, the shortest path can be analyzed.

3.2 Branch and Bound Algorithm (BBA) and Best First Search (BFS)

BBA is a global search method that offers searching for optimal results with minimal cost [14]. It is also known as the backtracking algorithm and is used in order to search for the solution of the optimization problem. The process of BBA starts with a branching and bounding process based on a set of active nodes and a broad search of these nodes. Previously, in MBT for SPL, BBA has been implemented for test case generation [5, 11]. However, this algorithm has been used to generate an abstract test suite that only satisfies all-state coverage [5]. The algorithm they proposed also computed the abstract test case by continue to loop until all remaining states have been visited in the test model. However, this algorithm is still in an early phase and the authors have highlighted issues of scalability, as the experimental results showed that random generation performed much better than BBA in the large test cases.

The integration between BBA and BFS helps to minimize the process of searching to produce test cases. We integrate BBA-BFS with FWA to obtain the full TCG flow. By having BFS in BBA, irrelevant branches that keep repeating can also be removed, so that only the applicable branch will be displayed as a result. The hybrid of BWA-BFS has been implemented in software testing. However, this algorithm is still in an early phase, but it has the potential to be one of the best heuristic search algorithms for test case generation.

4 Details of FWA-BBA-BFS

This algorithm starts with giving the definitions required. Based on the FWA results, it can produce a matrix of the test model. The information from FWA is imported into the BBA-BFS algorithm for the TCG process. The second step consists of the process of removing the irrelevant variables from the FWA matrix results.

Definition 5: Relevant variable

This refers to the input of FWA that can traversed all paths. To make it more precise, we define it as arcs, a = {vi, vn ∈ E}, that consist of values, {E = 1, 2,.., n}. An irrelevant variable is defined as {vi, vn ∈ E} that consists of values {E = 1, 2,.., n} where E !{INF | 0}.

After relevant variables have been selected, it goes to the next phase, which is the bounding phase. Here, vertices are added for a deep search of the path. Here, the permutation process starts. It consists of the selection, reduction and backtracking process starting from the initial states until the final state. From the initial state, one vertex is taken randomly to handle the permutations. In order to conduct the permutations, we have defined three types of coverage, which are all-state, all-transition and transition-pair coverage. Based on these three types of coverage, the algorithm will search the path to fulfil the coverage defined (Fig. 2).

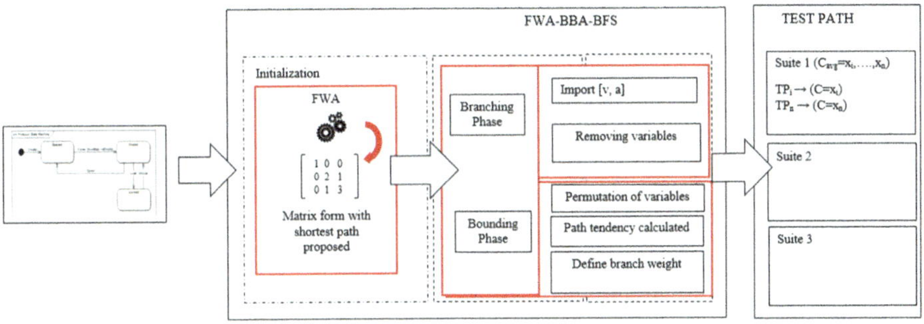

Fig. 2. Integration and hybrid of FWA-BBA-BFS

Definition 6: Permutations *P*, is conducted to obtain the queue list of the nodes. For example, we take one coverage defined, which is all-transition coverage, defined as AT = variable input V, in the space domain D, the list of nodes will be added in stack form $P = \{V_i \in D_i\} \in a \mid ia \, \Delta \, AT$. It produces the sequence of variables, V, based on active a or inactive *ia* types of variables. In the permutation process, the stack queue consist of:

Definition 7: Permutation queue consists of the list of test paths, TP, in the queue list. We define the permutations as $Queue = \{Q_{next} \in TP_i\}$ where Q_{next} refers to the next queue of the path. This process will end after a bounding search of neighbors' nodes reaches the final node. It will continue to add a new stack after the one branch has been considered. Next, the tendency of the path starts. For BBA-BFS, tendency refers to the position of test model branch so that it can provide the next searching direction. This process is controlled by the branch expression or predicate which can be positive or negative.

Definition 8: Path tendency PT refers to the attributes of the variable in the test model that fulfil the branching expression or predicate along the path. Weight values were classified as positive or negative: $PT \in \{w_{+ve}, w_{-ve}\}$. $+ve$ refers to higher values of the tendency which showed better results, while $-ve$ refers to cases where the lower value is better. In order to calculate the tendency values, the weight of each branch was calculated. In this algorithm, we refer to it as the cost of branching. Based on the weight of each branch, we added all the weights for each branch to get the total cost for each path. Furthermore, we also added a checking code, to prevent redundant generation. For this purpose *<arraylist>* will be used to insert the path nodes as integers. Thus, the issue of redundancy will be prevented which will accelerate the search process. The concept of BBA-BFS above is described based on the pseudocode shown in Algorithm 1.

5 Results

In order to observe the effectiveness of FWA-BBA-BFS, a set of experiment was carried out based on four case studies with different numbers of states in a statechart. Within the approach developed, the case study was selected based on FMs that describe the variability and commonality of the product line. The experiments were performed in the Windows 8.1 with 64 bits, Intel CORE i5 with 4.0 GB memory. The algorithms were implemented in Java based on the platform of NetBeans.

```
Algorithm 1: FWA-BBA-BFS
Input : matrix form of [v,a,w]
D_i : Space domain; B_b : k branch; B_b+1 : k branches along path
Output: Result [TP_1,...TP_n]
Step 1: Remove irrelevant
     if (variable >= 0)
              matrix[i][j] = 1;
     else if (variable <0 )
              matrix[i][j] = 0;
      else if (variable == "INF")
              matrix[i][j] = 0;
Step 2: Permutations of variables
Insert into queue int matrixTemp[][] = new int[v][a]
For i →1:|Q_i| do
      insertQueue (j > i; D_i ∈ Q_i)
    break; else
For (b_b, b_b+1) (a ∈ [1,k]) do
ArrayList<Integer> storage = new arrayList<Integer>(); Q_i.add(rank(b_b, b_b+1));
Step 3: Path tendency calculated
    PathTendency ← null; Foreach w_i
       If (w_i > 0)
         PathTendency = posINF;  w_i = w_+ve;
       Else if (wi < 0)
         PathTendency = NegINF;  w_i = w_ve;
  Return PathTendency;
End
```

5.1 Performance Evaluations

In order to evaluate our approach, four case studies were used as a benchmark for test case generation. Details of these case studies are shown in Table 1. The studies are based on different domains, which are a mobile phone, e-shop, alarm system and UTM Robokar. These case studies have different sized requirements in the test model.

Table 1. List of case studies.

Name	Types	Description	Total states	Total transition
Mobile phone [15]	Communication-based	To describe the mobile phone features	21	37
E-shop [16]	Electronic-based	To describe the online shop features	17	27
Alarm system [17]	User-based	To describe the functionalities of alarm system	5	15
Educational robotics [18]	Student learning-based	To describe the robotics of student learning process	45	78

5.2 Characterizing the Performance of Algorithms

To put this work into perspective, we have characterized the performance of the algorithm in terms of average time. In order to depict the variation of the results obtained, a box plot of the algorithm was constructed, as shown in Fig. 3, which is based on Table 2. The boxplot reveals the time and average time required to generate test cases, based on the algorithm used, in trying to fulfil the maximum coverage.

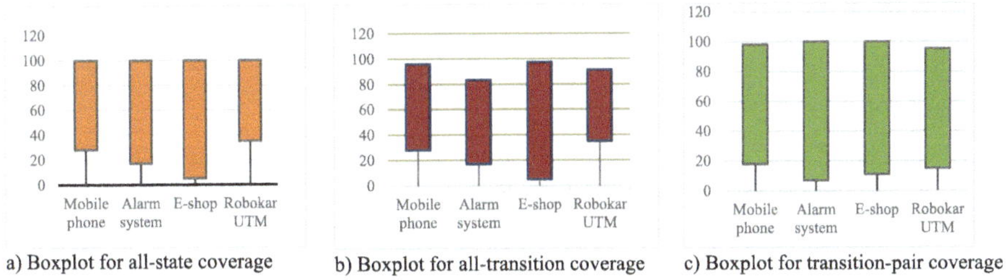

Fig. 3. Boxplot of coverage criteria

From Fig. 3, it can be seen that the proposed algorithm can achieve 100% for all-state coverage. For all-transition coverage, for the smallest case study, which was the e-shop, this algorithm covered 97.7% of transitions, whereas for the larger case study, it achieved only 87.2% coverage.

The transition pair coverage result shows that for two of the case studies, it covered 100% whereas mobile phone is 98.1% and Educational robotics is 95.5%. Thus, the results show that this algorithm can generate test cases with high effectiveness, based on the coverage criteria. Table 3 shows the details of 30 time runs divided into three parts which are the size of the minimum, maximum and average number of cases generated for FWA-BB-BFS and random-based search.

The reason for comparing our results with those of random search was that the previous work used random search with FWA and BBA for the test case generation [19]. Therefore, we wanted to compare our work with random in order to improve the

Table 2. Percentage of coverage criteria of proposed algorithm

Case study	FWA-BBA-BFS					
	Coverage					
	Best all-states	AVG coverage (%)	Best all-transition	AVG coverage (%)	Best transition pair	AVG coverage (%)
Mobile phone	15	100	48	96.1	18	98.1
Alarm system	17	100	14	90.5	7	100
E-shop	5	100	26	97.7	11	100
Educational robotics	45	100	74	87.2	15	95.5

Table 3. Size of test suite and execution time taken

Case study	Test suite size (AVG)	FWA-BBA-BFS	Random search
		Execution time (s)	Execution time (s)
Mobile phone	15	0.08	0.06
Alarm system	10	0.06	0.01
E-shop	8	0.05	0.01
Robokar UTM	20	0.08	0.06

performance of algorithm. From these results, it can be seen that our algorithm took 0.05 s to complete the execution whereas the random search took 0.01 s. This is due to the complexity of our algorithm that consists of a combination of three types of difference search techniques. How-ever, it was worthwhile, since the difference is not too large. For the largest sized case study, which is Educational robotics, we can see that FWA-BBA-BFS can challenge random generation, since the average value of the mutation score is 0.60, which is higher than that for the random search. The time for execution also longer for FWA-BBA-BFS, compared to random search.

6 Conclusion and Future Work

In this paper, we have described the impact of our proposed algorithm on test case generation. This algorithm is based on a heuristic search that can traverse all nodes, so that all states can be reached. The FWA-BBA-BFS approach has been evaluated in terms of its performance based on coverage, generation time, and size of test suite. In terms of these coverage criteria, it was found that FWA-BBA-BFS often produces a good coverage. As highlighted in the results section above, this algorithm can generate test cases by visiting each node and then the best nodes that are more promising will be discovered first.

Concerning our test results, FWA-BBA-BFS can generally cover all the nodes in a test suite of less than 30 cases, meaning that, this algorithm can search for the best results in minimal time. However, the results of the generation are also related to the design of the test model. The complexity of the test model used can also have an impact on generation. This due to the presence of more looping and sub-states in the test model, causing the process of generation to take longer to reach the final state. This issue can be simplified if the test model does not include too many looping states. In addition, based on the nature of the algorithms, the shortest path will be considered as the best path in the test model. The hybrid of BBA with BFS helps to generate the full path of generation. However, the proposed algorithm still requires improvement, since the starting point, which is the FWA, has removed the looping cycles of the test nodes. In addition, BBA has weaknesses, since only the weighted graph can be adopted to generate test cases. This caused difficulties for the tester to adopt the unweighted graph to conduct testing, since this algorithm has only capable to handle single types of test model.

For future work, it is intended to improve the optimization algorithm to maximize coverage, with minimal time generation and minimal size of test suite. Thus, we intend to extend this proposed work by introducing new algorithm to optimize the results.

Acknowledgement. This research is fully funded by Ministry of Higher Education Malaysia (MOHE) for FRGS Grant Vot No.5F117 and University Teknologi Malaysia for UTM-TDR Grant Vot No.06G23, which made this research endeavor possible.

References

1. Zhang, Y., Krinke, J., Petke, J., Harman, M., Langdon, W.B., Jia, Y.: Search based software engineering for software product line engineering. In: Proceedings of the 8th International Software Product Line Conference, vol. 1, pp. 5–18 (2014)
2. Model, C., Testing, B., Lines, S.P., Farrag, M.: Colored Model Based Testing for Software Product Lines (CMBT-SWPL), Technical University of Ilmenau (2010)
3. Ensan, F., Bagheri, E., Gasevic, D.: Evolutionary search-based test generation for software product line feature models. In: International Conference on Advanced Information Systems Engineering, pp. 613–628 (2012)
4. Cichos, H., Oster, S., Lochau, M., Schuerr, A.: Model-based coverage-driven test suite generation for software product lines. Model Driven Eng. Lang. Syst. **6981**, 425–439 (2011)
5. Devroey, X., Perrouin, G., Schobbens, P.-Y.: Abstract test case generation for behavioural testing of software product lines. In: 18th International Software Product Line Conference 2014, pp. 86–93 (2014)
6. Wang, S., Ali, S., Gotlieb, A.: Cost-effective test suite minimization in product lines using search techniques. J. Syst. Softw. **103**, 370–391 (2015)
7. Wang, S., Ali, S., Yue, T., Liaaen, M.: Using feature model to support model-based testing of product lines: an industrial case study. In: Proceedings of the International Symposium on the Physical and Failure Analysis of Integrated Circuits IPFA, pp. 75–84 (2013)
8. Sulaiman, R.A., Jawawi, D.A., Halim, S.A.: Coverage-based approach for model-based testing in Software Product Line. Int. J. Eng. Technol. **7**(4) (2018)

9. Devroey, X., Perrouin, G., Legay, A., Cordy, M., Schobbens, P., Heymans, P.: Coverage criteria for behavioural testing of software product lines. In: Proceedings of the 6th International Symposium on Leveraging Applications of Formal Methods, pp. 336–350 (2014, to appear)
10. Xing, Y., Gong, Y., Wang, Y., Zhang, X.: Path-wise test data generation based on heuristic look-ahead methods 2014 (2014)
11. Wang, Y.W., Xing, Y., Gong, Y.Z., Zhang, X.Z.: Optimized branch and bound for path-wise test data generation. Int. J. Comput. Commun. Control **9**(4), 497–509 (2014)
12. Floyd, R.W.: Algorithms. Commun. ACM **97**, 344–348 (1962)
13. Aini, A., Salehipour, A.: Speeding up the Floyd – Warshall algorithm for the cycled shortest path problem. Appl. Math. Lett. **25**(1), 1–5 (2012)
14. Hervieu, A., Baudry, B.: Pacogen : automatic generation of pairwise test configurations from feature models. In: International Symposium on Software Reliability Engineering, pp. 120–129 (2011)
15. Egyed, A., Segura, S., Lopez-Herrejon, R.E., Ruiz-Cortés, A., Parejo, J.A., Sánchez, A.B.: Multi-objective test case prioritization in highly configurable systems: a case study. J. Syst. Softw. **122**, 287–310 (2016)
16. Weißleder, S., Lackner, H.: Top-down and bottom-up approach for model-based testing of product lines. Electron. Proc. Theor. Comput. Sci. **111**(Mbt), 82–94 (2013)
17. Oster, S.: Feature model-based software product line testing, Technische Universität (2012)
18. Siti, N.M., Halim, S.A., Jawawi, D.N., Mamat, R.: Enhanced educational robotics feature model in software product line. Adv. Sci. Lett. **24**(10), 7251–7256 (2018)
19. Devroey, X.: Behavioural model based testing of software product lines. In: Software Product Lines Conference (SPLC 2014), pp. 1–8, August 2014

Occluded Face Detection, Face in Niqab Dataset

Abdulaziz Ali Saleh Alashbi(✉) and Mohd Shahrizal Sunar

Media and Game Innovation Centre of Excellence (MaGICX), Institute of Human Centre Engineering, Universiti Teknologi Malaysia, 81310 Skudai, Johor, Malaysia
{asaabdulaziz2, shahrizal}@utm.my

Abstract. The detection of faces with occlusion is one of the remaining challenges in the field of computer vision. Despite the high performance of current advanced face detection algorithms, occluded faces are under hot research and is still requires more investigation. There is a need of rich dataset for occluded faces to be used for enriching the training of cnn-deep learning models in order to boost the performance of face detection for occluded faces as in face in niqab. In this paper a dataset of occluded faces named faces in niqab is proposed. An experimental results indicated the poor performance of state-of-the-art face detection algorithms when tested in our dataset.

Keywords: Face detection · Object detection · Deep-learning · Computer vision · Artificial intelligence

1 Introduction

Face detection, a sub-domain of object detection is the most heavily investigated in the area of computer vision and still active area of research for more than two decades. It is the first step of all face application systems such as face recognition, face identification, surveillance and emotion recognition [1–3]. Huge improvement has been done since the rapid implementation of machine learning from the seminal work of Viola and Jones, 2001 [4].

A very good improvement has been done in this field, and achieved an average performance of 98% [5] in the unconstrained FDDB dataset. However detecting faces in certain scenarios is still worth studying as active research area, Occluded and partially occluded face detection is one of the remaining challenges in the field of computer vision. The occlusion problem of faces usually appears when covering faces either in partial or in total due to work requirement such as masks as in hospitals and also due to religious concern as in Muslim societies. A major requirement for Muslim women is to covering their face while being outdoors or when mixed with non-family members.

We used the word niqab to refer to the head covering worn by Muslim woman. Muslim women can often be distinguished by the way they dress. In general Muslim women cover their faces with niqab, burqa and\or hijab as in Fig. (2).

F. Saeed et al. (Eds.): IRICT 2019, AISC 1073, pp. 209–215, 2020.
https://doi.org/10.1007/978-3-030-33582-3_20

Fig. 1. Examples of Images from niqab dataset

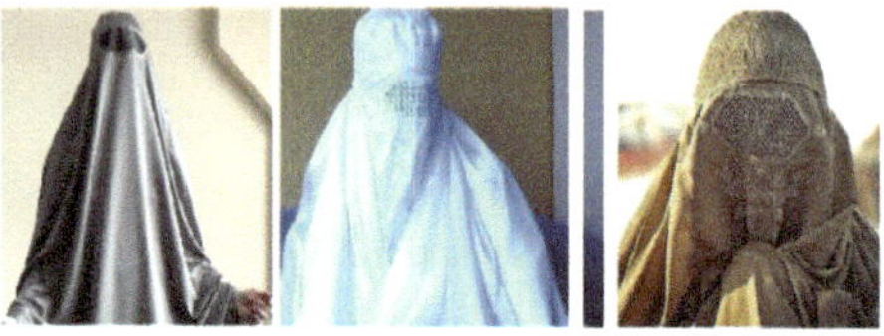

a) Burka full body covering

b) Khimar covering head and shoulders only eyes appears

c) Niqab a veil covering the face showing only eyes and some parts of forehead.

d) Cultural niqab covering wore in certain countries e.g. Yemen, Pakistan and Jordon

Fig. 2. Examples of different Muslim women covering their faces with different niqab styles

There is increasing concern that the improvement of better face detectors is now becoming more complex and too hard. In practice, the detection of heavily occluded faces [6] as in a woman with niqab covering face for instance is necessary and considered very helpful for several applications such as video surveillance and event analysis.

It is also highly demanded for security and monitoring concern and people counting applications. However the task is still challenging for most of state-of-the-art detectors [5, 7, 8], and no enough investigation done by researchers yet.

There are some reasons that could explain why state-of-the-art face detectors performance degrade dramatically when detecting faces covered with niqab. It is well known that huge number of images used for training cnn-deep-learning face detection it critical to boosting the performance of the detector. Several face detection datasets has been recently proposed to be used for training contains thousands of images with unconstrained faces [2, 9]. Although faces there are in different pose rotation and lighting environment and in different circumstances and with some partially occlusion. Unfortunately images that contain niqab faces are totally ignored and is not considered within these datasets. As a result face detectors certainly will degrade to detect faces with niqab due to the lack of sufficient features to be learned during training.

Almost state-of-the-art face detectors were trained using widerface [2] dataset which contains 18,839 images for training but doesn't contains faces covered with niqab except of some several photos that could be found in the surgeon occasions section which have less than 30 photos with faces covered with medical masks which share some similarity with faces in niqab.

For the aforementioned earlier and due to the lack of sufficient training dataset that is rich of faces in high degree of occlusion as in faces covered with niqab we proposed face detection datasets labeled as faces in niqab dataset. It has been classified in two categories one-face photos in which only one face per image and multiple-faces photos. We also examined two well-known face detectors in our dataset and reported the details in the experiment result section.

2 Related Studies

Human face was one of the early concern of computer vison since 1960s, [10] was an early reported work to address facial recognition under the domain of computer vision, an early work which tried segmenting visual system to sub problem [11].

We summarize some of the well-known face detection datasets in Table 1. AFW dataset [12] collected from internet using Flickr images. It has 205 images with 473 labelled faces. For each face, annotations include a rectangular bounding box, 6 landmarks and the pose angles [13], FDDB dataset contains the annotations for 5,171 faces in a set of 2,845 images [2]. PASCAL FACE [14] datasets are most widely used in face detection.

Caltech Web Faces collected from google image search and consists of 7,092 photos contain 10,524 faces in it [15]. Recently, IJB-A [16] is proposed for face detection and face recognition. It contains 24,327 images and 49,759 faces. MALF [17] is the first face detection dataset that supports fine-gained evaluation. MALF consists of 5,250 images and 11,931 faces collected from the internet. UCCS [18] dataset contain simple occlusions such as sunglasses, winter caps, fur jackets …etc. It contains 70,000 faces with high resolution images of average size of 5184 × 3456.

The FDDB dataset [2] has helped driving recent advances in face detection. However, it is collected from the Yahoo! news website which biases toward celebrity faces. The AFW and PASCAL FACE datasets contain only a few hundred images and has limited variations in face appearance and background clutters. The IJBA dataset has large quantity of labeled data; however, occlusion and pose are not annotated. The FDDB dataset labels fine-grained face attributes such as occlusion, pose and expression. The number of images and faces are relatively small.

Widerface Dataset [9] is one of the most widely used dataset for face detection for training and testing, It was proposed by The Chinese University of Hong Kong. It contains 32,203 images with 393,703 annotated face bounding boxes in unconstrained conditions such as occlusion facial expression, and lighting condition. It is noticed that none of the above mentioned datasets had included photos of occluded faces such as faces covered with niqab.

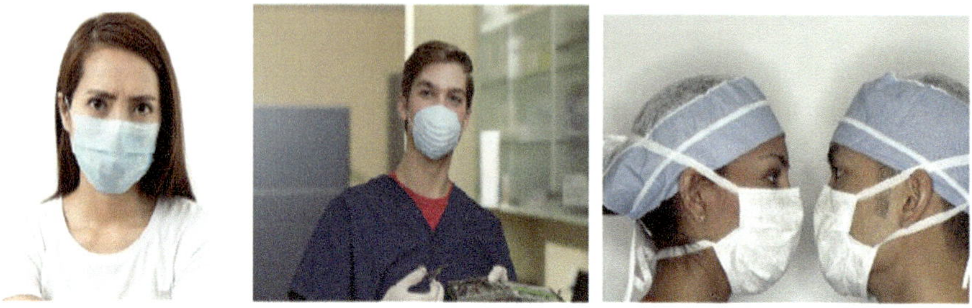

Fig. 3. Images from mafa dataset there is still parts of a face left uncovered

MAFA dataset [6] is the most recent dataset. It is more specific and is targeted to address the problem of occluded faces, it contains 30,811 images with 35,806 faces. The drawback in this dataset is that majority of photos are for faces covered with medical masks, with only few images containing faces covered with niqab. There is a clear difference between faces covered with masks as in Fig. (3) where only part of the nose and a mouth is covered whereas only eyes could be seen in faces covered in niqab as in Figs. (1) and (2) where only the eyes are left uncovered.

Table 1. Different face detection datasets properties

Dataset name	Properties		
	Number of images/faces	Date	Is occlusion considered?
AFW	205 images with 473 faces	2011	No
Caltech Web Faces	7,092 images with 10,524 faces	2005	No
FDDB	2845 images with 5171 faces	2010	–
Widerface	32,203 images with 393,703 faces	2016	Yes
MAFA	30,811 images with 35,806 faces	2017	Yes

3 Niqab Dataset

3.1 Dataset Collection

Niqab dataset Photos were collected from different internet resources using Google and Bing image search engine searching for content based photos that contains faces covered with niqab from internet and social network such as Instagram, Pinterest and Facebook. The search task resulted in more than 30 k photos manually filtered to ensure that only photos that contain at least one face covered with niqab is to be kept.

(a) Dataset Properties and Annotation

Niqab dataset consists of 10,000 photos with 12,000 faces. It is classified into two groups one group for only-one-face per image, the other group is for photos that contains more than one face per image.

Photos are labeled manually. Labelle-me annotation tool was used to draw rectangle for each face. All faces with niqab are labeled with respect to the context in which a face appears, this is a useful technique for training the detector utilizing the context within the image to lets cnn learns available features within the context [19, 20]. It is clearly illustrated in Fig. (4). This is important especially for occluded faces as the features gathered from faces are few due to the accusation so most of the face feature is unavailable. Therefore context can help a bet to gather more information about the face and its context.

The proposed niqab dataset can be used as an extended training for pre-trained face detection so that the new detector would be more robust on detecting heavily occluded face as in faces in niqab for instance. Another scenario is to be used as a benchmark for testing for evaluating face detection models.

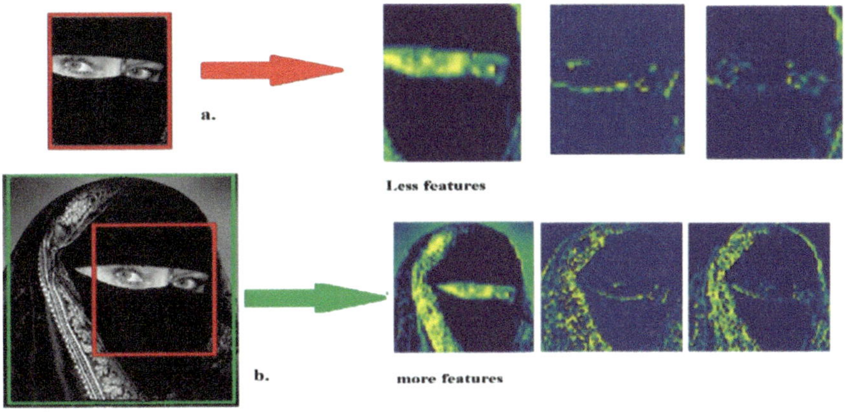

Fig. 4. Shows the importance of the labelling with context. **a.** The red rectangle labelling; face only with no context results in limited features. **b.** The green rectangle labelling; labelling with context gives more features for cnn to learn.

4 Experimental Results

We conducted an experiment of two state-of-the-art face detectors to find out how these face detectors perform in detecting faces in niqab. The experiment was limited to a sample of 466 images selected randomly from our niqab dataset due to limited time and technical issues in which we processed the calculation of the detected and undetected faces manually. A total number of 644 faces in the selected sample, each photo has only one occluded face. Figure (5) shows a set of some photos used for our experiment.

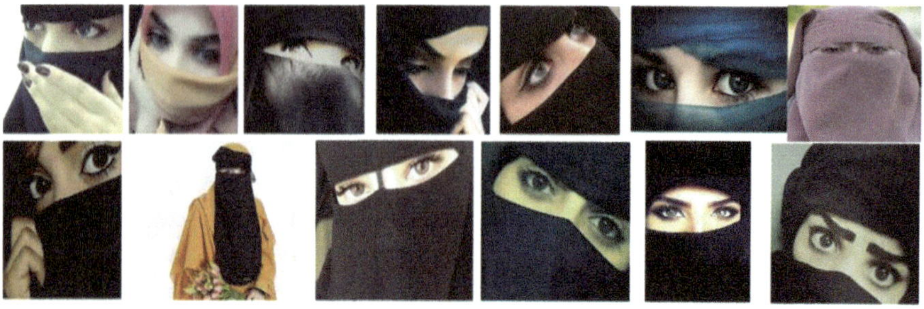

Fig. 5. Set of images from the selected sample used for the experiment

The first face detection algorithms which we used in this experiment was MTCNN [8] a well know face detection algorithm with outstanding performance in accuracy and real-time processing, we used the default threshold [0.5, 0.5, 0.7], and minimum face size = 24 as set by the author. The detection result was only 82 out of 466 faces were detected as true positive. The final result achieved by this algorithm was 94.2% for Precision and 17.7% for the Recall.

The second face detection algorithms was Mobielnet-ssd-based from google [21]. We also used the default settings. The detection result was just a little bit better than mtcnn. 97 out of 466 faces were detected successfully. The total results is 88.9% for Precision and 21.2% for the Recall.

Table 2 below gives the summary result obtained by both of the detectors. The accuracy result for both detectors is: 67.3% and 71.1% respectively.

It is clearly indicating that our proposed niqab dataset is a very challengeable dataset for the state- of-the-art face detectors. There is strong need for face detectors that can detect faces in heavily occluded situations as in faces in niqab.

Table 2. Experimental result on sample of dataset images

Method	FP	TP	FN	Precision	Recall
mtcnn face detector	5	82	379	94.2%	17.7%
Mobielnet-ssd face detector	12	97	355	88.9%	21.2%

5 Conclusion and Future Work

In this paper a dataset of occluded/covered faces named faces in niqab is proposed. It consists of ~ 10,000 images with approximately 12,000 faces. Two state-of-the-art face detection algorithms were tested in the proposed dataset and the results were presented and discussed. For our future work we are working on training face detector that can be able to perform better with high accuracy in detecting faces in niqab.

References

1. Alashbi, A.A.S., Sunar, M.S.B., AL-Nuzaili, Q.A.: Two stages haar-cascad face detection with reduced false positive. In: International Conference of Reliable Information and Communication Technology. Springer (2018)
2. Jain, V., Learned-Miller, E.G.: FDDB: a benchmark for face detection in unconstrained settings. UMass Amherst Technical report (2010)
3. Zhang, C., Zhang, Z.: A survey of recent advances in face detection. Technical report, Microsoft Research (2010)
4. Viola, P., Jones, M.: Rapid object detection using a boosted cascade of simple features. In: 2001 Proceedings of the 2001 IEEE Computer Society Conference on Computer Vision and Pattern Recognition, CVPR 2001. IEEE (2001)
5. Hu, P., Ramanan, D.: Finding tiny faces. In: 2017 IEEE Conference on Computer Vision and Pattern Recognition (CVPR). IEEE (2017)
6. Ge, S., et al.: Detecting masked faces in the wild with LLE-CNNS. In: Proceedings of the IEEE Conference on Computer Vision and Pattern Recognition (2017)
7. Ranjan, R., Patel, V.M., Chellappa, R.: Hyperface: a deep multi-task learning framework for face detection, landmark localization, pose estimation, and gender recognition. IEEE Trans. Pattern Anal. Mach. Intell. **41**(1), 121–135 (2019)
8. Zhang, K., et al.: Joint face detection and alignment using multitask cascaded convolutional networks. IEEE Signal Process. Lett. **23**(10), 1499–1503 (2016)
9. Yang, S., et al.: Wider face: a face detection benchmark. In: Proceedings of the IEEE Conference on Computer Vision and Pattern Recognition (2016)
10. Chan, H., Bledsoe, W.: A man-machine facial recognition system: some preliminary results. Panoramic Research Inc., Palo Alto (1965)
11. Papert, S.A.: The summer vision project (1966)
12. Köstinger, M., et al.: Annotated facial landmarks in the wild: a large-scale, real-world database for facial landmark localization. In: 2011 IEEE International Conference on Computer Vision Workshops (ICCV Workshops). IEEE (2011)
13. Zhu, X., Ramanan, D.: Face detection, pose estimation, and landmark localization in the wild. In: 2012 IEEE Conference on Computer Vision and Pattern Recognition (CVPR). IEEE (2012)
14. Yan, J., et al.: Face detection by structural models. Image Vis. Comput. **32**(10), 790–799 (2014)
15. Angelova, A., Abu-Mostafam, Y., Perona, P.: Pruning training sets for learning of object categories. In: 2005 IEEE Computer Society Conference on Computer Vision and Pattern Recognition (CVPR 2005). IEEE (2005)
16. Klare, B.F., et al.: Pushing the frontiers of unconstrained face detection and recognition: IARPA Janus Benchmark A. In: Proceedings of the IEEE Conference on Computer Vision and Pattern Recognition (2015)
17. Yang, B., et al.: Fine-grained evaluation on face detection in the wild. In: 2015 11th IEEE International Conference and Workshops on Automatic Face and Gesture Recognition (FG). IEEE (2015)
18. Günther, M., et al.: Unconstrained face detection and open-set face recognition challenge. In: 2017 IEEE International Joint Conference on Biometrics (IJCB). IEEE (2017)
19. Tang, X., et al.: Pyramidbox: a context-assisted single shot face detector. In: Proceedings of the European Conference on Computer Vision (ECCV) (2018)
20. Zhu, C., et al.: CMS-RCNN: contextual multi-scale region-based CNN for unconstrained face detection. In: Deep Learning for Biometrics, pp. 57–79. Springer (2017)
21. Howard, A.G., et al.: MobileNets: efficient convolutional neural networks for mobile vision applications. arXiv preprint arXiv:1704.04861 (2017)

Spin-Image Descriptors for Text-Independent Speaker Recognition

Suhaila N. Mohammed[1(✉)], Adnan J. Jabir[1], and Zaid Ali Abbas[2]

[1] Department of Computer Science, College of Science, University of Baghdad, Baghdad, Iraq
suhailan.mo@sc.uobaghdad.edu.iq,
Adnanjjabir@gmail.com

[2] Department of Electrical and Electronic Engineering, Faculty of Engineering, Universiti Putra Malaysia, 43300 Seri Kembangan, Selangor, Malaysia
zaidaliabbas@yahoo.com

Abstract. Building a system to identify individuals through their speech recording can find its application in diverse areas, such as telephone shopping, voice mail and security control. However, building such systems is a tricky task because of the vast range of differences in the human voice. Thus, selecting strong features becomes very crucial for the recognition system. Therefore, a speaker recognition system based on new spin-image descriptors (SISR) is proposed in this paper. In the proposed system, circular windows (spins) are extracted from the frequency domain of the spectrogram image of the sound, and then a run length matrix is built for each spin, to work as a base for feature extraction tasks. Five different descriptors are generated from the run length matrix within each spin and the final feature vector is then used to populate a deep belief network for classification purpose. The proposed SISR system is evaluated using the English language Speech Database for Speaker Recognition (ELSDSR) database. The experimental results were achieved with 96.46 accuracy; showing that the proposed SISR system outperforms those reported in the related current research work in terms of recognition accuracy.

Keywords: Deep belief network · Run length matrix · Speaker recognition · Speech spectrogram image · Spin-based descriptors

1 Introduction

The primary goal of building systems for speaker recognition is to extract and recognize the information contained in the speech signal for carrying speaker identity [1]. The development of a functional speaker recognition architecture has paved the way for many real-world applications, such as speech forensics, robotics, home control systems, voice mail and security control [2].

Speaker recognition techniques can be classified based on the type of spoken text into text-independent and text-dependent cases. Text-independent speaker recognition systems do not require the speaker to produce speech for the same piece of text in both training and testing steps, whereas in text-dependent speaker recognition, the spoken text in both training and testing must be same [3].

F. Saeed et al. (Eds.): IRICT 2019, AISC 1073, pp. 216–226, 2020.
https://doi.org/10.1007/978-3-030-33582-3_21

The human voice includes various discriminative attributes that can be used in speaker recognition [3]. However, one of the major difficulties in speaker recognition is selecting strong features as inputs to the classifier [4]. Different feature extraction methods are used to speaker recognition task including Linear Predictive Coefficients (LPC), Mel Frequency Cepstral Coefficients (MFCC) and others [3].

The spectrogram image of the speech can also be used to extract a speaker's discriminating features. A spectrogram describes the change in the spectral density of a signal with respect to time [5]. Neammalai et al. [6] applied the 2-D Fourier transform on the spectrogram image and the energy of the signal at the specific frequencies was calculated to form the feature vector which was then inputted to support the vector machine for classification. The method was tested on an audio database containing 510 instances each 1.5 s in length and the final accuracy was 93.68%. Nguyen et al. [7] used Scale-Invariant Feature Transform (SIFT) features which were extracted from a spectrogram image of the speech signal and classified with local naïve Bayes nearest neighbor. The proposed approach achieved classification results equal to 73%, 96%, 95%, 97%, and 97% accuracy on the ISOLET, English Isolated Digits, Vietnamese Places, Vietnamese Digits and JVPD databases, respectively. Solovyev et al. [8] considered different representations of sound commands, such as wave frames and spectrograms, which were used as an input to convolutional networks for classification; these were applied in TensorFlow speech recognition challenge organized by Google Brain team on the Kaggle platform. The accuracy achieved was 98.5% on a database consisting of 12 speakers.

The English Language Speech Database for Speaker Recognition (ELSDSR) has been widely used as a base to evaluate the speaker recognition systems. High levels of accuracy have been achieved by Saady et al. [9], using Zak transform (91.3% accuracy), Bora et al. [10], using LPC and MFCC techniques (92%) accuracy ratio, and recently, Soleymanpour et al. [11], using the Modified MFCC and neural network as a classifier (91.9% accuracy).

Although, the recent work achieved acceptable accuracy ratio, the identification of new strong features still needs further investigation to improve the recognition accuracy ratio. Thus, in this research, a new spin-image descriptor is generated for our proposed speaker recognition system (SISR), where the proposed spin-image descriptors are extracted from the spectrogram image of the voice signal, and then the deep belief networks are used as a classifier to achieve the speaker recognition task. The proposed SISR system is evaluated using the text-independent ELSDSR and its performance is compared against the current research work. The rest of the paper is organized as follows: the proposed system is described in Sect. 2. The experimental results based on the ELSDR dataset are illustrated in Sect. 3, while the conclusions and suggestions for future work are presented in Sect. 4.

2 Materials and Methods

The overall design of the proposed speaker recognition system encompasses four stages, as shown in Fig. 1, which are: (1) The speech spectrogram image generation stage, (2) the spectrogram image enhancement stage, (3) The feature vector extraction stage, and (4) the classification stage.

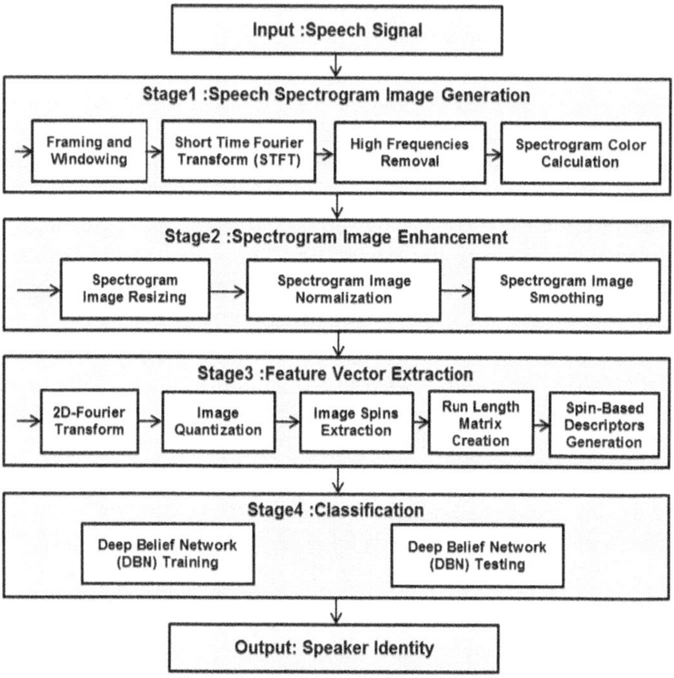

Fig. 1. The overall design of the proposed SISR system

The spectrogram image is generated in the first stage by applying a series of steps which include windowing, Short Time Fourier Transform (STFT), removal of high frequencies and calculation of color intensities. Before extracting the discriminating features from the generated spectrogram image, it is necessary to apply several image enhancement techniques to improve the image data by resizing the image and suppressing the undesired illumination and noise effects.

The goal of the feature extraction stage is to extract a set of strong discriminating features from the enhanced spectrogram image to represent a given speaker's identity. In this stage, the image is first transformed from the spatial domain to the frequency domain by applying 2D-Fourier transform. After that, spin-image-based descriptors are generated by computing the run length matrix for each spin of the transformed image. In the classification stage, the feature vectors extracted in the previous stage are populated into the deep belief neural network classifier, to train the classifier and update the network connection weights. The stable updated weights are then used in the testing step to perform the final recognition task.

2.1 Speech Spectrogram Image Generation

The first stage in the proposed system is speech spectrogram image generation. The most commonly used method for speech spectrogram construction is to represent the amplitude of a particular frequency at a given time with an intensity or color in the image. The following steps are applied to generate the spectrogram image for the speech signal:

Step 1: Speech Framing and Windowing

In this step, the speech is divided with respect to time into overlapped frames with length equal to 512 samples and overlap equal to 50% among successive frames. The windowing function is required to be applied on each frame to remove the disconnectivity effect among speech frames after the framing process [12]. Hamming window is used for this purpose.

Step 2: Short-Time Fourier Transformation

Since the spectrogram image is represented as a frequency versus time plot, each frame must firstly be transformed into the frequency domain by applying STFT, using the following 1D- Fourier transform function [13]:

$$F(v) = \frac{1}{n}\sum_{x=0}^{n-1} s(x)e^{-j2\pi\frac{(vx)}{n}} \tag{1}$$

where $j = \sqrt{-1}$. According to Euler's identity, the Fourier transform equation can be written as [13]:

$$F(v) = \frac{1}{n}\sum_{x=0}^{n-1} s(x)\left[\cos\left(\frac{2\pi}{n}(vx)\right) + j\,\sin\left(\frac{2\pi}{n}(vx)\right)\right] \tag{2}$$

with the real part, R(v), corresponding to the cosine term and the imaginary part, I(v), corresponding to the sine term. Thus, the magnitude of a complex spectral component can be represented as [13]:

$$\text{Magnitude} = |F(v)| = \sqrt{[R(v)]^2 + [I(v)]^2} \tag{3}$$

Step 3: Removal of High Frequencies

After applying STFT, smoothing can be achieved in the frequency domain by dropping out the high-frequency components and passing only the low-frequencies. A Gaussian low-pass filter which has the following transfer function is used for this purpose [14]:

$$H(v) = e^{-D^2(v)/2D_0^2} \tag{4}$$

where, $D(v) = [v - n/2]^2$ and D_0 is a specified distance from the origin of the transform.

Step 4: Spectrogram Color Calculation

Before calculating the intensity of the transformed amplitude, the log spectrogram (in decibels) is computed, because the human perception of sound is logarithmic. After that, the color intensity of each decibel value within each frame can be derived as follows [15]:

$$Color(v) = \left(\frac{255}{max - min}\right) * (10 * \log_{10}|\mathrm{F(v)}| - min) \tag{5}$$

where, min and max are the minimum and maximum decibel values, respectively.

Each frame in the speech will represent a column in the generated spectrogram image. Figure 2 shows the spectrogram image generated for a speech signal using grey-scale colors for two different speakers. As is clearly shown in the figure, there is a large amount of information laid out in the spectrogram image for the different speech signals, which can help in speaker-identity recognition.

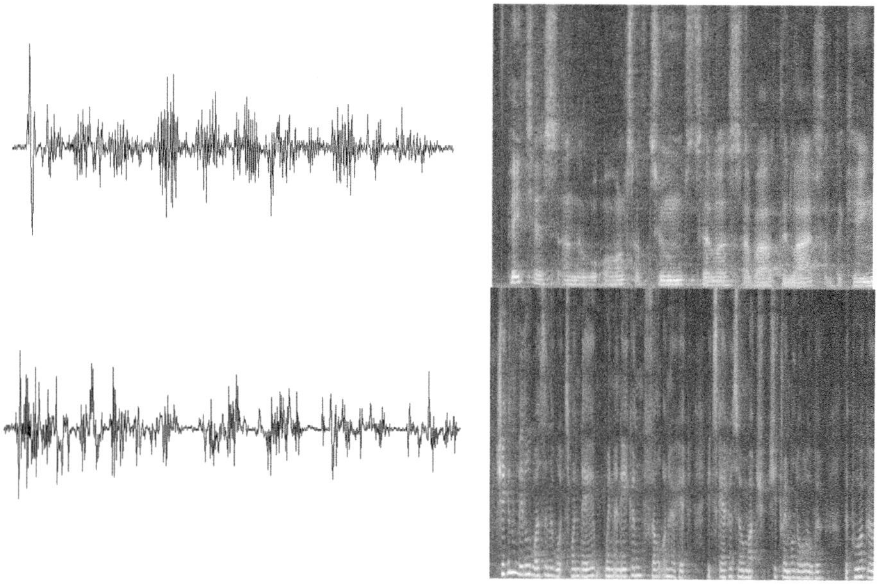

Fig. 2. The spectrogram images generated for the speech signal, using grey scale colors for two different speakers.

2.2 Spectrogram Image Enhancement

A sequence of image processing techniques is applied on the spectrogram image to make it more appropriate for the feature extraction stage [16]. Spectrogram image enhancement involves the following steps:

Step 1: Spectrogram Image Resizing

This step is required to scale down the generated spectrogram image, which leads to speeding up the next processing steps. Nearest neighbor interpolation is used for resizing purposes. In nearest neighbor interpolation, the new output pixel locations (x′, y′), can be calculated as [17]:

$$I_{new}(x, y) = \sum_{i=-k_s+1}^{k_s+1} \sum_{j=-k_s+1}^{k_s+1} I_{org}(i + x_o, j + y_o) h(i - F_x) h(j - F_y) \quad (6)$$

where, (x, y) are the new coordinates of the interpolated pixel, (x_o, y_o) are the original coordinates of the image pixel, $I_{new}(x, y)$ is the interpolated pixel value, $I_{org}(i + x_o, j + y_o)$ is the original pixel value, h represents the interpolation function, (i, j) denotes the boundary of the function (with kernel size = k_s) in the x-coordinate and y-coordinate, F_x represents the position of the interpolated pixel relative to its neighbors, and h is nearest neighbor interpolation kernel.

Step 2: Spectrogram Image Normalization

Image normalization step attempts to improve the contrast of the resized spectrogram by extending the color intensity range of the image and making full use of all possible intensities.

Step 3: Spectrogram Image Smoothing

The spectrogram image may include noisy pixels which degrade the image quality. The median filter is used to reduce noisy pixels that may exist in the normalized image, because it preserves useful detail in the image and can also preserve sharp edges. The median filter is applied 5 times with filter size equal to 3.

2.3 Feature Vector Extraction

To extract spin-image-based features from the enhanced spectrogram image, the following five steps are used:

Step 1: Time to Frequency Domain Transformation

A 2D- Fourier transform is firstly applied to transform the spectrogram image from the spatial domain to the frequency domain. The transformation step will help in focusing feature extraction work on the most important information (lying around the center of the image) and thus avoiding high dimensionality of features.

Step 2: Image Quantization

To reduce the number of possible intensities used to represent image colors and in turn reducing size of the run length matrix, uniform quantization is applied to the image using the following equation [13]:

$$I_q(x, y) = I(x, y)/Q_s \quad (7)$$

where, Q_s is the quantization step.

Step3: Image Spin Extraction

From the center of the image, N different spins (circular blocks) with radius r are extracted from the image, where each spin represents a 1-D vector of the spin pixels' values (pixels between two consecutive rings), as shown in Fig. 3.

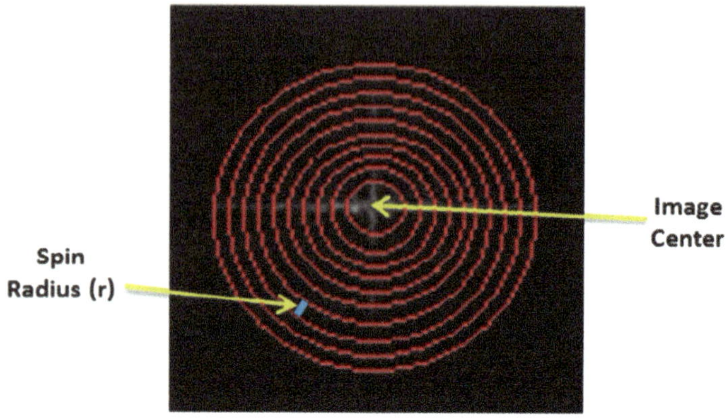

Fig. 3. Image spins

Step4: Run Length Matrix Creation

For each spin, a Run Length Matrix (RLM) is created. The RLM is considered as a base from which the descriptors are generated. RLM is a 2D array used to characterize the variance in the intensities of colors in the spin, such that, each entry in the array holds the number of runs of a given length and intensity when scanning the spin. The value at entry *(i, l)* in the RLM matrix, *h*, shows the number of runs of length *l* and of intensity *i* that appear in the spin. A joint conditional probability density function *p(i, l)* can be obtained from RLM by dividing each entry in the matrix by the sum of the entries.

Step 5: Spin-Based Descriptor Generation

The following five features are generated from RLM of each spin [18]:

1. Short-Run Emphasis (SRE): when the spin contains too many short runs, this feature will be high. SRE can be calculated as follows:

$$SRE = LRE = \sum_i \sum_l \frac{p(i,l)}{l^2} \quad (8)$$

2. Long-Run Emphasis (LRE): when the spin contains too many long runs, this feature will increase. LRE can be calculated as follows:

$$LRE = \sum_i \sum_l l^2 p(i,l) \quad (9)$$

3. Gray-Level Non-uniformity (GLN): when the number of runs of the same intensity increases, this feature will increase, implying a spin with highly varying intensities. The following equation is used to generate this feature:

$$GLN = \frac{\sum_i \left(\sum_l h(i,l)\right)^2}{\sum_i \sum_l h(i,l)} \tag{10}$$

4. Run-Length Non-uniformity (RLN): when the spin is highly varying and most likely has repeated patterns, this feature increases. The following equation is used to generate this feature:

$$RLN = \frac{\sum_l \left(\sum_i h(i,l)\right)^2}{\sum_i \sum_l h(i,l)} \tag{11}$$

5. Run Percentage (RP): This feature shows the percentage of runs in a spin when normalized with respect to the spin size (L). The smaller the percentage, the smaller the number of runs, which implies a spin with large homogeneous regions. The following equation is used to generate this feature:

$$RP = \frac{100}{L} \sum_i \sum_l h(i,l) \tag{12}$$

2.4 Classification Stage

A recent advance in the learning of deep artificial neural networks makes it possible to overcome the vanishing gradient problem existing on previous neural network models. This problem has been overcome using a pre-training step, where deep belief networks (DBNs) are formed by the stacked Restricted Boltzmann Machines (RBMs) performing unsupervised learning. Once a pre-training step is done, network weights are fine-tuned using regular error back propagation, while treating the network as a feed-forward net [19]. DBNs are probabilistic graphical models which have multiple hidden layers. RBM is a Boltzmann machine which is restricted to have only one hidden layer (H) and one visible layer (V) and also have no visible-visible and hidden-hidden connections [20].

The DBN classifier is used in the proposed system because of its strong classification capability. The classification stage includes a training step and a testing step. In the training step, the DBN is trained with the training dataset for the purpose of selecting the proper DBN architecture and adjusting the weights, which are then used in the testing step to guess the identity of an unknown speech signal.

2.5 Experimental Design

The performance of the proposed system was evaluated using the English Language Speech Database for Speaker Recognition (ELSDSR). ELSDSR includes voice messages from twelve males and ten females (twenty-two speakers in total), and the age of the speakers ranged from twenty-four to sixty-three. The corpus is divided into suggested training and test sets. Seven paragraphs of text were constructed and collected, which contain eleven sentences. For the training set, 154 (7 * 22) utterances are recorded; and for test set, 44 (2 * 22) utterances are provided [21]. In order to test the system performance, the training set specified in the database consists of 154 speech samples, which belong to 22 speakers, while the testing samples are assembled to test the proposed system performance. The number of hidden nodes used with DBNs is 10 and with 10000 epochs.

3 Experimental Results

Different values for N (number of spins) and r (radius) values were attempted and the accuracy achieved is recorded in Table 1. As shown in the table, the best accuracy was achieved for values of N = 10 and r = 6; thus, the total number of features adopted in the system is 60 features.

Table 1. The achieved accuracy with different values for r and N parameters

r	N									
	1	2	3	4	5	6	7	8	9	10
1	10.09	30.38	40.3	60.13	71.67	80.55	82.98	83.67	84.28	85.86
2	12.57	32.62	51.77	64.09	75.11	82.4	84.58	85.9	87.3	88.89
3	16.16	40.4	60.1	67.17	80.81	86.36	87.37	90.11	90.45	91.92
4	17.17	38.38	58.08	66.67	76.77	84.85	86.36	91.92	93.43	93.43
5	13.13	36.87	58.59	75.25	82.32	86.36	85.35	90.91	95.45	95.45
6	19.2	40.4	60.6	78.28	86.87	89.39	91.92	93.94	92.93	**96.46**
7	17.17	38.89	65.15	80.3	87.37	90.91	93.43	93.43	**96.46**	95.96

Table 2 shows a comparison with the results in some other published works on speaker recognition using the ELSDSR. The table clearly shows the superiority of the proposed system.

Table 2. A comparison between the proposed SISR system and current research works in speaker recognition on ELSDSR database

Authors	Methodology	Accuracy
Saady et al. [9]	Zak transform	91.3%
Bora et al. [10]	LPC and MFCC	92%
Soleymanpour et al. [11]	Modified MFCC and neural network	91.9%
The proposed SISR system	Image spins and run length matrix	**96.46%**

4 Conclusion and Future Work

New spin-image based descriptors for speaker recognition have been presented in this paper. The spectrogram image has been firstly transformed with 2D- Fourier transform in order to be used to derive the circular blocks (spins). These spins are used to create the run length matrix, which is utilized to extract the final descriptors that are finally populated into a deep belief network classifier. The experimental results with the text-independent ELSDSR showed that the proposed system gave accuracy equal to 96.46% with feature vector size equal to 60 features. To evaluate the performance of the proposed SISR system using different descriptors, other types of descriptors can be generated from the spectrogram, e.g. Local Binary Pattern (LBP) based descriptors.

References

1. Kekre, H.B., Kulkarni, V., Gaikar, P., Gupta, N.: Speaker identification using spectrograms of varying frame sizes. Int. J. Comput. Appl. **50**(20), 27–33 (2012)
2. Dhakal, P., Damacharla, P., Javaid, A.Y.: A near real-time automatic speaker recognition architecture for voice-based user interface. Mach. Learn. Knowl. Extr. **1**, 504–520 (2019)
3. Chauhan, T., Soni, H., Zafar, S.: A review of automatic speaker recognition system. Int. J. Soft Comput. Eng. (IJSCE) **3**(4), 132–135 (2013)
4. Fandrianto, A., Jin, A., Neelappa, A.: Speaker recognition using deep belief networks. [CS 229] Fall, 14 December 2012. http://cs229.stanford.edu/proj2012/JinFandriantoNeelappa-SpeakerRecognitionUsingDeepBeliefNetworks.pdf. Accessed 20 Apr 2019
5. Dennis, J., Dat, T.H., Li, H.: Spectrogram image feature for sound event classification in mismatched conditions. IEEE Signal Process. Lett. **18**(2), 130–133 (2011)
6. Neammalai, P, Phimoltares, S., Lursinsap, C.: Speech and music classification using hybrid form of spectrogram and fourier transformation. In: APSIPA (2014)
7. Nguyen, Q.T., Bui, T.D.: Speech classification using SIFT features on spectrogram images. Vietnam J. Comput. Sci. **3**, 247–257 (2016)
8. Radionov, A., Aliev, V., Shvets, A.A.: Deep learning approaches for understanding simple speech commands. arXiv:1810.02364v1 [cs.SD] (2018)
9. Saady, M.R., El-Borey, H., El-Dahshan, E.S.A., Yahia, S.: Stand-alone intelligent voice recognition system. J. Signal Inf. Process. **5**(04), 70–75 (2014)
10. Bora, A., Vajpai, J., Sanjay, G.: Speaker identification for biometric access control using hybrid features. Int. J. Comput. Sci. Eng. (IJCSE) **9**(11), 666–673 (2017)
11. Soleymanpour, H.M.M.: Text-independent speaker identification based on selection of the most similar feature vectors. Int. J. Speech Technol. **20**, 99–108 (2017)
12. Padmaja, J.N., Rao, R.R.: A comparative study of silence and non silence regions of speech signal using prosody features for emotion recognition. Indian J. Comput. Sci. Eng. (IJCSE) **7** (4), 153–161 (2016)
13. Umbaugh, S.E.: Digital Image Processing and Analysis. CRC Press, London (2010)
14. Makandar, A., Halalli, B.: Image enhancement techniques using highpass and lowpass filters. Int. J. Comput. Appl. **109**(14), 12–15 (2015)
15. Farina, A.: Methods. Springer, Dordrecht (2014)
16. Baraa, A.K., Abdullah, N.A.Z., Abood, Q.K.: Hand written signature verification based on geometric and grid features. Iraqi J. Sci. **56**(2C), 1799–1809 (2015)

17. Patel, V., Mistree, K.: A review on different image interpolation techniques for image enhancement. Int. J. Emerg. Technol. Adv. Eng. **3**(12), 129–133 (2013)
18. Goshtasby, A.A.: Image Registration Principles Tools and Methods. Springer, London (2012)
19. Bondarenko, A., Borisov, A.: Research on the classification ability of deep belief networks on small and medium datasets. Inf. Technol. Manag. Sci. **16**, 60–65 (2013)
20. Pezeshki, M., Gholami, S.: Distinction between features extracted using deep belief networks, pp. 1–4. arXiv:1312.6157v2 [cs.LG] (2014)
21. English language speech database for speaker recognition (ELSDSR). http://www.imm.dtu.dk/~lfen/elsdsr/index.php?page=index. Accessed 20 Mar 2019

Convergence-Based Task Scheduling Techniques in Cloud Computing: A Review

Ajoze Abdulraheem Zubair[1,2](✉), Shukor Bin Abd Razak[1], Md. Asri Bin Ngadi[1], Aliyu Ahmed[1], and Syed Hamid Hussain Madni[1]

[1] School of Computing, Faculty of Engineering, Universiti Teknologi Malaysia, 81310 Skudai, Johor Bahru, Malaysia
azajoze-2068@graduate.utm.my,
{shukorar,dr.asri}@utm.my,
ahmedaliyu8513@yahoo.com, madni4all@yahoo.com
[2] Kogi State Polytechnic, P.M.B 1101 Lokoja, Kogi state, Nigeria

Abstract. The cloud computing promises various benefits that are striking to establishments and consumers of their services. These benefits encourage more business establishments, institutes, and users in need of computing resources to move to the cloud because of efficient task scheduling. Task scheduling is a means by which the tasks or job specified by users are mapped to the resources that execute them. Task scheduling problems in cloud, has been considered as a hard Nondeterministic Polynomial time (Np-hard) optimization problem. Task Scheduling is use to map the task to the available cloud resources like server, CPU memory, storage, and bandwidth for better utilization of resource in cloud. Some of the problems in the task scheduling include load-balancing, low convergence issues, makespan, etc. Convergence in task scheduling signifies a point in the search space that optimize an objective function. The non-independent tasks has been scheduled based on some parameters which includes makespan, response time, throughput and cost. In this paper, an extensive review on existing convergence based task scheduling techniques was carried out spanning through 2015 to 2019. This review would provide clarity on the current trends in task scheduling techniques based on convergence issues and the problem solved. It is intended to contribute to the prevailing body of research and will assist the researchers to gain more knowledge on task scheduling in cloud based on convergence issues.

Keywords: Cloud computing · Task scheduling · Optimization · Convergence · Resource management · NP-hard problem

1 Introduction and Motivation

Cloud computing has been the latest technology delivered as utility computing after Cluster and Grid computing in a way similar to the delivery of traditional utilities such as water, electricity, gas, and telephony [1–3]. The NIST (National Institute of Standards and Technology) defines cloud computing as, a model for facilitating or enhancing convenient, on-demand network access to a common pool of configurable

F. Saeed et al. (Eds.): IRICT 2019, AISC 1073, pp. 227–234, 2020.
https://doi.org/10.1007/978-3-030-33582-3_22

computing resources such as applications, storage, networks, servers and services that can be speedily provisioned and released with insignificant management efforts or service provider [4, 5]. Cloud computing has been term as a paradigm that housed both deployment and delivery models as well as the five essential characteristics [6–9]. The cloud computing promises various benefits that are striking to establishments and consumers of their services. These benefits encourage more business establishments, institutes, and users in need of computing resources to migrate to the cloud.

For decades, one of the most persistent challenges in cloud computing is the resource management. It is an exercise that has to do with the procurement, release and controlling of resources. In cloud settings, resources are virtualized, and shared among multiple cloud users [10–12]. However, the results of task scheduling in cloud computing were not encouraging [13]. Some of the task scheduling challenges include, load balancing, energy efficient, resource utilization, execution time and response time (Fig. 1).

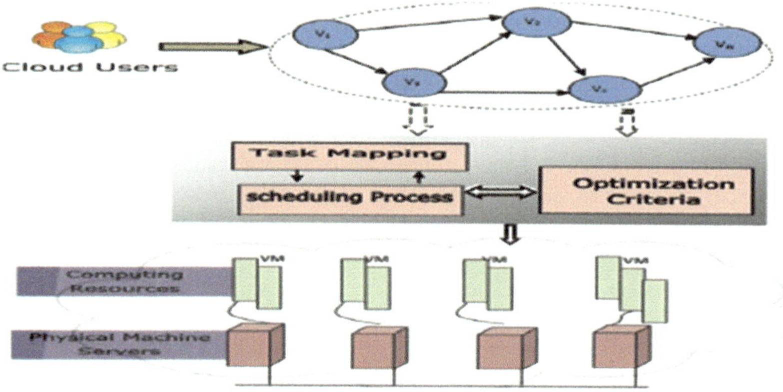

Fig. 1. Cloud task scheduling (Singh et al., 2017)

1.1 Problem Statement

Task scheduling in cloud computing has been considered as one of the most challenging issues in cloud resource management. It is worth to note that QOS-based optimization task scheduling is one of the important aspect of service level agreement (SLA) set by both cloud users and cloud service providers. From the literatures so considered, several QOS optimization techniques have been proposed to solve task-scheduling problems, which include convergence, imbalance and computational complexity among the selected nodes. For some time now, metaheuristic techniques which is classified as either bio-inspired and swarm intelligence have been deployed to solve task scheduling problems which is termed as NP-hard because it enforced some constraints. Most of the aforementioned techniques neither focus on how to attain global nor local solution. The poor performance or result obtained, suffered from either premature convergence or entrapment at local optimal.

1.2 Contribution of This Paper

The contributions of this review paper include-

- Stress the need to focus more on how to increase the speed of convergence in global search space.
- Enhance the metaheuristic technique by hybridizing with non-metaheuristic techniques to achieve a better solution

Section 2 is based on some related survey and review of researches in cloud task scheduling. Section 3 describe the detail review of literatures on convergence-based task scheduling techniques in cloud computing and its classification. Section 4 focuses on the open research issues and challenges in cloud task scheduling. Section 5 present the conclusion and recommendations.

2 Related Works

Task scheduling in cloud computing has been classified as an NP-hard problem because of the large-scale of heterogeneous workloads and resources which lead to a huge solution space and thus computational complexity. It is not seldom easy to find an optimal solution. It has been established that some of the existing techniques could not find feasible solution within a polynomial time to solve these complex problems. Metaheuristic-based techniques have proved to achieve near desired solution within a rational time. This paper therefore, tend to provide an extensive review and comparative analysis of different task scheduling techniques based on convergence issues in cloud computing environment.

A review has been conducted in the area of meta-heuristic scheduling approaches, which focus on three meta-heuristic techniques and two different algorithms. The technics includes Ant Colony Optimization (ACO), Genetic Algorithm (GA) and Particle Swarm Optimization (PSO), with other two while the two algorithms are League Championship Algorithm (LCA) and Bat Algorithm (BA) [14]. However, the review does not consider convergence issues [15]. Carried out a survey of league championship algorithm, which was based on a brief synopsis of LCA literatures in peer-review journals, conferences as well as book chapters. The survey highlighted some application of LCA such as solving a constrained optimization in mechanical engineering design, numerical function optimization and tackling a single machine scheduling nonlinear problem in Just-In-Time (JIT) system with batch delivery cost and different due dates. The paper equally highlighted some of the issues of the technique which include lack of generalization, robustness etc. However, the technique does not focus on convergence issues. A review on the existing load balancing techniques was conducted by [16]. The review focus on load balancing techniques that is prevalent in cloud computing which improves overall network signals, energy efficiency with enriched service delivery. The technique duel on cost minimization, reduction in response time, minimization of execution time and reduction of communication overhead using equitable load balancing techniques. However, the review does not in any way mention convergence issues.

[17], conducted a systematic review of resource allocation techniques which focus on the different resource schemes used by different researchers as well as the issues these techniques addressed and the metrics used to evaluate them. The review consider resource allocation as part of resource management, which enhanced the total performance of cloud infrastructure in terms of quality of service delivery as well as efficient resource utilization. However, the review does not focus on how to solve convergence issues. [18], conducted an extensive review of load balancing techniques. The review, which centered on the existing load balancing techniques and their associated issues and challenges such as geographical distribution of nodes, single point of failure, virtual machine migration, heterogeneous nodes, storage management, load-balancer scalability and algorithm computational complexity. Similarly, the review provides a comparative analysis of the articles used based on various metrics used by the techniques. These metrics include QoS metrics, single objective and multiple objectives metrics. However, the review does not consider convergence as an issue.

It is observed from the reviewed and surveyed papers that there is no iota of review about task scheduling techniques based on convergence issues.

3 Issues of Existing Convergence-Based Task Scheduling Techniques

The review of the existing techniques, which are mainly meta-heuristics, are classified into swarm intelligence and bio-inspired techniques as shown in Fig. 2 below.

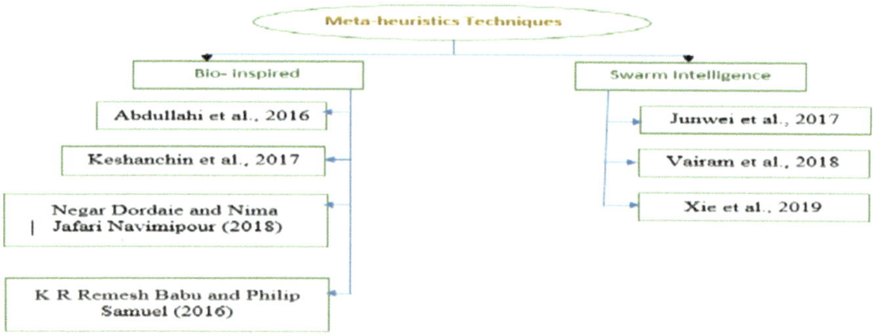

Fig. 2. The convergence techniques based literatures

3.1 Bio- Inspired Techniques

In [19] a hybridized symbiotic organism search technique for task scheduling in cloud computing location has been proposed. The technique employed simulated annealing (SA) with symbiotic organism search (SOS) to address the problem of task scheduling in cloud computing setting. Further, the combined technique optimize task scheduling by improving the convergence rate as well as quality of solution of SOS. Furthermore, a fitness function is proposed to equally address the problem of inefficient utilization of virtual machines (VMs).

Task scheduling in cloud computing environment using priority queues based on an improved genetic algorithm was propose by [20]. The propose technique named N-GA employs Genetic algorithm (GA) which is hybridized with a heuristic-based HEFT search, allocate subtasks to processors. Further, the projected specification, which are in the form of Linear Temporal Logic (LTL), were extracted. Furthermore, the combined techniques achieve a better execution time as well as minimization of makespan.

Task scheduling in cloud locations using a hybridized algorithm was propose by [21]. The techniques employed are Particle Swarm Optimization (PSO) and Hill Climbing Algorithm (HCA), which are meant to increase the convergent speed of candidate solutions in a population-based search space. Further, the hybrid techniques were able to accelerate search function to achieve a minimal job completion time.

An efficient load balancing techniques in cloud computing which was centered on an enhanced bee colony algorithm was proposed to solve the problems of imbalance among VMs, minimization of makespan as well as reducing the number of VM migration. Further, the prosed technique uses the foraging behaviour of honeybee for effective load balancing across the resource available. Furthermore, the task with smallest priority value are selected for migration to reduce the degree of imbalance among VMs. So tasks are not kept waiting unnecessarily before they are processed [22] (Table 1).

Table 1. Analysis of bio-inspired techniques in cloud task scheduling

Author	Focus of review	Techniques of review	Problem solved (metrics)	Nature of tasks	Environment	Remarks/limitations
Abdullahi et al., 2016	Metaheuristics	SOS	Makespan	Large-scale	Cloudsim	Much focus on the convergent capabilities of the technique employ
Keshanchin et al., 2017	Metaheuristics	N-GA	Task Execution time, Makespan	Workflows	C++ on Azure	No specific focus on convergent issues
Negar Dordaie and Nima Jafari Navimipour (2018)	Metaheuristics	PS&H	Makespan, Imbalance	Workflows	C++	Focus on convergence issues were not priority
K R Remesh Babu and Philip Samuel (2016)	Metaheuristics	EBC	Job completion time/Makespan	Heterogeneous cloudlets	Simulated Dynamic environment	No specific focus on convergent issues

3.2 Swarm Intelligence (SI)

An improved linear decreasing weight (LDW) with particle swarm optimization technique was proposed by [23]. The proposed technique, (LDW-PSO), prevent particle swarm optimization (PSO) technique falling into local convergence easily with minimum iteration or push. Further, the enormous complex problems were decompose into many smaller parts and distributed to several computing nodes for processing. Furthermore, the improved technique (LDW-PSO), tend to improve the computational defects in optimization and general performance.

A hybrid discrete particle swarm optimization technique for solving multiprocessor task scheduling problem was propose by [24]. The propose technique address the issue of premature convergence of Meta task in distributed system. Furthermore, the proposed work present a hybridization of HIDPSO with CSA namely Cyber HIDPSO (CHIDPSO) which will improve the diversity of the algorithm, such as minimization of makespan, mean flow time, cost reliability and resource utilization.

A unique directional and global convergent particle swarm optimization based on workflow scheduling in cloud-edge environment was proposed by [25]. The prose technique (DNCPSO) employs non-linear inertia weight with selection and mutation operations by directional search process, which reduces makespan, and cost drastically to obtain a feasible solution. Further, the scheduling problem of workflow applications provide a theoretical foundation for workflow scheduling strategy. Furthermore, the improved PSO enhanced the global and local search capabilities (Table 2).

Table 2. Analysis of Swarm Intelligence (SI) Techniques in cloud task scheduling

Author	Focus of review	Techniques of review	Problem solved (metrics)	Nature of Tasks	Environment	Remarks/limitations
Junwei et al., 2017	Metaheuristics	LDW-PSO	Improve local convergence	Independent	MATLAB 2010a GUI	Focus on convergence issues
Vairam et al., 2018	Metaheuristics	CHIDPSO	Makespan, mean flow time cost, reliability and resource utilization	Independent	JAVA Net Beans IDE	No specific focus on convergent issues
Xie et al., 2019	Metaheuristics	DNCPSO	Enhanced global and local search	Workflows	WorkflowSim 1.0	Much focus on the convergent capabilities of the techniques employ

4 Open Research Issues and Challenges in Cloud Task Scheduling

In the review literatures, most of the work done are based on hybridization of techniques to optimize the objective function. In providing the desired services to the consumers or cloud users, countless issues and challenges are encountered by the cloud providers. Consequently, an efficient scheduling technique needs to be provided by the cloud service providers that will satisfy both the cloud users and cloud service providers in terms of minimizing cost of processing task, reduce energy consumptions at cloud data centers, maximum utilization of cloud resource etc.

In task scheduling, the physical diverse resources are virtualize and mapped to various tasks, which facilitate the better performance of the cloud computing system by minimizing cost, execution time, makespan, throughput etc.

The challenges to task scheduling includes the diverse, dispersion and ambiguity of both workloads and resources in cloud computing locations. Therefore, it is robust to make cloud services and cloud-oriented applications more efficient by improving on the properties of the cloud locations.

5 Conclusion and Recommendations

There have not been any literature in cloud computing that consider convergence issues as among all the most important challenges in task scheduling. Most of the techniques employed dwell on either energy-based, cost-based, and load balancing based to enhance task scheduling. The solutions produced by these various existing meta-heuristic techniques can be enhance by considering hybrid approach in picking candidate solutions, final global optimal solution, and diversity of the solution or to modify the systems. These problems are therefore identify as open research issues that can be considered for future research.

Acknowledgment. This work was sponsored by the Nigerian Tertiary Education Trust Fund (TETFund) in collaboration with Kogi State Polytechnic Lokoja, Nigeria.

References

1. Rani, B.K., Rani, B.P., Babu, A.V.: Cloud computing and inter-clouds-types, topologies and research issues. Proc. Comput. Sci. **50**, 24–29 (2015)
2. Cheng, M., Li, J., Nazarian, S.: DRL-cloud: deep reinforcement learning-based resource provisioning and task scheduling for cloud service providers. In: Proceedings of Asia South Pacific Design Automation Conference ASP-DAC, January 2018, pp. 129–134 (2018)
3. Zhang, Q., Cheng, L., Boutaba, R.: Cloud computing : state-of-the-art and research challenges, pp. 7–18 (2010)
4. Mell, T., Grance, P.: The NIST Definition of Cloud Computing (2009)
5. Jarraya, Y., et al.: Securing the cloud, Ericsson Rev. English Ed., vol. 95, no. 2, pp. 38–47, 2017
6. Sasikala, P.: Research challenges and potential green technological applications in cloud computing. Int. J. Cloud Comput. **2**(1), 1–19 (2013)
7. Alkhater, N., Walters, R., Wills, G.: Telematics and informatics an empirical study of factors in fluencing cloud adoption among private sector organisations. Telemat. Inform. **35**(1), 38–54 (2018)
8. Rabai, L.B.A., Jouini, M., Ben Aissa, A., Mili, A.: A cybersecurity model in cloud computing environments. J. King Saud Univ.-Comput. Inf. Sci., **25**(1), 63–75 (2013)
9. Kratzke, N., Quint, P.: Understanding cloud-native applications after 10 years of cloud computing-a systematic mapping study. J. Syst. Softw. **126**, 1–16 (2017)

10. Arianyan, E., Taheri, H., Sharifian, S.: Novel energy and SLA efficient resource management heuristics for consolidation of virtual machines in cloud data centers. Comput. Electr. Eng. **47**, 222–240 (2015)
11. Zhou, J., Yao, X.: Multi-population parallel self-adaptive differential artificial bee colony algorithm with application in large-scale service composition for cloud manufacturing. Appl. Soft Comput. J. **56**, 379–397 (2017)
12. Singh, P., Dutta, M., Aggarwal, N.: A review of task scheduling based on meta-heuristics approach in cloud computing. Knowl. Inf. Syst. **52**(1), 1–51 (2017)
13. Achar, R., Thilagam, P.S., Shwetha, D., Pooja, H.: Optimal scheduling of computational task in cloud using virtual machine tree. In: 2012 Third International Conference Emerging Application Information Technology, pp. 143–146 (2012)
14. Kalra, M., Singh, S.: A review of metaheuristic scheduling techniques in cloud computing. Egypt. Inf. J. **16**(3), 275–295 (2015)
15. Abdulhamid, S.M., Latiff, M.S.A., Madni, S.H.H., Oluwafemi, O.: A survey of league championship algorithm: prospects and challenges. Indian J. Sci. Technol. **8**(February), 101–110 (2015)
16. Gabi, D., Samad, A., Zainal, A.: Systematic review on existing load balancing techniques in cloud computing. Int. J. Comput. Appl. **125**(9), 16–24 (2015)
17. Madni, S.H.H., Latiff, M.S.A., Coulibaly, Y., Abdulhamid, S.M.: Recent advancements in resource allocation techniques for cloud computing environment: a systematic review. Cluster Comput. **20**(3), 2489–2533 (2017)
18. Kumar, P., Kumar, R.: Issues and challenges of load balancing techniques in cloud computing. ACM Comput. Surv. **51**(6), 1–35 (2019)
19. Abdullahi, M., Ngadi, M.A.: Hybrid symbiotic organisms search optimization algorithm for scheduling of tasks on cloud computing environment. PLoS ONE **11**(6), 1–29 (2016)
20. Keshanchi, B., Souri, A., Navimipour, N.J.: An improved genetic algorithm for task scheduling in the cloud environments using the priority queues: formal verification, simulation, and statistical testing. J. Syst. Softw. **124**, 1–21 (2017)
21. Dordaie, N., Navimipour, N.J.: A hybrid particle swarm optimization and hill climbing algorithm for task scheduling in the cloud environments. ICT Express **4**(4), 199–202 (2018)
22. Snášel, V., Abraham, A., Krömer, P., Pant, M., Muda, A.K.: Innovations in bio-inspired computing and applications. In: Proceedings of the 6th international Conference on Innovations in Bio-inspired Computing and Applications (IBICA 2015), Kochi, India, 16–18 December 2015. Advances in Intelligent System and Computing, vol. 424 (2016)
23. Junwei, G., Shuo, S., Yiqiu, F.: Cloud resource scheduling algorithm based on improved LDW particle swarm optimization algorithm. In: Proceedings of 2017 *IEEE 3rd Information Technology* and *Mechatronics Engineering Conference* ITOEC 2017, January 2017, pp. 669–674 (2017)
24. Vairam, T., Sarathambekai, S., Umamaheswari, K.: Multiprocessor task scheduling problem using hybrid discrete particle swarm optimization. Sadhana - Acad. Proc. Eng. Sci. **43**(12), 1–13 (2018)
25. Xie, Y., et al.: A novel directional and non-local-convergent particle swarm optimization based workflow scheduling in cloud–edge environment. Future Gener. Comput. Syst. **97**, 361–378 (2019)

Imperative Selection Intensity of Parent Selection Operator in Evolutionary Algorithm Hybridization for Nurse Scheduling Problem

Huai Tein Lim[1(✉)], Irene-SeokChing Yong[2], and PehSang Ng[1]

[1] Department of Physical and Mathematical Science, Faculty of Science, Universiti Tunku Abdul Rahman, 31900 Kampar, Perak, Malaysia
limhuaitein@yahoo.com, psng@utar.edu.my
[2] University of Malaya, 50603 Kuala Lumpur, Malaysia
ireneching2017@um.edu.my

Abstract. Flexibility of shifts assignment in real-time condition is complex because it must consider multiple important aspects such as nurses' diverse requests and nurse ward coverage. In fact, creating nurses' work schedule is a time-consuming task and the created schedule may not be effective due to considerable dependence of the process on the head nurse's capability and working experiences. Thus, hospitals are becoming increasingly interested in the deployment of technology to solve the nurse scheduling problems. In the current research, three classifications of constraints namely the hard, semi-hard and soft constraints were implemented technically to refine undesirable work schedule in nurse scheduling. To deal with heavy constraints handling, this research implemented an enhancement of Evolutionary Algorithm with Discovery Rate Tournament parent selection operator (D_rT) to minimize constraints violations. Selection intensity resulted from hybridizing discovery rate of Cuckoo Search and tournament elements were used for exploration and exploitation. Correspondingly, three parent selections were tested, and D_rT parent selection was found to achieve the best accuracy which gives way to obtaining better quality schedule with lowest fitness value. In particular, the superiority of D_rT parent selection suggested that selecting elite parents and ensuring diverse characteristic of the selected parents in a population are especially useful in small-sized population.

Keywords: Evolutionary computation · Cuckoo search · Nurse scheduling problem · Parent selection · Selection intensity

1 Introduction

Unhealthy nurse workforce working environment in a hospital is not just contending with the issue of short supply of nurse workers. There are other equally challenging aspects to be considered in retaining the available scarce nurse workers. Among the notorious aspects that give rise to unhealthy nurse workforce working environment crisis are such as changes in the skill mix of nurses necessary to meet new service requirement [1], nurse burnout [2–4], personal pressure due to the ignorance of shifts

F. Saeed et al. (Eds.): IRICT 2019, AISC 1073, pp. 235–244, 2020.
https://doi.org/10.1007/978-3-030-33582-3_23

preferences and fairness [5, 6] and understaffing caused by uncertain absenteeism [7]. Under situations of unexpected circumstances or deficiency, a head nurse's fast and intellectual decision is critical to handle the real time working condition. A head nurse is also responsible to address subjective nurse preferences in situation where the automated nurse scheduling approaches are less effective. Hence, nurse scheduling research has been an ongoing effort since [8–14]. As a matter of fact, it is important that researchers keep working on automated nurse scheduling despite its complexity.

Generally, nurse scheduling deals with the allocation of a number of nurses to perform a set of shift duties within a time period, which is also subject to a number of constraints that need to be satisfied. Technically, nurse scheduling problem is a NP-hard problem which requires evolutionary computation heuristic [15, 16]. In evolutionary computation, exploration and exploitation of a search are contradictory. Therefore, population size, population selective pressure, population diversity, and randomization are carefully attended to balance the exploration and exploitation of search upon convergence issue. Thus, practically enhancing Evolutionary Algorithms (EA) in those aspects would be potential to address the nurse scheduling problem in real-time practice. This is the main aim of the study. In EA enhancement, parent selection operator in Evolutionary Algorithms shall be responsive to provide a diverse permutation space for crossover operators in order to preserve dissimilar individuals dwelling in a population. However, high selection pressure in a small population may cause great affection to premature convergence [17]. Thus, here goes a vivid sign of enhancing EA by studying the impact of elitism and dissimilarity between selected parents as the matter of controlling selective pressure.

In all, the research objective of this paper is to develop an efficient nurse scheduling model that could produce desirable nurse work schedule which satisfies nurse diverse requests in line with sustaining imperative nurse ward coverage. The function of the model is to minimize the penalty value of constraint violations while creating a schedule that satisfies all hard rules, semi-hard rules, and soft rules optimally within the capacity of nursing skills and staffing size. Generally, nurse schedule can be performed at two logical times: one before a schedule is put into place, and another during the implementation of the schedule. The former is known as offline schedule optimization which optimizes based on a set of assumptions to produce a final, static schedule. The latter is known as real-time schedule optimization which optimizes a schedule in progress (while it is being implemented) and allows dynamic updating of the schedule when more information becomes available to the optimization system [18]. In order to utilize the available nurses, this research takes schedule optimization into account, and the three classifications of constraints (i.e., hard, semi-hard, and soft) in nurse coverage are reviewed.

2 Nurse Coverage on Daily Basis

In this research, a series of interviews with the matron and head nurse of Cardiac Rehabilitation Ward (CRW) in Hospital Sultanah Bahiyah (HSB), Malaysia were conducted to obtain the data on daily nurse coverage in each shift. In addition, secondary data such as yearly record of implemented schedules, annual reports and official publication were also collected. As a conclusion, the numbers of nurses needed for operating CRW on a daily basis are shown in Table 1. The duties shifts include Morning Shift (M), Afternoon Shift (E), and Night Shift (N).

Table 1. Number of nurses needed in daily basis

CRW		Number of nurse needed
Total number of nurses		39
Senior nurse per shift		≥ 1
On duty shifts:	Morning shift (M)	5 to 7
	Evening shift (E)	5 to 7
	Night shift (N)	5 to 7

Next, a list of model constraints was considered where it was used to create a desirable schedule. Two types of constraint classifications, which are hard constraint and soft constraint, were implemented for this N-P hard problem as in [1, 6, 7]. The former was designed to conform to strict policies that determined the feasibility of a schedule. The latter was designed to conform to subjective personnel desires which were not necessarily be satisfied. To these extreme conditions, there is an awkward predicament which a semi-hard constraint can be relieved from the over strictness (i.e., hard constraint). In other words, the restriction was slightly loosened, yet it remained important. This is because the extreme of restriction in hard constraints may make the control over complexity of changes difficult in real time working environment. Such working principle was initially inspired by several earlier researches [19, 20], and it had also been implemented by a few researches thereafter in other applications [21, 22]. Therefore, additional classification with the three different groups of constraints (i.e., hard, semi-hard, and soft) were implemented in the current research to address nurse scheduling problem and to provide a better monitoring and control over the operation.

2.1 The Classification of Constraints

Constraints that are considered in the model are per listed below:

- 1 nurse only works 1 shift per day [hard constraint]
- Nurses are required to work 6 days a week [hard constraint]
- At least 6 nurses should be available for each Morning Shift and Evening Shift, meanwhile 5 nurses should be available for a Night Shift [hard constraint]
- At least 1 senior nurse is assigned to each work shift [hard constraint]
- Working more than 6 consecutive workdays is disallowed for each nurse [hard constraint]
- For each nurse, 2 off days are given to compensate for 3 consecutive Night Shift duties [hard constraint]
- A Night shift followed by a Morning Shift on the next day (N-M shift) is forbidden in any adjacent work shifts [hard constraint]
- The reassignment of shifts must adhere to the Forward Clockwise Direction rule (M < M/E/N/Off < N/E/Off < Off/N < Off) during the implementation [hard constraint]

- The range of 16%–21% of the total nurse is the tolerable coverage for each Morning Shift and Evening Shift [semi-hard constraint]
- Split off days or single work day is discouraged for nurses, unless the nurse has required the day off [semi-hard constraint]
- At least one weekend off is given to each nurse in a two-week schedule [semi-hard constraint]
- The ideal number assigned to each Morning Shift and Evening shift is 21% of total nurses, and a slightly smaller number, that is, 19% of total nurse, is ideally needed for a Night Shift for a ward [soft constraint]
- Consecutive off days arrangement for each nurse is preferable [soft constraint]
- Equivalent total number of M and E shifts ($[M + E]/2 \pm [1or2]$) is practiced for each nurse (by row) [soft constraint]
- Nurse request for off day should be approved, as it is his or her personal right [soft constraint]
- Stretch of consecutive off days is preferred for each nurse [soft constraint]

3 Construction of Evolutionary Algorithm

In the current research, Evolutionary Algorithm (EA) is implemented as the population-based heuristic with cuckoo search integrated at its parent selection operator. MATLAB R2010a language with Intel® Core™ i5-2410 M CPU @ 2.30 GHz and RAM 8 GB was used to program the EA. The detail of EA modelling is explained in the following sections.

Fundamentally, EA is an imitating evolutionary process which combines solutions in order to produce better solutions that eventually thrives on the survival of the fittest. The construction of EA involves parent selection, crossover, and mutation operators (see Fig. 1).

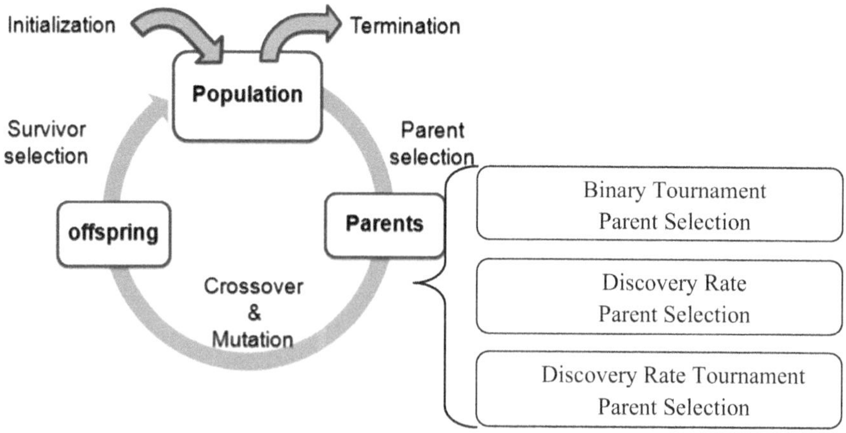

Fig. 1. Construction of EA with the proposed parent selection operators

The evaluation process yielded a multi-set of fitness values. This research work pinpoints the effectiveness of parent selection operator in EA. In this operator, EA does not create new individuals. Instead, the parent selection operator is concerned with the way of selecting some kinds of individuals from the initial population that can potentially produce a good child for the next generation. The selection work resembles a search path in a search space. In other words, parent selection is deliberately made considering population diversity, selective pressure, problem space or population size, and convergence. Logically, higher selection intensity can be an advantage for large problem spaces [23, 24]. However, small population size might lower the probability of enlarging population diversity [17]. Therefore, selection intensity is employed to compromise with the different level of population diversity. For the case of low diversity, ineffectively managing a small size of population and high selection pressure may lead to fast convergence (fast stuck in local optima).

However, much has been ignored about the mating strategy regarding the extent to which differences between selected parents would influence the outcome. In other words, it questions whether pair of diverse parents would have more potential to produce varying offspring than parents who look alike. This might be an important key of exploration in parent selection operator because it compromises flexible reproduction operators since most studies categorized a selection operator in an exploitation mode [25–27]. On the other hand, some studies overlooked this important aspect. They merely copied a number of individuals to a mating pool without considering any selection strategies [28, 29]. In fact, a parent selection operator may be responsive to provide some potential permutation space for reproduction operators. Therefore, handling diversity of population is a significant task which does not only generally preserve by replacement strategy, it indeed adapts to the population diversity by parent selection. Hence, controlling or managing a diversity search provides room to understand and enhance a parent selection operator.

Based on the discussion above, the current research chose Discovery Rate Tournament parent selection (D_rT) to execute elite selections as well as diverse selections simultaneously. Then, this D_rT parent selection was experimented with others two parent selection operators in EA to verify the performance of D_rT. These two parent selection operators are namely Discovery Rate parent selection (D_r) [17] and the well-known Binary Tournament parent selection (T) [30].

3.1 Binary Tournament Parent Selection

Binary Tournament parent selection (T) is a tourney of two or above randomly selected individuals from a population. The implementation of Binary Tournament selection is simple and does not involve fitness sorting. Although all individuals have the chance to be selected in order to preserve diversity, the tournament activity has its bias toward elite individual. In fact, tournament may influence the convergence speed. In this research, the classic Binary Tournament parent selection adopted from [30] was used to validate the performance of the newly proposed techniques.

3.2 Discovery Rate Parent Selection

In Cuckoo Search, discovery rate is the probability pa∈ [0, 1] when a host bird discovers alien eggs and subsequently build a completely new nest in a new location. Successful discovery of alien eggs happens when the Relative Difference (RD) is greater than the discovery rate [17]. Thus, Discovery Rate parent selection (Dr) is injecting relative difference into discovery concept in order to determine who shall be paired to proceed to the next recombination strategy. Through the alien egg discovery process, Discovery Rate parent selection (Dr) could study the impact of dissimilarity between selected parents.

The Discovery Rate parent selection (Dr) operator pinpoints the difference between selected parents in order to provide a diverse permutation space for reproduction operators. Approximately 40% of relative difference between two selected parents for a small population size is suggested by [17].

3.3 Discovery Rate Tournament Parent Selection

Besides maintaining exploration search of D_r parent selection [17] to produce Discovery Rate Tournament parent selection (D_rT), an additional focus of exploitation search was integrated into this selection intensity as well. In all, the characteristics of discovery rate tournament parent selection (D_rT) were shown as below.

(1) tournament that regards to elitism
(2) ensure a constant level of dissimilarity between the selected parents that regards to population diversity

The tournament was used to select the better fit individual as potential parents. The dissimilarity concept gives a practical advantage to D_r and D_rT parent selections. Both acclimatize to a population's condition regardless of the whole population's diversity. This condition reduced heavy computation. The detailed procedure of D_rT Parent Selection was shown in Fig. 2.

i. Randomly select two individuals from a population of size twelve
ii. Set a probability rate of discovery P_d and calculate the Relative Difference (RD), where the reference number falls on the individual who has bigger fitness value
iii. Discovery verification by comparing RD with P_d.
Discovered if RD $\geq P_d$, proceed to (iv)
Not discovered if RD $\leq P_d$, return the two individuals into the population and repeat (i) to (iii)
iv. In a tournament among the two, select the lower fit individual as parent 1. The other is sent back to the population
v. Repeat (i) to (iv) to get parent 2

Fig. 2. Procedure of discovery rate tournament parent selection

4 Experimental Result of Proposed Parent Selection

There was a need to determine the most suitable relative difference between parents in Discovery Rate parent selection series in order to proceed with further comparisons. Figure 3 presents the different levels of discovery rate d_r experimented by Discovery Rate Tournament Parent Selection.

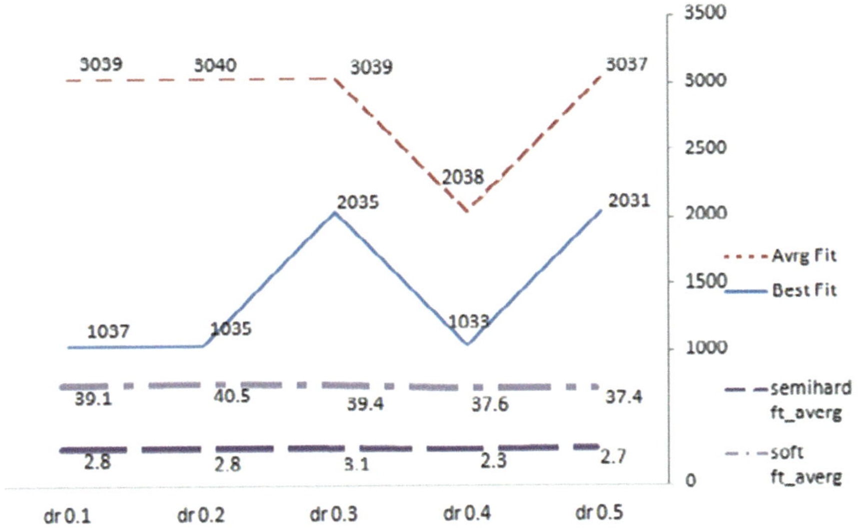

Fig. 3. The best fitness and average fitness at different levels of discovery rate d_r in D_rT parent selection

Selective pressure may supply or preserve population diversity. Relative difference rate is denoted as a selective pressure with the aim of determining an acceptable pressure during a selection. The fittest discovery rate d_r was 0.4 (see Fig. 3). This rate means that the difference of the selected parents (in terms of fitness) should be at least 40%. This level of pressure produced 1033 best fitness and 2038 average fitness over 20 runtimes. It resulted in violation of 2 semi-hard constraints and a penalty mark of 37.6 for soft constraints.

From the experiment, a suitable d_r may produce better fitness and diversity for the offspring. Approximately 40% of relative difference between the two selected parents for a small-sized population was suggested and applied to the experiments for comparing parent selection operators (in Sect. 4.1). In fact, d_r = 0.6 had produced fast convergence. The technique may be void because it was unable to counter the excessive high selective pressure. Hence, this outcome might be caused by the lower diversity in the population due to the small size of the population used. Thus, in the case of this research work, a high relative difference of 60% for selected parent was considerably negligible in this experiment.

4.1 Comparison of Parent Selection Operators

For a fair parent selection comparison, the experiment was implemented with the same Two-factor Blockwise crossover, directed mutation, steady-state replacement strategy, and fixed parameters (e.g., 100 Generations, 12 Population size, and 30 Runs) (Table 2).

Table 2. Output of EAs with three different parent selections

Parent selection	T	Dr	DrT
Best fitness	3040	3041	2033
Time (seconds)	129.77	135.23	153.07
Convergence level	15	93	39
Average fitness	5049	5050	5046
STD ('000)	1.78316	1.06143	1.55575
Feasible rate	12/30	10/30	10/30

In terms of best fitness, D_r parent selection produced mediocre results which were slightly inferior to T parent selection (i.e. 3041 − 3040, D_r was defeated by 1 penalty value of soft constraint violated or ((40– 41)/41) = 2.4% loss of soft constraint). One possible reason could be that the mating strategy which merely emphasized the diversity of parental gene alone was not enough. Therefore, as an advance of D_r parent selection operator, D_rT parent selection achieved the best accuracy by obtaining the lowest 5046 average fitness and 2033 best fitness. The created schedule (i.e. solution of 2033 best fitness) was able to grant all requested off days. The superiority of D_rT parent selection was suggestive of the importance of selecting elite parents and the diverse characteristic of the selected parents.

Nevertheless, one common drawback of these three different parent selection models was the lack of reliability. The small feasible rates which range from at least 33.3% (i.e., 10/30 * 100) to 40% (i.e., 12/30 * 100) indicate that T, Dr, and DrT were less reliable in generating feasible solution successfully in each run. This condition could be improved when suitable recombination operators were taken into further consideration.

5 Conclusion

A significant computerized scheduling system would reduce nurse's idle time and head nurse's schedule planning time. Hence, this research could aid in health care cost reduction (counted in term of time). Particularly, this model had included nurse fairness and preferences elements which were intended to reduce internal conflict amongst nurses. The model also enabled better chances of having significant Request Off day among nurses, equivalent chances of having significant weekend off day among nurses, and a balanced number of Morning Shift and Evening Shifts for each nurse. The fairness that is the average number of nurses for each type of shifts was also advocated.

Essentially, exploration and exploitation of EA are in a contradictory nature. Each of them yields to the opposite extreme act of convergence, that is, either it is fast or slow. Hence, EA balances between two. For this reason, EA hybridization in this research had carefully considered the population diversification, selective pressure, randomization principle, and convergence issue through the enhancement of parent selection operator and crossover operator.

Technically, Discovery Rate, Discovery Rate Tournament, and Tournament parent selection operators were studied for assessing population diversity in EA. To note, high selective pressure towards elite parents may result in premature convergence, and hence losing the population diversity, as demonstrated in EA with Tournament parent selection which causes great impact on fast convergence. Correspondingly, searching strategy that comprises more exploration could produce better performance.

References

1. Bard, J.F., Purnomo, H.W.: Short-term nurse scheduling in response to daily fluctuations in supply and demand. Health Care Manag. Sci. **8**, 315–324 (2005)
2. Kalisch, B.J., Aebersold, M.: Interruptions and multitasking in nursing care. Joint Comm. J. Qual. Patient Saf. **36**(3), 126–132 (2010)
3. Sangai, J., Bellabdaoui, A.: Workload balancing in nurse scheduling problem models and discussion. In: 2017 International Colloquium on Logistics and Supply Chain Management (LOGISTIQUA), pp. 82–87. IEEE, April 2017
4. Shahriari, M., Shamali, M., Yazdannik, A.: The relationship between fixed and rotating shifts with job burnout in nurses working in critical care areas. Iran. J. Nurs. Midwifery Res. **19**(4), 360–365 (2014)
5. Gormley, D.K.: Are we on the same page? staff nurse and manager perceptions of work environment, quality of care and anticipated nurse turnover. J. Nurs. Manag. **19**, 33–40 (2011)
6. Youssef, A., Senbel, S.: A bi-level heuristic solution for the nurse scheduling problem based on shift-swapping. In: 2018 IEEE 8th Annual Computing and Communication Workshop and Conference (CCWC), pp. 72–78. IEEE, January 2018
7. Clark, A.R., Walker, H.: Nurse rescheduling with shift preferences and minimal disruption. J. Appl. Oper. Res. **3**(3), 148–162 (2011)
8. Aickelin, U., Dowsland, K.A.: Exploiting problem structure in a genetic algorithm approach to a nurse rostering problem. J. Sched. **3**(3), 139–153 (2000)
9. Azaiez, M.N., Al Sharif, S.S.: A 0-1 goal programming model for nurse scheduling. Comput. Oper. Res. **32**, 491–507 (2005)
10. Glass, C.A., Knight, R.A.: The nurse rostering problem: a critical appraisal of the problem structure. Eur. J. Oper. Res. **202**(2), 379–389 (2010)
11. Van den Bergh, J., Beliën, J., De Bruecker, P., Demeulemeester, E., De Boeck, L.: Personnel scheduling: a literature review. Eur. J. Oper. Res. **226**(3), 367–385 (2013)
12. Wu, T.H., Yeh, J.Y., Lee, Y.M.: A particle swarm optimization approach with refinement procedure for nurse rostering problem. J. Comput. Oper. Res. **54**, 52–63 (2015)

13. Karmakar, S., Chakraborty, S., Chatterjee, T., Baidya, A., Acharyya, S.: Meta-heuristics for solving nurse scheduling problem: a comparative study. In: 2016 2nd International Conference on Advances in Computing, Communication, & Automation (ICACCA) (Fall), pp. 1–5. IEEE, September 2016
14. Bunton, J.D., Ernst, A.T., Krishnamoorthy, M.: An integer programming based ant colony optimisation method for nurse rostering. In: 2017 Federated Conference on Computer Science and Information Systems (FedCSIS), pp. 407–414. IEEE, September 2017
15. Tein, L.H., Ramli, R.: Recent advancements of nurse scheduling models and a potential path. In Proceedings of 6th IMT-GT Conference on Mathematics, Statistics and its Applications (ICMSA 2010), pp. 395–409, November 2010
16. Burke, E.K., Curtois, T.: New approaches to nurse rostering benchmark instances. Eur. J. Oper. Res. **237**(1), 71–81 (2014)
17. Lim, H.T., Ramli, R.: Enhancements of evolutionary algorithm for the complex requirements of a nurse scheduling problem. In: Proceedings of the 3rd International Conference on Quantitative Sciences and Its Applications (ICOQSIA 2014), vol. 1635, pp. 615–619. American Institute of Physics Conference Series (AIP) Publishing (2014)
18. Aggour, K.S., Moitra, A.: Advances in schedule optimization with genetic algorithms. GE Global Research, GRC111 (2003)
19. Muntz, A.H., Wang, K.: Workload model specifications and adaptive scheduling of semi-hard real-time controls. In: Proceedings of the First International Conference on Systems Integration, pp. 403–414. IEEE (1990)
20. Grandoni, F., Könemann, J., Panconesi, A., Sozio, M.: Primal-dual based distributed algorithms for vertex cover with semi-hard capacities. In: Proceedings of the Twenty-Fourth Annual ACM Symposium on Principles of Distributed Computing, pp. 118–125. ACM (2005)
21. Kelemen, A., Franklin, S., Liang, Y.L.: Constraint satisfaction in "conscious" software agents- a practical application. Appl. Artif. Intell. **19**, 491–514 (2005)
22. Abdallah, K.S., Jang, J.: An exact solution for vehicle routing problems with semi-hard resource constraints. Comput. Indu. Eng. **76**, 366–377 (2014)
23. Hutter, M., Legg, S.: Fitness uniform optimization. IEEE Trans. Evol. Comput. **10**(5), 568–589 (2006)
24. Kazimipour, B., Li, X., Qin, A.Q.: A review of population initialization techniques for evolutionary algorithms. In: IEEE Congress on Evolutionary Computation (CEC), pp. 2585–2592. IEEE (2014)
25. Ashlock, D.: Evolutionary Computation for Modeling and Optimization. Springer, USA (2005)
26. Al-Naqi, A., Erdogan, A.T., Arslan, T.: Fault tolerance through automatic cell isolation using three-dimensional cellular genetic algorithms. In: 2010 IEEE Congress on Evolutionary Computation (CEC), pp. 1–8. IEEE, New York (2010)
27. Veerapen, N., Maturana, J., Saubion, F.: An exploration-exploitation compromise-based adaptive operator selection for local search. In: Proceedings of the 2012 Genetic and Evolutionary Computation Conference (GECCO), pp. 1277–1284 (2012)
28. Tsai, C.C., Li, S.H.: A two-stage modeling with genetic algorithms for the nurse scheduling problem. Expert Syst. Appl. **36**(5), 9506–9512 (2009)
29. Yang, F.C., Wu, W.T.: A genetic algorithm-based method for creating impartial work schedules for nurses. Int. J. Electr. Bus. Manag. **10**(3), 182 (2012)
30. Burke, E.K., Smith, A.J.: Hybrid evolutionary techniques for the maintenance scheduling problem. IEEE Trans. Power Syst. **15**(1), 122–128 (2000)

Detection of Cirrhosis Through Ultrasound Imaging

Karan Aggarwal[1,2](✉), Manjit Singh Bhamrah[2], and Hardeep Singh Ryait[3]

[1] Electronics & Communication Engineering Department, M.M. Deemed to be University, Mullana 133207, Haryana, India
karan.170987@gmail.com

[2] Electronics & Communication Engineering Department, Punjabi University, Patiala 147002, Punjab, India
manjitsingh_manjit@rediffmail.com

[3] Department of Electronics & Communication Engineering, BBSBEC, Fatehgarh Sahib 140407, Punjab, India
hardeepsryait@gmail.com

Abstract. Cirrhosis is a liver disease that is considered to be among the most common diseases in healthcare. Due to its non-invasive nature, Ultrasound (US) imaging is a widely accepted technology for the diagnosis of this disease. This research work proposed a method for discriminating the cirrhotic liver from normal liver through US images. The liver US images were obtained from the radiologist. The radiologist also specified the Region of Interest (ROI) from these images and then the proposed method was applied on it. Two parameters were extracted from the US images through differences in intensity of neighboring pixels. Then these parameters can be used to train a classifier by which cirrhotic region of test patient can be detected. A 2-D array was created by difference in intensity of the neighboring pixels. From this array, two parameters were calculated. The decision was taken by checking these parameters. The validation of the proposed tool was done on 80 images of cirrhotic and 30 images of normal liver and classification accuracy of 98.18% was achieved. The result was also verified by the radiologist. The results verified its possibility and applicability for high performance cirrhotic liver discrimination.

Keywords: Cirrhotic liver · Ultrasound image · Intensity difference · Support vector machine · Region of interest

1 Introduction

World Health Organization (WHO) states that hepatitis C caused by hepatitis C virus is a liver disease. Chronic hepatitis C infection infected globally, approx 130–150 million people [1]. A significant number of people will develop liver cirrhosis due to poor medical attention. The last phase of chronic hepatopathies is considered to be cirrhosis which often moves towards hepatocellular carcinoma [2]. This disease is diagnosed by checking at the liver parenchyma's structure i.e. granular and the liver surface's aspects such as its roughness and contours [3].

F. Saeed et al. (Eds.): IRICT 2019, AISC 1073, pp. 245–258, 2020.
https://doi.org/10.1007/978-3-030-33582-3_24

Biopsy is observed as "golden standard" method which is used for diagnosing and treatment of all liver diseases [4]. It is not easily accepted by the patients due to in-vivo in nature. When all non-invasive techniques were observed out of these: US is the most general and maximum used imaging modality for diagnosing the liver disease because: (a) it is not expensive (b) no harmful radiation emitted from it (c) easily and widely availability of it (d) it has high sensitivity [5]. The main drawback of it is the dependency on the knowledge and experience of the operator i.e. Radiologist [6]. The additional techniques or systems are being developed as Computer Aided Diagnosis systems (CAD) which will reduce the dependency of the operator to get reproducible results [7, 8]. Therefore, developing CAD systems is so much important as it detects liver disease at an early stage. Due to this system, patients could save their life from unwanted nervousness, the probability of improvement increased and the cost required for offering such treatments for advanced liver diseases, reduced [9, 10].

Therefore, ultrasonography is the most preferred inspection technique used in CAD systems [11]. Here, the radiologist also described that the texture of the body tissues gave a significant visible belongings in classification of results in their radiological importance [12]. Texture of an image can be categorized by coarseness, roughness, granulation in surface, randomness and irregularity [13]. All these features were described by a spatial and intensity array of pixels in the textural US images [14, 15]. The normal liver has regular texture and cirrhosis liver has an irregular texture as shown in Fig. 1. This irregularity occurred due to sores, eruption as well as roughness on the surface of cirrhotic liver [16].

The scar tissue distorted the architecture of normal liver, which joined the perivenous and periportal areas through the group of connecting tissues [18]. This process may move towards to complete cirrhosis. Now, it is very difficult work to detect the change when ultrasound screening or scanning happens [19]. If a technique can be designed based on the theory of image processing, then early detection of disease is possible.

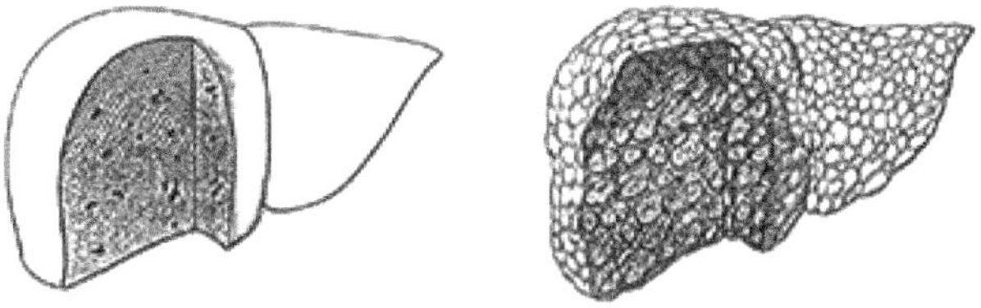

(a) (b)

Fig. 1. (a) Normal Liver (b) Cirrhotic Liver: given by NIDDKD, National Institutes of Health [17]

In this paper, a tool is designed for categorization of normal and cirrhotic liver through US images. In this, texture of US images is analyzed by proposed parameters and results are validated by support vector machine classifier. The results are also validated by the radiologist.

2 Literature Survey

Yeh et al. [20] presented a method to distinguish the images of liver non-steatosis and steatosis from high frequency ultrasound with SVM classifier. Two features were extracted i.e. non-separable wavelet transform features and GLCM features and accuracy 90.5% was obtained. Kadah et al. [21] proposed features to differentiate the fatty, normal and cirrhosis US images with KNN classifier. 100% sensitivity and 88.9% specificity were obtained. Graif et al. [22] presented an algorithm that determined the decrease in back scatters echo amplitude. This amplitude was the function of beam depth. This automatic algorithm is called far field slope algorithm. The US images of diffuse and normal liver diseases were non-compensated. 67% Sensitivity and 77% specificity were obtained by this algorithm. In radio frequency study [23], the extracted images were speckle and despeckle image with anatomic details. Bayes classifier was used on ten fatty liver and ten normal liver samples and obtained 95% accuracy.

Wu et al. [24] presented a system that was evolution based on hierarchical features of fusion system. In this system, KNN classifier was used and obtained 95.05% accuracy. Wu et al. [25] proposed optimization algorithm that designed the features. It helped to diagnose the automatic liver cirrhosis. The feature selection and extraction through US images were done by the self organization properties of genetic algorithm. An AdaBoost classifier is used to detect liver cirrhosis based on patient's database and US image. Murula et al. [26] proposed three parameters to distinguish the fibrosis and healthy subgroups in liver tissues. These three parameters were US attenuation, speed, coefficient and integrated backscatter coefficient (IBC). Then a discriminated analysis was done on these parameters to detect fibrotic groups. Lucieer et al. [27] proposed CAD system in which features were derived through US images of liver and accompanying spleens. The statistical analysis had been improved by incorporating all the old techniques i.e. dimension reduction, fractal dimension, non -parametric. Vermani et al. [28] presented computer aided diagnostic system in which B-mode US images were used. This system was evaluated on the images of hemangioma and metastatic carcinoma lesion [29]. The texture features were extracted through inside and outside regions of lesions. The optimal principal components were find out on the feature set by Principal component analysis [30, 31]. The components were classified by SVM.

Kyriacou et al. [32] presented 11 features to detect the normal, fatty and cirrhosis liver through US images. These features were Gray Level Difference Statistics (GLDS), Gray level Run Length Statistics (RUNL), Spatial Gray Level Dependence Matrices (SGLDM) and Fractal Dimension Texture Analysis (FDTA). The region of interest had the size of 32 $\times$ 32 pixels. The KNN classifier was used on the combination of FDTA and SGLDM features and obtained 82.2% accuracy. The author added one more dataset in this same study i.e. set of hepatoma images [33]. Then investigators got 80% accuracy with KNN classifier by using combination of three parameters i.e. RUNL, SGLDM and FDTA. With the use of neural network classifier, 82.67% accuracy was obtained.

Badawi et al. [34] presented eight features to distinguish between the normal, fatty and cirrhotic liver US images with fuzzy classifier. This study was evaluated on 140 cases of US liver images and obtained sensitivity 96% for the fatty liver classification.

Jiuqing Wan and Sirui Zhou proposed WPT of the cirrhotic and normal US liver images [35]. With the use of WPT, total 32 parameters were extracted. The dataset was of 390 normal and 200 cirrhotic livers US images. The data was classified with the use of SVM classifier and 85.79% accuracy obtained. Lee et al. [36] presented feature that is found on M-band wavelet transform. The cirrhotic, normal and hepatoma liver US images were used for this feature vector with hierarchical classifier. Accuracy of 96.7% was obtained in classification of normal and abnormal liver US images. Acharya et al. [37] proposed the features through the liver US images which was deploy on the combination of HOS, DWT and image texture features. Accuracy of 93.3% was achieved through these significant features by decision trees classifier for normal and FLD-affected abnormal livers classification.

Literature emphasizes that some other parameters are required to analyze the texture that creates the system more accurate and feasible.

3 Methodology

The US images were taken from radiology department of Rajindra Hospital, Patiala, Punjab under guidance of a skilled radiologist. Radiologist also specified the areas or regions in every US image which belongs to cirrhosis/normal as shown in Fig. 2. These areas are called Region of Interest (ROI). These selected regions were taken as templates of dimension 50×50 pixels from every US image of both cirrhotic/normal livers as in Fig. 2(a–d) and then proposed method was applied on it.

The total liver US images were 110. Out of these, 80 images of cirrhotic patients and 30 images of normal liver patients. The images were categorized in two categories i.e. cirrhotic and normal liver having dimension of 381 by 331 in JPG format. Every image was taken from one US instrument to abolish the property of alter in texture values and dimensions on the image. The difference in the dimensions of the images can be well observed if they were captured by different sources. The US probe having 15 cm depth and 5 MHz frequency are used for acquiring these images.

A novel tool is developed that will help to differentiate the US image of liver between cirrhotic and normal.

US image of normal liver has smooth surface or regular texture and cirrhotic liver has irregular texture or rough surface. Regular texture is nothing but the difference between neighborhood pixels intensity is very small. While irregular texture having intensity difference is high among neighboring pixels. To keep this in mind, an intensity difference technique is developed.

In this technique, the difference has been found out between the two neighboring pixels. To start with the first row in ROI, the difference of the first and the second pixel became the first element in 2-D array. Then, the difference of second and third pixel became the second element. Similarly, difference of third and forth pixel became third element and so on. Through this, the first row is created in the 2-D array. Similarly, the difference of the pixels from second row became the second row and third row and so on. Due to this, a 2-D array created.

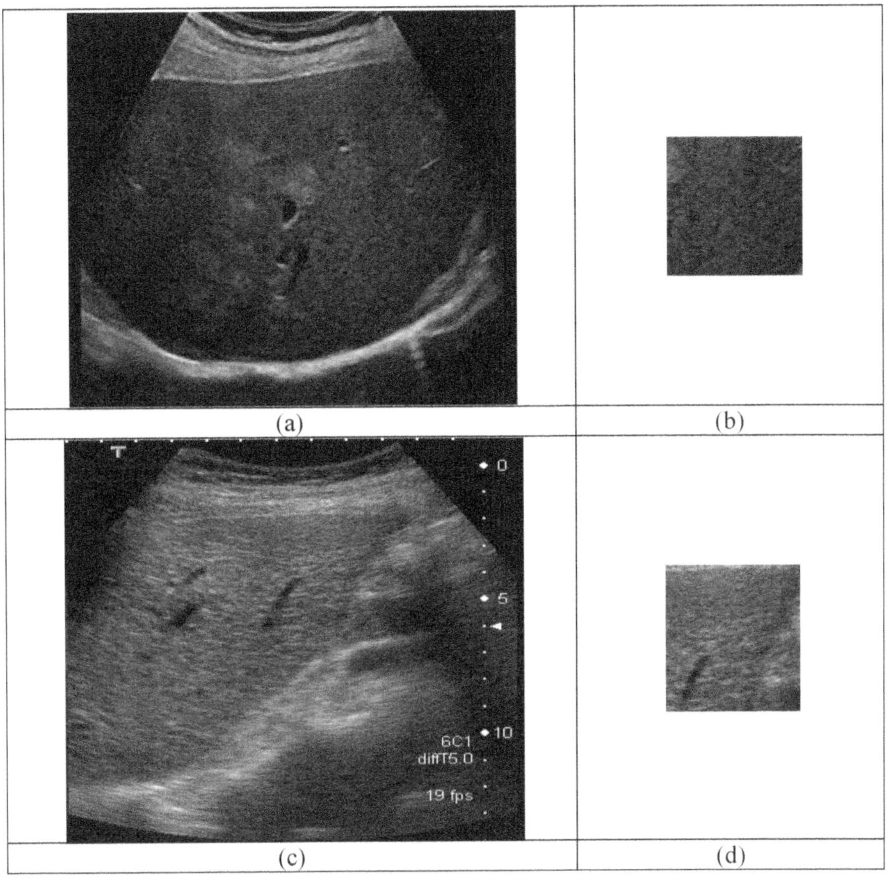

Fig. 2. (a) US image of normal liver (b) Region of Interest from normal liver (c) US image of cirrhotic liver (d) Region of Interest from cirrhotic liver

The mean and standard deviation are calculated from this array. The mean showed the average of all the pixel intensity in the ROI. These two values become the threshold values. The standard deviation showed the variation of the pixel intensity in the ROI. It means how much intensity varies. So, first threshold value is mean of the array and second one is standard deviation of the array.

The array elements are checked with these threshold values. With the use of first threshold value, all the array elements are checked and counted the array elements from the whole array which have value greater than threshold value. If the counted array elements are less in number, it depicts that difference between the pixels intensity is small then the image is uniform or having regular texture. While if number is high i.e. difference between the pixels is high then image is having irregular texture or rough surface. So, one parameter is extracted called array elements.

Now, value of the array elements is checked. If element value is higher than threshold value then replace the both pixels intensity values (minuend and subtrahend)

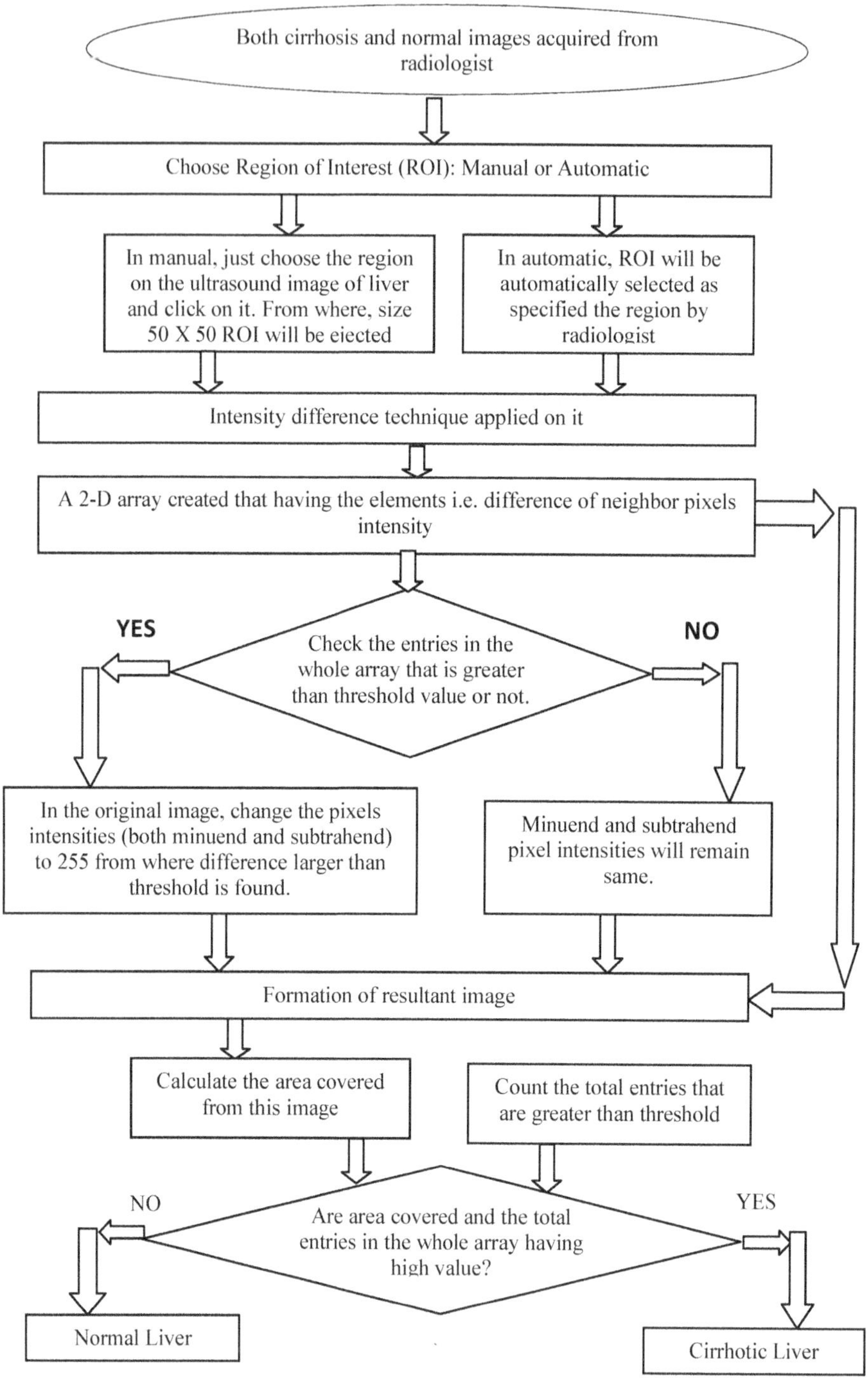

Fig. 3. Flow Chart of the proposed approach

by 255 in the original ROI image. But if it is lower then there is no change in both pixel intensities (minuend and subtrahend) as shown in Fig. 5. Due to this, resultant image is formed. From this image, second parameter is calculated called area covered.

By using these two parameters, judgment is taken that US image of liver is having regular or irregular texture i.e. normal and cirrhosis. The complete process is shown in Fig. 3.

To validate the results, support vector machine classifier is used. The leave-M-out cross validation was used. The two parameters i.e. array elements and area covered, were used with different threshold value for SVM classification.

Algorithm:

1. Image taken of both cirrhotic and normal liver.
2. ROI selection: manually and automatically
3. In manually, just select the region from the US image of liver. (Manual selection will be done under supervision of person that knows about the US images of liver as shown in Fig. 4(a)).
4. ROI will be automatically selected as specified the region by radiologist as shown in Fig. 4(b).
5. After ROI selection, intensity difference technique applied on it.
6. Due to this technique, a 2-D array created that having the elements i.e. difference of neighbor pixels intensity.
7. Count the elements in the array that is greater than threshold value. Threshold value is standard deviation and mean of the array.
8. If array element is higher than threshold then replace the both pixels intensity i.e. minuend and subtrahend by 255 in the original ROI image. But if it is lower then there is no change in the minuend and subtrahend pixel intensities in the original ROI image. So, the final image formed.
9. From the array, the total array elements are counted that are greater than 15 and from the resultant image, area covered is calculated.
10. By using this value and area, conclusion is taken that US image of liver is of cirrhosis and normal.

In this, there is an image button from where US image of liver are selected. US images of liver from radiologist act as input. There are 80 images of cirrhotic and 30 images of normal liver. Any image can select from these images that will act as input. There is also an image window that will show the selected image as shown in Fig. 4(a) and (b).

Then, next step is Region of Interest (ROI) selection from this US image. There is top down list which have two options; one is manual and other is automatic. From this list, any method can be selected.

In manual method, ROI selection is done by person itself. The person can just select appropriate region from the US images of liver with their knowledge. It might be defected. The person can just click on the region and 50×50 size ROI ejected from the image. In automatic method, region specified by the radiologist in the image that belongs to the cirrhotic is automatically ejected. There is also another ROI window that will show the selected ROI from the image.

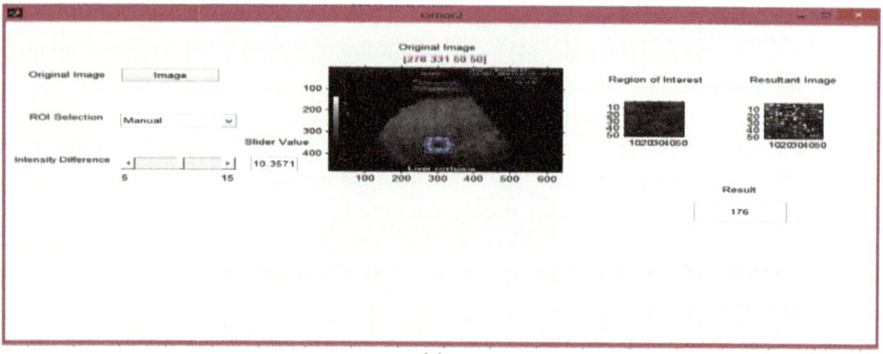

(a)

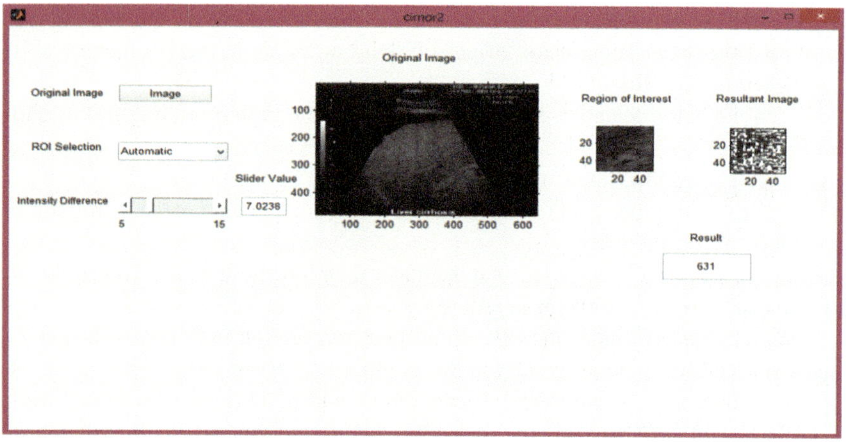

(b)

Fig. 4. (a) GUI for manual ROI selection (b) GUI for automatic selection

As the threshold value selected from this slider, intensity difference technique is applied on the ROI. This technique gives three things; one is an array having elements of difference between neighboring pixels, second is formation of new resultant image and third is sum value. There is also another resultant image window that will show the resultant image. This sum value box shows the sum value i.e. how many elements in the array are greater than threshold value.

In the resultant image, the both pixel intensities in the original ROI (minuend and subtrahend) become '255' which having the difference more than threshold value and remaining pixels will be same.

This '255' pixel intensity in resultant image depicts the cirrhotic area. In cirrhotic image, surface is rough or irregular i.e. difference between neighborhood pixels intensity will be large. So, almost full ROI will be enclosed by pixel intensity '255'. While in normal image, surface is regular i.e. difference between neighborhood pixels will be very low. So, pixel intensity '255' in the ROI will be very less. So, area covered as well as array elements becomes the most appropriate parameter to differentiate the liver that is cirrhotic and normal.

4 Results and Discussion

ROI of size 50 × 50 has been selected as templates. Therefore, one ROI was chosen from one US image i.e. total of 30 templates from 30 normal liver images and eighty templates were chosen from eighty cirrhosis liver images.

Through the intensity difference technique, an array of difference and a resultant image were generated as shown in Fig. 5.

This array contains the differences among the neighboring pixels. These differences represent the intensity variation from their adjacent pixel. If the difference is small, it means variation of intensity between adjacent pixels is small and vice-versa. If all the differences in the whole array are small then it shows that variation in intensity of pixels in the ROI image is small. If there is a small deviation in the pixel intensity in the ROI image then the image is smooth or surface is regular. While, the differences between the pixels in the array are high then it shows that variation of intensities in the ROI image is high. If there is high variation in pixels intensities in the ROI image then it shows that image is rough or surface is irregular. To check the differences, array was checked and counted the elements that having value greater than threshold value.

Threshold value is nothing but it creates a level. Below this level, image is regular and above this level, image is irregular. For that, two threshold values are taken from the array i.e. mean and standard deviation. The array elements that having value greater than mean of the array depict the irregularity of the surface and array elements that having value smaller than mean of the array depict the regularity in the surface.

The number of array elements that are greater than threshold value will represent that image is rough or smooth and image belongs to cirrhotic or normal. If the number is high, it means that image belongs to cirrhotic or vice versa.

In the final image, the 255 intensity value depicts that difference between neighboring pixels value are greater than threshold. In cirrhotic image, intensity differences between neighboring pixels are large. So, almost complete ROI is covered by 255 pixel intensity value. While in normal image, intensity differences between neighboring pixels are very small. So, 255 pixel intensity value in the ROI is very less.

In this paper, two parameters are used to differentiate cirrhotic liver and normal liver for 110 patients. Two parameters are area covered and array elements. These parameters are released for two different threshold values. However, amalgamation of one parameter for two thresholds moves to encouraging classification results. Parameters with lower value in each threshold have higher impact on normal liver side and vice-versa.

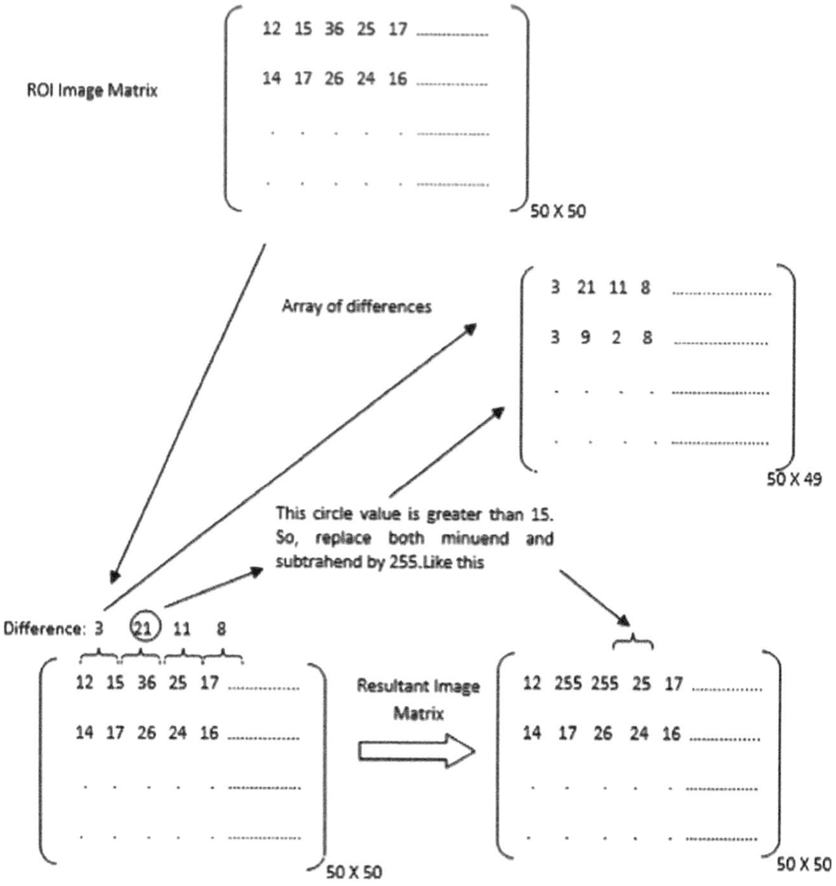

Fig. 5. Intensity Difference Technique

For statistical interpretation, widely accepted classifier is used i.e. Support Vector Machine (SVM) classifier. SVM considers the best hyperplane that have the largest margins between the two classes. The best hyperplane distinguish the data from one class to another. Margin means maximum thickness of the slab parallel to the hyperplane. In this, there are no data points inside.

Accuracy of this classifier will justify our methodology and decision.

To make sure our classification is unbiased toward parameters value through the cirrhotic ROIs, normal ROIs also incorporated in our examination. For 80 cirrhotic patients, 80 cirrhotic ROIs and for 30 normal patients, 30 normal ROIs were selected. The proposed technique in this research work is performed well to characterize the cirrhotic and normal liver.

Total 110 US liver images were used in experimentation. The area covered and array elements with two threshold values were taken from all images and then SVM applied on these values.

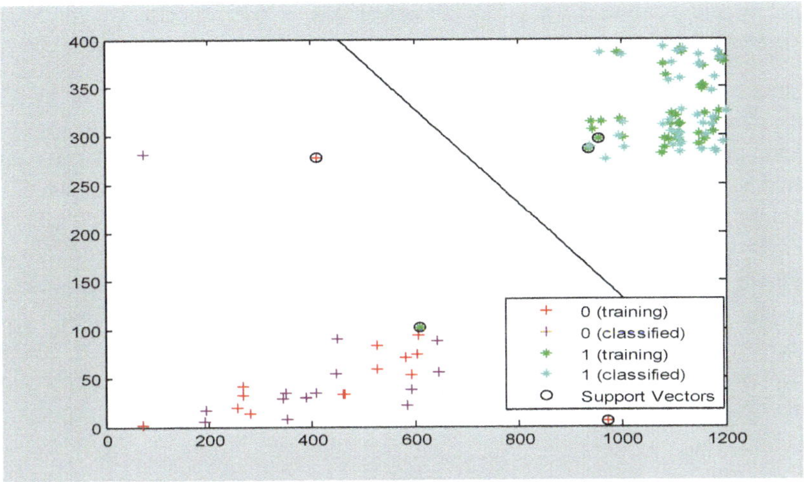

Fig. 6. SVM classification for array elements

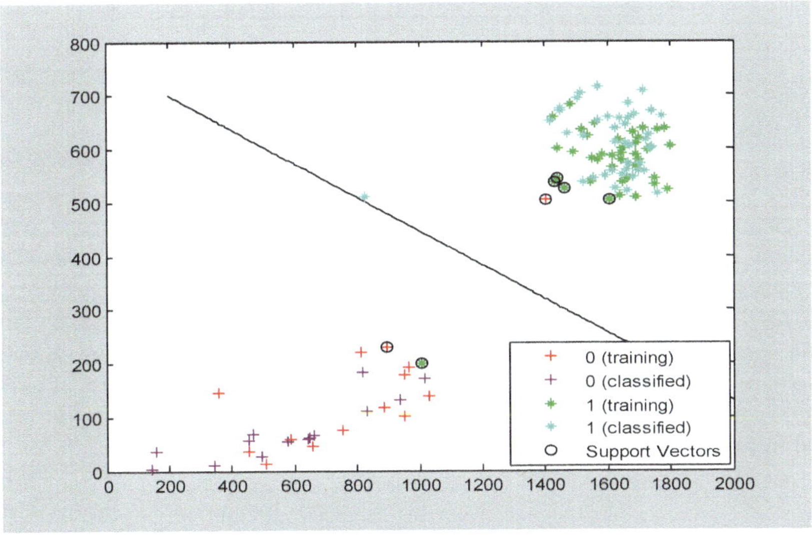

Fig. 7. SVM classification for area covered

For array elements:
In this classification as shown in Fig. 6, blue plus symbol '+' belongs to the cirrhotic liver and pink plus symbol '+' belongs to the normal liver. In this, x-axis is having area covered value from one threshold and similarly y-axis is having area covered value from another threshold. The line distinguishes the value of the cirrhotic and normal liver. 79 values from 80 cirrhotic values are classified accurately and 1 value is inaccurate classified. Similarly, 29 values from 30 normal values are classified

accurately and 1 value is inaccurate classified. So, 96.67% specificity, 98.75% sensitivity and 98.18% accuracy obtained as shown in Table 1.

For area covered:
In this classification as shown in Fig. 7, blue plus symbol '+' belongs to the cirrhotic liver and pink plus symbol '+' belongs to the normal liver. In this, x-axis is having area covered value from one threshold and similarly y-axis is having area covered value from another threshold. The line distinguishes the value of the cirrhotic and normal liver. 79 values from 80 cirrhotic values are classified accurately and 1 value is inaccurate classified. Similarly, 29 values from 30 normal values are classified accurately and 1 value is inaccurate classified. So, 96.67% specificity, 98.75% sensitivity and 98.18% accuracy obtained as shown in Table 1.

The classification accuracy obtained from this proposed technique is higher than the results obtained from previous methods as in the literature survey.

Table 1. Classification performance

	For counted array elements	For area covered
Accuracy	98.18%	98.18%
Sensitivity	98.75%	98.75%
Specificity	96.67%	96.67%

5 Conclusion

Liver US images have been used to characterize the cirrhotic and normal disease. In this research, a new method has been proposed that was based on the pixels' intensity variation. This technique provides two characterization parameters i.e. counted array elements and area covered. The parameters are extracted from the images which categorizes the images as cirrhotic/normal. The 96.67% specificity, 98.75% sensitivity and 98.18% accuracy obtained to detect the tested images. The results are also validated by radiologist. The results can be more accurate with knowledge of the stages of cirrhosis. It also reduces over diagnosis. The stages between cirrhotic and normal will be identified in future.

References

1. Virmani, J., Kumar, V., Kalra, N.: Prediction of cirrhosis based on singular value decomposition of gray level co-occurence matrix and a neural network classifier. In: Proceeding IEEE Conference Developments in E-systems Engineering, pp. 146–151 (2011)
2. Aggarwal, K., Bhamrah, M.S., Ryait, H.S.: The identification of liver cirrhosis with modified LBP grayscaling and Otsu binarization. SpringerPlus **5**, 1–15 (2016)
3. Masson, S.A., Nakib, A.: Real-time assessment of bone structure positions via ultrasound imaging. J. Real Time Image Process. **13**, 135–145 (2017)
4. Strauss, S., Gavish, E., Gottlieb, P.: Interobserver and intraobserver variability in the sonographic assessment of fatty liver. Am. J. Roentgenol. **189**, W320-3 (2007)

5. Doi, K.: Current status and future potential of computer-aided diagnosis in medical imaging. Br. J. Radiol. **78**, s3–s19 (2005)
6. Fujita, H., Uchiyama, Y., Nakagawa, T.: Computer-aided diagnosis: the emerging of three CAD systems induced by Japanese health care needs. Comput. Methods Programs Biomed. **92**, 238–248 (2008)
7. Hashem, A.M., Rasmy, M.E., Wahba, K.M.: Single stage and multistage classification models for the prediction of liver fibrosis degree in patients with chronic hepatitis C infection. Comput. Methods Programs Biomed. **105**, 194–209 (2012)
8. Polat, K., Günes, S.: A hybrid approach to medical decision support systems: combining feature selection, fuzzy weighted pre-processing and AIRS. Comput. Methods Programs Biomed. **88**, 164–174 (2007)
9. Sartakhti, J.S., Zangooei, M.H., Mozafari, K.: Hepatitis disease diagnosis using a novel hybrid method based on support vector machine and simulated annealing (SVM-SA). Comput. Methods Programs Biomed. **108**, 570–579 (2012)
10. Wang, Y., Ma, L., Liu, P.: Feature selection and syndrome prediction for liver cirrhosis in traditional Chinese medicine. Comput. Methods Programs Biomed. **95**, 249–257 (2009)
11. Adams, L.A., Angulo, P., Lindor, K.D.: Nonalcoholic fatty liver disease. Cana Med Assoc J. **172**, 899–905 (2005)
12. Hawlick, R.M.: Statistical and structural approaches to texture. Proc. IEEE **67**, 786–808 (1979)
13. Abramov, N., Fradkin, M., Rouet, L.: Configurable real-time motion estimation for medical imaging: application to X-ray and ultrasound. J. Real Time Image Process. **13**, 147–160 (2017)
14. Castellano, G., Bonilha, L., Li, L.M.: Texture analysis of medical images. Clin. Radiol. **59**, 1061–1069 (2004)
15. Filipczuk, P., Fevens, T., Krzyżak, A.: Computer-aided breast cancer diagnosis based on the analysis of cytological images of fine needle biopsies. IEEE Trans. Med. Imaging **32**, 2169–2178 (2013)
16. Chaieb, F., Said, T.B., Mabrouk, S.: Accelerated liver tumor segmentation in four-phase computed tomography images. J. Real Time Image Process. **13**, 121–133 (2017)
17. NIDDK (2010). http://digestive.niddk.nih.gov/ddiseases/pubs/cirrhosis_ez
18. Grangier, D., Bengio, S.: A discriminative kernel-based approach to rank images from text queries. IEEE Trans. Pattern Anal. Mach. Intell. **30**, 1371–1384 (2008)
19. Gulo, C.A.S.J., De Arruda, H.F., De Araujo, A.F.: Efficient parallelization on GPU of an image smoothing method based on a variational model. J. Real Time Image Process. **16**, 1–13 (2016)
20. Yeh, W.C., Jeng, Y.M., Li, C.H.: Liver steatosis classification using high-frequency ultrasound. Ultrasound Med. Biol. **31**, 599–605 (2005)
21. Kadah, Y.M., Farag, A.A., Zurada, J.M.: Classification algorithms for quantitative tissue characterization of diffuse liver disease from ultrasound images. IEEE Trans. Med. Imaging **15**, 466–478 (1996)
22. Graif, M., Yanuka, M., Baraz, M.: Quantitative estimation of attenuation in ultrasound video images: correlation with histology in diffuse liver disease. Invest. Radiol. **35**, 319–324 (2000)
23. Ribeiro, R., Sanches, J.: Fatty liver characterization and classification by ultrasound. In: Pattern Recognition Image Analysis. LNCS, vol. 5524, 354–361 (2009)
24. Wu, C.C., Lee, W.L., Chen, Y.C.: Evolution-based hierarchical feature fusion for ultrasonic liver tissue characterization. IEEE J. Bio. Health Inf. **17**, 967–976 (2013)
25. Wu, C.C., Lee, W.L., Chen, Y.C.: Ultrasonic liver tissue characterization by feature fusion. Expert Syst. Appl. **39**, 9389–9397 (2012)

26. Murala, S., Jonathan, Q.M.: Local mesh patterns versus local binary patterns: biomedical image indexing and retrieval. IEEE J. Bio. Health Inf. **18**, 929–938 (2014)
27. Lucieer, A., Stein, A., Fisher, P.: Multivariate texture-based segmentation of remotely sensed imagery for extraction of objects and their uncertainty. Int. J. Remote Sens. **26**, 2917–2936 (2005)
28. Virmani, J., Kumar, V., Kalra, N.: SVM-based characterization of liver ultrasound images using wavelet packet texture descriptors. J. Digit. Imaging **26**, 530–543 (2013)
29. Chaou, A.K., Mekhaldi, A., Teguar, M.: Elaboration of novel image processing algorithm for arcing discharges recognition on HV polluted insulator model. IEEE Trans. Dielectr. Electr. Insul. **22**, 990–999 (2015)
30. Heikkila, M., Pietikainen, M.: A texture-based method for modeling the background and detecting moving objects. IEEE Trans. Pattern Anal. Mach. Intell. **28**, 657–662 (2006)
31. Otsu, N.: A threshold selection method from gray-level histograms. IEEE Trans. Syst. Man Cybern. **9**, 62–66 (1979)
32. Kyriacou, E., Pavlopoulos, S., Konnis, G.: Computer assisted characterization of diffused liver disease using image texture analysis techniques on B-scan images. In: IEEE Nuclear Science Symposium, vol. 2, pp. 1479–1483 (1997)
33. Pavlopoulos, S., Kyriacou, E., Koutsouris, D.: Fuzzy neural network-based texture analysis of ultrasonic images. IEEE Eng. Med. Biol. Mag. **19**, 39–47 (2000)
34. Badawi, A.M., Derbala, A.S., Youssef, A.M.: Fuzzy logic algorithm for quantitative tissue characterization of diffuse liver diseases from ultrasound images. Int. J. Med. Inform. **55**, 135–147 (1999)
35. Jiuqing, W., Sirui, Z.: Features extraction based on wavelet packet transform for B-mode ultrasound liver images. In: 3rd International Congress on Image and Signal Processing, vol. 2, pp. 949–955 (2010)
36. Lee, W.L., Chen, Y.C., Hsieh, K.S.: Ultrasonic liver tissues classification by fractal feature vector based on m-band wavelet transform. IEEE Trans. Med. Imaging **22**, 382–392 (2003)
37. Acharya, U.R., Sree, S.V., Ribeiro, R.: Data mining framework for fatty liver disease classification in ultrasound: a hybrid feature extraction paradigm. Med. Phys. **39**, 4255–4264 (2012)

Methods to Improve Ranking Chemical Structures in Ligand-Based Virtual Screening

Mohammed Mumtaz Al-Dabbagh[1(✉)], Naomie Salim[2], and Faisal Saeed[3]

[1] Tishk International University, Erbil, Iraq
Mohamad.aldabagh@ishik.edu.iq
[2] Universiti Teknologi Malaysia, Skudia, Malaysia
[3] College of Computer Science and Engineering, Taibah University, Medina, Saudi Arabia

Abstract. One of the main tasks in chemoinformatics is searching for active chemical compounds in screening databases. The chemical databases can contain thousands or millions of chemical structures for screening. Therefore, there is an increasing need for computational methods that can help alleviate some challenges for saving time and cost in drug discover design. The ranking of chemical compounds can be accomplished according to their chances of clinical success by the computational tools. In this paper, the techniques that have been used to improve the ranking of chemical structures in similarity searching methods have been highlighted through two categories. Firstly, the taxonomy of using machine learning techniques in ranking chemical structures have been introduced. Secondly, we have discussed the alternative chemical ranking approaches that can be used instead of classical ranking criteria to enhance the performance of similarity searching methods.

Keywords: Molecular ranking · Marching learning techniques · Ranking chemical compounds · Ligand-based · Virtual screening · Alternative ranking techniques

1 Introduction

Virtual screening refers to the use of a computer-based method to process compounds from a library or database of compounds in order to identify and select the ones that are likely to possess a desired biological activity, such as the ability to inhibit the action of a particular therapeutic target. The selection of molecules with a virtual screening algorithm should yield a higher proportion of active compounds, as assessed by experiment, relative to a random selection of the same number of molecules [1, 2].

Machine learning is one of the subfields of computer science, which can play a significant role in chemo-informatics and gradually improve the solutions for a given problem. The umbrella of machine learning techniques evolved from several disciplines, which are pattern recognition, cognitive sciences, computer science and statistics. Pattern recognition can be considered to be one of the branches of machine learning which deals with soft computing methods for the classification or description

F. Saeed et al. (Eds.): IRICT 2019, AISC 1073, pp. 259–269, 2020.
https://doi.org/10.1007/978-3-030-33582-3_25

of observations. Cognitive sciences deal with the concepts of thinking and learning, which contributes to evolving machine learning techniques. Moreover, the methodological and technological approaches of statistics and computer science also support the development of machine learning methods.

Machine learning approaches are frequently utilized to facilitate description selection through predicting the most appropriate types of descriptors for a given search, design problem of a compound or classification purpose. Therefore, these approaches play an important role in establishing the relationship between chemical structures and compound properties, which include biological activities, chemical or physical properties [3]. In addition, machine learning methods can play an important role in ranking chemical structures in order to describe molecular similarity to reference structures. In this paper, the methods that can be used to enhance the chemical ranking have been highlighted through two categories. Firstly, the machine learning techniques that employed in various ways for ranking in LBVS, the common of these methods is used traditional ranking criteria. Hence, the paper has proposed group of alternative chemical ranking approaches in LBVS which can provide higher performance for ranking molecular compounds than classical ranking criteria [4].

2 Machine Learning Methods for Ranking Chemical Structures

Based on the use of machine learning methods we can classify the methods that have been extensively utilized in VS to discriminate active compounds from inactive ones, while assessing regression methods that have been extensively utilized in quantitative structure-activity relationship (QSAR) as well as quantitative structure-property relationship (QSPR) applications to predict the biological activity of compounds. Generally, in the training process of machine learning, known molecules, whether active/inactive, or biological activity develop decision rules that can be able to distinguish new molecules in the testing process. The end goal is to obtain active compounds at higher-ranking positions than inactive ones.

The popular approaches of classification methods are comprised of several techniques that have been recently investigated by researchers such as support vector machines (SVM), k-nearest neighbors (k-NN), decision trees (DT), naïve Bayesian methods (NBM) and artificial neural networks (ANN). On the other hand, the techniques that belong to regression methods are Partial Least Squares (PLS), Genetic Algorithm (GA), and Support Vector Regression (SVR). The following subsections report recent studies that have been presented by researchers.

2.1 Support Vector Machines (SVM)

The use of SVM in LBVS was reported for ranking library compounds according to their decreasing probability of activity [5, 6]. A new approach for ranking compounds by using a top-k ranking algorithm based on SVM was presented by Rathke [7], while [8] introduced a structure ranking approach based on kernel representation to minimise the ranking error.

The kernel functions for SVM that have been introduced were comprised of ligand and target kernels in order to capture various information for similarity assessment [9–11]. The graph kernels were employed to compute the similarity between labels graphs [12, 13]. Recently, [14] introduced a weighted voting approach based on multiple machine learning techniques to improve the quality of predictions. The combination between SVM-based and docking-based VS was introduced by [15] to retrieve novel inhibitors of c-Met from 18 million compounds.

2.2 Decision Trees (DT)

The role of DT is to rank the compounds that are more similar to reference molecules represented through their structure, where each molecular property connects with a leaf node, while the non-leaf nodes (i.e. root and internal nodes) assign with a molecular descriptor that uses a test condition for unknown compounds that are going through DT. The properties need to be sorted in decreasing order based on their importance according to queries. Therefore, several studies have employed the DT to discriminate chemical compounds into drug and non-drug [16, 17]. While the prediction of molecular properties presented by researchers [18] uses the model for the Prediction of the Blood-Brain Barrier Passage of Drugs. The prediction of ADME properties in drug discovery have been reported [19] for P-glycoprotein substrates [20], for blood–brain barrier permeation [21], and for solubility and permeability of drugs [22].

2.3 Naïve Bayesian Methods (NBM)

The use of Naïve Bayesian Methods (NBM) in LBVS, which is based on the mathematical Bayes theorem, was represented via the classification of compounds, and the predication of biological activity rather than physical-chemical properties. Some studies use NBM to predict the protein and bioactivity classification of molecules [23]. While predication for mechanism of phospholipidosis and toxicity of the compounds was introduced respectively [24]. On the other hand, the Bayesian network model is one of the applications of the Bayes techniques, which is frequently used in LBVS, whether in the machine learning world [25] or even as a similarity-based approach [26, 27].

2.4 K-Nearest Neighbors (K-NN))

The use of k-NN approach in LBVS was presented for classification and for the regression of compounds properties. [28] employed the k-NN technique with QSAR as a potential screening mechanism for larger libraries of target compounds, while the prediction for the k-NN method was based on the property of molecules reported by [29]. The k NN method was combined with QSAR methods such as artificial neural network (ANN) and Decision Forest (DF) and was applied to 3363 diverse compounds to enhance the ranking compounds [30]. The k-NN approach is used to distinguish between kinase inhibitors and other molecules [31]. Recently, the k-NN predication of enzyme has relied on multi-labels presented by [32].

2.5 Artificial Neural Networks (ANN)

The ANN techniques and their application has been introduced in QSAR [33], in pharmaceutical research [34] and in VS [35], while the use of techniques in regression models was presented by [36]. The role of self-organising maps for similarity in LBVS has been studied by researchers [37, 38]. Furthermore, the ANN integrates with Kohonen networks for the predication and identification of β-amyloid aggregation inhibitors [39].

2.6 Partial Least Squares (PLS)

The Partial Least Squares (PLS) is one of the statistical machine learning methods which is utilised in QSAR in order to predict the biological activities of compounds [40, 41] and develop a binary classification tool based on linear PLS-based prediction system by cytochrome P450 3A4 inhibition, while the build of computational models is based on Kernel partial least squares (K-PLS) and the electro-topological descriptors introduced [42]. Another approach was developed for the prediction of the hERG of low molecular weight compounds, which can be considered to be a major concern in the drug discovery process [43].

2.7 Genetic Algorithm (GA)

The use of GA approaches proved their efficiency when used in QSAR and 'drug-likeness' scenario during lead generation [44, 45]. The GA techniques used in docking generate conformers for a ligand via employed in the GOLD program [46] as well as in AutoDock [47]. Later [48] improved GA in the docking method based on various techniques such as entropy-based searching and multi-population genetic strategy.

2.8 Support Vector Regression (SVR)

Support Vector Regression (SVR) is one of regression techniques which is extensively used in QSAR and QSPR models to predict the structure features of how chemical compounds correspond to biological activities. The end goal is to predict compound activities or help the prediction via finding structural features then by ranking the compounds in decreasing order of activity. The non-linear chemical model reported [49] relies on the interpretation of the SVM and SVR models to achieve high performance for classification and regression rates. While the enhancement of ranking in docking used SVR methods that rely on two scoring functions (SVR-KB and SVR-EP) [50].

3 Alternative Ranking Approaches

The next group of techniques that can be used to improve ranking chemical structures is alternative ranking approaches. These alternative ranking approaches have been introduced and used in text information retrieval [51, 52]. Due to many similarities between text and chemical information retrieval, we thought that these alternative ranking approaches can be applied in virtual screening which can contribute to enhance the effectiveness of molecular similarity searching methods [53].

Alternative ranking approaches can be used instead of conventional ranking criteria of molecules, known as Probability Ranking Principle (PRP). The PRP is one of the most popular ranking techniques in virtual screening as well as text information retrieval. Faced with a collection of compounds, the user issues a query (i.e. target structure), and an ordered list of chemical compounds is returned. The principle of PRP depends on the compounds, which should be ranked in decreasing probability of relevance to the user's target structure.

The estimation of probability $P(\mu, \Re)$ can be achieved by using various similarity measure coefficients that are used to estimate the probability between pair of molecules [54]. The PRP can be prioritized the chemical structures to produce ranking list. The following equation represented the criteria of probability ranking principle:

$$\mu_i = \underset{\mu \in \mu_R \backslash \Re L}{\arg\, max}\, P(\mu, \Re) \tag{1}$$

where μ_i represents the ranking position for each molecule, μ and $\Re$ indicate database structures and reference structures respectively, $\Re L$ is the list of molecular structures that have been ranked, and a molecule μ belongs to the set of retrieved molecules μ_R for reference $\Re$ but not ranked yet ($\mu \in \mu_R \backslash \Re L$). The alternative ranking criteria can also prioritize the chemical structures to produce ranking list, but by using many other criteria not only depending on the similarity scores of molecules.

3.1 Interactive Probability Ranking Principle

The theoretical framework for extending the probability ranking principle (PRP) to the context of interactive text information retrieval known as Interactive Probability Ranking Principle (IPRP) was initially proposed by Fuhr [55]. Fuhr proposed to model ranking in Interactive information retrieval as a decision making process that considers two actors, which are a user and an information retrieval system. In this framework, the IPRP is derived as the optimum ordering of the choices presented in each situation, where the IPRP is not a sequential ranking algorithm because a ranking choice can be revised after an interaction has occurred. Recently, Zuccon et al. [56] further developed IPRP approach through the proposal of an effective instantiation of IPRP for a given situation and for each rank position. Formally, the ranking principle of an object (suppose it molecule) under the IPRP can be ranked as the following:

$$\mu_i = \underset{\mu \in \mu_R \backslash \Re L}{\arg\, max} \left[-P(\mu, \Re) \frac{\sum_{\mu' \in \Re L} P(\mu, \mu')}{|\Re L|} \right] \tag{2}$$

where $P(\mu, \mu')$ represents a similarity function between pair of molecules, while $P(\mu, \Re)$ represents a similarity function between library molecules and user target, $|\Re L|$ is the size of the list containing the documents that have been ranked so far (i.e. until the position $i - 1$, therefore $|\Re L| = i - 1$). The IPRP assumes that the user would examine each document in turn. Thus at each rank, the previous documents would influence the relevance of the subsequent documents [51].

3.2 Maximal Marginal Relevance

The Maximal Marginal Relevance (MMR) is one of the most popular methods for balancing relevance and diversity in text information retrieval. Carbonell and Goldstein [57] recognized that in specific search scenarios relevancy is not the only criteria that should be used to rank documents. They mixed between relevance and diversity estimations by means of a linear combination. As a result, they ended up to developing a ranking approach known as MMR. Formally, the MMR can rank the objects (we assumed as molecules μ) at position i by using the following criteria [51]:

$$\mu_i = \underset{\mu \in \mu_R \backslash \Re L}{\arg\, max} \left[\lambda P(\mu, \Re) - (1 - \lambda) \max_{\mu' \in \Re L} P(\mu, \mu') \right] \tag{3}$$

where $P(\mu, \Re)$ is similarity between database compounds μ and user reference structure $\Re$, is similarity function between pair of molecules, and λ is the hyper-parameter, which linearly combines query-similarity and molecular-similarity. The hyper-parameter can be inferred by the user, where the value of λ can be $0.5 < \lambda < 1$ $P(\mu, \mu')$. An number of studies in text information retrieval have attempted to modify and move away from the original ranking approach of Carbonell and Goldstein [57] in order to enhance the effectiveness. Leelanupab et al. [58] proposed to modify the MMR by considering average marginal relevance instead of maximal marginal relevance (i.e. the average similarity of documents $\underset{d' \in \Re L}{avg} P(d, d')$ used instead of the maximum similarity of documents $\max_{d' \in \Re L} P(d, d')$). He et al. [59] used cluster-based approaches for ranking diversification, which proved its efficiency in web diversity ranking and image diversity ranking. Recently, Santos et al. [60] have suggested that improvements in web search can be achieved when novelty-based approaches like MMR are deployed in combination with strategies that explicitly capture the coverage of the query's intents.

3.3 Portfolio Theory

Portfolio Theory (PT) that has been proposed for information retrieval can rank documents by balancing the probability estimates returned by a retrieval model with the variance of their estimates. However, under non-trivial conditions PT violates the original probability ranking principle. Wang and Zhu [61] have examined the assumption of certainty in the estimation of relevance brought forward by probability ranking principle, and argued against this assumption. Also, they suggested an analogy with financial models, and reported that the independence assumption is not realistic in many situations.

The concept of portfolio diversification with respect to risk and relations between stocks have been transposed to information retrieval giving rise to the Portfolio Theory ranking strategy [62]. The aim of the ranking strategy is to minimize the risk associated with ranking documents under uncertainty in their relevance estimates. This can be achieved by balancing the expected relevance value and its variance. The resulting ranking criterion combines different terms in order to produce the appropriate position of each object. Formally, for each ranking position i, molecules are selected according to the following equation [51]:

$$\mu_i = \arg\max_{\mu \in \mu_R \setminus \Re L} \left[P(\mu, \Re) - bw_\mu \sigma_\mu^2 - 2b \sum_{\mu \in \mu_R} w_\mu \sigma_\mu \sigma_{\mu'} P(\mu, \mu') \right] \quad (4)$$

where $P(\mu, \Re)$ is the estimated probability of relevance between database and reference structure, b represents the encoding of the risk propensity of the user, σ_μ^2 is variance associated to $P(\mu, \Re)$, and w_μ is a weight that expresses the importance of the rank position of μ and μ'. This weight term can be calculated by $w_{\mu_i} = \frac{1}{\log(1+i)}$, where i is the rank position of molecule μ_i. Wang and Zhu [61] demonstrated that the criterion is successful in ranking text documents in both ad-hoc and subtopic text retrieval scenario. Aly et al. [63] employed PT in the context of video story retrieval while a new version of PT ranking approach has developed by Zuccon et al. [64], which has been employed for different information retrieval systems

3.4 Quantum Probability Ranking Principle

The idea of using quantum theory in information retrieval was formally put forward by van Rijsbergen [65]. Recently, there has been a spate of work which aims to develop quantum inspired or quantum based information retrieval models [66–68]. These studies that have achieved in text information retrieval proved that the use of quantum can provide the best performance compared to other models. The employment of the quantum theory in ligand based virtual screening can provide better retrieval performance than other conventional models [4, 54], because there are many similarity aspects between the text and chemical information retrieval [53].

The ranking principle is derived by developing an analogy between the famous double-slit experiment and text document ranking. The original application demonstrated the wave theory of light, where it displays the fundamentally probabilistic nature of quantum mechanical phenomena [69]. Formally, the ranking of objects (let us assume as molecules) depending on the user needs can be achieved by using the quantum ranking criteria as the following [51]:

$$\mu_i = \arg\max_{\mu \in \mu_R \setminus \Re L} \left[P(\mu, \Re) - 2 \sum_{\mu' \in \Re L} \sqrt{P(\mu, \Re)}\, \sqrt{P(\mu', \Re)}\, P(\mu, \mu') \right] \quad (5)$$

where $P(\mu, \Re)$ represents the estimation of similarity between database and reference molecular structures, $P(\mu, \mu')$ represents the similarity between pair of molecules, which represents the correlation function. This ranking approach will be employed in this research for ranking the molecules in virtual screening.

4 Conclusion

Recently, the problems of ranking chemical compounds have gained more interest by researchers who are aiming to enhance the efficiency of VS. The machine learning techniques have been playing a vital role to develop the current approaches of various fields under the chemoinformatics filed. One of these fields is ranking chemical structure after high throughput screening is done. The improvement of chemical ranking techniques in LBVS can be achieved through various approaches whether by using machine learning techniques, or by using alternative chemical ranking approaches. The employment of machine learning techniques can be divided into two families: classification and regression. The methods that belong to the classification family are employed to distinguish active compounds from inactive ones, while the regression family applied applications to predict the biological activity of compounds based on certain features in molecules. Alternative ranking approaches for chemical structures have also been investigated in LBVS, which obtained better results than traditional PRP. Some of these alternatives' methods have been only employed in text information retrieval, which can open a way for researchers to apply it in chemoinformatics because many aspects are similar between the two areas. The end goal is to obtain active compounds at higher ranking positions than inactive ones, or certain activity compounds higher than other compounds according to reference structures.

References

1. Walters, W.P., Stahl, M.T., Murcko, M.A.: Virtual screening—an overview. Drug Discov. Today **3**(4), 160–178 (1998)
2. Johnson, M.A., Maggiora, G.M.: Concepts and Applications of Molecular Similarity. Wiley, New York (1990)
3. Gasteiger, J., Engel, T.: Chemoinformatics: A Textbook. Wiley (2006)
4. Al-Dabbagh, M.M., et al.: Quantum probability ranking principle for ligand-based virtual screening. J. Comput.-Aided Mol. Des. **31**(4), 365–378 (2017)
5. Jorissen, R.N., Gilson, M.K.: Virtual screening of molecular databases using a support vector machine. J. Chem. Inf. Model. **45**(3), 549–561 (2005)
6. Geppert, H., et al.: Support-vector-machine-based ranking significantly improves the effectiveness of similarity searching using 2D fingerprints and multiple reference compounds. J. Chem. Inf. Model. **48**(4), 742–746 (2008)
7. Rathke, F., et al.: StructRank: a new approach for ligand-based virtual screening. J. Chem. Inf. Model. **51**(1), 83–92 (2010)
8. Agarwal, S., Dugar, D., Sengupta, S.: Ranking chemical structures for drug discovery: a new machine learning approach. J. Chem. Inf. Model. **50**(5), 716–731 (2010)
9. Jacob, L., Vert, J.-P.: Protein-ligand interaction prediction: an improved chemogenomics approach. Bioinformatics **24**(19), 2149–2156 (2008)
10. Jacob, L., et al.: Virtual screening of GPCRs: an in silico chemogenomics approach. BMC Bioinf. **9**(1), 363 (2008)
11. Vert, J.-P., Jacob, L.: Machine learning for in silico virtual screening and chemical genomics: new strategies. Comb. Chem. High Throughput Screen. **11**(8), 677 (2008)

12. Rupp, M., Proschak, E., Schneider, G.: Kernel approach to molecular similarity based on iterative graph similarity. J. Chem. Inf. Model. **47**(6), 2280–2286 (2007)
13. Ralaivola, L., et al.: Graph kernels for chemical informatics. Neural Netw. **18**(8), 1093–1110 (2005)
14. Plewczynski, D.: Brainstorming: weighted voting prediction of inhibitors for protein targets. J. Mol. Model. **17**(9), 2133–2141 (2011)
15. Xie, Q.-Q., et al.: Combined SVM-based and docking-based virtual screening for retrieving novel inhibitors of c-Met. Eur. J. Med. Chem. **46**(9), 3675–3680 (2011)
16. Schneider, N., et al.: Gradual in silico filtering for druglike substances. J. Chem. Inf. Model. **48**(3), 613–628 (2008)
17. Klekota, J., Roth, F.P.: Chemical substructures that enrich for biological activity. Bioinformatics **24**(21), 2518–2525 (2008)
18. Deconinck, E., et al.: Classification tree models for the prediction of blood-brain barrier passage of drugs. J. Chem. Inf. Model. **46**(3), 1410–1419 (2006)
19. Hou, T., Wang, J., Li, Y.: ADME evaluation in drug discovery 8. The prediction of human intestinal absorption by a support vector machine. J. Chem. Inf. Model. **47**(6), 2408–2415 (2007)
20. de Cerqueira Lima, P., et al.: Combinatorial QSAR modeling of P-glycoprotein substrates. J. Chem. Inf. Model. **46**(3), 1245–1254 (2006)
21. Mente, S., Lombardo, F.: A recursive-partitioning model for blood–brain barrier permeation. J. Comput.-Aided Mol. Des. **19**(7), 465–481 (2005)
22. Lamanna, C., et al.: Straightforward recursive partitioning model for discarding insoluble compounds in the drug discovery process. J. Med. Chem. **51**(10), 2891–2897 (2008)
23. Koutsoukas, A., et al.: In silico target predictions: defining a benchmarking data set and comparison of performance of the multiclass naïve bayes and parzen-rosenblatt window. J. Chem. Inf. Model. **53**(8), 1957–1966 (2013)
24. Lowe, R., et al.: Predicting the mechanism of phospholipidosis. J. Cheminf. **4**, 2 (2012)
25. Wasserman, L.: Bayesian model selection and model averaging. J. Math. Psychol. **44**(1), 92–107 (2000)
26. Abdo, A., et al.: Ligand-based virtual screening using bayesian networks. J. Chem. Inf. Model. **50**(6), 1012–1020 (2010)
27. Ahmed, A., A. Abdo, and N. Salim, Ligand-based Virtual screening using Bayesian inference network and reweighted fragments. Sci. World J. (2012)
28. Kauffman, G.W., Jurs, P.C.: QSAR and k-nearest neighbor classification analysis of selective cyclooxygenase-2 inhibitors using topologically-based numerical descriptors. J. Chem. Inf. Comput. Sci. **41**(6), 1553–1560 (2001)
29. Konovalov, D.A., et al.: Benchmarking of QSAR models for blood-brain barrier permeation. J. Chem. Inf. Model. **47**(4), 1648–1656 (2007)
30. Votano, J.R., et al.: Three new consensus QSAR models for the prediction of Ames genotoxicity. Mutagenesis **19**(5), 365–377 (2004)
31. Briem, H., Günther, J.: Classifying "kinase inhibitor-likeness" by using machine-learning methods. ChemBioChem **6**(3), 558–566 (2005)
32. De Ferrari, L., et al.: EnzML: multi-label prediction of enzyme classes using InterPro signatures. BMC Bioinf. **13**(1), 61 (2012)
33. Patel, J., Chaudhari, C.: Introduction to the Artificial Neural Networks and their applications in QSAR studies. Altex **22**, 271 (2005)
34. Patel, J., Patel, L.: Artificial neural networks and their applications in pharmaceutical research. Pharmabuzz **2**, 8–17 (2007)
35. Selzer, P., Ertl, P.: Applications of self-organizing neural networks in virtual screening and diversity selection. J. Chem. Inf. Model. **46**(6), 2319–2323 (2006)

36. Hykin, S.: Neural Networks: A Comprehensive Foundation. Printice-Hall Inc., New Jersey (1999)
37. Hristozov, D., Oprea, T.I., Gasteiger, J.: Ligand-based virtual screening by novelty detection with self-organizing maps. J. Chem. Inf. Model. **47**(6), 2044–2062 (2007)
38. Bonachera, F., et al.: Using self-organizing maps to accelerate similarity search. Bioorg. Med. Chem. **20**(18), 5396–5409 (2012)
39. Afantitis, A., et al.: Ligand - based virtual screening procedure for the prediction and the identification of novel β-amyloid aggregation inhibitors using Kohonen maps and Counter propagation Artificial Neural Networks. Eur. J. Med. Chem. **46**(2), 497–508 (2011)
40. Hasegawa, K., Funatsu, K.: Partial least squares modeling and genetic algorithm optimization in quantitative structure-activity relationships. SAR QSAR Environ. Res. **11** (3–4), 189–209 (2000)
41. Zuegge, J., et al.: A fast virtual screening filter for cytochrome P450 3A4 inhibition liability of compound libraries. Quant. Struct.-Act. Relat. **21**(3), 249–256 (2002)
42. Wang, Y., Li, Y., Wang, B.: An in silico method for screening nicotine derivatives as cytochrome P450 2A6 selective inhibitors based on kernel partial least squares. Int. J. Mol. Sci. **8**(2), 166–179 (2007)
43. Roche, O., et al.: A virtual screening method for prediction of the HERG potassium channel liability of compound libraries. ChemBioChem **3**(5), 455–459 (2002)
44. Gillet, V.J., Willett, P., Bradshaw, J.: Identification of biological activity profiles using substructural analysis and genetic algorithms. J. Chem. Inf. Comput. Sci. **38**(2), 165–179 (1998)
45. Shi, L.M., et al.: Mining the NCI anticancer drug discovery databases: genetic function approximation for the QSAR study of anticancer ellipticine analogues. J. Chem. Inf. Comput. Sci. **38**(2), 189–199 (1998)
46. Jones, G., Willett, P., Glen, R.C.: Molecular recognition of receptor sites using a genetic algorithm with a description of desolvation. J. Mol. Biol. **245**(1), 43–53 (1995)
47. Morris, G.M., et al.: Automated docking using a Lamarckian genetic algorithm and an empirical binding free energy function. J. Comput. Chem. **19**(14), 1639–1662 (1998)
48. Kang, L., et al.: An improved adaptive genetic algorithm for protein–ligand docking. J. Comput.-Aided Mol. Des. **23**(1), 1–12 (2009)
49. Hasegawa, K., Funatsu, K.: Non-linear modeling and chemical interpretation with aid of support vector machine and regression. Curr. Comput.-Aided Drug Des. **6**(1), 24–36 (2010)
50. Li, L., Wang, B., Meroueh, S.O.: Support vector regression scoring of receptor-ligand complexes for rank-ordering and virtual screening of chemical libraries. J. Chem. Inf. Model. **51**(9), 2132–2138 (2011)
51. Zuccon, G.: Document ranking with quantum probabilities, in College of Science and Engineering, School of Computing Science, p. 222, University of Glasgow, UK (2012)
52. Zuccon, G., Azzopardi, L., Van Rijsbergen, C.K.: An analysis of ranking principles and retrieval strategies. In: Advances in Information Retrieval Theory, pp. 151–163. Springer (2011)
53. Willett, P.: Textual and chemical information processing: different domains but similar algorithms. Inf. Res. **5**(2) (2000)
54. Al-Dabbagh, M., et al.: A quantum-based similarity method in virtual screening. Molecules **20**(10), 18107 (2015)
55. Fuhr, N.: A probability ranking principle for interactive information retrieval. Inf. Retrieval **11**(3), 251–265 (2008)

56. Zuccon, G., Azzopardi, L., Rijsbergen, C.J.K.V.: The interactive PRP for diversifying document rankings. In: Proceedings of the 34th International ACM SIGIR Conference on Research and Development in Information Retrieval, pp. 1227–1228. ACM, Beijing, China (2011)
57. Carbonell, J., Goldstein, J.: The use of MMR, diversity-based reranking for reordering documents and producing summaries. In: Proceedings of the 21st Annual International ACM SIGIR Conference on Research and Development in Information Retrieval, pp. 335–336. ACM, Melbourne, Australia (1998)
58. Leelanupab, T., Zuccon, G., Jose, J.M.: When two is better than one: a study of ranking paradigms and their integrations for subtopic retrieval. In: Information Retrieval Technology, pp. 162–172. Springer (2010)
59. He, J., Meij, E., de Rijke, M.: Result diversification based on query-specific cluster ranking. J. Am. Soc. Inf. Sci. Technol. **62**(3), 550–571 (2011)
60. Santos, R.L., Macdonald, C., Ounis, I.: On the role of novelty for search result diversification. Inf. Retrieval **15**(5), 478–502 (2012)
61. Wang, J., Zhu, J.: Portfolio theory of information retrieval. In: Proceedings of the 32nd International ACM SIGIR Conference on Research and Development in Information Retrieval. ACM (2009)
62. Wang, J.: Mean-variance analysis: a new document ranking theory in information retrieval. In: Advances in Information Retrieval, pp. 4–16. Springer (2009)
63. Aly, R., et al., Beyond shot retrieval: searching for broadcast news items using language models of concepts. In: Advances in Information Retrieval, p. 241–252. Springer (2010)
64. Zuccon, G., Azzopardi, L., Rijsbergen, C.J.K.V.: Has portfolio theory got any principles?. In: Proceedings of the 33rd International ACM SIGIR Conference on Research and Development in Information Retrieval, pp. 755–756. ACM, Geneva, Switzerland (2010)
65. Rijsbergen, C.J.V.: The Geometry of Information Retrieval. Cambridge University Press, UK (2004)
66. Piwowarski, B., et al.: What can quantum theory bring to information retrieval. In: Proceedings of the 19th ACM International Conference on Information and Knowledge Management, pp. 59–68. ACM, Toronto, ON, Canada (2010)
67. Zuccon, G., Azzopardi, L.: Developing the quantum probability ranking principle (2010)
68. Arafat, S.: Foundations research in information retrieval inspired by quantum theory. University of Glasgow (2008)
69. Feynman, R.P.: The concept of probability in quantum mechanics. In: Proceedings of the Second Berkeley Symposium on Mathematical Statistics and Probability. University of California Press, Berkeley, California (1951)

Analysis on Misclassification in Existing Contraction of Fuzzy Min–Max Models

Essam Alhroob[1(✉)], Mohammed Falah Mohammed[1,2], Osama Nayel Al Sayaydeh[1], Fadhl Hujainah[1], and Ngahzaifa Ab Ghani[1]

[1] Faculty of Computer Systems and Software Engineering, Universiti Malaysia Pahang, 26300 Kuantan, Malaysia
essamalhroob@gmail.com

[2] Faculty of Science, Department of Computer Science, University of Zakho, Duhok, KRG, Iraq

Abstract. Fuzzy min–max (FMM) neural network is one of the most useful models for pattern classification. Various models have been introduced based on FMM model to improve the classification performance. However, the misclassification of the contraction process is a crucial issue that has to be handled in FMM models to improve classification accuracy. Hence, this research aims to analyse the existence and execution procedure of addressing the misclassification of the contraction in the current FMM models. In this manner, practitioners and researchers are aided in selecting the convenient model that can address the misclassification of the contraction and improve the performance of models in producing accurate classification results. A total of 15 existing FMM models are identified and analysed in terms of the contraction problem. Results reveal that only five models can address the contraction misclassification problem. However, these models suffer from serious limitations, including the inability to detect all overlap cases, and increasing the network structure complexity. A new model is thus needed to address the specified limitations for increasing the pattern classification accuracy.

Keywords: Patten classification · Misclassification · Fuzzy min- max · FMM models

1 Introduction

Artificial neural networks (ANNs) are mathematical models of artificial nodes that mimic the functions of biological neural networks, such as calculating and distributing information in the human brain [1, 2]. The first model of ANNs was proposed by McCulloch and Pitts in 1943 [3]. This model aimed to emulate the biological neural structure by formulating a mathematical model of biological neurons. It consists of a group of neurons and processes that are interconnected by weighed connections to form the network structure [4].

Many attempts have been made to develop an ANN architecture for the pattern recognition problem, including multilayer layer perceptron [5], hopfield neural network

F. Saeed et al. (Eds.): IRICT 2019, AISC 1073, pp. 270–278, 2020.
https://doi.org/10.1007/978-3-030-33582-3_26

[6, 7] and radial basis function [8]. In this regard, ANN has emerged as one of the most computerised approach used for pattern recognition [9]. The combination and application of the ANN with fuzzy set method have been used to reduce the restrictive assumptions of each existing ANN [10]. The primary aim of using the ANN along with fuzzy sets is to transform the input data into more meaningful outputs [11].

Fuzzy min–max (FMM) neural network is a neuro-fuzzy system that synthesises the role of a fuzzy set in neural networks for pattern classification [12]. The FMM model depends on the hyperbox, where hyperboxes are used to build the network and store knowledge. Each hyperbox has n-dimensions, which are specified by two corners called min (v) and max (w). Each hyperbox belongs to a single class [13]. The size of the hyperbox is limited by a parameter called expansion coefficient between 0 and 1. The learning process in FMM is used to find the suitable hyperboxes for each input sample by three-step processes: hyperbox expansion, hyperbox overlap test and hyperbox contraction. The FMM is an incremental learning type, where the data samples are performed in a sequential order. When an input sample is presented for the training, the network tries accommodating it in one of its existing hyperboxes that belongs to a similar class with the highest membership degree, and this process is known as hyperbox expansion. During the hyperbox expansion process, it is possible to create the overlap areas between the hyperboxes belonging to different classes. Thus, an overlap test process is used to identify whether any overlap exists between the hyperbox belonging to different classes, and this overlapping is solved using the contraction process [14].

The FMM learning stage is composed of the single-pass-through and online-adaptive system that is executed within a short timeline. Different models have been proposed based on the traditional FMM model to improve the classification performance. Examples are general FMM neural network (GFMM) [9], Inclusion/Exclusion classifier (EFC) [15], fuzzy min–max neural network classifier with compensatory neurons (FMCN) [16], data-core-based FMM neural network (DCFMN) [17] and enhanced fuzzy min–max neural network (EFMM) [18]. These models are indicated as FMM existing models [18, 19].

The misclassification of the contraction process is a primary challenge in FMM existing models [14, 16, 17]. Misclassification in the contraction during learning execution contributes to the inability of the FMM models to produce high classification accuracy [14]. However, the impact of misclassification of the contraction in FMM models has not been studied by many researchers [14, 19, 20]. Thus, the aim of the present research is to analyse and investigate the presence of the contraction misclassification problem in current FMM models. To this end, this research discusses and reveals the execution procedure of catering the contraction misclassification problem in FMM models.

As a comprehensive source for understanding the misclassification of the contraction in existing models of FMM, this study helps the researcher and practitioners who are working on pattern classification. It is also a useful guide for selecting the convenient model for addressing the misclassification issue of the contraction and improving the classification performance.

The rest of this paper is structured as follows. Section 2 illustrates the methodology that is constructed to conduct this research. Section 3 elaborates on and discusses the findings. Section 4 concludes this research and highlights future work.

2 Methodology

Methodology is an essential element that must be constructed to achieve the aim of a research [21, 22]. Thus, a methodology was designed to conduct this research and fulfil the defined objective. Figure 1 depicts the research methodology used, which comprises four concatenation phases: formulation of research questions, exploration of related research papers, analysis of the selected related research papers and results.

In the first phase, two research questions (Q) were specifically formulated based on the defined objective of this research. The constructed research questions are as follows:

Q1: What are the available FMM models that address the misclassification of the contraction?

Q2: How do the current FMM models (identified from Q1) address the misclassification of the contraction?

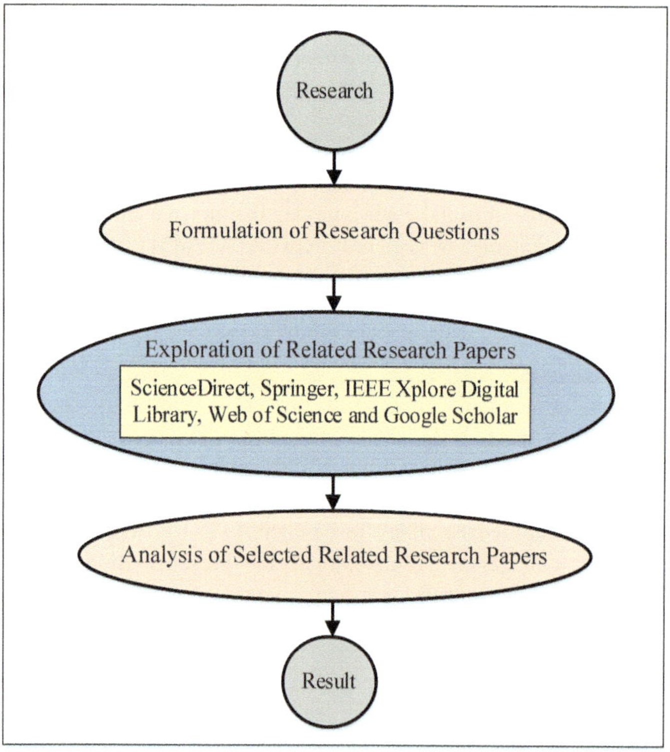

Fig. 1. Research methodology

Moreover, on the basis of the defined research questions, exploration of the related research papers was executed by performing an online search in five digital libraries:

ScienceDirect, Springer, IEEE Xplore Digital Library, Web of Science and Google scholar. These digital libraries are the most relevant to computation intelligence domain [22–24]. Additionally, these libraries present easy and powerful search engines, which are appropriate for automatic search of databases [25, 26].

A list of keywords was used in the search process. The keywords were specified based on the listed research questions of this work, including pattern classification, misclassification issue, OR error, OR problem, OR challenge in fuzzy min–max (FMM) pattern classification and FMM neural network, OR models and OR techniques.

Each collected research paper was then critically investigated by screening its title, keywords and abstract. Hence, studies that focus on the misclassification in FMM models' pattern classification and include at least an answer for the formulated research questions were included. The full content of each included related research paper was then studied and analysed to extract answers for the specified research questions.

3 Findings and Discussion

In this section, the findings of this research is elaborated and discussed. The findings reflect the outcome of answering each formulated research question. The critical analysis of the misclassification in the existing contraction in FMM models is conducted to obtain answers for the research questions. Each FMM model is analysed in terms of its ability to solve the misclassification issues for the contraction. Table 1 presents the analysis result of 15 existing FMM models derived from the selected studies. The result shows that 10 existing FMM models do not handle the misclassification issue accurately for the contraction. These models are FMM [27], Stochastic FMM [28], GFMM [9], WFMM [29], MFMM [30], MFMM-GA [31], EFMM [18], M-FMM [32], EFMM2 [33] and KnFMM [34]. The limitations of these models are related to their behaviour in inheriting the limitation of the contraction of the traditional FMM model, which was adopted in their learning process [19].

The contraction of these models induces the existence of membership confusion among the overlapped regions, thereby producing misclassification in the learning phase. Contraction also leads to loss of a part of the contracted hyperboxes' information. Hence, data distortion becomes a problem [14]. However, only five models can address the contraction misclassification problem, namely EFC [15], FMNC [16], DCFMN [17], RGFMM [35] and multi-level fuzzy min–max neural network (MLF) [36].

An inclusion/exclusion classifier (EFC) is proposed to execute the pattern classification for addressing the misclassification problem of FMM [15]. The EFC introduces two types of hyperboxes: inclusion and exclusion. Inclusion hyperboxes are used to indicate the input patterns of the same class, whereas exclusion hyperboxes refer to the overlap patterns. This model successfully addresses the misclassification by using the exclusion hyperboxes to represent an overlapped region instead of an FMM contraction.

Additionally, utilising the exclusion hyperboxes assists in reducing the learning algorithm from three processes (expansion, overlap test and contraction) to two processes (expansion and overlap test). However, Reference [14] reported that the number of inclusion/exclusion hyperboxes affect the model performance by increasing the network structure complexity.

Another model that can solve the misclassification issue is the FMCN. This model can perform the pattern classification process on the basis of the traditional FMM model [16]. In this model, the misclassification issue is handled by introducing compensatory neurons (CNs) for use in the learning phase instead of FMM hyperbox contraction. The CNs are composed of two sections: containment compensation neuron (CCN) and overlap compensation neuron (OCN). These sections are activated whenever the test sample falls into the overlap areas between hyperboxes from different classes. During the learning phase, one node is added in the compensation section for each overlapped region created. Two outputs relate to each OCN node. However, only one of these outputs is selected. Meanwhile, the CCN node has one output. The experimental result of Reference [16] showed that the classification performance of FMCN is better than that of traditional FMM. However, FMCN failed to use the suitable membership function for compensatory nodes [36]. Thus, the samples that fall in the overlapped areas between two different classes cannot be classified correctly [36]. Moreover, this model did not execute the expansion process in case any overlap is caused by the hyperbox that expanded with other existing hyperboxes of different classes. As a result, the network structure became complex because of the increasing number of nodes in the network hidden layer [14, 20].

Table 1. Analysis of existing of FMM models.

No.	Model name	References	Has misclassification issue in the contraction process
1.	GRFMM	[35]	No
2.	EFMM2	[33]	Yes
3.	EFMM	[18]	Yes
4.	DCFMN	[17]	No
5.	FMCN	[16]	No
6.	FMM	[27]	Yes
7.	GFMM	[9]	Yes
8.	KnFMM	[34]	Yes
9.	EFC	[15]	No
10.	MFMM	[30]	Yes
11.	MFMM-GA	[31]	Yes
12.	M-FMM	[32]	Yes
13.	MLF	[36]	No
14.	Stochastic FMM	[28]	Yes
15.	WFMM	[29]	Yes

A general reflex FMM neural network (GRFMM) was likewise proposed based on the GFMM model [35]. This model combines the algorithms of FMM classification and clustering in a unified framework along with the human reflex mechanism concept. The reflex mechanism is used to solve the overlap between the hyperboxes to avoid the misclassification problem instead of using FMM contraction. The reflex mechanism is comprised of compensatory neurons, which assist to approximate the complex data topology. Compensatory neurons are divided to OCNs and CCNs. During the training phase, compensatory neurons are added dynamically to the reflex mechanism. The experimental results of this model showed the efficiency of its performance on real datasets is better than that of the GFMM model. In addition, this model can specify the underlying data structure. However, the overlap test rules used in this model are insufficient to reveal all overlap cases [14, 20].

A DCFMN model for pattern classification was documented in [17]. The misclassification issue was the primary motivation for this model, which uses compensatory neurons to overlap a region of hyperboxes from different classes, such as FMCN. The difference between DCFMN and FMCN is that the former is composed of two types of neurons: overlapping neurons (OLNs) and classifying neurons (CNs). OLNs are used to address all types of overlap areas of different classes, whereas data patterns are classified by using CNs. A new membership function is used to consider the data characteristics and the impact of noise. The membership function contains three factors: geometric centre of the hyperbox, data core and noise. A new learning approach is introduced in this model as well, where the overlap test process starts to investigate the overlap after creating/expanding the hyperboxes for all the training data. Although this model can avoid the misclassification issue in its learning process, the overlap test rules used cannot discover all overlapped cases efficiently [14].

Furthermore, in [36], the MLF was introduced for pattern classification. This model successfully addresses the misclassification of the traditional FMM by using a multi-level structure. The MLF employs different small hyperboxes with separate classifiers in the network to handle the samples that fall in the overlapped area instead of using FMM contraction. During the training stage in the MLF model, each node is considered a separate classifier and has two segments: hyperboxes segment (HBS) and overlap boxes segment (OLS). HBS is used to create and adjust hyperboxes into the early training phase, while OLS is used to classify the samples that fall in the overlapped zones. The outputs of separate classifiers are integrated to form the ultimate output in the network. On the basis of the experiment conducted in [36], the misclassification of the MLF model is less than that of the traditional FMM model. However, the used overlap test rules are insufficient to specify all overlapping cases [14, 20]. This situation affects the classification accuracy in the MLF model [14].

4 Conclusion

The FMM neural network has recently emerged as one of the most useful neural networks for pattern classification purpose. The FMM has a number of important properties of the learning process, such as online learning, overlap classes, nonlinear separability, fast training and soft/hard decisions. However, the misclassification issue of contraction is considered a major challenge in existing FMM models. On this basis, this research presented a comprehensive investigation of the contraction misclassification problem in the FMM models.

To conduct this research, a research method was designed with four phases: formulation of research questions, extraction of research papers, studying and obtaining the answers from the collected related studies and result. A total of 15 FMM models were identified and analysed critically with respect to the contraction misclassification issue. The findings reveal that most FMM models cannot solve the misclassification problem of the contraction in their learning process.

Only five models can address the contraction misclassification problem, namely, EFC, FMNC, RGFMM, DCFMN and MLF. As revealed from the findings, these five models still have serious challenges with respect to network structure complexity and their inability to reveal overlapped cases. Thus, further work is required with the proposal of a new model that can address the misclassification issue. The propose model should have the ability to identify all the overlap regions between the hyperboxes of different classes and reduce the complexity of the network structure. Additionally, another future can be conducted to compare the accuracy of the identified FMM models in order to assist the research in identifying model can be considered as promising model in securing accurate classification results.

Acknowledgement. The researchers would like to express their sincere gratitude for the financial support of the Fundamental Research Grant Scheme (FRGS/1/2018/ICT02/UMP/02/15) and University Malaysia Pahang grant (RDU180369).

References

1. Al-Shayea, Q.K.: Artificial neural networks in medical diagnosis. Int. J. Comput. Sci. Issues **8**(2), 150–154 (2011)
2. AlRashidi, M., El-Hawary, M.: Applications of computational intelligence techniques for solving the revived optimal power flow problem. Electr. Power Syst. Res. **79**(4), 694–702 (2009)
3. McCulloch, W.S., Pitts, W.: A logical calculus of the ideas immanent in nervous activity. Bull. Math Biophys. **5**(4), 115–133 (1943)
4. Grossberg, S.: How does a brain build a cognitive code? Psychol. Rev. **87**(1), 1 (1980)
5. Simpson, P.K.: Artificial neural systems: foundations, paradigms, applications, and implementations. Pergamon (1990)
6. Sampson, G.: Parallel distributed processing: explorations in the microstructures of cognition, JSTOR (1987)
7. Hopfield, J.J.: Neurons with graded response have collective computational properties like those of two-state neurons. Proc. Nat. Acad. Sci. **81**(10), 3088–3092 (1984)
8. Broomhead, D.S., Lowe, D.: Radial basis functions, multi-variable functional interpolation and adaptive networks. Royal Signals and Radar Establishment Malvern (United Kingdom) 1988
9. Gabrys, B., Bargiela, A.: General fuzzy min-max neural network for clustering and classification. IEEE Trans. Neural Netw. **11**(3), 769–783 (2000)

10. Punitha, K., Latha, B.: Sampling imbalance dataset for software defect prediction using hybrid neuro-fuzzy systems with Naive Bayes classifier. Tehnički vjesnik **23**(6), 1795–1804 (2016)
11. Loron, A.S., Loron, M.S.: An integrated fuzzy analytic hierarchy process-fuzzy data envelopment analysis method for intelligent building assessment. Tehnicki vjesnik/Technical Gazette, vol. 22, no. 2 (2015)
12. Joshi, A., Ramakrishman, N., Houstis, E.N., Rice, J.R.: On neurobiological, neuro-fuzzy, machine learning, and statistical pattern recognition techniques. IEEE Trans. Neural Netw. **8** (1), 18–31 (1997)
13. Sayaydeha, O.N.A., Mohammad, M.F.: Diagnosis of the parkinson disease using enhanced fuzzy min-max neural network and OneR attribute evaluation method. In: 2019 International Conference on Advanced Science and Engineering (ICOASE), pp. 64–69 (2019)
14. Alhroob, E., Mohammed, M.F., Lim, C.P., Tao, H.: A critical review on selected fuzzy min-max neural networks and their significance and challenges in pattern classification. IEEE Access **7**, 56129–56146 (2019)
15. Bargiela, A., Pedrycz, W., Tanaka, M.: An inclusion/exclusion fuzzy hyperbox classifier. Int. J. Knowl.-based Intell. Eng. Syst. **8**(2), 91–98 (2004)
16. Nandedkar, A.V., Biswas, P.K.: A fuzzy min-max neural network classifier with compensatory neuron architecture. IEEE Trans. Neural Netw. **18**(1), 42–54 (2007)
17. Zhang, H., Liu, J., Ma, D., Wang, Z.: Data-core-based fuzzy min–max neural network for pattern classification. IEEE Trans. Neural Netw. **22**(12), 2339–2352 (2011)
18. Mohammed, M.F., Lim, C.P.: An enhanced fuzzy min–max neural network for pattern classification. IEEE Trans. Neural Netw. Learn. Syst. **26**(3), 417–429 (2015)
19. Alhroob, E., Ghani, N.A.: Fuzzy min-max classifier based on new membership function for pattern classification: a conceptual solution. In: 2018 8th IEEE International Conference on Control System, Computing and Engineering (ICCSCE), pp. 131–135 (2018)
20. Sayaydeh, O.N.A., Mohammed, M.F., Lim, C.P.: Survey of fuzzy min–max neural network for pattern classification variants and applications. IEEE Trans. Fuzzy Syst. **27**(4), 635–645 (2019)
21. Ahmed, M.M., Isa, N.A.M.: Knowledge base to fuzzy information granule: a review from the interpretability-accuracy perspective. Appl. Soft Comput. **54**, 121–140 (2017)
22. Hujainah, F., Bakar, R.B.A., Abdulgabber, M.A., Zamli, K.Z.: Software requirements prioritisation: a systematic literature review on significance, stakeholders, techniques and challenges. IEEE Access **6**, 71497–71523 (2018)
23. Hujainah, F., Bakar, R.B.A., Al-Haimi, B., Abdulgabber, M.A.: Investigation of stakeholder analysis in requirement prioritization techniques. Adv. Sci. Lett. **24**(10), 7227–7231 (2018)
24. Hujainah, F., Bakar, R.B.A., Al-Haimi, B., Nasser, A.B.: Analyzing requirement prioritization techniques based on the used aspects. Res. J. Appl. Sci. **11**(6), 327–332 (2016)
25. Hujainah, F., Abu Bakar, R.B., Al-haimi, B., Abdulgabber, M.A.: Stakeholder quantification and prioritisation research: a systematic literature review. Inf. Softw. Technol. **102**, 85–99 (2018)
26. Hujainah, F., Abu Bakar, R.B., Abdulgabber, M.A.: Investigation of requirements interdependencies in existing techniques of requirements prioritization. Tehnicki vjesnik **26**(4), 1186–1190 (2019)
27. Simpson, P.K.: Fuzzy min-max neural networks. I. classification. IEEE Trans. Neural Netw. **3**(5), 776–786 (1992)
28. Likas, A.: Reinforcement learning using the stochastic fuzzy min–max neural network. Neural Process. Lett. **13**(3), 213–220 (2001)

29. Kim, H.-J., Yang, H.-S.: A weighted fuzzy min-max neural network and its application to feature analysis. In: Wang, L., Chen, Y.O.K. (eds.) Advances in Natural Computation. LNCS, vol. 3612, pp. 1178–1181. Springer, Berlin (2005)
30. Quteishat, A.M., Lim, C.P.: A modified fuzzy min-max neural network and its application to fault classification. In: Soft Computing in Industrial Applications, pp. 179–188. Springer (2007)
31. Quteishat, A., Lim, C.P., Tan, K.S.: A modified fuzzy min-max neural network with a genetic-algorithm-based rule extractor for pattern classification. IEEE Trans. Syst. Man Cybern. - Part A: Syst. Hum. **40**(3), 641–650 (2010)
32. Shinde, S., Kulkarni, U.: Extracting classification rules from modified fuzzy min–max neural network for data with mixed attributes. Appl. Soft Comput. **40**, 364–378 (2016)
33. Mohammed, M.F., Lim, C.P.: A new hyperbox selection rule and a pruning strategy for the enhanced fuzzy min–max neural network. Neural Netw. **86**, 69–79 (2017)
34. Mohammed, M.F., Lim, C.P.: Improving the fuzzy min-max neural network with a K-nearest hyperbox expansion rule for pattern classification. Appl. Soft Comput. **52**, 135–145 (2017)
35. Nandedkar, A.V., Biswas, P.K.: A general reflex fuzzy min-max neural network. Eng. Lett. **14**(1), 195–205 (2007)
36. Davtalab, R., Dezfoulian, M.H., Mansoorizadeh, M.: Multi-level fuzzy min-max neural network classifier. IEEE Trans. Neural Netw. Learn. Syst. **25**(3), 470–482 (2014)

Modified Opposition Based Learning to Improve Harmony Search Variants Exploration

Alaa A. Alomoush[2(✉)], AbdulRahman A. Alsewari[1,2(✉)], Hammoudeh S. Alamri[3], Kamal Z. Zamli[2], Waleed Alomoush[3], and Mohammed I. Younis[4]

[1] IBM Centre of Excellence, Universiti Malaysia Pahang, Gambang, Malaysia
alsewari@ump.edu.my

[2] Faculty of Computing, College of Computing and Applied Sciences, Universiti Malaysia Pahang, Kuantan 26300, Pahang, Malaysia
allaa1030@gmail.com

[3] Computer Science Department, Imam Abdulrahman Bin Faisal University, Dammam, Saudi Arabia

[4] Department of Computer Engineering, College of Engineering, University of Baghdad, Baghdad, Iraq

Abstract. Harmony Search Algorithm (HS) is a well-known optimization algorithm with strong and robust exploitation process. HS such as many optimization algorithms suffers from a weak exploration and susceptible to fall in local optima. Owing to its weaknesses, many variants of HS were introduced in the last decade to improve its performance. The Opposition-based learning and its variants have been successfully employed to improve many optimization algorithms, including HS. Opposition-based learning variants enhanced the explorations and help optimization algorithms to avoid local optima falling. Thus, inspired by a new opposition-based learning variant named modified opposition-based learning (MOBL), this research employed the MOBL to improve five well-known variants of HS. The new improved variants are evaluated using nine classical benchmark function and compared with the original variants to evaluate the effectiveness of the proposed technique. The results show that MOBL improved the HS variants in term of exploration and convergence rate.

Keywords: Harmony search · Opposition based learning · Meta-heuristics · Evolutionary algorithms · Optimization

1 Introduction

The increasing number of complex systems and tools in our life requires more efficient optimization technique with lower cost. Hence these problems are difficult to be achieved using classical optimization algorithm because it requires much time and resource consumptions, the need for faster techniques such as metaheuristic has increased. Different metaheuristic algorithms are widely used to solve NP-problem, through obtaining a good result with sacrifices of finding optimal results, such as Genetic algorithm [1], Artificial Bee Colony [2], Firefly algorithm [3, 4] and harmony search Algorithm [5].

F. Saeed et al. (Eds.): IRICT 2019, AISC 1073, pp. 279–287, 2020.
https://doi.org/10.1007/978-3-030-33582-3_27

Harmony Search (HS) algorithm is a well-known metaheuristic algorithm introduced by Geem et al. [5]. HS has been implemented in many research fields, including Software testing [6–9], engineering [10], artificial intelligence [11, 12], and others [13, 14].

As many metaheuristics, the HS algorithm has some weakness, such as parameters-dependent, low convergence rate and falling into local optima. To solve these problems many researchers created new variants of HS with adaptive parameters selection or by modifying the original HS to improve its performance [15–19].

Opposition based learning is a new technique introduced by Tizhoosh [20] to improve algorithm exploration capabilities. Different variants have been introduced in the literature to improve this technique, and many algorithms adapted it, including HS variants and hybrid algorithms [21–23].

The motivation of the work is to improve some of the HS variants performance using a modified opposition technique (MOBL), introduced by previous work to improve the original variant of HS [18].

The MOBL aims to improve the HS [5] exploration and avoid falling in local optima. Since the MOBL has shown promising results, in this work, we have used it with some recent variants of HS to improve these variants performance. The five HS variants are: Global best harmony search (GHS 2008) [16], Novel best global harmony search (NGHS 2010) [24], Self-Adaptive Global Best Harmony Search Algorithm (SGHS2010) [25], Global Harmony Search with Generalized Opposition-Based Learning (GOGHS 2017) [23], and Hybrid Harmony Search and Simulated Annealing (HS-SA 2018) [26].

To evaluate the new technique with HS variants, we have used 9 classical benchmark functions as described in Table 1.

Table 1. Benchmark functions (D = 30).

Function	Function formula	Range
F1: Sphere	$\sum_{i=1}^{n} x_i^2$	−100, 100
F2: Schwefel's 2.22	$\sum_{i=1}^{D} \lvert X_i \rvert + \Pi_{i=1}^{D} = \lvert X_i \rvert$	−10, 10
F3: Step	$\sum_{i=1}^{D} (\lvert X_i + 0.5 \rvert)^2$	−100, 100
F4: Rosenbrock	$\sum_{i=1}^{D} 100 \times (X_i - X_{i-1}^2)^2 + (x_{i-1} - 1)^2$	−30, 30
F5: Schwefel's 2.26	$-\sum_{i=1}^{n} \left[x_i sin\left(\sqrt{\lvert x_i \rvert}\right)\right]$	−500, 500
F6: Rastrigin	$\sum_{i=1}^{D} (X_i^2 - 10cos(2\pi x_i) + 10)$	−5.12, 5.12
F7: Ackleys	$-20exp\left(-0.2\sqrt{\frac{1}{30}\sum_{i=1}^{D} x^2}\right) - exp\left(\sqrt{\frac{1}{30}\sum_{i=1}^{D} cos2x^2}\right) + 20 + e$	−32, 32
F8: Griewank	$\frac{1}{4000}\sum_{i=1}^{D} x^2 - \Pi_{i=1}^{D} cos\frac{x_i}{\sqrt{i}} + 1$	−600, 600
F9: Rotated hyper-ellipsoid	$\sum_{i=1}^{n}\left(\sum_{j=1}^{j=i} x_j\right)^2$	−100, 100

The following section will be organized as follow: HS and some of its variants will be described in Sect. 2, the proposed algorithm will be described in Sect. 3, results and discussion in will be described in Sect. 4, and finally, the conclusion will be described in Sect. 5.

2 HS and Its Variants

2.1 Original HS Structure

The HS mimics the creation of Jazz music, by implementing the tuning process of creating new music tune, similarly, the algorithm creates new possible value for specific objective functions. The original HS algorithm contains five main steps:

1: HS assign the values of its parameters (harmony memory consideration rate (HMCR), pitch adjustment rate (PAR), harmony memory size (HMS), and bandwidth (BW)).
2: Initializing population randomly within the objective function boundaries.
3: Improvising population will be accomplished by using the HS parameters (HMCR, PAR, and BW). This step contains two main parts:
   (a) Firstly, the algorithm will create a random value (rand) between (0, 1), and if (rand > HMCR), then a random value will be created using the following equation:

$$x_j^{'} = LB + R5 \times (UB - LB))$$

   (b) if (rand < HMCR), then a random HM will be chosen and another random value (rand2) will be created between (0, 1) and if (rand2 < PAR) then the value of the chosen HM will be modified using the following equation:

$$x_j^{'} = x_j^{'} \pm R4 \times bw$$

4: HM will be updated if the new value from step three is better than the worst value in HM.
5: Finally, the algorithm will stop if it meets the stopping criteria such as the maximum number of evaluations.

2.2 HS Variants

Since the introduction of the HS, it's grabbed many researchers' attention to solve different types of engineering and real-world problems. But to overcome the weakness in the original HS, many variants and hybrid algorithms have been introduced based on the HS. In this work, we will use five well-known variants of HS. The main goal is to improve these variants effectiveness.

Next, we will describe the five variants, and more detailed information's are provided in previous review paper [13]:

1: Global best harmony search [16]: in this variant, the authors modify the original HS by borrowing the idea of using best value to update the memory from another metaheuristic named particle swarm optimization [27]. Another modification by using updating formulas of PAR and BW through the search process as described in previous variants named improved harmony search algorithm [15].
2: Novel best global harmony search [24]: in this variants the author removed the PAR and HCMR parameters and modify the search process based on position updating and genetic mutation, to avoid falling in local optima and improve the algorithm performance.
3: Self-Adaptive Global Best Harmony Search Algorithm [25]: in this variant, the original GHS variant have been improved by adapting dynamic HCMR and PAR values, and BW value will be reduced through the iteration process.
4: Global Harmony Search with Generalized Opposition-Based Learning GOGHS [23]: in this variant, the concept of utilizing the best-obtained value to is used to update the HM, and the generalized opposition used to improve the algorithm performance and avoid falling in local optima.
5: Hybrid Harmony Search and Simulated Annealing [26]: a hybrid algorithm introduced in this work by implementing the temperature variable from another metaheuristic named Simulated Annealing [28] to improve the exploration and avoid falling into local optima.

3 Proposed Algorithms

Even though most of the HS variants have different modifications on the original HS but they still have the same updating process as the original HS as described in Algorithm 1 (between lines 12–14). In this work we will add the MOBL within the updating phase of each variant to improve their exploration.

The following section will describe the MOBL structure as part of the update process of all these variants including GOGHS, as we will replace the generalized opposition with the MOBL in this case for a fair comparison.

3.1 MOBL Structure

The MOBL introduced by previous work as part of a hybrid algorithm to improve the performance of the original HS algorithm. In this work, to improve the performance of the five variants of HS, the MOBL has been applied on each variant to improve its performance.

The general structure of MOBL described in Algorithm 2. The MOBL will be added within the HS updating process.

In Algorithm 1, $x\{d\}$ symbolize the new improvised vector, (rand) is a random number of $(0 \sim 1)$, d is the number of decisions variables, and $\bar{X}_i$ is the MOBL value. After the improvisation of HS breed a new value x_j, the MOBL will be used to modify it and create new value $\bar{X}_i$ in the update phase, between line 12 to14 of algorithm 1, and will replace it if it is better based on their objective function value.

Algorithm1: Modified opposition

1. $x\{d\} = \{x_1, x_2, \ldots\ldots x_d\}$
2. $rand = random\ value\ between\ (0\,,1)$
3. $for\ (i = 1\ to\ d\ )do$
4. $\bar{X}_\iota = -1 \times x\{i\} \times r;$
5. $if(f(\bar{x}) < f(x))$
6. $x = \bar{x}$

The main role of the MOBL is applied in the updating phase of each variant, to create a new possible solution. If the new value created using MOBL better than the obtained value from HS improvisation, the new value will be considered as the replacement, and after that, it will be compared with the worst value of the HM to replace it.

4 Results

To evaluate the proposed algorithms, we have used 9 standard benchmark function as described in tale 1. To evaluate the MOBL effectiveness on the HS variants performance, this research compares the original HS variants and these variants with the MOBL technique. All the variant applied based on their original description of their authors, except for two parameters, (harmony memory size = 5, and the maximum number of evaluation function = 50^4), we have used fixed values for all the variants for a fair comparison.

Tables 2 and 3 present the mean, standard deviation and running time of the obtained result of each HS variants verse its modified version using MOBL.

Table 2 present the results of three variants of HS, and its improved version using the proposed technique in this paper (the MOBL):

1: The first column has the results of the original HS.
2: The second column has the original HS with MOBL (HS-MOBL), which present better results than the original HS in the previous column.
3: The third column has the Global best harmony search (GHS).
4: The fourth column has the result of GHS with the MOBL as (GHS-MOBL), and we can see it has better results than the original GHS.
5: The fifth column has the results of a novel global best harmony search (NGHS).
6: The sixth column has the NGHS with MOBL as (NGHS-MOBL), and it provides better results than the original NGSH.

Table 3 presents two variants and one hybrid algorithm of the HS algorithm:

1: The first column has the results of a self-adaptive global harmony search (SGHS).
2: The second column has the SGHS with MOBL as (SGHS-MOBL), which improve the original SGHS performance.

3: The third column has the results of global harmony search with generalized opposition-based learning (GOGHS)
4: The fourth column has the results of GOGHS with MOBL as (GOGHS-MOBL), and since the original GOGHS used the generalized opposition, we have replaced it with our MOBL, which provide a better result as we can see in Table 3.
5: The fifth column provides the results of hybrid harmony search and simulated annealing (HS-SA).
6: The sixth column provides the results of HS-SA with MOBL as (HS-SA -MOBL), which provides better results than the original hybrid algorithm HS-SA.

As we can see in both Tables 2 and 3 the MOBL improved the performance of all variants and reduced the running time for most of the benchmark functions. The best overall results obtained by GOGHS after using MOBL.

Table 2. Mean and SD of the errors for the benchmark function optimization (D = 30).

F	Index						
		HS	HS-MOBL	GHS	GHS-MOBL	NGHS	NGHS-MOBL
F1	Mean	6.87	5.4E−10	0.01	3.9E−242	3.3E−15	6.6E−68
	SD	6.87	5.4E−10	0.01	3.9E−242	3.3E−15	6.6E−68
	Time	1.22	0.75	1.55	0.83	1.05	0.874
F2	Mean	0.11	3.7E−5	0.06	4.2E−126	4.5E−9	1.2E−36
	SD	0.11	3.7E−5	0.06	4.2E−126	4.5E−9	1.2E−36
	Time	1.18	0.83	1.42	0.78	1.02	0.938
F3	Mean	10.93	0.0	0.0	0.0	0.0	0.0
	SD	10.93	0.0	0.0	0.0	0.0	0.0
	Time	1.30	0.914	1.79	0.95	1.21	0.73
F4	Mean	266.39	28.28	86.05	28.62	14.66	24.80
	SD	266.39	28.28	86.05	28.62	14.66	24.80
	Time	1.14	0.69	1.35	0.7	1.01	0.58
F5	Mean	−12543	−12474	−12569	−12569	−12569	−12565
	SD	−12543	−12474	−12569	−12569	−12569	−12565
	Time	3.02	2.82	3.62	2.63	3.04	2.579
F6	Mean	0.85	7.8E−8	0.01	0.0	2.7E−12	0.0
	SD	0.85	7.8E−8	0.01	0.0	2.7E−12	0.0
	Time	2.91	2.5	3.19	2.42	2.86	2.210
F7	Mean	0.89	1.6E−5	0.02	4.4E−16	1.5E−8	4.4E−16
	SD	0.89	1.6E−5	0.02	4.4E−16	1.5E−8	4.4E−16
	Time	3.22	2.66	3.60	2.55	3.028	2.40
F8	Mean	1.08	8.01E−12	0.03	0.0	0.06	0.0
	SD	1.08	8.01E−12	0.03	0.0	0.06	0.0
	Time	3.41	2.87	3.63	3.0	3.23	3.35
F9	Mean	48.89	4.1E−9	0.48	0.0	2.5E−14	5.1E−69
	SD	48.89	4.1E−9	0.48	0.0	2.5E−14	5.1E−69
	Time	1.68	1.20	1.89	1.36	1.60	1.66

Table 3. Mean and SD of the errors for the benchmark function optimization (D = 30).

F	Index						
		SGHS	SGHS-MOBL	GOGHS	GOGHS - MOBL	HS-SA	HS-SA–MOBL
F1	Mean	1.613	0.0	4.9E−7	0.0	10.242	5.6E−14
	SD	1.613	0.0	4.9E−7	0.0	10.242	5.6E−14
	Time	2.188	1.513	1.198	0.833	0.124	0.344
F2	Mean	0.346	0.0	3.8E−5	0.0	0.851	2.0E−6
	SD	0.346	0.0	3.8E−5	0.0	0.851	2.0E−6
	Time	1.867	1.095	0.773	0.832	0.074	0.20
F3	Mean	1.266	0.0	0.1	0.0	11.766	0.0
	SD	1.266	0.0	0.1	0.0	11.766	0.0
	Time	1.966	1.130	0.821	0.957	0.173	0.344
F4	Mean	820	28.28	38.71	28.760	553.709	28.43
	SD	820	28.28	38.71	28.760	553.709	28.43
	Time	1.891	1.066	0.70	0.612	0.083	0.127
F5	Mean	−16238	−15243	−14272	−6779.76	−12542.17	−12475
	SD	−16238	−15243	−14272	−6779.76	12542.17	−12475
	Time	3.768	3.00	3.024	2.383	0.118	0.19
F6	Mean	0.984	1.0E−11	0.03	0.0	1.449	9.3E−10
	SD	0.984	1.0E−11	0.03	0.0	1.449	9.3E−10
	Time	3.796	3.207	2.07	1.854	0.107	0.15
F7	Mean	2.127	8.2E−8	6.8E−7	4.4E−16	1.610	2.0E−6
	SD	2.127	8.2E−8	6.8E−7	4.4E−16	1.610	2.0E−6
	Time	3.751	3.07	2.197	1.932	0.112	0.163
F8	Mean	0.327	2.5E−15	0.0	0.0	1.103	2.1E−12
	SD	0.327	2.5E−15	0.0	0.0	1.103	2.1E−12
	Time	4.109	3.415	2.6	2.108	0.112	0.172
F9	Mean	20.723	2.9E−13	4.7E−4	0.0	92.409	1.3E−10
	SD	20.723	2.9E−13	4.7E−4	0.0	92.409	1.3E−10
	Time	2.283	1.580	1.377	1.767	0.074	0.093

5 Conclusion

The current work attempts to explore a new modified version of opposition-based learning, to improve the HS algorithm exploration and some of its variants. A comparison between the variants before and after adding the MOBL made to provide an insight into the performance improvement. The HS variants with MOBL outperform the original HS variants, with more effective results. The GOGHS-MOBL variant provided the best-obtained results. For future work, we recommend applying the MOBL technique on other metaheuristics to improve their performance and solving some real-world problems.

Acknowledgment. This research is funded by, UMP (RDU190334): A Novel Hybrid Harmony Search Algorithm with Nomadic People Optimizer Algorithm for Global Optimization and Feature Selection, and (FRGS/1/2018/ICT05/UMP/02/1) (RDU190102): A Novel Hybrid Kidney-Inspired Algorithm for Global Optimization Enhance Kidney Algorithm for IoT Combinatorial Testing Problem.

References

1. Baker, J.E.: Adaptive selection methods for genetic algorithms. In: Proceedings of an International Conference on Genetic Algorithms and Their Applications. Hillsdale, New Jersey (1985)
2. Karaboga, D., Basturk, B.: Artificial bee colony (ABC) optimization algorithm for solving constrained optimization problems. In: International Fuzzy Systems Association World Congress. Springer (2007)
3. Yang, X.-S.: Firefly algorithm, stochastic test functions and design optimisation. arXiv preprint arXiv:1003.1409 (2010)
4. Alomoush, W., et al.: Firefly photinus search algorithm. J. King Saud Univ.-Comput. Inf. Sci., 1319–1578 (2018, in press)
5. Geem, Z.W., Kim, J.H., Loganathan, G.: A new heuristic optimization algorithm: harmony search. simulation **76**(2), 60–68 (2001)
6. Alsewari, A.A., Zamli, K.Z.: Interaction test data generation using harmony search algorithm. In: 2011 IEEE Symposium on Industrial Electronics and Applications (ISIEA). IEEE (2011)
7. Alsewari, A.R.A., Zamli, K.Z.: A harmony search based pairwise sampling strategy for combinatorial testing. Int. J. Phys. Sci. **7**(7), 1062–1072 (2012)
8. Alsewari, A.R.A., Zamli, K.Z.: Design and implementation of a harmony-search-based variable-strength t-way testing strategy with constraints support. Inf. Software Technol. **54** (6), 553–568 (2012)
9. Alsewari, A., Zamli, K., Al-Kazemi, B.: Generating t-way test suite in the presence of constraints. J. Eng. Technol. (JET) **6**(2), 52–66 (2015)
10. Nazari-Heris, M., et al.: Large-scale combined heat and power economic dispatch using a novel multi-player harmony search method. Appl. Therm. Eng. **154**, 493–504 (2019)
11. Cuevas, E., et al.: Circle detection by harmony search optimization. J. Intell. Robot. Syst. **66** (3), 359–376 (2012)
12. Alomoush, W., Norwawi, A.A.N., Alomari, Y.M., Albashish, D.: A survey: challenges of image segmentation based fuzzy C-means clustering algorithm. J. Theor. Appl. Inf. Technol. **96**(16), 18 (2018)
13. Ala'a, A., et al.: Comprehensive review of the development of the harmony search algorithm and its applications. IEEE Access **17**(7), 14233–14245 (2019)
14. Manjarres, D., et al.: A survey on applications of the harmony search algorithm. Eng. Appl. Artif. Intell. **26**(8), 1818–1831 (2013)
15. Mahdavi, M., Fesanghary, M., Damangir, E.: An improved harmony search algorithm for solving optimization problems. Appl. Math. Comput. **188**(2), 1567–1579 (2007)
16. Omran, M.G., Mahdavi, M.: Global-best harmony search. Appl. Math. Comput. **198**(2), 643–656 (2008)
17. Wang, C.-M., Huang, Y.-F.: Self-adaptive harmony search algorithm for optimization. Expert Syst. Appl. **37**(4), 2826–2837 (2010)

18. Alaa, A., Alomoush, A.A.A., et al.: Hybrid harmony search algorithm with grey wolf optimizer and modified opposition-based learning. IEEE Access **7**, 1–3 (2019)
19. Das, S., et al.: Exploratory power of the harmony search algorithm: analysis and improvements for global numerical optimization. IEEE Trans. Syst. Man Cybern. Part B (Cybern) **41**(1), 89–106 (2011)
20. Tizhoosh, H.R.: Opposition-based learning: a new scheme for machine intelligence. In: 2005 International Conference on Intelligent Agents, Web Technologies and Internet Commerce Computational Intelligence for Modelling, Control and Automation. IEEE (2005)
21. Gao, X., et al.: A hybrid optimization method of harmony search and opposition-based learning. Eng. Optim. **44**(8), 895–914 (2012)
22. Xiang, W.-L., et al.: An improved global-best harmony search algorithm for faster optimization. Expert Syst. Appl. **41**(13), 5788–5803 (2014)
23. Guo, Z., et al.: Global harmony search with generalized opposition-based learning. Soft. Comput. **21**(8), 2129–2137 (2017)
24. Zou, D., et al.: Novel global harmony search algorithm for unconstrained problems. Neurocomputing **73**(16), 3308–3318 (2010)
25. Pan, Q.-K., et al.: A self-adaptive global best harmony search algorithm for continuous optimization problems. Appl. Math. Comput. **216**(3), 830–848 (2010)
26. Assad, A., Deep, K.: A Hybrid Harmony search and Simulated Annealing algorithm for continuous optimization. Inf. Sci. **450**, 246–266 (2018)
27. Eberhart, R., Kennedy, J.: A new optimizer using particle swarm theory. In: 1995 Proceedings of the Sixth International Symposium on Micro Machine and Human Science, MHS 1995, pp. 39–43 (1995)
28. Kirkpatrick, S., Gelatt, C.D., Vecchi, M.P.: Optimization by simulated annealing. Science **220**(4598), 671–680 (1983)

Word Embedding-Based Biomedical Text Summarization

Oussama Rouane[1(✉)], Hacene Belhadef[1], and Mustapha Bouakkaz[2]

[1] University of Constantine 2 - Abdelhamid Mehri, Constantine, Algeria
{oussama.rouane,hacene.belhadef}@univ-constantine2.dz
[2] Computer Science Department, Faculty of Sciences,
University of Amar Telidgi, Laghouat, Algeria
m.bouakkaz@lagh-univ.dz

Abstract. In this paper, we have proposed a novel word embedding-based biomedical text summarizer. Biomedical words are represented by real dense vectors. Sentences are represented by summing-up the word vectors that contain. The PageRank algorithm is applied to rank sentences using the cosine similarity as a distance measure between sentences vectors. The top N highly ranked sentences are selected to build the summary. For the evaluation, we created a corpus of 200 biomedical papers downloaded from the Biomed Central full-text database. We used a pre-trained Word2vec model of word vectors generated from a combination of PubMed, PMC, and recent English Wikipedia dump texts. We compared our method with four other summarizers using: ROUGE-1, ROUGE-2, ROUGE-3, and ROUGE-SU4 metrics by evaluating the generated summaries with the abstracts of papers. Our summarizer achieved an improvement of 3.48%, 7.68%, 9.76%, and 3.47% respectively against the second-ranked summarizer.

Keywords: Biomedical text summarization · Word embedding · Word2vec · PageRank algorithm · ROUGE metrics

1 Introduction

Recent researches in the biomedical field have generated a massive amount of data in the web like texts, multimedia documents, and clinical health records etc. i.e., MEDLINE[1] contains about 28 million references of journal articles in life sciences that focus on biomedicine. To manage efficiently this knowledge, there is an increasing need to develop new techniques that allow researchers to access to the desired information in more efficiently and quickly manner.

Text summarization [1] is one of the most promising techniques that retains the main ideas from a massive quantity of textual data on the web.

The majority of text summarization systems are domain-independent; they could be applied to any document regardless of its type or domain. However, in the biomedical domain, these techniques are inefficient [18]. In order to manage this problem,

[1] https://www.nlm.nih.gov/bsd/pmresources.html.

F. Saeed et al. (Eds.): IRICT 2019, AISC 1073, pp. 288–297, 2020.
https://doi.org/10.1007/978-3-030-33582-3_28

researchers in this area have incorporated the semantic level to analyze these texts using domain-knowledge resources like: ontologies, thesaurus, and taxonomies [12], and then treating these texts using a conceptual representation retained from these resources (a set of concepts with their semantic relationships). These methods could be a solution to many problems in this domain. However, their major problem is that all the treatment are dependent on these resources. Moreover, they do not take into account the significance of the co-occurrence of words in the text that is proved in recent researches.

The recent apparition of the deep learning field led researchers to develop innovative techniques called "word embedding" which use neural networks to learn from a massive raw corpus vectors representations of text units such as words, sentences, and documents. These techniques are applied to many NLP tasks, and they achieved excellent results because they use both the global and the local co-occurrences of text units to learn vectors representations.

In this paper, we propose a word embedding-based biomedical text summarizer that incorporates one of these words embedding models called "Word2vec", and as the best of our knowledge, this current work is the first one that uses word embedding in biomedical text summarization. Firstly, each biomedical paper is split into sentences and tokenized into words. Words are represented by real-valued dense vectors learned using the word2vec function. Sentences are represented by summing-up the word vectors that contain. A graph is generated where nodes are represented by sentences vectors, and edges represent the cosine similarity between them. The PageRank algorithm is applied to the graph to rank sentences. Finally, the summary is generated by selecting the top N highly scored sentences and placed in the summary based on their original orders in the document.

The paper is organized as follow: in Sect. 2, we give a background of the domain and its related works. In Sect. 3, we present the architecture of the method. In Sect. 4, we present the evaluation process and the results of the experiments with a discussion, and finally, a conclusion and perspectives are given in the final section.

2 Background and Related Work

According to the definition in [9], a summary can be defined as: "a text that is produced from one or more texts, that contain a significant portion of the information in the original text(s), and that is no longer than half of the original text(s)".

Text summarization methods could be divided into many factors [1]. For example, they could be single or multiple, monolingual or multilingual, informative or indicative, generic or user-oriented, domain-independent or domain-specific, and extractive or abstractive.

In domain-independent text summarization, researchers have been proposed many methods that generate summaries using a direct representation of texts. For example, methods have been proposed based on a superficial analysis of the text like: the frequency of terms, the position of sentences, the presence of cue words or phrases, and sentence similarity to the titles [6]. These features also could be combined using a linear function to calculate a single score to each sentence or could be used for machine learning algorithms to determine the best subset of these features to generate concise summaries. Graph-based methods are also proposed in this field; they model a document as a graph where nodes represent text units and edges represent relationships between them. Then, many algorithms are used to determine the most salient units in these graphs for example, MEAD [20], TextRank [13], and LexRank [19].

The biomedical domain has its own properties that must be considered during the summarization task, for example, biomedical terms are often compounds, acronyms and abbreviations [5]. The variety of synonyms [16], and homonyms [23]; a lot of terms are not universal [8]. For this reason, researchers used biomedical knowledge-resources like UMLS[2], MeSH[3], and SNOMED[4]. These resources contain the majority of terms and concepts in this field with their definitions and the semantic relationships between them, by building a semantic representation model of the source text using these resources, biomedical texts become more understandable for machines, i.e., authors in [21] have combined lexical chains with biomedical concepts to construct summaries. Authors in [25] proposed a graphical representation of biomedical documents using the MeSH tree to construct summaries. Authors in [18] proposed a semantic-graph based approach using the UMLS metathesaurus. Authors in [22] combined clustering and frequent itemsets mining techniques with a conceptual representation of biomedical documents to enhance the quality of biomedical summaries.

These approaches proved their efficiency. However, analyzing biomedical texts using these knowledge-resources is a complicated process, because they range terms and concepts in a complex hierarchical structure which made them hard to maintain and do not have a consistent representation; they also neglect the co-occurrence of units in the whole document, which can give more information about the significance of these units.

Innovative techniques in the NLP field called: "word embedding" (Goldberg) are developed recently, these techniques learned a dense real valued vectors of words from a huge raw corpora. The main idea of these techniques is that the words that have the same meaning have similar vectors representations. There are different techniques to generate vector representation, e.g., dimensionality reduction [4] and probabilistic models [2]. In [14, 15], authors proposed a novel word embedding model based on neural networks called "Word2vec". This model gave a magic results, because it could predict word vectors with a simple algebraic operations like [King − Man + Woman = Queen] which is not the case in other older methods.

The Word2vec model uses two-layer neural network to learn word vectors, its input is words of the corpus and its output is a set of n-dimensional vectors. The goal of Word2vec is to group vectors of similar words in the same vector space.

Two variant of learning models are proposed and they can be used as part of the word2vec approach to learning word embedding:

2 https://www.nlm.nih.gov/research/umls/.

3 https://www.nlm.nih.gov/mesh/meshhome.html.

4 http://www.snomed.org/.

- Continuous Bag-of-Words (CBOW model).
- Skip-Gram Model.

The CBOW model learns vectors by predicting the current word based on its context. The continuous skip-gram models are learned by predicting the surrounding words given a current word. The context is defined by a window of neighboring words. The window's size is one of the parameters of the model (Fig. 1).

Word embedding techniques are used in many NLP tasks such as text classification, sentiment analysis, information retrieval… However, in the biomedical domain, there are some recent works i.e., biomedical concepts embedding [11], document retrieval [3], and question/answering summarization [7].

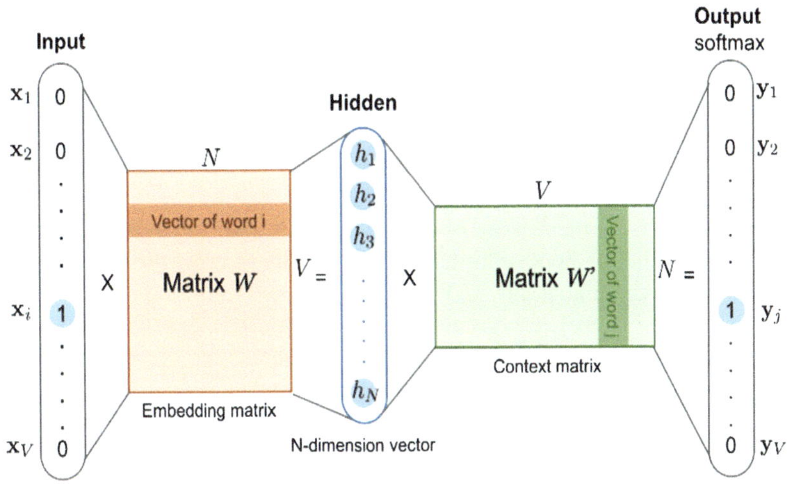

Fig. 1. CBOW and skip-gram models of Word2vec

3 Method

The main goal of this paper is to propose a novel biomedical text summarization method based on graph representation of distributed vectors. For this reason, the summarizer is composed of four steps: (1) document preprocessing, (2) graph construction, (5) sentence ranking and (6) summary generation as illustrated in Fig. 2.

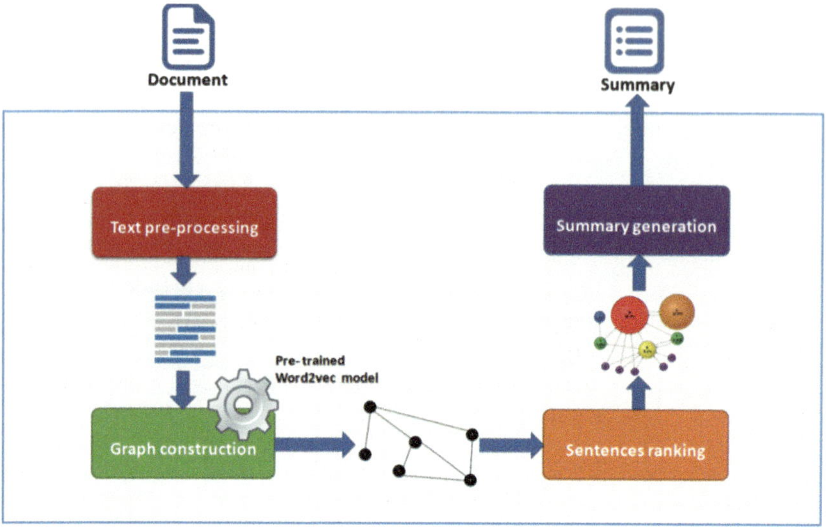

Fig. 2. The global architecture of the proposed summarizer.

3.1 Text Pre-processing

Firstly, we extract the main body of document. Then, we remove unnecessary parts from it i.e., the title, authors' information, abstract, keywords, the header of sections and subsections, the bibliography… We split the main body into sentences represented by the set $\{S_1, S_2 \ldots S_N\}$, each sentence is composed of a set of words $\{w_1, w_2 \ldots w_M\}$. In addition, we lemmatize words in the document, and we remove stop words related to the biomedical domain to reduce the vocabulary space (Fig. 3).

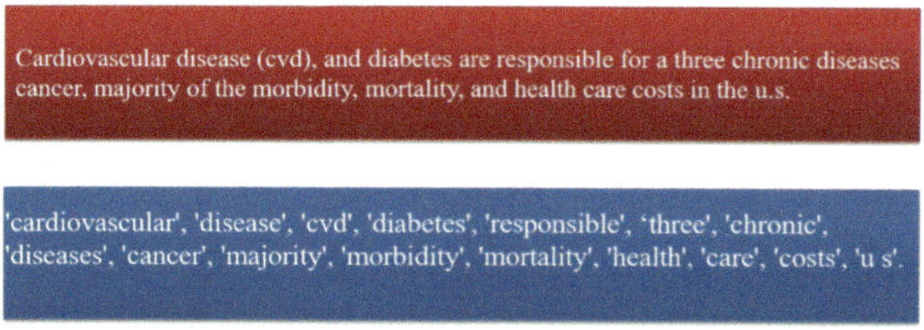

Fig. 3. Example of sentence pre-processing including: tokenization, lemmatization, and removing stop words.

3.2 Graph Construction

In the second step, we model the input document as a graph G = (V, E), where "V" is a set of nodes, i.e. the sentences of the document, and "E" is the set of edges representing the similarity values between them. Each sentence is composed of a set of words, and an n-dimensional embedding vector represents each word, as illustrated in the Fig. 4.

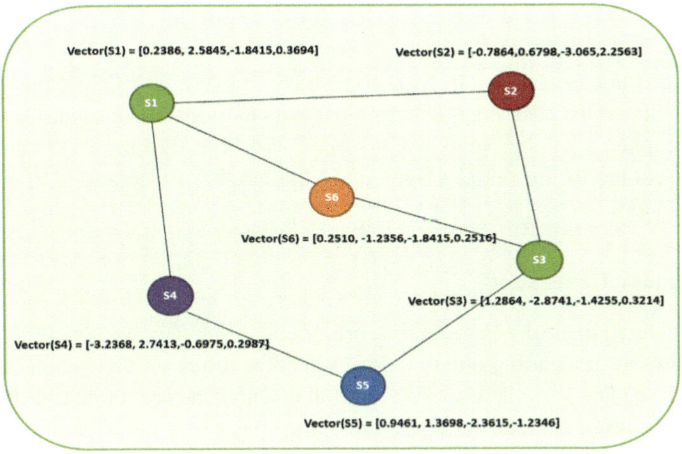

Fig. 4. Exampe of a document contains six sentences represented using a graph

We calculate the sentence vector as the element-wise sum of the distributed vectors of all words within this sentence as follow:

$$Vect(S_i) = \sum Vect(w_j) \;\; w_j \in S_i \tag{1}$$

We calculate the similarities between these sentences using the cosine similarity [24], given two sentences S_i and S_j represented by n-dimensional embedding vectors, the cosine similarity between them is defined as follow:

$$Similarity(S_i, S_j) \frac{\sum_{i,j=1}^{n} S_i \times S_j}{\sqrt{\sum_{i=1}^{n} S_i^2}\sqrt{\sum_{j=1}^{n} S_j^2}} \tag{2}$$

3.3 Sentences Ranking

Since we represented document as a graph of nodes (sentences vectors) and edges (cosine similarity between sentences vectors), we used the PageRank algorithm to rank nodes in the graph [17]. Here, G = (V, E) is a graph with "*V*" is the node set, "*E*" is the edge set, and N is the total number of nodes. Let $In(v_j)$ is the set of nodes that point to the node v_j, d is a dumping factor set to 0.85 by default, and $PR(v_j)$ is the PageRank value of the node v_j, the score of each node is calculated recursively using the formula as below:

$$PR(v_j) = \frac{1-d}{N} + d \times \sum_{v_i \in In(v_j)} \frac{PR(v_i)}{In(v_i)} \qquad (3)$$

An initial value is assigned to each node in the graph, after running the algorithm; a score is calculated for each node, which represents its importance within the graph, the algorithm iterates all nodes until it converges. We note that the final values are not affected by choice of the initial values; only the number of iterations to convergence may be different.

3.4 Summary Generation

After the calculation of the PageRank score of each sentence, we sort sentences based on their scores in descending order, and we select the top N scoring sentences, where N is a number specified by the user. We note that sentences are ranged in the summary based on their original order in the document to preserve their sequential meaning.

4 Experimental Results

4.1 Dataset

We constructed manually a corpus of 200 biomedical papers downloaded from the BioMed central full-text database[5]. We removed irrelevant units like graphics, tables, figures, captions, citation references, bibliography…We split each document into abstract and body section. We used the abstracts of papers (i.e., author's summaries) as reference summaries, and we used the body sections as input to our summarizer.

4.2 Pre-trained Word2vec Model

We used a pre-trained Word2vec model related to the biomedical domain, which is induced on a combination of PubMed, PMC, and texts extracted from a recent English Wikipedia dump[6]. The Word2vec function run using the skip-gram model with window equal to five, Subsampling threshold of 0.001, and hierarchical SoftMax training to create word vectors of dimension 200. The file size is 4 GB, and it contains about 5.5 billion word vectors.

4.3 Evaluation Metric

We used the ROUGE toolkit [10] to evaluate the performance of all tested methods. ROUGE compares the system generated summaries with model summaries and

[5] https://www.biomedcentral.com/.

[6] http://bio.nlplab.org.

evaluates the performance of the summarization method in term of how informative the generated summaries are. In this work, we used four metrics of ROUGE: ROUGE-1, ROUGE-2, ROUGE-3, and ROUGE-SU4. Both the measures estimate the shared content in terms of unigram, bigrams, three grams, and a maximum skip distance of four between bigrams. ROUGE scores have shown a significant correlation with experts' judgments in the Document Understanding Conference 2004.

4.4 Comparison Results

In this section, we compared the performance of the proposed summarizer with four summarizers, i.e., the Itemset-based summarizer, TextRank, SweSum, and Microsoft Auto summarize. The itemset-based summarizer [16] is a biomedical text summarization method that combines frequent itemset mining technique with a conceptual analysis of biomedical papers. TextRank [13] is a domain-independent text summarizer that represents sentences and their overlapping units as a graph, then ranking sentences using the PageRank algorithm. SweSum[7] is Swedish text summarizer developed for newspapers; it selects the most important sentences based on a linear combination of scores like sentences positions, sentences tags, and sentences containing numerical values. Microsoft auto summarize is a commercial application developed by Microsoft to summarize word documents, it determines the most important sentences based on the frequency of words that contains. The Table 1 shows the comparison results of our method against other summarizers in term of ROUGE-1, ROUGE-2, ROUGE-3, and ROUGE-SU4.

Table 1. The comparison results of our proposed method with other summarizers using the ROUGE package.

	ROUGE-1	ROUGE-2	ROUGE-3	ROUGE-SU4
Word embedding-based summarizer	**0.78674**	**0.32736**	**0.15620**	**0.37002**
Itemset-based summarizer	0.76026	0.30400	0.14230	0.35760
TextRank	0.75124	0.29303	0.13666	0.34219
SweSum	0.73604	0.27592	0.11931	0.32792
Microsoft auto-summarize	0.68012	0.23854	0.09302	0.28561

We can observe from the Table 1 that the proposed summarizer outperforms all tested summarizers in all ROUGE metrics. For example, it achieved a value of 0.32736 in ROUGE-2 while the second-ranked summarizer (itemset-based summarizer) achieved 0.30400 in the same metric. The improvement of our method in ROUGE-1, ROUGE-2, ROUGE-3, and ROUGE SU-4 are respectively: 3.48%, 7.68%, 9.76%, and 3.47%, which are considerable improvements. Textrank is classed third, fourth for SweSum and the worst is Microsoft auto summarize.

[7] http://swesum.nada.kth.se/index-eng.html.

The proposed method outperforms other test summarizers because it is based on the semantic meaning of words that are represented by dense distributed vectors and learned from huge corpora related to the biomedical domain. Therefore, it can capture the meaning of words, their relationships, and the semantic meaning of sentences more accurately than methods that use domain-knowledge resources, or domain-independent methods.

5 Conclusion and Future Work

We have proposed in the paper a new biomedical text summarizer based on word embedding, our method outperforms many tested summarizers, and the results are promising. Recently, the word embedding techniques could enhance largely the summarization tasks, especially in the biomedical domain since it can solve many problems in semantic based-methods. For future work, we will more focus on this domain to propose a new other summarization methods that are based on word embedding and deep learning techniques.

References

1. Afantenos, S., Karkaletsis, V., Stamatopoulos, P.: Summarization from medical documents: a survey. Artif. Intell. Med. **33**, 157–177 (2005). https://doi.org/10.1016/j.artmed.2004.07.017
2. Blei, D.M., Edu, B.B., Ng, A.Y., et al.: Latent Dirichlet allocation. J. Mach. Learn. Res. **3**, 993–1022 (2003). https://doi.org/10.1162/jmlr.2003.3.4-5.993
3. Brokos, G.-I., Malakasiotis, P., Androutsopoulos, I.: Using centroids of word embeddings and word mover's distance for biomedical document retrieval in question answering, pp. 114–118 (2016)
4. Deerwester, S., Furnas, G.W., Landauer, T.K., Harshman, R.: Indexing by latent semantic analysis. J. Am. Soc. Inf. Sci. **41**, 391–407 (1990). https://doi.org/10.1017/CBO9781107415324.004
5. Dzuganova, B.: English medical terminology – different ways of forming medical terms. JAHR – Eur. J. Bioeth. **4**, 55–69 (2013)
6. Edmundson, H.P.: New methods in automatic extracting. J. ACM **16**, 264–285 (1969). https://doi.org/10.1145/321510.321519
7. Fan, Q., Fang, Y.: An answer summarization method based on keyword extraction. In: BIO Web Conference, vol. 8, p. 03015 (2017). https://doi.org/10.1051/bioconf/20170803015
8. Friedman, C., Elhadad, N.: Natural language processing in health care and biomedicine. In: Biomedical Informatics, pp. 255–284. Springer, London (2014)
9. Hovy, E.: Automated text summarization. In: The Oxford Handbook of Computational Linguistics, pp 583–598. Oxford University Press (2005)
10. Lin, C.Y.: ROUGE: a package for automatic evaluation of summaries. In: Proceedings of Work Text Summarization Branches Out (WAS 2004), pp. 25–26 (2004)
11. Luo, X., Shah, S.: Concept embedding-based weighting scheme for biomedical text clustering and visualization. Appl. Inform. **5** (2018). https://doi.org/10.1186/s40535-018-0055-8

12. Menéndez, H.D., Plaza, L., Camacho, D.: A genetic graph-based clustering approach to biomedical summarization. In: Proceedings of 3rd International Conference on Web Intelligence, Mining and Semantics – WIMS (2013). https://doi.org/10.1145/2479787.2479807
13. Mihalcea, R., Tarau, P.: TextRank: bringing order into texts. Proc EMNLP **85**, 404–411 (2004). https://doi.org/10.3115/1219044.1219064
14. Mikolov, T., Chen, K., Corrado, G., Dean, J.: Distributed representations of words and phrases and their compositionality (2013a)
15. Mikolov, T., Yih, W.-T., Zweig, G.: Linguistic Regularities in continuous space word representations. association for computational linguistics (2013b)
16. Moradi, M., Ghadiri, N.: Quantifying the informativeness for biomedical literature summarization: an itemset mining method. Comput. Methods Program. Biomed. **146**, 77–89 (2017). https://doi.org/10.1016/j.cmpb.2017.05.011
17. Page, L., Brin, S., Motwani, R., Winograd, T.: The PageRank citation ranking: bringing order to the web (1998)
18. Plaza, L., Díaz, A., Gervás, P.: A semantic graph-based approach to biomedical summarisation. Artif. Intell. Med. **53**, 1–14 (2011). https://doi.org/10.1016/j.artmed.2011.06.005
19. Radev, D.R.: LexRank: graph-based lexical centrality as salience in text summarization. J. Artif. Intell. Res. **22**, 457–479 (2004). https://doi.org/10.1613/JAIR.1523
20. Radev, D.R., Jing, H., Budzikowska, M.: Centroid-based summarization of multiple documents: sentence extraction, utility-based evaluation, and user studies. Inf. Process. Manag. **40**(6), 919–938 (2000). https://doi.org/10.1016/j.ipm.2003.10.006. 40:10
21. Reeve, L., Han, H., Brooks, A.D.: BioChain. In: Proceedings of the 2006 ACM Symposium on Applied Computing - SAC 2006, p. 180. ACM Press, New York (2006)
22. Rouane, O., Belhadef, H., Bouakkaz, M.: Combine clustering and frequent itemsets mining to enhance biomedical text summarization. Expert Syst. Appl. **135** (2019). https://doi.org/10.1016/j.eswa.2019.06.002
23. Shortliffe, E.H., Cimino, J.J.: Biomedical Informatics: Computer Applications in Health Care and Biomedicine, 4th edn. Springer, London (2014)
24. Tan, P.-N., Steinbach, M., Kumar, V.: Introduction to Data Mining, 1 edn. Pearson, London (2006). ISBN-13: 978-0321321367, ISBN-10: 0321321367
25. Yoo, I., Hu, X., Song, I.Y.: A coherent graph-based semantic clustering and summarization approach for biomedical literature and a new summarization evaluation method. BMC Bioinform. **8**, 1–15 (2007). https://doi.org/10.1186/1471-2105-8-S9-S4

Discrete Particle Swarm Optimization Based Filter Feature Selection Technique for the Severity of Road Traffic Accident Prediction

Lawal Haruna[1,2](✉), Roselina Sallehuddin[1], and Haizan Mohammed Radzi[1]

[1] Computer Science Department, University Technology Malaysia, Skudai, Malaysia
lawalharunl@gmail.com, roselina@gmail.com, haizan@gmail.com

[2] Computer Science Department, Federal University Dutsinma, Dutsin-Ma, Nigeria

Abstract. The rate of traffic accident is increasing worldwide, which resulted to severe injuries, fatalities and economic loses. More than 1.25 million people lose their lives every year and 50 million are non-fatally wounded in road accident worldwide. The severity of road traffic accident are attributed to many factors which include human (age, gender), road (road alignment, roads surface), and environment (moisture, weather) among others. Nevertheless, some factors are more significant than others in increasing the severity level of injuries sustained by victims of accident. In previous studies, few feature selection techniques were used to get the most relevant and non-redundant features prior to prediction of severity of road traffic accident. Among the techniques of feature selection which includes filter, wrapper, embedded and hybrid, only filter technique has been considered. However, the filter feature selection techniques (filter correlation, mutual information) are inadequate at finding optimal features. Additionally, the redundant feature doesn't only affect the performance of the classification algorithm but also computational time. To this end, this paper, Proposed a hybrid discrete Particle Swarm Optimization with Fast Correlation Based Feature Selection (DPSO_FCBF) technique for selecting the most significant features for the prediction of severity of road traffic accident. In this technique, the discrete particle swarm optimization (DPSO) search for the optimum values and the fast correlation based feature selection (FCBF) evaluate the selected optimum values to select the most relevant features, then DPSO update the personal best (Pbest) and global best (Gbest) for the optimal solution. The proposed technique has been evaluated using benchmark road traffic accident dataset. The experimental result are compared with the result obtained from DPSO and FCBF individually, the proposed technique showed better result in terms of number of features selected, prediction accuracy and time complexity.

Keywords: Discrete particle swarm optimization · Feature selection · Prediction · Filter · Severity of road traffic accident

F. Saeed et al. (Eds.): IRICT 2019, AISC 1073, pp. 298–310, 2020.
https://doi.org/10.1007/978-3-030-33582-3_29

1 Introduction

Road traffic accident (RTA) becomes a major concern due to severe injuries, fatalities and economic losses. The cost of fatalities and injuries as a result of RTA has huge impact on society [1]. Globally, injuries sustained from RTA are reported to be the leading cause of deaths among younger people aged (15–29) old and are among the major three reasons of death among people aged (15–44) years [2]. An estimate of more than 1.25 million people lose their lives every year and fifty million 50 are non-fatally wounded in road accidents worldwide. The high number of deaths are mostly in low and middle-income nations where economic growth have been accompanied by increased number of vehicles and road traffic injuries [2]. Research revealed that in most of the countries in Africa, Latin America and Asia RTA becomes one of the major reasons of deaths in young and middle age adult between thirty and fourty nine (30–49) years [3], Jacobs *et al.* [4]. Further, the death as a result of accidents in lower and middle income nation are projected to rise by as much as 80% unless there are new methods to prevent incidence of accidents [2]. This pose a great challenge to various government and non-governmental organization in reducing the severity of RTA.

Most of the existing work on the severity of RTA focus on classification and prediction. However, the selection of input features in the severity of RTA is mostly done based on users' assumption or experience. On the other hand, feature selection is a very important method in building a reliable prediction strategies. By removing non-relevant and redundant features before prediction improved the prediction accuracy, save time and produced a more general classifier [5–9]. The redundant feature doesn't only affect the performance of the classification algorithm but also computational time [9]. Although feature selections has been used in some few studies in this domain [10–12]. Feature selection refers to the process of removing non-relevant and redundant features from the dataset [7]. Therefore, it becomes imperative to select the most significance features from the dataset using feature selection technique prior to prediction.

Several feature selection techniques are used to select the relevant and non-redundant features, the techniques are filter, wrapper, embedded and hybrid [13]. In filter feature selection, all features are graded based on a particular criteria. Thereafter, features with the uppermost grade are chosen [14]. The filter feature selection techniques are advantageous because they are easily scale to high dimensional datasets that are computationally simple and fast, it doesn't depend on the mining algorithm, feature selection needs to be performed only once, different classifiers can be evaluated and it tries to avoid over-fitting. On the other hand, the wrapper feature selection is a selection technique which is dependent on the learning algorithm, wrapper model used some predetermined learning algorithm to find out the relevant features. The wrapper may guarantee a good results [9]. However, wrapper is computationally expensive and has high risk of over fitting [9, 15]. [15], stated that some researchers avoided the use of wrapper approach because of high computational consumption of resources and high risk of overfitting.

On the other hand, embedded feature selection do selection in the course of modelling algorithm's implementation. These technique are embedded in the algorithm either as its normal or extended functionality. Common embedded techniques includes, decision tree algorithm such as CART, C4.5, random forest. While hybrid feature selection traditionally put together the best properties of filter and wrapper. First, filter technique in order to reduce the feature space dimension. Then, wrapper is engaged to discover the best candidate subset. Recently, numerous interesting optimization methods were proposed, such as: hybridant colony optimization, fuzzy-random-forest based feature selection, hybrid genetic algorithms [16]. In the prediction of severity of RTA more attention was given to classification, although the feature selection have been successfully employed to improve the performance of the classifier. However, among the feature selection technique wrapper, filter, embedded and hybrid, only filter (filter correlation, mutual information) are employed in the previous studies and these feature selection technique (filter correlation) are inadequate at finding optimal features [17]. To the best of authors' knowledge no any hybrid work on feature selection in this domain based on the available literature. This paper proposes the hybrid DPSO with FCBF technique to select significance features with the aim of improving the performance and reduce time taken in the prediction of severity of road traffic accident. The other part of the paper is organized as follows, Sect. 2 related work, Sect. 3 proposed technique, Sect. 4 experimental setup and result, Sect. 5 discussion and Sect. 6 conclusion.

2 Related Work

The effectiveness of the machine learning algorithm attached on the importance of the features selected. Selecting relevant and non-redundant features remain a challenging issue in data mining, as such several feature selection techniques have been proposed to select the most relevant and non-redundant features. Although no much work on feature selection in the prediction of severity of road traffic accident, below are the discussion of the few works relevant to this study.

[10] The authors used correlation coefficient and found that head on collision, highway width and type of vehicle at fault are the most significance factors.

[11] carried out a research to find relevant features and classification of road traffic accident pattern using correlation based feature selection, fast correlation based feature selection and mutual information for feature selection, and C4.5, CR-T, ID3, CS-CRT, CS-MC4, Naïve-bayes and random-tree for classification. The result shows an improvement in the performance of the classifiers.

[12] used correlation coefficient and the result shows high correlation of some variables to the target. These show the important of feature selection before prediction. However, the traditional filter techniques used in all the previous studies such as mutual information, correlation based feature selection (FCS) and fast correlation based feature selection (FCBF) are inadequate at finding optimal reductions [17], because they lack global search capability. In general, feature selections have been less studied in the prediction of severity of road traffic accident. This is because researchers pay more attention to classification and prediction, and majority select features based on assumption. Therefore, there is need to explore more feature selection techniques in the prediction of severity of RTA.

On the other hand, PSO has been used in other domain in selecting the most relevant features either as individual or hybrid, here we will consider the most relevant to our proposed technique. Some of the studies used hybrid techniques such as [17–19]. [18] Proposed (FCBF) combined with Genetic Algorithm (GA) and Particle Swarm Optimization (PSO), in the first part, the FCBF is applied to select the relevant and remove redundant features. The subset selected are now the input for GA. The selected features are set to one and other features are set randomly within 0 and 1. This same procedure is fallowed for the PSO. The wrapper approach is used for feature selection, where support vector machine SVM is used as fitness function and also for the classification. The performance of the two proposed methods shows that FCBF-PSO outperformed FCBF-GA in terms of accuracy and the number of features selected in most of the experiment carried out.

[19] Proposed correlation based feature selection (CFS) with Binary Particle swarm Optimization (BPSO) for gene selection and cancer classification. The proposed CFS-BPSO used wrapper approach, in the first place the CFS reduce the dimensionality of the data and the selected subset are used to select the optimal features using integrated BPSO with the help of Naïve-Bayes classifier. The result shows an improvement in the accuracy and low dimensional set of selected gene. All these hybrid methods none has proposed DPSO and FCBF for feature selection. In line with this, our proposed technique combines the high speed and efficiency of FCBF and global optimality of DPSO in selecting the significance features for the prediction of severity of road traffic accident and introducing a new approach different from other methods. Further, to the best of our knowledge this is the first time of using DPSO with FCBF in this domain based on the available literature.

3 Proposed Technique

In this part, we present our proposed feature selection technique for severity of road traffic accident prediction which follows DPSO based FCBF approach, the proposed technique is shown in Fig. 1. It comprises of two major part, the DPSO part and the FCBF part, the DPSO part is having two stages, the DPSO first stage initialize the parameters such as velocity, position and search for the optimal features, the position and velocity are updated upon each iteration, then the second stage of DPSO update the Pbest and Gbest to produce the selected optimal features after the FCBF evaluated the selected features. The FCBF part evaluate the selected features immediately the first stage of DPSO search for the optimal features, the next is for the FCBF to select the most relevant features and dispose the non-relevant and redundant features. There after the subset selected by FCBF are then send to DPSO as second stage to update the Gbest and Pbest. The DPSO_FCBF technique process is shown in Fig. 2.

3.1 Discrete Particle Swarm Optimization (DPSO)

The original Particle swarm optimization (PSO) was proposed in 1995 by Eberhart and Kennedy, the PSO is motivated by social and cooperative behavior exhibited by numerous animal colony, group or swarm such as flocking birds or schools of fish to fill their needs in the search space [18]. Then binary PSO was proposed in 1997 and later on the discrete particle swarm optimization (DPSO).

The algorithm is led by personal-experience (Pbest), overall- experience (Gbest) and the present movement of the particles to decide their next positions in the search space. The experiences are accelerated by two factors (cognitive memory and learning factor) c1 and c2 respectively, and two random numbers r1 and r2 generated between [0, 1] where the present movement is multiplied by an inertia factor w varying between 0 and 1 [wmin, wmax].

During the movement, the current position of particle i is representing $x_i = (x_{i1}, x_{i2}, \ldots\ldots x_{id}), i = 1, 2, \ldots\ldots, m$, d is the dimension of the search space. Each particle contain the current velocity of the particles; $v_i = (v_{i1}, v_{i2}, \ldots\ldots v_{id})$, $Pbest$, the personal best position of particle and, the global best position of the particle swarm. The PSO searches for the optimum solution by updating the position and the velocity of each particle based on the equations below:

$$x_{id}^{t+1} = x_{id}^{t} + v_{id}^{t+1} \tag{1}$$

$$v_{id}^{t+1} = w \times v_{id}^{t} + c1 \times r1 \times (p_{id} - x_{id}) + c2 \times r2 \times (p_{gd} - x_{id}) \tag{2}$$

The inertia weight (W), responsible for providing balance in-between the local and global search. The x_{id}^{t} and x_{id}^{t+1} are the present position and updated position for each iteration respectively. The v_{id}^{t} and v_{id}^{t+1} are the current velocity and updated velocity for each iteration respectively.

3.2 Fast Correlation Based Feature Selection (FCBF)

Is a multivariate filter feature selection technique which the attributes selection does not depend on the mining algorithm to be used to the best features and assess the relevance of features by looking only at the essential properties of the data [20, 21]. The correlation based feature selection has numerous advantages which includes Less execution/computational time [7, 18, 22–25], Very efficient [25], Reduction in computational cost [7, 18], improve the performance of learning algorithm [7, 23], attribute are treated independently [18, 23].

The two feature selection techniques are hybridized together so as to remove irrelevant and redundant features as they could lead to biased in result and longer training time. [20] The major goals of feature selection are to escape over-fitting, improve performance, and provide quicker and more effective models. The propose DPSO-_FCBF is expected to reduce the running time and increase accuracy of the classifier.

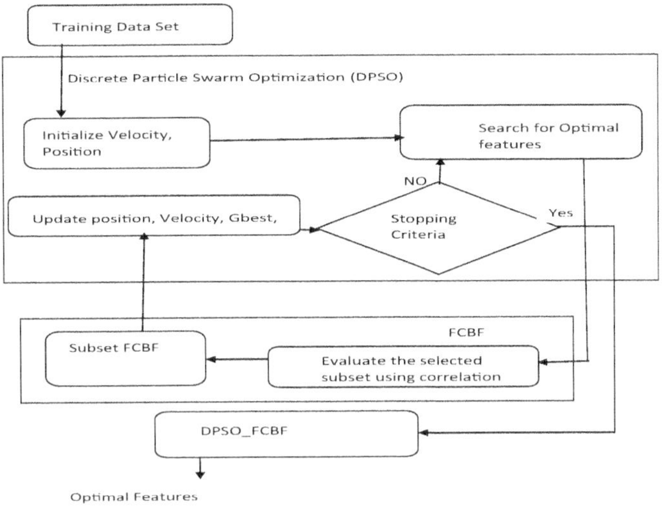

Fig. 1. Proposed DPSO_FCBF technique

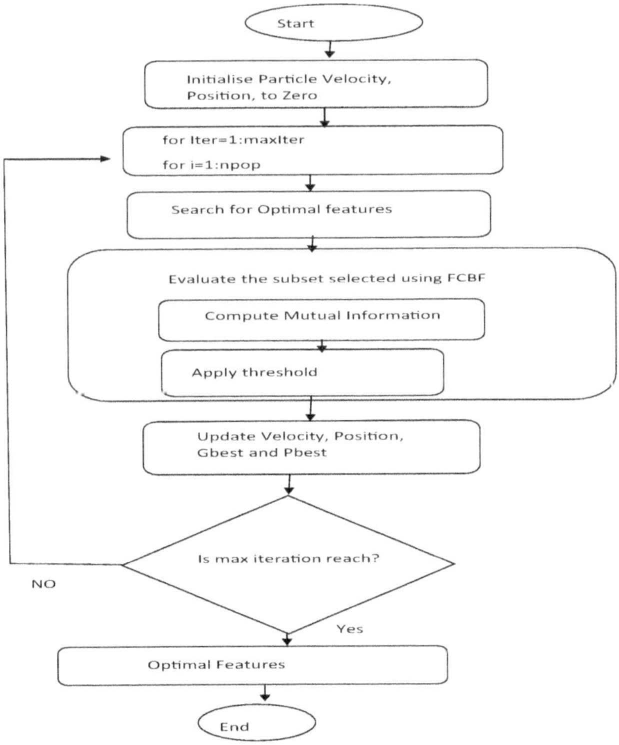

Fig. 2. DPSO_FCBF process

3.3 Dataset

For the purpose of this paper a benchmark dataset of road accident were obtained and used from [26], link https://data.sa.gov.au/data/dataset/road-crash-data, the data includes information related to driver characteristics, environmental and road factors from two files of crash and casualty information. The total of 6565 instances, 30 features and 3 classes of road accident related data for 2016. The features are 1. Und_Unit_Number, 2. Casuality_Number, 3. Casuality_Type, 4. Sex, 5. Age, 6. Position_Invehicle, 7. Seat_Belt, 8. Helmet, 9. Seats_Area, 10. Total_Unit, 11. Total_Cas, 12. Total_Fats, 13. Total_Si, 14. Total_Mi, 15. Month, 16. Day, 17. Area_Spead, 18. Position_Type, 19. Horizontal_Align, 20. Vertical_Align, 21. Road_Surface, 22. Moisture_Cond, 23. Weather_Cond, 24. Daylight, 25. Crash_Type, 26. Unit_Resp, 27. Entity_Code, 28. Traffic_Cntrls, 29. Dui_Involved, 30. Drugs_Involves. The three classes are; Major_Injury, Minor_Injury and fatality. See Tables 1 and 2 for the total number of features and instances, and selected features respectively. The repository has been used by [27, 28].

Table 1. Road accident dataset (Government of South Australia, 2016)

Dataset	Instances	Features	Classes
South Australia crash data	6565	30	3

Table 2. Selected features

Feature	Description	Feature	Description	Feature	Description
3	Casuality_Type	10	Total_Unit	17	Area_Spead
4	Sex	11	Total_Cas	18	Position_Type
5	Age	12	Total_Fats	19	Horizontal_Align
6	Position_Invehicle	13	Total Si	20	Vertical_Align
7	Seat_Belt	14	Total_Mi	21	Road_Surface
8	Helmet	15	Month		
9	Seats_Area	16	Day		

4 Experimental Setup and Result

The DPSO_FCBF proposed technique experiment setup is shown in Table 3 and Functional Link Artificial Neural Network Back Propagation FLANN BP is shown in Table 5, all were implemented on the MATLAB software. Irrelevant and redundant features were removed by the new (DPSO_FCBF) technique, 19 features were selected out of 30 features as the most significant features see Table 4. For comparison FCBF technique was used, where 27 features were selected see also Table 4. Then experiment were conducted using complete 30 features, 27 and 19 selected features for the prediction of severity of road traffic accident using FLANN model and BP learning algorithm, the result are shown in Tables 6, 7 and 8 respectively. The results shows a better accuracy and much reduction in training time in the result using DPSO_FCBF FLANN BP compared to before feature selection and using FCBF FLANN BP.

Table 3. Experimental parameters setting for DPSO_FCBF

Techniquetion	Popula	Inertia weight	Learning coefficient	Max iteration	Fitness
DPSO_FCBF	30	1	C1 = 2, C2 = 2	100	Sigmoid
FCBF					Threshold, Sigmoid

Table 4. Feature selected

Techniques	Selected features
DPSO_FCBF	3, 4, 5, 6, 7, 8, 9, 10, 11, 12, 13, 14, 15, 16, 17, 18, 19, 20, 21
FCBF	4, 5, 6, 7, 8, 9, 10, 11, 12, 13, 14, 15, 16, 17, 18, 19, 20, 21, 22, 23, 24, 25, 26, 27, 28, 29, 30

Table 5. Parameters for Functional link artificial neural network experiment

Parameters	FLANN BP	FCBF BP
Learning rate	0.2	0.2
Momentum	0.7	0.7
Maximum epoch	10000	10000

Table 6. Experiment before feature selection

Max. epoch	Training MSE	Testing MSE	Training accuracy%	Time
10000	0.049764	0.050761	90.0472	4858.1875
10000	0.048649	0.051477	90.2702	4790.7031
10000	0.049764	0.050761	90.0472	4755.672
10000	0.049764	0.050761	90.0472	4856.875
10000	0.049764	0.050761	90.0472	4754.425
10000	0.048649	0.051477	90.2702	4810.2656
10000	0.048649	0.051477	90.2702	4869.8125
10000	0.049764	0.050761	90.0472	4754.425

Table 7. Experiment using DPSO_FCBF FLANN BP

Epoch	Training MSE	Testing MSE	Training%	Testing%	Time
10000	0.048649	0.051477	90.2702	89.7045	4664.0625
10000	0.048649	0.051477	90.2702	89.7045	4609.7669
10000	0.048649	0.051477	90.2702	89.7045	4676.5156
10000	0.048649	0.051477	90.2702	89.7045	4644.3125
10000	0.048649	0.051477	90.2702	89.7045	4629.2431
10000	0.048649	0.051477	90.2702	89.7045	4677.4375
10000	0.04864	0.051477	90.2702	89.7045	4619.7813
10000	0.048649	0.051477	90.2702	89.7045	4680.3306

Table 8. Experiment using FCBF FLANN BP

Epoch	Training MSE	Testing MSE	Training%	Testing%	Time
10000	0.049055	0.052042	90.1889	89.5887	4764.1625
10000	0.049055	0.052042	90.1889	89.5887	4784.0922
10000	0.048751	0.051528	90.2498	89.6944	4699.6145
10000	0.048751	0.051528	90.2498	89.6944	4769.6125
10000	0.049055	0.052042	90.1889	89.5887	4800.1325
10000	0.048649	0.051477	90.2702	89.7045	4722.4322
10000	0.04864	0.051477	90.2702	89.7045	4781.7825
10000	0.048751	0.051528	90.2498	89.6944	4719.6112

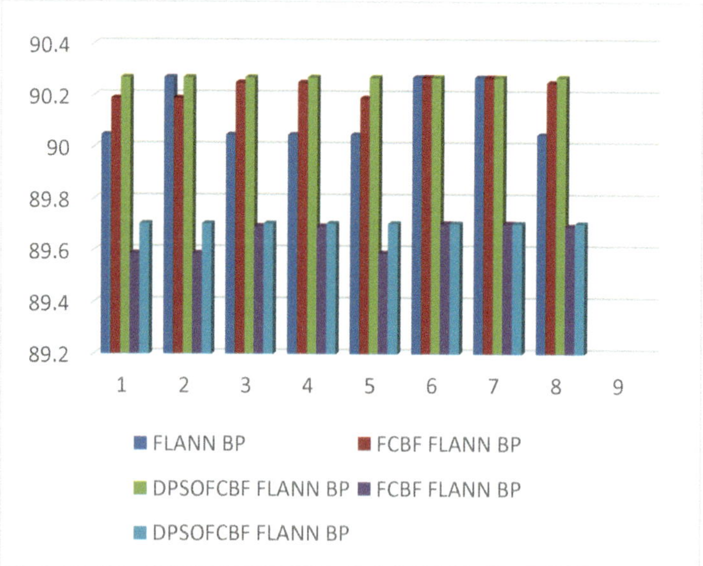

Fig. 3. Comparison of training and testing accuracy for FLANN BP, DPSO_FCBF FLANN BP and FCBF FLANN BP

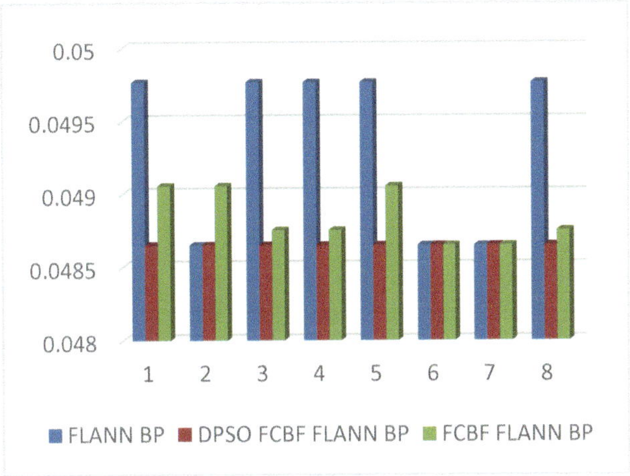

Fig. 4. Comparison of training MSE for FLANN BP, DPSO_FCBF FLANN BP and FCBF FLANN BP

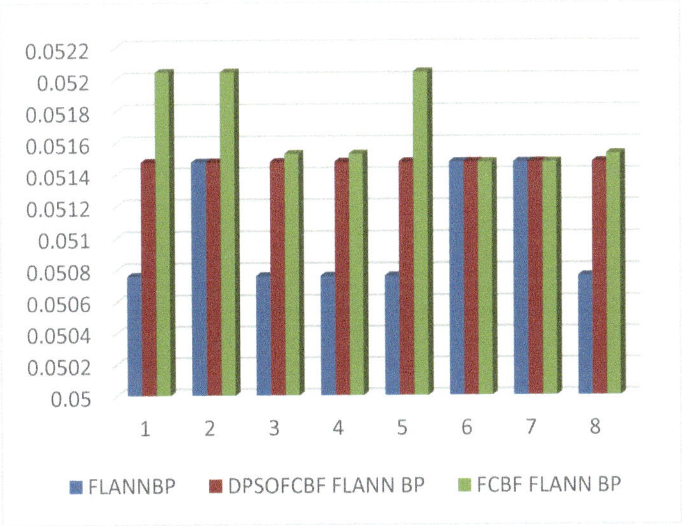

Fig. 5. Comparison of testing MSE for FLANN BP, DPSO_FCBF FLANN BP and FCBF FLANN BP

5 Discussion

In this paper, DPSO based Fast correlation based feature selection is proposed. In the proposed technique, the DPSO is used to search for optimal features and evaluate the features by using FCBF, the features that are more relevant are selected, subsequently update the optimal values. The sigmoid function is used as fitness function and threshold as stopping criteria for the FCBF.

Series of experiments has been conducted to obtained the best result, firstly, the feature selection experiment was conducted using new proposed technique of Discrete Particle Swarm Optimization with Fast Based Correlation Feature Selection (DPSO_FCBF) and secondly using FCBF, the data comprises of 6565 instances, the result shows that 19 and 27 features were selected out of 30 features as the most significance factors for the severity of road traffic accident as shown in Table 4.

Thirdly, experiments has been conducted for the prediction of Severity of road traffic accident using Functional Link Artificial Neural Network FLANN model and Back-propagation as learning algorithm before feature selection, the result is shown in Table 6. Fourthly, series of experiments was carried out for the prediction of severity of road traffic accident using FLANN model and backpropagation as learning algorithm after feature selections using the result obtained from DPSO_FCBF, FCBF techniques, and the result is shown in Tables 7 and 8 respectively. It has been observed that DPSO_FCBF FLANN BP outperformed FCBF FLANN and before feature selection in terms of accuracy and reduction in time in the prediction of severity of road traffic accident.

Figure 3 shows graph of the comparisons of prediction accuracy for training and testing for proposed technique and FCBF FLANN and FLANN. Where proposed technique performed better than FCBF FLANN and FLANN. Also Figs. 4 and 5 shows graph comparisons of the training and testing MSE for the proposed technique and FCBF FLANN and FLANN experiment, where it revealed that proposed technique has lower MSE for both training and testing compared to before feature selection and using FCBF FLANN BP.

5.1 Performance Measure

The performance evaluation used for evaluating the performance of the proposed DPSO_FCBF feature selection technique using FLANN and back propagation is mean square error (MSE), CPU processing time and percentage accuracy. The MSE works by calculating the error between the real values (train value) and the estimated values to measure the accuracy of the model. The MSE offers information on the short term performance of a model by permitting a term comparison of the actual difference between the predicted value and the measured value. The smaller value of MSE indicates better performance of the proposed model, CPU processing time is the interval time to execute an algorithm. In this study, MATLAB built-in stopwatch timer is used to record the processing time (includes training and testing) of the proposed model, with completion cycle of k-fold cross validation. The accuracy is the correct prediction percentage from a specified dataset that use to measure prediction accuracy, Eq. 4 shows the percentage accuracy of the proposed model.

$$\mathrm{MSE} = \frac{1}{N}\sum\nolimits_{j=1}^{N} (\mathrm{yj} - \mathrm{tj})^2 \tag{3}$$

$$\mathrm{Accuracy\%} = \frac{1}{N} \times \sum\nolimits_{j=1}^{N} \delta\mathrm{yj},\ \mathrm{tj} \times 100\% \tag{4}$$

Where yj = 1 or 0 according whether k = 1 or k $\neq$ 1 respectively; N is the size of the data set.

The hybrid DPSO_FCBF feature selection technique has significantly play a role in determining significant factors for the prediction of severity of RTA.

6 Conclusion

In this article, we proposed DPSO_FCBF feature selection technique implemented for severity of road traffic accident. In the proposed algorithm, the DPSO is used to search for optimal features and evaluate the features by using FCBF, the features that are more relevant are selected, subsequently update the optimal values. The sigmoid function is used as fitness function and threshold as stopping criteria for the FCBF. From the experiments conducted it can be concluded that the proposed technique has reduced the number of features selected by removing non-relevant and redundant features. The prediction result of the severity of road traffic accident using Functional Link Artificial Neural Network FLANN model and Backpropagation as learning algorithm shows that the performance in terms of accuracy has slightly improved and the time complexity has been remarkably improved compared to prediction before feature selection and using FCBF FLANN BP. The future work of this study is to improve the proposed technique for more accuracy and more reduction in computational time.

References

1. Jayasudha, K., Chandrasekar, C.: An overview of data mining in road traffic and accident analysis. J. Comput. Appl. **2**(4), 32–37 (2009)
2. World Health Organization: Global status report on road safety 2015. World Health Organization (2015)
3. Murray, C.J.L., Lopez, A.D.: Evidence-based health policy–lessons from the global burden of disease study. Science **274**(5288), 740–743 (1996)
4. Jacobs, G., Aeron-Thomas, A., Astrop, A.: Estimating global road fatalities (2000)
5. Tang, J., Salem, A., Huan, L.: Feature selection for classification: a review. Data Classif.: Algorithms Appl. 37 (2014)
6. Liu, H., Yu, L.: Toward integrating feature selection algorithms for classification and clustering. IEEE Trans. Knowl. Data Eng. **4**, 491–502 (2005)
7. Doshi, M.: Correlation based feature selection (CFS) technique to predict student Performance. Int. J. Comput. Netw. Commun. **6**(3), 197 (2014)
8. Archana, S.R.A., Thanabal, M.S.: Optimization algorithms for feature selection in classification: a survey. Int. J. Innov. Res. Comput. Commun. Eng. 4(2)
9. Liu, J., et al.: Feature Selection based on quality of information. Neurocomputing **222**, 11–22 (2017)
10. Moghaddam, F.R., Afandizadeh, Sh., Ziyadi, M.: Prediction of accident severity using artificial neural networks. Int. J. Civ. Eng. **9**(1), 41 (2011)
11. Shanthi, S., Ramani, R.G.: Feature relevance analysis and classification of road traffic accident data through data mining techniques. In: Proceeding of the World Congress on Engineering and Computer Science, vol. 1 (2012)

12. Mussone, L., Bassani, M., Masci, P.: Analysis of factors affecting the severity of crashes in urban road intersections. Accid. Anal. Prev. **103**, 112–122 (2017)
13. Hoque, N., Bhattacharyya, D.K., Kalita, J.K.: MIFS-ND: a mutual information-based feature selection method. Expert Syst. Appl. **41**(14), 6371–6385 (2014)
14. Miao, J., Niu, L.: A survey on feature selection. Proc. Comput. Sci. **91**, 919–926 (2016)
15. Bolón-Canedo, V., Sánchez-Maroño, N., Alonso-Betanzos, A.: Recent advances and emerging challenges of feature selection in the context of big data. Knowl.-Based Syst. **86**, 33–45 (2015)
16. Jović, A., Brkić, K., Bogunović, N.: A review of feature selection methods with applications. In: 2015 38th International Convention on Information and Communication Technology, Electronics and Microelectronics (MIPRO). IEEE (2015)
17. Wang, H., et al.: A correlation-based binary particle swarm optimization method for feature selection in human activity recognition. Int. J. Distrib. Sens. Netw. **14**(4), 155014771877278 (2018)
18. Djellali, H., et al.: Fast correlation based filter combined with genetic algorithm and particle swarm on feature selection. In: 2017 5th International Conference on Electrical Engineering-Boumerdes (ICEE-B). IEEE (2017)
19. Jain, I., Jain, V.K., Jain, R.: Correlation feature selection based improved-binary particle swarm optimization for gene selection and cancer classification. Appl. Soft Comput. **62**, 203–215 (2018)
20. Beniwal, S., Arora, J.: Classification and feature selection techniques in data mining. Int. J. Eng. Res. Technol. (IJERT) **1**(6), 1–6 (2012)
21. Hira, Z.M., Gillies, D.F.: A review of feature selection and feature extraction methods applied on microarray data. Adv. Bioinform. (2015)
22. Huang, C.-L., Tsai, C.-Y.: A hybrid SOFM-SVR with a filter-based feature selection for stock market forecasting. Expert Syst. Appl. **36**(2), 1529–1539 (2009)
23. Hall, M.A.: Correlation-based feature selection of discrete and numeric class machine learning (2000)
24. Michalak, K., Kwaśnicka, H.: Correlation-based feature selection strategy in classification problems. Int. J. Appl. Math. Comput. Sci. **16**, 503–511 (2006)
25. Senliol, B., et al.: Fast Correlation based filter (FCBF) with a different search strategy. In: 2008 23rd International Symposium on Computer and Information Sciences. IEEE (2008)
26. Government of South Australia "Road crash data". https://data.sa.gov.au/data/dataset/road-crash-data
27. Boufous, S., et al.: The impact of environmental, vehicle and driver characteristics on injury severity in older drivers hospitalized as a result of a traffic crash. J. Saf. Res. **39**(1), 65–72 (2008)
28. Connely, L.B., Supangan, R.: The economic cost of road traffic crashes: Australia, states and territories. Acci. Anal. Prevent. **38**(6), 1087–1093 (2006)

Classification of Mammogram Images Using Radial Basis Function Neural Network

Ashraf Osman Ibrahim[1,2,3](✉), Ali Ahmed[4], Aleya Abdu[3], Rahma Abd-alaziz[3], Mohamed Alhaj Alobeed[5], Abdulrazak Yahya Saleh[6], and Abubakar Elsafi[7]

[1] Faculty of Computer Science and Information Technology, Alzaiem Alazhari University, Khartoum North 13311, Sudan
ashrafosman2@gmail.com
[2] Arab Open University, Khartoum, Sudan
[3] Faculty of Computer Science, Future University, Khartoum, Sudan
[4] Department of Computer Science, Faculty of Computing and InformationTechnology, King Abdulaziz University, Rabigh 25732, Saudi Arabia
[5] Information Technology Department, Shendi University, Shendi, Sudan
[6] FSKPM Faculty, University Malaysia Sarawak (UNIMAS), 94300 Kota Samarahan, Sarawak, Malaysia
[7] Software Engineering Department, College of Computer Science and Engineering, University of Jeddah, P.O. Box 34, Jeddah 21959, Saudi Arabia

Abstract. Recently, computer aided diagnosis and image processing have received considerable attention from a number of researchers. Mammography is the most effective method for exposure of early breast cancer to increase the survival rate. This paper presents the classification method for mammogram Image using Radial Basis Function Network (RBF) technique. This method is focused on features extracted from the breast cancer mammogram image processing algorithms. The actual decision about the presence of the disease is then made by RBF network classifiers. We conducted this study in five stages; collecting images, Region of Interest (ROI), features extracting, classification and end with testing and evaluating. The experimental results shown the classification accuracy results of the RBF neural network 79.166% while MLP algorithm was 54.1667%, that illustrate the capability of the RBF network to obtain better classification accuracy results.

Keywords: Artificial Neural Networks · Radial Basis Function · Mammogram · Breast Cancer

1 Introduction

Breast cancer (BC) spreads around the world. The numbers of cases increased and there are million cases occurred annually. In addition, more than hundred thousand women die each year [1]. It is also expected that 1 in 30 will die from breast cancer [2]. Early diagnosis of the breast cancers by the radiologist reduces the death rate globally. There

F. Saeed et al. (Eds.): IRICT 2019, AISC 1073, pp. 311–320, 2020.
https://doi.org/10.1007/978-3-030-33582-3_30

are several methods exist for detecting breast cancers among which digital mammography is the most familiar and successful technique currently used by the radiology people [3].

Mammograms are collected from patients who are suspected for breast cancer mostly as full field mammograms, where the image detection and classification are high, due to the high image quality [4]. Mammography is one of the methods that can detect the BC disease earlier. It doesn't decrease the disease or stop it [5]. Regular screening can be a successful strategy to identify the early symptoms of breast cancer in mammographic images. This examination also ensures other pathologies detection suggesting cancer nature as being benign, malignant, or normal. The most important improvement is breast imaging which is possible due to the advancement in digital mammography [6].

Medical images classification can be successfully used for diagnostic and medicine teaching purposes. It is also used as a procedure to analyze the extracting models telling significant data classes. Many approaches have been formed to categorize masses to benign or malignant classes by different classification techniques [7].

In this paper, radial basis function network algorithm applied to classify mammography images to normal and abnormal mammograms. The aim of the study is to use RBF for mammograms images classification and to compare its performance with that of the Multilayer Perception Neural Network (MLP). For forecasting tumour, features are extracted from mammogram.

This paper is organized as thus: Sect. 2 reviews some related work available in the literature. Section 3 presents the proposed technique. Section 4 reports the experimental study. Finally, Sect. 5 concludes the paper.

2 Related Work

According to the previous studies, artificial neural networks (ANNs) have received more attentions of the researchers with the aim of improving the classification tasks. In this section, we highlight some studies done in medical classification using computational methods. In study [8], the researchers proposed a computer aided classification technique, they digitized mammography images using ANN and categorizing benign/malignant class on region of interest (ROI). The main mass that classifies mammographic characteristic is texture. The use of ANN classifier to exploits and to classify data as benign/malignant. Some statistical textural features extracted such as are mean, entropy, standard deviation, kurtosis sleekness and uniformity. The authors use their proposed method to increase classification method efficiency and to reduce the error rate resulting from false positive of malignancies. They proposed seven features with three layers ANN to classify marked regions into benign or malignant.

The researchers in [9] used ANN, Bayesian Network as a Hybrid method for predicting breast cancer prognosis. The hybridization combined both ANN and Bayesian Network. SEER data used with 9 variables as inputs for the proposed model. The ANN and hybrid method outperformed the Bayesian Network and help in decision making.

Another authors in [10], introduced three primary procedures: pre-processing, extraction of features, and classification. Back-Propagation Neural Network (BP) and Radial Basis Function Network (RBF) are used in the same research. It will be pre-processed first before the dataset to be entered the classifier. Then, with Gray-level-Co-occurrence Matrix (GLCM) the characteristics of each mammogram obtained. As standard input from the Mammographic Image Analysis Society (MIAS) database, normal and abnormal breast images are used. The computational experiments indicate that in performing classification of breast cancer, RBF network is better than BP network. Classification techniques are one of the greatest important and basic functions in machine learning and data mining [11]. Many of the researchers performed experiments on medical datasets, more recent research presented in [12] executed two class classification using the proposed model based on wavelet analysis utilizing Artificial Neural Network (ANN). Also, in [13], the researchers proposed an enhancements in pattern classification neural network training. They suggested training process motivated by neuron's biological met plasticity property and Shannon theory. Throughout training, the MLP updating the connection weights as learning process. This model training and measured using Breast Cancer data. Other study [14] introduced MLP hybrid with Differential Evolution (DE) method in multi-objective optimization problems for breast cancer diagnosis. Previous study [15] presented and considered at various clinical features for benign/malignant taxonomy and achieved statistical investigation tests. During the classification method, multiple linear and nonlinear models were implemented, including LDA, minimum square, K-nearest-neighbors, RBF and MLP. The authors recommended more optimizing for the model to improve the accuracy results.

Recently, [16] investigated about the ability of the Local Binary Pattern texture and Convolution Neural Network technique for automatic breast tumor images classification. Other recent study [17] proposed a new computational aided diagnosis technique of lung nodes. Recently some works used new techniques such as convolutional neural networks, deep learning and other new methods [18–21].

We can conclude from earlier works that the development of the medical images topics still towards and need more works in enhancing the methods and techniques to increase classification accuracy.

3 The Proposed Method

The proposed method **includes** four main steps as we can see in Fig. 1:

- Mammogram Database (MIAS dataset).
- Pre-Processing
- Features Extraction.
- Classification using RBF network.

The detail of these four steps is described below one by one.

First we collet MAIS data as appeared in the next section in detail.

The phases are used as thus; the model starting by mammogram images collection (MIAS dataset), pre-processing (equalization and image cropping founded ROI, extract the features using six functions (mean, Skewness, standard deviation, Contrast, Smoothness, Kurtosis). The classification process used RBF network as single classifier and last get classification output (Benign or Malignan) and evaluating the results.

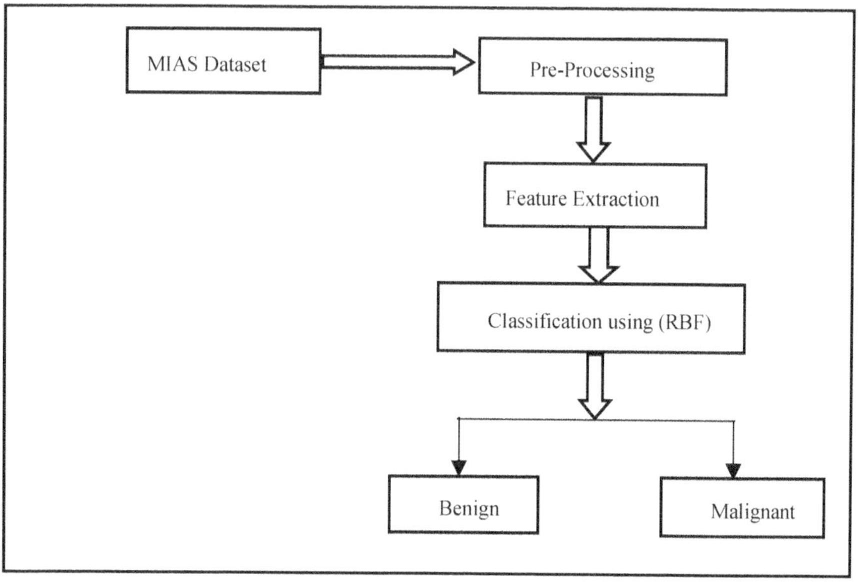

Fig. 1. The proposed method

3.1 Mammogram Database

In this section, we will give a brief account of dataset that used. We get the Mammographic Image Analysis (MIAS) data from online source [22]. Recently several researches used the same data for different reasons. The size of every mammography image is 1024 × 1024 pixels and 200 micron resolution. In addition, MIAS encloses a complete of 322 images of both breasts right and left of 119 cases.

3.1.1 Cropping Based on Region of Interest

Extracting the ROI is the next stage, ROI's are defined as regions characterized as districts containing object characterized objects of intrigued. In this stage we implemented crop method to each mammography image; a cropping process was utilized in order to select the attention parts of the image. The selected cropping area removed undesirable portions and keep the selected portions of the image which is resulting ROI as appeared in Fig. 2.

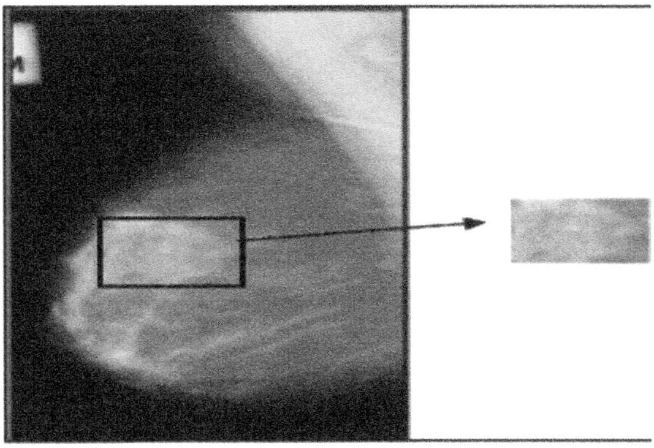

Fig. 2. Original mammogram image a ROI example of malignant tissue

3.2 Feature Extraction

Feature extraction considered as a key step of pattern classification. Basically, the accurate results of classifiers are based on robust features. In mammography images, the process of selecting ROI from (x, y) point and radius is actually the idea of MIAS data images. This phase implements Mean, Skewness, Standard Deviation, Smoothness, Contrast, and Kurtosis to cutting and selecting feature values from every mammography image.

3.3 Classification of Mammograms

After applying the previous steps resulting to numeric values with six attributes, numerical data were used for classification purpose. Two Artificial Neural Network techniques namely Radial Basis Function and Multilayer Perceptron implemented as classifiers and compare the result. Next sections, brief introduce of the both mentioned algorithms as individual classifier.

3.4 Radial Basis Function

Radial Basis Function Network Algorithm (RBF) is a type of neural network. It is a single layer of hidden unit of feed forward network. It is used as activation functions. The activation function is exactly a Gaussian function. It accomplishes classification process based on the input's match to samples from the training step. For classifying the inputs, the neurons calculate the Euclidean distance among inputs and prototype [23] as shown in Fig. 3.

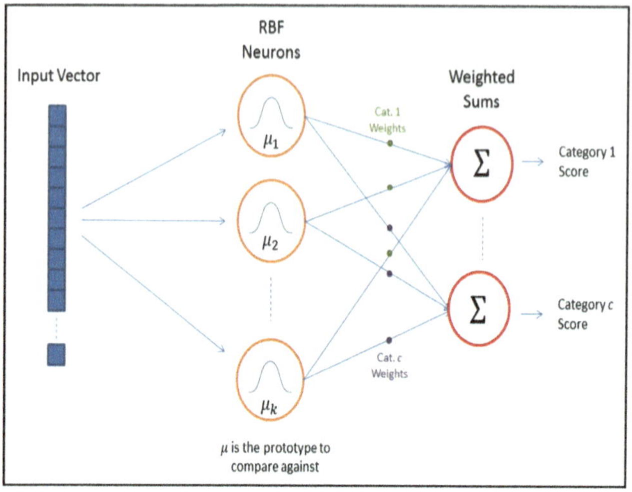

Fig. 3. Generic structure of RBF Network Architecture

3.5 Multilayer Perceptron Network (MPL)

MLP is a well-known type of artificial neural network, it is feed forward with multiple layers model. It read data from input layer and pass the signals to the output layer through hidden layer. It is similar to directed graph, each neuron connected to other via linked weights. The number of inputs determined by the input data and there is no fixed method to determine the number of neurons in hidden layer. For the output layers the number of the neurons depends on the classes of the problem applied. MLP uses type of learning known as supervised learning for training the network [24, 25].

3.6 Performance Measurements

For the performance measurement, in this paper statistical measures are used. Famous measure were used; sensitivity to identify the correct positive samples since it depends on the number of true positives and false negatives. Specificity is to predict the correct negative samples that depend on the number of true negatives and false positives. The accuracy is a measure to produce the level of accurate results of the proposed method. Equations 1, 2 and 3 indication the mathematical way of measures (accuracy, sensitivity and specificity) as follows:

$$Accuracy = \frac{TP + TN}{TP + TN + FP + FN} \tag{1}$$

$$Sensitivity = \frac{TP}{TP + FN} \tag{2}$$

$$Specifity = \frac{TN}{TN + FP} \tag{3}$$

Where, TP is true positive, FP is false positive, TN is true negative and FN is false negative.

4 Experimental Study

In this section, we will show experiments in the proposed model. MIAS database was used. The MIAS data was created to contain two experimental data on the similar mammogram images. In addition, the difference between them is that in the first data the images are divided into two classes: normal or abnormal. MIAS database is containing of 322 commented images. The abnormal cases/images in this data enclose with coordinates (x, y) and the radius (z). First feature extraction was used to and extracted six attributes. Next, we get a numeric data as a converted of the images. This paper apply two ANN classifiers (RBF network and MLP algorithm) with a simple use continues form of dataset. We divided our numeric data into 80% for training phase and 20% for testing phase. The results are offered in the coming section.

Table 1. Two classifiers results

Classifier	Training–testing	% Accuracy
RBF network	80–20%	79.17%
MLP		54.16%

After applying the two different classifiers of the artificial neural network, we calculated the overall Accuracy by using mathematical equations shown earlier (Eqs. 1–3). The results of RBF network classifier and multi-layer perceptron algorithm are presented in Table 1 and also Fig. 4.

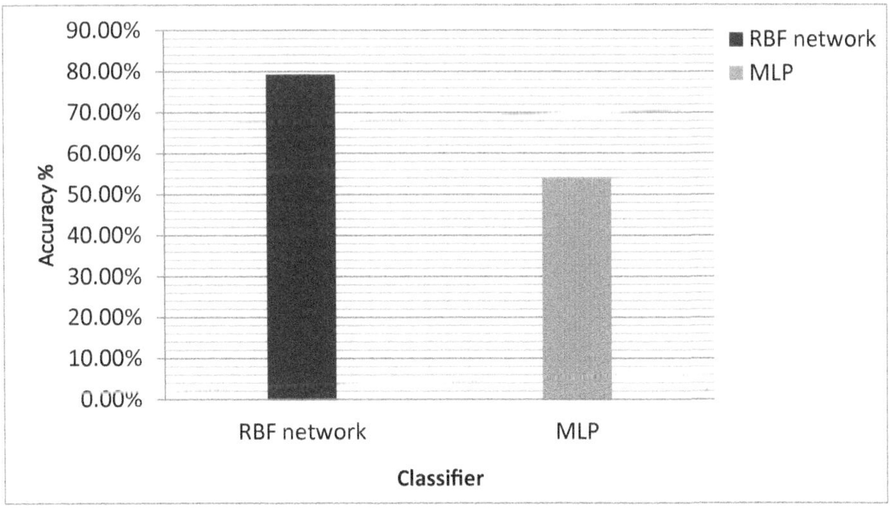

Fig. 4. Accuracy of the RBF and MLP network

As it can be noted from Table 2, the highest accuracy is achieved by RBF network. From this result, we can understand that classifier contain the best result of accuracy. The RBF network have less only one hidden layer compared to MLP network. In addition, RBF is easy to design and it gives stable and good generalization with robustness in performance as well. Therefore, the RBF outperform MLP. We can also see that the MLP classifier has the lowest accuracy value, which mean that classifier it is not the best depend from the accuracy result.

Table 2. Comparison of proposed and other methods accuracy.

Classifier	% Accuracy
RBF network	79.17%
MLP	54.16%
SVM [26]	53.16%
KNN [26]	55.20%

From Table 2, we can note that the proposed model achieved higher accuracy as single classifier compared with other methods in the literature. The comparison results include other two methods support vector machine (SVM) and K-nearest Neighbors (KNN) classifiers. The proposed model extracted the features from mammogram images using six functions that mentioned before and then classified the data using RBF network.

5 Conclusion

This paper introduced radial basis function network (RBF) technique used for classifying mammogram images. The use of artificial neural networks is justified by their own properties: learning ability, generalization and reminiscence, improving the identification of breast cancer classification. The mammograms image data used and crop ROI for all images to extract the features that help to classify the cancer using computational methods. The investigational results illustrate that the RBF algorithm have an ability to obtain better classification accuracy than MLP algorithm. For future work, we plan to propose optimization method to enhance the neural network structure and parameters. Since that will improve the classification accuracy of the mammogram images.

References

1. de Oliveira Martins, L., dos Santos, A.M., Silva, A.C., Paiva, A.C.: Classification of normal, benign and malignant tissues using co-occurrence matrix and Bayesian neural network in mammographic images. Paper Presented at the 2006 Ninth Brazilian Symposium on Neural Networks (SBRN 2006) (2006)
2. Loka, V.R.D., Putheti, S.: Classification of normal, benign and malignant tissues using fuzzy texton and support vector machine in mammographic images. Int. J. Comput. Appl. **82**(15), 36–39 (2013)

3. Punitha, S., Ravi, S., Devi, M.A., Vaishnavi, J.: Breast cancer detection using classification techniques in digital mammography: international science press. IJCTA **9**(7), 3123–3134 (2016)
4. Tang, J., Rangayyan, R.M., Xu, J., El Naqa, I., Yang, Y.: Computer-aided detection and diagnosis of breast cancer with mammography: recent advances. IEEE Trans. Inform. Technol. Biomed. **13**(2), 236–251 (2009)
5. Smith, R.A., Cokkinides, V., Eyre, H.J.: American cancer society guidelines for the early detection of cancer. CA: Cancer J. Clin. **56**(1), 11–25 (2006)
6. Jaffar, A.: Hybrid texture based classification of breast mammograms using adaboost classifier. Int. J. Adv. Comput. Sci. Appl. **8**(5), 321–327 (2017)
7. Choi, J.P., Han, T.H., Park, R.W.: A hybrid Bayesian network model for predicting breast cancer prognosis. J. Korean Soc. Med. Inform. **15**(1), 49–57 (2009)
8. Islam, M.J., Ahmadi, M., Sid-Ahmed, M.A.: An efficient automatic mass classification method in digitized mammograms using artificial neural network. arXiv preprint arXiv:1007.5129 (2010)
9. Pratiwi, M., Harefa, J., Nanda, S.: Mammograms classification using gray-level co-occurrence matrix and radial basis function neural network. Proc. Comput. Sci. **59**, 83–91 (2015)
10. Lee, Y.J., Mangasarian, O.L., Wolberg, W.H.: Survival-time classification of breast cancer patients (2008). http://www.cs.wisc.edu/dmi/annrev/rev0601/uj.ppt. [cited 2017]
11. Vani, G., Savitha, R., Sundararajan, N.: Classification of abnormalities in digitized mammograms using extreme learning machine. Paper Presented at the 2010 11th International Conference on Control Automation Robotics and Vision (2010)
12. Marcano-Cedeño, A., Quintanilla-Domínguez, J., Andina, D.: WBCD breast cancer database classification applying artificial metaplasticity neural network. Expert Syst. Appl. **38**(8), 9573–9579 (2011)
13. Mavroforakis, M.E, Georgiou, H.V, Cavouras, D., Dimitropoulos, N., Theodoridis, S.: Mammographic mass classification using textural features and descriptive diagnostic data. Paper Presented at the 2002 14th International Conference on Digital Signal Processing Proceedings, DSP 2002 (Cat. No. 02TH8628) (2002)
14. Ibrahim, A.O., Shamsuddin, S.M., yahya Saleh, A., Abdelmaboud, A., Ali, A.: Intelligent multi-objective classifier for breast cancer diagnosis based on multilayer perceptron neural network and differential evolution. In: 2015 International Conference on Computing, Control, Networking, Electronics and Embedded Systems Engineering (ICCNEEE), pp. 422–427. IEEE (2015)
15. Touahri, R., AzizI, N., Hammami, N.E., Aldwairi, M., Benaida, F.: Automated breast tumor diagnosis using local binary patterns (LBP) based on deep learning classification. Paper Presented at the 2019 International Conference on Computer and Information Sciences (ICCIS) (2019)
16. Mehre, S.A., Dhara, A.K., Garg, M., Kalra, N., Khandelwal, N., Mukhopadhyay, S.: Content-based image retrieval system for pulmonary nodules using optimal feature sets and class membership-based retrieval. J. Digit. Imaging **32**(3), 362–385 (2019)
17. Rasti, R., Teshnehlab, M., Phung, S.L.: Breast cancer diagnosis in DCE-MRI using mixture ensemble of convolutional neural networks. Pattern Recogn. **72**, 381–390 (2017)
18. Arevalo, J., González, F.A., Ramos-Pollán, R., Oliveira, J.L., Lopez, M.A.G.: Representation learning for mammography mass lesion classification with convolutional neural networks. Comput. Methods Program. Biomed. **127**, 248–257 (2016)
19. Mohamed, A.A., Berg, W.A., Peng, H., Luo, Y., Jankowitz, R.C., Wu, S.: A deep learning method for classifying mammographic breast density categories. Med. Phys. **45**(1), 314–321 (2018)

20. Yi, P.H., Lin, A., Wei, J., Yu, A.C., Sair, H.I., Hui, F.K., Harvey, S.C.: Deep-learning-based semantic labeling for 2d mammography and comparison of complexity for machine learning tasks. J. Digit. Imaging **32**(4), 565–570 (2019)
21. Ibrahim, A.O., Shamsuddin, S.M., Abraham, A., Qasem, S.N.: Adaptive memetic method of multi-objective genetic evolutionary algorithm for backpropagation neural network. Neural Comput. Appl. **31**(9), 4945–4962 (2019)
22. http://peipa.essex.ac.uk/info/mias.html. Accessed 2 Dec 2016
23. J. Suckling, J. Parker, D. Dance, S. Astley, I. Hutt, C. Boggis, et al.: The mammographic image analysis society digital mammogram database. In: Excerpt Medical International Congress Series, pp. 375–378 (1994). http://peipa.essex.ac.uk/info/mias.html
24. Rajini, N.H., Bhavani, R.: Classification of MRI brain images using k-nearest neighbor and artificial neural network. Paper Presented at the 2011 International Conference on Recent Trends in Information Technology (ICRTIT) (2011)
25. Kumar, G.J., Kumar, G.V.: Biological early brain cancer detection using artificial neural networks. Int. J. Comput. Sci. Eng. (IJCSE) **02**(08), 2721–2725 (2010)
26. Mohd, A.I., Mohamed Alhaj, A., Shahreen, K., Ashraf, O.I., Anik, H.A., Saima, A.L., Ali, A.: An enhancement of multi classifiers voting method for mammogram image based on image histogram equalization (2018)

Cycle Generative Adversarial Network for Unpaired Sketch-to-Character Translation

Leena Alsaati(✉) and Siti Zaiton Mohd Hashim(✉)

Universiti Teknologi Malaysia, Skudai, Malaysia
zeinij.leena@graduate.utm.my, sitizaiton@utm.my

Abstract. This research investigated the capabilities of Cycle-Consistent adversarial network (Cycle GAN) in the application of stick figure sketches to cartoon character translation. Few studies focused on the task of generating a variety of poses and facial expression of cartoon characters from simple sketches of stick figures, based on unpaired dataset samples. Furthermore, existing studies showed low performance in detecting rare pose features. In this research, two datasets have been created which consists of paired and unpaired images of manually drawn sketches and cartoon characters. The performance of Cycle GAN has been compared against a paired based model, Pix2Pix, by using qualitative and quantitative evaluation measurements. Results show that Pix2Pix outperforms Cycle GAN in accurately mapping cartoon characters to stick figures. Despite that, the Cycle GAN still managed to produce competing results.

Keywords: Cycle-consistent GAN · Pix2pix · Image translation

1 Introduction

Cartoon characters have been a part of us as early as when our eyes started observing the world. Drawing is used as a form of visual art in which the artist expresses their imagination though simple sketches to detailed illustrations, to present a picture, and using colours and shading to enhance its reality. Cartoons characters are featured in many areas such as cartoon production, media, and gamming.

Generative Adversarial Networks (GANs), a model that generates new images have shown exciting results. An example is the implementation of MIT's Nightmare Machine, where a GAN is used to generated horror themed houses and faces. Another implementation is the interactive image to image translation Pix2Pix demo, where it allows users to doodle objects and then produce real images. Training GANs to auto generate illustrations of characters from simple sketches, with defined features, can ease the production process for companies tremendously, as it lowers the production time by generating more characters in less time, and avoids repetitive tasks.

The objective of this research is to generate images of cartoon characters, from unpaired images of stick figure sketches and target cartoon characters, using Cycle GAN. It also compares its performance with synthesized images, from paired images of stick figures and cartoon characters, generated by Pix2Pix. Both models have been adopted as it has been proven to be a general-solution to many image translation tasks.

F. Saeed et al. (Eds.): IRICT 2019, AISC 1073, pp. 321–329, 2020.
https://doi.org/10.1007/978-3-030-33582-3_31

2 Related Work

2.1 Generative Adversarial Networks

GANs, proposed by Goodfellow et al. [1], is a neural network based on supervised generative models. The model consists of two networks: The generator and discriminator. A basic generator takes noise vector as input and generates fake images that resemble real images from the dataset. The discriminator receives both generated and real images, and outputs a probability of whether the given image is real or fake. The two networks work against each other in an adversarial manner, where the discriminator attempts to maximize the cost and the generator to minimize it.

2.2 Cycle-Consistent Adversarial Networks

Cycle GAN [2], is a extended version of Generative Adversarial Network that comprises of two generators and two discriminators. The first generator aims at translating from the input domain to the target domain, whereas the second generator translates from the target domain back to the input domain. The architecture of having two generators and discriminators aids in the cross-domain transfer task; which is transferring an image from the input to target domain without a one-to-one mapping between the input and target in the training dataset. In other words, the model uses an unpaired dataset. The power this unpaired based model, is the ability to generalize and apply a model to a variety of tasks.

2.3 Face and Character Generation

Most of the previous approaches were successful in producing good quality images of characters [3–5]. Furthermore, researchers have focused on the sketch to face translation task [6–9]. Kazemi et al. proposed a new framework [10], a conditional Cycle GAN, which is conditioned on soft biological traits as skin and hair. The aim of the work is to avoid the need of paired data during the training phase. Sketches of objects to generate real items or scenery painting has been done by [11], however, a paired dataset was used. Fukumoto et al. [12], proposed a modified version of StackGAN and using Pix2pix as a base model under StackGAN. The model successfully produces images of desired body poses from stick figure sketches. However, it suffers when assigned rare possess and generates blurred characters but still maintaining the desired shape. Based on literature review, no previous research has focused on specifying detailed facial character expressions from simple sketches with only basic face parts. Therefore, this research investigates the generation of cartoon images from sketches by defining both face expressions and body pose features.

3 Methodology

3.1 Dataset Creation

The dataset created for this research includes images of sketches as input for the model and target images of cartoon characters. The dataset creation process comprises of two parts: First, the sketches of stick figures were manually drawn to represent the facial and body pose features of a corresponding character image. Second, a cartoon character infographic template, created by vectorpocket, was used as the base to construct the images of the character. Using an editing software, parts of character were mixed and matched to produce different poses and facial expressions. The dataset consists a total of 211 images of size 256 × 256 pixels.

Defining Features. During the collection process, we take into consideration to cover certain common features. Such features include: Common leg possess: Walking, running, and sitting. Common arm gestures: Raising hands, bending arms, arms laid down. Face expressions: Happy, angry, sad, laughing, surprised, confused.

Defining Dataset Difficulty Rating. One of the issues with GANs presented is the difficulty of training a network using different types of datasets. According to [13], datasets can be rated: easy, moderate, and difficult according to specific properties. Such properties include framing, angles, number of objects in the dataset.

Table 1 shows the difficulty rating of datasets. To produce good quality images, we keep our datasets on a moderate difficulty rate by using similar images, common framing, but with varied rotation. A hard difficulty rating dataset was not covered since our scope is to entertain a single subject with no additional objects such as accessories, background furniture and scenes.

Table 1. Dataset difficulty ratings [13]

Difficulty	Description
Easy	Images are similar. Common framing, structure and features
Medium	Images share many similarities but framing and rotation varies
Hard	The dataset contains images of varying subjects, objects and angles

Dividing into Paired and Unpaired Datasets. To train the models, two datasets were prepared. The input and target images were organizing to match each model. A paired dataset was given for the Pix2Pix model, and unpaired dataset for the Cycle GAN. The paired dataset comprised of paired images, where each input has a corresponding target image. For the unpaired dataset, the paired correspondences between the input and target images does not exist.

3.2 Model Adoption

The experiment involved training an adopted Cycle GAN model [2]. Cycle GAN has been proven to accommodate a wide range of application. This includes and is not limited to image, audio, and video translations. Seeing the structure of how this model is able to find a relationship between two different distributions without the need of matching or paired data. More specifically for our dataset, it includes a set of sketches and cartoon characters which have different distributions. As a result, Cycle GAN was used in this research to tackle the task of image translation, more specifically the application of translating sketches to cartoon characters. In this case, this model makes the task of sketch to character translation more practical and flexible, as this allows the model to be flexible when handling any given sketch. To compare the results of Cycle GAN against a base model, the Conditional GAN (Pix2Pix) model [11] was used. It has also been proven to solve a wide variety of applications, however, unlike Cycle GAN, it requires a paired dataset which is our main point of comparison between the two models. The results of the two models are compared in the following section. Since GANs are computationally expensive to train on a standard CPU, a setup was made to train the model on a cloud platform, Floyd Hub, using a standard Tesla K80 GPU (12 GB Memory, 61 GB RAM, 100 GB SSD) (Fig. 1).

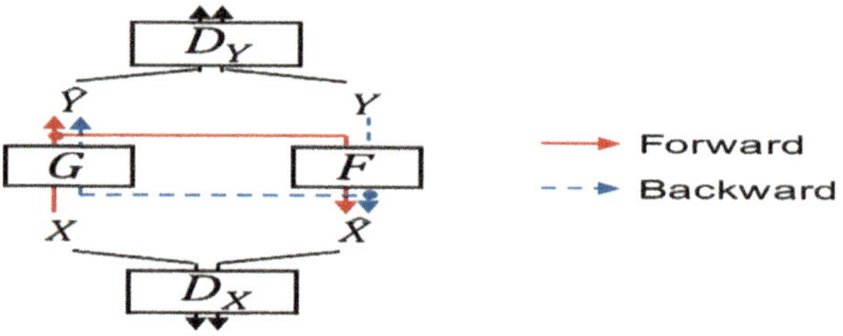

Fig. 1. CycleGAN architecture [14]. G and F are generators; D_X and D_Y are discriminators. X and Y are the real distributions, and $Y^\wedge$ and $X^\wedge$ represent the generated distributions, respectively.

3.3 Performance Evaluation

The quality evaluation of synthesized images is still an open and difficult task. In general, the main training goal of GANs is to generate believable data. The performance is usually estimated manually as Generative networks are frequently trained until the discriminator is unable to distinguish between real and fake data. There have been a few evaluation metrics that previous researchers have used in evaluating GANs; quantitative and qualitative methods. In this research, qualitative measurements such as human evaluation [2, 11] by evaluating the body poses [12] and facial features [10] are used. A criterion for measurement includes facial and body pose matching and direction, color, and sharpness, and missing parts. Furthermore, quantitative measurements such as accuracy percentage and loss function visualization are also used. Accuracy percentage calculates the percentage of number of good images generated divided by the total number of images. The loss function shows how the models are performing over 200 epochs.

4 Results and Discussion

Both models were trained for 200 epochs. The Cycle GAN took 14 h to train, and 29 s to test. The Pix2Pix took only an hour and 49 min to train, and one minute to test. The good and bad test results of Cycle GAN are shown in Figs. 2 and 3. The results show some mode collapse as CycleGAN mostly generates a single facial expression. Moreover, with some cases, it fails to match leg and arm poses. Despite that, it accurately maps many of the body poses and the quality of the images are sharp and not blurry. Measurements criteria is shown in Table 2 to highlight the accuracy of the main features in the cartoon character. Figures 6 and 7 show the Pix2Pix results.

Fig. 2. Good cases for Cycle GAN.

Fig. 3. Failure cases for Cycle GAN.

In Fig. 4, the loss function of the Cycle GAN is shown. Discriminator loss and generator loss correspond to the Adversarial Loss of domain B. Cycle loss refers to the

Cycle Consistent Loss of transferring from domain A to B and mapping back from domain B to A. This loss is important as it compensates for the paired image dataset. The Sketch Identity is an identity mapping, which encourages the translation to preserve colour and structure between the target and the synthesized images [2]. This helps to produce results while preserving initial features from the input images.

The blue line shows the discriminator loss function where it remains below 0.5, with some minor fluctuations in the beginning epochs, and reduces to almost zero near the end. The red line represents the generator loss where the loss values fluctuates throughout the end of the epochs. The fluctuations show that the generator is attempting to improve its loss, however, it does not converge to a certain value in the end. Since both models are competing, the fluctuations can be normal. Nevertheless, it seems that the discriminator has become stronger as it converged to zero by the 155 epochs. Despite the loss function, the generator still produced good results. The development of model during training is shown in Fig. 5.

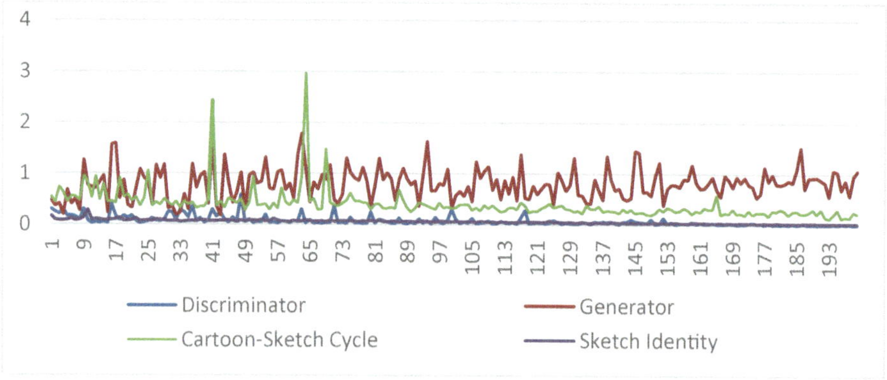

Fig. 4. Cycle GAN loss function.

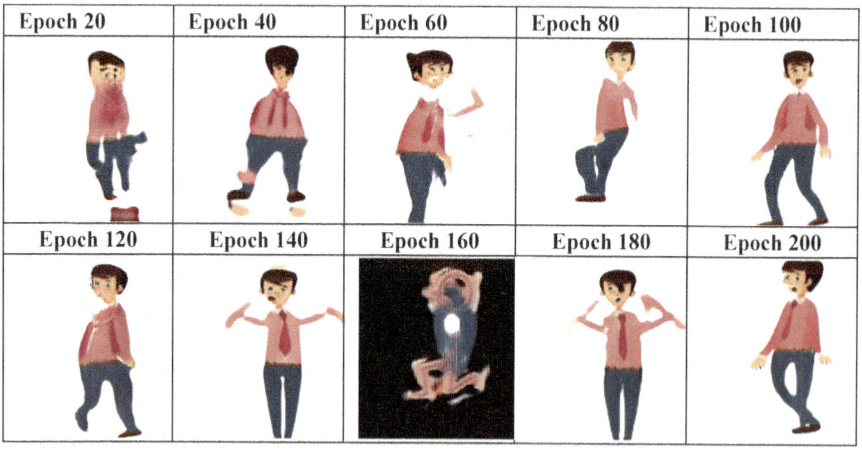

Fig. 5. Development of CycleGAN model at several epochs.

Fig. 6. Good cases for Pix2Px. Target column is the conditional attribute enforced on the model.

Fig. 7. Failure cases for Pix2Pix

Table 2. Observed feature percentage accuracy

Features	Cycle GAN	Pix2Pix
Color	100%	100%
Sharpness	100%	0%
Pose direction match	50%	100%
Leg pose match	90%	100%
Arm pose match	60%	100%
Facial match	30%	40%
Missing parts (face, hand, feet)	No (0%)	Yes (10%)

5 Conclusion

This research looked at the state-of-the-art methods, Cycle GAN and Pix2Pix, to synthesize images of cartoon characters from paired and unpaired training samples of stick figure sketches and cartoon characters. The experiment shows the adopted Cycle GAN produces results that compete with Pix2Pix that is based on paired training samples. The results of the Cycle GAN were much sharper, however, the Pix2Pix method produced cartoon characters that accurately maps the body poses and facial expressions of the stick figure sketches. Despite the need for further improvement made to the Cycle GAN to improve its mapping accuracy in the tasks of sketch to character translation, it has proved to work despite given an unpaired dataset.

References

1. Goodfellow, I.J., Pouget-abadie, J., Mirza, M., Xu, B., Warde-farley, D.: Generative adversarial nets, pp. 1–9 (2014)
2. Zhu, J., Park, T., Efros, A.A.: Unpaired image-to-image translation using cycle-consistent adversarial networks (2018)
3. Jin, Y., Zhang, J., Li, M., Tian, Y., Zhu, H.: Towards the high-quality anime characters generation with generative adversarial networks. In: Machine Learning for Creativity and Design Workshop at NIPS Work, Long Beach, California, USA, pp. 1–13 (2017)
4. Kataoka, Y., Matsubara, T., Uehara, K.: Automatic manga colorization with color style by generative adversarial nets. In: IEEE SNPD, Kanazawa, Japan, vol. 495, pp. 495–499 (2017)
5. Saito, M., Matsui, Y.: Illustration2Vec: a semantic vector representation of illustrations. In: SIGGRAPH Asia, no. 2 (2015)
6. Zhang, S., Ji, R., Hu, J., Lu, X., Li, X.: Face sketch synthesis by multidomain adversarial learning. IEEE Trans. Neural Netw. Learn. Syst. 1–10 (2018)
7. Huang, Y., Khan, S.M.: DyadGAN : generating facial expressions in dyadic interactions. In: IEEE Conference on Computer Vision and Pattern Recognition Workshops Work, DyadGAN, pp. 2259–2266 (2017)
8. Philip, C., Jong, L.H.: Face sketch synthesis using conditional adversarial networks. In: ICTC, pp. 373–378 (2017)

9. Wang, L., Sindagi, V.A., Patel, V.M.: High-quality facial photo-sketch synthesis using multi-adversarial networks. In: 13th IEEE International Conference on Automatic Face and Gesture Recognition High-Quality, pp. 83–90 (2018)
10. Royer, A., et al.: XGAN: unsupervised Image-to-image translation for many-to-many mappings, pp. 1–19 (2017)
11. Isola, P., Zhu, J.Y., Zhou, T., Efros, A.A.: Image-to-image translation with conditional adversarial networks. In: 30th IEEE Conference on Computer Vision and Pattern Recognition, CVPR 2017, pp. 5967–5976 (2017)
12. Fukumoto, Y., Shimizu, D., Shibata, C.: Generation of character illustrations from stick figures using a modification of generative adversarial network. In: 2018 IEEE 42nd Annual Computer Software and Applications Conference, vol. 01, pp. 183–186 (2018)
13. Selseng, S.: Guiding the training of generative adversarial networks (2017)
14. Fang, F., Yamagishi, J., Echizen, I., Lorenzo-trueba, J.: High-quality nonparallel voice conversion based on cycle-consistent adversarial network national. In: 2018 IEEE International Conference on Acoustics, Speech and Signal Processing, pp. 5279–5283 (2018)

Rabies Outbreak Prediction Using Deep Learning with Long Short-Term Memory

Abdulrazak Yahya Saleh[1(✉)], Shahrulnizam Anak Medang[1], and Ashraf Osman Ibrahim[2,3]

[1] FSKPM Faculty, Universiti Malaysia Sarawak (UNIMAS), 94300 Kota Samarahan, Sarawak, Malaysia
ysahabdulrazak@unimas.my

[2] Faculty of Computer Science and Information Technology, Alzaiem Alazhari University, Khartoum North 13311, Sudan

[3] Faculty of Computer Science, Future University, Khartoum, Sudan

Abstract. The purpose of this article is to evaluate the Long Short-Term Memory (LSTM) model performance for rabies outbreak prediction (ROP). Successful forecasting of the initial epidemic outbreaks can decrease the incidence of the ailment and save lives, but this type of research is costly, and an erroneous result can trigger false alarms, and the trustworthiness of the warning system will be at stake. As such, biosurveillance system developers are looking for highly sensitive outbreak prediction algorithms that will minimise the number of false alarms. Using the epidemiological data such as those of rabies to forecast novel and vital directions is a significant issue of public health, and it involves the collective attention of the machine learning (ML) communities. In this study, the data are obtained from HealthData.com and utilised for the performance evaluation of the LSTM algorithm. The algorithm performance is evaluated based on Root Mean Square Error (RMSE) and Accuracy, and compared with that of the traditional algorithm– the Autoregressive integrated moving average (ARIMA) model. The results from this research prove that a deep learning LSTM network can predict the disease prevalence, using the rabies datasets, with a good accuracy. The performance of the proposed model is evaluated by comparing with the ARIMA model. The LSTM model attains the best result with 97.10% accuracy, while the traditional ARIMA obtains 72.10%. Moreover, the LSTM model scores the lowest value of RMSE (2.04) compared with the ARIMA model which scores the highest (3.12). Through this study, it is obvious that the LSTM prediction model is an effective method for determining this viral disease, evidenced by a very low RMSE value and a high accuracy score.

Keywords: Long Short-Term Memory (LSTM) · Autoregressive Integrated Moving Average (ARIMA) · Machine Learning (ML) · Rabies Outbreak Prediction (ROP)

F. Saeed et al. (Eds.): IRICT 2019, AISC 1073, pp. 330–340, 2020.
https://doi.org/10.1007/978-3-030-33582-3_32

1 Introduction

Rabies is a well-known zoonotic disease. This disease has been classified under the Lyssavirus type of the Rhabdoviridae family, which infects warm-blooded creatures and causes lethal encephalitis [1]. A person bitten by a crazy infected dog has more than 99% chance of contracting the disease. Although rabies is preventable, the death toll is still high. It is estimated that 59,000 humans die from rabies every year on the planet Earth, with sub-Saharan Africa and Asia contributing to over 95% of these deaths [2]. According to a report of WHO [3], about 45% of the total deaths caused by rabies happen in all parts of the world. More than 3.3 billion persons are in danger of the rabies infection and it disproportionally affects the under-resourced and poorest populaces living in the lower-middle income nations; the infection rate is on the rise in a few of the countries. Ramos [4] explains that in addition to the undesirable effects on mortality, rabies has an adverse impact on the financial health of the affected populaces. Marí [5] says that about 75% of the uncontrollable rising numbers of maladies affecting humans nowadays originate from the animals, and 60% of these ailments are zoonotic in nature. Rabies is an exceptionally deadly disease that has been around for about 4000 years; it is caused by the bullet-shaped Lyssavirus that can infect all warm-blooded creatures with a mortality rate of nearly 100%, after the beginning of clinical signs or manifestations [6]. This illness presently kills in excess of 69,000 individuals, the vast majority of which are from Asia and Africa [7].

Epidemic prediction around the globe is an essential public health issue. Information about tracking, timely detection and forecasting of epidemics like outbreak time and peak intensity are vital for efficient health interventions. Wu mentions in [8] that epidemic control draws much attention and consideration of the government in every nation and the World Health Organization (WHO). Among the numerous issues are scourge control, and the epidemic prediction, which gauges the prevalence of the pandemic in specific locales. As indicated by Fricker [9], Bio-observation is the procedure of effectively assembling and breaking down information associated with human wellbeing and diseases; it provides early warning of an outbreak of a disease in a community and raises the general awareness of the human ailment in the populace. Outbreak forecasting algorithms carry out the job of computerised biosurveillance, analysing vast volumes of clinical information progressively and distinguishing the different phases of an ailment development. In the medicinal services industry, utilisation of time-series modelling and forecasting a future flare-up of certain uncontrollable illnesses and infection, which happen cyclically, is exceptionally critical. The determining of an ailment helps predict the course of an illness, cautions social insurance specialists about the potential health cost and alerts the relevant authorities to introduce control measures to avoid disease flare-ups. Zoonotic sicknesses are a group of uncontrollable diseases normally transmitted from animals to people. There are numerous methodologies that can be used to predict the spread of Zoonotic Diseases using Machine Learning techniques [10, 11]. However, there are limitations in conducting this kind of experiment when the actual datasets display irregular trends [12]. Therefore, forecasting different and crucial drifts of epidemiological data such as that of rabies is a significant public health issue.

One of the powerful methods which can be utilised for epidemic prediction is deep learning [13]. Deep learning is well known in the field of predicting the probability of event occurrences; it entails collecting huge amounts of data and uses them to forecast the outcome of the future [14]. It has been used to solve numerous problems in the world such as flood prediction and the malware detection [15, 16].

In this study, Deep Learning with Long Short-Term Memory is utilised to predict the rabies outbreaks by applying the time-series analysis. In this regard historical data are used to identify and validate the existing data patterns, which are then used to predict what will happen in the future. By utilising Deep Learning to predict rabies outbreaks, the local governments can put in place the necessary preventive measures and raise the awareness of the people about the impending danger so that they can take precautions and appropriate actions to protect themselves. As the saying goes, forewarned is forearmed, and the harm would be greatly reduced.

The rest of this article is organised in the following manner: Sect. 2 presents an explanation of the materials and methods, Long Short-Term Memory (LSTM), and the general framework of the methodology; while Sect. 3 analyses the findings; and finally, Sect. 4 delivers the conclusion and future work.

2 Materials and Methods

Based on the latest development in Deep Learning, a powerful new pattern of machine learning has emerged [17]. Deep Learning refers to a subset of machine learning that imitates the workings of the human brain in processing data and creating patterns for use in decision-making, by using 'deep' artificial neural systems, for example, recurrent neural networks (RNN), convolution neural networks (CNN) and deep neural networks (DNN). The area of Data Science has extended its boundaries into the sphere of healthcare and medicine. Although epidemiology and biostatistics have been in existence for several years, the current industrial flourishing in Deep Learning (DL) and Artificial Intelligence is a driving factor for innovative research in these fields. This study focuses on predicting the rabies outbreak and it is measured by the model performance (Accuracy). The proposed DL model performance is evaluated by comparing with that of ARIMA. The datasets of rabies are obtained from HealthData.com; it is an open-source data management platform from the United States, with 1000 samples, dataset characteristics (Univariate) and Features (2). The software used for ROP in the current research is Python because it is very flexible and popular among other programming languages.

2.1 Long Short-Term Memory (LSTM)

LSTM is a type of RNN; it is created to prevent the output of neural network for a certain input from exploding or decaying as it rounds over the feedback loops [18]. The pattern recognition of the Recurrent Networks is considered superior to those of other neural networks because the former allows the feedback of memory cells of the previous input, which are essential for settling frequent learning errands. LSTM networks deliver improved findings associated with other RNN models by mitigating the problem of vanishing gradient [19].

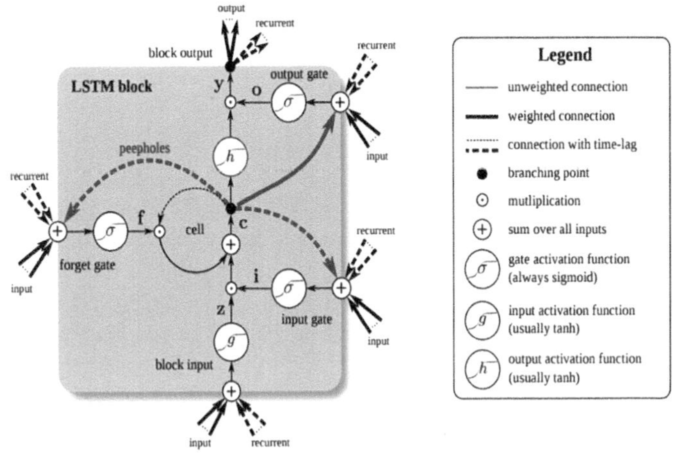

Fig. 1. Detailed diagram of a Long Short-Term Memory block

Figure 1 demonstrates the detailed diagram of the LSTM block utilised in the RNN hidden layers. The LSTM unit consists of four input weights and four recurrent weights. The LSTM is appropriate for forecasting the data of time series if there is a period venture with an arbitrary size [20]. It seems that the execution procedure of the epidemic forecasting models may be improved by employing LSTM and the data gathered for the research. An essential preferred standpoint of RNN is that the relevant data are accessible if they have been mapped into the IO successions. In any case, the gradient issue might arise as a result of the impact of a certain input on the hidden layer, which can be enlarged or reduced meaningfully throughout the circular connection. As certain input sources are overwritten, the affectability of the principal input diminishes after some time. Consequently, the system is "forgotten". The three gates are considered as non-linear additional units which control the cell activation. The forget gate duplicates the cell's past condition, whereas the input and output gates increase the cell's IO. The door enactment work f is a strategic sigmoid. The IO activation functions g and h of the cell, for the most part, utilise hyperbolic tangents or strategic sigmoid. Be that as it may, at times, h utilises the identity function. For whatever length of time that the forget gate is open and the input door is shut, the memory cell keeps on recalling the first input. Along these lines, LSTM is an algorithm that settles an issue in the conventional RNNs [21]. The conditions for forgetting, putting away, restoring, and outputting data in the cell appear as follows, separately:

$$f_t = \sigma(w_f.[h_{t-1}, x_t] + b_f) \tag{1}$$

$$i_t = \sigma(w_i.[h_{t-1}, x_t] + b_i) \tag{2}$$

$$\check{c}_t = tanh(w_c \cdot [h_{t-1}, x_t] + b_c) \tag{3}$$

$$c_t = f_t x c_{t-1} + i_t x \check{c}_t \tag{4}$$

$$o_t = \sigma(w_o.[h_{t-1}, x_t] + b_o) \tag{5}$$

$$h_t = o_t x \tanh(c_t) \tag{6}$$

After the data (x_t) are given to the LSTM cell in Eq. 1, function f_t controls the data to be forgotten in the cell layer. In Eqs. 2 and 3, data to be again kept in the cell layer are formed in i_t and $\check{c}_t$. In Eq. 4; the cell layer c_t is changed utilising *ft*, i_t, and $\check{c}_t$. In Eq. 5, the cell layer's information is utilised and h_t is the output. In Eq. 6, the cell state gets a value between −1 and 1 over the tanh function. The values of c_t and h_t are saved for the following repetition of LSTM.

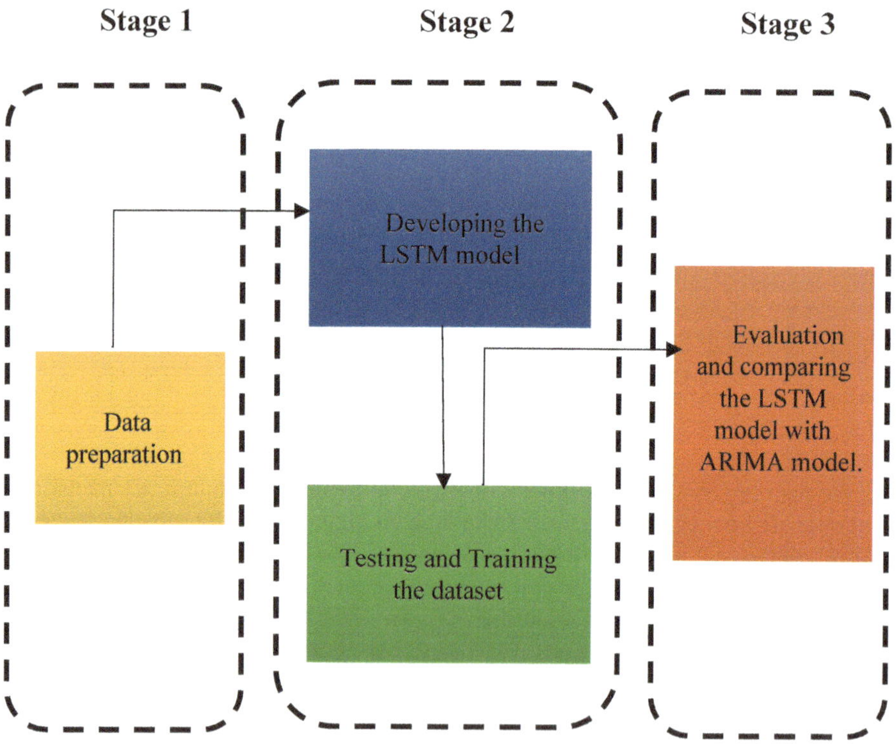

Fig. 2. Detailed diagram of an LSTM block

Figure 2 shows the general framework of the methodology. The first stage is obtaining the data samples for the experiment. The second stage is the description of the LSTM model, followed by testing and training the datasets. Finally, the last stage is evaluation and comparing the model with ARIMA. Data pre-processing is a data mining procedure which contains shifting CRUDE information to a reasonable formation. Genuine data are normally deficient, contradictory, and at the same time poor in some practices or points and might have many errors. Information preparation is an established approach for solving such problems. With information preparation, CRUDE information is ready for extra treatment. Information preparation is used in database-driven applications like client relationship and the executives, as well as standard-based applications (such as neural systems). Rouse in [22] states that data preparation is the way of collecting, joining, forming and categorising data so that they may be analysed as a key feature of data visualisation, examination and machine learning applications. One of the basic factors of the data preparation is that data processed for examination must be exact and expectable, so the aftereffects of BI and analytics applications will be considerable. Poor quality of data is a common problem, which may include missing values, mistakenness or other errors. Data collections should be put away in discrete samples or databases that have distinctive configurations to accommodate them. The ways of revising mistakes, providing confirmation and joining informational indexes are a main part of the data preparation process. Data preparation is, to a great extent, a robotised assignment, since it can take long stretches of work by the IT staff members or information experts to physically address each field in each document that will be utilised in an investigation, especially in a big data application. Machine learning algorithms can speed things up by looking at information fields, and naturally filling in clear qualities or renaming certain fields to guarantee consistency when information documents are being joined. The data for this research is collected from HealthData.com. It is an open-source data management platform from the United States with 1000 samples. The dataset characteristics are univariate and have two features. For an accurate performance comparison of diverse classifiers, it is essential to validate the data, and select the suitable performance metrics to be utilised. The classifiers from the data approximate the true features to build the models. The obtained data need to be split into training and testing datasets, in order to evaluate the constructed models. The training data are utilised to build the model, and the test data to assess it. It is assumed that the test data labels are recognised so that they can be put on the metrics of performance, which are accuracy and RMSE in this study.

3 Results

This section contains the results of the statistical analysis of LSTM for the rabies datasets. In this study, the LSTM model has experimented with a different number of epochs and different batch sizes. The findings of all datasets included are categorised according to Accuracy and RMSE.

3.1 Accuracy

This section explains the results and the LSTM model performance based on accuracy. In this research, the explained variance score is utilised to test the accuracy and the model performance. The described variance regression score function with the superlative probable score is 1.0; lower values indicate worse scenarios.

Table 1. Different epoch numbers with different batch size numbers (accuracy)

Number of epochs	Batch size				Accuracy (%)
	1	32	64	128	
50	✓				94.47
100	✓				97.3
120		✓			88
150			✓		84.3
200				✓	85.1

Each experiment shows a different outcome and we select the best result based on the accuracy as shown in Table 1. The best accuracy from these experiments is 97.3%, utilising 100 epochs and 1 batch size. Meanwhile, the lowest accuracy is 84.3%, utilising 150 epochs and 32 batch size.

3.2 Root Mean Square Error (RMSE)

The RMSE metric is utilised to show the variance between the actual and forecast values. A lower RMSE value shows a lesser variance between the actual and forecast results, which indicates an excellent forecasting performance.

Table 2. Different epochs number with different batch size number (RMSE (testing))

Number of epochs	Batch size				RMSE (Testing)
	1	32	64	128	
50	✓				2.76
100	✓				2.04
120		✓			4.82
150			✓		4.61
200				✓	4.58

Table 2 above presents the results of RMSE of the LSTM model. Lower values of RMSE indicate a better fit. Moreover, Table 2 shows that 100 epochs with batch size 1 produce the lowest RMSE value of 2.04; and 120 epochs with batch size 32 produce the highest RMSE value of 4.82.

Table 3 below shows a summary of the best results of the LSTM model. LSTM has a small RMSE value of (Testing) 2.04 and the highest accuracy of 97.30%.

Table 3. Summary of LSTM best results

Model	RMSE (test)	Accuracy (%)
LSTM	2.04	97.30

3.3 Comparative Study and Discussion

This study focuses on predicting the rabies outbreak and it is measured by the model performance (Accuracy). The performance of the proposed model is evaluated by comparing with that of ARIMA. The deep learning model (LSTM) displays extraordinary performance in comparison with the traditional method (ARIMA). The ARIMA model utilised for the study was noticed to be perfect when the incidences of the infectious diseases were steady and have zero growing or reducing drifts. But actual data may have drifts and abnormality. Consequently, deep learning might be a better analytical technique when analysing certain data and forecasting the upcoming conditions. Based on the findings of the earlier analyses, the model of deep learning adapts to growing and reducing drifts adequately well. Additionally, the LSTM model was noticed to be susceptible to reducing drifts and growing drifts, respectively. Figure 3 shows a comparative analysis of the LSTM model and ARIMA model respectively. Findings from this study evidently prove that the LSTM model is consistent in predicting the rabies outbreak. The LSTM model attains the highest accuracy of 97.30% compared with 72.10% by the ARIMA model.

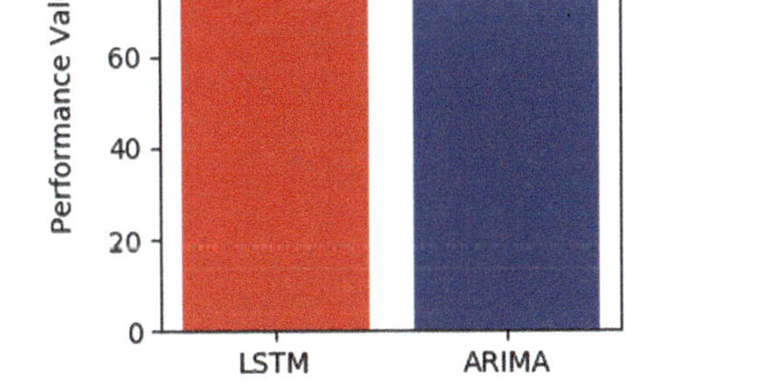

Fig. 3. Comparison of the accuracies of LSTM and ARIMA

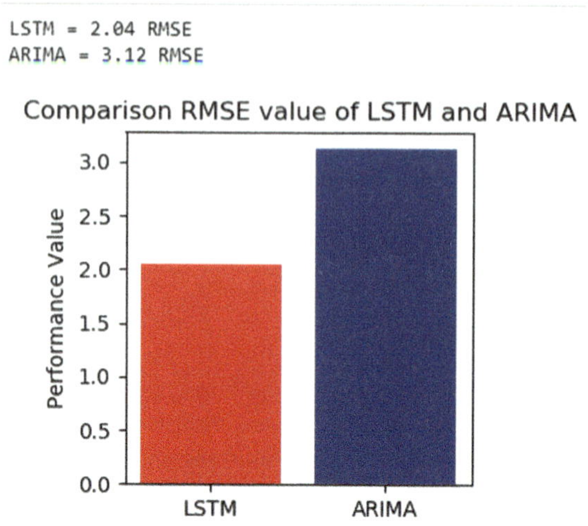

Fig. 4. Comparison of RMSE values of LSTM and ARIMA

In addition, Fig. 4 shows the comparative results of RMSE values between the LSTM and ARIMA models. The results show that LSTM has a low RMSE value of 2.04. Meanwhile, ARIMA obtains a high RMSE value of 3.118—higher than that of LSTM.

4 Conclusion

This study focuses on assessing the LSTM performance for forecasting the rabies outbreak and the results are compared with those of the ARIMA model. The results achieved by this study show that a deep learning LSTM network can predict an outbreak using a rabies dataset with a better accuracy. The LSTM model acquires the best accuracy results of 97.30%; meanwhile, the ARIMA model with a traditional network obtains 72.10%. As for the comparison of RMSE values, the LSTM model scores the lowest value of 2.04, and the ARIMA model scores the highest value of 3.12. From these results, the LSTM model is superior for forecasting the outbreak of this viral disease. The success of the LSTM model warrants it to be used for forecasting the outbreaks of other diseases like flu, H1N1, and Ebola or related problems. In future work, deeper architectures will be applied in order to achieve a better performance. More data are also required to improve the prediction performance. Finally, different machine learning algorithms should be utilised for the purpose of making more robust comparisons.

References

1. Hemachudha, T., Ugolini, G., Wacharapluesadee, S., Sungkarat, W., Shuangshoti, S., Laothamatas, J.: Human rabies: neuropathogenesis, diagnosis, and management. Lancet Neurol. **12**(5), 498–513 (2013)
2. Hampson, K., Coudeville, L., Lembo, T., Sambo, M., Kieffer, A., Attlan, M., et al.: Estimating the global burden of endemic canine rabies. PLoS Negl. Trop. Dis. **9**(4), e0003709 (2015). https://doi.org/10.1371/journal.pntd.0003709
3. Kole, A.K., Roy, R., Kole, D.C.: Human rabies in India: a problem needing more attention (2014)
4. Ramos, J.M., Melendez, N., Reyes, F., Gudiso, G., Biru, D., Fano, G., et al.: Epidemiology of animal bites and other potential rabies exposures and anti-rabies vaccine utilization in a rural area in Southern Ethiopia (2015)
5. Bueno-Marí, R., Almeida, A.P.G., Navarro, J.C.: Emerging zoonoses: eco-epidemiology, involved mechanisms and public health implications. Front. Publ. Health. **3**, 157 (2015)
6. Sparkes, J., Fleming, P.J.S., Ballard, G., Scott-Orr, H., Durr, S., Ward, M.P.: Canine rabies in Australia: a review of preparedness and research needs. Zoonoses Publ. Health **62**, 237 (2014)
7. Mähl, P., Cliquet, F., Guiot, A.L., Niin, E., Fournials, E., Saint-Jean, N., Aubert, M., Rupprecht, C.E., Gueguen, S.: Twenty-year experience of the oral rabies vaccine SAG2 in wildlife: a global review. Vet. Res. **45**(1), 77 (2014)
8. Wu, Y., Yang, Y., Nishiura, H., Saitoh, M.: Deep learning for epidemiological predictions. In: SIGIR, (2018). Ann. Agricul. Environ. Med. AAEM **22**(1), 76–79. https://doi.org/10.5604/12321966.1141372
9. Fricker, R.: Some methodological issues in biosurveillance. Stat. Med. **30**, 403–415 (2011)
10. H, Bamaiyi: 2015 outbreak of canine rabies in malaysia: review, analysis and perspectives. J. Vet. Adv. **5**(12), 1181 (2015). https://doi.org/10.5455/jva.19691231040000
11. He, J., Luo, L., Jin, R.G., Li, J.M.: The application of ARIMA in forecasting the cases of rabies in China different human groups. Zhonghua lao dong wei sheng zhi ye bing za zhi = Zhonghua laodong weisheng zhiyebing zazhi = Chin. J. Ind. Hygiene Occup. Dis. **36** (7), 512–515 (2018)
12. Chae, S., Kwon, S., Lee, D.: Predicting infectious disease using deep learning and big data. Int. J. Environ. Res. Public Health **15**(8), 1596 (2018). https://www.ncbi.nlm.nih.gov/pubmed/30060525
13. Wu, Y., Yang, Y., Nishiura, H., Saitoh, M.: Deep learning for epidemiological predictions. In: SIGIR (2018)
14. Zhang, Q., Yang, L.T., Chen, Z., Li, P.: A survey on deep learning for big data. Inform. Fusion **42**, 146–157 (2018)
15. Saleh, A.Y., Tei, R.: Flood prediction using seasonal autoregressive integrated moving average (SARIMA) model. Int. J. Innov. Technol. Explor. Eng. **8**(8), 1037–1042 (2019)
16. Saleh, A.Y., Francis, C.: A deep learning approach to Malware detection in android platform. Int. J. Innov. Technol. Explor. Eng. **8**(8), 1043–1048 (2019)
17. LeCun, Y., Bengio, Y., Hinton, G.: Deep learning; Nature Publishing Group, a division of Macmillan Publishers Limited, 28 May 2015. https://doi.org/10.1038/nature14539
18. Staudemeyer, R.C.: Evaluating performance of long short-term memory recurrent neural networks on intrusion detection data, October 2013
19. Brownlee, J.: Time series prediction with lstm recurrent neural networks in python with keras, p. 18 (2016). machinelearningmastery.com

20. Bao, W., Yue, J., Rao, Y.: A deep learning framework for financial time series using stacked autoencoders and long-short term memory. PLoS One **12**, e0180944 (2017). https://doi.org/10.1371/journal.pone.0180944
21. Alex, G.: Supervised Sequence Labelling with Recurrent Neural Networks; Studies in Computational Intelligence. Springer, Berlin (2012)
22. Rouse, M.: What is data preparation? - Definition from WhatIs.com, January. 2018. https://searchbusinessanalytics.techtarget.com/definition/data-preparation

Bioactivity Prediction Using Convolutional Neural Network

Hentabli Hamza[1](✉), Maged Nasser[1], Naomie Salim[1], and Faisal Saeed[2]

[1] School of Computing, Universiti Teknologi Malaysia, Johor Bahru, Malaysia
hentabli_hamza@yahoo.fr

[2] College of Computer Science and Engineering, Taibah University, Medina, Saudi Arabia

Abstract. According to the similar property principle, structurally similar compounds exhibit very similar properties as well as similar biological activities. Many researchers have applied this principle to discover novel drugs, thereby leading to the emergence of the prediction of the activities of compounds based on their chemical structure, since the toxic or biological properties of compounds are determined by their chemical structure, particularly, their substructures. The concept of functional groups (FGs) of connected atoms (small molecules) determining the properties and reactivity of the parent molecule forms the cornerstone of organic chemistry, medicinal chemistry, toxicity assessments and QSAR. This study introduced a novel predictive system, i.e., a convolutional neural network that enables the prediction of molecular bioactivities using a novel molecular matrix representation. The number of atoms in small molecules were investigated to determine its accuracy during the prediction of the activities of the orphan compounds. This approach was applied to popular datasets and the performance of this system was compared with three other classical ML algorithms. All the experiments indicated that the proposed model was able to provide an interesting prediction rate (accuracy of 90.21).

Keywords: Bioactive molecules · Activity prediction model · Convolutional neural network · Deep learning · Biological activities

1 Introduction

Biological systems function via the physical interactions that occur between the molecules. Hence, it is important to determine the molecular binding in order to understand biological systems and to discover novel drugs [1]. Pharmaceutical industries are devoting a lot of effort towards the discovery of novel drugs. These discoveries can improve our quality of life but, at the same time, they can also lead to many adverse effects [2]. Hence, pharmaceutical companies must ensure the safety of a drug during the research stage, as the observation of adverse effects during the late clinical phases may lead to heavy financial losses. However, despite the development of computational systems over the past 30 years, these systems have proven to be inaccurate in the prediction of molecular binding, and therefore, it is necessary to

F. Saeed et al. (Eds.): IRICT 2019, AISC 1073, pp. 341–351, 2020.
https://doi.org/10.1007/978-3-030-33582-3_33

conduct physical experiments to determine the binding [3]. The popular ML algorithms using the compound classification method for activity prediction (target) are the binary kernel discrimination (BKD) [4], Bayesian inference network for ligand-based virtual screening [5], Naivë Bayesian classifier (NBC) [6], artificial neural networks (ANNs) [7] and support vector machines (SVMs) [8]. The Bayesian belief network classifier is used for predicting ligand-based targets and their activities [1].

Deep learning is seen to dramatically improve advanced artificial intelligent tasks such as speech recognition, object detection and machine translation [9]. The deep architectural nature of this technique is useful for solving complex artificial intelligence-related problems [10]. In another study [11], a neural model was used to generate images with differing styles. The deep learning technology was used by [12] for a simultaneous analysis of sentiments from multiple modalities. This study introduced a novel approach, which is a convolutional neural network that predicts the bioactivity of molecules using small molecules as a unique molecular matrix representation. Also, it is a deep learning system that incorporates information on the substructures of the molecules for the activity prediction.

2 Materials and Methods

Initially, the data encoding, and input representation system are described along with the design of the deep convolutional network.

2.1 Datasets

The quality of the proposed prediction model was estimated using multiple datasets to validate the molecule classification based on the structure-activity relationship. This study used MDDR dataset (Table 1) which were described earlier and used by several studies for validating ligand-based virtual screening methods [1, 13]. this dataset comprised of 10 heterogeneous activity classes. The molecules in each class and the differences between the classes, which were computed based on the mean pairwise Tanimoto similarity that was estimated across all the molecular pairs within the class.

Table 1. MDDR activity classes dataset

Activity index	Activity class	Active molecules	Pairwise Similarity
09249	Muscarinic (M1) agonists	900	0.111
12455	NMDAreceptor antagonists	1400	0.098
12464	Nitricoxidesynthase inhibitor	505	0.102
31281	Dopaminehydroxylase inhibitors	106	0.125
43210	Aldose reductase inhibitors	957	0.119
71522	Reversetranscriptase inhibitors	700	0.103
75721	Aromatase inhibitors	636	0.110
78331	Cyclooxygenase inhibitors	636	0.108
78348	Phospholipase A2 inhibitors	617	0.123
78351	Lipoxygenase inhibitors	2111	0.113

2.2 Input Representation

The biological activity of a compound is an adverse property which affects its potential to be marketed as a drug. The toxic or biological properties of a compound are based on its chemical structure, particularly, its substructures. The concept of functional groups (FGs) of connected atoms (small molecules) determining the properties and reactivity of the parent molecule is the cornerstone of organic chemistry, medicinal chemistry, toxicity assessments and QSAR. However, surprisingly, little attention has been paid to the study of FGs from the perspective of cheminformatics. The majority of the theoretical studies utilized FGs as a basis for chemical ontologies, where FGs are "keys" that are used to hierarchically categorise molecules [14, 15].

A novel method, called the Mol2fgs (molecule to functional groups), was proposed. This technique can be used in cheminformatics for many problems related to the virtual screening, classification, biological activity prediction, similarity measurements and sub-structural search of molecules. Here, every compound was embedded within an $n \times n$ matrix, which characterised the various molecular properties. A distributed representation is viewed as a successful and popular ML approach [16]. This approach involves the encoding and storage of information within a system by interactions with other compounds. Deep learning possesses the ability to construct abstract features to help in predicting toxicity or biological activities.

The ChEMBL dataset [17] is an open, large-scale bioactivity database. The most recent version of the ChEMBL is comprised of more than 2.5 million compounds. The ChEMBL database consists of 63327 compounds, where the number of atoms in each compound is less than 15, such that they can be called small molecules [18], which are converted to ECFP4 by the Pipeline Pilot to the final size of 1024 [19], Table 2 describes the number of molecules in each set, from A to N, and the atom numbers in the ChEMBL dataset.

Table 2. Molecules number in each Sets of Atoms Number in ChEMBL dataset to select the small molecules

Sets	Atoms number	Molecules number	Sets	Atoms number	Molecules number
SetA	2	47	SetH	9	2068
SetB	3	82	SetI	10	3203
SetC	4	138	SetJ	11	5041
SetD	5	277	SetK	12	7401
SetE	6	489	SetL	13	9959
SetF	7	734	SetM	14	13865
SetG	8	1258	SetN	15	18765

In this study, the Mol2fgs was proposed to represent every molecule in the $n \times n$ matrix. This matrix was comprised of compounds displaying Tanimoto similarities to the small molecules selected from the ChEMBL dataset. The Tanimoto-based similarity searching (TAN) [20] technique applied the binary form of the Tanimoto coefficient to the binary data. A similarity score, S_{xy}, was used for computing the similarities between the 2 molecular ECFP4 fingerprints, i.e., X and Y, with a length of 1024, wherein 'A' represented the number of bits present in the X and Y fingerprints, 'B' represented the number of bits that were present only in X, while 'C' represented the number of bits present only in Y.

$$S_{xy} = A/(A + B + C) \tag{1}$$

Every row in the matrix was filled with the Tanimoto similarities to the small molecules selected from the ChEMBL dataset. This Mol2fgs representation helped in visualising and characterising every molecule in the matrix based on its interactions and similarities with the functional groups. Thereafter, this matrix described the toxic properties of the chemical compound. The Mol2fgs is a good tool for describing the computational toxicology as it constructs the abstract chemical features.

For studying the performance of the Mol2fgs representation, scatter graphs were also plotted using the 8294 molecules that had been categorised into 10 different biological activity classes in the MDDR dataset (Fig. 1). These scatter plots were used to determine the relationship between the different molecules within the same class based on their individual representations that were reduced to 3D structures using the principal component analysis (PCA) technique for representing their features. As can be easily observed in the figure, there was no overlapping in the Mol2fgs representation using just the small molecules with atom number 10 (Set K) selected from the ChEMBL dataset. Also, the biological activities of the molecules could be segregated.

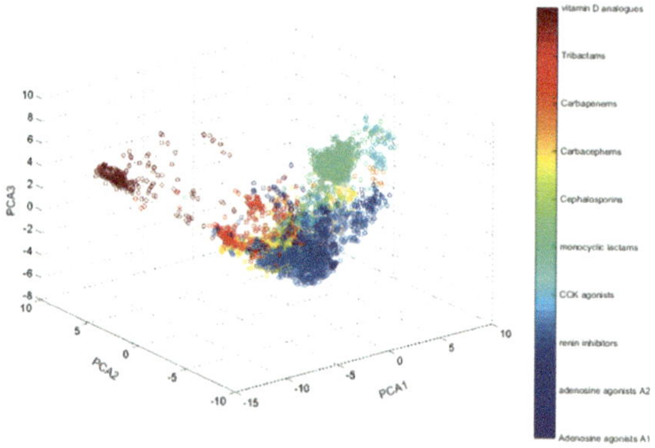

Fig. 1. D-scatter plots based on the Mol2fgs representation of 8294 different molecules selected from the 10 biological activity classes of the MDDR dataset using molecules with atom number 12 as small molecules.

2.3 Network Architecture

After collecting all the data, the various atom numbers in the case of the selected molecules from the ChEMBL dataset were investigated. The convolutional architecture with fully connected layers was considered as the default architecture. Based on many experiments result, the author used the Krizhevsky principles [21] as the best result to design the CNN-QSAR model configuration, through which the source code could be viewed [22]. The configuration followed the generic design described earlier [21]. Figure 2 illustrates the CNN-QSAR configuration.

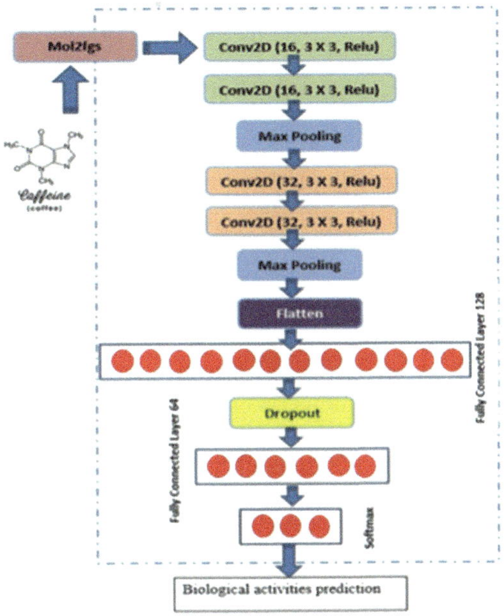

Fig. 2. The proposed CNN-QSAR configuration

Generally, the target prediction was carried out as follows: The problem involved predicting whether the given chemical compound, i, was active against the target, t. This information was encoded in a binary form, y_{it}, where $y_{it} = 1$ for an active compound, and $y_{it} = 0$ for a compound that was not active. During the training stage, a standard back-propagation algorithm was used for determining the CNN and minimizing the cross-entropy of the targets and the output layer activation.

2.4 Machine Learning Algorithms

The proposed technique was compared to 3 other available ML algorithms within the WEKA Workbench [23], i.e., the Naive Bayesian classifier (NaiveB) [24], SVM classifier (known as the LibSVM, LSVM) [25] and neural network classifier (RBFN) [26]. All the remaining parameters were kept as the default for every classifier in the WEKA Workbench.

3 Results and Discussion

The proposed code was implemented in Theano [27], which is a public deep learning software based on Keras [28]. The weights in the neural networks were initialised according to the Keras settings. The complete network was trained using the Dell Precision T1700 CPU system with a 14 GB memory and the professional-grade NVIDIA-Quadro discrete graphics. The training and testing of the deep network took 2 weeks.

3.1 Evaluation Measures

A 10-fold cross-validation technique was used to validate all the results of the proposed CNN-QSAR system. Similar to other prediction techniques, the accuracy (ACC) measurement was determined in order to quantify the performance of the different classification models. The accuracy refers to the general effectiveness of a model and is determined by the sum of the correct classifications divided by the total number of classifications. This phenomenon is described in Eq. 2 as:

$$ACC = (TP + TN)/\,(TP + TN + FP + FN) \tag{2}$$

where TP, TN, FP and FN are the number of true positives, true negatives, false positives and false negatives, respectively, with TP being the number of active molecules within the active set, and TN being the number of inactive molecules selected in the inactive set. Meanwhile, FP and FN refer to the number of active molecules present in the inactive set and the number of inactive molecules in the active set, respectively.

3.2 Results

In this study, the CNN-QSAR was proposed as a novel ligand-based activity prediction or target fishing method for unknown chemical compounds, the proposed CNN-QSAR technique was compared to 3 other ML algorithms present in the WEKA Workbench, i.e., NaiveB, LSVM, and RBFN, using optimal parameters.

This novel deep learning system was assessed for its computing prediction accuracy using the procedure described in Subsect. 2.3 for all the various CNN-QSAR molecule sets. All the results that were obtained from each set of atom numbers for A to N were compared using a box plot, which is a standardized way of displaying the distribution of results. This was Stage 1 of the results analysis, and it is described in further detail below. In Stage 2, the analysis results obtained from the 5 best CNN-QSAR situations were combined two by two and compared, based on their box plot diagram, with previous sets of models. In Stage 3, the prediction accuracy results for the 2 best CNN-QSAR sets obtained from Stage 2, which had been combined two by two and compared based on their box plot diagram with previous case models, were assessed. In Stage 4, the results for the best configuration obtained from the earlier 3 stages were compared with the 3 standard ML algorithms, i.e., NaiveB, LSVM, and RBFN.

Stage 1

Table 3 describes the sets of atom numbers for A to N in the ChEMBL dataset that was used for the proposed Mat2Fgs representation.

In Stage 1, the prediction accuracy values for 10 activity classes in the MDDR dataset were evaluated and compared. The comparison of the prediction accuracy for the Stage 1 experiments involving the CNN-QSAR sets A to N are described in Fig. 4.

As shown in Fig. 3, there was a clear difference in the mean values of the 14 situations. Also, the G, H, I, J and K situation models showed the best accuracy values, whereas the atom numbers for the molecules in these situations were 8, 9, 10, 11 and 12, respectively. These results indicated that the next stage was the analysis results obtained for the 5 situation models, G, H, I, J and K, which were combined two by two and compared based on their box plot diagram data with previous case models.

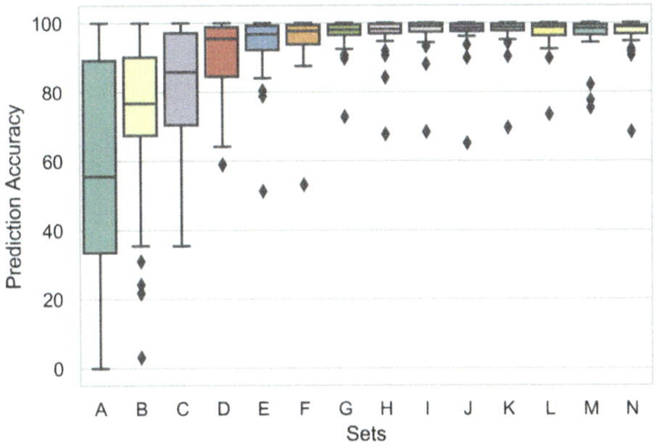

Fig. 3. Box plot diagram comparing the prediction accuracy values for the Stage 1 experiments involving the CNN-QSAR sets A to N.

Stage 2

In Stage 2, the prediction accuracy values of the 5 best CNN-QSAR situations obtained from the previous stage were analysed. They were combined two by two and compared with previous case models. A comparison of the prediction accuracy values for the Stage 2 experiments for 9 sets of models, G, H, I, J, K, G+H, H+I, I+J and J+K using a box plot diagram is described in Fig. 4.

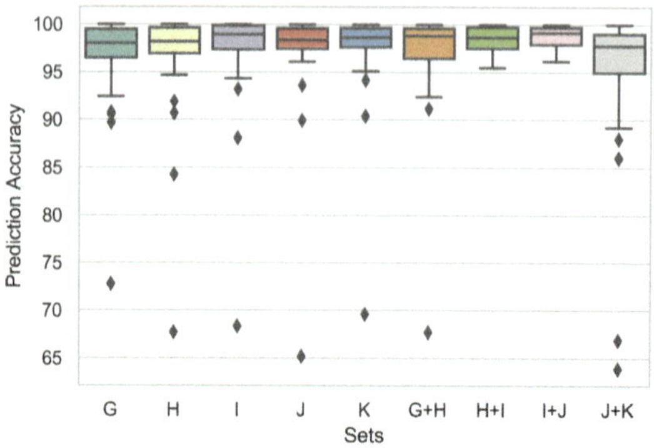

Fig. 4. Box plot diagram showing comparison of the prediction accuracy values for the Stage 2 experiments involving the CNN-QSAR situations of G, H, I, J, K, G+H, H+I, I+J and J+K

As shown in Fig. 4, there was a difference in the mean values for the 9 situations with regard to the 9 sets of atoms used in the Mol2Fgs representation. Also, the H+I and I+J sets of models showed the best accuracy value, whereas the atom numbers for the molecules in this situation were 9 + 10 and 10 + 11, respectively.

Stage 3

In Stage 3, the prediction accuracy values for the (SetH + SetI) and (SetI + SetJ) models and the combined (SetH + SetI + SetJ) models were analysed. The comparison of the prediction accuracy values for the 3 models using the box plot diagram are described in Fig. 5.

As shown in Fig. 5, the (SetH + SetI + SetJ) situation for the combined SetH, SetI and SetJ atoms used in the Mol2Fgs representation was better, gave the best prediction accuracy values and displayed very close mean accuracy values of 99.3%. This result indicated that the best set of small molecules that can be used for the new Mol2Fegs representation was the set of molecules with atom numbers 9, 10 and 11.

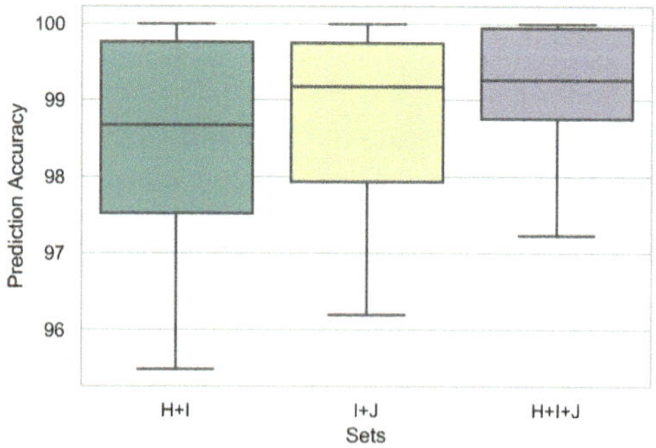

Fig. 5. Box plot diagram comparing the prediction accuracy values for the Stage 3 experiments involving the CNN-QSAR situation for the (SetH + SetI) and (SetI + SetJ) and the combined (SetH + SetI + SetJ) models.

Stage 4
In the final stage, the best model results obtained from the earlier 3 stages were compared with the 3 standard ML algorithms, i.e., NaiveB, LSVM, and RBFN. The comparison of the prediction accuracy values for the 4 models are described in the box plot diagram in Fig. 6.

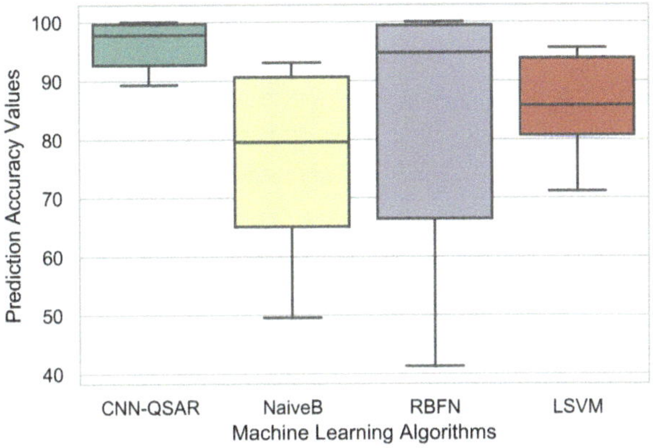

Fig. 6. Box plot diagram comparing the prediction accuracy values for the CNN-QSAR, NaiveB, RBFN and LSVM algorithms.

The box plot diagram in Fig. 6 presents the results of the comparison of the prediction accuracy values for the CNN-QSAR, NaiveB, RBFN and the LSVM algorithms. A larger variance was noted between the NaiveB and the RBFN ML algorithms compared to the CNN-QSAR method. This highlighted the diversity in the prediction accuracy noted in the algorithms with a median of 98.2%. Also, the CNN-QSAR method displayed a good prediction accuracy value of 99%, while the NaiveB and RBFN algorithms showed a mean prediction accuracy value of 80% and 93%, respectively. The box plot diagram in Fig. 6 indicated that there was a significant difference between the algorithms.

4 Conclusion

This study investigated the use of a deep learning convolutional network for predicting ligand-based targets and their activities. It was demonstrated that the new model provided an interesting result with regard to the prediction accuracy. A novel Mol2Fgs technique, which had less overlapping and was able to segregate the biological activities of the molecules, was also proposed. It was demonstrated that the small molecules were better at representing the functional groups for unknown molecules with atom numbers of between 9 to 11. The new CNN-QSAR technique was then applied to popular datasets and its performance was compared with 3 standard ML algorithms. All the experiments indicated that the CNN-QSAR algorithm exhibited interesting prediction rates (accuracy of 90.21%). Furthermore. Hence, the CNN-QSAR is presented as a stable and convenient activity prediction approach for unknown chemical compound targets. However, this area still needs to be explored further and better accuracy prediction techniques have to be developed for the highly diverse activity classes.

Acknowledgment. This work is supported by the Ministry of Higher Education (MOHE) and the Research Management Centre (RMC) at the Universiti Teknologi Malaysia (UTM) under the Research University Grant Category (VOT Q.J130000.2528.16H74 and R. J130000.7828.4F985).

References

1. Ammar, A., Leclère, V., Jacques, P., Salim, N., Pupin, M.: Prediction of new bioactive molecules using a Bayesian belief network. J. Chem. Inf. Model. **54**(1), 30–36 (2014)
2. Barakat, K.: Computer-aided drug design. J. Pharm. Care Heal. Syst. **1**(4), 1–2 (2014)
3. Kothiwale, S., Borza, C., Pozzi, A., Meiler, J.: Quantitative structure–activity relationship modeling of kinase selectivity profiles. Molecules **22**(9), 1–11 (2017)
4. Willett, P., Wilton, D., Hartzoulakis, B., Tang, R., Ford, J., Madge, D.: Prediction of ion channel activity using binary kernel discrimination. J. Chem. Inf. Model. **47**(5), 1961–1966 (2007)
5. Chen, B., Mueller, C., Willett, P.: Evaluation of a Bayesian inference network for ligand-based virtual screening. J. Cheminform. **1**(1), 1–10 (2009)
6. Xia, X., Maliski, E.G., Gallant, P., Rogers, D.: Classification of kinase inhibitors using a Bayesian model. J. Med. Chem. **47**, 4463–4470 (2004)
7. Winkler, D., Burden, F.R.: Application of neural networks to large dataset QSAR, virtual screening, and library design. Methods Mol. Biol. **201**, 325–367 (2002)
8. Kawai, K., Fujishima, S., Takahashi, Y.: Predictive activity profiling of drugs by topological-fragment-spectra-based support vector machines. J. Chem. Inf. Model. **48**(6), 1152–1160 (2008)
9. LeCun, Y., Yoshua, B., Geoffrey, H.: Deep learning. Nature **521**(7553), 436–444 (2015)
10. Bengio, Y.: Learning deep architectures for AI, vol. 2, no. 1. (2009)
11. Gatys, L. Ecker, A.S. Bethge, M.: A Neural algorithm of artistic style. arXiv Prepr., pp. 1–16 (2015)
12. Wang, H., Meghawat, A., Morency, L.P., Xing E.P.: Select-additive learning: improving cross-individual generalization in multimodal sentiment analysis, vol. 1 (2016)
13. Hentabli, H., Naomie, S., Saeed, F.: An activity prediction model using shape-based descriptor method. J. Teknol. **1**, 1–8 (2016)
14. Feldman, H.J., Dumontier, M., Ling, S., Haider, N., Hogue, C.W.V.: CO: a chemical ontology for identification of functional groups and semantic comparison of small molecules. FEBS Lett. **579**(21), 4685–4691 (2005)
15. Bobach, C., Böhme, T., Laube, U., Püschel, A., Weber, L.: Automated compound classification using a chemical ontology. J. Cheminform. **4**(12), 1–12 (2012)
16. Mikolov, T., Sutskever, I., Chen, K., Corrado, G., Dean, J.: Distributed representations of words and phrases and their compositionality, pp. 1–9 (2013)
17. Bento, A.P.: The ChEMBL bioactivity database: an update. Nucleic Acids Res. **42**(D1), D1083–D1090 (2014)

18. Lewis, R.A., Wood, D.: Modern 2D QSAR for drug discovery. Wiley Interdiscip. Rev. Comput. Mol. Sci. **4**(6), 505–522 (2014)
19. Saeed, F., Salim, N.: Using soft consensus clustering for combining multiple clusterings of chemical structures. J. Teknol. Sci. Eng. **63**(1), 9–11 (2013)
20. Hentabli, H., Saeed, F., Abdo, A., Salim, N.: A new graph-based molecular descriptor using the canonical representation of the molecule. Sci. World J. **2014** (2014)
21. Krizhevsky, A., Sutskever, I., Hinton, G.E.: ImageNet classification with deep convolutional neural networks. In: Advances in neural information processing systems, pp. 1–9 (2012)
22. Gupta, V.: Image classification using convolutional neural networks in keras (2017)
23. Witten, I.H., Frank, E., Hall, M.A., Pal, C.J.: Data Mining: Practical Machine Learning Tools and Techniques. Morgan Kaufmann, Burlington (2016)
24. John, G.H. Langley, P.: Estimating continuous distributions in Bayesian classifiers, pp. 338–345 (2013)
25. Chang, C.C.: LIBSVM: a library for support vector machines. ACM Trans. Intell. Syst. Technol. **2**, 271–2727 (2011)
26. Bugmann, G.: Normalized Gaussian radial basis function networks. Neurocomputing **20**(1–3), 97–110 (1998)
27. Bastien, F.: Theano: new features and speed improvements, pp. 1–10 (2012)
28. Chollet, F.: Keras documentation. Keras.Io (2015)

An Improved Jaya Algorithm-Based Strategy for T-Way Test Suite Generation

Abdullah B. Nasser(✉), Fadhl Hujainah, AbdulRahman A. Al-Sewari, and Kamal Z. Zamli

Faculty of Computer Systems and Software Engineering, Universiti Malaysia Pahang, 26300 Kuantan, Pahang, Malaysia
abdullahnasser@ump.edu.my

Abstract. In the field of software testing, several meta-heuristics algorithms have been successfully used for finding an optimized t-way test suite (where t refers to covering level). T-way testing strategies adopt the meta-heuristic algorithms to generate the smallest/optimal test suite. However, the existing t-way strategies' results show that no single strategy appears to be superior in all problems. The aim of this paper to propose a new variant of Jaya algorithm for generating t-way test suite called Improved Jaya Algorithm (IJA). In fact, the performance of meta-heuristic algorithms highly depends on the intensification and diversification capabilities. IJA enhances the intensification and diversification capabilities by introducing new operators search such lévy flight and mutation operator in Jaya Algorithm. Experimental results show that the IJA variant improves the results of original Jaya algorithm, also overcomes the problems of slow convergence of Jaya algorithm.

Keywords: T-way testing · Meta-heuristics · Jaya Algorithm · Improved Jaya algorithm

1 Introduction

Optimization is a mathematical method used for solving complex problems many disciplines, including engineering, biology, physics, business, and economics, by finding the best/optimal solution from the alternative set of solutions. Many meta-heuristic algorithms have been used successfully for finding an optimal or near optimal solution within reasonable time including Simulated Annealing (SA), Tabu Search (TS) Genetic Algorithm (GA), Particle Swarm Optimization (PSO), Differential Evolution (DE), Sine Cosine Algorithm (SCA), Flower Pollination Algorithm (FPA), Ant Colony Optimization (ACO), Jaya Algorithm (JA), Optimization Algorithm (TLBO) Cuckoo Search (CS) and Teaching–Learning-Based [1].

The performance of meta-heuristic methods highly depends on the intensification and diversification of the search process. Intensification explores the promising regions in the hope to find better solutions. On the other hand, diversification ensures that all regions have been visited, which allows the algorithm to jump out of local optimum [2].

F. Saeed et al. (Eds.): IRICT 2019, AISC 1073, pp. 352–361, 2020.
https://doi.org/10.1007/978-3-030-33582-3_34

In the field of software testing, several meta-heuristics algorithms have been successfully used for finding an optimized t-way test suite based on defined interaction strength (t) [3]. Nowadays, many researchers turn into t-way testing. T-way testing is a very efficient approach for minimizing the test cases based on the *t* interaction coverage. Finding an optimized set of test cases is a non-deterministic polynomial-time hard problem where increasing software inputs lead to an increase in computational time in an exponential manner and the complexity as well [4]. As a consequence, much existing research adopts meta-heuristic algorithms as the basis of their implementation including TS, SA, GA, ACA, PSO [5], HS [6] CS [7] and FPA [8].

This paper proposes a t-way test suite generation method, called IJA based on an improved of Jaya Algorithm [9] by enhanced its search capabilities. First, the proposed method utilizes lévy flight for enhanced its exploration part. In fact, lévy flight consists of random walks that are interspersed by long jumps which can help to get out of local minima [10]. Second, a mutation operator is added to improve the convergence rate which to measure how fast JA converge.

The rest of this paper is organized as follows. An overview of t-way testing will highlight in Sect. 2 while Sect. 3 reviews existing t-way strategies. Section 4 elaborates the description of the proposed strategy. Section 5 highlights and discusses the results. Lastly, Sect. 6 gives the conclusion and future work.

2 Overview of T-Way Testing

T-way testing ensures that every t-combinations (where t refers to interaction strength) of the inputs are covered in the final test suite. For mere understanding, consider the following insert page numbers dialog box in Microsoft word processor. In this example, three are parameters (with ignoring the buttons) which are:

1. Position has two values (top and bottom).
2. Alignment has three values (Left, center and Right).
3. Show-number-on-first-page has two values (Selected/Yes and Unelected/No) (Fig. 1).

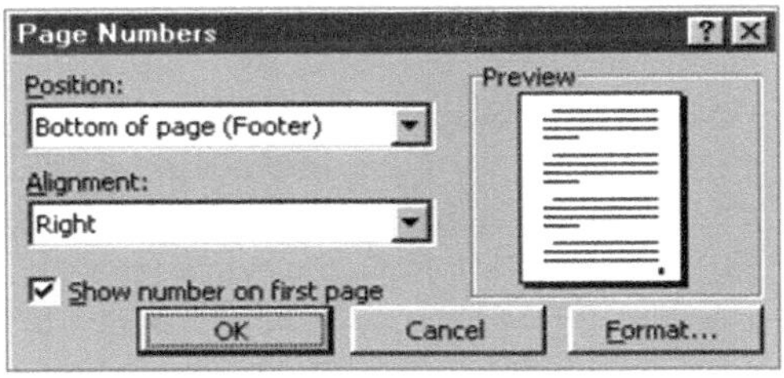

Fig. 1. Page numbers dialog box.

For testing such a simple dialog box, ideally, all the combinations of the inputs are considered. Here are 12 test cases need to be tested that can cover all the interactions.

However, by considering 2-way interaction, test suite size can be minimizing from 12 to 6 test cases. Two-way interaction testing ensures that every two-combination of parameters is covered in the final test suite at least one time. It should be noted that there is 50% reduction test suite size however the faults detection rate can be reached to 90%. Similar to two-way testing, three-way testing increase the faults detection reach to 99% [11].

3 Related Works

In the literature review, T-way Testing has been generally treated as an algebraic problem. In this approach, lightweight algebraic computations are adopted for generating the t-way test suite. Generating t-way test suite using the algebraic approach used two methods. The first method is based on mathematical functions where each cell of the test suite is computed directly based on the corresponding row and column using mathematical functions. Whereas the second one employs a recursive process to construct test suite by constructing a large test set from small test sets such as CA, MCA and TConfig, are limited to small configurations [12].

Computational approaches adopt a greedy algorithm to generate the test suite. Each iteration of the generation process tries to cover as many t-way combinations as possible, by generating one parameter at a time (OPAT) or one test at a time (OTAT). OPAT strategies start by generating the test suite for the smallest t-combinations. Then the test suite is extended horizontally by adding a parameter per iteration until all the parameters are covered. Examples of such approach are IPO, IPOG, IPOG-D, IPOF and IPAD2 [13].

OTAT strategies start generating one test case per iteration involved all the parameters that cover the maximum number of combinations. The iteration continues until all the t-combinations are covered. Due to its efficiency, there are many strategies that adopt OTAT techniques such as AETG [14], Jenny [15], TConfig [16], and WHITCH [17]. Recently, many OTAT based strategies adopt meta-heuristic algorithms for generating t-way test suite Applying a meta-heuristic algorithm, or in general search algorithms, in the field of testing also called Search-based software Testing (SBST).

The first publication on SBST was in 1976 however it was completely different from techniques that were used at the time. It was simply generating the test suite that consists of floating-point inputs [18]. Recently, the role of the search algorithm in SBST is to guide the search for a better solution within a limited time.

In the literature, there are many meta-heuristic algorithms has been used successfully for generating t-way test suite such as Genetic Algorithm (GA) [14], Ant Colony Algorithm (ACA) [19], Particle Swarm Optimization [20], Harmony Search [6], Flower Pollination Algorithm (FPA) [21], and Cuckoo Search (CS) [22].

Unlike the aforementioned strategies, new strategies based on more than single search engine methods have been proposed [3, 4, 23–25]. The new strategies attempt to improve the search process for generating test suite by adopting hybridization methods such as hybrid-FPA (eFPA) [3], Tabu search hyper-heuristic (HHH) [4], Q-learning sine-cosine [25], and fuzzy-TLBO (ATLBO) [23].

4 Proposed Strategy

4.1 Original Jaya Algorithm

Jaya Algorithm [9] is one of the population-based algorithms, designed for solving constrained and unconstrained problems. The main idea behind the algorithm is that any candidate solution should approach to the best solution and evade worse solution at the same time. Therefore, for updating current solution $X_{i,j,}$ Jaya Algorithm needs only to update the best and worse solutions then update the current solution using the following equation:

$$X'_{i,j} = X_{i,j} + Rnd_1\left(X_{best,j} - \left|X_{i,j}\right|\right) - Rnd_2\left(X_{worst,j} - \left|X_{i,j}\right|\right) \quad (1)$$

Where $X_{best,}$ X_{worst} are the best and worst obtained solutions, and $X'_{i,j}$ is the new candidate solution.

In term of parameter tuning, Jaya Algorithm is free parameters algorithm and need only to define common parameters such as population size and iteration number. Figure 2 show the summary of Jaya Algorithm.

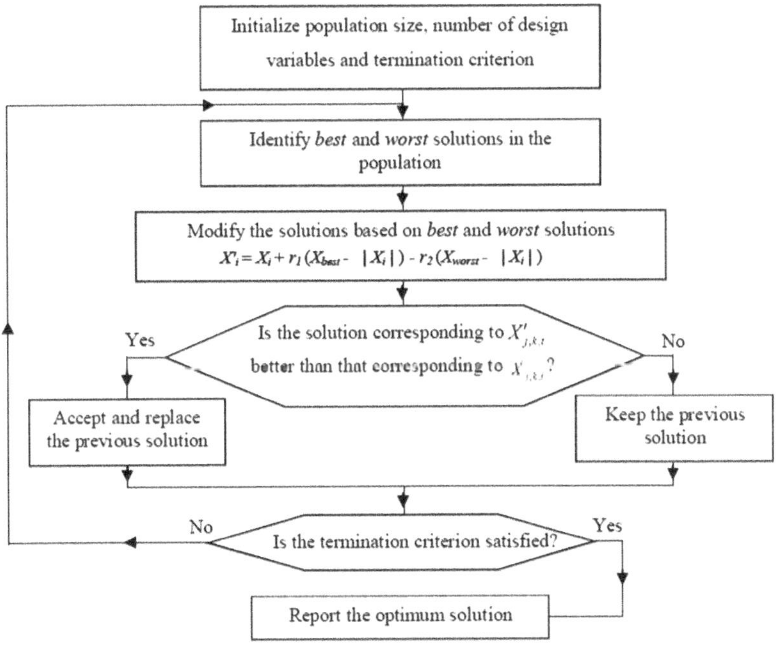

Fig. 2. Original Jaya Algorithm

4.2 Improved Jaya Algorithm Based Strategy for T-Way Test Suite Generation

Improved Jaya Algorithm (IJA) strategy is a t-way testing strategy for generating the test suite. It uses improved Jaya algorithm as core implementation for finding optimal test cases. The strategy starts generating all the t-combinations of the inputs, then using improved Jaya algorithm, IJA attempts to cover all those t-combinations using the smallest test suite.

As shown in Fig. 2, after generating all the t-combinations, improved Jaya algorithm generates its random population, then update the population using Eq. 1. To improve the diversification of generated solutions, lévy flight (i.e. Eq. 2) has been deployed on the population. Based on the current solution, a new solution is generated using lévy flight. After updating some solution using lévy flight, mutation operation is applied as well. The mutation operator used to improve convergence speed and maintains the diversity solution of the population from one generation to the next one (i.e. as one or more parts of the solution values are changed subtly).

To maintain the behavior of original Jaya algorithm a portion of the population is updated. Here, we applied elitism technique that used in GA, HS and CS, to ensure the high quality solution is passed to next generation; only a fraction of worse solutions are updated using lévy flight or mutation while the elite solutions (i.e. quality solutions) passed to next generation. In our proposed, we applied a simple mutation operator that randomly changes one parameter of the solution per iteration. Finally, all solutions of the population are evaluated based on fitness function, and the best solution, which cover the maximum number of t-combination, is selected to be added to the final test suite.

$$L\acute{e}vy(\lambda) = \frac{\lambda\Gamma(\lambda)\sin\left(\frac{\pi\lambda}{2}\right)}{\pi}\frac{1}{s^{1+\lambda}}, (s \gg s_0 > 0) \tag{2}$$

The recommended value of the distribution factor λ is selected from 0.3 to 1.99 [26]. $\Gamma(\lambda)$ is standard gamma function. This distribution is valid for large steps $s > 0$. Figure 2 illustrates the proposed improved Jaya algorithm-based strategy for t-way test suite generation (IJA) (Fig. 3).

```
Input:  p: Parameter number,
        V: = [v0..vj] parameter-values and
        t: interaction level
        P ∈ [0, 1];
Output: Final test suite F;
Let X be a set of population (candidate test cases);
Generate t-combination L based on P, V and t
Generate initial random population
while L is not empty do
   while stop criterion not meet do
      For each X_i of the population do
         Update X_i position using Eq. 1
      End For
   // Find a fraction of worse solutions and update them using lévy flight or mutation
      For each X_i of the population do
       If (random > p)
         If (random > 0.5)
            Generate X_{i+1} using lévy flight on X_i (Eq. 2)
         Else
            Apply mutation operator on X_i
         End if
       End if
      End for
      Evaluate the solutions
      Find the best solution
   End while
Add the best solution to F.
Remove covered t-combinations from L
End while
End-Procedure
```

Fig. 3. Improved Jaya algorithm pseudocode for t-way test suite generation

5 Results and Discussion

The proposed strategy is compared with existing t-way strategies such as PSO, CS and original Jaya using different t-way testing problems as shown in Table 1. In order to assess the enhancement of Jaya algorithm, the results of both the original Jaya algorithm and its proposed improvement are involved in the comparison. In addition, we compare the convergence rate of Jaya algorithm against the improved Jaya algorithm. Based on some existing works, the parameters of IJA are set at population size = 50 and the maximum number of improvements = 300 [6, 21, 24]. Different systems are used in this experiment as shown in Table 1. Systems configuration column describes the system such that y^x means the system consists x parameters each parameter has y values.

5.1 Performance Evaluation

In this section, IJA's results are compared with existing meta-heuristic strategies published in [5–7].

Table 1. Comparison with existing strategies

System#	Systems configuration	t	PSO	CS	Jaya	IJA
1	2^{10}	2	8	8	10	**8**
2	3^{10}	2	**17**	**17**	23	**17**
3	2^{10}	3	17	**16**	18	**16**
4	3^{6}	3	**42**	43	44	**42**
5	5^{7}	4	1209	1200	1195	**1178**
6	5^{8}	4	1417	1415	1345	**1331**
7	2^{10}	5	82	**79**	85	**79**
8	3^{7}	5	441	439	449	**435**
9	2^{10}	6	158	**157**	163	159
10	3^{7}	6	977	973	**971**	**971**
11	2^{10}	7	NA	NA	312	**297**
12	2^{10}	8	NA	NA	519	**502**
13	2^{10}	9	NA	NA	680	**584**
14	2^{10}	10	NA	NA	**1024**	**1024**

Table 1 presents the results of the comparison; each cell presents the best (i.e. smallest) test suite size obtained by t-way testing strategies while NA indicates that corresponding results not available. The results in Table 1 show that IJA performs better than PSO, CS, and original Jaya and obtains the best results (marks by bold font cell) while PSO obtains the worst results. Comparing only IJA and its counterpart, the results show that IJA performs much better than original Jaya, IJA outperforms original Jaya in almost all cases with exception of cases #10 and #14 where both strategies obtain the same results.

IJA strategy appears to obtains the optimum results in most of the cases owing to enhancement of its lévy flight which help improve the diversification and intensification since the lévy flight can be seen as random walk interrupted by long jump. Another point worth to highlight is that during the experiments, the results are improved after applying mutation operator, that's because the behaviour of mutation operator and how it works are totally different than how Jaya algorithm works. In mutation operator, during updating the solutions, only one input or more are changed while Jaya algorithm updates all inputs.

5.2 Convergence Rate Analysis

To assess the modification of IJA strategy, the convergence rates of both IJA and original Jaya strategies are studied. The convergence rates use to measure the speedup of meta-heuristic algorithm to reach the optimal solutions. We apply Jaya and IJA on two systems (Table 1) using different values of iteration (i.e. 5, 10, 20, 30, 40, 50, 100, 200, 300, 500, and 1000) (Table 2).

Table 2. Description of two problems

No.	Systems	t	System's description
Systems #1	10^5	2	5 parameters each with 10 values
Systems #2	5^2, 7^3, 3^1	3	6 parameters (i.e. 2 parameters with 5 values, 3 parameters with 2 values, and 1 parameter has 3 values

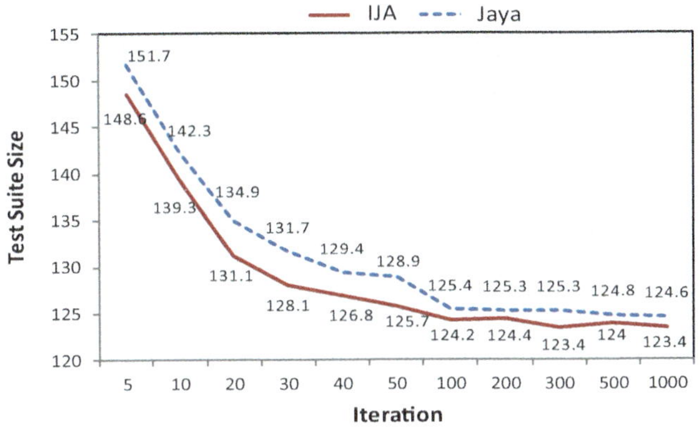

Fig. 4. Convergence rate of Jaya and IJA for system #1

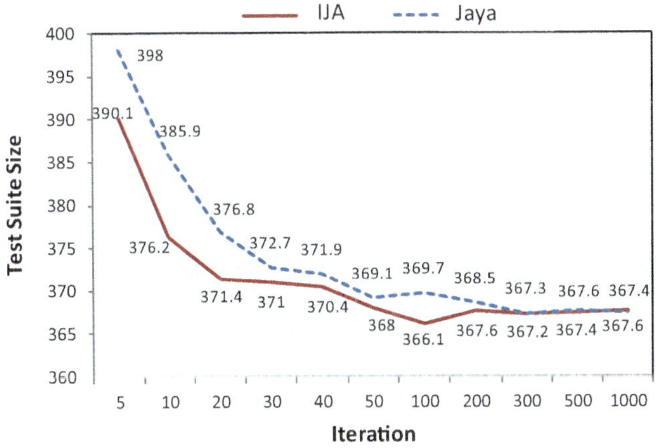

Fig. 5. Convergence rate of Jaya and IJA for system #2

Figures 4 and 5 show the convergence rate of Jaya and IJA for the two problems. As the figures show, convergence rates of IJA are faster than convergence rates of Jaya in the two problems. From the figures, we can see IJA reach to the optimal solution before the original Jaya due to its improvements by lévy flight and mutation operator.

6 Conclusion

In this paper, an improvement of Jaya algorithm for t-way test generation, called IJA, has been elaborated. IJA enhances the intensification and diversification capabilities by introducing new operators search such lévy flight and mutation operator into Jaya Algorithm. The experimental results show that introducing the new operators into IJA improves the performance of the original Jaya algorithm, also overcomes the problems of slow convergence of Jaya algorithm. As part of future work, we are planning to extend the experimental study by study the diversifications of IJA and include other existing t-way strategies such as eFPA, HHH, and ATLBO which consider as enhanced meta-heuristic based strategies.

Acknowledgements. The work reported in this paper is funded by Universiti Malaysia Pahang (UMP) grants "Optimization using Bee Colony" and "Prioritized t-way Test Suite Generation based on Chaotic Flower Pollination Algorithm". We thank UMP for the contribution and supports.

References

1. Yang, X.-S., Nature-Inspired Metaheuristic Algorithms. 2 edn. Luniver Press (2010)
2. Yang, X.S.: Nature-Inspired Optimization Algorithms. Elsevier, Amsterdam (2014)
3. Nasser, A.B., Zamli, K.Z., Alsewari, A.A., Ahmed, B.S.: Hybrid flower pollination algorithm strategies for t-way test suite generation. PLoS One **13**(5), e0195187 (2018)
4. Zamli, K.Z., Alkazemi, B.Y., Kendall, G.: A tabu search hyper-heuristic strategy for t-way test suite generation. Appl. Soft Comput. **44**, 57–74 (2016)
5. Ahmed, B.S., Zamli, K.Z., Lim, C.P.: Constructing a t-way interaction test suite using the particle swarm optimization approach. Int. J. Innov. Comput. Inform. Control **8**(1), 431–452 (2012)
6. Alsewari, A.R.A., Zamli, K.Z.: Design and implementation of a harmony-search-based variable-strength t-way testing strategy with constraints support. Inf. Softw. Technol. **54**(6), 553–568 (2012). (in English)
7. Ahmed, B.S., Abdulsamad, T.S., Potrus, M.Y.: Achievement of minimized combinatorial test suite for configuration-aware software functional testing using the cuckoo search algorithm. Inf. Softw. Technol. **66**, 13–29 (2015). (in English)
8. Nasser, A.B., Alsewari, A.A., Tairan, N.M., Zamli, K.Z.: Pairwise test data generation based on flower pollination algorithm. Malays. J. Comput. Sci. **30**(3), 242–257 (2017)
9. Rao, R.: Jaya: a simple and new optimization algorithm for solving constrained and unconstrained optimization problems. Int. J. Ind. Eng. Comput. **7**(1), 19–34 (2016)
10. Yang, X.-S., Deb, S.: Cuckoo search via lévy flights. In: World Congress on Nature and Biologically Inspired Computing, pp. 210–214. IEEE (2009)

11. Kuhn, D.R., Kacker, R.N., Lei, Y.: Practical combinatorial testing. National Institute of Standards and Technology (NIST) Special Publication, vol. 800, p. 142 (2010)
12. Hartman, A., Raskin, L.: Problems and algorithms for covering arrays. Discrete Math. **284** (1), 149–156 (2004)
13. Lei, Y., Kacker, R., Kuhn, D.R., Okun, V., Lawrence, J.: IPOG: a general strategy for t-way software testing. In: 14th Annual IEEE International Conference and Workshops on the Engineering of Computer-Based Systems, ECBS 2007, pp. 549–556. IEEE (2007)
14. Cohen, D.M., Dalal, S.R., Fredman, M.L., Patton, G.C.: The AETG system: an approach to testing based on combinatorial design. IEEE Trans. Softw. Eng. **23**(7), 437–444 (1997)
15. Jenkins, B.: Jenny download page (2003). http://www.burtleburtle.net/bob/math. Accessed 16 Dec 2014
16. Williams, A.: TConfig download page. University of Ottawa (2008). http://www.site.uottawa.ca/~awilliam/[Accessed Accessed 23 Dec 2014
17. Hartman, A., Klinger, T., Raskin, L.: IBM intelligent test case handler. Discrete Math. **284** (1), 149–156 (2010)
18. Miller, W., Spooner, D.L.: Automatic generation of floating-point test data. IEEE Trans. Softw. Eng. **3**, 223–226 (1976)
19. Shiba, T., Tsuchiya, T., Kikuno, T.: Using artificial life techniques to generate test cases for combinatorial testing. In: Proceedings of the 28th Annual International Computer Software and Applications Conference, COMPSAC 2004, pp. 72–77. IEEE (2004)
20. Ahmed, B.S., Zamli, K.Z.: A review of covering arrays and their application to software testing. J. Comput. Sci. **7**(9), 1375–1385 (2011)
21. Nasser, A.B., Zamli, K.Z., Alsewari, A.A., Ahmed, B.S.: An elitist-flower pollination-based strategy for constructing sequence and sequence-less t-way test suite. Int. J. Bio-Inspired Comput. **12**(2), 115–127 (2018)
22. Nasser, A.B., Alsewari, A.R.A., Zamli, K.Z.: PairCS: a new approach of pairwise testing based on cuckoo search algorithm. Presented at the SOFTEC Asia 2015, Kuantan, Malaysia (2015)
23. Zamli, K.Z., Din, F., Baharom, S., Ahmed, B.S.: Fuzzy adaptive teaching learning-based optimization strategy for the problem of generating mixed strength t-way test suites. Eng. Appl. Artif. Intell. **59**, 35–50 (2017)
24. Zamli, K.Z., Din, F., Kendall, G., Ahmed, B.S.: An experimental study of hyper-heuristic selection and acceptance mechanism for combinatorial t-way test suite generation. Inf. Sci. **399**, 121–153 (2017)
25. Zamli, K.Z., Din, F., Ahmed, B.S., Bures, M.: A hybrid Q-learning sine-cosine-based strategy for addressing the combinatorial test suite minimization problem. PLoS One **13**(5), e0195675 (2018)
26. Pavlyukevich, I.: Lévy flights, non-local search and simulated annealing. J. Comput. Phys. **226**(2), 1830–1844 (2007)

Artificial Intelligence Techniques for Predicting the Flashover Voltage on Polluted Cup-Pin Insulators

Ali. A. Salem[1], R. Abd-Rahman[1(✉)], Samir A. Al-Gailani[2], M. S. Kamarudin[1], N. A. Othman[1], and N. A. M. Jamail[1]

[1] University Tun Hussein Onn Malaysia (UTHM), 86400 Batu Pahat, Johor, Malaysia
rahisham@uthm.edu.my

[2] School of Electric and Electronic Engineering, Universiti Sains Malaysia, Gelugor, Malaysia

Abstract. In this paper, flashover characteristics of polluted cup-pin insulators have investigated by means of laboratory test and a mathematical model. Data from experimental works combined with the model results from a new mathematical modelling are used to derive algorithm for Artificial Neural Network (ANN) and Adaptive Neuro-fuzzy Inference System (ANFIS) for determining the critical Flashover characteristics (current and voltage). Series of laboratory testing and measurement are carried for 1:1, 1:5, 1/10 and 1:15 ratios of top to bottom surface salt deposit density on cup and pin polluted insulators (T/B). The new model was derived based on dimensional analysis approach of the parameters which commonly effect the phenomenon of pollution flashover of insulators. This model was developed by establishment the relationship between flashover current and voltage, arc constant, arc length and layer pollution conductivity of insulator. The constant of arc A and n is determined using genetic algorithm. Comparative studies have evidently shown that the proposed AI-based technique gives the satisfactory results compared to the analytical model and test data with the coefficient of determination R-Square value more than 96%.

Keywords: Outdoor insulators · Artificial Neural Network · Flashover · Mathematical model

1 Introduction

The exposure of insulator surface to various conditions of environmental and pollution depositions is inevitable in almost all energy systems [1]. Flashover on outdoor insulator occurs due to increase of electrical stresses on the transmission line combined with the high pollution level [2]. Critical voltage of flashover on polluted outdoor insulator surface is one of the important parameters for accessing reliability of a power system. Flashover prediction especially using numerical approaches have been always become a topic of interest among researchers [3]. Artificial intelligence for example can be used to analyses degree of pollution level, tracking/erosion and also for estimation

F. Saeed et al. (Eds.): IRICT 2019, AISC 1073, pp. 362–372, 2020.
https://doi.org/10.1007/978-3-030-33582-3_35

of the flashover voltage. In this work, data from the experimental works of uniform and non-uniformly polluted insulators combined with the proposed mathematical model are utilized to develop an enhanced flashover parameters prediction algorithm via Artificial Neural Network (ANN) and Adaptive Neuro-fuzzy Inference System (ANFIS). Flashover voltages of polluted insulator for 1:1, 1:5, 1/10 and 1:15 ratios of T/B surface ESDD on cup-pin insulators (porcelain and glass) are investigated.

2 Methods

2.1 The Proposed Mathematical Model

The process of contaminated insulators flashover has been carefully studied by several researchers. In this work, dimensional analysis method (DAM) has been used to establish the flashover current/voltage mathematical model on the polluted insulators surface. From previous studies, it has been noticed that the common variables which effect the pollution flashover are current (I), pollution severity in terms of pollution layer resistance (*Rp*), constant of arc (A) and humidity (H). The relationship between these variables can be written as,

$$I = f(Rp, A, H) \tag{1}$$

The fundamental dimensions (mass-*M*, length-*L*, time-*T* and current-*I*) matrix of the above variables using DAM and with exponents of the respective variables can be confined to:

	I	*Rp*	*A*	*H*
M	0	1	1	1
L	0	1	1	-3
T	0	-3	-3	1
I	1	-2	n-1	0

	x1	*x2*	*x3*	*x4*
	I	*Rp*	*A*	*H*
M	0	1	1	1
L	0	1	1	-3
T	0	-3	-3	1
I	1	-2	n-1	0

By solve the above matrix using homogeneous linear algebraic equations, the value x2, x3 and x4 in terms of x1 can be given as,

$$x2 = -\frac{x1}{n+1}, \quad x3 = \frac{x1}{n+1}, \quad x4 = 0$$

By assigning x1 = 1, the matrix can be rewritten as,

	x1	x2	x3	x4
	I	R_P	A	H
π	1	$-\frac{1}{n+1}$	$\frac{1}{n+1}$	0

The general equation can be expressed as:

$$\pi = I_C (R_P)^{-\frac{1}{n+1}} (A)^{\frac{1}{n+1}} \tag{2}$$

By applying the Buckingham's, π theorem [5], Eq. (2) can be re-expressed as,

$$I_C = K (R_P)^{-\frac{1}{n+1}} (A)^{\frac{1}{n+1}} \tag{3}$$

Where I_C is the critical current in mA, K is the dimensional constant determined from the experimental flashover voltages, A and n are the Ayrton's constants of arc and R_P is the contamination layer resistance per unit length in kΩ/cm. The resistance can be written as [4],

$$R_P = \frac{F}{\sigma(L-x)}, \quad F = \int_0^l \frac{1}{\pi.D(l)} dl, \quad x = \frac{L}{n+1} \tag{4}$$

Where F is form factor, σ is conductivity of pollution layer, L is leakage distance and x is length of arc. Substituting for R_P from (4), the new expression for the critical current (I_C) is given by,

$$I_C = K \left(\frac{F}{\sigma(L-x)} \right)^{-\frac{1}{n+1}} (A)^{\frac{1}{n+1}} \tag{5}$$

According on [6], the relationship between conductivity of pollution layer and Equivalent salt deposit density (ESDD) given by,

$$\sigma = (369.05 \cdot ESDD + 0.42) \cdot 10^{-6} \tag{6}$$

Using (6) for conductivity, the critical current proposed in final expressed for the single disc insulators is written as,

$$I_C = K1 \left(\frac{F}{[(369.05 \cdot ESDD + 0.42) \cdot 10^{-6}] * (L-x)} \right)^{-\frac{1}{n+1}} (A)^{\frac{1}{n+1}} \tag{7}$$

In non-uniform polluted insulator, the top and bottom sides were polluted separately in terms of different equivalent salt deposit density (ESDDT and ESDDB), the average equivalent salt deposit density ESDD should be satisfied by:

$$ESDD = \frac{ESDD_T \cdot A_T + ESDD_B \cdot A_B}{A_T + A_B} \tag{8}$$

$$\frac{T}{B} = \frac{ESDD_T}{ESDD_B} \tag{9}$$

Where $ESDD_T$ represents the equivalent salt deposit density of the top side, $ESDD_B$ represents the equivalent salt deposit density of bottom side, AT is the top side surface area, AB is the bottom side surface area and r is the ratio of pollution degree in the top side to bottom side on the non-uniform polluted insulator. The critical current in (7) under non-uniform condition is given as,

$$I_C = K1\left(\frac{F}{\left[\left(369.05\left(\frac{ESDD_T \cdot A_T + ESDD_B \cdot A_B}{A_T + A_B}\right) + 0.42\right) \cdot 10^{-6}\right] * (L - x)}\right)^{-\frac{1}{n+1}} (A)^{\frac{1}{n+1}} \tag{10}$$

To determine dimensional constant, the K1 and K2 against ESDD are graphed and

2.2 The Proposed Model for Flashover Voltage

According on Obenaus, the flashover voltage given by,

$$U_F = xAI_C^{-n} + R_p I_C (L - x) \tag{11}$$

Substituting for *Ic* from (10), the flashover voltage proposed model in final expressed in uniform is

$$\begin{aligned} U_F = & xA\left\{K1\left(\frac{F}{[(369.05 \cdot ESDD + 0.42) \cdot 10^{-6}] * (L - x)}\right)^{-\frac{1}{n+1}} (A)^{\frac{1}{n+1}}\right\}^{-n} \\ & + R_p\left\{K1\left(\frac{F}{[(369.05 \cdot ESDD + 0.42) \cdot 10^{-6}] * (L - x)}\right)^{-\frac{1}{n+1}} (A)^{\frac{1}{n+1}}\right\}(L - x) \end{aligned} \tag{12}$$

In non-uniform case is

$$\begin{aligned} U_F = & xA\left\{K1\left(\frac{F}{\left[\left(369.05\left(\frac{ESDD_T \cdot A_T + ESDD_B \cdot A_B}{A_T + A_B}\right) + 0.42\right) \cdot 10^{-6}\right] * (L - x)}\right)^{-\frac{1}{n+1}} (A)^{\frac{1}{n+1}}\right\}^{-n} \\ & + R_p\left\{K1\left(\frac{F}{\left[\left(369.05\left(\frac{ESDD_T \cdot A_T + ESDD_B \cdot A_B}{A_T + A_B}\right) + 0.42\right) \cdot 10^{-6}\right] * (L - x)}\right)^{-\frac{1}{n+1}} (A)^{\frac{1}{n-1}}\right\}(L - x) \end{aligned} \tag{13}$$

Further, the flashover voltage and critical current have been determined for two insulators (glass and porcelain) based on (12) and is compared with experimental results and predicted results estimated using ANN and ANFIS.

In order to determine K1, the relation between K1 and ESDD is graphed, the best curve fitting was given in the following expressed,

$$K1 = k\, e^{-0.57\ ESDD} \qquad \text{For uniform pollution} \qquad (14)$$

$$K1 = k\, e^{-0.14\left(\frac{T}{B}\right)} \qquad \text{For non-uniform pollution} \qquad (15)$$

Where k is a constant depend on the insulator type, the value of k for glass and porcelain insulators used in this work are 4.71 and 7.93 in uniformly pollution and 9.66 and 9.32 in non-uniformly pollution respectively.

2.3 Experimental Design

2.3.1 Experiment Facilities

The experimental procedure were carried out in a different multifunctional artificial chamber approximately 1.2 m × 1.0 m and circuit arrangement is illustrated in Fig. 1. The voltage was measured through AC divider of capacitive voltage while leakage current measured through shunt resistance of 10 k ohm. The data acquisition monitoring system is consisting of personal computer (PC), DAQ card and LabVIEW software.

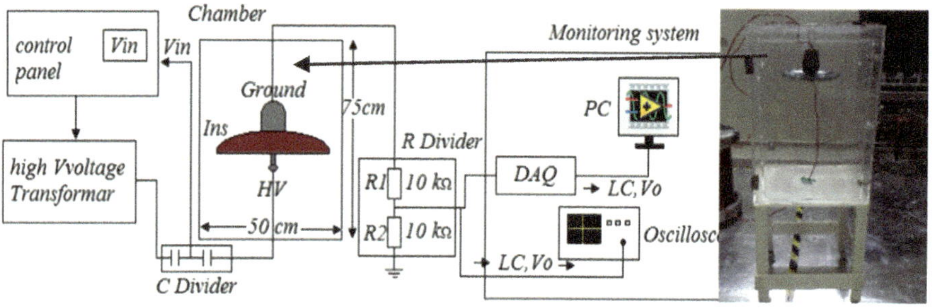

Fig. 1. Schematic view of measurement circuit and test setup.

2.3.2 Test Specimens

The insulators models are porcelain (XP-70) and glass (LXY-70). The clean and polluted insulators were tested under different voltage sources. The structure schematic diagram and parameters of structural of samples are shown in Table 1 in which H is the insulator height, D is the insulator diameter and L is the leakage distance.

Table 1. Dimensions of the tested insulators

	Type	Materials	H/cm	D/cm	L/cm	F
	LXY- 70	Glass	14.6	25.5	31	0.89
	XP-70	Porcelain	14.6	25.5	29.5	0.73 6

2.3.3 Test Procedure

Prior to the test, the insulator was first carefully cleaned using alcohol and water with conductivity lower than 10 μS/cm. Pollution suspension is prepared by mixing appropriate ratio kaolin with Sodium Chloride (NaCl) in water to achieve the desired conductivity level [7]. The salinity of the pollution layer on the insulator is then determined using a conductivity meter. The contamination was applied according to solid layer method using brush before allowed to dry naturally for 24 h. A sprayer is used to wet the polluted insulator sample under normal ambient condition with temperature of 25 °C and humidity of 85–90%.

2.4 Determination of Arc Constants Using Genetic Algorithm

The voltage across the arc have expressed in terms of Ayrton's equation as [8]:

$$E_{arc} = AI_{arc}^{-n} \tag{16}$$

Where A and n are the arc constants. Here, E_{arc} represents the arc voltage gradient and I represent arc current. The flashover voltage against ESDD test results can be used to compute the constants of arc mathematically for different ESDD using genetic algorithm technique. The data of experimental arc voltage and arc current at different ESDD obtained from two practical insulators are substituted in (16) resulting in to total of 46 equations for each insulator type. These equations have only unknown's arc constants 'A' (V/cm) and 'n'. To optimize the arc constants, these constants A and n can be determined by minimizing the following objective function:

$$\sum_{i=1}^{46} \left[E_{arc} - \mathrm{f}\left(N \cdot I_{arc}^{-n}\right)\right] \tag{17}$$

To compute the constants of arc, the function given in (17) is defined with constraints with the help of previous researcher's arc constants [9]. The constraints defined are

$$500 \leq \mathrm{N} \geq 30 \quad \text{and} \quad 0.2 \leq \mathrm{N} \geq 1.1 \tag{18}$$

To get the arc constants A and n, 46 equations formulated are optimised using 'GA' function from the Optimization tool box in MATLAB. The result of the optimization has given value of arc constants (A = 62.53 and n = 0.73) for porcelain insulator, and (A = 75.93 and n = 0.68) for glass insulator.

2.5 Derivation of Artificial Intelligence (AI)

The severity of deposited pollution on the surface of insulators under uniform and non-uniform condition and the insulators dimension were studies in attempt to derive prediction of insulators flashover voltage using ANN and ANFIS. The ANN and ANFIS were simulated using MATLAB R2019a Toolbox. The four variables representing the ESDD and form factor (F) and leakage distance (L) instead of pollution layer resistance (*Rp*) and constant of arc (A) were used as inputs of ANN and insulator flashover voltage and critical current as the outputs. The structure of ANN and ANFIS showed in Fig. 2.

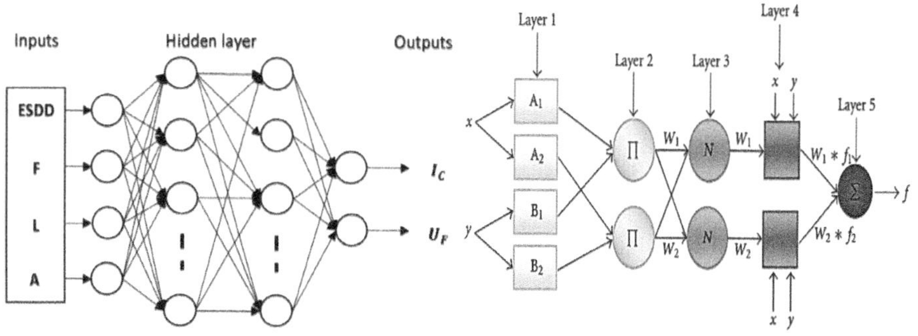

Fig. 2. Equivalent ANN and ANFIS architecture

3 Results and Discussion

3.1 Critical Current and Flashover Voltage U_{FO} for Polluted Insulators

Figures 3 and 4 show the plot of the critical current and flashover voltage in correlation with equivalent salt deposit density ESDD obtained the proposed model under uniform and non-uniform pollution condition. The laboratory tests results of critical current and flashover voltage are shown in Table 2. It can be observed that the uneven contamination degree has incredible effect on the critical current and flashover voltage. The flashover voltage increases with an increase in the pollution ratio in the area near the high voltage terminal on the bottom surface of the insulator for both insulators (Fig. 5). On the other hand, with decrease of pollution on the bottom, the maximum of flashover voltage of uniformly polluted insulator increases by 43.57% and 37.81% for the glass porcelain insulator respectively.

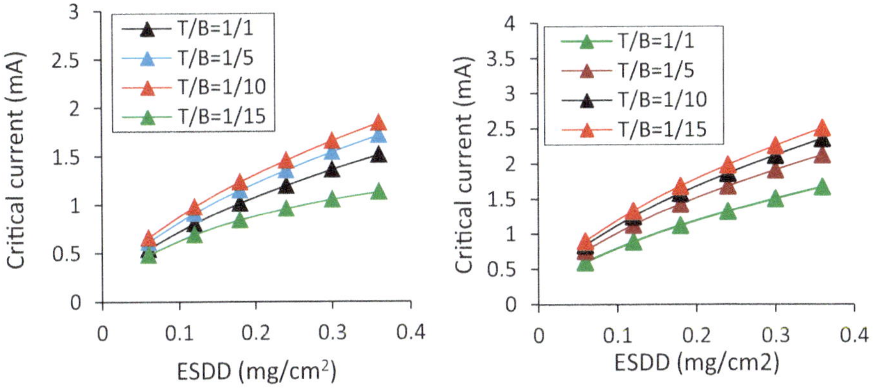

(a) (b)

Fig. 3. Critical current and ESDD with different non-uniform level pollutions (a) glass and (b) porcelain

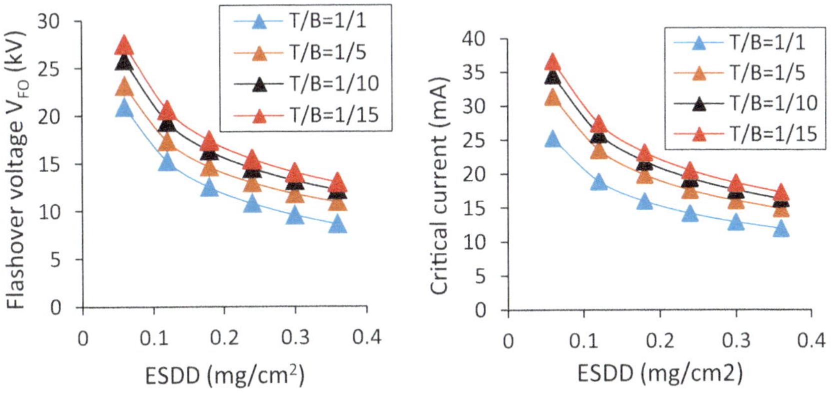

Fig. 4. Flashover voltage and ESDD with different non-uniform level pollutions (a) glass and (b) porcelain

Table 2. Laboratory test results of critical current and flashover voltage of glass and porcelain insulators

Insulator type	ESDD mg/cm^2	T/B = 1/1		T/B = 1/5		T/B = 1/10		T/B = 1/15	
		Ic	Vc	Ic	Vc	Ic	Vc	Ic	Vc
A	0.06	0.44	21.96	0.51	22.96	0.62	25.98	0.72	27.54
	0.12	0.62	16.08	0.78	17.08	0.95	19.08	0.98	20.08
	0.18	0.81	12.22	1.00	15.22	1.15	15.22	1.28	17.22
	0.24	0.99	10.10	1.26	12.99	1.34	13.99	1.40	15.99
	0.3	1.14	8.98	1.37	11.98	1.54	12.98	1.73	13.98
	0.36	1.21	8.03	1.55	10.31	1.73	12.31	1.77	13.23

(continued)

Table 2. *(continued)*

Insulator type	ESDD mg/cm^2	T/B = 1/1		T/B = 1/5		T/B = 1/10		T/B = 1/15	
		Ic	Vc	Ic	Vc	Ic	Vc	Ic	Vc
B	0.06	0.61	25.82	0.81	32.56	0.86	35.13	1.02	37.26
	0.12	0.88	18.26	1.17	24.13	1.29	26.08	1.36	28.12
	0.18	1.05	15.01	1.40	19.22	1.47	20.22	1.81	22.65
	0.24	1.41	13.55	1.74	16.99	1.90	18.10	1.85	19.68
	0.3	1.54	12.87	1.85	15.98	2.24	17.98	2.31	18.06
	0.36	1.60	11.58	2.06	13.77	2.32	16.03	2.48	17.51

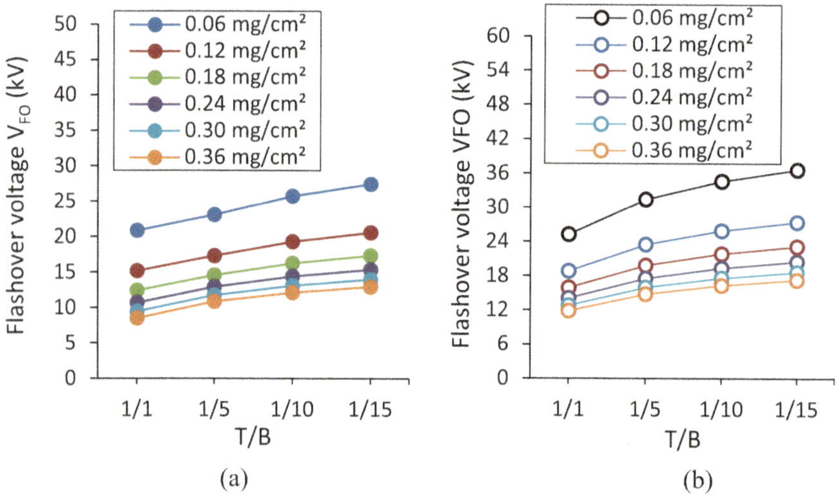

Fig. 5. Flashover voltage against T/B with different non-uniform level pollutions. (a) glass and (b) porcelain

3.2 Validation of Model and Artificial Intelligence Results

Employment of the arc constants values in flashover voltage model (Eq. 13) allows us to calculate of the flashover voltage model. The ANN, ANFIS models of glass and porcelain insulators was carried out based to the properties of insulators in pollution conditions. The comparison of the ANN, ANFIS, mathematical model based on DAM and the laboratory test results under different non-uniform pollution condition is given in Figs. 6 and 7. It is clear from Figs. 6 and 7 that the simulation process based on AI techniques have been given the accurate prediction of critical current and flashover voltage under different pollution levels.

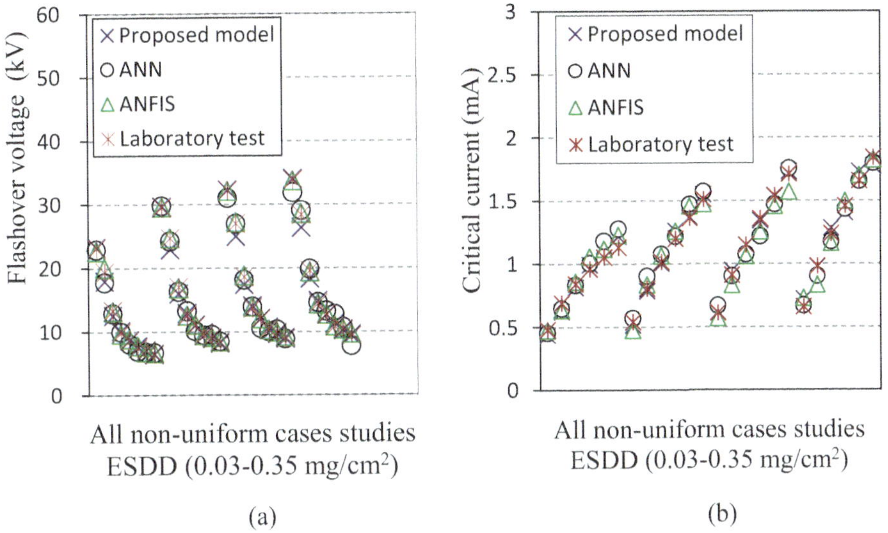

Fig. 6. Comparison between ANN, ANFIS, mathmatical model based on DAM and tests results for all non-uniform pollution cases of glass insulator: (a) Flashover voltage and (b) Critical current

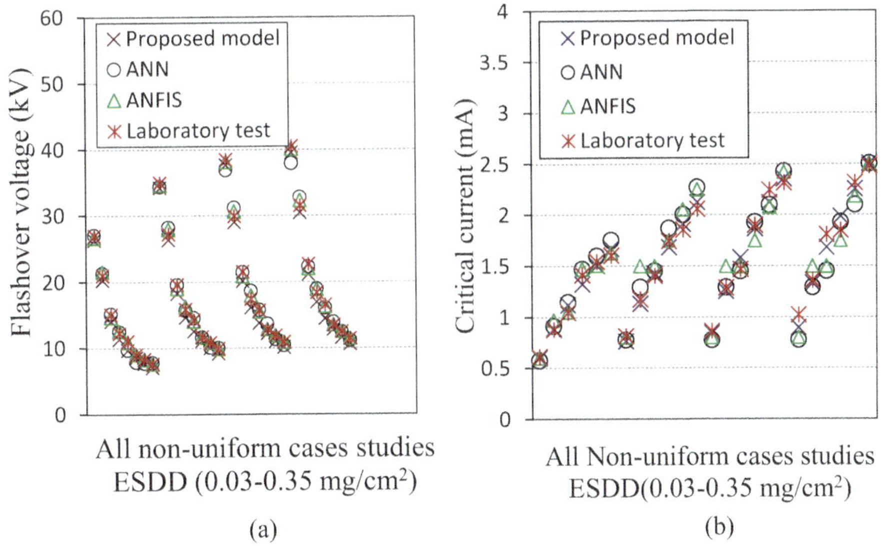

Fig. 7. Comparison between ANN, ANFIS, mathmatical model based on DAM and test results for all non-uniform poullation cases of porcelain insulator: (a) Flashover voltage and (b) Critical current

4 Conclusion

This paper has proposed a new mathematical model based on Dimensional Analysis Method (DAM) for uniform and non-uniform polluted insulators. The results obtained from experimental test was used to achieve the suitable values of arc constants A and n using genetic algorithm optimization. Two artificial intelligence techniques were designed for the estimation of critical current and flashover voltage in artificially non-uniform polluted glass and porcelain insulators. The mathematical model based to the pollution characteristics of the insulator, ESDD and constant of arc was applied and validated using experimental flashover voltage results. Therefore, the determination of the constants of arc allows the simulation of the critical conditions for the flashover voltage. Several of non-uniform pollution cases 1:1, 1/5, 1/10 and 1/15 have considered in this work, these pollution conditions have a great influence on the U_{FO}. An artificial neural network has been applied to the measurement results and a good agreement has been achieved with the high determination of coefficients (more than 96.5%).

References

1. Salem, A.A., Abd-rahman, R.: A review of the dynamic modelling of pollution flashover on high voltage outdoor insulators. J. Phys. Conf. Ser. **1049** (2018)
2. Ahmad, H., Abd-Rahman, R., Ahmad, M.H.: Evaluation of transmission line insulator for I-Type string insulator design. In: PECON 2016 – 2016 IEEE 6th International Conference on Power and Energy, pp. 50–53 (2017)
3. Salem, A.A., et al.: Proposal of a dynamic numerical approach in predicting flashover critical voltage, vol. 10, no. 1, pp. 51–58 (2019)
4. Salem, A.A., et al.: The effect of insulator geometrical profile on electric field distributions, vol. 14, no. 2, pp. 618–627 (2019)
5. Salem, A.A., Rahman, R.A., Kamarudin, M.S., Othman, N. A.: Factors and models of pollution flashover on high voltage outdoor insulators: review. In: 2017 IEEE Conference on Energy Conversion (CENCON), pp. 241–246 (2017)
6. IEC Std. 60507: Artificial pollution tests on high-voltage insulators to be used on AC systems (1991)
7. Dixit, P., Krishnan, V., Nagabhushana, G.R.: Studies on pollution performance of ceramic insulators under AC excitation. In: Proceedings of 16th International Symposium on High voltage Engineering (ISH 2009), 28th August 2009, Cape town, South Africa, paper NoE-35, pp. 1331–1336 (2009)
8. Wilkins, R.: Flashover voltage of high voltage insulator with uniform surface pollution films. Proc IEE **116**(3), 457–465 (1969)
9. Suresh, A.G., Dixit, P.: ANN model to predict critical flashover voltages of polluted porcelain disc insulators, vol. 12, no. 11, pp. 2942–2951 (2017)

Heart Disease Diagnosis Using Diverse Neural Network Categories

Mostafa Ibrahem Hassan[1,2](✉), Ahmed Hamza Osman[3](✉), and Eltahir Mohamed Hussein[4](✉)

[1] Administration of IT, Faculty of Igraa for Computer Studies, International University of Africa, Khartoum, Sudan
halange@iua.edu.sd

[2] Sudan Academic for Graduate Studies, Africa City of Technology, Khartoum, Sudan

[3] Department of Information System, Faculty of Computing and Information Technology, King Abdulaziz University, Jeddah, Kingdom of Saudi Arabia
ahoahmad@kau.edu.sa

[4] Department of Information System, Faculty of Engineering, Sudan University of Science and Technology, Khartoum, Sudan
altahir33@yahoo.com

Abstract. This study indicated an intelligent diagnosis method about the heart disease based on the diversity of artificial neural networks algorithms. The major objective was to improve the heart disease diagnosis accuracy, and reducing the miss-diagnosis results. One of the disadvantages of the current heart disease diagnosis methods that the diagnosis is expensive and not accurate enough to confirm the diagnosis of the heart disease. Our study will help the medical doctors to make appropriate diagnosis process and thus take specific treatment of the heart diseases easily. Our suggested method used Heart Diseases Database (Cleveland database) for training and testing phases. We used five neural networks: Quick, RBFN, Multiple, Prune, and Exhaustive. Our results revealed that the dynamic neural network algorithm was the best method diagnosing heart diseases accurately. We recommend using Dynamic neural network algorithm as the routine and sensitive method for diagnosis of heart disease easily and specifically.

Keywords: Neural networks · Heart disease · Cleveland

1 Introduction

The heart muscle is the most active part of the human body and is surrounded by three main arteries which supply food and oxygen, it begins to work after about 20 days of pregnancy [1], pumping 7,000 L of blood a day. There are different types of heart diseases like coronary artery diseases, congenital heart defects … etc. The cardiac arrhythmias are irregularity and disturbance in the heartbeat, which become clear by irregularity and abnormality, too fast (tachycardia) or abnormally too slow (bradycardia) [2], it's a change in the electrical system [3], which leads to the inability of the heart to pump blood efficiently. Studies have shown that there are about 5.6 million

F. Saeed et al. (Eds.): IRICT 2019, AISC 1073, pp. 373–385, 2020.
https://doi.org/10.1007/978-3-030-33582-3_36

patients in the United States, and it is expected that number of patients may exceed 15.9 million patients by 2050 [4], The World Health Organization (WHO) forecasts that the death toll from heart diseases could reach 23.6 million deaths worldwide [5]. Heart diseases cause more than 26% of deaths in UK; in 1961 there are 320,000 cases of deaths attributed by CVD [6].

Our methodology is based on using data mining techniques to improve heart diseases diagnosis. We used different neural networks algorithms (Dynamic, Quick, RBFN, Multiple, Prune, and Exhaustive Prune), for training and testing the neural networks, That was done by using the heart diseases database (Cleveland Database). This dataset is divided into different groups (70%, 60%, and 50%) for training and (30%, 40%, 50%) for testing respectively. Through this methodology the obtained results proved that the dynamic neural has a best accuracy in diagnosis of the heart diseases, for helping cardiologist that improve diagnosing these diseases. Also we compared our results with the previous studies used which used the same database, and we found this methodology surpassed all of them [9] and [13].

The related works of this study are presented in the next section: related works, Sect. 3 reviewed the neural network algorithms. Sections 4 discussed research methodology framework, while Sects. 5 described the dataset, and experimental results and discussion section is in 6. The last section demonstrated conclusion and future work.

2 Related Works

D'essai proposed efficient approaches for the heart diseases prediction according to the Probabilistic Neural Network (PNN) technique. The dataset clustered into different clusters using k-means algorithm. The dataset after clustering was evaluated in the existing algorithms such as Naive Bayesian, decision tree; the accuracy of the new approach achieved was 94.6% in predictions [10]. Shouman et al. Suggested a new application by using k-Nearest-Neighbors (k-NN) to diagnose a heart diseases. The results of the study showed that k-NN application has the higher accuracy compared to other determining algorithms like neural network. Voting is a process of integrating more several different classifiers. K-Nearest-Neighbors achieved 97.4% accuracy without voting and 92.7% with voting [11]. Purushothaman et al. testified the various classification techniques for the prediction of the heart diseases with different datamining algorithms. The accuracy of a single model consists of decisions tree; artificial neural network and the Naive Bayesian which are 76%, 85% and 69% respectively. The new suggested hybrid approaches show that the accuracy is 96%; therefore, the proposed hybrid model has produced better and more reliable and effective results when compared to the other data mining algorithms used [12]. Chandna used the k-nearest neighbor's algorithm to create a hybrid approach to integrate learning algorithm with the features selection technology. The proposal methodology used information obtained from the Neuro-Fuzzy Adaptive Inference System (ANFIS). ANFIS is a common mechanism between neural networks and the inference system, where the accuracy of the new approach is 98.24% [13]. Pravabathi et al. presented an overview paper showed a number of research and scientific papers using DNFS (decision tree based on neural fuzzy system). They concluded that the decision tree and naive bayesian made the best accuracy in the diagnosis of heart diseases amounted to 95% [14]. Ajam presented a model for diagnosing the level of heart diseases using back propagations neural network, where dataset was divided as: 60% for network training, 20% for network validation and 20% for testing process. The Network accuracy achieved for heart diseases is 88% [15]. Rabbi et al. evaluated the popular data mining classification techniques in the heart diseases using heart diseases dataset (Cleveland database). The dataset was tested using support vector machine (SVM), k-nearest neighbor (KNN) and artificial neural network (ANN) algorithms. The experimental result showed that the classification accuracy using SVM achieved (85.1852%) while the result of using KNN (82.963%) and the accuracy of using artificial neural network achieved (73.3333%) [16]. Mitra and Samantha presented a new way to classify arrhythmia after 100 simulations. The backpropagation algorithm used based was on the heart diseases database (Cleveland database), showed positive results with accuracy of 87.71% [17]. Padma, Latha and Jayakumar designed a support system to analyze the ECG using LVG. The training was conducted using four ECG data models obtained from the ECG simulator, including heart failure, irregular heartbeats, shrinkage, contraction and muscle infarction. LVG achieved high performance with a full total accuracy of 95.5% when 200 samples were trained and tested [8]. The study discussed the classification of the diseases in both normal and abnormal situations. The UCI ECG signals were used to train and test neural network models as well as basic Backpropagation Algorithm in the analysis and diagnosis of arrhythmia. The performance of the classification was evaluated using several measures such as Mean Squared Error (MSE), Accuracy Receiver Operating Characteristics (ROC) and Area Under Curve (AUC), gained better results in terms of the classification, reaching 86.67%, 93.75% and 93.1% respectively [9]. Özbay et al. presented a comparison between neural networks and fuzzy logic in the classification of arrhythmias. Data were taken from MIT - BIH ECG. 10 different training models were used, representing 40 males and 52 females between the ages of 19 and 39. These data were classified by C (MEANS Clustering), which reduced the number of sectors by aggregating similar sectors in training data. After the tests, the results showed that the multilayered neural network worked better and faster [18].

3 The Neural Networks Algorithms

Neural Network algorithms are considered as the popular machine learning techniques used among the current data mining models. In general the neural network is a kind of mathematical algorithms that solve a number of problems, and stemmed from the idea of neural network in humans organized in layers of neurons: the first one is the input layer and the last one is the output layer, and are a layers between them. The linking between a neuron layers considered with a weight value. The neurons multiply each input value from the previous layer neurons with the weight of the connecting with these neurons, and then multiply all the multiplication outputs. The results reflect the different conversion by neuron type. The output is considered a neuron that is transferred to the neurons of the subsequent layer. We used main six different neural network techniques; these were RBFN, Quick, Dynamic, Multiple, Prune and Exhaustive Prune neural network. In this research, the neural network was trained to identify the symptoms of the heart diseases, and then tested on the diseases to prove the good result of identification through different results obtained on training phase.

3.1 RBF Neural Network

Radial Basis Function Network is a special type of neural network. It has three layers: input layer, hidden layer and output layer. The hidden layer or the receptor layer includes a number of neuron that represent clusters at input patterns. These clusters are built on radial basis functions or functions to the distance between the centers of RBF's vector values entered (Fig. 1).

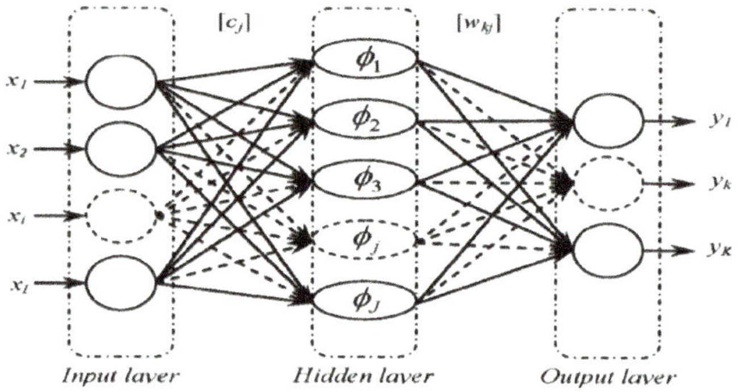

Fig. 1. RBFN neural network algorithm [19]

3.2 Quick Neural Network

In this type of neural network, a single layer network is used. In a normal way, it contains one hidden layer with a number of neurons, which is determined by Max (3, (n0 + ni)/20). The reverse diffusion method is a method for finding the small value of the total error box of the output value calculated by the network based on the supervised learning method (Fig. 2).

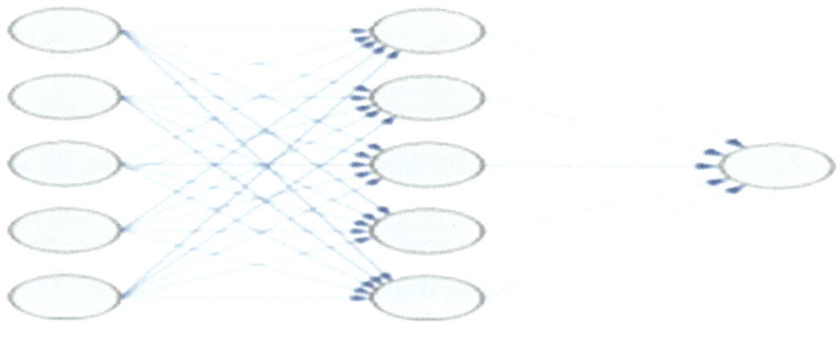

Fig. 2. Quick neural network algorithm [20]

The training process of this network aims to create a balance between the ability of the network to correctly respond to the input samples used in the training process (stored by the network) and the ability of the network to give a good response to similar income but not identical to that used in the training.

3.3 Dynamic Neural Network

The idea of this network is changing the topography during the training process by adding a number of neurons to increase the accuracy of the network. There are two phases to this type of training: topology and final network training to reach the best result. The network is built in two hidden layers, each layer with a number of neurons. The initial network training is usually done in one session by creating two copies of the primary network left and right of the grid. The network on the right side adds one neuron to the second hidden layer. The network parameters are trained in one cycle and specify the overall error for any network and calculates the sum of (p*j), where the (j) represents the output and (p) represents the number of records recorded in the session. If the left network has a simple error, so it is saved, and then one neuron is added to the right of the grid for the first hidden layer. If the right network has a small error rate, the left network is changed by a copy of the right network and neuron will be added to the second hidden layer of the right network. Then the two networks are trained for another cycle and the training process is repeated by increasing the courses to meet the required (Fig. 3).

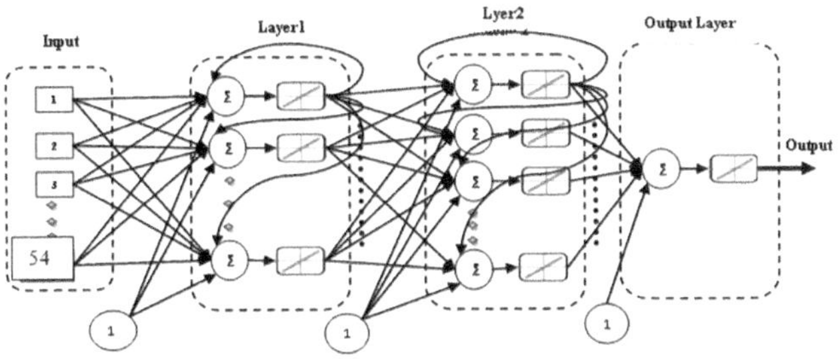

Fig. 3. Dynamic neural network algorithm [21]

3.4 Multiple Neural Networks

This type of algorithm creates a number of neural networks and trains them in parallel, where the highest resolution network returns the best result (the final value). The topography of the network is configured in the simple model where a single layer network is created for a number of hidden units starting from the number 3 to the number of neurons input, if it is less than 12 and cannot in any case be greater than 60 neurons. The grid generates any number of neurons inputs in sequence 3, 4, 7, 12 and so on, an increase for any step twice larger than before, for each and every single layer network. A set of two-tier networks will be created. The first layer has the same number of neurons hidden in the single-layer network, and the number of neurons in the second layer varies and varies between networks. The network generates any number of neurons in the binary layer in the order of 2, 5, 10, 17 and etc. until the neurons reach the initial number in the hidden layer (Fig. 4).

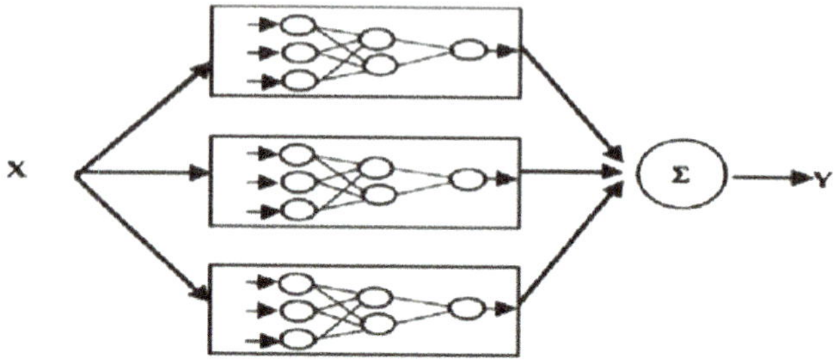

Fig. 4. Multiple neural networks algorithm [22]

3.5 Prune Neural Network

The Prune technique starts with a large network and evaluates the network by deleting the non-useful (active) neuron from input, layer and hidden. Prune's procedures are done in two stages: elimination of unwanted layers and elimination of neurons input (Fig. 5).

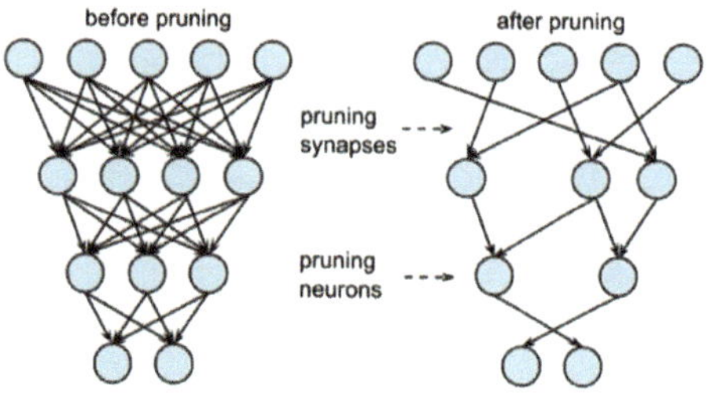

Fig. 5. Prune neural networks algorithm [23]

3.6 Exhaustive Prune

The idea of this network is deleting the weakest neurons in the hidden layers in the network, after checked by training the network; the network starts with wide numbers of neurons and eliminate the weakest of them for arriving to the best accuracy. Although this method gives acceptable accuracy but has disadvantaged that is taking long time in training and testing process.

4 Research Methodology Framework

The methodology of this research was divided into several phases starting with the dataset collection and ending with prediction results diagnosis. This dataset known as Heart Diseases Database (Cleveland database) was obtained from University of California which contains 75 attributes. The actual attributes used were 14. Attributes were trained and tested in different neural networks used in this research, firstly we pre-processing dataset by converting the dataset into CSV files, outlier removal, and missing value amending. Secondly we divided the heart dataset into three groups (50%, 60%, and 70%) for learning and (50%, 40%, 30%) for testing respectively. Thirdly we applied each segment into neural network algorithms such as Dynamic, Multiple, Prune, and RBFN algorithm). Finally by using the testing group we diagnosed the heart diseases cases (Fig. 6).

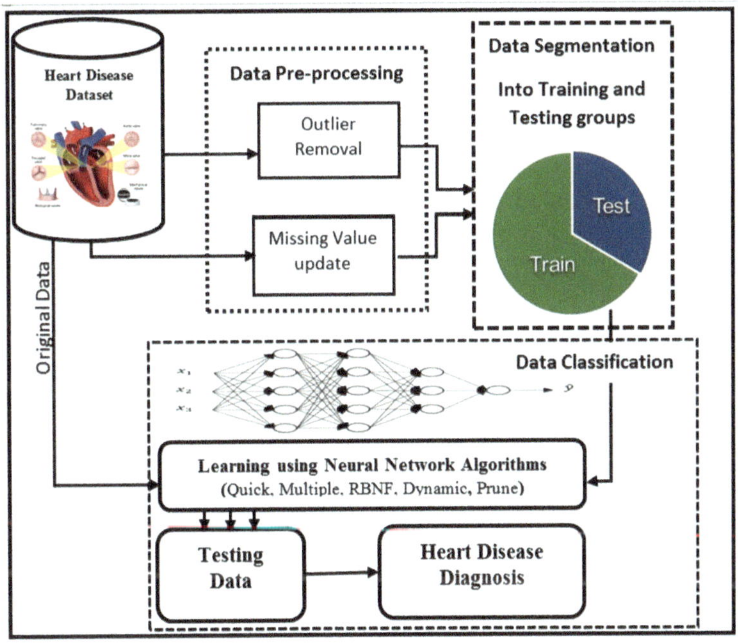

Fig. 6. Research methodology framework

The framework of this methodology contained a number of necessary and important steps from the beginning of selection the database and finally to the diagnosis of heart diseases using artificial neural networks:

Firstly, Database: we used heart diseases database from University of California, which was considered as one of the best heart diseases databases open and available for researchers. Many studies used this database, which contained 303 records of heart patients and 14 features.

Secondly, Data Pre-processing: The initial processing of the data contained two parts, the first part was the disposal of the abnormal values, which were not present in this database where there were no abnormal values, and in the second part, empty fields were filled where some of the fields were blanks and have no values. Therefore, a number of null values were treated in this study and compensated with appropriate values, by calculating the average of attributes values.

Thirdly, Data Segmentation: the data were divided into three main groups. Each group was divided into two categories, for the first group (70% training, 30% test), the second group (60% training, 40% test) and third group (50% training, 50% test).

Fourthly, Data classification: in this stage we used the three groups of databases for training and testing the neural networks used in this study, so that each neural network was trained and tested in order to obtain best results.

Fifthly, Evaluation: the end we gathered all the results obtained from the different neural networks for evaluation and comparison of the previous studies results by using same database to determine the best method to diagnose heart diseases.

5 Dataset

So, we used heart Diseases Database (Cleveland database) which included 75 attributes of Heart diseases that collected and gathered by the Dr. Robert Detrano, a cardiologist from the Cleveland Clinic Foundation and V.A. Medical Center [24]. Actually 14 attributes from the dataset were used for training of neural networks. The data contain 303 records (sample data) represented the information of the patients of the heart diseases (Table 1).

Table 1. Summary of the attributes information

No	Attribute no	Attribute details
1	#3	Age in years
2	#4	Male or Female
3	#9	Type of Chest pain
4	#10	Blood pressure
5	#12	Serum cholesterol
6	#16	Fasting blood sugar >120 mg/dl
7	#19	Electrocardiographic
8	#32	The maximum of heart rate
9	#38	Exercises induced angina
10	#40	ST depression
11	#41	Slope of peak exercise ST
12	#44	Number of major vessels
13	#51	Thal
14	#58	Diagnosis of heart diseases by diameter narrowing (the predicted attribute)

The dataset contains 14 attributes mentioned above were the predication of the field referred to diagnose heart diseases by checking the diameter narrowing. We applied the dataset group into the network by training part of the dataset to learn algorithm and testing to evaluate the neural network to achieve the desired accuracy [7].

6 Results and Discussion

In our study we performed different experiments using six neural networks (Multiple, Dynamic, Quick, RBFN, Prune and Prune Executive), also we used Heart Diseases Database (Cleveland database) for training neural networks to improve the heart diseases diagnosis accuracy, the dataset was classified into three groups: the first group the dataset was divided into 70% for training and 30% for testing, in the second group the dataset was divided into 50% for training and 50% for testing, and in the third group the dataset was divided into 60% for training and 40% for testing, after that we performed every neural network on these three groups of dataset separately (Fig. 7).

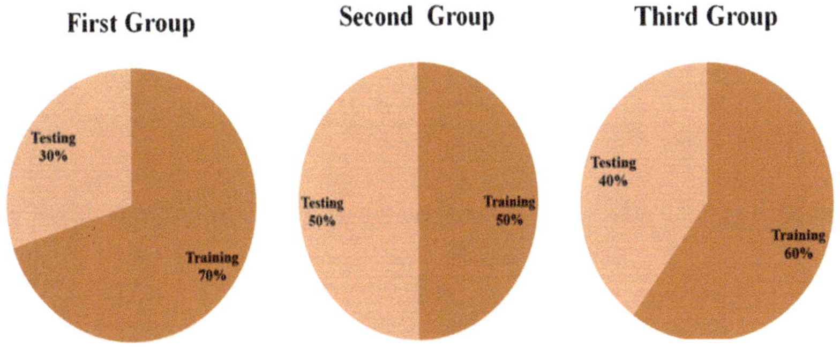

Fig. 7. Dataset classes groups

After segmentation of dataset into different groups we determined the predication field for training and testing neural networks, one by one in each of dataset groups. In the experimental design we have found different results (Table 2).

In the First Group we trained (Multiple, Dynamic, Quick, RBFN, Prune and Exe-Prune) (on the 70% of dataset size and testing on 30% of it. We found the multiple network achieved best accuracy 99.873 in training; also we found multiple network achieved the best result 99.727% in testing. In the second group the dataset was divided into 50% for training and 50% for testing process, we found the Dynamic neural network got the best accuracy 97.5552% in training and also best accuracy in testing process 99.875%. In the third dataset group we divided the dataset into 60% for training and 40% for testing process. After training and testing different neural networks we found the Dynamic network achieved best accuracy in training process 99.526% and also the dynamic network got the best result in testing process 99.698% (Table 3).

Table 2. Training and testing neural networks with dataset groups

Dataset groups	Neural network	Training result	Testing result
First group Training 70% Testing 30%	Multiple	99.873	99.727
	Dynamic	99.856	99.161
	Quick	90.829	92.847
	RBFN	83.769	86.806
	Prune	86.584	82.991
	Exe-Prune	87.307	86.067
Second group Training 50% Testing 50%	Multiple	97.5552	99.274
	Dynamic	99.797	99.875
	Quick	90.411	90.411
	RBFN	80.886	82.448
	Prune	87.137	87.395
	Exe-Prune	87.576	83.851
Third group Training 60% Testing 40%	Multiple	97.108	99.698
	Dynamic	99.526	99.862
	Quick	90.016	92.176
	RBFN	84.632	82.723
	Prune	87.457	84.96
	Exe-Prune	86.355	85.104

Table 3. Best result achieved by training and testing neural networks with dataset groups.

	Dataset groups	Neural network	Best result	Error
Training	*First group* Training 70%	Multiple	99.873	0.127
	Second group Training 50%	Dynamic	99.797	0.203
	Third group Training 60%	Dynamic	99.526	0.474
Testing	*First group* Testing 30%	Multiple	99.727	0.273
	Second group Testing 50%	Dynamic	99.875	0.125
	Third group Testing 40%	Dynamic	99.862	0.138

In the training stage of the first group (70% of dataset), we found that multiple neural network got best result (99.873%) by Error 0.127%. In the second group (training 50% of dataset), we found the Dynamic neural network got best result (99.797%) by Error 0.203%. In the third group by training 60% of dataset we found the Dynamic got the best result (99.526%) by Error 0.474%.

In the testing stage, of the first group (30% of dataset), we found the multiple neural network got best result (99.727%) by error 0.273%. In the second group of dataset by testing 50%, we found the dynamic neural network got the best result (99.875%) by error 0.125%. In the third group of dataset by testing the neural networks on 40%, we found the dynamic neural networks got the best result (99.862%) by error 0.138%.

We compared several studies which used the current database to improve the diagnosis of heart diseases, using different techniques (Table 4).

Table 4. Comparing with previous studies used heart diseases database

Study	Method	Tool	Accuracy
Artificial neural network based cardiac arrhythmia diseases diagnosis	Classification of the diseases in both normal and abnormal situations by evaluated the performance of the classification by using (MSE), (ROC) and (AUC)	Basic backpropagation algorithm	86.67%, 93.75% and 93.1% Respectively
Cardiac arrhythmia classification using neural networks with selected features	Classify arrhythmia	Backpropagation algorithm	877.1%
Diagnosis of heart diseases using data mining algorithm	Hybrid algorithms	k-nearest neighbor's ANFIS (neural network and fuzzy inference system)	98.24%
Our methodology	Training 60% - Testing 40%	Dynamic neural network	99.862%

Our study revealed that, the best result that was achieved by the Dynamic Neural Network is 99.862% in 60% training and 40% for testing, similar as the study of Hybrid algorithms using k-nearest neighbor's ANFIS (neural network and fuzzy inference system) which indicated high accuracy 98.24%. Our results was better than the study conducted for the classification of the diseases in normal and abnormal situations by evaluation the performance of the classification using (MSE), (ROC) and (AUC) and Basic Backpropagation Algorithm where achieved accuracy: 86.67%, 93.75% and 93.1% Respectively. Also our result was superior the classify arrhythmia method that using backpropagation algorithm where its accuracy reached 877.1%. [9, 13, 17].

7 Conclusion and Future Work

The heart diseases have become more common and widespread in the world today that causes many deaths. The use of data mining and its techniques represents opportunities for cardiologist to diagnose diseases early and help patients effectively. This study proved that the use of the Multiple and Dynamic Neural Networks confirmed the diagnosis accurately which avoid misdiagnosis. More studies can be conducted by Dynamic Neural Network in the future. Furthermore, a combination of different data mining techniques such as the k-means and Decision Tree algorithms will be applied with Dynamic Neural Network to enhance more accuracy.

References

1. Severs, N.J.: The cardiac muscle cell. BioEssays **22**(2), 188–199 (2000)
2. Kastor, J.A.: Cardiac Arrhythmias, Encyclopedia of Life Sciences. Macmillan Publishers Ltd., Nature Publishing Group. e LS (2001)
3. National Heart Foundation of Australia: Understanding abnormal heart rhythm, Heart Foundation, pp. 1–9 (2017)
4. Curran, T., Sheppard, G.: Cardiac arrhythmias: mechanisums of arrhythmias – atrial, ventricular, conduction and ST changes, canterbury district health board, pp. 1–74 (2011)
5. Hakim, A.N.: Predicting heart attacks in patients using artificial intelligence methods. Mod. Appl. Sci. **10**(3), 66–78 (2016)
6. UK Factsheet - British Heart Foundation. https://www.bhf.org.uk/what-we-do/our-research/heart-statistics. Accessed 25 July 2019
7. UCI Machine Learning Repository - Center for Machine Learning and Intelligent Systems. http://archive.ics.uci.edu/ml/datasets/heart+Diseases. Accessed 25 July 2019
8. Padma, T., Latha, M., Jayakumar, K.: Decision making algorithm through LVQ neural network for ECG arrhythmias. In: 13th International Conference on Biomedical Engineering, vol. 23, pp. 85–88. Springer, Heidelberg (2009)
9. Jadhav, S.M., Nalbalwar, S.L., Ghatol, A.A.: Artificial neural network based cardiac arrhythmia diseases diagnosis. In: International Conference on Process Automation, Control and Computing, pp. 1–6. IEEE Xplore,12 August 2011
10. Dessai, I.S.F.: Intelligent heart disease prediction system using probabilistic neural network. Int. J. Adv. Comput. Theory Eng. (IJACTE) **2**(3), 2319–2526 (2013)
11. Shouman, M., Turner, T., Stocker, R.: Applying k-nearest neighbors in diagnosing heart diseases patients. Int. J. Inform. Educ. Technol. **2**(3), 220–223 (2012)
12. Purusothaman, G., Krishnakumari, P.: A survey of data mining techniques on risk prediction: heart diseases. Indian J. Sci. Technol. **8**(12), 1–5 (2015)
13. Chandna, D.: Diagnosis of heart diseases using data mining algorithm. Int. J. Comput. Sci. Inform. Technol. (IJCSIT) **5**(2), 1678–1680 (2014)
14. Prabhavathi, S., Chitra, D.M.: Analysis and prediction of various heart diseases using DNFS techniques. Int. J. Innov. Sci. Eng. Res. (IJISER) **2**(1), 1–7 (2016)
15. Ajam, N.: Heart diseases diagnoses using artificial neural network, network and complex systems, vol. 5, no. 4, pp. 7–11 (2015). ISSN 2224-610X (Paper), ISSN 2225-0603(Online)
16. Rabbi, M.F., Uddin, M.P., Ali, M.A., Kibria, M.F., Afjal, M.I., Islam, M.S., Nitu, A.M.: Performance evaluation of data mining classification techniques for heart diseases prediction. Am. J. Eng. Res. (AJER), **7**(2), 278–283

17. Mitra, M., Samanta, R.: Cardiac arrhythmia classification using neural networks with selected features. Proc. Technol. **10**, 76–84 (2013)
18. Özbay, Y., Ceylan, R., Karlik, B.: A fuzzy clustering neural network architecture for classification of ECG arrhythmias. Comput. Biol. Med. **36**(4), 376–388 (2006)
19. http://www.it.hiof.no/prosjekter/hoit/html/nr2_02/age.html
20. SOURCEFORGE. http://fann.sourceforge.net/report/report.html. Accessed 25 July 2019
21. Researchgate. https://www.researchgate.net/figure/Dynamic-Neural-Network-structure-with-3-layers-and-one-output_fig2_271423140. Accessed 25 July 2019
22. Researchgate. https://www.researchgate.net/figure/Combining-multiple-neural-networks_fig1_220871290. Accessed 25 July 2019
23. O'REILY. https://www.oreilly.com/ideas/compressing-and-regularizing-deep-neural-networks. Accessed 25 July 2019
24. UCI Machine Learning Repository - Center for Machine Learning and Intelligent Systems. https://archive.ics.uci.edu/ml/machine-learning-databases/heart-disease/. Accessed 25 July 2019

Solving the Minimum Dominating Set Problem of Partitioned Graphs Using a Hybrid Bat Algorithm

Saad Adnan Abed[1(✉)] and Helmi Md. Rais[2]

[1] High Performance Cloud Computing Center,
Universiti Teknologi PETRONAS, 32610 Seri Iskandar, Perak, Malaysia
saad.adnan_g03549@utp.edu.my

[2] Department of Computer and Information Sciences,
Universiti Teknologi PETRONAS, 32610 Seri Iskandar, Perak, Malaysia
helmim@utp.edu.my

Abstract. The minimum dominating set (MDS) is the problem of finding the minimum number of nodes that have connections to all other nodes in a given graph. This problem belongs to the NP-complete complexity class that cannot be solved exactly in a polynomial time. Hence, we deployed a stochastic method to estimate good solutions for the MDS in a reasonable time. However, the problem space of the MDS problem grows exponentially with respect to the graph size. Therefore, our proposed method partitions the given graph to sub-graphs that can be tackled independently, which reduces the computational time of finding the MDS solution. This paper investigates the swarm intelligence behaviour represented by a population-based approach called the bat algorithm (BA) to find the smallest set of nodes that dominate the graph. The BA explores a wide area of the search space; thus, it is capable of the diversification procedure. However, population-based algorithms are not good in exploiting the search space in comparison to single-solution based methods, therefore, we included the Simulated annealing (SA) algorithm to balance between exploitation and exploration in order to reach a best possible solution. To analyse the performance of the proposed partitioning scheme, we experimented with the hybrid algorithm on the graph, with and without partitioning. The gained results showed significant speed up when the partitioning scheme was applied.

Keywords: Minimum dominating set · Graph partitioning · Simulated annealing · Hybrid methods · Bat algorithm

1 Introduction

The minimum dominating set (MDS) is a graph problem that consists of finding minimum number of nodes that have connections to all the other nodes. The complexity of MDS is classified as an NP-hard [1], which cannot be solved exactly in polynomial time. The MDS is one of the optimisation problems that have wide applications in wireless networking [2], network clustering [3], and social networks [4]. In the study of Mohanty et al. the authors proposed a greedy construction procedure for

F. Saeed et al. (Eds.): IRICT 2019, AISC 1073, pp. 386–395, 2020.
https://doi.org/10.1007/978-3-030-33582-3_37

estimating the minimum connected dominating set (MCDS) [5]. This procedure consists of three phases which are constructing a pseudo dominating set, constructing a Steiner Tree, and filtering. The proposed method in this paper reduces the graph size by decomposing it before searching for the MDS.

To find the MDS, it is required to search the problem space intensively. However, this would require a prolonged processing time when big graph instances are considered. Hence, partitioning the graph to subsets can simplify the search process, where this has been studied for some earlier optimisation problems [6, 7].

In computational intelligence, there are some recent classes known as swarm intelligence (SI) which are commonly used for wide domains. SI is the branch of methods that have the collective behaviour of decentralised, self-organized systems, natural or artificial. Typically, SI systems consist of a group of agents that can communicate with each other by acting on their local environments. These agents follow very simple rules, however, such agents can lead the search to the emergence of very complicated global behaviour, far beyond the capability of a single individual agent [8]. Just to name a few instances of natural-inspired SI, it includes, bird flocking, fish schooling [9] and ant foraging [10].

Practically, population-based metaheuristic algorithms are effective in exploring the problem space of a given problem. Recently, a mapping of the microbats echolocation behaviour to algorithmic steps named the bat algorithm (BA) was presented by Yang to solve continuous optimization problems [11]. The BA attempts to increase the diversity of the solutions based on the frequency tuning technique inspired from the pulse emission rates of the bats when searching for their prey. In addition, BA try to exploit the problem space effectively using automatic zooming, which is basically implemented based on random walks. The BA has been applied efficiently for solving complex optimisation problems and has received extensive attention.

However, considering a binary problem like MDS requires an intensive local search to yield the best possible solution [12]. Hence, in this study we investigated some partitioning techniques to simplify the instances of the given problem. Also, we made use of the intensive local search inside the BA for achieving effective solutions for the partitioned MDS (PMDS).

This paper is organised into five sections including the current one which has given an introduction about our research. Section 2 describes the works related to our method and to the tackled problem. Whilst, Sect. 3 describes in detail our proposed method. The penultimate section presents and discusses the experimental results. Finally, Sect. 5 gives a conclusion for the proposed model.

2 Related Work

he problem of figuring out the minimum number of nodes in a graph that have connections to all the other nodes is classified as the NP-hard problem [1]. Several researches have been devoted to solving the MDS considering the computational complexity of the problem. Big graph instances of the MDS pose a huge problem space and thus, approximation and heuristic methods are invested.

Hence, many research studies have trended in the use of the stochastic approaches for solving the MDS problem and its variants. In particular, some meta-heuristic techniques, such as genetic algorithm [13], memetic algorithm [14], ant colony algorithm [15], and simulated annealing (SA) [12] have been used to address the problem of MDS. The SA attempts to skip the potential local optima by considering worse movements sometimes. Consequently, the SA is a general case of the Hill Climbing (HC) method which accepts the best solutions only.

In general, a dominating set in the graph theory has three variants, namely the minimum dominating set (MDS) minimum connected dominating set (MCDS), and minimum weight dominating set (MWDS). Recently, these variants have been solved effectively based on meta-heuristic approaches [12, 14, 16]. Finding the smallest dominating set is the main objective of the MDS problem, however, several studies considered the processing time of finding the MDS. Lin et al. developed a hybrid method that consists of greedy a construction procedure and some stochastic operators to find the best quality solution for the MWDS in the shortest time [14]. In a different manner, Giap and Ha attempted to accelerate the process of solving the MDS problem by parallelising the genetic algorithm [17]. Hence, they studied a parallel scheme that suited the genetic algorithm operators to yield the desired solutions for the MDS problem.

Intuitively, processing small instance of a given problem can be solved accurately and rapidly in comparison to big instances of the same problem. Accordingly, some studies trended to partition the problem instances into smaller one to simplify the solving procedure. For example, particular instances of the vehicle routing problem (VRP) had been partitioned into sectors and solved in a parallel scheme [6]. Similarly, Reimann et al. proposed decomposing the whole problem of the VRP into a number of smaller sets of geographically close tours [18].

3 Proposed Method

The proposed model of this study is based on the theoretical analysis from the literature review of the MDS problem. The designed model in our study implemented a partitioning of the graph instances since the graph nodes are uniformly distributed. Decomposition of a graph problem into sub-graphs was applied successfully for the vehicle routing problem [6], which motivated us to partition the MDS graph instances. The proposed model in this study tackled the MDS problem based on the following steps:

1. Reading the graph instances.
2. Partitioning the given graph into sub-graphs.
3. Applying HBA for each sub-graph.
4. Aggregating and filtering the solutions of the tackled sub-graphs.

The first stage of the proposed model was performed to get the graph instances which were generated in a random uniform distribution based on the study of Bettstetter in 2002 [19]. The next stage was to decompose the graph nodes into sub-graphs based on the Cartesian coordinates, where a given graph is partitioned into four subsets (see Fig. 1). Since the graph nodes are distributed uniformly on a square area, the decomposition procedure partitioned the graph into sub-graphs of equivalent areas. The nodes in each sub-graph is obtained based on their coordinates, thus a node belong to a graph that does not exceeds its area. Practically, partitioning the graph into more subsets would lead to a higher demand for the filtering procedure, and thus, it would require more processing time. On the other hand, using fewer subsets would make the parallel algorithm proposed in this study and the sequential one exploit roughly the same processing time. Each of these subsets was tackled independently by the proposed hybrid bat algorithm (HBA). Thereafter, the solution of each sub-graph was aggregated with other solutions of the same graph instance to form a solution for the complete graph. Usually, the produced solution tended to include redundant nodes because of the partitioning scheme. Hence, the solution was filtered by examining the feasibility of the solution when a dominating node was dropped from it. If the solution was a dominating set after the node's exclusion, then this node would be considered redundant and would be deleted from the final solution.

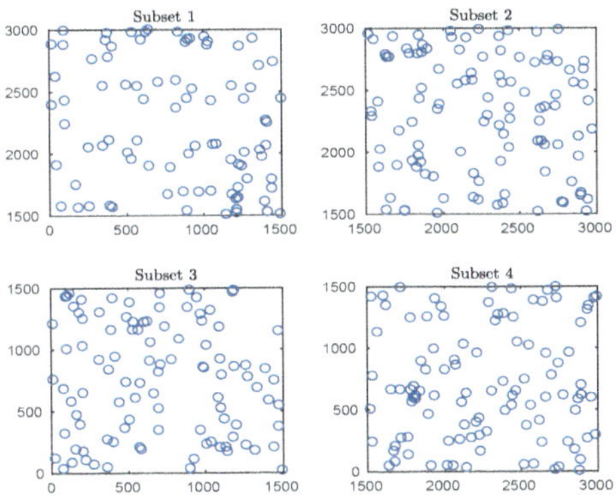

Fig. 1. Partitioned graph instance of 400 nodes scattered on area of 3000 × 3000 unit.

The Bat algorithm (BA) is a bio-inspired algorithm that mimics the micro-bats behaviour [11]. The BA controls the search process using three parameters, namely, frequency, pulse rate, and loudness. Practically, the frequency falls in a specific range $[f_{min}, f_{max}]$ corresponding to a wavelength range $[\lambda_{min}, \lambda_{max}]$. In this study, we implemented the default setting of the BA which was proposed to set the frequency in the range [0, 100]. The pulse rate fell between zero where there were no pulses at all and one as a maximum rate of the pulses.

In the BA, the pulse rate and the loudness behave dynamically, where they change their values in each iteration. Once a bat finds its prey, the loudness is decreased, whilst the pulse rate is increased. The loudness can be set at any value that suits the problem. Since the BA simulates natural bats, a set of rules were defined to update a bat's velocity v_i^t and position x_i^t at iteration t. Amongst the deployed bats, there existed a current best bat x*. Thus, the aforementioned rules were mapped to the updating equations as follows:

$$f_i = f_{min} + (f_{max} - f_{min})\beta \quad (1)$$

$$v_i^t = v_i^{t-1} + (x_i^t - x^*)f_i \quad (2)$$

$$x_i^t = x_i^{t-1} + v_i^t \quad (3)$$

where β is a uniformly distributed number falling in the period [0, 1]. In Eq. 2, x* represents the global best solution which is indicated by comparing all the solutions.

Generally, the BA has similar features to the global search of the particle swarm algorithm, as well as to the local search of the simulated annealing. Where, the BA performs the local search step using a random walk as follows:

$$x' = x + rand\ A^t \quad (4)$$

where ϵ is a random number falls between −1 and 1, and A^t is the bats' average loudness at the current iteration.

The proposed model in this study attempted to search the partitioned graph effectively. However, population-based metaheuristics are incompetent in exploiting the search space as excellently as the single-based algorithms. Hence, this study has provided a balanced search method that is capable on exploiting and exploring the problem space. Though the typical BA includes a local search which is a random walk, it is not sufficient to exploit the local neighborhood of the MDS problem. Thus, in this study the BA incorporated an intensive local search procedure represented by the simulated annealing (SA) algorithm [20]. The SA has been selected in this work due to its excellent search ability in various domains. The SA mimics the physical process of heating a material and then gradually dropping the temperature to decrease the defects, thus minimising the system's energy. The SA always accepts the best solution and it considers the worse moves regarding to the following criterion:

$$exp^{\left(\frac{-\Delta S}{T}\right)} > R$$

where ΔS is the cost difference between the new and current solutions, whilst T represents the temperature of the current iteration, and R is a random number between one and zero.

SA is a meta heuristic method which is based on a single solution, that starts the search by using an initial solution then updates it successively. In each iteration, the SA permutes the current solution to move to the surrounding areas of the problem space looking for the best possible solution. In this work, the movement to a new solution was obtained using a flip as a neighbourhood structure. The flip works on including an uncovered node to the DS or excluding a node from it. This neighbourhood structure guarantees the change in the DS size, as it would be decreased if a node was excluded from the set. Also, the DS size may be increased when a node is included, however, the uncovered nodes would be decreased and thus the objective value would be minimised.

In this paper, the stopping criterion of the SA is the frozen temperature, which is the state of no further improvement being gained. The final temperature is reached using a geometric cooling scheduler which decreases the system temperature gradually. The scheduler is performed every 16 iterations by multiplying the current temperature by a predefined cooling factor (α) which is set to 0.95. The initial and final temperatures are set to 1 and 1e-3 respectively. These values were adopted from the study of the SA on graph partitioning [21]. However, setting the final temperature to lower values (prolonging the search process) would not improve the solution quality for the partitioned instances.

In the context of a local search process, our proposed method performs it within a specific probability called the local search rate (lr) which has been set experimentally to 20%. Limiting the local search procedure to a lower ratio would not give enough space to exploit the problem space. On the other hand, allowing a lot of local search would lead to premature convergence to a local optimum. Hence, the experiment shows that 20% is a suitable ratio of performing the local search procedure for the MDS problem.

4 Experimental Results

The proposed model in this paper was aimed at finding a good quality MDS solution for a given graph in short time. Typically, the MDS can be determined for small graph instances rapidly. However, big instance would require a long processing time to identify the MDS. Hence, the quartering scheme in this study was employed to simplify the search process for big MDS graphs. In such graphs, each vertex is the centre of a circle of diameter r. Two vertices, v and u, are connected if their circles are intersected at any point. Due to the most applications of MDS arising in the network field, r was considered the coverage range of each node in the network. In this study, we tackled four MDS benchmark graph instances generated using random uniform distribution. To show the gained speed up we compare the gained result with HBA of non-partitioned graph. Based on the directions in the study of Bettstetters [19], we used the random uniform distribution to generate the graph instances. In each graph instance, all the nodes had the same transmission range. For the purpose of experimenting the algorithm stability, we ran each graph instance for ten times then we reported the statistics of the overall executions as shown in Table 1. This table shows a comparison between the multiple HBAs (HBA-PMDS) that were used for the partitioned graph and the HBA (HBA-MDS) that tackled the whole graph at once. Also, Table 1 presents the experimental results of the basic BA to compare it with the proposed HBA. In this study, the partitioning scheme has been designed for big graph instance, and thus we experiment the proposed algorithm for the graphs that had 200 nodes or more.

The results reported in the Table 1 were based on four metrics which were minimum cost (Min), average cost (Mean), standard deviation (Std), and running time in seconds (Ts). The minimum cost reports the quality of the best solution gained through the runs. Whilst, the average cost shows the median quality of the solutions gained during the experimental runs. Finally, the standard deviation shows the amount of variation amongst the yielded solutions. The values in the minimum cost field shows the size of the dominating set for the tackled graph. Notably, the obtained dominating set was smaller in higher connectivity graphs. Comparing the HBA-MDS to the HBA-PMDS, it is worth to note that the latter was faster by at least 50% for 200 nodes graph and more faster for bigger graphs. Also, the HBA-MDS and HBA-PMDS outperform

Table 1. The experimental results of the proposed algorithm.

Graph size			MDS-BA				MDS-HBA				PMDS-HBA			
Area	|V|	R	Min.	Avg.	Std.	T(s)	Min.	Avg.	Std.	T(S)	Min	Avg	Std	T(s)
3000X3000	400	210	76	77.4	1.07	171.21	70	70.5	0.52	290.73	69	69.1	0.31	46.22
	400	220	73	74.3	1.33	132.47	68	68.4	0.51	257.81	67	67.3	0.48	41.12
	400	230	69	70.4	1.26	115.02	61	61.8	1.03	251.66	60	60.1	0.31	40.03
	400	240	61	62	0.82	89.23	56	56.3	0.94	170.06	56	56	0	37.58
2500X2500	350	200	62	63.3	1.15	124.38	57	57.5	0.84	183.61	55	56	0.66	40.25
	350	210	58	59.7	1.49	117.25	53	53.6	1.34	189.08	51	52.5	0.70	41.59
	350	220	52	53.2	1.13	110.94	48	49.5	0.87	190.36	47	47	0	36.35
	350	230	49	50.1	0.74	93.41	46	47.7	0.67	138.48	46	46	0	28.57
2000X2000	2000	180	50	51.9	1.19	105.67	46	46.6	1.07	165.27	43	43.2	0.63	32.02
	2000	190	48	49.3	1.15	89.28	43	43.3	0.67	107.5	40	40.9	0.31	23.19
	2000	200	45	46.6	1.07	77.64	39	39.2	0.42	81.10	37	37.4	0.69	26.45
	2000	210	40	41.4	0.84	72.93	35	36	0.47	80.21	34	34	0	22.14
	2000	220	38	38.8	0.78	59.41	32	32.4	0.51	70.32	31	31.9	0.31	22.52
1500X1500	250	130	54	55.2	1.03	52.83	51	51.3	0.94	69.72	48	49	0.47	35.91
	250	140	46	47.3	0.95	47.35	42	43.1	0.56	61.69	40	40	0	29.26
	250	150	40	41	0.81	40.87	37	37.3	0.48	54.31	37	37	0	23.73
	250	160	36	36.8	0.79	38.94	34	34.1	0.31	48.88	33	33	0	31.30
1000X1000	200	100	39	40.1	0.99	40.81	35	33.6	0.55	53.06	33	33.3	0.44	39.23
	200	110	36	37.4	0.84	37.54	30	30.7	0.44	47	30	30.4	0.50	38.02
	200	120	31	31.9	0.73	34.97	27	27.2	0.37	44.69	26	26.2	0.40	38.13
	200	130	28	28.8	0.78	25.83	23	23.4	0.49	30.06	23	23	0	39.56
	200	140	25	26.4	0.69	18.05	21	21.2	0.48	21.76	19	19	0	38.34
	200	150	20	21.2	0.63	15.29	18	18.03	0.18	20	18	18	0	40.22
	200	160	18	18.9	0.57	13.94	16	16.1	0.25	15.58	15	15	0	39.98
700X700	200	70	37	37.7	0.82	35.84	34	34.4	0.76	64.11	33	33.4	0.49	43.26
	200	80	32	32.6	0.69	28.03	29	29.3	0.44	57.39	28	28.2	0.46	32.10
	200	90	28	28.2	0.42	19.47	24	24.1	0.25	30.43	23	23	0	33.25
	200	100	22	22.1	0.31	16.28	19	19.9	0.25	18.68	17	17	0	28.42
	200	110	19	19.3	0.48	13.96	17	17.8	0.49	18.29	15	15	0	27.12
	200	120	16	16.4	0.51	11.42	15	15.63	0.36	15.58	14	14	0	27.38

the basic BA significantly. Though the basic BA was inferior to the HBA-MDS, the BA was slightly faster but it was not as stable as the HBA-MDS. The basic BA performed faster as it does not invoke local search procedure which saves the time. However, the basic BA yielded higher variations amongst the runs than the proposed HBAs. This due to the premature convergence to a local optimum solution in each run.

To show the performance of the proposed algorithm in comparison to the related works, we provide an experimental comparison as shown in Fig. 2. Since the proposed algorithm in this work is dedicated for big graph instances, then the comparison in Fig. 2 is for graphs of an order equivalent to 250 or higher. In Fig. 2, the HBA-PMDS is compared with two greedy methods [22], and a simulated annealing algorithm called SAMDS [12]. This comparison was based on the DS size over various node's ranges. The higher range refers to the dense connectivity and thus, the DS size is smaller. The proposed HBA-PMDS outperformed the other algorithms through all the graph instances. Particularly, the HBA-PMDS is significantly better for the higher ranges which are 240, 230, 220, and 160 in Fig. 2a, b, c and d respectively.

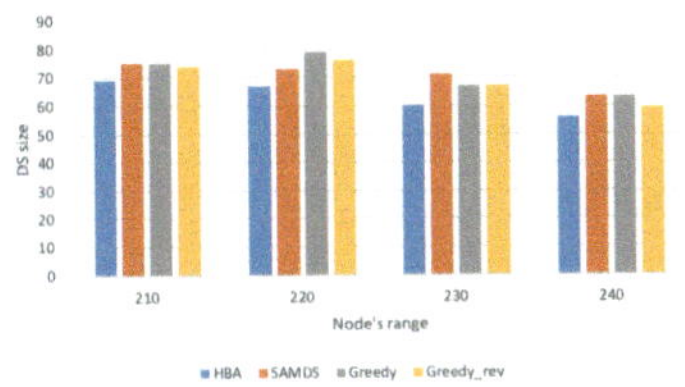

(a) Graph instance of 400 nodes distributed over an area of 3000X3000 unit

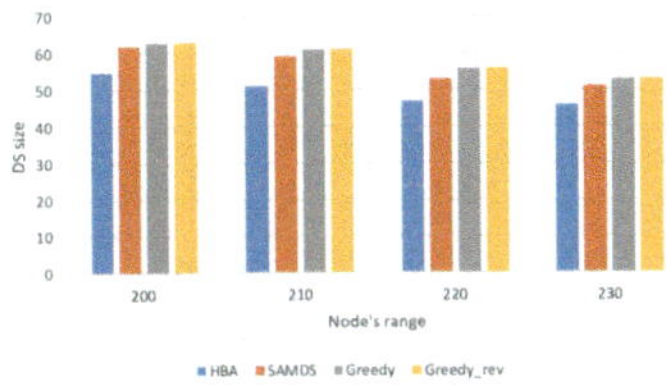

(b) Graph instance of 350 nodes distributed over an area of 2500X2500 unit

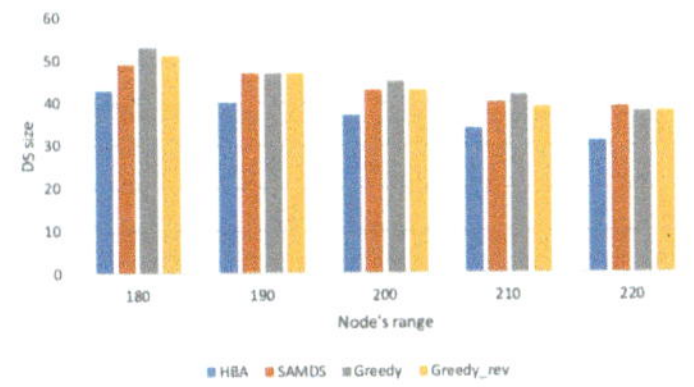

(c) Graph instance of 300 nodes distributed over an area of 2000X2000 unit

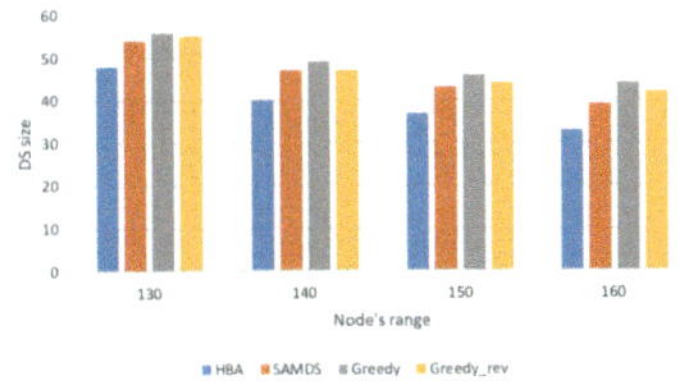

(d) Graph instance of 250 nodes distributed over an area of 1500X1500 unit

Fig. 2. Performance comparison between the proposed HBA and the SAMDS, Greedy, and Greedy_rev.

5 Conclusion

This study has employed stochastic approach to solve the MDS problem in a reasonable time. Generally, stochastic approaches are classified into two groups, i.e., single-based and population-based methods. Each of these types has a specific search strategy that characterises it from the others. Population-based methods are effective in exploring the search space, whilst single-based methods are effective in exploiting the search space. Hence, we make use of both features to orient the search towards the potential best solutions. The proposed method in this study has been aimed at gaining a good quality solution in a short time. Hence, we divided the given graph into subsets and then tackle each subset using the hybrid algorithm. Overall, the proposed partitioning technique gained a high speed up over the same algorithm without partitioning. In future studies, we would investigate more possible partitioning techniques if they could produce higher quality solutions in reasonable time.

References

1. Johnson, D.S., Garey, M.R.: Computers and Intractability: A Guide to the Theory of NP-Completeness. WH Freeman, San Francisco (1979)
2. Vaishnav, Z.B., Sajja, P.S.: Knowledge-based approach for word sense disambiguation using genetic algorithm for Gujarati. In: Information and Communication Technology for Intelligent Systems, pp. 485–494. Springer, Singapore (2019)
3. Oprişa, C., Cabău, G., Pal, G.S.: Multi-centroid cluster analysis in malware research, pp. 94–103. Springer (2018)
4. Ran, Y., Zhang, Z., Du, H., Zhu, Y.: Approximation algorithm for partial positive influence problem in social network. J. Combin. Optim. **33**, 791–802 (2017)
5. Mohanty, J.P., Mandal, C., Reade, C.: Distributed construction of minimum connected dominating set in wireless sensor network using two-hop information. Comput. Netw. **123**, 137–152 (2017)
6. Taillard, É.: Parallel iterative search methods for vehicle routing problems. Networks **23**, 661–673 (1993)
7. Walshaw, C.: A multilevel approach to the travelling salesman problem. Oper. Res. **50**, 862–877 (2002)
8. Eberhart, R.C., Shi, Y., Kennedy, J.: Swarm intelligence. Elsevier, Amsterdam (2001)
9. Eberhart, R., Kennedy, J.: A new optimizer using particle swarm theory. In: Proceedings of the Sixth International Symposium on Micro Machine and Human Science, MHS 1995, pp. 39–43. (1995)
10. Dorigo, M., Maniezzo, V., Colorni, A.: Ant system: optimization by a colony of cooperating agents. IEEE Trans. Syst. Man Cybern Part B (Cybern.) **26**, 29–41 (1996)
11. Yang, X.-S.: A new metaheuristic bat-inspired algorithm. In: González, J.R., Pelta, D.A., Cruz, C., Terrazas, G., Krasnogor, N. (eds.) Nature Inspired Cooperative Strategies for Optimization (NICSO 2010), pp. 65–74. Springer, Heidelberg (2010)
12. Hedar, A.-R., Ismail, R.: Simulated annealing with stochastic local search for minimum dominating set problem. Int. J. Mach. Learn. Cybern. **3**, 97–109 (2012)
13. Hedar, A.-R., Ismail, R.: Hybrid genetic algorithm for minimum dominating set problem, pp. 457–467. Springer, Heidelberg (2010)
14. Lin, G., Zhu, W., Ali, M.M.: An Effective Hybrid Memetic Algorithm for the Minimum Weight Dominating Set Problem. IEEE Trans. Evol. Comput. **20**, 892–907 (2016)
15. Ho, C.K., Singh, Y.P., Ewe, H.T.: An enhanced ant colony optimization metaheuristic for the minimum dominating set problem. Appl. Artif. Intell. **20**, 881–903 (2006)
16. Jovanovic, R., Tuba, M.: Ant colony optimization algorithm with pheromone correction strategy for the minimum connected dominating set problem. Comput. Sci. Inf. Syst. **10**, 133–149 (2013)
17. Giap, C.N., Ha, D.T.: Parallel genetic algorithm for minimum dominating set problem. In: 2014 International Conference on Computing, Management and Telecommunications (ComManTel), pp. 165–169 (2014)
18. Reimann, M., Doerner, K., Hartl, R.F.: D-ants: savings based ants divide and conquer the vehicle routing problem. Comput. Oper. Res. **31**, 563–591 (2004)
19. Bettstetter, C.: On the minimum node degree and connectivity of a wireless multihop network. In: Proceedings of the 3rd ACM International Symposium on Mobile Ad Hoc Networking & Amp; Computing, pp. 80–91. ACM, Lausanne (2002)
20. Kirkpatrick, S., Gelatt, C.D., Vecchi, M.P.: Optimization by simulated annealing. Science **220**, 671–680 (1983)

21. Johnson, D.S., Aragon, C.R., McGeoch, L.A., Schevon, C.: Optimization by simulated annealing: an experimental evaluation; part I, graph partitioning. Oper. Res. **37**, 865–892 (1989)
22. Sanchis, L.A.: Experimental analysis of heuristic algorithms for the dominating set problem. Algorithmica **33**, 3–18 (2002)

A Voice Morphing Model Based on the Gaussian Mixture Model and Generative Topographic Mapping

Murad A. Rassam[1,2(✉)], Rasha Almekhlafi[2], Eman Alosaily[2], Haneen Hassan[2], Reem Hassan[2], Eman Saeed[2], and Elham Alqershi[2]

[1] Information Technology Department, College of Computer, Qassim University, Buraidah, Kingdom of Saudi Arabia

[2] Faculty of Engineering and Information Technology, Taiz University, 6803 Taiz, Yemen

murad.rassam@ieee.org

Abstract. In this paper, a new model for voice morphing is proposed. The spectral characteristics of a source speaker's speech have been transferred to speech as it was spoken by another designated target speaker. The proposed model performs a phoneme segmentation of the voice signal and then transforms the spectral characteristics of each segment using a Linear Prediction model. The spectral features extracted using the Linear Prediction Coding (LPC) technique are aligned using the Dynamic Time Wrapping (DTW). The Generative Topographic Mapping (GTM) method was used for modeling the LPC features. Then, the transformation is achieved using the Gaussian Mixture Model (GMM). The transformed code-books are finally converted to prediction coefficients, and the excitation signal is filtered in order to synthesis the speech. A correlation test is performed between the source, and target signals showed a high correlation. The results reveal that the proposed model is promising in terms of recognizing full sentences in addition to individual words.

Keywords: Voice morphing · DTW · GTM · GMM · Signal processing · Correlation

1 Introduction

Voice is a series of waves and vibrations that moves and travels through transmission media (air, solid, liquid). The human voice is more complicated than we think as the scientists say, it is more unique than a fingerprint [1]. Voice processing concerned with voice signals, and those signals can be processed [2]. To do that, signals are usually processed in a digitized form due to that every person has unique vocal qualities. These qualities are essentially caused by people 's vocal tract, that construct a transfer function which maps between the input signal and the output signal. Therefore, voice processing can be considered as a branch of digital signal processing applied to voice signals.

Voice morphing is a technique that modifies a speech pattern of a source speaker to different speech pattern that impersonates another speaker or to obscure the identity of the source speaker [3]. Voice morphing offers a wide range of applications such as

F. Saeed et al. (Eds.): IRICT 2019, AISC 1073, pp. 396–406, 2020.
https://doi.org/10.1007/978-3-030-33582-3_38

Text-to-speech, voice disguise, and low bandwidth speech encoding, among many others [5]. Text-to-speech adaptation also called voice morphing system, is one of the applications of voice morphing technology that gained the interest of many researchers and developers. Voice morphing system is usually trained on small amounts of voices data in order to form new voices that have a lower cost than that of the original speakers [4]. It can also be trained based on previous records of person's voices to create a new speech in a situation where the original speaker is not available. Another use of voice morphing is to hide the identity of the speakers during transferring their speeches [5].

There are many voice morphing models have been suggested in the literature. Most of these solutions used signal processing techniques and machine learning to address the issue of the voice morphing. However, the extant voice morphing models suffer from several problems such as low quality, low robustness, and difficult training process. To the best of our knowledge, most of the existing works morphed only one word and have not considered a full sentence.

In this paper, a full sentence based voice morphing model is proposed. The voice morphing problem is formulated as a machine learning problem that aims at learning a transformation function that maps the speech of source speaker to speech of a target speaker. The proposed voice morphing model consists of three phases: the speech analysis and features modeling phase, the conversion phase, and the synthesis phase. In the first phase, the speech signals of the source and target speakers are analyzed, and the voice characteristics are extracted. Formant frequencies which were commonly used for voice conversion [6] have also been used in this study. Formant frequencies are essentially the characteristic resonant frequencies of the vocal tract that can [7]. The extraction process is performed by means of a mathematical optimization technique that is very popular in the speech processing world, namely the linear prediction coding technique (LPC) [12]. In the second phase, the Generative Topographic Mapping (GTM) [14] was used to represent the acoustic features of the sound signal and transformed into those of the target speaker. In the third phase, the linear prediction filter excites the glottal excitation (the LPC residual) of the source speaker in order to produce the desired speech.

The rest of this paper is organized as follows. Section 2 demonstrated the key related studies of voice conversion and morphing. Section 3 describes the proposed model and details the different stages involved in the design of the model. Section 4 shows the results of the experimental results of the proposed model on TIMIT dataset. It further reports the evaluation process of the model and discusses the results. Section 5 concludes the paper.

2 Related Studies

Voice morphing systems have received much interests from researchers and industry in the recent few years. Machine learning and different signal processing techniques were commonly used to address the issues of the voice morphing systems. For example, Carlo Drioli [8] proposed Radial Basis Function Networks (RBFN) as a special case of feed-forward neural network for the modeling of the spectral changes related to the control of important sound parameters, such as pitch or intensity. The identification process of such conversion functions was based on learning the shape of the conversion from a few couples of target spectra in a dataset. Authors in [9] presented a new multi-scale voice-morphing algorithm. The algorithm performs the morphing at different sub-bands by using the theory of wavelets and models the spectral conversion using the RBFN. The results obtained on the TIMIT [10] speech dataset demonstrated the effective transformation of the speaker identity. The experimental results showed that the conversion is successful in terms of producing speech that can be recognized as the target speaker, although the speech signals sounded muffled. Furthermore, it is reported that most distortion occurred in the unvoiced parts of the signals. According to [10], the muffled effect, as well as the distortion, could be due to the removal of some of the highest frequency components of the speech signals. Therefore, the authors suggested that methods should include all frequency levels in the morphing process. Songar and Harita [11] proposed a model for voice conversion based on Pitch Synchronization Over-Lap Add algorithm (PSOLA). In this model, PSOLA works directly on the signal waveform without any sort of modeling and therefore does not lose any detail of the signal. The design of the PSOLA algorithm was based on the decomposition of a signal into a series of elementary waveforms in such a way that each waveform represents one of the successive pitch periods of the signal and the sum (overlap-add) of them to reconstitute the signal. The proposed voice morphing methods in the literature suffer from several problems such as low quality, low robustness, and difficult training process. In addition, to the best of our knowledge, most of the existing works morphed only one word and have not considered a full sentence.

In this paper, a new model that considers the morphing process for a full sentence is proposed. The proposed voice morphing model consists of three phases: the speech analysis and features modeling phase, the conversion phase, and the synthesis phase. The following section describes the proposed model indetatiles.

3 Description of the Proposed Model

The proposed Voice morphing model consists of three phases: the speech analysis and features modeling phase, the conversion phase, and the synthesis phase. Figure 1 shows the flow chart of the proposed model. In most approaches of voice morphing, the speech signals need to be segmented before morphing can take place. Therefore, the training speech from source and target speakers was first segmented into phonemes [13]. Phonemes are the smallest units of sound that comprise the words. A word is represented by a sequence of phonemes so-called phonetic spelling.

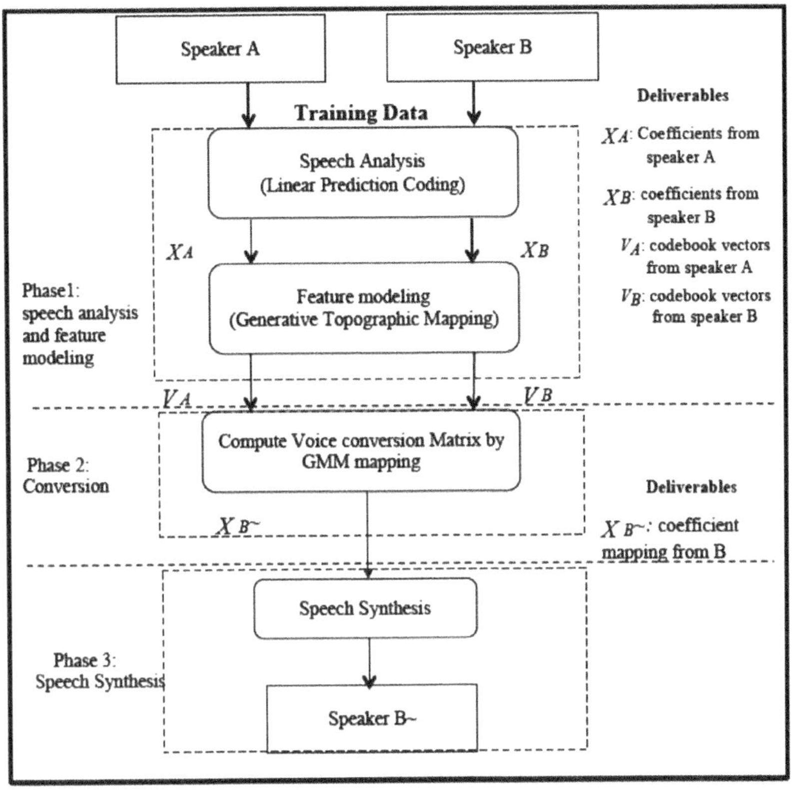

Fig. 1. The proposed model

3.1 Speech Analysis and Features Modeling Phase

In this phase, a mathematical optimization technique, namely the Linear Prediction Coding (LPC) technique [12] was used to analyze the source and target speech signals so as to extract their voice characteristics. Linear Prediction Coding represents the speech signal as a linear combination of past values of a hypothetical input to a system whose output is the given signal. A set of predictors coefficients are obtained by minimizing the sum of the squared differences between the actual speech samples and the linearly predicted samples. As a result, the acoustic space for each speaker is defined by these high-dimensional predictor coefficients and relate closely to formant frequencies. Furthermore, the LPC residual was used to approximate the glottal excitation signal.

In order to boost the robustness of the conversion, Generative Topographic Mapping (GTM) was used to represent the high dimensional speech spaces as a continuous probability density. GTM creates high-dimensional data variables from a smaller number of hidden or latent variables. This relationship between the latent and the data space is a non-linear through using a non-linear parameterized mapping from the latent space to a corresponding point in the data space.

Speech Analysis

In the speech analysis phase, the Linear Prediction Coding was used to the phonemes. As shown in Fig. 2, the speech analysis phase comprises two steps: converting the sound sequence values to zero-mean white noise process and phoneme segmentation.

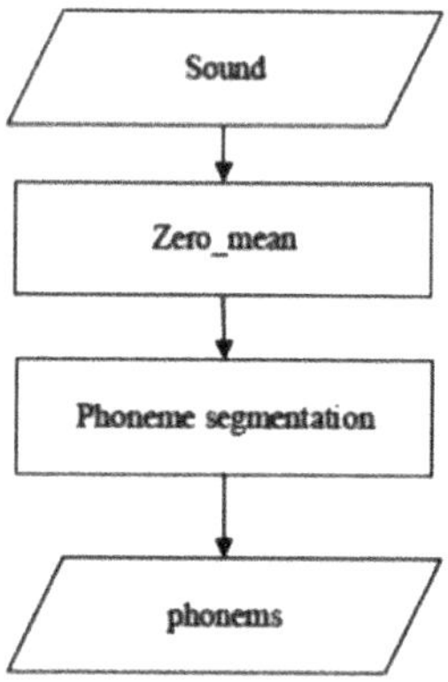

Fig. 2. Speech analysis diagram.

Features Modeling

The features are modeled in a parametric constrained Gaussian Mixture Model that is topologically orientated using the Generative Topographic Mapping (GTM). Figure 3 shows the steps of the Feature Modeling phase.

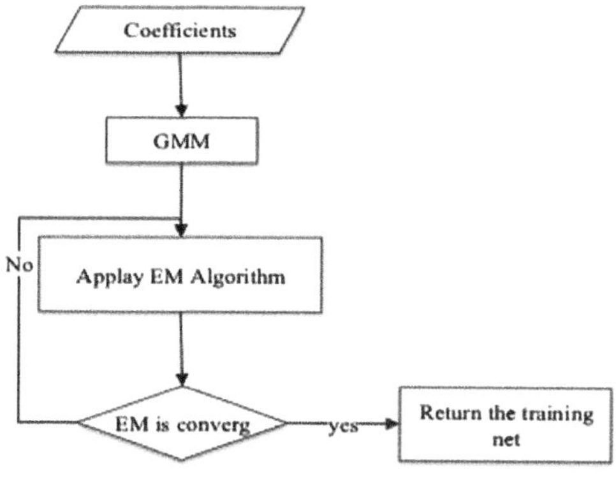

Fig. 3. Features modeling diagram

3.2 Conversion Phase

In this phase, the GTM was used to represent the acoustic features of the sound signal and then transformed to those of the target speaker using GMM mapping. That is, a one-to-one relationship between the codebook entries of the source speaker and the target speaker has been created. Then, each transformed codebook was used to construct a different linear prediction filter.

The Mapping Process

For each test phoneme, the posterior probability distribution of each frame is calculated, and the closest codebook vector from the source codebook was determined. Then, the codebooks were transformed using the codebook mapping method. The transformed codebooks were utilized to learn the prediction coefficients so as to be used for the filtering of the excitation signal. The mapping process between the code vector of the source speaker and the target speaker is shown in Fig. 4.

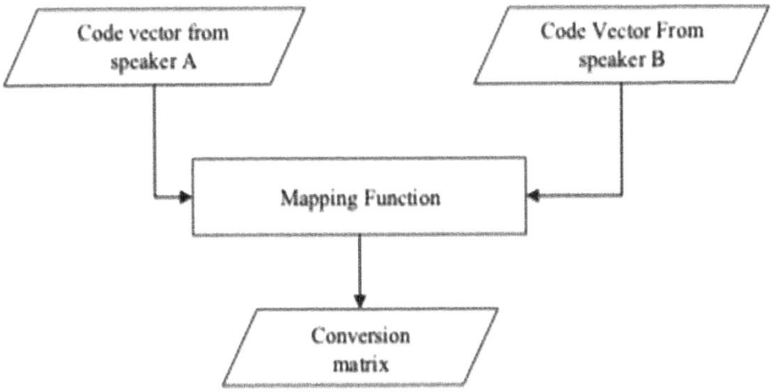

Fig. 4. Mapping process

3.3 Synthesis Phase

In the synthesis phase, the desired speech was produced using the trained linear prediction filter. The morphed speech is synthesized from the linear prediction parameters. The system used had the same parametric representation as that used in extracting the features of the original speech signals, as shown in Fig. 5.

The excitation signal impulses along with the morphed LPC coefficients were used as control parameters that were used to supply the synthesizer. Because the speech signal was analyzed frame-by-frame it was also synthesized frame-by frame. The used speech synthesis technique was essentially a filtering technique where the morphed LPC coefficients act as a filter to the excitation signal, and the output of this is the synthesized speech signal.

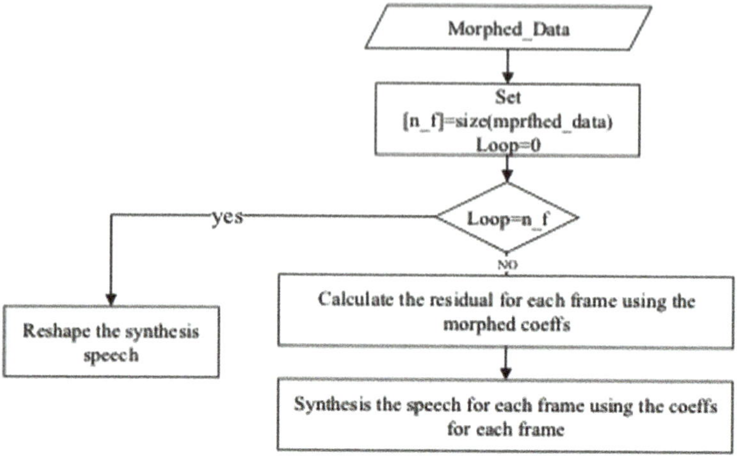

Fig. 5. The speech synthesis process

4 Results and Evaluation

The results of the three phases are described in the following subsections, followed by the evaluation of the proposed Model.

4.1 Phase 1: Speech Analysis and Features Modeling

The phonemes segments of the training data speech of source and target speakers was firstly extracted and then stored in the phoneme table. Figure 6 shows a sample of Target phoneme table.

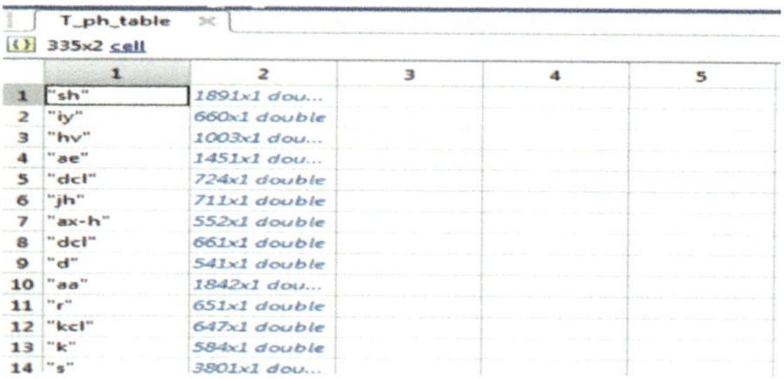
T_ph_table

335x2 cell

	1	2	3	4	5
1	"sh"	1891x1 dou...			
2	"iy"	660x1 double			
3	"hv"	1003x1 dou...			
4	"ae"	1451x1 dou...			
5	"dcl"	724x1 double			
6	"jh"	711x1 double			
7	"ax-h"	552x1 double			
8	"dcl"	661x1 double			
9	"d"	541x1 double			
10	"aa"	1842x1 dou...			
11	"r"	651x1 double			
12	"kcl"	647x1 double			
13	"k"	584x1 double			
14	"s"	3801x1 dou...			

Fig. 6. A sample of a target phoneme

Features Extraction Using LPC

The phonemes that were extracted from the training speech was time-aligned. The spectral characteristics of source and target speaker vocal tract characteristics were represented by 18th order LPC analysis in a frame-by-frame basis. Next, the linear prediction coefficients were obtained by the Generative Topographic Mapping as described earlier. Figure 7 shows the coefficients produced from the LPC analysis for the first three frames.

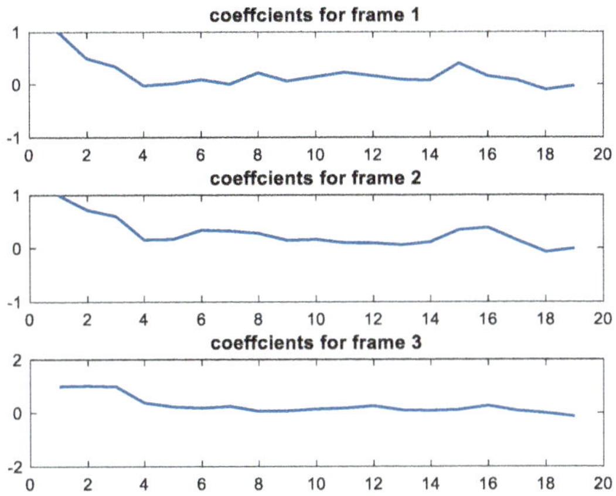

Fig. 7. Plot of coefficients for three frames

Features Modeling

Variable number of Mixtures were used for designing the model varies from 2-64 depending on the amount of training data. The spherical covariance matrix was used for the covariance type. The output of the GTM Model is a NET variable containing six fields as follows.

- type = 'gtm'
- nin = dimension of data space
- dimlatent = dimension of latent space
- rbfnet = RBF network data structure
- gmmnet = GMM data structure
- X = sample of latent points.

4.2 Phase 2: GMM Mapping

GMM mapping was applied with the help of netlab toolbox [15]. The GMM model is containing the following fields.

- mix.type = 'gmm';
- mix.covar_type – 'spherical';
- mix.nin = dimention of the data;
- mix.ncentres = number of mixture components
- mix.priors = a vector of weights in each elements.
- mix.covars = covariances of Gaussians.
- mix.centres = means of Gaussians: stored as rows of a matrix;
- mix.nwts = total number of numerical elements in the mix.

Only the centers of the GMM model were mapped. The mapped GMM Centers are then transformed to prediction coefficients, which will then filter the excitation signal in order to synthesize speech.

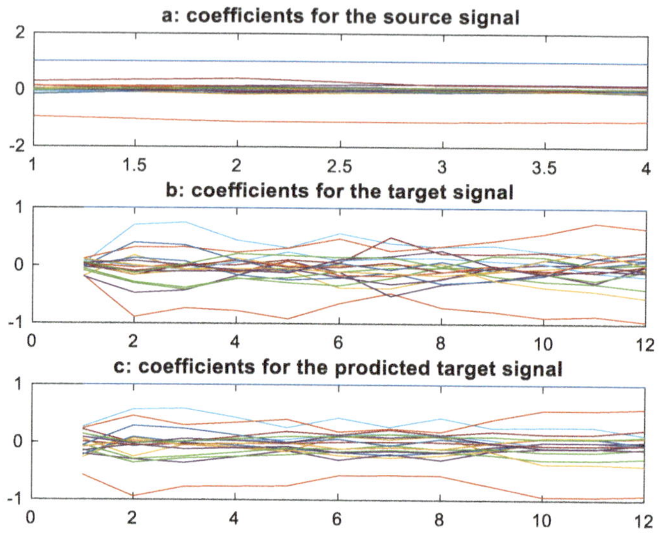

Fig. 8. Coefficients for the source, target, and the predicted target.

Figure 8a shows the coefficients for the source phoneme. Figure 8b shows the coefficients for the target phoneme. Figure 8c shows the coefficients for predicted target phoneme. As can be noticed from Fig. 8, the predicted target coefficients are similar to the target coefficients but not exactly the same.

4.3 Phase 3: Speech Synthesis

Figure 9 shows the predicted target wave after performing GMM mapping, then synthesize the coefficients produced from GMM. As observed, the predicted target is much closer to the original target. Figure 9(a) shows the original source wave after Dynamic Time Wrapping. Figure 9(b) is the original target wave after Dynamic Time Wrapping. Figure 9(c) shows the Predicted target wave synthesized using the coefficients produced from the GMM.

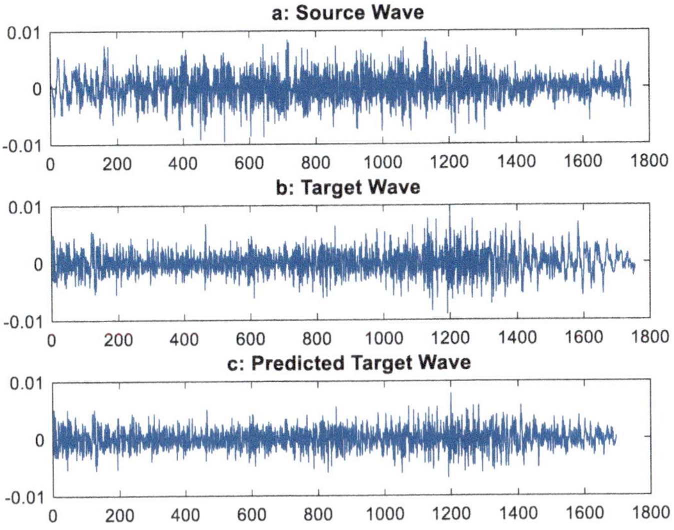

Fig. 9. A plot of source, target, and predicted target signal.

4.4 Evaluation

One method used for measuring the similarities between two signals is to apply the correlation function provided by Matlab. If the correlations near to 1, then the correlation between the two signals is high. Random phonemes were chosen for the evaluation. The correlation coefficient among the targets and predicted phonemes was used to evaluate the similarity of the predicted and targeted phonemes. Table 1 shows the correlation results between the target and the predicted target are much closer to 1, which indicates the high similarity between the target and the predicted phonemes. Table 1 also shows that the correlation between the source and the predict phonemes is low, which indicates the difference between the source and the newly created phonemes.

Table 1. Correlation evaluation

Phonemes	Correlation (Between the target and the predicted target signals)	Correlation (Between the source and the predicted target signals)
"s"	0.9203	0.6559
"w"	0.9390	0.7465
"ay"	0.9864	0.5959
"l"	0.9788	0.7603
"ah"	0.9767	0.6925
"z"	0.9243	0.7095
"f"	0.8347	0.5457
"aw"	0.9894	0.7109
"th"	0.9345	0.6628
"n"	0.9625	0.6185
"hh"	0.8307	0.6941
"g"	0.9407	0.6240

5 Conclusion

In this study, a new model for voice morphing is proposed. The aim of the proposed model was to transfer the spectral characteristics of a source speaker's speech to speech as it was spoken by another designated target speaker. The Generative Topographic Mapping (GTM) method. The Generative Topographic Mapping (GTM) method was used for modeling the LPC features. Then, the transformation is achieved using the Gaussian Mixture Model (GMM). The proposed model has the ability to deal with a full-sentence morphing. The results show that there is a high similarity between the predicted and the target morphs, which implies the high accuracy of the prediction. In future work, we plan to design a package for voice morphing system using different techniques for android-based end users.

References

1. Hutchinson, M.: Methods for voice conversion (2012)
2. Saundade, M., Kurle, P.: Speech recognition using digital signal processing. Int. J. Electron. Commun. Soft Comput. Sci. Eng. **2**, 31 (2013)
3. Orphanidou, C., et al.: Voice morphing using the generative topographic mapping (2003)
4. Kain, A., Macon, M.W.: Spectral voice conversion for text-to-speech synthesis (1998)
5. Mccree, A.: Low-Bit-Rate Speech Coding. Information Systems Technology Group, MIT Lincoln Laboratory (2008)
6. Abe, M., Nakamura, S., Shikano, K., Kuwabara, H.: Voice conversion through vector quantization. In: Proceedings of IEEE ICASSP (1988)
7. Rabiner, L.R., Schafer, R.W.: Digital Processing of Speech Signals. Prentice-Hall Signal Processing Series (1978)
8. Drioli, C.: Radial basis function networks for conversion of sound spectra. EURASIP J. Appl. Signal Process. **2001**, 36–44 (2001)
9. Orphanidou, C., Moroz, I.M., Roberts, S.J.: Wavelet-based voice morphing (2004)
10. Garofolo, J.S.: TIMIT Acoustic-Phonetic Continuous Speech Corpus LDC93S1. Web Download. Linguistic Data Consortium, Philadelphia (1993)
11. Songar, A., Harita, M.B.: MATLAB based voice conversion model using PSOLA algorithm. Int. J. Digit. Appl. Contemp. Res. **1**, 2319–4863 (2013)
12. Makhoul, J.: Linear prediction: a tutorial review. Proc. IEEE **64**, 561–580 (1975)
13. Hosom, J.-P.: Automatic time alignment of phonemes using acoustic-phonetic information, May 2000
14. Markus, J.F.: GTM: the generative topographic mapping, April 1998
15. Netlab Toolbox. http://www1.aston.ac.uk/eas/research/groups/ncrg/resources/netlab/

A Semantic Taxonomy for Weighting Assumptions to Reduce Feature Selection from Social Media and Forum Posts

Ali Muttaleb Hasan(✉), Taha Hussein Rassem, Noorhuzaimi Mohd Noor, and Ahmed Muttaleb Hasan

Faculty of Computing (FKOM), University Malaysia Pahang, 26300 Gambang, Kuantan, Pahang DarulMakmur, Malaysia
alimatlab65@yahoo.com,
{tahahussein,nhuzaimi}@ump.edu.my,
ahmed.matlab11@gmail.com

Abstract. Numerous researchers have worked on the knowledge-based semantics of words to clarify the ambiguity of (https://github.com/alimuttaleb/Ali-Muttaleb/blob/master/Synonym.txt) synonyms in various natural-language processing fields, such as Wikipedia, websites, and social networks. This paper attempts to clarify ambiguities in the lexical semantics of taxonomy in social media. It proposes a new knowledge-based semantic representation approach that can handle ambiguity and high dimensionality issues in text mining. The proposed approach consists of two main components, namely, a feature-based method for incorporating the relationships between lexical sources and a topic-based reduction method to overcome high dimensionality issues. These components help weight and reduce the relevant features of a concept. The proposed approach captures further lexical semantic similarity between words. It also evaluates the use of (https://wordnet.princeton.edu) WordNet 3.1 in text clustering and constant weighting assumption in the feature-based method used to select concepts/words from social media. To address ambiguity, the semantics of concepts with small feature subset size reduction are represented, and the performance of the semantic similarity measurement is improved. The proposed method evaluates word semantic similarity using the (https://github.com/alimuttaleb/semantictaxonomy/blob/master/mc30.txt) MC30 dataset in WordNet and obtains the following results for semantic representation: r = 0.82, p = 0.81, m = 0.81, and nz = 0.96.

Keywords: Semantic taxonomy · Feature-based method · Semantic representation · Feature selection · Gloss · Social media · MC30

1 Introduction

Social media is a collection of online communications channels dedicated to community-based input, interaction, content sharing, and collaboration. Social media posts generally contain a message alongside user details and a time stamp [1]. A thread comprises a collection of posts defined by a title with a description that summarizes the

F. Saeed et al. (Eds.): IRICT 2019, AISC 1073, pp. 407–419, 2020.
https://doi.org/10.1007/978-3-030-33582-3_39

intended discussion and a starter. The thread has a tree-like structure, containing all the posts from the oldest to the most recent. A thread may contain any number of posts and even multiple posts from the same member. A member can check the complete discussion that occurred within the thread. Many subtopic discussions may evolve from a single thread through members' posts. Such posts may contain information that is unrelated to the thread, which may arouse the interest of other members and start a new discussion on the post [2]. Various discussions on different topics can run in several threads at any time. Members with similar interests become involved in discussions by forming groups. Members of one group can participate in the discussion of any other group at any time without restriction [3]. Previous studies focused on measuring semantic relatedness on the basis of the distance between two lexical concepts in natural language processing (NLP) [4], the semantic similarity between concepts that serve as their taxonomical resemblance, the semantic representation formulated for each concept/word on the basis of semantic features in lexical resources like WordNet [5], and Wikipedia as the semi-structured resource [6, 7]. Semantic representations based on these language sources are called knowledge-based approaches, which map words in a natural language text into corresponding units (concepts) in the knowledge source. However, previous experimental results are inconclusive. Several studies [8–13] demonstrated that employing lexical sources is helpful for document clustering, whereas other works [14, 15] showed that the conventional text clustering technique without lexical sources outperforms knowledge-based techniques. This discrepancy is due to the lack of uniformity in comparing the previous experimental results in terms of knowledge sources. The present work fundamentally focuses on WordNet, though the results from using politics on social media show that any other knowledge structure could be used. To address the problems deduced from the semantic taxonomy of concepts, the features selected to represent the lexical source are based on WordNet 3.1. The semantic taxonomy on the semantic representation redundancy is also used to reduce the irrelevant features for dictionaries, such as WordNet 3.1. Certain words suffer from semantic issues due to the relationships between these words. This problem is caused by the use of dictionaries in the semantic representation of texts. Therefore, to solve the issue, the features are combined rather than separated. The choice of features is based on weightings, which are in turn considered to be based on extraction. Words with weighting are more important, as the weighting process mainly depends on the taxonomy for each feature. The weighting of the concept or synsest is determined through the taxonomy of the parent word, the number of children it has, and its depth. These weightings and their relationships have not been addressed in previous studies. Moreover, the features of semantic relationships are unimportant and irrelevant. For example, the relationship of the hyponym of *car* is that it is a one word that means a vehicle, which indicates a lack of semantics of each individual feature in the representation. This paper aims to investigate the application of knowledge in lexical sources for text mining tasks. Its specific goals are as follows: to propose weighting semantic taxonomy on the semantic representation approach for handling the constant weighting assumption in the feature selections of a feature-based method with the synsest of taxonomy, non-taxonomy[1], and glosses[2]; and to reduce the features to overcome the high dimensionality in the document of semantic representation. All the proponents have an equal voice across the paper. Forum discussions are topic-centered and grow only when members reply. When users post something on social media, numerous words are ambiguous and require analysis. Figure 1 shows an example of the synonyms and ambiguity between words.

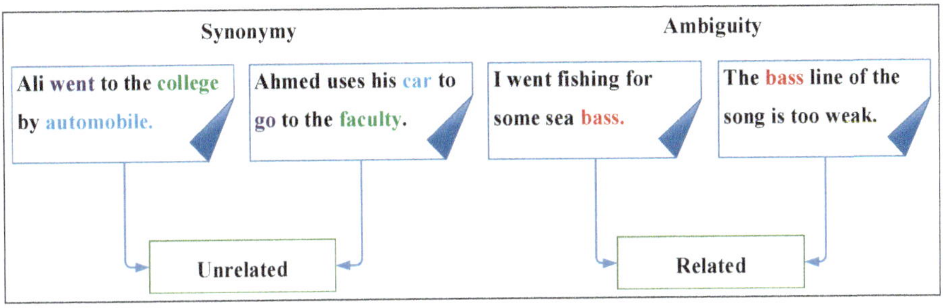

Fig. 1. Example of synonym and ambiguity issues.

The rest of this paper is organized as follows. Section 2 explores the related literature and discusses the present work. Section 3 presents the semantic relationship between two words along with various proposed semantic representations. The new weighting of the taxonomy is discussed in Sect. 4, whereas the experiments and results are presented in Sect. 5. Finally, the conclusions of this paper are given in Sect. 6.

2 Related Works

The current research framework stems from the use of WordNet 3.1, which is similar to previous studies on semantic representation and similarity measures. Semantic representation approaches, including distribution- and knowledge-based approaches, semantic similarity measurement, and knower-based text mining, are reviewed comprehensively. From the literature review, semantic information can be organized to provide efficient semantic measures by introducing a new approach for representation [16–18], proposing new measures [19–21], and combining semantic relations [6] to improve the performance of measuring semantic relations. Existing studies tend to examine such relations by using the comparative evaluation among approaches, measures [4], or between lexical sources [22, 23].

2.1 Lexical Source and Semantic Relationships in Social Media

Several approaches have been introduced to handle the Word Sense Disambiguation (WSD) using semantic similarity measurement among closely related words [24].

[1] https://github.com/alimuttaleb/Ali-Muttaleb/blob/master/Non-taxonomy

[2] https://github.com/alimuttaleb/Ali-Muttaleb/blob/master/Glosses

Synonyms and ambiguous words may present several meanings, and WSD is used to identify the correct one. Therefore, WSD aims to select the correct meaning for a word with several meanings in a sentence.

2.1.1 Detecting and Capturing the Similarity in Clustering

The semantic representation method has advantages and disadvantages. As previously mentioned in the Introduction, the strengths of this method in detecting structural similarity in the semantic representation of data lack one key. Therefore, a semantic representation method was introduced on the basis of a statistical approach to overcome this shortcoming. This approach combines statistics and knowledge to build a semantic profile of words. However, the experimental results have been inconclusive, thus far. Several studies [9, 10] demonstrated that using lexical sources is helpful for document clustering, whereas other works [15] showed that the conventional text clustering technique without lexical sources outperforms knowledge-based techniques.

2.1.2 Semantic Taxonomy of Structural Similarity Detection

The semantic textual similarity (STS) used WordNet 3.1 on the basis of lexical source to combine features from the weighting assumption and to reduce irrelevant features from social media posts. In addition, the co-occurrence association model assumes that every single vector inherently captures all text features that are necessary for computing similarity. Existing measures typically preclude similarity features beyond content, hence implying that similarity can be computed by exclusively comparing the text content.

2.1.3 The Semantic Relationship Between Two Words

The semantic relationship, which is based on the weighting of features to capture efficient words using vectors, is determined by measuring the concepts between two words using semantic representation. Most methods focus on the semantic similarity/relatedness measurement in the line combing similarity of two words. However, the measurement of semantic relatedness tends to be a combing method of SS and SR measurements. The lexicon contains several semantic features, such as synonyms, 'is-a' relations, and textual definitions. Thus, the main challenge is how to select informative features that improve the semantic representation of concepts. As such, exploiting the features from statistical methods seems fair. However, several experimental studies introduced the combination of numerous semantic features for measuring relatedness, such as text mining tasks, thereby requiring special investigation to select features [25]. Therefore, semantics is represented as a vector. However, the semantic relatedness between two texts basically measures the similarity between two vectors using the cosine similarity, after which the description of semantic relatedness ends. Low cosine distance implies great semantic relatedness to make the similarity between two words zero. The feature-based method is used in the current work to combine the features with synsest of taxonomy, non-taxonomy, and glosses, as well as to reduce these features. For the taxonomy of weighting for relevant features, we introduce our proposed method where important words have more weight.

3 The Proposed Semantic Representation

This section analyzes the problems that arise from using the feature-based method. Dictionaries such as WordNet3.1 suffer from semantic representation for the relation between words and also lack semantics for each individual feature. That will be handled by this research to addresses such issues and describes the overview of the semantic representation used for solving issues in different directions. This figure shows the collected data by inserting words in WordNet and analyzing them to determine their similarities. Moreover, we analyze texts and reduce the irrelevant features by determining relevant words and extracting their features from weights to find the cosine between similar words and clustering with the LDA and LSA model, to reduce the annoying features of the text. We use WSD to test the best words in the text and extract them in the form of text clustering to be understood by machine learning (Fig. 2).

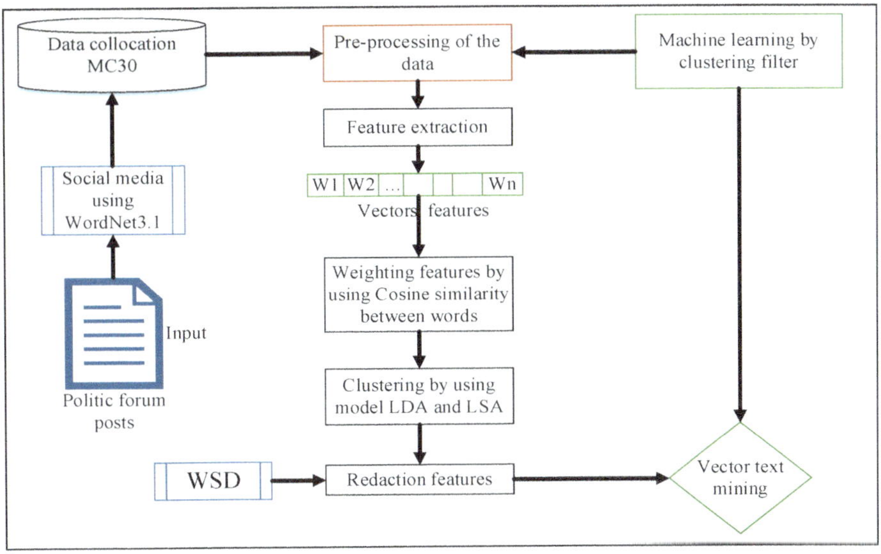

Fig. 2. The proposed work of semantic representation to reduce the features from Social Media.

3.1 Hyperspace Analogue to Language (HAL)

The measure of the hyperspace analogue to language (HAL) is given using a context window of words with size m in the corpus. For each hyperspace analog measure over a pair of words, the context window m is weighted depending on the distance and frequency of words in the context. This model is combined as follows:

$$HAL(w, w_1) = \sum_{i=1}^{m} f(w, w_1, i)(m - i + 1) \tag{1}$$

The proposed $f(w, w_l, i)$ is defined as the frequency with which the word w appears as the neighbor of word w_l at a distance of i. The definition of *HAL* is that the vector created for each word to be weighted depends on the distance and the frequency of words in the context and its column and row vectors, consequently containing the context words. For example, the vector of the text mining for "*The fundamental notion of the utterance society*" utilizes a *four-word moving window* after eliminating the stop words (shown in Fig. 3) from the sentence. Table 1 shows the matrix of words frequency co-occurrence.

Table 1. Matrix frequency co-occurrence of words

Context	Fundamental	Notion	Utterance	Society
Fundamental	5	4	0	0
Notion	4	0	0	0
Utterance	5	3	0	5
Society	4	2	5	0

n_{ij} = number of examples common to clusters i and j
a_i = Sum of contingency cells in row i
b_j = Sum of contingency cells in column j
n_{ij} = number of example common to cluster i and cluster j
a_i = Sum of contingency cells in row i
b_j = Sum of contingency cells in column j

$$\sum ij \binom{n^{ij}}{4} = \binom{5}{4} + \binom{4}{4} + \binom{0}{4} + \binom{0}{4} = 2.25 \tag{2}$$

$$\sum ij \binom{n^{ij}}{4} = \binom{5}{4} + \binom{2}{4} + \binom{5}{4} + \binom{0}{4} = 3 \tag{3}$$

The counting process is completed by weighting the taxonomy on the occurrence. For instance, if e occurs closely to c, the weighting obtained is 5; if e is separated from c by a word, the weighting obtained is 4, and so forth with the next down by 1 distance weight among 5 neighbors. Table 1 shows the computations of occurrence on semantic representation.

3.2 Correlated Occurrence Analogue to Lexical Semantic

This technique typically considers each word as an independent lexical unit regardless of its synonyms and ambiguity. For example, a *car* is not related to *automobile*. The lexical semantic has dozens of senses, which considerably reinforce lexical ambiguity. On the basis of the lexical source, a model is proposed as a weighting word in semantic representation, where the occurrence model is represented as a word matrix with size W. This model is used to compute the weight of each word. As an instance in the big corpus of text mining, several words, such as *Car, Vehicle,* and *Motor,* can be expected to appear near each other. This similarly applies to words, such as *Iraq, Yemen,* and *Egypt*. In the correlated occurrence, this text matrix that represents unique words n is illustrated as a column and row. The matrix has a size $m \times m$. For every word, c will count the number of times every word e occurs in proximity to c.

$$Corr(i,j) = \frac{w_{ij} \times Sum - \sum_{L=1}^{N} f_{iL} \sum_{k=1}^{N} f_{kj}}{\sqrt{\sum_{L=1}^{N} f_{iL} \times \left(Sum - \sum_{L=1}^{N} f_{iL}\right) \times \sum_{k=1}^{N} f_{kj} \times \left(Sum - \sum_{k=1}^{N} f_{kj}\right)}} \tag{4}$$

Table 2. Co-occurrence square matrix of text mining

	Car	Vehicle	Motor	Iraq	Yemen	Egypt
Car	**4**	**3**	0	0	0	0
Vehicle	3	0	**5**	0	0	0
Motor	0	**5**	**4**	0	0	0
Iraq	0	0	0	3	4	0
Yemen	0	0	0	4	0	5
Egypt	0	0	0	0	5	3

Table 2 shows an example for the results of words occurrence based on the weighting taxonomy. From the table, we can clearly see that the columns named *Car, Vehicle,* and *Motor* have similar rows. Similarly, columns named *Iraq, Yemen,* and *Egypt* have the same rows. The distance measure between two words as a vector becomes a proximity for the similarity of meanings of word representation via vectors. The distance-measure is the cosine of the angle between two vectors. Following the above example, words with small cosine distances have similar meanings. *For instance, Vehicle* does not seem directly related to *Car,* but both have a shared meaning with *Motor*. As a result, we can easily group words by their meanings, even by indirect association. Although, identical vectors with an angle of 0 have a cosine distance of 1, unrelated vectors would be orthogonal with an angle of 90° and the cosine value will be 0. For easier computation of cosine distances, matrices are often normalized over the rows to the unit longitude of 1 of the row's vectors.

3.3 Co-occurrence Association Model

On the basis of occurrences and co-occurrences, association measures have been introduced to quantify the degree of the relationship between two words [26]. Words with good associated value in the semantic representation to a given word are considered similar. Association measures have employed different techniques of text mining. Thus, this section describes how this measure is used to determine the weight of each word in the semantic taxonomy of structural similarity based on semantic representation. Table 3 is constructed using the statistics of two words.

Table 3. The contingency table for two words W_1 and W_2

	$\breve{w}1$	$\breve{w}2$	
W_1	a	b	$X_1 = a + b$
W_2	c	d	$X_2 = c + d$
	$Y_1 = a + c$	$Y_2 = b + d$	$N = a + b + c + d$

$\breve{W}1$ is the context without the word W_1.

$\breve{W}2$ is the context without the word W_2.

a: The number of times the word W_1 appears with the word W_2 in the same context.

b: The number of times the word W_1 appears without the word W_2 in the same context.

c: The number of times the word W_2 appears without the word W_1 in the same context.

d: The number of contexts in the corpus in which both words do not appear.

3.4 Semantic Similarity Retrieval Model

Various semantic similarity methods have been devised and tested on WordNet 3.1 to determine semantic similarities among terms. Semantic similarity methods can be classified into four distinct categories, namely, edge counting, information content, hybrid, single-ontology and cross-ontology methods. Most semantic similarity methods assign high similarity to terms with low hierarchy and small distance (regarding path length). However, single ontology methods can be of good use where every term is represented by its tree hierarchy. A term is usually referred to as a stemmed (nonstop) word. Terms that are frequent and/or infrequent are removed, and every term is represented by a computed weight. This weight is a function of the occurrence frequency. Here, the inverse-document frequency model (tf_i. idf_i) is used for weight calculation with a user-defined threshold *t*. The purpose is to assign high weights to terms that are semantically similar in the query, whereas the weights of non-similar terms do not change. Representation is performed as follows: $d_i = tf_i . idf_i$.

3.5 Lexical Source of WordNet 3.1 from Social Media

In the lexical source, most of the words used WordNet as the source. In our work, the lexical source is used on the basis of the *Constant Weighting Assumption* to overcome the features. Lexical relations hold between semantically related word forms, whereas semantic relations hold between word meanings. These latter relations include hypernym/hyponym (superordinate/subordinate), entailment, antonym, cause, entailment, participle of verb, pertainymy, similarity, verb group the attribute, domain of synset (*topic, region, and usage*), and the member of this domain used meronym/holonym. The knowledge of lexical source is based on the text mining method using WordNet 3.1 to represent the semantic textual similarity of documents from social media. As mentioned in Fig. 3, the dataset and preprocess words are collected. The feature's extraction from the posting is then analyzed such that for each word, weighting is converted into vectors after reducing irrelevant features and high dimensionality using the cosine similarity between words.

3.5.1 Preprocessing

The preprocessing step consists of the following stages: stop-words, stemming, and normalization.

3.5.2 Stop-Words

First, the stop-words in each document are applied to segment the text into the basic linguistic units called removing [27, 28]. Figure 3 shows the stop of words.

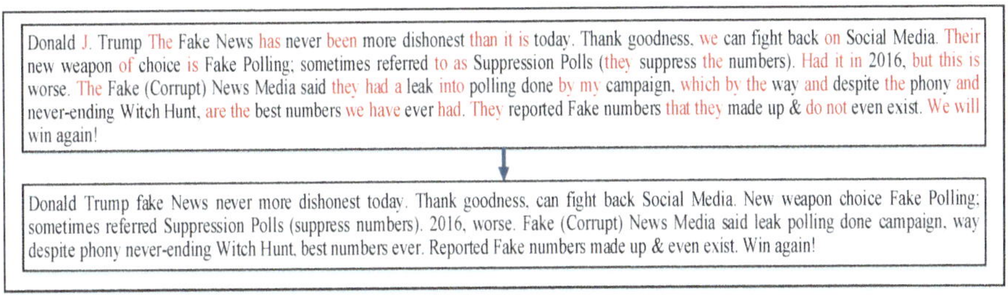

Fig. 3. The step of stop-words

3.5.3 Stemming

The second function is stemming, which excludes the stem from the document before representing the semantics. Stemming occurs such as in *argue, argued, argues, arguing,* and *argus* to the stem *argu*, which may lead to poor semantic representation of words. For example, the text "*It was pointless to argue the matter*" contains the stemming word *argue* that is a descriptive term in a text and has to be excluded [29, 30].

3.5.4 Normalization

The third function is normalization that excludes normalized words from the document before semantic representation. For example, context-independent normalization includes removal of non-alphanumeric characters or diacritical marks and regular expressions. [31, 32].

4 A New Weighting of the Taxonomy

The 'is-a' relations are used to represent the knowledge in the lexical source as the semantic taxonomy. The semantic taxonomy is a network among concepts in the lexicon source, in which the nodes represent the concepts and the edges represent the hypernyms[3] and hyponyms relations. The hypernyms and hyponyms are a relation that organized nominal and verbal synset as the inheritance semantic taxonomy.

[3] https://github.com/alimuttaleb/Ali-Muttaleb/blob/master/Hypernym

4.1 Reduction Features of Constant Weighting Assumption

As previously mentioned, the feature-based method focuses purely on representing the semantics of concepts. However, this method makes a tacit assumption that all features in the semantic representation of the specific concept have the same relevance. For instance, in the semantic of *car*, this method considers the features (*Vehicle, convertible, accelerator, train,* and *cable car*) to have the same semantic associations with the concept *car*. Therefore, this assumption is based on semantic features. However, several features are more related to the given concept than others are. Consequently, this assumption may make the feature-based method unsuitable for representing the semantics of texts in the text mining tasks in contrast to employing external lexical sources. With the lexicon containing several semantic features, such as synonyms, 'is-a' relations, and textual definitions, our main challenge is how to select informative features that improve the semantic representation of concepts.

5 Experiments and Results

5.1 Experiments

This section describes the methodology of experiment, that is, how to reduce the features to overcome limitations, such as synonym, ambiguity, and high dimensionality of the text mining tasks. The LSA and LDA models can be considered as dimensionality reduction models. In the LDA model, the semantic representation is constructed as a word of documents matrix, in which rows represent the words, whereas the columns represent documents that are given in the corpus, such as WordNet 3.1. The WordNet of the semantic taxonomy shows a good correlation for the SS datasets, such as the MC30 dataset, to make the similarity zero between words. Table 4 shows that the results are high in features that are based on selections, such as taxonomy, non-taxonomy, and glosses using our proposed method on the MC30 dataset, especially with the measure nz. The table shows the obtained correlation averages of $r = 0.829$, $\rho = 0.81$, $\mu = 0.819$, and $nz = 0.967$, which are good results were obtained by our proposed method tested on the MC30 dataset.

Table 4. The good results of Non-zero between two words by using MC30.

Correlation averages				
Features	r	ρ	μ	NZ
Taxonomy	0.85	0.784	0.815	0.933
Non-taxonomy	0.661	0.747	0.702	0.833
Gloss	0.508	0.629	0.562	0.167
Proposed method	**0.829**	**0.81**	**0.819**	**0.967**

In Fig. 4, the comparison of the semantic taxonomy in the WordNet 3.1 of the proposed method was achieved depending on the measures proposed by previous studies that achieved the highest μ correlations in the MC30 datasets. These results can be attributed to the specific relations. Table 4 shows that the proposed method achieved the highest μ correlation of 0.82 compared with the highest value of previous studies, which was 0.79 [33]. The results of MC30 are visualized in Fig. 4. We can obtain the best results with new techniques like nz.

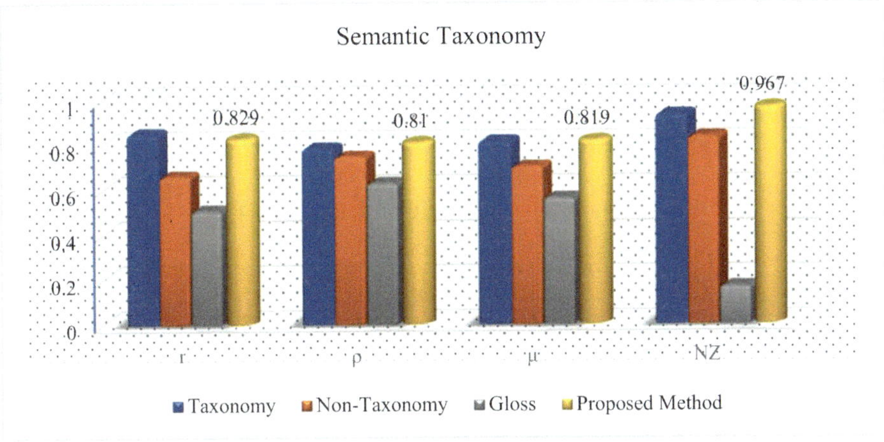

Fig. 4. The semantic taxonomy of Non-Zero similarity between words.

6 Conclusion

By using a technical model and findings from previous studies, we present a solution to problems based on feature selections regarding the ambiguity of words posted on social media. Our scheme generally achieved significant results and considerable improvement in comparison with standard schemes. Although the results we obtained are very close to standard methods and schemes of term weighting, we achieved the highest μ correlations in the MC30 datasets. Dictionaries such as WordNet 3.1 suffer from semantics for each relationship. For example, the *hypernym* and *hyponym* relations based on the taxonomy the *Car* give a word such as *Vehicle*, which indicates a lack of semantics of each individual feature semantic representation. However, the description of *Vehicle* is not represented, if the features of the semantic relationships are irrelevant, such as if the definition of *Car* contains this word. In addition, the semantic relationships in WordNet are insufficient and unbalanced features. Based on these problems, the goal of this paper is to determine how to weight the features to overcome the main issue of reducing the similarity between two words based on our non-zero measurement in the proposed method. Thus, feature taxonomy is further used to obtain the accuracy by non-zero 0.96. Our approach is useful to obtaining precise predetermined knowledge about which features of a certain data are the most suitable to handle and reduce the ambiguity in social media posts through people posting. The method does not have a very high computation demand and provides the highest μ correlations when compared with previously derived methods. We are currently conducting further research and studies working on the potential application of our method on the English language using WordNet 3.1 versions.

Acknowledgment. This work is supported by the University Malaysia Pahang (UMP) via Research Grant UMP RDU1803141.

References

1. Kaplan, A.M., Haenlein, M.: Users of the world, unite! The challenges and opportunities of Social Media. Bus. Horiz. **53**(1), 59–68 (2010)
2. Balog, K., Mishne, G., De Rijke, M.: Why are they excited?: Identifying and explaining spikes in blog mood levels. In: Proceedings of the Eleventh Conference of the European Chapter of the Association for Computational Linguistics: Posters & Demonstrations. Association for Computational Linguistics (2006)
3. Zhu, G., Iglesias, C.A.: Exploiting semantic similarity for named entity disambiguation in knowledge graphs. Expert Syst. Appl. **101**, 8–24 (2018)
4. Budanitsky, A., Hirst, G.: Evaluating wordnet-based measures of lexical semantic relatedness. Comput. Linguist. **32**(1), 13–47 (2006)
5. Agirre, E., et al.: A study on similarity and relatedness using distributional and wordnet-based approaches. In: Proceedings of Human Language Technologies: The 2009 Annual Conference of the North American Chapter of the Association for Computational Linguistics. Association for Computational Linguistics (2009)
6. Witten, I.H., Milne, D.N.: An effective, low-cost measure of semantic relatedness obtained from Wikipedia links (2008)
7. Gabrilovich, E., Markovitch, S.: Computing semantic relatedness using wikipedia-based explicit semantic analysis. In: IJcAI (2007)
8. Fodeh, S., Punch, B., Tan, P.-N.: On ontology-driven document clustering using core semantic features. Knowl. Inf. Syst. **28**(2), 395–421 (2011)
9. Wei, T., et al.: A semantic approach for text clustering using WordNet and lexical chains. Expert Syst. Appl. **42**(4), 2264–2275 (2015)
10. Saif, A., Ab Aziz, M.J., Omar, N.: Reducing explicit semantic representation vectors using Latent Dirichlet allocation. Knowl.-Based Syst. **100**, 145–159 (2016)
11. Saif, A., Ab Aziz, M.J., Omar, N.: Evaluating knowledge-based semantic measures on Arabic. Int. J. Commun. Antenna Propag. **4**(5), 180–194 (2014)
12. Saif, A., Ab Aziz, M.J., Omar, N.: Mapping Arabic WordNet synsets to Wikipedia articles using monolingual and bilingual features. Natural Language Engineering. **23**(1), 53–91 (2017)
13. Saif, A., et al.: Weighting-based semantic similarity measure based on topological parameters in semantic taxonomy. Nat. Lang. Eng. **24**(6), 861–886 (2018)
14. Fodeh, S.J., Punch, W.F., Tan, P.-N.: Combining statistics and semantics via ensemble model for document clustering. In: Proceedings of the 2009 ACM Symposium on Applied Computing. ACM (2009)
15. AlAgha, I., Nafee, R.: Investigating the efficiency of WordNet as background knowledge for document clustering. J. Eng. Res. Technol. **2**(2) (2016)
16. Gabrilovich, E., Markovitch, S.: Wikipedia-based semantic interpretation for natural language processing. J. Artif. Intell. Res. **34**, 443–498 (2009)

17. Hassan, S.: Measuring semantic relatedness using salient encyclopedic concepts. University of North Texas (2011)
18. Taieb, M.A.H., Aouicha, M.B., Hamadou, A.B.: Computing semantic relatedness using Wikipedia features. Knowl.-Based Syst. **50**, 260–278 (2013)
19. Batet, M., Sánchez, D., Valls, A.: An ontology-based measure to compute semantic similarity in biomedicine. J. Biomed. Inform. **44**(1), 118–125 (2011)
20. Sánchez, D., Batet, M., Isern, D.: Ontology-based information content computation. Knowl.-Based Syst. **24**(2), 297–303 (2011)
21. Zhou, Z., Wang, Y., Gu, J.: A new model of information content for semantic similarity in WordNet. In: 2008 Second International Conference on Future Generation Communication and Networking Symposia, FGCNS 2008. IEEE (2008)
22. Zesch, T.: Study of semantic relatedness of words using collaboratively constructed semantic resources. Technische Universität (2010)
23. Zesch, T., Gurevych, I., Mühlhäuser, M.: Comparing Wikipedia and German WordNet by evaluating semantic relatedness on multiple datasets. In: Human Language Technologies 2007: The Conference of the North American Chapter of the Association for Computational Linguistics; Companion Volume, Short Papers. Association for Computational Linguistics (2007)
24. Al-Tashi, Q., Hasan, A.M.: Word sense disambiguation: a review. Southern Connecticut State University, Hilton C. Buley Library 1, 2, pp. 20–458 (2019)
25. Rassem, T.H., et al.: Restoring the missing features of the corrupted speech using linear interpolation methods. In: 2017 AIP Conference Proceedings. AIP Publishing (2017)
26. Pecina, P.: Lexical association measures and collocation extraction. Lang. Resour. Eval. **44** (1–2), 137–158 (2010)
27. Hasan, A.M., Zakaria, L.Q.: Question classification using support vector machine and pattern matching. J. Theor. Appl. Inf. Technol. **87**(2) (2016)
28. Omar, N., Al-Tashi, Q.: Arabic nested noun compound extraction based on linguistic features and statistical measures. GEMA Online® J. Lang. Stud. **18**(2) (2018)
29. Hasan, A.M., Rassem, T.H., Noorhuzaimi, M.: Combined support vector machine and pattern matching for arabic islamic hadith question classification system. In: International Conference of Reliable Information and Communication Technology. Springer (2018)
30. Al-Tashi, Q., et al.: Binary optimization using hybrid grey wolf optimization for feature selection. IEEE Access **7**, 39496–39508 (2019)
31. Al-Tashi, Q., Rais, H., Jadid, S.: Feature selection method based on grey wolf optimization for coronary artery disease classification. In: International Conference of Reliable Information and Communication Technology. Springer (2018)
32. Hasan, A.M., Rassem, T.H., Karimah, M.: Pattern-matching based for Arabic question answering: a challenge perspective. Adv. Sci. Lett. **24**(10), 7655–7661 (2018)
33. Aouicha, M.B., Taieb, M.A.H., Hamadou, A.B.: Taxonomy-based information content and wordnet-wiktionary-wikipedia glosses for semantic relatedness. Appl. Intell. **45**(2), 475–511 (2016)

Content-Based Scientific Figure Plagiarism Detection Using Semantic Mapping

Taiseer Abdalla Elfadil Eisa[1(✉)], Naomie Salim[2], and Abdelzahir Abdelmaboud[3]

[1] College of Science and Arts - Girls Section, King Khalid University, Mahayil, Asir, Saudi Arabia
Teisa@kku.edu.sa
[2] Faculty of Computer Science and Information Systems, Universiti Teknologi Malaysia, 81310 Skudai, Johor, Malaysia
[3] College of Science and Arts King, Khalid University, Mahayil, Asir, Saudi Arabia
aelnour@kku.edu.sa

Abstract. Plagiarism is to steal others' work using their words directly or indirectly without a credit citation. Copying others' ideas is another type of plagiarism that may occur in many areas but the most serious one is the academic plagiarism. Academic misconduct forms high-profile plagiarism cases at universities. Therefore, technical solutions are strictly demanded for automatic idea plagiarism detection. Detection of figure plagiarism is a challenge field of research because not only the text analytics but also graphic features are analyzed. This paper investigates the issue of idea and figure plagiarism and proposes a detection method which copes with text and structure change. The procedure depends on finding similar semantic meanings between figures by applying image processing and semantic mapping techniques.

Keywords: Plagiarism detection · Figure plagiarism detection · Semantic meanings · Scientific publications

1 Introduction

There are many definitions for plagiarism, but all around one important point, to steal the works of others without credit situation. It can be some words or large paragraph; however, the most serious case is to plagiarize the results or the contributions from another author and insert it into one's work without citations. As plagiarism is categorized into two general types plagiarism of text and plagiarism of ideas [1, 2]. To understand what the plagiarism is and how can be detected, first must distinguish what is the difference between literal plagiarism and idea plagiarism. Literal plagiarism concerns with words which is serious in the humanities and literature fields. For instance, a poem's main axis is the words and the expression of the text and changing the words means a new poem; here the wordings are the most important part of the poem [3].

F. Saeed et al. (Eds.): IRICT 2019, AISC 1073, pp. 420–427, 2020.
https://doi.org/10.1007/978-3-030-33582-3_40

In contrast, idea plagiarism usually happens in the science fields, where the basics are facts, theories and scientific methods, but not words [4]. Recently, a lot of research handled text plagiarism [5, 6] but in comparison less attention is given to idea, image and figure plagiarism in scientific publications [7, 8]. However, some research detect text plagiarism used Semantic Role Labeling techniques (SRL) based semantic meaning methods to detect plagiarism [9]. While logical tree-structured representation was proposed in [10] as a framework, namely Intelligent Plagiarism Reasoner in Scientific Publications. The idea relied on semantic-based methods to detect different types of plagiarism in scientific publications at different levels from sentence level to the document level.

Few researches deeply investigated the problem of idea plagiarism. One of the good studies proposed by [11] investigated ontology based solution to detect text and idea plagiarism through section base importance. It investigated the use of ontology and WordNet in detecting paraphrasing and idea plagiarism in research publications using natural language processing techniques. The compression of the concepts within documents directed towards idea plagiarism percentage to be achieved using offline ontology; another design of a method to detect idea plagiarism using ontology Learning and ontology mapping presented by [12]. This method used ontology learner and ontology mapper to learn ontologies from blocks of text and mapper to map between similar concepts, the drawback of their methods is that the accuracy and performance of the system depend on the accuracy of the ontology.

Another different view to detect plagiarism using ontology is proposed in [13]. The study presents probabilistic ontology and reasoning methods to detect duplicate publication and plagiarism based on the holistic approach. Remarkably, the system returns a statistical rate for plagiarism cases rather than detecting the current plagiarism case.

2 Related Studies

In a related work, recently some techniques to detect plagiarism in non-textual elements of scientific articles such as bar chart, flow chart and diagram figure. Al-Dabbagh in [14] proposed a technique that can detect plagiarism in bar charts. They developed a method for identifying the intimate relationship between text and graphical components using word 2-gram and Euclidean distance methods to accurately detect similarities in the contents of bar charts through the help of threshold values. The limitation of their work is that they method can only detect exact copy plagiarism of bar charts.

Rabiu and Salim [15] proposed an approach for detecting figures plagiarism. In their approach, they compared the effectiveness of textual and structural representations of techniques to support figure plagiarism detection. A textual comparison method was designed to detect exact copy plagiarism of figure using Jaccard similarity measure while the structural comparison method was designed to compare the text within the components as well as the relationship between the components of the figures using graph edit distance measure. The limitation of their method is that it was evaluated with small set of figures. In addition the method could not detect plagiarism in case of plagiarism of change text or synonymies replacement.

While Arrish and colleagues in [16] proposed a technique for detecting flow-chart figure plagiarism centering on shape-based image processing and multimedia retrievals. The limitation of this work that the proposed technique is incapable of detecting plagiarism from flow-charts that possess dynamic or complex characteristics. Table 1 gives the summary of existing work in plagiarism detection.

Table 1. Existing work in semantic and idea plagiarism detection

	Techniques	Limitations
Text base plagiarism detection	Semantic Role Labeling techniques [9]	Only can detect exact copy and synonymies replacement plagiarism
	Intelligent Plagiarism Reasoner in Scientific Publications [10]	No evaluation for framework
	Ontology based solution to detect text and idea plagiarism [12]	This technique was not extended by introducing concept maps for text plagiarism detection and did not consider the topical organization of the research publications
Non-text plagiarism detection	Probabilistic ontology and reasoning methods to detect duplicate publication and plagiarism based on the holistic approach [13]	No evaluation to the method
	textual and structural approaches to detecting figure plagiarism in scientific publications [15]	Manually extraction of the features. Detect only copy and paste plagiarism
	detecting flow-chart figure plagiarism centering on shape-based [16]	Ignored the text inside the figure and the flow between components.
	Intelligent Bar Chart Plagiarism Detection in Documents [14]	Detects only copy and paste plagiarism

Therefore, proposed work, aim to apply a method capable of detecting and measuring plagiarism percentages in figures. The method aims to combine basic features with dynamic or modified features using new feature representation. Such representative features are meant to produce better understanding of the contents of figures which will enable accurate semantic comparisons between the suspected and original figures.

The objective of this study is to investigate the issues idea and figure plagiarism and proposes a procedure to detect figure plagiarism of others' ideas in the scientific publications. The procedure depend on extracting figure features from the image figure and store it as text file to be compared semantically based on the extracted features. The paper is organized as follows: Sect. 3 provides a discussion of idea plagiarism in scientific publications, Sect. 4 presents the proposed methodology, and Sect. 5 concludes the paper.

3 Plagiarism of Idea in Scientific Publications

Plagiarism in a scientific publication is different from plagiarism in general, since scientific publications have certain rules [17]. Usually, the author distinguishes between two types of plagiarism in scientific writing, which are plagiarism of data and plagiarism of text, which can be further classified as word plagiarism and idea plagiarism, the first one when the writer plagiarize data as tables or figures with or without some modifications, and insert it in his work without credit citation. This is the most serious case and is a clearly theft. When such cases are discovered, severe penalties should be carried.

Although text plagiarism is considered less serious than data plagiarism, but that depends on the plagiarize sections. For instance, The JIDC's editors do not accept any degree of plagiarism, when the section describes research using the same techniques from another paper. While the Office of Research Integrity in the USA overlooked some degree of plagiarism in the text because it is less dangerous as the plagiarism of data [3, 17, 18].

The true measure for the originality of a scientific article is the content; therefore, propose method to compare between figures based on it is content, to understand the content of the figure as human do using image processing and computer vision techniques. By using the benefit of image processing and computer vision (opencv) can understand the image and extract the most important features of the figure, represent it as areal text and store the extraction in textual file. The detection procedure then measure the similarities between figures based on the extracted feature.

Idea plagiarism occurs when an original idea is stolen and even with a complete change in original text or structure; therefore, this kind of plagiarism cannot be detected as usual plagiarism cases using traditional approaches. In fact, proposed extraction method provides all features of the figure in textual format. It can be used to find the semantic similarity between the figure after extracting the similar feature and comparing between figures based on similar/dissimilar features to detect each type of figure plagiarism.

4 Proposed Method to Detect Plagiarism of Figure

To detect idea plagiarism of figure, the important aspect is to understand the meaning of the figure to be able to compare between them. Understanding the content of the figure in terms of text and structure as human do can be achieved by extracting the textual and structural features. This will help to find a high similarity score between source and suspicious figure, whereby deeply look into the detection of plagiarism in scientific figure elements. Here, the assumption was made to consider monolingual setting for English language in both suspicious and collected figures.

According to the plagiarism taxonomy [1]; semantic based meaning, section based important and context based adaptations might be used to investigate idea plagiarism. The context based adaptation is accordingly chosen to carry comprehensive figure plagiarism. The proposed methods provide new representation to the content of the figure by extracting the figure's features from image and store them based on the type of the features known. Then semantic mapping techniques to compute and identify plagiarism of figures, depending on the semantic compressions for context between candidate and suspicious figure on the concept level. Within the figure, features are extracted and the representation of existing concepts and the relations between them are achieved.

The main objective is to investigate how to extract the important features of figure and how to compare between figures to measure plagiarism. Accordingly, three main steps are proposed to detect plagiarism of figure; which are feature extraction of the scientific figure, applying semantic mapping to extract similar concepts/features between figures, and finally, calculating the degree of similarities to measure figure plagiarism.

4.1 Extraction of Figure Features

The difficult task is to extract the features from the image of figure, (Fig. 1) illustrates the main procedure to extract feature. The figure shows the process of feature extraction from figure using image processing technique and computer vision.

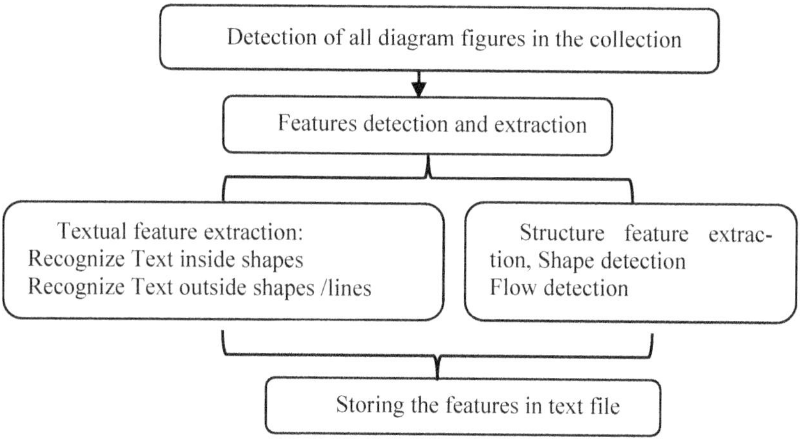

Fig. 1. Process of feature extraction from figure

The purpose here is to extract the most important features of the figure. Features are classified based on their type to textual feature and structural features. Textual features describe any text within the figure, not only, but can label the text based on it is position on the figure, such as the text inside or near the shape of the component or surrounded the figure, to recognize the text inside the figure OCR techniques can be used, OCR have wildly used in information extraction and information retrieval [19] use OCR to extract the text in Bar-chart image for plagiarism detection.

Structural features such as type of the shape of the components e.g. rectangle, diamond or decision and the flow between shapes (components) such as next or decision link based on the type of the flow, to extract the structural features methods from image processing and computer vision can be applied. Algorithm 1 shows the steps of the proposed method for the shape feature extraction.

Algorithm 1. Method steps of shape features based extraction

Input: Image of the figure
Output: Figure text descriptions
Step 1: Pre-processing
    Get gray-scale image
    Get binary image format.
    Thinning
Step 2: Shape data extraction and labelling
    Edge detection
    Obtain the contour of each shape inside the figure.
    Detect the quadrilateral shape forms
    Get the type of quadrilaterals
    Text recognition using OCR
Step 3: Edge data extraction
    Get image between shape-I and shape-j
    Get lines between shape-I and shape-j
    Label the Flow based on:
        Edge directivity
        The number of flows between two shapes
        Shape type
Step 4: Storing the features in out text file

The last step of the extraction process is to store the extracted features in text file format for further processing in the next step to compare between figures based on the extracted features.

4.2 Semantic Mapping

Semantic mapping is used to find the semantic relationship between different figures. Thus, in the semantic level, semantic mapping relates the associated concepts/features between figures. To measure the similarities between figures, semantic mapping is needed to extract the common features between figures. However, the key function in semantic mapping is the possibility to calculate similarity among textual and structural features.

Different methods for semantic mapping include: Probabilistic method [20], statistical method [21], machine learning method [22], ontology matching [23, 24], information flow method [25], fuzzy concept mapping method [26] In proposed method, predicate argument structure will be used to measure the similarities between the predicates of figure, Compute the similarity scores, which is predicate of each figure in the sematic mapper will be compared against the predicate of the source figure.

5 Conclusion and Future Work

In this study the problem of idea and figure plagiarism detection in scientific publications was investigated, and a proposed method is presented using feature extraction and semantic mapping techniques. This method copes with text and structure change. The procedure depends on similar semantic meaning between figures, which is considered for future work to present results and the evaluations of the proposed methods.

References

1. Alzahrani, S.M., Salim, N., Abraham, A.: Understanding plagiarism linguistic patterns, textual features, and detection methods. IEEE Trans. Syst. Man Cybern. Part C (Appl. Rev.) **42**(2), 133–149 (2012)
2. Sergey, B., et al.: Embedding plagiarism detection mechanisms into learning management systems. In: Scholarly Ethics and Publishing: Breakthroughs in Research and Practice, A. Information Resources Management, Hershey, PA, USA, pp. 216–231. IGI Global (2019)
3. Bouville, M.: Plagiarism: words and ideas. Sci. Eng. Ethics **14**(3), 311–322 (2008)
4. Habibzadeh, F., Shashok, K.: Plagiarism in scientific writing: words or ideas? Croat. Med. J. **52**(4), 576–577 (2011)
5. Vani, K., Gupta, D.: Unmasking text plagiarism using syntactic-semantic based natural language processing techniques: comparisons, analysis and challenges. Inf. Process. Manag. **54**(3), 408–432 (2018)
6. Agarwal, B., et al.: A deep network model for paraphrase detection in short text messages. Inf. Process. Manag. **54**(6), 922–937 (2018)
7. Meuschke, N., et al.: HyPlag: a hybrid approach to academic plagiarism detection. In: The 41st International ACM SIGIR Conference on Research & Development in Information Retrieval, Ann Arbor, MI, USA, pp. 1321–1324. ACM (2018)
8. Meuschke, N., et al.: An adaptive image-based plagiarism detection approach. In: Proceedings of the 18th ACM/IEEE on Joint Conference on Digital Libraries, Fort Worth, Texas, USA, pp. 131–140. ACM (2018)
9. Osman, A.H., et al.: An improved plagiarism detection scheme based on semantic role labeling. Appl. Soft Comput. **12**(5), 1493–1502 (2012)
10. Alzahrani, S., et al.: iPlag: intelligent plagiarism reasoner in scientific publications. In: 2011 World Congress on Information and Communication Technologies (WICT) (2011)
11. Deepika, J., et al.: A knowledge based approach to detection of idea plagiarism in online research publications. Int. J. Internet Distrib. Comput. Syst. **1**(2), 51–61 (2011)
12. Shenoy, M.K., Shet, K., Acharya, U.D.: Semantic plagiarism detection system using ontology mapping. Adv. Comput. **3**(3), 59–62 (2012)
13. Foudeh, P., Salim, N.: A holistic approach to duplicate publication and plagiarism detection using probabilistic ontologies. Springer, Heidelberg, pp. 566–574 (2012)
14. Al-Dabbagh, M.M., et al.: Intelligent bar chart plagiarism detection in documents. Sci. World J. **2014**, 1–11 (2014)
15. Rabiu, I., Salim, N.: Textual and structural approaches to detecting figure plagiarism in scientific publications. J. Theor. Appl. Inf. Technol. **70**(2), 356–371 (2014)
16. Arrish, S., et al.: Shape-based plagiarism detection for flowchart figures in texts. Int. J. Comput. Sci. Inf. Technol. (IJCSIT) **6**(1), 113–124 (2014)
17. Mason, P.R.: Plagiarism in Scientific Publications Editorial Article (2009)

18. Habibzadeh, F., Shashok, K.: Plagiarism in scientific writing: words or ideas? PMC3160704 (2011)
19. Al-Dabbagh, M.M., Salim, N., Rehman, A., et al.: Intelligent bar chart plagiarism detection in documents. Sci. World J. **2014**, 612787 (2014)
20. Mitra, P., Noy, N., Jaiswal, A.: OMEN: a probabilistic ontology mapping tool. In: Gil, Y., et al. (eds.) The Semantic Web – ISWC 2005, pp. 537–547. Springer, Heidelberg (2005)
21. Wang, P., Xu, B.: Debugging ontology mappings: a static approach. Comput. Inform. **27**, 21–36 (2008)
22. Shahri, S., Jamil, H.: An extendable meta-learning algorithm for ontology mapping. In: Andreasen, T., et al. (eds.) Flexible Query Answering Systems, pp. 418–430. Springer, Heidelberg (2009)
23. Albagli, S., Ben-Eliyahu-Zohary, R., Shimony, S.E.: Markov network based ontology matching. J. Comput. Syst. Sci. **78**(1), 105–118 (2012)
24. Rubiolo, M., et al.: Knowledge discovery through ontology matching: an approach based on an Artificial Neural Network model. Inf. Sci. **194**, 107–119 (2012)
25. Kalfoglou, Y., Schorlemmer, M.: IF-Map: an ontology-mapping method based on information-flow theory. In: Spaccapietra, S., March, S., Aberer, K. (eds.) Journal on Data Semantics I, pp. 98–127. Springer, Heidelberg (2003)
26. Doussot, D., et al.: Using fuzzy conceptual graphs to map ontologies. In: Meersman, R., Tari, Z. (eds.) On the Move to Meaningful Internet Systems 2006: CoopIS, DOA, GADA, and ODBASE, pp. 891–900. Springer, Heidelberg (2006)

TwitterBERT: Framework for Twitter Sentiment Analysis Based on Pre-trained Language Model Representations

Noureddine Azzouza[1(✉)], Karima Akli-Astouati[1], and Roliana Ibrahim[2]

[1] FEI - Department of Computer Science, RIIMA Laboratory, University of Science and Technology Houari Boumediene, Bab ezzouar, Algiers, Algeria
n_azzouza@esi.dz, kakli@usthb.dz

[2] School of Computing, Faculty of Engineering, Universiti Teknologi Malaysia (UTM), 81310 Johor Bahru, Johor, Malaysia
roliana@utm.my

Abstract. Sentiment analysis has been a topic of discussion in the exploration domain of language understanding. Yet, the neural networks deployed in it are deficient to some extent. Currently, the majority of the studies proceeds on identifying the sentiments by focusing on vocabulary and syntax. Moreover, the task is recognised in Natural Language Processing (NLP) and, for calculating the noteworthy and exceptional results, Convolutional Neural Networks (CNN) and Recurrent Neural Networks (RNN) have been employed. In this study, we propose a four-phase framework for Twitter Sentiment Analysis. This setup is based on the pre-trained Bidirectional Encoder Representations from Transformers (BERT) model as an encoder for generating sentence depictions. For more effective utilisation of this model, we deploy various classification models. Additionally, we concatenate pre-trained representations of word embeddings with BERT representation method to enhance sentiment classification. Experimental results show better implementation when it is evaluated against the baseline framework on all datasets. For example, our best model attains an F1-score of 71.82% on the SemEval 2017 dataset. A comparative analysis on experimental results offers some recommendations on choosing pre-training steps to obtain improved results. The outcomes of the experiment confirm the effectiveness of our system.

Keywords: Twitter Sentiment Analysis · Word embedding · CNN · LSTM · BERT

1 Introduction

Sentiment analysis has emerged as a key field of research in Natural Language Processing due to the spread of user-generated content through online blogs, reviews, social media and forums. Several techniques have been recommended in the literature which aim to distill sentiment information from text, enabling service providers and users to make opinion-oriented decisions.

F. Saeed et al. (Eds.): IRICT 2019, AISC 1073, pp. 428–437, 2020.
https://doi.org/10.1007/978-3-030-33582-3_41

The effective utilisation of neural networks in various applications [1, 2] and the accessibility to huge volumes of labelled data have triggered greater emphasis on sentiment classification. Supervised models are trained on documents [3], sentences [4], or phrases [5]. We propose Twitter Sentiment Analysis (TSA) as it is an interpersonal interaction site which is advancing with the usage of short forms. Furthermore, the language diversity makes it a more difficult task [6].

Pre-trained representations of words obtained by LMs (Language Models) have become widely known recently in the field of NLP (Natural Language Processing). Pre-trained LMs encrypt contextual data and high-level characteristics of semantics, language and modelling syntax, producing advanced results across a broad range of tasks, like named entity recognition [7], machine translation [8] as well as text classification [9]. Such representations can be highly effective for learning general language representations by controlling large quantities of untagged data, e.g. the pre-trained BERT (Bidirectional Encoder Representations from Transformers) [10]. It has been demonstrated that BERT can be adjusted to create advanced models for several NLP tasks, like question answering as well as machine translation.

In this study, we propose TwitterBERT, a four-phase setup with several variations of classification models. This model based on BERT is expanded by including an intermediate training phase (S) following the preliminary phase (P) on very large amount of Twitter data. For greater utilisation of the model in Twitter Sentiment Analysis, a variety of classification models are deployed for the fine-tuning phase (F) like BLSTM and CNN. For more improvement, pre-trained word embeddings are deployed in the fourth phase of blending. The key contributions of this study are mentioned below:

1. Presenting a framework derived from pre-trained BERT (Bidirectional Encoder Representations from Transformers) with an additional step of pre-training.
2. Carrying out a refinement on pre-trained language models by deploying a variety of classification models such as CNN, Linear, Bi-LSTM and Bi-GRU. These are mounted on top of the language model's final layer.
3. Intermingling the prior model with pre-trained word embedding models on various word embedding variants such as GloVe, Word2Vec, and FastText. We seek to deploy Twitter-based embeddings for improved representations.

The remaining paper is structured as follows. Section 2 carries on with the associated work. The TwitterBERT framework for Twitter Sentiment Analysis is provided in Sect. 3 along with Twitter dataset explanation. Section 4 presents the experiment and the discussion about the results obtained. The conclusion is offered in Sect. 5.

2 Related Work

2.1 Twitter Sentiment Analysis

TSA (Twitter Sentiment Analysis) is introduced by [11] as an approach based on the lexicon through a machine learning technique for categorising the sentiments of the tweets. Their contribution is to induce a tweet corpus based on emoticon as weakly tagged training data. [12] built an advanced model by obtaining rich manual features, that include traditional and diverse sentiment lexicon features. There are several dictionaries which have been formulated either semi-automatically or automatically, such as SentiWordNet [13] and WordNet-Affect [14]. The efficacy of the lexicon-based sentiment technique is signified by the coverage and precision of the dictionary [15].

Of late, TSA that use word embeddings has demonstrated large abilities. The experiments of [16] on extensively used datasets show a substantial improvement. These experiments are based on doc2vec and word2vec feature vectors. The work explained in [17] proposes a sentiment-based and a standard embedding related to a variety of pooling tasks to extract the sentiment from the tweets. [18] is a remarkable work, where both sentiment-based and standard embeddings are combined and utilised with several pooling functions, with the aim to obtain target-oriented sentiment of the comments in Twitter.

Deep learning approaches have begun to modify models for NLP. This tremendous work includes vector representation of words using neural networks and categorisation [6]. Various processes have been proposed effectively for TSA especially for semantic modelling. For example, the convolutional neural network contains several layers with word vector acting as input as described in [19]. For the representation of word distribution, a Bi-LSTM network model is utilised to carry out multitask sentiment assessment which is able to handle syntactic and semantics of the text proposed by [20]. A model is built on CNN–Bi-LSTM for statement-level sentiment classification, a strategy of divide-and-conquer which first classifies statements into various types, then carries out sentiment assessment separately on each statement as explained in [21].

2.2 Recent Advances in Natural Language Embeddings

Pre-trained word embedding approaches like GloVe and word2vec were comprehensively deployed in NLP for initialising the first layer of a neural network prior to training for a fresh task. These are superficial representations—not with respect to context—learned on the base of word concurrency though not sequential context [22]. CoVe [23]—Contextual Word Vectors—are a kind of word embeddings with which an encoder is trained in an attentional seq-to-seq machine translation model. It is pre-trained on huge parallel datasets and employs character n-gram embeddings along with word embeddings.

[9] presents ULMFiT which contains 3 steps to attain good outcomes on downstream language categorisation tasks. Initially, a general LM (language model) is pre-trained on Wikipedia text. Further, target task LM is adjusted. Finally, a fine-tuning is done where pre-trained LM is augmented with a couple of standard feed-forward layers and a softmax normalisation at the end for estimating a target label distribution. This leads us to ELMO (Embeddings from Language Model) which was proposed by [7] which learns context-based word representations through pre-training an LM in an unsupervised manner. The basis for ELMo is the bidirectional Language Model (biLM).

Similar to the concept of ELMO, OpenAI GPT (Generative Pre-training Transformer) [24] extends the unsupervised LM (language model) to quite a larger scale by getting it trained on a huge free text dataset. GPT is a multi-layer decoder transformer which adjusts the same base model for each of the end tasks. BERT (Bidirectional Encoder Representations from Transformers) [10] is a direct successor of [24]. It is pre-trained on a big LM on free dataset and then adjusted on particular tasks without customised network frameworks. The biggest enhancement of BERT is making training bi-directional. The model is trained to predict both left and right context using multi-layer bidirectional encoder transformer architecture.

3 Materials and Methods

3.1 Framework Overview

Figure 1 depicts the pipeline of the four-phase TwitterBERT setup. At first, the files of datasets pre-processed, which includes discarding redundant data (such as, URLs, HTML tags and strings going beyond maximum length allowed) and obtaining the structured contents. The tokenizer lowercases the inputs and divides these words into most common n-grams present in the pre-training dataset.

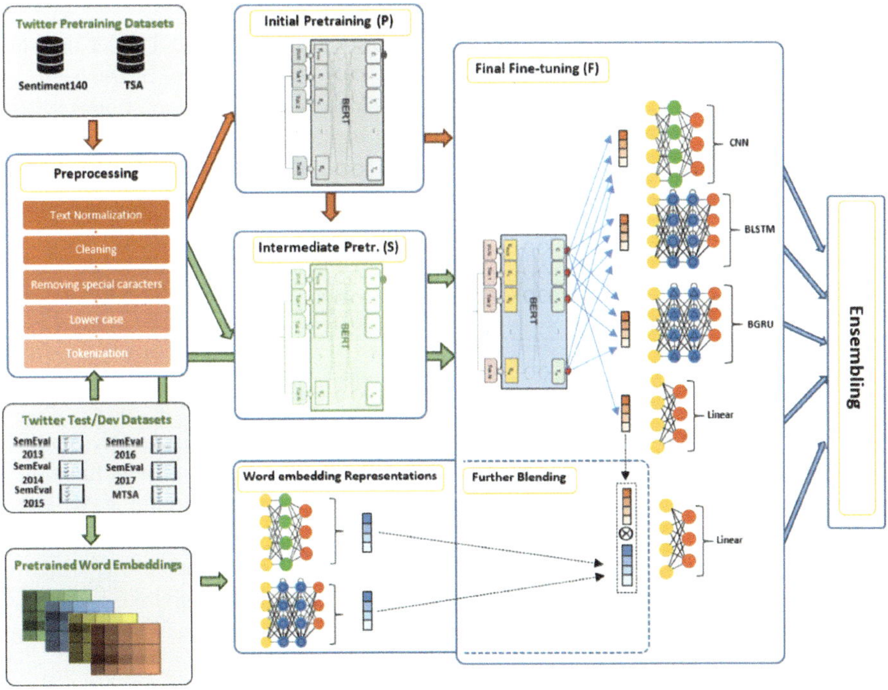

Fig. 1. General framework and pipeline of our system for Twitter Sentiment Analysis.

The initial phase of TwitterBERT architecture is the "**Initial pre-training (P)**" phase which is an unsupervised pre-training with LM (language modelling) using a huge tweets dataset. Then, further pre-training on the already pre-trained BERT model is performed for special datasets. This task is known as the **"intermediate pre-training (S)"** phase. Further, the "**Final fine-tuning (F)**" is carried out on top of the earlier pre-trained models. To completely explore the data produced by the encoder, we pile a series of various task modules on the encoder as per various classifiers. For better representation, classifiers based on multiple word embeddings are utilised to further improve the predictions by combining the 2 pipelines. This "**Further Blending**" will generate multiple combinations depending on the employed embedding or the model of classification. Lastly, we assemble those models by a voting approach as final prediction by the system. The various specified phases and classifiers will be outlined below.

3.2 Four-Stage Framework

Initial pre-training (P): The first phase follows the process of the BERT model [10]. The transformer parameters are learned using 2 unsupervised prediction functions: next sentence prediction and masked language modelling. This stage enables the model to encapsulate general knowledge regarding the language. We train the LM on 2 datasets: TSA [25] over 1 million steps and Sentiment 140 [11] over 1.5 million step. Table 1 depicts the particulars of these datasets.

Intermediate pre-training (S): After BERT's pre-training on huge Twitter datasets, the resultant models will then be pre-trained on special test datasets. In this step, we gather over 60 K tweets from each of the test datasets and only the content of the tweets is exploited. There is no need for labelled data.

Final fine-tuning (F): In this phase, we adjust the pre-trained LMs (language models) on the specified datasets, with a simple adjustment, according to which all the layers of the model are trained concurrently. Naturally, we could utilise the sentence encoding directly for classification, but we went beyond that and we employed 4 types of models that will be piled over the encoder. These models are outlined below:

1. ***Linear:*** This refers to the default BERT model with respect to sequence classification. For a completely connected layer for performing softmax, the final hidden state of the first word ([CLS]) from BERT functions as the input.
2. ***CNN:*** Convolutional Neural Network (CNN) model has a considerable ease for implementation and is especially fast for training. Moreover, the softmax function is selected as activation function, while the max pooling is used to combine the output of the convolutional layers.
3. ***BLSTM:*** A bidirectional LSTM with the ultimate hidden state of all the words in the sequence from BERT is keyed to the bi-LSTM. The stacked final LSTM hidden state of the sequence is associated with an entirely connected layer to carry out softmax.
4. ***BGRU:*** This model has originated from the previous one, but GRU cells are used instead of LSTM cells.

Based on the phase used in subscript, our generated model (M) is named. Essentially, $\mathbf{M_f}$, $\mathbf{M_{pf}}$ and $\mathbf{M_{psf}}$ are obtained with different variants of fine-tuning classifier. Table 2 presents the outcomes for these models.

Further Blending: The concatenation method has been introduced, which is inspired from the relationship of the classification task and language modelling. Independent model classifiers and pre-trained language models are used for both sentence representations. On the basis of pre-trained word embeddings, the representation of the encoded sentence in the final layer of the language model is concatenated with the classification model's hidden state. In this manner, both the representations are used. We recommend two models: the BLSTM model and a CNN classifier. According to word embeddings, we deployed six (6) pre-trained word embeddings (Table 1): Word2Vec (Twitter [26]), GloVe (crawl [27] and Twitter [27]), FastText(crawl [28]) and ConceptNet (Numbernatch [29]). These models will be known as MCE (pre-trained language Models with Concatenated Embedding). The results are given in Fig. 2.

3.3 Datasets

The datasets deployed for testing are SemEval 2013 [30], SemEval 2014 [31], SemEval 2015 [32], SemEval 2016 [33], SemEval 2017 [34] and MTSA [35]. In the initial pre-training task, the TSA [25] and Sentiment 140 [11] datasets are used for pre-training the language model. These datasets are presented in brief in Table 1.

Table 1. Statistics of the studied datasets.

Dataset	Positive	Negative	Neutral	Total
Sentiment 140	800000	800000	0	1600000
TSA	371341	179050	629566	1179957
SemEval 2013	1479	560	1513	3552
SemEval 2014	982	202	669	1853
SemEval 2015	1043	366	987	2396
SemEval 2016	7059	3231	10342	20632
SemEval 2017	2375	3972	5937	12284
MTSA	1187	1038	4186	6411

4 Experimental Results

The experiments conducted are described in this section. Six different datasets are used for every performance experiment, which is extensively used by the Sentiment Analysis community. $BERT_{Base}$ is used (12-layer transformer blocks, 768 hidden-size, and 110 M parameters) in all the experiments. The macro averaged F1-Score is the metric used.

Table 2. Macro averaged F-Score of the language models classifiers (Mf, Mpf, and Mpsf)

Dataset		SemEval 2013	SemEval 2014	SemEval 2015	SemEval 2016	SemEval 2017	MTSA
Baseline		68.49	70.86	64.84	63.30	68.50	65.80
M_f	M_f-L	69.02	71.23	66.21	65.08	69.10	66.31
	M_f-CNN	53.23	62.56	54.17	50.98	55.67	52.18
	M_f-BLSTM	58.31	65.19	56.64	52.70	59.90	55.72
	M_f-BGRU	58.76	66.37	57.13	52.36	59.71	54.16
M_{pf}	M_{pf}-L	68.78	**71.96**	**66.76**	66.10	69.23	65.91
	M_{pf}-CNN	53.82	63.42	52.54	50.64	54.10	51.86
	M_{pf}-BLSTM	59.25	66.41	55.91	53.27	58.89	54.34
	M_{pf}-BGRU	58.94	65.63	56.07	53.61	57.97	55.68
M_{psf}	M_{psf}-L	**69.86**	70.23	65.82	**66.24**	**69.81**	**67.16**
	M_{psf}-CNN	54.43	63.45	53.72	51.29	55.41	52.45
	M_{psf}-BLSTM	60.26	66.31	56.36	52.87	59.16	54.84
	M_{psf}-BGRU	59.34	66.87	56.14	53.05	58.62	55.10

Table 2 represents the results that indicate very competitive models are the ones that use initial pre-training stage with linear classifier. Unexpectedly, BLSTM and CNN classifier exhibit a low F1-score over BERT encoder.

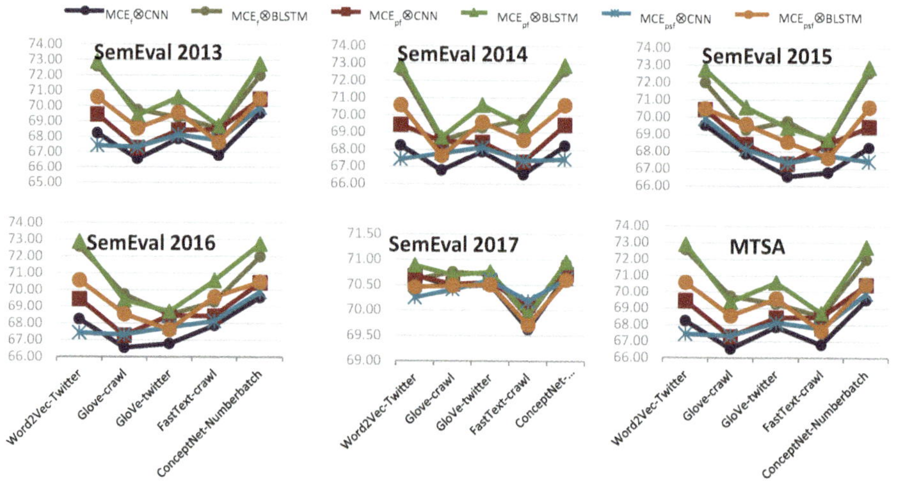

Fig. 2. Performance of different models including concatenated embeddings.

The final model is an aggregate of the models that have the best performance. Using a specific threshold value on F1-score metrics for the ensemble, a minimum of five models are filtered and derived for every classifier. At the end, an efficient integration strategy is employed to generate a robust sentiment analysis. In order to fuse the results of the various models, the voting method is used (Table 3).

Table 3. Results for ensembling all models.

Dataset	SemEval 2013	SemEval 2014	SemEval 2015	SemEval 2016	SemEval 2017	MTSA
Ensemble (All)	72.61	72.48	68.23	68.10	71.82	69.26

According to Table 2, it can be seen that M_{psf}-L exhibits the maximum F1-score. Compared to the models with fine-tuning only, the models with pre-training S- or -P have better capabilities. It is observed that there is no noticeable improvement in LM performance with CCN layer or BLSTM/BGRU layer stacking of the transformer. With regard to the Models with Concatenated Embeddings (MCE), it is found that the majority of the recommended models attain better performance individually, compared to the baselines. Additionally, it is observed that with BLSTM training of the embeddings-based models, the performance of MCE⊗BLSTM exceed that of the MCE⊗CNN models. As per the expectations, the pre-training (P) improves the results of models. Moreover, it can be concluded that the use of Twitter-based word embeddings such as Word2Vec-Twitter [26] improve the performance more than Wikipedia or crawl embeddings. With regards to the ensemble, this approach results in marginally improved performance compared to the individual models. It can be observed that the addition of the MCE_{psf}⊗BLSTM to the ensemble leads to the most improvement.

5 Conclusion

TwitterBERT: A four-stage framework for Twitter Sentiment Analysis based on pre-trained Language Model has been elaborated in this study. Highly competitive results were obtained for all the studies' datasets. The proposed framework is dependent on pre-trained language models' representations that are derived from Transformers (BERT). It is shown that the application of large Twitter data for pre-training can lead to improved performance. Further analysis of concatenation of other pre-trained word embeddings indicate that Twitter-based embeddings exhibit enhanced performance when concatenated with BERT encoder representation. In future endeavours, the objective will be to experiment with more pre-trained language models. For improved representation of tweets, an LM hierarchical architecture could be studied.

References

1. Bahdanau, D., Cho, K., Bengio, Y.: Neural machine translation by jointly learning to align and translate. In: 3rd International Conference on Learning Representations ICLR 2015 (2015)
2. Le, Q., Mikolov, T.: Distributed representations of sentences and documents. In: 31st International Conference on Machine Learning, vol. 32, pp. II–1188–II–1196 (2014)
3. Yang, Z., Yang, D., Dyer, C., He, X., Smola, A., Hovy, E.H.: Hierarchical attention networks for document classification. In: HLT-NAACL 2016, pp. 1480–1489 (2016)
4. Jianqiang, Z., Xiaolin, G., Xuejun, Z.: Deep convolution neural networks for Twitter sentiment analysis. IEEE Access **6**, 23253–23260 (2018)
5. Zhang, L., Zhou, Y., Duan, X., Chen, R.: A hierarchical multi-input and output Bi-GRU model for sentiment analysis on customer reviews. In: IOP Conference Series: Materials Science and Engineering, vol. 322, no. 6 (2018)
6. Abid, F., Alam, M., Yasir, M., Li, C.: Sentiment analysis through recurrent variants latterly on convolutional neural network of Twitter. Future Gener. Comput. Syst. **95**, 292–308 (2019)
7. Peters, M., et al.: Deep contextualized word representations. In: NAACL-HLT 2018, pp. 2227–2237 (2018)
8. Ramachandran, P., Liu, P., Le, Q.: Unsupervised pretraining for sequence to sequence learning. In: EMNLP 2017, pp. 383–391 (2017)
9. Howard, J., Ruder, S.: Universal language model fine-tuning for text classification. In: ACL, no. 1, pp. 328–339 (2018)
10. Devlin, J., Chang, M.W., Lee, K., Toutanova, K.: BERT: pre-training of deep bi-directional transformers for language understanding. In: NAACL-HLT 2019, pp. 4171–4186 (2019)
11. Go, A., Bhayani, R., Huang, L.: Twitter sentiment classification using distant supervision. pdf. CS224 N Project Report (2009)
12. Kiritchenko, S., Zhu, X., Mohammad, S.M.: Sentiment analysis of short informal texts. J. Artif. Intell. Res. **50**, 723–762 (2014)
13. Baccianella, S., Esuli, A., Sebastiani, F.: SentiWordNet 3.0: an enhanced lexical resource for sentiment analysis and opinion mining. In: Proceedings of Lr, vol. 10, pp. 2200–2204 (2010)
14. Strapparava, C., Valitutti, A.: WordNet-affect: an affective extension of WordNet. In: Proceedings of 4th International Conference on Language Resources and Evaluation, vol. 4, pp. 1083–1086 (2004)
15. Giahanou, A., Crestani, F.: Like it or not: a survey of Twitter sentiment analysis methods. ACM Comput. Surv. **49**, 28:1–28:41 (2016)
16. Arslan, Y., Küçük, D., Birturk, A.: Twitter sentiment analysis experiments using word embeddings on datasets of various scales. In: NLDB 2018, pp. 40–47 (2018)
17. Vo, D.T., Zhang, Y.: Target-dependent Twitter sentiment classification with rich automatic features. In: IJCAI 2015, pp. 1347–1353 (2015)
18. Zhang, P., He, Z.: Using data-driven feature enrichment of text representation and ensemble technique for sentence-level polarity classification. J. Inf. Sci. **41**(4), 531–549 (2015)
19. Liao, S., Wang, J., Yu, R., Sato, K., Cheng, Z.: CNN for situations understanding based on sentiment analysis of Twitter data. Procedia Comput. Sci. **111**(2015), 376–381 (2017)
20. Balikas, G., Moura, S., Amini, M.-R.: Multitask learning for fine-grained twitter sentiment analysis. In: SIGIR 2017, pp. 1005–1008 (2017)
21. Chen, T., Xu, R., He, Y., Wang, X.: Improving sentiment analysis via sentence type classification using BiLSTM-CRF and CNN. Expert Syst. Appl. **72**, 221–230 (2017)

22. Kamkarhaghighi, M., Makrehchi, M.: Content Tree Word Embedding for document representation. Expert Syst. Appl. **90**, 241–249 (2017)
23. McCann, B., Bradbury, J., Xiong, C., Socher, R.: Learned in translation: contextualized word vectors. In: NIPS 2017, pp. 6297–6308 (2017)
24. Radford, A., Salimans, T.: Improving language understanding by generative pre-training. Technical report, OpenAI, pp. 1–12 (2018)
25. Vadicamo, L., et al.: Cross-media learning for image sentiment analysis in the wild. In: ICCV Workshops 2017, pp. 308–317 (2017)
26. Godin, F., Vandersmissen, B., De Neve, W., Van de Walle, R.: Named entity recognition for Twitter microposts using distributed word representations. In: NUT@IJCNLP 2015, pp. 146–153 (2015)
27. Pennington, J., Socher, R., Manning, C.D.: GloVe: global vectors for word representation. In: EMNLP 2014, pp. 1532–1543 (2014)
28. Mikolov, T., Grave, E., Bojanowski, P., Puhrsch, C., Joulin, A.: Advances in pre-training distributed word representations. In: LREC 2018 (2018)
29. Speer, R., Chin, J., Havasi, C.: Conceptnet 5.5: an open multilingual graph of general knowledge. In: AAAI 2017, pp. 4444–4451 (2017)
30. Nakov, P., Rosenthal, S., Kozareva, Z., Stoyanov, V., Ritter, A., Wilson, T.: SemEval-2013 Task 2: sentiment analysis in Twitter. In: SemEval@NAACL-HLT 2013, pp. 312–320 (2013)
31. Rosenthal, S., Ritter, A., Nakov, P., Stoyanov, V.: SemEval-2014 Task 9: sentiment analysis in Twitter. In: SemEval@COLING, pp. 73–80 (2014)
32. Rosenthal, S., Nakov, P., Kiritchenko, S., Mohammad, S., Ritter, A., Stoyanov, V.: SemEval-2015 Task 10: sentiment analysis in Twitter. In: SemEval@NAACL-HLT 2015, pp. 451–463 (2015)
33. Nakov, P., Ritter, A., Rosenthal, S., Sebastiani, F., Stoyanov, V.: SemEval-2016 Task 4: sentiment analysis in Twitter. In: SemEval@NAACL-HLT 2016, pp. 1–18 (2016)
34. Rosenthal, S., Farra, N., Nakov, P.: SemEval-2017 Task 4: sentiment analysis in Twitter. In: SemEval@ACL 2017, pp. 502–518 (2017)
35. Kenyon-Dean, K., et al.: Sentiment Analysis: It's Complicated! In: NAACL-HLT 2018, pp. 1886–1895 (2018)

Internet of Things (IoT)

Evaluation of SRAM PUF Characteristics and Generation of Stable Bits for IoT Security

Pyi Phyo Aung(✉), Koichiro Mashiko, Nordinah Binti Ismail, and Ooi Chia Yee

Malaysia-Japan International Institute of Technology, Universiti Teknologi Malaysia, Kuala Lumpur, Malaysia
phyo.aung@graduate.utm.my,
{k.mashiko,nordinah.kl,ooichiayee}@utm.my

Abstract. Better security is essential for IoT devices these days. Device authentication of IoT devices can be done by using Static Random-Access Memory Physical Unclonable Functions (SRAM PUF). However, SRAM PUF has poor stability and relatively high error rate. Temporal Majority Voting (TMV) and other Error Correction Codes (ECCs) has been used to improve SRAM PUF performance. But they require a lot of processing time and hardware resources. Most of the microcontrollers used in IoT devices do not have that. Still, those methods cannot produce sufficiently stable bits of SRAM PUF. The data remanence nature of SRAM cells can be utilized to generate much more stable SRAM PUF with low error rates. In this paper, we made use of both TMV and data remanence to obtain SRAM PUF characteristics of microcontrollers used in IoT devices. The characteristics of SRAM PUF such as biasness, uniqueness and stability have been analyzed and investigated for better understanding of SRAM PUF on different chips. Moreover, by using the data remanence method, we managed to obtain 128 bits of SRAM PUF from 512 bits of initial SRAM values with the error rate of 3.77×10^{-8} and the stability of 99.983% which can be implemented on simple microcontrollers.

Keywords: Physical Unclonable Function · SRAM PUF · Security

1 Introduction

Data security is very important. Insecure system in data access is a serious risk which can cause a lot of problems nowadays [1]. This is especially true for IoT devices since most IoT devices sense and store confidential data of their users [2]. Moreover, a lot of IoT devices are based on small and inexpensive processing chips without consideration of security challenges [3]. Current technologies in security and device authentication use private-key and public key cryptography. In such systems, the keys are generated by the computation of the host devices and the security protocols must be stored in a memory. However, almost all IoT devices usually do not have enough computing power to generate keys on their own [4].

To achieve a more secure device authentication system, Physically Unclonable Functions (PUFs) are suggested. PUFs are digital fingerprints of semiconductor devices

F. Saeed et al. (Eds.): IRICT 2019, AISC 1073, pp. 441–450, 2020.
https://doi.org/10.1007/978-3-030-33582-3_42

which can be used for security, authentication and attestation purposes [5]. PUFs are based on physical imperfections and variations which are naturally occurring during the manufacturing of semiconductor devices. They are unpredictable and practically impossible to duplicate. They provide a good measure of security without using so much resources and energy from the host device. There are several types of PUFs such as Ring-Oscillator PUF, Arbiter PUF, SRAM PUF, etc. [6].

1.1 SRAM PUF

All microcontrollers used in IoT systems today have on-chip SRAM. SRAM PUFs can make use of the existing SRAMs in such microcontrollers as a unique ID or key [7]. Although semiconductor manufacturing process is precisely controlled, parameters such as threshold voltage, mobility, capacitance, resistance, etc., of these transistors distribute within some specified range [8]. A typical SRAM cell consists of six transistors. The schematic of an SRAM cell is shown in Fig. 1. When the SRAM is powered up, the mismatch between transistors M_2 and M_4 in each cell produces random and unpredictable data of '0' or '1' [9]. Since SRAM PUF uses only the initial value of SRAM, using SRAM PUF for security and key generation can be implemented easily on a microcontroller without affecting the memory performance of the system.

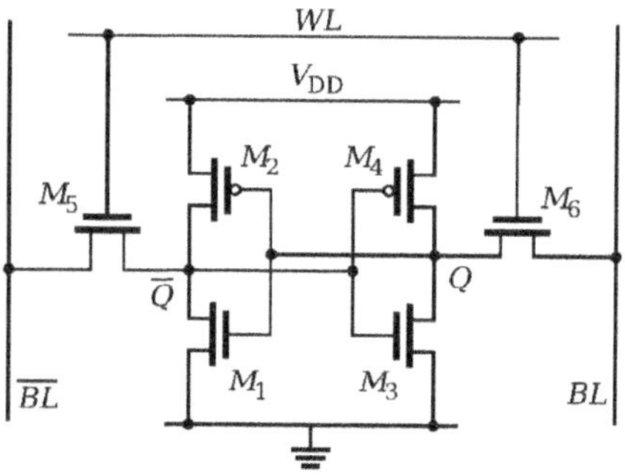

Fig. 1. Schematic of a typical six-transistor SRAM cell

However, since SRAM PUF is a type of weak PUF, the start-up values of SRAM do not always give the exact same PUF values. This is due to the nature of static memory and the voltage and current behavior of the CMOS SRAM chips [9]. However, more research is needed to be done to investigate the SRAM PUF characteristics in order to obtain more reliable SRAM PUFs.

2 Related Works

Using the initial start-up values of SRAM as digital fingerprint was first proposed in 2007 by Holcomb et al. [8]. They used the uninitialized values of CMOS SRAM and obtained 128 bits of random number for cryptography key generation from 256 bytes.

So far, there are several researches focused on the Error Rate and usage of SRAM PUF as security and authentication protocols. Böhm et al. studied the SRAM PUF on microcontrollers and used an ECC to reduce the error rate to lower than 7.0×10^{-7} [9]. Vijayakumar et al. proposed to provide more reliable SRAM PUF using modified Temporal Majority Voting (TMV) [10]. Their technique provided a simulation result of 128-bit key with error rate of 10^{-6}. However, no experimental result was provided in their paper. Many other researchers have obtained fairly reliable authentication keys for selected platforms [11–13].

However, TMV and other ECCs requires large computation power and long test time. In addition, they focused primarily on the individual error rate and not on the overall stability of the system. M. Liu et al. proposed a new method which utilizes the data remanence nature of SRAM cells. They wrote all '0's (or all '1's) to the entire array of SRAM cells. Then they read the data again after powering down for a short period of time to determine the strongest '1's (or '0's) [14].

In this paper, both TMV and data remanence methods are utilized not only to find and reduce the Error Rate, but also to analyze and improve other important characteristics of SRM PUF such as Internal Humming Distance, Uniqueness, Biasness or Uniformity and Stability of different microcontrollers used in IoT applications.

3 Evaluation of SRAM PUF Characteristics

3.1 Analytic Techniques

The startup values of SRAM from multiple various microcontrollers are first obtained from challenge value C using a specialized program for the specific PUF function P. The obtained PUF value or the response value R is then stored in computer for further data analysis [15]. Therefore, the PUF function can be represented with a Challenge-Response Pairs (CPRs) as follows:

$$P(C) = R \quad (1)$$

Each of the obtained PUF response values R will be compared and analyzed with the most represented values via Temporal Majority Voting (TMV). The difference in each bit of the R is called the Intra-distance D_{intra} [16]. In other words, D_{intra} is the error bit of the response value R and it can be used to get the error rate E of the SRAM PUF. The error rate can be represented as the fraction of the number of different bits over total number of bits n in percentage.

$$E = \frac{D_{intra}}{n} \times 100\% \quad (2)$$

The uniqueness of the PUF can be represented by the value of Inter-distance D_{inter}. The inter-distance D_{inter} is the difference between two PUF Responses of equal number of bits from different, but the same type of, chips [16]. That can be calculated from the response values R obtained from two or more chips using the same challenge C. The formula to calculate D_{inter} is as follows.

$$D_{inter}(R_P, R_Q) = \Delta(R_P, R_Q) \tag{3}$$

Similar to error rate, the inter-distance of the PUF function can also be represented using the fraction over the number of bits n [28]. The fractional inter-distance I can be calculated using the following formula.

$$I(R_P, R_Q) = \frac{\Delta(R_P, R_Q)}{n} \tag{4}$$

Another useful attribute of PUF is the biasness towards either 0 or 1 which is the ratio between the number of 0 s and 1 s in a stream of bits. This is also known as Fractional Hamming Weight W. It can be calculated using the following formula.

$$W(R) = \frac{\#(\mathrm{i} : R_i \neq 0)}{n} \tag{5}$$

Stability S of the SRAM PUF can be evaluated from the fraction between the number of stable bits (the bit that never changes its value during all iterations), and the total number of bits. The stability can be represented as follows.

$$S = \frac{\#(\mathrm{i} : E_i = 0)}{n} \tag{6}$$

The evaluation of SRAM PUF characteristics is illustrated in Fig. 2. In this paper, the error rate E, the fractional inter-distance I and the bias W will be analyzed to find the optimal condition for each of the systems. The results will be used for comparison and benchmarking of the error rate reduction using data remanence algorithm discussed in this paper.

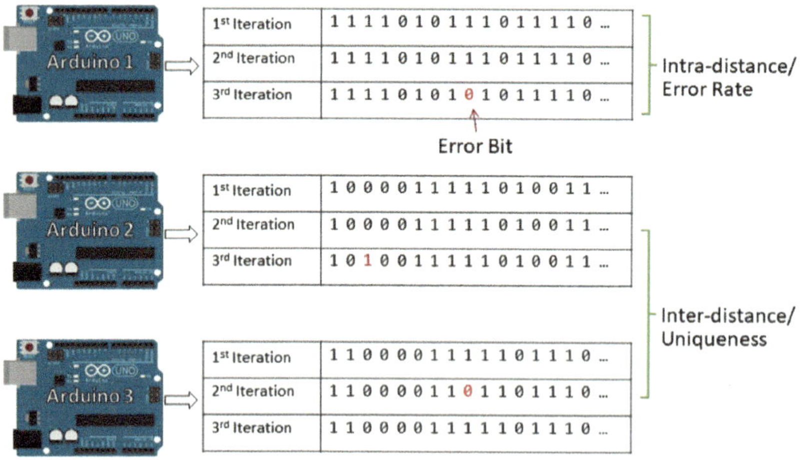

Fig. 2. Illustration of calculating SRAM PUF characteristics for each Microcontroller

3.2 Evaluation Results

The uninitialized SRAM PUFs from various microcontrollers have been extracted and the SRAM Characteristics discussed in Sect. 3.2 have been evaluated by using TMV. 256 Bytes of SRAM PUF responses for each of the microcontrollers were recorded for more than 100 iterations to calculate the error rate, biasness, uniqueness and the stability. The experiment was done under room temperature and the following microcontrollers has been analyzed. The results of the analysis are listed in Table 1.

1. Six AtMega328P Microcontrollers from Arduino UNO boards,
2. Four AtMega328 SMD Microcontrollers from Arduino Nano Boards,
3. Five STM32F103 Microcontrollers from STM32 Blue-Pill Development Boards,
4. Three ESP8266 Microcontrollers from ESP8266 Node MCU.

Table 1. Comparison of SRAM PUF characteristics on different Microcontrollers

		AtMega328	$AtMega328_{SMD}$	STM32	ESP8266
Error rate	E_{AVG}	1.817%	1.768%	3.338%	2.342%
	E_{Max}	2.393%	2.589%	5.273%	6.457%
	E_{Min}	0.270%	1.054%	1.758%	1.078%
Biasness	W_{Avg}	62.071%	61.097%	50.885%	50.002%
	W_{Max}	62.824%	61.914%	52.930%	51.549%
	W_{Min}	61.084%	60.302%	48.828%	49.061%
$Uniqueness_{Avg}$		40.881%	41.996%	48.367%	50.287%
$Stable\ bits_{Avg}$		90.004%	91.504%	78.532%	81.447%

Those microcontrollers are selected because they are used in a lot of IoT applications in the market. They also have relatively similar in SRAM size. Both AtMega328 Microcontrollers have similar characteristics although their physical size differs a lot. Although the SRAM PUF of two AtMega328 Microcontrollers are relatively more stable, their biasness is not very well uniformed. With the average biasness of more than 60%, the SRAM PUF of the AtMega328 microcontrollers are biased towards '1'. However, those of STM32 and ESP8266 are well uniformed with average biasness of around 50% while their stability is significantly less.

All in all, the SRAM PUFs of all the platforms can produce promising results to be used for unique security key generation. The biasness and the occurrence of error bits are normally distributed throughout the iterations. Figures 3 and 4 illustrate the distribution of Error Rate Biasness for an AtMega328 Microcontroller respectively.

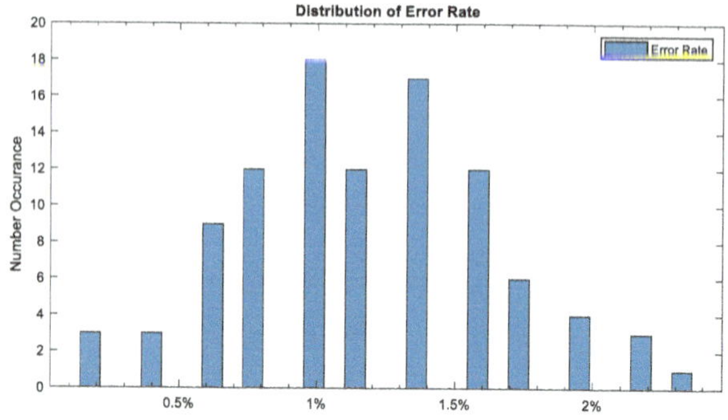

Fig. 3. Distribution of error rate on AtMega 328 Microcontroller

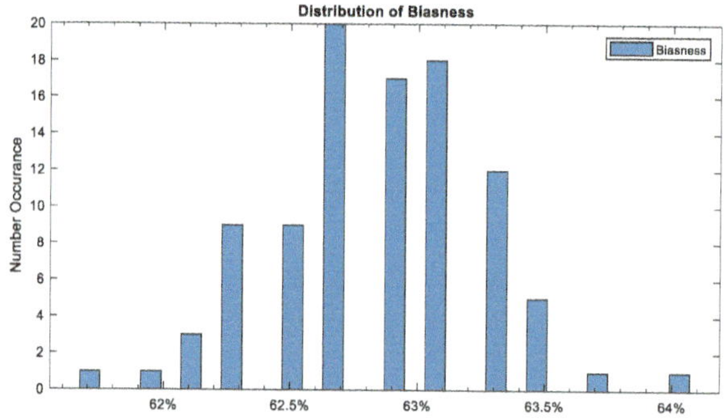

Fig. 4. Distribution of biassness on AtMega 328 Microcontroller

Figure 5 plots the Error Rate for first 100 iterations of an AtMega238 Microcontroller. The error rate is ranging from 0.2% to 2.4% but it never goes to zero. It means that there are always a few bits flipping for each iteration and therefore the stability of 100% could not be achieved using the TMV method alone.

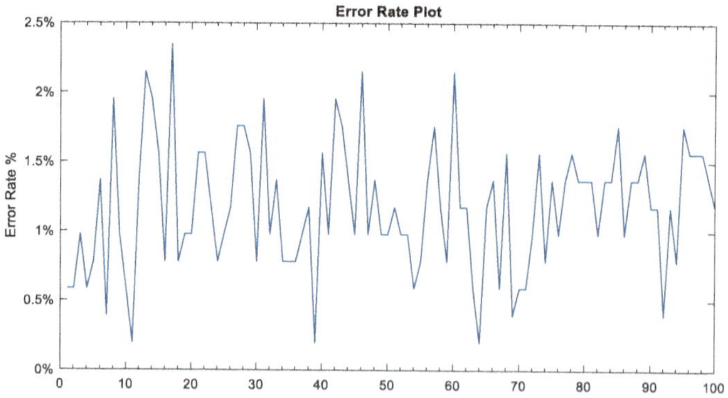

Fig. 5. Error rate of an AtMega328 Microcontroller for one hundred iteration

4 Obtaining Stable SRAM Bits

4.1 Data Remanence Algorithm

One of the characteristics of SRAM PUF is the remanence of data in the memory after power down for a short period of time [14]. This is due to the internal capacitance of the transistors in each memory cell. This phenomenon can be utilized to obtain a more reliable PUF data by setting and resetting of the whole memory array before reading the initial SRAM values. To obtain strong 0 SRAM cell, all the SRAM cells are written the value of '1's at first. And then, the SRAM cells will be powered down for the time T. The SRAM is then powered up again and the data is read. By doing so, only the strongest '0' cells will flip to the value of '0' in the data stream.

The strong 1 SRAM cells can also be identified using similar method by setting all the cells to 0s. The process has to be repeated several times to obtain the most reliable results. This method can reduce the testing time to identify strong 0 and strong 1 SRAM cells compared to ordinary means of extracting SRAM PUF such as using Temporal Majority Voting. Figure 6(a) shows the block diagram of the system that has been used to obtain the strong '1' and strong '0' SRAM PUF. And Fig. 6(b) illustrates the algorithm to obtain those values.

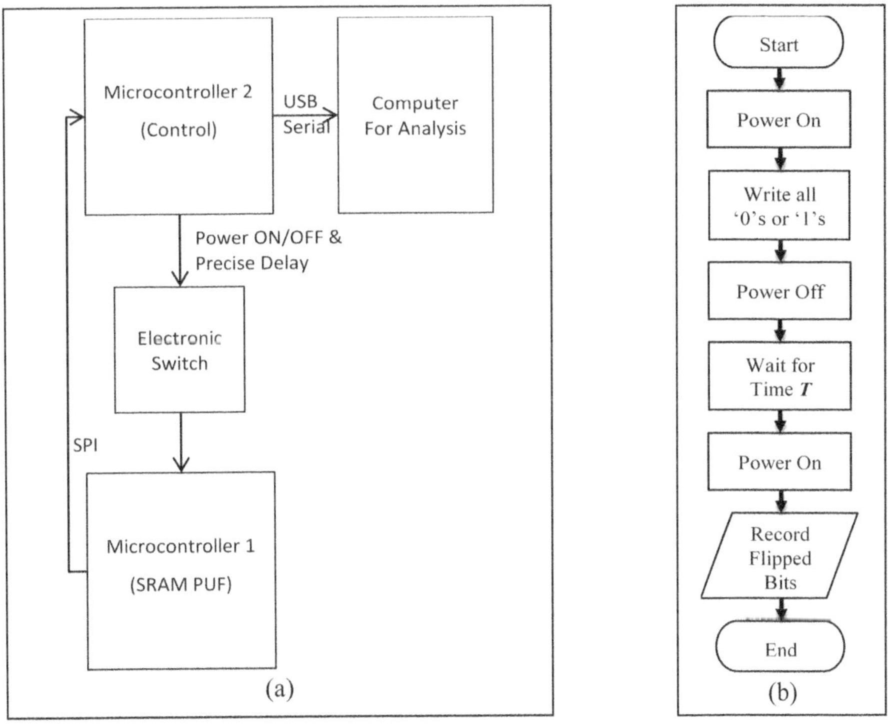

Fig. 6. (a) Block diagram of the system for obtaining Strong '0's and Strong '1's using two microcontrollers, (b) Algorithm used for obtaining the Strong '0' and Strong '1' bits

4.2 Obtaining Strong '0's and Strong '1's from SRAM PUF

In order to use data remanence method for stable SRAM PUF, the power on time and the power off time must be controlled precisely and independently. So, an additional microcontroller was used between the microcontroller under test and the computer to obtain Strong '0's and Strong '1's. 512 bits (64 Bytes) of initial SRAM values from four AtMega328 Microcontrollers were recorded under different power down times *T*s. Twenty iterations have been done for each value of *T* and the data from a microcontroller is shown in Fig. 7. At the *T* of around 280 ms, some of the bits start flipping and they can be considered as the strongest bits. The memory addresses of both strong '1's and strong '0' for various power down times are recorded and compared with the initial SRAM values for later calculations.

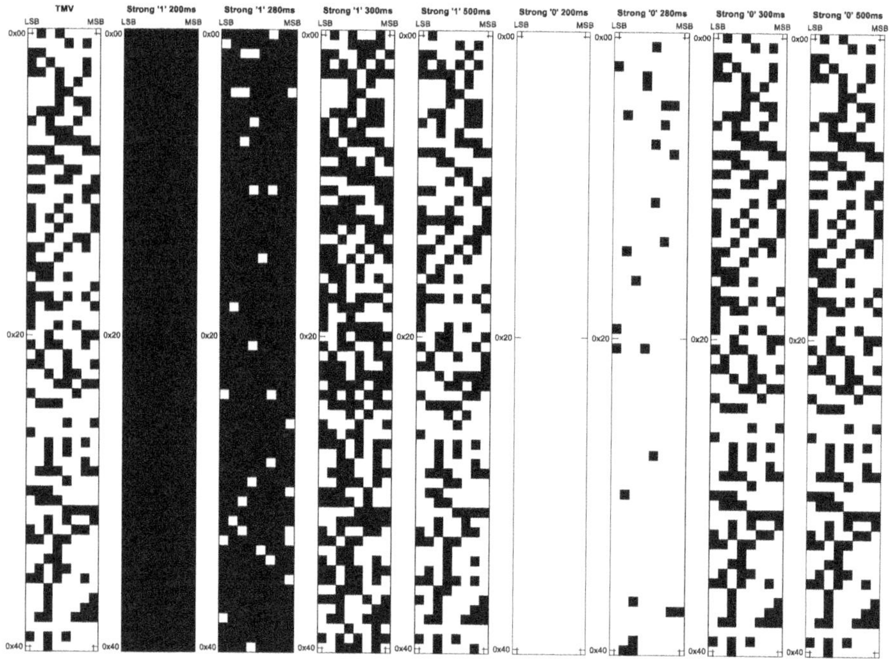

Fig. 7. SRAM PUF data at various power down times

To calculate the error rate and stability of the SRAM PUF, 128 selected strong bits, i.e. the bits which flips, out of 512 bits were recorded with their memory cell addresses at the power down time *T* of 300 ms. They are then compared with the initial SRAM values of the respective microcontrollers on those specific addresses. Similar steps have been done for both strong '1' and strong '0' bits at $T = 300$ ms and $T = 500$ ms respectively. 192 bits were used for the *T* value of 500 ms. The results are summarized in Table 2.

The individual strong '1' and strong '0' tests have the best results at $T = 300$ ms. When both of 128 strong '1' bits and 128 strong '0' bits were used at the same time, we managed to obtain the error rate of 3.77×10^{-8} (3.77×10^{-6} %) and the stability of more than 99.98%. This method used fewer number of iterations and consuming less time than TMV method and it is clear that this method can achieve significant error rate reduction and stability improvements over the TMV method.

Table 2. Summary of SRAM characteristics under various power down times

	TMV	Strong '1' T = 300 ms	Strong '1' T = 500 ms	Strong '0' T = 300 ms	Strong '0' T = 500 ms	Strong '1' & Strong '0' T = 300 ms
Error rate	1.817%	0.0138%	0.3105%	0.0293%	0.546%	3.77×10^{-6} %
Biasness	62.071%	38.086%	46.680%	66.992%	65.039%	*NA*
Stability	90.004%	99.805%	97.852%	99.707%	97.266%	99.983%

5 Conclusion

Although the SRAM PUF has the potential to be used for device authentication and security purposes, the performance of the SRAM PUF is yet to be investigated intensely for it to be used as a reliable security protocol. The various characteristics of SRAM PUF from different microcontrollers, which are being used in IoT applications these days, have been evaluated.

The SRAM PUFs on microcontrollers have comparatively high error rate and instabilities. But those errors can be reduced significantly by utilizing data remanence algorithm that chooses the preferable memory cell to achieve the desirable performance. We set up the evaluation system using two microcontroller boards to choose strong 1 s and 0 s conveniently. As a result, the error rate of less than 10^{-7} and the stability of more than 99.9% have been achieved. This result shows that this remanence algorithm has a potential to achieve secure device authentication for IoT devices with microcontroller and on-chip/off-chip SRAMs while eliminating the need for complicated ECC algorithm.

References

1. Cilardo, A., Barbareschi, M., Mazzeo, A.: Secure distribution infrastructure for hardware digital contents. IET Comput. Digit. Tech. **8**(6), 300–310 (2014)
2. Kemp, S.: Digital in 2017: Global Overview, 24 January 2017. https://wearesocial.com/special-reports/digital-in-2017-global-overview. Accessed 20 Oct 2018
3. Aman, M.N., Chua, K.C., Sikdar, B.: Position paper: physical unclonable functions for IoT security. In: Proceedings of the 2nd ACM International Workshop on IoT Privacy, Trust, and Security, Xi'an, China, pp. 10–13. ACM (2016)

4. Noh, J., et al.: Secure key exchange scheme for WPA/WPA2-PSK using public key cryptography. In: 2016 IEEE International Conference on Consumer Electronics-Asia (ICCE-Asia) (2016)
5. Chatterjee, U., Chakraborty, R.S., Mukhopadhyay, D.: A PUF-based secure communication protocol for IoT. ACM Trans. Embed. Comput. Syst. **16**(3), 1–25 (2017)
6. Maes, R.: Physically unclonable functions: concept and constructions. In: Maes, R. (ed.) Physically Unclonable Functions: Constructions, Properties and Applications, pp. 11–48. Springer, Heidelberg (2013)
7. Sklavos, N.: Securing communication devices via physical unclonable functions (PUFs). In: Reimer, H., Pohlmann, N., Schneider, W. (eds.) ISSE 2013 Securing Electronic Business Processes: Highlights of the Information Security Solutions Europe 2013 Conference, pp. 253–261. Springer Fachmedien Wiesbaden, Wiesbaden (2013)
8. Holcomb, D.E., Burleson, W.P., Fu, K.: Initial SRAM state as a fingerprint and source of true random numbers for RFID tags (2007)
9. Böhm, C., Hofer, M., Pribyl, W.: A microcontroller SRAM-PUF. In: 2011 5th International Conference on Network and System Security (2011)
10. Vijayakumar, A., Patil, V., Kundu, S.: On improving reliability of SRAM-based physically unclonable functions. J. Low Power Electron. Appl. **7**(1) (2017)
11. Lipps, C., et al.: Proof of concept for IoT device authentication based on SRAM PUFs using ATMEGA 2560-MCU. In: 2018 1st International Conference on Data Intelligence and Security (ICDIS) (2018)
12. Xu, X., et al.: Reliable physical unclonable functions using data retention voltage of SRAM cells. IEEE Trans. Comput.-Aided Des. Integr. Circuits Syst. **34**(6), 903–914 (2015)
13. Das, J., et al.: MRAM PUF: a novel geometry based magnetic PUF with integrated CMOS. IEEE Trans. Nanotechnol. **14**(3), 436–443 (2015)
14. Liu, M., Zhou, C., Tang, Q., Parhi, K.K., Kim, C.H.: A data remanence-based approach to generate 100% stable keys from an SRAM physical unclonable function. In: 2017IEEE/ACM International Symposium on Low Power Electronics and Design (ISLPED), Taipei, pp. 1–6 (2017)
15. Deutschmann, M., Iriskic, L., Lattacher, S.L., Münzer, M., Tomshchuk, O.: A PUF based hardware authentication scheme for embedded devices, pp. 1–18. Technikon (2017). http://www.technikon.com
16. Deutschmann, M., Iriskic, L., Lattacher, S.L., Münzer, M., Stornig, F., Tomashchuk, O.: Research on the applications of physically unclonable functions within the Internet of Things. Technikon, pp. 1–9 (2018). http://www.technikon.com

Generic 5G Infrastructure for IoT Ecosystem

Saeed Khorashadizadeh[1(✉)], Adeyemi Richard Ikuesan[2], and Victor R. Kebande[3]

[1] Faculty of Computing, Universiti Teknologi Malaysia, Skudai, Malaysia
saeedkh@live.com

[2] Department of Cybersecurity and Networking, School of Information Technology, Community College of Qatar, Doha, Qatar
Richard.ikuesan@ccq.edu.qa

[3] IoTaP Research Center, Department of Computer Science & MT, Malmö Universitet, Malmö, Sweden
Victor.kebande@mau.se

Abstract. While the Internet of Things (IoT) is still gaining rapid adoption in an upward trajectory means across many smart areas in recent years, still, there is a need to develop a scalable ecosystem that is able to support future IoT implementations, given the heterogeneity and increased information flow among IoT devices. The lack of effective interoperability, availability, reliability, and performance in IoT are a few challenges that hinder the effectiveness of IoT deployment and communication, and this has acted as a stumbling block for the optimisation of IoT platforms. That notwithstanding, the advent of the Fifth Generation (5G) networks, has seen a significant shift on how the IoT-paradigm operates. This article introduces a discussion on the generic 5G infrastructure for IoT environment that can support future deployments. The article begins by conducting a technical review of the evolution of the First generation (1G) through 5G cellular networks. Thereafter, an illustration of the Fourth Generation (4G) architecture, stating its features as a precedent that shows to what extent 5G network may thrive or support IoT ecosystems, is given. Furthermore, the paper explores both the architectural requirements and futuristic vision of 5G infrastructure in the perspective of IoT applications. Most importantly, the paper has also explored the IoT market share and forecast on growth and how it affects different industrial sectors. The authors believe that the conclusions that have been made in this paper will act as a pacesetter and give a direction worth exploring once the 5G infrastructure for IoT ecosystems is implemented.

Keywords: IoT · Ecosystem · Industrial Internet of Things · Fourth Generation (4G) · Fifth Generation (5G) · Smart city · Healthcare · Telehealth · Telemedicine

1 Introduction

The prevalence of outstanding innovations and increased desire to unify and connect everything has meant that a rich variety of Information and Communication Technology (ICT) constitute the Internet of Things (IoT) platform. This composition combines diverse heterogenous dependencies in order to establish the possibilities of

F. Saeed et al. (Eds.): IRICT 2019, AISC 1073, pp. 451–462, 2020.
https://doi.org/10.1007/978-3-030-33582-3_43

interconnectedness in almost all 'things', everywhere and at any time. Importantly, such connectivity allows IoT devices to be able to receive, transmit, analyse and transcribe data into knowledge, actions, and decisions successfully. This need of constantly coming up with the next effective 'thing' in a connected atmosphere has also seen an influx of ICT-based technologies, that have been integrated with IoT as a result. Examples of such technologies that have seem massive integration as a result include cloud computing, big data analytics, Artificial Intelligence (AI), edge computing, and fog computing. Furthermore, the integration of these technologies has created vast opportunities for individuals and enterprises, which has also led to the creation of useful use-cases and applications that have been able to take advantage of the usefulness that comes along [1].

Worldwide spending of IoT is projected to shift given how fast different technologies are being integrated to IoT. For example, the International Data Corporation (IDC) [1], projects that spending on IoT will reach $745 Billion by 2019, while Gartner forecasts that AI will be the driving factor behind IoT and by the year 2020, there is an estimation that over 25 billion devices will be connected. Notably, the success and proliferation of IoT has been attributed to miniaturisation and pervasiveness [5] of the Internet and its effectiveness to human endeavors. It is worth noting that at the heart of successful IoT implementations, data is considered as a very important commodity. Therefore, in order to transform data from an IoT environment into useful information, it is important to ensure 'things' interoperability and the safety of the sensing environment. This requires proper network communication and coordination of devices in order to achieve a more effective environment. Through these IoT-based applications, technology providers may be able to empower the lives of the end-users in diverse areas.

The main objective of the paper is to propose a generic 5G infrastructure for IoT ecosystems which is pertinent to the device proliferation that is being experienced presently. Whilst existing studies have provided technical and empirical detail of IoT devices, platforms, as well as application, to the best of the authors' knowledge, this study is the first to provide a baseline for understanding the market implication of the 5G network in IoT domain. Furthermore, this study provides a holistic perspective through which the impact of 5G can be quantified in relation to the previous generations of networks; i.e., 1G, 2G, 3G and the 4G.

The rest of the paper is organised as follows: Sect. 2 gives a review on wireless technologies in IoT. This is followed by Sect. 3 that gives a study on IoT market share and future growth. Section 4 gives a discussion of the study while Sect. 5 gives a conclusion.

2 Literature Review

Though various wired technologies had come into play in order to establish the connectivity between IoT devices with a promising level of reliability and security, it is important to note that such networks have in most cases been unable to address several issues in IoT networks. For example, issues like mobility and flexible connection of every device from everywhere at any time, have had some concerns which have been least or not addressed. Hence, that presents no choice but to shift to wireless communications. Technologies like Wi-Fi, Bluetooth, IEEE 802.15.4, 4G and 5G can be used to connect to devices in IoT networks. Therefore, this section concentrates on discussing the following: Wireless network challenges in IoT, evolution of IG through 5G, 4G and 5G respectively.

2.1 Wireless Networks Challenges in IoT

Common wireless connectivity concerns in heterogenous environments that affects IoT networks include, but are not limited to, interoperability, self-management, maintainability, signaling, bandwidth and power consumption, which is further summarised in Table 1 [2]. Table 1 shows the existing wireless network challenges in IoT and a description of each challenge is given too.

Table 1. Wireless networks challenges in IoT.

Challenge	Description
Interoperability	Ability and flexibility to connect and communicate with different network interfaces
Self-management	Human independent, ability to adapt to failures and changes
Maintainability	Ability to identify various services due to its level of strength and status
Signaling	Reliability in the bidirectional stream of data
Bandwidth	Sufficient bandwidth and practical bandwidth management
Power consumption	Minimal battery consumption in wireless communication

2.2 Evolution of 1G Through 5G Cellular Networks

In the last few decades, several cellular networks have been developed, one after the other, in order to deliver better speed, latency, frequency, and technology. This has been done to improve communication and more importantly, the focus has mainly been on speed and the technologies used. For example, speed and technology, has evolved remarkably from the First Generation (1G) with 2 Kbps which is relatively slow, to the Fourth Generation (4G) with the speed of 1 Gbps. However, 4G has other multiple technologies such as Wi-Max and Long-Term Evolution (LTE). Additionally, significant features have been developed upon the release of the 4G relative to its predecessors. Moreover, the 3G also has comparable features. These features include being able to adjust mobile web access, enhancing IP telephony, gaming services, high-definition mobile TV, video conferencing, 3D television, and cloud computing. A detailed comparison of the generations of cellular network is outside the scope of this article; however, the main characteristics of each generation is summarised in Table 2.

Given the enormous advantages of a fast 5G cellular network over other cellular networks, consideration would be given to 5G. Nonetheless, for a better understanding

Table 2. A comparison of cellular network generations.

Generation	Deployment	Bandwidth	Key improvement
1G	1970–80	2 Kbps	Foundational structure and legacy architecture
2G	1990–04	64 Kbps	Introducing data services
3G	2004–10	2 Mbps	Mobile broadband access
4G	2010–Present	1 Gbps	WLAN integration
5G	~2020	>1 Gbps	Low-latency, high availability

of the magnitude of the development in the 5G, first, a closer look at 4G network architecture and structure is required. This is therefore presented in the next section.

2.3 Fourth Generation (4G)

Fourth Generation (4G) or often called Long-Term Evolution (LTE) is a huge enhancement as compared to the earlier generations in almost every aspect. 4G offers new business opportunities to IoT-based technologies leveraging its high level of connectivity to the Internet. The native all-IP architecture of 4G makes it ideal for IoT applications, and the built-in security along with robust and scalable traffic management capabilities that is able to provide ubiquitous services. This service can be expanded multimedia services as well as a low-cost per-bit, which makes it the best candidate for an IoT network infrastructure [6].

From a network architecture point of view, the 4G network is very similar to the earlier generation, 3G. However, there are some major improvements that are worth further emphasis. Circuit-switching is removed in 4G, but, the backward compatibility is maintained by segmenting voice data into packets and routing them through the IP backbone using (Voice Over IP) VOIP technology [3, 4]. Another main enhancement in 4G networks is the integration of Wireless LANs (WLAN) into the mobile network, in which flexibility, scalability, and simplicity increased remarkably. As a result, enterprises can scale their network with simpler network architecture, less cost, and faster data transfer rate.

The advantage of using the 4G network is not stopped in its architecture only, as larger bandwidth means better Quality of Service (QoS). As a result, several new services become available as users can have high-resolution data transfer, high-quality video-conferencing, and the capability of creating ad-hoc network structure. 4G networks also support IPv6, which significantly increase the capacity of hundreds of device connectivity [8].

However, there are some issues and limitations in 4G networks which limits it applicability as a networking-platform for a fully IoT infrastructure deployment. The 4G network consists of various wireless technologies and such diversity despite the advantages, creates multiple issues too. First, the interconnection of such heterogeneous networks is very challenging as each wireless network has different protocols and architectures for transport, routing, mobility management, and so forth. Second, high QoS is not guaranteed by transferring data in multiple wireless network infrastructures. Finally, maintaining a high-level of security by monitoring, analysing, and software patches in mixed embedded wireless technologies in the 4G network are indeed very difficult. As a result, the key requirements for IoT implementation such as interoperability, high availability, low-latency, etc. are not addressed well in 4G cellular networks.

2.4 Fifth Generation (5G)

Fifth Generation (5G) started in the late 2010 with the aim to provide far better levels of connectivity and coverage and to focus on World-Wireless World Wide Web (WWWW). The 5G network is an all-IP based architecture, which promised that all information and services are carried by a single network transparent. Hence, a user terminal and a number of independent autonomous Radio Access Technologies (RAT) [5] supply services through Cloud Computing Resources (CRR), which provides on-demand network access to configurable computing resources in 5G platforms [9].

In addition to the IP based architecture, 5G networks are needed to support eight major requirements [9]. These requirements include $\approx$10 Gbps data rates, $\leq$1 ms round trip latency, high bandwidth, facility to connect a massive number of devices, having an availability of 99.999%, providing 100% coverage, a significant reduction in energy usage by 90% and finally high battery life. And 5G networks also must be able to address scalability by Network Function Virtualisation (NFV), network virtualisation capability, and improve complex network management by mobility control, access control, resource efficient network virtualisation and big data analysis to support smart services.

In the context of industry and society, 5G is projected to provide an age of boundless connectivity and intelligent automation by providing sophisticated integration of multiple technologies like cloud computing with the delivery of vast communication services. Additionally, 5G will develop under the umbrella of ubiquitous internet access vision to create the networks and platforms to digitise economies, societies, and to address the five mobile industry objectives that are listed next [10]:

1. Provide limitless connectivity for all: 5G will co-exist with 4G networks and other technologies to deliver high speed, reliable and secure experience to provide massive use-cases for development.
2. Deliver prime economic future networks: To develop and implement IoT solutions, 5G vision is to deliver not only far better network performance but also significantly cost-effective approaches. This will rely on both licensed and unlicensed solutions.
3. Speed up the digital transformation of industry verticals: Providing the networks and infrastructures to operate the digitisation and automation of industrial applications.
4. Revolutionise the mobile broadband experience: Providing a massive fast speed with the less than 10 ms latency.
5. The novelty of IoT use-cases and critical communications services: 5G will establish intelligent IoT nodes for a great number of frameworks and also provide infrastructures for the adoption of critical communication services.

3 IoT Ecosystem Applications

Advancements in technology have paved the way for utilising IoT in a vast range of domains. Every single day, a new case study or a new product is introduced either by individuals or industries, that illustrates what we have to do to enhance the quality of our lives. According to [7], IoT will connect more than 20 billion devices to the current internet infrastructure by 2020. That will significantly revolutionise the way people live, work, travel, and organisations interact in the near future. Additionally, the market growth, which started at around 15% in 2016 will be estimated to reach more than 40% in 2020, (See Fig. 1). From Fig. 2, it is evident that the global IoT market share will be dominated mainly by Smart Cities (26%), the Industrial IoT (24%), and healthcare (20%).

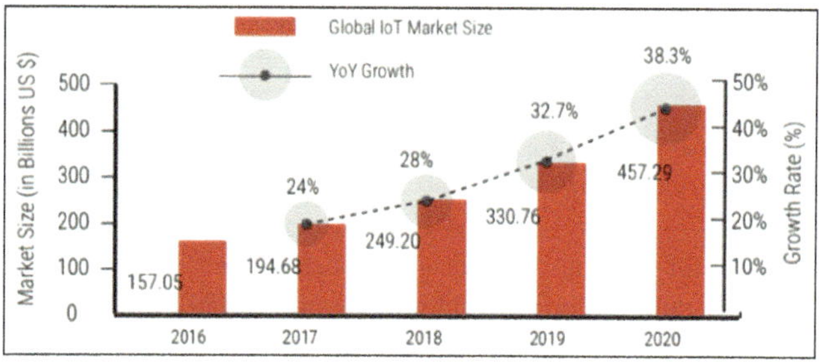

Fig. 1. Market size and growth forecast [7]

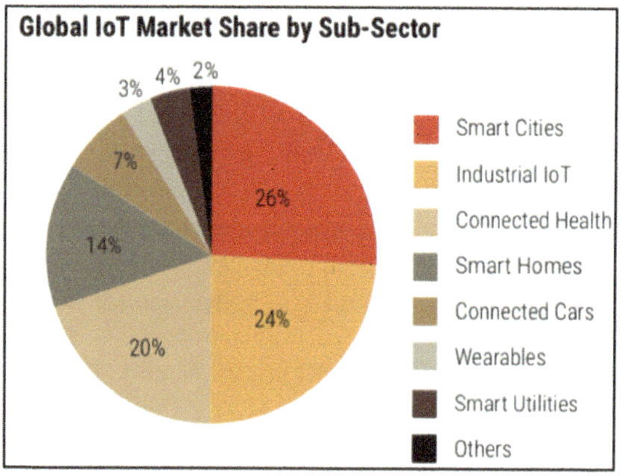

Fig. 2. Global IoT market share [7]

3.1 Smart Cities

Several studies and technologies have been developed to enhance the life experience in cities around the world in the most intuitive and creative ways. Create urban societies with an emphasis on efficiency and performance in energy, transportation, and utilities. In that essence, several uses-cases and scenarios have arisen under the context of the smart city such as [11]: Smart parking, smart traffic management, smart street lighting, and smart waste management, etc., which intend to provide better cost-effective welfare communities, (see Table 3).

Table 3. Smart city applications and use cases.

Application	Use case
Public services	Citizen services, tourist services, public transportation, information services
Transportation	Smart roads, connected vehicle, smart parking, smart traffic, noise, and pollution
Public safety	Smart lighting, environment, asset tracking, video surveillance, emergency response
Sustainability	Environmental monitoring, smart waste management, smart energy, smart metering, smart water
Integrated smart functions	Smart care, smart education, smart governance, smart planning, smart/open data
Infrastructure	Smart infrastructure, structural health, smart irrigation

Such a futuristic perspective of the smart city is indeed demanding a high level of consideration and expectation of the future generation of cellular networks, 5G, which elaborately described in the previous section. Without discounting the aforementioned, there is no doubt that 5G networks with the expected key features will make the adaptation of IoT technology on a large scale [13]:

- Massive device connectivity means connectivity of homes, streets, traffic junctions, public places, railway stations, airports, etc., which potentially support smart traffic systems, smart homes, public safety, and security for the smart cities.
- Large bandwidth indeed will deliver both upload and download of video-rich services to support various uses cases in smart cities. For example, by processing the raw video footages from the street junctions by special video analytics, we would be able to detect, classify, and track cars, motors, and pedestrians.
- Low latency in 5G networks potentially delivers the possibility of driverless cars for example. High availability of 5G networks not only create a high level of mobility environments for cars and fast trains but also provide high-reliability requirements for driverless cars and traffic monitoring.

3.2 Industrial Internet of Things (IIoT)

The Industrial Internet of Things (IIoT) aims to provide a new spectrum in automation, industrial control, intelligent manufacturing, smart industry, asset performance management, and optimisation. In addition, on-demand service model in IIoT is projected to create both a new way of servicing customers and new revenue models for enterprises [11].

Improved operational efficiency and productivity, creation of new business opportunities, reduced downtime, and maximising asset utilisation are some of the top objectives behind the development of IIoT on a large scale [15]. In that essence, the use cases such as smart factory applications and smart warehousing, goods and transportation monitoring, connected logistics, manufacturing equipment monitoring, asset tracking, and smart logistics, etc. are introduced and possible challenges and solutions are highlighted along the way. That leads us to the Fourth Industrial Revolution (4IR) since the initial Industrial Revolution of the 18th century [15], which will emerge by arising technologies such as robotics, AI, IIoT, 5G cellular networks, quantum computing, biotechnology, etc.

In the scope of this paper, the authors take a closer look at five core characteristics of 5G networks that make 4IR is an absolute reality in the near future [12].

1. **Flexibility**
   a. *Increased customisation:* With fast and reliable communication between machinery, sensors and computing systems, we can have a smoother adjustment for the production line while meeting consumer demands.
   b. *Supply chain:* Having servers in the edge cloud, communicating with sensors and machines, we will be able to automate several systems and processes, which would lead to complete control from material handling to production and inventory management and as the result, reduce assembly time and enhance overall productivity.
   c. *Assembly flexibility:* Wireless connectivity of 5G makes it possible to reconfigure the production systems to cope with changing product lines with far more flexibility than wired networks would ever provide.
2. **Efficiently**
   a. *Increased value:* Not only the technical operational work becomes more efficient, but also a study shows that the total cost of ownership of factory operating by wireless connectivity is 5 times lower than the wired.
   b. *Reconfiguring on the fly:* At any point in time, wireless connectivity makes it possible to reconfigure the machines with ease and help for future-proof of the production line.
   c. *Improved reliability:* Promising five-nines reliability (99.9999%) by 5G networks means you can have a peace in mind for even critical operational lines and with the flexibility to increase or decrease the bandwidth to match the required demand.

3. **Productivity**
   a. *Improved productivity:* 5G can improve the productivity with automated information and knowledge sharing and with low latency and high reliability we can mitigate errors and provide new digital services as added value for customers.
   b. *Streamlined maintenance:* Faster and safer maintenance is now possible by having off-site specialists to communicate with local engineers in real time, using virtual or augmented reality to supervise the maintenance operations.
   c. *Guaranteed production:* Isolating critical operations on a local private network to mitigate the loss of connectivity to external networks will ensure production is maintained.
4. **Safety and Security**
   a. *Emergency shutdown:* High reliability and low latency of 5G connectivity mean immediate intrusion detection enables you to shut down machinery quickly to avoid possible equipment damage and ensuring worker safety.
   b. *Network security:* Having servers in the edge cloud on your physical premises to maintain and secure easier and traffic management to isolate critical traffics from the other traffics.
5. **Technology and Services**
   a. *Technology:* By having the 5G infrastructure, we able to have advanced analytics to control and manage the entire network through Software Defined Network (SDN) and Network Functions virtualisation (NFV).
   b. *Services:* Accelerating network services to provide less complex design and deployment along with new network business models.

3.3 Healthcare

While visiting clinics and hospitals for city residences when a patient gets ill, it seems very common and acceptable, but for patients who are living in rural areas and require to travel long distance is very challenging and life-threatening in some cases. With the advent of telehealth and remote home monitoring systems, though, we could bring doctors and specialists to patients. However, the complexity of such equipment and infrastructure are the major challenges in the healthcare industry. But, 5G technology with promising the delivery of high reliability, availability, large bandwidth, and low latency seems to have the potential solutions for all these requirements.

In [14] five factors are highlighted that 5G will transform the healthcare industry significantly:

1. *High speed:* Transmitting large image files generated by MRIs and other medical equipment to specialists for review.
2. *Low-latency and large bandwidth:* Expanding telemedicine, in which requires high-quality real-time videos to handle telemedicine apportionments.
3. *High availability:* Overcome the language barriers for patients by communicating in other languages would increase extremely the availability of doctors and specialists.

4. *High availability and low-latency:* Real-time remote monitoring and gathering data can be used to improve personalised and preventive care by utilising IoT devices, sensors and wearables, which the latter significantly becomes popular in recent years.
5. *Embedded technologies:* Without any doubt, Big data and Artificial Intelligence (AI) play a major role in future healthcare where AI will determine potential diagnoses and decide on the best treatment plan for patients and early interventions when necessary.

In another study [14] healthcare is categorised in different use cases as is shown in Table 4, and the impact of a 5G network is analysed for each group individually. In which, patient and hospital applications will develop and empower under unique features of 5G networks as described before and AI and IoT help to analyse data and managing medical data records to generate valuable information that could contribute not only to market plans, sales development for players but also to product creation and medical service improvements.

Table 4. Healthcare major use-cases.

Use case	Description
Patient applications	Patient-centered applications used outside hospital areas
Hospital applications	Applications used in hospitals as medical equipment
Medical data management	Systems to manage and analyse patient records and other medical data
Other	Applications with the aim to facilitate quick access to medical assistance

Through enabling all these technologies, the overall healthcare systems will provide a cost-effective high quality of care for all patients regardless of their location and the level of accessibilities.

4 Discussion

The pervasive nature of IoT applications and networks is gaining wider utilisation and adoption in the human day-to-day activity. This growing trend present myriad opportunities for developing nations, particularly in Africa. However, such growing opportunities presents a medium for security concerns. Security and privacy issues are major challenges in the IoT space and this has also been identified in the study on Virtual Internet of Threats (VIoT) introspection, which has identified the security and privacy issues as a result of sensor proliferation. Ranging from the application of IoT in transportation such as the intelligent transport services, and intelligent traffic controller, to smart health as depicted in Fig. 3; security and privacy present a ranging challenge in IoT platforms. Whilst the inherent vulnerabilities in the typical 1G, 2G, 3G, and 4G have been widely studied, the challenges introduced by the speed of 5G presents a

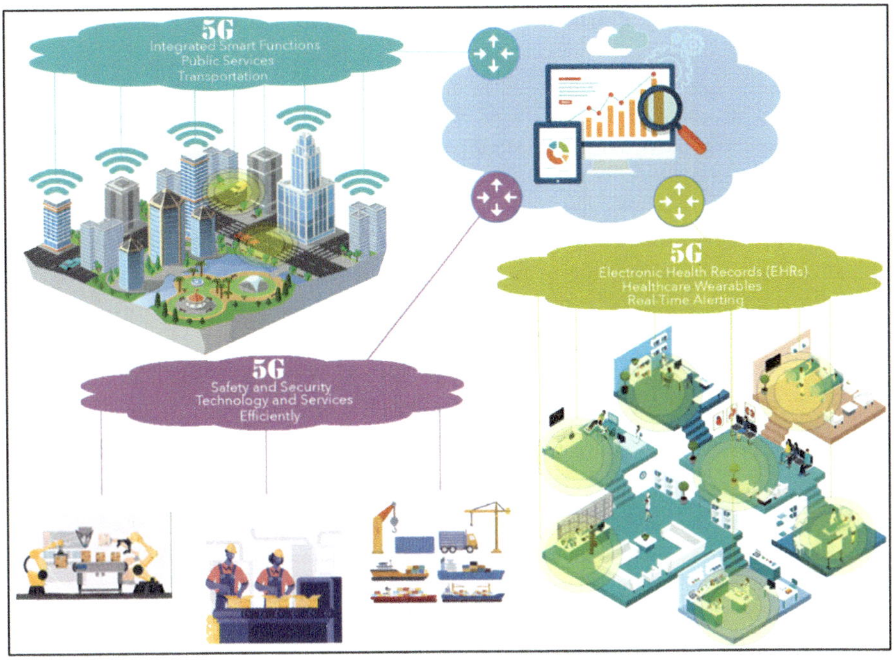

Fig. 3. Sample of IoT Application in a 5G Network

higher threat factor for the IoT platform. For instance, the speed of data transmission in an IoT platform could be leveraged for data exfiltration in any IoT-platform, without a 'noticeable' degradation in data rate. Therefore, greater emphasis will be required to address the potential security challenges in the 5G enabled connectedness. The advantages of the 5G cellular network can span beyond the immediate application of human day-to-day activities. On the IIoT platform, the integration of 5G cellular network in industrial settings such as Supervisory Control and Data Acquisition (SCADA) systems, Cyber-Physical Systems (CPS), critical infrastructure systems, and Programmable Logic Controller (PLC) platforms further opens the spectrum of application of IoT for exploration. This is particularly enhanced with the growing trend of software-based networking platform; the Software Defined Networks (SDN) for instance paves way for enterprise network connectivity. Whilst high speed could enable remote and faster system shutdown in an attack scenario, 5G could potentially enhance the propagation of malicious and subtle attacks. It is the author's opinion that the generic 5G infrastructure meant for IoT ecosystems presents a more open approach that allows formidable exploration of core issues and challenges that could help to harden IoT from the aforementioned threats. Furthermore, the 5G infrastructure provides a platform for an effective implementation of IoT related communications, and the IoT ecosystem in general.

5 Conclusion and Future Work

The authors have walked through the very first generation of cellular networks to the latest technology, 5G, by conducting a review and doing a brief comparison of the evolution from 1G to 5G. Furthermore, the study has given exploratory insights into 4G features and architectures. With that knowledge in mind, the study further highlighted the key components of 5G technology and compare them with its predecessors. In addition, the study evaluated the current and future trend of IoT applications, analysed the case studies, and the impacts of 5G cellular networks in the smart city, industry, and healthcare sectors under the context of IoT utilisations.

Future work is aimed at creating an SDN-based 5G Testbed for IoT ecosystems that will allow us to simulate live attacks. Through this, the authors will be able to propose mitigating solutions with a degree of certainty.

References

1. What is the Internet of Things? Internet of Things definitions. In: i-SCOOP. https://www.i-scoop.eu/internet-of-things. Accessed 10 Mar 2019
2. Samuel, S.S.I.: A review of connectivity challenges in IoT-smart home. In: 2016 3rd MEC International Conference on Big Data Smart City, ICBDSC 2016, pp. 364–367 (2016). https://doi.org/10.1109/ICBDSC.2016.7460395
3. Singh, R.K., Singh, R.: 4G LTE cellular technology: network architecture and mobile standards. Int. J. Emerg. Res. Manag. Technol. **5**, 1–6 (2016). https://doi.org/10.23956/ijermt/2016
4. 4G. Wikipedia. https://en.wikipedia.org/wiki/4G. Accessed 15 Mar 2019
5. Agiwal, M., Roy, A., Saxena, N.: Next generation 5G wireless networks: a comprehensive survey. IEEE Commun. Surv. Tutor. **18**, 1617–1655 (2016). https://doi.org/10.1109/COMST.2016.2532458
6. Vora, M.L.J.: Evolution of mobile generation technology: 1G to 5G and review of upcoming wireless technology 5G. Sci. J. Impact Factor **2**(10), 281–291 (2015)
7. GrowthEnabler: Market pulse report, Internet of Things (IoT): discover key trends & insights on disruptive technologies & IoT innovations (2017)
8. Smart cities: acceleration, technology, cases and evolutions in the smart city. i-SCOOP. https://www.i-scoop.eu/smart-cities-smart-city/. Accessed 10 Mar 2019
9. Rao, S.K., Prasad, R.: Impact of 5G technologies on smart city implementation. Wirel. Pers. Commun. **100**, 161–176 (2018). https://doi.org/10.1007/s11277-018-5618-4
10. Obiodu, E., Giles, M.: 5G era: age of boundless connectivity and intelligent automation. GSM Assoc. (2017)
11. The Industrial Internet of Things (IIoT): The business guide to Industrial IoT. https://www.i-scoop.eu/internet-of-things-guide/industrial-internet-things-iiot-saving-costs-innovation/. Accessed 10 Mar 2019
12. 5G Connected Industries. https://networks.nokia.com/5g/use-cases/industry/infographic. Accessed 10 Mar 2019
13. Five ways 5G transforms healthcare. https://www.business.att.com/learn/updates/how-5g-will-transform-the-healthcare-industry.html. Accessed 10 Mar 2019
14. 5G Healthcare. https://www.ericsson.com/en/networks/trending/insights-and-reports/5g-healthcare. Accessed 10 Mar 2019
15. Fourth Industrial Revolution. https://en.wikipedia.org/wiki/Fourth_Industrial_Revolution. Accessed 10 Mar 2019

Sensor Network in Automated Hand Hygiene Systems Using IoT for Public Building

Michael O. Omoyibo, Tawfik Al-Hadhrami(✉), Funminiyi Olajide, Ahmad Lotfi, and Ahmed M. Elmisery

School of Science and Technology, Nottingham Trent University, Nottingham NG11 8NS, UK
Michael.omoyibo@my.ntu.ac.uk,
{Tawfik.Al-adhrami,Funminiyi.Olajide,Ahmad.Lotfi,
Ahmed.Elmesiry}@ntu.ac.uk

Abstract. The implementation of Sensor Network is a vital area of research in achieving accurate data transfer for automated hand hygiene systems. Sensing devices is also called host or end systems, they provide a service to collate data and can access the internet. The investigation of the entire sensors which include Infrared, water, ultrasonic, distance and motion sensor are based on the Internet Protocol Stack. Internet Protocol Stack consist of five layers such as physical, link, network, transport and the application layer, these layers work together to transmit data across the network. Public building is considered vital to be protected from the spread of infection through poor hand hygiene. In this sense an automated system is required to encouraging people to mitigate the spread of infections in such environment. In this context, this paper proposes a Sensor Network Automated Hand Hygiene Systems in public building (SN-AHHS). Sensor Network Automated Hand Hygiene Systems was designed to prevent spread of infection occurring in public building facilities by prompting users. The designed system uses smart devices (sensors and actuators) for SN-AHHS system. Internet of Things (IoT) platform is employed to process and analyze the collected data from these smart devices in the facility. SN-AHHS has been tested and evaluated, the results show that all component can communicate effectively which is vital to activate the require functionality to alert the user to perform good hand hygiene, it means SN-AHHS is activated when the user step into the facility.

Keywords: Smart dispenser · IoT platform · Touch free facility · Sensors and actuators · Short- and long-range communication technology

1 Introduction

Infection Control is a broad subject of its own, and it covers various subheading, including Healthcare-associated infection (HAI), which is one of the significant causes of mortality [1, 3]. The NHS had a couple of strategies to reduce the risk of patients contracting these infections [2].

Because this can cause the hospital paying a lot of fines and could result in the death of patients should the patient be infected by these bacteria [1]. Currently, in

F. Saeed et al. (Eds.): IRICT 2019, AISC 1073, pp. 463–476, 2020.
https://doi.org/10.1007/978-3-030-33582-3_44

public building around the country, toilets facility has facility seat, Sink and Dispenser these components are required for the users to wash their hands after using the facility. And it has been proven that regular hand wash is enough to reduce the incidence of diseases, thus improving the quality of life [19]. In most case, the user's compliance with washing hands is low, and this is a risk of been infected by bacteria [1]. UV lamps are another approach focused on hand-wash timing using a video monitoring system. The UV lamp could not produce the effect of hand hygiene [19]. Many of the current hand hygiene technology has based their concept of monitoring on the Point of Care (PoC) for every patient, and this means a sensor is placed on the wardroom to communicate with worker badge and remind worker to wash hands when entry and leaving the room [19]. The drawback to PoC is that it required a lot of medals, and it doesn't address any staff members. This proposed system, Sensor Network Automated Hand Hygiene Systems in public building (SN-AHHS) is designed to prevent the spread of infection in public building like healthcare and food environment. The system is activated when the user steps into the facility and this fitted in strategic point around the building. SN-AHHS provide additional feature such as touch-free facility toilet system. The rest of the paper prearranged into the following section. Section 2 reviews related literature on smart toilet facility around healthcare and food environment. Section 3 presents details of the designed stage for the proposed system. Section 4 presents a comprehensive evaluation and discussion for the proposed method, which includes functional testing and analytical results. Section 5 present summary/conclusion of the paper and highlighted future works, lastly the References.

2 Related Work

There are numbers of hand hygiene systems which have been developed in past years to help reduce the spread of infection in healthcare environment. These systems range from the most common alcohol gel, soap dispenser to more complex systems. Where tag is used to detect compliance to hand hygiene (HH), or if a staff have applied alcohol gel on their hands to prevent the spread of bacteria [4]. Another investigation in [6], the system uses sensors to detect alcohol, where staff are required to wear tag/sensor, if the tag sense alcohol it changes color from green to red, but if it senses no alcohol the color remain green. The draw back to is the system could only detect compliance from staff who wear the tags, therefore the application of this system is limited. Another investigation presented in [5], uses the wearable, passive RFID wristband, wall mounted dispenser and a software to measure hand hygiene compliance rate. This system consists of three parts, which include passive RFID tagged wearable wristbands, RFID reader embedded wall mounted gel dispensers and the software tool to see the hand hygiene compliance numerical results. And uses the ZigBee IEEE 802.15.4 protocol to transfer data to the server. Further study presented in [7], this system uses three different components to measure real-time hand hygiene compliance. These components include infrared emitters marking at location where hand hygiene is required. A wearable electronics device to monitors communicating with dispensers and decoding the signal from monitored area, and instrumented stationary and wearable dispensers. Another study presented in [8], this system investigates staff compliance of hang hygiene using passive RFID tag attached to staff members' hospital ID badge with RFID reader placed in the patient room. In all related work, there is not an automated system prompting the hygiene to users, instead monitors the staff if they are compliance with applying the alcohol gel on their hands. The propose Sensor Network Automated Hand Hygiene Systems (SN-AHHS) is a smart system, it is designed to mitigate the spread of infection in public building. SN-AHHS is meant to focus to encourage users to perform good hand hygiene after using the facility. The system consists of smart soap dispenser, touch free facility seat, an LCD screen, buzzer to alert users to use the dispenser, an LED to lit-up facility environment and an IoT platform to analyzed collected data from the sensors. The paper aims to provide solution via SN-AHHS system to promote the fight against poor hand hygiene in public building e.g. hospital, school and restaurant.

3 SN-AHHS Architectures and Design

This section focusses on developing the system by looking at the design, planning and evaluation. It's also covered the aim and objective, the functional and non-functional requirement, the functionality of the component. The SN-AHHS system consist of smart devices such as water sensor, Infrared sensor, ultrasonic and ping sensors. More to that is the 2.2" TFT LCD, Buzzer, jumpers wire, mega bread board, an Arduino microcontroller, smart dispenser and a battery, the system uses the Internet Protocol Stack to transmit data. The components are put into two categories, one gets the data and the other output the action which is the actuators. Each of the devices are installed around the facility and interconnected by short rang communication medium-Bluetooth. For example, the distance sensor is installed on the facility sealed seat. This is used to measure the distance of a person to the facility sealed seat, which senses the person and open the sealed seat ready for use. The same sensor activates the Infrared sensor to flush the water container, at this point the smart dispenser is activated by the Infrared sensor. The dispenser buzzer/alarm is turn on, and it display pictogram of hand dispensing soap on the LCD. This help to guide the user through a process of hand hygiene. The paper is arranged to highlight details of the devices as it progresses into the system.

3.1 Functional Requirement

In Software Engineering Paradigm Software Development Life Cycle (SDLC) consist of 7 phases, which include Requirement Collection and Analysis as one of the phases. This section focusses on what the system would do or how the system would react when the system is install in the facility, and when users interact with it. The following bullet point is meant to describe the functionality of the system and when user engages the facility.

- The system is meant to turn on the light in the facility area when user step into the facility.

- The facility seat is meant to open ready for use when user stand between 3–20 cm to the facility seat.
- The system is meant to activate the flush when facility seat sensor detects no user's presence.
- The system is meant to record the flush count in the IoT platform.
- The flush sensor should activate the smart dispenser to display pictogram of hand dispensing soap and hand hygiene on LCD TFT 2.2" screen, while alarm remain on.
- The dispenser should turn off the alarm when it dispenses soap.
- The system is meant to record the dispense count in the IoT platform.
- The system should indicate if power level is low.
- The system should indicate if washing liquid is low.

3.2 Non-functional Requirement

This section focusses on the process that should be made available for the system to deliver its potentials. These include things like smart alert via it IoT Platform. The following bullet point show few non-functional requirements in the system.

- The system should display alert message via its IoT platform if the power is running low and require changes to the power.
- The system should request for change to washing liquid if it falls below require level.
- The system should send alert messages of software update if newer version is available.

3.3 Software for SN-AHHS Architecture

The software was developed in C/C++ Language using Arduino IDE [9]. Furthermore, variable are names assigned to memory in the computer, so each device is called by name to provide a memory space to store data of the device [18]. There are different data type in software development, data type can also be called variable type, some data type has more memory space than others for example integer and string. Therefore, in order to link this software to the system this require the microcontroller, the microcontroller consists of pin positions, each position is call by name, which is same as the variable [9].

This paragraph focusses on the setup API. The aim of the setup API as the name implies. Its setup the variable to received and to distribute data using the command words INPUT, OUTPUT, LOW and HIGH with the API digitalWrite and pinMode. The LOW and HIGH keywords mean off/on, for example when use on a device it switches the device on and off. This paragraph focusses on the loop API, the loop API contains most of the code, however part of code that control the facility seat would be discussed in flowchart, however there are more section of the code that is not discuss in this paper. Because same principle can be applied to other section of the code. Figure 1 show the Loop API flowchart. This section of the code controls the facility sealed seat to open and close, such as if the user is 3- 20 cm away from the facility seat, it triggers the facility seat to open. And when the user moves away from the facility seat the

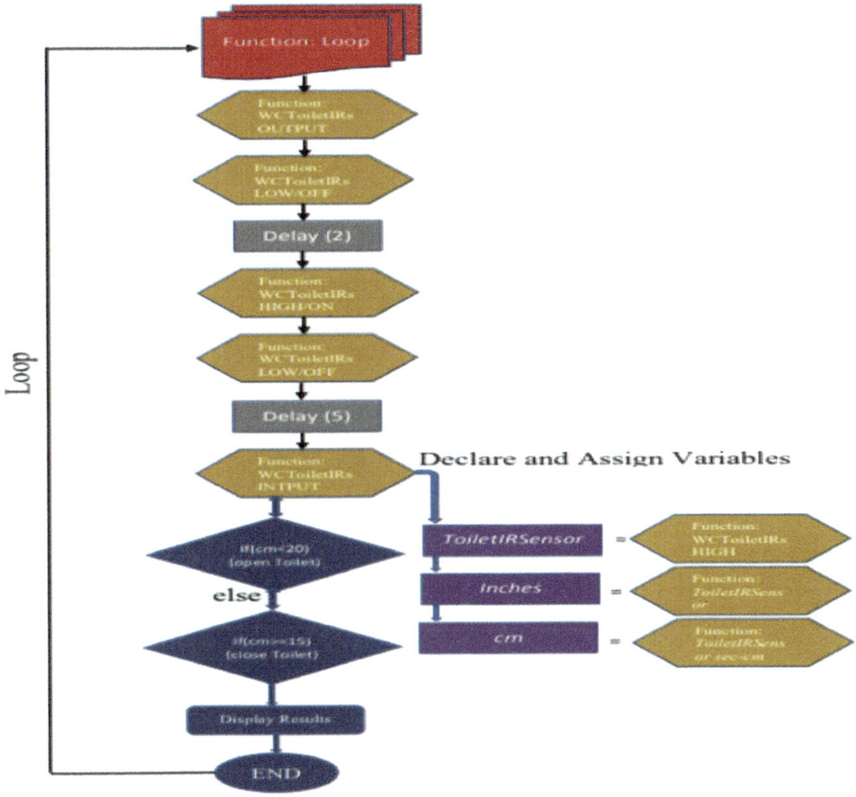

Fig. 1. Loop function flowchart

sealed seat return to its original position. This system uses a smart mechanism to open and close the facility seat. The loop API contained section of the code with their required API, this section is assigned to open and close the facility seat as show in Fig. 2.

Fig. 2. Picture for the SN-AHHS experiment

3.4 Hardware for SN-AHHS Architecture

Figure 3 show the top front view of the system; this section focuses on the dispenser which is one of the key components of the system. The dispenser is trigger after the user finish using the facility seat, and the dispenser further turn on the buzzer/alarm, the dispenser incorporates the 2.2TFT LCD Display. Soap are release via a built in Infrared sensor in the dispenser, this in return switch off the buzzer. The dispenser is on the right-hand side in Fig. 2.

Figure 3 is the zoom top view of the system, showing the component position as a circuit. Which includes ultrasonic sensor, mega bread board, water sensor, wires, distance sensor and the Infrared sensor.

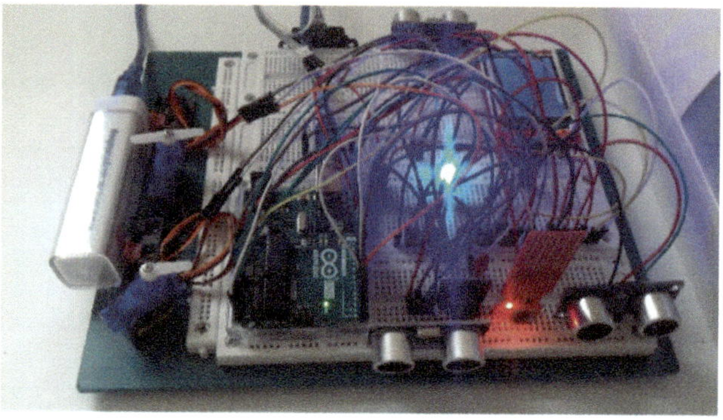

Fig. 3. Picture for the SN-AHHS experiment

3.5 SN-AHHS Algorithm

The SN-AHHS system incorporate a flush counter device (water sensor) to collate numbers of flush. A smart soap dispenser that displays pictogram of dispensing soap and hand hygiene on its TFT 2.2" LCD screen. And Buzzer/Alarm as an actuator, this is activated if user ignore hand hygiene. The buzzer is turn off if the user dispense soap from the smart dispenser. Figure 6 displays the image of the system in the facility and how the component is link in order to provide service as describe in Sect. 3 above. The system is centrally controlled by the microcontroller, the component sends data to the board-microcontroller via communication link (wire) [18] and that data is returned to the required devices to activate the next device for action. These data are collated and stored in a local database via communication link employing Internet Protocol Stack, then further analyzed in the IoT Platform.

Internet Protocol Stack consist of five layers such as physical, link, network, transport and the application layer, each of these layers act on the data and provide a service to the next layer. The Application Layer Protocol is responsible for the HTTP and the SMTP, the HTTP is a web protocol, this help to display the require information on the screen of both source and destination end systems, while SMTP responsible for messages [18]. The Internet's, Transport Layer transports application messages between application endpoints, using either of the transport protocol TCP or UDP [18]. The Network layer is responsible for moving network packets know as datagrams from one host to other, this layer uses the Internet Protocol IP address to determine the destination of the packet [18]. The Link Layer and the Physical Layer are responsible for handling communication over a specific link, they are implemented in a network interface for example the Ethernet or the Wi-Fi associated with a given link [18]. To move a packet from one node (host or router) to the next node in the route the network layer relies on the link layer, at each node the network layer passes the datagram down to the link layer [18]. Figures 4 and 5 shows a hybrid topology and SN-AHHA architecture for the SN-AHHS system respectively.

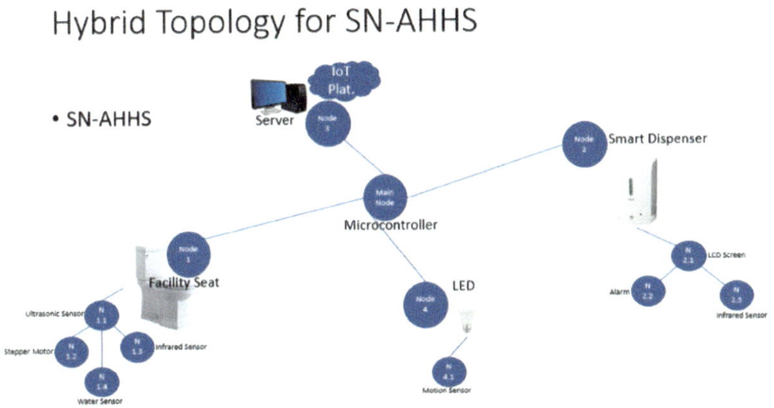

Fig. 4. Hybrid topology for SN-AHHS

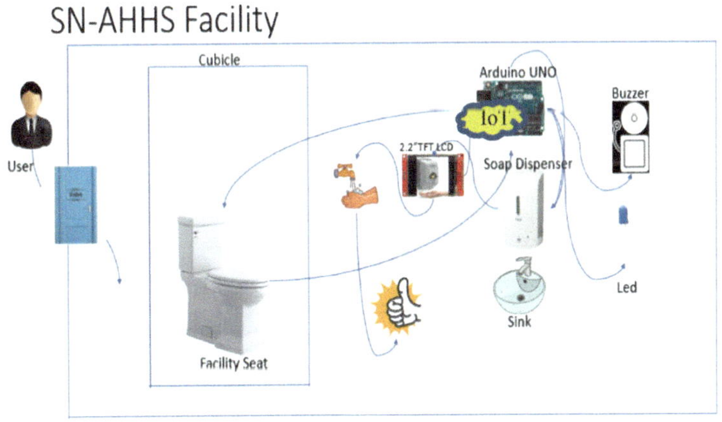

Fig. 5. SN-AHHS system architecture

Figure 6 shows a flowchart describing how the system deliver it functions, the algorithm is developed and run in a loop. The process starts and end with one user and then ready to repeat the same process with the new user. In this algorithm (problem solving operations) the motion sensor activates the lighting in the facility. The ping sensor open the facility seat while user stay in the require distance. The ping sensor is synchronized with the Infrared sensor to activate the flush for the facility seat. The Infrared sensor further activates the smart dispenser, the dispenser consists of alarm and LCD screen which are activated with the dispenser. The dispenser remains activated on till user engage with it to deactivate alarm, at this point the system return to its original state and wait for the next user. The Sensor Network Automated Hand Hygiene Systems (SN-AHHS) experimental components. In other to provide the require service the project need couple of items which include Software and Hardware for the experiment.

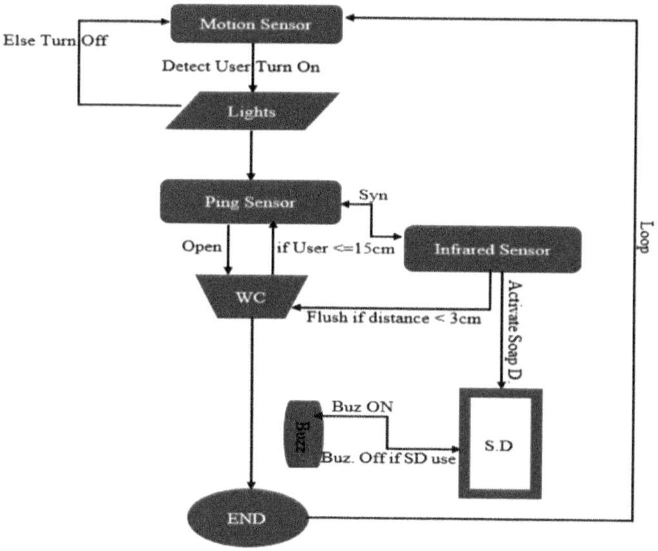

Fig. 6. SN-AHHS flowchart

3.6 SN-AHHS Circuit

Figure 7 describe SN-AHHS Circuit. The circuit consists of the microcontroller, bread board and the required component, it is presented as a schematics [16]. Figure 8 shows the pins position in the microcontroller and the component that is attached to the pin. This pin enables the system to collate data as an input and distribute these data around the system to trigger response as an output via the required component. For example, the Buzzer is trigger if the user ignores performing hand hygiene.

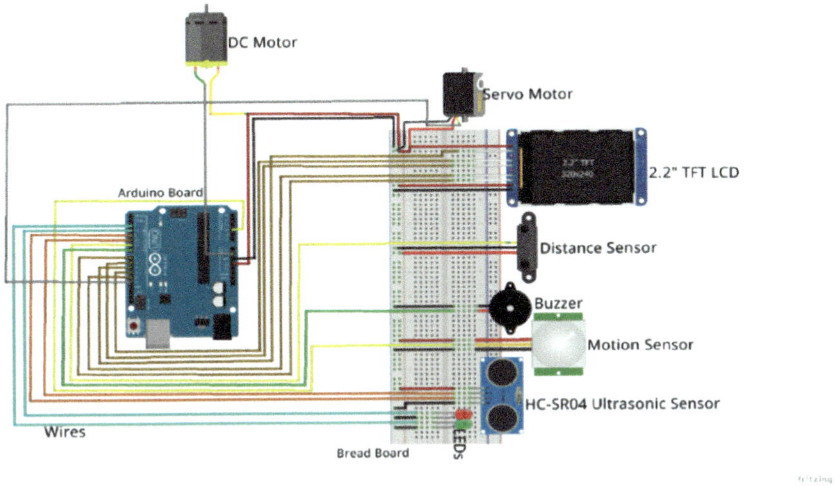

Fig. 7. IoT-ICSA schematic

4 Evolution and Discussion

Sensor Network Automated Hand Hygiene Systems (SN-AHHS) aimed to provide a service in public building to stop the spread of bacterial. When people engage with the facility, they may ignore to perform hand hygiene after using the facility. The system has employed the use of smart devices/sensors, microcontroller, 2.2" TFT LCD Display, IDEs and more for this purpose. And in this regard an experiment for this system has been stage in the previous subheading. The result shows that the system components can communicate employing Internet Protocol Stack shown in Table 1. The system can prompt people through its LCD screen to perform hand hygiene and remind people by its alarm in the smart dispenser if they forget. The next paragraph is set to run a simple functionality test to confirm if the system achieved it aims.

4.1 Functional Testing

This form of testing is done to check if the system follows the requirements provided by the client, and to produce the require results. The functional testing results in class diagram for the Sensor Network Automated Hand Hygiene Systems (SN-AHHS). The diagram presents the visual image of the SN-AHHS component and provides the link between all section of the system. The SN-AHHS consist of the base class, four dependent class and two subclasses to the Soap Dispenser class. Each class respond to the require trigger as user engage the system. The test was run to confirm the connection among components. These classes represent Smart Devices, for example the infrared sensor control the open/close of the facility seat and activate the distance sensor to provide flush. And the distance sensor further activates the dispenser to display pictogram for hand hygiene and trigger its alarm.

4.2 Analytical Results

This section focuses to analyze the results based on data collated via sensors for the SN-AHHS system. The data are analyzed and presented in various types of graph shapes. Figure 8 show the results analyzed using serial plot via Arduino IDE. The 4-little square on the right top corner represent 4 sensors, the blue square is facility seat sensor, the red represent the LCD sensor, the green represent the LED sensor and the orange represent the Flush sensor. The Y axis analyze the data when the system is in use or idle, and the X axis show duration (time) the system has remained idle or has been in use. The blue line show if the system is interacting with user, for example if the blue line stays between 300 and 400 the system is not in use, but if the line goes from 400 to 600+ this confirm the facility seat is in use, and if the blue line remain in 600+ position, it confirm the system is in used and duration (time) the user stays on the facility seat. The same apply to the red, green and orange lines but between 0 and 100, 0 represent that the system has been trigger while 100 means the system is idle. The flush sensor (orange line) is trigger when the blue line return to 300/400 position, the flush sensor is activated, the data is recorded as 1 count, Fig. 10 show the movement of the orange line drop from 94 to 0. At this point the green line remain activated because it lit up the facility environment. The red line (LCD sensor) is triggered by the flush sensor to display pictogram and activate the dispenser to turn on the alarm/buzzer. The graph shows the movement of the red line drop from 95 to 0. The process continues as a loop.

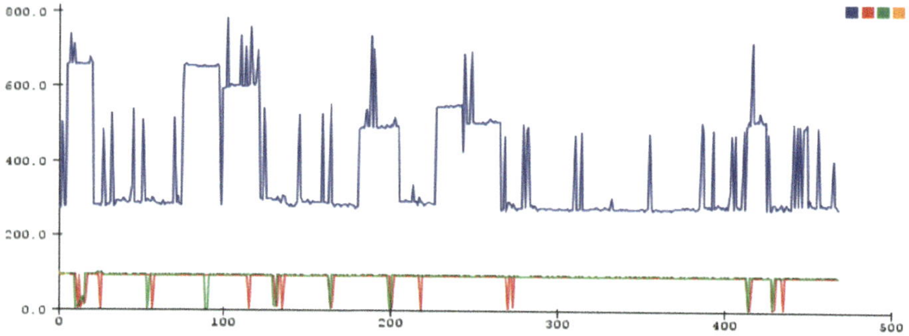

Fig. 8. Serial plots analysis for SN-AHHS

Collated Data from SN-AHHS Sensor

Figure 9 represent the collated data via facility seat (Infrared sensor), Flush (Distance sensor) Dispenser (Infrared sensor) and Light (Ultrasonic) sensors. The system work in a loop so the sensors are designed to continually gardening data around the facility environment until a user approaching the facility. Figure 9 show repetition and progression of numbers using a line graph. The high input value of the facility seat sensor confirms the facility seat is in use, and the flush sensor record every time a flush is activated by the user. This applied to the dispenser sensor, as presented in Fig. 9, the light sensor is activated as user step into the facility-toilet environment. These collected data are extracted from among the data generated by the SN-AHHS system.

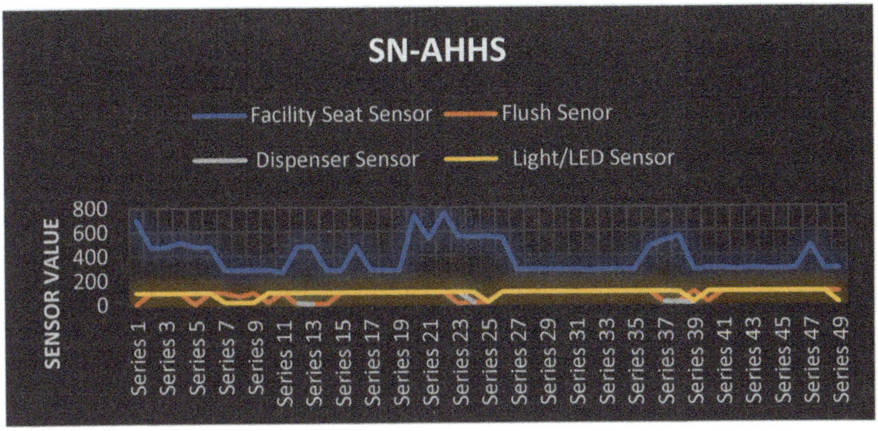

Fig. 9. Collated data from SN-AHHS

Figure 10 the blue line presents the facility seat, the value 500–700 means the facility seat is in use. And the orange line show numbers of flush, so the gray line is for soap dispenser, in Fig. 10 clearly shows what series a flush or dispenser is activated. For example, a flush is activated in series 14, while soap is dispensed in series 13.

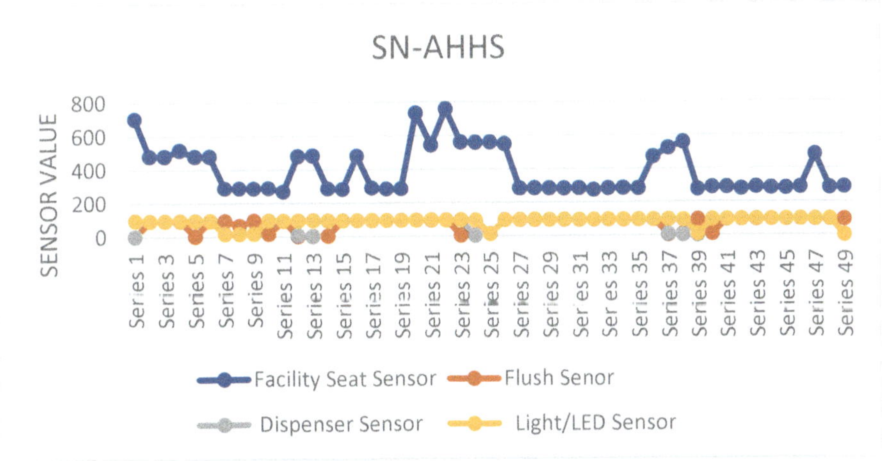

Fig. 10. Collated data from SN-AHHS

SN-AHHS Smart Device Analysis

This section provides a visual data analysis in scatter line graph of the SN-AHHS system in Fig. 11. The information presented represent the interaction between components of the system. In no. 1 on the X axis, the value of facility seat Sensor is 700, this means a user is very close, in this case the sealed seat opens ready for use. The light sensor is 95, this value activates the system to lit up the environment, the flush and

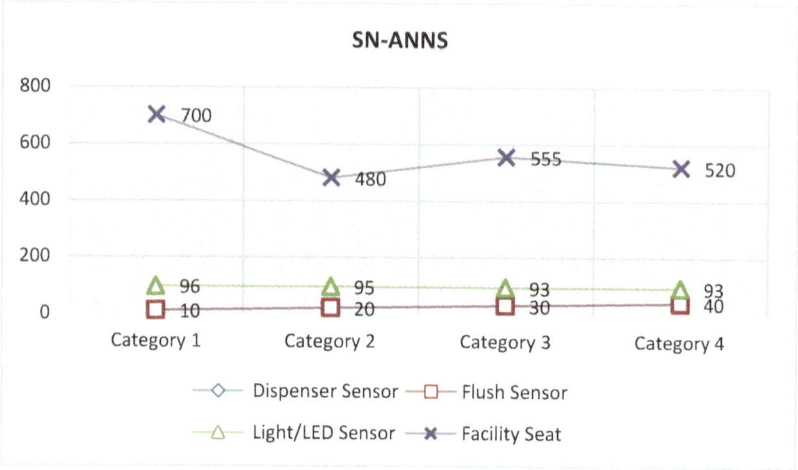

Fig. 11. SN-AHHS smart device analysis

dispenser sensor takes a scale of 1:10, this is meant to make better visual presentation. So, for every activation of the facility seat sensor, 1 flush is performed, and the dispenser is activated.

SN-AHHS Facility Seat Analysis

Figure 12 at category 2, the facility seat (Sensor) sensor is 480, this means the user is less close to the than the previous section above. The flush and dispenser sensor recorded 2 activations, this number is addition to the previous count, this means 1 in the category 1 and 1 in category 2.

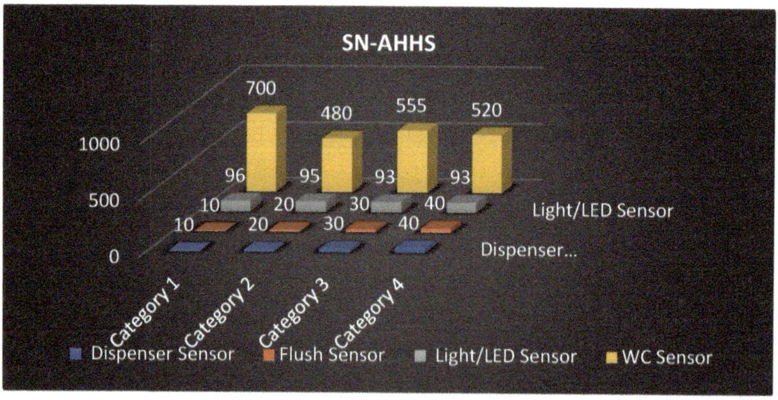

Fig. 12. SN-AHHS facility seat analysis

SN-AHHS Analysis Progression for Smart Devices

Figure 13 present a progression of the system smart devices, as it interacts with the user. This section compares the System Determinant against X axis the Dispenser Progression Y axis. The system is first activated by the light sensor when the user enters the facility-toilet environment, this has very less impact on the system. And the next activation is the facility seat sensor, this further trigger other components, which include the soap dispenser.

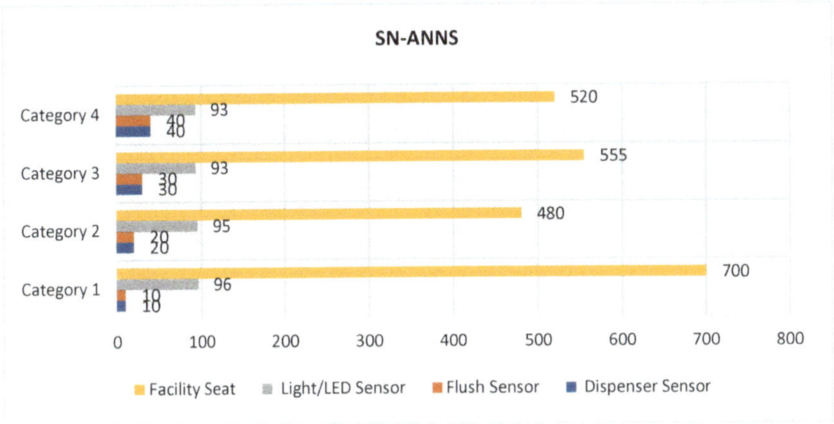

Fig. 13. SN-AHHS analysis progression

5 Conclusion

In this paper the SN-AHHS architecture is presented, an experiment copy was developed as reference point to the propose system. This work is considered as the first phase of publication to the project, the subject Internet of Things has not fully addressed in this publication, it is meant to be address in the next paper. The results for this system was presented in the result section, in summary the result is as follow: The system responds to user, the soap dispenser communicates with the facility seat and trigger the Buzzer and the buzzer continues buzzing until the user dispense soap and perform hand hygiene, the comprehensive results is in the analytical result section above. Further investigation will be carrying out to study the quality of services in terms of packet delivery ratio and time, speed, and sensors accuracy.

References

1. World Health Origination (WHO): Hand Hygiene in Outpatient and Home-based Care and Long-term Care Facilities (2012). https://apps.who.int/iris/bitstream/handle/10665/78060/9789241503372_eng.pdf;jsenid=E839BAC6060B76EC8FD4006086AD2C1B?sequence=1
2. NHS: Infection Prevention Control (2018). https://www.nottinghamshirehealthcare.nhs.uk/infection-prevention-and-control/

3. Lash, M.T., et al.: A large-scale exploration of factors affecting hand hygiene compliance using linear predictive model. In: 2017 IEEE International Conference on Health Informatics (ICHI), USA, pp. 66–73 (2017)
4. Bal, M., Abrishambaf, R.: A system for monitoring hand hygiene compliance based-on Internet of Things. In: 2017 IEEE International Conference on Industrial Technology (ICIT), USA, pp. 1348–1353 (2017)
5. Akif Meydanci, M.: RFID based hand hygiene compliance monitory station. In: 2013 IEEE International Conference on Control System, Computing and Engineering, Mindeb, Malaysia, pp. 573–576 (2013)
6. Simmonds, B., Granado-Villar, D.: Cost effectiveness of an electronic hand hygiene monitoring system (EHHMS) in the prevention of healthcare-associated infections. Am. J. Infect. Control **40**(5), e117–e118 (2012)
7. Levchenko, A., Boscart, V., Fernie, G.: Hand hygiene monitoring and real-time prompting system. In: 2012 IEEE International Systems Conference SysCon 2012 (2012)
8. Johnson, R., Tsouri, G., Walsh, E.: Continuous and automated measuring of compliance of hand-hygiene procedures using finite state machines and RFID. IEEE Instrum. Meas. Mag. **15**(2), 8–12 (2012)
9. Arduino: Library (2018). https://www.arduino.cc/en/Reference/Libraries. Accessed 06 Sept 2018
10. Zanella, A., Bui, N., Castellani, A., Vangelista, L., Zorzi, M.: Internet of Things for smart cities. IEEE Internet Things J. **1**(1), 22–32 (2014)
11. Kulkarni, C., Karhade, H., Gupta, S., Bhende, P., Bhandare, S.: Health companion device using IoT and wearable computing. In: 2016 International Conference on Internet of Things and Applications (IOTA) (2016)
12. Suthokumar, G., Tharmakulasingam, J., Johnnirajh, A., Subendran, P., Dias, D.: Smart building—towards the insight remotely. In: 2016 International Conference on Internet of Things and Applications (IOTA) (2016)
13. Google: 2.2" TFT LCD Display (2018). https://www.google.co.uk/search?safe=active&biw=1242&bih=602&tbm=isch&sa=1&ei=3dV-W5PyDYrWgAbM_a6IDA&q=tft+displays+2.2&oq=tft+displays+2.2&gs_l=img.3…707519764.707537673.0.707539181.32.15.0.0.0.0.421.1093.5j2j4-1.8.0….0…1c.1.64.img..26.4.391.0..0j0i67k1j0i30k1j0i8i30k1.0.idryJcOthEw#imgrc=sbt-7_T1qFqS-M. Accessed 05 Sept 2018
14. Pande, H., Kharat, M.U.: Modified WiseMAC protocol for energy efficient wireless sensor networks with better throughput. In: Internet of things Application IoT. 2005 IEEE/SP 13th Workshop on Statistical Signal Processing, pp. 615 – 618 (2005/2016)
15. Domingo, M.C.: An over view of the Internet of Things for people with disability. J. Netw. Comput. Appl. 584–596 (2012). www.elsevier.com/locate/jnca. Accessed 29 Oct 2011
16. Stalling, W.: Fritzing electronics. In: 2016 Wireless Communications and Networks, 2nd edn. Pearson Education, USA. Schematics Presentation (2005). http://fritzing.org/download/. Accessed 05 Sept 2018
17. Deitel, H.M.: C++ How to Program, 5th edn. Pearson Education, In, USA (2005)
18. Haidegger, T., Varga, V., Lehotsky, A., Rona, P., Pethes, R., Szeremy, P., Szilagyi, L., Ferenci, T., Kovacs, L.: Information technology tools employed in infection control. In: 2015 16th IEEE International Symposium on Computational Intelligence and Informatics (CINTI) (2015)

Security Attacks in IEEE 802.15.4: A Review Disassociation Procedure

Abdullah A. Alabdulatif(✉)

Computer Department, College of Sciences and Arts, Qassim University, Al-Rass, Saudi Arabia
A.Alabdulatif@qu.edu.sa

Abstract. The specification of Physical layer PHY and MAC layer for Low Rate Wireless Personal Area network (LR-WPAN) is provided in the IEEE 802.15.4 standard. This standard offers four services of security to protect the frames during the transmission process. Disassociation is one of features provided by the MAC layer in IEEE 802.15.4 that is sent by devices to leave the network or the coordinator PAN to force the associated device to leave the network. Wherefore, Disassociation procedure should initiate and finish between legitimate nodes to avoid DoS attacks that can affect services of the network. The aim of this paper is to analyze and review Disassociation procedure to verify the process is secure and free of weaknesses that can threaten the security in WPAN. In the analysis of Disassociation procedure, it was found that the disassociation process can face a new potential attack. This potential attack can be executed in two different scenarios based on the target node.

Keywords: IoT security · Disassociation procedure · IEEE 802.15.4 standard

1 Introduction

Currently, researchers and developers in wireless environments face one of the most difficult challenges; how to enhance security and protect wireless networks. The Eavesdropping attack is a potential attack faced by wireless networks from an attacker with only a little effort and suitable radio receivers. Thus, it is difficult to detect or prevent such attack on wireless networks [1, 2]. The IEEE organizations, who are responsible for wireless networks standards, update their standards periodically to avoid any weaknesses and improve security in these networks. However, researchers and developers, through their study and analysis of the security of WLANs standards, identify new vulnerabilities and, then propose ideas to stop attackers form abusing these vulnerabilities.

The approval on the 1st version of the IEEE 802.15.4 standard was in 2003; the standard describes the specifications of the PHY layer and MAC layer for LR-WPAN [3]. The wireless connection requires the low cost of network connectivity. Moreover, the throughput capacity is needed by some applications with limited energy, thus, the IEEE 802.15.4 standard includes these requirements. The IEEE 802.15.4 standard is used in different applications that work in different fields, such as industrial, healthcare, residential, and the military, which require a number of features in the network such as

F. Saeed et al. (Eds.): IRICT 2019, AISC 1073, pp. 477–485, 2020.
https://doi.org/10.1007/978-3-030-33582-3_45

low data rate and the relaxed quality of service (QoS) [4]. There are a number of reasons for failure and congestion of the network such as error or latency, for example, the attack's behavior on WLANs when launching attacks such as spyware on a secret transmission process or injecting forged frames of battlefield monitoring applications [4]. Both the PHY layer and the MAC layer in IEEE 802.15.4 standard can be a target for various potential attacks. The IEEE 802.15.4 standard has some weaknesses that can be exploited to attack the network, for instance the open and unprotected environment of the wireless network makes it easy to access the range of the network compared with the difficulty of access in the wired network. Thus, the attacker's goals and the type of attack determine the level of effect this attack has on the network.

There are a number of frame types that transfer between the nodes in the IEEE 802.15.4 standard and the PAN coordinator, Beacon, Command, Data, or Acknowledgement frames. Each of these frame types has a different format that corresponds to the task it is going to achieve over the network [5, 6]. The IEEE 802.15.4 standard is a wireless network, which means the nodes exchange the frames over radio waves. Therefore, an attacker will be able to monitor, receive, or intercept these frames over the network [7, 8]. Thus, launching a spoof attack can be easily achieved by an attacker in order to collect information about the network, which leads to an active attack launch against the network such as denial of service (DoS).

There are a variety of objectives or motives for launching a DoS attack that are achieved by the concerted efforts of one or more people to prevent services from being provided efficiently or at all, temporarily or forever [9]. The attacker can launch the DoS attack by flooding massive unlawful data into the network, thus denying a legal client access to the resources in the network [8]. The Radio Jamming attack is an example of a DoS attack against the PHY layer in IEEE 802.15.4. This attacker aims to increase the Signal Noise Ratio (SNR) through the intentional emission of radio signals in ongoing radio communications. Thus, the signal is distorted before it is received into the node in the network and establishes a DoS attack [10]. Moreover, the MAC layer in the IEEE 802.15.4 standard faces different types of attacks such as DoS against data transmissions [9], DoS against GTS requests [9], and DoS against CAP maintenance [11], among others. All of these attacks can be either a cause or effect of a DoS attack.

Disassociation process is used to achieve Disassociation for one of the nodes in the Network. The process can be initiated either by a legitimate node or the coordinator of PAN to disconnect the node within the network in the securely. However, the attacker can also use Disassociation process to launch DoS attack against the legitimate nodes in the network which forces the legitimate nodes to leave the network or disconnects the nodes in the network. Nzabahimana [12] has presented a DoS attack against Disassociation process called Rogue Access Point. In the Rogue Access Point attack, the attacker has to install a rogue access point in the range of the network. Then, the network has two access points, one is that installed by the attacker and the other is a legitimate access point. Next, the attacker aims to remove legitimate access points through disconnecting IoT network from it. Then, each device in the network will be reconnected with an access point installed by the attacker. As a result, the whole of the network is connected to rogue access point and controlled by attacker [12].

As number of potential attacks have been identified by the researchers against different packets used by the IEEE 802.15.4 standards. This paper will focus on the study and analysis of Disassociation packet and Disassociation process in order to identify any weakness in them. So, the aim is to identify and explain the possibility of a new scenario followed by an attacker to launch any type of attack against Disassociation packet and Disassociation process.

In this paper, a brief overview of the IEEE 802.15.4 standard was presented. In Sect. 2, the four security services of MAC sublayer in IEEE 802.15.4 standard are presented. The security packet in the IEEE 802.15.4 standard is reviewed, focusing on the disassociation packet in Sect. 3. In Sect. 4, the new potential attacks against the Disassociation packet are identified through two scenarios. Section 5 follows with the conclusion.

2 Security Services in IEEE 802.15.4 Networks

The IEEE 802.15.4 provides four security services, access control, message integrity, message confidentiality, and replay protection. The security services are applied by the MAC sublayer in the IEEE 802.15.4 standard, which is responsible for checking the incoming and outgoing frames in the higher layers when demanded [3].

2.1 Access Control

This service blocks unauthorized nodes from gaining access to the network and to be an authorized node. Thus, any malicious or fake frames received from unauthorized nodes have to be detected and rejected by nodes in the network [13, 14].

2.2 Message Integrity

This service detects any tampering with the messages during the transmit process over the network between authorized nodes. Any tampering in a message has to detect by a recipient of the message for rejection and discard the message. The message authentication code (MAC) is the mechanism used in the IEEE 802.15.4 standard to provide message authentication and integrity for messages. Both participants are able to compute the MAC using a shared secret cryptographic key and comparing results to determine whether to accept or reject the message [13, 14].

2.3 Message Confidentiality

This service hides the contents of a message; therefore, the message is unreadable for any unauthorized nodes when transmitted over the network. The IEEE 802.15.4 standard utilizes the encryption technique and the Nonce technique to provide message confidentiality. Thus, unauthorized nodes are unable to recover any information from an encrypted message. Additionally, they are unable to infer any of the content of the encrypted message. Moreover, Nonce technique is used to generate a new nonce without repetition every time calls to the encryption algorithm occur. Thus, it adds some changes to the encryption procedure which are then reflected on the set of messages [13, 14].

2.4 Replay Protection

The adversary is able to eavesdrop on the wireless network environment; thus, it is easy to capture any legitimate message from the network. Then, the adversary resends the captured message later in the network to launch a replay attack. The IEEE 802.15.4 standard uses the sequence number technique to prevent this attack. Therefore, each packet should assign a sequence number which increases with each packet sent over the network. Thus, the recipient is able to reject any packet when it finds the sequence number of packets is equal to or smaller than the current sequence number [13, 14].

3 The Security Packet in the IEEE 802.15.4 Networks

The MAC layer is responsible for security in the IEEE 802.15.4 standard under the control of the application. Therefore, the security requirements are set by the application through specifying control parameters in each packet, and the security in the IEEE 802.15.4 standard is by default disabled when the parameters are not set by the application. The acknowledgment packets are unsupported in the security in the IEEE 802.15.4 standard, so the security is not applied to acknowledgment packets, whereas remaining of packet types are supported in the security optionally through integrity and confidentiality protection of data field in the packet [15].

There are a number of security suites supported in the IEEE 802.15.4 standard. An application uses these security suites to select and control the security protection types that are required to the transmitted data based on the security properties and protection levels needed. The first suite is null, in which no level of security is provided. The second type is the AES-CTR suite which provides encryption only, whereas authentication is provided by the AES-CBC-MAC suite. The AES-CCM suite offers both encryption and authentication, which means it a high- level of security, providing both integrity and encryption of data.

3.1 Disassociation Packet

Disassociation is one of features provided by the MAC sublayer in IEEE 802.15.4. The coordinator in PAN or any associated devices can be used to start the disassociation process. If the PAN coordinator wish an association device to leave the PAN, then it sends a Disassociation Notification command to the association device. Conversely, if an association device desires to leave the PAN, then the device sends a Disassociation Notification command to the coordinator in the PAN [3].

Figure 1 shows the general format for the MAC frame. The MAC header in the frame appears in a fixed order. However, the frames may be differentiated because some fields may not be contained in all frames based on the type of frame and the values of fields. For example, the format of the frame control field includes the information to define the frame type, addressing fields, and other control flags. The Security Enabled field is one of the control flags used to control the protection of the MAC frame. Thus, setting the value of the Security Enabled field to one means the MAC sublayer provides the protection for the frame; otherwise, the frame does not protect by setting the value to zero. Therefore, in the case of the Security Enabled field set to one, it means it presents the Auxiliary Security Header field in the MHR [3].

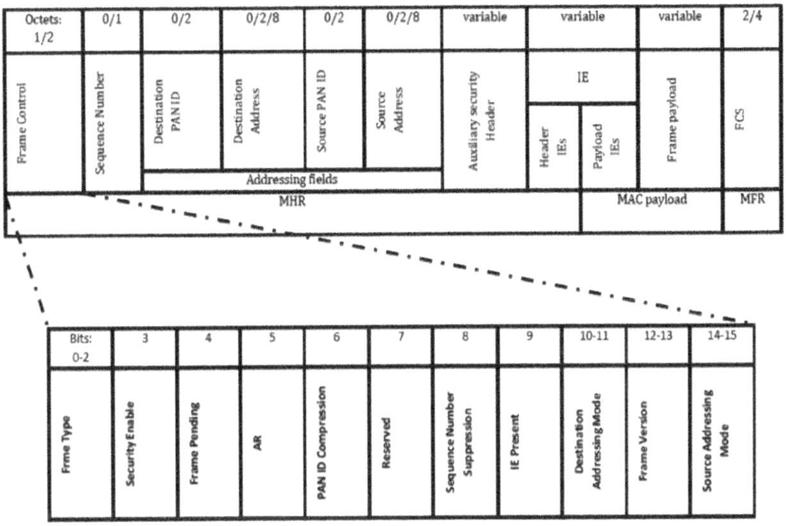

Fig. 1. General MAC frame format.

4 Potential Attacks Against Disassociation Packet

This section provides a study and analysis of the new potential attacks against the Disassociation packet in the IEEE 802.15.4 standard. These new potential attacks explain two different scenarios that could happen to the Disassociation packet in the IEEE 802.15.4 standard when the attacker launches a DoS attack. The Disassociation packet attack is classified as an active attack and it requires collecting some information about the network in order to commence the attack. A successful sniffing attack is the first step to launch Disassociation packet attacks. Its objective is to monitor and look for any useful information from packets during the transition process between the nodes that can be used during the active attack. The sniffing attack is a passive attack whose only aim is to obtain any important information in preparation for the attack of the nodes in the network. Furthermore, it does not require any action from the attacker to change or modify the contents of the packets of data through the transfer process between the nodes [16].

The disassociation process in the IEEE 802.15.4 standard can start from either the PAN coordinator or the associated node through sending the disassociation packet. If the PAN coordinator desires to discontinue with an association node in the network, then the disassociation process has to be initiated by the PAN coordinator though sending a disassociation packet to the target association node. After that, the Acknowledgment packet should be sent from the association node to the coordinator PAN, which means the disassociation process has been successfully completed. In some situations, the disassociation packet does not receive the association node because the PAN coordinator fails to extract and send the packet. Therefore, the PAN coordinator assumes the node to be disassociated. In the other scenario, the PAN coordinator receives the disassociation packet sent from the association node in the network when the association node desires to leave the network, as shown in Fig. 2. After that, the PAN coordinator has to send the Acknowledgment packet to the association node, which means the disassociation process has been successful. In some situations, the Acknowledgment packet is sent from the PAN coordinator, but does not receive the target node for some reasons linked to in the network, so the node considers itself disassociated.

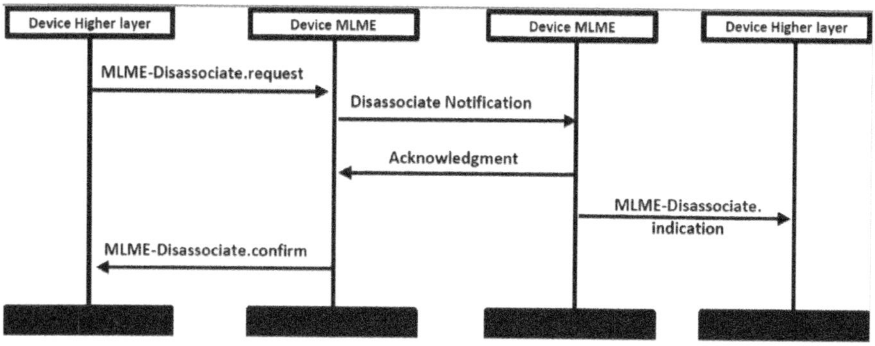

Fig. 2. Message sequence chart of disassociation

A potential attack can be launched against the MAC layer in the IEEE 802.15.4 standard exploiting the weakness of the association and disassociation processes. The processes of association and disassociation can be used to attack the MAC layer in order to oblige either the PAN coordinator or a target association node to execute the process of disassociation that causes the DoS attack. The DoS attack can be launched against the MAC layer through the weakness identified in the disassociation packet in two different two scenarios based on the targeted node and the attacker connectivity node as follows.

4.1 First Scenario

Figure 3 shows that the target node is any node trying to associate with PAN, the coordinator of the network, and the attacked node to either connect or disconnect with the PAN coordinator. Firstly, the attacker uses the sniffing attack to look for an association packet request over the network exchange between the unassociated node and the PAN coordinator to execute the association process. If an association packet request is detected by the attacker over the network, then a fake disassociation packet is created by an attacker with a destination address as the target node address who sent the association packet request and the source address is the PAN coordinator address. After that, the attacker sends a disassociation packet with the necessary information to the node that has to reply to the PAN coordinator an Acknowledgment packet for completing the disassociation process successfully. In this situation, the attacker switches between the address's source and destination in the association packet to be the address of the disassociation packet. Thus, the destination address in the association packet would be the source address in the disassociation packet and vice versa. As a result, a valid disassociation packet will be delivered to the node from the attacker, and the node thinks the coordinator PAN wishes the node to disassociate and leave the network. Therefore, it sends an Acknowledgment packet to coordinator PAN and disassociation with the network.

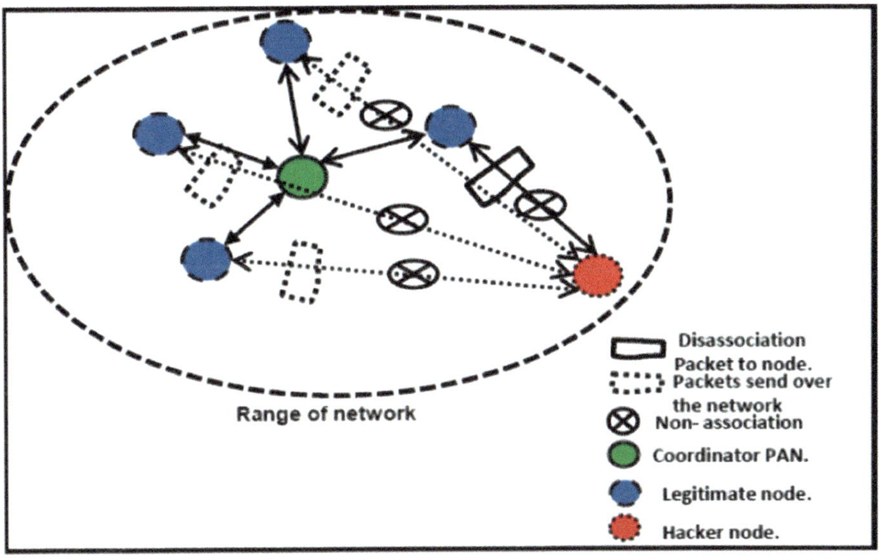

Fig. 3. Disassociation packet attack scenario 1 disassociation.

4.2 Second Scenario

Figure 4 shows that the association packet response, which is the reply from the PAN coordinator to the node to successfully complete the association process, is the target packet for the attacker. Regardless, the attacked node either connects or disconnects with the PAN coordinator. Firstly, the attacker uses the sniffing attack to look for an association packet response over the network exchange between the unassociated node and the PAN coordinator to execute the association process. If there is an association packet response detected by the attacker over the network, then a fake disassociation packet created by the attacker with a destination address of the PAN coordinator's address who sent the association packet response and source address is the node address. After that, the attacker sends a disassociation packet with the necessary information to the PAN coordinator which has to reply to the node an Acknowledgment packet for completing the disassociation process successfully. In this situation, the attacker switches between address's source and destination in the association packet to be the address of the disassociation packet. Thus, the destination address in the association packet will be the source address in the disassociation packet, and vice versa. As a result, a valid disassociation packet will receive the PAN coordinator from the attacker. Thus, the PAN coordinator thinks the node wishes to disassociate and leave the network. Therefore, an Acknowledgment is sent packet to the node and assumes node disassociation from the network.

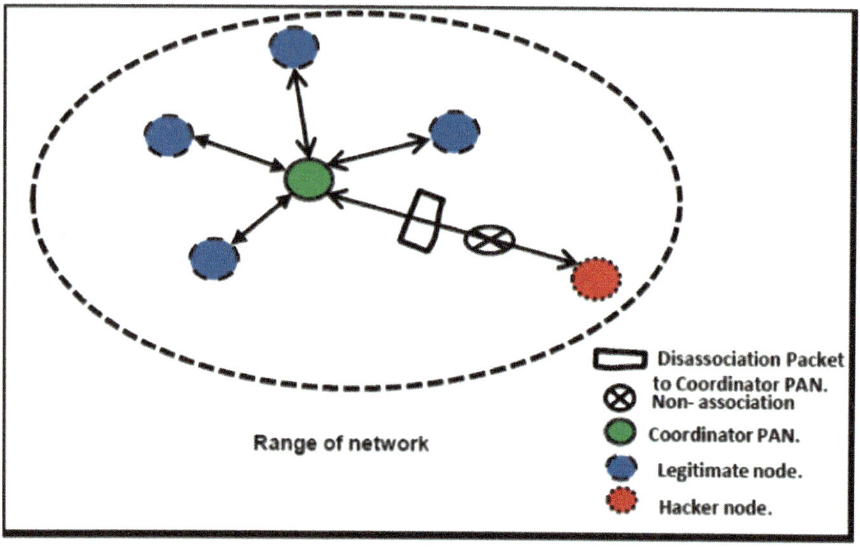

Fig. 4. Disassociation packet attack scenario 2 disassociation.

5 Conclusion

The IEEE 802.15.4 standard presents the specifications for the physical layer and MAC layer for Low Rate Wireless Personal Area network. This work aims to study and analyze the security in IEEE 802.15.4 standard. Also, it focuses on evaluating security of the disassociation process. New potential attacks have been identified against disassociation process in IEEE 802.15.4 standard. The attacker can exploit the weakness in the disassociation process to force the legal node in the network to dissociate. Moreover, the attacker's node can launch the attack in both cases to either associate or dissociate with the network. However, success in launching the attack is based on the sniffing attack and the level of security in the network. Additionally, the attack node has to look for any association packet sent over the network to start executing the attack because the association packet, either request or respond, has acquires sensitive information necessary to launch the attack.

Acknowledgments. This work is fully supported by Qassim University by the contract no. (2337-alrasscac-2018-1-14-S).

References

1. Alabdulatif, A., Ma, X.: Analysing the EAP-TLS handshake and the 4-way handshake of the 802.11i standard. Int. J. Inf. Secur. Res. (IJISR) **3**(34), 439–448 (2013)
2. Alabdulatif, A., Ma, X., Nolle, L.: Analysing and attacking the 4-way handshake of ieee 802.11i standard. In: 8th International Conference for Internet Technology and Secured Transactions (ICITST), pp. 382–387. IEEE (2013)
3. Committee, L.S., et al.: IEEE std 802.15.4-2015 (revision of IEEE std 802.15.4-2011). IEEE standard for low-rate wireless personal area networks (LR-WPANS). IEEE Computer Society (2016)
4. Amin, Y.M., Abdel-Hamid, A.T.: A comprehensive taxonomy and analysis of IEEE 802.15.4 attacks. J. Electr. Comput. Eng. **2016**, 4 (2016)
5. Elijah, O., Rahman, T.A., Orikumhi, I., Leow, C.Y., Hindia, M.N.: An overview of Internet of Things (IoT) and data analytics in agriculture: benefits and challenges. IEEE Internet Things J. **5**(5), 3758–3773 (2018)
6. Hussain, I., Negi, M.C., Pandey, N.: Security in ZIGBEE using steganography for IoT communications. In: System Performance and Management Analytics, pp. 217–227. Springer (2019)
7. Cheema, R., Bansal, D., Sofat, S.: Deauthentication/disassociation attack: implementation and security in wireless mesh networks. Int. J. Comput. Appl. **23**(7), 7–15 (2011)
8. Razzaq, M.A., Gill, S.H., Qureshi, M.A., Ullah, S.: Security issues in the Internet of Things (IoT): a comprehensive study. Int. J. Adv. Comput. Sci. Appl. **8**(6), 383 (2017)
9. Frustaci, M., Pace, P., Aloi, G., Fortino, G.: Evaluating critical security issues of the IoT world: present and future challenges. IEEE Internet Things J. **5**(4), 2483–2495 (2017)
10. Amoozadeh, M.: Towards Robust and Secure Collaborative Driving and Interactive Traffic Intersections. University of California, Davis (2018)
11. Amin, Y.M., Abdel-Hamid, A.T.: A simulation model of IEEE 802.15.4 GTS mechanism and GTS attacks in OMNET ++/MiXiM + NERA. Comput. Inf. Sci. **11**(1), 78–89 (2018)
12. Nzabahimana, J.P.: Analysis of security and privacy challenges in Internet of Things. In: 2018 IEEE 9th International Conference on Dependable Systems, Services and Technologies (DESSERT), pp. 175–178. IEEE (2018)
13. Yang, Y., Wu, L., Yin, G., Li, L., Zhao, H.: A survey on security and privacy issues in Internet-of-Things. IEEE Internet Things J. **4**(5), 1250–1258 (2017)
14. Sicari, S., Rizzardi, A., Grieco, L.A., Coen-Porisini, A.: Security, privacy and trust in Internet of Things: the road ahead. Comput. Netw. **76**, 146–164 (2015)
15. Frustaci, M., Pace, P., Aloi, G.: Securing the IoT world: issues and perspectives. In: 2017 IEEE Conference on Standards for Communications and Networking (CSCN), pp. 246–251. IEEE (2017)
16. Perlman, R., Kaufman, C., Speciner, M.: Network Security: Private Communication in a Public World. Pearson Education India (2016)

Traditional Versus Decentralized Access Control for Internet of Things (IoT): Survey

Mohammed Saghir[1(✉)], Bassam Ahmed H. Abu Al Khair[1], Jamil Hamodi[2], and Nibras Abdullah[3]

[1] Computer Science Department, University of Science and Technology, Sana'a, Yemen
msfat7070@gmail.com, bassamAhmed2020@gmail.com
[2] Computer Science Department, Hodeida University, Hodeida, Yemen
Jamil@gmail.com
[3] National Advanced IPv6 Center, University Saints Malaysia (USM), 11800 George Town, Penang, Malaysia
nibras@usm.my

Abstract. Internet of Things concept has started from the stage of early construction and increased with full speed to become a part of the Internet in the future. In our world, there are billions of an internet device has the technical challenges to manage them. Although Internet access management techniques exist, they are using 3rd party or centralized techniques to offer a set of technical constraints to manage them globally. The Block-Chain (BC), a notebook technology, offers a solution to secure trust in decentralized systems, and combines the trust, consistency and use of the shared Ledger in a public network that could be useful for building confidence in Internet objects. However, the use of BC with Internet things by itself involves a lot of challenges and has become a hot topic. Therefore, a methodical review of literature is necessary to know what is the proposed on this subject. This paper discusses a survey of key issues and design factors in IoT security that need to be considered for Access Control (AC) model designers as the IoT security requirements. In addition, we offer a broad survey of modern AC frameworks in IoT object research. We classify the proposed mechanisms into two categories: traditional and decentralized. Besides, we are discussing how the BC can be a key enabling factor for the access management model designed in IoT security.

Keywords: IoT security · BC · IoT protocols · Network security · Smart contracts · AC

1 Introduction

The IoT is a concept that refers to serial practices to connect an enormous number of devices to communication networks [1]. There are several types of systems in the IoT field, such as Smart Home, Smart City, and Healthcare System. While IoT can give us a lot of gains, it increases the risk of being exposed to various security and privacy threats; some of these threats are very difficult and may be new to the Internet provider. Before the advent of IoT, data disclosure, and denial of service was the most reported

F. Saeed et al. (Eds.): IRICT 2019, AISC 1073, pp. 486–494, 2020.
https://doi.org/10.1007/978-3-030-33582-3_46

security threat. In IT security threats there is not only information theft or denial of service. But there is a possibility that there are threats associated with devices, such as physical security. In addition, many devices in the internet have a limited capacity which adds to the burden of use such as the low amount of storage and power, as well as the challenges for the existing management of AC systems using the centralization of management control techniques (such as 3rd-party), to compound the challenges in IoT objects.

IoT and BC have introduced new concepts and technologies. They have transformed concepts and created new capabilities into their own system. The BC is creating applications with possibilities to share the main properties of them based on their relationship and utilize the decentralized characteristics of BC for use in IoT objects. In fact, the BC is a breakthrough technology that radically modifies our concept of central management for authorization to access control and representation of digital notebooks at the core of financial systems based on decentralized such as Bitcoin and, very rapidly, there are many other systems using decentralized models such as Storj, a p2p cloud network to storage [2].

In this paper, we present a review and compare four main traditional architectures commonly employed; XACML [3], OAuth [4], UMA [5] and FedCAC [6] in the IoT AC field, and study the work proposed and describe the different authorization platform solutions existing in the IoT field (Sect. 2.1). As well, we offer to study the works proposed in the decentralization of architecture in IoT and authorization of architectural solutions existing in IoT. There are more works to present the solution for the decentralization of architecture in the Internet of Things and a detailed explanation will be provided in the Sect. 2.2 for example: FairAccess [2], ControlChain [7], Management hub [8], TrustChain [9], IoTChain [10], and BlendCAC [11].

2 Review of the Literature of AC Methods

The Many AC architectures have been suggested in the survey to treat the security problems in IoT. The most recent and relevant are listed below, and we categorize them based on their access to two categories: traditional architecture and decentralized engineering.

2.1 Traditional Architectures

In this section, we study the works proposed in the traditional architectures in IoT, describe the various authorization platform solutions existing in IoT area:

First, the Extensible AC Markup Language (XACML) [3] is an XML-based language and offered by standardized the Organization for the Advancement of Structured Information Standards (OASIS) that is used for making authorization in AC and describes both. In addition to doing the following two functions: an AC policy language such as Attribute Based AC (ABAC) and AC decisions such as request and response. An overall study on XACML can be found in [12]. In this direction, the public authorization platform for the IoTs is suggested in [13]. It provides precise and malleable AC management for devices with limited memory and low power. When considering the current Internet standards and suggestion of AC solutions like XACML and SAML [14], the local AC decision is implemented by the local PEP (Policy Enforcement Point) object, considering its local terms and taking the authority over the XACML commitment. Because of the hard work of assessing the XACML security strategy for devices that have high resource constraints, most decision-making to grant access is an external PDP. Therefore, the solution remains to do not achieve lightness (very heavy) to be included in the device side.

Second, the AC method is based on Open Authorization (OAuth) is a framework for controlling access for customers who access resources on web servers. The OAuth1.0 [15] Protocol and OAuth 2.0 Framework [4] was created to solve privacy and AC problems for a range of wide online applications and licensing that enables users to grant 3rd-party (RP) applications that act as an associated party (RP) and access to protected resources (hosted on a specific service) that play the role of identity provider (IdP) and 3rd-party credentials. An overall study of OAUTH can be found in [4]. There are already a lot of OAuth depending on user accounts provided by main service providers (social networking) such as Microsoft, Facebook, Google, etc. This framework is productive when the number of objects and links is small, but it cannot be completely managed when this number grows.

Third, the AC solution based on UMA [5] is characterizes how it can be used in the Protocol of Web Authorization (OAuth) with user managed access (UMA) in an Internet area; it presents various scenarios with a Vehicle, Automotive Door, etc. These scenarios include web applications, smart phones and Internet objects. UMA is a relatively new control of access, which is less commonly used in Internet things, but it provides interesting features, such as user mode and offline playback. Therefore, many IETF efforts [5] are being run to adapt UMA with IoTs. To achieve the required and appropriate licensing mechanisms, an overall study of diverse security protocols was conducted. The UMA protocol that integrates the OAuth protocol has been identified as an appropriate mechanism for the perceived authorization structure and the UMA application specification is specified in SWoT [12]. Shortage of detail is one of the main disadvantages of OAuth and the personal UMA file and the technical details or implementation are not clear and are still missing.

Fourth, the Federated Capability-based AC model (FedCAC) [6]: Recently, the authors proposed a new framework for addressing challenges like granularity, dynamicity, and scalability in secure AC for IoTs. By authorizing some of the identity authentication and job authorization to the center delegator, he called the centralized policy decision making center (PDC) to reduce the overhead node traffic.

The FedCAC framework suggested distributing some jobs of validation to local nodes to be lighter and context awareness authorized. By the share of secure access authorization tasks between smart nodes in IoTs that allows node-to-node communication, which supports scalability and interoperability in IoTs network. However, it is basically still having weaknesses in using the centralized authentication control scheme that includes single-point of failure and traffic overload.

2.2 Decentralization Methods

In this section, we study the proposed works in the decentralized architecture of the IoT and the half of the various architectural solutions existing for authorization in the field of Internet of things. In this subject, the authors [2] propose a new framework for controlling IoT access based on BC technology. Presentations include a reference model of the proposed framework within the objectives, models, architecture and mechanism of the Internet. In addition, FairAccess gives an authorized management framework and an alias for maintaining privacy that enables users to own and control their data. To implement the FairAccess form, for use and modify the BC to the decentralized AC manager, through providing new kinds of transactions which provide award access, and delegate and revoke access.

In the FairAccess work [2] when the time token expired or cancelled, the subject wants to connect with the owner or give it other tokens and the new token required at least two blocks in BC, so it is costly, and that needs more time in access grants and they only support token-based authorizations. But in the work [7] the authors offer a new framework called ControlChain. In ControlChain, each identity can be associated with any other identity and can give parameters and properties to those associated with using the low processor profile in the authorization process for each identity. Although ControlChain has a complete architectural layer, it's decentralized and resilient. Allow work offline. It does not include the ControlChain architecture of Internet devices and objects in the BC.

In comparison with ControlChain [7] the author presents in [8] a new architecture for controlling Internet of Things objects to solve a challenge that stores BC information stored in Internet of Things objects that is due to their restricted nature. Architecture provides decentralized management of the AC system and is connected to geographically distributed sensor networks. Methods are based on BC technology during the implementation of AC policies. By adopting BC, this removes centralized access control management. Architecture [8] has a fully distributed management system for controlling access to Internet objects based on BC technology. Most Internet devices will not be able to store BC information because of their restricted nature. Thus, this structure does not contain Internet devices in BC objects, but rather recognizes a node called a management hub that requests AC information from BC on behalf of IoT devices. It is divided into six different components such as managers, wireless sensor networks, smart nodes, BC network, proxy nodes and management centers. It supports proof of concept. Therefore, the management center node functions as a single central point server in the AC and there may be heavy traffic congestion when there are frequent access requests or updates.

In this work [10] the author proposed a platform to supply an end to end (E2E) solution for the security of managers of secure access to IoT devices named IoTChain and combine it with the Object Security Architecture for IoT (OSCAR) and the Authorization and Authentication for Constrained Environments (ACE) Framework. With Replace with the only trusted authorization server in the ACE framework with an unreliable authorization of the BC to make AC resources strong and flexible and may maintain privacy and require that the attacker control at least 51% of the BC before the attacker can obtain illegal tokens. Furthermore, OSCAR uses a self-healing group key distribution system to enable effective multicast streaming of IoT resources. However, the IoTChain [10] platform does not adequately deal with the expansion of ban strings in case of use where IoT coefficients are very frequent, but they seem to trust E2E connections to IoT objects as transaction exporters to BC.

Another work in [9] Authors offer TrustChain. A TrustChain can create reliable transactions between strangers without centralized control. This allows new areas to use BC with a focus on building confidence among individuals. Their innovative approach provides scalability, openness and Sybil-resistance with the replacement of PoW with a mechanism for validating and correcting transactions. A TrustChain is a leakproof data structure without permission to store agent transaction logs (agent records). They create a constant series of interactions required temporarily for each agent. It is parallel in nature and creates each agent in its own configuration block. TrustChain includes a new Sybil-resistant algorithm called NetFlow to determine an agent's confidence in an online community. NetFlow ensures that agents who get resources from the community also contribute to return. There is evidence that the irrefutable history of historical transactions provides security and scalability without the need for universal consensus. The major contribution of the work [9] is to extend the architecture of the most general BC developed so far. It offers a way to provide distributed trust, free of any gatekeeper, while still offering strict limits on the profitability of the Sybil attack. Also this work [9] was limited to a model resistant to the Sybil attack and did not address the rest of the types of potential attack (e.g. replay, man-in-the-middle, ballot-stuffing, slander, eclipse, etc.) on the loss of a trusted identity between the callers to the system and focus on the challenging problem of Sybil-resilient calculation of trust scores and operational systems built upon TrustChain.

Finally, inspired by intelligent nodes and BC technology, a decentralized framework for AC to IoTs, authors [11], BlendCAC, an AC based on the decentralized ability based on BC for Internet IoTs. BlendCAC aims to effectively monitor access to devices, services and information in a wide range of Internet systems. Using to the BC network, it is proposed to create a mechanism to delegate the ability to deploy access permissions. Proposes a strong identity-based identity management strategy that benefits from a smart contract for recording, licensing and cancellation of access authorization. BlendCAC is the different scheme discussed above, with providing granularity, scalability and dynamicity for AC strategies in IoTs. By embedding AC policy into a smart contract, published via the BlendCAC scheme, IoT devices (or users in IoT objects) are the master of controlling their resources (data, devices, or services) rather than being supervised by a central authority. Enforcing authorization and accessing correct authentication among a large number of distributed edge devices allows for more coordination on edge networks.

3 Comparison Based on the Challenges for IoT Mechanism

In this section we will comparison for the two categories that are traditional models and decentralization architectures with criteria evaluated based on the challenges of support for the IoT mechanism such as: Scalability, Fault tolerant, no 3rd-party, new authorization, get authorization, integrated relationship, Compatibility and Low object overhead, those are challenges for the IoT mechanism.

Scalability: The new architecture is decentralized such as, FairAccess, ControlChain etc. Decentralized architectures are more scalable methods than the others (OAuth, UMA, and XACML). Because the architectures designed to make decentralized principles were in mind from start to work. However, all entities in the management hub architecture are part of the BC technology except Internet objects and the management center (hub) nodes. The BlendCAC uses a large number of distributed edge devices to force delegation and access to the correct check to allow for more coordination on edge networks.

Fault Tolerant: This standard assesses the impact of hardware or communication connections. Of course, decentralization is more tolerant with the error (fault tolerant) of those central. However, FairAccess requires that the resource owner publish a token every time that access is requested. This gives ControlChain a slightly better evaluation, but not more than management hub is allowed to fault tolerant, because this architecture does not include IoT devices and BC objects, instead, it identifies a new node called the administrative hub that requests AC information from the BC on behalf of Internet devices.

No 3rd Parties: Using this criterion for evaluating the dependence on 3rd parties to prevent the detection of censorship, fraud and interference. Whereas, the centralized architectures depend on 3rd parties when using server control for making to authenticate the notarized access and, thus, only all decentralized architectures receive positive evaluations.

New Authorization: In the new authorization standard, we can evaluate the lead time to create or change a mandate (latency). The central architecture, XACML, OAuth, and UMA have low latency in these activities because their database update is straightforward. Thus, they received a positive assessment. For FairAccess, ControlChain, and TrustChain, the full analysis of this standard depends on the speed of blocks deployment and, essentially, the type of proof used in blocks mining. Since the most common and well-known type of proof of action and this type of proof imposes considerable time on latency and provides a negative assessment. It is important to note, for example, that FairAccess has a central symbolic resource (owner) similar to central architecture, but the contract cannot be modified with a token published on BC. Although, the management hub architectures propose the management hub node to manage and block the dissemination by using proof-of-stack. However, it has low latencies in these activities but not more than the centralized architectures. Because any node may be having more than one of the management hubs to manage.

Get Authorization: This criterion evaluates the response time for authorization. XACML, OAuth, UMA, ControlChain, and the models using Administration Centers such as management hubs, contain proxies in real-time. FairAccess has a greater response time because after an indirect request to the resource owner, two consecutive pieces are extracted before granting access. Thus, it is the only one receiving a negative assessment.

Integrated Relationship: The work [7] is contains a structure designed to allow direct relationships with rules. Thus, the only positive evaluation was. As well as to the innovative TrustChain approach to scalability, openness and Sybil resistance with the replacement of PoW technicality to determine the validity and integrity of transactions. And, the management center (hub) has a communication between administrators and IoT objects through the BC network allowing communication across the platform.

Compatibility: This standard evaluates architectural compatibility with many of the models currently used in Internet objects. XACML, TrustChain, Management hub, ControlChain, IoTChain and BlendCAC correspond to a large amount of forms and has received a positive evaluation. Because the Node in a BC network must include a copy of the BC not in IoT devices, that is present in the Management hub, it has allowed being more compatible more than previous architecture. But, OAuth, UMA, and FairAccess can work with virtually any model in the background; however, they only work with access codes in the foreground and are therefore negatively evaluated.

Low Object Overhead: This standard evaluates the amount of excessive loading of the object through the authorization process. In central architecture, each authorization process is delegated to a robust entity by design. However, nothing prevents Fair-Access and ControlChain from authorizing their authorization process as well. In fact, ControlChain can facilitate the automation of this process when necessary. The device, in Relationship BC, can search for other devices for the same owner and ask for support in the authorization process. Therefore, all structures were evaluated positively except for the management center that requests AC information from the BC on behalf of IoT devices that have been positively evaluated double. To conclude, we summarize the existing work in this area that offers the solutions for challenges in IoT AC management in line with the objectives of the comparison in Table 1.

Table 1. Architecture comparison based on in the challenges of IoT.

AC Arch.	Scalability	Fault tolerant	No 3rd parties	New authorization	Get authorization	Integrated relationship	Compatibility	Low object overhead
XACML	No	No	No	Yes	Yes	No	Yes	Yes
OAuth	No	No	No	Yes	Yes	No	No	Yes
UMA	No	No	No	Yes	Yes	No	No	Yes
FedCAC	Yes	Yes (partially)	No	Yes	Yes	Yes	Yes	Yes (partially)
FairAccess	Yes	Yes (partially)	Yes	No	No	No	No	Yes
ControlChain	Yes	Yes	Yes	No	No	Yes	Yes	Yes
Management hub	Yes	Yes (complete)	Yes	No	No	Yes	Yes	Yes (partially)
TrustChain	Yes	Yes	Yes	No	No	Yes	Yes	Yes
IOTChain	No	Yes	Yes	No	No	Yes	Yes	Yes (partially)
BlendCAC	Yes	Yes	Yes	No	Yes	Yes	Yes	Yes

4 Conclusion

To our knowledge, many of the major studies published by many IoT researchers in their studies have focused on the development of a governance control structure that meets all the new problems and challenges raised by lot, but few of them deal with management and express control policies. In accessing the IoT scenario in this section, we have scanned the most important works in this area that have been overlooked by many previous works. By surveying existing structures used to control Internet access in the field using traditional structures and decentralized structures. At present, this field is in a very nascent stage. The infrastructure structure shows security and maturity signs. However, much must happen in the areas of Internet applications, objects and communication technologies. These fields will surely mature and affect human life in ways unimaginable over the next decade. From the context of IoT technology, BC based smart contracts are expected to play a major role in the management of secure for IoT devices and control them and most importantly secure them. So, we discuss and summarize some of the core features of BC that can be very useful for the Internet of things in general, and IoT security in particular. Although, IoTChain [10] suggests an unreliable IoT structure where IoT objects register themselves in a series of keys to store, organize, and share data flows safely without requiring them to provide a trusted middleware. However, the IoTChain [10] work does not adequately deal with the expansion of the prohibition strings in case of use where IoT coefficients are very frequent but trust in communication from E2E for IoT objects as transaction issuers to BC.

The BC has provided a solution to these challenges readily and safely. The BC has been widely used to provide reliable and documented identity, property tracking and control of products, commodities and assets. Methods such as TrustChain [9] suggest enabling authenticated transactions using BC while maintaining transaction integrity in distributed systems. IOT objects are no exclusion. BC can be used to record and give objects connected to IOT objects, with a host of features and complex relationships that can be downloaded and stored in a distributed group ledger. BC also provides reliable decentralized management, governance and movement at every point in the supply list and the life cycle of an Internet device. Supply list can include many agents, such as manufacturer, vendor, supplier, distributor, shipper, installer, owner, repair, reinstallation, etc. And, in BlendCAC [11] Encapsulation of AC policies in intelligent nodes, deployed through this platform, Internet objects and devices (or Internet user objects) are the main masters to control their resources (data, devices or services) rather than supervised by the Central Authority. BlendCAC uses many distributed edge devices to force delegation and access to the correct check to allow for more coordination on edge networks. From our review, we note that BC has interesting features required for a wide range of IoT systems, especially in managing access control.

References

1. Siddiqa, A., Shah, M.A., Khattak, H.A., Akhunzada, A., Ali, I., Razak, Z.B., Gani, A.: Social internet of vehicles. Complexity, adaptivity, issues and beyond. IEEE Access **6**, 62089–62106 (2018)
2. Ouaddah, A., Abou Elkalam, A., Ait Ouahman, A.: FairAccess: a new blockchain based access control framework for the Internet of Things. Secur. Commun. Netw. **9**(18), 5943–5964 (2016)
3. Webfarmr.eu: XACML 3.0.: enhancements, Nano scales Res. Lett. **6**(1), 297 (2011). https://doi.org/10.1186/1556-276x-6-297, http://www.ncbi.nlm.nih.gov/pubmed/21711787
4. Hardt, D.: The OAuth 2.0 authorization framework (2012). http://tools.ietf.org/html/rfc6749.html
5. Tschofenig, H., Maler, E., Wahlstroem, E., Erdtman, S.: Authentication and authorization for constrained environments using OAuth and UMA. IETF Internet Draft (2015). draft-maler-ace-oauth-uma-00.txt
6. Xu, R., Chen, Y., Blasch, E., Chen, G.: A federated capability-based access control mechanism for Internet of Things (IoTs). In: 2017 Sensors and Systems for Space Applications XI, vol. 10641, p. 106410U. International Society for Optics and Photonics, May 2018
7. Pinno, O.J.A., Gregio, A.R.A., De Bona, L.C.: ControlChain: blockchain as a central enabler for access control authorizations in the IoT. In: GLOBECOM 2017-2017 IEEE Global Communications Conference, December 2017, pp. 1–6. IEEE (2017)
8. Novo, O.: Blockchain meets IoT: an architecture for scalable access management in IoT. IEEE Internet Things J. **5**(2), 1184–1195 (2018)
9. Otte, P., de Vos, M., Pouwelse, J.: TrustChain: a Sybil-resistant scalable blockchain. Future Gener. Comput. Syst. (2017)
10. Alphandy, O., Amoretti, M., Claeysy, T., Dall'Asta, S., Duday, A., Ferrari, G., Rousseauy, F., Tourancheauy, B., Veltri, L., Zanichelli, F.: Department of Engineering and Architecture, University of Parma, Italy University of Grenoble Alpes, CNRS, Grenoble INP, LIG, F-38000 Grenoble, France. IoTChain: A BC Security Architecture for the Internet of Things (2018). https://hal.archives-ouvertes.fr/hal-01705455/document
11. Xu, R., Chen, Y., Blasch, E., Chen, G.: BlendCAC: a blockchain-enabled decentralized capability-based AC for IoTs. arXiv:1804.09267v1 [cs.NI], 24 April 2018. https://arxiv.org/abs/1804.09267, https://doi.org/10.1007/978-3-662-48000-7_29
12. El-Aziz, A.A., Kannan, A.: A comprehensive presentation to XACML (2013)
13. Seitz, L., Selander, G., Gehrmann, C.: Authorization framework for the Internet-of-Things. In: 2013 IEEE 14th International Symposium on a World of Wireless, Mobile and Multimedia Networks (WoWMoM), pp. 1–6. IEEE, June 2013
14. Cantor, S.M., Kemp, I.J., Philpott, N.R., Maler, E.: Assertions and protocols for the oasis security assertion markup language. OASIS Standard, pp. 1–86, March 2005
15. Hammer-Lahav, E.: The oauth 1.0 protocol (2010)

Intelligent Communication Systems

Antenna Design Using UWB Configuration for GPR Scanning Applications

Jawad Ali[1,2](✉), Noorsaliza Abdullah[1], Asim Ali Khan[2], Roshayati Yahya[1], Muzammil Jusoh[3], and Ezri Mohd[1]

[1] Department of Communication Engineering, Faculty of Electrical and Electronic Engineering, Universiti Tun Hussein Onn Malaysia (UTHM), Parit Raja, Johor, Malaysia
[2] Department of Electrical and Computer Engineering, COMSATS University Islamabad (CUI), Lahore, Pakistan
[3] School of Computer and Communication Engineering, Universiti Malaysia Perlis (UniMAP), Kangar, Perlis, Malaysia
engrjawad.ali@outlook.com

Abstract. Ultra-wideband (UWB) technology is gaining more attention, as the wireless communication research is progressing with every passing day. This radio technology offers a greater bandwidth with nominal characteristics such as low power, system compactness, and cost-effectiveness. In 1887, first experiment was conducted on UWB for electromagnetic wave production [1]. However, the technology was restricted to military applications only. Thereafter, with the increasing demand of UWB spectrum, US states department Federal Communication Commission (FCC) localized the bandwidth of 7.5 GHz from 3.1–10.6 GHz for the commercial and experimental purpose. Since then, this UWB technology was also used in the development of antenna design, because antennas are the front-end components of the radio frequency (RF) system used in many applications. Therefore, the recent development of RF system is focusing mainly on the search and rescue applications for victim detection behind the obstacle, to make UWB a significant technology. The aim of writing this research article is to highlight the characteristic configuration of a tilted arc-shaped compact size planar antenna design with the frequency ranging from 2.8–15.6 GHz and a gain of around 6 dB for human detection applications. To fulfill the purpose, an antenna is carried out under different configurations for the development of a 2D image in the form of radar cross section (RCS). Thereafter, the analysis is performed to validate the human skin detection capability of the designed antennae. Therefore, a suitable antenna configuration possess proper results is proposed with the help of experimental modeling to proceed further.

Keywords: Ultra-wideband antenna · GPR · UWB applications · Antennae

1 Introduction

Antenna devices are known for the conversion of electrical energy into radio waves and vice versa. It is considered to be the front-end component device for radio frequency (RF) systems as a useful transmitter and/or receiver. However, the antenna became an

F. Saeed et al. (Eds.): IRICT 2019, AISC 1073, pp. 497–510, 2020.
https://doi.org/10.1007/978-3-030-33582-3_47

important component for electrical devices in late 18's, especially in wireless communication field; scientist named Henrich Hertz was the first person to experiment the radio waves [2].

Ultra-wideband (UWB) technology now a day's is gaining more attention mainly in terms of antenna technology, which can be useful for different applications especially in the field of communication engineering. The technology mainly incorporates a system possess very large operating bandwidth; from the study the bandwidth in this case should be greater than 500 MHz with less power consumption characteristics [2]. The localization of free UWB spectrum by Federal Communications Commission (FCC) in 2002, opened the new gates for research and development in different technologies mainly those with greater spectrum. The frequency band allocation for UWB applications varies from 3.1 GHz to 10.6 GHz which incorporates a 7.5 GHz of bandwidth. Based on the study, UWB has several advantages which are not limited to a larger bandwidth and minimum power consumption. UWB also has low cost of implementation, high data rate because of the greater spectrum, and is also capable of obstacle penetration in many GPR applications [1] studied earlier. Since, UWB has a wide range of frequencies due to which it is useful for many narrowband applications, prominent in the field of communication engineering main of which lies in Wireless Local Area Network (WLAN) with frequencies i.e. 3.6 GHz, 4.9 GHz, 5 GHz and 5.9 GHz, Worldwide Interoperability for Microwave Access (WiMAX) possess 3.5 GHz, and other wireless related applications [3]. Further to this, UWB also has contributions in the field of radar dominantly for ground penetrating radars (GPR), imaging and positioning radars [4, 5].

Ground Penetrating Radar; more specifically Surface Penetrating Radar (SPR) [6] is becoming one of the top positioned ground probing techniques among all existing geophysical methods. The basics of aforementioned technology is related to the reflections and scattering of the radio waves [7] generated using different wireless components for localization, identification, and imagination of the targets buried in the soil as well as under various guises [8]. Thus, the aim of writing this article is to discuss the antenna design for GPR applications. Therefore, the working principle of GPR technology is shown in Fig. 1, mainly to identify the target behind the object. The mentioned figure clearly illustrated that the transmitter antenna transmits a radio wave which probes into the ground surface or rubble and strikes the target to be detected. The target then have different responses to the received energy, some of the signals are reflected back which is being received by the receiver in order to detect the buried target while other is being absorbed by the target and scatters into different direction which is known as the scattered energy, this energy cannot be detected by the receiver. However, it can be extracted from the literature study [7] that there is also another kind of interference or cluttering called refracted energy from the other buried objects as shown in Fig. 1, which may also cause a disturbance in the signal received by the receiver. Moreover, the receiver also has the effect of the transmitted signal which it receives directly from the transmitter. To deal with all these unwanted signal and to achieve the accuracy in results, simple filtering techniques can be useful. Nowadays the performance in the field of GPR is more likely linked with ultra-wideband (UWB) technology. Various GPR applications were reported in the field of engineering and technology for different military as well as civilian purpose. It mainly involves the detection of land mines, imaging through/behind the wall, also in the methodologies for remote sensing in non-destructive interference testing of a concrete and rescue of people buried under-debris/in opaque environment [9]. The detection of buried victims under the debris is becoming more and more important task of the recent era, the main reason of this is the mega constructions which on the one hand is fascinating the humans with their huge structures. While on the other hand, it is very dangerous to risk the life of the people, when they are exposed to natural disasters such as earthquakes and floods.

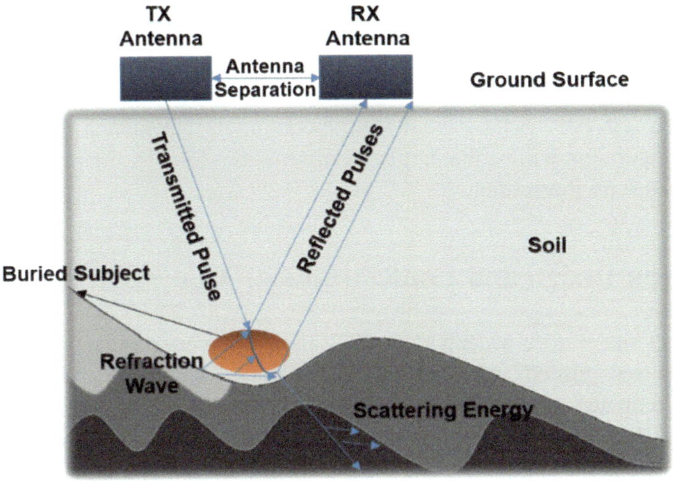

Fig. 1. Working principle of GPR technique.

Thus for the above-mentioned reasons, the study on the implementation of UWB antennas using GPR technique for detection and identification of the buried victims is carried out. Moreover, many UWB antenna designs were reported in the study previously conducted by different researchers. Refering to the literature study mentioned in [10] one can identify the appropriate reason for the choice antenna type. In addition to this, the antenna designed on a low frequency band was studied to improve the penetration depth mainly. Whereas the antenna designed on high frequency band was performed to attain better imaging resolution for GPR technologies. Thereafter, the antennas such as bow-tie, spiral, horn, planar etc. were designed and analyzed under different simulation environment [10]. These reported designs were analyzed for bandwidth, gain and directive properties, which are the basic requirement of antenna components for GPR applications [11]. From the analysis, it has been concluded that planar antenna can be carried out to further the study, as horn antennas are usually of very large size and other antennas are more complex to develop compared to planar antennas, which are very simple and have ease of fabrication.

Therefore, this article is about the design of a tilted arc-shaped UWB antenna with operating range from 2.8 GHz to 15.6 GHz under different configurations i.e. Monostatic and Bistatic, in order to validate the antenna for scanning purpose. Since, the UWB band has been adapted with the help of literature study on previously developed antennas for human detection behind the obstacle [10], where the designs were configured with very large bandwidth and gain. The reason for choosing UWB is that the low-frequency components will help in increasing penetration depth because of greater wavelength, which causes a number of waves to pass through the surface. While the reason for choosing greater bandwidth is mainly related to resolution imaging for human detection, the greater is the bandwidth the more information can be carried. Another technique called defected ground structure (DGS) [5] is also the part of this study for antenna characteristics such as gain and bandwidth achievement. Consequently, the article is divided into two main parts where antenna design and its results will be discussed. Followed by the design, modeling of the antenna for monostatic and bistatic configuration will be the part of the discussion. Finally, the proposed work will be concluded with the results.

2 Antenna Design and Configurational Analysis

The compact shape ultra-wideband (UWB) planar antenna designed with the integration of defected ground structure (DGS) method is simulated using commercially available Computer Aided Design (CAD) simulator, known as Computer Simulation Technology (CST) under Microwave Studio (MWS) environment. Whereby, different antenna configurations, as mentioned earlier were modeled, while simulations were carried out to perform the experiments for human detection behind the obstacle. To perform the in-depth analysis of a proposed antenna design as well as the modeling of the antenna under different configurations, this section is categorized into sub-sections. Where the designing of a proposed arc shape ultra-wideband antenna which is configurable for human detection behind the obstacle is the part of discussion under Sect. 2.1. Thereafter, Sect. 2.2 will discuss the simulation-based experimental modeling for scanning behind the wall. The proposed configuration constitutes a planar antenna to transmit/receive the signal; an obstacle such as a concrete wall for signal penetration and a human skin model representing the target to be detected using CST microwave studio. Finally, a MATLAB is programmed to develop the 2D image of the simulated configurations in the form of Radar Cross Section (RCS) for the received reflections to analyze the reflection strength and antenna scanning capability.

2.1 Planar Antenna Design

The implementation of an ultra-wideband antenna design for GPR technology and for the validation of human detection capability behind the obstacle, a tilted arc-shaped planar antenna is proposed. The design is carried out using Rogers RT-5880 dielectric substrate possess $\varepsilon r = 2.2$ having defected ground structure characteristics to match the bandwidth and gain of the antenna. As mentioned earlier that the reason for choosing planar antenna is its design simplicity as well as the leverage of involved parameters to achieve the required results. Basically, a planar antenna consists of three main components; a radiating patch for signal propagation through the surface; a substrate material with some dielectric constant value to reduce the fringing effect of the radiating patch during excitation; and a ground to neutralize the effect. To achieve the tilted arc-shaped radiating patch, the antenna was first carried out with an elliptical shape patch radiator. Whereby, a partially circular slot is then introduced towards the minor axis of an elliptical patch at the place where the current is minimum. Thereafter, an integrated circular slot is further optimized by shifting it to different positions for the tuning of scattering parameter mainly S_{11}. Therefore, the resultant of optimization is a tilted arc-shaped planar antenna. The designed UWB antenna is shown in Fig. 2, where the size in terms of antenna compactness is demonstrated by the dimensions of a substrate. However, the compactness of UWB antenna has the challenge of bandwidth matching at lower frequency [11, 12] over the band from 3–10 GHz. The term bandwidth match, illustrating the cut-off value for the antenna which in this is below −10 dB for the range of frequencies. To achieve the bandwidth matching over the entire UWB band while maintaining the size compactness, a modified half grounded method is introduced in the design. In this method, instead of entire copper coated ground, a half grounding is etched on one side of the substrate. The gain acquired as a result of aforementioned half grounded structure is low, however, to validate antenna for GPR applications there must be a nominal gain, which in this case is standardize to 6 dB [5]. Thus, to achieve the targeted gain, a DGS method [5, 13] is used based on which the slots were added at the different positions on the ground surface as a result of an accumulated process where slots were created one after the other. The design prototype of the proposed antenna including the patch as well as ground with DGS is presented in Fig. 2.

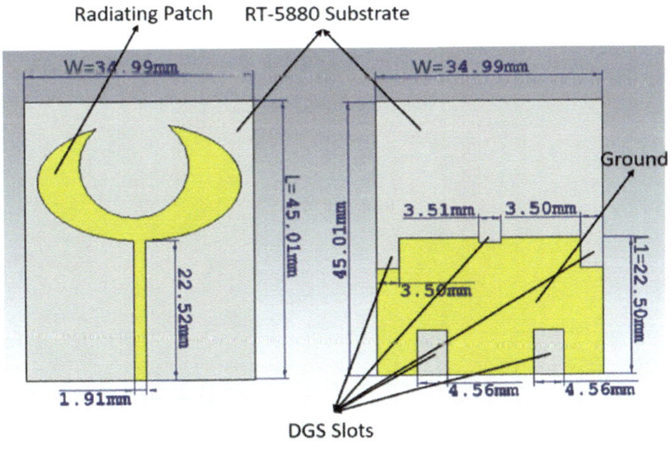

Fig. 2. Design of tilted arc antenna: Top view (left); Bottom view (right).

The slots shown in Fig. 2 are the combination of even and odd dimensions, acquired as a resultant of parameteric study. The summary of antenna in terms of the frequency band, antenna impedance, materials used in the design, dimensions of the substrate as well as ground, with the major and minor axis of the tilted arc is mentioned in Table 1. Moreover, the cut-off frequencies to analyze the far-field of the antenna are also the part of Table 1.

Table 1. Parameters and dimensions of the designed antenna.

Parameter	Material	Values
Major arc axis (mm)	Copper	31
Minor arc axis (mm)	Copper	19
Substrate L × W (mm)	RT-5880	45.01 × 34.99
Ground L1 × W (mm)	Copper	22.5 × 34.99
Frequency band (GHz)	–	2.8 to 15.6
Impedance (Ohm)	–	50
Cut-off frequencies (GHz)	–	3, 6.5, 8
Center frequency (GHz)	–	9.2
Bandwidth (GHz)	–	12.8

As mentioned earlier, the complete design discussed in this section is carried out for simulation using CST Microwave studio environment. Thereafter, a successful simulation of the proposed antenna was carried for fabrication. The fabricated antenna as shown in Fig. 3, was then carried out for the measurement. Whereby, the antenna is connected to a Vector Network Analyzer (VNA), an equipment to measure the return loss in terms of S_{11}, which shows the aptitude of the antenna [2]. However, one can manipulate S_{11} as the signal sent from the antenna in the form of the wave using port 1 of VNA, which is then received by the antenna using the same port as a result of mode switching characteristic of VNA. Therefore, the antenna in this case is measured up to the maximum available frequency of the VNA.

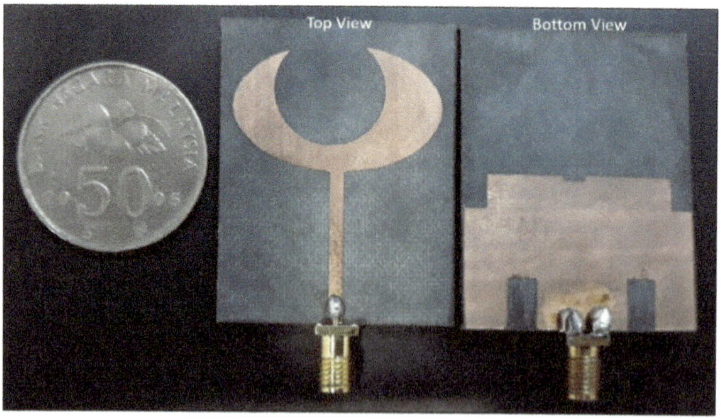

Fig. 3. Fabricated antenna: Top view (left); Bottom view (right).

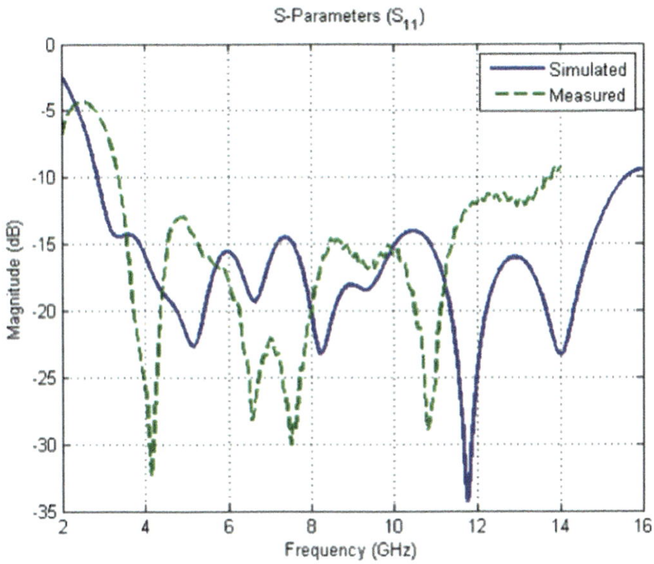

(a)

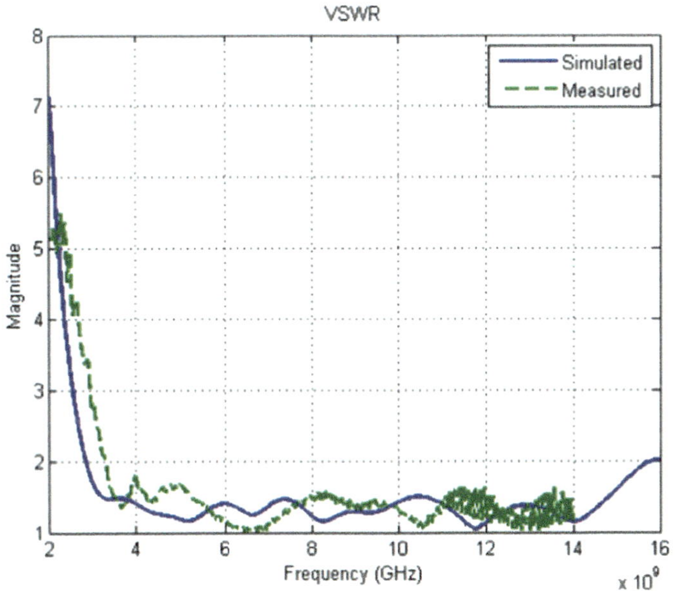

(b)

Fig. 4. RT-5880 based antenna (a) Return loss (b) VSWR.

However, the measured results illustrating a good agreement between simulation and measurement as shown in Fig. 4(a). As a result of which it can be stated, that the proposed antenna has the ability to perform operation from 2.8 GHz to 15.6 GHz,

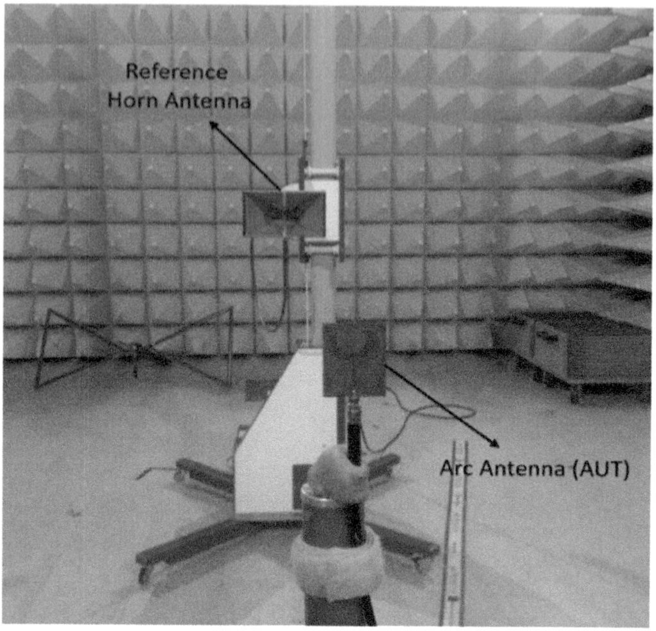

Fig. 5. Measurement set-up for antenna radiation.

since all the frequencies within the mentioned range are below −10 dB. Thereafter, the Voltage Standing Wave Ratio (VSWR) of the antenna was measured as shown in Fig. 4(b). Whereby, the VSWR is basically representing the reflected power of an antenna which should be small for better antenna performance. Therefore, Fig. 4(b) shows that the simulated and measured results showing a good agreement between the two. Which also provided the information that the antenna is operating between 2.8 GHz to 15.6 GHz without having a single interference during its path.

Furthermore, the antenna is carried out to measure the radiation pattern and gain in an anechoic chamber available at UTHM. The Antenna under test (AUT) is placed in front of reference horn antenna in an anechoic chamber for the radiation pattern measurement. The complete measurement set-up is demonstrated in Fig. 5. Whereby, the obtained results illustrated in Fig. 6 is validating the measured and simulated patterns of antenna radiation. Thereafter, the gain transfer method [14–16] is used to measure the gain, which in this case is 6.05 dB. Hence, it can be concluded that these obtained characteristics of the antenna are suitable to carry out the antenna for scanning applications especially for human detection behind the obstacle.

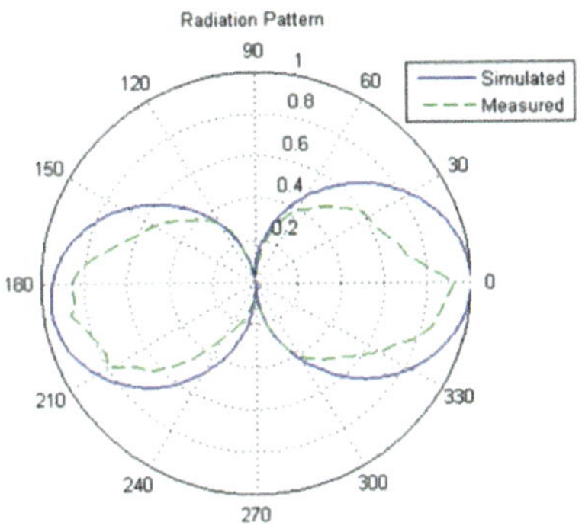

RT-5880 based antenna at 3GHz

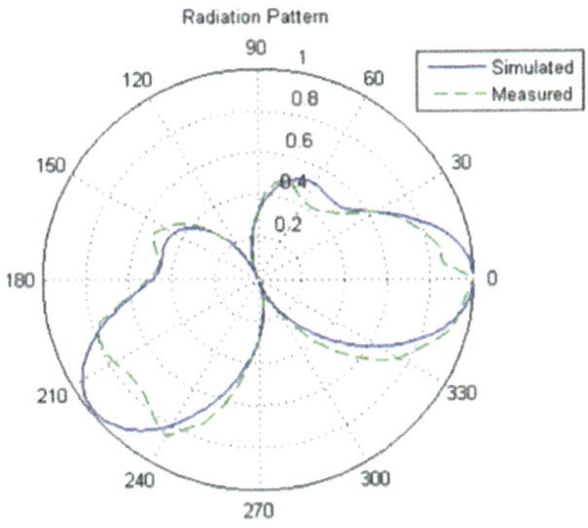

RT-5880 based antenna at 6.5GHz

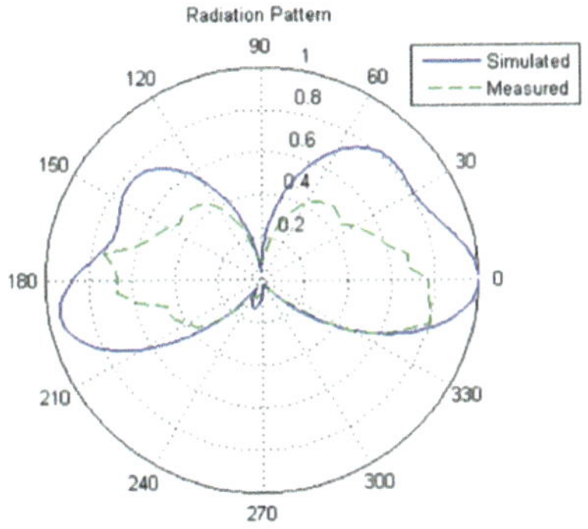

RT-5880 based antenna at 8 GHz

Fig. 6. Measured antenna radiation pattern.

2.2 Antenna Configurations Test for Scanning/Detection

The analysis of antenna presented in Sect. 2.1, provided the information about antenna bandwidth, and gain. Consequently, this section carried out a discussion on antenna modeling using different configurations for the detection of a targeted objects behind the obstacle. Therefore, in this study, a targeted object is human skin as it is the first and foremost thing which can be in contact with an electromagnetic waves during transmission, even if the buried person is not in the conscious state. While a concrete wall is considered as an obstacle, whereby, the simulation-based experimental set-up was carried out. During the process, a wave signal is transmitted using arc shape antenna, the signal penetrates through the obstacle which in this case is a concrete wall, to hit the target. Therefore, the target has different responses to the received signal, among which some of the energy is reflected back from the target towards the receiving antenna. Using the aforementioned theory, a monostatic configuration of the antenna is carried out to observe the detection capability of an antenna behind the wall. Whereby, the term monostatic is associated with the technology, when a single antenna can be used as a Transmitter (Tx) and a Receiver (Rx) [17] by switching the mode between the two. The scenario for human detection using monostatic configuration of the antenna with entire set-up is illustrated in Fig. 7. From the figure, it can be seen that the monostatic antenna placed at a certain distance from the concrete wall is used to transmit and/or to receive the signal. In the meantime, a targeted skin model of a human in the form of a rectangular block is placed behind the wall. The distance of the antenna to the wall is measured as Wall_D whereas, the distance between the wall and target is determined as Target_D, these defined distances can be varied to measure the penetration depth. The dielectric value of concrete wall is εr = 5.46 [18] as it does remained constant over the entire UWB frequency band. At the same time, it has been observed during the experiment that the dielectric constant for the skin is varied [19] at different UWB frequencies with the value appears to be εr = 26.31–36. Following the above mentioned monostatic configuration, the antenna has been simulated at different horizontal positions in front of the wall and data is being collected for each antenna simulation. The collected data is then processed using MATLAB based signal processing algorithm in which matrices have been used to develop the 2D imaging in the form of RCS and to validate the strength of a signal reflected from human skin, which then verify the detection capabilities of the designed antenna.

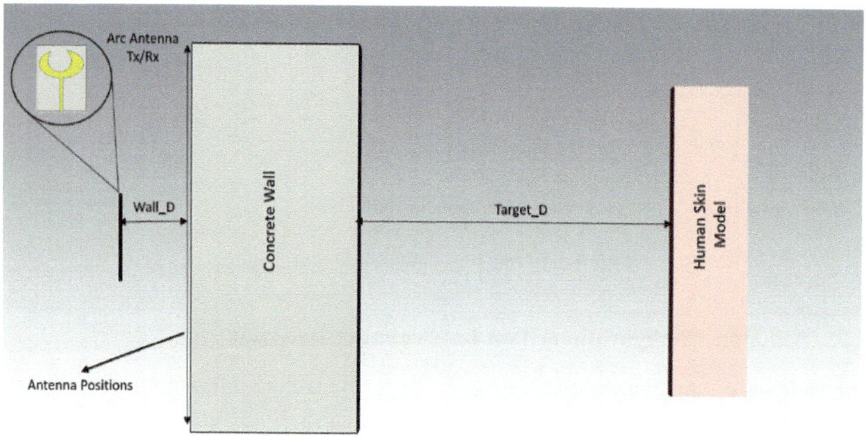

Fig. 7. Monostatic antenna configuration side view.

The 2D image developed as a result of monostatic based antenna configuration is presented in Fig. 8. From the figure, only a few traces have been observed and the detected target is demonstrating its presence in the encircled area. However, the traces are not very prominent, the main reason for which is the use of single antenna for transmission and reception at the same time. Thus, with the compact size single antenna, it is very difficult to receive the proper signal mainly due to the angle of reflection [20], as it is not necessary to receive the entire reflected signal in the same direction from where it was transmitted.

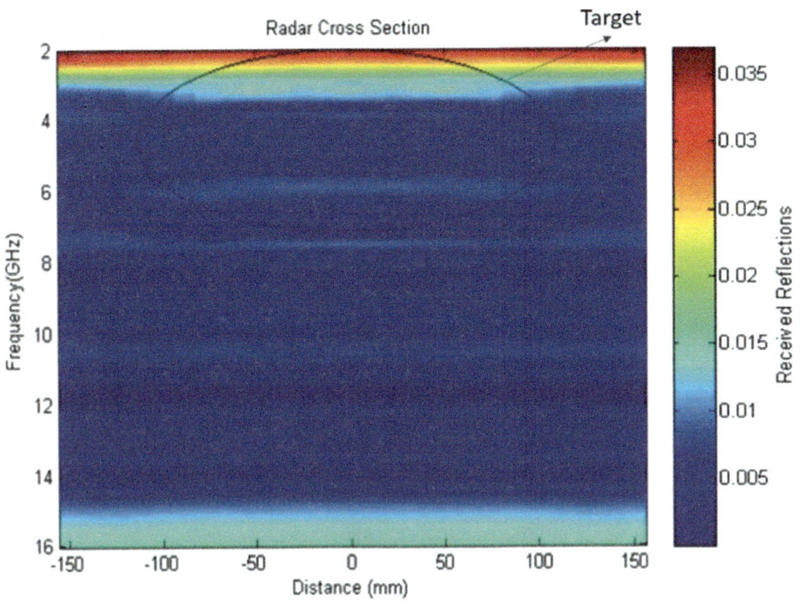

Fig. 8. RCS of monostatic antenna configuration using RT-5880.

Subsequently, to validate the antenna detection capability, another configuration named bistatic is experimented, as the traces of the object in case of monostatic configuration was not sufficiently significant. Therefore, the term bistatic is associated with the technology where two antennas can be used instead of the single antenna. Following the configuration, each antenna is dedicated for a single operation i.e. one antenna for transmission, while the other antenna for reception purpose, placed at the fixed distance from each other [21]. The proposed set-up for bistatic configuration is illustrated in Fig. 9.

Thus, it can be seen from Fig. 9 that two antennas were used by placing apart at a fixed distance from one another. Once again, the distance of the antenna to the concrete wall is measured as Wall_D, whereas the distance between the wall and targeted skin can be determined as Target_D. Thereafter, the proposed configuration is carried out for the experimental simulation in which both antennas placed at a fixed distance have been shifted to different horizontal positions in front of the wall to collect the data for each simulation. The collected data is once again processed using a MATLAB based algorithm to visualize the 2D image in the form of RCS for the validation of the received reflection's strength. The developed image for bistatic configuration demonstrates the significant results as can be seen from Fig. 10. The detected human skin in the figure is showing a clear traces encircled in the red and yellow color. While the red color outside the encircled area is the direct reflection of the wall on the antenna. The colors mentioned in Fig. 10 is illustrating the received reflection's strength. Thus, the bistatic configuration is validating the human skin scanning capability of the designed UWB antenna with prominent reflection strength.

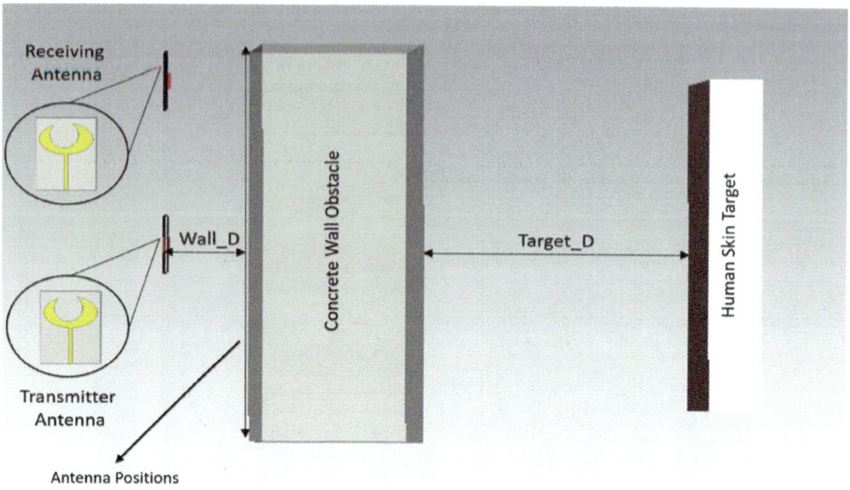

Fig. 9. Bistatic antenna configuration top view.

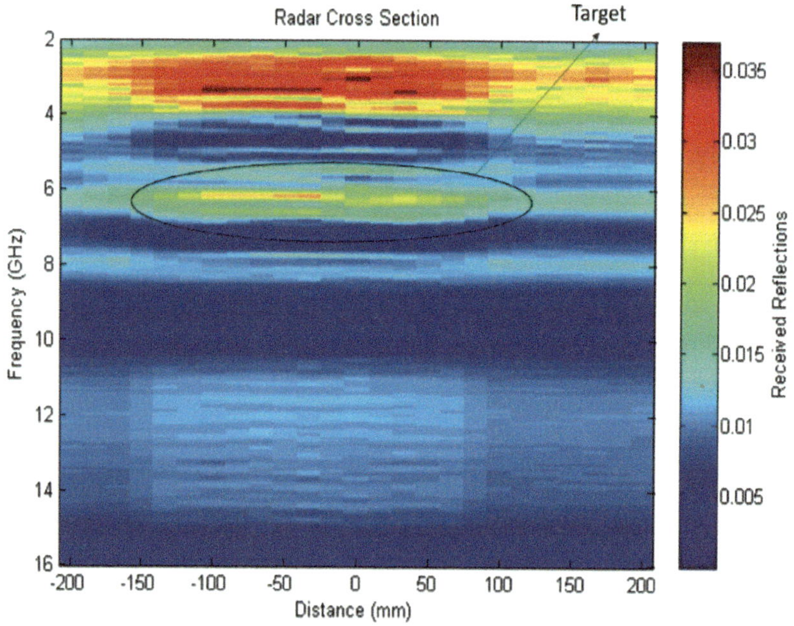

Fig. 10. RCS of bistatic horizontal antenna configuration.

3 Conclusion

This paper presented an insight on UWB antenna design and its use in scanning applications specifically for human detection behind the obstacle and/or under the rubble. A tilted arc-shaped antenna is proposed during the study, which is then carried out to undergo through different simulation and measurement processes for design validation purpose. The acquired bandwidth for the proposed antenna ranges from 2.8–15.6 GHz with 6.05 dB of gain, which is very helpful in human detection technology. Furthermore, the antenna was carried for different configurations i.e. monostatic as well as bistatic, to validate the detection capability of the designed antenna. The purpose of the analysis for the construction of a 2D image from the reflections is fulfilled using MATLAB, while, CST Microwave studio based simulation tool for the design. Thus, from the developed study it can be concluded that the proposed tilted arc-shaped antenna has the ability to detect the buried victims using bistatic configuration.

Acknowledgement. The authors would like to express the gratitude to ORICC, Universiti Tun Hussein Onn Malaysia (UTHM) for supporting this work under TIER 1 research grant VOT U860.

References

1. Valderas, D., et al.: Ultrawideband Antennas: Design and Applications. Imperial College Press, London (2010)
2. Balanis, C.A.: Antenna Theory Analysis and Design. Wiley, Hoboken (2005)
3. Zarrabi, F.B., Mansouri, Z., Gandji, N.P., Kuhestani, H.: Triple-notch UWB monopole antenna with fractal Koch and T-shaped stub. AEU-Int. J. Electron. Commun. **70**(1), 64–69 (2016)
4. Elboushi, A., Sebak, A.R., Denidni, T.: Through wall gap detection using monostatic Radar. Appl. Comput. Electromagn. Soc. J. **28**(5), 411–418 (2013)
5. Ali, J., Yahya, R., Abdullah, N., Sapuan, S.Z.: Ultra-wideband monostatic antenna for behind the wall detection. Int. J. Electr. Comput. Eng. **7**(6), 2936–2941 (2017)
6. Kota, S., Gunavathi, N.: Design and analysis of TEM horn antenna for GPR applications. In: 2017 Third IEEE International Conference on Sensing, Signal Processing and Security, pp. 15–22 (2017)
7. Jianjun, X., Ling, H.: A target imaging method of multiple-input-multiple-output ground penetrating radar-based on direction of arrival estimation. Curr. Sci. **114**(5), 00113891 (2018)
8. Serhir, M., Lesselier, D.: Wideband reflector-backed folded bowtie antenna for ground penetrating radar. IEEE Trans. Antennas Propag. **66**(3), 1056–1063 (2018)
9. Ahmed, A., Zhang, Y., Burns, D., Huston, D., Xia, T.: Design of UWB antenna for air-coupled impulse ground-penetrating Radar. IEEE Geosci. Remote Sens. Lett. **13**(1), 92–96 (2016)
10. Ali, J., Abdullah, N., Ismail, M.Y., Mohd, E., Shah, S.M.: Ultra-wideband antenna design for GPR applications: a review. Int. J. Adv. Comput. Sci. Appl. **8**(7), 392–400 (2017)
11. Ali, J., Abdullah, N., Yahya, R., Naeem, T.: Monostatic ultra-wideband GPR antenna for through wall detection. In: 2017 EPJ Web of Conferences, vol. 162, pp. 010761-4 (2017)

12. Koziel, S., Bekasiewicz, A.: A structure and simulation-driven design of compact CPW-fed UWB antenna. IEEE Antennas Wirel. Propag. Lett. **15**, 750–753 (2016)
13. Pandhare, R.A., Zade, P.L., Abegaonkar, M.P.: Miniaturized microstrip antenna array using defected ground structure with enhanced performance. Eng. Sci. Technol. Int. J. **19**(3), 1360–1367 (2016)
14. Gain Transfer Method. http://www.measurementest.com/2010/09/how-to-measure-antenna-gain-part-1-gain_08.html. Accessed 01 Mar 2019
15. Gain Transfer Method. http://www.diamondeng.net/library/GainXfer.pdf. Accessed 29 Dec 2018
16. Brochu, C.J., Morin, G.A., Moffat, J.W.: Gain measurement of a cavity-backed spiral antenna from 4 to 18 GHz using the three-antenna method. In: Defense Research Establishment Ottawa, Ontario, vol. 52, pp. 1–52 (1998)
17. Etellisi, E.A., Elmansouri, M.A., Filipovic, D.S.: Wideband monostatic simultaneous transmit and receive (STAR) antenna. IEEE Trans. Antennas Propag. **64**(1), 6–15 (2016)
18. Ahmad, F., Amin, M.G., Kassam, S.A.: Synthetic aperture beamformer for imaging through a dielectric wall. IEEE Trans. Aerosp. Electron. Syst. **41**(1), 271–283 (2005)
19. Tamyis, N.M., Ghodgaonkar, D.K., Taib, M.N., Wui, W.T.: Dielectric properties of human skin in vivo in the frequency range 20–38 GHz for 42 Healthy volunteers. In: Proceedings of the 28th URSI General Assembly (2005)
20. Jol, H.M.: Ground Penetrating Radar: Theory and Applications. Elsevier Science, Amsterdam (2009)
21. Kang, W., Kim, C.R., Kim, J.H., Park, S.G., Cho, S.J., Son, J.S., Kim, K.W.: A study of antenna configuration for bistatic ground-penetrating radar. In: 2016 IEEE 16th International Conference on Ground Penetrating Radar (GPR), pp. 1–4 (2016)

A Robust Hybrid Model Based on Kalman-SVM for Bus Arrival Time Prediction

Abdirahman Osman Hashi[1,2(✉)], Siti Zaiton Mohd Hashim[2(✉)], Toni Anwar[2(✉)], and Abdullahi Ahmed[1(✉)]

[1] Faculty of Computing, SIMAD University, Mogadishu, Somalia
Wadani12727@gmail.com, aaayare@simad.edu.so
[2] Department of Computer Science, Faculty of Computing, Universiti Teknologi Malaysia, 81310 Skudai, Johor, Malaysia
sitizaiton@utm.my, toni_anwar@yahoo.com

Abstract. Transport management is a system that allows travelers to plan, schedule and manage their journey from takeoff till arrival. Passengers involved in the journey either use private or public vehicles and they always estimate their arrival-time. As the total vehicles on the road increases, the possibility of traffic congestion would also increase which is becoming a major problem for traffic planning and scheduling. To address the issue of congestion, many scholars' opted for solutions such as building a proper infrastructure and developing an accurate travel time prediction models such as historical datasets and statistical models. However, these solutions are not efficient from an economic perspective. Therefore, the key goal of this research paper is to propose a new robust model that could forecast the arrival-time of buses by using support vector machine and Kalman filtering algorithm. The experimental results showed that the prediction model produced an outstanding performance in terms of prediction accuracy over the available historical traffic data (achieved) models using Genetic Algorithm – Support Vector machine (GA-SVM) and Artificial - Neural - Networks (ANN). Based on the analysis, the performance of this prediction model is 87.63% better than ANN which was 83.5%. The proposed model has contributed to the field of study by introducing an integrated process for estimation bus arrival-time prediction in Artificial, data mining and Transportation Engineering field.

Keywords: Bus arrival time prediction · Support vector machine · Kalman Filtering · Machine learning

1 Introduction

The development of public transportation system and growth number of passenger's causes to gain for some technologies to be adopted, those technologies includes bus-tracking, location and also communication while it becomes more notorious [1]. To help passengers, it is a good assistant by providing a reliable service in the public

F. Saeed et al. (Eds.): IRICT 2019, AISC 1073, pp. 511–519, 2020.
https://doi.org/10.1007/978-3-030-33582-3_48

transportation; so that, these technologies should play an essential role by predicting the arrival-time at stops in easier way and make the information available [2].

Moreover, it helps also by the passengers to be scheduling for their journey an efficient way as long as they get the available accurate for the bus arrival-time prediction, and this should be helpful to passengers for efficient journey preparing and avoiding delay times [2, 9]. For the implementing an effective tour planning means, a powerful prediction of bus arrival-times has been gaining more significance [3]. The availability of some technologies that includes a Vehicle-location-information (VLI), as a part of Advance-Travel-Information (ATIS) takes part to enhance the performance precision of the bus arrival-time [4, 11].

In the recent decades, due to the rapid enhancement of tracking technologies have been widely gained to be implemented on various sectors among private and public vehicle, and this is going to generate a huge data of vehicles such that includes speed and location by managing the fleets and other transportation application [5]. For such technology includes wireless communication technologies such as GPS [6]. Hence, the data on vehicle are usually regarded in shape of the trajectories and also made a significant way for real time approximation of arrival-times [3, 10]. Although the research about bus arrival-time prediction is still in progress, many researchers are proposed by different prediction frameworks to predict the arrival-time of buses such as exploiting gathered trajectory bus data that has been developed progressively [7].

However, dude the complexity and many factors that effect in order to forecast the arrival-time of buses; it is essential for the consideration of traffic condition, dwell time at stops, to become the prediction accurately and mannerly of time[8, 12]. Therefore, developing a model could forecast the arrival-time of buses by using a combined model based on Kalman Filtering Algorithm with Support vector machine is the objective of this research. This paper is organized to be four sections. The upcoming section demonstrates the proposed methodology through which this framework is going to develop. The third section describes the result and discussions of the proposed framework, followed by the designed experimental setup, as ordered according to their sequences. Finally, the fourth section describes the conclusions from our results and outlines the future work.

2 Methodology

The main focus of the research is towards developing a good accuracy prediction model, and therefore research methodologies are developed to achieve the research's objectives. Since the research methodology is a guidance which describes how to conduct the systems objective. And also defines the way in which the data are collected in a research project.

2.1 Research Framework

In this framework, firstly some parameters initially is made and these parameters will be used by SVM to do estimation. The result of estimation is contained some errors that controlled by the condition in GA. If the condition inside GA is not satisfied then this process will be repeated. The condition can be based on several iterations or a threshold for error measurement that we must pass it on. Of course, sometimes it can be a combination of both. After finding expected parameters, there are two phases must be done. Phase 1 includes learning process that trained SVM model based on the obtained parameters will create a model and pass the model to the Kalman Filtering Algorithm. Phase 2 contains testing process that uses a new unseen data set, which is called Test Set or Test Data Set, by evaluating the performance accuracy for the prediction (Fig. 1).

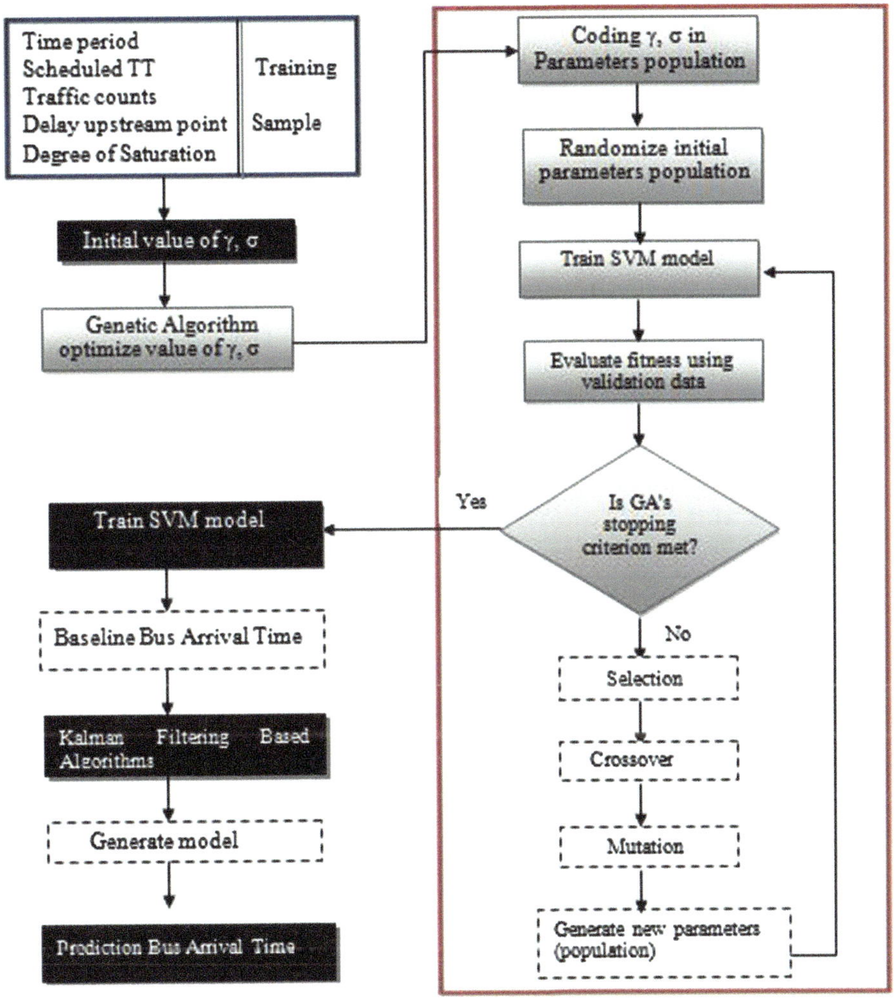

Fig. 1. Reseach framework

2.2 SVM and Kalman Filtering Algorithm

Basically, a route is divided into segments in each way or direction, while starting and finishing at two following time-point stops will be haven at each segment. It is supposed that the bus arrival or departure times and real information on positing for the vehicle at specific point could be handle by AVL. To forecast the arrival-time of a buses with along particular segment at instant i + 1, the arrival-time will be estimated based on different types of independent variables or explanatory variables that is used for this research as follow: Time period, Delay Degree of Saturation, Traffic Counts, Scheduled Travel Time and Delay at the upstream timing point. After employed Genetic algorithm (GA) to perform feature optimization (feature selection) so that only the best independent parameters are feeds into the support vector machine (SVM) algorithm.

This step is a very important pre-processing step because we assumed not all the features in the traffic data are important in traffic prediction. The normal genetic algorithm works by creating a random population; the default population size for GA is 100, for every generation. In each generation, the algorithm produces a new better solution based on the fitness function by performing mutations, crossover and selection. This better solution produce during the current iteration are considered to be the parents of the next generation.

The final solution generated by GA is a set of features that can best predict traffics using support vector machine (SVM). After that, SVM plays an essential role for predicting the base line travel times on the basic historical trips occurring data at a giving time-of-day, travel time on the current segment and route segment. Whereas, Kalman Algorithm provides the latest travel times as a dynamically.

3 Results and Discussion

The proposed model which is GA-SVM with Kalman Filter is to predict bus arrival time by using set of independent features idientified. As we explained, generally there are three key components done in this research: First using the SVM with Kalman to build a good regression model that forecasts the bus travel moment. Whereas second and third conducting GA parameters optimization to enhance SVM with Kalman precision and time efficiency and enhancing the accuracy performance of the bus arrival-time.

In the subsequent section, the experiment procedure, results and analysis is simplified to include the fore mentioned components: SVM with Kalman has successfully proved its aptitude in generating a good prediction model. With the appropriate kernel and hyper parameters setting, it achieves a better accuracy than the standard solution, such as Artificial Neural Network.

As shown in the performance, the RMSEs for SVM with Kalman are outperformed than those achieved by Artificial Neural Network, demonstrating the superiority of SVM with Kalman over its benchmark. The following figures are the results obtained from each algorithm and comparison between the algorithms. Each figure or table is followed with accompanying explanation.

3.1 Dataset Description

In this study the dataset used, is obtained from Melbourne, Australia[6]. The buses that operate Route 246 in inner Melbourne were recorded this dataset. Schematically the dataset is divided into four sections which are almost in the same length and it is around 8 km for long as illustrates in Fig. 2.

Fig. 2. A graphic appearance of the study test bed [6].

Bus that runs on this route is starting from 10 min in off-peak hours to around 30 min in the peak hours. Without any dedicated lanes for their operation, the buses in this route operate in a mixed traffic. The data collection was happening for 6 months, during that time all the travel time was recorded. Nevertheless, in February 2007, the dataset was provided for the purpose of research, and the dataset contains 1,800 travel-time that will be matching to each part point while each section point have two consecutive timing point.

This journey consists of four sections that are defined by five timing point stops as shown Fig. 3. In fact, there are 30 min difference between the soonest arrival bus and late one on this journey.

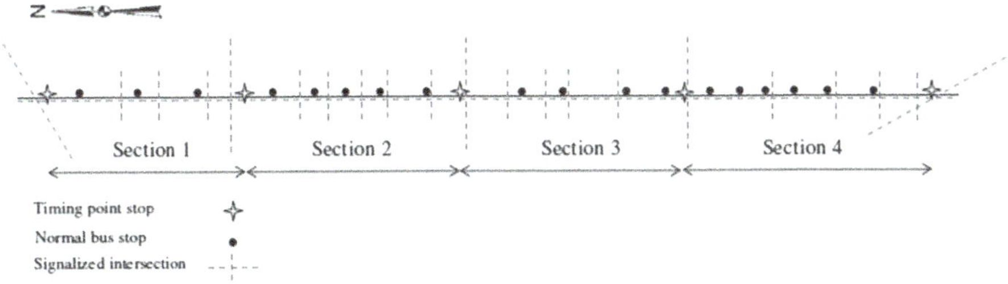

Fig. 3. Signalized intersection and normal bus stops.

3.2 Predicted Travel Time After Implemented SVM with Kalman

To do the travel time variability, firstly we optimized the selected variables using GA in matlab. In this research, 100 is assigned in the population size; whereas crossover and the mutation probability were setted to the 0.7 and the 0.6, respectively. This upcoming demonstrates that value that predicts the model and the actual value. RMSE is a repeatedly used to measure of the differences among the forecasted value by our model or an approximator and the observed actually value. We found that 87.63% is the performance of this prediction and this will neglect the ANN performance that is 83.5% (Fig. 4).

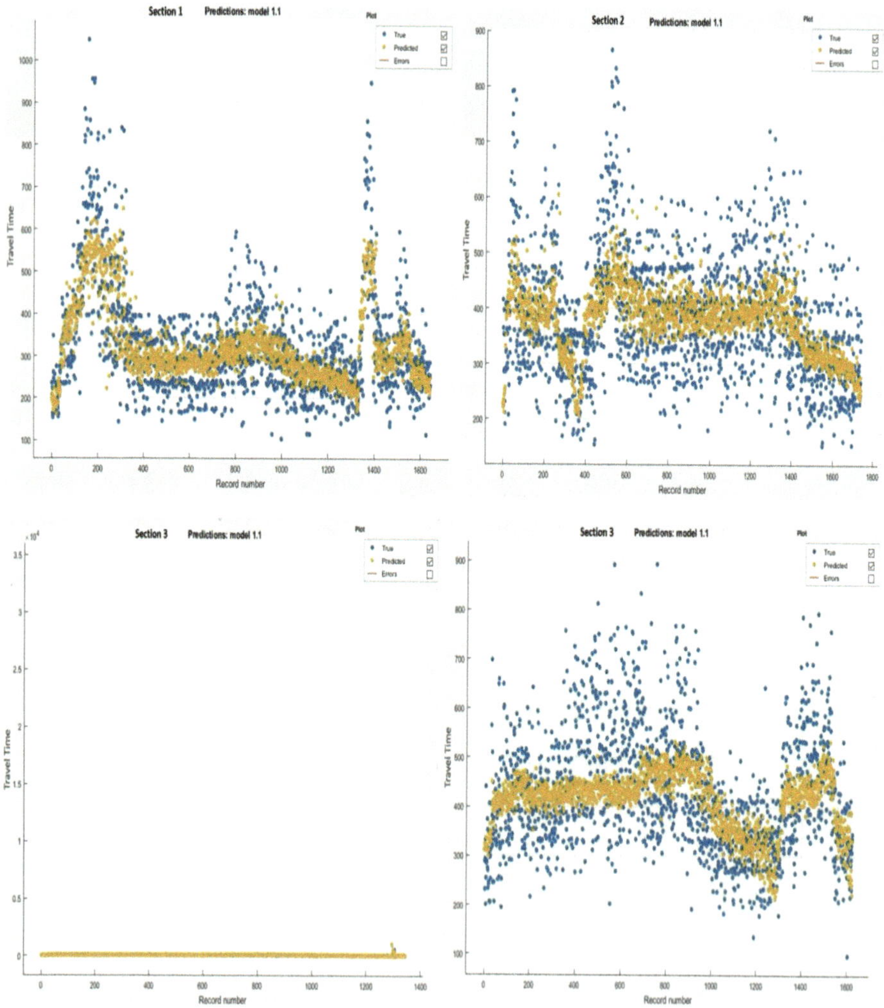

Fig. 4. The travel time observations along with the prediction

Before we implemented Kalman, we observed that the variation of travel time in each particular period of day is vary from each other. As shown in Table 1.

Table 1. Expected Travel time Produced by SVM with Kalman Filtering (seconds)

Section	Day off peak	Morning peak	Inter peak	Afternoon peak	Late night	Whole day
Clifton Hill to Johnston	154	194	179	168	115	154
Johnston to Bridge	184	167	209	191	179	167
Bridge to Toorak	132	108	114	105	97	97
Toorak to St Kilda	180	211	203	192	197	180

From Table 1 it can be seen that the variation of travel time in each particular period of day is vary from each other. For instance, in test section (Bridge to Toorak), minimum travel time in Late Night off Peak (7 pm to 12 am) is 97 which is below 100 s, however for the rest periods is almost above 100 s. On the other hand, the bus from (Toorak to St Kilda) has the maximum travel time for the Morning peak and Inter peak over the 200 s whereas the rest periods is almost lower than 200 s. Moreover, for maximum travel time, it is obviously shown that there is more than 200 s differences between inter peak and morning peaks. Hence we proposed prediction for each period to ensure the accuracy for travel moment. By Kalman, it could be seen that the observed data that has improved the degree of saturation passed the line while green is on.

Before we implemented Kalman filtering, we trained SVM model to test the expected travel time without the kalman. The observed data shows that the travel time from stop to another is variant as Table 2 illustrates. From Table 2 it can be seen that the variation of travel time in each particular period of day is vary from each other. For instance, in test section (Clifton Hill to Johnston), maximum travel time in Morning Peak (7 pm to 12 am) is 1049 which is upper 1000 s, however for the rest periods is almost above 200 s.

Table 2. Expected Travel time Produced by SVM alone (seconds)

Section	Day off peak	Morning peak	Inter peak	Afternoon peak	Late night	Whole day
Clifton Hill to Johnston	436	1049	576	596	458	1049
Johnston to Bridge	521	866	600	721	648	866
Bridge to Toorak	402	574	482	565	483	574
Toorak to St Kilda	698	659	900	893	643	900

Whereas the (Bridge to Toorak), is the minimum travel time for the whole day specially the day off the peak. If we take look (Johnston to Bridge and Toorak to St Kilda), the travel time is almost closed to each other or for the whole day for 866 s and 900 s respectively, whereas the difference is less than 40 s. Hence, such improvement can decrease to improve the performance of the accuracy for the prediction. Because of lack the dynamic travel adjustment that SVM will not do to predict by it is model.

3.3 Model Evaluation

For the model performance, we compare the three models those are ANN, GA-SVM and GA-SVM- Kalman and here is there result of the performed algorithms:

Table 3 presents the performance of the proposed method against the ANN model. Obviously the performance of GA-SVM with kalman is much better than ANN. As it can be seen, the poorest performance is belonged to peak times (AM and PM), however, our method could improve it from 140 s to 119 s for AM peak and from 108 s to 81 s for PM peak. From this result, GA-SVM-Kalman could improve the error rate more than 20 s in each. We had 4 and 6 s improvement for inter peak and off peak respectively. Moreover, we achieve 78 s against 90 s in all day prediction model.

Table 3. Performance of ANN versus GA-SVM-Kalman (RMSE in seconds)

Model	AM peak	Inter peak	PM peak	Off-peak	All day
ANN	140	65	108	59	90
GA-SVM	132	66	96	58	86
GA-SVM-KALMAN	119	61	81	53	78

4 Conclusions

This paper applies GA-SVM with Kalman Filter to forecast arrival-time of buses foundation of basis independent features including flow traffic data. As we mentioned, generally there are three key components that is done in this research: Using the SVM with Kalman to build a good regression-model that forecast the arrival time of buses. By conducting the parameters of SVM, search automatization experiment to improve SVM with Kalman time competence, accuracy and enhancing the accuracy performance of the bus arrival-time. In this paper, the SVM with Kalman has successfully proved its aptitude in generating a good prediction model. With the appropriate kernel and hyper parameters setting, it achieves a better accuracy than the standard solution, such as Artificial Neural Network. As mentioned in our performance, the RMSEs for SVM with Kalman are lower than those achieved by Artificial Neural Network, demonstrating the superiority of SVM with Kalman over its benchmark.

As result, this research work has explored a new robust effective way arrival-time of bus forecast nation by combining different arrival-time bus models which is support vector machine with kalman filtering algorithm. The experimental results and findings of this research reveal the capabilities of support vector machine and kalman filtering algorithm in solving the accuracy of bus arrival - timeprediction. The algorithm developed in this research is proofed to be an effective in finding the appropriate features of bus arrival-time while enhanced the accuracy of arrival-time in term of buses with less computational expenses.

In future research, other evolutionary algorithms such as PSO being one of the Meta Heuristic methods that can be implemented and inexpensive because of its limited CPU speed and memory requirements and also unlike GA, the PSO is unique in terms of information sharing mechanism. Information to others, in PSO, are only taken from the 'best' particles. The information sharing mechanism is one-way, and best solution is targeted by the evolution. ANFIS could be used in modeling as well, to achieve better performance including less computational complexity and high accuracy prediction for data which has a natural temporal ordering.

Acknowledgments. The authors would like to express their cordial thanks to University Technologi Malaysia (UTM), for the Research University Grant no. PY/2017/00570. The authors would also like to acknowledge VicRoads for supplying the SCATS data for this research as we are also grateful to Soft Computing Research Group (SCRG) and UTM-Big Data Centre for their support and incisive comments in making this research a success.

References

1. Balasubramanian, P., Rao, K.R.: An adaptive long-term bus arrival time prediction model with cyclic variations. J. Public Transp. **18**(1), 6 (2015)
2. Yang, M., Chen, C., Wang, L., Yan, X., Zhou, L.: Bus arrival time prediction using support vector machine with genetic algorithm. Neural Netw. World **26**(3), 205 (2016)
3. Zadeh, S.M.T.S., Anwar, T., Basirat, M.: A survey on application of artificial intelligence for bus arrival time prediction. J. Theoret. Appl. Inf. Technol. **46**(1) (2012)
4. Zaki, M., Ashour, I., Zorkany, M., Hesham, B.: Online bus arrival time prediction using hybrid neural network and Kalman filter techniques. Int. J. Mod. Eng. Res. **3**(4), 2035–2041 (2013)
5. Amita, J., Singh, J.S., Kumar, G.P.: Prediction of bus travel time using artificial neural network. Int. J. Traffic Transp. Eng. **1**(5), 410–424 (2015)
6. Mazloumi, E., Currie, G., Rose, G.: Using GPS data to gain insight into public transport travel time variability. J. Transp. Eng. **136**(7), 623–631 (2010)
7. Ganesh, K., Thrivikraman, M., Kuri, J., Dagale, H., Sudhakar, G., Sanyal, S.: Implementation of a real time passenger information system. arXiv preprint arXiv:1206.0447 (2012)
8. Altinkaya, M., Zontul, M.: Urban bus arrival time prediction: a review of computational models. Int. J. Recent Technol. Eng. (IJRTE) **2**(4), 164–169 (2013)
9. Lee, W.C., Si, W., Chen, L.J., Chen, M.C.: HTTP: a new framework for bus travel time prediction based on historical trajectories. In: Proceedings of the 20th International Conference on Advances in Geographic Information Systems, pp. 279–288. ACM (2012)
10. Dong, J., Zou, L., Zhang, Y.: Mixed model for prediction of bus arrival times. In: 2013 IEEE Congress on Evolutionary Computation (CEC), pp. 2918–2923. IEEE (2013)
11. Zegeye, K.G.: A dynamic prediction of trip time for transit vehicles in Brazil using GPS data. Master thesis, Department of Civil Engineering & Management, University of Twente, Enschede, The Netherlands (2010)
12. Wang, L., Zuo, Z., Fu, J.: Bus arrival time prediction using RBF neural networks adjusted by online data. Procedia-Soc. Behav. Sci. **138**, 67–75 (2014)

Future Internet Architectures

Muhammad Ali Naeem, Shahrudin Awang Nor, and Suhaidi Hassan(✉)

InterNetworks Research Laboratory, School of Computing, University Utara Malaysia (UUM), 06010 Sintok, Malaysia

Abstract. The tremendous growth of today's Internet traffic has launched new challenges in present IP Internet architecture. Similar content called again and again that creates a large amount of Internet traffic. There is a need to implement an efficient architecture to handle this problem. It is necessary to reduce the content redundancy to make Internet architecture efficient. Now the new research is introducing about the converting of IP host-based architecture Information-based. The Internet tradition has changed from IP to content dissemination. Host-to-host infrastructures tend to vanish to construct way for one-to-many or many-to-many sharing and reclamation of information objects with a growing request for an enhanced network design. Clients just desire to know about the content's availability and how can get it quickly where they wished as well as do not care about the physical location. While inventive solutions have proposed successfully and proved to address new Internet paradigm and these solutions provide a lot of benefits in terms of security, mobility, scalability and manageability.

Keywords: Information centric networking · Content Centric Network · Caching

1 Introduction

The Internet played an immense role in our lives. Since the 60s to 70s keep the aim of connecting a few of the machines. The Internet is always working on the top of the protocols stack TCP/IP [1]. The present Internet paradigm is a host-to-host model, based on host-centric communication that depends on both the sender and receivers where the data is an exchange between end-host. The sender is completely in charge of the communication only the data is sent which the sender wished. The host-oriented architecture is complicated and exactly matches the early Internet. A transmission is apparent for changing information. Actually, Internet applications are the end-to-end transmission, for example, emails, messages, chat. Based on the previous forecasts, Internet traffic experienced rapid growth in the past several years. According to the Cisco's VNI report, the global IP traffic has increased eight times in past five years, and the CAGR (Compound annual growth rate), is expected to be 35% during 2015–2019. Most of the traffic is related to content retrieval applications. For example, Video on Demand (VoD), YouTube. It is expected that video traffic alone will report for 86% of all the IP traffic in 2020 [2].

The requirement is increasing for bandwidth [3–9] and user-generated content (UGC), time-shift TV and high definition VoD, traffic is generated by receiver-driven

F. Saeed et al. (Eds.): IRICT 2019, AISC 1073, pp. 520–532, 2020.
https://doi.org/10.1007/978-3-030-33582-3_49

approaches that will continue its high growth rate. Knowledge can never stop, and the study has become a new Internet paradigm. So, the idea of content-centric networking offered a new model for communication that describes what data transmitted on the network rather than what network components are transmitting the data. The ICN aim is to shift the current complicated Internet paradigm to a simple and generic one. The basic networking unit is no longer the identified node (servers, routers, terminals). The ICN networking activities are a reality based on the named content objectives. ICN is a receiver-driven networking model, where end-users only express their interest in a given content. The entire network is in charge of routing the named content based only on the content names towards the best content container and delivering the contents through the reverse paths to the end-users. Due to the increasing of the network traffic, the current Internet is not sufficient to control that type of traffic [10].

Therefore, the current host-centric Internet paradigm is converting to the content-centric paradigm. In ICN, the content is the major component instead of hosts. The ICN creates a great interest in the Internet research community and this led to a high increment in the interest of ICN approaches. The marvelous progress of the Internet and the productions of novel applications to accomplish developing needs have particular growth to the new needs of the architecture, such as sustenance for accessible content distribution, security, mobility. Nevertheless, the Internet hasn't ever fabricated to face these types of requirements and in order to support it, a circumstance of functionality coverings is appearing, for example, Mobile IP. That covering increased the overall complexity of architectures and provides only temporal results [11]. Several presents and appearing issues still not faced satisfactorily by the existing Internet. This rose up the demand for a new technological approach for efficient Internet architecture and covers its actual need [12].

There is a research society requires that having recognized the limits of the present Internet is deliberating the important needs and goals of the Future Internet, and suggesting new architectures and prototypes face them. In this perspective, ICN developed as a favorable applicant for the Future Internet architecture. The fact is that the Internet is progressively recycled for information distribution, relatively less use for end-to-end communication between end hosts. ICN objective is to provide present and future requirements enhanced than the current Internet architecture [13]. ICN provides in-network caching (or storage, more generally) to facilitate multicast mechanisms, thus, it deployed an efficient and appropriate supply of desired information to the users. Therefore, ICN focuses on information and its distribution. There are many projects exist to implement the idea of ICN.

ICN implementation was pioneered at TRIAD, which makes a base for ICN research. [14]. Content-Centric Networking (CCN) [15], now that is known as Named Data Networking (NDN) architecture [16]. Publish-Subscribe Internet Routing Paradigm (PSIRP) [17], currently known as Publish-Subscribe Internet Technology (PURSUIT) project. The network of Information (NetInf) also developed for the Future Internet (4WARD) project [18], now its implementation is (SAIL) project. The architectures of these ICN projects are somehow different with respect to their implementation details. The main goal behind these architectures is, compare the results of these architectures to achieve high performance to implement the idea of ICN.

2 Survey and Issues

The existing Internet was initially designed as military experiments keeping in mind the target as sharing of resources. Because of the expensive cost and scarcity of resources, only a little number of consumers had access to it [19]. Nowadays, the situation has changed, the cost of Internet-capable devices has dropped and the number of persons accessing it has exceeded 3 billion, as shown in Fig. 1. Today there exist over one device per capita and this number is expected to grow to nearly three devices in 2019. However, the number of users is not the only aspect that has brought change to the Internet. Available and transferred contents on the Internet have skyrocketed as well. Google has already indexed 1 trillion web pages. Every 5 min, 4.5 Petabytes cross the Internet. Smartphones transfer 1.6 Exabytes monthly. These numbers are expected to keep growing. In 2019, networks will deliver 13.5 Petabytes every 5 min and Smartphones will transfer 10 Exabytes monthly. The Video-on-Demand (VoD) traffic will be roughly equivalent to 4 billion DVDs per month. Unfortunately, the evolution of the Internet architecture has not followed the growing patterns of Internet users and transferred data. The original Internet architecture has been altered with incremental changes to provide all the services that the Internet providers today. The first Internet prototype, Advanced Research Project Agency Network (ARPANET), connected just one specific type of computers/micro-computers. Later on, other types of networks such as satellite and mobile radio network were interconnected through ARPANET. To handle this interconnection, the TCP/IP stack was designed.

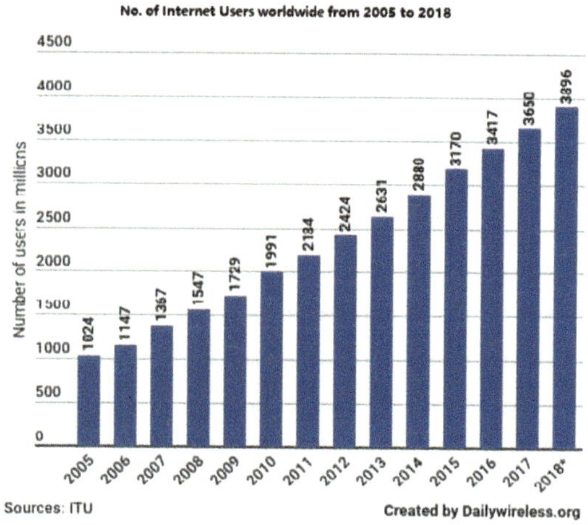

Fig. 1. Number of Internet Users

In 1981, the Domain Name System (DNS) was created to reduce the complexity to find hosts into this network of networks. Lately, the emergence of the World Wide Web (WWW) permitted Internet to reach 600 million users in the late 2,000 s and this number has kept growing since then. In less than a century, the Internet evolved from serving a few selected groups of scientists to reach more than 3 billion users. It moved from serving non-profit scientific and military activities to be an important part of the world economy. Indeed, Internet accounts for 3.4% of GDP in both G8 and BRICS countries [20]. It adapted from the exchange of static files and messages to serve hours of digital video and multi-party video conferences. Nevertheless, the communication paradigm has remained unchanged and the network operation is still led by an end-to-end model. The current communication paradigm of the Internet presents many limitations. First, the Internet architecture can offer multiple copies of content, but these copies are not linked together. In order to share resources of the network, additional overlay mechanisms have been proposed over the TCP/IP stack. The fast and reliable content transfer requires application-specific mechanisms such as Content Delivery Networks (CDN) [21] or Peer to Peer (P2P) services. Second, security is achieved by third-party applications and services. Trust in retrieved content is not easy to achieve and most of the connections rely on untrustworthy locations.

Third, the fact of retrieving content is strongly related to its location. Every Internet packet is addressed following source and destination addresses. Every time content is requested, addresses are needed. Our requests are strongly linked to an address, signaling a particular location, when users do not care about the location but get the content as fast as possible. The direct manner to resolve this problem is to replace were with what [22]. Host-to-host communication was an abstraction to solve the problems of the 60s. We argue that a paradigm based on named content is better for today's communication. Based on this premise, Information-Centric Networks (ICNs) was proposed. ICN architectures are clean-state designs of the Internet, where the content is located at the center of the scene. In contrast to the current Internet where packets are addressed according to source and destination nodes (i.e. IP addresses), ICN architectures address content according to content name at the network layer.

Every packet is named and this name serves to route contents. The aim of ICN is to develop architectures for content delivery. Several architectures have been proposed for ICN such as CCN, FP-7 PURSUIT, and SAIL NetInf. These architectures have been defined giving high priority to satisfy end-users expectations and prioritizing efficient content distribution. Among these new architectures, CCN has attracted considerable attention from the research community. The main issue impeding deployment is that one or more of these actors need to be given incentives to start deployment of the core ICN functionality. This functionality boils down to two main network components: in-network storage for caching of information objects, and the name resolution/routing service. We have explained and motivated the ICN approach to the network of the future. Based on the designs proposed in the research community, we have discussed the major trade-offs [23]. The different properties and design choices of the analyzed approaches according to the previous discussion. However, remaining challenges that need to be addressed for the ICN ideas to become deployed and used on a wider scale.

The number of NDOs are so large than the number of hosts in the current Internet, which means the ICN routing and name resolution system has a harder job than today's global IP routing and DNS name resolution. It remains to be shown that the proposed mean for aggregation of routing information and scalability of name resolution works in practice. Requests for content are visible to the ICN network, resulting in a possibly worse privacy situation than exists today. On the other hand, it might not be possible to relate a request to a particular person. The privacy issues need to be investigated in more detail in order to understand the full consequences and find means to mitigate them [24]. Ubiquitous caching probably does not sound too appealing for some content owners, who fear that their content can be illegally spread. The combination of technical mechanisms and new laws and regulation provide an acceptable solution. The incentives for all involved players have to be clearly communicated to advance deployment. It can be supported by standardizing central ICN aspects (e.g., naming). Deployment is also simplified if the ICN architecture allows growth from the edges (i.e., incremental deployment is supported) [25].

3 ICN Approaches

3.1 Content-Centric Networking (CCN)

CCN (Content-Centric Networking) [10] it is the initial project that was implemented at Palo Alto Research Center (PARC) on the basis of information-centric. CCN is the most attractive ICN project (Jacobson et al. 2009). The target of CCN is to transform the location-based Internet design to the named information-based. The unit of information is denoted as content in CCN that has a hierarchical naming structure in prefix form. For example, ccnx:/parc.org/video/. Two primitives Interest packets and Data packets are used to make communication possible. Interest packets denote the client's requests data packets show the response from the source [20]. Interest packet matches with a data packet in the form of prefix order when a hit occurred. CCN nodes are divided into three sections that are: Content Store (CS), Pending Interest table (PIT) and Forwarding Information Base (FIB). CS has the ability to store the transmitted content at the delivery path. PIT manages all the requests and specified path for them. FIB indicate the right path for interest packet to forward the appropriate source also from source to the requester. The working of CCN architecture is illustrated in Fig. 2. When clients send their requests for required contents then the requests forwarded to the node that is nearest to the requested user. Then, the request is compared with the existing contents in CS. If the request does not match with any of the present entry the CS forward the request to the sources. When the requested content is found then the content is sent back to the requester by locally cached content as well as a copy also stores at the nodes that appears along the delivery path back to the requesters.

3.2 Network of Information (NetInf)

The NetInf is ICN architecture to implement the future Internet paradigm. It is a part of 4WARD project [26] that derives from the Scalable and Adaptive Internet Solutions (SAIL). It combines the Information Object and bit-level object into Named Data Object NDO. NDOs have three qualities identifier (unique name), metadata (semantics) and security (hash functions). The bit-level object is the real body of the content which contains the binary data of the content itself. The flat naming structure is used in NetInf in the form of P:L. Where P is the public-key of the content provider and L is the content-label provided by the content suppliers. NDOs are registered in Name

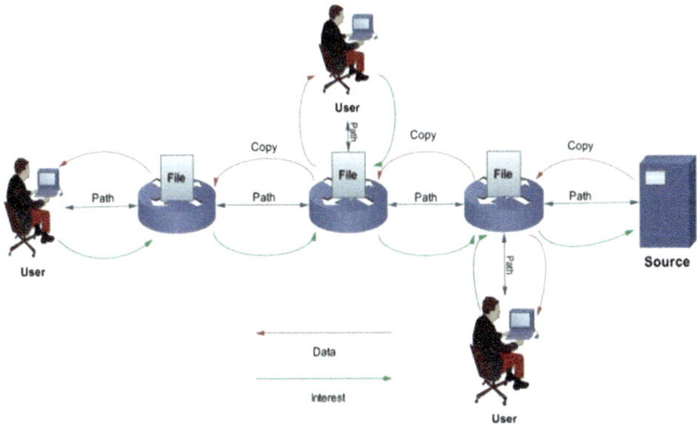

Fig. 2. Content Centric Networking

Resolution System (NRS) with their locations when the publishers publish theirs. All client requests received by NRS, then NRS perform matching [23] between the requests and existing NDOs.

NRS send the request to the best appropriate source if there exist any of the entry related to the request. Figure 3 explains how the client's requests forward in the network according to NRS. In this architecture, all the contents have to register in NRS and the NRS continuously update. NRS performs matching between the existing entries and requests. If there exists any of the entry related to request the NRS forward the request towards the source. Then the source will response to the requester using the reverse path or follows the direct path as shown in the figure.

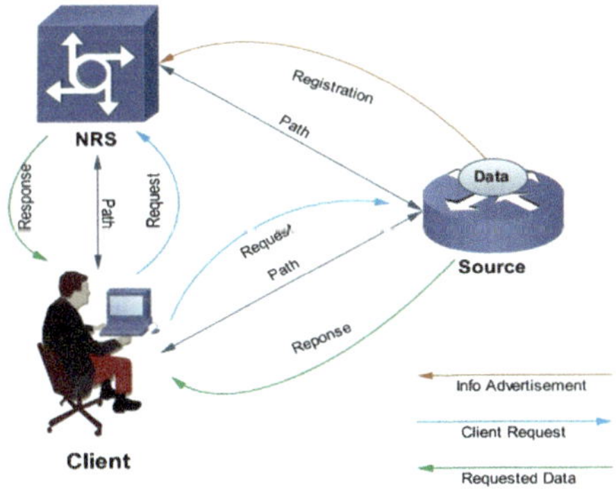

Fig. 3. Network of Information

3.3 Data-Oriented Network Architecture (DONA)

Data-Oriented Network Architecture (DONA) is a new hygienic ICN architecture and it is location independent. DONA is the first complete ICN architecture [27] that was established at (Koponen et al. 2007) to converts the host-host based network into the data-centric. Hierarchical names are used in DONA and the unit of information is represented by Named Data Object (NDO). Each information object uniquely identified the name with a public-private key as P:L. where P represented the public key defined by the publisher and L is the label assigned by the publisher. Routing of NDOs is done by using Resolution Handlers (RHs). Each node contains a registered information table that stores three types of information. Name of NDO P:L, next hope RH to know about the content availability and third is next hope distance. Client's request and data from the source is forwarded using two approaches: "FIND" and "REGISTER". When an RH receives user request (find) then each request is compared with existing entries of registration table, if the request does not match with any existing entry then the request is forwarded to the corresponding next-hop RH as shown in Fig. 4. The data is sent back to the requester by followed the reverse path when the required content is found as well as cached a copy of requested data at the return path or route through the direct path as shown in the figure. If the registration table contains more than one entry then request is sent to the nearest source of the required content [28].

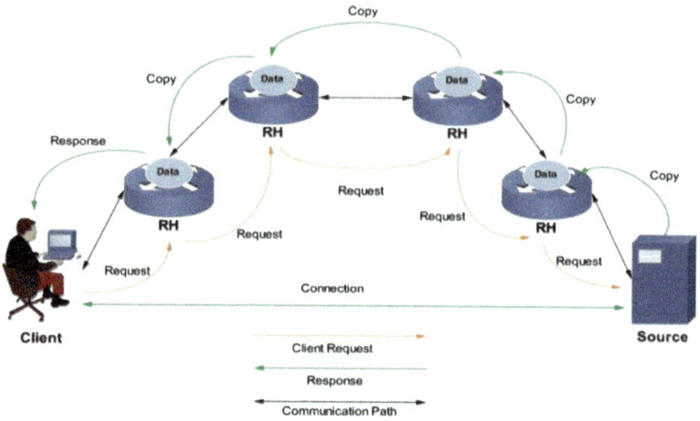

Fig. 4. Data-Oriented Network Architecture

3.4 Publish-Subscribe Internet Routing Paradigm (PSIRP)

The development of Publish-Subscribe Internet Routing Paradigm (PSIRP) is based on published-subscribe architecture. PSRIP (Lagutin et al., 2012) is derived from Publish-Subscribe Internet Technology (PURSUIT) [29]. Both architectures are the part of European project FP7 that was started in 2008 and ended in 2012. These projects were introduced to implement and validate the aim of information-centric network. In PSRIP, Rendezvous system is used to handle routing and naming requirements (e.g. scope and uniqueness). The rendezvous system is divided into several RendezVous Nodes (RN) and these nodes are interconnected to make communication possible. All the contents are published using two types of identifiers that are known as Rendezvous identifier (Rid) and the scope identifier (Sid). These are used to define the unique name and the global scope of contents. The Scope identification Structure provides in hierarchically organized data and the identifiers express unique label as well as scope for all contents.

The RN is responsible for performing matching between the publisher's content and the subscription request (client request). After matching the Forwarding Node (FN) is responsible to forward the content towards the subscriber from the content container. FN consists of the bloom filters. Bloom filters are used to detect a specific interface by which the requested content needs to forward.

The subscription request accumulates the return path in the bloom filter and creates the Fid. By using the Fid the required content to return to the subscriber. Figure 5 explains the rendezvous architecture. The client rendezvous and source rendezvous are interconnected to communicate with each other. The client has to send the request to related rendezvous. If the request match with any of the existing entry then the request is sent to the source that is mentioned in rendezvous. A copy of the requested data is copied at the delivery path from source to the requester as shown in Fig. 5. It is clear that from the table all the ICN architectures are content-oriented rather than location. All the ICN architectures have their own properties as shown in below Table 1.

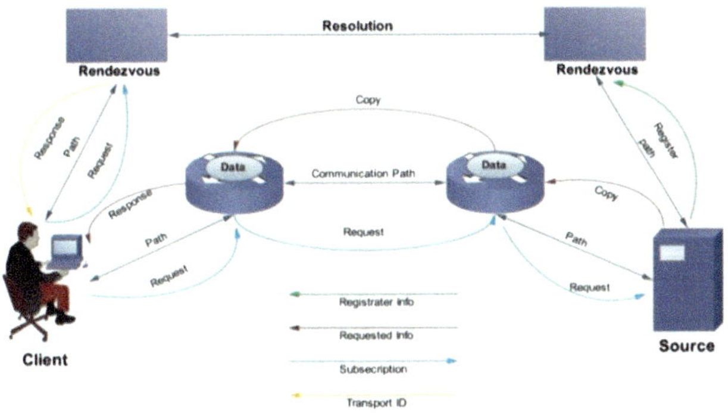

Fig. 5. Publish-Subscribe Internet Routing Paradigm

Table 1. ICN architectures

ICN Architectures' Names	CCN	DONA	NefInf	PSRIP
Naming structure	Hierarchical	Flat	Flat	Flat
Naming strength	Signature, external security	Signature, KPI	Signature, Hash function, KPI	Signature, KPI
Routing of client's request	Named-based	Named-based using RH	Named-based NRS (Hybrid)	Named-based using Rendezvous
Routing of data object	Reverse path as router mentioned	Reverse path or direct path	Routing using bloom filters	Reverse path or direct path
Human understandable	Readable	Unreadable	Unreadable	Unreadable
Data object granularity	Packets	NDOs	NDOs	NDOs

4 Overview of ICN

Information-centric networking is one of the most significant researches related to the future Internet. The basic concept of ICN is to switch the address-based Internet to the named information-based paradigm. Recently most of the projects are done to present the idea of ICN. The content name-based routing is firstly introduced in the TRIAD project [30]. After that, most of the researchers show their interests in this new Internet paradigm and proposed several different ICN architectures. In ICN, the unit of transmitting data is present in the form of content objects. All ICN architecture is divided into several parts in which important parts are: Naming, Routing, Security, Caching, and Mobility.

4.1 ICN Naming

One of the most important and fundamental modules of ICN is naming. Name is only the identifier by which the transmission of NDOs takes place. Therefore, the information is independent of its location because of unique names. According to ICN architectures, naming is divided into two different schemes. Each naming scheme has its advantages.

Hierarchical Naming Scheme. The structure of the hierarchical naming scheme is alike to the present URLs and in the form of the prefix-rooted tree-like manner. This naming scheme can perform shortest and longest match by aggregating the routing information in prefix order to improve the system scalability. Mostly hierarchical names are human-readable that help the users to write it manually. For example ccnx:/parc.org/video/widget1/version2/chunk2. This shows that an NDO can consist of more than one version and chunks.

Flat Naming Scheme. In this scheme, a special signature P:L provided by its publisher to identify the uniqueness of each content. Where P is the public key cryptographic hash of the publisher and L is label provided by the publishers. These names are not in human-readable form. For example: 1DB76EB8DFD6B0b92A293AADC8.

4.2 ICN Routing

Routing is the core part of all ICN architectures. It is used to transmit data from one place to another in a network. ICN provides named content-based routing rather than location-based routing. The routing is used for two purposes when a user request is created. First, for user request from the user to the source and the second one is routing from the source to the requester (content delivery path). In ICN, information retrieval mechanism has two main sections. First is the discovery of required information and the second is delivery of that discovered information [31]. The discovery of contents is related to client request how a client request can reach the accurate source and the delivery is related to the rules that are used to propagate the data in a network. In ICN routing, the publishers publish their contents by advertising in the network. According to different architectures, ICN routing is divided into two main components [32], these are name-based routing and name resolution routing.

Name Based Routing. In name-based routing, publishers spread the registration messages in the network about their NDOs and these messages saved at each node [33]. The client request directly sent to the network. Each router has a specific information entity (registration table) about the published NDOs and all the incoming requests are compared with the existing information. If any of the network nodes has required data the node will respond by using the reverse path to the client. Name Based Routing used in Content-Centric Networking (CCN).

Name Resolution Routing. In this mechanism, the data is spin by using an extra system known as Name Resolution System (NRS) and the registration information about the publishers NDOs saved at NRS. The working is same as current DNS. By using this system all the information about the published NDOs is saved in NRS [34]. Therefore the client requests have to go to NRS first then the NRS will send the client request to the corresponding data source. Then the required data is sent to the client by using the reverse path or it can be a direct path from source to the client. Name Resolution Routing is used in Network of Information (NetInf).

4.3 ICN Security

In early Internet, security was not a big issue researcher's focus on the delivery of information rather than the information security. Now a day's peoples have faced many security problems. For example, Denial of services (DoS) and the receiver's security depend on sender [35]. To solve these kinds of problems IPSec and DNSSec protocols were introduced in IP Internet. But these protocols cannot fulfill the security issues of host-host Internet architecture. Therefore ICN decupled sender and the receiver and it enclosed a specific type of information in one arbitrary object that has its own security features given by the publishers (makers). Several kinds of security features are integrated with information objects (NDOs). For example, some publishers used a specific type of signatures with the name of their contents (e.g. /aueb.gr/ai/main.html/aueb.gr/ai/main.html) only authorized people can decrypt the signature. Several ICN architectures are established to implement different security features according to their architectural design.

4.4 ICN Caching

The cache is one of the basic components of ICN that is integrated to implement the main idea of the ICN network paradigm. Caching is the temporary storage that is provided to all ICN nodes to store the transmitted data for a specific time to increase the transmission efficiency of the network. Mostly caching is used to store a copy of incoming data from source to the requester. Caching also decreases the propagation delay and improve the performance of network utilization [10]. Caching makes ICN Internet paradigm differs from the presently running Internet paradigm. Cache makes possible to store transmitted data near to the user that decrease the response time with efficient resources utilization and data transmission service. Client request for a specific piece of information (NDO) can satisfy any of the nodes having the corresponding data in its cache. According to the ICN architectures caching is divided in two main schemes. These are On Path Caching and Off Path Caching.

On Path Caching. On path caching is a plain approach of placing data on the network path. In this, the node responds to the requester using the local copy of stored data. When an ICN node received a request for a specific NDO so, the request is compared with storage data of that node. If there exists a copy of required data the node will send it directly back to the requester [10]. Almost all ICN architectures support on path caching.

Off Path Caching. In Of Path Caching, there is a need to integrate a special entity with a network known as name resolution system based on data replication [10]. All NDOs have to register in NRS first then the user can request. All the publishers registered their contents in NRS. Therefore the client request first needs to go to NRS then NRS compare with existing entries and send it to the corresponding node (source). Then the node sends a copy of the required data back to the requester by following the same path.

4.5 ICN Mobility

Current Internet is focused on end-to-end communication. It supports fixed location (host) with a specific type of IP address. If a person wants to connect with the Internet it needs to have an IP that will show the user's location. It is necessary to host to stick a particular location to communicate with other hosts [36]. If a user wants to communicate during his moving so he needs to change the IP because his location changes continuously so it is very difficult to communicate. For example mail. Now a day's, multihoming (connecting a host or a computer network to more than one network) is an essential issue because choosing a data path to handle the transmission of information among the several connections (hosts) is very difficult. So it highly damages the data flow and increases the network traffic that creates delay and bandwidth problem. Now the host-based Internet paradigm is converting to content-based because the present network paradigm is insufficient to solve these kinds of problems competently. Content oriented network paradigm is independent of physical locations and it is easy to communicate using information-based network architecture because users can send their requests from anywhere to any source. There is no need to make connections between hosts [37]. In ICN all nodes have the ability to store information (NDOs) so, if a node disconnect from the network users can request to any other node to get their required information. ICN mobility is divided into three main sections mobility of clients, mobility of contents (NDOs) and mobility of network. During client mobility it is not necessary to stick a fixed location he can establish several connections with the network for communication and he can also retrieve a copy of desired data. During content mobility, the information about the content stored in nodes and at NRS needs to be updated but a copy of the same data can exist more than one source. For example, original content at the server and network node have a copy of that content. During the mobility of the network, all the information about the publisher's content needs to be updating.

References

1. Alberti, A.M., Casaroli, M.A.F., Singh, D., da Rosa Righi, R.: Naming and name resolution in the future internet introducing the NovaGenesis approach. Future Gener. Comput. Syst. **67**, 163–179 (2017)
2. Cisco. Cisco Visual Networking Index: Forecast and Methodology 2014–2019 White Paper. Cisco, Technical report (2015)
3. Fazea, Y.: Numerical simulation of helical structure mode-division multiplexing with nonconcentric ring vortices. Opt. Commun. **437**, 303–311 (2019)
4. Fazea, Y.: Mode division multiplexing and dense WDM-PON for Fiber-to-the-Home. Optik **183**, 994–998 (2019)
5. Fazea, Y., Alobaedy, M.M., Ibraheem, Z.T.: Performance of a direct-detection spot mode division multiplexing in multimode fiber. J. Opt. Commun. **40**, 161–166 (2019)
6. Fazea, Y., Amphawan, A.: 5 × 5 25 Gbit/s WDM-MDM. J. Opt. Commun. **36**, 327–333 (2015)
7. Fazea, Y., Mezhuyev, V.: Selective mode excitation techniques for mode-division multiplexing: a critical review. Opt. Fiber Technol. **45**, 280–288 (2018)
8. Fazea, Y., Amphawan, A.: 32 channel DQPSK DWDM-PON for local area network using dispersion compensation fiber. In: EPJ Web of Conferences, vol. 162, pp. 1–3 (2017)
9. Hassan, S., Fazea, Y., Habbal, A., Ibrahim, H.: Twisted Laguerre-Gaussian mode division multiplexing to support blockchain applications. In: Region 10 Conference, TENCON 2017–2017, pp. 2421–2425. IEEE (2017)
10. Naeem, M.A., Nor, S.A., Hassan, S., Kim, B.-S.: Performances of probabilistic caching strategies in content centric networking. IEEE Access **6**, 58807–58825 (2018)
11. Naeem, M.A., Ali, R., Kim, B.-S., Nor, S.A., Hassan, S.: A periodic caching strategy solution for the smart city in information-centric Internet of Things. Sustainability **10**, 1–16 (2018)
12. CV Networking. CiscoGlobal Cloud Index: forecast and methodology, 2015–2020. White paper by Cisco (2017)
13. Paschos, G.S., Iosifidis, G., Tao, M., Towsley, D., Caire, G.: The role of caching in future communication systems and networks. arXiv preprint arXiv:1805.11721 (2018)
14. Ahlgren, B., Dannewitz, C., Imbrenda, C., Kutscher, D., Ohlman, B.: A survey of information-centric networking. IEEE Commun. Mag. **50**, 26–36 (2012)
15. Arianfar, S., Nikander, P., Ott, J.: On content-centric router design and implications. In: Proceedings of the Re-Architecting the Internet Workshop, p. 5 (2010)
16. Chai, W.K., He, D., Psaras, I., Pavlou, G.: Cache "less for more" in information-centric networks (extended version). Comput. Commun. **36**, 758–770 (2013)
17. Cui, X.-D., Huang, T., Jiang, L., Li, L., Chen, J.-Y., Liu, Y.-J.: Design of in-network caching scheme in CCN based on grey relational analysis. J. China Univ. Posts Telecommun. **21**, 1–8 (2014)
18. Dräxler, M., Karl, H.: Efficiency of on-path and off-path caching strategies in information centric networks. In: 2012 IEEE International Conference on Green Computing and Communications (GreenCom), pp. 581–587 (2012)
19. Fang, C., Yao, H., Wang, Z., Wu, W., Jin, X., Yu, F.R.: A survey of mobile information-centric networking: research issues and challenges. IEEE Commun. Surv. Tutor. **20**, 2353–2371 (2018)
20. Sarno, L., Tsiakas, I., Ulloa, B.: What drives international portfolio flows? J. Int. Money Finance **60**, 53–72 (2016)

21. Singhal, R., Geyer, E.K., Makholm, H., Mortensen, C.W.: Marker based reporting system for hybrid content delivery network and peer to peer network, ed. Google Patents (2019)
22. Lu, Z., Wang, Y., Yang, Y.R.: An analysis and comparison of CDN-P2P-hybrid content delivery system and model. JCM **7**, 232–245 (2012)
23. Naeem, M.A., Nor, S.A., Hassan, S., Kim, B.-S.: Compound popular content caching strategy in named data networking. Electronics **8**, 771 (2019)
24. Muhammad Ali Naeem, S.A.N., Hassan, S.: IP-internet data dissemination challenges and future research directions. J. Adv. Res. Dyn. Control Syst. **10**, 1596–1606 (2018)
25. Peng, C., Han, Q.-L., Yue, D.: Communication-delay-distribution-dependent decentralized control for large-scale systems with IP-based communication networks. IEEE Trans. Control Syst. Technol. **21**, 820–830 (2013)
26. Qazi, F., Khalid, O., Rais, R.N.B., Khan, I.A.: Optimal content caching in content-centric networks. Wirel. Commun. Mob. Comput. **2019**, 15 (2019)
27. Saucez, D., Schmidt, T., Kutscher, D., Eum, S., Psaras, I., Corujo, D., et al.: Information-Centric Networking (ICN) Research Challenges (2016)
28. Vasilakos, A.V., Li, Z., Simon, G., You, W.: Information centric network: research challenges and opportunities. J. Netw. Comput. Appl. **52**, 1–10 (2015)
29. Xylomenos, G., Ververidis, C.N., Siris, V.A., Fotiou, N., Tsilopoulos, C., Vasilakos, X., et al.: A survey of information-centric networking research. IEEE Commun. Surv. Tutor. **16**, 1024–1049 (2014)
30. Yan, H., Gao, D., Su, W., Foh, C.H., Zhang, H., Vasilakos, A.V.: Caching strategy based on hierarchical cluster for named data networking. IEEE Access **5**, 8433–8443 (2017)
31. Saha, S., Lukyanenko, A., Ylä-Jääski, A.: Cooperative caching through routing control in information-centric networks. In: Proceedings IEEE INFOCOM 2013, pp. 100–104 (2013)
32. Barakabitze, A.A., Xiaoheng, T., Tan, G.: A survey on naming, name resolution and data routing in information centric networking (ICN). Int. J. Adv. Res. Comput. Commun. Eng. **3**, 8322–8330 (2014)
33. Bari, M.F., Chowdhury, S.R., Ahmed, R., Boutaba, R., Mathieu, B.: A survey of naming and routing in information-centric networks. IEEE Commun. Mag. **50**, 44–53 (2012)
34. Azgin, A., Ravindran, R.: Hash-based overlay routing architecture for information centric networks, ed. Google Patents (2018)
35. Fu, X., Kutscher, D., Misra, S., Li, R.: Information-centric networking security. IEEE Commun. Mag. **56**, 60–61 (2018)
36. Adhatarao, S., Arumaithurai, M., Kutscher, D., Fu, X.: NeMoI: network mobility in ICN. In: International Conference on Communication Systems and Networks, pp. 220–244 (2018)
37. Kitagawa, T., Ala, S., Eum, S., Murata, M.: Mobility-controlled flying routers for information-centric networking. In: 2018 15th IEEE Annual Consumer Communications & Networking Conference (CCNC), pp. 1–2 (2018)

Compute and Data Grids Simulation Tools: A Comparative Analysis

S. M. Argungu[1,2], Suki Arif[1(✉)], and Mohd. Hasbullah Omar[1]

[1] InterNetWorks Research Laboratory, School of Computing, Universiti Utara Malaysia, 06010 Sintok, Kedah, Malaysia
argungu@internetworks.my, {suki1207,mhomar}@uum.edu.my

[2] Department of Computer Science, University of Science and Technology, Aliero, Kebbi, Nigeria

Abstract. The ever increasing need for grid systems in scientific, business and "what if" real world types of applications, coupled with the dynamic nature of computing infrastructure, has necessitated the need for clear distinctions between the various types of grid systems. This is particularly important when evaluating real-life applications using simulation environments. The very knowledge of what simulation environment to use in evaluating the performances of different types of grid models will go a long way in helping to arrive at a true representation of the system studied. Equally important is the need to identify distinctively the different real life works scenarios, in which these systems are applied. The grid industry is endowed with powerful simulation tools to enable researchers evaluate their designs prior to actual implementations. However, often researchers get busy developing extended versions of these simulators, at the expense of the precious time needed to solve the problem at hand, which is partly due to the wrong choice of simulation environment. This study is inspired by the need to compare and contrast between the two major grid types (Computational and Data Grids) in terms of areas of applications and the simulation environment appropriate for performance evaluation relating to each of these grid systems. This will help researchers to making an informed decision while considering simulation environments to be used in their projects, and help identify the relevant as well as suitable measurable metrics. In addition, the research findings will help to eliminate ambiguity while testing real life applications, and reduces inconsistency in the obtained results.

Keywords: Data Grids · Compute Grids · Simulation Simulators

1 Introduction

More often than not, confusion abounds when trying to differentiate between the computational grids (CG) and data grids (DG). This is because the former and later both emerged from the same pool of existing technologies such as the distributed computing, world-wide-web and the internet services, various cryptographic and virtualization technologies [2, 3]. Different writers have varying views about the discrepancies between the computational grids and data grids, most of which succeeded in

F. Saeed et al. (Eds.): IRICT 2019, AISC 1073, pp. 533–544, 2020.
https://doi.org/10.1007/978-3-030-33582-3_50

providing only a superficial difference between the two concepts without going into details. In other words, when looking at the two concepts from the infrastructural point of view, the distinction between data grids and computational grids may not be obvious. This is because; both are seen to have been evolved from similar if not the same set of computing infrastructure. In other words, their core components comprise of the same computing and data resources (CPU, Processors, Memory, Storage facilities, Software, data and Networking facilities) [1, 2].

However, Confusions usually amount when discussing issues such as Grid Application Platforms, Grid Performance Evaluation Metrics and the types of real life applications scenarios to which they are applied. Also, parameters such as "Load Balancing, Job Scheduling, Resource Discovery, Resource Allocation and Data Replication introduce more confusions, and unless explicit reference is made to a particular type of grid system, the reader may be wondering which of the grid's category? Even when explicit reference is made to a particular type of grid system, the choice of evaluation metrics that are suitable for the simulation environment relevant to the grid system in question, may present ambiguity. In addition to exploring the various motivations for deployment of the grid infrastructure and the underlying architectures, this paper attempts to use some important measureable metrics in drawing the margin between Data Grids (DG) and Computational Grids (CG). The rest of this paper is organized as follows. Section 2 - presents an overview and classification of grid systems, Sect. 3 - Relationship of Data Grids with Compute Grids, Sect. 4 - Major grid projects and deployment focus, Sect. 5 - Characteristics Features of the grid, Sect. 6 - Grid system Simulators/Application areas, and Sect. 7 - Conclusions and onward look.

2 Overview of Grid Systems

Grid systems in general, leverage on efficient utilization of a pool of heterogeneous systems with optimal workload management, exploiting an enterprise's entire computational resources (servers, networks, storage or information); acting together to create one or more large pools of computing resources [2]. Grid Computing (GC) is the ability to utilize the ideal CPU cycles and storage facilities from millions of computer systems across a worldwide network to function as a single powerful mainframe computer. These systems come together to provide inexpensive, accessible pool that could be harnessed by users irrespective of their locations. This is similar to the way power companies and their users share the electrical grids. The Grids technology can be termed as goal-motivated and exciting global efforts to develop a platform in which individual users can access computers, databases, and experimental facilities simply and transparently, without having to consider where those resources are located. It is a model for allowing organizations to use a large number of computing resources on demand irrespective of their administrative domains. Generally, Grids cannot be looked at as a revolutionary technology; instead, they have evolved from prevailing technologies such as the distributed computing, world-wide-web, the internet services, various cryptographic and virtualization technologies [2, 3]. Virtualization happens to be the strongest factor that strengthens the motivation for the advancement of the grid systems [3, 13]. Virtualization can be defined as the seamless integration of geographically dispersed and heterogeneous systems, which enables grid users to access the services of the grid in a transparent way without concern for the location of the computing resources. To the users' perspective, all they have to do is submit jobs to the grids via a single point of entry or node. It then remains duty bound to the grid system to locate/find the desirable/available computing resources, which can serve the needs of the users [3]. So far, grids have been severally explained as systems of heterogeneous nature, which exhibit some characteristics taxonomy. In the next section, we shall briefly look at the classifications of grid systems.

2.1 Classifications of the Grid Systems

Grid systems have been classified according to the following classes namely; *Computational Grids, Data grids, Data-centric Grids, Services Grids, Access Grids, Application Service Grids, Interaction Grids, Knowledge Grids and Utility Grids* [2, 5, 11].

Computational Grid (e.g., TeraGrid, China Grid and APACGrid) combines the computational power of distributed resources including clusters, desktops, and supercomputers to provide services for high performance computing systems. They connect machines of an organizational domain by utilizing their idled CPU-cycles to provide higher computational capacity than the capacity of any individual machine in the system. Here, the focus is on optimization of applications' execution time that requires an excessive amount of computer processing cycles.

Data Grid (e.g., LHCGrid, GriPhyN) is patronized by data-intensive tasks that consist of dispersed data warehouses providing the facilities to store massive amount of data that can be accessed, relocated, and managed as if they were small files. Data Grids (DG) provide hardware and software infrastructure for synthesizing new information from existing data repositories distributed over a wide area network. Here, the focus is based on solving large-scale data management problems.

Data-centric Grid (DCG) is a system of grids that provide access and computational facilities to large distributed data repositories. The difference between Data Grids (DG) and Data-centric Grid (DCG) is that in the latter, heavy computations are moved onto data rather than data to computation. Distributed Data Mining (DDM) is a typical real life application that employs the services of Data-centric Grids.

Services Grids (SG) refer to systems that convey services that are not provided by any single local machine within the grid, and is further sub-divided into three subgroups, namely;

On-demand, which aggregate resources with a view to providing new services.
Collaborative, connecting users and applications via a virtual workspace, and Multimedia, providing infrastructure for real-time multimedia applications.
Application Service Grid (ASG). For example, NetSolve and GridSolve provide access to distant applications and libraries that are hosted on various data centers and Computational Grids (CG).
Interaction Grids (IG). Provides visualization abilities between participants by according users with interactive and collaborative services.
Knowledge Grids (KG). Focused on providing business analytical services by the means of knowledge discovery, acquisition, management, and processing.

Utility Grids (UG). Provide all the typical Grid services such as data services, computing services, discovery and allocation of resources, Quality of Service and contract management.

Classifying grids in-terms of their characteristic services as explained in [2] may not suffice to differentiate between the computational and data grids concepts. A more formidable classification may be to look at the grid systems in terms of the resource types as explained in [5], which portrays grids as comprising of two broad resource types, namely, *hardware* and *software* resources. The hardware resources are further subdivided into computing, storage and network resources. On the other hand, the software resources are further subdivided into data, application software and utility software. Thus, by deployment, CG focuses more on providing access to shared computing resources (processor, hard disk, memory, network bandwidth, etc.), of which data may be involved as a constituent (resource) of the storage facility. The grids classification advanced by [11] boiled down to two major categories when subjected to the resources types. In other words, the Application Service and Utility Grids are software biased and thus can be placed under DG. On the other hand, the Interaction and Knowledge Grids can be placed under CG since they are hardware biased. The section that follows, explains further the difference between Compute and Data Grids systems.

3 Interactions of Data Grids with Compute Grids

Now, if Grids were conceived in terms of layered architecture and following the categorical placement of DGs and CGs respectively under software and hardware, then it could as well be said that DG is a layer running beneath and supporting the CG? An attempt will be made to answer this question in this section. From the early network computers, it was obvious that these machines were dependent on other powerful computers called servers for their data and/or computational needs. It was obvious also that there was a direct link between the machine called the *client* and the one called *server*. Thus, the server becomes over burden when more clients get connected to it, with a resultant increase in the work load. In other words, the server was laden with more requests from the clients and eventually a bottleneck would occur. This client-server relationship could also be seen in the World-Wide-Web and the Internet models, a limitation that was promised to be resolved by the Peer-to-Peer (P2P) Network technology [34]. The P2P posed another challenge by introducing communication overheads into the systems when every client is treated as a server [34].

In the same vein, early Grid Computing (GC) model faced similar challenges as experienced in the previous technologies, in which case some client machines established a one-to-one communication link with the data server. The one-to-one communication posed a scalability issue as computational nodes were distributed all over the network, which also adversely affects the delivery of data items over the network. To resolve the problem of data delivery, another layer was added to the Compute Grid model known as the Data Grid Layer (DGL). The DGL facilitates management of Computational Grid's data requirement by providing Data Caching, Data Replication, Data Delivery and Data Movement according to the application's needs. In addition, the DGL helps facilitate data connections with both local and upstream data servers that are distributed over the network globally. Figures 1 and 2 shows the layered architecture for compute and data grids, respectively. Figure 3 is Data Grid Layer added beneath Compute Grid [5].

Grid Applications Science, Engineering, Commercial applications, Web portals
Grid Programming environments and tools Languages, interfaces, libraries, compilers, parallelization tools
User-level middleware-resource aggregators Resource management and scheduling services
Core grid middleware Job submission, storage access, info services, trading accounting
Security infrastructure Single-sign-on, authentication, secure communication
Grid fabric PCs, workstations, clusters, networks, software, databases, devises

Fig. 1. Compute Grid Layered Architecture [34]

Application	Discipline-Specific Data Grid Application
Collective (App)	Coherency control, replica selection, task management, virtual data catalog, virtual data code catalog …
Collective (Generic)	Replica catalog, replica management, co-allocation, certificate authorities, metadata catalogs,
Reso	Access to data, access to computers, access to network
Connect	Communication, service discovery (DNS), authentication, authorization, delegation
Fabric	Storage systems, clusters, networks, network caches …

Fig. 2. DG layered Architecture [34]

The following Fig. 3 depicts integration of DGL's facilitating the management of CG services, with a resultant load balancing on the data centers

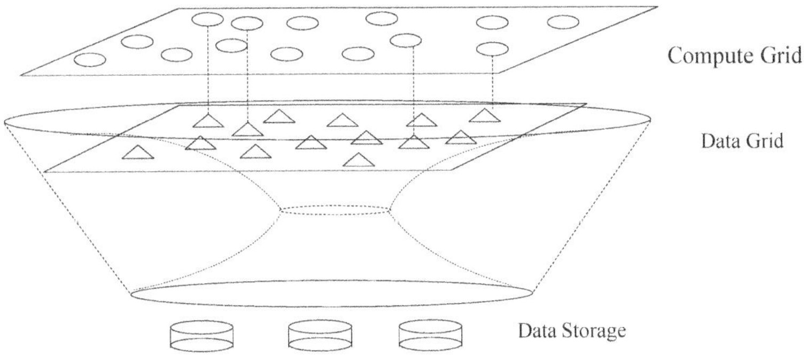

Fig. 3. Data grid layer added beneath compute grid [5]

3.1 Motivations for Data Grids

Some of the major motivations for integrating DG layer beneath the CD layer are explained by the research work in [23]. Thus, DG systems relieve CG systems from the burden of data management. DG Coordinate compute and data layer, which include streamlining access to data centers, data synchronization, managing data connections, etc. In addition, DG facilitates task and resource management - managing scheduled tasks to facilitate faster computational activities with minimum data blockage. DG help in sharing of data, distribution, and location information - thus facilitating data mobility, caching, scheduling of computing tasks. DG and CG layers work together to provide Data Regionalization including synchronization, distribution, transaction ability, fault tolerance, data availability, data management, and distribution.

4 Some Major Grid Projects and Their Deployment Focus

The following Table 1 summarized some of the major grids projects developed over the years with their deployment motivation and application areas.

Table 1. Some major grid projects and their deployment focus

Grid Project	Developer	Deployment focus	Grid type
Open Source Grid	American based Grid Infrastructure	Scientific research applications	Computational Grid
Tera Grid	High-end computational resources based in the USA	Scientific research applications	Compute and Data Grid
BeInGrid Business Experiments in Grid	European Union	Scientific and business research applications	Services Grid
EGEE: Enabling Grids for E-Science	European Commission	HEP (High Energy Physics), Biomedical Sciences Research	Data Grid
Grid5000	National Grid Project of France	A platform for Academic Research, Supports Open Science, Parallel & Distributed Computing including Cloud, HPC, and Big Data	Compute and Data Grid
Grid PP: Grid for particle Physics	European Union	Part of the project EGEE, Supports HEP experiments	Compute and Data Grid
		Extends the LHC experiments to other communities	
Cross-grid	European Union	Used in HEP of CERN, Biomedical Research, Earth observation, climatology & Nanotechnology	Interaction Grid
Data Grid	European Union	Used in HEP of CERN & Biomedical Research	Data Grid
LHC Grid: Large Hydron Collider	Europe, America, and Asia	Management and Distribution of LHC data from CERN to some major computer centers in Europe, North America, and Asia	Data Grid
GriPhyN: Grid Physics Network	NSF-funded project	Supports Large-scale Data management in Physics experiments such as HEP, Astronomy, and Gravitational Wave Physics	Data Grid

5 Characteristic Features of the Grid Systems

Computational Grids (CDs) allow users to take a computational task and optionally split it into multiple parts; execute them on different grid nodes in parallel [23]. Apparently, this strategy tends to speed up the computational tasks since resources from different grid nodes could be used in parallel.

5.1 Compute Grids Characteristics

Some of the most important Computational Grid characteristics features include: (i) Automatic Deployment-Allows automatic deployment of classes and resources onto the grid without any extra steps from the user. (i) Topology Resolution-Allows computing nodes provisioning based on a given node characteristics or configurations specified by users. (iii) Collision Resolution- CGs allow users to monitor and control the order by which their jobs are executed, Jobs to be rejected, and how many jobs can be executed in parallel. (iv) Load Balancing- CGs enable system loads to be properly balanced within the grid, made possible by making use of load balancing policies. (v) Fail-over- Jobs automatically fail over onto other nodes in case of failure or node crash. (vi) Chec-points- Long running jobs should be able to periodically store their intermediate state; useful for fail-overs. (vii) Grid Events-A mechanism that query all events that happened on remote grid nodes during grid task execution. (viii) Node Metrics-Ability to provide dynamic grid metrics for all grid nodes such as vital node statistics from CPU load to average job execution time; useful for load balancing. (ix) Plugability-Ability to blend into any environments. For instance, if running on top of G-Lite, a CG should totally reuse G-Lite communications and discovery protocols. (x) Data Grid Integration- Ability to natively integrate with DGs. More often businesses will need both CG and DG features working within the same application.

5.2 Data Grids Characteristics

Data Grids allow users to distribute their data across the grid environment [23]. Some of the important DG characteristics features include: (i) Data Replication-Ability to replicate data within the grid; consumes resources, however. (ii) Data Invalidation-Ability to load data on demand; changes to data on one of the nodes will prompt purging or terminating of the same on all other nodes. (iii) Distributed Transactions-A Data Coherency policy is making sure that cache updates are working just like database updates; any update failure will trigger rolling back of the whole transaction. (iv) Data Backups-. Ability to assign some nodes to be used for data backups; useful for fail over. (v) Data Affinity/Partitioning-Feature that enables splitting or partitioning of whole data set into multiple subsets and assigns every subset to a grid node for execution.

6 Grid Systems Simulation Tools and Areas of Applications

In this section, Compute and Data Grids will be analyzed in two folds. However, space may not allow us to show the detailed analysis here. In Table 2, CG and DG have been compared in terms of existing grid simulation tools and their design motivations.

Table 2. Comparison of CG Vs DG based on simulation tools and design motivation

Simulation tool	Design motivation	Grid platform
GridSim	Allows modeling and simulation of objects in parallel and distributed computing (PDC) systems [18]	Computational Grid
SimGrid	A Scientific instrument to study the behavior of large-scale distributed systems such as Grids, Clouds, HPC or P2P systems	Computational and Data Grids
Bricks	Resource Scheduling tool for compute and data grids	Computational and Data Grids
GangSim	Grid scheduling studies, capable of supporting studies for controlled resource sharing	Computational Grid
OptoSim	Grid simulator designed to test dynamic scheduling and replication strategies used in optimizing the efficiency of a Grid [19]	Data Grid
G3S	Grid Security services Simulator	Compute Grid
CloudSim	Designed to support modeling and simulation of large scale Cloud computing data centers [20]	Cloud Data Centers
GreenCloud	Green Cloud [41] Provides a simulation environment for energy-aware cloud computing data centers	Cloud Data Centers
PlanetLab	PlanetLab [36] is an open platform for planetary scale services that emulate Network, CPUs and applications services. Each user can get a slice of the platform, multiple competing services run in parallel on shared nodes	Computational Grids
EmuLab	Emulab [37] is an emulator maintained by the University of Utah. It is an emulator that runs on over 200 nodes, most of them PC's. It uses the same script language as the ns [32] network simulator which is a widely used simulation tool for network topologies. The Emulab software enables studying networked and distributed Systems by transparently Integrating and supporting a variety of different experimental measurements: emulation, simulation, and live-Internet as well as virtualized emulation	Distributed Systems
SimBOINC	SimBOINC [39] is a simulator based on the SimGrid toolkit designed for the simulation of scheduling strategies in heterogeneous and volatile desktop grids and volunteer computing systems. Users can dedicate their idle computing power to one of several scientific projects producing real research data.	Computational Grids and Data Grids
HuskySim	HuskySim [40] is a discrete event Grid simulator toolkit developed in JAVA. It enables simulation of both static and dynamic job scheduling strategies to different processors and a hybrid approach combining static and dynamic techniques	Computational Grids
HuskySim	HuskySim [40] is a discrete event grid simulator developed in JAVA. It enables simulation of both static and dynamic job scheduling strategies	Computational Grids

Table 3. Comparison of CG Vs DG based on performance metrics

Research approach	Grid platform	Grid simulator	Measureable metrics
Novel Job Scheduling for Improving Data Grid's Performance [24]	Data Grid	OptorSim	• Access time • Job throughput • Network utilization • Storage utilization
Scheduling, Dynamic Data Replication Strategy [25]	Data Grid	OptorSim	• Access Latency • Network Utilization
Enhanced, Adaptive Scoring Job Scheduling Algorithm [26]	Compute + Data Grid	GridSim	• Storage Utilization • Transmission Power • Processing Power • Job Completion Time • Job Failure Rate
Dynamic Replication Strategy [27]	Data grid	OptorSim	• Network Utilization • Storage Usage • Job Throughput
Scheduling Reliability, Time, Cost & Constraints [28]	Compute Grid	OptorSim	• Mean Job Time • Total Job Time • Storage Element Size
Integrating Scheduling & Replication with Performance Guarantee [29]	Data Grid	OptorSim	• Access Time
Heuristic algorithm for scheduling [30]	Compute Grid	GridSim	• Mean Job Time • Resource Utilization
Hybrid batch job scheduling [31]	Compute Grid	GridSim	• Mean Job Time • Resource Utilization
Data Replication Approach Consistency Guarantee [32]	Data Grid	GridSim	• Bandwidth Usage • Response Time • System Load, • System Availability
Hybrid Approach for Monitoring & Scheduling [33]	Compute Grid	GridSim	• Mean Job Time • Resource Utilization

Table 3 gives a summary of a critical review of previous researches and the corresponding performance evaluation metrics.

7 Conclusions and Future Work

Looking at the grid systems from the architectural point of view, compute and data grids exhibit slightly different architectural constituents. In terms of design principles/motivation, the two concepts differ largely. The design of Computational Grids has been motivated by the desperate need for robust computing infrastructure to function as and/or replace the traditional centralized supercomputing infrastructure so that it can cope with current enabling technologies such as massive MIMO [42–48] to realize 5G and other applications. Whereas in the case of data grids, the design had been strongly motivated by the need to create databases using dedicated Grid Database

Management Software (GDBMS). The GDBMS enables efficient database management services such as data storage, replication, retrieval, movement, relocation, supported by robust storage facilities. Although these could not be achieved without the use of computing facilities, yet the emphasis is on data and databases. In addition, data grids and compute grids have been explicitly categorized under software and hardware, respectfully.

It has also been shown that data grid is layered beneath the compute grid to support the activities of the later. In terms of business and scientific applications, both Compute and Data Grids have been seen to play major roles. However, data grids prove to be more relevant when it comes to data intensive applications as is the case with the LHC, Biomedical, and High Energy Physics experiments. Previous researches carried out have shown significant convergence in the use of the various measurable metrics. However, from Table 4, GridSim and SimGrid simulators are visibly more active in terms of evaluating computing resources such as the processor, memory, and transmission power. On the other hand, OptorSim is more conversant with storage element and network utilization, as these entities are biased towards activities relating to data and databases. Thus, Data Replication functionality happens to be one of the unique characteristics of Data Grids.

To the depth of this research, this is the first time the two major grids types are compared in a single research based design motivation, areas of applications and performance metrics. In the future, this work will be extended to compare other types of grid systems in terms of the layered architecture, design motivations, supported applications, and simulation tools. In addition, attempt will be made to investigate the performance of CG and DG using the most pronounced measurable metrics such as Access Time, Job Throughput, Network and Storage Utilization in both GridSim and OptorSim simulation environments.

References

1. Jacob, B., Brown, M., Fukui, K., Trivedi, N.: Introduction to Grid Computing. IBM International Technical Support Organization, USA, pp. 1–5 (2005)
2. Magoules, F., et al.: Introduction to Grid Computing, pp. 1–5. CRC Press, Taylor and Francis Group, London (2009)
3. Magoules, F., et al.: Introduction to Grid Computing, pp. 10–14. CRC Press, Taylor and Francis Group, London (2010)
4. Vaithiya, S., et al.: Ontology based resource discovery mechanism for mobile grid environment. In: Proceedings of IEEE, pp. 154–159 (2013). https://doi.org/10.1109/adcons
5. Qureshi, M.B., et al.: Survey of Grid Resource Allocation Mechanisms. Springer, Dordrecht
6. Magoules, F., et al.: Grid Resource Management, pp. 1–17, 61–95. CRC Press, Taylor and Francis Group, London (2010)
7. Buyya, R.: Economic-based distributed resource management and scheduling for grid computing. Ph.D. thesis, School of Computer Science and Software Engineering, Monash University, Melbourne, Australia, p. 180 (2002)
8. Ranjan, R., Harwood, A., Buyya, R.: A taxonomy of peer-to-peer based complex queries: a grid perspective. abs/cs/0610163 (2006)

9. Khan, S., Ahmad, I.: A cooperative game theoretical technique for joint optimization of energy consumption and response time in computational grids. IEEE Trans. Parallel Distrib. Syst. **20**(3), 346–360 (2009)
10. Khan, S.U.: A game theoretical energy efficient resource allocation technique for large distributed computing systems. In: International Conference on Parallel and Distributed Processing Techniques and Applications (PDPTA), Las Vegas, pp. 48–54 (2009)
11. Yeo, C.S., Buyya, R., Assunção, M.D., Yu, J., Sulistio, A., Venugopal, S., Placek, M.: Utility computing on global grids (chap. 143). In: Bidgoli, H. (ed.) The Handbook of Computer Networks. Wiley, New York (2007). ISBN 978-0-471-78461-6
12. Dharma, T.N., et al.: Data replication in data intensive scientific applications with performance guarantee. IEEE Trans. Parallel Distrib. Syst. **22**(8) (2011)
13. Ciubăncan, M., Dulea, M.: Implementing advanced data flow and storage management solutions within a multi-VO grid site. In: 2017 16th RoEduNet Conference on Networking in Education and Research (RoEduNet), pp. 1–4. IEEE, September 2017
14. http://en.wikipedia.org/wiki/File:Data_Grid_Multiple_Topologies_1.Jpg. Accessed 21 Jan 2015
15. http://www.usatlas.bnl.gov/computing/grid/griphyn/
16. http://www.realitygrid.org/publications.shtml
17. http://www.nesc.ac.uk/action/projects/project_action.cfm%3Ftitle=236=>crossgrid
18. http://www.buyya.com/gridsim/
19. http://www.gridpp.ac.uk/demos/optorsimapplet/
20. http://www.cloudbus.org/cloudsim/
21. http://www.sdsc.edu/srb/index.php/DGMS
22. http://it.toolbox.com/blogs/technews/grid-computing-advantagesand-disadvantages
23. http://java.dzone.com/articles/compute-grids-vs-data-grids
24. Mansouri, N., Dastghaibyfard, G., Horri, A.: A novel job scheduling algorithm for improving data grid's performance. In: 2011 International Conference on P2P, Parallel, Grid, Cloud and Internet Computing, pp. 142–147 (2011)
25. Zarina, M., Ahmad, F., bin Mohd Rose, A.N., Nordin, M., Deris, M.M.: Job scheduling for dynamic data replication strategy in heterogeneous federation data grid systems. In: 2013 2nd International Conference on Informatics and Applications (ICIA), pp. 203–206 (2013)
26. Aparnaa, S.K., Kousalya, K.: An enhanced adaptive scoring job scheduling algorithm for minimizing job failure in the heterogeneous grid network. In: 2014 International Conference on Recent Trends in Information Technology (ICRTIT), pp. 1–6 (2014)
27. Yeo, C.S., Buyya, R., Assuncao, M.D., Yu, J., Sulistio, A., Venugopal, S., Placek, M.: Utility computing on global grids. In: Bidgoli, H. (ed.) Handbook of Computer Networks. Wiley, New York (2006)
28. Maheshbhai, L.A.: Job scheduling based on reliability, time and cost constraints under grid environment. In: 2011 Nirma University International Conference on Engineering (NUiCONE), pp. 1–5, 8–10 December 2011
29. Lakshmi, R., Tang, B.: Integrating scheduling and replication in data grids with performance guarantee. In: IEEE Globecom 2011 Proceedings (2011)
30. Wang, J., Wu, G., Zhang, B., Hu, X.: A heuristic algorithm for scheduling on grid computing environment. In: 2012 Seventh ChinaGrid Annual Conference, pp. 36–42 (2012)
31. Zahedani, S.D., Dastghaibyfard, G.: A hybrid batch job scheduling algorithm for grid environment. In: 2014 4th International Conference on Computer and Knowledge Engineering (ICCKE), pp. 763–768 (2014)
32. Abawajy, J.H., Deris, M.M.: Data replication approach with consistency guarantee for data grid. IEEE Trans. Comput. **63**(12), 2975–2987 (2014)

33. Rajkumar, S., Rajkumar, N., Suresh, V.G.: Hybrid approach for monitoring and scheduling the job in heterogeneous system. In: 2014 International Conference on Information Communication & Embedded Systems (ICICES), pp. 1–5, 27–28 (2014)
34. Foster, I.: The anatomy of the grid: enabling scalable virtual organizations. In: 2001 Proceedings of First IEEE/ACM International Symposium on Cluster Computing and the Grid, pp. 6, 7 (2001)
35. Song, H.J., Liu, X., Jakobsen, D., Bhagwan, R., Zhang, X., Taura, K., Chien, A.: The microgrid: a scientific tool for modeling computational grids. In: ACM/IEEE 2000 Conference on Supercomputing, p. 53. IEEE, November 2000
36. http://drjslab.org/2014/05/27/list-of-simulator-for-cloud-and-grid-computing/
37. Stoller, M.H.R.R.L., Duerig, J., Guruprasad, S., Stack, T., Webb, K., Lepreau, J.: Large-scale virtualization in the Emulab network testbed. In: USENIX Annual Technical Conference, pp. 113–128, June 2008
38. Cutler, C., Hibler, M., Eide, E., Ricci, R.: Trusted disk loading in the Emulab network testbed. In: CSET, August 2010
39. SIMBoinc Home Page. http://simboinc.gforge.inria.fr/
40. Kerasha, M.A., Greenshields, I.: Huskysim: a simulation toolkit for application scheduling in computational grids. In: Proceedings of the 13th International World Wide Web Conference on Alternate Track Papers & Posters (WWW Alt. 2004), pp. 380–381. ACM, New York (2004)
41. Kliazovich, D., Bouvry, P., Audzevich, Y., Khan, S.U.: GreenCloud: a packet-level simulator of energy-aware cloud computing data centers. In: 2010 IEEE Global Telecommunications Conference (GLOBECOM 2010), pp. 1–5. IEEE, December 2010
42. Fazea, Y.: Numerical simulation of helical structure mode-division multiplexing with nonconcentric ring vortices. Opt. Commun. **437**, 303–311 (2019)
43. Fazea, Y.: Mode division multiplexing and dense WDM-PON for Fiber-to-the-Home. Optik **183**, 994–998 (2019)
44. Fazea, Y., Alobaedy, M.M., Ibraheem, Z.T.: Performance of a direct-detection spot mode division multiplexing in multimode fiber. J. Opt. Commun. **40**, 161–166 (2019)
45. Ibraheem, Z.T., Rahman, M.M., Fazea, Y., Ahmed, K.K.: PAPR reduction in OFDM signal by incorporating Mu-Law companding approach into enhanced PTS scheme. J. Opt. Commun.
46. Fazea, Y., Amphawan, A.: 32 channel DQPSK DWDM-PON for local area network using dispersion compensation fiber. In: EPJ Web of Conferences, vol. 162, p. 01016. EDP Sciences (2017)
47. Fazea, Y., Mezhuyev, V.: Selective mode excitation techniques for mode-division multiplexing: a critical review. Opt. Fiber Technol. **45**, 280–288 (2018)
48. Fazea, Y., Sajat, M.S., Ahmad, A., Alobaedy, M.M.: Channel optimization in mode division multiplexing using neural networks. In: 2018 IEEE 14th International Colloquium on Signal Processing & Its Applications (CSPA), pp. 173–175. IEEE (2018)

Resource Discovery Mechanisms in Shared Computing Infrastructure: A Survey

Mowafaq Salem Alzboon[1(✉)], M. Mahmuddin[2], and Suki Arif[2]

[1] Jadara University, P.O. Box 150459, Irbid 13115, Jordan
malzboon@jadara.edu.jo
[2] InterNetWorks Research Laboratory, School of Computing, Universiti Utara Malaysia, 06010 Sintok, Kedah, Darul Aman, Malaysia

Abstract. The increased number of computers, enlarged network bandwidth, more powerful computers, consumed resources and the acceptance of the Internet has driven the ongoing demand for new and better ways to execute huge problems in shared computing infrastructure such as grid and P2P computing. Resource discovery is extremely significant and challenging issues in Grid and P2P. It is complex because resources are heterogeneous, dynamic, geographically distributed, manageability and autonomous. A well-organized resource discovery mechanism is the fundamental requirements for resource sharing infrastructure, as it supports and facilities the resource administration and scheduling of applications. In computing systems like Grid, provide many resources across multiple administrative domains for the achievement of the goal. The major challenge faced by the system is to find the right resources on the network. In this paper, we study the resource discovery mechanisms that have been used in distributed computing systems so far. The mechanism for discovery of resource is divided into different types: centralized, decentralized, and hierarchical. We review the main developments in these categories and outline the new challenges. This paper also provides a discussion of the differences between the mechanisms considered for scalability, dynamism, reliability and real consultations as well as directions for future research.

Keywords: Grid computing · Resource discovery · Reliability

1 Introduction

World Wide Web (WWW) was the aspirant networking development of the Defense Advanced Research Project Agency (DARPA) [1]. Initially, it was used just by a small number of military sections in the United States to exchange information. Next, in a short period of time, the project became so popular and was utilized by many to form a network of computers. This expansion of WWW was achieved through the educational and research institutions which further expanded to other domains. Presently, the Internet has turned out to be a part of humanity hi-tech everyday life where a majority of individuals cannot imagine life without the Internet. The Internet has grown into being a major information highway of the current world as it is used in browsing and exchanging information, publishing e-commerce and various kinds of multimedia [2].

F. Saeed et al. (Eds.): IRICT 2019, AISC 1073, pp. 545–556, 2020.
https://doi.org/10.1007/978-3-030-33582-3_51

The Internet has become one of the members of the commodity list in the computing world. As a result of this, new research has emerged and attracted the attention of people around the globe to the use idle computers that are connected to the Internet [3]. Computers are described as idle when any program is not employing them or when a computer is connected to the Internet, but it is not executing any work or computation [4]. This area has offered researchers the room to explore idle computers within the Internet for improved utilization using Shared Computing Infrastructures (SCI) which is a type of distributed computing that allows large-scale resource sharing system through the Internet. It makes it possible to share files, software and hardware resources (e.g., collaborative computing, peer to peer computing and grid computing). The Grid is one of the significant developments in the field of engineering and computer science and provides a clear future in the global use of various optimal distributed resources (software and hardware). Grid computing and peer-to-peer computing are both hot topics at present. The convergence of the two systems is increasingly visible, and Open Grid Services Architecture (OGSA) provides a framework for integrating grid and peer to peer.

Grids are systems that manage and share resources, and also solve problems in heterogeneous dynamic environments [5]. A Grid is publicly shared, and this includes a system formed from different resources, shared by multiple administrative or Virtual Organizations (VO) which offer access to useful resources based on an agreed set of rules. The fundamental concept of Grid networking is not new, and there are several meanings in this present day for Grid, but the foundational concept of Grid computing revolves around "aggregating geographically dispersed computing, storage, and network resources, which are coordinated with the aim of delivering higher service quality, improved performance, more optimal utilization, and easy data access" [6]. SCI facilitates VOs to share data and applications in a heterogeneous, open environment [13]. The central idea of SCI is to launch like environment, over sharing vicinity, integrating heterogeneous elements such as a server, storage, and computer network parts in a scalable, wide area across computing infrastructure.

The Grid middleware which harmonizes share parts and components are equivalent to the Operating System (OS) in a standalone computer medium [7]. A shared computing infrastructures like Grid and P2P computing is a large-scale, heterogeneous, dynamic collection of independent systems which are geographically distributed and interconnected with high-speed networks to meet the ever-increasing demand for high-performance computing. A resource information service, which collects resource information and provides resource search functionality for locating desired resources, is a crucial component in shared computing infrastructures.

A typical example of shared computing infrastructures consists of a central computer with a centric database called Grid Resource Information Service (GRIS) where the information about the available resources is updated and stored. Moreover, a significant number of computers called workers which occasionally keep posting their resource condition among a centric database or GRIS in the central computer. User submits the jobs to be performed by the central computer that in turn generate some resource query to a centric database for the suitable resources then a central computer directs the workers to carry out the jobs and finally collects the results. Commonly, a Grid computing consist of three main parts one central computer typically as a server, which holds the complete administrative responsibilities of the system, A network of computers running special Grid computing network software and middleware that is a collection of computer software. Grid computing utilizes middleware to coordinate disparate IT resources across a network and a Grid middleware distribution can be described as a software stack or a collection of cooperating components, protocols and services which will give users the ability to access the resource distribution of a Grid. Examples of a Grid middleware includes Advanced Resource Connector (ARC), European Middleware Initiative (EMI), gLite, Globus toolkit, GridWay, Open middleware infrastructure Institute United Kingdom (The OMII-UK distribution), Oracle Grid Engine, Alchemy, and Legion [8–14].

2 Centralized Resource Discovery Mechanisms

Central architecture which discovers resources by using selected group of controllers in a system is known as the centralized mechanism [15]. The information about different available services can be obtained from the server in these systems. In the centralized mechanism, the request is generated by the entity for a certain service and is forwarded to the server. The server searches the required resources and then provides theses resource to the concerned entity. They can search the required resources in less time as compared to other mechanisms. This mechanism faces bottleneck when the number of nodes in a system goes beyond several hundred because this architect assigned set of controllers to process all queries. Some centralized mechanisms in a Grid, which are used in determining available resources, are now discussed.

In 1997, a new system was proposed by [16] for the management of resources in Grid systems known as Meta-computing Directory Service (MDS). Application Programming Interface (API) and data representation are used in Meta-computing Directory Service where the central servers maintain the computing service on Lightweight Directory Access Protocol (LDAP). In the LDAP servers, specialized data structures such as MDS entries are used to represent the resources. LDAP handles the maintenance of resource information and queries response. This method is useful in providing dynamic, distributed and diverse information of the resources but MDS is of centralized nature, so it has a single point of failure. Due to this reason, MDS-2 was designed and developed with the enhanced hierarchical manner by [17] in 2001.

In 2003, [18] presented Grid Market Directory (GMD), a web-based service, which uses XML format for providing services to clients and users. As compared to MDS, GMD provides efficient services and supports the enabling of Grid economy. GMD Query Web Page (QWP) and GMD Portal Manager are the two most important components of GMD. GMD-QWS helps the discovery of resources which are necessary to meet the requirements of queries by using web service processing. The GMD Portal Manager provided the management of publication and registration of resources on a web service and in a large-scale Grid system, the system suffers bottleneck as it uses central query processing. Likewise, the system suffers a singular failure point from the central resource repository database.

As a result of the dynamicity requiring manual updates, the system does not seem appropriate for Grid environments with dynamic attributes. Although range attributes and multi-attribute queries are supported in the repository system, dynamic attribute queries cannot be utilized in the system as it is not supported. Another observation is that the response time increases when there is a growth in workload. A centralized mechanism for discovery of resource, in web-based Grids, was proposed by [19] in 2007. This system consists of Universal Description Discovery and Integration (UDDI) [25], Simple Object Access Protocol (SOAP) [20], Hypertext Transfer Protocol (HTTP) [19] and Grid Web Services Description Language (GWSDL) [13].

In a service-oriented architecture, for the discovering and publishing of the software components, UDDI [21] is considered as standard and in this mechanism, UDDI is used for the Grid services discovery and likewise holding of key values of Grid resource information. Grid services are described in an extended edition by using GWSDL. In a Grid environment, the web service communication is established by SOAP where the interface to obtain responses and sending requests is provided by HTTP. The web service was run by using the central database and servers which face single point of failure and bottleneck. The repository database was used to hold resource information which does not support dynamic attribute query even though it supports multi-attribute and range queries. By using these given modifications, UDDI can be used for resource discovery in a Grid service. The security is high as it is operated under a central system but dynamic attribute queries and bottleneck are not supported while still possessing very low scalability. In a Grid, the adaptive resource selection and resource discovery were presented by [22] in 2010. The authors discussed three different models; Adaptive push-pull model, Adaptive pull model, and Adaptive push model. Adaptive Pull Model refers to the model in which one node is made as a coordinator in a Grid environment, with only one daemon running.

Dynamic information such as load on CPU, RAM, and CPU speed are collected and pulled by the daemon, and it has a single point of failure due to huge traffic. However, with an increase in traffic, the time will also increase to pull the process of resource discovery query. Now in the process where the master daemon is working by the use of the coordinator node while other nodes use a local daemon for gathering information and push them to the node is the operation process for the adaptive push model. To record resource activity, the database is maintained by the centralized coordinator. The coordinator and master daemon face huge traffic while having a singular failure point. The third model is a combination of both models described above known as the adaptive push-pull model. This model is made up three layers where the first layer (Layer-1) is the Main Coordinator, the second layer (Layer-2) are the Aggregators and the third layer (Layer-3) consist of the Grid nodes. To collect information about the state of the system such as the RAM, the load on CPU or CPU speed, a daemon is run by each Grid node and this information will be pushed toward aggregator instead of sending it to the main coordinator.

The aggregator maintains the database of sub-resources and then the daemon will respond to the queries by pulling information from the aggregator. This will decrease the traffic but does not remove the single point of failure. This Grid is inefficient at large scale. In the central mechanism, the information is updated and stored on central servers. It increases the traffic to the network which does not support dynamic attribute queries, although most mechanisms support multi-attribute queries. Due to centralized control of information, security and authorization issue are handled easily, but there is still the disadvantage of low scalability. As a result of the enormous traffic of nodes, the system suffers the bottleneck problem, and because it operates on a central database, parallel access is limited thus high workload decrease time efficiency [10]. These drawbacks were removed in the decentralized mechanism which changes the database to avoid failure. Client/server structural design which discovers resources by using a selected group of controllers in a system is known as the centralized mechanism [10, 13].

The information about different available resources can be obtained from the server in these systems in this mechanism; the request is generated by the entity for a certain resource and is the outcome is forwarded to the servers. Then a search is performed by the server on the required resources which then simultaneously makes provision of these resources to the concerned entity. This mechanism face bottleneck when the number of nodes in a system goes beyond several hundred because in this architecture a set of controllers is assigned to process all queries. It likewise increases the traffic to the network where dynamic attribute queries are not supported, though a significant number of mechanisms support multi-attribute queries. Due to centralized control of information, security and authorization issue are handled easily, but the low scalability disadvantage is still present. The system still continues to suffer from bottle-neck problem as a result of the massive traffic of messages. Decrease in time efficiency is still a resultant problem of the parallel access to the central database due to high workload [10, 13].

3 Decentralized Resource Discovery Mechanisms

It has been clearly discussed that the centralized mechanism does not support large-scale networks [23, 24]. Therefore, replacement by a decentralized mechanism is expedient in which the central server is removed, and all nodes work for the discovery of resource in the extensive network. In a decentralized mechanism, there is liberty for Grid resources as they can leave or join at any time, likewise disconnecting and connecting to the routers is also allowed [25–30]. These mechanisms support load balancing, fault tolerance, high scalability and removal of Single-Point-of-Failure (SPF) [34], although the systems still do not support dynamic query attributes. The system requires an information update for resource periodically, and network traffic is very high [28, 31]. In 2001, in [32] presented a method for the discovery of resource which is fully decentralized and in these systems, resources are placed on nodes within the network. The architecture of the mechanism is smooth, configured and decentralized. In this mechanism, the local node receives the request from the user, and this request is moved forward from one node to other within the network. Each node will check the availability of the source at their end and if not found move it to other nodes. The process continues till the source is found, and the request is transferred to the node using one of these: "best neighbor," "experience based + best neighbor, "random", or "experience based + random." However, amongst these four, the "experience based + random algorithm" performs best. In this mechanism the resource is found on a priority basis and when the first resource is identified it is transferred to the requestor and the search stops. The flooding in this system reduces scalability but it is inefficient in terms of time to response to vital systems because there is a possibility of false positive error. Despite all these problems, this system is useful for balancing the load. In 2004, [13, 33] proposed a system in which attributes are used to discover resources. The nodes with similar resources in this system make a group and form information communities by the discovery mechanism. The formation of communities and easy availability of information increases the performance but at the same time, overhead maintenance is also increased. The introduction of decentralized mechanism with flat network structure for the discovery of resource was presented by [34] in 2005, and the algorithm was quite different from the one used previously. To forward the request, "experience based + random" algorithm is used. In addition to the reserve nodes, some additional matching can also be obtained on demand from the requestor. This mechanism is controlled in such a way that only one resource is provided to a request even if the number of available resources is more than required. The efficiency of the selection process reduces the identification time required for the discovery of resource. Although when a large scale is considered, the results are less efficient, and this mechanism also supports false positive error.

The researchers in [23] introduced a new system in 2007, which is based on a combination of keywords and clusters of nodes. The information of requested resource is transmitted by the help of multi keywords which are received by under-standing of the required resource. These key words are matched at nodes side which increases the speed as the request is forwarded to the focused clusters. In Knowledge Grid, by helping retrieval of knowledge based on multi-key words, this mechanism fits to realize sharing of knowledge and knowledge integration. This whole mechanism shows that the network size, memory size and the hopes where the message routes do not matter. This was referred to as a distributed Grid environment. In 2012, [35] presented Learning Automata based Resource Discovery (LARD) which improves the hit ratio average, hop count average and also controlled the overhead of message. In a decentralized mechanism, Grid resources can leave or join at any time as afore-mentioned, likewise disconnecting and connecting to the routers is also allowed. These mechanisms do not support balancing of load, fault tolerance, high scalability, single point of failure and dynamic attribute of the query. These mechanisms require an update of information for resource periodically.

4 Hierarchical Resource Discovery Mechanisms

Like the Globus resource discovery mechanism, the information of resources is indexed, updated and sorted under a set of hierarchical nodes [36]. The connectivity of the distributed system is provided when the proximity clusters of the resources are built at the closer level of the resources. In these approaches, since servers have been organized hierarchically, the queries are also processed hierarchically in such a way that each server is responsible for partitions of information resource site [37]. Due to the use of multilayer architecture, these mechanisms guarantee low search delay and have static partitioning scheme [38], although some servers face more load than others. Since it takes a long time for resource information to reach the upper node, they do not work well when the value of attributes changes rapidly.

Furthermore, the traffic and request of routes are reduced when a fault is faced by using multi layers architecture. Also, in comparison to the mechanism based on Data Structure Tree (DST), the fault tolerance of services is improved more than 28% by proposed mechanism [52, 53]. However, the hierarchical mechanism is not efficient for large-scale Grid as the tree root represent an SPF. Hierarchical mechanisms are more scalable than centralized systems. They provide few proximity query capabilities that are raised from the system's design and reduce the network traffic. The drawback of this system is that queries are not answered at the lower level, and it cannot provide fast response to the queries. At times, for small and medium sized Grids, the hierarchical mechanisms are practical and for very large Grids super peer mechanisms are more effective. Sufficient time is required to update resource information from the leaf to the root, and SPF occurs in tree root system which minimizes the system fault tolerating feature. MDS-2 was introduced in 2001. It provides Grid Resource Information Service (GRIS), a configurable information provider component and Grid Index Information Service (GIIS), a directory component. In MDS-2, GRIS service is responsible for providing dynamic and static resource information, and when using GIIS, the resource information is stored in directories. As per the local and virtual organization-specific policies, the information provider registers itself to a directory and in its registered directory, the provider then updates its resource status [3, 4, 39, 40].

The provider is deleted from its directory after reported unavailable for a predefined period or if the provider of the resource is incapable of updating its resource information. Consideration is made in the results obtained for discovery queries that are relevant till the time provider exists in a directory. In MDS-2, from a rich variety of system components, LDAP used a uniform means for storing system information. It is also used for query processing and constructing a consistent namespace for resource information across the system. The use of Grid Security Infrastructure (GSI) credentials provides support for secure data access. It also provides high scalability and multi-attribute queries. However, if the system is used for long period without restarting, the server causes a significant problem in load and memory usage. Although in different clients, the time to answer a query also varies. A new Grid resource discovery mechanism and job submission system were proposed by [13, 23] in 2005. This is also considered as a complete Grid management middleware.

In this hierarchical topology, the Grid environment is divided into VO where each VO consists of slave nodes and master nodes. In the Grid system, the slave nodes retrieve resource information from each machine and the master nodes then update the resource information. Each master machine describes the slave machine using XML format with the IPs and names of slaves are available in the XML file. By the preparation of a configuration resource which includes the requested resource information, the resource discovery is realized. There is a decrease in the occurrence of a possible bottleneck problem because it is a hierarchical web service based system. Because the task of the discovery of resources is being handled by the master machines, the majority of the slave machines become invisible within the system after the failure of a master machine. Also, as a result of resource discovery fulfilling its task with no hashing function involved, this mechanism supports all multi-attribute, range and dynamic attribute queries. However, the proposed mechanism is not efficient for the large-scale Grid. Also, at the tree root, single point of failure is faced by this system.

Hierarchical mechanisms are more scalable than centralized systems. The drawback of this system is that queries are not answered on the lower level, and it cannot provide fast response to the queries. At times for small and medium sized Grids, the hierarchical mechanisms are practical and for very large Grids super peer mechanisms are more effective. Sufficient time is required to update resource information from the nodes of the leaf to the node in the root and since single failure point occurs in the system of the tree root, hence decreasing fault tolerating feature of the system [4, 39, 41–43].

5 Mechanisms Comparison

Based on the characteristics and network structure, the resource discovery mechanisms have been categorized as follows; hierarchical, P2P, decentralized and centralized mechanism. In centralized mechanisms, the authorization and security issues are handled easily through central system administration, and it supports range and multi-attribute queries. These mechanisms have an SPF, though not scalable, reduces overall performance and have a high computational overhead. Node dynamicity is ignored in these mechanisms, and dynamic attribute queries are also not supported because of periodic updates. Likewise, due to high workload and limited parallel access to the system, the response time could be very high.

In the centralized mechanism, a central database is a main drawback, and the global database must be replicated to overcome this drawback. Decentralized mechanisms are based on this idea. The problem of centralized mechanisms is removed in decentralized mechanisms as it supports multi-attribute and ranges queries since they are resolved without any hashing. These mechanisms scale well, tolerate the failure of distributed operations, balance loads and tolerate node dynamicity. Although it manages the network architecture, additional overhead is created. Structured P2P mechanisms, unstructured P2P mechanisms, Hybrid P2P mechanisms and super to peer mechanisms are the four categories of P2P mechanisms. Regarding query correctness and SPF, P2P mechanisms that are unstructured are very reliable and robust. Also, they support multi-attribute, range and dynamic attribute queries and can tolerate node dynamicity. Since both resources and indexing nodes are distributed to the network, these approaches can handle easily dynamicity of the Grid [44–47].

However, to deliver service to requester they cannot give a definite assurance of the hops number taken. By using of TTL, they also suffer from false positive errors. Due to TTL field, these methods cannot assure that outcomes will become backed if they are available in the network. The network wide broadcast storm also affects them. Despite their disadvantages, these methods are familiar with the Internet. Regarding traffic load, structured P2P RDM are scalable. To maintain their fixed structure, they require self-organizing mechanisms that are strong. In highly dynamic systems, membership management hinders their usability. These mechanisms adopt rich resource description, and the implementation is difficult. Dynamic attribute queries are not fully supported due to periodic updates. In these mechanisms, robustness level is very low, and periodic updates are needed. Lose of queries in the network may also occur due to the failure of a node as the queries are spread to the system by subsequent a definite pathway in the topological configuration, results in the SPF. By using this method the false positive errors is removed. In short, P2P mechanisms suffer from security issues and response time but are reliable regarding query correctness. P2P mechanisms also tolerate node dynamicity and provide high scalability.

For Grid systems in which frequent updates and scores of requests exist, hierarchical mechanisms do not scale well. In tree root system, SPF decreases the system fault tolerating feature. On comparing with the centralized mechanism, these approaches offer better failure tolerance and capability due to different hierarchical servers. As a result of system design, these mechanisms provide some proximity to search abilities and can reduce the network traffic. These mechanisms are less reliable regarding SPF but are reliable regarding query correctness. Dynamic attribute queries are not supported because of periodic updates but the range, and multi-attribute queries are supported [48–50].

6 Conclusion and Future Work

Different techniques for discovery of resources using Grid are discussed in detail. Furthermore, this paper describes the taxonomy of mechanism in the discovery of Grid resources. These mechanisms are divided into four structures namely centralized, decentralized, and hierarchical. These structures are discussed, and a detailed comparison is provided. Based on findings, the centralized mechanisms are not scalable, do not support dynamic attribute queries, and also suffer from SPF. The failure of the system at a single point is not observed in mechanisms which are decentralized. In decentralized mechanisms, loads are balanced, and it also covers the problem of failure due to distributed operations. The problem of scalability and failure of the system also faced in hierarchical mechanisms. Future research work should be based on developing a mechanism which covers all the problems faced by the different mechanisms discussed above to cope with the current massive need for 5G and its enabling technology [51–56].

References

1. Ridgway, R.: Microwave photonics programs at DARPA. J. Lightwave Technol. **32**(20), 3428–3439 (2014)
2. Strawn, G.: Masterminds of the arpanet. IT Prof. **16**, 66–68 (2014)
3. Alzboon, M.M.M.S., Arif, A.S.: Towards self-resource discovery and selection models in grid computing. ARPN J. Eng. Appl. Sci. **11**(10), 6269–6274 (2016)
4. Arif, S., Alzboon, M.S., Mahmuddin, M.: Distributed quadtree overlay for resource discovery in shared computing infrastructure. Adv. Sci. Lett. **23**(6), 5397–5401 (2017)
5. Kosińska, J., Kosiński, J., Zieliński, K.: Virtual grid-new paradigm of system resources dynamic organization. Comput. Inform. **30**, 749–760 (2012)
6. Bindu, P.L.H., Venkatesan, R., Ramalakshmi, K.: Perspective study on resource level load balancing in grid computing environments. In: 2011 3rd International Conference on Electronics Computer Technology, pp. 321–325, April 2011

7. Al-Namari, M.A., Mansoor, A.M., Idris, M.Y.: A brief survey on 5G wireless mobile network. Int. J. Adv. Comput. Sci. Appl. **8**(11), 52–59 (2017)
8. Liu, H., et al.: Study on virtualization-based simulation grid. In: 2010 International Conference on Measuring Technology and Mechatronics Automation, pp. 685–689, March 2010
9. Shen, H., Lin, Y., Li, T.: Combining efficiency, fidelity and flexibility in resource information services (2013)
10. Zarrin, J., Aguiar, R.L., Barraca, J.P.: Resource discovery for distributed computing systems: a comprehensive survey. J. Parallel Distrib. Comput. (2017)
11. Junior, A.M.G., Sato, L.M., Massetto, F.I.: A parallel application programming and processing environment proposal for grid computing. In: 2012 IEEE 15th International Conference on Computational Science and Engineering, pp. 154–161, December 2012
12. Brocco, A., Malatras, A., Hirsbrunner, B.: Enabling efficient information discovery in a self-structured grid. Future Gener. Comput. Syst. **26**(6), 838–846 (2010)
13. Jafari Navimipour, N., Masoud Rahmani, A., Habibizad Navin, A., Hosseinzadeh, M.: Resource discovery mechanisms in grid systems: a survey. J. Netw. Comput. Appl. **41**, 389–410 (2014)
14. Kouicem, D.E., Bouabdallah, A., Lakhlef, H.: Internet of things security: a top-down survey. Comput. Netw. **141**, 199–221 (2018)
15. Yu, J., Buyya, R.: A taxonomy of workflow management systems for grid computing. J. Grid Comput. **3**(3–4), 171–200 (2006)
16. Fitzgerald, S., et al.: A directory service for configuring high-performance distributed computations. In: Proceedings of the Sixth IEEE International Symposium on High Performance Distributed Computing (1997)
17. Czajkowski, K., Fitzgerald, S.: Grid information services for distributed resource sharing. In: Proceeding of 10th IEEE International Symposium on High-Performance Distributed Computing, pp. 181–194 (2001)
18. Yu, J., Venugopal, S., Buyya, R.: Grid market directory: a web services based grid service publication directory. arXiv Prepr. arXiv:cs/0302006, pp. 1–9 (2003)
19. Kaur, D., Sengupta, J.: Resource discovery in web-services based grids. World Acad. Sci. Eng Technol. **31**, 284–288 (2007)
20. Cokuslu, D., Hameurlain, A., Erciyes, K.: Grid resource discovery based on web services. In: 2009 International Conference for Internet Technology and Secured Transactions, ICITST 2009, pp. 1–6 (2009)
21. Benson, E.: Evaluation of UDDI as a provider of resource discovery services for OGSA-based Grids. In: Proceedings 20th IEEE International Parallel & Distributed Processing Symposium (2006)
22. Kovvur, R.M.R., Kadappa, V., Ramachandram, S., Govardhan, A.: Adaptive resource discovery models and resource selection in grids. In: 2010 1st International Conference on Parallel, Distributed and Grid Computing, PDGC - 2010, pp. 95–100 (2010)
23. Jeyabharathi, C., Pethalakshmi, A.: New approaches with chord in efficient P2P grid resource discovery. arXiv Prepr. arXiv:1401.2008, vol. 41, no. 8, pp. 3831–3849, June 2014
24. Azab, A.A., Kholidy, H.A.: An adaptive decentralized scheduling mechanism for peer-to-peer Desktop Grids. In: 2008 International Conference on Computer Engineering & Systems, pp. 364–371, November 2008
25. Le, N.T., Hossain, M.A., Islam, A., Kim, D.Y., Choi, Y.J., Jang, Y.M.: Survey of promising technologies for 5G networks. Mob. Inf. Syst. **2016**, 25 (2016)
26. Nithya, K.A., Rajaduari, R., Ganesan, M.: A survey on structural health monitoring based on internet of things. Int. J. Pure Appl. Math. **117**(19), 389–393 (2017)

27. Ray, P.P.: A survey on Internet of Things architectures. J. King Saud Univ. Comput. Inf. Sci. **30**(3), 291–319 (2018)
28. Akpakwu, G.A., Silva, B.J., Hancke, G.P., Abu-Mahfouz, A.M.: A survey on 5G networks for the Internet of Things: communication technologies and challenges. IEEE Access **6**(c), 3619–3647 (2017)
29. Condoluci, M., Lema, M.A., Mahmoodi, T., Dohler, M.: 5G IoT industry verticals and network requirements, no. 2010, pp. 148–175 (2017)
30. Chowdhury, N., Boutaba, R.: Network virtualization: state of the art and research challenges. Commun. Mag **3**(July), 20–26 (2009)
31. Chowdhury, N., Boutaba, R.: A survey of network virtualization. Comput. Netw., 1–29 (2010)
32. Iamnitchi, A., Foster, I.: Resource discovery in large resource-sharing environments (2003)
33. Vashisht, P., Sharma, A.: Decentralized P2P grid resources discovery model in LC-Trie structured overlay. In: 2010 First International Conference on Parallel, Distributed and Grid Computing (PDGC 2010), pp. 330–333, October 2010
34. Tangpongprasit, S., Katagiri, T., Kise, K.: A time-to-live based reservation algorithm on fully decentralized resource discovery in Grid computing. Parallel Comput. (2005)
35. Fattahi, S.M., Charkari, N.M.: Distributed resource discovery in grid with efficient range query. In: 2009 14th International CSI Computer Conference, pp. 335–340, October 2009
36. Ebadi, S., Khanli, L.M.: A new distributed and hierarchical mechanism for service discovery in a grid environment. Future Gener. Comput. Syst. **27**(6), 836–842 (2011)
37. Ge, X., Jia, Z.: Research on spatial knowledge service composition in grid environment. In: 2009 Second International Conference on Intelligent Computation Technology and Automation, pp. 259–262 (2009)
38. Yin, Y., Cui, H., Chen, X.: The grid resource discovery method based on hierarchical model. Inf. Technol. J. (2007)
39. SalemAlzboon, M., Arif, S., Mahmuddin, M., Dakkak, O.: Peer to peer resource discovery mechanisms in grid computing : a critical review. In: The 4th International Conference on Internet Applications, Protocols and Services (NETAPPS2015), pp. 48–54 (2015)
40. Mahmuddin, M., Alzboon, M.S., Arif, S.: Dynamic network topology for resource discovery in shared computing infrastructure. Adv. Sci. Lett. **23**(6), 5402–5405 (2017)
41. Jamal, A.A., Teahan, W.J.: Alpha multipliers breadth-first search technique for resource discovery in unstructured peer-to-peer networks. Int. J. Adv. Sci. Eng. Inf. Technol. **7**(4), 1403–1412 (2017)
42. Sehra, S.S., Singh, J., Rai, H.S.: Dynamic scheduling in grid environment with the improvement of fault tolerant level. Indian J. Sci. Technol. **8**(November), 507–515 (2015)
43. Nagarajan, V., Maluk Mohamed, M.A.: A decentralized two phase resource discovery model for peer-to-peer grid environments, vol. VII, no. Ii (2016)
44. Francesca, S., Carlo, C.G., Luca, D.N., Rocco, F., Marco, R.: Comparison of low-complexity algorithms for real-time QRS detection using standard ECG database. Int. J. Adv. Sci. Eng. Inf. Technol. **8**(2), 307–314 (2018)
45. Kasim, S., Hassan, R., Zaini, N.H., Ahmad, A.S., Ramli, A.A., Saedudin, R.R.: A study on facial expression recognition using Local Binary Pattern. Int. J. Adv. Sci. Eng. Inf. Technol. 7(5), 1621–1626 (2017)
46. Ahad, M.A., Biswas, R.: PPS-ADS: a framework for privacy-preserved and secured distributed system architecture for handling big data. Int. J. Adv. Sci. Eng. Inf. Technol. **8**(4), 1333–1342 (2018)
47. Wattana, M., Siriluk, B., Khotwit, S.: Counting and separating damaged seeds of soybean seeds using image processing. Adv. Sci. Eng. Inf. Technol. **8**(4), 1366–1371 (2018)

48. Myung, R., Yu, H., Lee, D.: Optimizing parallelism of big data analytics at distributed computing system. Int. J. Adv. Sci. Eng. Inf. Technol. **7**(5), 1716 (2017)
49. Hassan, M.H., Mostafa, S.A., Budiyono, A., Mustapha, A., Shamini, S.: A hybrid algorithm for improving the quality of service in MANET. Int. J. Adv. Sci. Eng. Inf. Technol. **8**(4), 1218–1225 (2018)
50. Teo, J.: Analyzing the scalability performance of Crossover-first and self- adaptive Differential Evolution algorithms for complex numerical optimization. Int. J. Adv. Sci. Eng. Inf. Technol. **7**(5), 1847–1852 (2017)
51. Fazea, Y.: Numerical simulation of helical structure mode-division multiplexing with nonconcentric ring vortices. Opt. Commun. **437**, 303–311 (2019)
52. Fazea, Y.: Mode division multiplexing and dense WDM-PON for Fiber-to-the-Home. Optik **183**, 994–998 (2019)
53. Hassan, S., Fazea, Y., Habbal, A., Ibrahim, H.: Twisted Laguerre-Gaussian mode division multiplexing to support blockchain applications. In: Region 10 Conference, TENCON 2017–2017. IEEE, pp. 2421–2425 (2017)
54. Fazea, Y., Alobaedy, M.M., Ibraheem, Z.T.: Performance of a direct-detection spot mode division multiplexing in multimode fiber. J. Opt. Commun. **40**, 161–166 (2019)
55. Fazea, Y., Mezhuyev, V.: Selective mode excitation techniques for mode-division multiplexing: A critical review. Opt. Fiber Technol. **45**, 280–288 (2018)
56. Fazea, Y., Sajat, M.S., Ahmad, A., Alobaedy, M.M.: Channel optimization in mode division multiplexing using neural networks. In: 2018 IEEE 14th International Colloquium on Signal Processing & Its Applications (CSPA), pp. 173–175 (2018)

Improving QoS for Non-trivial Applications in Grid Computing

Omar Dakkak[1], Shahrudin Awang Nor[2], Suki Arif[2], and Yousef Fazea[2(✉)]

[1] Computer Engineering Department, Faculty of Engineering, Karabük Üniversitesi, 78050 Karabük, Turkey
omardakkak@karabuk.edu.tr

[2] InterNetWorks Research Laboratory, School of Computing, Universiti Utara Malaysia, 06010 Sintok, Kedah, Malaysia
yosiffz@uum.edu.my

Abstract. Classical scheduling mechanisms don't satisfy the requirements for the end user, especially if the number of the jobs has increased massively in grid computing environment. To meet the expectations for non-trivial applications, the efficiency of the system has to be improved and the resources have to be ultimately utilized. Thus, backfilling technique becomes highly required due to its efficiency in exploiting the resources by filling the gaps that was created in the scheduler by short jobs. There are two well-known mechanisms, which are Extensible Argonne Scheduling System (EASY) and Conservative Backfilling (CONS). EASY is very aggressive and well uses the resources, however it causes a delay for the jobs ahead in the queue, while CONS solve this issue at the expense of system efficiency. In addition, and to further improve the scheduling quality, schedule-based approach has to be implemented. This approach provides information for the incoming job parameters and the resources capabilities; thus, the mechanism schedules the jobs in advance. This approach has shown a significant improvement compared with queue-based approach. In this paper, a new mechanism is proposed, namely Swift Gap. This mechanism implements schedule-based approach and applies multi-level scheduling method. In the first stage, the mechanism finds the right place for the newly arrival job, while in the second stage it manipulates the jobs' positions for further optimization. Moreover, this paper introduces the completion time scheme. This scheme minimizes both start time and processing time. The evaluation has shown the significant impact of Swift Gap alongside the completion time rule compared to CONS and EASY.

Keywords: Scheduling · Grid · Swift Gap · Completion time scheme

1 Introduction

Due to the dynamicity of grid computing, scheduling the jobs becomes challenging, particularly when the number of the jobs increases [1]. The classical scheduling mechanisms suffer from lack of flexibility when it allocates the jobs to the available resources. For instance Shortest Job First (SJF), which gives the privilege to the short

F. Saeed et al. (Eds.): IRICT 2019, AISC 1073, pp. 557–568, 2020.
https://doi.org/10.1007/978-3-030-33582-3_52

jobs at the expense of the long ones [2, 3]. Even though most of the jobs in High Performance Computing application workloads (HPC) are very short ones, Still SJF performance is questionable when this mechanism runs in real grid system.

In a real grid system and HPC workload such as 5G and its enabling technology [4–8], when the system is running for months or years, small jobs have to be processed faster, but not at the expense of the long jobs. Even though 10% only of the workloads are long jobs [9], but this ratio cannot be neglected. Moreover, traditional scheduling mechanisms cannot deal with the fragmentations that created in the queue due to different arrival times for the jobs. Theses fragmentations are a CPU idle times which can be exploited if a proper policy is applied. Thus and when the number of the gaps (fragmentations) becomes high, traditional mechanisms performance is inefficient due to poor utilization of available resources [10].

As a result, backfilling policy becomes required for such a system. The backfilling policy has no order or fixed rule to schedule the jobs, it simply backfills the short jobs in the gaps in order to reduce the waiting time for the whole jobs in the scheduler. The backfilling was described as something for nothing, a benefit without a tradeoff [9].

In order to reduce the waiting time for the jobs which need to be processed, available resources have to be fully utilized and therefore schedule-based approach has to be considered. In the queue-based system, the scheduling is executed blindly. The queue-system doesn't require any information about the incoming jobs; this might cause a delay for the rest of the jobs ahead in the queue when backfilling is applied. This can be justified due to the unawareness of the jobs execution time, the available resources and the size of the jobs. When such important information is missing, applying backfilling policy will be non-efficacious [11].

The schedule-based approach is of flaws which exist in the queue-based system approach. It requests information related to the incoming jobs and the available resources. This approach has information about the job execution time, the number of the jobs, the number of available resources and the computational power of the resources. Originally, this approach was primarily designed for the static grid environment, later on it was developed to adapt the real dynamic systems in grid computing environment [11]. The execution time for the jobs is provided by the user, using "real time estimates". To achieve the best results, it was found that real time estimates provided by the user have to be multiplied by a substantial factor [12, 13].

In the following sections, two of the most well-known backfilling mechanisms are reviewed. Our new mechanism called Swift Gap with the proposed completion time scheme is introduced in Sect. 3. Finally, an evaluation for Swift gap using real workloads that captured from HPC applications is performed.

2 Related Work

This section reviews comprehensively two of related work in scheduling mechanisms in grid computing environment. Since, the proposed mechanism in this study is based on backfilling techniques, the presented review highlights the mechanisms that share the same aspect and technique.

2.1 Extensible Argonne Scheduling System (EASY)

EASY was developed by Lifka [14] for IBM SP system. Eventually, EASY was combined with IBMs Load-Leveler administration software. EASY was well known as efficient and simple scheduler. EASY uses queues-based approach. EASY gives the short jobs highly advantage by backfilling them aggressively even though the possibility of delaying the subsequent jobs ahead in the queue.

EASY schedules the jobs based on their arrival time to the system. This mechanism takes scheduling decisions when a new job arrives to the system or when a running time for a current job is over. EASY keeps searching for the shortest job that can be processed on the idle resource at the present time for backfilling purpose, provided that it will not cause a delay for the first job in the queue. This can be achieved by checking if the backfilled job fits the available resources in terms of running time and number of processors required.

2.2 Conservative Backfilling (CONS)

Unlike EASY, CONS, which is proposed by Mualem and Feitelson in [13], doesn't cause any delay for the ahead jobs in the queue to backfill the short jobs. CONS schedules the jobs using schedule-based approach, thus it has the ability to set the schedule in advance for the incoming jobs. As mentioned before, schedule-based approach enables the implemented mechanism to predict the number of the jobs and the execution time for the jobs almost precisely. This information is provided by the users, or it can be retrieved from a historical database for a certain workload.

However, the authors of the Maui scheduler [15] proved that this claim is incorrect. They showed that if a job (J2) is going to be backfilled, it can cause a pseudo delay to the job (J1) ahead if the Job (J0) was suddenly terminated during its processing time. This case seemingly occurs only when a job (J0) ends before its full requested runtime has elapsed. Figure 1 presents this case.

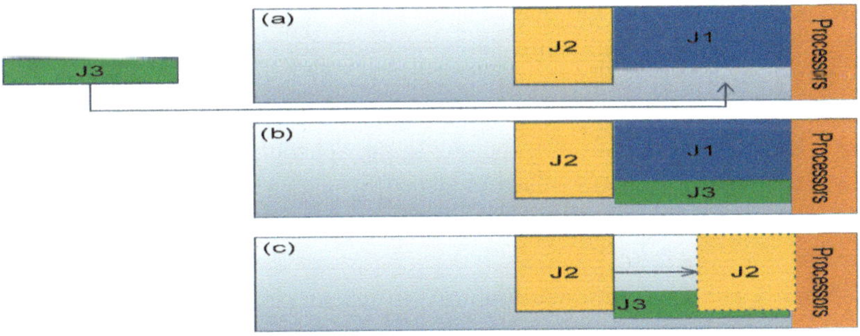

Fig. 1. Pseudo delay in CONS.

Figure 1 highlights the criticism of CONS scheduling performance. Obviously, when the Job (J0) is terminated for some reason; Job (J1) will try to fill the generated gap. However, Job (J2) prevents this opportunity by moving ahead of Job (J1) because the backfilling decision was already considered.

Although the aforementioned criticism is valid, nevertheless this might be an advantage stands up for CONS rather than a disadvantage. Because permitting job (J1) to be processed in this case forces the scheduler to re-backfill Job (J2), this will consume extra computational time which cannot be neglected, especially if this case occurred frequently.

3 Swift Gap Mechanism

Swift Gap mechanism implements schedule-based approach in order to schedule the jobs in advance. Swift Gap applies multi-scheduling level. Swift Gap is a combination of two mechanisms; the first mechanism is Best Gap [16] while the second one is Tabu Search [17]. The idea is to build the initial solution first and later on, Tabu Search optimizes the initial solution (periodically) by manipulating the positions of the jobs among the machines [18].

Swift Gap applies high hybridization level between Best Gap and Tabu Search. In this type of hybridization, method (M1) runs, and once it is over, method (M2) starts. This, hybridization scheme is also known by loosely coupled, since there is no interacting between both mechanism's structures. Figure 2 presents the hybridization scheme of Swift Gap mechanism.

3.1 Stage One: Building Initial Solution

In the first stage and for each newly job arrives to the schedule $(j1, j2, \ldots jn)$ a machine will be selected $(m1, m2, \ldots mn)$. The machine schedule will be scanned in order to find a suitable gap fits the job. If multi gaps are detected, the earliest gap will be selected. The movement of the job will be evaluated based on *acceptation criteria*.

Decision (*Current*) refers to the current position for the job, while Decision (*Next*) refers to the new proposed position for the job. If Scheduler (*Next*) is better than Scheduler (*Current*), the proposed move will be accepted and the solution will be included into Scheduler (*Current*); otherwise the Scheduler (*Current*) will be applied. Finally, Scheduler (*Current*) will be introduced as initial solution to be optimized in the next stage.

3.2 Stage Two: The Optimization

In this stage, the job will be moved (manipulated) among the resources, in order to make this job proceed earlier if possible. A machine $(m1, m2, \ldots mn)$ will be selected to scan its schedule for a job $(j1, j2, \ldots jn)$. The selected job will be included in Tabu-list Job. This list saves the optimized job in order to avoid any duplicated selection for the

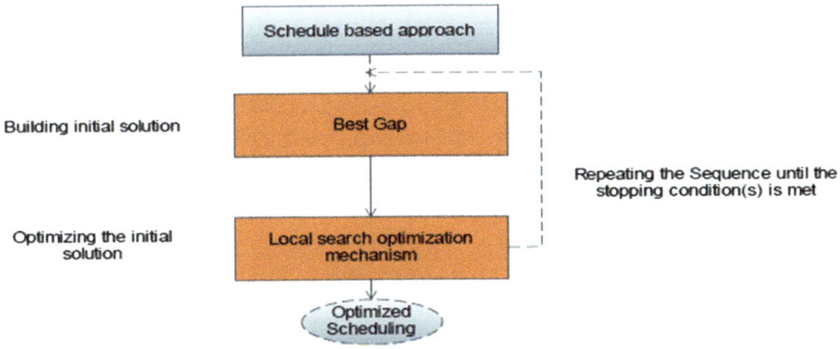

Fig. 2. Swift Gap Hybridization Scheme.

same job in a short time. Once a suitable gap is located, the evaluation starts based on the *acceptation criteria.*

If Decision (*Next*) is better than Decision (*Current*), the move will be accepted and included tobScheduler (*Current*). Otherwise the movement will be rejected and Scheduler (*Current*) will be executed. When the schedule of the machine contains only jobs that belong to Tabu-list Job, it means this machine is fully used and will be ruled out from the next cycle.

The stopping condition for the optimization is the number of requested iterations or when time allocated for each iteration elapsed. If one of the previous conditions is validated, the optimization stage has to be concluded. Pseudo code for Swift Gap presents the aforementioned steps.

Algorithm 1 Stage One

```
slot [A]= slot [B]=0, i =0, j=0;
select Job [job1,...job_n]; Select machine [machine_1,...machine_m];
for i,j=1 to n,m do
      if selected machine is suitable to perform job and gap found then
   evaluate the solution based on acceptance criterion;
      if new solution is good move then
place the job into the gap and include into the initial solution slot [A];
      end if
      else
      reject the move and include it in initial solution slot [A];
      end if
end for
return initial solution (slot [A]);
```

Algorithm 2 Stage Two

```
initial solution , slot [A]= slot [B]=0, insert iteration and time limit;
select random Job [job1,...job_n] belongs to machine [m];
while (i < iterations number) do
    i=i+1;
if all jobs tested, then tabujob-list = 0 then
add selected job to tabujob-list;
end if
if acceptance criterion = true then
remove job from slot [A];
move job into earliest suitable gap and compress slot [A];
include solution into slot [A];
else
keep the job and include in slot [A];
    end if
if all jobs in machine schedule belong to tabujob-list then
    machine = used;
machine selection next cycle = false;
end if
tabujob-list = full; remove oldest job;
  if optimization limit exceeded then
stop;
end if
end while
return slot [A];
```

3.3 Completion Time Scheme

Our aimed acceptance criterion is the completion time, reducing the completion time will minimize the time spent by the jobs in the scheduler and this in turns will improve the Quality of Service (QoS) for the end user. In order to reduce the completion time, the start time for the jobs and/or processing time have to be reduced. This can be observed based on the following expression [19, 20]:

$$Cj = Sj + Pj \tag{1}$$

To minimize the completion time, the best start time has to be selected, and the fastest machine available has to be chosen to process the job. The selection adapts the weight function that enables the mechanism to define which value is best for the job. The selection is between the *Current* position for the job and the *Next* (proposed) position for the job. The following expression presents the weight function [21]:

$$Weight_{Metric} = \frac{A - B}{A} \tag{2}$$

$$Weight_{Metric} = \frac{B - A}{A} \tag{3}$$

Switching between (A and B) in the numerator enables the mechanism to evaluate the move for different criteria at the same if it has to, i.e. the less start time indicates improvement in the scheduling performance, while the less machine speed considered as deterioration in the QoS for the end user. Based on this argument, the proposed weight functions in purpose of best start time and the fastest machine selection are:

$$Weight_{start\ time} = \frac{Current - Next}{Current} \tag{4}$$

$$Weight_{machine\ speed} = \frac{Next - Current}{Current} \tag{5}$$

Therefore, the Total weight as below:

$$Total_{Weight} = Weight_{start\ time} + Weight_{machine\ speed} \tag{6}$$

The implementation of the completion time scheme in Swift Gap is executed in two stages. The first one for building initial scheduler, in this stage only the start time weight is implemented. Whereas in the second stage both start time machine speed weight are implemented simultaneously.

The reason behind implementing the start time only in the first stage belongs to the limitation of Best Gap. Best Gap only deals with the newly arrival jobs. Best Gap priority is to find a suitable place within the cluster only, while Tabu Search has the ability to move the job among the clusters. Usually, the capabilities of the resources are the same in one cluster, and thus, implementing the machine speed weight increases the time complexity for the proposed mechanism without any improvement in the total performance.

Algorithm 3 Completion Time Scheme

1. get the start time and the weigh machine usage for Slot (A);
2. get the start time and the weigh machine usage for Slot (B);
3. weight $_{\text{start time}} = \frac{As-Bs}{As}$
4. weight $_{\text{weigh machine usage}} = \frac{Bu-Au}{Au}$
5. total weight = weight $_{\text{start time}}$ + weight $_{\text{weigh machine usage}}$
6. if total weight ≤ 0 then
7. reject the move and select Slot (A) as a solution;
8. else
9. accept the move and select Slot (B) as a solution;
10. end if

4 Swift Gap Evaluation

To study the impact of Swift Gap with the completion time scheme, we will compare its performance against CONS and EASY. The measured performance metrics are:

Slowdown: This metric refers to how many times the job was delayed. This metric has no unit. The numerator is normalized by the denominator and both units are measured in seconds [22]:

$$sld = \sum_{j=1}^{j} \frac{Completion\,Time - Submission\,Time}{Completion\,Time - Start\,Time} \tag{7}$$

Bounded Slowdown: the bounded slowdown eliminates the effect of the very short jobs by multiplying the denominator with a threshold ξ. Several studies have recommended to consider $\xi = 10$ [23].

$$B\_sld = \sum_{j=1}^{j} \frac{Completion\,Time - Submission\,Time}{10(Completion\,Time - Start\,Time)} \tag{8}$$

Tardiness: this metric refers to the time that jobs have spent to be accomplished corresponding to a certain due date. The tardiness can be measured by the following expression [19]:

$$Tardiness = \sum_{j=1}^{j} \max(0, Cj - di) \tag{9}$$

Where, Cj is the completion time and di due date time.

Waiting Time: is the time that jobs spent in the scheduler from the submission time till the processing time. Waiting time expression is given as follows [23]:

$$Tw = Start\,time - Submission\,Time \tag{10}$$

Response Time (Flow time): is the aggregate of running time and waiting time.

$$Resp\,Time = Tr + Tw \tag{11}$$

Where, Tr is the running time [23].

4.1 Workload and Simulation Properties

The simulated workload is "*hpc2n*"; this workload has 202871 as maximum records (jobs number). This is a cleaned version of the total log. This workload consists of one cluster, namely Seth. Seth has 240 AMD Athlon MP2000+ processors, running at 1.667 GHz. The total system peak performance is 800 Gflops. The total RAM is

1048576 GB. The machines are linked by a 3D SCI interconnect organized as a $4 \times 5 \times 6$ grid, empowered by rapid network connection speed.

Table. 1 Below shows the simulation settings

Simulation settings	*Info and values*
Simulator	Alea 3.1
Number of simulated jobs	[5000–110000]
Simulation time	3000
Simulated resources	CPU
Runtime estimates	True
Runtime estimates factor	2
Iterations	100
Fairness	Enabled
ε	10
Workload	HPC2 N
Simulator OS	Windows 7

4.2 Simulation Results and Discussion

In this simulation, the implementation covers only the start time rule. The implementation of machine speed was ruled out. This is due to the properties of the workload (hpc2n). As mentioned above, this workload contains only one cluster (Seth). Subsequently, applying this rule is meaningless, since the computational power is the same within Seth.

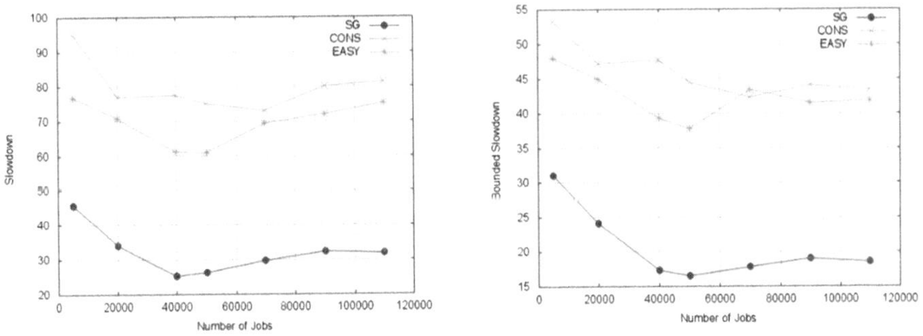

Fig. 3. Swift Gap v CONS and EASY (Slowdown and Bounded Slowdown)

To clarify the significant impact of reducing the completion time, following debate proves how completion time is involved in each performance metric which simulated in this study. In the *slowdown expression*, when the start time is reduced in the denominator, the completion time in both denominator and numerator will be reduced as demonstrated in Fig. 3.

If the start time was reduced by value equal to (x), thus the completion time will be reduced by value equal to (x) as well, subsequently:

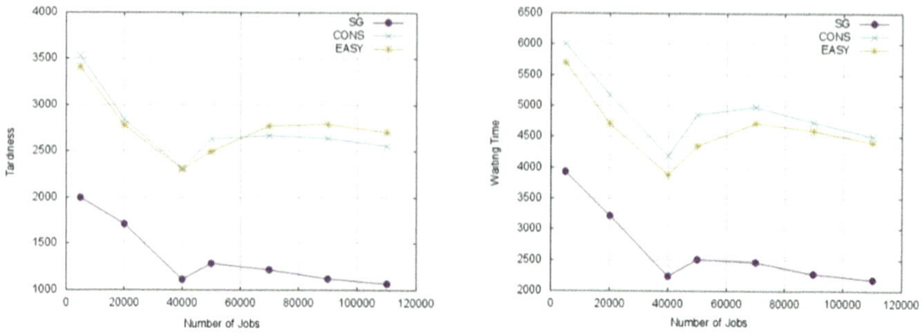

Fig. 4. Swift Gap v CONS and EASY (Tardiness and Waiting Time)

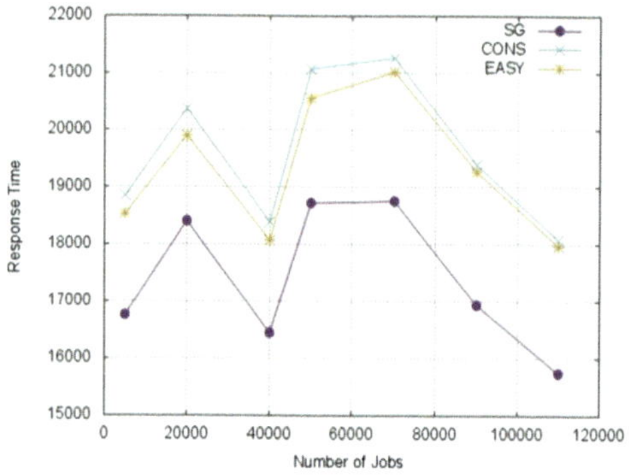

Fig. 5. Swift Gap v CONS and EASY (Response Time)

$$
\begin{aligned}
sld(new) &= \sum_{j=1}^{j} \frac{(Completion\ Time - x) - Submission\ Time}{(Completion\ Time - x) - (Start\ Time - x)} < sld(old) \\
&= \sum_{j=1}^{j} \frac{Completion\ Time - Submission\ Time}{Completion\ Time - Start\ Time}
\end{aligned}
\tag{12}
$$

The same argue is applicable for *Bounded Slowdown*, since both metrics have the same expression. For the *Tardiness* presented in Fig. 4, when completion time scheme is applied, the completion time *Cj* will be reduced (due to reducing the start time and machine speed). Let's assume that the completion time value is equal to (x), which means:

$$
\begin{aligned}
Tardiness(new) &= \sum_{j=1}^{j} \max(0, (Cj - x) - di) < Tardiness(old) \\
&= \sum_{j=1}^{j} max(0, Cj - di)
\end{aligned}
\tag{13}
$$

For the *Waiting Time* (Fig. 4), our completion time scheme will reduce the time for the jobs in the scheduler. Assuming the starting time is reduced by (x) due to the implementation of the completion time, which means:

$$\begin{aligned} Tw(new) &= (Start\,time - x) - Submission\,Time < Tw(old) \\ &= Start\,time - Submission\,Time \end{aligned} \quad (14)$$

Finally, the new *response time* after the completion time scheme implementation has to be improved as shown in Fig. 5. Since the *waiting time* is reduced as shown previously. Thus, when the *waiting time* is reduced by (x) value, the new *response time* will be:

$$\begin{aligned} Resp\,Time(new) &= (Tr) + (Tw - x) < Resp\,Time(old) \\ &= Tr + Tw \end{aligned} \quad (15)$$

5 Conclusion and Future Work

This paper presents a new mechanism namely "Swift Gap". Swift Gap is a combination of two mechanisms by applying multi-scheduling level. In the first stage, the mechanism has to find a right place for the newly arrival jobs to create an initial solution. Based on the previous stage, Swift Gap optimizes the initial solution by applying local search meta-heuristic algorithm.

Moreover, the completion time scheme was applied to further improve the performance. The implemented completion time scheme involves two factors, which are the start time rule and machine speed rule. Our evaluation has shown that Swift Gap alongside with the completion time scheme have a significant impact on the measured performance metrics. The simulated workload (hpc2n) has shown that Swift Gap outperforms both CONS and EASY even though this workload consists of one cluster, which eliminates the effect of the proposed machine speed rule.

For future work, our simulation will cover more workloads which are valid for grid and cloud computing environment [24] for further evaluation. In addition, more analysis will be provided against other scheduling mechanisms such as Flexible CONS, SJF and CONS integrated with Tabu Search (CONS-TS).

References

1. Klusacck, D., Rudová, H.: Improving QoS in computational Grids through schedule-based approach. In: Scheduling and Planning Applications Workshop at the Eighteenth International Conference on Automated Planning and Scheduling (ICAPS 2008), Sydney, Australia (2008)
2. Azmi, Z.R.M., et al.: Scheduling grid jobs using priority rule algorithms and gap filling techniques. Int. J. Adv. Sci. Technol. **37**, 61–76 (2011)
3. Sajat, M.S., et al.: A critical review on energy-efficient medium access control for wireless and mobile sensor networks. J. Telecommun. Electron. Comput. Eng. (JTEC) **8**(10), 89–94 (2016)

4. Fareed, A., et al.: Channel impulse response equalization based on genetic algorithm in mode division multiplexing. J. Telecommun. Electron. Comput. Eng. (JTEC) **10**(2–4), 149–154 (2018)
5. Fazea, Y.: Numerical simulation of helical structure mode-division multiplexing with nonconcentric ring vortices. Opt. Commun. **437**, 303–311 (2019)
6. Fazea, Y.: Mode division multiplexing and dense WDM-PON for Fiber-to-the-Home. Optik **183**, 994–998 (2019)
7. Fazea, Y., Alobaedy, M.M., Ibraheem, Z.T.: Performance of a direct-detection spot mode division multiplexing in multimode fiber. J. Opt. Commun. **40**, 161–166 (2019)
8. Fazea, Y., Amphawan, A.: 5 × 5 25 Gbit/s WDM-MDM. J. Opt. Commun. **36**(4), 327–333 (2015)
9. Lee, C.B.: On the user-scheduler relationship in high-performance computing (2009)
10. Dakkak, O., Nor, S.A., Arif, S.: Analyzing the QoS criteria from end user's perspective in computational grid environment. In: TENCON 2017–2017 IEEE Region 10 Conference. IEEE (2017)
11. Klusacek, D.: Dealing with uncertainties in grids through the event-based scheduling approach. In: Fourth Doctoral Workshop on Mathematical and Engineering Methods in Computer Science (MEMICS 2008) (2008)
12. Zotkin, D., Keleher, P.J.: Job-length estimation and performance in backfilling schedulers. In: 1999 Proceedings of the Eighth International Symposium High Performance Distributed Computing. IEEE (1999)
13. Mu'alem, A.W., Feitelson, D.G.: Utilization, predictability, workloads, and user runtime estimates in scheduling the IBM SP2 with backfilling. IEEE Trans. Parallel Distrib. Syst. **12** (6), 529–543 (2001)
14. Lifka, D.A.: The ANL/IBM SP scheduling system. In: Workshop on Job Scheduling Strategies for Parallel Processing. Springer (1995)
15. Jackson, D., Snell, Q., Clement, M.: Core algorithms of the Maui scheduler. In: Workshop on Job Scheduling Strategies for Parallel Processing. Springer (2001)
16. Chlumsky, V., Klusácek, D., Ruda, M.: The extension of torque scheduler allowing the use of planning and optimization in grids. Comput. Sci. **13**(2), 5–19 (2012)
17. Glover, F., Laguna, M.: Tabu Search*. Springer (2013)
18. Dakkak, O., Nor, S.A., Arif, S.: Proposed algorithm for scheduling in computational grid using backfilling and optimization techniques. J. Telecommun. Electron. Comput. Eng. (JTEC) **8**(10), 133–138 (2016)
19. Rizal, Z., et al.: Combinatorial Rules Approach to Improve Priority Rules Scheduler in Grid Computing Environment (2012)
20. Fibich, P., Matyska, L., Rudová, H.: Model of grid scheduling problem. In: Exploring Planning and Scheduling for Web Services, Grid and Autonomic Computing, pp. 17–24 (2005)
21. Xhafa, F., Abraham, A.: Computational models and heuristic methods for Grid scheduling problems. Future Gener. Comput. Syst. **26**(4), 608–621 (2010)
22. Gritsenko, A.V., et al.: Decomposition analysis and machine learning in a workflow-forecast approach to the task scheduling problem for high-loaded distributed systems. Mod. Appl. Sci. **9**(5), 38 (2015)
23. Feitelson, D.G.: Metrics for parallel job scheduling and their convergence. In: Workshop on Job Scheduling Strategies for Parallel Processing. Springer (2001)
24. Dakkak, O., et al.: From grids to clouds: recap on challenges and solutions. In: AIP Conference Proceedings. AIP Publishing (2018)

The Role of Management Techniques for High-Performance Pending Interest Table: A Survey

Raaid Alubady[1], Suhaidi Hassan[2](✉), and Adib Habbal[3]

[1] Information Networks Department, College of Information Technology, University of Babylon, Babylon, Iraq
alubadyraaid@itnet.uobabylon.edu.iq
[2] InterNetWorks Research Laboratory, School of Computing, Universiti Utara Malaysia (UUM), 06010 Sintok, Malaysia
[3] Computer Engineering Department, Faculty of Engineering, Karabuk University, Karabük, Turkey

Abstract. Most of the services used by Internet consumers such as social network platforms, video-on-demand, on-line gaming, web Media, and IP Television which are content-centric in nature; meaning they focus on named content objects instead of being focused on the host-location. In this context, many projects around named data propose redesigning and developing the communication of Internet-based on named data. NDN (Named Data Networking) is an ideal solution to achieve efficient data sharing and retrieval since NDN focuses on the contents themselves regardless of their sources. The focus of this survey is a unique characteristic presented by NDN; PIT (Pending Interest table). PIT is part of three fundamental data structures newly introduced in the NDN router to enable full functionality of NDN. NDN router depends on reverse paths in PIT to return back Data packets to consumers. Accordingly, the PIT may present stringent restrictions in terms of scalability, for-warding, and management. The challenging task is the design of a scalable and manageable PIT because it requires per-packet updating and controlling the impact of increasing Interest packets with the highest Interest lifetime of PIT. Therefore, this survey describes into greater detail the background and several important previous researches related to issues of PIT which is PIT management based on PIT placement, and replacement, PIT implementation as a data structure, and Adaptive Interest Life-time. Thus, would assist in defining the general framework of this survey.

Keywords: Named Data Networking · Pending Interest Table · Interest lifetime · Data structure · And replacement policy

1 Introduction

In recent times, the focus has been on the applicability, satisfaction and also the ubiquity of application on virtually every single connected device. Examples are seen in user generated content [1] and provided services such as Flicker and YouTube as well as social networks, video on-demand, online game, web on-demand, and IPTV or

F. Saeed et al. (Eds.): IRICT 2019, AISC 1073, pp. 569–582, 2020.
https://doi.org/10.1007/978-3-030-33582-3_53

in another word triple services (audio, video, data) using massive MIMO technology [2–7], which place content at a very center of the picture, thereby enabling users not to only consume the generated contents by a small set of well-identified providers, but also in producing and sharing their own content. These thus, become customer contents. Online social networking is also guided much and oriented towards content generation, access and sharing [8]. Therefore, it makes enabling wide sharing and access to the contents to be one of the main goals of the conventional Internet [9]. Also, according to [10], in the current Internet, the endpoint-centric connections model doesn't cope with the overwhelming utilization anymore. In today's practice on the Internet, huge traffic is indeed noticed which are driven by large information generation and the resulting services proportional to the information produced. Thus, information ranging from simple rich site summary feed aggregators in providing multimedia streaming through summary services, including non-static Web content and user-generated content. To overcome these challenges, NDN (Named Data Networking) that also called content | data-centric network paradigm was presented over few years to target a clean slate architecture redesign by placing data (content) in the foreground, at the heart of network environment. This new generation of the Internet (*i.e.*, NDN) [11–13] introduced a re-directional modification in the Internet communications. NDN emphasis on information accessing with lesser concern on location through its new database approach, allows networks to actively deliver data and content with wider categorization. While the information plays the key role in an NDN.

2 Named Data Networking

NDN is a project under a special program of FIA (Future Internet Architecture) based on CCN (Content-Centric Networking) novelty. The studies directions for this research work are fully focused on forwarding and routing improvements for CCN [11]. As part of the goals of NDN architecture are to completely redesign and develop the communications of the Internet by replacing Internet Protocol with content chunks (data) as a universal components of transport [14]. NDN is simple and efficient communication transactions, which are driven by consumers who broadcast Interest packet (requests) to request for data by his name, regardless of the IP addresses of the nodes supplying the data. Interests' packets are forwarded by inter-mediate nodes upstream to providers that are any node storing or owing the requested content. Providers simply respond with the asked for Data packets, which goes through the way back to the subscribers [15]. In NDN, named-object is hierarchically deployed similar to a tree structured concept organized in a lexicographic ordered tree. Bearing this ordering, the leaves match up to the content of interest, and each node from the connected leaf represents a common name prefix related to some collection and chunk of content objects. While addressing contents by means of this naming tree, the NDN thus dynamically supports the management of its contents by authorizing content users in making their respective requests on the basis of the prefix appended to the names as distinct for each content yet to be created [10]. Every NDN node has three types of data structures for processing of packets forwarding. They are the CS (Content Store). FIB (Forwarding Information Base) and PIT (Pending Interest Table) [11, 14, 15].

3 Pending Interest Table

PIT is a fundamental data structures newly that only introduced in the NDN router design to enable a full functionality of NDN [16]. Forwarding in NDN thought to be tasteful as a NDN router keeps up the PIT entries for each Interest packet. Every PIT record has five fields (see Fig. 1) [17]. The PIT in NDN router does not only the requested Interests that requires to be forwarded for up streaming to a data source but it saves the face in which the Interest is received. At every point of data traversal, an entry is saved and recorded to the PIT. The cached or saved Interest is in turn deleted upon upstream data fulfillment through delivery or when the PIT entry Lifetime expires. An important aspect of the PIT forwarding is the fact that when multiple requests amounting to the same Interest names are aggregated which saves duplicating forwarding. In this case, only the instances of the same interest are stored. The resulting saved faces are further seen as an aid to instigate the forwarding of the returned data to the request subscribers [16].

Name	List of Incoming Face (s)		List of Outgoing Face(s)			Lifetime	Forwarding Strategy
Content name	Interface ID	Arrival time	Interface ID	Send time	Retransmission Number	Entry expire time	Strategy name

Fig. 1. PIT entry fields

4 PIT Research Issues

There are yet many challenges and solutions that need to be proffered in the development and deployment aspects of PIT. Thus, this calls for in-depth investigation in terms of PIT since it has become one of the most significant and highly important research areas in the field of content-centric. The authors in [18] highlights some open issues in PIT, which is identified and desirable properties for PIT.

i. ***Lookup operates:*** since the PIT increases when subscribers sending their Interest packet and decreasing when Data packets arrive at the NDN routers. In case of the number of PIT entries increases, these required access speed to PIT structure and the possibilities available in this meaning by using current memory techniques, that make the PIT size representable as the main bottleneck that affects the whole NDN infrastructure. Because of the PIT is very dynamic for all incoming Interest packets and matching Data packets, a special process (lookup process) must be happening in the PIT which should be performed faster.

ii. ***Filtering:*** one of these issues, malicious users can create artificial Interest packets in order of filling the PIT on NDN routers. Hence, it is implementing a Distributed Denial of Service attack. This type of attack can possibly be implemented by distributing the generated request packets that include valid destination prefixes without existing resource names, in order that the NDN routers correctly forward Interest packets and keep new entries inside the table. Nonetheless, replies never come back. Another issue of security in NDN architectures is the vulnerability of NDN router to the PIT pollution attacks. This type of attack includes sending random Interest packets for content as a way to modify large values for the lifetime field.

iii. ***Entries situation:*** PIT should provide low-routing overhead, metadata updates avoid congestion, low-latency entry operations addition and deletion. The interesting question at this point, how to determine that the Interest packet deletion must be expiry-time based on some hybrid or explicit schemes.

iv. ***Interest lifetime:*** other issues are, when the dropping of both the Data and Interest packets occur on the way as a result of the network impairments, deciding where the NDN router or the endpoint could be, setting the timeout in duration and retransmitting the Interest packet after a timeout remain the key issues that could affect the performance of the entire network. In addition, the impact is noticed directly on the occupancy of the PIT. However, there is no unanimous agreement on how often should this periodic retransmission be done.

v. ***PIT overflow:*** the PIT is very dynamic due to the high-speed packet arrive rate to the PIT. Thus, for all incoming request packet and matching Data packet, hence a special process must be happening in this table. These processes should have to be performed faster to avoid this table may be overflowed, which cause the delay and packet loss for these packets because the PIT receives and remove the Interest packets exponentially. It is not easy to predict the tables are full. If the table is overflowing, subscribers' requests will be discarded from the NDN routers and based on this, subscribers will experience an increasing Interest retransmitting rate that will lead to a complete collapse of the whole network.

vi. ***PIT replacement:*** cache memory remains one of the most studied topics in computer architecture research. With the ever-increasing speed added between processor and memory, this emphasizes the need for a more efficient memory hierarchy for better network and content manageability. When PIT sets become full and a new Interest packet of memory needs to be placed into the PIT, there should be a provision for replacement based on the built-in algorithm for content placement. Replacement policy, therefore, is one of the most crucial factors of determining the effectiveness of cache memory.

5 PIT Management Techniques

PIT management techniques can be classified into four categories: adaptive Interest Lifetime, PIT placement, PIT implementation (data structure), and PIT replacement. Figure 2 shows the main PIT management techniques based on previous studies.

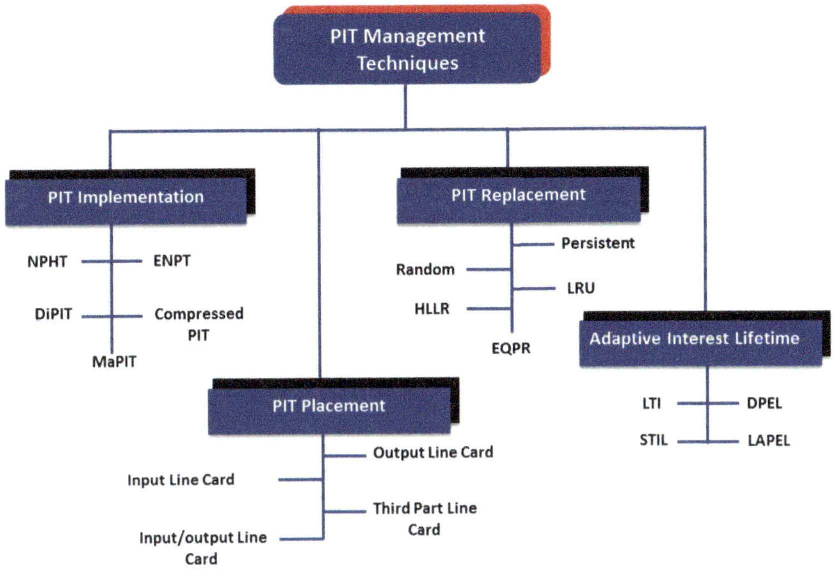

Fig. 2. PIT management techniques

5.1 PIT Implementation Approaches

Accordingly, the PIT may present stringent restrictions in terms of scalability. The challenging task is the design of a scalable PIT because it requires per packet updating and longer naming. To address the problem, several PIT approaches have seen proposed. In this section many of PIT implementation approaches are described include: NPHT, ENPT, DiPIT, Compressed PIT, and MaPIT.

- **Name Prefix Hash Table (NPHT)**

NPHT [11] is the default approach in which FIB and PIT share the hash table logical. Two types of structures are indexed; they are Propagation Entries and Forwarding Information Entries. In PIT, Interest packets are presented in form of Propagation Entries. Each packet has a unique nonce field. Propagating Hash Table (PHT) is established to store all the nonce fields. Because of the uniqueness propriety of nonce fields, they used as keys in PHT. The benefit of this design is to prevent loops in the network due to its impact on the operational flows in CCNx. NPHT approach is effective during ordinary operation because the main idea is to implement PIT. The hash table is capable of doing a quick lookup operation, but it is suffering from the following drawbacks. Because of the PIT in CCN is based on exact matching. Each entry must be memorized, which may be leads to large memory requirements. This situation is getting worse if the number of entries has increased because it results in increasing the delay time experienced by a consumer. In the case, the NPHT approach will be not efficient for scalability issue.

- **Encoded Name Prefix Trie (ENPT)**

This approach denoted by ENPT [19] relies on encoding the NCE (Name Component Encoding) to shrink the PIT size by speeding up the performance of insert operation, looking up operation, and delete operation. Name Component Encoding stores the objects as character strings to be searched within the table. Also, for emerging the names with the same prefix, this approach used Name Prefix Trie (NPT) strategy. Because of nature of NDN, the names will be organized in a hierarchical way. This contributes to more organized and manageable PIT. However, shrinking the PIT size in this way is not sufficient for faster PIT access. Therefore, the names are associated with codes (integer values) so that the PIT operations such as lookup, delete, and insert will be faster. ENPT reaps some benefits from both access frequencies considerably advanced fulfilling the access speed necessary for PIT, in this case the memory utilization can be decreased. According to the authors, the achieved results demonstrate that resultant PIT consists of 1.5 M entries, and frequencies for insert, delete, and lookups are 0.9 M/s, 1.4 M/s, and 0.9 M/s, respectively.

- **Distributed Bloom Filter Based PIT (DiPIT)**

DiPIT approach in [20] introduced as implementation of PIT. It performs better on PIT tables in terms of space, speed, and cost of data retrieval. For DiPIT, BFs (Bloom Filters) are an important part because it is faster updated and lookup operations. Furthermore, these are also very effective for memory utilization, but the ratio of "false positive" can be increased in this regard. DiPIT-based design, it builds one PIT*i* on each CCN face. In the DiPIT design, BFs are used to construct PIT*i* tables. Moreover, all faces share an additional BF. The idea behind this design is that each PIT*i* works independently and the footprints of the Interest packets are recorded in a CBF (Counting Bloom Filter) that are arriving from the perspective face. In this way, Data packets attained all PIT*i* tables are parallel, checked and then forwarded on the faces if there is any matching footprint in the PIT*i*. In the end, DiPIT approach can increase throughput for an incoming packet, and to reduce the memory usage (up to 63%) in the NDN router. However, this approach has a drawback since it does not filter Interest duplications for the reason that the same Interest packet arriving at different faces are recorded individually at different PIT*i*. Moreover, DiPIT did not mention about the impact CBF at PIT when the false negative is occurring if any of counters decrease to zero, hence it can lead to packet loss and redundant traffic.

- **Compressed PIT**

Compressed PIT approach [21] is a introduced a PIT compression with UBF (United Bloom Filter) to compress PIT effectively. For compressing PIT with BFs (Bloom Filters), the key is a face and the value is a name set. In the transposed PIT entries, it is associated to each face, there is one BF that exemplifies the name set. The essential form of BF is not suitable for this compression because the deletion of elements is not supported. However, BF is utilized to represent every one of the possible extensions of BF. For every received Interest packet, CCN router checks the Interest name, in case it is matched in any of the Bf*s*, the Interest packet will not be forwarded. On the other hand, for every Data packet, which is received from a face, CCN router will check the

Interest name in all available BFs. The packets will be forwarded to the faces with BFs that contain the name. As a result, Compress PIT with BF is using an extension of BF namely UBF to eradicate the demand of remove to avoid such unexpected anomalies. Hence, the communication can easily be recovered from the errors through retransmission. The most important advantage of UBF, the density can be controlled whereas there is no incorrect deletion.

- **MaPIT**

To enhance the implementation of PIT to fulfill the existing network requirements and memory technology, Li *at el.,* [22] designed a data structure MBF (Mapping Bloom Filter) and an improved PIT called MaPIT to meet the PIT requirements and make its usage of the existing memory chip. MBF included of two methods; PS (Packet Store) and IT (Index Table). IT is deployed on the on-chip memory to get into the PS. Recorded contents will be designated for dispersed locations on the PS that is fixed cache in the off-chip memory. IT consists of two structures, i.e., a regular BF and MA (Mapping Array). While using the BF, the objects are checked if they are in available in the MBF or not. To reach the PS, the MA value is utilized as being the offset address of the PS. MaPIT consists of two storages, which are off-chip memory and an on-chip memory. Off-chip is the memory where CBF is working, while on-chip memory is the one where the IT of MBF is working as a summary. Therefore, MaPIT approach presented to improve the implementation of PIT. This approach can be minimized to the consumption of memory to 2.097 MB, lower the probability of false positive is to be under 1% for 2000000 names. Furthermore, it is implemented on SRAM memory as on-chip memory to satisfy today's memory technology as well as network requirements. Hence, the proceeding algorithm has been required an additional much more storage space since many additional components included.

5.2 PIT Placement Schemes

In NDN router design, there are three placements to handle this issue: input line, output line, input/output lines and third part schemes.

- **Input Line-Card Scheme**

Originally proposed in [20, 21], the studies indicated that a PIT must be composed in all input line-cards. Consequently, a requested or sent Interest packet creates an entry only in the PIT of the line-card when received. In the event that the corresponding Data packet is received at the output line-card, the data are broadcasted to all input line-cards for the PIT which further indicates a decision of the Data being forwarded or otherwise. The above action enables multipath placement, however lacking loop detection to correct Interest aggregation, as each PIT is only aware of the local list faces and *list_nonces*. Particularly, this placement requires a number of lookups of the PIT for the subsequent returning of Data which could also result into a bottleneck.

- **Output Line-Card Scheme**

Furthermore as proposed in [19], it also submitted that an optimal positioning of the PIT must be on all output line-card. In doing this, the Interest packet should create the

PIT entry in the output line-card forwarding, and the selection should be done using the LPM of the FIB. Using the approach, the aggregation of interests is received at several line-cards however this also proves some limitations multipath cases. The scenario works in the fashion of receiving Interests handled at a line-card *(m)* been forwarded to two distinct output line-cards *(n)* and *(r)*, Upon satisfaction, the Data returning route is used through forwarding twice the line-card *(m)*; accordingly, the resulting Interest creates two entries in (PIT_n), and (PIT_r). The challenge therefore remains that the line-card *(n)* and *(r)* is thus, unable to detect if Data packet were earlier received at the other line-card. Additionally, assuming another Interest is received at line-card *(m)* is sent to line-card *(n)* with another Interest being a different Interest from the initial, for the same Data at line-card *(z)* but sent to line-card *(r)*. This implies, the initial received Data on line-card *(m)* or *(n)*, should satisfy both Interests; in practice therefore, two Data are needed for this PIT's placement.

- **Input/Output Line-Card Scheme**

Originally, this placement was proposed in [19] but squashed in favor of the output-only placement. This implies that it is important to place the PIT in both input and output line-card positions. In like manner, Interest makes a PIT entry for both the input line-card at the point that it is received and at the output line-card where it ought to be sent. Contrasted with the output just for the purpose of placement, it has two advantages: the absence of the superfluous look-ups in the FIB and copied packets the multipath. In any case, the input-output likewise experiences the last multipath issue as it additionally needs both the input and output Data so as to serve Interests acquired from various line-cards, which might forwarded to output line-cards. Additionally, loops prove difficult to be identified for the same reasons as above.

- **Third Part Line-Card Scheme**

This can be traced to the study proposed in [23], it designates that a PIT must be placed at all input line-card as in the input-only placement. In an event that an Interest is posed for content (*C*) at the line-card, thus it is "delegated" to a third party line-card just as the name implies. The resulting third party line-card is hereby selected as the $K = contentIDmodN$, where N is the number of line-cards in the router and *contentID* is the hash (e.g., CRC-32) bearing the content name, or $H(C)$. Consequently, the content aggregation is done for all entries of (*C*) to the PIT completely independent on the line-card bearing the Interest received for (*C*). From the output point, no PIT is required; when the Data is acknowledged through receipt, the output line-card then identifies (K) through the operations of a modulo evaluation as: $H(C)$ mod (N). Executing operations and other actions enables the placement for both loop detection and multipath as the third-party line-card to act as the entry aggregation point. For instance, when two Data are expected at two different output line-cards the initial received Data received is thus forwarded to a third-party line-card where it consumes each pending Interest. The same dissemination pattern is followed at the instance of the Data received as the second part. It is then forwarded to a third-party line-card.

5.3 PIT Replacement Policies

Replacement policy in PIT, is one of the important factors that determines the effectiveness of cache [24]. This section describes some possible PIT replacement policies i.e. Persistent, Random, LRU [17, 25], HLLR [26] and EQPR [27].

- **Persistent Replacement Policy**

Persistent policy is assumed as the default replacement police in NDN which adds to speeding up operations thereby reducing the complexity in implementation [17, 28]. It has been proposed as a solution to coordinate PIT in Interest interactions. A Persistent replacement policy requires that mostly new Interest packet should be selected and not to be rejected from its allocated memory space when there isn't free space found in its PIT. As a result, the memory space may be well underutilized. Thus, this may result in relatively poor performance. Therefore, this policy presents a major drawback since its reject any incoming Interest even with has less lifetime or a high popular entry in PIT. Although the Persistent policy is a simple to implement and easy to operate since it is rejected the new entries when the PIT size reached its limit.

- **Random Replacement Policy**

Just as the name implies, this policy randomly selects a candidate data and discards it in order to present free space when necessary. This policy unlike the previous does not require holding any information about the access in history. The discarding of Interest are randomly initiated. Random policy seems pretty easy in implementation due to the advantage of a pseudo-random counter for the whole cache operation. It however, consumes few resources but performs badly due to its non-usage-based dimension. Its performance relies solely on real randomness of the sequence. Although it is simple to implement in hardware. However, it less efficient because some entries may accumulate large request counts or may have little Interest lifetime as a factor is replacing and it might be the newly entry created one to be evicted from PIT.

- **Least Recently Used (LRU) Replacement Policy**

LRU policy is the most popular policy that is attributed to its ease of implementation and good performance. It is based on the least recently used cache replacement rule. LRU policy faces keeping both more frequent entries used in PIT and quick adaptation to the potential changes in entry popularity. In the analysis of LRU policy, due to the recently in use, the entry which with the highest access is most likely to be accessed again in the near future, and the entry that has been "least recently used" would be replaced by the PIT controller when the PIT demands a new entry adding. LRU is selected as a common one for NDN because of its low complexity, although its ability to identify the popular content is poor. Moreover, without consideration of factors, for example, the number of entries to keep track of increases. LRU becomes more expensive if one wants to ensure that the policy always discards the least recently used entry as well as it's harder to implement, slower.

- **Highest Lifetime Least Request (HLLR) Replacement Policy**

The HLLR is NDN replacement policy type which is used to manage memory (i.e., PIT) with methods employed within the NDN nodes. The standard fundamentals of this policy include the PIT keeping track of the number of times an Interest is requested in memory by assigning a counter on every face that is requested as Interest from the PIT. Each time a reference is made to an Interest, the counter is incremented by one. When the PIT gets full and has a new Interest waiting to be added to the PIT, a search operation is initiated for the Interest with the least faces of frequency, highest lifetime among all the entries present in the PIT in order to be used in determining which entry is the most suitable to be replaced. Hence, HLLR replaces the entry that has the minimum number of requests in one entry with the highest entry lifetime value. HLLR policy had tested using ndnSIM simulator and evaluated the results with some existing PIT replacement policies, i.e. Persistent, LRU and Random. The obtained simulation results showed that the HLLR policy can be approximate and even outperform LRU policy with much lower complexity, for a wide range of cache organizations.

- **EQPR Replacement Policy**

EQPR policy is proposed an as a novel PIT replacement policy. This policy using two factors: Estimated Round Trip Time and priority value for determining which existing entry inside PIT will be evicted. This aim towards in order to achieve Quality of Service for the valuable consumers who pay an additional cost for better service. The authors' model has classified Interest packets into two kinds: Prioritized Interest packet and Non-prioritized Interest packet. In order to make EQPR policy take the reservation decision and take replacement decision after the PIT is overflow; by making the node to keep track of the number of replaced PIT entries after each time interval. Also, by considering a threshold value which is denoted as the number of replaced entries. When an Interest packet arrives at the PIT, it checks the threshold value in case it is lower than the number of replaced PIT entries value, here it starts reserving the number of entries for prioritized interest packets, which also depend on the prioritized traffic. For that, EQPR policy take decision by making only prioritized Interest packets are allowed and non-prioritized Interest packets are lost. Thus, NACK packet with label 'PIT full' will be sent to downstream nodes. In this case, victim entry among all non-prioritized entries in the PIT that satisfies PEL < RTT need to be found. The EQPR policy was evaluated using ndnSIM simulator. The simulation results showed that prioritized traffic is always satisfied and total satisfaction rate remains the same.

5.4 Adaptive Interest Lifetime Techniques

In NDN, there are several adaptive Lifetime schemes have seen proposed such as DPEL, LTI, DPEL, STIL and LAPEL. This section shall carefully review each of the lifetime schemes in wider perspectives.

- **Long-Term Interest (LTI)**

Yao *et al.*, [29], design and implemented the LTI (Long-Term Interest) mechanism for real time applications based on the open source code. LTI enables a longer Lifetime for interest packets, and keep-alive interest packets that are used to keep the long-term interest by extending the interest packet with one more optional selector type. This thus, indicates the interest type. The value of the selector type can be long-term interest, keep-alive interest, normal interest, or unsubscribe interest. The LTI type indicates that the interest packet will be stored in the PIT for a longer time and will not be removed immediately after the requested Data packet is received. The keep-alive interest is used to keep the LTI alive for one more Lifetime. Normal interest is set as the default type. The unsubscribe Interest type is reserved for consumers to explicitly cancel the LTI. The *type* selector is optional. If it is not specified, the default type normal Interest is used. Thus, the extension to NDN packets is backwardly compatible with the original Interest packets. The Data packet is the same as the original. A content consumer sends an LTI packet with the channel name to request the real-time Data packet. This type of element is added as an optional selector. Then, the consumer periodically sends a keep-alive interest packet with the same content name to keep the LTI packet alive. The keep-alive Interest period can be set to the default value or set manually. The tests in live media stream and audio conference show that LTI can achieve better performance for real time applications.

- **Dynamic PIT Entry Lifetime (DPEL)**

Bouk *et al.*, [30] proposed a new scheme named DPEL which dynamically adjusts the lifetime of Interest in PIT for vehicular NDN. In DPEL scheme, the lifetime of every Interest packet is needed to be calculated at every traversed hop in a distributed fashion in order to acquire the resulting lifetime in an alleviated form if the numbers of relay nodes and hops are increasing. Since the network is always deployed as a highly non-static environment, in an event that Interests are not satisfied, the data provider and forwarding seem hard to assume that the Data shall traverse on the same reverse route. Hence, the resulting profession of the PIT by that Interest is termed ineffective and thus deletion operation should be engaged rather than setting the PIT to normal or static Lifetime for all Interest. In DPEL, the PEL. In DPEL, the PEL of new PIT entry is dynamically adjusted in the PIT of each intermediate node that forwards Interest. The Interest's PEL at a successive forwarder is computed using Interest's hop-count and forwarder node's own ISR (Interest Satisfaction Rate) information. It is evident from the results that any forwarding node with higher ISR has a very small average PEL because it frequently purges PIT entries after receiving the requested data packets. As a result, a forwarding node with higher ISR does not need to store PIT entry for a longer time. The result that DPEL mitigates approximately 40%, 55%, and 77% of PEL and number of PIT entries. Thus it's ensure the minimum number of PIT entries.

- **Smart Threshold Interest Lifetime (STIL)**

Alubady *et al.*, [24, 31, 32] designed and implemented a new mechanism named STIL that manage the Interest lifetime at content PIT in NDN router, which adjusts the Interest packet lifetimes as a function of the network load. Thus, NDN router grants larger lifetime values for incoming Interest packets in case of PIT is not full and when the low traffic. On the contrary, NDN router grants shorter lifetime values for incoming Interest packets in order to avoid overstaying of PIT entries during heavy network load and the PIT is full as the main goal for this mechanism. Basically, STIL has two options for adaptation Interest packet lifetime: the first option is to change incoming Interest packet lifetime based on given face received maximum number of Interest packet and recorded in the 'In-face list field', whereas the second option that is to change incoming Interest packet lifetime based on all entries in the PIT. The analysis results show that in NDN router when coupled with STIL, the performance can successfully stabilize the PIT table and improve to manage the PIT in order to avoid the Interest to stay in PIT a long time when it is full.

- **Limit Based Adaptive PIT Entry Lifetime (LAPEL)**

LAPEL scheme was proposed by Bouk *et al.,* [33] that adaptively estimates the PIT entry lifetime at Interest packet's forwarding for vehicular Named Data Networks. This scheme estimates the PIT entry lifetime based on parameters of the lower layer, and it does not have a constant factor. Thus, is made more adaptive comparing with the other proposed schemes. In LAPEL scheme, the vehicle generates Interest packets for requesting a content (data) with the broadcast limit plus retry count, and at each successive Interest packet's forwarders' PIT, the LAPEL just decays the PIT entry lifetime until they reach the broadcast boundary. So, LAPEL scheme behavior is entirely in line with the Interest-Data packets' response, which varies with the change in the maximum hop limit. The Simulation results obtained for validating and evaluating the proposed LAPEL scheme show that LAPEL scheme can reduce the PIT size as 8.4%; also has 52% fewer PELs, and achieves 4.5% more Interest satisfaction ratios, for any simulated TTL and Interest rates.

6 Conclusion

In the structure of the NDN architecture, the reference to the data and information is replaced with unique identification using a name similar to the current URLs; however, readable and strongly keyed to provide unique characteristics. Names are carefully deployed using strong key concepts such as hash functions to attain name schemes. Structurally, every NDN router or cache-able device possesses the distinct data structures of CS, FIB and PIT. With the PIT being the structure that decides what face of the data is needed to routed and saved, replaced to give way to incoming interests and the lifetime computation of the residency of the interest after their receipt. Furthermore, the router can control magnitude and volumes of the Data flooded into the network using its PIT size for the decision of accommodating and regulating the inputs. Data caching, on the other hand, provides the network with copies of the requested and granted interests and data in the NDN. This architecture, therefore, controls network congestions and minimize the amount of packet loss and drops. However, there exist yet some magnanimous researches a challenge that requires solutions to sustain adequate network functioning and thus, opens a wide opportunity for researchers to explore to come up with solutions. Therefore, this survey we have broadly classified PIT management into four classes, PIT data structure, PIT replacement and placement and adaptive Interest lifetime.

References

1. Cha, M., Kwak, H., Rodriguez, P., Ahn, Y.Y., Moon, S.: Analyzing the video popularity characteristics of large-scale user generated content systems. IEEE/ACM Trans. Netw. **17**(5), 1357–1370 (2009)
2. Fazea, Y.: Numerical simulation of helical structure mode-division multiplexing with nonconcentric ring vortices. Opt. Commun. **437**, 303–311 (2019)
3. Fazea, Y.: Mode division multiplexing and dense WDM-PON for Fiber-to-the-Home. Optik (Stuttg) **183**, 994–998 (2019)
4. Fazea, Y., Alobaedy, M.M., Ibraheem, Z.T.: Performance of a direct-detection spot mode division multiplexing in multimode fiber. J. Opt. Commun. **40**, 161–166 (2019)
5. Fazea, Y., Mezhuyev, V.: Selective mode excitation techniques for mode-division multiplexing: a critical review. Opt. Fiber Technol. **45**, 280–288 (2018)
6. Fazea, Y., Sajat, M.S., Ahmad, A., Alobaedy, M.M.: Channel optimization in mode division multiplexing using neural networks. In: 2018 IEEE 14th International Colloquium on Signal Processing & Its Applications (CSPA), pp. 173–175 (2018)
7. Ibraheem, Z.T., Rahman, M.M., Fazea, Y., Ahmed, K.K.: PAPR reduction in OFDM signal by incorporating Mu-Law companding approach into enhanced PTS scheme. J. Opt. Commun.
8. Passarella, A.: A survey on content-centric technologies for the current internet: CDN and P2P solutions. Comput. Commun. **35**(1), 1–32 (2012)
9. Ejaz, W., Azam, M.A., Saadat, S., Iqbal, F., Hanan, A.: Unmanned aerial vehicles enabled IoT platform for disaster management. Energies **12**(14), 1–19 (2019)
10. Mathieu, B., Truong, P., Peltier, J.-F., You, W., Simon, G.: Information-centric networking: current research activities and challenges. In: Hassnaa, M., Sherali, Z. (eds.) Media Networks - Architectures, Applications, and Standards, pp. 141–184 (2012)
11. Jacobson, V., Smetters, D., Briggs, N., Thornton, J., Plass, M., Braynard, R.: Networking named content. In: Proceedings of the 5th International Conference on Emerging Networking Experiments and Technologies, pp. 1–12 (2009)
12. Saxena, D., Raychoudhury, V., Neeraj, S., Becker, C., Cao, J.: Named data networking: a survey. Comput. Sci. Rev. **19**, 15–55 (2016)
13. Xylomenos, G., et al.: A survey of information-centric networking research. IEEE Commun. Surv. Tutor. **16**(2), 1–25 (2014)
14. Bari, M.F., Chowdhury, S.R., Ahmed, R., Boutaba, R., Mathieu, B.: A survey of naming and routing in information-centric networks. IEEE Commun. Mag. **50**(12), 44–53 (2012)
15. Amadeo, M., Molinaro, A., Ruggeri, G.: E-CHANET: routing, forwarding and transport in information-centric multihop wireless networks. Comput. Commun. **36**(7), 792–803 (2013)
16. Abu, A.J., Bensaou, B., Wang, J.M.: Interest packets retransmission in lossy CCN networks and its impact on network performance. In: Proceedings of the 1st International Conference on Information-Centric Networking - INC 2014, pp. 167–176 (2014)
17. Afanasyev, A., Moiseenko, I., Zhang, L.: ndnSIM: NDN Simulator for NS-3, University of California, Los Angeles. Technical report NDN-0005, California, Los Angeles, pp. 1–7 (2012)
18. Alubady, R., Hassan, S., Habbal, A.: A taxonomy of pending interest table implementation approaches in named data networking. J. Theoret. Appl. Inf. Technol. **91**(2), 411–423 (2016)
19. Dai, H., Liu, B., Chen, Y., Wang, Y.: On pending interest table in named data networking. In: Proceedings of the Eighth ACM/IEEE Symposium on Architectures for Networking and Communications Systems - ANCS 2012, pp. 211–222 (2012)

20. You, W., Mathieu, B., Truong, P., Peltier, J.-F., Simon, G.: DiPIT: a distributed bloom-filter based pit table for CCN nodes. In: 2012 21st International Conference on Computer Communications and Networks (ICCCN), pp. 1–7 (2012)
21. Li, Z., Bi, J., Wang, S., Jiang, X.: Compression of pending interest table with bloom filter in content centric network. In: CFI 2012: Proceedings of the 7th International Conference on Future Internet Technologies, pp. 1–4 (2012)
22. Li, Z., Liu, K., Zhao, Y., Ma, Y.: MaPIT: an enhanced pending interest table for NDN with mapping bloom filter. IEEE Commun. Lett. **18**(11), 1915–1918 (2014)
23. Matteo, V., Diego, P., Linguaglossa, L.: On the design and implementation of a wire-speed pending interest table. In: 2013 IEEE Conference on Computer Communications Workshops (INFOCOM WKSHPS), vol. 13, pp. 369–374 (2013)
24. Alubady, R., Hassan, S., Habbal, A.: Pending interest table control management in Named Data Network. J. Netw. Comput. Appl. **111**, 99–116 (2018)
25. Hassan, S., Alubady, R., Habbal, A.: Performance evaluation of the replacement policies for pending interest table. J. Telecommun. Electron. Comput. Eng. **8**(10), 125–131 (2016)
26. Alubady, R., Hassan, S., Habbal, A.: HLLR : highest lifetime least request policy for high performance pending interest table. In: 2016 IEEE Conference on Open Systems (ICOS), Langkawi, Malaysia, 10–12 October 2016, HLLR, pp. 42–47 (2016)
27. Buragohain, M., Gudipudi, P., Anwer, Z., Nandi, S.: EQPR : enhancing QoS in named data networking using priority and RTT driven PIT replacement policy. In: ICC 2019 - 2019 IEEE International Conference on Communications (ICC), pp. 1–7 (2019)
28. Mastorakis, S., Afanasyev, A., Moiseenko, I., Zhang, L.: ndnSIM 2.0 : a new version of the NDN simulator for NS-3. University of California, Los Angeles. Technical report NDN-0028, California, Los Angeles, pp. 1–8 (2015)
29. Yao, C., Fan, L., Yan, Z., Xiang, Y.: Long-term interest for realtime applications in the named data network. In: Proceedings of ACM AsiaFI12, pp. 1–8 (2012)
30. Bouk, S.H., Ahmed, S.H., Yaqub, M.A., Kim, D., Gerla, M.: DPEL: dynamic PIT entry lifetime in vehicular named data networks. IEEE Commun. Lett. **20**(2), 336–339 (2016)
31. Alubady, R., Hassan, S., Habbal, A.: Adaptive interest lifetime in named data networking to support disaster area. J. Telecommun. Electron. Comput. Eng. **10**(2–4), 29–34 (2018)
32. Alubady, R., Hassan, S., Habbal, A.: Adaptive interest packet lifetime due to pending interest table overflow. Adv. Sci. Lett., 1–5 (2016)
33. Bouk, S.H., Ahmed, S.H., Kim, D., Park, K.J., Eun, Y., Lloret, J.: LAPEL: hop limit based adaptive PIT entry lifetime for vehicular named data networks. IEEE Trans. Veh. Technol. **67**(7), 5546–5557 (2018)

Software Defined Network Partitioning with Graph Partitioning Algorithms

Shivaleela Arlimatti[1(✉)], Walid Elbrieki[2], Suhaidi Hassan[2], and Adib Habbal[2,3]

[1] Shaikh College of Engineering and Technology, Belgaum, India
sarlimatti@gmail.com
[2] InterNetWorks Research Laboratory, School of Computing, Universiti Utara Malaysia, Changlun, Malaysia
[3] Computer Engineering Department, Faculty of Engineering, Karabuk University, 78050 Karabuk, Turkey

Abstract. Software Defined Networks is an emerging paradigm in Internet communication world that increases the flexibility of today's networks by decoupling control plane and data plane of the network devices. The fundamental aim is to centralize the control and reduce the complexity of the networks. The communication medium between control and data plane is through OpenFlow protocol, an open standard network protocol designed to manage the network traffic by software programs. To increase the scalability and flexibility of controllers the OpenFlow controllers are distributed based on location and network types. However, most critical issue is minimizing the communication cost between the controller domains. In this paper, two graph partitioning algorithms Fiduccia-Matthyses algorithm and Kernighan-Lin algorithm are used to minimize the communication cost between distributed OpenFlow controller domains. The implementation of the algorithms is under Matlab simulation environment. The methodology used for the proposed algorithms is to interchange the elements from one domain to other domain to calculate the gain. The simulated results show that Kernighan-Lin algorithm minimizes more communication cost rather than the Fiduccia-Matthyses algorithm.

Keywords: Fiduccia-Matthyses · Kernighan Lin · Communication cost · OpenFlow

1 Introduction

The current economic slowdown is affecting the technology, industries and their market, networking being one of them. The economic-technological trends of networking and computing domains are converging. Conventionally, network operators configure network policies separately for individual network devices for efficient deployment of network applications. With the limited tools, they manually transform high-level policies to low-level commands for configuration of network conditions.

Considering all these factors, future network infrastructure needs simple, cost-effective and dynamic network management systems that rely more on software rather than hardware. The main point in current networks is the vertical integration of basic

F. Saeed et al. (Eds.): IRICT 2019, AISC 1073, pp. 583–593, 2020.
https://doi.org/10.1007/978-3-030-33582-3_54

network planes, such as control plane and data plane. Both, the control plane that decides to handle the network traffic, and the data plane which forwards the traffic based on the decision of the control plane, are tightly coupled. This tight coupling of control plane and data plane minimizes the flexibility and scalability of the network, which is the chief motivating factor for researchers to innovate new technology with low-cost optimization of the network. The most promising approaches in recent years are the still emerging paradigms, the Software Defined Networks (SDN) [1]. SDN is an emerging network approach that aims to design the network with software programs. SDN is a programmable network; behavior of network elements and software programs handle their flow of control. These software programs operate independently of network hardware. Having centralized program makes the whole network run efficiently and effectively [2]. This programmability and automation greatly reduce network operators' time, by provisioning and maintaining the network device-by-device, thereby effectively increasing agility.

The whole paper is organized as, Sect. 1 introduces the conventional networks as well as SDN. Section 2 explains the overall concept of SDN. Section 3 gives with the Architecture of SDN, Sect. 4 comes up with OpenFlow and Sect. 5 show the implementation of partitioning algorithm with KL as well as FM. The paper is concluded in Sect. 6.

1.1 From Conventional to Software Defined Networks

The Internet is an infrastructure, which is becoming increasingly difficult to change its protocols, physical infrastructure and its performance. Moreover, Internet applications are Internet Protocol (IP)-based and because of its huge size, it is difficult to maintain and manage especially with the introduction of 5G and its enabling technology such as SDN controller and massive MIMO [3–9]. With limited tools and huge infrastructure, network operators manually configure high-level policies to low-level commands to change network events and applications. However, the technology introduces an increase in cost and complexity, and is facing significant issues such as low scalability and flexibility. In addition to this, the horizontal integration of control and data plane makes it difficult to configure each device independently [10].

The network intelligence in SDN is logically centralized and forwarding devices are programmed by this centralized controller to forward packets from source to destination. SDN supports for flow-based rather than destination-based forwarding, where flow is a sequence of packets moving from a source to its destination [2]. It unifies the behaviors of network devices such as switches, routers and firewalls [11]. Figure 1 shows a comparison of traditional network and SDN. The traditional network comprises of data and control plane in one device, whereas in SDN control is separated and placed outside.

1.2 Limitations of Conventional Networking Technology

In conventional networks all the devices have the pre-installed software, installed by operators. These devices are operated for manual-network configuration, managing the

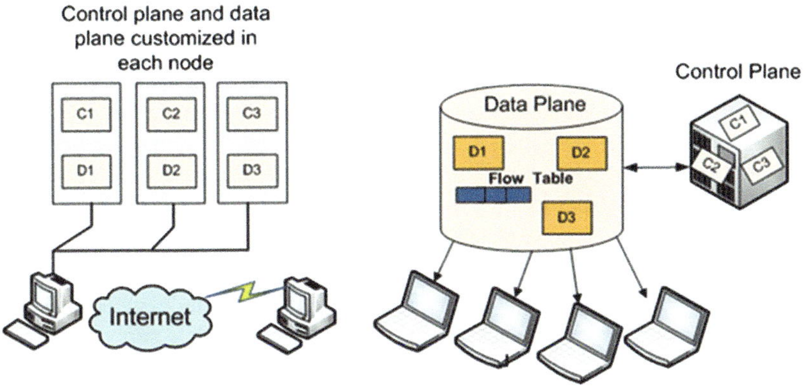

Fig. 1. Traditional network vs software defined networks

interfaces and are configured separately [11]. The architecture of traditional network device is shown in Fig. 2. The figure illustrates each device is comprised of all the planes such as data plane, control plane and application plane.

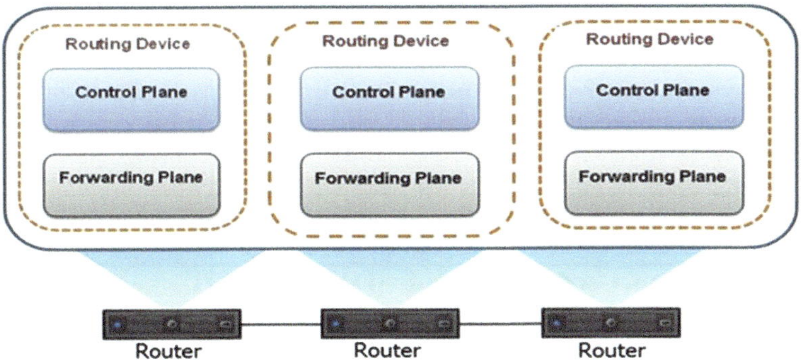

Fig. 2. Conventional router in traditional network

The main limitations of conventional networks are networks complexity, inconsistent policies, closed systems and physical reallocation of hosts in data centers. The goal of network operators is to design a protocol that is used for managing and maintaining the complex networks (due to their independent devices). Host virtualization is drastically increasing in data centers. This process is lengthy, expensive and quite difficult to obtain communication and cooperation between device vendors and network operators. It is even difficult to set network parameter policies.

2 Software Defined Networks

SDN is an emerging network approach of computer networks that allow network admins to handle network amenities through abstraction of lower level functionality. The traditional networks are ill-suited to support the scalable, dynamic computing and storage needs of today's computing environments such as changing traffic patterns, the consumerization of IT, the rise of cloud services, Big data handling, complexity that leads to stasis, inability to scale and vendor. On the contrary, SDN architecture is dynamic, cost-effective, manageable and adaptable, making the network ideal for high bandwidth and allowing directly programmable, agile, centrally managed, programmatically configured, open standards based and vendor neutral by supporting the nature of today's network applications. The SDN architecture decouples or disassociates the network control functions that makes decisions about where the traffic is sent from the forwarding functions that forward traffic to the selected destination enabling the control programmable directly and underlying forwarding devices to be abstracted for network services and applications dependence [12].

SDN network technology has garnered much attention today for its ability to address issues of programmability in networking and enable faster and easier network innovation. The idea of the programmable network has been conceived to facilitate network evolution. SDN provides more flexibility and new features to the network in the area of network applications. These applications may include traffic engineering and network management, security, routing optimization and NFV Lara 2014, [13]. These new features or services (which are added to SDN) are managed by only controllers which are easily programmable. In SDN, network intelligence is logically centralized in controllers that are software based. Google network deployed SDN with OpenFlow Wide Area Network (WAN), working as a production network. Google internal global network interconnects with its own data center network worldwide [14].

3 Architectural Components of SDN

The previous Section discussed the technical definition of SDN. In this section, architectural components of SDN are introduced. A simplified view of SDN architecture is shown in the Fig. 3. The clean separation of control, data and application planes are discussed in the subsections below.

3.1 SDN Application Plane

SDN applications are software programs that directly and programmatically communicate with their network requirements to controllers through northbound APIs. They have an abstracted view of the network for their internal decision-making. SDN application consists of more than one northbound interface with control logic. Network applications are the brains of the network as they implement the control logic, which is translated into commands to be installed on the forwarding devices from data-plane.

The SDN application contains applications designed by the user to fulfill their requirements through programmable control logic. These applications can access and

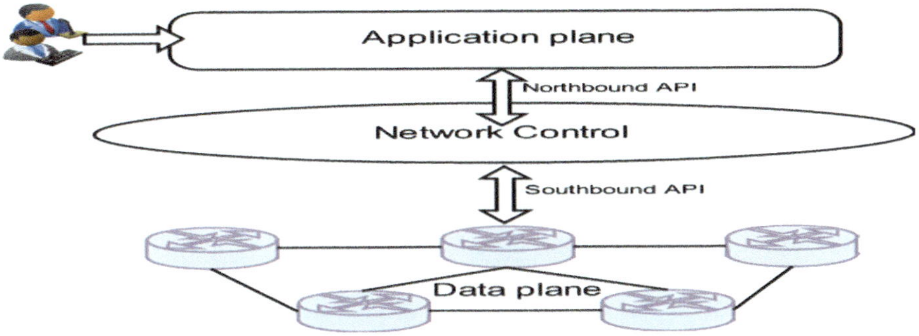

Fig. 3. Architecture of Software Defined Networks

control forwarding devices at data-plane. Such applications include load balancing [1, 15], seamless mobility [16], dynamic access control [17] and NFV [18]. In spite of this, a variety of use cases can be grouped into five application categories: traffic engineering, mobility and wireless, monitoring and measurement, data center networking, and security and dependability [15].

3.2 SDN Controller

The control plane is a bridge between infrastructure layer and application layer. Controller is a component in SDN, where the network complexity resides. It is a critical element, which supports to generate network configuration based on policies set by the network administrator.

There are three layers for the control platform such as the application, orchestration and service: SDN orchestration is defined as "the ability to program automated behavior in a network to co-ordinate the required networking to support applications and services". The orchestration starts with consumer service orders and consumer driven actions such as the ordering a service from the website [18, 19].

3.3 SDN Data Plane

The SDN data path, also known as infrastructure layer, consists of network devices such as switches and routers. SDN usually considers switches as simple forwarding devices. Nowadays virtualized networks and data center infrastructure are using these software switches and are emerging as promising solutions to data center networks because data centers contain more virtual access ports than physical access ports [20].

The data plane elements are specialized in data forwarding based on forwarding rules or flow rules. These rules are divided into two categories: a. processing forwarding rules and, b. installing forwarding rules [2], based on the current task of the rule. These flow table will have flow rules that tell how the packet is handled. An OpenFlow forwarding switch is based on flow tables, with each entry in the flow table consisting of three fields [21] as shown in the Fig. 4. The three fields of the flow tables are, namely: A matching rule: field RULE, in which the packet header is matched with

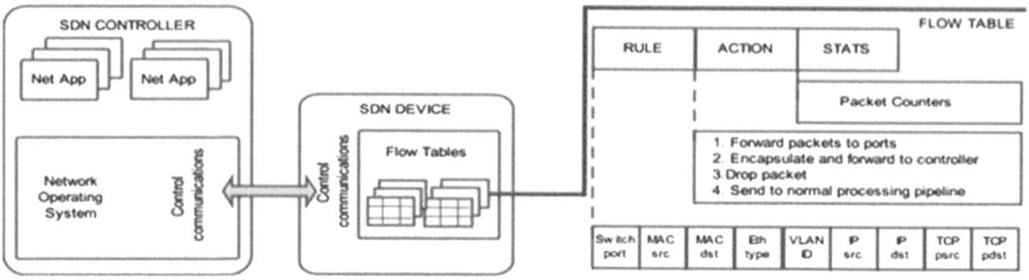

Fig. 4. SDN with Flow Tables.

the flow table entry. Set of instructions: field ACTION, if the matching is successful according to the flow table entry, then the packet is sent to next destination address, otherwise, the packet is sent back to the controller for new rule installation. When a packet enters the data plane element, then lookup process is carried out to find the match. If no match is found, then default rule is installed on the packet. The majority of the rules follow the table sequence number and the row order in the table. Forwarding rules are flexible enough to take any action on the arrived packet.

4 OpenFlow

The SDN is a new emerging architecture in which control plane and data plane are decoupled, and controllers are programmable, which is basic nature of SDN. In this architecture, network intelligence is centralized and controls the whole network. The Network appears as a single logical switch to the applications. By centralizing control plane, network managers have the flexibility to manage, configure and optimize network resources through SDN programs [10]. SDN architecture supports API's that implement common network services, which include access control, multicast, routing, bandwidth, management, storage optimization, traffic engineering, energy usage and different types of policy management.

The logical OpenFlow switch consists of flow tables and group tables as shown in the Fig. 5. The two tables perform packet lookups and forwarding, and an attachment to external controllers called OpenFlow protocol. Controllers can update, add and delete flow entries from flow table by proactively or reactively. ForCES and OpenFlow both the protocols push towards the new requirements of programmability for the benefit of innovative services. As mentioned earlier, OpenFlow is the southbound interface that connects control and forwarding plane elements. This is one of the major barriers to the acceptance of new networking technology.

5 Graph Partitioning Algorithms

The graph partition is defined by dividing the graph into sub graphs, in such a manner that all the sub graphs are of the same size and have a few connections between the sub graphs for the purpose of computing, task distribution and parallel processing. Graph

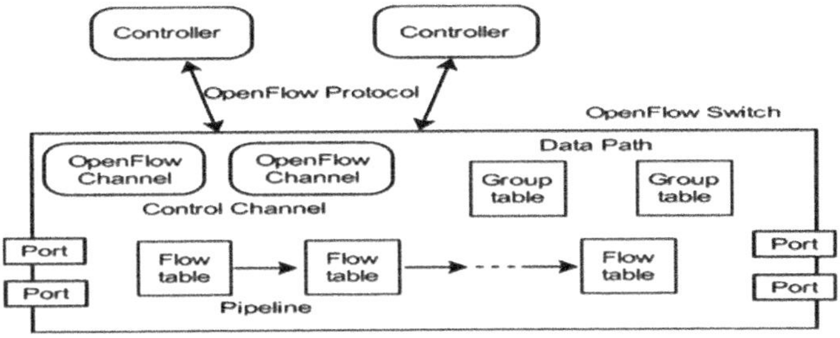

Fig. 5. Architecture of OpenFlow Switch

partition is mathematically represented as a graph G = (V, E) with vertices V and edges E, such that it partitions the graph G into smaller components with some specific properties.

$$\text{Let } S = S_1, S_2, S_3, \ldots .S_n, \tag{1}$$

Where S are the switches and n are total number of switches, Here, the number of switches is referred to as nodes for graph partitioning.

The links connecting to switches are given with the set

$$E = e1, e2, e3, \ldots .e\,m, \tag{2}$$

where m is the total number of links.

Number of controllers considered in the network graph are two and given as

$$C = C_1, C_2 \tag{3}$$

and each controller is associated with n/2 number of switches.

Kernighan-Lin algorithm divides a weighted graph with n nodes, into two equal parts, each of the size $n/2$. This partitioning minimizes the sum of the weights of two parts or groups. Kernighan-Lin is a heuristic algorithm for solving the problem of graph partitioning.

The Pseudo code for Simple partitioning algorithm is given as follows:

```
Algorithm 0.1: Pseudo Code for Simple Partitioning Algorithm
1: Divide the graph into two equal parts
2: Repeat until no nodes left
3: Select ai ∈ A, bi ∈ B
4: Swap a i and b i
5: Let Ci be the cost of the partition after swapping ai, bi
6: Return (A,B )
7: While cost continues to be reduced
8: Call using the returned partition for new starting point.
```

Moving a node S_1 from domain C_1 to domain C_2 would increase the value of the cut set by I a and decrease it by Ea. Where E a is external communication cost and I a is internal communication cost. Therefore, cost profit after moving S_1 from C_1 to C_2 is given by,

$$D_a = E_a - I_a \tag{4}$$

If $D_a > 0$ then gain is positive and $D_a < 0$ then the gain is negative. To maintain balanced partition, the elements of the networks are swapped from one domain to another domain. If two elements $S_1 \in C_1$ and $S_1 \in C_2$ are swapped the reduction in the cost is given by,

$$\mathrm{Gain}_{s1s2} = D_{s1} + D_{S2} - 2C_{s1}\,C_{s2} \tag{5}$$

Where C_{s1}, C_{s2} indicates there is an edge from S_1 to S_2.

The initial-cutsize and the final-cutsize is calculated as,

$$\text{Initial - cutsize} = \sum (C_{s1si} + C_{s2si} + C_{s3si}) \tag{6}$$

Where i = n/2 to n

In this section, Matlab has been used for building the mathematical models and performing experiments. This research took cut in cost size as the parameter for the simulation. To validate the proposed mechanism this research considered 4 switches to 12 switches for experimentation. Simulation was executed 5 times and each iteration was repeated for 6 runs.

Figure 6a shows the graph of total cut size of traditional distribution of controller mechanism and KL mechanism. The horizontal axis represents the number of switches participating in the program that are from 4 to 12. The vertical axis represents total cut size obtained before applying the algorithm and after applying the algorithm. In both the cases, cut size is observed increasing gradually as the numbers of switches increases. This is because of minimum communication cost between the domains. KL algorithm reduces the communication cost by swapping the elements from one domain to other. The swapping minimizes the communication cost by reducing inter and intra

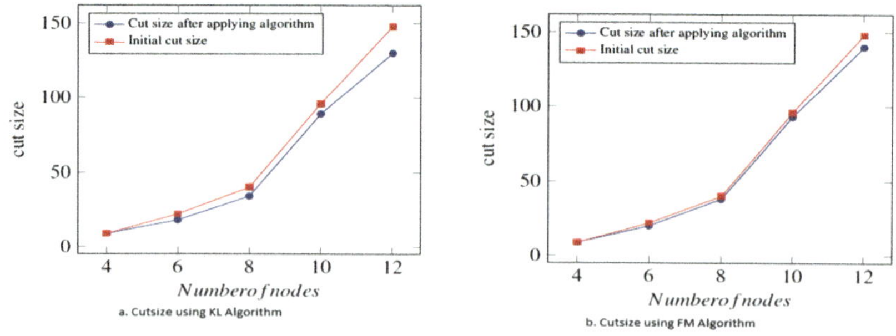

Fig. 6. a. Cutsize using KL algorithm b. Cutsize using FM algorithm

communication cost. Figure 6b shows the graph after applying FM algorithm. The custsize increases as the number of switches are increased.

Both the algorithm mechanisms are compared in Fig. 7. In both the cases cut size is noted to be increasing; however, KL algorithm gives better results than FM algorithm. At the beginning, all three are at the same point because a number of switches are less but as the traffic switch number increases, the difference between the graph lines is seen widening.

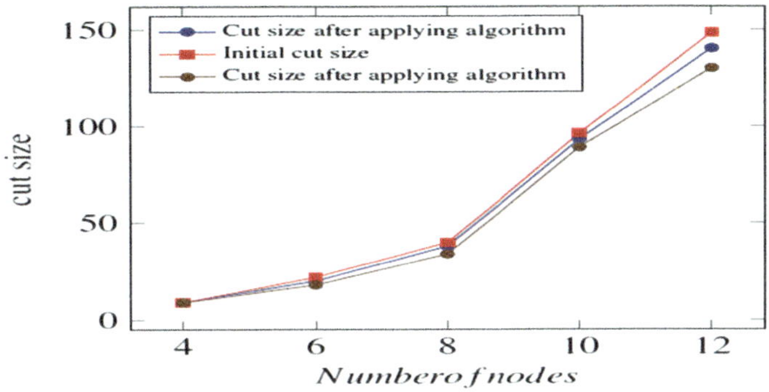

Fig. 7. Comparison of Cutsize using KL and FM algorithm

6 Conclusion

In this paper, we discussed two graph-partitioning algorithms, KL algorithm and FM algorithm. The system model for both the algorithms are discussed to show the similarity and difference. Both algorithms perform well in partitioning the network graph. However, KL algorithm gives the better results than FM algorithm this is because, in KL partitioning the elements of the domains are swapped and the gain is calculated for swapping the elements. The swapping interchanges the inter domain link cost to intra domain link cost and intra domain to inter domain. Whereas in FM graph partitioning algorithm elements are moved from one domain to other domain and the gain is calculated for the same. This gives imbalance in the graph weights. Our simulation results based on both the algorithms, show that the KL algorithm outperforms and is more robust in terms of parameter such as cut size cost. Even though this work focuses on inter domain communication cost minimization problem, the solution has wide application range. The algorithms are can be simulated in real simulation environment.

References

1. Al-Najjar, A., Layeghy, S., Portmann, M.: Pushing SDN to the end-host, network load balancing using OpenFlow. In: IEEE International Conference on Pervasive Computing and Communication Workshops (2016)
2. Hassas Yeganeh, S., Ganjali, Y.: Kandoo: a framework for efficient and scalable offloading of control applications. In: Proceedings of the First Workshop on Hot Topics in Software Defined Networks, pp. 19–24, August 2012
3. Fazea, Y.: Numerical simulation of helical structure mode-division multiplexing with nonconcentric ring vortices. Opt. Commun. **437**, 303–311 (2019)
4. Fazea, Y.: Mode division multiplexing and dense WDM-PON for Fiber-to-the-Home. Optik **183**, 994–998 (2019)
5. Fazea, Y., Alobaedy, M.M., Ibraheem, Z.T.: Performance of a direct-detection spot mode division multiplexing in multimode fiber. J. Opt. Commun. **40**, 161–166 (2019)
6. Fazea, Y., Amphawan, A., Qtaish, O.: Mode division multiplexing of helical-phased spot mode and donut mode in multimode fiber interconnects. In: 2017 IEEE Symposium on Computer Applications & Industrial Electronics (ISCAIE), pp. 200–205 (2017)
7. Fazea, Y., Mezhuyev, V.: Selective mode excitation techniques for mode-division multiplexing: a critical review. Opt. Fiber Technol. **45**, 280–288 (2018)
8. Fazea, Y., Sajat, M.S., Ahmad, A., Alobaedy, M.M.: Channel optimization in mode division multiplexing using neural networks. In: 2018 IEEE 14th International Colloquium on Signal Processing & Its Applications (CSPA), pp. 173–175 (2018)
9. Hassan, S., Fazea, Y., Habbal, A., Ibrahim, H.: Twisted Laguerre-Gaussian mode division multiplexing to support blockchain applications. In: Region 10 Conference, TENCON 2017–2017. IEEE, pp. 2421–2425 (2017)
10. Bari, M.F., Roy, A.R., Chowdhury, S.R., Zhang, Q., Zhani, M.F., Ahmed, R., Boutaba, R.: Dynamic controller provisioning in software defined networks. In: CNSM, pp. 18–25 (2013)
11. Casado, M., Freedman, M.J., Pettit, J., Luo, J., Gude, N., McKeown, N., Shenker, S.: Rethinking enterprise network control. IEEE/ACM Trans. Netw. (ToN) **17**(4), 270–1283 (2009)
12. Foster, N., Guha, A., Reitblatt, M., Story, A., Freedman, M.J., Katta, N.P., Monsanto, C., Reich, J., Rexford, J., Schlesinger, C., Walker, D.: Languages for software-defined networks. IEEE Commun. Mag. **51**(2), 128–134 (2013)
13. Casado, M., Freedman, M.J., Pettit, J., Luo, J., McKeown, N., Shenker, S.: Ethane: taking control of the enterprise. In: ACM SIGCOMM Computer Communication Review, vol. 37, no. 4, pp. 1–12. ACM, August 2007
14. Das, S., Parulkar, G., McKeown, N.: Unifying packet and circuit switched networks with OpenFlow (2009)
15. Feldmann, A.E., Foschini, L.: Balanced partitions of trees and applications. Algorithmica **71** (2), 354–376 (2015)
16. Liyanage, M., Gurtov, A., Ylianttila, M. (eds.): Software Defined Mobile Networks (SDMN): Beyond LTE Network Architecture. Wiley, Hoboken (2015)
17. Cui, L., Yu, F.R., Yan, Q.: When big data meets software-defined networking: SDN for big data and big data for SDN. IEEE Netw. **30**(1), 58–65 (2016)
18. Dabkiewicz, S., van der Pol, R., van Malenstein, G.: OpenFlow network virtualization with FlowVisor. research project 2, University of Amsterdam (2012)

19. Heller, B., Sherwood, R., McKeown, N.: The controller placement problem. In: Proceedings of the First Workshop on Hot Topics in Software Defined Networks, pp. 7–12. ACM, August 2012
20. Casado, M., Garfinkel, T., Akella, A., Freedman, M.J., Boneh, D., McKeown, N., Shenker, S.: SANE: a protection architecture for enterprise networks. In: USENIX Security Symposium 2006, vol. 49, p. 50, August, 2006
21. Han, B., Gopalakrishnan, V., Ji, L., Lee, S.: Network function virtualization: challenges and opportunities for innovations. IEEE Commun. Mag. **53**(2), 90–97 (2015)

Organizing Named Data Objects in Distributed Name Resolution System for Information-Centric Networks

Walid Elbrieki[1(✉)], Suhaidi Hassan[1], Shivaleela Arlimatti[2], and Adib Habbal[3]

[1] InterNetWorks Research Laboratory, School of Computing, Universiti Utara Malaysia, 06010 Sintok, Kedah, Malaysia
walid@internetworks.my
[2] Shaikh College of Engineering and Technology, Belgaum, India
[3] Computer Engineering Department, Faculty of Engineering, Karabuk University, 78050 Karabuk, Turkey

Abstract. The Information-centric Networks (ICN), an important research direction of the future Internet architecture, has gained lot of attention from the research community. The aim is to improve the current Internet with a new architecture where the design is completely based on information instead of the host. The information is named and is called as Named Data Object (NDO), utilized for data registration and name resolution, and the system that translates the object identifiers to network address is known as Name Resolution System (NRS). The random NRS distribution and network segregation are important and challenging issues in increasing NDO registration and storage. It is challenging for a single NRS to handle more than 10^{15} expected NDO's with low latency and good throughput, along with the issues interest flooding and interest congestion. To overcome these problems NRS is distributed based on the location of routers. The mechanism is called Distributed Name Resolution Mechanism (DNRM). The NDO storage needs to be organized to increase the scalability of NRS. Distributed hash table and Bloom Filter have their own problems such as high latency in individual element searching and member deletion respectively. To overcome the above-mentioned issues the Balance Binary Tree (BBT) data structure is introduced to manage large NDO storage. The implementation results reduce the end-to-end delay by increasing network throughput. The contribution of this study is significant in promoting the use of NRS in ICN for handling the heterogeneity of the future Internet.

Keywords: Name Resolution System · Named Data Objects · Balanced Binary Tree · Information-Centric Networks

1 Introduction

Current Internet architecture was introduced in the 1960s. The main intention behind the promulgation was to interconnect computing resources across a distributed geographical area. This led to benefits such as resource sharing, storing and transformation

F. Saeed et al. (Eds.): IRICT 2019, AISC 1073, pp. 594–603, 2020.
https://doi.org/10.1007/978-3-030-33582-3_55

of information among hosts. These hosts largely from different geographical areas and could communicate to one another by establishing connections between them. The resulted communication model gives conversation between two machines, one for resource utilization and the other for access provision. In this regard, Internet Protocol (IP) packets are having two identifiers, for source host address and destination host address. This thus forms the Internet traffic between these two hosts. According to the studies [1–3], users are tending their interests in the Internet for "what content?" as against "where is the content". However, the current situation is still driven on the host-centrism even as the need for the users is far tending to the content-centrism. An attempt of partly dissociating the host-to-host semantics will be to bend into the Internet of content. It could, be the "what" content being replaced with "where". Recently, Internet usage has been moved from host-oriented model to a content-oriented model. The shortcomings of traditional Internet architecture motivated the research community to a new architectural submission of the future Internet known as an ICN. The ICN technology is one of the alternatives for the future Internet predicted challenges [4]. The main aim of emerging technology is to shift the host-oriented communication to a content-centric [1]. Research in this area is still in its infancy as numerous challenges are to be addressed such as name resolution, name-based routing, security, caching and content dissemination [5].

The NRS is beneficial in the application of route-by-name. To deal with substantial number of NDOs Yuncheng *et al.* in [6] evaluated time and space complexity of storage data structure collected from traffic trace of campus backbone. They proposed Distributed Resolution Services (DRS) to time and space complexity model.

In this paper, the study shows the distribution of NRS and organizing the contents of NRS that is NDO in the form of Balanced Binary Tree (BBT). NRS is an element of an ICN that translates the object identifiers into network addresses and supports clients in searching NDO using names. The mechanisms proposed in this paper are specifically designed to improve the performance and efficiency of an ICN. The technology has close relation with Internet of Things (IoT), Software Defined Networks (SDN) and mangy more in field of security and privacy [7–9] (Fig. 1).

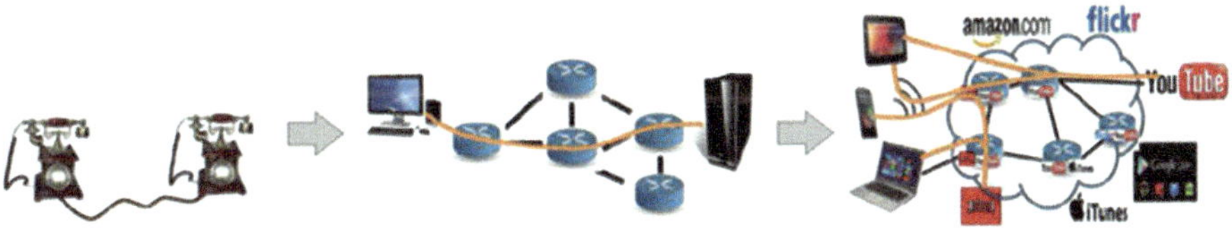

Fig. 1. Evolution of Information-Centric Networks

The whole paper is organized as, Section one introduces the concept and evolution of ICN. Section two explains the concept of ICN, along with Interest and data transmission in ICN and explains Distributed NRS. Sections three gives with the name registrations in ICN whereas section four comes up with implementation of DNRS mechanism and Section five show the implementation of BBT based NDO storage mechanism. Section six elaborates the results and discussion and the paper is concluded in Section seven.

2 Information-Centric Networks

This is an approach that evolves the Internet infrastructure away from its host-centric paradigm to a content-centric paradigm based on end-to-end principle and perceptual connectivity, to a network architecture where the center point is named information (data or content).

Nowadays the Internet is transforming from an academic significance to a global infrastructure for large-scale information of over a billion connected devices using massive MIMO [1–6] and other enabling technologies. The annual data transfer is in Exabyte [17] and indexed web pages approximately to the mark of trillion [18]. Even though information dissemination is the primary interest of customers, the operation of the Internet is on host centric communication model. This leads the networking community to explore new architectures for the future Internet, which shows its direction towards ICN [19]. The overview is shown in Fig. 2.

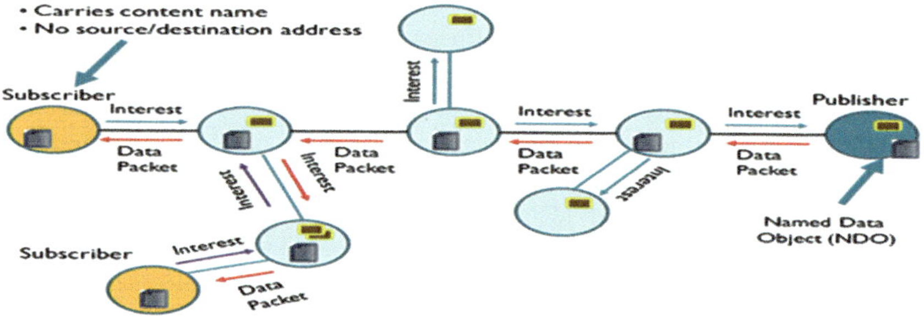

Fig. 2. Overview of Information-Centric Networks.

2.1 Interest and Data Transmission in ICN

NDO packet forwarding and routing in the network is handled by publishers and subscribers. When the subscriber receives the message as a source address and content name from NRS, the subscriber sends the request message to the source. This message is transmitted through the routing protocol (e.g. Open Shortest Path First (OSPF) routing protocol). When the publisher receives the message, the data is sent back to the subscriber from the same path traveled by the request message. While transmitting back, the message travels through routers. Each router when gets data, publishing function must be executed. As shown in Fig. 3, the interest and data are transmitted in the network.

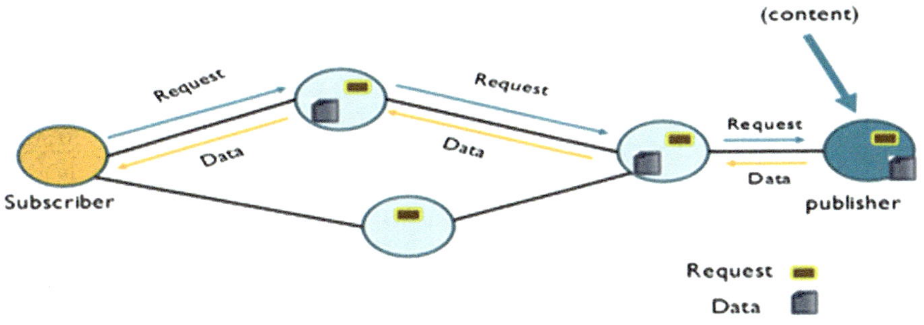

Fig. 3. Interest and data transmission in ICN

2.2 Distributed NRS

The main objective of NRS is to design a system for name resolution with the objectives like, low latency, optimal network utilization, scalability, scoping and agility. NRS stores the bindings between object IDs with its corresponding locators, in which object copies are available. However, the NRS is independent of routing and forwarding. The centralized or single NRS in the network has drawbacks such as,

- Single Point of Failure (SPoF)
- Scalability
- Overloading
- Congestion in interest and data messages
- Interest Flooding on the storage of NDOs

To overcome all these issues, the NRS is distributed. Distributed NRSs are allocated with routers based on the Nearest Neighbor algorithm. The details and mathematical model of distribution of NRS explained in our paper [20]. The conceptual diagram of distributed NRS is as shown in Fig. 4 below:

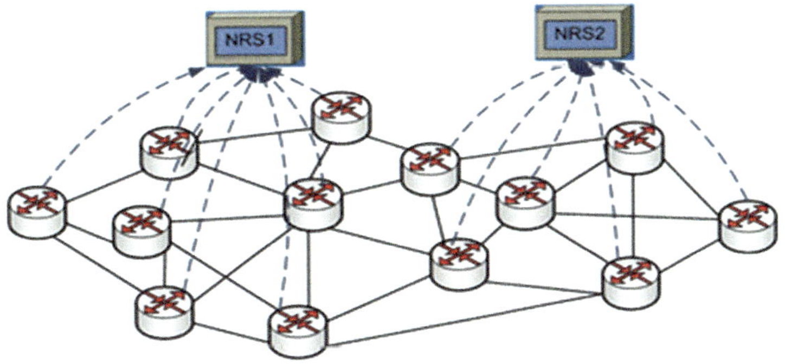

Fig. 4. Distributed name resolution system

The purpose of NRS distribution is to handle storage system. The balanced binary search tree is a self-balancing tree data structure, handles the NRS better than

Distributed Hash Table (DHT) and Bloom Filter (BF). In this paper, we construct a BBT-NDO storage NRS which consists of a Balanced Binary Tree data structure. The details and mathematical model of distribution of NRS explained in our paper [21]. The network of BBT-NDO storage is defined by a balanced binary tree and parent-child relationship with individual NDOs and balancing operation of the tree with each insertion of NDO.

We divided the network into several domains as a prerequisite for this mechanism. Each domain has its own NRS to store the NDOs. All the resolution domains are interconnected. In our work, we consider one NRS for better understanding and set of operations on it. Balanced Binary Tree data structure conceived to efficiently perform all data structure basic operations on large data sets [21, 22].

3 Name Registration in NRS

Assume a simple network with n number of publishers. All n numbers of publishers will publish their NDOs in one NRS. There are two NRS in the network and nodes grouping is very much necessary for dividing the network in two parts. The two NRS in the network as shown in Fig. 5.

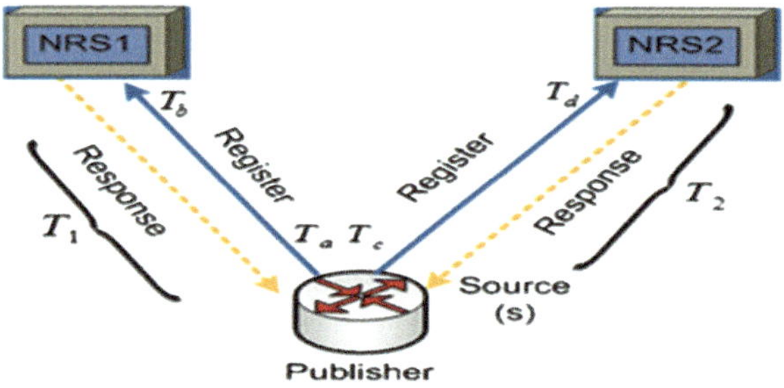

Fig. 5. Name registration in NRS

Let P_x be publisher, registers NDO x into NRS1 and NRS2 both. Let T_1 be a time taken by the P_x to register in NRS1 and T_2 be the time taken by P_x to register in NRS2.

```
Step 1: Subscriber Si sends an interest  Ia to NRSi Where Ia be-
longs Sa.
Step 2: For every NRSi
Step 3: If interest matching is successful,
the NRSi provides locator address to the requested subscriber.
Step 4: else if Interest message Iais forwarded to NRSj
Then NRSj sends locator address to the requested subscriber
Step 6: else
  The subscriber sends a request message to locator for origi-
nal data
```

The NDO and the requests sent by the subscribers are mapped in this stage. As stated above in this paper, data objects are stored in the form of balanced binary tree data structure. The Searching Time is faster in our mechanism as compared to Distributed Hash Table and Bloom Filters. Our simulation result for the searching time of NDO is compared with Distributed Hash Table searching time and for which we have outstanding results. The searching time graph is shown in the Fig. 6. In general, DHT is the best NRS till date. However, it takes more time to search one element in the whole list of NDOs. Searching a group of elements in DHT is always better. The graph shows that the DHT searching time for very high compared to BBT searching time. In BBT each iteration is reduced by its half of the searching time.

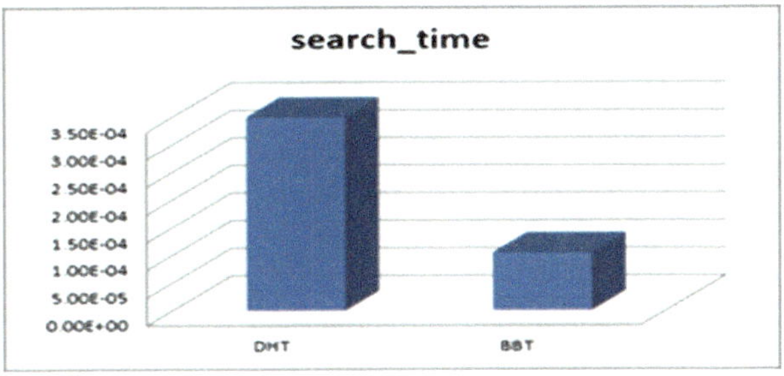

Fig. 6. NDO search time

4 DNRS Mechanism Implementation

Figure 10 shows the average throughput of proposed DNRM mechanism, traditional distribution of NRS and single NRS network over random topology. The horizontal axis represents the number of publishers from 10 to 50 numbers. The vertical axis represents the total throughput in bits/second. Initially, for 10 numbers of publishers, the traditional distribution of NRS throughput was more as compared to our mechanism, interestingly DNRM throughput increases gradually as the number of publishers are increased. Single NRS and Traditional distribution throughput over the same speed followed a similar trend, but with lower values. However, proposed mechanism is considerably increasing in the later stages. The increase in the total throughput is because of more NRSs in the network, which reduces the searching time and processing time, and leads to increase in the total throughput (Fig. 7).

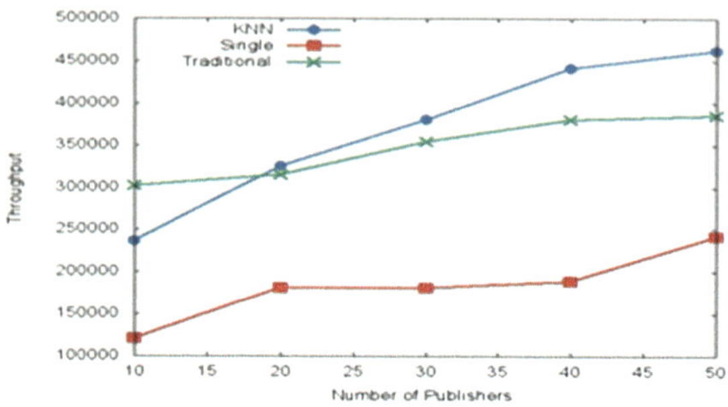

Fig. 7. Throughput in random topology

Figure 8 presents average end-to-end delay in traditional distribution of NRS, DNRM distribution and single NRS with several publishers. However, the DNRM line shows less delay as compared to other mechanisms. This is because DNRM minimizes the time of transmitting the data from sender to receiver by decreasing the end-to-end delay. On the other hand, the traditional distribution shows more delay even though the NRSs are distributed. The reason for this is, in our mechanism the logical distance between the publisher and the NRS and subscriber to NRS is less, minimizes the overall delay in the network.

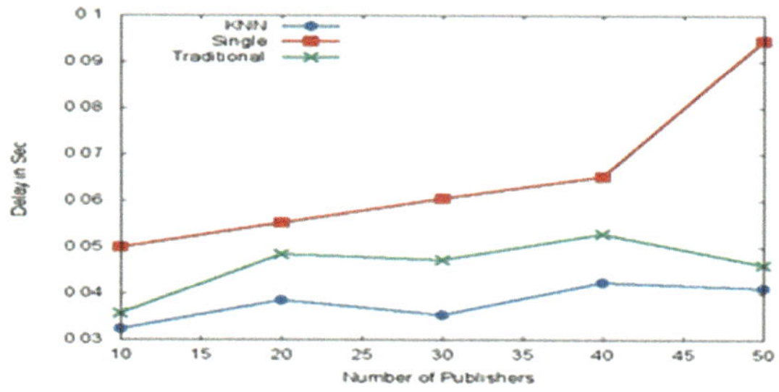

Fig. 8. Delay in random topology

5 BBT Based NDO Storage Mechanism

The main application of ICN paradigm is, scalability of NRS which must increase as the Internet scales heavily. The NDO population is huge and increases significantly day by day. The approximate indexable web estimation is 10^{12} web pages [23, 24] and it is going to increase, NRS with efficient storage structure is required to support for forwarding and routing of huge data. Envision of Internet of Everything expected to be 50 billion interconnected devices. Managing all these becomes challenging.

Figure 9 illustrates the total throughput of distributed hash table and balanced binary tree in bits/second. The simulation results are collected on power law topology with different numbers of publishers. Initially, the total throughput is equal for both the mechanisms. The distributed hash table throughput was fluctuating in between. However, balanced binary tree throughput was gradually increasing as the number of publishers is increased.

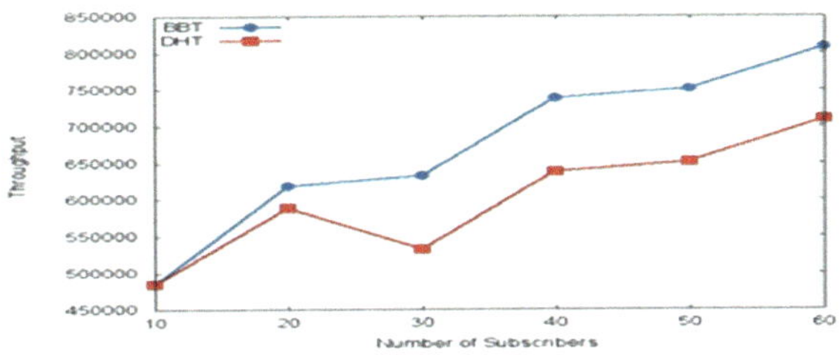

Fig. 9. Throughput in BBT

From Fig. 10, the average end-to-end delay differences of distributed hash table and balanced binary tree mechanism for a different number of publishers. Although both mechanisms showed a gradual rise in the delay as the numbers of publishers are increased, distributed hash table mechanism seemed to be higher than a balanced binary tree. In brief, balanced binary tree achieves lower delay than distributed hash table.

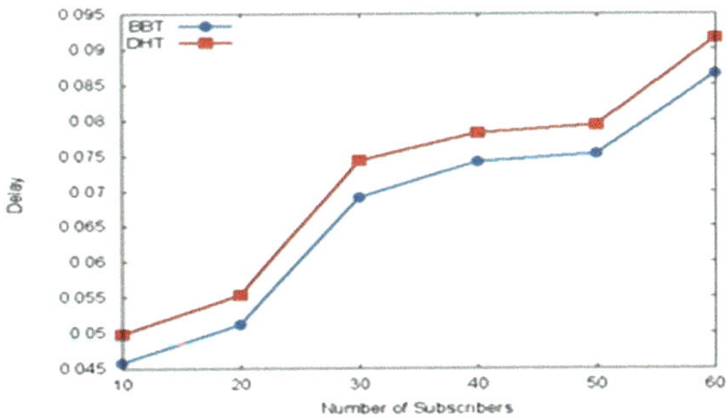

Fig. 10. End-to-End delay in BBT

6 Results and Discussion

For every source and destination pair, our simulation mechanism calculates average end-to-end delay. The delay is minimized with our mechanism because the logical distance or the cost between each pair of nodes is minimized between routers or publishers or subscribers and NRS. The previous mechanisms proposed only the distribution of NRS either based network size or location. Our mechanism logical distance between the NRS and the underlying network devices is reduced to increase the total throughput and average end-to-end delay.

The more interesting results can be seen in storage mechanism i.e. BBT based NDO storage mechanism. In this case end-to-end delay, is increasing as the input increases but there is a significant difference between BBT based NDO and DHT mechanisms. The results of BBT based NDO storage is better than DHT. It is due to the organization of the NDOs in the form of a balanced binary tree.

The performance metrics total throughput is calculated in the distributed NRS network with a systematic arrangement of NDOs. The throughput of the network increases with our mechanisms with respect to publishers publishing the NDOs and subscribers searching for data with the interest message. In both the mechanisms, the throughput is higher than the compared mechanisms. This is due to DNRM reduces the communication cost between the routers and NRS so by allowing more registration or interests in less time, in addition, BBT based NDO mechanism arranges the NDOs in the form of tree data structure so that searching time minimizes. These are some of the reasons that help to increase the network total throughput.

7 Conclusion

This paper discussed the introduction of DNRM and BBT based NDO mechanism and evaluation of both the mechanisms used for the performance of NRS in ICN. The first mechanism, distributed NRS was developed using the nearest neighbor algorithm. This is the extension of traditional distributed NRS, minimizes publishing time of the NDOs in the network by partitioning the network. In the second mechanism, the balanced binary tree is used for storing the NDOs. This storage data structure reduces the searching time by half of its original time. Both the mechanisms are implemented in an OMNeT++ simulation environment with INET framework, total throughput and average end-to-end delay over distinct types of topological scenarios. The evaluation was accomplished with the examination of NRS in ICN performance with end-to-end delay and throughput as compared to normal distributed NRS with previous work. In the second mechanism, searching time of distributed hash table and balanced binary tree are compared. Finally, the obtained results show that distributed NRS with balanced binary tree data structure outperforms. Moreover, these two mechanisms have been proven to be the best mechanisms for segregating the network and storage of NDOs.

The proposed mechanisms focused on distributing the NRS and storage utilization. Although DNRM and BBT based NDO mechanism were evaluated through a reliable simulator, it would be interesting to evaluate those mechanisms using real data in a real environment, a good extension to the life of this research.

References

1. Ahlgren, B., Dannewitz, C., Imbrenda, C., Kutscher, D., Ohlman, B.: A survey of information-centric networking. IEEE Commun. Mag. **50**, 26–36 (2012)
2. Barabási, A.L., Albert, R.: Emergence of scaling in random networks. Science **286**, 509–512 (1999)
3. Zhang, D., Liu, H.: Routing and name resolution in information-centric networks. In: 2013 22nd International Conference on Computer Communications and Networks (ICCCN) (2013)

4. Pan, J., Paul, S., Jain, R.: A survey of the research on future internet architectures. IEEE Commun. Mag. **49**, 26–36 (2011)
5. Bari, M.F., Chowdhury, S.R., Ahmed, R., Boutaba, R., Mathieu, B.: A survey of naming and routing in information-centric networks. IEEE Commun. Mag. **50**, 44–53 (2012)
6. Kutscher, D., Eum, S., Pentikousis, K., Psaras, I., Corujo, D., Saucez, D., Schmidt, T., Waehlisch, M.: ICN research challenges. Work Prog. (2014)
7. Benedetti, P., Ventrella, A.V., Piro, G., Grieco, L.A.: An SDN-aided information centric networking approach to publish-subscribe with mobile consumer. IEEE (2019)
8. Frey, M., Gündoğan, C., Kietzmann, P., Lenders, M., Petersen, H., Schmidt, T.C., Juraschek, F., Wählisch, M.: Security for the industrial IoT: the case for information-centric networking. IEEE (2019)
9. Shen, Q., Wu, J., Li, J.: Edge learning based green content distribution for information-centric internet of things. IEEE (2019)
10. Fazea, Y.: Numerical simulation of helical structure mode-division multiplexing with nonconcentric ring vortices. Opt. Commun. **437**, 303–311 (2019)
11. Fazea, Y.: Mode division multiplexing and dense WDM-PON for Fiber-to-the-Home. Optik **183**, 994–998 (2019)
12. Fazea, Y., Alobaedy, M.M., Ibraheem, Z.T.: Performance of a direct-detection spot mode division multiplexing in multimode fiber. J. Opt. Commun. **40**, 161–166 (2019)
13. Fazea, Y., Amphawan, A.: 32 channel DQPSK DWDM-PON for local area network using dispersion compensation fiber. In: EPJ Web of Conferences, vol. 162, p. 01016. EDP Sciences (2017)
14. Fazea, Y., Sajat, M.S., Ahmad, A., Alobaedy, M.M.: Channel optimization in mode division multiplexing using neural networks. In: 2018 IEEE 14th International Colloquium on Signal Processing & Its Applications (CSPA), pp. 173–175. IEEE (2018)
15. Ibraheem, Z.T., Rahman, M.M., Fazea, Y., Ahmed, K.K.: PAPR reduction in OFDM signal by incorporating Mu-Law companding approach into enhanced PTS scheme. J. Opt. Commun.
16. Fazea, Y., Mezhuyev, V.: Selective mode excitation techniques for mode-division multiplexing: a critical review. Opt. Fiber Technol. **45**, 280–288 (2018)
17. Xylomenos, G., Ververidis, C.N., Siris, V.A., Fotiou, N., Tsilopoulos, C., Vasilakos, X., Katsaros, K.V., Polyzos, G.C.: A survey of information-centric networking research. IEEE Commun. Surv. Tutor. **16**, 1024–1049 (2014)
18. D'Ambrosio, M., Dannewitz, C., Karl, H., Vercellone, V.: MDHT: a hierarchical name resolution service for information-centric networks. In: Proceedings of the ACM SIGCOMM Workshop on Information-Centric Networking (2011)
19. Pavlou, G.: Keynote 2: information-centric networking: overview, current state and key challenges. In: 2011 IEEE Symposium on Computers and Communications (ISCC) (2011)
20. Elbreiki, W., Arlimatti, S., Hassan, S., Habbal, A., Elshaikh, M.: A distributed name resolution system in information centric networks. AIP Publishing
21. Hassan, S., Elbreiki, W., Arlimatt, S., Habbal, A.: Named data object organization in distributed name resolution system for information centric network environment, pp. 1936–7317 (2016)
22. Manku, G.S.: Balanced binary trees for ID management and load balance in distributed hash tables (2004)
23. Takamizawa, H., Saji, K., Aritsugi, M.: A replica management protocol in a binary balanced tree structure-based P2P network. JCP **4**, 631–640 (2009)
24. Google. We knew the web was big (2008). https://googleblog.blogspot.my/2008/07/we-knew-web-was-big.html

Scheduling Criteria Evaluation with Longer Job First in Information Centric Network

Ibrahim Abdullahi[1(✉)], A. Suki M. Arif[2], and Yousef Fazea[2]

[1] Department of Mathematics/Computer Science, Ibrahim Badamasi Babangida University, Lapai, Niger State, Nigeria
ibrojay01@ibbu.edu.ng

[2] InterNetWorks Research Laboratory, School of Computing, Universiti Utara Malaysia, 06010 Sintok, Kedah Darul Aman, Malaysia

Abstract. Schedulers are always important for good system performance in Operating systems. In Operating system concepts, the better and flexible a scheduler, the more likely the system attain optimization through maximum Central Processing Unit (CPU) utilization, maximum throughput, minimum turn-around time, minimum waiting time and less Context switch. Scheduling practice has also seemed to be of essence to the Internet. Packets are schedule to cross hops systematically to enhance quality of service (QoS) and minimize excessive bandwidth utilization. This report presents some scheduling algorithms and suggests an extension of the Longest Job First (LJF) algorithm to enhance system performance in multiprogramming domain and Internet. Cache improvement is shown by tapping the advantages of the LJF in Information Centric Networking. This was achieved by introducing a new technique called Combinational Burst Time (CBT) and Combined Weighted Average (Cwa). Monte Carlo simulation using discrete Poisson for the process burst time generation was demonstrated to achieve some of the optimization criteria in CPU Scheduling.

Keywords: CPU scheduling · Optimization · Combinational Burst Time CBT · Longest Job First · Caching · Operating system · Information-Centric Networking

1 Introduction

Process is defined as any program in execution [1]. It can also be referred to as an individualistic unit of a program. Processes exist in different forms in computer system viz. co-operating process, dis-joint, interacting processes etc. Depending on the type of process, characteristics are placed in the Process Control Block (PCB Multiprocessing environment has played vital role in increasing the performance such as Multi-input-multi-output [2–6] which involves more than a process in the system, therefore the need for good scheduling algorithm is a requisite. The scheduler handles how the processes will be admitted into the CPU and various processes admitted will be placed in the queues. The Schedulers are charged with ordering these processes based on the operating system design and specification. Therefore, every process shall be placed in a

F. Saeed et al. (Eds.): IRICT 2019, AISC 1073, pp. 604–614, 2020.
https://doi.org/10.1007/978-3-030-33582-3_56

ready queue before the scheduler decides upon its execution exercise. In a new paradigm known as Information-Centric Networking (ICN) [7, 8] scheduling is seen as a key to making caching appropriate by engaging the mid-node serving a requested interest (information) to the subscriber. ICN is a paradigm that aids in making information readily available to a user in a content-centric approach in lieu of the traditional Internet host-centric concept.

2 Introduction

Scheduling is the method, actions by which processes, threads (light weight processes) and data flows are given access to the system resources (Process time, CPU, communication bandwidth) which is usually done to balance a system and achieve best quality of service (QoS) [9, 10]. It also refers to the series of actions performed on a system to distribute processes to resources using long-term, mid-term and the short-term schedulers. For the ICN paradigm, schedulers could be seen as the function of forwarding interest to requesters through special feature known as the Forwarding Information Base (FIB) [8, 11].

2.1 Scheduler

Scheduler is the module that moves jobs/processes from queue to queue, see Fig. 1. Queuing diagram of process scheduling. The scheduler evaluation can be measured using the following standard metrics:

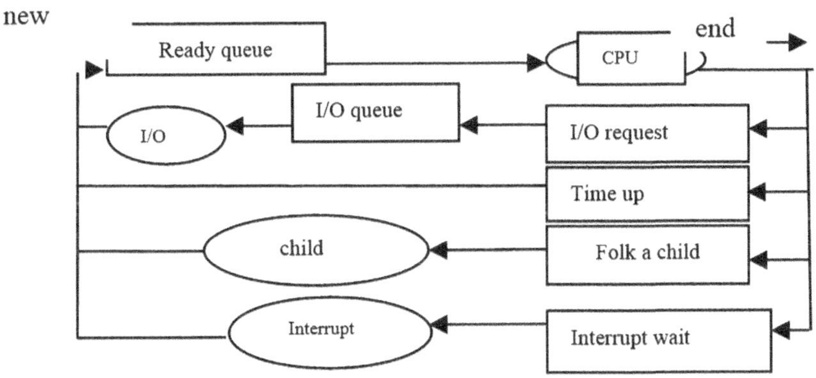

Fig. 1. Queuing diagram of a process scheduling [1]

- Throughput: the total number of processes that completes their execution per unit time.
- CPU utilization: Percentage of time the CPU was kept busy processing.
- Fairness/waiting time: the total time a process waits in the system due to arrival, suspension or pre-emption.

- Latency: this defines the total response time and the overall turn-around time of a process in a system.

The process on creation is referred to as "new"; it gets into the "ready queue" as shown on Fig. 1, before it is scheduled to the CPU. Processes can be moved to the I/O queue on request of an I/O operation or expired time allotted.

Admission/Long Term Scheduler
The long-term scheduler decides which job/process is to be brought to the main memory from a lower level memory location. In modern operating system machines such as Windows, there is always a high priority to real-time processes than others. Common examples of the long-term schedulers are implemented in large scale systems such as batch processing, computer clusters and super computers

Middle/Midterm Scheduler
From the concept of memory, it is understood that processes can be moved temporarily in and out of the CPU without getting to the waiting phase. This is only possible when the process does not request and I/O. the processes are forcibly (pre-empted) taken out by the user based on the designs or priority (see Fig. 2. Midterm scheduler); the moving of the processes in and out in this form is known as "swapping in and swapping out" also referred to as "paging in and paging out". Midterm scheduler is the module which handles the paging in and out of processes.

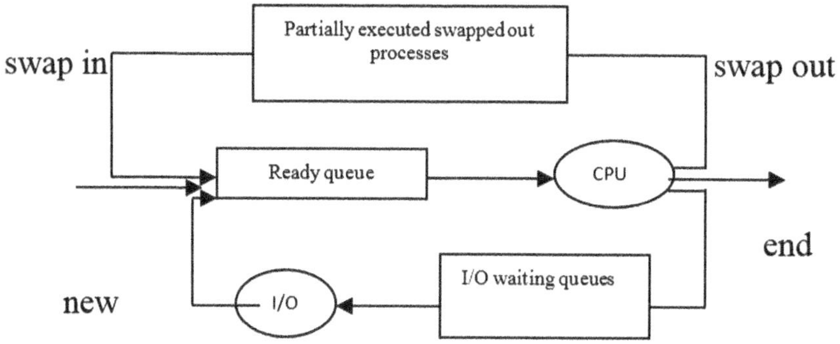

Fig. 2. Midterm schedular [1]

CPU/Short Term Scheduler
This scheduler does more of its work in the ready queue. It selects based on some conditions (CPU scheduling) which order and the process that should be assigned the CPU. Its response time is usually faster than long term schedulers. It performs based on clock interrupts, I/O interrupts and system calls. Short term scheduler handles pre-emption depending on the algorithm, priority or time slice.

3 Scheduling Algorithm

3.1 First Come First Served (FCFS)

This is the most popular type of scheduling algorithm, just as the name implies the first process that arrives to the queue is the first to be processed and assigned the CPU. This is implemented in grocery stores, packet-switching in computer networks, gas stations etc. FCFS advantages include high turnaround time, high response time, minimal scheduling overhead since context switch is absent. In Networks this can be seen as a fair share and delivery of packets

3.2 Shortest Job First (SJF)

It differs from the FCFS through its characteristics of serving the process with the lowest execution time (burst time) ahead of the any process with a higher execution time. The algorithm is referred to as selfish algorithm since longer jobs waits indefinitely for shorter jobs regardless of the time of arrival [12]. The SJF exists in two forms viz. SJF with pre-emption and SJF without pre-emption. Its distinct characteristics include better throughput, maximum CPU usage, low average waiting time but a disadvantage of leading processes to starvation (indefinite waits of the long jobs).

3.3 Priority Scheduling

This is a scheduling algorithm that allocates CPU to process according to some numbers known as 'priority mark score'; mostly the lower the values imply higher priority. Its characteristics include: all evaluation metric depends on priority order costly to implement due to the large number of context switches involved, could also lead to starvation.

3.4 Round Robin

A process scheduling algorithm that operates in cyclic manner and order. It operates based on a fixed time slice called "quantum" which every process uses and move out of the CPU for the next process to use. Each process in the queue is allocated the CPU through the amount of quantum set by the operating system designer [13]. It could be generally referred to as a starvation free algorithm as all processes are considered simultaneously. Round Robin properties include: Only a process run at time, processes give up CPU when quantum is reached, all processes are assumed to be treated equally with fairness, but it is uniformly inefficient.

3.5 Longest Job First Extended with CBT

A new technique that will extend the already existing LJF will be called Combinational Burst Time (CBT) [10], to enhance system performance by achieving some of the optimization criteria in system scheduling. These criteria are: Better CPU utilization, maximum throughput, minimal response time, minimal overall average waiting time, and lesser context switches [10].

4 ICN Overview

Information-centric network needs caching to deliver and service the components of CS, PIT and FIB. This content store provides the mechanism for saving data, interest as information. This is used by the ICN architecture to manage the lodgement of the contents. PIT on the other hand is a unique table that keeps track of all events on nodes. A router node has a PIT that records all the details of the served objects/interests. When request is served to the nodes, PIT records the interest by keeping a record of it being served. In an event that the interest was not found in the CS, PIT saves the interest as a pending data. Once the interest moves in the network, the status will change. FIB fashion as a router in the traditional Internet. It is charged with the operational responsibility of forwarding interest to the nodes requested. This is however possible when the interests exist on the network.

From Fig. 3, a subscriber sends out an interest into the network. The interest is then moved into the first router which has the aforementioned components (CS, PIT and FIB). Each taking care of its functionality as the interest traverses into the network. The most important point to note here is the ability of the nodes caching this interest. The publisher on the other end sends out a data which to the user known as interest. Naming the interest and resolving it becomes the challenge of ICN. It is needed by the network NRS to translate the interest and data to match up. If the interest is matched with the data, the PIT records the interest and sends the resulting mapping to the FIB.

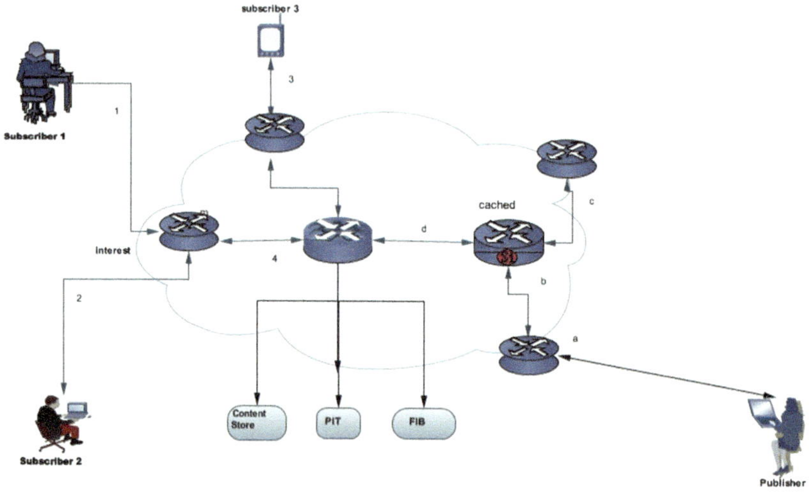

Fig. 3. ICN overview

Additionally, caching is the major factor in ICN with the advantage of once a node caches a result, there will be no need to visit the sole publisher when similar interests are needed. This is seen as a major advantage of reducing as a major advantage of reducing excessive bandwidth consumption, lesser delay, lower energy requirement and overall QoS.

5 Cache Scheduling

Scheduling as shown throughout this report has become so vital in every operation. For ICN achieving adequate scheduling is most of essence. Several cache deployment and placement policies have been used in web caching, data caching, memory caching etc. For ICN, proxy cache, reverse proxy, adaptive, transparent, active caching are seen as possibilities that could mitigate the wastage and redundancy of information as complained in the studies by Psaras *et al.* Zhang and Wang in different studies [16-18] and [24]. Caching can be defined as the predictor storage or saving of information/interest data to serve as when needed with lower delay and fast delivery. It usually places contents in its limited memory based on some itemized strategies. These contents are place based on its recency, or frequency of usage or referral.

5.1 Proxy Cache

Is a cache deployment method that tends to benefit the user of subscriber faster. In this scheme, the cache-enabled router or node is placed a bit closer in distance than the publisher point. Here scheduling can be said to be shorter job first in lieu of the LJF. Advantage of this practice serves the subsequent requester on-path closer to the previous subscriber.

From Fig. 4, using the ICN technique at all nodes, the cache positioning is done using the proxy scheme. Here the cache edge is closer to the user as against a different scheme such as the reverse proxy (described below). The advantage is that a user or group of user requires less energy to get a subscribed information from the publisher. However, the disadvantage is seen on the issue of data corruption, inconsistency from a neighbour fetch.

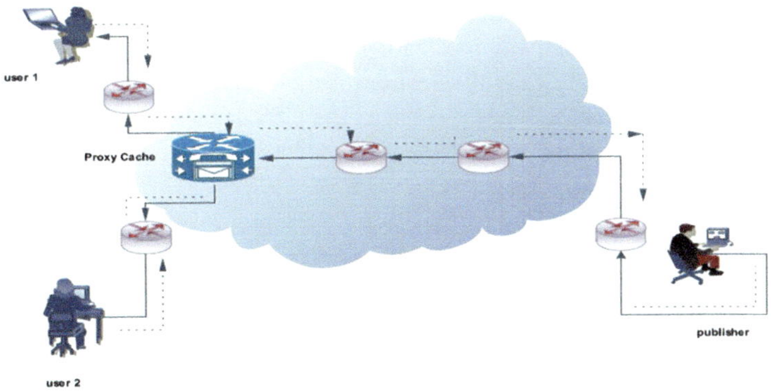

Fig. 4. Proxy Cache

5.2 Reverse Proxy Cache

Just as the name implies, it is similar to the proxy with only distinction in the positioning and placement of the cache-enabled router closer in distance to the publisher. The LJF is needed to access the far position of the cache. The reverse proxy is seeming popular in practice as the position is seen as a buffer that reduces the streaming, high accessing and reduced queue problems as shown in Fig. 5.

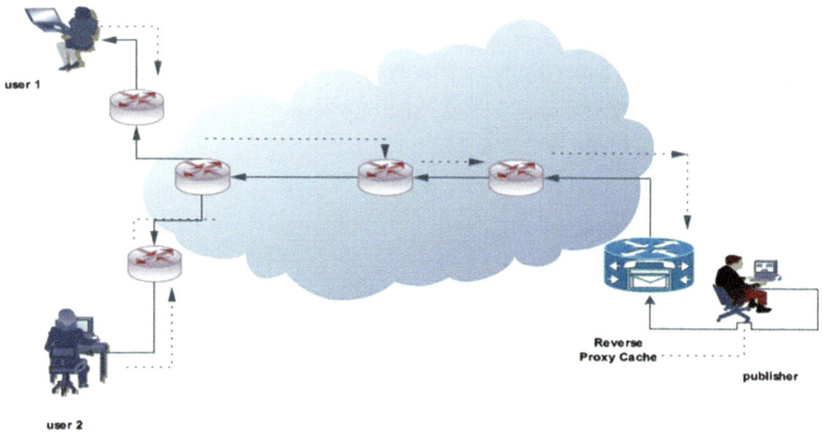

Fig. 5. Reverse proxy

The user here is placed a bit far from the cache making its scheduling LJF adoptable to pull the content published. Its advantage is the high chance of always fetching a reliable content at refreshes from the publisher. Once an update is posed to the network by the publisher, the update is circulated faster. Here the proposed LJF +CBT would fashion in the way of creating higher memory spaces on-path to enable the data transfer. Once the careful selection of the cache-enabled node is designated, the LJF will be to access the longer distance. The CBT is therefore the average number of spaces in memory for the travel to fetch the data.

5.3 Transparent Cache

In transparent cache, the cache placement and positioning are determined and scheduled by an administrator. ICN of PSRIP feature this approach using an NRS-like operation. The cache coordination aids in load balancing, upward stream minimization and a good schedule to avoid bottleneck clustering as shown in Fig. 6. Collaborating caching has the ability to mitigate the starvation problem seen in CPU scheduling.

The advantages of vending and subscribing the proposed LJF+CBT helps improve the overall network and work performance similar to the Shared Memory Architecture (SMA) in computer architecture field.

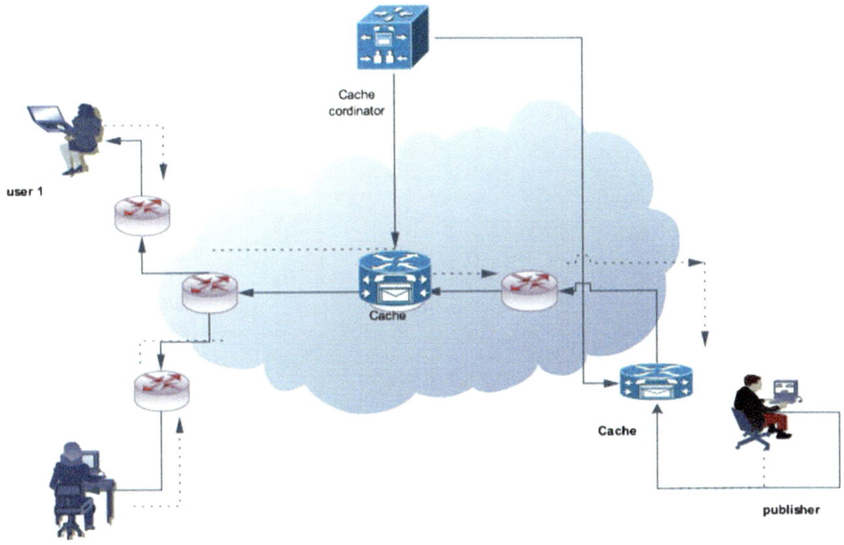

Fig. 6. Transparent Cache

5.4 Adaptive Cache

The scheduling in adaptive cache looks opportunistic and predictor. The cache schedule and placement are similar to the mesh topology architecture of the network where almost the entirety of the nodes is cache enabled. The servicing and forwarding of this placement scheme use the round-robin nature for placement. However, some points of intelligence are always seen as the scheduling is grouped into chunks and served from the popular groups as autonomous nodes as shown in Fig. 7.

5.5 Active Cache

In active cache, the cache is placed dynamically and also allowing for personalization of the content. This practice here is seen as the LJF+CBT approach. The cache placement point is designated and chosen by a node user make it proxy, or reverse proxy depending on the user's flexibility [14].

The figure above shows the flexibility of the node and the user deciding the optimal cache placement point for the content. This serves the advantage of vending into LJF for servicing and minimizing bandwidth consumption when requests are sent at later time. The proposed concept of scheduling in [15] could also be used to improve the LJF+CBT as shown in Fog computing.

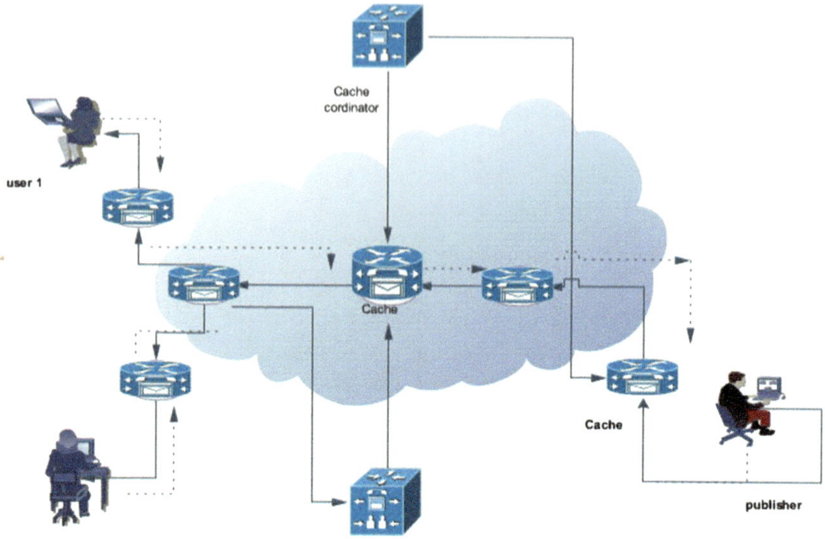

Fig. 7. Adaptive Cache

6 LJF+CBT Shift with ICN

ICN is therefore an envisioned replacement to the traditional Internet that solves the huge amount of multimedia messages and file dominated on today's Internet.

However, without adequate caching and scheduling, the dream will look unfeasible (Fig. 8).

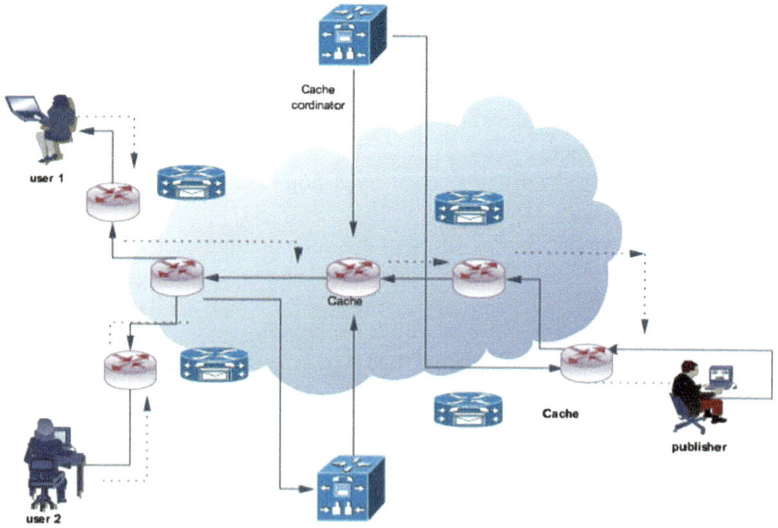

Fig. 8. Active Cache

7 Conclusion

In conclusion, based on the stated assumptions, processes were termed to be of the same nature as the processes were not differentiated as CPU bound or I/O bounds. This makes the proposed LJF+CBT applicable and more flexible to use on any process category type for easy combination of the burst times. The results obtained has shown that lower burst time processes are starved in this scheduling algorithm thereby increasing the overall average waiting time of the problem as shown in the ordinary LJF. This contributes to the reason why most scheduling practices avoid the LJF algorithm. In practice, a new machine performs better in terms of efficiency measure when newly branded; the LJF+CBT would be helpful in solving high tedious and time demanding processing in multi-tasking systems which have set due dates. In environments where processes executive concurrently but have a general due date, LJF +CBT will perform better. Therefore, the results obtained from the simulation proved that the processes are admitted into queues based on the phases. The result provided the LJF+CBT winning over the ordinary LJF by having lesser values in waiting times and turnaround time as 62.80 over 88.53 and 73.93 over 99.67 respectively. LJF+CBT will thus be helpful in video gaming, mobile application resource management through multi-tasking to avoid convoy and starvation problems if extended. Additionally, LJF +CBT demonstrate the potential of in-path aching in ICN thereby reducing the excessive bandwidth consumption through the path predictor calculation using the placement schemes of proxy, reverse proxy, adaptive, active and transparent forms of cache deployment. Further research is encouraged to suit and solve the problem of CPU starvation and information redundancy in the router nodes of ICN.

References

1. Galvin, P.B., Gagne, G., Silberschatz, A.: Operating System Concepts. Wiley, Hoboken (2003)
2. Fazea, Y., Amphawan, A.: 5 × 5 25 Gbit/s WDM-MDM. J. Opt. Commun. **36**, 327–333 (2015)
3. Hassan, S., Fazea, Y., Habbal, A., Ibrahim, H.: Twisted Laguerre-Gaussian mode division multiplexing to support blockchain applications. In: Region 10 Conference, TENCON 2017–2017, pp. 2421–2425. IEEE (2017)
4. Fazea, Y., Amphawan, A.: 32 channel DQPSK DWDM-PON for local area network using dispersion compensation fiber. In: EPJ Web of Conferences, vol. 162, pp. 1–3 (2017)
5. Fazea, Y.: Numerical simulation of helical structure mode-division multiplexing with nonconcentric ring vortices. Opt. Commun. **437**, 303–311 (2019)
6. Fazea, Y., Mezhuyev, V.: Selective mode excitation techniques for mode-division multiplexing: a critical review. Opt. Fiber Technol. **45**, 280–288 (2018)
7. Jacobson, V., Smetters, D.K., Thornton, J.D., Plass, M.F., Briggs, N.H., Braynard, R.L.: Networking named content. In: Proceedings of the 5th International Conference on Emerging Networking Experiments and Technologies, pp. 1–12 (2009)
8. Ahlgren, B., Dannewitz, C., Imbrenda, C., Kutscher, D., Ohlman, B.: A survey of information-centric networking. IEEE Commun. Mag. **50**, 26–36 (2012)

9. Biersack, E.W., Schroeder, B., Urvoy-Keller, G.: Scheduling in practice. ACM SIG METRICS Perform. Eval. Rev. **34**, 21–28 (2007)
10. Abdullahi, I., Junaidu, S.: Empirical framework to mitigate problems in longer job first scheduling algorithm LJF+CBT. Int. J. Comput. Appl. **75** (2013)
11. Ahlgren, B., D'Ambrosio, M., Marchisio, M., Marsh, I., Dannewitz, C., Ohlman, B., et al.: Design considerations for a network of information. In: Proceedings of the 2008 ACM CoNEXT Conference, p. 66 (2008)
12. Fluhrer, S., Mantin, I., Shamir, A.: Weaknesses in the key scheduling algorithm of RC4. In: International Workshop on Selected Areas in Cryptography, pp. 1–24 (2001)
13. Mohanty, R., Das, M., Prasanna, M.L.: Design and performance evaluation of a new proposed fittest job first dynamic round robin (FJFDRR) scheduling algorithm. arXiv preprint arXiv:1109.3075 (2011)
14. Sourlas, V., Gkatzikis, L., Flegkas, P., Tassiulas, L.: Distributed cache management in information-centric networks. IEEE Trans. Netw. Serv. Manag. **10**, 286–299 (2013)
15. Abdullahi, I., Arif, S., Hassan, S.: Ubiquitous shift with information centric network caching using fog computing. In: Computational Intelligence in Information Systems, pp. 327–335. Springer (2015)

An Evaluation of Performance of Location-Based and Location-Free Routing Protocols in Underwater Sensor Networks

Nasarudin Ismail[1(✉)] and Mohd Murtadha Mohamad[2]

[1] Pusat Teknologi Maklumat, Universiti Tun Hussein Onn Malaysia, 86400 Parit Raja, Batu Pahat, Johor, Malaysia
nasar@uthm.edu.my

[2] School of Computing, Faculty of Engineering, Universiti Teknologi Malaysia (UTM), 81310 Skudai, Johor, Malaysia

Abstract. Nowadays, both academia and industries has been actively conducting research and development of Underwater Wireless Sensor Networks (UWSNs) to support various widely available application such as oil/gas monitoring system, tsunami monitoring, disaster prevention and environment monitoring. However, developing an efficient communication in UWSNs is a demanding task due to the irregular characteristic of underwater environment. Due to this reason, the research in network technology of UWSNs has been underway to overcome the lack of available network technology in UWSNs. For this research, we are going to evaluate the performance of the existing proposed UWSNs routing protocols under different network densities. The objective of this research is to compare the performance of location based and location free UWSNs routing protocols under several different dense network condition. We have selected several UWSNs routing protocol location-free like Underwater Flooding (UWFlooding) and Depth Based Routing (DBR) and for location based like Vector-Based Forwarding (VBF) Protocol and Hop-by-Hop Vector-Based Forwarding (HH-VBF). The results clearly shown that location-based is outperform the location-free routing protocols in most test. However due to the impractical use of finding location of nodes in water environment, location-free is still the best solution for UWSNs.

Keywords: UWSNs · Routing protocol · Flooding · Location-based · Location-free · Aqua-Sim

1 Introduction

Currently, human knowledge about underwater environment is still lack in behind in comparison with land exploration. Exploration of knowledge about the land and its structure are remarkably a success due to the technological advancement in Wireless Sensors Network (WSN). As such, researchers are eagerly wanted to explore in an underwater environment with the same technology which can called as Underwater Sensor Networks (UWSNs) [1]. However, due to its unique features in underwater

F. Saeed et al. (Eds.): IRICT 2019, AISC 1073, pp. 615–624, 2020.
https://doi.org/10.1007/978-3-030-33582-3_57

environments such as high water pressure, wide area and environmentally vicious conditions, the use of UWSN is the most appropriate way by using un-manned discovery in that environment [2].

There are many differences between Terrestrial Wireless Sensor Network (TWSNs) for land and UWSNs. Due to the differences, UWSNs could not use the same existing established technology design for TWSNs. Routing protocol design is one of the hottest research topics at UWSN that can ensure the transmission of packet data at UWSN from source nodes to destination nodes is reliable and effective. In UWSNs, this routing protocol design is one of the key issues in UWSNs [3–5].

At UWSNs, most of the sensor's node is powered by a battery that has limited capacity and is difficult to charge and replace [1]. As a result of the use of acoustic waves for communication, where acoustic use experienced high latency, high error probability and low bandwidth, sensor nodes in UWSNs suffer more energy usage compared to TWSNs sensor nodes [6, 7]. For that reason, to extend the life span of the network at UWSNs, the design of energy-efficient protocols should be considered without sacrificing network performance such as by minimizing energy consumption during communication [6, 8, 9]. Communication in UWSNs is generally sending and receiving packets, either a control pack or a data packet, then the routing protocol in the network layer is an important protocol in delivering packets from the source to the destination. To reduce energy consumption in UWSNs, there are several proposed technique in routing protocol such as clustering, reducing flooding transmission, mobicast, limited retransmission and intelligent algorithm like reinforcement learning to preserve the battery power in sensory node [6].

Opportunistic Routing (OR) approach is the most applicable routing technique to use in UWSNs versus end-to-end routing like in TWSNs due to highly dynamic condition in underwater environment [10, 11]. However, OR approach which are flooding the network with redundant packets could lead to higher energy consumption because of higher network traffic [10, 12]. Mostly OR algorithm in UWSNs, based on their approach to find the next forwarder node can be divided into 2 categories, location based and location free [13]. In location based OR category, the protocol is using the node location information during the network implementation. While, for location free OR category, it uses other information like depth information instead using location information.

The objective of this research is to compare the performance of existing proposed location free and location based OR under different network density condition. Even though, we already know that getting location information is not very possible in underwater environment due to inapplicable GPS signal in water, but this research wants to compare how actually these two categories of OR protocols perform in the same network condition in simulation environment.

The rest of this paper is organized as follows. Section 2 discusses the related work with location free and location-based OR. Section 3 includes the simulation environment and parameters. Section 4 includes the discussion and results of the simulation experiments and the last section is the conclusion for future research.

2 Related Works

OR has been revealed as a promising routing approach for UWSNs instead of using end to end approach. OR can be divided into two categories, known as location-based and location-free protocols. Location-based protocols like VBF [14], HH-VBF [15], GEDAR [16] and FBR [17] uses the full location information of the nodes while for location-free protocols like DBR [18], EEDBR [19], WDFAD-DBR [20] and UWFlooding uses the local information like depth, residual energy information without knowing the exact location of the nodes. For each routing protocol mention above, the details implementation of the routing protocols is explain in the following subsections.

2.1 Location-Based Protocols

VBF [14]: is an OR geographical location information routing protocol based on mobility and does not require the state information of the sensor nodes. VBF is proposed as a solution to two important problems in underwater environment, namely the continual movement of ordinary nodes by water current and energy efficiency. In VBF, each source node creates its own pipeline towards the sink and embeds its sink location, and its own location as a forwarder node in the packet and broadcasts this packet. As this protocol belongs to the receiver-based subcategory, all the sensor nodes are responsible in identifying their eligibility to forward the data. After receiving a packet, the sensor node located nearer than the radius of the vector can forward the packet by updating the forwarder node information; otherwise, it will simply discard the packet.

Through the node forwarding mechanism only few nodes are involved in the packet forwarding which focuses on the scalability and robustness of the VBF. As only few nodes are involved in forwarding the packet, the network traffic and energy consumption decrease significantly.

HH-VBF [15]**:** is an enhanced version of VBF proposed to improve the probability of communication void in sparse environments in order to increase successful data delivery. In addition, it needs a smaller radius than VBF in finding eligible forwarder nodes while causing an improvement in the packet delivery ratio. HH-VBF uses multiple virtual pipes from the source to the destination nodes. In HH-VBF the vector is computed on per-hop basis i.e. each forwarding node computes a vector from itself towards the sink. As this HH-VBF is also receiver- based, the eligible sensor node for forwarding is determined by the sensor node based on the radius range computed by the previous forwarder node.

GEDAR [16]**:** is uses an OR algorithm forwarding strategy in which sensor nodes forward the data packets to multiple sinks in their transmission range. If there are no neighbor inside their progression area, the depth of a sensor node is adjust to new depths to locate at least one forwarder candidate node. GEDAR uses a beacon messages to find a next hop forwarder candidate node which have a smallest distance than current node to the sink. By using the OR algorithm strategy, GEDAR achieves high packet delivery ratio, avoids void holes, and also provides the recovery mechanism (depth adjustment) to recover data from a void region.

FBR [17]**:** in this FBR algorithm, each node should know its own position and the location of sink. Before any communication occurs, sender and next forwarder would communicate to each other. In this algorithm, to reduce the energy consumption, it uses different power level for each different radius for routing procedure. The performance of FBR is very sensitive to the angle Ø toward the sink because if the angle is too large or too small, it would affect the performance in sparse or dense deployment.

2.2 Location-Free Protocols

DBR [18]**:** is the first location-free OR routing protocol for UWSNs. The purpose of this study is to design a location-free routing protocol algorithm to address dynamic network topology issues in UWSNs with appropriate energy use. For this purpose, each sensor node is equipped with a cheap pressure sensor to calculate their local depth. In order to address the forwarding area problem, each sender node would employ the depth information to identify its progression area headed for the sink. For this purpose, the source node would embed its local depth information into each packet during the forwarding and later broadcast the packet. Once the receiver node receives the packet, it will extract the sender node depth information from the data packet, and compare it with its own depth. If the receiver node is shallower than sender node, it will simply remove the packet. In contrast, the receiver node is a qualified for next hop node and should participate in the data packet forwarding toward the sink.

This process would be repeated until the data arrives at the sink. In order to avoid the collision, this protocol uses packet holding time based on depth information. A highest ranking node would forward the packet immediately while other would wait until their holding time expired or remove the packet once receive same packet.

EEDBR [19]**:** aim to balance the energy of nodes and prolong the lifetime of network. This protocol would select the next candidate node based on their depths and residual energy. So only the smaller depths node would be chosen as the next candidate node. This protocol comprises of two phase, acquisition phase and data forwarding. The acquisition phase would collect all the neighbor node depths and residual energy and data forwarding phase would choose only node with the smaller depth and high residual energy. The packet holding time would be computed based on node residual energy. However for this protocol, the acquisition phase should run several time to update the node residual energy, this could lead to high network overhead.

WDFAD-DBR [20]**:** is proposed to overcome the communication void problem in UWSNs. It is designed to improve the communication in sparse and dense condition by advancing sense of void holes and dividing the progression region based on network density and channel condition. This routing protocol employs the multi-sink approach and makes forwarding decision based on the differences of the weightage depth difference of two-hop nodes. WDFAD-DBR is receiver-based protocol and use holding time before forward the data. It also use history information on neighbours to improve the network performance and at the same time to minimize the load of network.

However, the mechanism proposes in this protocol is not suitable for data sensitive application and it will use a very complex algorithm to find the two-hop depth information.

UWFlooding [21]**:** is the example of flooding approach in UWSNs which introduce in Aqua-Sim NS2 based simulation tools. It performs without any consideration of the node information. It just forward the packet in multiple path and duplicate it from the source until the packet arrives at the sink. In theory, this example flooding approach would have many problems and could increase the network overhead when there is an increase in density of node.

3 Simulation of Location-Based and Location-Free OR Using Aqua-Sim

3.1 Simulation Setup

The underwater sensor nodes are placed at random location and some selected set of underwater sensor nodes would start CBR traffic and send packets to the one sink node located at the surface of the water. Most of the parameters of the simulations are used as default of Aqua-Sim. There are 4 UWSNs routing protocols used in this simulation, UWFlooding (just example underwater flooding approach in Aqua-Sim), VBF [14], HH-VBF [15] and DBR [18]. While for some of the important parameters that are used in this simulation are given in the following Table 1. The simulations will be repeated for several times and only the average values of results are taken into account. The randomness of the topology and network conditions, traffic start and stop time are controlled by seeding the random number generator with respect to the run number.

Table 1. Some important parameters of the simulation

Parameters	Value
Area	500 m * 500 m * 500 m
Channel	Acoustic waves
No underwater nodes	20, 40, 60, 80, 100
Number of sinks	One at the surface
Movement model	Static model
Routing protocols	UWFlooding, DBR, VBF, HH-VBF
Propagation model	UnderwaterPropagation
Physical model	UnderwaterPhy
Mac protocol	Broadcast Mac
Traffic type	CBR
CBR flows	3
Initial energy nodes	1000 J
Transmission range	250 m
Energy consumption	P_{tx} = 2 W, P_{rx} = 0.1 W, P_{idle} = 10 mW
Geometric spreading	2 (Spherical)
Acoustic signal speed	1500 m/s
Avg. data packet size	<200 bytes
Data transfer rate	10 kbps
Simulation time	500 s
Runs	5

3.2 Metrics Used for Evaluating the Performance

In order to evaluate the performance of UWSN network routing protocols, the following metrics are considered:

Packet Delivery Ratio - PDR (%): Packet delivery ratio is the ratio of the number of packets successfully received by the sink node at the surface of water to the total number of packets sent into the network by the UWSN sensor nodes.

Cumulative Delay (milliseconds): The average time interval between the generation of a packet in a source UWSN node and the successfully delivery of the packet at the UWSN sink node. It counts all possible delays that can occur in the source and all intermediate nodes, including queuing time, packet transmission and propagation, and retransmissions at the MAC layer. In this work, we are finding the average of the cumulative end to end delay of all the data packets. It is measured in milli-seconds.

Number of Dropped Data Packets at Application Layer (Nos): It is actually not dropped packets, these are packets that cannot reach the sink because of the overhead, delay and other reasons. In other words, it is equal to the number of sent data packets minus the number of received data packets at sink.

Energy Consumption (Jules): The average of the total consumed energy of all the nodes of the network. It is measured in Jules.

4 Results and Discussion

We simulate different density underwater networks by deploying different number of underwater nodes on the same fixed, three dimensional area 500 m × 500 m × 500 m. We used three CBR traffic flows and they will try to send a maximum of 300 packets in all different dense network conditions. To simulate different network density scenario, the number of underwater sensor nodes is changed from 20 nodes to 120 in steps of 20.

4.1 Analysis of the Results

The following Fig. 1 shows the performance in terms of PDR with respect to different number of underwater nodes. As shown in the graph, the performance of location free routing protocols is poor. The performance of the location-based routing protocols VBF and HH-VBF seems to be almost equal and better than two location free algorithms. Even though the performance of all the protocols is reduced with the increase of network density, a high performance is achieved at an appropriate network density. For example, in this small topographical area, if we place 60 nodes to cover that area, the two location-based routing protocols VBF and HH-VBF are giving better performance in terms of PDR.

The following Fig. 2 shows the performance in terms of cumulative delay with respect to different number of underwater nodes. The performance of the location-based routing protocols HH-VBF seems poorer than the normal VBF. While for

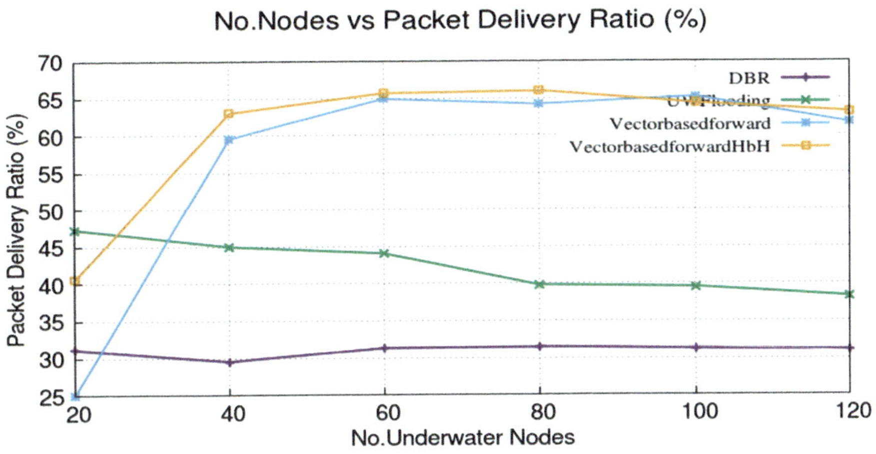

Fig. 1. Number of Nodes vs Packet Delivery Ratio (%)

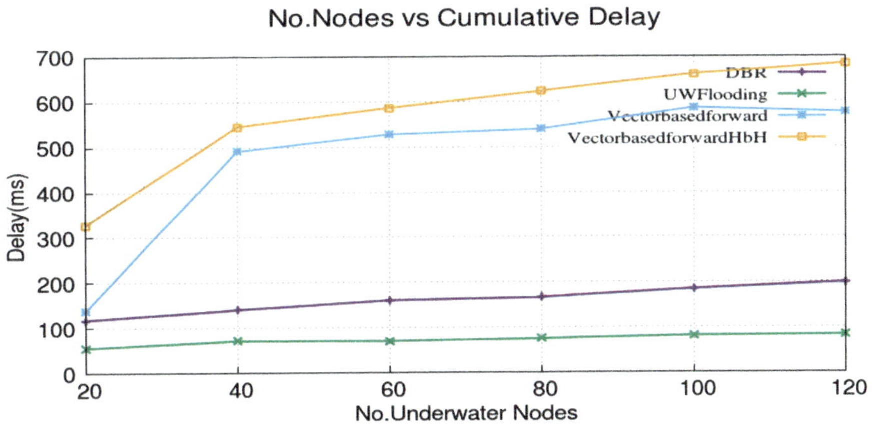

Fig. 2. Number of Nodes vs Cumulative Delay

location-free routing protocols both of them UWFlooding and DBR perform better than both of location-based. Due to the use of flooding approach, UWFlooding is the best performance is this cumulative delay simulation due to its vigorous nature of forwarding packets at each hop.

The following Fig. 3 shows the number of data packets dropped at application layer with respect to different number of underwater nodes. As shown in the graph, the performance of location-free routing protocols is depleted than the location-based routing protocols. Among of the location free routing protocols, the performance of UWFlooding is better compare to DBR. The performance of the two location-based routing protocols in terms of dropped data packets seems to be almost equal and better than location-free routing protocols even at high node densities.

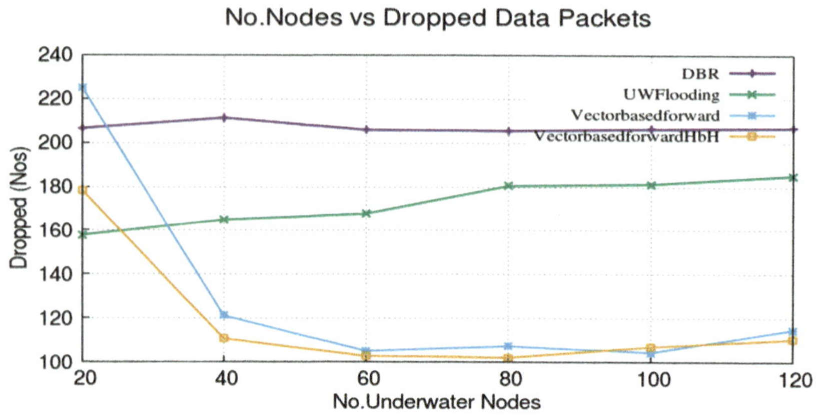

Fig. 3. Number of Nodes vs Dropped Data Packets

The following Fig. 4 shows the performance in terms of energy consumption with respect to different number of underwater nodes. The performance of the two location based routing protocols seems to be equivalent and better than the other compared location free routing algorithms. As shown in the graph, obviously, the energy consumption is increasing due to the increase in network density

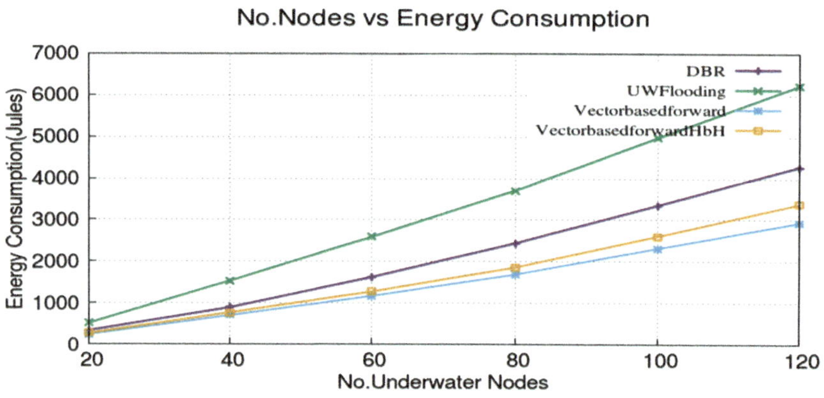

Fig. 4. Number of Nodes vs Energy Consumption

5 Conclusion

This evaluation and its results clearly shows that the performance of the location-free and location-based routing protocols in UWSNs. The result indicates that location-based routing protocols as expected perform better than the location-free routing protocols. But, in practical situation and real world, many problem will arise in resolving location of the underwater nodes and make them impractical to apply them on UWSNs scenario. As we already know that GPS signal is unavailable in underwater because high frequencies signal face the quick absorption problem in water environment [2].

For location-free routing protocols, even though the DBR shows less overheads, the overall performance of DBR is obviously poor than UWFlooding. Between the two location-free routing protocols, UWFlooding performs better than DBR in term of the cumulative delay, dropped packets and PDR but having issue on energy consumption due to its flooding approach which increase network overhead. It is giving much hope on doing a much better location-free routing protocol for UWSNs by reducing the overheads realized in UWFlooding. So in the future we will try to address these issues and will propose a new location-free routing protocol UWSNs.

References

1. Akyildiz, I.F., Pompili, D., Melodia, T.: Underwater acoustic sensor networks: research challenges. Ad Hoc Netw. **3**(3), 257–279 (2005)
2. Ayaz, M., Baig, I., Abdullah, A., Faye, I.: A survey on routing techniques in underwater wireless sensor networks. J. Netw. Comput. Appl. **34**(6), 1908–1927 (2011). https://doi.org/10.1016/j.jnca.2011.06.009
3. Climent, S., Sanchez, A., Capella, J.V., Meratnia, N., Serrano, J.J.: Underwater acoustic wireless sensor networks: advances and future trends in physical, MAC and routing layers. Sensors (Basel) **14**(1), 795–833 (2014)
4. Li, N., Martínez, J.-F., Meneses Chaus, J., Eckert, M.: A survey on underwater acoustic sensor network routing protocols. Sensors **16**(3), 414 (2016). http://www.mdpi.com/1424-8220/16/3/414
5. Ismail, N., Mohamad, M.M.: Review on energy efficient opportunistic routing protocol for underwater wireless sensor networks. KSII Trans. Internet Inf. Syst. **12**(7), 3064–3094 (2018). http://itiis.org/digital-library/manuscript/2058
6. Zenia, N.Z., Aseeri, M., Ahmed, M.R., Chowdhury, Z.I., Shamim Kaiser, M.: Energy-efficiency and reliability in MAC and routing protocols for underwater wireless sensor network: a survey. J. Netw. Comput. Appl. **71**, 72–85 (2016). http://linkinghub.elsevier.com/retrieve/pii/S1084804516301230
7. Awan, K.M., Shah, P.A., Iqbal, K., Gillani, S., Ahmad, W., Nam, Y.: Underwater wireless sensor networks: a review of recent issues and challenges. Wirel. Commun. Mob. Comput. **2019**, 1–20 (2019). https://www.hindawi.com/journals/wcmc/2019/6470359/
8. Ovaliadis, K., Savage, N., Kanakaris, V.: Energy efficiency in underwater sensor networks: a research review. Sci. Technol. **3**(1), 151–156 (2010). http://www.jestr.org/downloads/volume3/fulltext272010.pdf
9. Ahmed, M., Salleh, M., Channa, M.I., Rohani, M.F.L.: Energy efficient routing protocols for underwater wireless sensor networks: a review. TELKOMNIKA (Telecommun. Comput. Electron. Control) **15**(1), 212 (2017). http://journal.uad.ac.id/index.php/TELKOMNIKA/article/view/4706
10. Darehshoorzadeh, A., Boukerche, A.: Underwater sensor networks: a new challenge for opportunistic routing protocols. IEEE Commun. Mag. **53**(11), 98–107 (2015)
11. Ghoreyshi, S., Shahrabi, A., Boutaleb, T.: A novel cooperative opportunistic routing scheme for underwater sensor networks. Sensors **16**(3), 297 (2016). http://www.mdpi.com/1424-8220/16/3/297

12. Coutinho, R.W.L., Boukerche, A., Vieira, L.F.M., Loureiro, A.A.F.: Design guidelines for opportunistic routing in underwater networks. IEEE Commun. Mag. **54**(2), 40–48 (2016). http://ieeexplore.ieee.org/lpdocs/epic03/wrapper.htm?arnumber=7402259
13. Kheirabadi, M.T., Mohamad, M.M.: Greedy routing in underwater acoustic sensor networks: a survey. Int. J. Distrib. Sens. Netw. **2013**(9), 21 (2013)
14. Xie, P., Cui, J.H., Lao, L.: VBF: vector-based forwarding protocol for underwater sensor networks. Lecture Notes in Computer Science (including Subseries Lecture Notes in Artificial Intelligence and Lecture Notes in Bioinformatics). LNCS, vol. 3976, pp. 1216–1221 (2006)
15. Nicolaou, N., See, A., Xie, P., Cui, J.-H., Maggiorini, D.: Improving the robustness of location-based routing for underwater sensor networks. In: OCEANS 2007 – Europe, pp. 1–6. IEEE (2007). http://ieeexplore.ieee.org/document/4302470/
16. Coutinho, R.W.L., Boukerche, A., Vieira, L.F.M., Loureiro, A.A.F.: GEDAR: geographic and opportunistic routing protocol with Depth Adjustment for mobile underwater sensor networks. In: 2014 IEEE International Conference on Communications (ICC), pp. 251–256. IEEE (2014). http://ieeexplore.ieee.org/lpdocs/epic03/wrapper.htm?arnumber=6883327
17. Jornet, J.M., Stojanovic, M., Zorzi, M.: Focused beam routing protocol for underwater acoustic networks. In: Proceedings of Third ACM International Workshop on Wireless Network Testbeds, Experimental Evaluation & Characterization, WuWNeT 2008, p. 75 (2008). http://portal.acm.org/citation.cfm?doid=1410107.1410121
18. Yan, H., Shi, Z.J., Cui, J.H.: DBR: depth-based routing for underwater sensor networks. Lecture Notes in Computer Science (including Subseries Lecture Notes in Artificial Intelligence and Lecture Notes in Bioinformatics). LNCS, vol. 4982, pp. 72–86 (2008)
19. Wahid, A., Dongkyun, K.: Analyzing routing protocols for underwater wireless sensor networks. Int. J. Commun. Netw. Inf. Secur. **2**(3), 253–261 (2010)
20. Yu, H., Yao, N., Wang, T., Li, G., Gao, Z., Tan, G.: WDFAD-DBR: weighting depth and forwarding area division DBR routing protocol for UASNs. Ad Hoc Netw. **37**, 256–282 (2016)
21. Xie, P.X.P., Zhou, Z.Z.Z., Peng, Z.P.Z., Yan, H.Y.H., Hu, T.H.T., Cui, J.-H.C.J.-H., et al.: Aqua-Sim: an NS-2 based simulator for underwater sensor networks. In: Ocean 2009, MTS/IEEE Biloxi - Marine Technology for Our Future: Global and Local Challenges, pp. 1–7 (2009)

Development of WDM System in Optical Amplifiers by Manipulating Fiber Length and Bandwidth for Telecommunication System

Roby Ikhsan[1], Romi F. Syahputra[1], Suhardi[1], Saktioto[1(✉)], Nor Ain Husein[2], and Okfalisa[3]

[1] Department of Physics, Universitas Riau, Jl. HR Soebrantas KM. 12, 5, Pekanbaru 28293, Riau, Indonesia
ikhsanroby7@gmail.com, romifadlisyahputra@yahoo.com, salimihamzah@gmail.com, saktioto@yahoo.com
[2] Jabatan Fizik, Fakulti Sains, Universiti Teknologi Malaysia, 81310 Skudai, Johor, Malaysia
norainh@utm.my
[3] UIN Sultan Syarif Kasim, Jl. HR Soebrantas KM. 15, Pekanbaru 28293, Riau, Indonesia
okfalisa@gmail.com

Abstract. The use of fiber optic for telecommunication system has widely developed because of limitations on communication ranges and methods for information demands. In the application of optical fiber communications, signal strength is reduced due to attenuation, absorption and dispersion effects of source, structure, and geometry of fiber, even for single mode fiber, these problems still exist. In order to investigate these phenomena, this paper designs and operates a simple optical design consisting of wavelength division multiplexing (WDM) which is able to multiplex various wavelength sources to one fiber optic by using various source wavelengths. This mechanism allows a bidirectional communication with several or more fibers without any interference. The output will describe the bit error rate (BER) and Q-factor and using two samples of Semiconductor Optical Amplifier (SOA) and Fiber Raman Amplifier (FRA). The simulation evaluates these amplifiers by manipulating the frequency of bandwidth until 50 GHz, a low BER corresponding to large Q-factor is reached. The BER value for SOA device is 7.59×10^{-16} and 1.54×10^{-27} for SOA device respectively for a bandwidth of 50 GHz. These data depict that SOA is able to launch the wavelength to high performance than that of FRA device. Both optical amplifiers describe better performances in term of the value of BER and Q-factor having much smaller than 10^{-12} and larger than that of 6 respectively for a distance of 120 km for SOA and 100 km for FRA.

Keywords: Optical fiber · Optical circuit · WDM · FRA · SOA

F. Saeed et al. (Eds.): IRICT 2019, AISC 1073, pp. 625–633, 2020.
https://doi.org/10.1007/978-3-030-33582-3_58

1 Introduction

A high speed telecommunication system is a key success in transferring information around the world, whether the space is separated by miles away. One of the information carriers is an electromagnetic wave which has various frequencies. Electromagnetic wave applied in optical circuit and communication systems uses high carrier frequencies source ($\sim$100 Tbit/s) in the visible or near-infrared wavelength region [1]. Those sources of carrier frequency bandwidth are particularly much smaller by five orders of the value ($\sim$1 GHz) [2]. Optical fiber circuit and communication system are able to deal with optical fiber as media for carrying, transferring and filtering data and information from one place to another. Optical fiber system generally consists of transmitting, receiving and filtering devices which are able to convert electrical to the light signal. This fiber cable also carries the visible and invisible light and a receiver device which admits the spectral signal back to change into an electrical signal.

There has been a massive need to advance the communication system with the swift growth in telecommunication systems such as fast-signal internet device and error linkage system which meet the first need and commit for the service in term of a provider [3, 4]. These service and development have produced an increased high concern in applying a wavelength division multiplexing (WDM) for advanced optical communication and network system.

WDM component has the function and capability of transmitting into a single fiber optics and multiple numbers of waves with various frequencies source. Kinds of data and information channels in term of wavelength and intensity can be added to a different location by applying an add/drop multiplexer system. That adding system can improve and fit the flexibility of the WDM system [4–6]. Sending multiple signals is needed to gain a high bandwidth in optical communication. WDM systems come to fulfill this necessity. However, this communication still faces a lot of obstacle such as wavelength and refractive indices resulting a dispersion problem [5–8]. Dispersion results in the widening signal and influences the source wavelength so that the intensity losses will be obtained from the original data at the receiver end.

There are a lot of techniques which have developed to play as dispersion compensation such as using FBG [7–11], chirp grating [12], artificial neural network [13], optoelectronic method [14], dispersion compensate fiber [15], and orthogonal frequency division multiplexing [16]. Another technique is using an optical amplifier. Optical amplifier are widely used and become component in telecommunication networks. These are hoped to have a better service and maintenance towards all system of optical circuit by applying optical amplifiers corresponding to another optical component, and of course, this requires efficient and effective optical amplifiers because of large power losses as far as the optical circuit operates. This circuit and link suffer from attenuation and dispersion. The attenuation from geometry effect produces intensity signal losses and restricts the transmission ranges. Nevertheless, the beam source input will broaden as a result of the dispersion. It will lead to interference, limitation in the fiber signal and increase bit error rate (BER) [17]. Here, an optical amplifier is required to improve and compensate for the intensity losses. Nowadays, kinds of optical amplifier technologies that have developed with distinguished properties. SOA has been used to compensate dispersion [18, 19]. In this paper, two amplifiers are chosen to investigate the performance of amplification ranges i.e., SOA and FRA. They will be compared at various bandwidth and length of the fiber. The other one, Erbium-Doped is not considered in this result, where it is underway to model for different purposes.

2 Design and Operation

The circuit is dealt with basic optical communication having components which consist of transmission link, filter, transmitter, and a receiver device [20]. A design is set by preparing the power input and several optical components to deliver, change and increase the signal where it is found reduced. The system launches the power bringing data and information and transmits it using electromagnetic wave from the source to the receiver through optical fiber as shown in Fig. 1. The optical circuit contains four transmitters which use equal channel spacing of 50 GHz. This circuit is operated by simulating the power input and output. The design and operation for simulation are described as follows.

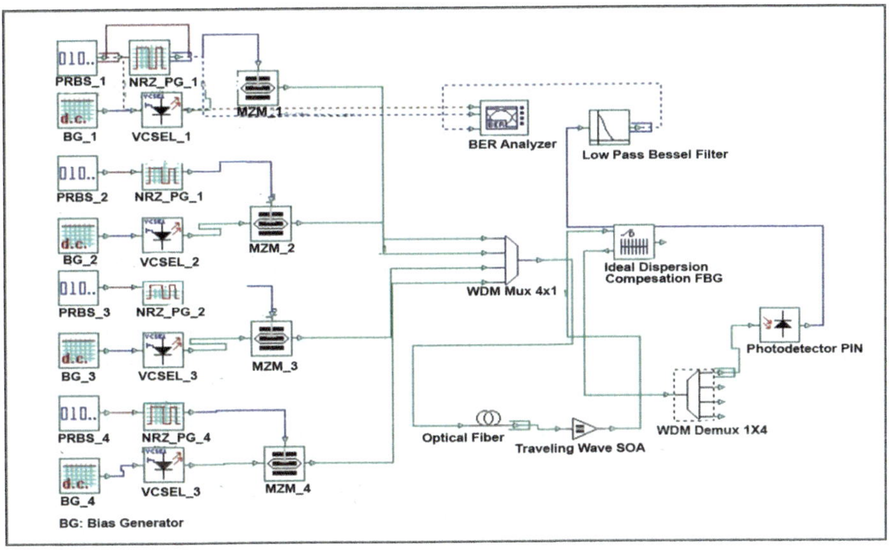

Fig. 1. WDM system using SOA

The electromagnetic input containing electrical signal data is shown by 0's and 1's generated by a non-return-zero (NRZ) pseudo-random binary sequence (PRBS) [21]. This signal source will be then modulated with a visible or invisible laser source generated by Vertical Cavity Surface Emitting Laser (VCSEL) through Mach-Zehnder Modulator (MZM). VCSEL's supply input signal starting from 193.1 THz to 193.25 THz of frequency and temperature of 20 °C. MZM will mix the input of electrical signal and produce optical signal as the output which delivered into the multiplexer.

Single mode fiber is applied for fewer intensity losses. It can also result greater data transfer with low dispersion and enable to operate in long optical fiber system. Therefore it is compatible to deal with a transmission link. This transmission link circuit using SMF, the operation circuit system results in a dispersion of about 16.75 ps/km/nm and attenuation of 0.3 dB/km. The 50 km of fiber length draw on various bandwidth simulations.

This simulation uses FBG functioning the wavelength filter and transmission as dispersion compensation and amplifier as fiber linear loss compensation (attenuation). The optical circuit system is set for 50 km fiber optics length which is combined with SOA by injecting electrical current of 0.042 A, and followed by the attenuation of the fiber is0.3 dB/km. The four-channel frequencies are 193.1 THz, 193.15 THz, 193.2 THz and 193.25 THz as shown in Fig. 1. Raman Amplification specifies a length of 10 km and pumps laser frequency of 980 nm as depicted in Fig. 2.

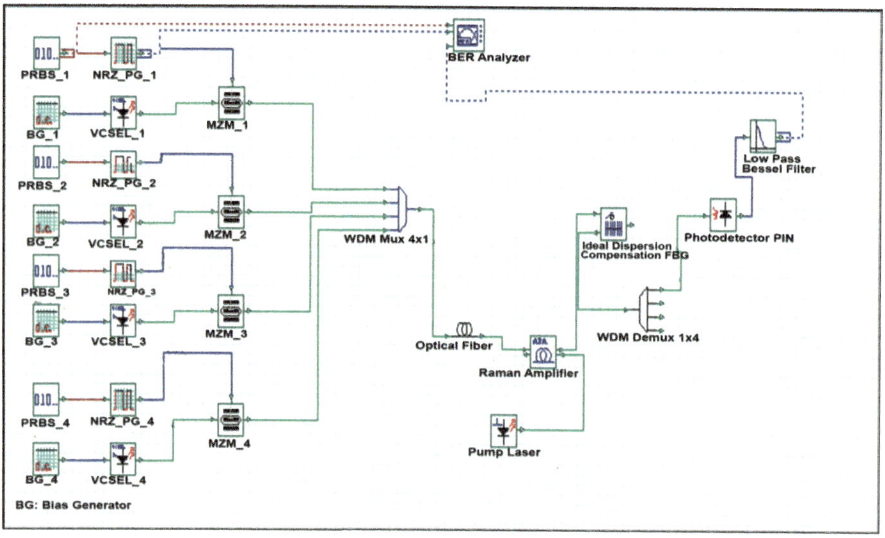

Fig. 2. WDM system using FRA.

For block receiver component, PIN photo-detector is set as a sensor of the optical signal and change the wavelength source to electrical signal and deliver it to low pass filter by using Bessel filter function which is filtering high frequency carrier signal and transmitting the low frequency signal The BER analyzer is then applied to summarize BER and quality factor (Q-factor) of the optical design and operation circuit system.

3 Result and Discussion

The optical circuit proposed in Figs. 1 and 2 are design and operation of the amplifier system having a frequency of 10 GHz bandwidth and they produce the output results for the Q-factor is 5.44 for SOA and 7.07 for FRA respectively. These outputs are also detected by eye diagrams as shown in Fig. 3.

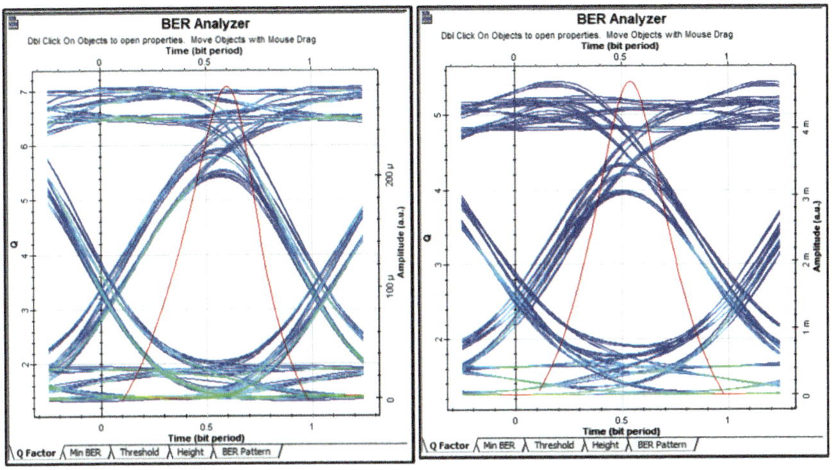

Fig. 3. Eye diagram of FRA (left) and SOA (right) for 10 GHz bandwidth

The increment of the bandwidth until 20 GHz will also produce the Q-factor increased, and BER is decreased and of course, it is required for optical transmission. However, when the bandwidth is higher than 20 GHz, the Q-factor and BER tend to be stable. Furthermore, BER, Q-factor value and the eye diagrams for various values is shown in Fig. 4. Figure 4 describes the eye diagrams look slightly smooth although the 10 Ghz is better than 20 GHz. The small slope of the eye diagram indicates that the jitter is not significant effect. The value of noise which can be tolerated is rather high, it means the small value of noise occurs hence the small distortion occurs [22]. The formed eye diagram is not surprising where the losses occur, much power is wasted and the value of noise allowed is still small. This shows that the gradient resulted produces the sensitivity of the jitter is large.

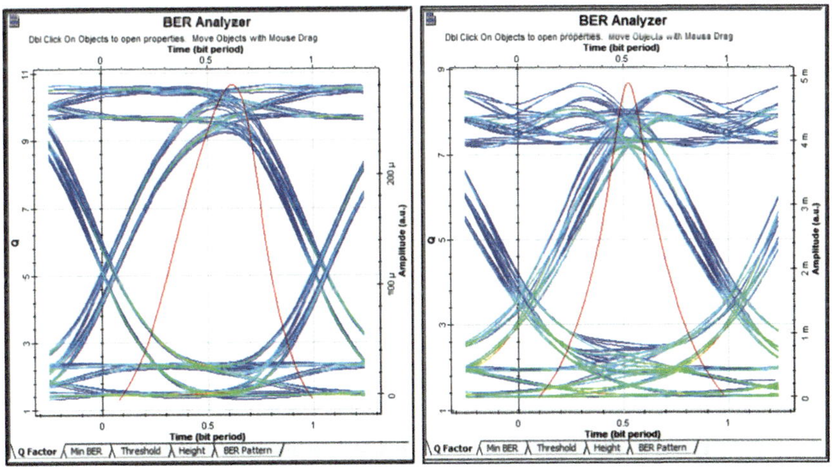

Fig. 4. Eye diagram of FRA (left) and SOA (right) for 20 GHz bandwidth

These eye diagrams can produce interferences by effects of multipath where the signal wave from the input source, mainly as power is delivered to a sensor by two or more paths, with certain condition, the components of the electromagnetic wave can interfere. The wave signal propagates where the modulation is formed by the various signal patterns such as sine and cosine inputs at a precise function of position and time. Both diagrams are geometrically a better performance even though it seems that there is a slight distortion because of dispersion condition as depicted at the certain lines shown by 'left arm' and 'right arm' in the picture. Of course, there is also noise occurs such as top line (noise "1") compared to down line (noise "0").

As shown in the eye diagram, the opening of eye as it is set at the middle region, the performance of FRA is large compared to SOA device at set at the center. The number of disturbance can be allowed by the wave described as a vertical length of the diagram is high and also a horizontal length of the diagram is large. The length of horizon describing the time function to the waveform informs that the sensitivity to timing error is small, meaning the pattern is good. The time jitter is also tiny as described by the bit period of value of 1 and 0. In conclusion, in accordance with the graph and geometry analysis, FRA is better result than that of SOA.

From Table 1, it can be observed both SOA and FRA when the increment of bandwidth, the value of Q-factor goes up but otherwise bit error rate goes down. If there is no SOA component, then Q-factor is zero and BER is 1. More details information of Q-factor is depicted in Fig. 5.

Table 1. Q-factor and BER by varying the bandwidth applying SOA and FRA

No	Bandwidth (GHz)	Q-Factor		BER	
		FRA	*SOA*	*FRA*	*SOA*
1	10	7.07	5.44	7.26×10^{-13}	2.45×10^{-08}
2	20	10.65	8.66	8.60×10^{-27}	2.25×10^{-18}
3	30	11.16	8.48	3.16×10^{-29}	1.12×10^{-17}
4	40	11.05	8.63	1.04×10^{-28}	3.02×10^{-18}
5	50	10.81	7.98	1.54×10^{-27}	7.59×10^{-16}

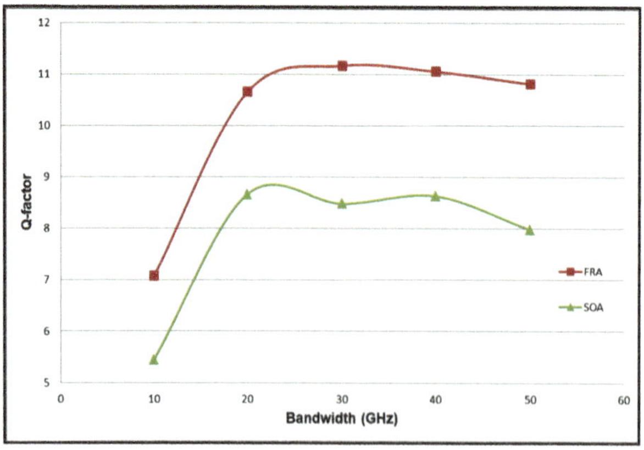

Fig. 5. Q-factor for FRA and SOA

As seen in Table 2, it is found that when the increment of fiber length until 40 km, the value of Q-factor will be high but opposite to BER. However, when fiber length is above 40 km, it will be *vice versa* for amplifier quality. The Q-factor is also shown in Fig. 6. The propagation of signals due to the higher frequency used to result in the quality of wave reduced corresponding to energy decreased. This also occurs when the length of fiber is long [23], then the power losses are affected by the geometry and structure of fibers which imply to the SOA and FRA. As depicted in Fig. 6, FRA quality factor goes down if the fiber length increases. It is not surprised since the FRA is much affected by the frequency then the power is much lower [24]. However, this does not explain that FRA performance is not good enough since the wave propagation is not readily reached due to attenuation.

Table 2. Optical amplifier quality for SOA and FRA

No	Fiber length (km)	Q-Factor		BER	
		FRA	SOA	FRA	SOA
1	20	9.29	7.83	7.23×10^{-21}	2.51×10^{-15}
2	40	13.43	9.42	2.08×10^{-41}	2.05×10^{-21}
3	60	9.7	7.79	1.52×10^{-22}	3.28×10^{-15}
4	80	8.15	7.97	1.78×10^{-16}	7.76×10^{-16}
5	100	6.05	8.24	7.30×10^{-10}	8.87×10^{-17}
6	120	2.96	9.23	1.52×10^{-03}	1.03×10^{-20}
7	140	0	7.86	1	1.96×10^{-15}

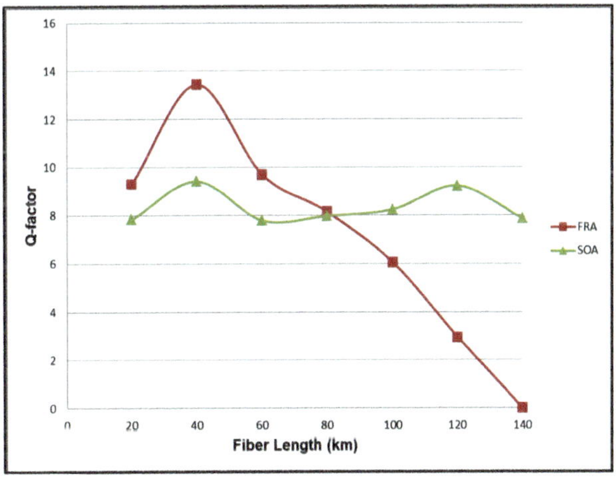

Fig. 6. Q-factor for various fiber length

4 Conclusion

This article can be concluded that the WDM system for SOA and FRA has been successfully designed, operated and compared. The optical circuit is set for 50 km length injecting electrical current of 0.042A for SOA and is found that the intensity loss of the optical fiber is about 0.3 dB/km. The four-channel frequencies are set on the above THz regions which relate to channel spacing of 50 GHz. For 20 GHz bandwidth, it is obtained that the BER is low and Q-factor is high. Beyond the signal frequency of 20 GHz, Q-factor and BER value have nearly linear profile. Also, based on geometry analysis of eye diagram, FRA device has a better performance compared to SOA device having the highest Q-factor and lowest BER value. Nonetheless, FRA device is actually not as stable as the SOA device.

References

1. Agrell, E., Karlsson, M., Chraplyvy, A.R., Richardson, D.J., Krummrich, P.M., Winzer, P., et al.: Roadmap of optical communications. J. Opt. **18**(6), 063002 (2016)
2. Agrawal, G.P.: Fiber-Optic Communication Systems, 4th edn. Wiley, New York (2010)
3. Panda, T.K., Mishra, P., Patra K.C., Barapanda, N.K.: Investigation and performance analysis of WDM System implementing FBG at different grating length and datarate for long haul optical communication. In: 2017 IEEE International Conference on Power, Control, Signals and Instrumentation Engineering (ICPCSI), pp. 483–488. IEEE New York (2017)
4. Collings, B.C., Mitchell, M.L., Boivin, L., Knox, W.H.: A 1021 channel WDM system. IEEE Photonics Technol. Lett. **12**(7), 906–908 (2000)
5. Keiser, E.G.: A review of WDM technology and applications. Optical Fiber Technology **5** (1), 3–39 (1999)
6. McCoy, A.D., Thomsen, B.C., Ibsen, M., Richardson, D.J.: Filtering effects in a spectrum-sliced WDM system using SOA-based noise reduction. IEEE Photonics Technol. Lett. **16**(2), 680–682 (2004)
7. Othman, M.A., Ismail, M.M., Sulaiman, H.A., Misran, M.H., Said, M.A.M., Rahim, Y.A., ChePee, A.N., Motsidi, M.R.: An analysis of 10 Gbits/s optical transmission system using fiber Bragg grating (FBG). IOSR J. Eng. (IOSRJEN) **2**(7), 55–61 (2012)
8. Prashad, B., Mallick, B., Parida, A.K.: Fiber Bragg grating as a dispersion compensator in an optical transmission system using optisystem software. Int. Res. J. Eng. Technol. (IRJET) **2** (6), 9–14 (2014)
9. Kumar, K., Jaiswal, A.K., Kumar, M., Agrawal, N.: Performance analysis of dispersion compensation using fiber Bragg grating (FBG) in optical communication. Int. J. Curr. Eng. Technol. **4**(3), 1527–1531 (2014)
10. Litchinitser, N.M., Patterson, D.B.: Analysis of fiber Bragg gratings for dispersion compensation in reflective and transmissive geometries. J. Lightwave Technol. **15**(8), 1323–1328 (1997)
11. Sharma, A., Singh, S., Sharma, B.: Investigations on dispersion compensation using fiber Braggs grating. Int. J. Comput. Appl. **73**(2), 34–43 (2013)
12. Mohammadi, S.O., Mozzaffari, S., Shahidi, M.M.: Simulation of a transmission system to compensate dispersion in an optical fiber by chirp gratings. Int. J. Phys. Sci. **6**(32), 7354–7360 (2011)

13. Warm, S., Bunge, C.-A., Wuth, T., Petermann, K.: Electronic dispersion precompensation with a 10 Gb/s directly modulated laser. IEEE Photonics Technol. Lett. **21**(15), 1090–1092 (2009)
14. Foo, B., Corcoran, B.: Lowery, a: optoelectronic method for inline compensation of XPM in long-haul optical links. Opt. Express **23**(2), 859–872 (2015)
15. Nielsen, L.G., Wandel, M., Kristensen, P., Jorgensen, C., Jorgensen, L.V., Edvold, B., Palsdottir, B., Jakobsen, D.: Dispersion-compensating fibers. J. Lightwave Technol. **23**(11), 3566–3579 (2005)
16. Lowery, A.J., Armstrong, J.: Orthogonal-frequency-division multiplexing for dispersion compensation of long-haul optical systems. Opt. Express **14**(6), 2079–2084 (2006)
17. Golani, O., Elson, D., Lavery, D., Galdino, L., Killey, R., Bayvel, P., Shtaif, M.: Experimental characterization of nonlinear interference noise as a process of intersymbol interference. Opt. Lett. **43**(5), 1123–1126 (2018)
18. Hossain, M.S., Howlader, S., Basak, R.: Investigating the Q-factor and BER of a WDM system in optical fiber communication network by using SOA. Int. J. Innov. Sci. Res. **13**(1), 315–322 (2015)
19. Vedala, G., Hameed, M.A., RongqingHui, R.: Digital compensation of SSBI in direct detection multicarrier system with SOA nonlinearities. IEEE Photonics Technol. Lett. **29**(4), 369–372 (2017)
20. Kikuchi, K.: Fundamentals of coherent optical fiber communications. J. Lightwave Technol. **34**(1), 157–179 (2016)
21. Aldouri, M.Y., Aljunid, S.A., Ahmad, R.B., Fadhil, H.A.: Bit error rate (BER) performance of return-to-zero and non-return-to-zero data signals optical code division multiple access (OCDMA) system based on and detection scheme in fiber-to-the-home (FTTH) networks. Opt. Appl. **41**(1), 173–181 (2011)
22. Kim, J.-H., Lim, J.-H., Kim, B., Sim, J.-Y., Park, H.-J.: An adaptive equalizer for high-speed receiver using a CDR-assisted all-digital jitter measurement. J. Semicond. Technol. Sci. **15** (2), 155–167 (2015)
23. Marhic, M.E., Andrekson, P.A., Petropoulos, P., Radic, S., Peucheret, C., Jazayerifar, M.: Fiber optical parametric amplifiers in optical communication systems. Laser Photonics Rev. **9**(1), 50–74 (2015)
24. Clivati, C., Bolognini, G., Calonico, D., Faralli, S., Mura, A., Levi, F.: In-field Raman amplification on coherent optical fiber links for frequency metrology. Opt. Express **23**(8), 10604–10615 (2015)

5G Channel Propagation at 28 GHz in Indoor Environment

Ahmed M. Al-Samman[1(✉)], Tharek Abdul. Rahman[1], and Tawfik Al-Hadhrami[2(✉)]

[1] Wireless Communication Center, Universiti Teknologi Malaysia, Skudai, Malaysia
ahmedsecure99@gmail.com
[2] School of Science and Technology, Nottingham Trent University, Nottingham, UK
tawfik.al-hadhrami@ntu.ac.uk

Abstract. The propagation characteristics for the 5G wideband channel based on the path loss model and time dispersion parameters have been presented in this paper. The power degradation has been investigated based on path loss, while the RMS delay spread is used to investigate the time dispersion of 5G channel at 28 GHz frequency for Indoor environment. Extensive measurements have been conducted using wideband channel sounder within 1 GHz bandwidth. Omnidirectional transmitter and highly-directional receiver horn antennas are used for co- and cross-polarized antenna configurations. Results shown that the path loss exponent is 0.9 for Vertical-Vertical (V-V) and 1.8 Vertical-Horizontal (V-H) Polarizations. The mean value for the attenuation factor due to the cross-polarization antenna configuration is 11.9 dB, which represents the discrimination factor (XPD). The time dispersion results found that the Root Mean Square (RMS) delay spread values vary between 0.3 ns to 11 ns and between 0.3 ns to 6.4 ns for Vertical-Vertical (V-V) and Vertical-Horizontal (V-H) Polarizations, correspondingly.

Keywords: 28 GHz · Path loss · RMS delay spread · 5G mobile system · Polarizations

1 Introduction

Nowadays, the demand for wireless technology applications is increasing rapidly which requires a huge capacity. An advanced wireless communication technology must be adopted for the next generation of 5G networks to adapt to the demand for future applications. A novel band is required for the 5G wireless networks due to the overcrowding of the frequency below 6 GHz band [1] and the narrowness of its bandwidth. The utilization of millimeter wave (mm-wave) band considers to achieve lower latency and enable large capacity comparing with conventional communication below 6 GHz [2]. To support high data rates and dynamic spatial frequency reuse, the mm-wave systems are expected to become the core part of 5G mobile networks to support multi-gigabit communication applications such as high definition television (HDTV) and

F. Saeed et al. (Eds.): IRICT 2019, AISC 1073, pp. 634–642, 2020.
https://doi.org/10.1007/978-3-030-33582-3_59

ultra-high definition video (UHDV) [3]. Recently, most research has gained attention on mm-wave bands such as 28 GHz, 38 GHz, 60 GHz and 70 GHz [4–6].

In literature, many measurements have been conducted at different mm-wave bands in indoor and outdoor environments [3, 7–12]. There are different eventualities and environments particularly in tropical region i.e. Malaya that channel modeling is required although most of the recent studies in this field.

In this work, a wideband channel propagation characteristic is investigated in indoor environment in tropical region i.e. Malaysia. The measurements have been conducted at 28 GHz mm-wave for line of sight (LOS) scenario within 1 GHz bandwidth. The close-in reference (CI) path loss model is used to model the signal degradation on 5G wireless network for co- and cross-polarized antenna configurations. The close- in reference distance with discrimination factor (CIX) path loss model is simplified by using proposed cross-polarization model (XPL) to estimate the XPD factor, where XPD factor represents the attenuation from cross-polarization antenna configuration. The time dispersion factors are investigated based on the root mean square (RMS) delay spread.

The rest of paper is organized as follows. Section 2 describes the measurement campaign; Sect. 3 presents path loss model analysis. The time dispersion parameters are given in Sect. 4. The paper is concluded in Sect. 5.

2 Measurement Campaign

The measurements were conducted at 28 GHz using 5G channel sounder. To produce the wideband sounding signal, the Arbitrary Waveform Generator (AWG) has been used as transmitter (Tx). High speed digitizer with 1 GHz bandwidth interleaving mode was used at the receiver (Rx) for sounding signal acquisition with 1 ns multipath resolution. The details of the measurement equipment can be found in [13, 14].

The measurements were conducted in the corridor at the level fifteen of the School of Engineering (Research Floor) in UTM KL campus. The span of the corridor is 2.4 m × 40 m and the celling height is 2.7 m. It has plywood and glass doorways, and the partitions are produced from concrete, glass and gypsum board. The floor ground is roofed with glazed ceramic tiles, and the ceiling is created from fiberglass materials. The measurement environment is shown in Fig. 1. The Tx antenna was fixed during the measurements while the Rx antenna was moved along a corridor. To emulate indoor hotspot in the wall, the Tx antenna was placed 1.7 m above the floor. The Rx antenna was placed 1.5 m above the floor as the typical handset level heights. The received signal was recorded at each 1 m till Tx-Rx separation distance 40 m. The measurements were conducted by using LOS scenario using omnidirectional antenna vertical polarized and wide beam width horn antenna vertically and horizontally polarized for co- polarization (V-V) and cross- polarization (V-H) configurations. Table 1 presents the measurement parameters. The details of the measurement equipment can be found in [13, 14].

The measurements were conducted in the corridor at the level fifteen of the School of Engineering (Research Floor) in UTM KL campus. The span of the corridor is 2.4 m × 40 m and the celling height is 2.7 m. It has plywood and glass doorways, and

Fig. 1. Measurement environment.

Table 1. Measurement parameters

Carrier frequency	28 GHz
AWCS signal	10th order PRBS (length = 1024)
AWCS chip rate	1000 Mcpa
AWCS chip width	1 ns
Digitizer sampling rate	3.2 Gigasamples/second
RF BW	1 GHz
Rx LO power	10 dBm
Transmitted power	0 dBm
Gain of Tx antenna	3 dBi
Rx antenna Gain (Vertical Polarization, Horizontal Polarization)	11.6 dBi, 13.1 dBi
Rx Azimuth/Elevation HPBW	37.6°, 44.8°
Height of Tx/Rx antenna	1.7, 15 m
Polarization of Tx	Vertical
Polarization of Rx	Vertical/Horizontal

the partitions are produced from concrete, glass and gypsum board. The floor ground is roofed with glazed ceramic tiles, and the ceiling is created from fiberglass materials. The measurement environment is shown in Fig. 1. The Tx antenna was fixed during the measurements while the Rx antenna was moved along a corridor. To emulate indoor hotspot in the wall, the Tx antenna was placed 1.7 m above the floor. The Rx antenna was placed 1.5 m above the floor as the typical handset level heights. The received signal was recorded at each 1 m till Tx-Rx separation distance 40 m. The measurements were conducted by using LOS scenario using omnidirectional antenna vertical polarized and wide beam width horn antenna vertically and horizontally polarized for co- polarization (V-V) and cross- polarization (V-H) configurations. Table 1 presents the measurement parameters.

3 Path Loss Analysis

The physically-based CI path loss model is used in this work and defined as [3, 15]:

$$P_{loss}^{CI}(f,d)[dB] = P_{loss}(f,d_0) + 10n\log_{10}(^{d}/_{d_0}) + X_\sigma \tag{1}$$

The $P_{loss}(f,d)$ segment is the path loss at different frequencies with various Tx-Rx separation distance. The $P_{loss}(f,d_0)$ segment is the path loss in dB at close-in (CI) distance, d_0, of 1 m and X_σ is a zero-mean Gaussian distributed random variable with standard deviation σ dB (shadowing effect). The minimum mean square error (MMSE) is used to calculate the path-loss exponent (n) and the standard deviation.

Due to the cross-polarization, the attenuation could be calculated at each Tx-Rx separation distance for each frequency as:

$$XPL(f,d)[dB] = P_{loss(V-V)} - P_{loss(V-H)} \tag{2}$$

where $P_{loss(V-V)}$ and $P_{loss(V-H)}$ component denote to the co- and cross-polarization path loss, correspondingly. By performing the average for all XPL values over the distance at carrier frequency, the cross-polarization discrimination (*XPD*) factor was calculated using Eq. (2).

Joining the *XPD* and CI path loss model to produce new case of cross-polarization propagation which is known by “close-in reference distance with XPD (CIX) path loss model” [10]:

$$P_{loss}^{CIX}(f,d)[dB] = P_{loss}(f,d_0) + 10n\log_{10}(^{d}/_{d_0}) + XPD + X_\sigma^{CIX} \tag{3}$$

where n is the path loss exponent for V-V polarization that is determined from measurements by using (1), XPD is the V-H polarization attenuation factor in dB and X_σ^{CIX} is the random variable for CIX model with zero-mean Gaussian and standard deviation in dB.

Figure 2 shown, the path loss and the best fit for CI and CIX model along the corridor. In both instances, inverse relationship has been generated between the path-loss and (Tx–Rx) separation distance. It can be shown that the path loss exponent (n = 0.9) is low, indicating that the multipath components add up constructively from both side walls (as a waveguide effect) along the corridor. Using the CI model for V-H antenna configuration, the path loss exponent is 1.8, as shown in Fig. 3, indicating that the signal drop for cross-polarization is double of co-polarization. The CIX model is used for cross-polarization, where the XPD factor is 11.9 dB, as shown in Fig. 2. It can also be shown that the standard deviation for V-H is greater than the standard deviation for V-V by 1.2 dB, indicating that more signal fluctuation for large-scale at V-H. In both co- and cross-polarization, the standard deviation values are small, as shown in Fig. 2, implying large-scale signal fluctuations is not significant in the indoor corridor environment for wideband propagation at 28 GHz.

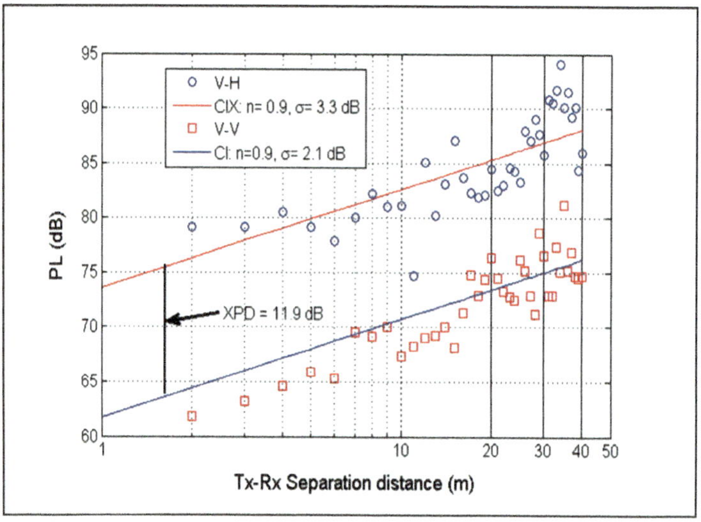

Fig. 2. CIX path loss models for the indoor channel at 28 GHz band for V-V and V-H antenna configurations.

The cross-polarization attenuation factor is investigated using XPL proposed model of (3), as shown in Fig. 4. Due to the Vertical-Horizontal antenna configuration in different location, the XPL attenuation factor has supplementary loss of signal as shown in Fig. 4. It varies between 6.3 dB to 18.5 dB along the corridor. The 11.9 dB value is the XPD discrimination factor which is the important value. It does not mean anything that the XPD factor value using the XPL model is totally matching with the XPD value that is calculated using the CIX path loss model. This implies, using the XPL model of (3), the XPD factor could be calculated without using CIX model to avoid the MMSE computation complexity.

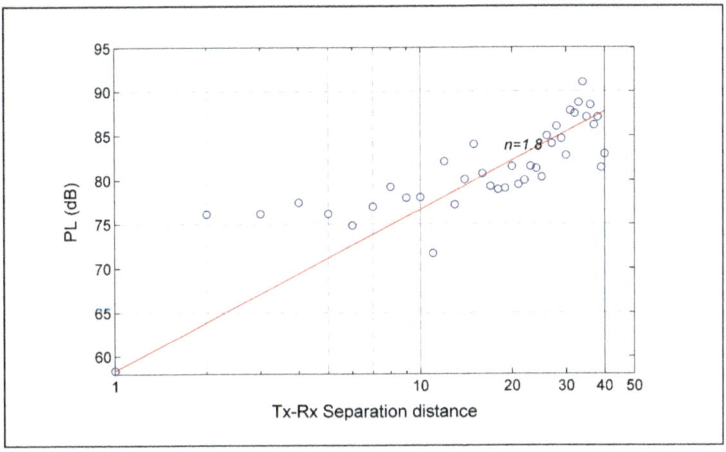

Fig. 3. Path loss models for the indoor channel at 28 GHz band for V-H antenna configuration.

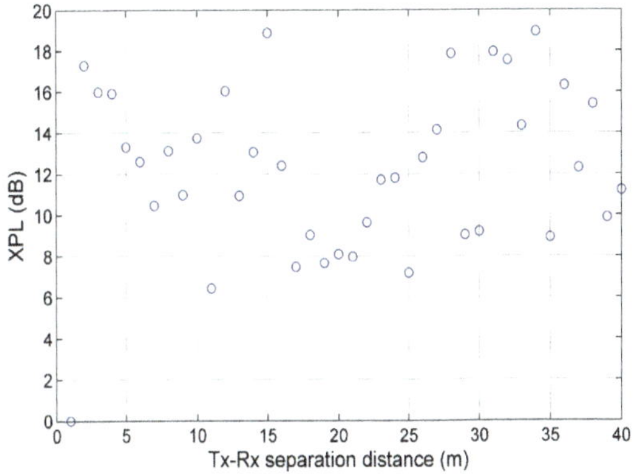

Fig. 4. Attenuation factor (XPL) of cross-polarization versus Tx-Rx separation distance for indoor channels at 28 GHz.

Finally, low correlation effects for the XPL attenuation has been shown in Fig. 4 with measured frequency and Tx-Rx separation distance.

4 Time Dispersion Analysis

The delay spread for the Root Mean Square (RMS) is used to analyze the time dispersion [16]. It has been reported by the second central moment of the power delay profile (PDP) as [17]:

$$\tau_{rms} = \sqrt{\overline{\tau^2} - \overline{(\tau_m)}^2} \tag{4}$$

where $\overline{\tau^2}$ is the second moment of the PDP and it is given as:

$$\overline{\tau^2} = \frac{\sum_k p(\tau_k) \cdot (\tau_k)^2}{\sum_k p(\tau_k)} \tag{5}$$

and $\overline{\tau_m}$ is the mean excess delay, also given as:

$$\overline{\tau_m} = \frac{\sum_k p(\tau_k) \cdot \tau_k}{\sum_k p(\tau_k)} \tag{6}$$

with p and τ are defined as the power and delay of the k*th* path, respectively. A threshold of 10 dB below the strongest path was chosen to avoid the effect of noise on the statistics of multipath arrival times.

Figure 5 shows the scatter plot for RMS delay spread versus Tx-Rx separation distance for V-V and V-H antenna configurations. It can be shown that the delay spread values vary between 0.3 ns to 11 ns with the mean value of 4.3 ns for V-V antenna polarization. For V-H antenna polarization, the values of delay spread vary between 0.3 ns to 6.4 ns with the mean value of 2.3 ns. It is noted that the correlation is low between RMS delay spared, Tx-Rx separation distance, and antenna polarization. This indicates that the RMS delay spread depends on the number and power of multipath components beside the excess delay (the time arrival of each multipath component

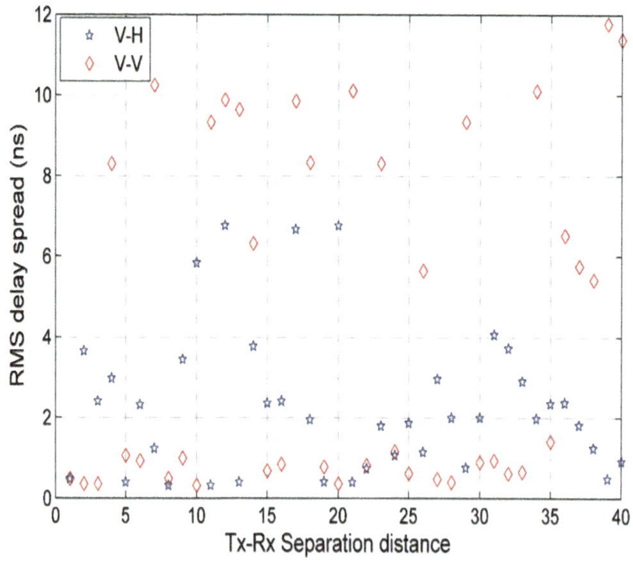

Fig. 5 RMS delay spread versus Frequency for V-V and V-H polarization measurement for indoor channels.

relative to the timely arrival of the first path). It can be concluded that the time dispersion characteristics of the channel such as delay spread and a number of multipath components are considerably similar over the frequency bands below and above 6 GHz, which is encouraging for extending the existing channel propagation models to the frequency range for the 5G system.

5 Conclusion

Path loss models and time dispersion for millimeter-wave at 28 GHz for indoor corridor environment are presented. The cross-polarization effect is studied using simple proposed model. The mean value of cross-polarization attenuation along corridor is 11.3 dB. The path loss exponent is 0.9 for Vertical-Vertical (V-V) and 1.8 Vertical-Horizontal (V-H) antenna configurations. In addition, the RMS delay spread mean values are 4.3 ns Vertical-Vertical (V-V) and 2.3 ns for Vertical-Horizontal (V-H) antenna configurations. The low values of path loss exponent and RMS delay spread in this environment indicate that the 28 GHz frequency with co- and cross-polarization antenna configurations is a suitable candidate for the environments that have the same structure of the study case.

Acknowledgment. The authors would like to thank the Research Management Centre (RMC) at Universiti Teknologi Malaysia for funding this work under Vot.04E21. Also, the authors would like to acknowledge the UTM research grant (Vot 4J218), Universiti Teknologi Malaysia.

References

1. Al-Samman, A.M., Rahman, T.A., Nunoo, S., Chude-Okonkwo, U.A., Ngah, R., Shaddad, R.Q., Zahedi, Y.: Experimental characterization and analysis for ultra wideband outdoor channel. Wirel. Pers. Commun. **83**(4), 3103–3118 (2015)
2. Kovalchukov, R., Moltchanov, D., Samuylov, A., Ometov, A., Andreev, S., Koucheryavy, Y., Samouylov, K.: Evaluating SIR in 3D mm wave deployments: direct modeling and feasible approximations. IEEE Trans. Wirel. Commun. **18**(2), 879–896 (2018)
3. Rappaport, T.S., MacCartney, G.R., Samimi, M.K., Sun, S.: Wideband millimeter-wave propagation measurements and channel models for future wireless communication system design. IEEE Trans. Commun. **63**(9), 3029–3056 (2015)
4. Al-Samman, A., Al-Hadhrami, T., Daho, A., Hindia, M.H.D., Azmi, M., Dimyati, K., Alazab, M.: Comparative study of indoor propagation model below and above 6 GHz for 5G wireless networks. Electronics **8**(1), 44 (2019)
5. Al-Samman, A., Rahman, T., Hindia, M., Daho, A., Hanafi, E.: Path loss model for outdoor parking environments at 28 GHz and 38 GHz for 5G wireless networks. Symmetry (Basel) **10**(12), 672 (2018)
6. Haneda, K., Jarvelainen, J., Karttunen, A., Kyro, M., Putkonen, J.: Indoor short-range radio propagation measurements at 60 and 70 GHz. In: The 8th European Conference on Antennas and Propagation (EuCAP 2014), pp. 634–638 (2014)
7. Murdock, J.N., Rappaport, T.S., Heath Jr., R.W., Daniels, R.C.: Millimeter Wave Wireless Communications, 1st edn. Prentice Hall, New Jersey (2015)

8. Deng, S., Samimi, M.K., Rappaport, T.S.: 28 GHz and 73 GHz millimeter-wave indoor propagation measurements and path loss models. In: 2015 IEEE International Conference on Communication Workshop (ICCW), pp. 1244–1250. IEEE, June 2015
9. MacCartney, G.R., Deng, S., Rappaport, T.S.: Indoor office plan environment and layout-based mm wave path loss models for 28 GHz and 73 GHz. In: 2016 IEEE 83rd Vehicular Technology Conference (VTC Spring), pp. 1–6, May 2016
10. Maccartney, G.R., Rappaport, T.S., Sun, S., Deng, S.: Indoor office wideband millimeter-wave propagation measurements and channel models at 28 and 73 GHz for ultra-dense 5G wireless networks. IEEE Access **3**, 2388–2424 (2015)
11. Al-Samman, A.M., Rahman, T.A., Azmi, M. H.: Indoor corridor wideband radio propagation measurements and channel models for 5G millimeter- wave wireless communications at 19 GHz, 28 GHz and 38 GHz Bands. Wirel. Commun. Mob. Comput. **2018** (2018)
12. Wang, C., Bian, J., Sun, J., Zhang, W.: A survey of 5G channel measurements and models. IEEE Commun. Surv. Tutorials **20**(4), 3142–3168 (2018)
13. Brochure, S.: Keysight technologies 5G channel sounding, reference solution. http://about.keysight.com/en/newsroom/pr/2015/30jul-em15109.shtml
14. Al-Samman, A.M., Rahman, T.A., Azmi, M.H., Hindia, M.N., Khan, I., Hanafi, E.: Statistical modelling and characterization of experimental mm-wave indoor channels for future 5G wireless communication networks. PLoS ONE **11**(9), e0163034 (2016)
15. Al-Samman, A.M., Rahman, T.A., Azmi, M.H., Hindia, M.N.: Large-scale path loss models and time dispersion in an outdoor line-of-sight environment for 5G wireless communication. AEU Int. J. Electron. Commun. Press **70**(11), 1515–1521 (2016)
16. Al-Samman, A.M., Rahman, T.A.: Experimental characterization of multipath channels for ultra-wideband systems in indoor environment based on time dispersion parameters. Wirel. Pers. Commun. **95**(2), 1713–1724 (2016)
17. Rappaport, T.S.: Wireless Communication Principles and Practice. Prentice-Hall Inc., New Jersey (2002)

Design Specification of Context Cognitive Trust Evaluation Model for V2V Communication in IoV

Abdul Rehman(✉) and Mohd Fadzil Bin Hassan

Centre for Research and Data Science (CeRDaS), Computer and Information Science Department, Universiti Teknologi PETRONAS, 32610 Seri Iskandar, Perak Darul Ridzuan, Malaysia
{abdul_18000023,mfadzil_hassan}@utp.edu.my

Abstract. VANET (Vehicle Ad Hoc Network) is now making its shift towards IoV (Internet of Vehicle). On road smart vehicles are expected to be the part of vehicle network in near future. Information integrity during V2V (Vehicle to Vehicle) communication is one of the key barriers in the implementation of vehicle networks. Vehicle networks face many drastic changes in the network topology during a short period of time. Absolute central network availability is difficult in vehicular networks that makes V2V communication significantly important. Trustworthiness plays a very important role to ensure integrity. Many models of trust evaluation have been proposed with a common trait of fixed nature. As possibility of scenario change is high for every other event so the fixed nature of previous models seems impractical. Our work aims to solve the problem by introducing context awareness between the available information and trust evaluation. It is expected that the proposed solution will overcome the limitations of current models and provide a hybrid adaptive model. Our model is based on context scenario adaptation rather than depending on a limited set of situations.

Keywords: VANET · Trust evaluation · IoV

1 Introduction

Over the past decade several new technologies are converging into the VANETs. The emergence of IoT (Internet of Things) has attracted researcher's attention to the unification of VANETs with the IoT [1–3] which formed a new dimension of research as IoV. New generations vehicles are being equipped with technology gadgets [4]. A new paradigm of technology and innovation of vehicular networks started with IoV and SIoV (Social Internet of Vehicles) [5, 6]. Compared to VANET, IoV has a very wide vision where all the cars are attached via the internet. IoV is emerging as an important part of the smart and intelligent cities with future vision of metropolitan cities [3, 6]. While there are still issues of implementation such as safety, policy, uniform infrastructure, social ethics and others, the world looks forward to partial or complete implementation of vehicle networks.

F. Saeed et al. (Eds.): IRICT 2019, AISC 1073, pp. 643–652, 2020.
https://doi.org/10.1007/978-3-030-33582-3_60

Distinctive characteristics of vehicular network splits it from other ad hoc networks [7]. The wireless nature of the network in VANETs/IoV makes it more vulnerable against security threats [8]. Malicious vehicles/nodes can manage and divert traffic according to a systemic plan that may lead to disaster. Any bad activity may lead to traffic congestion and mass accidents which is a potential threat to human life.

A big picture of the IoV shown in Fig. 1. Internet is the backbone of IoV, where all the data is synchronized with cloud. Cloud provide service to IoV using cellular communication. All the vehicles are equipped with OBU (On Board Unit) which refers to all sensors and devices related to communication in network. RSU (Road Side Units) is the main part of VANET infrastructure where all vehicles can connect and communicate. Cloud serves as a centralized IoV communication hub.

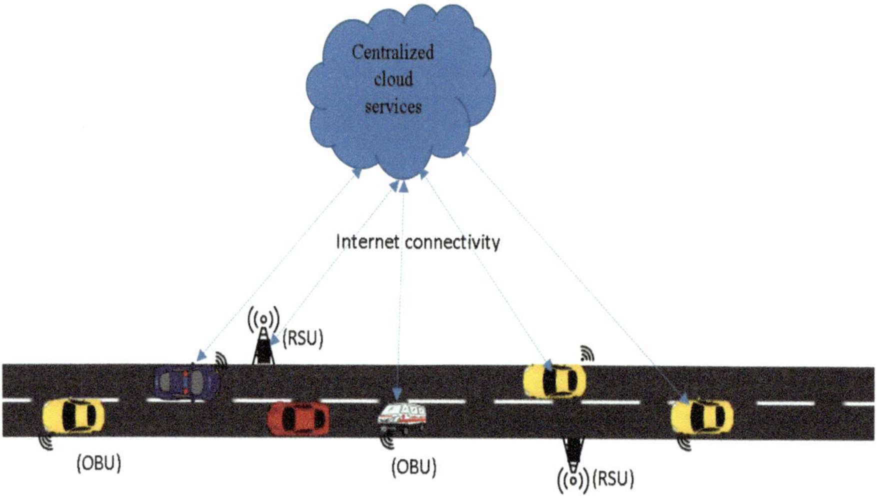

Fig. 1. An overview of IoV

Due to its mobility nature, the vehicle networks have no geographical limitations. In high mobility most of the times node is disconnected from centralized network, in this situation V2V communication is very important. Unavailability of centralized network is a common issue lies in vehicular networks where vehicles must be part of network which is unsupervised. Relaying on centralized network in such dynamic environment also creates the problem of SPOF (Single Point of Failure) phenomena that means if internet connectivity is compromised by any mean will lead to network disruption.

Considering security domain in vehicular network, trust among the nodes is one of the biggest challenge to overcome [8–10]. The exchange of critical message such as accident report, road blockage, traffic congestion, natural disaster and many others requires integrity to be assured before any action is taken. Selfish and malicious vehicles can divert traffic to desired routs.

All models have one thing in common that they have chosen fixed parameters or scenarios that are unsuitable for dynamic vehicle networks. Most models have tried to overcome certain issue in limited experimental environment that only works well if there are no major changes. As we already know the nature of vehicular network is dynamic, so the available models are unable to provide solution in an adaptive manner.

Building the context is a solution that seems to fit best in this problem domain. Our proposed model for trust evaluation is focused on introducing context awareness to address the highlighted issue. For each critical message received, the scenario-based context will be built by the model. In addition, the model is expected to adapt the appropriate trust calculation module for each scenario.

The information communication in vehicular networks can broadly classified into two categories [5] namely, critical messages; messages related to traffic management, accident report, road blockage, diversion, driving conditions and traffic congestion. Second category is, infotainment; such as music, weather update, city guide, social communication, personal information and others. Our work is focused critical messages only.

2 Related Work

Security has always been the concern of ad hoc networks, especially where the breach of information can lead to a threat to human life. As researchers in their work [6] gave a broad overview of the IoV and discuss different aspects of upcoming challenges security is one of them. In another work [20] authors gave basic idea of IoV vision, they also took security as high concern for IoV future. For the development of the IoV there is great need for intelligent systems [6]. Existing solutions in VANETs/IoV for trust evaluation and management do not provide ultimate solution therefore require improvement [13, 21, 22]. Some key aspects discussed below.

2.1 Trust

Trustworthiness is the degree of legitimacy of a node in distributed systems such as VANETs, ITS (Intelligent Transport System) and others. Other than secure delivery of data from node to node the trustworthiness of received contents is essential [13, 14]. Trust is the basic and critical aspect of VANET/ITS (intelligent transport systems) [14, 22]. Trust management is one of the key security challenges in vehicular networks due to its dynamicity in all aspects [4, 9, 13, 14]. The trust evaluation framework must be decentralized [4]. The trust model must calculate the trust level in short time[11]. According to studies [5, 12] on VANETs the trust between the nodes can be classified in two basic types; trust related to data and trust related to node.

2.2 Trust Measuring Types

Different techniques that are used to evaluate the trust can be classified in three types [8–10, 13–15] as shown in Fig. 2.

Fig. 2. Trust measuring types

(i) Entity centric, these types of models are based on the node trust credit building over the time. A proposed fuzzy logic-based trust model [14]. The experience is a key attribute in the models whereas in VANETs node interaction experience is unlikely to take into consideration while calculating trust in real time. A model presented by[16] work on prior experience and suggestion from neighbors and CA (certificate authority) are used to build vehicle trust. Some other entity centric approaches[17, 18] use mix of CA and cryptographic authentication. The main issue with these types of models is, they assume once the node is authenticated it cannot perform malicious activity.

(ii) Data centric, these types of models are based on the data (message) to evaluate the trust. When we talk about data centric models one basic model [19], where researchers worked on scalability. They assigned different roles to different vehicles and on other hand also considered experience. In studies researchers [4, 13, 20] worked on neighbor nodes provide their opinion about trustworthiness of certain event. The trust is calculated on majority estimation.

Combining both the data and entity properties to evaluate the trustworthiness of the message received. A utility theory based model [10] using experience proposed for evaluating trust. One of the popular model presented by [7] using a hybrid approach which is based on combination of Bayesian rule to find the trust and Dempster Shafer theories for handling uncertainty.

Problem with majority of models is that they are limited to work under inflexible and predefined information and scenarios. There is a requirement for such system that can handle change at real time.

2.3 V2V

All the components of IoV infrastructure communicate with each other such as vehicle to infrastructure, vehicle to RSU and others. Our focus is V2 V communication. In vehicular networks there can be a situation where no connection is available with centralized infrastructure[9]. IoV requires efficient and secure wireless solution with limited connectivity resources. In a futuristic study [21] they discuss that uninterrupted internet access seems to be costly and will be rising issue in future that increase the importance of node to node communication without involvement of infrastructure. They further discuss importance of proximity in IoT based networks but at the same time they highlight the need for trust building between the nodes. The studies show there is requirement for node to node communication without centralized network.

2.4 Context Awareness

Context awareness can enable adaptation [22]. The author in a [23] on ad hoc networks explains that VANET trust management schemes must be adaptive with respect to the environment. The fixed security infrastructure lacks arability in such type of networks. The use of ontology for context building show the capabilities of ontological inference technique. Furthermore, they also incorporate limited resource availability in their model.

AI is mostly be used to develop context [24, 25] using different techniques such as rule based, formal logic and others. Formal logic an AI technique can also be used to formulate context [24]. Some context models are based on social sciences phenomena of activity theory, which build a context around certain activity such as [24] their model was based on services that to execute action.

3 Proposed Trust Model

An overview of the proposed trust evaluation model is illustrated in Fig. 3. The context building module is main core of our work that is based on AI with cognitive nature. Third module is mapping the context with trust calculation modules and last one is for evaluating the trust value by using different modules. All the components are discussed below in detailed methodology. In contrast with the previous approaches our model is of a flexible nature that can be observed in Fig. 3. The Fig. 4 shows the low-level view of presented model. Our model is focused to take the advantage of maximum information available and build a context around it.

3.1 Parameter Module

The parameter module is responsible for managing all available information from sensory data, received message extraction and historical information related to different entities. Data classification and prioritization is the next responsibility of this module. Due to diverse nature of data it is classified by multiclass, vector weight and linear classification depending upon the requirement. The deliverable of this layer to the context layer are "cues".

3.2 Context Module

This module has main responsibility to perform the cognitive context building around certain set of available information (cues). It has three sub modules, first context management module responsible for handling received cues from parameter layer, second context inference engine which is mainly responsible for inference and third context information builder which builds the context information to be fed in next layer.

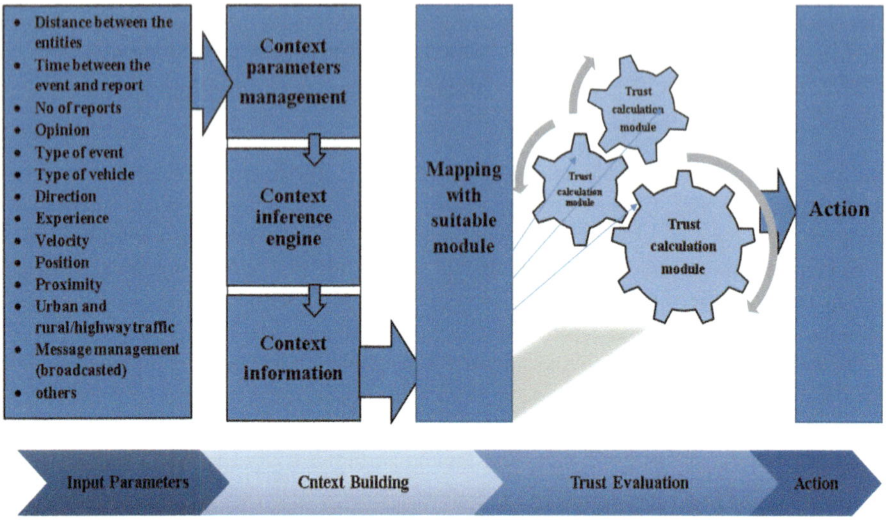

Fig. 3. Trust evaluation model

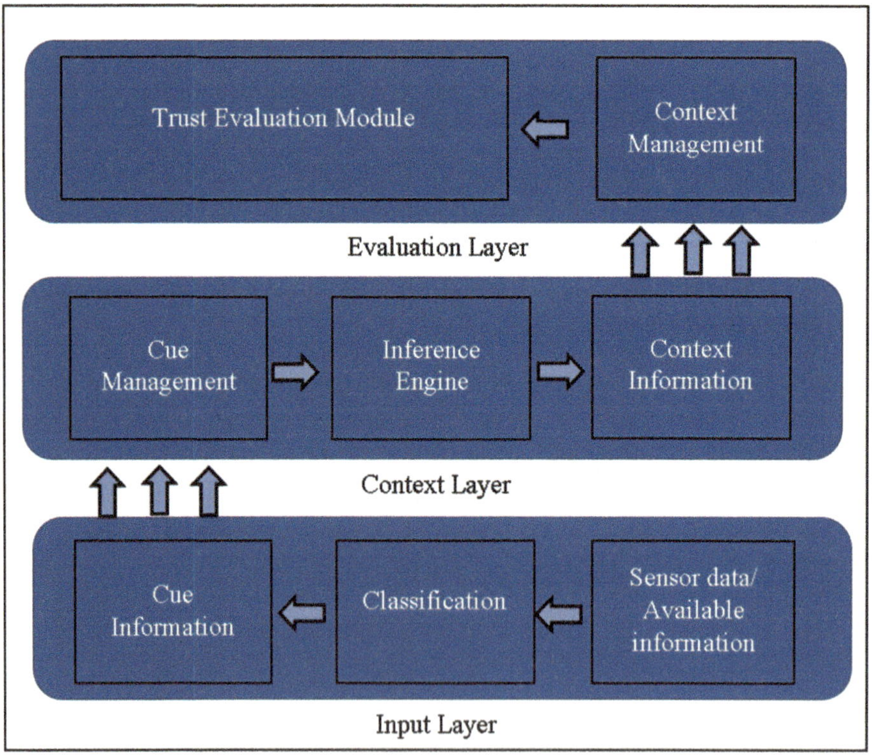

Fig. 4. Low level view of model

Inference Engine Architecture

An architecture of inference engine is designed as preliminary work. Our model is based on AI method for building the context. As per the nature of our work we are expected to use formal logic and ontology techniques to build context. On other hand our work also requires handling uncertainty for dealing with uncertainty approaches like probability and fuzzy logic. Our proposed model for inference engine architecture is shown in Fig. 5. Our proposed inference engine will use rule repository that is designed using ontology, formal logic and probability. The facts are the perimeter data that will be used to infer the context scenario.

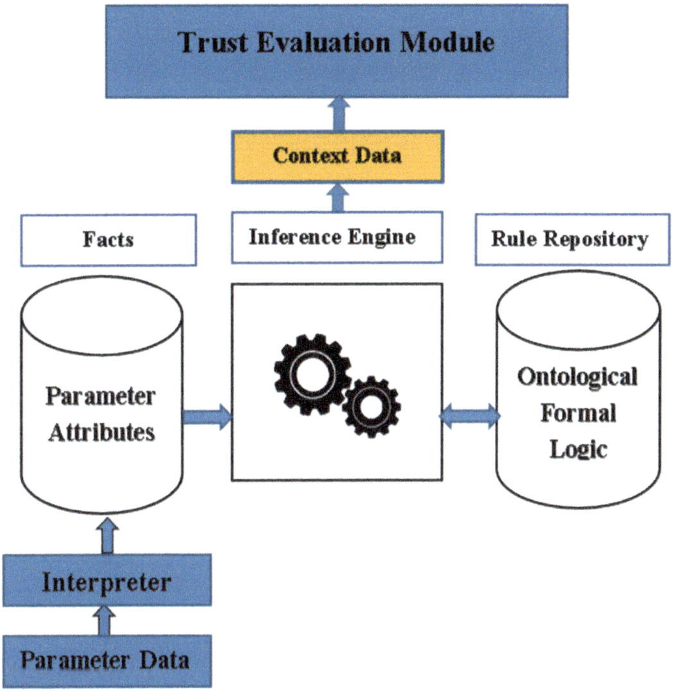

Fig. 5. Context inference engine

3.3 Trust Evaluation Module

The trust evaluation is finally responsible for computing the level of trust for certain event under all available information. This module is flexible to adapt the context information received from context layer. Based on context information one or multiple sub modules are adapted to construct the level of trust.

4 Design Analysis and Discussion

The model is intended to use all available resources to infer and attain the highest level of trust. Following are some scenario-based analysis to further elaborate our model.

Scenario 1 Heavy Traffic: In certain event where, multiple reports are available for single events and the number of reporting cars are multiple, the model can infer and verify received information. The opinion of neighbor nodes play an important role in achieving maximum trust. These events are likely to occur in urban traffic. Most of the present models may work well in such scenario due to sufficient available information.

Scenario 2 Moderate Traffic: In events with moderate number of nodes, enough information is available, in these types of situations our model takes into consideration all available information. Unlike other models which only consider certain set of parameters. It is also notable in these scenarios present models may work well only if available information matches their selective parameters but badly fail when event is random.

Scenario 3 Low Traffic: Events with fewer nodes are the most critical in term of trust evaluation due to less availability of information. These types of events occur in rural areas, highways and low rush hours. In these situations, level of uncertainty is very high. Our model handles these scenarios very well by utilizing maximum available perimeters such as direction of reporter, speed, proximity and others.

Based on above discussion an experiment is conducted while taking three parameters (i) Number of reporting cars (ii) Opinion of neighbor nodes (iii) Experience (previous trust level). Opinion and experience have ability to handle null value 0, in most of the other models the null value is ignored which is very essential for new vehicles to enter in network. The context scenario "Cs" is built around above-mentioned parameters and their ranging values. The context Cs = {Cs1, Cs2, Cs3 … Csn} is fed into most suitable trust evaluation module "TE" where TE = {TE1, TE2, TE3… TEn}. Initially system is tested against limited scope. The preliminary results show there is potential for system to be tested against complete set of parameters and automated inferred context using context inference engine.

5 Conclusion and Future Work

The aim of this study is to provide a model for evaluating message trust in vehicle networks. It has been found that the present evaluation models lack flexibility in terms of the available information, thus requiring an adaptive context aware model. In addition, the current models work in fixed or limited number of scenarios and take a limited number of parameters that are not suitable for vehicle networks. Our contribution will certainly benefit the upcoming era of IoV, IoT, ITS and smart city projects.

Our future direction is to classify the parameters, create low level model of inference engine along with the trust evaluation mechanism and finally simulating the whole model for evaluation.

References

1. Alam, K.M., Saini, M., El Saddik, A.: Toward social internet of vehicles: concept, architecture, and applications. IEEE access **3**, 343–357 (2015)
2. Xie, Y., Su, X., He, Y., Chen, X., Cai, G., Xu, B., et al.: Stm32-based vehicle data acquisition system for internet-of-vehicles. In: 2017 IEEE/ACIS 16th International Conference on Computer and Information Science (ICIS), pp. 895–898 (2017)
3. Yang, F., Wang, S., Li, J., Liu, Z., Sun, Q.: An overview of internet of vehicles. China Commun. **11**, 1–15 (2014)
4. Sheet, D.K., Kaiwartya, O., Abdullah, A.H., Cao, Y., Hassan, A.N., Kumar, S.: Location information verification using transferable belief model for geographic routing in VANETs. IET Intell. Transp. Syst. **11**, 53–60 (2017)
5. Butt, T.A., Iqbal, R., Shah, S.C., Umar, T.: Social internet of vehicles: architecture and enabling technologies. Comput. Electr. Eng. **69**, 68–84 (2018)
6. Contreras-Castillo, J., Zeadally, S., Guerrero-Ibañez, J.A.: Internet of vehicles: architecture, protocols, and security. IEEE Int. Things J. **5**, 3701–3709 (2018)
7. Wei, Z., Yu, F.R., Boukerche, A.: Trust based security enhancements for vehicular ad hocnetworks. In: Proceedings of the fourth ACM international symposium on Development and analysis of intelligent vehicular networks and applications, pp. 103–109 (2014)
8. Hasrouny, H., Samhat, A.E., Bassil, C., Laouiti, A.: VANet security challenges and solutions: a survey. Veh. Commun. **7**, 7–20 (2017)
9. Saini, M., Alelaiwi, A., Saddik, A.E.: How close are we to realizing a pragmatic VANET solution? a meta-survey. ACM Comput. Surv. (CSUR) **48**, 29 (2015)
10. Yao, X., Zhang, X., Ning, H., Li, P.: Using trust model to ensure reliable data acquisition in VANETs. Ad Hoc Netw. **55**, 107–118 (2017)
11. Gazdar, T., Belghith, A., Abutair, H.: An enhanced distributed trust computing protocol for VANETs. IEEE Access **6**, 380–392 (2018)
12. Bujari, A., Palazzi, C.E., Vitella, A.: Broadcasting messages in the internet of vehicles. In: 2018 IEEE 29th Annual International Symposium on Personal, Indoor and Mobile Radio Communications (PIMRC), pp. 58–62. IEEE (2018)
13. Raya, M., Hubaux, J.-P.: Securing vehicular ad hoc networks. J. Comput. Secur. **15**, 39–68 (2007)
14. Soleymani, S.A., Abdullah, A.H., Zareei, M., Anisi, M.H., Vargas-Rosales, C., Khan, M.K., et al.: A secure trust model based on fuzzy logic in vehicular ad hoc networks with fog computing. IEEE Access **5**, 15619–15629 (2017)
15. Sumithra, S., Vadivel, R.: An overview of various trust models for vanet security establishment. In: 2018 9th International Conference on Computing, Communication and Networking Technologies (ICCCNT), pp. 1–7. IEEE (2018)
16. Mármol, F.G., Pérez, G.M.: TRIP, a trust and reputation infrastructure-based proposal for vehicular ad hoc networks. J. Netw. Comput. Appl. **35**, 934–941 (2012)
17. Li, X., Liu, J., Li, X., Sun, W.: RGTE: a reputation-based global trust establishment in VANETs. In: 2013 5th International Conference on Intelligent Networking and Collaborative Systems, pp. 210–214. IEEE (2013)
18. Cui, J., Wu, D., Zhang, J., Xu, Y., Zhong, H.: An efficient authentication scheme based on semi-trusted authority in VANETs. IEEE Trans. Veh. Technol. **68**(3), 2972–2986 (2019)
19. Chen, C., Zhang, J., Cohen, R., Ho, P.H.: A trust-based message propagation and evaluation framework in vanets. In: Proceedings of the International Conference on Information Technology Convergence and Services (2010)

20. Minhas, U.F., Zhang, J., Tran, T., Cohen, R.: Towards expanded trust management for agents in vehicular ad-hoc networks. Int. J. Comput. Intell. Theory Prac. (IJCITP) **5**, 03–15 (2010)
21. Want, R., Schilit, B.N., Jenson, S.: Enabling the internet of things. Computer, 28–35 (2015)
22. Schmidt, A., Beigl, M., Gellersen, H.-W.: There is more to context than location. Comput. Graph. **23**, 893–901 (1999)
23. Pathan, A.S.K.: Security of Self-organizing Networks: MANET, WSN, WMN, VANET. CRC Press, Boca Raton (2016)
24. Kofod-Petersen, A., Cassens, J.: Using activity theory to model context awareness. In: International Workshop on Modeling and Retrieval of Context, pp. 1–17 (2005)
25. Schmidt, A.: Ubiquitous computing-computing in context (2003)

Small and Bandwidth Efficient Multi-band Microstrip Patch Antennas for Future 5G Communications

Abdulguddoos S. A. Gaid(✉), Osaid A. S. Qaid(✉), Moheeb A. A. Ameer(✉), Fadi F. M. Qaid(✉), and Belal S. A. Ahmed(✉)

Department of Communication and Computer Engineering, Faculty of Engineering, Taiz University, Taiz, Yemen
asry2000@hotmail.com, osaid.ucv2016@gmail.com, malborihy.5@gmail.com, fadifuadm995@gmail.com, b736397022@gmail.com

Abstract. In this paper, two multi-band microstrip patch antennas are proposed for 5G mobile devices. The proposed antennas have low profile structure. The first antenna is tri-band circular-shaped integrated on FR-4 epoxy with overall dimensions of $8 \times 7.6 \times 0.508$ mm^3. It operates at 40/50/64 GHz with a maximum gain of 5.19/5.23/8.269 dB respectively. The obtained bandwidths of this antenna are 2/9/5.83 GHz at 40/50/64 GHz respectively. The second one is a rectangular-shaped dual-band antenna printed on Rogers RT5880 with overall dimensions of $10 \times 10.22 \times 0.78$ mm^3. It operates at 28/39 GHz with a maximum gain of 7.73/7.02 dB respectively, and the achievable bandwidths are 2.19/2.84 GHz at 28/39 GHz respectively. These designs are very compact, directive and bandwidth efficient (greater than 5% of the center frequency). These characteristics make them suitable for mobile devices where the space is a major issue.

Keywords: 5G · Compact microstrip antennas · Multi-band microstrip antenna · Circular microstrip antenna

1 Introduction

Nowadays, mobile communication networks are attracting a considerable interest and represent the main direction in future developments. This interest arises due to the growing demand for higher data rates, higher capacities by the emergence of new electronic devices and increased number of internet users. It is expected that the fifth generation (5G) wireless communication will provide higher data consumption rates and will enable connectivity between machines, devices, people and so on. And there will be connection between everything, anytime and anywhere. To support these higher data rates, the large available spectrum in the millimeter band should be utilized [1]. Unfortunately, signals in this band suffer from a severe path loss. To compensate for the path loss, the transmitting/receiving antennas are required to be directive and bandwidth efficient. Microstrip has been considered the preferred candidate for the new

F. Saeed et al. (Eds.): IRICT 2019, AISC 1073, pp. 653–662, 2020.
https://doi.org/10.1007/978-3-030-33582-3_61

communication systems due to its good characteristics, such as low profile, small size, low cost, ease of fabrication and simplicity of integrating into circuit boards [2, 3]. However, microstrip antennas have poor gain and are bandwidth inefficient. Over the past years, engineers and researchers have proposed numerous techniques and ideas to improve the bandwidth and the gain of microstrip antennas so that it can fulfill the requirements of the new wireless networks [4, 5].

Many studies on mm-wave Microstrip antennas were conducted to improve their gain and bandwidth. For instance, Al-alem and Kishk [6] proposed a single-band microstrip antenna operating at 60 GHz, implemented on Rogers 3003 substrate. The patch dimensions of the proposed antenna is 6.81×6 mm^2. A maximum gain of 11 dB was obtained while a bandwidth is 7 GHz. Khattak et al. [7] proposed a dual- band elliptical microstrip patch antenna operating at 28/45 GHz. This antenna is printed on Rogers RT5880 with a radius of 2 mm and overall dimensions of $6 \times 6 \times 0.508$ mm^3. The maximum gain of 7.6 dB and 7.2 dB were obtained at 28 GHz & 45 GHz respectively. The maximum gain averaged over the two center frequencies is about 7.4 dB. In addition, the bandwidth provided by this antenna is 1.3 GHz & 1 GHz at 28 GHz and 45 GHz respectively, and the total bandwidth is 2.3 GHz. A single band antenna at 38 GHz, implemented on FR-4 epoxy substrate with thickness of 1.6 mm was proposed by Seker et al. [8]. The patch size is 5.9×4.4 mm^2 and has a maximum gain of 1.26 dB and bandwidth of one GHz. Authors in [9] designed a dual-band antenna at 38/54 GHz printed on Rogers RT5880 with thickness of 0.508 mm. The patch has a size of 2×2 mm^2. This antenna has a gain of 6.9 dB & 7.4 dB (an average of 7.15 dB) and a bandwidth of 1.94, 2 GHz at 38 GHz & 54 GHz respectively. Furthermore, a tri-band antenna at 50/57/63 GHz, printed on Rogers RO3003 with a thickness of 0.51 mm was proposed in [10]. The patch size is 5×5 mm^2. The maximum gain is 5.66 dB and the bandwidth is 1.2 GHz, 1 GHz and 2 GHz (total bandwidth is 4.2 GHz) at 50/57/63 GHz respectively. Paper [11] presents a dual-band microstrip antenna at 28/38 GHz; implemented on Rogers RT5880 substrate with a thickness of 1.575 mm. This patch has a size of 20×20 mm^2, achieving a maximum gain of 5.75 dB & 7.23 dB (averaged max gain is 6.49 dB). The obtained bandwidth is 4.7 GHz & 3.6 GHz (total bandwidth is 8.3 GHz) at 28/38 GHz respectively. A maximum gain of 3.3 dB at 28/31.45/34.6 GHz was achieved by the tri-band antenna proposed in [12]. This antenna was implemented on RT5880 substrate with a thickness of 1.6 mm and a patch size of 4×3 mm^2. The obtained bandwidth is 1.37 GHz, 0.11 GHz and 2.4 GHz (total bandwidth is 3.88 GHz) at 28/31.45/34.6 GHz respectively.

The above discussion on the reported results is summarized in Table 1 with a comparison to the performance of our proposed antennas. The proposed antennas in this work are better in terms of gain, bandwidth and size as can be clearly noticed.

This paper is organized as follows: Sect. 2 illustrates the design process of the two proposed microstrip patches, while our simulation results are presented and discussed in Sect. 3. Finally, in Sect. 4 the conclusion of the paper is presented.

2 Design Procedure

We start the design process by carefully selecting the dielectric substrate material based on the required performance. Type, thickness, dielectric constant and the loss tangent are important parameters that play a crucial role in deciding the bandwidth, the gain, the efficiency and volume of the antenna. The following sections will illustrate the design procedure of each antenna.

Table 1. Comparative analysis of different antenna designs for millimeter wave range

Ref.	Patch size (mm)	Frequency (GHz)	S11 (dB)	Gain (dB)	Average Gain (dB)	Bandwidth (GHz)	Total bandwidth (GHz)
[6]	6.81 × 6	60	−12.3	11	11	7	7
[7]	Radius = 2	28	−40	7.6	7.405	1.3	2.3
		45	−14	7.21		1	
[8]	5.9 × 4.4	38	−24.3	1.26	1.26	1.021	1.021
[9]	2 × 2	38	−15.5	6.9	7.15	1.94	3.94
		54	−12	7.4		2	
[10]	5 × 5	50	−15	5.66	5.66	1.2	4.2
		57	−22			1	
		63	−20			2	
[11]	20 × 20	28	−40	5.75	6.49	4.7	8.3
		38	−32.5	7.23		3.6	
[12]	4 × 3	28	−12.5	3.3	3.3	1.37	3.88
		31.45	−16.6			0.11	
		34.6	−15.7			2.4	
Design I	Radius = 1.8	40	−31.839	5.19	6.23	2	16.83
		50	−22.47	5.23		9	
		64	−22.27	8.269		5.83	
Design II	3.21 × 4.24	28	−34.49	7.738	7.34	2.2	5.03
		39	−18.444	7.028		2.83	

2.1 Design I

To design the circular patch, we have to calculate the radius of the patch using the following steps:

Step 1. Since the dimension of the patch is treated as a circular loop, the actual radius a of the patch is given by [13].

$$a = \frac{F}{\left\{1 + \frac{2h}{\pi\varepsilon_r F}\left[\ln\left(\frac{\pi F}{2h}\right) + 1.7726\right]\right\}^{1/2}} \tag{1}$$

Where

$$F = \frac{8.791 \times 10^9}{f_r\sqrt{\varepsilon_r}}$$

h is the substrate thickness, f_r is the resonance frequency in hertz and ε_r is the dielectric constant of the substrate.

Step 2. Determine the effective radius a_e of the microstrip antenna by applying Eq. (2)

$$a_e = a\left\{1 + \frac{2h}{\pi\varepsilon_r F}\left[\ln\left(\frac{\pi F}{2h}\right) + 1.7726\right]\right\}^{1/2} \tag{2}$$

Step 3. Determine the resonant frequency $(f_r)_{110}$ for the dominant TM^Z_{110} which is given by [11].

$$(f_r)_{110} = \frac{1.8412 v_0}{2\pi a_e\sqrt{\varepsilon_r}} \tag{3}$$

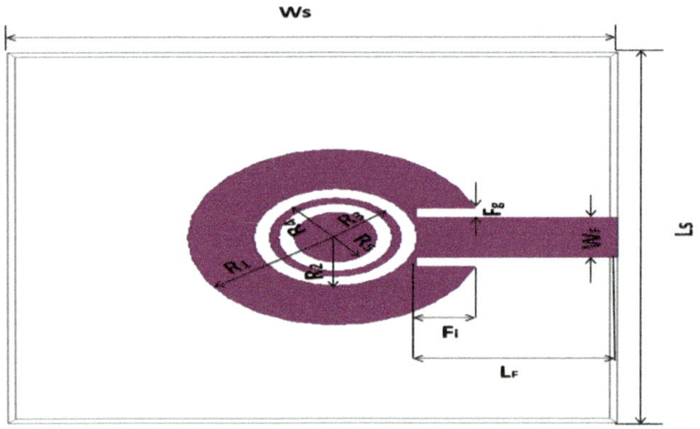

Fig. 1. Proposed antenna design geometry

After that, the antenna is simulated using a simulation tool such as HFSS simulator to get the optimum characteristics. Based on the parametric study and simulation results, a 0.508 mm thickness, low cost FR-4 glass epoxy substrate with $\varepsilon_r = 4.4$, is chosen for the antenna design. The initial antenna radius, which equals to 0.7 mm at 38 GHz is obtained from Eq. (1). A full electromagnetic simulation and optimization of the antenna is then performed using HFSS.13.

Table 2. Optimized geometrical dimensions of the proposed antenna.

Parameter	Value (mm)
Ls	7.6
Ws	8
R1	1.8
R2	1
R3	0.8
R4	0.7
R5	0.5
F_i	0.82
L_F	2.45
W_F	0.8
Fg	0.2
H	0.508

A top view of the antenna is shown in Fig. 1. The patch is fed by a 50 Ω microstrip line through an inset to improve the matching of the line to the patch. Two annular slots are etched on the patch to improve its performance. Table 2 summarizes the optimized values of the proposed antenna parameters.

2.2 Design II

For the rectangular patch antenna, we first calculate its length and width following the steps:

Step 1. Apply Eq. (4) to determine the width of the rectangular patch.

$$\mathrm{W} = \frac{c}{2f_r}\sqrt{\frac{2}{\varepsilon_r + 1}} \tag{4}$$

where c is the free space light velocity.

Step 2. Apply Eq. (5) to determine the effective dielectric constant, $\in_{eff}$ of the microstrip antenna.

$$\in_{eff} = \frac{\varepsilon_r + 1}{2} + \frac{\varepsilon_r - 1}{2\sqrt{1 + \frac{12h}{w}}} \tag{5}$$

Step 3. Apply Eq. (6) to determine the extended incremental length of the patch, ΔL.

$$\Delta L = 0.421h\frac{\left(\in_{eff} + 0.3\right)\left(\frac{W}{h} + 0.264\right)}{\left(\in_{eff} - 0.258\right)\left(\frac{W}{h} + 0.8\right)} \tag{6}$$

The actual length of the patch, L, can now be determined from Eq. (7).

$$L = \frac{\lambda}{2} - 2\Delta L = \frac{c}{2f_r\sqrt{\varepsilon_{eff}}} - 2\Delta L \tag{7}$$

Finally, the effective length L_e is determined by Eq. (8).

$$L_e = L + 2\Delta L = \frac{\lambda}{2} \tag{8}$$

The structure of the proposed multi-band antenna is shown in Fig. 2. It is printed on RT/Duroid5880 substrate with a thickness $h = 0.78$ mm and a dielectric constant $\varepsilon_r = 2.2$. The size of this patch is 4.24×3.21 mm^2.

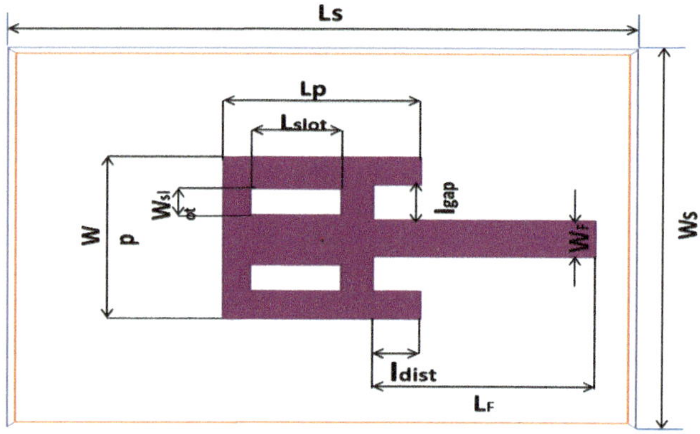

Fig. 2. The proposed slotted rectangular patch antenna geometry

The design and optimization of theoretical parameters were achieved by HFSS. After many optimizations and miniaturizations, the final optimized dimensions of the designed structure are listed in Table 3.

Table 3. Optimized geometrical dimensions of the proposed antenna

Parameter	Value (mm)
Ws	10
Ls	10.22
Wp	4.24
Lp	3.21
W_F	0.975
L_F	3.6
I_{dist}	0.77
I_{gap}	0.924
Wslot	0.7
L_{slot}	1.45
h	0.78

3 Simulation Results and Discussions

3.1 Design I

The results of the circular patch microstrip antenna with annular slots are given below.

The bandwidth and simulated return loss versus frequency are illustrated in Fig. 3 (a). A return loss magnitudes of −31.8, −22.5 and −22.3 dB are obtained with total bandwidth under −10 dB of 2 GHz, 9 GHz and 5.83 GHz at 40, 50 and 64 GHz respectively. The total bandwidth of the three bands is 16.83 GHz. The VSWR versus frequency is illustrated in Fig. 3(b). A VSWR of 1.05, 1.16 and 1.16 are obtained at 40, 50 and 64 GHz respectively.

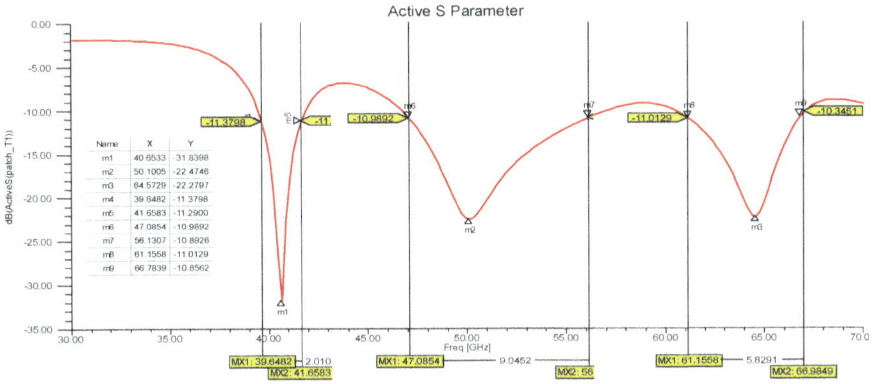

(a)

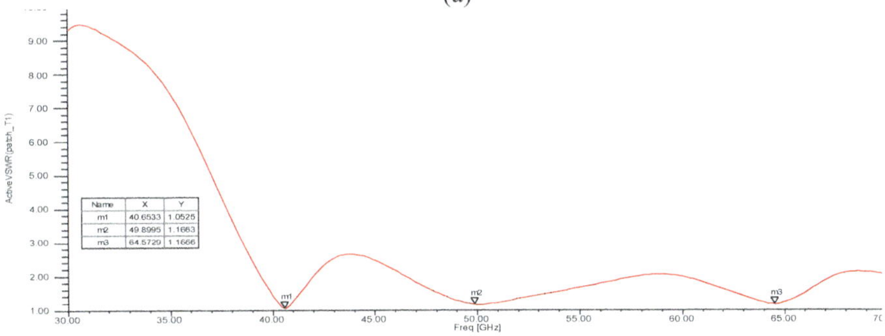

(b)

Fig. 3. Design I simulated (a) Return loss (b) VSWR.

Figure 4 shows the 3D gain of the proposed antenna. The maximum gain of 5.19, 5.23 and 8.26 dB are obtained at 40, 50 and 64 GHz respectively. The averaged maximum gain over the three resonance frequencies is 6.23 dB.

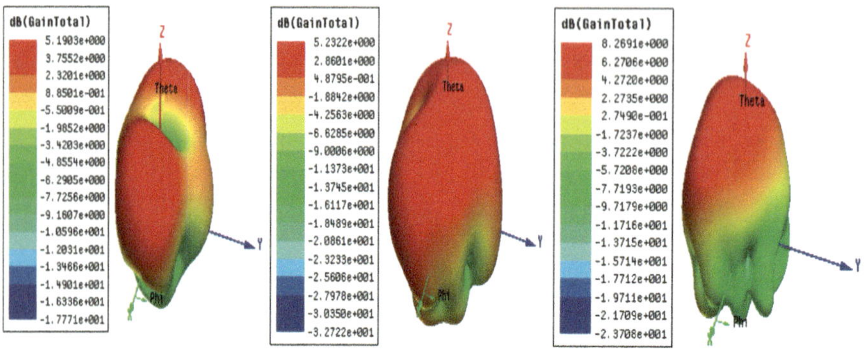

(a) (b) (c)

Fig. 4. Design I simulated 3D gain at (a) 40 GHz (b) 50 GHz (c) 64 GHz.

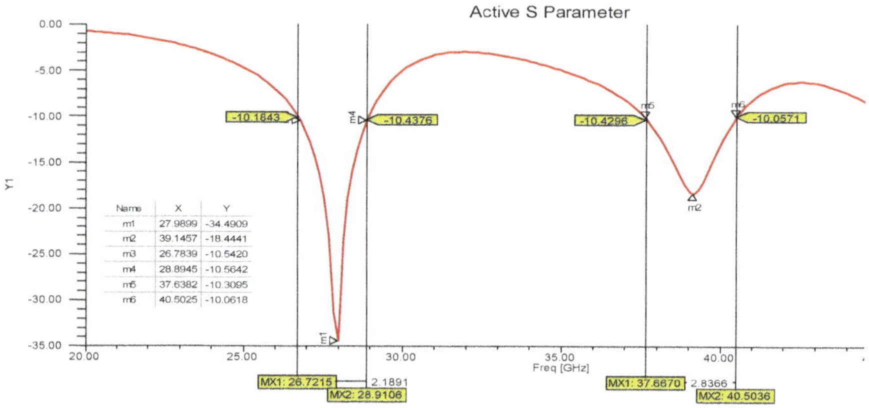

Name	X	Y
m1	27.9899	-34.4909
m2	39.1457	-18.4441
m3	26.7839	-10.5420
m4	28.8945	-10.5642
m5	37.6382	-10.3095
m6	40.5025	-10.0618

(a)

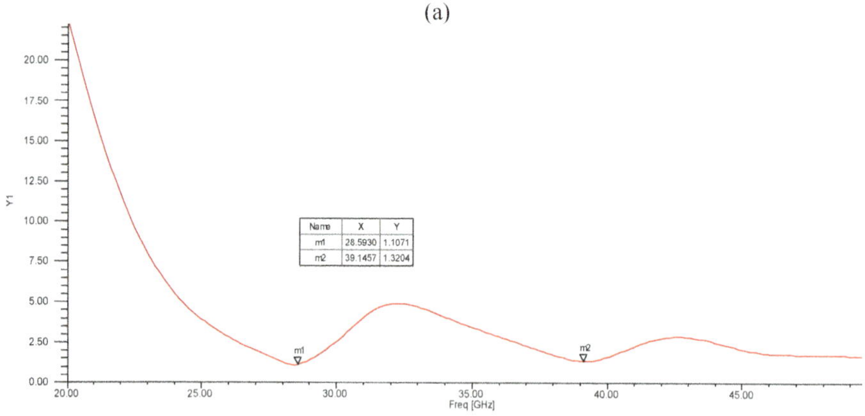

Name	X	Y
m1	28.5930	1.1071
m2	39.1457	1.3204

(b)

Fig. 5. Design II simulated (a) Return loss parameter (b) VSWR

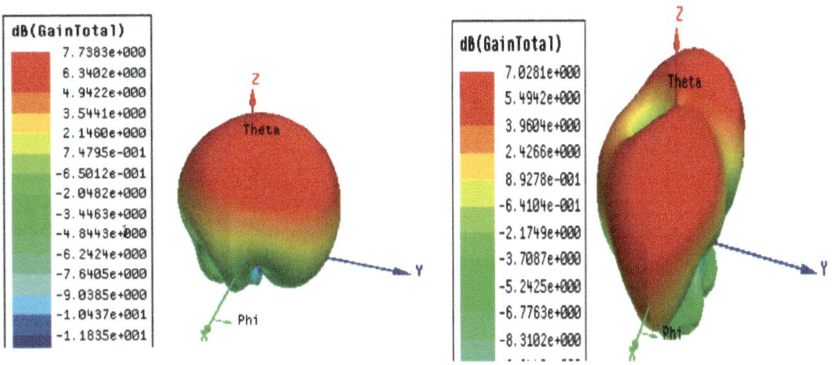

(a) (b)

Fig. 6. Design II simulated 3D gain at (a) 28 GHz (b) 39 GHz

3.2 Design II

The bandwidth and simulated return loss of the rectangular antenna versus frequency are illustrated in Fig. 5(a). A return loss magnitude of −34.5 and −18.4 dB are obtained with a total bandwidth under −10 dB of 2.19 and 2.84 GHz at 28 and 39 GHz respectively. The total bandwidth of the two bands is 5.03 GHz. The VSWR versus frequency performance is illustrated in Fig. 5(b). A VSWR of 1.1 and 1.3 is obtained at 28 and 39 GHz respectively.

Figure 6 shows the 3D gain of the proposed rectangular antenna. The maximum gain of 7.73 and 7.02 dB is obtained at 28 and 39 GHz respectively with averaged max gain which equals to 7.34 dB.

4 Conclusions

In this research, two small microstrip patch antennas for 5G wireless standards have been proposed, designed and analyzed. The structures of these antennas are very low profile, high gain and bandwidth efficient. The total bandwidth provided by the tri-band circular antenna is 16.83 GHz while the maximum averaged gain over the resonance frequencies is 6.23 dB. The dual-band rectangular antenna provides a bandwidth of 5.03 GHz and 7.34 dB of maximum gain averaged over the two resonance frequencies. These proposed antennas are suitable for 5G standards due to their compactness and high performance.

References

1. Niu, Y., Li, Y., Jin, D., Su, L., Vasilakos, A.V.: A survey of millimeter wave communications (mmwave) for 5G: opportunities and challenges. Wirel. Netw. **21**(8), 2657–2676 (2015)
2. Roh, W., Seol, J.Y., Park, J., Lee, B., Lee, J., Kim, Y., Cho, J., Cheun, K., Aryanfar, F.: Millimeter-wave beamforming as an enabling technology for 5G cellular communications: Theoretical feasibility and prototype results. IEEE Commun. Mag. **52**(2), 106–113 (2014)
3. Saed, M.A.: Reconfigurable broadband microstrip antenna fed by a coplanar waveguide. Prog. Electromagnet. Res. **55**, 227–239 (2005)
4. Cao, W., Zhang, B., Liu, A., Yu, T., Geo, D., Wei, Y.: Gain enhancement for broadband periodic endfire antenna by using split-ring resonator structures. IEEE Trans. Antennas Propag. **60**(7), 3513–3516 (2012)
5. Levine, E., Malamud, G., Shtrikman, S., Treves, D.: A study of microstrip array antennas with the feed network. IEEE Trans. Antennas Propag. **37**(4), 426–434 (1989)
6. Al-Alem, Y., Kishk, A.A.: Simple high gain 60 GHz antenna. In: International Symposium on Antennas and Propagation & USNC/URSI National Radio Science Meeting 2018, pp. 1693–1694. IEEE (2018)
7. Khattak, M.I., Sohail, A., Khan, U., Barki, Z., Witjaksono, G.: Elliptical slot circular patch antenna array with dual band behaviour for future 5G mobile communication networks. Prog. Electromagnet. Res. **89**, 133–147 (2019)
8. Şeker, C., Güneşer, M.T.: A single band antenna design for future millimeter wave wireless communication at 38 Ghz. Eur. J. Eng. Formal Sci. **2**(2), 34–38 (2018)
9. Imran, D., Farooqi, M., Khattak, M., Ullah, Z., Khan, M., Khattak, M., Dar, H.: Millimeter wave microstrip patch antenna for 5G mobile communication. In: International Conference on Engineering and Emerging Technologies (ICEET) 2018, pp. 1–6. IEEE (2018)
10. Firdausi, A., Hakim, G., Alaydrus, M.: Designing a tri-band microstrip antenna for targetting 5G broadband communications. In: MATEC Web of Conferences 2018, vol. 218, pp. 03015. EDP Sciences (2018)
11. Rahayu, Y., Hidayat, M.I.: Design of 28/38 GHz dual-band triangular-shaped slot microstrip antenna array for 5G applications. In: 2nd International Conference on Telematics and Future Generation Networks (TAFGEN) 2018, pp. 93–97. IEEE (2018)
12. Sumi, M., Hirasawa, K., Shi, S.: Two rectangular loops fed in series for broadband circular polarization and impedance matching. IEEE Trans. Antennas Propag. **52**(2), 551–554 (2004)
13. Balanis, C.A.: Antenna Theory: Analysis and Design, 4th edn. Wiley, Hoboken (2016)

Compact and Bandwidth Efficient Multi-band Microstrip Patch Antennas for 5G Applications

Abdulguddoos S. A. Gaid(✉), Amjad M. H. Alhakimi(✉), Osama Y. A. Sae'ed(✉), Mohammed S. Alasadee(✉), and Ali A. Ali(✉)

Department of Communication and Computer Engineering, Faculty of Engineering, Taiz University, Taiz, Yemen
asry2000@hotmail.com, amjad770497211@gmail.com, osama12.12.12c@gmail.com, moalasadee@gmail.com, alialzuriqi1989@gmail.com

Abstract. This paper proposes two compact and bandwidth efficient multi-band rectangular microstrip patch antennas for future mm-w 5G wireless communications. The first proposed one is a dual-band antenna with dimensioning to $7.68 \times 9.18 \times 0.78\ \text{mm}^3$. This antenna provides approximately 2.086 GHz/7.143 GHz of bandwidth and produces a maximum gain of 8.13 dBi/6.33 dBi at 28 GHz/38 GHz respectively. The second antenna is a tri-band resonating at 28 GHz, 45 GHz and 60 GHz. This antenna has a small size of 7.2×8.8 $0.78\ \text{mm}^3$. The first band is 3.543 GHz wide with a maximum achievable gain of 5.19 dBi. The second band has a bandwidth of 4.687 GHz and a maximum gain of 5.85 dBi. Moreover, a maximum gain of 6.07 dBi is obtained at the third band, which has a bandwidth of 5.926 GHz. The proposed antennas are low profile and compact enough to be integrated into handheld mobile units. Furthermore, the antennas provide enough bandwidth (>5% of the center frequency) and an acceptable gain which are required by the mm-w 5G wireless systems.

Keywords: Multi-band microstrip patch antenna · Mm-w 5G · Compact antennas

1 Introduction

The last few decades have witnessed a rapid development and fast growth in the mobile and wireless communications industry, evolving from the first generation (1G), which was an analog voice-based technology, to the second-generation (2G) digital technology then later on to the third generation (3G), and nowadays the fourth generation (4G, LTE) [1]. The ever-increasing demand for higher data rates due to increasing number of users of computers, smart phones and tablets connected to the internet, thrust for web browsing and video streaming, made it necessary to move forward a further step to a new generation, the fifth generation networks (5G) [2].

5G networks are characterized by their high data rates and high capacity [3]. These demands are realized by using antenna arrays consisting of microstrip patches at base stations and on the terminals [4, 5]. On the other hand, high data rates are achievable by shifting to the millimeter band at which a large spectrum is available. However, the

F. Saeed et al. (Eds.): IRICT 2019, AISC 1073, pp. 663–672, 2020.
https://doi.org/10.1007/978-3-030-33582-3_62

frequencies in the millimeter band suffer from a severe path loss [6]. To overcome these impairments, small size, high bandwidth and high gain antennas are required [7, 8]. Microstrip patches are planar, low profile and adaptable; however, microstrip patches are inherited by their low bandwidth and low radiation gain [9].

During the last decades, engineers and researchers carried intensive research to overcome these shortcomings, i.e., its small bandwidth and poor gain [10]. We will here survey some literature to see the efforts in improving the bandwidth, the gain and the size of the microstrip patch antennas to suit the requirements of the 5G new technology.

For example, Rahayu and Hidayat [11] designed a dual-band antenna at 28/38 GHz whose patch size is $3.2 \times 3\ \text{mm}^2$. The maximum gain achieved is 5.75 dBi and 7.23 dBi at 28 GHz & 38 GHz respectively. Hence, the maximum gain averaged over the two resonance frequencies is 6.49 dBi. In addition, the total bandwidth (<-10 dB) provided at both bands are 4.864 GHz and 3.602 GHz with a total of 8.466 GHz. Abdelaziz and Hamad [12] proposed a tri-band antenna operating at 10/28/38 GHz with a patch size of $9.7 \times 9.9\ \text{mm}^2$. The total achieved bandwidth is 2.604 GHz whereas the maximum gain averaged over the three center frequencies is 8.19 dBi. In addition, a dual-band antenna operating at 28/38 GHz with a patch sizing to a $3.4 \times 4.127\ \text{mm}^2$ was proposed by Kamal and Sarwar [13]. The maximum gain averaged over the two bands is 6.265 dBi and the total bandwidth provided is 1.38 GHz. Finally, Khattak et al. [14] proposed a dual band circular patch antenna with a radius of 2 mm resonating at 28/45 GHz. the maximum gain averaged over the two band is 7.405 dBi while the total bandwidth offered is 2.3 GHz.

Table 1 summarizes the results discussed above and compares them with the outcomes of this work. We can clearly see that the proposed antennas in this research managed to tradeoff between the patch size, bandwidth and radiation gain. For instance, in [12] the achieved gain is slightly better than the gain of our proposed patches; however, the proposed patches in this work are much smaller and superior in terms of bandwidth. To summarize, our proposed antennas here are compact, band-width efficient and directive antennas.

This paper is organized as follows: Sect. 2 illustrates the design process of the two proposed microstrip patches, while our simulation results are presented and discussed in Sect. 3. Finally, in Sect. 4 the conclusion of the paper is presented.

2 The Design Procedure

The first step in the designing process is to select carefully the substrate material based on the desired performance required. Type, thickness, dielectric constant and the loss tangent are important parameters that play a crucial role in deciding the bandwidth, the gain, the efficiency and also the volume of the antenna [12]. In this work, Rogers RT/duroid 5880 substrate whose thickness is 0.78 mm, dielectric constant is $\varepsilon r = 2.2$ and with a loss tangent of $tan\delta = 0.0009$, is used. Additionally, the radiator patches and the ground planes of both antennas are made of a copper sheet with a thickness of 0.035 mm. According to the desired resonance frequency, the patches' lengths and widths are first roughly obtained from applying the transmission line model as stated below [15]. The width is calculated first using Eq. (1) and subsequently apply Eqs. (2) to (5) to find the approximate length of the patch.

Table 1. A comparison between this work and some of the recently published work.

Ref.	Antenna size mm^3	Resonant frequencies GHz	S_{11} dB	BW GHz	Gain dBi	Total BW GHz	Average gain dBi
[11]	20 × 20 × 1.575	28	−40.64	4.864	5.75	8.466	6.49
		38	−32.66	3.602	7.23		
[12]	20 × 16.5 × 0.508	10	−25.8	0.343	5.67	2.604	8.19
		28	−24.9	0.761	9.33		
		38	−32.9	1.5	9.57		
[13]	——	28	−19.3	0.9	7.03	1.38	6.265
		38	−18.7	0.48	5.5		
[14]	6 × 6×0.508	28	−40	1.3	7.6	2.3	7.405
		45	−14	1	7.21		
Dual-band microstrip patch	7.68 × 9.18 × 0.78	28	−33.87	2.087	8.13	9.2107	7.232
		38	−31.87	7.143	6.334		
Tri-band microstrip patch	7.2 × 8.8 × 0.78	28	−24.1	3.543	5.19	14.156	5.703
		45	−18.14	4.687	5.85		
		60	−31.2	5.926	6.07		

The following step is then to optimize these calculated length and width so that the best results are obtained. The optimization process is performed using any one of the available commercial high frequency simulation software's such as the HFSS simulator, which is used in this work. The optimized dimensions of the two proposed patch antennas are summarized in Tables 2 and 3.

$$w = \frac{c}{2f_0\sqrt{\frac{\varepsilon_r+1}{2}}} \tag{1}$$

Where εr is the dielectric constant of the Rogers substrate, f_0 is the center resonance frequency and c is the speed of light.

Equation (2) is used to calculate the patch's length. After all, the effective length, the length extension and the effective dielectric constant are calculated first applying Eq. (3) through Eq. (5).

$$L = L_{eff} - 2\Delta L \tag{2}$$

Where L_{eff} is the effective length and ΔL is the length extension. These parameters are calculated using the following equations:

$$L_{eff} = \frac{c}{2f_0\sqrt{\varepsilon_{reff}}} \tag{3}$$

Table 2. Optimal parameters of the proposed rectangular dual-band antenna.

Parameters	Symbols	Values (mm)
Patch's width	Wp	4.5
Patch's length	Lp	3
Substrate's length	Ls	7.68
Substrate's width	Ws	9.18
Substrate's thickness	h	0.78
Circle radius	Re	0.4
Half-circle radius	Rm	0.5
Gap	Wg	0.1
Insertion	Yg	0.58
Feeder's length	Lf	1.85
Feeder's width	Wf	0.9
Position of circle (ref. to position half circle)	x, y	−0.6, −1.4
Position of half-circle	x, y	0,0

Table 3. Optimal parameters of the proposed rectangular tri-band antenna

Parameters	Symbols	Values (mm)
Patch's width	Wp	4.308
Patch's length	Lp	3.155
Substrate's length	Ls	7.2
Substrate's width	Ws	8.8
Substrate's thickness	h	0.78
Circle radius	R1	0.4
Slot length	Lm	2.6
Gap	Wg	0.1
Insertion	Yg	0.5
Feeder's length	Lf	1.85
Slot width	Wm	0.1
Ground width	Wgr	4.4

Where ε_{reff} is the effective dielectric constant, which is determined by:

$$\varepsilon_{reff} = \frac{\varepsilon_r + 1}{2} + \frac{\varepsilon_r - 1}{2}\left(1 + 12\left(\frac{h}{w}\right)\right)^{-\frac{1}{2}} \tag{4}$$

The extension length, ΔL is obtained from the following formula:

$$\Delta L = 0.412h \frac{\left(\varepsilon_{reff} + 0.3\right)\left(\frac{w}{h} + 0.264\right)}{\left(\varepsilon_{reff} - 0.258\right)\left(\frac{w}{h} + 0.8\right)} \tag{5}$$

Where w is the patch width obtained from eq. (1) and h is the dielectric substrate thickness.

2.1 The Dual-band Rectangular Microstrip Patch Antenna

Figure 1 shows a top and bottom views of the dual-band rectangular patch antenna. Two circular and one-half circular slots are etched on the patch to improve the return loss, gain and bandwidth [12, 14]. The radiating patch size is $3 \times 4.5\ mm^2$ and the overall volume of the antenna is $7.68 \times 9.18 \times 0.78\ mm^3$. This antenna is a dual-band resonating at 28/38 GHz. A 50 Ω microstrip line of 1.85 mm length and 0.9 mm width is utilized to feed the radiating patch; using inset technique to improve the matching between the line and the patch. Table 2 summarizes the optimized dimensions of this rectangular dual-band antenna.

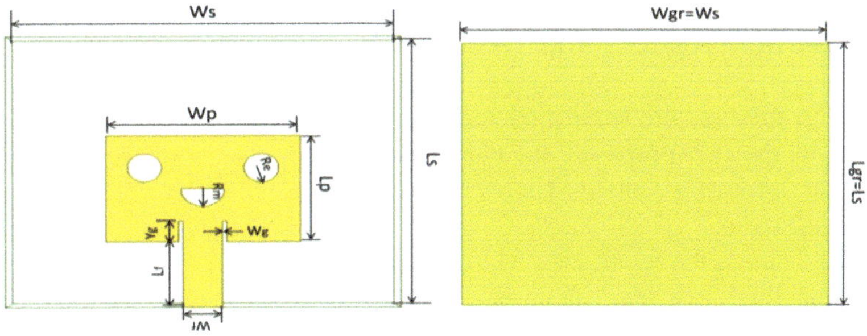

(a) (b)

Fig. 1. The dual-band patch antenna (a) Top view (b) Bottom view.

2.2 The Tri-band Rectangular Microstrip Patch Antenna

Figure 2 shows the geometrical configuration of the proposed tri-band microstrip patch antenna. The overall volume of the antenna is $7.2 \times 8.8 \times 0.78\ mm^3$. To improve the return loss and bandwidth of the proposed antenna, a partial ground is performed. According to [16], applying the partial ground technique enhances the antenna bandwidth by about 30%. The impedance matching is also improved when comparing with the full ground. Besides that, the patch is fed by a 50 Ω microstrip line, which is a 1.85 mm long and 1 mm wide. The inset feed technique is applied in order to get the best match between the feeding line and the radiating patch. Slots are etched on the radiating patch to improve the antenna performance. The optimized dimensions of this antenna are summarized in Table 3.

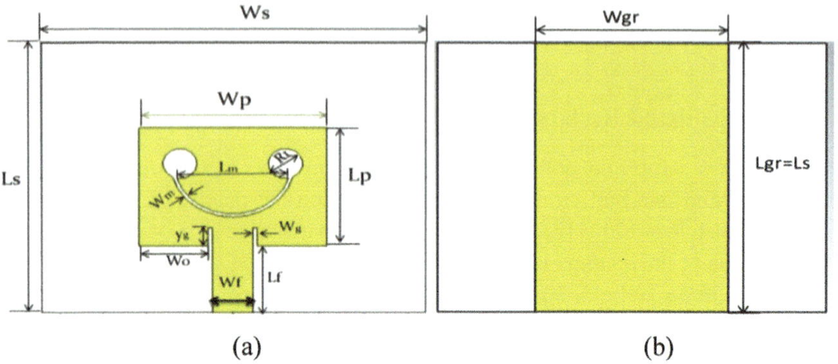

Fig. 2. Geometry of the tri-band microstrip antenna (a) Radiator patch view (b) Partial ground view.

3 Simulation Results and Discussions

This section presents the simulation results we have obtained and interpret. The performance of the two proposed antennas are simulated and optimized by using two commercial software's namely, High Frequency Structural Simulator HFSS and CST Microwave Studio.

Figure 3 shows the return loss, S11 of the dual-band antenna versus frequency. It is clearly seen that, at a frequency very close to 28 GHz, the value of S11 is about −33.8771 dB, and at 38 GHz the value of the return loss is −31.8737 dB. The bandwidth (<−10 dB) provided by this antenna is 2.0678 GHz (7.385%), around 28 GHz, extending from 26.6125 GHz to 28.6992 GHz. The second band is 7.1429 GHz (18.79%), extending from 35.9259 GHz to 43.0688 GHz. The total bandwidth offered by this proposed antenna is about 9.2107 GHz, which is much better than the bandwidth offered by antennas proposed in [11–14]. The VSWR against the frequency is shown in Fig. 4. The value of this parameter is less than 2 throughout the −10 dB bandwidth. At 28/38 GHz, the value of VSWR is 1.04 and 1.05 respectively. Figure 5 shows that the proposed antennas is directive where the maximum gain is about 8.13 dBi at 28 GHz and about 6.334 dBi at 38 GHz. The maximum gain averaged over the two center frequencies is about 7.232 dBi, which indicates its good directivity.

The return loss performance of the proposed tri-band microstrip patch is shown in Fig. 6. The value of the maximum return loss at the first band is −24.1 dB, whereas it is −18.17 dB and −31.17 dB at the 2nd and 3rd bands respectively. The first band centered at 28 GHz has a bandwidth (<−10 dB) of 3.5433 GHz (12.66%), extending from 26.526 to 30.069 GHz. A bandwidth of 4.6872 GHz (10.42%) in the second band is centered at 45 GHz, extending from 42.732 to 47.419 GHz. Moreover, a bandwidth of 5.9265 GHz (9.88%) is available in the third band, centered at 60 GHz, extending from 57.365 to 63.291 GHz. Total bandwidth offered by this antenna is 14.157 GHz, which indicates the broadband nature of the proposed antenna.

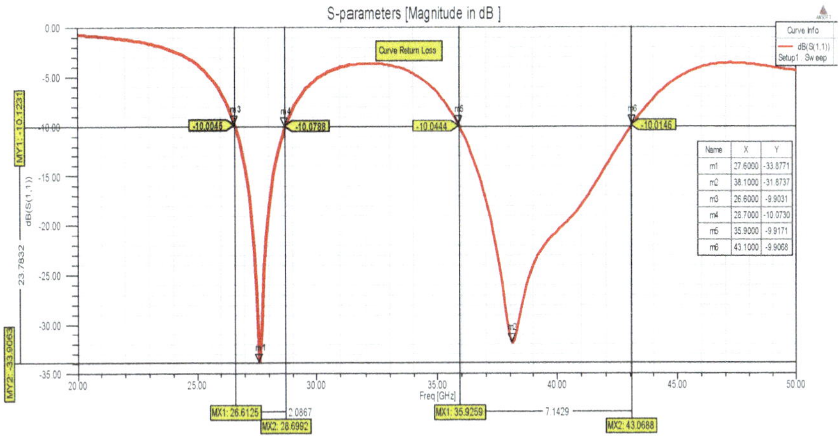

Fig. 3. Simulated return loss (S11) Vs frequency of the dual-band microstrip patch antenna.

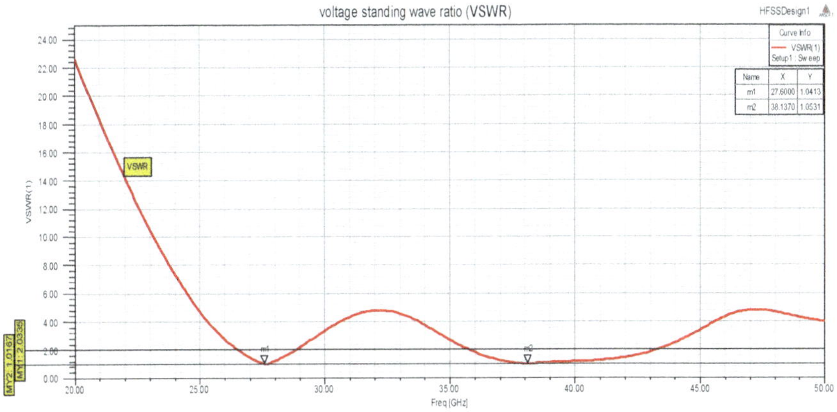

Fig. 4. Simulated VSWR Vs frequency of the dual-band microstrip patch antenna.

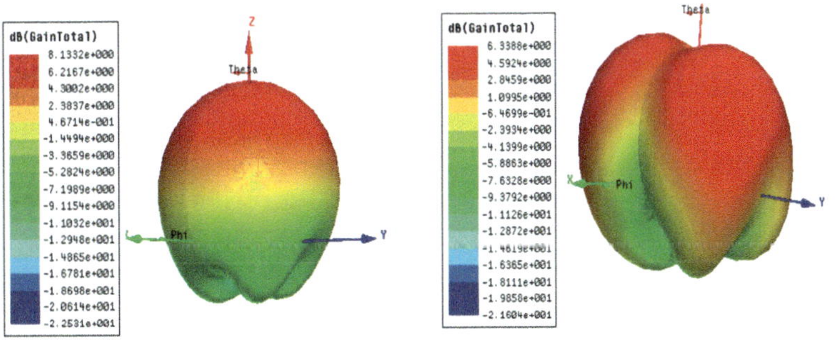

(a) (b)

Fig. 5. Simulated 3D gain of the dual-band microstrip patch antenna at (a) 28 GHz (b) 38 GHz.

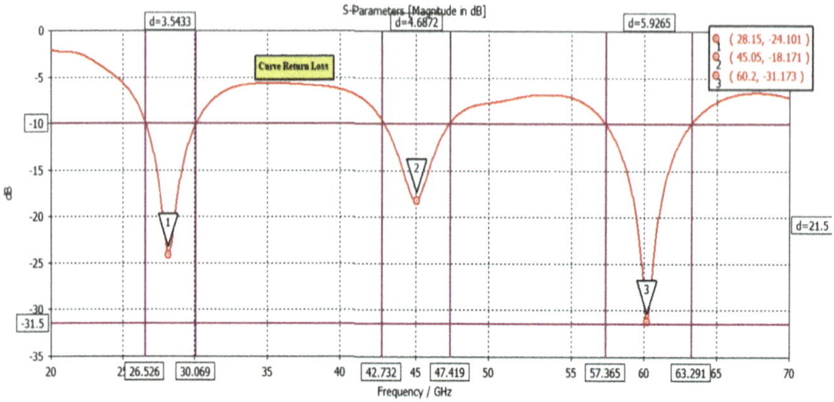

Fig. 6. Simulated return loss (S11) Vs frequency of the tri-band microstrip patch antenna.

Figure 7 shows the VSWR characteristics of the tri-band microstrip patch. The value of VSWR is 1.14, 1.28 and 1.06 at 28 GHz, 45 GHz and 60 GHz respectively. Generally speaking, the value of this parameter is less than 2 throughout the bandwidth of the antenna, revealing the good matching between the feeding microstrip line and the radiating patch.

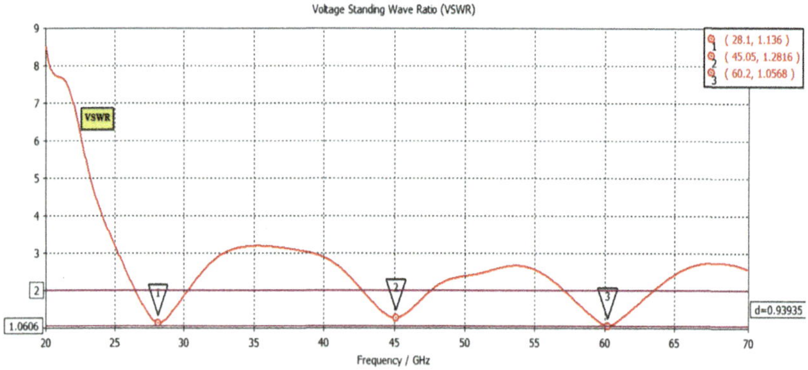

Fig. 7. Simulated VSWR Vs frequency of the tri-band microstrip patch antenna.

Figure 8 shows the gain of the tri-band antenna at the three bands. Figure 8(a) shows the gain at 28 GHz, where it is seen that a maximum gain of 5.19 dBi is provided. Figure 8(b) shows the gain at 45 GHz at which the maximum gain achieved is 5.85 dBi. Furthermore, a 6.07 dBi of maximum gain is realized as shown in Fig. 8 (c). Therefore, the maximum gain averaged over the resonance frequencies is about 5.7 dBi, meaning that the patch is directive enough.

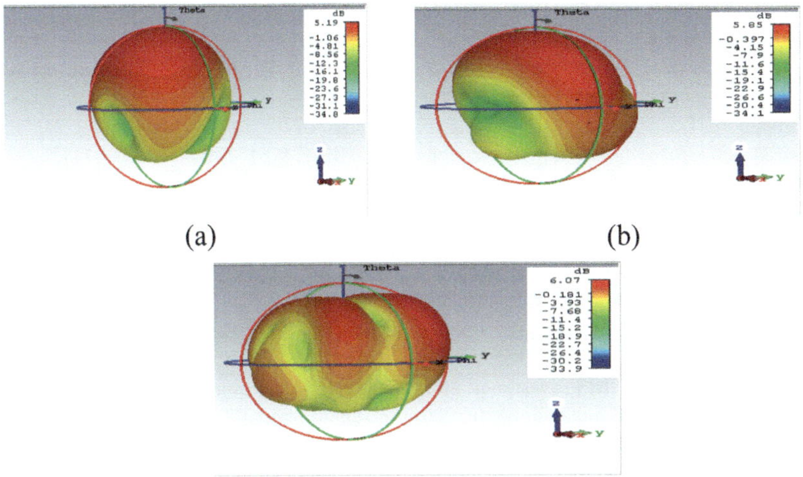

(a) (b)

(c)

Fig. 8. Simulated 3D gains of the tri-band microstrip patch antenna at (a) 28 GHz (b) 45 GHz, and (c) 60 GHz

4 Conclusions

In this paper, two rectangular multi-band microstrip patch antennas have been proposed, designed and analyzed. The two antennas are characterized by their low profile, compactness, good directivity and bandwidth efficiency. The dual-band antenna achieves a total bandwidth of 9.2107 GHz and realizes a maximum averaged gain of 7.232 dBi. The tri-band, however, achieves a total bandwidth of 14.157 GHz and a maximum gain of 5.7 dBi averaged over the three center frequencies. The proposed antennas are suitable for 5G terminals due to their small size and good performance.

References

1. Rappaport, T.S., Sun, S., Mayzus, R., Zhao, H., Azar, Y., Wang, K., Wong, G.N., Schulz, J.K., Samimi, M., Gutierrez, F.: Millimeter wave mobile communications for 5G cellular: It will work. IEEE Access **1**, 335–349 (2013)
2. Rao, R.G.S., Sai, R.: 5G–Introduction & future of mobile broadband communication redefined. Int. J. Electron. Commun. Instrum. Eng. Res. Dev. **3**(4), 119–124 (2013)
3. Ahmad, W., Khan, W.T.: Small form factor dual band (28/38 GHz) PIFA antenna for 5G applications. In: MTT-S International Conference on Microwaves for Intelligent Mobility (ICMIM) 2017, pp. 21–24. IEEE (2017)
4. Gampala, G., Reddy, C.: Design of millimeter wave antenna arrays for 5G cellular applications using FEKO. In: IEEE/ACES International Conference on Wireless Information Technology and Systems (ICWITS) and Applied Computational Electromagnetics (ACES) 2016, pp. 1–2. IEEE (2016)

5. Ying, Z.: Antennas in Cellular Phones for Mobile Communications. Proc. IEEE **100**(7), 2286–2296 (2012)
6. Chen, S., Zhao, J.: The requirements challenges and technologies for 5G of terrestrial mobile telecommunication. IEEE Commun. Mag. **52**(5), 36–43 (2014)
7. Rappaport, T.S., Murdock, J.N., Gutierrez, F.: State of the art in 60-GHz integrated circuits and systems for wireless communications. Proc. IEEE **9**(8), 1390–1436 (2011)
8. Kaur, N., Malhotra, S.: A review on significance of design parameters of microstrip patch antennas. In: 5th International Conference on Wireless Networks and Embedded Systems (WECON) 2016, pp. 1–6. IEEE (2016)
9. Singh, I., Tripathi, V.S.: Microstrip patch antenna and its applications: a survey. Int. J. Comp. Tech. Appl. **2**(5), 1595–1599 (2011)
10. Summakieh, M.A., Mokayef, M.: Single wideband microstrip patch antenna for 5G wireless communication application. IJRDO J. Electr. Electron. Eng. **1**(4), 7–8 (2016)
11. Rahayu, Y., Hidayat, M.I.: Design of 28/38 GHz dual-band triangular-shaped slot microstrip antenna array for 5G applications. In: 2018 2nd International Conference on Telematics and Future Generation Networks (TAFGEN), pp. 93–97. IEEE (2018)
12. Abdelaziz, A., Hamad, E.K.: Design of a compact high gain microstrip patch antenna for tri-band 5 G wireless communication. Frequenz **73**(1–2), 45–52 (2019)
13. Kamal, M.S., Islam, M.J., Uddin, M.J., Imran, A.: Design of a tri-band microstrip patch antenna for 5G application. In: International Conference on Computer, Communication, Chemical, Material and Electronic Engineering (IC4ME2) 2018, pp. 1–3. IEEE (2018)
14. Khattak, M.I., Sohail, A., Khan, U., Barki, Z., Witjaksono, G.: Elliptical slot circular patch antenna array with dual band behaviour for future 5G mobile communication networks. Prog. Electromagnet. Res. **89**, 133–147 (2019)
15. Balanis, C.A.: Antenna Theory Analysis and Design, 4th edn. Wiley, Hoboken (2016)
16. Rashid, S., Chakrabarty, C.K.: Bandwidth enhanced rectangular patch antenna using partial ground plane method for WLAN applications. Putrajaya Campus: University Tenaga Nasional, pp. 8–9 (2015)

Towards the Development of a Smart Energy Grid

Moamin A. Mahmoud[1](✉), Alicia Y. C. Tang[1], Andino Maseleno[1], Fung-Cheng Lim[2], Hairoladenan Kasim[2], and Christine Yong[3]

[1] Institute of Informatics and Computing in Energy, Universiti Tenaga Nasional, 43000 Kajang, Selangor, Malaysia
{moamin,aliciat,andino}@uniten.edu.my

[2] College of Computing and Informatics, Universiti Tenaga Nasional, 43000 Kajang, Selangor, Malaysia
{fclim,hairol}@uniten.edu.my

[3] ICT Division, Tenaga Nasional Berhad, 50470 Kuala Lumpur, Malaysia
christine@tnb.com.my

Abstract. Current electricity transmission and distribution networks can classified as conventional electricity networks because they have not been able to provide excellent services and present real-time data. This network has not been able to provide reliability, safety and efficiency in supplying electrical energy even not yet have the flexibility to be integrated with the generation of renewable energy or micro-grid. In this paper, a comprehensive review has been conducted to map the literature studies to a coherent taxonomy of Smart Energy grid. The fundamentals of smart grid presented in this paper can benefit readers who wish to embark in smart energy grid research and applications development.

Keywords: Smart energy grid · Power system · Renewable energy · Internet of Things

1 Introduction

Smart energy grid is a modern electric grid concept that has a high degree of flexibility, accessibility and efficiency [1]. Smart energy grid network involves digital sensors, smart meter, online monitoring, automation equipment and two-way communication system that allows an operator and a consumer to interact with each other in order to improve the reliability in services compared with the existing power systems [2]. Smart grid is an integrated and interdisciplinary power grid technology concept that is still being developed [3]. The Telegestore project is the first power grid to use smart meters on its 27 million subscribers, which are connected using low-bandwidth communication technology and Power Line Communication [4].

Smart Energy Grid consists of three main elements, namely information technology, telecommunication, and electric power [5]. All these elements work together to enable a two-way communication between the electricity provider and the consumer. With Smart grid, the transfer of electrical energy happens not only from the provider of

F. Saeed et al. (Eds.): IRICT 2019, AISC 1073, pp. 673–682, 2020.
https://doi.org/10.1007/978-3-030-33582-3_63

electricity to the consumer, but also vice versa. If it turns out consumers that have solar cells can generate electrical energy from sunlight, thus, when the electrical energy from the solar cells is greater than the needs of consumers, the consumer can send electrical energy to the existing grid. Consumers gain money from the utility company for that matter.

While the smart grid according to the Department of Energy (DOE) USA is a power system based on sensing technology, communication, digital control, information technology (IT) and other field equipment that serves to coordinate the existing processes in the power grid making it more effective and dynamic in its management. The feasibility and reliability of the power supply is one of the vital parameters in energy distribution especially for cross-application or operator. The information available in each generator, transmission and distribution area is usually only for each local power grid and system data is not based on real time data. This is the challenge for the future, so that the power grid can be more qualified with a high level of reliability as well safe from interference. The inefficiency of existing systems is one of the things which encourages the presence of smart grid paradigm.

To provide a clear insight to this problem and support researchers, we need a background understanding of various aspects related to smart energy grid. In this paper, a comprehensive review is conducted to map the literature studies to a coherent taxonomy of smart energy grid. To do so, we review four main elements constitute smart energy grid which are Features, Components, Architecture, and Areas.

2 Smart Energy Grid

Smart Grid is a modern electric grid concept whose planning, development and research have developed rapidly [6, 7]. Some countries in the USA and Europe have been very intensive to realize the technology. In the USA the smart grid implementation is coordinated by the Department of Energy (DOE) together with EPRI (Electric Power Research Institute) with its project called "Intelligent Grid" [8]. This project develops communication process between power grid with computer to improve reliability of power system and service to consumer. In addition DOE also affiliates with the industry sector through the Gridwise program. The focus of this program is to define the communication design and smart grid standard, simulation and analysis tools, smart technology, test and demo plant infrastructure, legal protection and market framework. Meanwhile, Smart Grid European Technology Platform (ETP) has a vision that power system networks in Europe must be flexible in meeting consumer needs, accessible, reliable and economical [2]. To realize that vision the main requirements include: (a) creating a technical solution that can be applied in a cost effective way so as to accept the integration of all sources of electrical energy; (b) harmonizing regulations and facilitating cross-commercial structures in Europe in the service of electrical energy; (c) issuing technical regulations and standards; (d) developing an IT-based system; (e) ensuring the successful integration of the old system design with the new system. Until now, there is no mutual decision on the definition of smart grid.

Each country and research institute of the world defines smart grid differently but generally has the same framework. According to NIST smart grid can be defined as a network of electrical power systems that use two-way information technology, secure cyber communications technology and intelligence computing integrated across the spectrum of electrical energy systems ranging from generators to consumers. The anticipated benefits and requirements of SG are the following [9]: (1) Improving power reliability and quality; (2) Optimizing facility utilization and averting construction of back-up (peak load) power plants; (3) Enhancing capacity and efficiency of existing electric power networks; (4) Improving resilience to disruption; (5) Enabling predictive maintenance and self-healing responses to system disturbances; (6) Facilitating expanded deployment of renewable energy sources; (7) Accommodating distributed power sources; (8) Automating maintenance and operation; (9) Reducing greenhouse gas emissions by enabling electric vehicles and new power sources; (10) Reducing oil consumption by reducing the need for inefficient generation during peak usage periods; (11) Presenting opportunities to improve grid security; (12) Enabling transition to plug-in electric vehicles and new energy storage options; (13) Increasing consumer choice; (14) Enabling new products, services, and markets.

3 Features

Smart Grid is a modern electrical energy network that intelligently integrates power grids with communications devices that support electricity generation and distribution networks to be more attractive, communicative and qualified. The term grid is used for an electricity system that may support all or some of the following four operations: electricity generation, electricity transmission, electricity distribution, and electricity control. Smart grid will be a main key in the transformation process in the electrical energy sector because of several benefits, among others.

Safe and Reliable: High-level cyber security becomes one of the most important parameters for safe and reliable smart grid [10].

Self Healing: Self-healing is a term used to describe the ability of the power grid to detect, to predict, to anticipate and to respond to disturbances that occur in the system quickly based on data or information transmitted by the sensors that have been installed [11].

Efficient and Smart: Each domain in the smart grid that includes generators, transmission, distribution and consumer will be equipped with Advanced Sensor Infrastructures (ASI) that serves to provide specific parameters data in the smart grid network in real time desired [12, 13].

Accommodating: Smart grids will be more flexible to be integrated with power generation from renewable energy such as solar energy generation and wind energy in the form of micro grid [14, 15]. Smart grid is also more accommodating to scattered small scale power plants (distributed generation and micro grid) [16].

Environmentally Friendly: It is possible that the integration of large scale renewable energy-based micro power generation makes smart grid very environmentally friendly

[17]. Utilization of renewable energy will reduce the impact of greenhouse gases thereby driving the decline of global warming (global warming).

Priority Quality and Stability: Smart grid will be more focused in maintaining the quality of power (power quality) in the supply of electrical energy. Problems that harm consumers and providers of electric energy can be reduced to the fullest. The problem usually occurs because of unstable voltage, rise and fall of frequency and the emergence of harmonics.

4 Components

A smart grid intelligently integrates the activities of all users/customers (power plants, customers) in order to provide/to produce electricity efficiently, sustainably, economically and securely. Smart grid uses innovative products and together serve with monitoring, control, communication and self-healing technologies to [2]:

- Facilitate better relationships and operations of all generators and technologies
- Provide flexibility to customers and to determine part of optimization of the operation of the system
- Provide customers with information and options for supply,
- Significantly reduce the environmental impact of the entire power supply system.
- Improve reliability and safety of supply.

Smart grid is the management of electrical energy with integrated concepts and reduces the dependence on natural resources derived from fossils. This integration enables communication and information exchange between electricity generation, distribution and consumption to make smarter decisions about energy consumption and production. In the smart grid there are three elements of technology that is electric power technology, information, and telecommunications [5]. The three elements are integrated with each other, allowing two-way communication between power companies such as Tenaga National Berhad and consumers. The transfer of electrical energy in the smart grid is unlike conventional systems that are only one way but can also be done otherwise. If it turns out consumers have their own electrical energy sources such as solar panels that can generate electrical energy from sunlight, when the electrical energy generated is excessive, the consumer can send the resulting electrical energy to the existing grid. From that consumers are not only pay utility bills alone, but also can make money from generated electricity.

The smart grid consists of communication networks, advanced sensors and control tools that serve to monitor the running of the system [18]. The received power current and power consumption used will be in data and control in real time, hence from that power current need to be controlled in order to get optimal and efficient network performance. Therefore, it is necessary to set up a device on the network that can monitor the flow of two-way communication and information between the power plant, the control base and the consumer. Regulatory methods in this system are obtained based on data collected on the control base that receives data from sensors that monitor energy consumption in real time, weather conditions, operating status, equipment conditions and energy availability of power plants and energy owned by consumers. The data then will be used to predict energy and energy needs that will be channeled to consumers. In addition, the data obtained will be used as a control grid which requires more energy. On smart grid if the source of electrical energy from one power plan cannot distribute electrical energy, then the source of electrical energy can be diverted and distributed from other sources. Smart grid parts consist of an integrated communication system, modern hardware, modern control & instrumentation (I & C) and smart software parts.

5 Architecture

In the application of a smart grid system usually consists of users, bulk generation, network, communication.

Users (Customer): Smart consumption will permit reaction to interface need between distribution management and building automation [18]. Local production is currently not a major component, but local production is expected to be the future referrer of the smart grid requirement. Smart home is home equipped with automation systems. The home automation system incorporates a variety of controls for lighting, shutter, temperature control and other equipment to enable efficient, economical and increased the comfort energy use.

Bulk Generation: Smart generation will include the increased use of power electronics in order to control harmonics, fluctuating generation failures from renewable energy as well as the need for increased flexibility of fossil power plants due to increased fluidation from renewable energy sources [19].

Network (Transmission Dan Distribution): Automation of substation and protection is the backbone for the operation of a secure transmission network. Power Quality and Power Monitoring Systems act similarly to a company's quality management system [20]. They are free from operating systems, controls and management and supervise all electrical activities and equipment in the same network. Therefore such a system can be used as an early warning system and must analyze failure and find suitable grounds.

Communication: Communication as a whole is the backbone of the smart grid. Only by exchanging information at the syntactic and semantic levels of smart grid benefits can be achieved [21]. Security of critical infrastructure has always been a major issue. But the smart grid solution will encounter a huge increase in data exchange for observation capabilities and also for control. Therefore, the security of this data exchange and the components behind it will have a better impact.

6 Areas

Smart grid technology areas (each with individual technology) across a whole network range from multiple generation through transmission and distribution to various types of electricity customers. A number of technologies are actively utilized and considered established for their development and application, while others still require further development and demos. A fully optimized electrical system will be deployed across all technological areas. But not all technologies need to be installed to improve the intelligence of the network.

Monitoring and Control Area: Real time monitoring and display of power system components and performance, along the connection and in large geographic areas, helps operators to understand and to optimize the behavior and performance of power system components. Advanced operating equipment system avoids blackout and facilitates the integration of renewable energy sources [22].

Integration of Information and Communication Technology: Communications infrastructure needs to be prepared using either a private network (radio network) or public network operator (internet, cellular, cable or telephone) communications utility [23]. Along with communication devices, computing, control system software and software, it is necessary to plan enterprise resources to support the exchange of two-way information between stakeholders enabling the use and management of a more efficient network.

Integration of Renewable Energy and Distributed Generator: The integration of large-scale renewable technology, distributed energy resources which cover large scale in transmission, medium scale in distribution level and small scale in commercial or settlement can be a challenge for the delivery and control of these resources and for the operation of electrical systems. Energy storage systems, both electric and thermal (thermal), can alleviate problems such as production coupling and energy delivery. Smart grids can help by automating control of generation and demand (in addition to demand response forms) to ensure supply and demand balance [24].

Improved Transmission Application: There are several technologies and applications for transmission system. Flexible AC transmission system (FACTS) is used to improve the control of the transmission network and maximize the power transfer capability [25]. The deployment of this technology on the line can improve efficiency and delay additional investment needs. The DC high-voltage technology (HVDC) is used to connect offshore wind power generators and centralized solar power plants with large electrical environments, with loss reductions system and improved control systems, enabling efficient use of far energy source from the load center [26]. The use of high Temperature Superconductors (HTS) can significantly reduce transmission losses and allow for higher-performance current delimiters, despite the debate over market-ready technology.

Distribution Network Management: Distribution and sub-station sensing and automation can reduce shutdown time and repair, maintain voltage levels and improve asset management. Advanced distribution automation processes real-time information from sensors and meters for fault location, automatic reconfiguration of feeder, stress and reactive power optimization, or for control of distributed generation.

Advanced Metering Infrastructure (AMI): Advanced metering system which consists of electronic state of the art/digital hardware and software, which combines data measurement intervals with continued long distance communication available. The system allows detailed measurement, time-based information and the collection and transmission of information to various parties. AMI usually refers to an overall measurement system and collection that includes a meter at a customer's location, a communications network between customers and service providers, such as gas, electricity, or water utilities, and receipt of data and management systems that make information available to service providers.

Infrastructure for Electric Car Battery Charging: The electric vehicle battery charging infrastructure handles billing, scheduling and other smart features for smart charging (the vehicle grid) during low energy demand. In the long term, it is conceivable that large installation filling will provide additional power system services such as capacity reserves, peak load cuts and vehicle filling regulations on the network.

System on the Customer Side: Customer-side System: which is used to help industry to manage electricity consumption, service level and housing, including energy management systems, energy storage devices, smart equipment and distributed generator. Improved energy efficiency and peak load demand reduction can be accelerated from home with display or energy dashboards, smart equipment and local energy storage. The demand response includes user manual and automatic user response, price-responsive equipment and thermostats connected to an energy management system or controlled by signal from the operator or utility system.

7 Future Directions

A. IoT Technology

IoT technology is widely used for traffic monitoring, industrial processes, the military and the environment. The use of Internet of Things (IoT) in the power system network is the result of the development of information and communication technology [27, 28]. Currently the implementation of information and communication technology in the power grid is still very limited even in some place is still not exist. So that the level of automatic is low. This has an impact on weak governance information on the power grid, especially the simultaneous transmission and distribution networks also affects the decline in service to consumers. The current electrical network still has many loopholes or weaknesses in the network data communication system. For example in monitoring the quality of power in the network distribution, the operator must make a field visit to monitor the desired network so that the data obtained is not real-time.

B. Agent Technology

One of the potential methods towards smart grid is the applications of agents in smart grid operations [29–32]. Agents are designed to handle resources autonomously [34–38] using agent-based modeling and simulation [39–41], which offer flexibility for each resource to collaborate/coordinate/negotiate with each other to achieve the desired goal efficiently [42–44]. It was reported by [29] that any planned feature of smart grid can be brought into reality by applying intelligent agents as the distributed control building blocks.

8 Conclusion

The fundamentals of smart grid presented in this paper can benefit readers who wish to embark in smart grid research and applications development. Several features of smart grids are discussing which are, safe and reliable, self-healing, efficient and smart, accommodating, environmentally friendly, priority quality and stability. specifications. While the specifications are discussed from three angles Components, Architecture, and Areas. The presented components are integrated communication system, modern hardware, modern control and instrumentation, smart software. While the reviewed architecture is Users, Bulk generation, Network, and Communication. Finally the discussed areas are monitoring and control, integration of information and communication technology, integration of renewable energy and distributed generator, integration of renewable energy and, distributed generator, improved transmission application, distribution network management, advanced metering infrastructure (ami), infrastructure for electric car battery charging, system on the customer side. With Smart Grid technology, consumers will have full control to manage their electrical energy consumption. Sensor technology and automatic control in the Smart Grid allow the automatic enabling of consumer electrical appliances by taking into account the number of available electrical energies.

Acknowledgments. This work is sponsored by Tenaga Nasional Berhad (TNB) under TNB R&D Seeding Fund Scheme No. TC-RD-18-19. We also gratefully appreciate Universiti Tenaga Nasional & Uniten R&D for securing and managing the fund.

References

1. Aktas, A., Erhan, K., Ozdemir, S., Ozdemir, E.: Experimental investigation of a new smart energy management algorithm for a hybrid energy storage system in smart grid applications. Electr. Power Syst. Res. **144**, 185–196 (2017)
2. Gungor, V.C., Sahin, D., Kocak, T., Ergut, S., Buccella, C., Cecati, C., Hancke, G.P.: Smart grid technologies: communication technologies and standards. IEEE Trans. Ind. Inf. **7**(4), 529–539 (2011)
3. Lund, H., Werner, S., Wiltshire, R., Svendsen, S., Thorsen, J.E., Hvelplund, F., Mathiesen, B.V.: 4th Generation District Heating (4GDH): Integrating smart thermal grids into future sustainable energy systems. Energy **68**, 1–11 (2014)
4. Botte, B., Cannatelli, V., Rogai, S.: The Telegestore project in ENEL's metering system. In: CIRED 2005–18th International Conference and Exhibition on Electricity Distribution, pp. 1–4. IET, June 2005
5. Yu, X., Xue, Y.: Smart grids: a cyber–physical systems perspective. Proc. IEEE **104**(5), 1058–1070 (2016)
6. Amin, S.M., Wollenberg, B.F.: Toward a smart grid: power delivery for the 21st century. IEEE Power Energy Mag. **3**(5), 34–41 (2005)
7. Gungor, V.C., Lu, B., Hancke, G.P.: Opportunities and challenges of wireless sensor networks in smart grid. IEEE Trans. Ind. Electron. **57**(10), 3557–3564 (2010)
8. Haase, P.: Intelligrid: a smart network of power. EPRI J., 28–32 (2009)

9. National Institute of Standards and Technology. NIST framework and roadmap for smart grid interoperability standards, release 1.0 (2010). http://www.nist.gov/publicaffairs/releases/upload/smartgridinteroperabilityfinal.pdf
10. Gharavi, H., Ghafurian, R. (Eds.). (2011). Smart grid: The electric energy system of the future
11. Farhangi, H.: The path of the smart grid. IEEE power Energy Mag. **8**(1), 18–28 (2010)
12. Momoh, J.A.: Smart grid design for efficient and flexible power networks operation and control. In: 2009 IEEE/PESPower Systems Conference and Exposition, (PSCE 2009), pp. 1–8. IEEE, March 2009
13. Bu, S., Yu, F.R., Cai, Y., Liu, X.P.: When the smart grid meets energy-efficient communications: green wireless cellular networks powered by the smart grid. IEEE Trans. Wirel. Commun. **11**(8), 3014–3024 (2012)
14. Wang, C., Li, P.: Development and challenges of distributed generation, the micro-grid and smart distribution system. Autom. Electric Power Syst. **2**(004) (2010)
15. Kanchev, H., Lu, D., Colas, F., Lazarov, V., Francois, B.: Energy management and operational planning of a microgrid with a PV-based active generator for smart grid applications. IEEE Trans. Ind. Electron. **58**(10), 4583–4592 (2011)
16. Battaglini, A., Lilliestam, J., Haas, A., Patt, A.: Development of supersmart grids for a more efficient utilisation of electricity from renewable sources. J. Clean. Prod. **17**(10), 911–918 (2009)
17. Aghaei, J., Alizadeh, M.I.: Demand response in smart electricity grids equipped with renewable energy sources: a review. Renew. Sustain. Energy Rev. **18**, 64–72 (2013)
18. Siano, P.: Demand response and smart grids—a survey. Renew. Sustain. Energy Rev. **30**, 461–478 (2014)
19. Liang, X.: Emerging power quality challenges due to integration of renewable energy sources. IEEE Trans. Ind. Appl. **53**(2), 855–866 (2016)
20. Kayastha, N., Niyato, D., Hossain, E., Han, Z.: Smart grid sensor data collection, communication, and networking: a tutorial. Wirel. Commun. Mob. Comput. **14**(11), 1055–1087 (2014)
21. Naumann, A., Bielchev, I., Voropai, N., Styczynski, Z.: Smart grid automation using IEC 61850 and CIM standards. Control Eng. Pract. **25**, 102–111 (2014)
22. Ashok, A., Hahn, A., Govindarasu, M.: Cyber-physical security of wide-area monitoring, protection and control in a smart grid environment. J. Adv. Res. **5**(4), 481–489 (2014)
23. Rohjans, S., Dänekas, C., Uslar, M.: Requirements for smart grid ICT-architectures. In: 2012 3rd IEEE PES International Conference and Exhibition on Innovative Smart Grid Technologies (ISGT Europe), pp. 1–8. IEEE, October 2012
24. Ehsan, A., Yang, Q.: Optimal integration and planning of renewable distributed generation in the power distribution networks: a review of analytical techniques. Appl. Energy **210**, 44–59 (2018)
25. Albatsh, F.M., Mekhilef, S., Ahmad, S., Mokhlis, H., Hassan, M.A.: Enhancing power transfer capability through flexible AC transmission system devices: a review. Front. Inf. Technol. Electron. Eng. **16**(8), 658–678 (2015)
26. Flynn, D., Rather, Z., Ardal, A., D'Arco, S., Hansen, A.D., Cutululis, N.A., Smith, C.: Technical impacts of high penetration levels of wind power on power system stability. Wiley Interdisc. Rev. Energy Environ. **6**(2), e216 (2017)
27. Al-Turjman, F.: 5G-enabled devices and smart-spaces in social-IoT: an overview. Future Gener. Comput. Syst. **92**, 732–744 (2019)
28. Saleem, Y., Crespi, N., Rehmani, M.H., Copeland, R.: Internet of things-aided Smart Grid: technologies, architectures, applications, prototypes, and future research directions. IEEE Access **7**, 62962–63003 (2019)

29. Malik, F.H., Lehtonen, M.: A review: agents in smart grids. Electr. Power Syst. Res. **131**, 71–79 (2016)
30. Mahmoud, M.A., Ramli, R., Azman, F., Grace, J.: A development methodology framework of smart manufacturing systems (Industry 4.0). MySEC (2018)
31. Ahmad, A., Zaliman, M., Yusof, M., Ahmad, M.S., Ahmed, M., Mustapha, A.: Resolving conflicts between personal and normative goals in normative agent systems. In: 2011 7th International Conference on Information Technology in Asia 12 Jul 2011, pp. 1–6. IEEE (2011)
32. Ahmed, M., Ahmad, M.S., Yusoff, M.Z.: Modeling agent-based collaborative process. In: International Conference on Computational Collective Intelligence 10 Nov 2010, pp. 296–305. Springer, Heidelberg (2010)
33. Jassim, O.A., Mahmoud, M.A., Ahmad, M.S.: A multi-agent framework for research supervision management. In: Distributed Computing and Artificial Intelligence, 12th International Conference, pp. 129–136. Springer, Cham (2015)
34. Mahmoud, M.A., Ahmad, M.S., Ahmad, A., Yusoff, M.Z., Mustapha, A.: A norms mining approach to norms detection in multi-agent systems. In: 2012 International Conference on Computer & Information Science (ICCIS) 12 June 2012, vol. 1, pp. 458–463. IEEE (2012)
35. Ahmed, M., Ahmad, M.S., Yusoff, M.Z.: Mitigating human-human collaboration problems using software agents. In: KES International Symposium on Agent and Multi-Agent Systems: Technologies and Applications, 23 June 2010, pp. 203–212. Springer, Heidelberg (2010)
36. Mahmoud, M.A., Ahmad, M.S., Yusoff, M.Z.: A norm assimilation approach for multi-agent systems in heterogeneous communities. In: Asian Conference on Intelligent Information and Database Systems 14 March 2016, pp. 354–363. Springer, Heidelberg (2016)
37. Mahmoud, M.A., Ahmad, M.S., Yusoff, M.Z., Idrus, A.: Automated multi-agent negotiation framework for the construction domain. In: Distributed Computing and Artificial Intelligence, 12th International Conference 2015, pp. 203–210. Springer, Cham (2015)
38. Mahmoud, M.A., Mustapha, A., Ahmad, M.S., Ahmad, A., Yusoff, M.Z., Hamid, N.H.: Potential norms detection in social agent societies. In: Distributed Computing and Artificial Intelligence 2013, pp. 419–428. Springer, Cham (2013)
39. Mahmoud, M., Ahmad, M.S., Yusoff, M.Z.: Development and implementation of a technique for norms-adaptable agents in open multi-agent communities. J. Syst. Sci. Complex. **29**(6), 1519–1537 (2016)
40. Mostafa, S.A., Ahmad, M.S., Annamalai, M., Ahmad, A., Gunasekaran, S.S.: A dynamically adjustable autonomic agent framework. In: Advances in Information Systems and Technologies 2013, pp. 631–642. Springer, Heidelberg (2013)
41. Mostafa, S.A., Darman, R., Khaleefah, S.H., Mustapha, A., Abdullah, N., Hafit, H.: A general framework for formulating adjustable autonomy of multi-agent systems by fuzzy logic. In: KES International Symposium on Agent and Multi-Agent Systems: Technologies and Applications 20 June 2018, pp. 23–33. Springer, Cham (2018)
42. Mostafa, S.A., Ahmad, M.S., Ahmad, A., Annamalai, M., Gunasekaran, S.S.: A flexible human-agent interaction model for supervised autonomous systems. In: 2016 2nd International Symposium on Agent, Multi-Agent Systems and Robotics (ISAMSR) 23 August 2016, pp. 106–111. IEEE (2016)
43. Mostafa, S.A., Ahmad, M.S., Annamalai, M., Ahmad, A., Gunasekaran, S.S.: A conceptual model of layered adjustable autonomy. In: Advances in Information Systems and Technologies 2013, pp. 619–630. Springer, Heidelberg (2013)
44. Mostafa, S.A., Ahmad, M.S., Ahmad, A., Annamalai, M., Mustapha, A.: A dynamic measurement of agent autonomy in the layered adjustable autonomy model. In: Recent Developments in Computational Collective Intelligence 2014, pp. 25–35. Springer, Cham (2014)

A Survey of Geocast Routing Protocols in Opportunistic Networks

Aliyu M. Abali[1(✉)], Norafida Bte Ithnin[1], Tekenate Amah Ebibio[1], Muhammad Dawood[1], and Wadzani A. Gadzama[2]

[1] School of Computing, Faculty of Engineering, Universiti Tekenologi Malaysia, Skudai, Malaysia
aliyuabali@fud.edu.ng, afida@utm.my, mastangalbaloshi@gmail.com, amatek008@gmail.com

[2] Federal Polytechnic Mubi, Mubi, Nigeria
ask4gadzama@gmail.com

Abstract. Geocasting supports communication in a challenged environment by addressing various routing issues to a wide range of applications it exploits intermittent connectivity among mobile devices to support opportunistic message transfer over a short-range radio transmission, Variety of surveys have been carried out in the literature concerning general OppNets. The main purpose of this paper is to provide a survey of OppNet geocast routing protocols. It further highlights Geocast OppNet characteristics and envisioned application areas different from general OppNets. Additionally, this paper studies existing geocasting routing solutions based on message endpoint, forwarding utility computation approach, determining the relay node, routing decision approach and message dissemination. Finally, we have identified current challenges and future research direction for routing in OppNet geocast.

Keywords: Routing · Opportunistic networks · OppNet geocast

1 Introduction

Opportunistic networks (OppNets) attempts to bridge the gap between human activity and the behaviour of network elements by opportunistically exploiting mobility characteristics to support data forwarding [1]. In OppNets, the mobility of nodes is considered as an opportunity. Contrary to this, it is perceived as a factor that negatively impacts the performance of the network in traditional networks as the communication success is subject to the availability of route between endpoints. In recent years, the world has witnessed substantial growth in mobile technologies, smart off the shelf mobile devices have increased penetration and became available to the teaming world population at affordable price. According to a survey conducted by We Are Social and Hootsuite, the number of mobile phone users is increasing with 5.135 billion, up 4 percent year-on-year, a smartphone is considered as a preferred choice for personal users connecting to the internet, it constitutes a greater share of internet traffic that surpass all other devices combined. traditional MANETs differs in many ways from OppNet the TCP/IP and MANETs architectures are not designed to operate in an

F. Saeed et al. (Eds.): IRICT 2019, AISC 1073, pp. 683–694, 2020.
https://doi.org/10.1007/978-3-030-33582-3_64

environment that is characterised with frequent network partitions, long delay paths and limited node's resources [2, 3]. Thus, the concept of DTN/OppNet is primarily designed to tolerate long delays and intermittent connectivity between end nodes. Routing in OppNet is based on store-carry-forward (SCF) communication paradigm in which the source node store a data bundle in its memory, physically carry it from one point to another as the device goes about its daily activities, the data-bundle is forwarded to multiple relay nodes through a wireless communication interface such as Bluetooth and Wi-Fi. Eventually, the generated message can be delivered to the destination node. The opportunistic network is regarded as human-centric; it exploits human movement to aid data forwarding [4]. Opportunistic networking is motivated by increased mobile device usage and human social interaction coupled with the key technological development in wireless communication. Geographic routing in the form of unicast multicast and anycast is well investigated in OppNets. However, it becomes more desirable to send a message to a defined geographic location/area, geocasting in OppNets is addressing such problem; to send a message to all nodes currently located in a geographic location/area.

Geocasting has directed researchers to focus on location-based content sharing, i.e. content is not associated with a node but a geographic location e.g. floating content [5], however, the challenge is more associated with routing to the destination location than floating the content within a defined area. Geocasting offers new services and applications in the real-world implementation of opportunistic networks; sending an emergency notification for evacuation in a disaster-affected region, location-targeted advertisement, finding nearby friends or publish-subscribe service. Opportunistic network geocast can play a major role in disaster scenarios, e.g. in the aftermath of a natural disaster (earthquake or tsunami) existing infrastructure is partially or completely destroyed rendering mobile networks inaccessible. This kind of situation warrants all forms of improvising means of communication. Establishing a connection between mobile devices opportunistically without requiring infrastructure could play an important role in sending rescue notifications. Location targeted advertisement is also feasible with Geocast in scenarios where a large number of users gather in a small location in which cellular service degrades with an increase in the number of users in a small location. Another possible application of Geocasting is a service discovery confined within a geographic location, this applies to government censorship in restricting a broadcast of a news item.

Routing in OppNet geocast can be classified into; inter-cast and intra-cast routing. Inter-cast routing is aimed to forward the message to destination cast while intra-cast used flooding approach of disseminating message within a cast, there are also routing approaches that enhance the delivery performance of intra-cast routing classify message recipients, for instance, to address only nodes with a given a buffer requirement and energy threshold.

So far, a large portion of survey on general OppNet has been proposed. Similarly, research has been conducted for geocast MANET protocols that focus mainly on networks with a predefined route. However, little attention is given to opportunistic geocast solutions. To the best of authors' knowledge, researchers are yet to carry out an investigation that leverages the research issues in OppNet geocast. The research in [6] highlighted some application of geocast in OppNet scenarios. But the outcome of the work targeted general OppNet routing solutions. Therefore, this work highlights the state-of-the-art survey for OppNet geocast routing protocols with a holistic view of different research areas in movement prediction methods, message endpoint identifications, forwarding utility computations, forwarding algorithms and message dissemination. Each of these areas has been investigated with examples and finally, the illustrated different areas of research and contributions of existing works can provide researchers with opportunities to further address research challenges in OppNet geocasting.

The article is structured as follows, the next section discusses the background, envisioned applications and unique characteristics of OppNet geocast routing. Section 3 describes a taxonomy describing a routing flow of OppNet geocast routing. Section 4 describes the current and potential directions of OppNet geocast routing. The last section concludes the paper.

2 Background on Opportunistic Geocast Routing

The idea of sending a message selectively to a location defined by x and y coordinate is introduced by [7], the work envisioned the use of GPS for location-dependent services, although, it is specifically designed to address routing and addressing method to integrate $x\,and\,y$ coordinates into the IP (Internet Protocol) to facilitates location-dependent services. However, the widespread portable handheld devices with sensing capabilities have open more research issues that are largely connected to location-dependent services. The application of geocast service (Emergency notification, advertisement, military tactical manoeuvres etc.) is more practical and economical in the mobile opportunistic network because it does not require pre-installed infrastructure to support its communication.

2.1 Envisioned Applications of Geocast Routing

There are various real-world applications of geocasting that targets geographical area/location instead of sending message to an individual device. Geocasting can support communication in critical situations. Emergency notification such as natural and manmade disasters, advertisement in a large gathering such as a sporting event, geographically restricted service discovery, data offloading to compliment the service of cellular networks. Major real-world applications of OppNet geocasting are outlined below:

- *Emergency notification in pre-disaster and post-disaster situations*. Study shows that every second natural disaster affects one person in the world (IDMS, 2017), this is caused by extreme environmental and climate changes that are continuously disrupting the natural ecosystems, severe natural disasters may likely to occur continuously which results in a huge impact on people and their welfare especially in developing economies. Using technology to support communication in such an environment will improve public service effectiveness. Although, rapid deployment of communication networks in the aftermath of disaster for data communication between all responding entities (rescue command and control Centre, casualty collection area etc.) is difficult [8].

- *Advertisement (Location Targeted Advertisement).* OppNet geocasting can be used for an advertisement that targets a specific location. In some cases a large number of users are gathered in a small area such as a football stadium, the cellular network is not designed to support communication in such scenarios [9]. Which degrades mobile network reception [10], although, the rollout of 5G networks is coming with significant improvement in data capacity and throughput, experts predict 10 times data transmission speed compared with the existing 4G and support of content sharing, 5G is specifically designed to support IoT and Big Data with a huge cost of infrastructure replacement. Opportunistic advertisement, *when people connect, they influence and persuade.* Placing OppNets in IoT will influence many business opportunities [11].
- *Geographically restricted service discovery.* In OppNet geocasting, mobile devices can request restricted services. Context-based data could be subscribed as well (e.g. customised news and advertisement). Floating content – social media plays a significant role in connecting people from in and around the world with online social networks such as Facebook, twitter etc. however is not always desirable to rely on infrastructure services for location-aware services [12] due to location privacy, content privacy, connectivity, geographic validity, temporal validity, and user identification.
- *Data offloading-Compliments the services of Cellular Networks.* The growing popularity of smartphones applications users and the need to share contents (video, audio) and access to cloud-based services have increased the traffic of cellular networks, mobile cellular networks cannot support the unprecedented overload on the existing infrastructure, upgrading such infrastructure is capital intensive. In addition, regulations have forced the mobile operators to utilise only a small portion of the overall licensed radio spectrum, which is equally expensive. Data offloading is considered as a complementary wireless technology to transfer data originally targeted to flow through the cellular network to improve some key performance indicators [13]. Due to these challenges experts result in finding ways of adopting various schemes to solve the problem by sharing the limited spectrum. [14, 15] envisioned data offloading in opportunistic mobile networks to relieve the burden of mobile data traffic of cellular networks.

2.2 Unique Characteristics of OppNet Geocast Routing

Mobile devices are versatile and can contribute differently to various wireless communication domains. OppNet geocasting preferably exploit spatial and temporal aspects of user movement. GPS-enabled devices to establish its location, use of the identified location to address a message and short-range communication between devices within a communication range. The main characteristics of geocasting in OppNets are as follows.

- **High Degree of Spatial and Temporal Aspects of User Movement.** Researchers give more attention to important mobility metrics, like contact time and inter-contact time between nodes. Attention is also given to delay and network capacity, the model has two parameters to measure the delay incurred in the network [16]. The spatial mobility characteristics are mostly ignored in the design of communication protocols for mobile partitioned networks, this information is vital and improves the efficiency and effectiveness of protocols in a partitioned network especially when mobility pattern is repetitive over a period [17]. However, in opportunistic networks both spatial and temporal aspects of nodes' mobility need to be considered [18]. The spatial characteristics depict particles movement in physical space has attracted the attention of researchers in different field of studies such as physical, geographical and social sciences. Spatial locality suggests that people move and spend most of their time of the day within a geographical region, while regularity in human movement indicates that people visit one or few places and have a tendency of revisiting the places in the future.
- **Messages are Sent Explicitly to Geographic Location.** Inexpensive portable handheld devices are equipped with global positioning system (GPS) that enable devices to establish their coordinates and eventually location-based applications and services of user devices were introduced for various communication schemes, for instance in ad-hoc networks, geometric ad-hoc routing protocols [19] is proposed to enhance efficiency and scalability, furthermore geocast routing solutions have gain researchers attraction in recent years [20], but mobile device penetration has introduced the new paradigm shift in geocast communication service and solutions. OppNet geocast differs in some aspects with general OppNets for instance, messages are destined to a given geographical area [18, 21, 22] while in general OppNets routing solutions that uses node's identification as message destination [23, 24].
- **Nodes Address Corresponds to Their Geographic Location.** Location as message Endpoint Identifier (EID) Geocast message must contain information about destination as its endpoint identifier [18]. Casts are pre-defined and known to all networking devices, cast identifier may be included in the message otherwise message destination address can be defined as one of the following; cycle [21, 25], cycle with a radii [5], polygon coordinates [18], squire grid [26, 27], a message must include one of the above cast definition, a forwarding node always compares the received message ID with its own location.

3 A Taxonomy of OppNet Geocast Routing Protocols

Due to the unique aforementioned characteristics, routing messages from a source to destinations that are identified by a specified geographic location follows the flow described in Fig. 1. The taxonomy consists of five phases: (i) acquiring location information on-the-go; (ii) defining the destination cast as well as determining whether the current position is located within it; (iii) computing a forwarding utility to quantify the routing contribution that can be made by nodes encountered outside the destination cast; (iv) selecting a relay based on the forwarding utility; and (v) disseminating the message among nodes encountered within the destination cast. This section summarises these phases and identifies related research contributions.

3.1 Predicting Node Movement

Two tasks are involved in predicting node movement [25, 28]. The first task (i.e., collecting mobility traces) has to do with sampling and maintaining node location, while the second task deals with extracting information relevant to a given message from the available movement history.

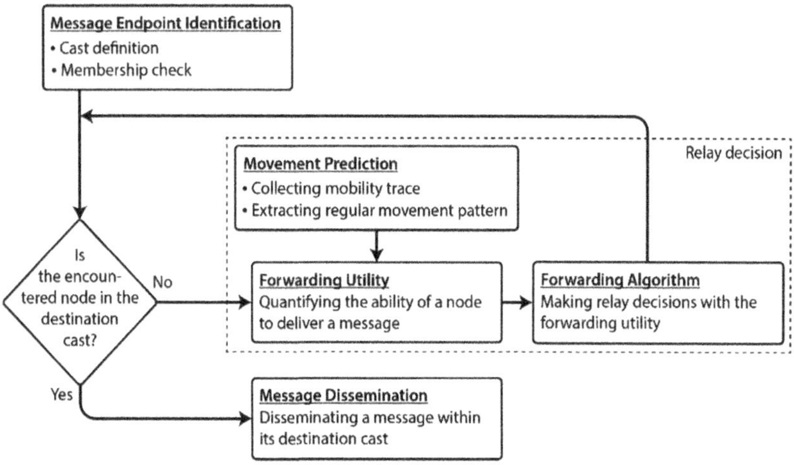

Fig. 1. Taxonomy oppnet geocast routing protocols

- **Location Sampling.** Node location may either be sampled periodically or triggered by an external event. In periodic sampling, the choice of sampling interval presents a trade-off between accuracy and energy consumption: fine-grained intervals provide a more accurate movement history and higher energy consumption, while larger intervals consume less energy but may fail to record important locations. With an event-triggered approach, the location is sampled only when a node is encountered. While this records only locations in which encounter opportunities are available, some of the encounters may not be regular. Furthermore, a high population in urban areas may lead to energy consumption concerns. [25] provides a detail location sampling of nodes, while in [27] and [18] information about location sampling is not specified. LoSeRO and DTN-FLOW [26, 29] collect GPS traces at regular interval and logged into a file to determine a correlation of a number of visits between nodes to a given location.
- **Location Information Maintenance.** The authors in [26] define a place as a collection of nearby locations in the network. The geographic area of the network is partitioned by a grid of square cells, and each cell is considered as a place. Every

node then maintains a list of its most visited places, namely its MobyZone. In [25] a location is obtained either by Global Positioning System (GPS) or Geolocation API. In LOOP, the overall mobility record of a device is in the format $r_i = \langle d_i, t_i, l_i \rangle$, the equation indicates l_i is a location visited by a node on a day d_i and a location l_i.

3.2 Identifying Message Endpoint

A message in OppNet is created along with the Endpoint Identifier specifying the destination address of the message. Unlike unicasting and multicasting, the naming approach in geocasting depends on the geographical location/area and the membership of the location. A location could be a point, polygon, grid, circle etc. Membership of a node is subject to time and space. However, this is different from unicasting and multicasting, message destination is assigned to a node with its Endpoint Identifier (EID) [18].

- **Cast Definition:** In geocasting, the message destination is defined as a cast, a cast is a geographical area considered as a destination of a transmitted message. All network users reside within a defined geographical area is a potential recipient of a message, in other words, a user is a recipient if a message destination address corresponds with a user current location. A cast can be defined as a point, circle, polygon, grid, etc. floating content defines cast as a circle with a user defines the centre and two radii r and a [5]. Similarly [21, 30] defines cast as a circle with user defines centre and radius r. [26] and [27] uses a square grid with a pre-define grid. Authors in [29] defines cast as a Landmark sub-area (a pre-defined area) Louvre [31]. Authors [25] defines cast as point, a user defines the point. [18, 32] defines cast a polygon, a user defines the cast as a Point of Polygon, casts are defined as a set of pairs of two dimensional coordinates, an approach that is used widely in computer graphics and image processing and the strategies (Point in Polygon) are used to identify the position of point inside or outside the polygon, these strategies include Crossing Test, Angle Summation Test and Triangle Test [33].
- **Membership Check:** Node is identified with a location(s) to qualify as a recipient of a message, the node checks whether it is a member of a location i.e. belongs to a cast. Frequently visited places also known as Point of Interests (PoIs) are locations nodes often visit. [18, 32] check nodes currently located in Point in Polygon. Floating Content aggregates all nodes currently located in the radius as a criterion to fulfil its membership to the cast and a valid recipient of a message. Authors in [26, 29] determine the membership of nodes in the cast as a Visit-cast. LOOP and the works in [21, 30] is applicable to any (Area-cast, Home-cast, Visit-cast) membership criteria.

3.3 Computing the Forwarding Utility

Forwarding utilities in OppNet geocast routing are usually probabilistic values that are computed when nodes encounter each other. Hence, larger values indicate that the encountered node can make more contribution towards delivering the message in concern. The forwarding utility in [12] is derived from the number of visits that the encountered node has made to the destination cast of the message. However, it may be challenging to find a node that often visits the destination cast if the message is generated far away from the destination cast. The work in [21] is another approach that divides the geographic area of the network into non-overlapping cells. A suitable relay node is selected based on a factor known as the Expected Visiting Rate (EVR), which illustrates the frequency at which it visits the destination cast. While these factors enable the forwarding utility to better indicate suitable relay nodes, the division of network area into cells is unclear [18] as different sizes would affect network performance. Authors in [25] propose a more sophisticated approach by predicting future location visits from locally obtained mobility records in the form of day type, time slot and visited a location. Two forwarding utilities, namely original–landmark metric and current–landmark metric, are presented, either of which can be used to determine a more suitable relay node.

3.4 Forwarding Algorithm

Opportunistic network paradigm was proposed for challenged environment i.e. environment where end-to-end route between source and destination does not exist. The forwarding algorithm is designed for a message carrier to always identify the best node with higher chance to relay a message to its destination. Different approaches are proposed for relay node selection; Utility-based forwarding algorithms require a node to maintain an up to date utility metric to suitable relay node [34]. The forwarding decision is made based on the nodes' forwarding utility metric. Authors in [35] utilise contextual information using Kalman filter, Seek and Focus in [36] is a timer-based used to select a node with recent encounter time with a destination, [21] utilises Expected Visiting Rate to select relay node. [25] Offers two strategies for relay decisions i.e. flexible comparison and adaptive relay strategies. Visit patterns is used to select a relay node as well, [37] network area is divided into regions and each user maintain a probability of visits to each of those regions. Similarly, [38] divided the network location into an equal-sized square grid, nodes maintain a list of the frequently visited region to determine their contribution towards message forwarding.

3.5 Message Dissemination

Message dissemination within the cast is a typical floating content issue as best described in [5, 12] this work offers an opportunity to enable localised information sharing without the need for infrastructure, data collection or central control. Ultimately it offers some degree of privacy by not allowing remote access to content within a defined geographical location. The works in [18, 32] offer privacy-aware message dissemination. DA-GSAF (Direction-Aware Geocasting Spray and Flood) is an extended GSAF, the algorithm offers location-based routing without sharing user context information in the network. DA-GSAF has two phases, the first phase applies the following conditions; if there is a copy of bundle left to distribute another copy, if the current encountered node is not heading towards the destination location and if the current encountered node is moving towards the message destination location. Flooding a message is resource consuming but could guarantee optimum message delivery ratio [39].

4 Current State and Potential Research Direction in OppNet Geocasting

Geocasting plays an important role in scenarios where there is partial or no access to communication infrastructure due to large scale disasters, terrorist attack or government censorship. The concept of smart cities, which is fuelled by a human population increasingly moving towards urban areas can stand to benefit from OppNet geocasting. Determining nodes that can contribute to the delivery of a message requires a general understanding of user movement. Fortunately, human movement is regular to a certain degree that it can be predicted. Routing schemes are thereby designed to learn the user's movement patterns from mobility traces collected with the device on the go. The spatiotemporal properties of human movement, which refers to the tendency for people to move within a local region and visit only one or few places regularly, can be exploited to make geocast routing decisions. The following are the summary of the current state and future direction of OppNet geocasting.

- Geocast routing follows a generic approach for providing solutions to routing issues, they are not specifically designed for smart city scenarios with a very high rising node population. The message transmission overhead and overhead of metadata have been addressed in some existing works. However, an increase in node population in smart city environment results in excessive information overhead in which existing solutions cannot handle.
- High membership check overhead. Membership check overhead incurred by the polygon approach increases with the number of points in the message EID. Even if the issue of usability is resolved via a user-friendly interface (e.g., one that allows users to select destination casts from a list of location names and then converts the selection to the appropriate set of coordinates for the message EID).
- Poor usability in cast definition. From the users' perspective, defining casts via the polygon approach brings about usability issues. In a post-disaster scenario, for instance, users heard about the affected area through the media and may only be aware of the names of specific streets or suburbs that were mentioned. Hence, in real-world, users may not often be provided with detailed knowledge about the range of the affected area.
- Consequently, nodes whose movement history suggest that they will be moving closer to the destination cast or they often visit the destination cast are considered suitable relays in existing works, However, visiting only a few places also denotes that nodes tend to stay in these locations for relatively long periods of time [40, 41]. Therefore the history of node's encounter alone cannot ensure delivery of message within its lifetime.
- Acquiring movement traces through GPS for forwarding utility computation depletes the resource constraint mobile devices. Investigating the impact of collecting movement traces on mobile nodes in a real world implementation may bring about improvement in OppNet routing solutions.

5 Conclusion

In this paper, we have presented a survey of OppNet geocast routing protocols. The routing protocols differ in how the regular movement pattern is extracted from movement traces, the definition of the targeted message destination location, forwarding utility computation, and routing approach namely; inter-cast (routing between locations) and intra-cast (routing within the location). Accordingly relay decisions that play a fundamental role in ensuring the message is delivered to the intended destination location is highlighted and discussed. Finally, message dissemination within the defined geographical area has been studied and compared. The major outcomes of this survey are twofold, firstly, the taxonomy that illustrates the flow of OppNet geocast routing. Secondly, the current state and future direction of OppNet geocast. The taxonomy identifies challenges related to each aspect of routing in OppNet geocast and the future directions highlight routing challenges as well as techniques to define a message destination cast (location) i.e. the concept of message addressing and End Point Identification.

References

1. Conti, M., Giordano, S., May, M., Passarella, A.: From opportunistic networks to opportunistic computing. IEEE Commun. Mag. **48**(6), 126–139 (2010)
2. Fall, K.: A delay-tolerant network architecture for challenged internets. In: Proceedings of the 2003 Conference on Applications, Technology, Architectures and Protocols Computer Communications - SIGCOMM 2003, pp. 27–34 (2003)
3. Fall, K., Farrell, S.: DTN: an architectural retrospective. IEEE J. Sel. Areas Commun. **26**(5), 828–836 (2008)
4. Conti, M., Kumar, M.: Opportunities in opportunistic computing. Computer **43**, 42–50 (2010)
5. Ott, J., Hyytiä, E., Lassila, P., Vaegs, T., Kangasharju, J.: Floating content: information sharing in urban areas. In: 2011 IEEE International Conference on Pervasive Computing and Communications PerCom, pp. 136–146. IEEE (2011)
6. Amah, T.E., Kamat, M., Moreira, W., Bakar, K.A., Mandala, S., Batista, M.A.: Towards next-generation routing protocols for pocket switched networks. J. Netw. Comput. Appl. **70**, 51–88 (2016)
7. Navas, J.C., Imielinski, T.: GeoCast - geographic addressing and routing. In: Proceedings of the 3rd Annual ACM/IEEE International Conference Mobile Computing Network (MobiCom 1997), pp. 66–76 (1997)
8. Manoj, B.S., Baker, A.H.: Communication challenges in emergency response. Commun. ACM **50**(3), 51–53 (2007)
9. Shafiq, M.Z., Ji, L., Liu, A.X., Pang, J., Venkataraman, S., Wang, J.: Characterizing and optimizing cellular network performance during crowded events. IEEE/ACM Trans. Netw. **24**, 1308–1321 (2016)
10. Wakeman, I., Naicken, S., Rimmer, J., Chalmers, D., Fisher, C.: The fans united will always be connected: Building a practical DTN in a football stadium. In: Lecture Notes of the Institute for Computer Sciences, Social-Informatics and Telecommunications Engineering, LNICST (2014)

11. Guo, B., Yu, Z., Zhou, X., Zhang, D.: Opportunistic IoT: exploring the social side of the internet of things. In: Proceedings of the 2012 IEEE 16th International Conference Computer Supported Cooperative Work in Design CSCWD 2012, pp. 925–929 (2012)
12. Ott, J., Hyyti, E., Lassila, P., Kangasharju, J., Santra, S.: Floating content for probabilistic information sharing. Pervasive Mob. Comput. **7**(6), 671–689 (2011)
13. Rebecchi, F., et al.: Survey data offloading techniques in cellular networks : a survey. IEEE Commun. Surv. Tutor. **17**(2), 580–603 (2014)
14. Han, B., Hui, P., Kumar, V.S.A., Marathe, M.V., Pei, G., Srinivasan, A.: Cellular traffic offloading through opportunistic communications (2010)
15. Han, B., Hui, P., Kumar, V.S.A., Marathe, M.V., Shao, J., Srinivasan, A.: Mobile data offloading through opportunistic communications and social participation. IEEE Trans. Mob. Comput. **11**, 821–834 (2012)
16. Groenevelt, R., Nain, P., Koole, G.: The message delay in mobile ad hoc networks. Perform. Eval. **62**(1–4), 210–228 (2005)
17. Piorkowski, M.: Mobility-centric geocasting for mobile partitioned networks. In: Proceedings of the International Conference on Network Protocols, ICNP, pp. 228–237 (2008)
18. Rajaei, A., Chalmers, D., Wakeman, I., Parisis, G.: Efficient geocasting in opportunistic networks. Comput. Commun. **127**, 105–121 (2018)
19. Kuhn, F., Wattenhofer, R., Zhang, Y., Zollinger, A.: Geometric ad-hoc routing: of theory and practice. In: Proceedings of the 22nd Annual Symposium on Principles Distributed Computing (PODC) (2003)
20. Maihöfer, C., Daimlerchrysler, A.C.: Geocast routing protocols. Delab. Csd. Auth. Gr. **6**(2), 32–42 (2004)
21. Ma, Y., Jamalipour, A.: Opportunistic geocast in large scale intermittently connected mobile ad hoc networks. In: The 17th Asia Pacific Conference on Communications, pp. 445–449. IEEE (2011)
22. Costantino, G., Maiti, R.R., Martinelli, F., Santi, P.: LoSeRO: a locality sensitive routing protocol in opportunistic networks. In: Proceedings of the 31st Annual ACM Symposium Applied Computing - SAC 2016, vol. 14, no. 8, pp. 644–650 (2016)
23. Spyropoulos, T., Psounis, K., Raghavendra, C.S.: Spray and focus: efficient mobility-assisted routing for heterogeneous and correlated mobility. In: Proceedings of the Fifth Annual IEEE International Conference Pervasive Computing and Communications Workshops (PerCom Work 2007), pp. 79–85 (2007)
24. Vahdat, A., Becker, D., et al.: Epidemic routing for partially connected ad hoc - google scholar (2000)
25. Lu, S., Liu, Y., Liu, Y., Kumar, M.: LOOP: a location based routing scheme for opportunistic networks. In: MASS 2012 – 9th IEEE International Conference on Mobile Ad-Hoc and Sensor Systems, pp. 118–126 (2012)
26. Costantino, G., Maiti, R.R., Martinelli, F., Santi, P.: LoSeRO. In: Proceedings 31st Annual ACM Symposium Applied Computing - SAC 2016, pp. 644–650 (2016)
27. Lu, S., Liu, Y.: Geoopp: geocasting for opportunistic networks. In: 2014 IEEE Wireless Communications and Networking Conference (WCNC), vol. 3, pp. 2582–2587. IEEE (2014)
28. Yuan, Q., Cardei, I., Wu, J.: Predict and relay: an efficient routing in disruption-tolerant networks. In: Proceedings tenth ACM International Symposium Mobile ad hoc Network Computing - MobiHoc 2009 (2009)
29. Chen, K., Shen, H.: DTN-FLOW: inter-landmark data flow for high-throughput routing in DTNs. IEEE/ACM Trans. Netw. **23**(1), 212–226 (2015)
30. Ma, Y., Jamalipour, A.: Opportunistic geocast in disruption-tolerant networks. In: GLOBECOM - IEEE Glob. Telecommunications Conference (2011)

31. Lee, K.C., Le, M., Härri, J., Gerla, M.: LOUVRE: landmark overlays for urban vehicular routing environments. In: IEEE Vehicular Technology Conference (2008)
32. Singh, J., Dhurandher, S.K., Woungang, I., Takizawa, M.: Centrality based geocasting for opportunistic networks, vol. 1. Springer, Heidelberg (2019)
33. Eric, H.: Point in polygon strategies. In: Graphics Gems IV (1994)
34. Wang, T., Cao, Y., Zhou, Y., Li, P.: A survey on geographic routing protocols in delay/disruption tolerant networks. Int. J. Distrib. Sens. Netw. **2016** (2016)
35. Musolesi, M., Mascolo, C.: CAR: context-aware adaptive routing for delay-tolerant mobile networks. IEEE Trans. Mob. Comput. **8**(2), 246–260 (2009)
36. Spyropoulos, T., Psounis, K., Raghavendra, C.S.: Efficient routing in intermittently connected mobile networks: the single-copy case. IEEE/ACM Trans. Netw. **16**(1), 63–76 (2008)
37. Burns, B., Brock, O., Levine, B.N., Burns, B., Levine, B.N.: MV routing and capacity building in disruption tolerant networks. In: Proceedings of the IEEE 24th Annual Jt. Conference IEEE Computing Communications Society, pp. 398–408 (2005)
38. Aviv, A.J., Sherr, M., Blaze, M., Smith, J.M.: Privacy-aware message exchanges for geographically routed human movement networks. In: Lecture Notes in Computer Science (including subseries Lecture Notes in Artificial Intelligence and Lecture Notes in Bioinformatics) (2012)
39. Zhang, X., Neglia, G., Kurose, J., Towsley, D.: Performance modeling of epidemic routing. Comput. Netw. **51**, 2867–2891 (2007)
40. Hasan, S., Schneider, C.M., Ukkusuri, S.V., González, M.C.: Spatiotemporal Patterns of Urban Human Mobility. J. Stat. Phys. **151**, 304–318 (2013)
41. Zhang, L., Cai, Z., Lu, J., Wang, X.: Spacial mobility prediction based routing scheme in delay/disruption-tolerant networks. In: Proceedings 2014 International Conference Identification Information Knowledge Internet of Things, IIKI 2014, pp. 274–279 (2014)

Hybrid Storage Management Method for Video-on-Demand Server

Ola A. Al-wesabi[1,2(✉)], Nibras Abdullah[2,3], and Putra Sumari[1]

[1] School of Computer Science, Universiti Sains Malaysia (USM), 11800 Gelugor, Penang, Malaysia
ola_osabi@yahoo.com, putras@usm.my

[2] Faculty of Computer Science and Engineering, Hodeidah University, Hodeidah, Yemen
neprasf@gmail.com

[3] National Advanced IPv6 Center, Universiti Sains Malaysia (USM), 11800 Gelugor, Penang, Malaysia

Abstract. One of the important components of video-on-demand (VOD) systems is the hybrid storage server. This component consists of use hard disk drives (HDDs), solid-state drives (SSDs), and RAM, collectively fulfilling the requirement of simultaneous fast data access and large data distribution to numerous users. However, current hybrid storage systems still pose numerous challenges. The integration and roles of the HDD, SSD, and RAM are relatively weak in terms of optimizing fast access prior to streaming to a large number of simultaneous users. The HDD and SSD exhibit poor data layout and streaming controller in supporting the production of a high number of simultaneous streams. This paper proposes (1) the flash cache hybrid storage system (FCHSS) VOD servers. The FCHSS architecture has no RAM, which is removed and replaced by a flash-based SSD. (2) The new data layout stores thousands of video segments in the HDD and SSD. The streaming management scheme, namely, flash cache-data streaming controller (FC-DSC), is also proposed to support FCHSS. The proposed VOD server-based FCHSS with the FC-DSC shows a 73.53% and 23.71% enhancement in average total response time, and a 35.63% and 298.91% enhancement in throughputs for all request sizes compared with the average of the total response time and throughputs of the hybrid storage system HSS VOD server- and the VOD server-based feedback-based adaptive data migration (FADM).

Keywords: Flash cache hybrid storage system (FCHSS) · Flash cache-data streaming controller (FC-DSC) · Hard disk drives (HDDs) · I/O response time · Solid-state-drive (SSD) · Throughput · Video on demand (VOD)

1 Introduction

A video-on-demand (VOD) system allows users to access any video at any time. This system enables users to view videos in the real-time streaming mode for extended durations. Given that the main function of a storage system is to store video data, a brief discussion about video data is given [1].

F. Saeed et al. (Eds.): IRICT 2019, AISC 1073, pp. 695–704, 2020.
https://doi.org/10.1007/978-3-030-33582-3_65

Video data are continuous media data that change over time. In other words, the time at which the data are presented influences the validity of the information. Therefore, processing of such video data is time sensitive [2]. In addition, a video contains a set of sequentially shown frames. Normally, video data involves three requirements: size, real time, and throughput.

A VOD server contains a large number of large-sized videos. Thus, storing data in the main memory is impossible. For this reason, most video data are usually fetched and read from a secondary storage device. Therefore, the storage subsystem has to achieve the demands imposed by multimedia data [3, 4].

Given that video data are continuous media data presented as sequences of frames, failure to meet the real-time constraint causes a hiccup in video presentations. On the server side, the storage subsystem should be able to satisfy real-time requirements. In general, playing back video data involves retrieving data from hybrid storage devices (HDDs and SSD); however, owing to the weak collaboration among storage devices (e.g., HDDs) still have high seek time and high rotational latency even if disk-scheduling algorithms and admission controllers are employed to reduce seek time and rotational latency [5, 6].

In this paper, the design of the proposed enhanced hybrid storage server for a VOD system is presented. The aims of this study are to improve the hybrid storage for the VOD servers. Second, this research proposes a video streaming management scheme required to manage user requests and helps enhance the performance of hybrid VOD servers.

In the subsequent sections of paper, the related works of hybrid storage VOD server, the proposed methods, experimental design, evaluation results and conclusion will be discussed in detail.

2 Related Studies

The most related studies on employing a hybrid storage (HDD/SSD) to enhance the performance of the VOD storage server are discussed as follows:

Ling et al. proposed a feedback-based adaptive data migration (FADM) method for VOD hybrid storage devices (HDD/SSD) [7]. FADM consists of a hybrid storage, a replacement manager (RM), a data migration manager (DMM), and a threshold manager (TM). The hybrid storage consists of HDDs and the cache. The cache in FADM consists of two parts: C is cache device (RAM) and Cs is SSD. The most popular segments are placed in SSD for better performance. This research concentrated on controlling the write load of SSDs because of the limitation of the number of erase cycles of SSDs. The rule of moving data between HDDs and SSDs in real time was adopted. This method allows segments in HDD to be moved into SSD when its popularity is higher than that of the least popular video in SSD by DMM. In this study, the authors mentioned that the size of C represents 5% of the total size of all videos in blocks, whereas the size of Cs equals 10% of the total cache size. However, the procedures of DMM, RM, and TM after serving each request increase the overhead processing time. Also, the architecture caches a part of the videos in the local storage and performs the migration based on the updated popularity of the requested video blocks. This does not guarantee the continuity of streaming videos.

In previous work [6], the HSS VOD server was proposed. In this proposed VOD storage server system, SSD is used as disk storage in the VOD server to provide increased I/O bandwidth with additional flexibility, such as the capability to specify different I/O priorities for different volumes in the VOD storage system. The HSS uses SSD-based storage as an extension of the permanent HDDs that hold the most popular videos. Thus, many popular videos can be served quickly and simultaneously. Unpopular videos are also considered, with the RAM used as a buffer cache to the requested videos that are located in the HDDs.

The proposed FCHSS architecture has no RAM, which is removed and replaced by a flash-based SSD. The flash-based SSD is used instead of the RAM as a cache for the HDD to achieve fast data access and service to a large number of simultaneous users.

3 Methods

The FCHSS consists of a central server, a VOD unit cluster server, hybrid unit (HU) components, and a proxy server. The unit cluster is deployed in a local area network together with a number of users. The enhanced architecture comprises an extra component called the HU. The HU consists of HDD, SSD, and The flash-based SSD. The three storage components constitute the mechanism of the video being streamed to the users. Figure 1 presents the proposed FCHSS VOD server with its software architecture as an enhanced HU for VOD systems.

The inclusion of a HU in the system delivers several features.

1. The HU is attached only to the cluster server because this server is closest to the users and is the busiest server in terms of streaming to many users. If the request video is located at cluster server, the content is directly streamed to the user. Otherwise, it has to pulls the video from the proxy or central server.
2. The SSD stores the top 10% popular videos. Popular videos constitute more than 80% of the total views [8, 9]. Hence, 10% of the total number of videos that have the highest popularity are stored in SSD provides the highest read speed and minimize start-up latency.
3. The HDD is assisted by a flash-based SSD with a small size (16 GB) as a buffer cache for videos. Thus, the requested videos are fetched segment by segment from HDD and sent to flash-based SSD. In addition, certain old videos are deleted to accommodate new videos when no available space exists in the flash-based SSD.
4. The SSD operates independently and without assistance from the flash-based SSD.

The structure of the proposed FCHSS server consists of the following:

- Hybrid storage (HDD/SSD and flash-based SSD)
- Software architecture that consists of the flash cache-media management system (FC-MMS) component, flash cache-controller (FC-C) component, flash cache-data stream controller (FC-DSC) scheme (FC-DSC-HDD and FC-DSC-SSD components), index table (MDMT), and data layout scheme component.

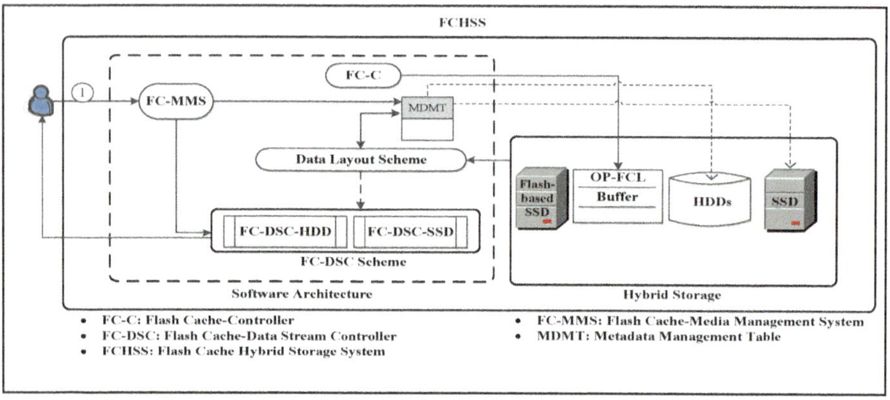

Fig. 1. Proposed FCHSS VOD server with its software architecture.

3.1 FC-MMS Component

When requested videos that hit the FCHSS VOD server are not located in its hybrid storage devices, the FCHSS VOD server retrieves the content from the central server or the nearest proxy server. Then, FC-MSS locates the requested content in HDD and transmits it to the client. FC-MMS includes information about the physical address of the new video inside MDMT. FC-MSS manages the requests between the elements in the FCHSS VOD server.

MDMT stores the information of the video files located inside the FCHSS VOD server. This information determines the place of the videos in one of the hybrid storages (HDD/SSD), in which 10% of popular videos are located in SSD. Moreover, MDMT stores the specifications related to video files, such as video size, status, and physical address. MDMT is located in the SSD-based storage for rapid accessing and reading, flexible control, and easy recovery during power failures. The flowchart of the FC-MMS procedures when the requests of users hit the FCHSS VOD server is presented in Fig. 2. The figure also shows the decision controller process, in which the FC-MMS scheme is executed to serve ongoing requests.

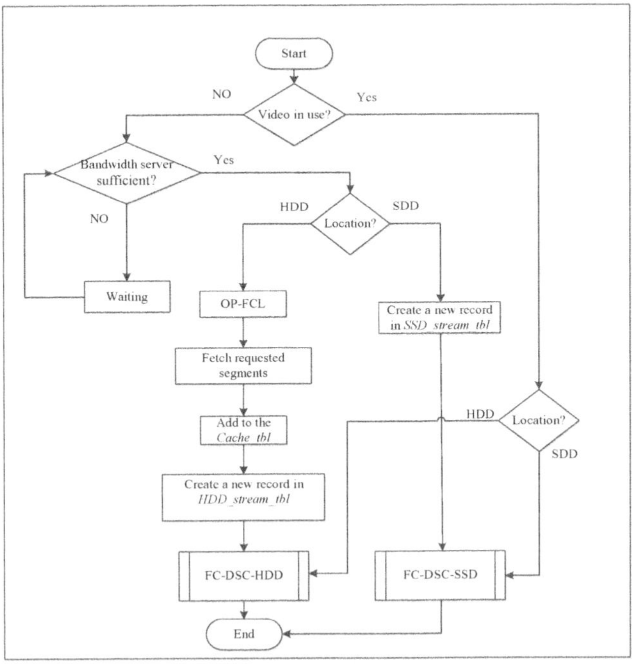

Fig. 2. FC-MMS process.

3.2 FC-C Component

In the FCHSS VOD server, flash-based SSD is used as a cache instead of RAM. The replacement of RAM with flash-based SSD requires careful use of the latter because of the differences between the two architectures. As a result of the efficient management of metadata and of the handling of discarded RAM data, flash-based SSD requires an intelligent design to overcome these challenges and avoid a reduction in performance. Although the workload of the VOD application mostly includes reading, flash-based SSD as cache must maintain written data workloads for buffer caching with protected writing for better utilization.

For the optimal use of flash-based SSD, an existing intelligent technique, namely OP-FCL [10], is adopted to maximize performance by controlling the read and write processes for efficient management of garbage collection.

The SSD-based cache buffers video requests for videos stored in HDDs. The requested video data are fetched from HDD segment by segment to accommodate the small capacity of SSD, which is used as a buffer cache. Moreover, OP-FCL dominates and manages the blocks in flash-based SSD to avoid the negative effect of several operations, such as erase-before-write and garbage collection. OP-FCL improves the overall performance of flash-based SSD because of the intensive write workload that results from the nature of the caching process. The use of the OP-FCL technique depends on the performance criteria of the work of the FC-DSC scheme.

3.3 FC-DSC Scheme

FC-DSC scheme in the FCHSS VOD server utilizes multicasting in streaming video data to users. The flow of the FC-DSC scheme is related to the data layout scheme. In the video layout scheme, each video is partitioned into N segments, in which the segment size is constant S_{seg}. Every segment consists of a number of blocks with a block size of 512 KB. The total number of segments for each video is measured by dividing the length of the video by the segment size.

Notably, FC-DSC scheme keeps a record of each streaming to manage and exploit this streaming to other users' requests for the same video file. The information of the streaming records is kept in a streaming table (SSD_stream_tbl and HDD_stream_tbl). The streaming table consists of seven parameters. These parameters are defined as follows:

- *Stream_ID*: It pertains to the unique number identification for each new stream.
- *List_of_users*: This consists of a list that contains of a set of *user_id*. Basically, it pertains to users that share and join this stream.
- *Video_Name*: It indicates the video file title.
- *Seg_no*: It denotes the segment number of the current streaming that is continuously updated within the allotted time.
- *Address*: It refers to the physical address of the first block of the segment number.
- *Rem_b*: It refers to the value of the last segment number of the remaining segments. This column includes a list of the last segment numbers of the remaining segments for each user that joins this streaming having the same index.
- Tot_{seg}: It represents the total number of segments of the video files.

FC-DSC-SSD Component

The FC-DSC-SSD scheme handles the requests for video files stored in the SSD-based storage in the FCHSS server. The operations of the algorithm begin by looking for the segment number in the streaming table. The rows of the same video file in the streaming table are arranged in ascending order according to segment numbers. This ranking ensures that a new request that arrives at the streaming table is selected with the same required segment number or a close larger one. After determining the same segment number or a close larger one as the demanded segment, the *user_id* joins the list of users for the current row stream. This operation is sufficient if the segment number is the same. Otherwise, a new stream is created for the remaining segments. The FC-DSC-SSD algorithm generates a new stream when the required segment is larger than the segment number of the current stream and then begins streaming to the user (FC-stream-SSD algorithm).

FC-DSC-HDD Component

The procedures start by looking for the demanded segment number in the segment number (*seg_no*) column in the streaming table (*HDD_stream_tbl*). If the demanded segment number has the same or a larger segment number in the streaming table for the same video file, then the *user_id* is added to the list of users sharing the current stream. When remaining segments exist (demanded segment number < *seg_no* for the same video) and the demanded segment is available in the cache table, the location address of the segment is assigned to the newly created streaming in the streaming table for the remaining segments. Otherwise, the algorithm executes OP-FCL and fetches the requested segment from HDD to the SSD-based cache (SSD_{cache}) and assigns the location address of the fetched segment to the newly created streaming in the streaming table. Then, buffering streaming is initialized to the user by the stream-FC-HDD algorithm.

When the demanded segment number is larger than the segment number in the streaming table for the same video file, a new stream should be created in the streaming table. In this case, the address from the cache table should be obtained while the segment is still available in the buffer. Otherwise, OP-FCL should be applied, and the segment must be fetched from HDD to the SSD-based cache. The stream-FC-HDD algorithm is executed to start buffering streaming to the user.

4 Experimental Design

Modified DiskSim4.0 (FCHSS simulator) assists in designing the FCHSS VOD server [hybrid storage (HDDs and SSD) with another small SSD used as a cache for HDDs]. Different scenarios are executed to measure the performance of the proposed FCHSS VOD server architecture. The essential performance metrics are average response time, and throughput for the SSD-based cache device.

4.1 Simulation Setup and Parameters

The simulation of the FCHSS VOD storage server is modeled by the modified DiskSim4.0 (FCHSS simulator), which depends on the hybrid storage (HDD and SSD) with a small-capacity flash-based SSD as a cache for HDD. The capacity of the SSD-based storage is 128 GB, and the capacity of the SSD-based cache is 16 GB. Both types of SSD are MLC. In the FCHSS simulator, SSD emulates a typical MLC SSD whose average read and write latencies are 50 and 900 μs/page, respectively, with an erase latency of 2000 μs/block; the number of pages per block is 128 [11, 12].

Meanwhile, the type of HDD used in the FCHSS simulator is a validated Seagate Cheetah 15 K.5 model ST3146855FC with 146 GB capacity and an IO throughput of up to 200 MB/s [13, 14]. The number of HDDs is 16, which are striped as an array by RAID 0 of the storage server. The total capacity of the HDDs is 2336 GB.

To simulate the FCHSS VOD storage server, all video files are assumed to have the same constant bit rate. The typical constant bit rate for good-quality video streaming over the Internet is 245 KB/s. Furthermore, the video files are assumed to have the same length of 90 min, which is the usual length of movies [15]. Therefore, the size of the video file is approximately equal to 1292 MB. In this simulation, the number of video files is 1,200 distinct videos. Ten percent of the most popular video files are located in the SSD-based storage, whereas the rest of the video files (a larger number) are stored in HDDs.

The measurement evaluation for the design of the proposed FCHSS VOD server should follow the arrival of the requests to the multimedia server. DiskSim4.0 can internally generate synthetic workload or use external traces. In the previous HSS VOD server, synthetic traces are extracted in distinct files during simulations to generate synthetic VOD server trace models. The same synthetic traces that are used in the HSS simulator are used as external traces called log files for the FCHSS simulator. Each log file contains various scales of intensive workload. The validated disk parameters and the requests from log files are fed into the FCHSS simulator, which adopts our modular design for our proposed FCHSS VOD server architecture.

4.2 Simulation of the FADM VOD Server

The size of RAM and SSD in FADM is related to the total number of blocks for all video files. In the FADM study, the size of the RAM represents 5% of the total size of the entire videos in blocks, whereas the SSD size accounts for 10% of the total cache size. In this simulation setup, the number of video files is 1200, and the size of the blocks for each video is 2,584 divided equally into 23 segments. Thus, the size of the RAM is 155,040 blocks, and the SSD size is 15,504 blocks.

5 Simulation Results and Discussion

These simulations are applied to show the effect of the number of requests on the performance of the hybrid storage in the FCHSS server. These experiments are executed also to compare it with the previously the hybrid storage HSS VOD server and the VOD server-based FADM in terms of average response time, and throughput.

Nine log files (synthetic VOD server traces) are extracted. Each log file contains various scales of intensive workload. The validated disk parameters and the requests from the log files are fed into the modified FCHSS simulator that adopts the modular design for the proposed FCHSS architecture.

In this experiment, the number of requests starts from 1000 requests to 256,000, where the number of requests increases by 2x. Figure 3 easily shows the difference between the total time taken by the existing works to serve the request dispatcher and the proposed hybrid storage FCHSS VOD server. The proposed VOD server-based FCHSS that utilizes SSD as buffer cache in its architecture enhances the average response times for all various scales of intensive workload by 72.55% and 23.71% compared with the average of the total response times taken by VOD server-based FADM and the previous proposed VOD server-based HSS that uses RAM as a buffer cache, respectively.

As shown in Fig. 4, when the number of requests exceeds 1,000 requests, the aggregate throughput drops significantly until the number of requests is 4,000. When the number of the requests exceeds 4,000 requests, the aggregate throughput will drop slightly. This outcome is due to the HDD generates a random read I/O pattern to storage, which reduces the scalability of serving the number of concurrent video streams that the disk device can serve, when HDD serves a large number of video streams simultaneously. In addition, the existing HSS server stores the entire video file in the SSD-based storage that helps in benefiting from the increased IOPS provided by the SSD. Thus, the HSS VOD server outperforms the VOD server-based FADM, in which the throughput is improved by 194.11%. However, the proposed FCHSS VOD server outperforms the VOD HSS server- and the VOD server-based FADM, enhancing the throughputs by 35.63% and 298.91%, respectively, because of the use of SSD in its architecture at both storage and cache level.

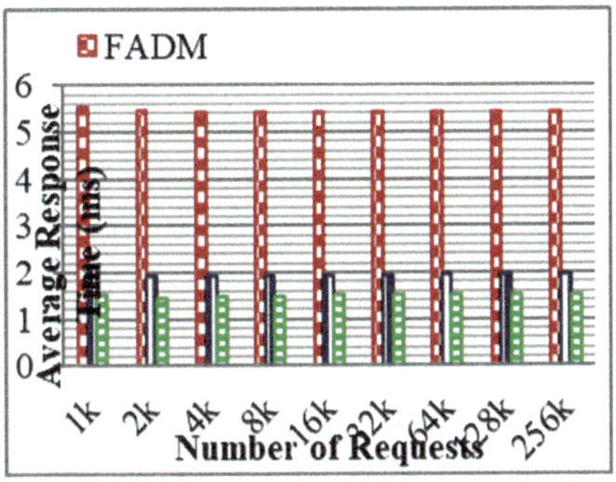

Fig. 3. Comparison between the FCHSS, the HSS, and the FADM in terms of average response time with various numbers of requests.

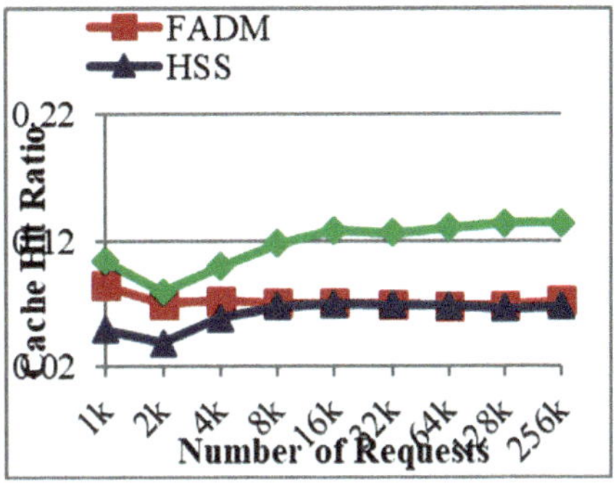

Fig. 4. Comparison between the FCHSS, the HSS, and the FADM in terms of throughput with various numbers of requests.

6 Conclusions

In a VOD system, the most important component is the server. The server stores a large number of video files and distributes these files from the disk storages to a large number of users. Data storage and distribution is a complicated task because of real-time requirements and large size of video data. These characteristics require servers to be able to quickly retrieve videos and distribute them to a large number of simultaneous users in real time. In this paper, the design of the proposed FCHSS VOD server structure is provided. Two main components (FC-MMS and FC-DSC) are proposed to improve the performance of the hybrid storage in the VOD FCHSS server. FC-MMS engages with the other components in the FCHSS VOD server, handles the user requests that reach the server, and directs the requests to the FC-DSC component according to the demand video location in the hybrid storage. FC-DSC is formulated to manage and control data streaming depending on the requests in the FCHSS VOD server. However, the experimental results highlight the performance improvement of the proposed VOD FCHSS server relative to the existing VOD HSS server and the VOD server-based FADM.

References

1. Al-wesabi, O.A., Sumari, P., Al-wesabi, M.A.: Proxy caching strategies for multimedia distribution. Int. J. Adv. Comput. Technol.(IJACT) **8**(2), 74–87 (2016)
2. Al-wesabi, O.A., Abdullah, N., Sumari, P.: Data stream management system : for video on demand hybrid storage server. Int. J. Intell. Syst. Technol. Appl. (IJISTA) **18**(5), 470–493 (2019)
3. Al-wesabi, O.A., Sumari, P., Al-wesabi, M.A.: Efficient architecture for large-scale video on demand storage server. In: International Conference on Control Systems, Computing and Engineering (ICCSCE), pp. 395–400. IEEE (2015)
4. Al-wesabi, O.A., Sumari, P.: Streaming management scheme for hybrid storage video on demand server. Adv. Sci. Lett. **22**(10), 2695–2699 (2016)
5. Al-wesabi, O.A., Sumari, P.: Hybrid storage architecture for video on demand server. In: 4th International Conference on Software Engineering and Computer Systems (ICSECS), Malaysia, pp. 6–10. IEEE, Kuantan (2015)
6. Al-wesabi, O.A., Abdullah, N., Sumari, P.: On the design of video on demand server-based hybrid storage system. In: Recent Trends in Information and Communication Technology. IRICT 2017, Lecture Notes on Data Engineering and Communications Technologies, vol. 5, pp. 306–315. Springer, Heidelberg (2017)
7. Ling, Q., Xu, L., Yan, J., Zhang, Y., Li, F.: A Feedback-based adaptive data migration method for hybrid storage VOD caching systems. Multimedia Tools Appl. **75**(1), 165–180 (2016)
8. Xu, K., Liu, X., Ma, Z., Zhong, Y., Chen, W.: Exploring the policy selection of the P2P VoD system: a simulation-based research. Peer-to-Peer Netw. Appl. **8**(3), 459–473 (2015)
9. Ma, C., Yan, Z., Chen, C.W.: Forecasting initial popularity of just-uploaded user-generated videos. In: International Conference on Image Processing (ICIP), pp. 474–478. IEEE (2016)
10. Oh, Y., Choi, J., Lee, D., Noh, S.H.: Caching less for better performance: balancing cache size and update cost of flash memory cache in hybrid storage systems. In: Proceedings of the 10th USENIX Conference on File and Storage Technologies (FAST), vol. 12, San Jose, CA, USA (2012)
11. Kan, L.V., Jang, Y.H.: U.S. Patent Application No. 15/631,773 (2018)
12. Hachiya, S., Johguchi, K., Miyaji, K., Takeuchi, K.: Hybrid triple-level-cell/multi-level-cell NAND flash storage array with Chip Exchangeable Method. Jpn. J. Appl. Phys. **53**(4S), 04EE04 (2014)
13. Wu, Q.M., Xie, K., Zhu, M.F., Xiao, L.M., Ruan, L.: DMFSsim: a distributed metadata file system simulator. Appl. Mech. Mater. **241**, 1556–1561 (2013)
14. Yin, S., Li, X., Li, K., Huang, J., Ruan, X., Zhu, X., Cao, W., Qin, X.: REED: a reliaenergy-efficient RAID. In: 44th International Conference on Parallel Processing (ICPP), Beijing, China, pp. 649–658. IEEE (2015)
15. Uyangoda, L., Ahangama, S., Ranasinghe, T.: User profile feature-based approach to address the cold start problem in collaborative filtering for personalized movie recommendation. arXiv preprint arXiv:1906.00365 (2019)

New Architecture Design of Cloud Computing Using Software Defined Networking and Network Function Virtualization Technology

Abdullah Ahmed Bahashwan, Mohammed Anbar, and Nibras Abdullah

National Advanced IPv6 Centre (Nav6), Universiti Sains Malaysia (USM), 11800 Gelugor, Penang, Malaysia
bahashwan@student.usm.my, {anbar,nibras}@usm.my

Abstract. Cloud computing is a new technology model that consists of a pool of configurable resource and multiple data centers in distributed areas. It offers a service over the internet for users using virtual data, and who share resources for storing and accessing data from anywhere. Cloud computing supports big-scale platforms that consist of thousands of servers and hundreds of switches. Currently, the cloud is suffering from many issues in term of management, security and other essential aspects. This paper discusses integrating Software Defined Networking (SDN), with Network Function Virtualization (NFV) based on cloud computing, to eliminate some of the existing security issues faced by cloud computing. This research paper proposes a new architecture design by taking advantage of the NFV technology to build a virtual security mechanism such as Network Intrusion Detection System (NIDS).

Keywords: Software-defined networking (SDN) · Network Function Virtualization (NFV) Based in Cloud Computing

1 Introduction

Cloud computing is a model that is available everywhere; in this way it is more convenient that users can access the network and use a shared pool of configurable computing resources. For in-state services, systems, applications and storage, all these can be speedily supported and managed with minimum effort or controlled by a service provider [1, 2]. The cloud computing consists of five crucial features; measured service, broad network access, on-demand self-service, rapid elasticity and resource pooling. Similarly, there are three unique types of cloud computing service models, namely; Infrastructure as a Service (IaaS), Software as a Service (SaaS) and Platform as a Service (PaaS) [3, 4].

Furthermore, the fundamental of cloud computing is data center structures, which are maintained and managed by universal operators such as Google, Amazon and Microsoft. All these cloud computing systems ensure the smooth running of thousands of servers. The demands of cloud computing dramatically increase, while at the same

F. Saeed et al. (Eds.): IRICT 2019, AISC 1073, pp. 705–713, 2020.
https://doi.org/10.1007/978-3-030-33582-3_66

time, the companies are progressively using crucially enhanced computing environments to fulfil customers' requirements. Meanwhile, there are more technologies that are being used, including remote data storage devices, cloud-based networks, virtualized desktop and servers. These technologies require more computing resources, planning and maintenance [5].

In addition, since the cloud computing is rapidly growing, companies and service providers have to come out with smart solutions to help the data center reduce capital expenditures, as well as increasing network performance, accuracy of critical information systems, and to accommodate these demands. Nonetheless, there is quick movement driven by the many organizations and academia to design a control paradigm to enable higher management and flexibility. This model is known as Software Defined Networks (SDNs). It is a new networking architecture that separates data forwarding functions from control functions through a centralized controller. This innovative model or architecture is designed for next-generation networks that have been rapidly highlighted with OpenFlow (OF) protocols [6].

Software-Defined Networking (SDN) architecture can overcome the drawbacks and complexities of traditional cloud computing. Precisely, SDN enables enterprise networks to reduce operational cost and flexibility in network management services, and with minimal efforts [7]. In the following subsections, a brief review of SDN is presented in 1.1. On the other hand, 1.2 discusses SDN based cloud computing, while Network Function Virtualization (NFV) is described in 1.3. Section 2 provides related studies of using SDN with NFV. Our proposed architecture is discussed in Sect. 3. Finally, a conclusion is provided in Sect. 4.

1.1 Software-Defined Network (SDN)

SDN model is a novel technology of networking which uses a centralized controller that organizes the whole network, manages network flows with minimal effort to make the network more scalable and agile [8]. To clarify, the primary purpose of Software-defined networking is to enhance network security. The difference between traditional network and network using SDN is illustrated in Fig. 1 [9].

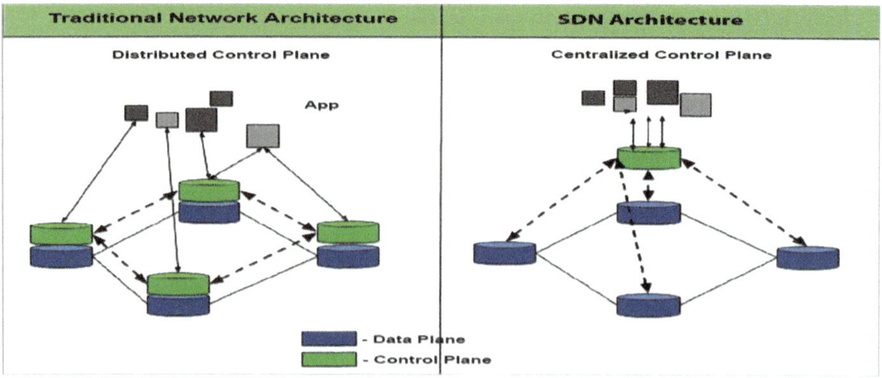

Fig. 1. Traditional network vs SDN network [10]

In traditional networking paradigms, the data plane and control plane are located within every device in the system. For instance, network switches make their own decisions locally based on the information received from neighboring devices. In this manner, the process of selecting the past path is a bit slow, sometimes resulting in increased network complications. In contrast, SDN is a new paradigm that separates the control and forwarding planes through an integrated interface [11].

Similarly, the controller is based on a programmable software that gives it the ability to control the entire network, then execute uncomplicated decisions by sending control packets to the forwarding devices, using SDN protocols through the southbound API. SDN architecture consists of three layers. First, the Application layer which consists of programs that offer communication with SDN via Application Programming Interface (API). Second, the Control layer which has the SDN controller, and thirdly, the Infrastructure layer that consists of SDN switching devices as shown in Fig. 2 below. The recent layers are connected using two interfaces; the Northbound API that allows the application layer to interact with the controller, and the Southbound API which serves as an interface between the SDN controller and network resources [12].

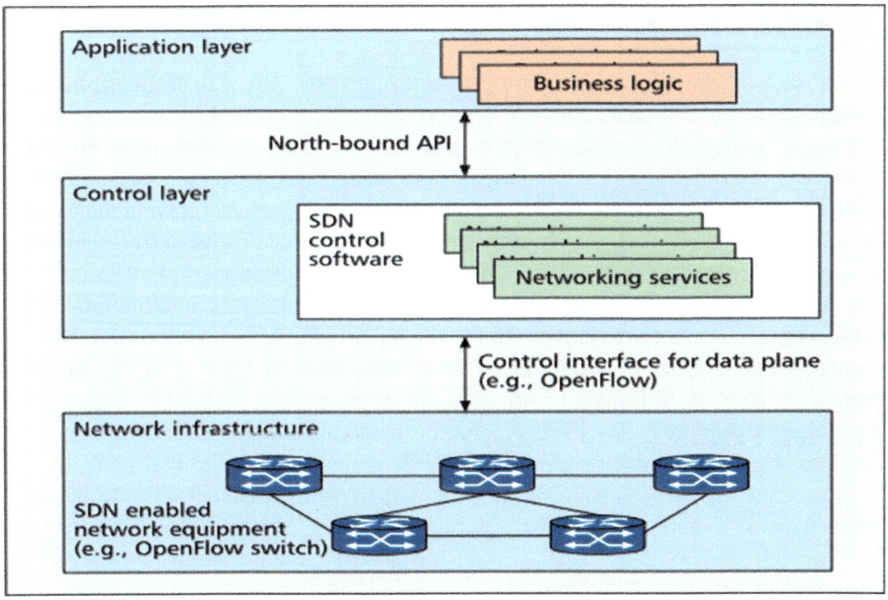

Fig. 2. SDN architecture.

Furthermore, OpenFlow is the most common controller that is used in SDN. OpenFlow allows the controller to communicate directly between itself and the forwarding plane devices, like switches using OpenFlow protocol. Additionally, Open flow plays a significant role in maintaining rules and getting remarkably higher performance functions [13]. Overall, there are several controllers announced and advances constructed, for example, Open-Daylight [14], RYU [15], Floodlight [16] and Nox [17].

1.2 SDN Based Cloud Computing

Cloud computing is the primary infrastructure which includes various applications such as cloud services. Relatively, cloud computing is an excellent building that associates a significant number of servers and switches connected with high-speed communication links. Additionally, creating a cloud computing that is effective, controllable, securable and scalable, can be done by deploying Software Defined Network within Cloud computing [18]. Also, SDN architecture offers and supports (APIs) which facilitates network services such as traffic engineering [19]. Similarly, integrating SDN with cloud computing offers various advantages including;

- Extremely high performance, high level of services, and enhanced resource utilization.
- The real physical network topology is dissociated from the virtual network topology.
- Deploying SDN to the cloud computing will result in reducing efforts of controlling network complexity.
- Elastic Cloud computing can fulfill demand requirements in a short time.

1.3 Network Function Virtualization

Virtualization technologies give the opportunity to come out with Network Function Virtualizations to shift the complete class of network devices functions, to become virtualized without the need to deploy hardware appliances in the networks. NFV allows network functions to be used as software that can be executed in one or multiple VMs. NFV is an innovation that is implemented in a system and plays a vital role in guaranteeing high network provenance, flexibility, lower cost, agile infrastructure and higher level of management services. Henceforth, NFV is working in both control and data planes in network infrastructure. This new technology is contributing to the communication operators to join up diverse kinds of hardware network devices such as switches, routers, firewalls and Intrusion Detection Systems (IDS). Using this means, it enables the network to function without hardware equipment. The NFV infrastructure is working to support the hypervisor, which is responsible for controlling the network functions. One of the advantages of using NFV is to reduce the capital cost [9].

Nowadays, the NFV is conducted to the markets with four types of software-based virtual network functions. The first is virtual switching. Virtualized network security devices such as IDS and Firewalls, and virtualized network services such as load balancers, traffic analysis and network monitoring utilization include any other applications in the network. In this case, it is a big challenge. Every type of the above-mentioned features requires a suitable isolation layer and a comprehensive controller for the NFV. To overcome this issue, the combination of SDN and NFV can solve this matter in an integrated cloud computing platform [20].

2 Related Studies

Cloud computing uses virtualization as a crucial and critical function, and there are two significant innovations to overcome those circumstances, in other words, SDN and NFV based on cloud computing. In this proposed architecture, leveraging these two technologies to the ability of having a scalable and elastic network, that fulfills the demands of getting virtual network infrastructure, and virtual network functions to be integrated into cloud computing in a manner that reduces the capital cost and operational cost. This section is discussing different researches done over the last few years. Authors' designs aim to come out with an efficient architecture of cloud computing.

According to Suh et al. [21] they propose a firewall software application that runs over the SDN OpenFlow controller, in order to show that most of the firewall functionalities are converted to be a software application, instead of dedicated physical devices. In addition, for the experiment, they use POX OpenFlow controller and the SDN network topology while deploying VirtualBox and Mininet. However, the proposed firewall software was deployed by using sets of manual rules that were configured to firewall application. So, the drawback of this approach is in using the deep packet inspection, which inspects all the network traffic without infiltration, and this is not suitable for highspeed network.

According to Vilalta et al. [22], the proposed generic architecture for SDN and NFV services is deployed over multidomain transport networks and distributed data centers. They present two use-cases of possible NFV services: a virtual Path Computation Element (vPCE) and the deployment of virtual SDN controllers (vSDN), on top of the virtualized transport networks.

According to Buyya et al. [9] they did their research on cloud computing and distribution systems (CLOUDS). They showed how network clouds can work together with architectural frameworks and principles, while hosting SDNs and NFVs for geographically distributed multi-cloud computing environments. The findings of their research illustrate that challenges of the programmable network cloud can be managed using the SDNs and NFVs. For them, it will not only increase network performance, but it will also reduce the operational cost and electricity consumption among others.

From the recent researches, it is possible to conclude that there is a clear momentum for exploiting the proposed novelty in a combination of SDN and NFV to enhance network services, and to further allow efficient security mechanisms to fulfill network demands. However, in this proposal architectural design will deploy an SDN controller for centralized network management and using the NFV to build a virtual Network Intrusion Detection System which is not yet proposed before. It is a step forward for the provisioning of network functions and a great opportunity to shift a hardware security device into the virtual layer. In the following sections, Sect. 3 underlines the proposed architecture of this research paper.

3 Proposed Architecture

This section is focusing on the proposed architecture to solve the cloud computing challenges. These limitations include manageability, Flexibility, security, high level of performance, and reducing capital cost using the two innovations SDN and NFV based in cloud computing. Figure 3 below illustrates an abstract overview of our methodology and stages.

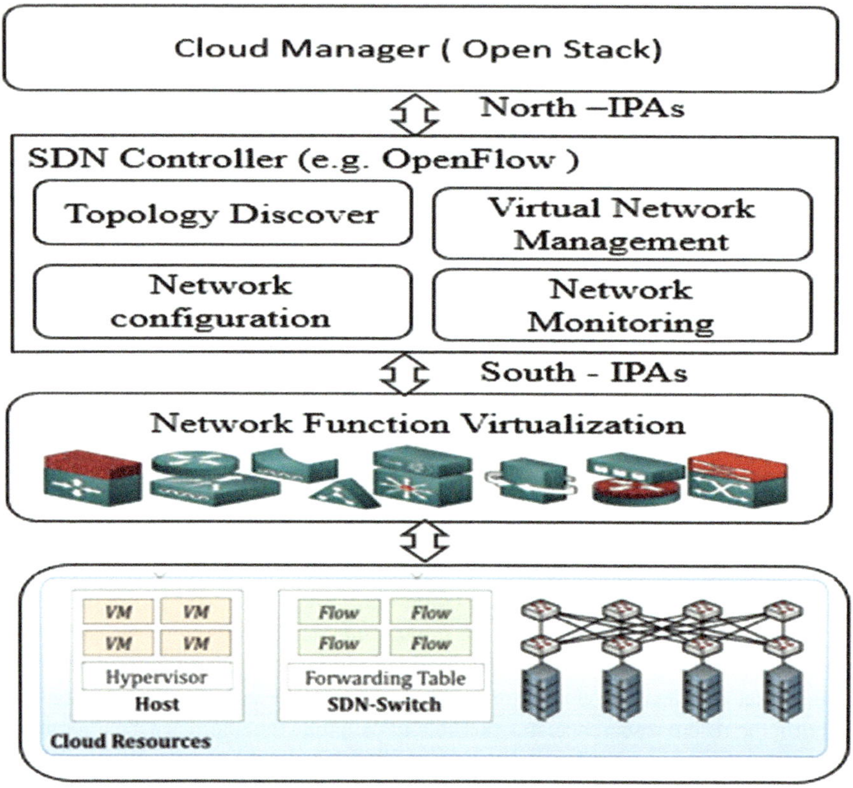

Fig. 3. Proposed architecture

The proposed architecture consists of four stages. The first stage focuses on the north API which contains the cloud directory. The proposed architecture carefully uses OpenStack since it is widely known as an open-source. The main function of OpenStack is controlling the cloud resources. Thus, the OpenStack plays a significant role in terms of managing IP addresses, advancing network services, compatibility with SDN technology, and controlling traffic. The second stage is concerned with the network Operating System (OS), which contains a set of applications used to power the network services. For instance, security, routing and managing of network traffic. The third stage is the NFV technology, the main goal of this stage is to shift the traditional devices such as Network Intrusion detection systems (IDS), into virtual software applications to be installed in south API for further security, and to filter the traffic inbound and outbound to the cloud or SDN controller. The primary goal of NIDS is to detect the network traffic and identify any suspicious forms that may break into the network. The last stage is the cloud resources, which contains the SDN Switches that have the forwarding table and retrieves the commands from the controller. In summary, SDN and NFV are complementary technologies, and each one can leverage off the other to be able to have a scalable and elastic network, as well as to get a virtual network infrastructure that is integrated into cloud computing, in such a way as to reduce both capital costs and operational costs.

4 Conclusion

Currently, the cloud is more complicated with the number of customers rapidly increasing every day. Furthermore, the emergence of new technologies like IoT technology has also led to even more complicated operations. The traditional Cloud is a high scale network which suffers from many challenges and limitations, such as, achieving a high level of performance in case of traffic congestion, achieving a high level of sufficient bandwidth, and tolerating a large number of VMs in the large-scale data center. The second challenge is getting high scalability and elastic network manageability by getting a full abstraction of the entire system. Other challenges include inadequate flexibility to accommodate several different kinds of hardware and software at the same time, security issues, and cost of operations.

The main goal of SDN is to simplify the network complexity and architecture by decoupling the control and data planes, and having abstraction view of the whole network, while at the same time successfully connecting large numbers of servers with high-speed communication links. SDN is an innovative technology for the near future for all network types. This paper attempts to solve data center challenges by integrating SDN and NFV to extract more efficient features for cloud computing, and to come out with an innovation architecture platform. This is possible because a marriage of the two leads to more efficient management of cloud computing by producing outcomes such as NIDS. Moreover, in this research, the focus is on these technologies to have a scalable and elastic network. Another key function is to reduce the capital and operational costs in cloud computing. Other functions range from increasing flexibility and functionality, to enhance overall network security and performance, to increase proper management of virtual data with or without hardware.

Acknowledgment. This research was supported by the Bridging Research Grant, Universiti Sains Malaysia(USM)No: 304/PNAV/6316271

References

1. Dillon, T., Wu, C., Chang, E.: Cloud computing: issues and challenges. In: 2010 24th IEEE International Conference on Advanced Information Networking and Applications, pp. 27–33 (2010)
2. Alanazi, S.T., Anbar, M., Karuppayah, S., Al-Ani, A. K., Sanjalawe, Y.K.: Detection techniques for DDoS attacks in cloud environment. In: Intelligent and Interactive Computing, pp. 337–354. Springer, Singapore (2019)
3. Amairah, A., Al-tamimi, B.N., Anbar, M., Aloufi, K.: Cloud computing and internet of things integration systems: a review. In: International Conference of Reliable Information and Communication Technology, pp. 406–414. Springer, Cham (2018)
4. Alanazi, S.T., Anbar, M., Karuppayah, S., Al-Ani, A.K., Sanjalawe, Y.K.: Detection techniques for DDoS attacks in cloud environment. In: Intelligent and Interactive Computing, pp. 337–354. Springer, Singapore (2019)
5. Karakus, M., Durresi, A.: A survey: control plane scalability issues and approaches in software-defined networking (SDN). Comput. Netw. **112**, 279–293 (2017)
6. Ishimori, A., Farias, F., Cerqueira, E., Abelém, A.: Control of multiple packet schedulers for improving QoS on OpenFlow/SDN networking. In: 2013 Second European Workshop on Software Defined Networks, pp. 81–86. IEEE (2013)
7. Wang, B., Zheng, Y., Lou, W., Hou, Y.T.: DDoS attack protection in the era of cloud computing and software-defined networking. Comput. Netw. **81**, 308–319 (2015)
8. AL-Adaileh, M. A., Anbar, M., Chong, Y.W., Al-Ani, A.: Proposed statistical-based approach for detecting distribute denial of service against the controller of software defined network (SADDCS). In: MATEC Web of Conferences, vol. 218, p. 02012. EDP Sciences (2018)
9. Buyya, R., Son, J.: Software-defined multi-cloud computing: a vision, architectural elements, and future directions. In: International Conference on Computational Science and Its Applications, pp. 3–18. Springer, Heidelberg (2018)
10. A primer on software defined networking (SDN) and OpenFlow standard. https://opensourceforu.com/2017/10/primer-software-defined-networking-sdn.openflow-standard/. Accessed 06 Dec 2018
11. Denazis, S., Haleplidis, E., Salim, J.H., Koufopavlou, O., Meyer, D., Pentikousis, K.: Software-defined networking (SDN): layers and architecture terminology (2015)
12. Zubaydi, H.D., Anbar, M., Wey, C.Y.: Review on detection techniques against DDoS attacks on a software-defined networking controller. In: 2017 Palestinian International Conference on Information and Communication Technology (PICICT), pp. 10–16. IEEE (2017)
13. Son, J., Buyya, R.: A taxonomy of software-defined networking (sdn)-enabled cloud computing. ACM Comput. Surv. (CSUR) **51**(3), 59 (2018)
14. Platform Overview – OpenDaylight. https://www.opendaylight.org/what-we-do/odlplatform-overview. Accessed 06 Dec 2018
15. Build the future of Open Infrastructure. https://www.openstack.org/. Accessed 06 Dec 2018
16. Floodlight OpenFlow Controller -Project Floodlight. http://www.projectfloodlight.org/floodlight/. Accessed 06 Dec 2018
17. Gude, N., Koponen, T., Pettit, J., Pfaff, B., Casado, M., McKeown, N., Shenker, S.: NOX: towards an operating system for networks. ACM SIGCOMM Comput. Commun. Rev. **38** (3), 105–110 (2008)
18. Azodolmolky, S., Wieder, P., Yahyapour, R.: Cloud computing networking: Challenges and opportunities for innovations. IEEE Commun. Mag. **51**(7), 54–62 (2013)

19. Cziva, R., Jouët, S., Stapleton, D., Tso, F.P., Pezaros, D.P.: SDN-based virtual machine management for cloud data centers. IEEE Trans. Netw. Serv. Manag. **13**(2), 212–225 (2016)
20. Duan, Q., Ansari, N., Toy, M.: Software-defined network virtualization: an architectural framework for integrating SDN and NFV for service provisioning in future networks. IEEE Netw. **30**(5), 10–16 (2016)
21. Suh, M., Park, S.H., Lee, B., Yang, S.: Building firewall over the software-defined network controller. In: 16th International Conference on Advanced Communication Technology, pp. 744–748. IEEE (2014)
22. Vilalta, R., Mayoral, A., Muñoz, R., Casellas, R., Martínez, R.: The SDN/NFV cloud computing platform and transport network of the ADRENALINE testbed. In: Proceedings of the 2015 1st IEEE Conference on Network Softwarization (NetSoft), pp. 1–5. IEEE (2015)

Movement Pattern Extraction Method in OppNet Geocast Routing

Aliyu M. Abali[1(✉)], Norafida Bte Ithnin[1], Muhammad Dawood[1], Tekenate Amah Ebibio[1], Wadzani A. Gadzama[2], and Fuad A. Ghaleb[1]

[1] School of Computing, Faculty of Engineering, Universiti Tekenologi Malaysia, Johor Bahru, Malaysia
aliyuabali@fud.edu.ng, afida@utm.my, mastangalbaloshi@gmail.com, amatek008@gmail.com, fuadeng@gmail.com

[2] Federal Polytechnic Mubi, Yola, Nigeria
ask4gadzama@gmail.com

Abstract. OppNets is a subset of MANET in which there is no point to point connection between devices. Despite challenges of intermittent connectivity and unpredictable mobility characteristics, mobile nodes need to communicate and share valuable information without infrastructure. The routing strategy in OppNet follows a store-carry-forward paradigm, in which a node store messages in its buffer, carry the messages as it moves about its daily activities and opportunistically relay the messages one after another upon meeting their target destinations. This paper presents a movement pattern extraction, a new approach for extracting regular movement pattern of nodes from existing traces to predict a node future location. Human movement is predictable with a high degree of certainty that people visit certain locations and have the opportunity to meet other people on a regular basis. The proposed method incorporate message lifetime and previous history of encounter to extract locations from the GPS trace with respect to message lifetime using Neural Network to predict future location visit of nodes. Prediction method can be used in geocast routing schemes to determine the node's contribution in forwarding a message to a destination location. Experiment result indicates that the proposed method achieves significant performance over existing methods in terms of accuracy and stability in prediction.

Keywords: OppNets · Geocast · Movement pattern extraction

1 Introduction

The idea of geocasting was first investigated under UUMAP project [1] that bind together IP addresses to geographic locations, a database with an array of geographic locations of internet host were stored and maintained. The authors in [2] presents the first packet routing algorithms to a geographic location with cartesian routing. Cartesian routing uses longitude and latitude based addresses. Each node in the network can determine its geographic address and the nearby router that is within a communication

F. Saeed et al. (Eds.): IRICT 2019, AISC 1073, pp. 714–723, 2020.
https://doi.org/10.1007/978-3-030-33582-3_67

range. Geocasting is widely studied in the context of MANETs and VANETs. OppNet Geocasting is another variant of Mobile Ad Hoc Networks that take advantage of node movement to establish communication between mobile devices without the need of infrastructure. In OppNets data is transmitted according to store carry forward mechanism, intermittent connectivity, movement of handheld devices carried by individuals and undefined topology of the network have made a remarkable achievement in the real-world communication applications that is inherently delay–tolerant and does not require a priori route for data transmission. Such application fields include data offloading in mobile networks [3], emergency notification [4], information sharing [5], vehicular networks [6] etc.

This paper presents a Message Lifetime Awareness Movement Prediction, a method for movement pattern extraction that exploits partial regularity of human movement. Human movement is regular in nature and exhibits a high degree of repetition, they tend to visit certain places regularly and have an encounter with other people regularly as they carry out their daily activities [7]. Accurate prediction of the node's future visit to locations from the existing history of visits called movement traces can improve relay node selection for message forwarding in location-based routing strategies.

In line with the above, this research address inaccuracy in prediction of node's future visits to locations by effectively incorporating message lifetime awareness in movement pattern extraction from the history of encounters to predict node future location with high accuracy and consistency. The movement pattern mining is formulated based on Neural Network to learn large-scale GPS traces and extract only locations without zero remaining TTL thus the overhead is reduced, and this can have a significant impact on routing performance of (city-wide) large scale scenarios.

The sections of the paper are organized as follows; related work is presented in Sect. 2, Sect. 3 present the detail proposed movement pattern extraction, Sect. 4 present the experimental result and the last section, Sect. 5 present the conclusion and future work.

2 Related Work

Geocast in OppNets have been studied in MV algorithms presented in [8] the networked region is divided into sub-regions each user maintains a frequency of visit probabilities to each sub-region, each divided area has a dedicated node. A message is forwarded from a mobile node to a stationary destination node. EVR (Expected Visiting Rate) is proposed in [9] for delay tolerant networks with the purpose of addressing routing decision by an intermediate node. EVR addressed exponential decay to allow relay node to determine its future visit to locations to make a routing decision. The algorithm in [10] exploits human movement characteristics particularly the regularity embedded in human daily movement to predict node future visit to locations, based on the prediction, the nodes' ability to deliver a message to its destined region is quantified using a defined metrics of distance between current node location, destination of the message and current time slot. Consequently, more weight can be allocated to the latest time slot. Authors in [11] applies a greedy approach for routing a message to the destination location and in a second phase floods a message to all nodes currently located in a geographical area. Similar to [11, 12] is a centrality based OppNets geocast, the scheme combines unicasting and flooding based approaches, a message is first forwarded to a cast destination using a unicasting scheme and centrality degree is used as a function to intelligently flood the entire network area with the message. In [13] identified the challenges with forwarding message from node to node with the probability of meeting a node that could deliver the message to the destination or location, however, chances are less to get the required number of nodes to perform forwarding strategies especially in a sparse environment thus leading to insufficient throughput. LoSeRO [14] is a multicasting geocast routing protocol that explores user point of interests (POIs) to route message to a destination location, Nodes utilize the history of past movement to independently build its movement profile called MobyZone. The MobyZone of all users forms the basis of opportunistic communication networks. GSAF is proposed in [15], the scheme allows users to arbitrary define a cast (A cast is a set of coordinates in a two-dimensional space (the network). The coordinates define the edges of the cast). The sender defines the cast and routes it together with the message, thus the encountered node only checks the message header to identify whether they reside within the cast.

3 Proposed Movement Pattern Extraction Method

This section details the proposed method of extracting regular movement pattern of nodes from available movement traces to predict the future movement of nodes to provide accurate location information to improve message delivery, reduce prediction inaccuracies and computation overhead. This work incorporates message lifetime awareness in a method of extraction with the goal of finding a suitable relay that can route the message from a source node to a destination location. In order to achieve this goal a movement prediction method that incorporate message lifetime awareness is necessary. Therefore, a message Lifetime Aware Movement Prediction-MLAMP is proposed to ensure simplicity in mobility pattern extraction (only messages with remaining TTL are considered for computation) for easy computation to reduce overhead in exchanging information during the forwarding process. Movement model is designed to extract regular movement pattern that is relevant to nodes in order to improve routing decisions.

3.1 Movement Prediction

MLAMP relies on GPS traces to improve accuracy in selecting the intermediary node for message forwarding, devising an accurate ability of intermediary node is a key to routing decision. Each node maintains a previous history of encounter for subsequent predictions, the history table is maintained and updated periodically for prediction of a future visit to a location. Message Lifetime-Aware Movement Prediction (MLAMP) is a method of collecting mobility trace and movement pattern extraction. Collecting mobility trace in a location table and extracting only movement pattern of nodes based on the remaining lifetime of the message. Human movement exhibits some degree of repetition, introducing the concept of spatial locality that stated that there is high repetitive of visiting one location or few [16] and travel only for short distances exploiting node regular movement to design movement pattern method that conforms with ability of a node to contribute in forwarding message to destination. To collect Mobility Trace. The GPS is sampled at regular intervals and the recorded location is stored in a location table. The location table consists of cells that hold visited locations for a number of days, and each day consists of several time slots of equal intervals known as periods. As shown in Eq. A, the location table can be illustrated as a matrix of $m+1$ rows and $n+1$ columns, which represent periods and days, respectively. $l_{i,j}$, a recorded location is stored in the cell defined by i, the corresponding period, and j, the corresponding day. For example, $l_{1,0}$ represents the GPS location sampled in the second period of the first day.

$$L = \begin{bmatrix} l_{0,0} & l_{0,1} & \cdots & l_{0,n} \\ l_{1,0} & l_{1,1} & \cdots & l_{1,n} \\ \vdots & \vdots & \ddots & \vdots \\ l_{m,0} & l_{m,1} & \cdots & l_{m,n} \end{bmatrix} \quad \text{(A)}$$

The periods in the location table represent actual time values (*i.e.*, 12am to 1am, 1am to 2am, …, 11pm to 12am) and the days represent days of the week (*i.e.*, Monday, Tuesday, …, Sunday). In real-world implementation, the length of each period should be enough to capture changes in user location (*e.g.*, 1 h or less) but not too small to occupy excessive memory or require the GPS to be sampled too frequently (*e.g.*, 1 s). Upon initialization, locations are first stored in the first column (*i.e.*, from $l_{0,0}$ to $l_{m,0}$), then stored in the next column (*i.e.*, from $l_{0,1}$ to $l_{m,1}$), and so on until the last column (*i.e.*, from $l_{0,n}$ to $l_{m,n}$). Since this work rely on the regularity embedded in human movement, it is interested to focus on only information of the most recent week and consider older information to be stale. Hence, when the location table is full, the process restarts from the first column by replacing old records with new ones.

Figure 1, which shows the locations visited by Node A on the first day, is used to provide further explanation on the operation of the location table. Assuming the user leaves home to work at 8am, arrives at the office by 9am, leaves the office by 6pm, and finally arrives at home by 8pm, then: the locations recorded from $l_{0,0}$ to $l_{7,0}$ and $l_{20,0}$ to $l_{23,0}$ correspond to where the user lives; the locations from $l_{9,0}$ to $l_{17,0}$ represent where the user works; and the other locations (in $l_{8,0}$, $l_{18,0}$ and $l_{19,0}$) are recorded as the user commutes between these two places or visits other points-of-interest (PoIs). On the next day, the same process continues from $l_{0,1}$ to $l_{23,1}$. The location table eventually gets full at $l_{23,6}$, but this does not affect its operation because of locations in the cells of the first day (*i.e.*, $l_{0,0}$ to $l_{23,0}$) are replaced with locations recorded on the next day. This way, the regularity embedded in the node's daily movement can be captured and any changes in the usual pattern can be identified as it becomes more regular (*e.g.*, the user could decide to temporarily relocate to a friend's place or taking a new job in a different location). The term "steps" is used to denote the position of a cell with respect to the current cell. To demonstrate this, consider a full location table of 24 periods and 7 days. If the index of the current period is 22 and that of the present day is 6, then:

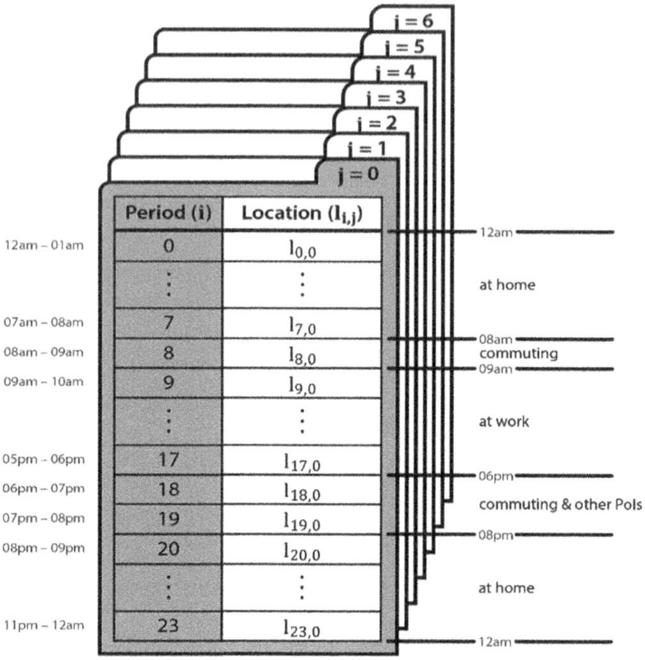

Fig. 1. The location table of Node A

1. The index of the period 5 steps after the current period is 3, while the index of the day 1 step after the present day is 0.
2. The location in the period 5 steps after the current period is given as $l_{3,0}$.
3. The current record is $l_{22,6}$, while the location in the period 5 steps after the current period is given as $l_{3,0}$.

It can be observed that, for the location 5 steps after the current record, i is 3 instead of being the sum of 22 and 5, while j changes from 6 to 0. This is due to the cyclic operation of the location table, whereby the last period of the present day is followed by the first period of the next day. Hence, given the index of the current period as p and that of the present day as d, the location a step(s) after the current record is given as $l_{i,j}$, where i and j are given by Eqs. A and B, respectively, $m+1$ is the number of periods in a day and $n+1$ is the number of days in a week. The value of $m+1$ can be obtained with Eq. C, where 86,400 is the number of seconds in a day and Δt is the length of a period in seconds.

$$i = (p + a)\ mod\ (m + 1) \tag{B}$$

$$j = \left(d + \left\lfloor \frac{p + a}{m + 1} \right\rfloor\right)\ mod\ (n + 1) \tag{C}$$

$$m + 1 = 86,400/\Delta t \tag{D}$$

3.2 Movement Pattern Extraction

Determining the delivery contribution that can be made by carrying a message requires the node to first extract its movement pattern from the location table. In order to describe the method of extracting movement pattern, the location table is define as L, a sequence of daily location sequences as shown in Eq. D, where i and j are given by Eqs. E and F, respectively. The sequence is written in terms of its elements, $l_{i,j}$ such that: a, the inner index, has values running from 0, the minimum, to m, the maximum; and b, the outer index has values running from 0, the minimum, to n, the maximum.

$$L = \left\{ \left\{ l_{i,j} \right\}_{a=0}^{m} \right\}_{b=0}^{n} \tag{E}$$

$$i = a \; mod \; (m+1) \tag{F}$$

$$j = \left(b + \left\lfloor \frac{a}{m+1} \right\rfloor \right) mod \; (n+1) \tag{G}$$

L begins with the first daily location sequence by initializing the outer index to its minimum value and defines each element by incrementing the inner index. So: for the first element of the sequence, $a = 0$ and $b = 0$; for the next element, $a = 1$ and $b = 0$; and so, on until $a = m$ and $b = 0$, *i.e.*, the last element of the first daily location sequence. Then L continues with the second daily location sequence by incrementing the outer index and defining its elements by resetting the inner index and incrementing it. So: for the first element of the second daily location sequence, $a = 0$ and $b = 1$; for the next element, $a = 1$ and $b = 1$; and so, on until $a = m$ and $b = 1$. The process continues until the table is full, *i.e.*, when $a = m$ and $b = n$.

In order to determine the delivery contribution that can be made by carrying a message, only locations in the periods during which the message TTL is not equal to zero are considered. Hence, given a message with T seconds remaining TTL, the extracted movement pattern can be written as $L(r)$, a submatrix of the location table matrix as shown in Eq. G, where r is the number of periods in T (*cf.*, Eq. H), p is the index of the current period, d is the index of the present day, and i and j are given by Eqs. A and B, respectively.

$$L(r) = \left\{ \left\{ l_{i,j} \right\}_{a=p}^{p+r-1} \right\}_{b=d}^{d+n} \tag{H}$$

$$r = T/\Delta t \tag{I}$$

To demonstrate the proposed method of extraction, consider L_A to be Node A's full location table of 24 periods and 7 days. As shown in Eq. I, the values of m and n will be 23 and 6, respectively. Also, consider the index of the current period to be 22 and that of the present day to be 6, *i.e.*, $p = 22$ and $d = 6$. Given a message that has a remaining TTL of 18,000 s, the extracted movement pattern can be written as $L_A(5)$ in Eq. J, since $\Delta t = 3600$ and Eq. H evaluates to $r = 5$.

$$L_A = \left\{ \left\{ l_{i,j} \right\}_{a=0}^{23} \right\}_{b=0}^{6} = \begin{bmatrix} l_{0,0} & l_{0,1} & l_{0,2} & \cdots & l_{0,6} \\ l_{1,0} & l_{1,1} & l_{1,2} & \cdots & l_{1,6} \\ l_{2,0} & l_{2,1} & l_{2,2} & \cdots & l_{2,6} \\ \vdots & \vdots & \vdots & \ddots & \vdots \\ l_{23,0} & l_{23,1} & l_{23,2} & \cdots & l_{23,6} \end{bmatrix} \tag{J}$$

4 Experimental Result

The performance evaluation of this research is carried out in MATLAB to determine the accuracy of our method and compare with LOOP [10] and LoSeRO [14]. In the simulation, Neural Network is used to learn and extract regular movement pattern from a sample of a large simulated dataset generated to depict a movement of nodes for a typical work.

There are no large publicly available traces of human movement with handheld devices collected in a city environment that conform to a working day movement in OppNet. From Table 1, a total of 141595 data is sampled. Accuracy, Precision, Recall, and F-measure are adopted for this model as evaluation metrics, moreover, in this study, the effectiveness of the proposed Message Lifetime Aware Movement Prediction (MLAMP) is compared with the existing baselines (LOOP & LoSeRO) relevant to this work. Performance of Neural Network NN for a given set of tests is carried out. T is denoted as a test set, T $= \{x_i, TP_i \mid 1 < i < p\}$, where x_i is the new test instance recorded in the form of <*daytype, timeslot*> and the predicted set of locations is denoted as TP_i. The prediction set of labels is evaluated for new instance in terms of Accuracy, Precision, Recall, F-measure. User device stores a movement pattern that is extracted in the form of $<x_i, TP_i>$ for each test instance x_i, the real set of location is denoted by R_i and the predicted set of locations is denoted by TP_i.

As shown in Table 1, the total sample is 141595, the total sample for TP, FP, and FN is 53405, 3023, 82412 and 2755 respectively. Similarly, the prediction model achieved an average of 0.04, 0.05, 0.96, 0.95, 0.95 and 0.95 for FPR, FNR, Acc., Prec., Recall and F-mea. respectively. As reported in Table 1, the False Positive Rate for all scenarios 0.04 maximum and 0.03 minimum, the False Negative Rate for all scenarios is 0.05.

Accuracy is considered as an important metric for MLAMP, as indicated in Fig. 2, the highest value of accuracy recorded is above 0.96 and it can be concluded that the prediction of future location can be achieved with high accuracy. The performance of MLAMP in terms of F-Measure is highly stable (95%) and reliable throughout the 60 days given in the scenario.

Table 1. Result of performance measures of all scenarios

Scen. No.	TP	FP	TN	FN	FPR	FNR	Acc.	Prec.	Recall	F-mea.
1	678	37	1065	42	0.03	0.06	0.96	0.94	0.95	0.94
2	1372	73	2115	68	0.03	0.05	0.96	0.95	0.95	0.95
3	2059	112	3169	101	0.03	0.05	0.96	0.95	0.95	0.95
4	2744	154	4222	136	0.04	0.05	0.96	0.95	0.95	0.95
5	3429	196	5275	171	0.04	0.05	0.96	0.95	0.95	0.95
6	4117	239	6324	203	0.04	0.05	0.96	0.95	0.95	0.95
7	4799	280	7381	241	0.04	0.05	0.96	0.95	0.94	0.95
8	5475	320	8445	285	0.04	0.05	0.96	0.95	0.94	0.95
9	6155	352	9513	325	0.04	0.05	0.96	0.95	0.95	0.95
10	6843	389	10568	357	0.04	0.05	0.96	0.95	0.95	0.95
11	7523	420	11637	397	0.03	0.05	0.96	0.95	0.95	0.95
12	8211	451	12698	429	0.03	0.05	0.96	0.95	0.95	0.95
Total sample/ measure	53405	3023	82412	2755	0.04	0.05	0.96	0.95	0.95	0.95

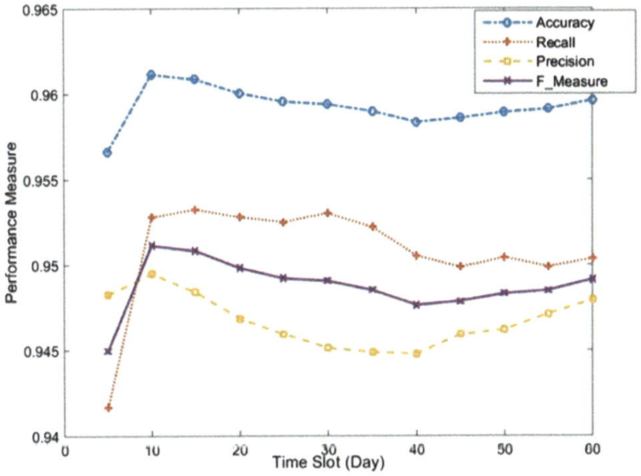

Fig. 2. Performance measures (accuracy, recall, precision & F-M) of different time slots

Comprehensive analysis of MLAMP against two existing works is carried out. LoSeRO and LOOP are the state of the art OppNets routing protocols that evaluate movement prediction methods, these works implemented movement prediction in their routing schemes. The proposed movement prediction method, MLAMP is compared with LoSeRO and LOOP as illustrated in Fig. 3, they recorded 72% and 51% respectively in terms of performance prediction. MLAMP is stable with high stability for 12 instances (days) as compared to the existing works. In contrast to MLAMP the fluctuant prediction performance of LoSeRO and LOOP is due to the unnecessary computation and location update that may incur more overhead when frequency of encounter high in an increased network environment.

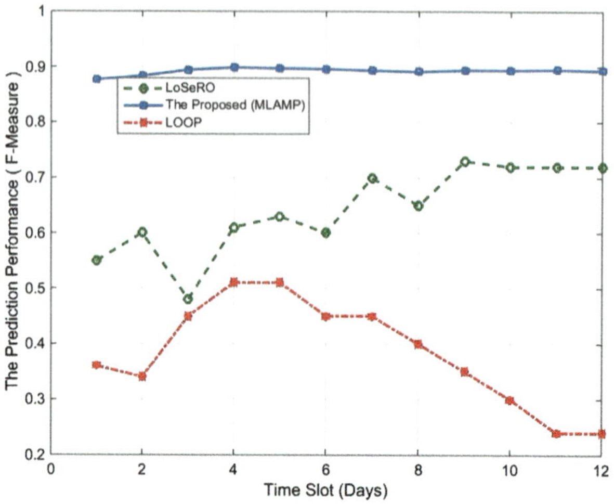

Fig. 3. Prediction performance comparison of MLAMP and baseline (LOOP, LoSeRO)

5 Conclusion

In this paper, a new movement prediction method is proposed for OppNet geocast routing. It indicates that the proposed method can effectively be used for accurate prediction of the node's future location to improve the routing performance of geocast protocols. The proposed solution presented movement pattern extraction that incorporates message lifetime awareness in the process of extracting locations to be visited so that the regular visit of node to a location can be used to determine its contribution in message forwarding. The performance of the proposed method is evaluated in terms of prediction accuracy and compare with well-known existing geocast movement prediction methods. The experiment results show that our proposed method outperformed LOOP and LoSeRO in terms of location prediction performance with very high stability.

In the future, the work can be extended to include illustrations with use cases on how a node running MLAMP decide to select a best forwarder between two nodes considering their forwarding utility function to extract movement pattern in respect of message lifetime.

References

1. The UUCP Mapping Project: UUCP mapping project (1985). http://www.uucp.org/index.shtml
2. Finn, G.G.: Routing and addressing problems in large metropolitan-scale internetworks (1987)
3. Han, B., et al.: Cellular traffic offloading through opportunistic communications. In: Proceedings of the 5th ACM Workshop on Challenged Networks - CHANTS 2010, p. 31 (2010)

4. Martín-Campillo, A., Crowcroft, J., Yoneki, E., Martí, R.: Evaluating opportunistic networks in disaster scenarios. J. Netw. Comput. Appl. **36**(2), 870–880 (2013)
5. Boldrini, C., Conti, M., Passarella, A.: Impact of social mobility on routing protocols for opportunistic networks. In: 2007 IEEE International Symposium on a World of Wireless, Mobile and Multimedia Networks, WOWMOM (2007)
6. Farahmand, F., Patel, A.N., Jue, J.P., Soares, V.G., Rodrigues, J.J.: Vehicular wireless burst switching network: enhancing rural connectivity. In: 2008 IEEE Globecom Work, GLOBECOM 2008 (2008)
7. González, M.C., Hidalgo, C.A., Barabási, A.L.: Understanding individual human mobility patterns. Nature **453**(7196), 779–782 (2008)
8. Burns, B., Brock, O., Levine, B.N., Burns, B., Levine, B.N.: MV routing and capacity building in disruption tolerant networks. In: Proceedings of IEEE 24th Annual Joint Conference of the IEEE Computer and Communications Society, pp. 398–408 (2005)
9. Ma, Y., Jamalipour, A.: Opportunistic geocast in disruption-tolerant networks. In: IEEE Global Telecommunications Conference - GLOBECOM (2011)
10. Lu, S., Liu, Y., Liu, Y., Kumar, M.: LOOP: a location based routing scheme for opportunistic networks. In: 9th IEEE International Conference on Mobile Ad-Hoc and Sensor Systems (MASS 2012), pp. 118–126 (2012)
11. Lu, S., Liu, Y.: Geoopp: geocasting for opportunistic networks. In: IEEE Wireless Communications and Networking Conference (WCNC), vol. 3, pp. 2582–2587 (2014)
12. Singh, J., Dhurandher, S.K., Woungang, I., Takizawa, M.: Centrality based geocasting for opportunistic networks, vol. 1. Springer, Cham (2019)
13. Chen, K., Shen, H.: DTN-FLOW: inter-landmark data flow for high-throughput routing in DTNs. IEEE/ACM Trans. Netw. **23**(1), 212–226 (2015)
14. Costantino, G., Maiti, R.R., Martinelli, F., Santi, P.: LoSeRO: a locality sensitive routing protocol in opportunistic networks. In: Proceedings of the 31st Annual ACM Symposium on Applied Computing - SAC 2016, vol. 14, no. 8, pp. 644–650 (2016)
15. Rajaei, A., Chalmers, D., Wakeman, I., Parisis, G.: Efficient geocasting in opportunistic networks. Comput. Commun. **127**, 105–121 (2018)
16. Wang, S., Liu, M., Cheng, X., Li, Z., Huang, J., Chen, B.: HERO - a home based routing in pocket switched networks. Lecture Notes in Computer Science (including subseries Lecture Notes in Artificial Intelligence and Lecture Notes in Bioinformatics), LNCS, vol. 7405, pp. 20–30 (2012)

Advances in Information Security

A Framework for Privacy and Security Model Based on Agents in E-Health Care Systems

Mohammed Ateeq Alanezi and Z. Faizal Khan(✉)

College of Computing and Information Technology, Shaqra University, Shaqraa, Kingdom of Saudi Arabia
{alanezi.mohd, faizalkhan}@su.edu.sa

Abstract. Usage of various systems for Electronic Healthcare Record (EHR) is increased in recent years. Majority of the hospitals and other laboratories requires health data sets during the treatment process. For enhancing the standard of the services provided by these health centers, the electronic health records should be shared so that it can be used by various users based on their requirements. Various researches have been done in the past for enhancing the security and privacy of these EHR data. Even though, it is not completely achieved. In this paper, a novel framework for maintaining the security and privacy of the EHR data based on agent based system is proposed. The agent based system consists of various intelligent agents such as the user interface agent, authentication agent, connection establishment agent and the connection management agent with distinct and expedient working features. These intelligent agents make ease of use, effective communication between patients/users and the E-service providers. This system monitors the data flow and focuses on various security and privacy concerns that the e-healthcare environment is facing.

Keywords: Information Technology · Electronic health records · Agent based systems · Agents · Privacy and security

1 Introduction

The influence of Information Technology (IT) has induced and combined numerous electronic health information from various places such as laboratories of medical research, hospitals and health care firms. This led to the formation of a novel concept called electronic health (e-Health). This can be defined as the application of IT based technologies and the practices of e-commerce for the entire sharing and processing, of the health information. This information is called as the EHR. The sensors based medical systems can be controlled by various controls in order to record of patient endlessly throughout the medical process. Hence, users like medical staff can access these collected health records since it contains most of the confidential and sensitive health data. Different methodologies has been proposed earlier in for maintaining the privacy and security of the EHR [1, 2]. But, these methods need more security in order to distribute the health data. The E healthcare Systems are real-time and has patient information which is in digital format. These are maintained by licenced persons. These

F. Saeed et al. (Eds.): IRICT 2019, AISC 1073, pp. 727–733, 2020.
https://doi.org/10.1007/978-3-030-33582-3_68

data sets where formed by acquiring various data from different patients. In these EHRs, the authorised persons can be the patients or the doctors. The data present in the servers can be available in local or cloud based which stores, analyse the stored health data [3]. The components which are present in the networks can be the inter connector between the patients and the medical staff for enhancing the broadcasting and distribution of data [4]. Though there are many benefits in these systems, more treats are there in terms of the security and the privacy for the data used in it. These security threats are caused because of their design [20]. These threats can be classified into various categories such as the data collection level [5–10], transmission level [11–14], and storage level [15–19] which are described more clearly in Sect. 3. Due to these threats in security and privacy of the EHR data, some users are not ready to use these applications. Hence, it is necessary to make sure that the users should be ready to use the system without any hesitation. Therefore, it is important to propose a system for maintaining the privacy and security in the EHR data.

In this paper, a detailed survey is done for analysing the security and privacy threats present in the healthcare system. Then, a novel agent based system for providing the security and privacy for EHR data is also proposed. The agent based system consists of various intelligent agents such as the user interface agent, authentication agent, connection establishment agent and the connection management agent with distinct and expedient working features. These intelligent agents make ease of use, effective communication between patients/users and the E-service providers. Various functions such as interface framing, user registration, user authentication, connection establishment and connection maintenance where carried out by the agent based system.

The remainder of this paper is as follows. Section 2 presents the various types of attacks and the methods proposed earlier to avoid it. Section 3 presents the intelligent agents and the proposed agent based method for providing the privacy and security in the E healthcare systems. Conclusions and the future work are depicted in Sect. 4.

2 Related Studies

Obtaining security and privacy in e-Health data is a crucial part in maintaining the objectives of this advanced methodology. There is a possibility of different forms of attacks while making the health related data to be shared among the users and the medical persons [11]. Hence, majority of the healthcare institutions have already developed the advanced framework for making the data as highly secured. Various guidelines have been proposed earlier for maintaining the requirements of privacy and security in the field of e-Health. Generally, the healthcare systems can be easily accessed by intruders. This sort of unauthorised access affects the overall performance of healthcare systems [4]. Networks present in the hospitals [3, 19] can be easily accessed by the hackers. Jamming based attack, Collision based attack, De-synchronization based attacks, spoofing and selective forwarding based attacks are the common types of attacks that is done so far during the stage data collection. Threats such as altering the overall information, embedding unwanted data, can be done at the level of data collection. Various security models have been proposed by earlier researchers in order to avoid these types of attacks at the stage of data collection.

Shin et al. [12] studied the different types of security models proposed already for the health- care platform and tried to find out how to avoid the leakage of information in this health- care platform. The framed various requirements of the security in order to make sure that the overall security and privacy in electronic health are up to the standard. As a solution to the above mentioned problem, the authors used a modified security model called as the Role Based Access Control (RBAC). They proposed a novel architecture by using the improved RBAC model. This model is proposed for maintaining the privacy and security in the health care data mainly at the stage of transferring the data between the user and the health care provider. Their algorithm can be applied in smart devices also. The results obtained revealed that their proposed algorithm provides more security and privacy in securing the health care data.

Morchon and Wehrle in [18] proposed a novel modular based privacy and security mechanism for the healthcare applications. Their algorithm is a modified version of the already available RBAC system. This system is developed because of two main conditions. First one is to assign and share the access control policies of mechanisms proposed for access control to the entire nodes of a sensor. Secondly, for storing the data which are related to the user such as the location, time, information about the user's health. It can be classified in to critical condition, emergency condition or normal condition. Various attacks such as altering the patient's medical information, intruding in the middle, data tampering, scrambling, signalling, unfairness in resource allocation, modification of message, data Interception can be done in the networks. Other threats such as spying of the data, altering the overall information, interrupting in between the communication, sending additional and unwanted signals to block the source of the message where caused by the attackers at the data transmission level. It is easy for an unauthorised developer to construct the systems for intruding or changing the data of the patient through the wireless based technology. Hence, a mechanism for controlling the overall transfer should be there whenever the patient's information is transferred [8].

Khennou et al. [20] proposed a novel framework for secure authentication and transmission of health data using advanced encryption techniques. This methodology is proposed for the security of two purposed such as the data and Channel. The security of channel is provided by utilizing the SSL on the HTTP layer present in the networks, while the data security is provided on the SOAP layer which is present above the HTTP. The authors proposed that their methodology can be used along with multi-factor authentication in order to provide authentication in the health data. Kahani et al. [3] framed a new scheme of access control in health data. Their methodology is based on a zero-knowledge protocol for verifying and maintaining the user's identity. Their framework combines the public key and a secret session key which is generated by Derive Unique Key Per-Transaction (DUKPT) scheme in order to establish secure communication between different health entities.

Gajanayake et al. [17] proposed new mechanism access control for enabling the privacy and security of the e-Health data. Their method is a combination of three security based models. Such as the Discretionary Access Control, Mandatory Access Control and the Role Based Access Control [16] in order to form a novel architecture. Their methodology allows the user as well as the healthcare providers to access the data in a privileged manner. The mail disadvantage of this methodology is it can be used only with in one type of network model while accessing the health data.

Access level based attacks can be a threat for the storage. It can change the health information of a patient, changing the overall configuration of system or servers which is involved in the monitoring purposes. These attacks can be done in the Interference of Patient's Information, accessing of Medical Information of Patient without approval etc. can be done in the storage level of patient's medical records. Various issues based on the hardware and software present in maintaining the patient's medical record can cause an intervention in the healthcare system. Also it is also possible that the health care system can be easily accessed by various types of threats such as viruses, Trojans, and spyware etc. [19].

Sun and Wang [2] proposed an improved access control usage and model. Main task of this method is to give the privilege to access private data more securely. It consists of six important components such as, attributes of subject, object attributes, rights, authorizations, obligations, and conditions. In these attributes, the authorizations, obligations, and conditions are components of usage control decisions which are further applied to determine whether a user is allowed to access the data. The presence of other attributes paves the way for solving certain shortcomings that have been common in access controls. The main disadvantage of this model is that it represents only a first step for authorization model.

2.1 Motivation of the Proposed Framework

Challenges faced by the conventional learning techniques are as follows:

- Various methods were proposed earlier for improving the EHR. Majority of the EHR systems are still in process, but with poor accuracy.
- Artificial Intelligence (AI) algorithm based methods was utilized for improving the accuracy of the HER model. Though, it was highly scalable, but the performance was poor in terms of accuracy.
- The methods based on fuzzy systems were proposed later on for the health records. The errors where evaluated. The method was more efficient and robust, but resulted in poor accuracy.

In order to rectify the drawbacks present in the existing methodologies, this framework is proposed.

3 Intelligent Agents

An agent is an autonomous and flexible computer based system that can receive input and output from the environment. There are many other advantages of the agents. For example, agents can be given knowledge to do specific task, collaborate with each other, and extendible which are very useful when developing secure systems. An agent can represent a user and do tasks on behalf of the user. Moreover, agent systems are extendible in nature. A new agent can be created instantly and added to the existing system in order to represent a new user such as a network monitoring agent, without changing the entire system.

4 Proposed Framework

An agent based system for providing effective and secure E Health security services has been proposed in this section. The reason behind this proposed framework is to design a secured methodology for E-Health. The proposed framework is controlled by the patient who is considered a major part of the E-health system. Architecture of the proposed model is shown Fig. 1. The agent based system proposed in this work has two types of intelligent agents such as the user interface agent, authentication agent. These agents make ease of use, effective communication between patients/users and the E-service providers. All the policies were defined by the user interface agent. A health database server is used to store all the username and the passwords used for the health care data. An authentication agent is used to validate the user. It is also used to verify the user credentials present in the health database.

The proposed framework consists of various intelligent agents for the user interface, authentication, establishment of connection and its management with respective features for working. These agents make effective communication between the patients as users and the providers of E-service. Various functions such as interface framing, user registration, user authentication, connection establishment and connection maintenance where carried out by this agent based system.

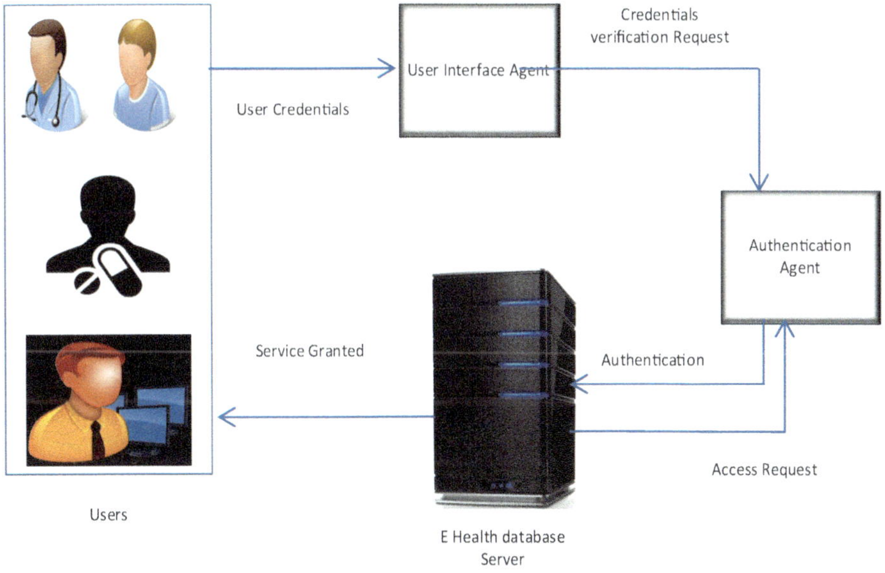

Fig. 1. Architecture of the proposed security model

The information phase present in the health database server consists of the basic and all the necessary information and data regarding the patient. Information such as location of the patient and time of the data collected, Age, Status, Blood Group, Height, Weight, BMI, Heart rate, Blood Pressure, Sugar level, Operation history etc. where present in the information server.

4.1 Working Principle

The proposed System accepts the service request from the customer through the user interface agent. It accepts the username and password of users. The User interface agent is connected through a website or a mobile based application. The authentication agent validates the username and password of the customer. After authenticating the user, a connection is established between the user and the server side by the connection establishment agent. This paves the way for effective and secure access of the e health care data. The information about the established connection will be also stored in the database so that it can be to use for further references. After establishing the connection, the authentication agent grants the permission for users to access the credentials of the person who requested the data.

4.2 Research Findings

- The proposed framework based on agent based system can provide an effective and secure E Health security services
- This framework is controlled by the patient who is considered a major part of the E-health system
- The two types of intelligent agents such as the user interface agent, authentication agent make ease of use, effective communication between patients/users and the E-service providers
- The User interface agent is connected through a website or a mobile based application so that it will be easy to handle for the users.

5 Conclusion and Future Work

Patient's health information should be kept secure in medical based servers so that the healthcare specialists and the doctors can access and use it at the time of treatments. In order to make sure the data is secure a novel agent based technology is proposed in this paper. Attacks at these health data servers can alter the information, modify the overall data and make unauthorised person as an accessed user. The proposed security model aims to avoid these using the intelligent agent based systems. This model is novel since it uses the intelligent agents for the entire process of providing access control and sharing. The proposed model consists of agents, each of which is in charge for different task. This method is a simple and efficient access control mechanism based on the agents functionalities. Future work can be the implementation of the proposed model with more agents in real-time health datasets and also to calculate its efficiency and accuracy.

References

1. Barua, M., et al.: Peace: an efficient and secure patient-centric access control scheme for eHealth care system. In: 2011 IEEE Conference Computer Communications Workshops (INFOCOM WKSHPS) (2011)
2. Fernández-Alemán, J.L., et al.: Security and privacy in electronic health records: a systematic literature review. J. Biomed. Inform. **46**(3), 541–562 (2013)

3. Kahani, N., Elgazzar, K., Cordy, J.R.: Authentication and access control in e-Health systems in the cloud. In: 2016 IEEE 2nd International Conference on Big Data Security on Cloud (BigDataSecurity), IEEE International Conference on High Performance and Smart Computing (HPSC), and IEEE International Conference on Intelligent Data and Security (IDS), pp. 13–23 (2016)
4. Shinde, S.S., Patil, D.: Review on security and privacy for mobile healthcare networks: from a quality of protection perspective. Int. J. Eng. Res.-Online Peer Reviewed Int. J. **3**(6) (2015)
5. Habib, K., Torjusen, A., Leister, W.: Security analysis of a patient monitoring system for the Internet of Things in eHealth. In: Proceedings of the International Conference on eHealth, Telemedicine, and Social Medicine (eTELEMED 2015) (2015)
6. Sankaranarayanan, S., Ganesan, S.: Applications of intelligent agents in health sector-a review. Int. J. E-Health Med. Commun. **7**(1), 1–30 (2016)
7. Sun, W., Cai, Z., Li, Y., Liu, F., Fang, S., Wang, G.: Security and privacy in the medical Internet of Things: a review. Secur. Commun. Netw.-Hindawi **28** (2018)
8. Manirabona, A., Fourati, L.C., Boudjit, S.: Investigation on healthcare monitoring systems: innovative services and applications. Int. J. E-Health Med. Commun. **8**(1) (2017)
9. Han, S., Zhao, S., Li, Q., Ju, C.-H., Zhou, W.: PPM-HDA: privacy-preserving and multifunctional health data aggregation with fault tolerance. IEEE Trans. Inf. Forensics Secur. **11**(9), 1940–1955 (2016)
10. Tang, J., Liu, A., Zhao, M., Wang, T.: An aggregate signature based trust routing for data gathering in sensor networks. Secur. Commun. Netw. **2018**(6328504), 30 (2018)
11. Sun, W., Cai, Z., Liu, F., et al.: A survey of data mining technology on electronic medical records. In: Proceedings of the International Conference on E-Health Networking, Application and Services, pp. 1–6 (2017)
12. Shin, M., Jeon, H., Ju, Y., Lee, B., Jeong, S.: Constructing RBAC based security model in u-healthcare service platform. Sci. World J. **2005**, 1–13 (2014)
13. Chandrasekhar, S., Ibrahim, A., Singhal, M.: A novel access control protocol using proxy signatures for cloud-based health information exchange. Comput. Secur. **67**, 73–88 (2017)
14. Bonab, T.H., Masdari, M.: Security attacks in wireless body area networks: challenges and issues. Acad. R. Des Sci. D OutreMer Bull. Des Seances **4**(4), 100–107 (2015)
15. Santos-Pereira, C., et al.: A secure RBAC mobile agent access control model for healthcare institutions. In: Proceedings of the 26th IEEE International Symposium on Computer-Based Medical Systems. IEEE (2013)
16. Azeez, N., Ademolu, O.: CyberProtector: identifying compromised URLs in electronic mails with Bayesian classification. In: International Conference Computational Science and Computational Intelligence (CSCI), Las Vegas, NV, USA, pp. 959–965 (2016)
17. Gajanayake, R., Iannella, R., Sahama, T.: Privacy oriented access control for electronic health records. e-Journal Health Inform. 8(2), 175–186 (2014)
18. MuthamilSelvan, T., Balamurugan, B.: Comparative performance analysis of various classifiers for cloud e-health users. Int. J. E-Health Med. Commun. **10**(2), 86–101 (2019)
19. Wellington, K.: Cyberattacks on medical devices and hospital networks: legal gaps and regulatory solutions. Santa Clara High Tech. LJ **30**, 139 (2013)
20. Khennou, F., Chaoui, N.E.H., Khamlichi, Y.I.: A mogration methodology from legacy to new electronic health record based open EHR. Int. J. E-Health Med. Commun. **10**(1), 55–75 (2019)

Normal Profile Updating Method for Enhanced Packet Header Anomaly Detection

Walid Mohamed Alsharafi[1,2], Mohd Nizam Omar[1(✉)], Nashwan Ahmed Al-Majmar[2], and Yousef Fazea[1]

[1] InterNetworks Research Laboratory, School of Computing, Universiti Utara Malaysia UUM, 06010 Sintok, Malaysia
niezam@uum.edu.my

[2] Department of Computer Science and Information Technology, Faculty of Science, Ibb University, Ibb, Yemen

Abstract. There is a significant need for various Intrusion Detection Systems (IDS) methods for packet behavior anomaly detection, due to the consistent exposure of packets to frequent intrusion threats. Thus, Packet Header Anomaly Detection (PHAD) considered as one of many significant approaches that is used for detecting threats on network packet. However, this approach still suffers from high generation of false alarm rate. This paper investigates a Normal Profile Updating Method (NPUM) for enhancing the PHAD based IDS model. This method updates normal profile of anomaly IDS using further processing of both the normal and abnormal data identified by anomaly detector. Simulation experiments and DARPA intrusion detection evaluation data sets are used for testing the proposed method. Results show that the proposed method can reduce the false positive alarms and improve the performance in terms of accuracy of detection. The major contributions of this research include the design of an enhanced PHAD-based IDS. This would contribute toward the enhanced IDSs to strengthen network security.

Keywords: IDS · PHAD · Anomaly detection · Normal profile · False alarm

1 Introduction

Digital security is the arrangement of advances and procedures that is designed to ensure and protect that computers, networks, systems, and data from attack, unauthorized access, change, or destruction [1]. It includes detecting and preventing attacks or intrusions. Up to date, detection threats on network and information systems are still a significant research issue. IDS characterize and recognizes potential dangers within a computer or a network. The signature and anomaly IDSs are the two major technologies which are generally utilized in the plan and advancement of the IDS model.

PHAD is an anomaly IDS model proposed by Mahoney and Chan [2, 3]. PHAD model is different than the other conventional network-based anomaly IDS model in that it models protocols rather than user behaviors [4] which means if there is anything

F. Saeed et al. (Eds.): IRICT 2019, AISC 1073, pp. 734–747, 2020.
https://doi.org/10.1007/978-3-030-33582-3_69

wrong with the packet header values of any of the specified fields, there anomaly will be detected [5–7]. PHAD approach have been fairly successful in detecting intrusion threats on network packets [8–11]. However, PHAD based IDS instants models still suffer due to generation of high false alarm rate [10, 11].

PbPHAD model proposed in [10] still one of the researches work that effectively enhanced the standard of PHAD model that was first described in [3]. PbPHAD model became a reference model for several recent works which are [11–15]. In these works, authors proposed different Anomaly Detection methods based on the PbPHAD model however they do not examine the normal profile direction in their improvements. The main purpose of this paper is to investigate a normal profile updating method for enhancing the PbPHAD model in terms of reducing the false alarms. Thus, we are exploiting the PbPHAD model to investigate the ability extent of our proposed NPUM method to reduce the false positives in PHAD based IDS models.

In this work, we firstly redevelop the PbPHAD model and apply it on two selected hosts (Host I. 127.016.114.050 and Host II. 127.016.112.050), from the MIT Lincoln Lab 1999 off-line intrusion detection evaluation data set [16], to get the number of false positives packets from the detected anomalous packets. We then reapply the PbPHAD model on the same selected hosts using, however, the proposed NPUM method. Finally, we compare the results of applying the PbPHAD model before and after using the proposed method to evaluate the efficiency of our method in reducing the false positives.

The rest of the paper is organized as follows. Section 2 overviews the related works. Section 3 provides an overview of the reference PbPHAD model. The proposed NPUM method is introduced in Sect. 4. Section 5 compares and discusses the results of the experiments. Section 6 offers conclusions and future work.

2 Related Work

According to the data type of network traffic, network IDSs could be split into three classes, which are system calls, TCP/IP connection descriptions and packets level [12]. The proposed IDS in this research belongs to the later level, i.e. packets level, because it is focusing on the anomalous behavior of the packet headers.

The idea of PHAD was first portrayed in [2, 3]. The authors presented a model that is able to learn the normal range of values for 33 fields of the Ethernet, IP, TCP, UDP and ICMP protocols. PHAD examines packet header information rather than the usual IP addresses and port numbers. One of burdens of PHAD is equipped for examining only a field at a time. This can lead to high probability of not able to detect Denial of Service (DoS) types of attacks [11]. Another disadvantage of PHAD is that it uses a general statistical model for all values in the packet headers for all protocols [10].

Protocol based PHAD (PbPHAD) model, proposed in [10], was drawn from a standard PHAD. It differs, however, from PHAD in using only 3 specific protocols, which are TCP, UDP and ICMP, because of their unique behavior when communicating among hosts, client and servers depending on the purpose and application used for a particular session. Although PbPHAD model outperforms the results from PHAD and DARPA best system and it has been demonstrated as a very promising model to be used for an anomaly-based IDS, the percentage of false positives is quite big for the detected anomalous packets based on the PbPHAD model alone [10]. Therefore, there is still room for further improvement for this model [14, 15]. As such, in this paper, we examine our proposed NPUM method to reduce the false positive on the PbPHAD model.

3 Overview of the PbPHAD Model

PbPHAD model [10] is intended to detect and recognize the typical anomalous behavior of network traffic packets. In addition, it is used to distinguish the level of maliciousness from a set of detected anomalous packets that are identified from the aggregated statistically modeled. The studies on PbPHAD structured it depend on three specific protocols, which are TCP, UDP and ICMP, on account of their unique behavior when communicating among hosts, client and servers depending on the purpose and application used for a specific session. In a view of this, a progressively precise factual model with better granularity, which represents the three picked protocols, can be built for detecting the anomalous behavior of the testing data [10]. For each protocol, if each field is indexed as i, $i = 1, 2, \ldots, n$, the PbPHAD structured is built depend on the proportion of the normal number of distinct field values in the training data, Ri, against the absolute number of packets associated with each protocol, Ni. The proportion, Pi = Ri/Ni represents the likelihood of the network seeing normal field values in a packet. Hence, the likelihood of inconsistencies will be 1 − Pi for each relating field. Each packet header field containing values not found in the normal profile will be assigned a score of 1 − Pi and will be summed up to give the total value for that particular packet as shown in relation 1 by [10].

$$Score_{packet} = \sum_{i=1}^{n} (1 - p_i), \qquad i = 1, 2, \ldots, n \tag{1}$$

As the value of Ri varies greatly, authors of PbPHAD model use log ratio (i.e. *log (Ri/Ni)*) in the PbPHAD model. In Tables 1 and 2 we formed PbPHAD statistical model for the two selected hosts Host I and Host II respectively. Here, the values of columns TCP, UDP and ICMP are calculated based on: *Relative Percentage Ratio of 1 − log(Ri/Ni)* to give the total probability of 1 for each protocol [10].

Table 1. PbPHAD statistical model for Host I incoming packets

No.	Field name	R	N	Anomaly score		
				TCP	UDP	ICMP
1	icmpchecksum	2	72	0.0000000	0.0000000	0.0397776
2	icmpcode	1	72	0.0000000	0.0000000	0.0444618
3	icmptype	2	72	0.0000000	0.0000000	0.0397776
4	udpchecksum	2	23475	0.0000000	0.0700562	0.0000000
5	udplen	42	23475	0.0000000	0.0517845	0.0000000
6	udpdestport	1027	23475	0.0000000	0.0325994	0.0000000
7	udpsrcport	3	23475	0.0000000	0.0676228	0.0000000
8	tcpoption	2	439879	0.0631072	0.0000000	0.0000000
9	tcpurgptr	1	439879	0.0661025	0.0000000	0.0000000
10	tcpchecksum	2	439879	0.0631072	0.0000000	0.0000000
11	tcpwindowsize	1098	439879	0.0358479	0.0000000	0.0000000
12	tcpflag	9	439879	0.0566076	0.0000000	0.0000000
13	tcpheaderlen	3	439879	0.0613550	0.0000000	0.0000000
14	tcpack	235828	439879	0.0126441	0.0000000	0.0000000
15	tcpseq	232808	439879	0.0126998	0.0000000	0.0000000
16	tcpdestport	468	439879	0.0395330	0.0000000	0.0000000
17	tcpsrcport	5545	439879	0.0288499	0.0000000	0.0000000
18	ipdest	1	463426	0.0663278	0.0921168	0.1037266
19	ipsrc	28	463426	0.0519283	0.0721186	0.0812080
20	ipchecksum	1	463426	0.0663278	0.0921168	0.1037266
21	ipprotocol	3	463426	0.0615804	0.0855235	0.0963023
22	ipttl	1	463426	0.0663278	0.0921168	0.1037266
23	ipfragptr	2	463426	0.0633325	0.0879569	0.0990424
24	ipfragid	65475	463426	0.0184069	0.0255637	0.0287856
25	iplength	702	463426	0.0380062	0.0527835	0.0594359
26	iptos	3	463426	0.0615804	0.0855235	0.0963023
27	ipheaderlength	1	463426	0.0663278	0.0921168	0.1037266
Total		**543060**		**1.000**	**1.000**	**1.000**

Table 2. PbPHAD statistical model for Host II incoming packets

No.	Field name	R	N	Anomaly score		
				TCP	UDP	ICMP
1	icmpchecksum	2	50	0.0000000	0.0000000	0.0364165
2	icmpcode	1	50	0.0000000	0.0000000	0.0409881
3	icmptype	2	50	0.0000000	0.0000000	0.0364165
4	udpchecksum	2	33575	0.0000000	0.0702221	0.0000000
5	udplen	46	33575	0.0000000	0.0519209	0.0000000
6	udpdestport	4067	33575	0.0000000	0.0257604	0.0000000
7	udpsrcport	3	33575	0.0000000	0.0678555	0.0000000
8	tcpoption	2	722877	0.0625475	0.0000000	0.0000000
9	tcpurgptr	1	722877	0.0654186	0.0000000	0.0000000
10	tcpchecksum	2	722877	0.0625475	0.0000000	0.0000000
11	tcpwindowsize	912	722877	0.0371875	0.0000000	0.0000000
12	tcpflag	8	722877	0.0568054	0.0000000	0.0000000
13	tcpheaderlen	3	722877	0.0608681	0.0000000	0.0000000
14	tcpack	384656	722877	0.0121508	0.0000000	0.0000000
15	tcpseq	383431	722877	0.0121640	0.0000000	0.0000000
16	tcpdestport	620	722877	0.0387861	0.0000000	0.0000000
17	tcpsrcport	1553	722877	0.0349827	0.0000000	0.0000000
18	ipdest	1	756502	0.0656070	0.0924489	0.1044660
19	ipsrc	28	756502	0.0518046	0.0729996	0.0824882
20	ipchecksum	1	756502	0.0656070	0.0924489	0.1044660
21	ipprotocol	3	756502	0.0610564	0.0860366	0.0972198
22	ipttl	1	756502	0.0656070	0.0924489	0.1044660
23	ipfragptr	2	756502	0.0627359	0.0884032	0.0998940
24	ipfragid	65536	756502	0.0196696	0.0277170	0.0313198
25	iplength	825	756502	0.0377911	0.0532527	0.0601747
26	iptos	3	756502	0.0610564	0.0860366	0.0972198
27	ipheaderlength	1	756502	0.0656070	0.0924489	0.1044660
Total		**841712**		**1.000**	**1.000**	**1.000**

4 Normal Profile Updating Method (NPUM) for Enhancing PHAD Model

The main idea of the proposed NPUM method is to investigate the ability of anomaly detector to re-process the abnormal data (i.e. anomalous data) identified by that anomaly detector itself for refining these abnormal data from normal data before flag them as the end anomalous detected data.

4.1 Architecture of NPUM Method

Figure 1 shows the proposed architecture where one same anomaly detector, which is PbPHAD, is applied in two modules of the architecture which are the anomaly detection module and the anomalous refining module. Next, the operation of the aforesaid two modules is described.

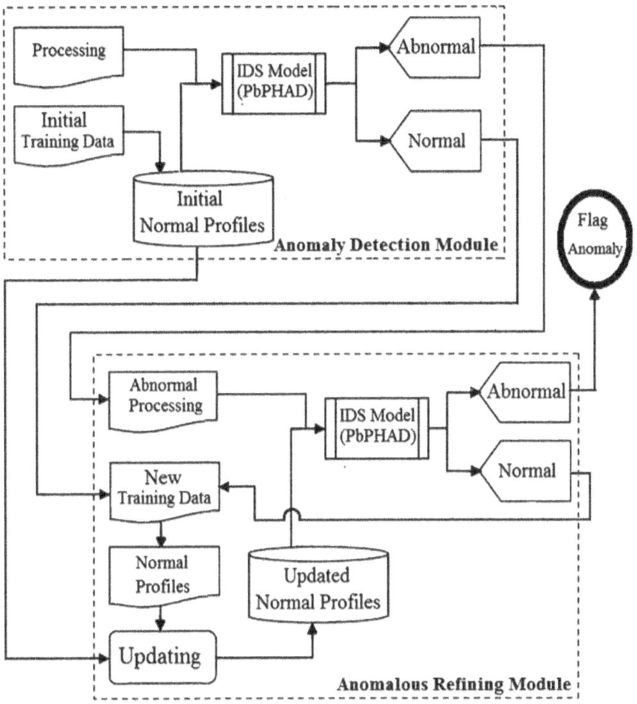

Fig. 1. NPUM architecture for false positives reduction in anomaly detection

Anomaly Detection Module. This module operates as simple anomaly detection system. Firstly, using training data, static unchanged normal profiles are established by an anomaly detector component (i.e. the PbPHAD model). Then, the anomaly detector uses the established normal profiles for processing data to identify which are normal and which are abnormal. Each of the normal data, the abnormal data, and the established normal profiles will be dealt as output data from this module and as input data for the next module.

Anomalous Refining Module. The main idea of operation of this module can be divided into two main steps: (1) creating the updated normal profiles using the normal data that output by the previous module, (2) the abnormal data, that output by the previous module too, will be refined from normal data. Thus, in the first step the normal data will be re-used for creating new training data. In the second step the updated normal profiles will be used for further re-processing the abnormal data to refine them from the normal one.

4.2 Simulation Design

For the simulation, firstly the PbPHAD model is redeveloped using the Structured Query Language (SQL Server), which also used as a database management system (DBMS) because it has the adequate power to be utilized to deal with huge number of packet records in the data set. Then, the design in Fig. 1 is regarded and the process flow in Fig. 2 is followed for the simulations.

Figure 2 shows the process flow of building the PbPHAD model (depicted by the components in stages 1, 2 and 3) and the enhancement to the process flow (depicted by the components in stages 4 and 5) in order to implement the proposed NPUM method. The process flow can be divided into 5 stages as follows:

Stage 1. Data Preparation (Steps a and b). In this stage, training data, testing data and attack identification file were downloaded from MIT Lincoln Lab web site [15]. The attack identification file available in the text format while the raw data of training and testing data are in compressed tcpdump format and therefore they were converted to comma separated values (*.csv) files and exported into the database.

Stage 2. Building the Normal Profile (Step c). In this stage, a normal profile table for each of the selected hosts was created to using the training data taken from week 3 for building a normal profile table for each of the selected hosts. Distinct values for each of the 27 fields in the TCP/IP packet were inserted into normal profile table to be used in the experiment to detect anomalous packet header field values.

Stage 3. Re-preparation the Data Set (Step e). In this step the result of Stage 3 (step d) of the original PbPHAD process flow was used to sort the testing data into two categories: ‘Anomalous’ data and ‘Normal’ data (non-anomalous). The new appeared normal data is combined with the old training data to build new updated training data. On the other hand, the Anomalous data is used as new testing data. So the result of this stage is a new data set that is needed at the next stage.

Stage 4. Re-applying the PbPHAD Model to the Re-prepared Data Set (Steps f and g). In this Stage the new data set resulted from the previous stage is used to repeat the stages 2 and 3 which include re-building the normal profile and re-running the experiment.

Stage 5. Results Composition Process (Step h). In this Stage a process of composition, which is explained next, is used to output the ‘Final Result’ i.e. the final detected anomalous packets.

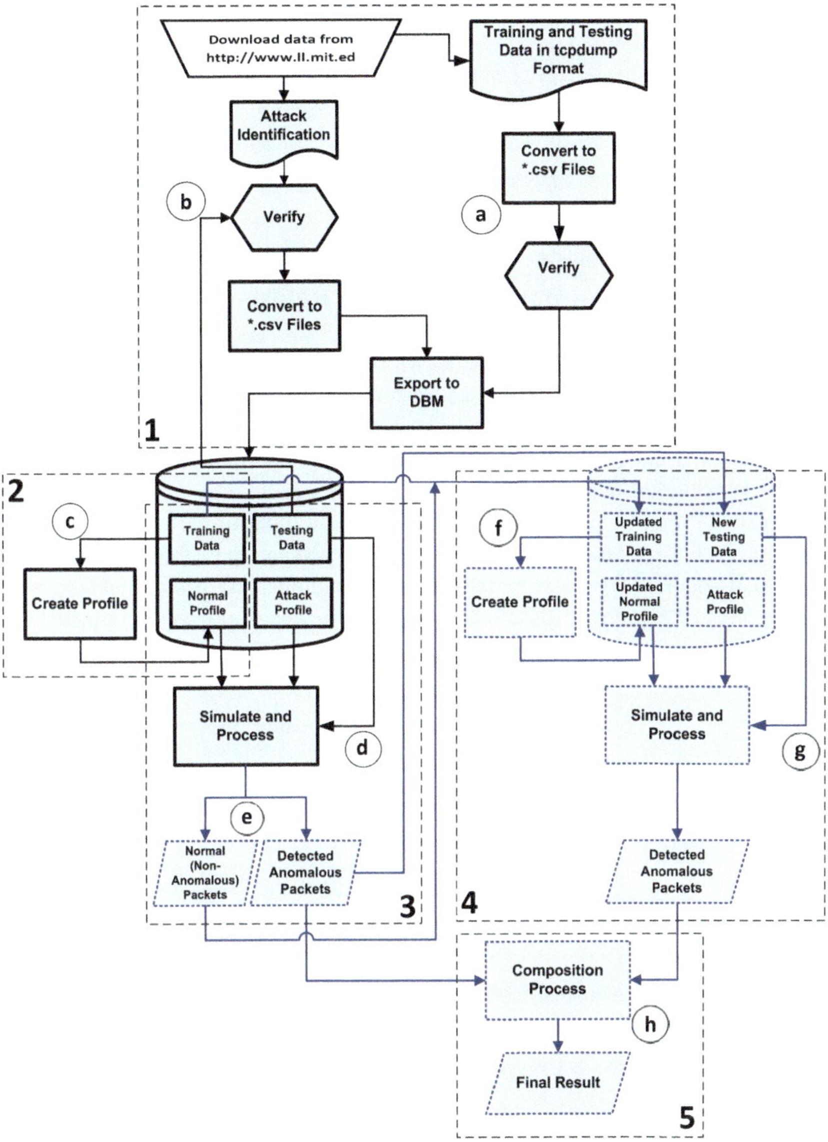

Fig. 2. Enhanced PbPHAD process flow for NPUM

4.3 Composition Process

It is the composition between the two counterparts of "Detected Anomalous Packets" in Stage 3 and Stage 4 based on calculation the average values of anomalous scores of the packets. The resulted averages then compared with the same certain preset threshold that was used in the above explained Stage 3 'Running the Experiment'. If the average value surpassed the threshold, it is captured into a "Collected Detected Anomalous

Packets table" which can be compared with the attack instances embedded in the data set to get the corresponding new false positive table which it can be compared with the first one to get the false positive reducing percentage. The enhanced PbPHAD model was tested on the 2 weeks of the inside testing incoming data for both selected hosts; Host I which comprises of 1,594,463 packets and Host II which comprises of 1,523,935 packets.

5 Experimental Results on the 1999 DARPA IDS Evaluation Data Set and Comparison with the Reference Model

Result of all detected anomalous packets which had surpassed certain preset thresholds of the anomaly score. The detected anomalous packets represent about 79% of more than one million packets, i.e. 1,241,573/1,594,463 packets and 1,211,744/1,523,935 packets, which is the total incoming test data for the selected hosts Host I and Host II respectively including all false positives.

If given the number of False Positive (FP), the number of False Negative (FN), the number of True Positive (TP), the number of True Negative (TN), the total number of actual positives Pos = TP + FN and the total number of actual negatives Neg = FP + TN then, the following six main performance measures were adopted for evaluation of the proposed method:

$$True\ Positive\ Rate\ (TPR) = \frac{TP}{Pos} = \frac{TP}{TP + FN} \tag{2}$$

$$False\ Positive\ Rate\ (FPR) = \frac{FP}{Neg} = \frac{FP}{FP + TN} \tag{3}$$

$$True\ Negative\ Rate\ (TNR) = \frac{TN}{Neg} = \frac{TN}{FP + TN} \tag{4}$$

$$False\ Negative\ Rate\ (FNR) = \frac{FN}{Pos} = \frac{FN}{TP + FN} \tag{5}$$

$$Precision = \frac{TP}{TP + FP} \tag{6}$$

$$Accuracy = \frac{TP + TN}{Pos + Neg} \tag{7}$$

5.1 Results for the Selected Host I

Figures 3 and 4 show the packet level results of testing data processed by the enhanced PbPHAD model for Host I after applying the proposed NPUM method and compare it with the counterpart results processed by the reference model [9].

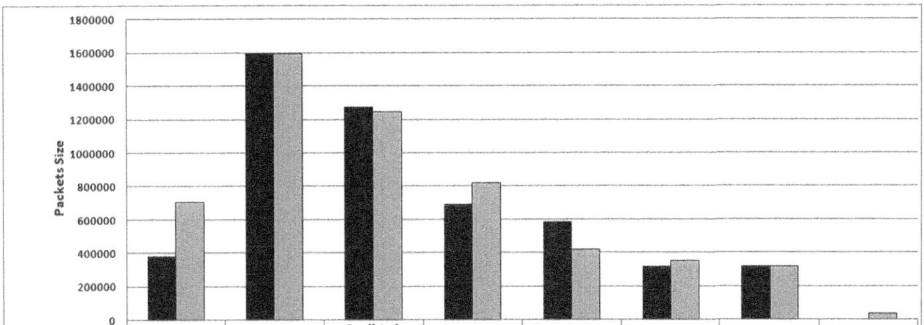

	Training Data	Testing Data	Predicted Anomalous Packets	True Positives (TP)	False Positives (FP)	Predicted Normal Packets	True Negatives (TN)	False Negatives (FN)
PbPHAD	383681	1594463	1273475	691198	582277	320988	317103	3885
Enhanced PbPHAD	704669	1594463	1241573	820275	421298	352890	318082	34808

Fig. 3. Result comparison on testing data processed using DARPA 1999 dataset for Host I

Figure 3 compares the results from simulations of both the proposed enhanced PbPHAD model and the reference model for the selected host. The figure shows the total amount of packets of training and testing data processed for anomaly detection. The figure also shows the distribution of the packets in the testing data by the four detection classes; TP, FP, TN, and FN. Figure 3 demonstrates that compared to the reference model, the proposed model shows increments of 19%, 0.31% and 796% on the number of TP, TN, and FN respectively while it shows a decrement of 28% on the number of FP.

Likewise, Fig. 4 compares the results from simulations of both the proposed enhanced PbPHAD model and the reference model for the selected Host I. However comparison here focusing on the rates of the detection classes (i.e. TPR rate, FPR, TNR, and FNR) as well as Precision and Accuracy. The rate results shows that compared to the reference model, the proposed model shows decrements of 4% and 8% on the TPR and FPR respectively while it shows increments of 8%, 4%, 12%, and 8% on the TNR, FNR, Precision rate and Accuracy respectively.

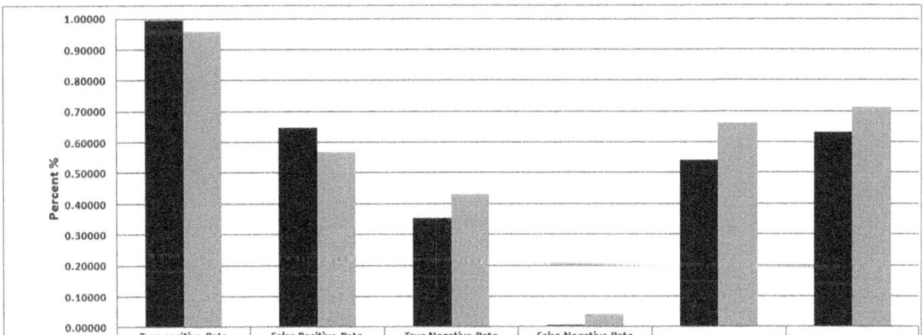

	True positive Rate (TPR)	False Positive Rate (FPR)	True Negative Rate (TNR)	False Negative Rate (FNR)	Precision	Accuracy
PbPHAD	0.99441	0.64742	0.35258	0.00559	0.54277	0.63238
Enhanced PbPHAD	0.95929	0.56980	0.43020	0.04071	0.66067	0.71394

Fig. 4. Performance comparison for Host I

5.2 Results for the Selected Host II

Figures 5 and 6 show the packet level results of testing data processed by the enhanced PbPHAD model for Host II after applying the proposed NPUM method and compare it with the counterpart results processed by the reference model which is the PbPHAD model [9].

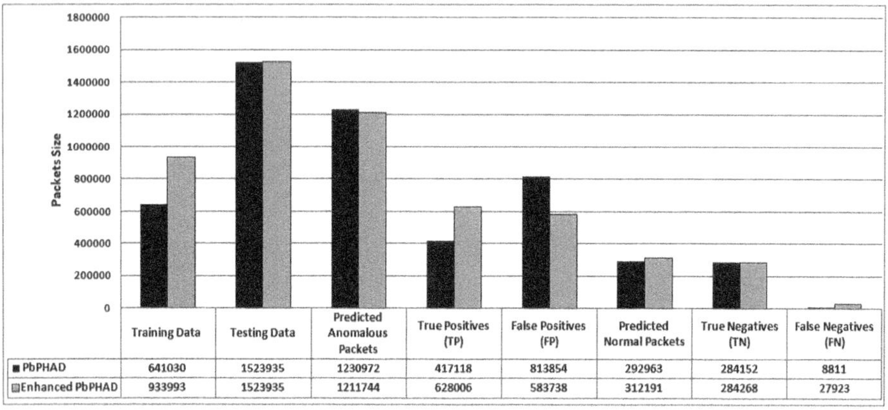

Fig. 5. Result comparison on testing data processed using DARPA 1999 dataset for Host II

Figure 5 compares the results from simulations of both the proposed enhanced PbPHAD model and the reference model for the selected host. The figure shows the total amount of packets of training and testing data processed for anomaly detection. The figure also shows the distribution of the packets in the testing data by the four detection classes; TP, FP, TN, and FN. Figure 5 demonstrates that compared to the reference model, the proposed model shows increments of 51%, 0.04% and 217% on the number of TP, TN, and FN respectively while it shows a decrement of 28% on the number of FP.

Likewise, Fig. 6 compares the results from simulations of both the proposed enhanced PbPHAD model and the reference model for the selected Host II. However comparison here focusing on the rates of the detection classes (i.e. TPR rate, FPR, TNR, and FNR) as well as Precision and Accuracy. The rate results shows that compared to the reference model, the proposed model shows decrements of 2% and 7% on the TPR and FPR respectively while it shows increments of 7%, 2%, 18%, and 14% on the TNR, FNR, Precision and Accuracy respectively.

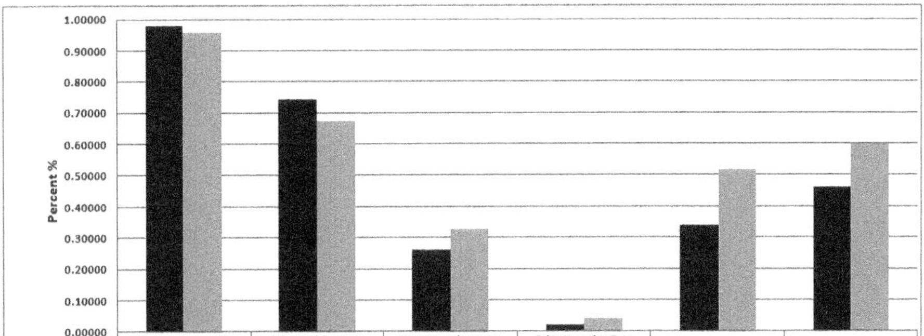

	True positive Rate (TPR)	False Positive Rate (FPR)	True Negative Rate (TNR)	False Negative Rate (FNR)	Precision	Accuracy
PbPHAD	0.97931	0.74121	0.25879	0.02069	0.33885	0.46017
Enhanced PbPHAD	0.95743	0.67250	0.32750	0.04257	0.51827	0.59863

Fig. 6. Performance comparison for Host II

5.3 Result Discussion

The proposed, in this paper, NPUM method has been demonstrated to be a promising method for enhancing PHAD model regarding reducing the false positive alarms by further processing of both the normal and abnormal data identified by this model. To summarize, the NPUM method has offered an enhanced PbPHAD model which, in comparison with the reference model, has produced the following results worthy of note:

- The enhanced PbPHAD model achieved an increment on the number of true positive packets (16% for Host I and 34% for Host II) and it achieved a decrement of 28% on the number of false positive packets for both selected hosts. This is considered to be the most important result of the proposed method.
- The enhanced model is better than the reference model by 12% on the precision rate and better by 8% on the accuracy rate for the Host I. Likewise, for the Host II, the enhanced model is better by 18% on the precision rate and better by 14% on the accuracy rate.
- Although the enhanced model shows a modest increment of 4% and 2% on the false negative rate for the selected Host I and Host II respectively, it shows, however, a notable increment of 796% and 217% on the number of false negative packets for the selected Host I and Host II respectively. This is a weakness because of trade-off between false negative and false positive.

6 Conclusions

From above mentioned result discussion, the proposed method shows a modest decrement in false positive rate, it presents a good decrement in number of false positive packets while it keeps high attack detection rate as in the reference model. Hence, the proposed method demonstrate high ability of reuse and refine testing date resulted after applying anomaly detector for further processing to update the normal profile that in turn can be used in refining the abnormal data and thus in reducing the false positives. Next, in the future work, we will investigate the possibility of incorporating our detector generating algorithm proposed in [16] for designing an effective Host based Anomaly Detection model inspired by artificial immune system (AIS) that could overcome the above-mentioned weakness and insufficiency of the proposed enhanced PbPHAD model.

References

1. Buczak, A.L., Guven, E.: A survey of data mining and machine learning methods for cyber security intrusion detection. IEEE Commun. Surv. Tutorials **18**(2), 1153–1176 (2016)
2. Mahoney, M.V.: Network traffic anomaly detection based on packet bytes. In: Proceedings of the 2003 ACM Symposium on Applied Computing, pp. 346–350. ACM (2003)
3. Mahoney, M.V., Chan, P.K.: PHAD: packet header anomaly detection for identifying hostile network traffic (2001)
4. Aydın, M.A., Zaim, A.H., Ceylan, K.G.: A hybrid intrusion detection system design for computer network security. Comput. Electr. Eng. **35**(3), 517–526 (2009)
5. Garg, A., Maheshwari, P.: PHAD: packet header anomaly detection. In: 2016 10th International Conference on Intelligent Systems and Control (ISCO), pp. 1–5. IEEE (2016)
6. Deka, R.K., Kalita, K.P., Bhattacharya, D.K., Kalita, J.K.: Network defense: approaches, methods and techniques. J. Netw. Comput. Appl. **57**, 71–84 (2015)
7. Al-Safwani, N., Fazea, Y., Ibrahim, H.: ISCP: In-depth model for selecting critical security controls. Comput. Secur. **77**, 565–577 (2018)
8. Elbasiony, R.M., Sallam, E.A., Eltobely, T.E., Fahmy, M.M.: A hybrid network intrusion detection framework based on random forests and weighted k-means. Ain Shams Eng. J. **4** (4), 753–762 (2013)
9. Lee, K.-C., Chang, J., Chen, M.-S.: PAID: packet analysis for anomaly intrusion detection. In: Washio, T., Suzuki, E., Ting, K.M., Inokuchi, A. (eds.) PAKDD 2008. LNCS (LNAI), vol. 5012, pp. 626–633. Springer, Heidelberg (2008). https://doi.org/10.1007/978-3-540-68125-0_58
10. Shamsuddin, S.B., Woodward, M.E.: Modeling protocol-based packet header anomaly detector for network and host intrusion detection systems. In: Bao, F., Ling, S., Okamoto, T., Wang, H., Xing, C. (eds.) CANS 2007. LNCS, vol. 4856, pp. 209–227. Springer, Heidelberg (2007). https://doi.org/10.1007/978-3-540-76969-9_14
11. Yassin, W., Udzir, N.I., Abdullah, A., Abdullah, M.T., Muda, Z., Zulzalil, H.: Packet header anomaly detection using statistical analysis. In: de la Puerta, J.G., et al. (eds.) International Joint Conference SOCO 2014-CISIS 2014-ICEUTE 2014. AISC, vol. 299, pp. 473–482. Springer, Cham (2014). https://doi.org/10.1007/978-3-319-07995-0_47
12. Kamarudin, M.H., Maple, C., Watson, T., Sohrabi, S.N.: A new unified intrusion anomaly detection in identifying unseen web attacks. Secur. Commun. Netw. **2017**(2539034), 1–18 (2017)
13. Cao, X., Chen, B., Li, H., Fu, Y.: Packet header anomaly detection using Bayesian topic models (2016). http://eprint.iacr.org/2016/040.pdf
14. Mahboubian, M., Udzir, N.I.: A naturally inspired statistical intrusion detection model. Int. J. Comput. Theor. Eng. **5**(3), 578 (2013)

15. Kamarudin, M.H., Maple, C., Watson, T., Sohrabi S.N.: Packet header intrusion detection with binary logistic regression approach in detecting R2L and U2R attacks. In: 2015 4th International Conference on Cyber Security, Cyber Warfare, and Digital Forensic, pp. 101–106 (2015)
16. Massachusetts Institute of Technology: DARPA intrusion detection scenario specific datasets. Lincoln Laboratory (1999). https://www.ll.mit.edu/ideval/data/1999data.html
17. Alsharafi, W.M., Omar, M.N.: A detector generating algorithm for intrusion detection inspired by AIS. ARPN J. Eng. Appl. Sci. **10**(2) (2015). ISSN-1819-6608

Hybrid Solution for Privacy-Preserving Data Mining on the Cloud Computing

Huda Osman[1,2(✉)], Mohd Aizaini Maarof[1], and Maheyzah Md Siraj[1]

[1] Faculty of Engineering, School of Computing, Universiti Teknologi Malaysia (UTM), 81310 Johor Bahru, Malaysia
omohuda2@graduate.utm.my

[2] Faculty of Computer Science and Information Technology, University of Kassala, Kassala, Sudan

Abstract. The continuous evolution of the cloud has attracted many enterprises to outsource their data to the cloud to accomplish data mining tasks and other services. So that data owners can enjoy the benefits of the cloud without fear from violations of data privacy, PPDM approaches came to protect the privacy of data while preserving the usefulness of data. This article reviews the most popular models of PPDM over the cloud along with their strengths and the weaknesses to conclude the research gap. Some of the current PPDM models still vulnerable to various types of attacks like the K-anonymity model whereby surface from background knowledge attack, homogeneity attack, similarity attack and probabilistic inference attack, others of them consume great computational complexity, for example, the methods which depended on cryptography. The hybrid solution was proposed to protect the data privacy and overcome the current problems of PPDM include identify and attributes disclosures whereby preservation the privacy of data before to outsourced on the cloud is the main focus. K-Anonymization is combined with homomorphic encryption to rid from their limitations and take their advantage together to enhance data privacy-preserving and maintain the data utility of outsourced data.

Keywords: Privacy-Preserving · Data mining outsourcing · Cloud data privacy

1 Introduction

Lately, cloud computing has revolutionized in institutions and organizations whereby their data-processing mechanism, especially in the method of data storing, access and processing including data mining and analysis [1]. The great benefits of cloud computing are alluring institutions and organizations to outsource their data and computational operations to the cloud where it is characterized under reduced costs, flexibility, reduced administrative burden and auto scalability [2, 3]. The outsourced data includes private and sensitive information, like financial records in the business or banking datasets, patients' illness or symptoms in medical datasets and so on [4]. Therefore, there is a need for techniques and methods to preserve the privacy of private data when deployed to the cloud for mining operations or other operations see Fig. 1, Privacy-Preserving Data Mining (PPDM) techniques in the cloud aim to preserve the

F. Saeed et al. (Eds.): IRICT 2019, AISC 1073, pp. 748–758, 2020.
https://doi.org/10.1007/978-3-030-33582-3_70

privacy of private data when mining it on the cloud while maintaining the utility of data so that data mining techniques can produce accurate results. Cloud computing became a great field to be focusing on data mining after it penetrated into many business domains and scientific computing. Accomplish data mining in the cloud enables the organizations from centralization management of data and software efficiently. PPDM aims to protect private information from a violation. Many studies explained that attackers often reveal information from third-party services or third-party clouds [5]. For example, one of the recent security breaches In April 2016, The National Electoral Institute of Mexico has been the victim of a violation of more than 93 million voter registration records, where data was stored on an insecure cloud server of Amazon. The an-other breakthrough was in 2012 for Dropbox, the attackers stole 68 million user accounts in order to obtain cash values of it by Bitcoins (BTC) [6]. The privacy and security threats like leakage of data, service or session hijack, insecure API, malicious insiders and insecure data deletion over the cloud, should be controlled proactively [7]. A fresh security violation in Google displayed that any administrator of the server has access to the secret information can misuse it easily. The worst trouble is that the administrator of the honest-but-curious server can violate privacy without being revealed [8].

2 Related Works

The common PPDM methods over the cloud can be divided into three wide groups are anonymization-based methods, cryptography-based methods and hybrid-based methods.

2.1 Anonymization-Based Methods

Anonymization is often used to concealment the identity of data owners and/or sensitive information, it one of the most widely used privacy-preserving methods. It makes the data or individual's information vague and unknown to maintain privacy (see Fig. 2) is performed by implementing different privacy models try to preserve the identity of the owner by implementing one or more of the data sanitizing operations namely generalization, suppression, anatomization, perturbation [9, 10]. K-anonymity,

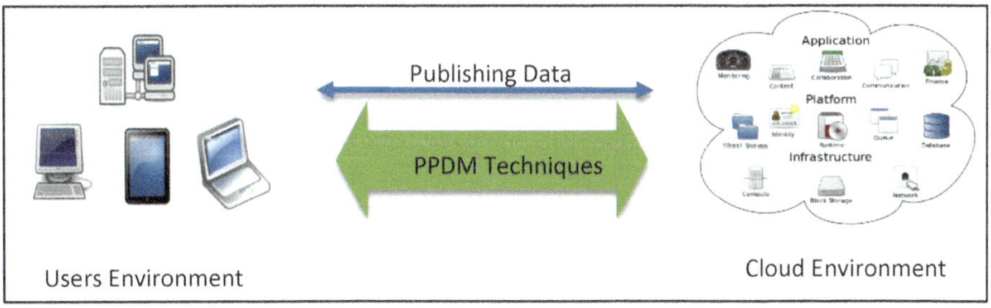

Fig. 1. Privacy-preserving data mining in the cloud

L-diversity and T-closeness are the most common privacy models of anonymization these privacy models using generalization to preserve the privacy, generalization firstly introduced by [11]. Generalization is preserving the truthfulness of generalized data, in contrast, it causes great information loss when the values are more generalized, and information loss adversely affects data utility which in turn reduces the accuracy of data mining. Chakravorty et al. [12] utilized the MapReduce paradigm to construct a novel iterative scalable k-anonymity model to privacy-preserving of individuals data when it sharing in an intercloud. [13] Anonymized microdata before publishing it to analysis and mining processes at service providers on the cloud through k-anonymity model, to ensure data anonymization they have anonymized different QIs for various service providers on cloud-based on their different background knowledge.

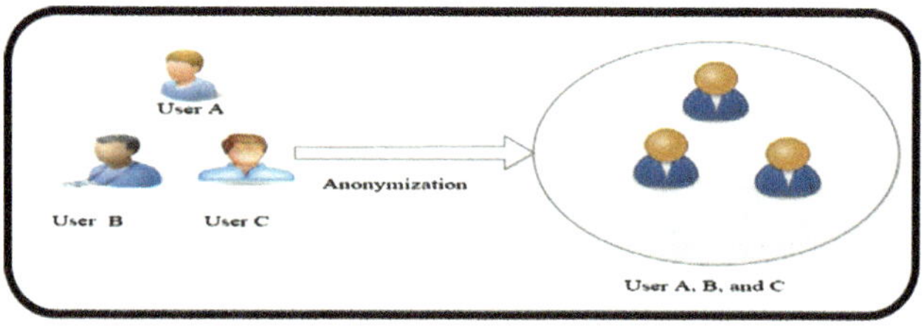

Fig. 2. Anonymization generates vagueness

K-Anonymity Model

K-anonymity is a basic privacy model and can be considered as a standard for preserving privacy through anonymization [14, 9], it has a simple definition and so many algorithms available for implementation, therefore, we find it attracts many researchers, unfortunately, it still vulnerable to various kinds of attacks such as background knowledge attack, homogeneity attack, similarity attack and probabilistic inference attack (also known as skewness attack) [15]. The anonymized table is satisfying the K-anonymity condition if and only if every record is indistinguishable from at least k-1 other records [16]. The k-anonymity model is a secure versus record linkage attack. In contrast, this model is unsafe against attribute linkage attack (also famous by attribute disclosure) because it has not any restrictions on the values sensitive attribute and has lack diversity among sensitive attributes (SA) values of Quasi-Identifiers group (QI-group) [17].

L-Diversity Model

Since the k-anonymity model was identified with a number of flaws, an upgraded privacy model was designed using a disparate concept named L-diversity [18]. The SA values of QI-group eventually diversity after utilization of the L-diversity approach. The altered microdata table is said to fulfill the L-diversity approach if and only if L "well-presented" values for SAs are accommodated in each QI-group. However, this model faces skewness attack and similarity attack skewness.

T-Closeness Model

This model needed the dispersal of a SA in any QI-group to be near to the distribution of this SA in the whole table. To evaluate the distance between the two distributions, the Earth Mover Distance (EMD) metric was utilized. In any case whereby the distance of distribution between SA in certain QI-group and this SA in the whole table not increase than a threshold t, then this QI-group is satisfying t-closeness. Issues own of SAs disclosures may be solved via the t-closeness model. T-closeness tends to destroy the linkage among QIDs and SAs, it is possibly leading to the destruction utility of data. Nevertheless, this model may not be suitable with a varied series of tables including the ones with SA's numerical as the property is necessitated via a computational process [19, 20].

2.2 Cryptography-Based Methods

Encryption techniques are very desirable because they preserve data privacy and confidentiality while preserving the utility of data but it requires a more computational process that negatively affects performance [4], there are many researchers used encryption methods to privacy-preserving. Deshmukh et al., [21] utilized the homomorphic encryption technique to preserve the privacy of patients on the cloud at diagnosing patient sickness, its improved diagnosis accuracy. Samanthula et al. [1] evolved a secure k-NN classifier protocol as a new protocol on encrypted data within the semi-honest model to privacy-preserving at applied classification method (i.e. at data mining process) over the cloud. Save data confidentiality, cover patterns of data access and keep input inquiry privacy of the user. Later [22] introduce a system utilized privacy-preserving k-NN classification protocol on encrypted data to preserve the privacy of customer's information during a data mining operation in the cloud.

Homomorphic Encryption

Homomorphic encryption is defined as an encryption way that permits to conducting computation processes on encoded data [21, 23]. With regards to this, after enciphered data are generated, the decryption results of these data are equal to the outcome of the same procedures that were implemented on plaintext [24]. Homomorphic encryption is the best-suited method for privacy preservation as the results for analysis and data mining. We observed that although the technique is successful in the privacy preservation and reverse process it has high time complexity and implementation complexity [25, 26].

2.3 Hybrid-Based Methods

Newly there are some of the researchers focusing on introducing mixed techniques to satisfy privacy requirements and achieved strong privacy. Table 1 summarizes more recent works that used hybridization to produce strong data privacy over the cloud.

Table 1. Comparison of hybrid-based Methods.

Authors	Issues	Techniques	Strengths	Weakness
Li et al. [27]	Privacy	Homomorphic encryption and a secure comparison	Safeguard data owner's privacy	Run time is high
Aldeen Yousra and Mazleena [5]	Privacy	K-anonymity, L-diversity and (a, k)-anonymity	Outperformed on L-diversity, K-anonymity (α, k)-anonymity	All techniques based on anonymization
Aldeen and Salleh [28]	Privacy and utility	K-anonymity with data relocation	Improved the utility	The expense of data truthfulness
Kohlmayer et al. [29]	Privacy	K-anonymity with (SMC)	Improvement the privacy	Computational complexity
Yang et al. [30]	Privacy	Cryptography, statistical analysis and anonymization	Just encrypted EIs	Optimal privacy not accomplished

3 The Current Research Gap and Problem Analysis

The PPDM methods have been developed and improved by many researchers as discussed in Sect. 3. However, these methods still have shortcomings are suffer from poor data utility or/and vulnerable to inference by various attacks [10]. Table 2 provides a summary of the weaknesses currently present in PPDM methods. When data is released to the cloud for mining operations, there are concerns about privacy violations that must be considered in order to ensure that data privacy is maintained. A lot of opportunities are existing to mishandle sensitive information if leaked or revealed [31]. The Privacy violation can be done by attribute disclosure, identity disclosure or/and membership disclosure [32, 20]. Hence, the dataset must be protected from various data disclosures before being released to the cloud to ensure that data privacy is not compromised. Some researchers in part 3.3 develop hybrid methods to obtain strong privacy, [5] and [28] combined anonymization techniques to accomplish development privacy-preserving as a weakness for the previous studies, whole the incorporated techniques are based on anonymization only. The techniques which rely on anonymization are based primarily on the processes of generalization and suppression that cause great loss of data. On [29] and [27] computational complexity is so high result in cryptography processes. While [30] was merge anonymization with the cryptography method, the privacy status was still not at the optimum level.

Table 2. Summary of current shortcomings in PPDM methods.

PPDM techniques	Current shortcomings
Anonymization-based techniques	Exposing to some types of attacks Utility issue
Cryptography-based techniques	Large computational complexity

4 The Proposed Solution for Privacy-Preserving Data Mining Over the Cloud

The proposed Hybrid Anonymization Cryptography (HAC) model is based on the k-anonymity technique, identification technique and homomorphic encryption, where it combines the three techniques to take advantage of their respective advantages to provide strong privacy while not sacrificing data utility. This work will use Cloud-RSA to implement homomorphic encryption and develops the Heuristic K-anonymity technique to implement dataset anonymization.

Proposed HAC Model

This section defines the proposal to build a HAC model that will achieve the preservation of dataset privacy. It contains the steps of the solution at a higher level. The detailed implementation steps of HAC is included in the next paper.

Phases of HAC Processing. To accomplish privacy-preserving of a dataset and simultaneously maintain data utility, the next four phases are required in HAC, for more clarification see Figs. 3, 4 and 5.

1. Dataset partition vertically: The data owner utilizes this component to fulfill the privacy of preserved data before publishing. The main table T is split into three parts which are explicit identifier attributes (EIs), quasi-identifiers attributes (QIs) and sensitive attributes (SAs) like income information, medical information or data about the location of individuals and so on.
2. Determine QIs by identification method: QIs Identification technique first construct attribute graph based on relations between the attributes then the QIs is acquired by searching at cut-vertex in attribute graph which contains all potential relationships among externally available information and published data. In other words, the QIs mentioned in the first phase are redefined by the QIs Identification method to avoid experimental selection that may adversely affect the effectiveness of the K-anonymity method and thus the utility of the anonymous data.
3. Data anonymization: In this phase, data is anonymized by the K-anonymity technique. In fact, only EIs and QIs are anonymized. EIs are suppressed because they recognize the individual explicitly. If the EIs attributes are redefined and thus violate the privacy of individual so easily, for this reason, should be get rid of any EIs attribute. QIs are generalized after been identified by the QIs identification method in the first stage.

4. Encrypt the dataset: in this phase only SAs is encrypted by homomorphic encryption this will ensure that privacy is maintained; If the intruder succeeds in identifying the record by linking QIs with available external information or from his background (background knowledge attack), he will not be able to identify the encoded sensitive attributes because homomorphic encryption allows performing operations on encrypted data. This will be enabling the service provider on the cloud to perform the mining operations (such as classification) on data without having the key to decrypt the data. The only party that is capable to decipher the encrypted data to see results of mining is the data owner.

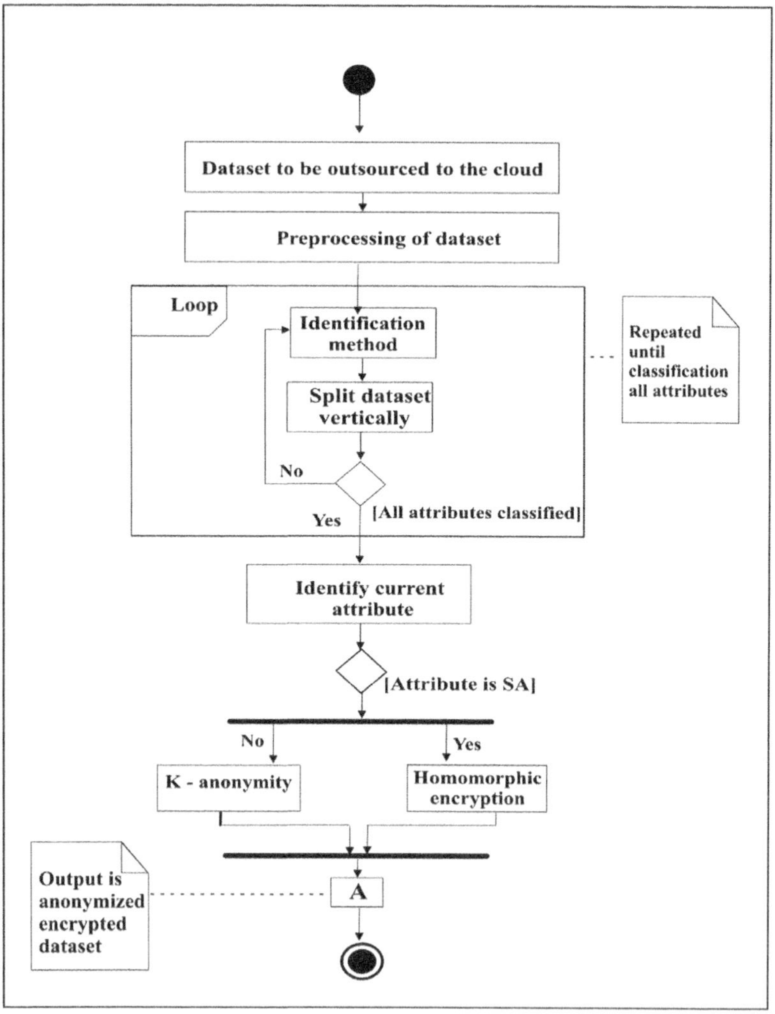

Fig. 3. Activity diagram of HAC

```
INPUT: Quasi_Identifiers, Diversion _Method, K
For i=1 to i ≤ K  do
    X=Find_Index_of_Min (Diversion _Level[])
    IF (Diversion _Method(x)==INCREMENTAL_GENERALIZATION)
        IF (Quasi_Identifiers == numbers)
            Num_Generalization(x);
        ELSE
            Union_Generalization(x);
    ELSE (Diversion _Method(x)==INCREMENTAL_SUPPRESSION)
            Incremental_Suppression(x);
OUTPUT: K_Anonymized_Table
```

Fig. 4. Pseudocode for developed Heuristic K-anonymity technique

Expected Deliverables of HAC Model

When we combine the homomorphic encryption with the k-anonymity technique, we can overcome the flaws of k-anonymity and the attacks on it. In contrast, we will reduce the high computational of homomorphic encryption by anonymization QIs via k-anonymity. Encrypting the SAs through homomorphic encryption will ensure that there is no similarity or homogeneity in SAs even if the distribution of SAs values is skewed. If we consider that the attacker has information about his victim on QIs attributes, he can't know the SAs of the victim. Thus, identify and attributes disclosures can be prevented, different types of attack, namely the homogeneity attack, background knowledge attack, similarity attack and probabilistic inference attack can be overcome. When using basic types of encryption to encrypt sensitive attributes, problems such as that of the attacker may be eligible to decrypt the encrypted database with one of the known decryption methods. Therefore, if homomorphic encryption is used, data mining operations can be performed on encrypted sensitive attributes without requiring decryption. Thus, ensuring the confidentiality and integrity of sensitive data.

Cloud-RSA Key Generation

INPUT: Two large distinct primes p and q of the same bits long.

- Compute $N = pq$ *and* $\emptyset(\mathrm{N}) = (p-1)(q-1)$.
- Select an integer $\boldsymbol{e}$ that is relatively prime to Ø (N).
- Determine the integer $\boldsymbol{d}$ satisfying ed $\equiv 1\ (mod\ \emptyset(\mathrm{N}))$.

OUTPUT: (ek, pk)

The evaluation key is ek = (N); the private key is $pk = (N, e, d)$.

Cloud-RSA Encryption

INPUT: pk, plaintext = $m_1, \ldots, m_k \in \mathbb{Z}$N (i.e. SAs in dataset).

- Compute *the ciphertexts* $c_i \equiv m_i^e\ (mod\ N)$ *where* $1 \le i \le k$.

OUTPUT: $c_i \equiv E\ (pk, m_i)$

Cloud-RSA Evaluation

INPUT: ek, ciphertexts = $c_1, \ldots, c_k \in \mathbb{Z}$N (i.e. after encrypted SAs in dataset).

- Compute $C \equiv c_1 \times \ldots \times c_k\ (mod\ N)$

OUTPUT: $\boldsymbol{C}$. // the evaluation process will execute by the service provider on the cloud.

Cloud-RSA Decryption

INPUT: pk, and a ciphertext C.

- Compute $C^d\ (mod\ N) \equiv m_1 \times \ldots \times m_k$

OUTPUT: $m_1 \times \ldots \times m_k = D\ (pk, C)$

Fig. 5. Cloud-RSA algorithms

5 Analysis and Discussion

Through theoretical comparison for our proposed solution with [5, 27] and [30], note that [5] merge anonymization methods to improve data privacy, this gives good performance in terms of execution time thus computational complexity compared with cryptography methods. On the other hand, the generalization that used as base-operation in anonymization leads to big information loss, in addition to that the privacy level introduces by anonymization methods not perfect; where are still vulnerable to some types of attacks as mentioned previously. Our proposed solution uses limited generalization by k-anonymity to reduce information loss and strengthening the level of privacy preservation provided by k-anonymity via combining it with homomorphic encryption.

The proposed solution reduced much of the computational complexity by using homomorphic encryption only for the sensitive attributes, other than that in related works such [21, 22] and [27] which using encryption techniques where the entire dataset is encrypted before being released to the cloud. The homomorphic encryption also plays a large role in reducing arithmetic complexity by eliminating the burden of the decryption process at the service provider on the cloud. The researchers in [30] introduce a hybrid solution similar to this work but the privacy-preserving level was still not at the optimum level. Whereby they encrypted EIs, anonymized QIs while SAs left as plain text. If the anonymized QIs are linked to SAs this will lead to attribute and identity disclosures. This problem cannot be found in the solution presented in this work, where EIs are suppressed, even if the anonymized QIs are linked to SAs the attacker can't predict anything because SAs are homomorphically encrypted. Table 3 gives a summary of this discussion.

Table 3. The comparison of our expected deliverables versus the related works.

PPDM techniques	Privacy level	Data utility	Computational burden
Anonymization-based techniques	Middle	Low	Low
Cryptography-based techniques	High	High	High
Hybrid techniques	Middle	Low	Middle
Our hybrid solution	High	High	Middle/low

6 Conclusion

More study about the privacy of data owing to identify disclosure and attributes disclosure at outsourced data on the cloud to mining is needed. This article reviewed most popular PPDM models on the cloud along with summarized their shortcoming to deduce the current research gap. Later, it suggested a hybrid model address the current problems of PPDM including identity and attributes disclosures whereby maintaining the privacy of data prior to outsourced on the cloud is the main focus. The proposed model merges homomorphic encryption, K-anonymity method and identification method to take advantage of their combined advantages to develop strong privacy while maintaining data utility. The resulting anonymous encrypted dataset will be immune to attacks associated with identity and attributes disclosures. For future work, the proposed model will be implemented in real Adult and Banking Marketing datasets (2014) from UCI, then outsourced them to execute classification methods on the cloud. The performance results would then be investigated further.

Acknowledgments. We would like to thank Ministry of Higher Education and University of Kassala – SUDAN for funding this work.

References

1. Samanthula, B.K., Elmehdwi, Y., Jiang, W.: k-nearest neighbor classification over semantically secure encrypted relational data. IEEE Trans. Knowl. Data Eng. **27**(5), 1261–1273 (2015)
2. Suresh, S., Huang, H., Kim, H.J.: Scheduling in compute cloud with multiple data banks using divisible load paradigm. IEEE Trans. Aerosp. Electron. Syst. **51**(2), 1288–1297 (2015)
3. Zhang, L., Yan, Y., Wang, W., Hao, X.: Finding quasi-identifiers for k-anonymity model by the set of cut-vertex (2018)
4. Zhang, H., Zhou, Z., Ye, L., Du, X.: Towards privacy preserving publishing of set-valued data on hybrid cloud (2018). https://ieeexplore.ieee.org
5. Aldeen Yousra, S., Mazleena, S.: A new heuristic anonymization technique for privacy preserved datasets publication on cloud computing. J. Phys: Conf. Ser. **1003**, 012030 (2018)
6. Storagecraft Home: 7 Most Infamous Cloud Security Breaches (2016). https://blog.storagecraft.com/7-infamous-cloud-security-breaches/. Accessed 12 May 2019
7. Islam, S., Ouedraogo, M., Kalloniatis, C., Mouratidis, S., Gritzalis, S.: Assurance of security and privacy requirements for cloud deployment models (2018). https://ieeexplore.ieee.org
8. Chen, B., Cheung, P., Cheung, P., Kwok, Y.: CypherDB: a novel architecture for outsourcing secure database processing (2018). https://ieeexplore.ieee.org
9. Mendes, R., Ao, J.O.: Privacy-preserving data mining : methods, metrics, and applications. **5** (2017)
10. Reddy, S.R., Raju, K.V.S.V., Valli Kumari, V.: Personalized privacy preserving incremental data dissemination through optimal generalization. J. Eng. Appl. Sci. **13**(11), 4205–4216 (2018)
11. Samarati, P., Sweeney, L.: Generalizing data to provide anonymity when disclosing information (1998)
12. Chakravorty, A., Wlodarczyk, T., Rong, C.: Privacy preserving data analytics for smart homes (2013)
13. Wang, J., Zhao, Y., Jiang, S., Le, J.: Providing privacy preserving in cloud computing. In: 3rd International Conference on Human System Interaction, HSI 2010 - Conference Proceedings (2010)
14. Mortazavi, R., Jalili, S.: Preference-based anonymization of numerical datasets by multi-objective microaggregation. Inf. Fusion **25**, 85–104 (2015)
15. Nayahi, J.J.V., Kavitha, V.: Privacy and utility preserving data clustering for data anonymization and distribution on Hadoop. Future Gener. Comput. Syst. **74**, 393–408 (2017)
16. Rebollo-Monedero, D., Forné, J., Soriano, M., Allepuz, J.P.: p-Probabilistic k-anonymous microaggregation for the anonymization of surveys with uncertain participation. Inf. Sci. **382**, 388–414 (2017)

17. Jain, P., Gyanchandani, M., Khare, N.: Improved k-anonymity privacy-preserving algorithm using Madhya Pradesh State election commission big data. In: Studies in Computational Intelligence, vol. 771, pp. 1–10. Springer (2019)
18. Machanavajjhala, A., Gehrke, J., Kifer, D.: ℓ - diversity : privacy beyond k-anonymity, vol. V, pp. 1–47 (2006)
19. Domingo-Ferrer, J., Soria-Comas, J.: From t-closeness to differential privacy and vice versa in data anonymization. Knowl.-Based Syst. **74**, 151–158 (2015)
20. Abdelhameed, S.A., Moussa, S.M., Khalifa, M.E.: Privacy-preserving tabular data publishing: a comprehensive evaluation from web to cloud. Comput. Secur. **72**, 74–95 (2018)
21. Mayil, S., Vanitha, M.: A survey on privacy preserving data mining techniques for clinical decision support system **5**(5), 6054–6056 (2016)
22. Uttarwar, N., Pradhan, M.A.: K-NN data classification technique using semantic search on encrypted relational data base. In: Proceedings - 2nd International Conference on Computing, Communication, Control and Automation, ICCUBEA 2016 (2017)
23. Omer, M.Z., Gao, H., Mustafa, N.: Privacy-preserving of SVM over vertically partitioned with imputing missing data. Distrib. Parallel Databases **35**(3–4), 363–382 (2017)
24. Selvaraj, B., Periyasamy, S.: A review of recent advances in Privacy preservation in health care data publishing. Int. J. Pharma Biosci. **7**(4), 33–41 (2016)
25. Kaur, A., Sofat, S.: A proposed hybrid approach for privacy preserving data mining (2016)
26. González-Serrano, F.-J., Amor-Martín, A., Casamayón-Antón, J., Madrid, S.: Supervised machine learning using encrypted training data. Int. J. Inf. Secur. **17**, 365–377 (2018)
27. Li, L., Lu, R., Choo, K.K.R., Datta, A., Shao, J.: Privacy-preserving-outsourced association rule mining on vertically partitioned databases. IEEE Trans. Inf. Forensics Secur. **11**(8), 1547–1861 (2016)
28. Aldeen, Y.A.A.S., Salleh, M.: A Hybrid K-anonymity Data Relocation Technique for Privacy Preserved Data Mining in Cloud Computing. J. Internet Comput. Serv. **17**(5), 51–58 (2016)
29. Kohlmayer, F., Prasser, F., Eckert, C., Kuhn, K.A.: A flexible approach to distributed data anonymization. J. Biomed. Inform. **50**, 62–76 (2014)
30. Yang, J.-J., Li, J.-Q., Niu, Y.: A hybrid solution for privacy preserving medical data sharing in the cloud environment. Future Gener. Comput. Syst. **43–44**, 74–86 (2015)
31. Taric, G.J., Poovammal, E.: A survey on privacy preserving data mining techniques. IOSR J. Comput. Eng. Vers. III **17**(5), 2278–2661 (2017)
32. Fung, B.C.M., Wang, K., Chen, R., Yu, P.S.: Privacy-preserving data publishing. ACM Comput. Surv. **42**(4), 1–53 (2010)

Detecting False Messages in the Smartphone Fault Reporting System

Sharmiladevi Rajoo[1], Pritheega Magalingam[1(✉)], Ganthan Narayana Samy[1], Nurazean Maarop[1], Norbik Bashah Idris[2], Bharanidharan Shanmugam[3], and Sundaresan Perumal[4]

[1] Razak Faculty of Technology and Informatics, Universiti Teknologi Malaysia, Skudai, Malaysia
sharmila2207@yahoo.com, {mpritheega.kl, ganthan.kl, nurazean.kl}@utm.my

[2] International Islamic Universiti Malaysia, Gombak, Malaysia
norbik@iium.edu.my

[3] School of Engineering and Information Technology, Charles Darwin University, Casuarina, Australia
bharanidharan.shanmugam@cdu.edu.au

[4] Faculty of Science and Technology, Universiti Sains Islam Malaysia, Gombak, Malaysia
sundresan.p@usim.edu.my

Abstract. The emergence of the Internet of Things (IoT) in Smart City allows mobile application developers to develop reporting services with an aim for local citizens to interact with municipalities regarding city issues in an efficient manner. However, the credibility of the messages sent rise as a great challenge when users intentionally send false reports through the application. In this research, an evidence detection framework is developed and divided into three parts that are a data source, IoT device's false text classification engine and output. Text-oriented digital evidence from an IoT mobile reporting service is analyzed to identify suitable text classifier and to build this framework. The Agile model that consists of define, design, build and test is used for the development of the false text classification engine. Focus given on text-based data that does not include encrypted messages. Our proposed framework able to achieve 97% of accuracy and showed the highest detection rate using SVM compared to other classifiers. The result shows that the proposed framework is able to aid digital forensic evidence experts in their initial investigation on detecting false report of a mobile reporting service application in the IoT environment.

Keywords: Internet of Things · Smartphone · Application · Reporting services · Smart City · Text classifiers

1 Introduction

The Internet of Things (IoT) is the new future of the Internet. It offers individuals, communities and the societies to connect and obtain services over the Internet anytime and anywhere. Furthermore, the IoT also improves interactions between people and

F. Saeed et al. (Eds.): IRICT 2019, AISC 1073, pp. 759–768, 2020.
https://doi.org/10.1007/978-3-030-33582-3_71

devices. In the context of Smart City, the integration of IoT technologies allows municipalities to manage their infrastructure efficiently and to offer interactive reporting services. As the smartphones become more affordable and it is the easiest way to connect, many town councils have utilized this tool to deal with city issues, making city life smarter. Mobile reporting services allow citizens to report issues with data captured through their smartphones [1]. There are a few applications such as Find It, Fix It (USA) [2], Improve My City (Greece) [3] and CitiAct (Malaysia) [4] that have been developed for citizens to participate and report information like faulty traffic lights, vandalism, crime, natural disasters and other city-related issues.

Aggarwal and Abdelkader [5] mentioned in their paper several challenges in mobile reporting services. This includes user privacy and trustworthiness of the reports. Sensor data in mobile devices such as the GPS collects private information of the users such as their daily routes and this can invade their privacy if it is not preserved. On the other hand, the credibility of the report is also one of the major concerns as users may intentionally send misleading information to the system. Thus, false reports such as offensive or threatening messages should be identified and filtered out before it is sent to the town councils. In this paper, we focus on the trustworthiness issue because false information affects the response team in the aspect of time, energy and other resources. This research takes the false reports as digital evidence of illegal mobile reporting.

Researchers in the past have been concentrating primarily on users' reputation and behavior to detect false information in the reporting services [6–12]. The researchers have evaluated trustworthiness of a report by evaluating the user's reputation [6]. The challenge of this technique is that a user with high reputation might also able to send false reports deliberately to sabotage the system. On the other hand, the people who uses the fault reporting system is public and the reputation or the behavior of the people cannot be detected pre-hand. However, elements in the text can also be considered as one of the methods to identify false reports because certain keywords in the text can actually determine whether the report is true or false.

Thus, the motive of this research is to propose an evidence detection framework for mobile reporting services using a text classifier technique. Text classifier has been widely used for spam detection and sentiment analysis [13]. Besides that, it is also applied in the field of digital forensics to analyze text-oriented digital evidence based on keyword search analysis to aid the experts in their investigation [14]. An analysis has been done to compare the performance of Naïve Bayes (NB), Support Vector Machines (SVM), Decision Tree (DT) and K-Nearest Neighbours (KNN) in terms of its accuracy and robustness. This research resolves the false report recognition problem in the CitiAct application by introducing a suitable method that could achieve 97% of accuracy and the phases involved to analyze large text from the application.

2 Related Works

2.1 False Text Detection in Reporting Services

Firoozjaei et al. [15] proposed a score based framework called False Request Detection Algorithm (FRDA) to detect false emergency requests. According to the author, false alarms can occur where reports are sent on good intents such as misjudgments and based on humanity or kind intents. FRDA evaluates the trustworthiness of the callers based on the callers' reported location and their trust score history. A caller with high trust score is more trustworthy compared to a caller with less trust score. This approach needs the trust score beforehand. Secondly, Kantarci et al.[6] proposed a reputation-based framework named Trustworthy Sensing for Crowd Management (TSCM) where the system validates users with high reputations has a higher probability of sending true reports, whereas users with low reputations will be validated with a lower probability of sending true reports. As discussed in the introduction, reputation and behavior itself cannot be used to predict the authenticity of the reports. Besides that, researchers have also proposed several algorithms to identify false reports by analyzing user's behaviour in reporting the information. Wang et al. [7] stated that different degree of confidence may affect the correctness of the data. Therefore, they proposed a confidence-aware truth estimation algorithm using Maximum Likelihood Estimation (MLE) approach by measuring the degree of users' confidence in their report. Next, Marshall et al. [8] came up with another algorithm using Maximum Likelihood Estimation (MLE) approach called emotional-aware truth discovery algorithm. They indicated that emotional reported information (claims) should not be classified as true or false claims because it can lead to inaccuracy in truth discovery results. Furthermore, Marshall et al. [9] also stated that factors such as time, location and scale influence the degree of hardness of claims. Hence, they proposed a hardness-aware truth discovery algorithm.

Some applications implement a credit-based reward to attract the user to participate and provide information. Users will get some incentives if they actively participate using the application. As a result, users misuse it by sending a large number of spam reports to increase their participation and get rewarded. Ghosh et al. [10] proposed a probabilistic-based approach to identify and filter spam reports through automated confidence assignment. The proposed approach is then validated by Barnwal et al. [11] using Waze traffic alerts. Prandi et al. [12] developed a trustworthiness model for a system called Mpass (mobile Pervasive Accessibility Social Sensing) system. This system evaluates the trustworthiness of the information collected by combining the accuracy of sensors, user's source credibility and the authoritative reports coming from authoritative data sources. Hence, a Mpass trustworthiness model is developed and conducted an agent-based simulation to assess the model.

In recent research by Bhatti et al. [16] aimed to propose an online rescue operation that detects accident and sends notification. The IoT physical system that relies on four sensor inputs to detect road accidents and is connected to medical rescue system for an immediate medical assistance. They claim that using the sensor inputs, the system able to reduce false positive rates in the reporting part. However, the proposed system needs to be tested in a real time scenario to confirm the reduction in false positive rates. Based on the previous studies, researchers have mainly focused on users' reputation and behavior in identifying the trustworthiness and usefulness of the report. The textual or the content of the report should also be taken into consideration in identifying true and false reports. Text classifier technique has been widely used in various fields mainly in spam filtering and sentiment analysis to analyze large amount data gathered from different sources. It is also applied in the field of digital forensics such as keyword analysis, detecting deceptive criminal identities, identifying groups of criminals who are engaging in various illegal activities, criminal profiling and evidence discovery. Al-Zaidy et al. [17] stated that digital evidence is usually in the form of textual data such as emails, chat messages from mobile phones and social media and web pages. These textual data are unstructured and due to that, investigators had to conduct their investigations in a tedious manner as they had to extract information manually. With text classifier technique, it aids the investigators to detect and extract useful information or digital evidence.

Application of Text (keyword) Classifiers in Various Domains. Criminals often hide their evidence in a digital device by changing the file type. It is necessary for the digital forensic investigator to identify the file type and classify them as legitimate and illegitimate file types [18]. As the usage of Internet and email continue to grow, the number of spam emails continues to increase rapidly. Lakshmi and Radha [19] compared two different classifiers techniques which are the Support Vector Machine (SVM) and K-Nearest Neighbours (KNN) to investigate the best classifier in classifying spam messages. They collected 20 datasets for the experiment and features such as header, subject and body were used to evaluate the performance of the classifiers. The analysis was carried out using a data mining software called Tanagra. The performances of the classifiers were evaluated based on their ease of learning, error rate and predictive accuracy. The results showed that compared to KNN, SVM consumed more time to learn and train the data. However, SVM has the least error rate and achieved a higher predictive accuracy of 77% compared to KNN which is 75%.

Likewise, authors Trivedi and Dey [20] also conducted a comparative analysis on three classifiers: Naïve Bayes (NB), Support Vector Machine (SVM) and Decision Tree (DT) to evaluate their performances in spam email classification. The performances of each classifier were measured based on their accuracy and false positive rate (FP rate). FP rate is the rate of misclassification of legitimate emails. Based on the overall results, SVM achieved an accuracy of 97.8%, followed by NB with 97.6% and DT with 96.6%. In terms of FP rate, SVM has the lowest FP rate compared to NB and DT.

Application of Text Classifier in Sentiment Analysis. Mobile network traffic was treated as text and important semantic features were extracted from it using N-gram method from the text of the traffic flows. Later the researchers have used SVM classifier to take the important features as its feed for malware detection [21]. Gautam and Yadav conducted sentiment analysis on customers' review about their satisfaction of the purchased product [22]. They did a comparative analysis of Support Vector Machine (SVM) and Naïve Bayes (NB). The experiment was implemented in Python and the results showed that SVM achieved a better result than NB with an accuracy of 88.2% while NB achieved an accuracy of 85.5%. Based on the comparisons that have been discussed above, we found that SVM performed better in all the experiments compared to the other three classifiers. Therefore, for our research, we apply SVM text classifier for our proposed framework because SVM does not only have higher accuracy, it is also robust, suitable for binary classification and it is easy to implement. Besides that, to the best of our knowledge, SVM is not used for false report identification in the IoT environment.

3 Methodology

This part explains the method that we have used to build the classifier. Agile model is used in this study for the development of the false text classification engine. It is a simple implementation and iteratively enhances the evolving versions until the full system is implemented. This model was chosen because it is adaptive and easy to understand [23]. Based on activities in Table 1, first, we identified the requirements of tools and its packages needed to build the classifier. Then, an algorithm to extract the suitable features is designed. The algorithm is designed by first gathering all the mechanics on how the CitiAct application works. Understanding the nature of the report generated from the CitiAct is essential. Then the data preparation steps were identified. Following to that, suitable classifiers' procedure was studied and its comparison process were designed. After forming the algorithm, it is developed and finally the source code is tested where the performance is compared with other classifiers.

Real time dataset is collected from CitiAct database. Four attributes are used from the dataset; report text, location, address and type of report. The tool that will be used for programming and text classification is RStudio [24]. The advantage of RStudio is that it is flexible and generates a high quality analysis. Continuous literature review and documentation have been done on each part to update the recent work and compare our result with the previous work.

Table 1. Activities conducted in each agile model phase.

Phases	Activities
Define	The requirement that is needed for this experiment is the installation of SVM packages in RStudio
Design	An algorithm to detect false texts using SVM text classification is designed
Build	The algorithm is built and developed using R studio
Test	The source code is tested in R and its performance measure is evaluated

4 Evidence Detection Framework

The report sent by public through the CitiAct application is an evidence that we need to obtain in order to investigate whether it is a true or untruthful report. Our proposed False Report Evidence Detection Framework applies SVM text classifier technique to identify true and false reports (see Fig. 1). The description of the framework and the experiment steps are explained below.

4.1 Data Preparation

First, the dataset that contains a list of reports has to be labelled as F (represents false report) or T (represents true report). It will then undergo preprocessing by tokenizing the stream of a text document into words (tokens), removing stop words such as "a", "the", "and", "of" and "while" and converting words into their root forms. Next, features are extracted by identifying distinctive keywords that occur frequently in the dataset.

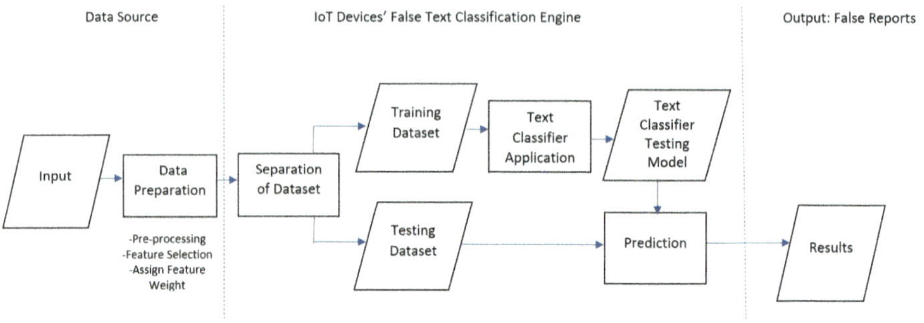

Fig. 1. False report evidence detection framework.

The selected features (keywords) are given weight based on a number of times the terms are used in the dataset. Table 2 shows the first 6 reports in the dataset.

Table 2. Example of a labelled and preprocessed dataset.

Doc	Data	Label
1	Two potholes at the junction	+1
2	Garbage unattended	+1
3	Faulty traffic light	+1
4	S@& toys on sale	-1
5	Pointless app	-1
6	Potholes near playground	+1

Doc	Pre-Processed Data	Label
1	Two, potholes, junction	+1
2	Garbage, unattended	+1
3	Faulty, traffic, light	+1
4	Toys, sale	-1
5	Pointless, app	-1
6	Potholes, near, playground	+1

Once the preprocessing is completed, the next step is extracting features and generating a list of terms from each document. Each term is given an index and it is sorted in alphabetical order.

IoT Devices' False Text Classification Engine. The prepared dataset is sent for text classification analysis that undergoes a series of steps. The first step is the separation of dataset. The dataset is separated into training and testing dataset with the proportion of 70:30. The training dataset will have the reports labelled as T (True) or F (False). This is for SVM to learn the properties of true and false reports. The testing dataset is used for prediction and therefore this portion of the dataset will not be labelled. Instead, SVM will try to predict and categorize the reports into their respective classes. Then the data are converted into term matrix which is in the form of a:b; a is the index of the term and b is the number of times the term occurs in that particular document. Table 3 shows some examples of the calculated term matrix.

Table 3. Calculated term matrix.

Doc	Data	Matrix	Label
1	Two, potholes, junction	13:1 9:1 4:1	+1
2	Garbage, unattended	3:1 14:1	+1
3	Faulty, traffic, light	2:1 12:1 5:1	+1
4	Toys, sale	11:1 10:1	−1
5	Pointless, app	8:1 1:1	−1
6	Potholes, near, playground	9:1 6:1 7:1	+1

In the text classifier application, support vector machine is applied and trained to learn the characteristics or features of true and false data from the training dataset. It computes and assigns a vector to each data. Hence, the data will be separated into two classes; class 1 and 2 where they represent true and false data respectively. The text classifier testing model is built when an optimal hyperplane is generated by identifying the support vectors and maximizing the margin. This separates two classes (True and False). In this paper, we calculated the equal margins between class 1 and class2. True class is denoted as (+1) and the false class is denoted as (−1). Then the SVM is trained and separates the reports into two classes. Class 1 (indicating True reports) and Class 2 (indicating False reports) as shown in Table 4. Thus, this model is used for predicting testing datasets.

Table 4. Classes assigned by SVM.

Doc	Data	Matrix	Label
1	Two, potholes, junction	13:1 9:1 4:1	1
2	Garbage, unattended	3:1 14:1	1
3	Faulty, traffic, light	2:1 12:1 5:1	1
4	Toys, sale	11:1 10:1	2
5	Pointless, app	8:1 1:1	2
6	Potholes, near, playground	9:1 6:1 7:1	1

In the prediction part, the text classifier testing model is then used with the testing dataset for prediction. The number of reports that are correctly predicted can be determined by referring to the confusion matrix [25]. Finally, a list of false texts from the mobile reporting services in the IoT environment that are extracted by the false text classification engine are displayed.

5 Results and Discussion

Naïve Bayes (NB), Support Vector Machines (SVM), Decision Tree (DT) and K-Nearest Neighbours (KNN) are most commonly used classifiers and among these four, SVM has proven to be the most popular because of its good performance and higher accuracy [26, 27]. Therefore, we conducted an experiment to test our framework by applying SVM text classifier and the results are compared with NB, DT and KNN.

SVM produced consistent results with the highest accuracy of 89% and 97% with 500 and 1000 dataset respectively and also with lowest error rate although the time taken to train and predict the data is higher compared to the other three classifiers. This could be because as the dimension gets larger, SVM consumes more time to train and test data. However, SVM works well with larger datasets and able to identify more distinct features to produce good classification results. On the other hand, with the shortest amount of time taken to train and test data, KNN also performed well by achieving the second highest accuracy of 85% in the first experiment 500 dataset were added.

NB produced the lowest result with an accuracy of 70% with 500 datasets but has outperformed DT and KNN with 1000 dataset by achieving the second highest accuracy of 96%. DT performed relatively well with the 500 datasets and achieved the third highest accuracy of 83% in a shorter amount of time but produced the lowest results with 1000 datasets. This is due to overfitting issue that causes a high error in generalization. On the whole, based on the results produced these experiments, SVM is found to be the most accurate and robust text classifier compared to NB, DT and KNN. Table 5 shows results of the four classifiers and the highlighted row indicates that the performance levels of text classifiers are determined by the results of the accuracy.

Table 5. Text classifiers results

Algorithm	SVM		NB		DT		KNN	
No. of datasets	500	1000	500	1000	500	1000	500	1000
Training time	0.13	0.15	0.05	0.07	0.06	0.10	0.04	0.05
Testing time	0.09	0.12	0.05	0.06	0.04	0.06	0.03	0.05
Precision	0.88	0.96	0.78	0.95	0.85	0.90	0.84	0.92
Recall	0.91	0.97	0.75	0.96	0.81	0.91	0.85	0.94
F1-Measure	0.89	0.96	0.76	0.95	0.83	0.90	0.84	0.93
Accuracy	**0.89**	**0.97**	0.77	0.96	0.83	0.91	0.85	0.93
Error rate	0.1	0.03	0.23	0.04	0.17	0.09	0.15	0.07

Comparing with the previous similar field of research such as application of SVM classifier on spam messages prediction [19], mobile filetype identification and classifying it as legitimate or illegitimate filetypes [18, 28], spam email classification [20], it is found that SVM has outperformed other classification techniques and has the lowest false positive rates. Following to this, SVM is used in our proposed framework and is applied on larger dataset to detect false reports. The framework manages to identify 85% of the false reports from the large dataset.

6 Conclusion and Future Work

The proposed work can be part of digital investigation on criminal cases such misuse of online application or security threats; transmitting false information in the IoT environment. Thus, our proposed framework can also be applied in the digital investigation area as it can help investigators to obtain important information from the unstructured data.

The proposed false report detection algorithm is yet to be tested on the real application to investigate the real time data and to obtain the performance results. The features used in this research is only from the static dataset that were collected from the CitiAct application. At the moment, the proposed framework is only able to read English language based text reports. CitiAct may receive messages both in English and Malay language. There are possibilities to enhance the algorithm to read both the languages. Based on these limitations, the algorithm can be tested with larger datasets and by adding variety of features. Since this algorithm is first being developed for CitiAct application in Malaysia, it should be further enhanced to read other languages and enable this false report identification algorithm to be applied to any other types of smartphone applications in the IoT environment.

Acknowledgement. This work is supported in part by Redtone IOT Sdn. Bhd., Malaysia. We thank Redtone IoT staff members who provided insight and data that greatly assisted the research.

References

1. Walravens, N.: Mobile city applications for Brussels citizens: Smart City trends, challenges and a reality check. Telematics Inform. **32**(2), 282–299 (2015)
2. Cook, J.: Tell the City of Seattle about potholes and graffiti with new 'Find It, Fix It' app (2013). https://www.geekwire.com/2013/city-seattle-potholes-graffiti-find-fix-it-app/. Accessed 15 July 2019
3. (Greece), T.I.T.I.-M.G. Improve My City. http://smartcityapps.urenio.org/improve-my-city_en.html. Accessed 15 July 2019
4. REDtoneIOTSdn.Bhd., CitiAct – Your Key in Building the Next Smart City Applications (2016)
5. Aggarwal, C.C., Abdelzaher, T.: Social sensing. In: Managing and Mining Sensor Data, pp. 237–297. Springer, Heidelberg (2013)
6. Kantarci, B., Mouftah, H.T.: Trustworthy sensing for public safety in cloud-centric internet of things. IEEE Internet Things J. **1**(4), 360–368 (2014)
7. Wang, D., Huang, C.: Confidence-aware truth estimation in social sensing applications. In: 2015 12th Annual IEEE International Conference on Sensing, Communication, and Networking (SECON). IEEE (2015)
8. Marshall, J., Wang, D.: Towards emotional-aware truth discovery in social sensing applications. In: 2016 IEEE International Conference on Smart Computing (SMARTCOMP). IEEE (2016)
9. Marshall, J., Syed, M., Wang, D.: Hardness-aware truth discovery in social sensing applications. In: 2016 International Conference on Distributed Computing in Sensor Systems (DCOSS). IEEE (2016)

10. Ghosh, N., et al.: A probabilistic approach for filtering out spam reports in a vehicular participatory sensing system. In: 2016 8th International Conference on Communication Systems and Networks (COMSNETS). IEEE (2016)
11. Barnwal, R.P., et al.: Enhancing reliability of vehicular participatory sensing network: a bayesian approach. In: 2016 IEEE International Conference on Smart Computing (SMARTCOMP). IEEE (2016)
12. Prandi, C., et al.: A trustworthiness model for crowdsourced and crowdsensed data. In: 2015 IEEE Trustcom/BigDataSE/ISPA. IEEE (2015)
13. Pandey, U., Chakravarty, S.: A survey on text classification techniques for e-mail filtering. In: 2010 Second International Conference on Machine Learning and Computing (ICMLC). IEEE (2010)
14. Tayal, D.K., et al.: Crime detection and criminal identification in India using data mining techniques. AI Soc. **30**(1), 117–127 (2015)
15. Firoozjaei, M.D., J. Park, and H. Kim. Detecting false emergency requests using callers' reporting behaviors and locations. In: 2016 30th International Conference on Advanced Information Networking and Applications Workshops (WAINA). IEEE (2016)
16. Bhatti, F., et al.: A novel internet of things-enabled accident detection and reporting system for smart city environments. Sensors **19**(9), 2071 (2019)
17. Al-Zaidy, R., et al.: Mining criminal networks from unstructured text documents. Digit. Invest. **8**(3–4), 147–160 (2012)
18. Marturana, F., et al.: A quantitative approach to triaging in mobile forensics. In: 2011 IEEE 10th International Conference on Trust, Security and Privacy in Computing and Communications (TrustCom). IEEE (2011)
19. Lakshmi, R.D., Radha, N.: Spam classification using supervised learning techniques. In: Proceedings of the 1st Amrita ACM-W Celebration on Women in Computing in India. ACM (2010)
20. Trivedi, S.K., Dey, S.: Interaction between feature subset selection techniques and machine learning classifiers for detecting unsolicited emails. ACM SIGAPP Appl. Comput. Rev. **14** (1), 53–61 (2014)
21. Wang, S., et al.: Detecting android malware leveraging text semantics of network flows. IEEE Trans. Inf. Forensics Secur. **13**(5), 1096–1109 (2018)
22. Gautam, G., Yadav, D.: Sentiment analysis of twitter data using machine learning approaches and semantic analysis. In: 2014 Seventh International Conference on Contemporary computing (IC3). IEEE (2014)
23. Munassar, N.M.A., Govardhan, A.: A comparison between five models of software engineering. IJCSI **5**, 95–101 (2010)
24. Torgo, L.: Data mining with R. Learning by case studies. University of Porto, LIACC-FEP (2003). http://www.dcc.fc.up.pt/~ltorgo/DataMiningWithR/. Accessed 1 July 2019
25. Panigrahi, P.K.: A comparative study of supervised machine learning techniques for spam e-mail filtering. In: 2012 Fourth International Conference on Computational Intelligence and Communication Networks. IEEE (2012)
26. Brindha, S., Prabha, K., Sukumaran, S.: A survey on classification techniques for text mining. In: 2016 3rd International Conference on Advanced Computing and Communication Systems (ICACCS). IEEE (2016)
27. Rajoo, S., et al.: A comparative study of text classifier for mobile crowdsensing applications. Adv. Sci. Lett. **24**(1), 686–689 (2018)
28. Gopal, S., et al.: Statistical learning for file-type identification. In: 2011 10th International Conference on Machine Learning and Applications and Workshops (ICMLA). IEEE (2011)

Local Descriptor and Feature Selection Based Palmprint Recognition System

Chérif Taouche[1,2](✉) and Hacene Belhadef[2]

[1] RELA(CS)² Laboratory, University of Oum El-Bouaghi, 04000 Oum El-Bouaghi, Algeria
ch_taouche@yahoo.fr

[2] NTIC Faculty, University of Constantine 2, 25016 Constantine, Algeria
hacene.belhadef@univ-constantine2.dz

Abstract. In this paper, we present a system based on a feature selection approach for solving the recognition problem in a multimodal biometric system combining both left and right palmprints of the same subject. In particular, the fusion of two or more traits, at feature-level, results in a long feature vector that needs large storage space, makes the execution time of the recognition task very long, and may include redundant and irrelevant features that can affect the recognition accuracy. To overcome these problems, feature selection is performed using genetic algorithms (GAs) and backtracking search algorithm for a comparison purpose. The experimental results show the usefulness of feature selection, especially the use of genetic algorithms, on the robustness of the multimodal biometric system as regards the feature vector length and run-time reduction, and the significant increase of the recognition rate.

Keywords: Multimodal biometrics · Feature selection · Local descriptor · Palmprint recognition · Feature-level fusion · Evolutionary algorithms

1 Introduction

As known, unimodal biometric systems proved their effectiveness in the recognition field. However, they still have some problems that make them less safe and less accurate, such as identity theft (spoofing), non-universality, noisy data, partial occlusion, variation in exposure and variation in lighting [1]. To solve some of or all these drawbacks, if it's possible, with the aim of increasing the security level and recognition accuracy, multi-biometrics systems based authentication is one of the possible promising and effective alternatives. The aforementioned drawbacks can be overcome, using multi-biometrics systems, by the fusion of more than one trait belonging to the same subject in a recognition system. This is because of the combination of two or more modalities cause it difficult to spoofing multiple biometric modalities of a genuine subject. Moreover, multi-biometrics authentication overcomes the problem of non-universality because of the utilization of multiple biometric modalities ensure the population coverage sufficiently [2]. However, the fusion of two or more traits results in a long feature vector which needs a large storage space, makes the execution time of the recognition task very long, and may include redundant and irrelevant features

F. Saeed et al. (Eds.): IRICT 2019, AISC 1073, pp. 769–778, 2020.
https://doi.org/10.1007/978-3-030-33582-3_72

which can affect the recognition accuracy by decreasing it. Therefore, feature selection which picks out the major or significant features is a good solution to overcome these problems.

Several biometric modalities are used in the recognition of individuals. Among these modalities, we find the palmprint one. By its comparison with other biometric traits, the important properties of the biometric characteristics (stability, individuality, universality, and collectability) are satisfied by the palmprint modality. Moreover, it has the following characteristics: non-intrusive, relatively stable features, very easy data collection, and unique features. The palmprint modality is highly efficient, even with the use of low-resolution images provided by a low-cost device. Hence, we can say that the palmprint recognition represents a promising, very interesting, and a fertile area of research.

Many are the researches that have been presented on multi-biometrics systems fusing the palmprint trait and other biometric traits [2–13].

According to the state-of-the-art study on multi-biometric systems based on palm prints, not many are the researches that have been conducted on the fusion of right and left palmprints, although they are complementary, even in the real life, and richer in discriminant information as mentioned above. Thus, in this paper, we present a multi-biometrics system fusing the right and left palmprint traits of the same subject. The fusion is performed at feature level followed by a feature selection technique, the matching and decision steps.

The extraction of features from the right and left palmprints is performed utilizing two techniques. These are based on the introduction of the multi-scale block principle on the local directional pattern (LDP) [14] descriptor and its variant called enhanced local directional pattern (ELDP) [15]. The produced descriptors are as follows: The multi-block local directional pattern (MBLDP) and the multi-block enhanced local directional pattern (MBELDP). Thus, the obtained descriptors perform the calculation operations on the average values of the different blocks in contrast to the original ones that perform it on the pixel values. Hence, macrostructures and microstructures can be captured by the resulting descriptors in a gradient space. Moreover, as the gradient space is robust against noise and illumination changes, these descriptors are able to have the following characteristics: complete image representation, production of a very few feature number, High resistance to changes of illumination and robustness to noise. Thus, these properties allow the produced descriptors to give higher performance and this is in terms of the execution time and recognition accuracy.

Once extraction of features is performed from the right and left palmprint modalities by utilization of the aforementioned descriptors, the fusion of the two obtained feature vectors is done in a serial manner that means simply concatenating them in a single global feature vector. The reason behind choosing it, as known, is that the fusion at feature-level presents richer pieces of information compared to the decision and matching score-level fusion which allows it to be able obtaining higher performance. After the fusion task, feature selection is then applied to the new constructed global feature vector. The reason behind applying feature selection task is to reduce the feature vector size, decreasing the execution time, and increasing the recognition accuracy by selecting the significant features and neglecting the irrelevant and redundant ones. As a feature selection technique, two algorithms are used in a comparison purpose: the well-known Genetic Algorithms (GAs) and the recently devised Backtracking Search Algorithm (BSA) [16]. Then, the Chi-square similarity measure is used in the matching step. Finally, intensive experiments were conducted using the public available IITD database.

The rest of the paper is as follows. Section 2 presents the local descriptors used for feature extraction. Section 3 describes the feature fusion and the feature selection techniques used in this paper. In Sect. 4, the different steps of the proposed method are explained. Section 5 presents the used database, the experiments setting, and the obtained results with in-depth discussion. Finally, a conclusion of the paper is provided in Sect. 6.

2 Feature Extraction Step Using Local Descriptors

In this section, local descriptors proposed and used in the paper for feature extraction are described.

2.1 MBLDP Descriptor

Once the average values of the different blocks of the original image are calculated, the pixels of the less size produced image are subject to kirsch mask in 8 directions in a 3*3 neighborhood. This task allows the production of an eight-bit binary code which is assigned to each pixel of the new image. The code is obtained by comparison of the eight relative edge response values $|m_i|$ (i = 0, 1, 2 … 7). So, the bits b_i that corresponds to the k most top edge response values are defined to the '1' value in contrast to the resting (8 – k) bits that are defined to '0'. The code of the MBLDP descriptor is calculated by Eq. (1) as follows:

$$MBLDP_k = \sum\nolimits_{i=0}^{7} b_i(m_i - m_k)2^i, b_i(x) = \begin{cases} 1, x \geq 0 \\ 0, x < 0 \end{cases} \quad (1)$$

with, m_k is the edge response value that corresponds to the k-th most top significant directional-response.

2.2 MBELDP Descriptor

In contrast to the MBLDP descriptor that exploits the absolute values to generate the code, MBELDP considers the sign of the original edge response values. Thus, the MBELDP local descriptor exploits the index number i corresponding to the first and second maximum response values. Hence, the MBELDP code corresponding to each pixel in the new image is produced by the use of a double-digit octal number $(ab)_8$ that we can convert it into a decimal one by using the Eq. (2) as follows:

$$MBELDP\, Decimal\, code = a \times 8^1 + b \times 8^0 \quad (2)$$

with, a and b represents the decimal index numbers corresponding, respectively, to the first and second maximum original edge response values m_a and m_b.

3 Feature Fusion and Feature Selection Steps

3.1 Feature Fusion Step

After the extraction of features from the left and right palmprints with the use of one among the presented descriptors, the resulting two feature vectors are combined (or fused) in a single vector with a serial manner. The latter is a simple concatenation of the produced feature vectors.

3.2 Feature Selection Step

The well-known population-based Genetic Algorithms and Backtracking Search Algorithm are successfully introduced, widely used in feature selection, and given very good and high results [9, 17]. In contrast to GAs, BSA is based on two populations, the current and the old population. Obviously, the binary GAs and binary BSA are used in feature selection.

In order to make the feature selection step, either with binary GAs or with binary BSA, we represent an individual of a population by a binary bit string. The latter is of fixed length and consisting of a series of 0's and 1's. The number of features in the global vector resulted from the combination of the two feature vectors, obtained from the left and right palmprints, represents the binary bit string length (L). In this binary string, the value '0' means that the associated feature is neglected, while the '1' indicates that is considered in the following step. The initial population, composed of N individual each of which is represented by a string of L bits, is generated in a random manner. A fitness function is necessary to evaluate an individual or a candidate solution. Maximizing the recognition rate is the purpose of this function (see Eq. (3)). So, to calculate the recognition rate (fitness function), Chi-square similarity is used. The latter is measured between each test image in the probe set and all the training samples in the gallery set. Then, for each test image, the distance with the lowest value is considered to determine if this probe image and the corresponding gallery one are from the same subject (class) and increment or not the acceptance number. Finally, the recognition rate is calculated by the division of the acceptance number by the test images total number.

$$Recg.Rat = \frac{Nb_{acc}}{Nb_{tes}} \tag{3}$$

with, Nb_{tes} represents the images number in the probe database while Nb_{acc} is the number of accepted images.

Finally, we have chosen the control parameters for the binary GAs like those in [1]. Also, for binary BSA, the control parameters were chosen as those of [9]. But, in exception, according to the tests performed in [9] we have set the population size at 40 for the two binary algorithms. These parameters are illustrated in the Table 1.

4 The Proposed System

In this section, we cite the different steps that compose the presented multi-biometrics system in this paper.

1. Without using any preprocessing step, ROI's of the right and left palmprint images are used, in contrast to the conducted research in [9].
2. Features are extracted from the right and left palmprint images utilizing the descriptors proposed in this paper.
3. Feature vectors obtained at the precedent step are fused in a serial manner. In other words, the two vectors are concatenated and a single global feature vector is produced like in [2].
4. Using either binary GAs or binary BSA, feature selection is performed to augment the recognition accuracy, decreasing the execution time, and also reducing the size of the feature vector.
5. For matching, the Chi-square similarity measure is applied between the training and testing vectors produced at the precedent step (feature selection).

5 Experimental Results and Discussion

In this section, we present the used database, the experimental results of unimodal and multi-modal systems followed by an analysis of the obtained results.

5.1 Experimentations

In the conducted experimentations, we have used the contactless IITD palmprint V1 database which is publicly available [18].This database is constructed by the acquisition of seven samples of right and left hands from 235 persons. Regions of interest (ROI's) of size 150 × 150 pixels of all palmprint images are also available in the database. The palmprint images of the IITD database presents serious or important variations on translation, pose, rotation and projection. Hence, palmprint images in the aforementioned database are more close to the real-applications than those of the rest databases like the PolyU one [5].

In order to ensuring a maximum of variations between the images for evaluating the proposed approach, we selected a sub-database composed of 96 individuals each of which have twelve well extracted images ROI's (six right palmprints and six left palmprints). We have chosen six samples per subject because of that the number of subjects with 7 sample images in the database is very small. So, the considered sub-database is composed of 1152 palmprint ROI's (Right = 96 × 6 = 576, Left = 96 * 6 = 576).

To perform the experimentations, two image sets are constructed, the training set and the test set, by considering one sample per person as a training image in contrast to other researches [2, 9] that considered two or more images. The utilization of more than one training image per person gives high recognition values which not effectively reflect the performance of the system. Hence, the training set contains one image from each person and the test set contains the remaining samples.

For proving the performance and the effectiveness of the proposed approach, experiments are performed for different unimodal and multi-biometrics systems. First, concerning the unimodal systems, experiments are separately performed on the right and left palmprint sets. This is because of the utilized database is composed of the right and left palmprints for each person. Hence, training and test sets are separately constructed for right and left palmprints. Thus, the training set of each category is composed of 96 samples and the test one includes the remaining 480 images. Second, for the multi-biometrics systems that combine right and left palmprints, training and test sets are formed. Hence, for each person, the training set includes one sample from each palmprint category (96 right + 96 left = 192 images) and the test one includes the remaining images (480 right + 480 left = 960 images).

Intensive experiments were conducted for the two categories of biometrics systems, unimodal and multimodal, without of applying any preprocessing task in contrast to some researches [9]. So, the experiments are conducted utilizing some local descriptors (LBP, ELDP, and LDP) in addition to those proposed in this paper (MBELDP, MBLDP). Also, these experiments are performed with the use and without utilizing feature selection.

For the unimodal systems, the experiments are conducted for each palmprint category separately. The resulting image, produced by applying a descriptor on each image, is decomposed into many non-overlapping equal blocks. From each constructed block, a feature histogram is extracted. Then, a large global histogram is produced by concatenation of the obtained feature histograms. For matching, the Chi-square similarity measure is applied between the new constructed global histograms from each test image and all the training palmprint images. The experimental results are shown in Table 2.

For the multi-biometrics systems, the experiments are performed on both right and left palmprint images in parallel. The two resulting images, produced by applying a descriptor on both right and left images, are decomposed into many non-overlapping equal blocks. Then, from each constructed block, a feature histogram is extracted. Then, a large global histogram is produced by concatenation of the obtained feature histograms for each palmprint category (right and left). The resulting global histograms of the right and left palmprints are fused by concatenating them in a serial manner resulting in a high dimensional global feature vector. Then, for the reduction of the global feature vector size and improvement of the recognition accuracy, feature selection using binary GAs or binary BSA is applied. Finally, the Chi-square similarity measure is applied between the resulting test feature vectors and all training ones obtained in the same manner. The experimental results are shown in Table 3.

5.2 Discussion

Looking at the results in Table 2, we see that for all used descriptors and in all cases, the unimodal biometric systems based on the right palmprint images are more accurate than those based on the left ones. Moreover, the recognition accuracy of the presented descriptors (MBELDP, MBLDP) are higher than those of other ones (ELDP, LDP, and

LBP). Hence, the presented descriptors increased the recognition rates of the original ones for the right and left palmprints. Furthermore, there is a remarkable improvement in the obtained results, when feature selection is applied, for all descriptors which can go up to more than 5% in some cases. However, in the most cases, recognition rates obtained using binary GAs as a feature selection are higher than those of the binary BSA. This leads us to conclude that binary GAs are more reliable and more suitable than binary BSA for features selecting in palmprint recognition purpose.

Table 1. Different control parameters values for feature selection using GAs and BSA

Control parameters	GAs	BSA
Binary bit string size	Determined in execution by the feature vector size	Determined in execution by the feature vector size
Population size	40	40
Generations	50	50
Crossover probability	0.9	random
Mutation probability	0.01	random
Fitness function	Determined by Eq. (3)	Determined by Eq. (3)
The amplitude control: F	Not applicable	1

Table 2. Results of unimodal systems based on left palmprints and right palmprints with and without feature selection (F.S.)

	Left palmprint			Right palmprint		
Descriptor	Without F.S. (%)	F.S. with GAs (%)	F.S. with BSA (%)	Without F.S. (%)	F.S. with GAs (%)	F.S. with BSA (%)
LDP	71.45	77.29	74.58	80.20	84.37	82.91
ELDP	68.33	73.95	72.50	78.33	83.54	81.87
LBP	78.33	83.54	81.87	85.41	87.29	87.08
MBLDP	85.83	87.70	87.29	88.75	90.20	90.00
MBELDP	85.00	87.08	87.29	88.54	88.96	88.95

From the experimental results summarized in Table 3, in contrast to the results of the unimodal systems shown in Table 2, we see that there is a remarkable improvement of the recognition rates obtained by the multi-biometrics systems based on the combination of the right and left palmprints without utilization of feature selection for all descriptors tested in this paper. Moreover, there is a remarkable improvement of the recognition rates when feature selection is applied to the resulting global feature vectors, which is not less than 1% and up to 7.08%, for all multi-biometrics systems based on the different local descriptors including MBELDP and MBLDP. Thus, the proposed multi-biometrics systems based on the fusion at feature level using the presented descriptors with the utilization of a feature selection technique, especially the binary GAs, overcomes the state-of-the-art ones investigated in this article (ELDP, LDP, and LBP). Furthermore, the obtained recognition rates in the experiments based on feature selection technique using binary GAs are remarkably higher than those based on binary BSA in almost all cases with the same parameters. So, the superiority and suitability of the binary GAs than the binary BSA for feature selection in the palmprints recognition purpose is confirmed once again.

Looking at the results in Table 3, we see that the multi-biometrics systems based on the fusion of the right and left palmprints with a binary GAs-based feature selection technique utilizing MBELDP and MBLDP presented the highest values, 97.08% and 96.66% as recognition accuracy, respectively. Hence, these results indicate the effectiveness and the superiority of the MBELDP-based approach than it's of the MBLDP.

Table 3. Results of multimodal systems based on left and right palmprints with and without feature selection (F.S.)

Descriptor	Without feature selection (%)	Feature selection with GAs (%)	Feature selection with BSA (%)
LDP	85.00	92.08	88.95
ELDP	85.62	89.37	87.70
LBP	92.91	95.41	94.79
MBLDP	95.20	96.66	96.25
MBELDP	94.37	97.08	96.66

Table 4. Examples of execution time and feature vector length of multimodal systems based on some descriptors with and without feature selection

Descriptor	Without feature selection (GAs)		With feature selection (GAs)	
	Time	Vector length	Time	Vector length
LDP	3.9000	450	1.6848	225
MB-ELDP	0.5148	44	0.5304	32

Results shown in Table 4 confirm that the feature selection reduces the size of the feature vector considerably which can go up to 50%. Also, feature selection reduces the execution time of the system and can go up to more than 2 times less. This advantage is guaranteed much more for feature vectors whose size is very large (e.g. for LDP descriptor) and not guaranteed for those with a very small size (e.g. for MB-ELDP descriptor).

Last, it should be noted that the experiments were we have done using MATLAB 2013a software and a personal computer with the following characteristics:

- Core ™ i3-4005U CPU (1.70 GHz)
- RAM of 4 GB
- 64 bits professional version of Windows 7 operating system.

6 Conclusion

In this present paper, the importance of the use of feature selection in biometrical systems has been investigated. For this purpose, we proposed a multi-biometrics system based on combining right and left palmprint images of an individual for recognition purpose by exploiting some presented local descriptors based on the multi-block principle (MBLDP and MBELDP). After applying one from these descriptors to the right and left palmprints, feature extraction is performed and fusion of the two resulted feature vectors were done at feature-level in a serial manner that means a simple concatenation which produces a single large global vector. After the fusion task, feature selection, especially the evolutionary algorithms based techniques, was applied to the new constructed global feature vector. The reason behind applying feature selection task was to reduce the feature vector size, decreasing the execution time, and increasing the recognition accuracy by selecting the significant features and neglecting the irrelevant and redundant ones. Among these, we have chosen genetic algorithms and backtracking search algorithm. Intensive experiments were conducted using a sub-database that was constructed from the known IITD database in order to test the robustness and strengths of the proposed approach that was focused on the introduction of feature selection. The obtained experimental results show that the use of the feature selection is very important and significantly improved the recognition rates of all the tested biometrical system including the proposed multi-biometrics system using MBELDP and MBLDP local descriptors. Moreover, the proposed multi-biometrics systems outperformed all unimodal and multimodal biometric systems based on the remaining local descriptors, and this is whether with or without applying the feature selection technique.

As a future work, our effort will focus on improving the proposed approach by considering other feature selection methods and a preprocessing step like in [9].

References

1. Hanmandlu, M.: Multimodal biometric system built on the new entropy function for feature extraction and the refined scores as a classifier. Expert Syst. Appl. **42**(7), 3702–3723 (2015)
2. Hezil, N., Boukrouche, A.: Multimodal biometric recognition using human ear and palmprint. IET Biom. **6**(5), 351–359 (2017)
3. Xu, Y., Fan, Z.Z., Qiu, M.N., Zhang, D., Yang, J.Y.: A sparse representation method of bimodal biometrics and palmprint recognition experiments. Neurocomputing **103**, 164–171 (2013)
4. Xu, Y., Zhu, Q., Zhang, D., Yang, J.Y.: Combine crossing matching scores with conventional matching scores for bimodal biometrics and face and palmprint recognition experiments. Neurocomputing **74**(18), 3946–3952 (2011)
5. Xu, Y., Fei, L., Zhang, D.: Combining left and right palmprint image for more accurate personal identification. IEEE Trans. Image Process. **24**(2), 549–559 (2015)
6. Zhang, D., Guo, Z., Liu, G., Zhang, L., Liu, Y., Zuo, W.: Online joint palmprint and palmvein verification. Expert Syst. Appl. **38**(3), 2621–2631 (2011)
7. Dai, J., Zhou, J.: Multi-feature based high-resolution palmprint recognition. IEEE Trans. Pattern Anal. Mach. Intell. **33**(5), 945–957 (2011)

8. Morales, A., Ferrer, M.A., Kumar, A.: Towards contactless palmprint authentication. IET Comput. Vis. **5**(6), 407–416 (2011)
9. Farmanbar, M., Toygar, Ö.: Feature selection for the fusion of face and palmprint biometrics. SIViP **10**(5), 951–958 (2016)
10. Saini, N., Sinha, A.: Face and palmprint multimodal biometric systems using Gabor-Wigner transform as feature extraction. Pattern Anal. Appl. **18**(4), 921–932 (2015)
11. Ghulam Mohi-ud-Din, S., Bin Mansoor, A., Masood, H., Mumtaz, M.: Personal identification using feature and score level fusion of palm- and fingerprints. SIViP **5**(4), 477–483 (2011)
12. Raghavendra, R., Dorizzi, B., Rao, A., Kumar, G.H.: Designing efficient fusion schemes for multimodal biometric systems using face and palmprint. Pattern Recogn. **44**(5), 1076–1088 (2011)
13. Charfi, N., Trichili, H., Alimi, A.M., Solaiman, B.: Bimodal biometric system for hand shape and palmprint recognition based on SIFT sparse representation. Multimed. Tools Appl. **76** (20), 20457–20482 (2017)
14. Jabid, T., Kabir, M.H., Chae, O.: Robust facial expression recognition based on local directional pattern. ETRI J. **32**(5), 784–794 (2010)
15. Zhong, F.J., Zhang, J.S.: Face recognition with enhanced local directional patterns. Neurocomputing **119**, 375–384 (2013)
16. Civicioglu, P.: Backtracking search optimization algorithm for numerical optimization problems. Appl. Math. Comput. **219**(15), 8121–8144 (2013)
17. Altun, A.A., Kocer, H.E., Allahverdi, N.: Genetic algorithm based feature selection level fusion using fingerprint and iris biometrics. Int. J. Pattern Recogn. Artif Intell. **22**(3), 585–600 (2008)
18. Kumar, A.: Incorporating cohort information for reliable palmprint authentication. In: 6th International Proceedings on ICVGIP'08, Bhubneshwar, India, pp. 583–590 (2008)

A Harmony Search-Based Feature Selection Technique for Cloud Intrusion Detection

Widad Mirghani Makki, Maheyzah M.D. Siraj, and Nurudeen Mahmud Ibrahim(✉)

Faculty of Engineering, School of Computing, Universiti Teknologi Malaysia, 81310 Johor Bahru, Johor, Malaysia
wida25@hotmail.com, Maheyzah@utm.my, imnurudeen2@live.utm.my

Abstract. Recently cloud computing has enjoyed widespread patronage due to its economy of scale and flexibility. However Cloud computing is confronted with security challenges. Intrusion detection can be used to protect computer resources from unauthorized access. However the presence of insignificant features in Intrusion Detection dataset may have a negative effect on the accuracy of Intrusion Detection System (IDS). Feature selection is utilized to remove noisy and insignificant attribute to improve IDS performance. However, existing feature selection techniques proposed for cloud IDS cannot guarantee optimal performance. Therefore, this research article proposes a Harmony Search based feature selection technique to improve the performance of cloud IDS. The attributes selected were assessed using Random Forest classifier and experimental results of the Harmony Search based technique achieved an attack detection rate of 79% and a false alarm rate of 0.012%. In addition performance comparison shows that the proposed Harmony search outperforms existing feature selection technique proposed for cloud IDS.

Keywords: Feature selection · Intrusion detection · Harmony Search · Cloud computing

1 Introduction

According to Mell and Grance [1] cloud computing is defined as a computing model that provides subscription-based internet access to computing facilities which is offered readily with little management effort or intervention of service provider. Cloud computing has three service models namely: software-as-a-service (SaaS), platforms-as-a-service (PaaS) and Infrastructure-as-a-service (IaaS) and four modes of service deployment namely: private, public, hybrid and community cloud [2]. Inspite of many advantages of cloud computing, it is still confronted by security challenges. Despite the growing enlightenment on network security, existing security measures cannot guarantee adequate security against threats such as DoS and malware. Therefore developing more effective security solutions is paramount. The traditional security solutions such as cryptography, authentication and firewall is unable to provide enough security owing to the polymorph nature of intrusion [1]. Therefore IDS is necessary to provide

F. Saeed et al. (Eds.): IRICT 2019, AISC 1073, pp. 779–788, 2020.
https://doi.org/10.1007/978-3-030-33582-3_73

additional security layer. Intrusion detection is an approach used to monitor the activities occurring in a computer system for evidence of cyber-attacks [3]. The detection technique for IDS can be categorized as signature-based and anomaly detection. The signature-based technique matches incoming traffic with known attack signature [4] while anomaly detection creates a normal profile which is subsequently used to detect anomaly.

The accuracy of an IDS is hampered by noisy and redundant attributes. Consequently discarding the noisy features is crucial to enhancing the accuracy of an IDS [5]. The speed and accuracy of a classification algorithm is vital for the efficacy of an IDS [6]. Data that is composed of fewer features provides more insight about the data processes and facilitates the extraction of knowledge [7]. Feature selection involves the elimination of irrelevant attributes in a dataset in order to improve the accuracy of the classifier. The two main classification of feature selection technique are the wrapper and the filter approach. The filter approach utilizes statistical technique to rank features and then eliminates features that are beyond a threshold while the wrapper approach employs machine learning techniques to assess the quality of features [8]. Various techniques have been proposed for selecting features for cloud IDS; these techniques are discussed as follows: In [9] a method for feature selection using four filters was used to select features to enhance attack detection in cloud computing. However, the disadvantage of the filter technique is that it can discard relevant attributes with less information in isolation but more informative when combined with other features [10]. In [11] Rough Set Theory was used as an attribute selection approach for choosing relevant features from KDD cup dataset. Particle Swarm Optimization was employed to select features for cloud intrusion detection [12]. In [6] feature selection using Genetic Algorithm was proposed. In [13] feature selection using Ant Colony Optimization for enhancing cloud security was proposed. However these techniques uses stochastic search technique that cannot guarantee optimality [14]. Therefore, the research paper proposes a Harmony Search based feature selection approach to improve the accuracy of intrusion detection in cloud computing. The remainder of the paper is structured as follows: Sect. 2 reviews the related works, the Harmony Search approach is presented in Sect. 3, the experimental set up is presented in Sect. 4, Sect. 5 discusses the result while the conclusion of the paper is presented in Sect. 6.

2 Related Works

The major classification of feature selection approach is the filter and wrapper approach. Under the filter approach the following research works have been conducted. In [9] an Ensemble Multi Filter Feature Selection (EMFFS) was used to select features to enhance the detection of DDoS attack in Cloud computing. EMFFS is comprised of Information Gain, Gain Ratio, Relief and Chi-squared. The authors employed the EMFFS to select 13 attributes from NSL-KDD dataset. In [15] Information Gain was utilised to select 8 attributes from NSL-KDD dataset. In [16] Gain Ratio was employed to select 35 attributes from NSL-KDD dataset. Both authors showed that feature selection improved the detection accuracy of IDS. However these methods are based on filter approach, and the filter approach has no ideal mechanism for selecting the size of attributes and the filter approach may discard important attributes with less information in isolation but more informative when in combination with other features. In [11] Rough Set Theory was used as an attribute selection approach for choosing relevant features from KDD cup dataset. Particle Swarm Optimization was employed to select features for cloud intrusion detection in Ref [12]. In [6] feature selection using GA was employed to reduce the unnecessary features for cloud IDS. In [13] feature selection using Ant Colony Optimization was proposed for cloud IDS. However these techniques uses stochastic search technique that cannot guarantee optimality [14]. Therefore a more effective feature selection technique is required to enhance the accuracy of cloud IDS.

3 Harmony Search Technique

Harmony Search (HS) is regarded as an evolutionary algorithm inspired by the improvisation process of music players [17]. The main concept of Harmony Search is to use the synchronization process adopted by music players when producing a song for achieving a best solution for optimization problems. Fundamentally, any possible outcome of the optimization solution is modelled in form of a harmony and the optimization parameters are treated as musical notes. The optimal harmony (solution) is chosen as the harmony with the maximum of some criteria used for the optimization. To use HS for feature selection, the basic steps in the algorithm is illustrated as follows:

Step 1: The first step involves the initialization of the algorithm parameters and the optimization task.
Step 2: The second step initializes the Harmony Memory (HM).
Step 3: A New Harmony Memory is improvised in the third step.
Step 4: In the fourth step the HM is updated; in the event that the New Harmony performs better than the worst harmony in the HM, the New Harmony is included in HM and the worst Harmony is discarded from HM.
Step 5: If the termination criteria is not fulfilled, go to step 3.

Figure 1 shows the steps of the Harmony Search (HS) feature selection. The first and second step involves the initialization of the parameters and Harmony Memory. This offers a domain note to each attribute selector. Features selectors may choose similar sets or empty sets. Then new features are arbitrarily selected and evaluated from the available features to form a new subset during the improvisation phase. The subsequent phase involves the updating of the Harmony Memory, during this phase a comparison of the evaluation score of the fitness function is performed. If the newly obtained subset achieves higher score than the existing subset in the Harmony Memory, then the new subset is replaced with the existing subset in the Harmony Memory, otherwise this subset is discarded. During the iteration phase, the process is repeated until a termination criteria is fulfilled. The termination criteria is when a maximum number of iteration is achieved. The final feature subset is the subset with the highest accuracy.

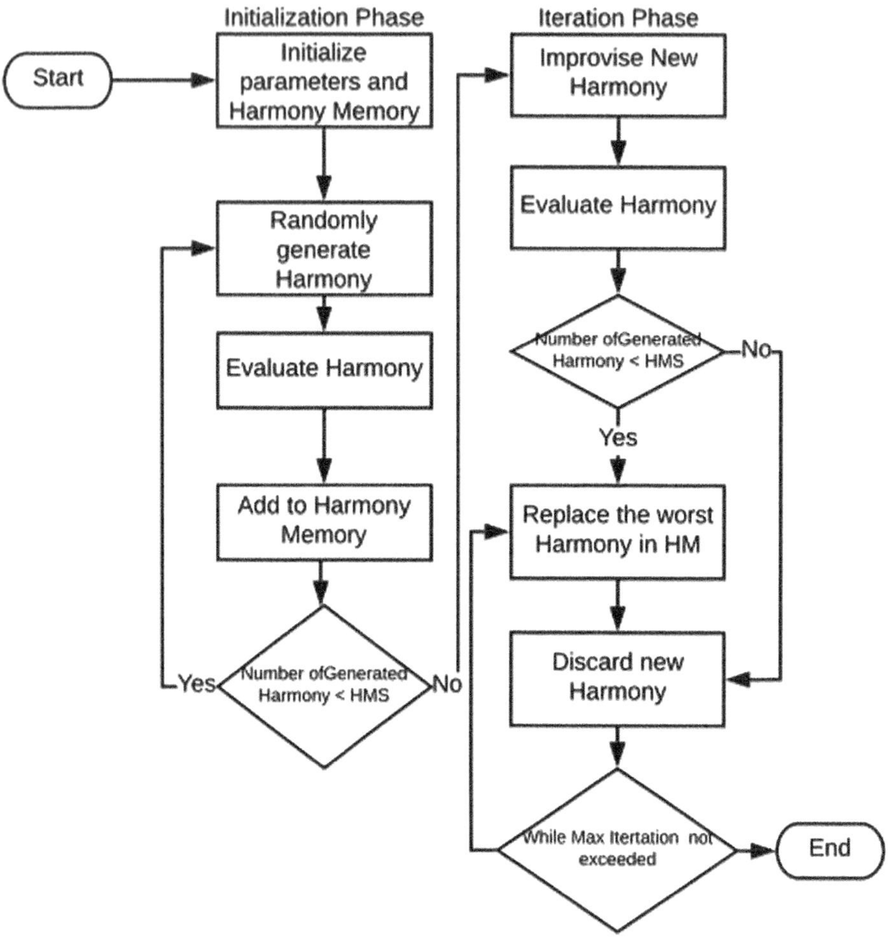

Fig. 1. Harmony search feature selection

The aforementioned steps are discussed in more details as follows:

i. Parameter Setting for the Optimization Task

To demonstrate the mechanism of HS attribute selection, the problem is expressed as an optimization task as shown in Eq. (1).

$$\max f(x)\, subject\, to\, x^j \in X_j \forall j = 1, 2, \ldots, N \quad (1)$$

$f(x)$ denotes the fitness function; the fitness function used for the proposed feature selection task is correlation based feature selection [18]. Upon selection of the set of features by the HS algorithm, the features are assessed using the fitness function. x^j denotes the variables in the dataset, while X_j denotes its possible set of values and N represents the number of variables.

ii. Harmony Memory Initialization

The initialization of the memory with an arbitrary generated values is performed in the second step. Every harmony possesses its own value used for the optimization of the fitness function.

iii. Generating a New Harmony from HM

The third step involves the derivation of a new vector of HM depending on the availability of memory, pitch adjustment and randomization. It is feasible to choose the new values by utilizing the parameters of HMCR which falls in the range of 0 and 1.

$$\bar{X}^J = \begin{cases} \bar{x}^j \in \{x_1^j, x_2^j, x_N^j\} \; with\, probability\, HMCR \\ \bar{X}^j \in X \quad\quad with\, probability\,(1 - HMCR) \end{cases} \tag{2}$$

The HMCR is the likelihood of choosing an attribute from the HM repository, while (1-HMCR) is the likelihood of arbitrarily choosing a probable value not limited to the value stored in the HM. Furthermore every part j of the New Harmony vector $\bar{x}$ is investigated to determine whether or not it pitch adjustment is necessary. The pitch adjustment decision for $\bar{x}$ is shown in Eq. (3).

$$\bar{x} = \begin{cases} Yes\, having\, probability\, PAR \\ No\, having\, probaility\,(1 - PAR) \end{cases} \tag{3}$$

Adjustment of pitch for every instrument is utilized to enhance the solution (selected features) and avoid local optima. The procedure involves adjusting the neighboring values of some decision variable in the harmony. If the decision for the pitch adjustment for a variable x^j is yes, x^j is replaced using the equation as follows:

$$\bar{x}^j = \bar{x}^j + rb \tag{4}$$

Where b determines an arbitrary distance bandwidth and r is a uniform distribution ranging from 0 and 1.

iv. Harmony Update

The fourth step checks the current Harmony to ascertain whether or not it is better than the previous Harmony, then the current Harmony is replaced with the previous Harmony.

v. Termination Condition

In the final phase the Harmony Search technique terminates having achieved the termination condition, (maximum iteration exceeded) otherwise, step 3 and 4 are replaced to create a New Harmony search again.

4 Dataset and Experimental Setup

The dataset used to validate the Harmony Search is NSL-KDD. NSL-KDD is an improved version of KDD-Cup 99 and commonly used as a benchmark dataset to test IDS algorithms both for conventional network and for cloud computing such as in reference [9]. NSL-KDD is composed of selected instances from KDDCup 99 that is made up of 41 attributes and a class label. It consists of both training and test set. The training dataset has 24 attack types while the test datasets has 14 new attack types. The classification of the intrusions are: Probe, DoS, U2R and R2L. The entire data composition consists of 67,343 normal instances, 45,927 DoS, 11656 probe, 995 R2L, and 52 U2R while the test set contains 9711 normal, 7456 DoS, 2421 probe, 2756 R2L and 200 U2R [9] (Table 1).

Table 1. Features in NSL-KDD dataset

Feature number	Description	Type	Feature number	Description	type
1	duration	Numeric	22	is_guest_login	Numeric
2	Protocol_type	Symbolic	23	count	Numeric
3	service	Symbolic	24	srv_count	Numeric
4	flag	Symbolic	25	serror_count	Numeric
5	src_bytes	Numeric	26	srv_serror_rate	Numeric
6	dst_bytes	Numeric	27	rerror_rate	Numeric
7	land	Numeric	28	srv_error_rate	Numeric
8	wrong_fragment	Numeric	29	same_srv_rate	Numeric
9	urgent	Numeric	30	diff_srv_rate	Numeric
10	hot	Numeric	31	srv_diff_host_rate	Numeric
11	num_failed_login	Numeric	32	dst_host_count	Numeric
12	logged_in	Numeric	33	dst_host_srv_count	Numeric
13	num_compromised	Numeric	34	dst_host_same_srv_rate	Numeric
14	root_shell	Numeric	35	dst_host_diff_srv_rate	Numeric
15	su_attempted	Numeric	36	dst_host_same_srv_host_rate	Numeric
16	num_root	Numeric	37	dst_host_srv_diff_host_rate	Numeric
17	num_file_creation	Numeric	38	dst_host_serror_rate	Numeric
18	num_shell	Numeric	39	dst_host_srv_serror_rate	Numeric
19	num_access_file	Numeric	40	dst_host_rerror_rate	Numeric
20	num_out_of_bound_cmd	Numeric	41	dst_host_srv_rerror	Numeric
21	is_hot_login	Numeric			

5 Experimental Result and Analysis

Result analysis for the attribute selected is provided in this section. The attribute chosen by the HS algorithm were tested by using Random Forest classification algorithm [20]. Table 2 depicts the attribute selected using Harmony Search algorithm. From Table 2 only 18 features were selected from the 41 features using Harmony Search. Table 3 depicts the outcome of the classification for the selected attribute using Harmony search and Random Forest classification algorithm. The result in Table 3 shows that with Harmony Search feature selection, a detection rate of 99.2% was achieved for DoS attack, 96.2% detection rate was achieved for Normal class, 97.6% detection rate was achieved for probe attack, 89.5% detection rate was achieved for R2L and 12.4% detection rate was achieved for U2R attack. In terms of False Positive Rate the proposed HS algorithm recorded a false positive rate of 0.003% for DoS attack, a false positive rate of 0.03% was achieved for the Normal data, for probe attack a false positive rate of 0.003% was achieved for R2L attack a false positive rate of 0.002% was achieved and for U2R attack a false positive rate of 0.02 was achieved. Analysis of the result shows that HS algorithm has a good detection rate for DoS, Normal, Probe, R2L; however for U2R attack the detection rate was relatively lower with the Random Forest algorithm achieving only 12.4%. For the false positive rate the HS algorithm with Random Forest achieves a low false positive rate for all the classes.

Table 2. Features selected by Harmony Search

Algorithm	Selected attribute	Number of attribute
Harmony search	2,4,8,11,12,23,25,26,29,30,31,32,33,35,37,38,39,41	18

Table 3. Classification result for selected features

Class	Detection rate (%)	False positive rate (%)
DoS	99.2	0.003
Normal	96.2	0.03
Probe	97.6	0.003
R2L	89.5	0.022
U2R	12.4	0.002

5.1 Comparison with Related Work

Table 4 and Figs. 2 and 3 shows the comparison result between the proposed Harmony Search (HS) versus Genetic Algorithm (GA) [6] and Particle Swarm Optimization (PSO) [12]. From these results it can be observed that the detection rate for HS outperforms GA and PSO for all classes of attacks except U2R attacks that has a relatively lower detection rate compare to PSO. Precisely, PSO achieved a detection rate of 15% while HS achieved a detection rate of 12.4% and GA achieved 10%. In terms of false positive rate HS achieved a lower false positive rate compared to GA and PSO for all the attack classes. The higher performance of HS can be attributed to the fact that GA and PSO are stochastic heuristic based technique that cannot guarantee optimality. Also the proposed HS uses correlation based feature selection as a fitness function which has the advantage of choosing only attributes that are highly linked to a class and discriminates features that have high correlation with one another.

Table 4. Comparison of proposed HS vs GA and PSO.

Classes	Detection rate (%)			False positive rate (%)		
	HS	GA	PSO	HS	GA	PSO
DoS	99.2	92.4	96.3	0.3	1.5	7
Normal	96.2	93.2	95.1	3	9.3	4.6
Probe	97.6	92.5	96.8	0.3	0.4	3
R2L	89.5	75.2	88.8	2.2	3.8	2.4
U2R	12.4	10.0	15.0	0.2	1	0.2
Average	79	73	78.4	1.2	3.2	3.4

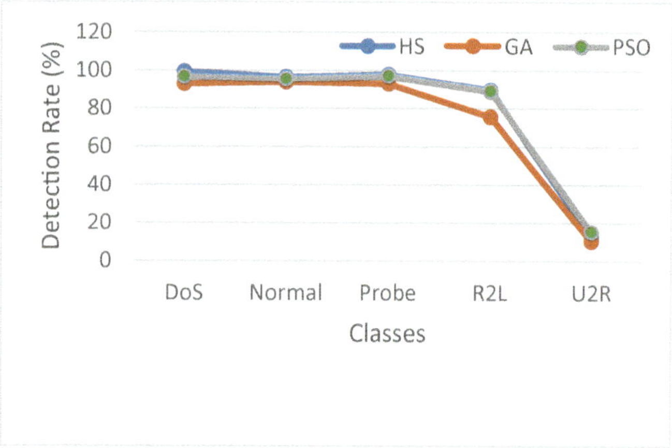

Fig. 2. Comparison of detection rate between HS vs GA and PSO

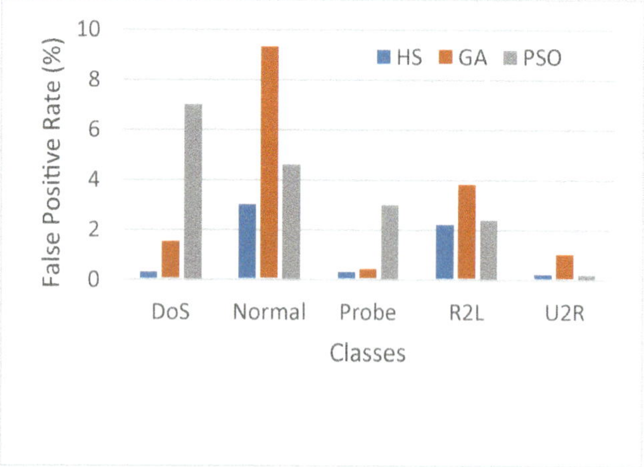

Fig. 3. Comparison of false positive rate between HS vs GA and PSO

6 Conclusions

Cloud computing is a computing model that provides computing resource as a service via the internet. Cloud computing has numerous benefits however it has many security issues. IDS are software's employed to detect computer misuse. However IDS is confronted with poor accuracy due to insignificant and noisy attributes. Therefore this research paper proposes a feature selection approach using Harmony Search to improve the detection accuracy of cloud intrusion detection. Using Harmony Search algorithm the features in NSL-KDD dataset is reduced to 18 features and the Random Forest algorithm was used as the classification algorithm to evaluate the features selected by the Harmony Search algorithm. The experimental evaluation result shows that the features selected by Harmony Search algorithm achieves good classification result. In addition it outperforms existing feature selection techniques proposed for cloud IDS.

Acknowledgement. We would like to thank Malaysia's Ministry of Higher Education (MOHE) and Universiti Teknologi Malaysia (UTM) for funding this work through research grant with vote number (0G73).

References

1. Mell, P., Grance, T.: The NIST definition of cloud computing. http://faculty.winthrop.edu/domanm/csci411/Handouts/NIST.pdf. Accessed Feb 2016
2. Khorshed, M.T., Ali, A.S., Wasimi, S.A.: A survey on gaps, threat remediation challenges and some thoughts for proactive attack detection in cloud computing. Future Gener. Comput. Syst. **28**, 833–851 (2012)
3. Scarfone, K., Mell, P.: Guide to intrusion detection and prevention systems (idps). NIST special publication 800-94 (2007)
4. Liao, H.-J., Lin, C.-H.R., Lin, Y.-C., Tung, K.-Y.: Intrusion detection system: a comprehensive review. J. Netw. Comput. Appl. **36**(1), 16–24 (2014)
5. Aghdam, M.H., Kabiri, P.: Feature selection for intrusion detection system using ant colony optimization. Int. J. Netw. Secur. **18**(3), 420–432 (2016)
6. Kannan, A., Maguire, G.Q., Sharma, A., Schoo, P.: Genetic algorithm based feature selection algorithm for effective intrusion detection in cloud networks. In: Proceedings of the 2012 IEEE 12th International Conference on Data Mining Workshops (ICDMW), Brussels, Belgium, 10 December 2012, pp. 416–423. IEEE (2012)
7. Kang, S.-H., Kim, K.J.: A feature selection approach to find optimal feature subsets for the network intrusion detection system. Cluster Comput., 1–9 (2016)
8. Hira, Z.M., Gillies, D.F.: A review of feature selection and feature extraction methods applied on microarray data. Advances in bioinformatics (2015). https://www.hindawi.com/journals/abi/2015/198363/abs/. Accessed January 2018
9. Osanaiye, O., Cai, H., Choo, K.-K.R., Dehghantanha, A., Xu, Z., Dlodlo, M.: Ensemble-based multi-filter feature selection method for DDoS detection in cloud computing. EURASIP J. Wirel. Commun. Netw. **1**, 130 (2016)
10. Chandrashekar, G., Sahin, F.: A survey on feature selection methods. Comput. Electr. Eng. **40**(1), 16–28 (2014)

11. Muthurajkumar, S., Kulothungan, K., Vijayalakshmi, M., Jaisankar, N., Kannan, A.: A rough set based feature selection algorithm for effective intrusion detection in cloud model. In: Proceedings of the International Conference on Advances in Communication, Network and Computing, pp. 8–13 (2013)
12. Zhou, L.-H., Liu, Y.-H., Chen, G.-L.: A feature selection algorithm to intrusion detection based on cloud model and multi-objective particle swarm optimization. In: Proceedings of the 2011 Fourth International Symposium on Computational Intelligence and Design (ISCID), Hangzhou, China, 28–30 October 2011, pp. 182–185. IEEE (2011)
13. Ibrahim, N.M., Zainal, A.: A feature selection technique for Cloud IDS using Ant Colony Optimization and Decision Tree. Adv. Sci. Lett. **23**(9), 9163–9169 (2017)
14. Jensen, R., Shen, Q.: Fuzzy-rough data reduction with ant colony optimization. Fuzzy Sets Syst. **149**(1), 5–20 (2005)
15. Aljawarneh, S., Aldwairi, M., Yassein, M.B.: Anomaly-based intrusion detection system through feature selection analysis and building hybrid efficient model. J. Comput. Sci. **25**, 152–160 (2018)
16. Pham, N.T., Foo, E., Suriadi, S., Jeffrey, H., Lahza, H.F.M.: Improving performance of intrusion detection system using ensemble methods and feature selection. In: Proceedings of the Australasian Computer Science Week Multiconference, Brisbane, Australia, 29 January–2 February 2018, p. 2. ACM (2018)
17. Geem, Z.W.: Novel derivative of harmony search algorithm for discrete design variables. Appl. Math. Comput. **199**, 223–230 (2008)
18. Hall, M.A., Smith, L.A.: Feature subset selection: a correlation based filter approach. In: Proceedings of the 1997 International Conference on Neural Information Processing and Intelligent Information Systems, pp. 855–858. Springer, Berlin (1997)
19. Singh, D., Patel, D., Borisaniya, B., Modi, C.: Collaborative ids framework for cloud. Int. J. Netw. Secur. **18**(4), 699–709 (2016)
20. Breiman, L.: Random forests. Mach. Learn. **45**(1), 5–32 (2001)

Security Assessment Model to Analysis DOS Attacks in WSN

Abdulaziz Aborujilah[1(✉)], Rasheed Mohammad Nassr[1], Tawfik Al-Hadhrami[2], Mohd Nizam Husen[1], Nor Azlina Ali[1], AbdulAleem Al-Othmani[1], Nur Syahela[1], and Hideya Ochiai[3]

[1] University Kuala Lumpur, 50250 Kuala Lumpur, Malaysia
abdulazizsaleh@unikl.edu.my
[2] Nottingham Trent University, Nottingham NG1 4FQ, UK
[3] Graduate School of Information Science and Technology, University of Tokyo, Tokyo, Japan

Abstract. Internet of Things (IoT) technology is getting more importance as many organizations from different sectors are moving toward adopting this technology. It introduces an efficient ways of collecting, processing and distributing data. Wireless sensor networks (WSNs) play a vital role in the expanding growth of internet of things (IoT). However, DoS (denial-of-service) attacks are a major threat of WSNs. In this study, graph theory based model has been proposed to analyse the security of WSN under DoS attacks. The proposed model describes and assesses the security of WSN when it encounter DoS attacks by finding the attacks success probability, attack cost, attack impact, mean-time-to-compromise, attack risk and return-on-attack values. The impact of mitigation methods toward strengthening the WSN security has been described. smart home scenario WSN has been tacking as an example. The results showed that the capability of the proposed model to analysis the effects of DoS attacks in WSN. Also it showed the impact of mitigation methods in improving WSN security.

Keywords: IoT security assessment · WSN · DoS attacks

1 Introduction

Network technology continuous growing has evolved new technologies such as IoT and cloud computing networks [1]. IoT is emerging technology by where wide range of physical objects such as home appliances and vehicles are communicating with each other and with human beings via the internet. It is unfortunate that IoT technology is a primary target for viruses and malicious attacks which cause huge damage on it [2]. WSN is widely used as communication network in IoT applications [3]. Zigbee is a WSN protocol which implements an IEEE 802.15.4-based specification [4]. This protocol is mainly designed for low power consumption, low data rate, low cost and two-way wireless communications devises. It is used to connect smart home appliances objects such as lighting control environment monitoring systems air conditioners, refrigerators and washing machines and controls systems such as smart door lock,

F. Saeed et al. (Eds.): IRICT 2019, AISC 1073, pp. 789–800, 2020.
https://doi.org/10.1007/978-3-030-33582-3_74

surveillance camera [2]. Attackers can exploit Zigbee network to gain unauthorised access to smart home networks by flooding the network bandwidth and misusing computing power resources. Graph theory is an approach which used to analyse the interactive behaviour of attackers and defenders. The authors in [5] explain how graph theory might be used to formulate the interaction between a network administrator and attacker. The attacker might follow a multi-stage plan to achieve his/her attack goals while the administrator might stope the attacks by following a multi-stage plan prevention method, this interaction can be depicted in a graph model [6]. Fault tree analysis (FTA) is a graph based method that is used to analyse different vulnerable scenarios of computer systems and how I can be mitigated methods. FAT is used to formulate the behaviours of different computer systems such as data centres, power plants, web shops and airplanes systems.

This paper focuses on analysing the security behaviour of WSN that is infected by DoS attacks. This work adapt the security framework proposed by Ge et al. [2] and propose FAT based model that present the internal and external attacks of WSN.

The remaining of the paper is organised as follows. Section 2 presents the related works of IoT based security models. Section 3 show the proposed WSN architecture. Section 4 present the FAT model. In Sect. 5, Security Assessment Implementation is explained. Then in Sect. 6 the security assessment models results are presented. Section 7 discusses the results and Sect. 8 concludes the work.

2 Related Works

WSN is enables the smart object to communicate with each other. IoT application such as smart houses, smart buildings, smart parking mainly really on WSN. WSN consist of several sensors (SNode) that gather the data and send it to sink node [7]. Sink analyses and synthesizes the collected data and make them ready for the main purpose of WSN application [8] DOS (denial of service) attacks are mainly target WSN security. Gavrić et al. [9] examined common DoS attacks in WSN and mentioned that such attacks can take place in applications, networks, medium access and physical layers of WSN.

In this section describes and discuss the taxonomies of DOS attacks of WSN. It includes DoS attacks in application layers such as sensor overloading attacks; network layer such as sinkhole attack; medium access layer such as jamming attack and physical layer such as Collision attack. The taxonomy of DOS attacks in WSN is shown in Fig. 1.

Sensors overloading is an attack that targets end user applications. A malicious node continuously regenerates dummy messages and forwards it to the sink node every 100 ms for 2 min. This attack causes the network to be freezed. The infects the Zigbee network by pluging a Zigbee device that is prepared with DoS software [9]. The sinkhole attack occurs when malicious node operates as router to pass the packets to their destination. Malicious node uses a strong transmitter with greater computational and communication power to attract other nodes to send their packets through it. This might motive Zigbee network nodes to send more packets thorough this malicious node (router) [10].

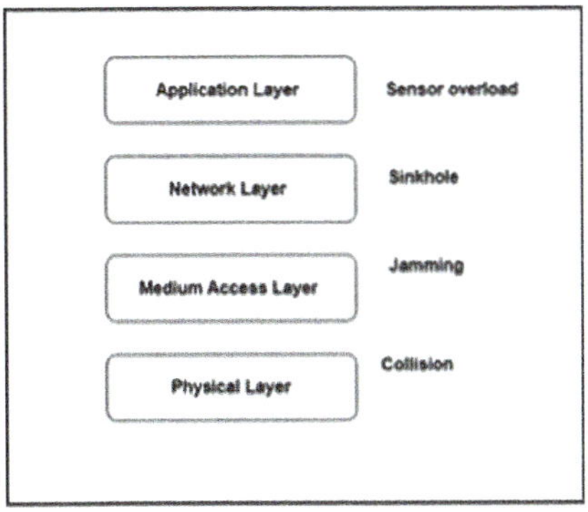

Fig. 1. Example of DoS attacks

Collision DoS attacks occur when two nodes, normal and malicious send their packets simultaneously on the same frequency. Moreover, the malicious node intercepts the packet generated from the legitimate node and collision which might cause loss of packets. An example of DoS attacks is sinkhole attacks that redirects the packets from their wanted destination to an different destination. Figure 2 shows collision DoS attacks (change base station to sink).

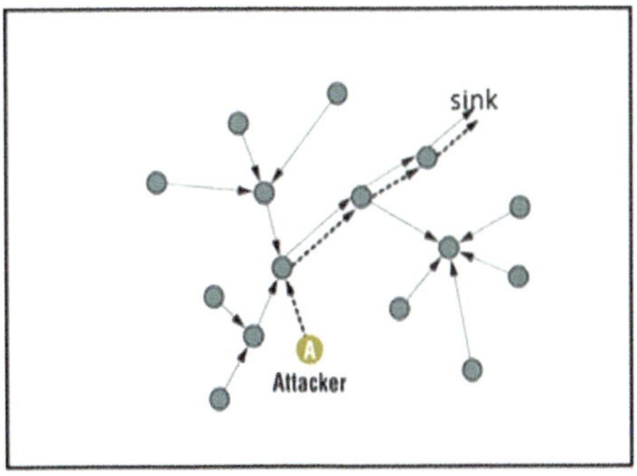

Fig. 2. Collision DoS attacks

Jamming DoS attack is a physical layer attack in which the attacker attempts to make WSN resources unavailable. It happens when the attacker monitor data transferring within the network upon which he/she intentionally generates a jam signal. The random jammer alternates the time between sleeping and jamming modes and repeats this process until the radio signals freeze in the "sleeping" mode [11].

In [2], IoT security models are classified into the security models in IoT into security framework, game based security modelling and adaptive security models.

Cao, et al. [12] proposed a method to detect and localize energy depletion attack in Zigbee networks. This method focuses on determining the severity level of the attacks on Zigbee networks. For example ghost attack. This attack leverages the IEEE 802.15.4

security suites vulnerabilities to duplicate the nodes energy and cause DoS attacks. An analytical model to quantify the impact of the induced DoS attack was proposed. Then a three-phases algorithm was presented to detect and localise the attacker. This depends on analysing network flow variations. This detection method is evaluated through a computer simulation that shows the correctness of the proposed detection method. Finally, a real experiment was conducted to validate the effectiveness of the detection method.

In [13] they have proposed a framework to secure the smart home that allows users to balance between performance, security, and overhead. The efficacy of this work was evaluated by implementing it in sinkhole and distributed DoS attacks. The evaluation results showed that the attacks were detected successfully. It also showed positive improvement in lowering overhead and detection latency in comparison with the existing systems.

Ge [2] framework presents how graph based approach are used to visualize the interaction behaviour between the attackers and defenders. It segregates the network environment into upper, middle and lower layers. The upper layer represents the topology of the possible attacks which are visualised in the attacks graph (AG) struture. The middle layer represents network level information. The lower layer represent the nodes layer (vulnerabilities levels) which are visualised in the attack tree (AT) structure.

Ge [2] security formwork consists of five phases: (1) data processing, (2) security model generation, (3) security visualisation, (4) security analysis, and (5) model updates.

In phase 1, the security workers observe the IoT system scenario and prepare system information and security metrics to construct an IoT network. In phase 2, the security model generator receives the input form phase 1 and generate security matric values. In phase 3, using graph and tree structures, attack paths are visualised with three levels of details. Attacks graph (AG) represents the upper and middle layers. The upper layer represent the possible attacks paths while the middle layer represents the network level. The lower layer represents the nodes layer (vulnerability layer). In phase 4, IoT network information along with security matrices are fed into the security evaluator (SHARPE) software that computes the security analysis results based on Flute tree reliability analysis. In phase 5, defence strategies are proposed and model inputs (security metrics) are updated accordingly. Then, phases 1 to 4 are repeated and security matrices are recomputed and security of IoT network are reanalysed. The security decision maker can use the output of the security model analysis to find the most effective defence strategies to protect the IoT network.

Referring to Fig. 3, we described use IoT1 = (S1, T1, V1) vector as an input for the IoT generation phase. It consist of 5-tuble GSM (U, M, L, CU, M, CM, L) where U is the attacks graph model for the upper layer, M is attacks graph for the middle layer and L is set of attacks tree of V (the lower layer). CU, M refers to the relationship between the higher and middle layers, and CM, L formulates relationship between middle and lower layers, CM, L = $\{(s, t) | s \in S, t \in s \text{ nodes}\} \subseteq S \times T$. Each node is associated with one or more vulnerabilities and has a corresponding AT in the lower layer. When partial mapping CM, L: $T \rightarrow L$ gives the associated AT. More details of this security framework are explained in [2].

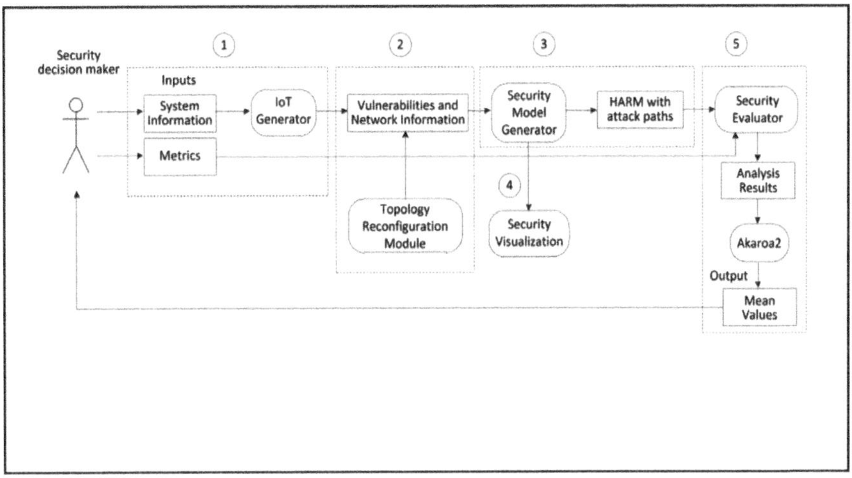

Fig. 3. Mengmeng et al.'s security formwork

This study will extend the framework that has been done suggested in [2] through implementing it in analysis DoS attacks in WSN and smart home scenario is used as an example. The proposed security model will show capability of using graph theory to figure out how the model the paths of DoS attacks in WSN.

3 The Proposed Assessment Model

The main goal of the proposed model to analyse and synthesis the DoS attacks effects in WSN. Also it give a better understanding of DoS attacks in the WSN network by identifying the possible paths that the attackers might pursue to run DoS attacks. Moreover, it assists information security workers to find the most efficient countermeasures to mitigate the damage caused by DoS attacks. Smart home scenario was selected as an example for WSN applications. The smart home seniors consists of IoT objects that are connected via the Zigbee network. The Zigbee network connects to the main smart home router through Wi-Fi connection. Figure 4 shows a smart home layout. Smart homes are subject to internal and external attackers. In the case of the internal attacks, the attacker plugs a malicious node into the Zigbee network and lunches the DoS attacks at the physical, medium access control, and network layers. Whereas the outsider attacker can lunch DoS attacks at physical and medium access control only.

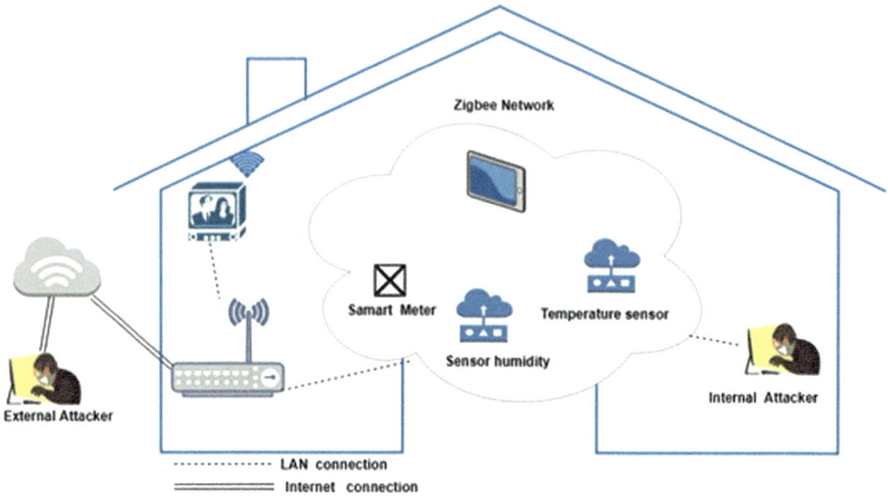

Fig. 4. Smart home scenario

4 Fault Tree Design

By referring to seniors of smart home which has been described in the previous section. The fault tree model was proposed. The fault tree visualizes the logic of DoS attacks that take place in WSN in form of reliability tree. The main goal of this security model is to provide a better understanding of DoS attacks in the Zigbee network and to identify the paths the attackers might pursue to run DoS attacks. This will assist information security workers to find the most efficient countermeasures to mitigate the damage caused by DoS attacsk. Figure 5 shows the proposed security model.

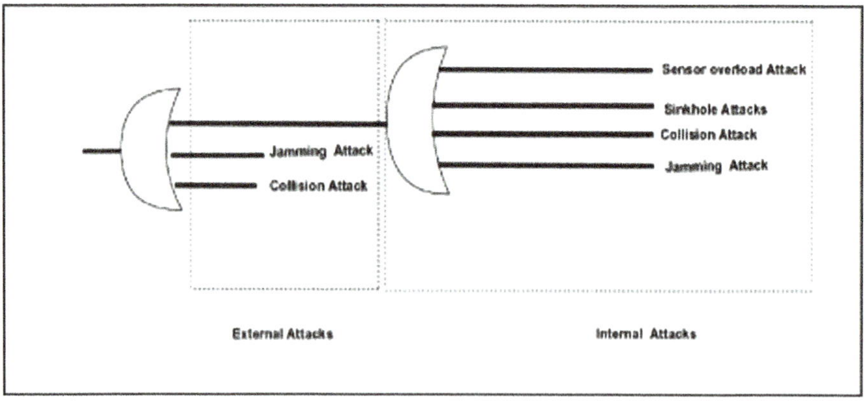

Fig. 5. Fault tree of DoS attacks in WSN

5 Security Assessment Implementation

Smart home architecture used in this study consists of the Zigbee network connected to the internet via a Wi-Fi connection. The behaviour of the Zigbee network under DoS attacks is analysed through by finding the values of attack success probability asp_v. Algorithm 1 shows how asp_v is calculated. Medium time to compromise of (mttc) is calculated based on algorithm 4. Compromise rate cr_v values are either extricated from Ge et al., [2] work or are assumed based on the researcher experience as there is no authenticated source of cr_v. Attacks cost is calculated based on Ge et al., framework and attack impact is calculated based on algorithm as Fig. 7 show.

```
Data: AP and asp_t (t ∈ ap)
Result: ASP
begin
    H ← {t|t ∈ ap for some ap ∈ AP}
    Construct a directed graph graph with node set H
    for each attack path (t_1, ...., t_n) ∈ AP do
        for each i ∈ {2, ..., n} do
            Include edge (t_{i-1}, t_i) with value 1 − asp_{t_i} in graph
    end
    ASP ← 1 − CalculateProbability(graph)
end
```

Fig. 6. Algorithm of attack ASP calculation

$$ac_b = \begin{cases} \sum_{a \in c(b)} ac_a, & b \in B \\ & g(b) = AND \\ \min_{a \in c(b)} ac_a, & b \in B \\ & g(b) = OR \end{cases}$$

$$ac_t = ac_{root}$$

$$ac_{ap} = \sum ac_t, \quad ap \in AP$$

Fig. 7. Algorithm of attack acv calculation

$$aim_b = \begin{cases} \sum_{a \in c(b)} aim_a, & b \in B \\ & g(b) = AND \\ \max_{a \in c(b)} aim_a, & b \in B \\ & g(b) = OR \end{cases}$$

$$aim_t = aim_{root}$$

$$aim_{ap} = \sum_{t \in ap} aim_t, \quad ap \in AP$$

Fig. 8. Algorithm of aimv calculation

Data: at_t and cr_v $(v \in A)$
Result: $mttc_t$
begin
Create a tree *tree* with structure at_t and values cr_v as leaves
$mttc_t \leftarrow CalculateMean(tree)$
end

Fig. 9. Algorithm of $mttc_t$ calculation

In algorithm 1, a reliability graph model is used to calculate the probability that no attacks paths can reach its goal. The result is subtracted from 1 to find the ASP value. SHARPE software [14] is used to create the reliability graph model and calculate ASP values (Figs. 6, 8 and 9).

6 Results

Table 1 shows the security matric values of each vulnerability in the Zigbee network. For the matric values of sensory overload attack, we estimate the values based on how easy the attacks can be implemented and the availability of the required tools to run the attacks. It is not difficult for internal attackers to plugin his/her computer into smart home network then plugin XBee device into the computer. Then The XTCU free application provided by Digi can be used to send flood the sensor connected to the Zigbee network with a huge number of dummy messages which may cause the ZigBee network to freeze or block the communication between sensors. Sinkhole attacks matric values are extracted from Gen et al., work. In contrast jamming attacks can be generated by insider or outsider attackers, so the attacker needs to place an outcast device close to the network so that he/she can change the power spectral density to cause a jamming attack (PSD) [11, 15] and [9]. Implementing jamming attacks is not easy because it requires special hardware. Collision attacks take place when two sensors try to send packets simultaneously. Loss packets or sum control error appear in the transmitted messages.

Table 1. Metric values of the vulnerabilities

	asp_v	ac_v	aim_v	cr_v
Sensor overload	1	3.0	10	0.05
Sinkhole (tab 1)	0.8	3.0	4.0	0.012
Sinkhole (tab 2)	0.8	3.0	4.0	0.012
Sinkhole (tab 3)	0.8	3.0	10	0.012
Jamming (internal attacks)	0.6	6.0	8	0.02
Jamming (external attacks)	0.6	6.0	8	0.02
Collision	0.6	5.0	6	0.05
Sinkhole (tab 2)	0.8	3.0	4.0	0.012
Sinkhole (tab 3)				

7 Discussion

Based on the proposed security model of DoS attacks in the Zigbee network and by referring to the Zigbee network security analysis results, jamming and collision DoS attacks are the most likely attacks that threat the Zigbee network. This is because both can be implemented by internal or/and external attackers and can cause more damage relative to other DoS attacks. However, in accordance with study of described in [9] jamming attacks is most common attacks in WSN. Protecting the Zigbee network against jamming attacks will have more impact on improving the Zigbee network in general. Jamming defence techniques that can be classified as passive and active. Passive techniques can only detect jamming attacks once it occurs while the active technique can help prevent the attack. We modify the vulnerability matric table based on the proposed defence techniques. Then we reconstruct the IoT security analysis accordingly. Table 2 shows security analysis of the Zigbee network with and without defence techniques.

Table 2. WSN security analysis

	asp(ap)	ac(ap)	aim(ap)	mttc(ap)	(ap)	roa(ap)
No defense	0.11	29	50	178.57	37.6	10.44
Multipath routing technique	0.04	23	24	208.33	34.4	9.05
Network configuration SDN technique	0.11	23	24	178.57	40.8	9.90

Table 2 shows that by deploying a defence technique, MTTC of Zigbee network equals 208.33. In contrast, mttc decreases to 178.57 in case of no defence technique or by using multipath routing technique. [2] network configuration SDN technique [16] only detection technique is used. Similarly, deploying prevention technique decreases asp(ap) from 0.11 to 0.04, R(ap) from 40.8 to 34.4 and roa(ap) from 9.90 to 9.05.

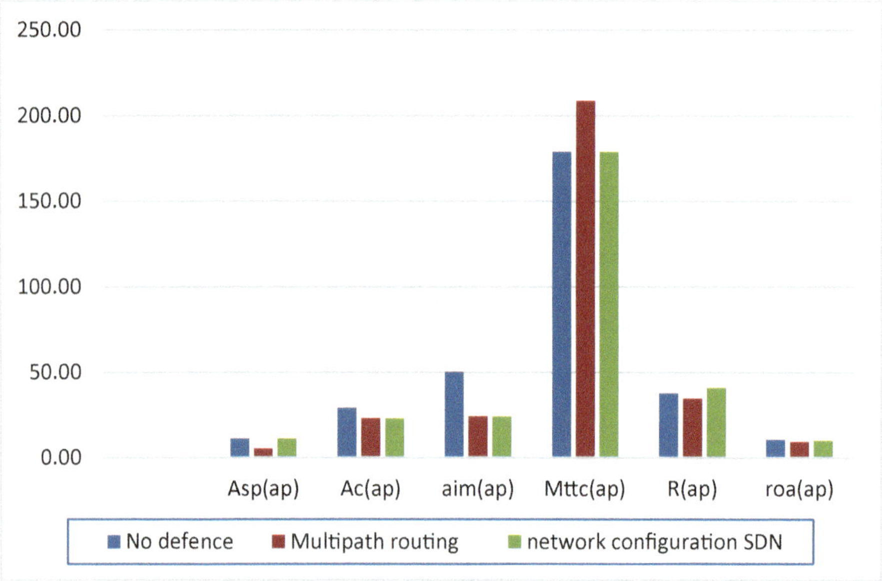

Fig. 10. DoS defense methods comparison

These finding show that Flute tree theory is an applicable modelling method to understand and analyse IoT objects in the Zigbee network. It also a reliable method that can assist decision makers in the security sector to analyse weakness in their IoT networks and assess different defence strategies to select the most efficient strategy (Fig. 10).

8 Conclusion

IoT security modelling is a challenging task since as it is made up of a large number of heterogeneous and mobile devices. DoS attacks often target IoT networks. In this paper, we proposed an extension on the security framework proposed by Mengmeng [2]. We experimented with modelling DoS attacks on smart home technology. A smart home design was proposed to show the possible internal and external DoS attacks. The metric values for each vulnerability that consist of the values of Asp(ap), Ac(ap), aim (ap), R(ap) and roa(ap) were calculated. The proposed security model has been validated with and without defence methods and MTTC was calculated to show the extent to which the proposed defence method can improve the reliability of the Zigbee network. This work contributes to existing knowledge of computer system security modelling by providing new scenarios to implement Mengmeng et al.'s security framework [2] on DoS attacks prevention. The present study lays the groundwork for future research to propose and implement a more accurate and reliable security framework to analyse the security behaviour of IoT objects. This study was limited by the absence of authenticated source for IoT vulnerabilities such as severity level and expectancy rate of the vulnerabilities. Further research is required to determine other important parameters which need to be consider in IoT network security assessment such as human intervention and common practice mistakes.

References

1. Hoque, N., Bhattacharyya, D.K., Kalita, J.K.: Botnet in DDoS attacks: trends and challenges. IEEE Commun. Surv. Tutorials **17**(4), 2242–2270 (2015)
2. Ge, M., Hong, J.B., Guttmann, W., Kim, D.S.: A framework for automating security analysis of the internet of things. J. Netw. Comput. Appl. **83**, 12–27 (2017)
3. Lazarescu, M.T.: Design of a WSN platform for long-term environmental monitoring for IoT applications. IEEE J. Emerg. Sel. Top. Circ. Syst. **3**(1), 45–54 (2013)
4. Kinney, P.: Zigbee technology: wireless control that simply works. Commun. Des. Conf. **2**, 1–7 (2003)
5. Durkota, K., Lisý, V., Bošanský, B., Kiekintveld, C.: Optimal network security hardening using attack graph games. In: Twenty-Fourth International Joint Conference on Artificial Intelligence (2015)
6. Ruijters, E., Stoelinga, M.: Fault tree analysis: a survey of the state-of-the-art in modeling, analysis and tools. Comput. Sci. Rev. **15**, 29–62 (2015)
7. Cayirci, E.: Wireless sensor networks. In: Wireless Information Highways, IGI Global, pp. 273–301 (2005)
8. Brodsy, J., McConnell, A.: Jamming and interference induced denial-of-service attacks on IEEE 802.15. 4-based wireless networks. In: Proceedings of the SCADA Security Scientific Symposium, pp. 1–2 (2009)
9. Brodsky, J., McConnell, A.: Jamming and interference induced denial-of-service attacks. In: IEEE 802.15. 4-Based Wireless Networks, Proc. SCADA Security Scientific Symp. 2009 (S4), pp. 1–2 (2009)
10. Mpitziopoulos, A., Gavalas, D., Konstantopoulos, C., Pantziou, G.: A survey on jamming attacks and countermeasures in WSNs. IEEE Commun. Surv. Tutorials **11**(4), 42–56 (2009)

11. Radmand, P., et al.: ZigBee/ZigBee PRO security assessment based on compromised cryptographic keys. In: 2010 International Conference on P2P, Parallel, Grid, Cloud and Internet Computing, pp. 465–470 (2010)
12. Cao, X., Shila, D.M., Cheng, Y., Yang, Z., Zhou, Y., Chen, J.: Ghost-in-zigbee: Energy depletion attack on zigbee-based wireless networks. IEEE Internet Things J. **3**(5), 816–829 (2016)
13. Sisodia, D., Mergendahl, S., Li, J., Cam, H.: Securing the smart home via a two-mode security framework. In: International Conference on Security and Privacy in Communication Systems, pp. 22–42 (2018)
14. Sahner, R.A., Trivedi, K., Puliafito, A.: Performance and Reliability Analysis of Computer Systems: An Example-Based Approach Using the SHARPE Software Package. Springer, Cham (2012)
15. Egli, P., Netmodule, A.G.: Susceptibility of Wireless Devices to Denial of Service Attacks. Technical White Paper. Netmodule AG, Bern (2006)
16. Scott-Hayward, S., Callaghan, G.O., Sezer, S.: SDN security: a survey. In: 2013 IEEE SDN For Future Networks and Services (SDN4FNS), pp. 1–7 (2013)

Information Security Research for Instant Messaging Service in Taiwan – Build a Private Instant Messaging

Weng Chia-Cheng[1](✉) and Chen Ching-Wen[2](✉)

[1] College of Management, National Kaohsiung University of Science and Technology (NKUST), Kaohsiung, Taiwan
0228904@nkust.edu.tw

[2] Department of Information Management, National Kaohsiung University of Science and Technology (NKUST), Kaohsiung, Taiwan
chingwen@nkust.edu.tw

Abstract. Instant Messaging (IM) is a service for online communication over the Internet, providing communication between individuals and communities using text, stickers, files and videos. In the early days, Users in Taiwan use Instant messaging services such as MSN and Skype. But now a large number of users in Taiwan rely on the LINE. The use of LINE has expanded from individuals to government units and even military units. In September 2014, Korea strengthened its surveillance of Kakao Talk and LINE on the grounds to prevent the spread of anti-government speech on the Internet and accessing user's private communication content. And so far, LINE has not yet planned to establish a dedicated computer server in Taiwan. In order to prevent the communication content of users who use LINE in Taiwan to cause other Countries who is not necessary to illegally obtain it by means of side recording, back door, Trojan horse, etc., but she can be monitored, read and analyzed by legal means to pose a threat to the Information security of Taiwan. In order to enhance the awareness of information security in military units in Taiwan. This study provides how to build a private Instant messaging service and database in her own computer room. Provide reference to the Information security strategy in Taiwan that use Instant messaging service on the premise of "Information security is National security".

Keywords: Instant messaging · LINE · Information security is National security

1 Introduction

Instant messaging (IM) is an online communication service over the Internet. It provides communication between individuals and communities to send text, stickers, files and video. The Instant messaging used by Taiwan in the early days are MSN, Skype and so on. They provided the function of VoIP which made everyone want to get rid of the nightmare of high telephone charges and use it madly. Today, LINE which provides functions such as Line@, message announcements, and message retrieval is favored by

F. Saeed et al. (Eds.): IRICT 2019, AISC 1073, pp. 801–809, 2020.
https://doi.org/10.1007/978-3-030-33582-3_75

Taiwan people too. In June 2018, LINE counted nearly 21 million Taiwanese users, accounting for 89% of Taiwan's total population of 23.58 million [1] (see Fig. 1).

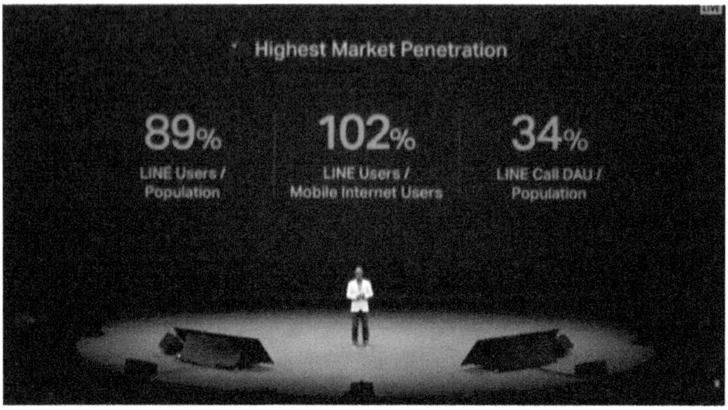

Fig. 1. More than 89% of the population use LINE in Taiwan, Image Source: iThome [1].

This shows that the use of LINE is already a new culture in Taiwan, from personal communication to government units, and even military units are also.

1.1 Research Motivation

LINE is a subsidiary of Korea's Naver Group in Japan. Its servers are built in Korea and Japan. In September 2014, in order to prevent the Internet from spreading rumors, the Korea prosecutor strengthened the monitoring of Kakao Talk, LINE and the user's chat history [2]. But as of today, LINE has not announced plans to build servers locally in Taiwan [3]. In other words, the daily communication content of nearly 21 million Taiwanese users is all in the servers of Korea and Japan.

At present, the research on Information security of Instant messaging services, general academic research only ends with the discussion of network security. that is,

1. When you download an app after registering your account information, you must provide the user's photo, permission to GPS and camera features before installing [4].
2. Explore the use of side records, backdoors, Trojans, etc. to steal personal information.
3. Explore the various ways of Cyber warfare.

Tactics and weapons are focused on military journals that although academic research is complete, there is little literature in military journals, but this study originality is an information weapon. We must know that military Information security should take precedence over other Information security.

On the other hand, the research on the Information security of instant messaging service is very complete. In fact, the Information security problem of the server is the real problem involving National security [5]. Because the local government can use the investigation of drugs, money laundering, fraud, etc. as a reason, or according to the Telecommunications Act, Information security Act, that can directly monitor, read, and analyze all the information in the servers. Just like the self-built mail server and website, I introduced how to build a IM service by myself. It's like your company's important data, but the database is just as dangerous in someone else's computer room. This study mean is that IM's services and databases should be placed in own computer room, especially military units.

In Taiwan today, there are 89% of the total population is LINE users, who include military units. Can we not worry about military secret information being directly monitored, read and analyzed?

1.2 Research Purposes

Of course, this is the fact that we can't control LINE servers in Korea and Japan. However, in order to prevent the user's communication content from being directly monitored, reviewed and analyzed by other Countries, that is a problem of Information security hazards in Taiwan. This study has an in-depth understanding of the technology and construction of Instant messaging systems, so that let Taiwan use the Instant messaging service under the premise of National security.

We need to have the ability to establish a private Instant messaging system, so the system must use its own equipment. A secure Instant messaging system can be guaranteed as long as the equipment complies with ISO 27001, ISO 27017, ISO 27018 Safety certification.

2 Literature Discussion

2.1 Cyber Warfare

The rapid development of Information & Internet technology has had a major impact on individuals, businesses and government organizations. Of course, military operations are no exception. The type of future warfare is dominated by largescale Cyber warfare, replacing the traditional way of launching missiles [6].

Therefore, "Cyber warfare" have gradually become a kind of military weapon. The enemy can weaken military combat power through the network such as communication, traffic, transportation and other important infrastructure [7].

2.2 Communication Content

The communication content is basically the same as the item recorded by the phone, but it will be clearer which number to record, at which time, from which cell site, which member to call and additional files for various media. Even government agencies are the same, and instant messaging users cannot directly access the Internet service provider's communications. For example, the police units must apply to the Internet service provider in accordance with the act in order to handle the case. When you are

"Eavesdropped" (communication monitoring) by the police unit, someone will record the contents of each of your calls and record the evidence.

3 Research Method

This research consists in the construction of private Instant messaging, which is a private SaaS, and therefore uses the system development research methodology proposed by Nunamaker, Chen and Purdin is cited in the paper by Shaw Ruei-Siang [8]. The system development research method is a problem-solving method based on system theory and system analysis as the working mode. It provides an integrated structure. In other words, the system development research method can be interpreted as: "A method and technique for seeking solutions to problems" [8]. Therefore, the construction of the entire system needs to go through four stages: System analysis, System planning, System production and Testing, evaluation, correction.

3.1 System Analysis

1. **About Security:** To build a private Instant messaging system, you must first have a basic information equipment of the server. In Taiwan, government units or military units that have strong demand for Instant messaging and have confidentiality requirements for communication content, most of them have passed ISO27001, ISO27017, ISO27018 Information security certification [9].
2. **About Server:** Preparing a 2U or 4U server is enough to build a private Instant messaging system for yourself. The firewall setting TCP Port 80 & 443 is open to the Internet, allowing the Internet to connect to the Instant messaging service. If VoIP is required, the firewall will then open TCP/UDP Port 5060 & 5061. The application can be the same server as the database, or it can be set up independently. A high-capacity storage is a must. It is strongly recommended to use an SSD hard disk to speed up access.
3. **About Operating System:** Linux is used in the operating system, such as RedHat, Debian and SUSE can be selected. If you use high performance server, HPUX, AIX and Solaris operating system can also be set smoothly.
4. **About Language:** The programming language adopts Java JDK 8 or upper. The main consideration is that JAVA can be cross-platform and easy to transplant. Web site uses Apache Tomcat or JBOSS, but it is necessary to proxy 8080 & 8443 port to 80 & 443 port to facilitate presentation in various media.
5. **About Database:** Use MySQL, MariaDB or PostgreSQL as the database. Oracle is a good choice if you use a mainframe or high performance server.

3.2 System Planning

1. Just like the web server, when the mobile phone is online, it will contact the AP server. The DB server will record the IP address and connection time of the mobile phone (see Fig. 2).

2. The A mobile phone first completes the text, stickers, video, file or other information, and transmits it to the AP server through the Telecommunication network. All messages will be stored in the DB server. The AP server will check the status of the B mobile phone online, and if it is online, it will send a notification (except text message) to the B mobile phone (see Fig. 2).
3. When the B mobile phone reads the notification of the A mobile phone, the A mobile phone will display "Readed". If the B mobile phone wants to open a message such as a picture, video or file, the AP server will transmit the message from the DB server to the B mobile phone (see Fig. 2).
4. Hypothesis that the Storage(File system) capacity of the DB server is large enough, or that Storage is built with Hadoop (Big Data Framework), The message can be saved for a long time (see Fig. 2).
5. Related applications include VoIP, Conference system, Notebook, Keep, etc., which can be expanded according to requirements (see Fig. 2).

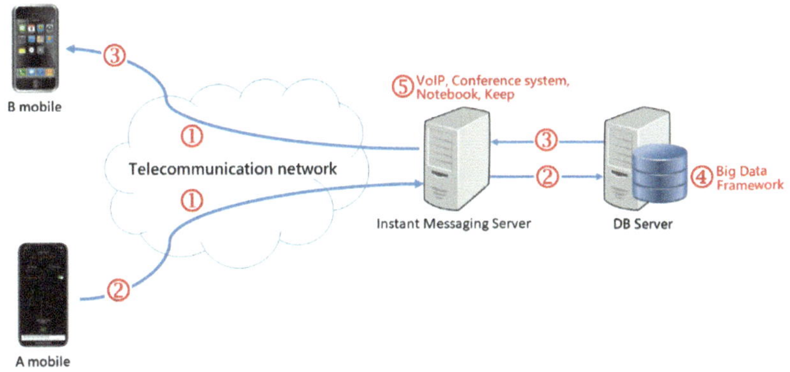

Fig. 2. Instant messaging system architecture, Image Source: This study

3.3 System Construction

1. Write the program in Java JDK 11 language in RedHat (see Fig. 3).

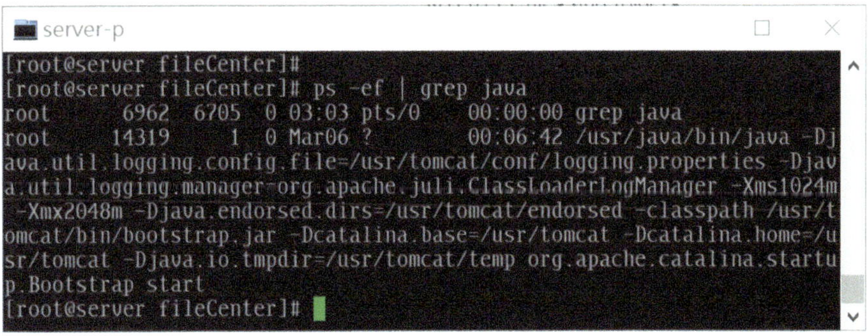

Fig. 3. Developing the Instant messaging system in Java, Image Source: This study

2. A flow chart to illustrate the developer steps that have been followed in this study. (see Fig. 4).

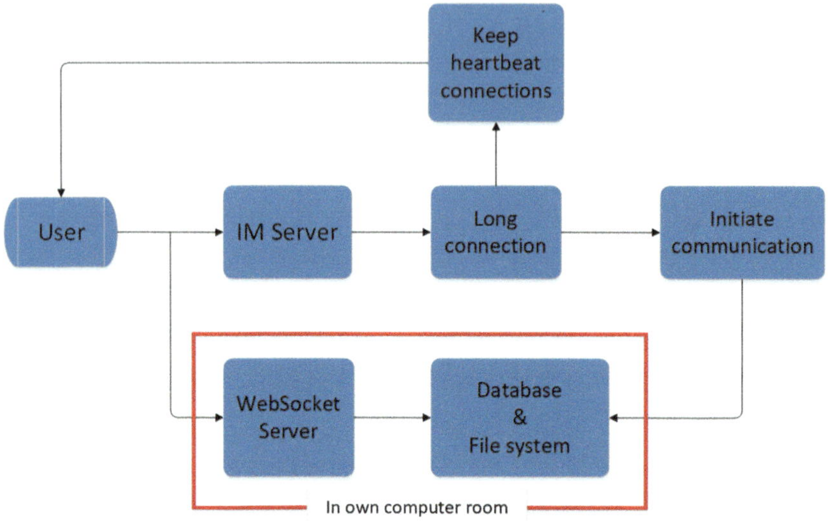

Fig. 4. The developer steps that have been followed in this study, Image Source:: This study

3. The user interface is modeled after the LINE interface, including textures, emoticons, etc., and supports the Chinese and English languages (see Fig. 5).

Fig. 5. The user interface, Image Source: This study

4. Text, stickers, audio and video files can be transfer, or stored in Notepad, Keep, and Postscript (see Fig. 6).

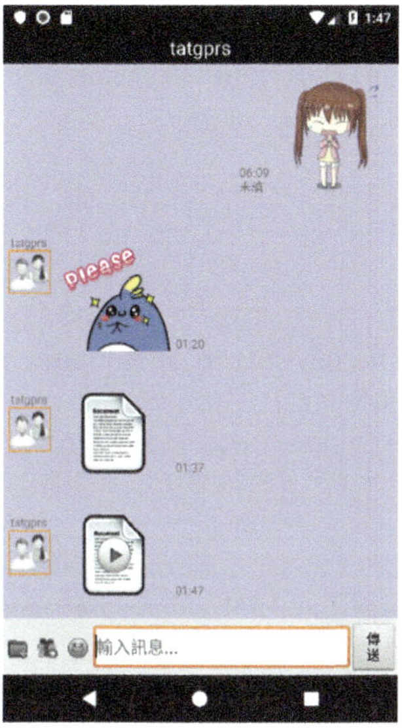

Fig. 6. Send text, stickers, audio and video files, Image Source: This study

5. All communication messages, text, stickers, videos or data files are saved in the database server. (see Fig. 7).

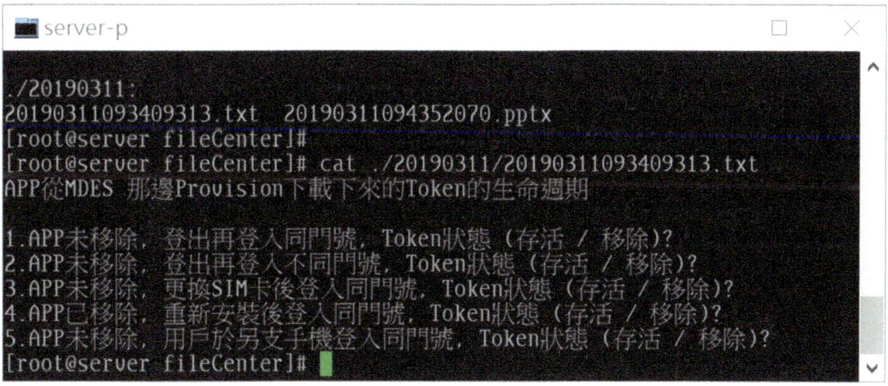

Fig. 7. All files are saved in the database server, Image Source: This study

3.4 Testing, Evaluation and Correction

1. The market share of iOS and Android has reached more than 90% for mobile operating systems. The evaluation prioritizes the development of two versions of

the iOS and Android apps. After numerous revisions and corrections, both versions of the app can be successfully executed on the mobile phone.
2. Mobile phones using the Android operating system are varied, so try to find different brands of mobile phones for testing. Currently, the common mobile phones on the market can execute the App smoothly.
3. All data is transmitted in TCP Port 443, and files encoded in UTF 8 can be displayed normally on the mobile phone.

3.5 Compare with LINE

- The same functions

1. You can send it to your friends via the smartphone app.
2. You can chat with your friends 1:1, or you can communicate with the group.
3. You can send text, stickers, audio and video files.

- The difference functions

1. LINE is a public SaaS, but the services of this study is private. [10]
2. We are unable to manage the LINE database, but this study will manage the database by myself. (see Fig. 8).
3. Have a professional information staff to build a Private Instant Messaging, or to order out.

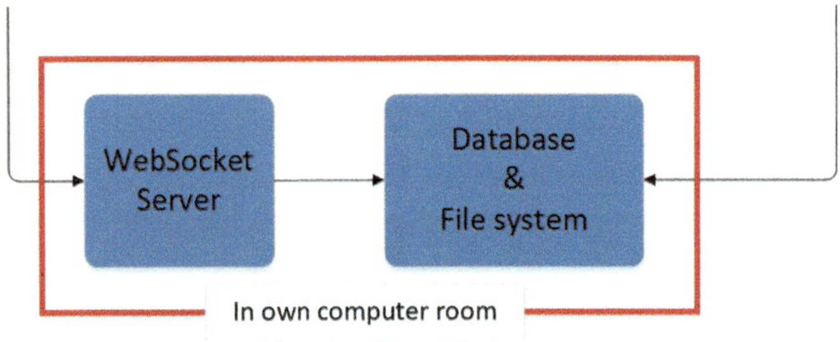

Fig. 8. The proposed solution compared with the LINE, Image Source: This study

4 Conclusion

- The non-feeling threat is more serious than the feeling threat

When you register your account, you must provide personal information and permission to read the internal information of the phone, that it is easy to prevent the feeling threat. Instead, local government can monitor, read, and analyze the user's information through legal means. The non-feeling threat of completely revealing the communication content of individuals, government units and even military units without any security measures is a very serious harm to National security.

- Build a private Instant messaging service to protect Information security

For the confidential content of communication, this study advocates try not to use public SaaS, such as LINE. Because the server is not in our Country. Confidential communications of military units must be secured by Information security through private Instant messaging services.

- Easy build a private Instant messaging service

To build a private Instant messaging services will be as easy as for building a private email service or website.

References

1. iThome. https://www.ithome.com.tw/news/124203. Accessed 26 June 2018
2. Liberty Times Net. https://news.ltn.com.tw/news/world/paper/821286. 14 Oct 2014
3. iThome. https://www.ithome.com.tw/news/129579. Accessed 25 Mar 2019
4. Tseng Po-Yuan, F., Hsieh Chih-Yuan, S.: Discussion on the influence of baoxin dream phenomenon on military security from the perspective of risk management. Army Bimonthly **53**(551), 113–128 (2017)
5. National Security Council, F.: National Qualification Security Strategy Report - Information security is National security. 1st edn. National Security Council, Taiwan (2018)
6. Chen Liang-Chu, F., Fan Chung-Ping, S., Hsieh Chua-Jung, T.: The study on exploring the topics of cyber warfare security and management: using growing hierarchical self-organizing map. J. Inf. Manag. **54**(557), 99–128 (2016)
7. Liu Ting-Chu, F.: Discussion on constructing key factors of information security protection of national Army Cloud Data Center. Army Bimonthly **53**(551), 92–110 (2018)
8. Shaw Ruei-Siang, F., Tsao chin-Feng, S., Chang Hui-Hsin, T.: A study of interface integrated campus cloud service management systems. J. Inf. Manag. **22**(4), 475–496 (2015)
9. National Information & Communication Security Taskforce. https://nicst.ey.gov.tw/Page/7CBD7E79D558D47C/45d1f417-ca0c-4b30-aafa-ffeb9070a257. Accessed 22 May 2014
10. Wang Chien-Min, F., Wu Li-Chun, S, Li Hung-Wei, T.: LINE instant messaging software protocol and security analysis. In: (CISC 2014) Cryptology and Information Security Conference (2014)

A Model of Information Security Policy Compliance for Public Universities: A Conceptual Model

Angraini[1,4](✉), Rose Alinda Alias[2], and Okfalisa[3]

[1] School of Computing, Faculty Engineering, Universiti Teknologi Malaysia, 81310 Skudai, Johor, Malaysia
Angraini@uin-suska.ac.id
[2] Department of Information System, Azman Hashim International Business School, Universiti Teknologi Malaysia, 81310 Skudai, Johor, Malaysia
[3] Department of Informatics Engineering, Faculty Science and Technology, Universitas Islam Negeri Sultan Syarif Kasim, Pekanbaru, Riau, Indonesia
[4] Department of Information Engineering, Faculty Science and Technology, Universitas Islam Negeri Sultan Syarif Kasim, Pekanbaru, Riau, Indonesia

Abstract. The university is an organization that manages much public information, and therefore, information security policies are developed to ensure data security. However, during implementation still founded disobey behavior user and has an impact on data security. The previous research has been conducted to find influencing factor user comply with information security, although some model and theories still limited to implementation. There is a lack of researchers combine behavioral theory and organizational theory to develop models and previous model inadequate to universities that have unique characteristics. This study aims to explore and identify factors that influence information security compliance and continue to develop conceptual models for assessing information security policies. This conceptual model creates based on a systematic literature review and preliminary study. The results in the conceptual model found several variables, namely habits, attitudes, moral beliefs, self-efficacy from behavioral theories and human culture, commitment, rewards, costs can be used to evaluate user compliance with information security policies. Conceptual will be tested further to contribute to help universities to ensure and assess users to comply with information security policies.

Keywords: Information security · Security policy · Compliance · User behavior

1 Introduction

This study addresses the issue of compliance with information security policies such as individual frequently trespass policies [1]. Employees unrealized non-compliance with information security lead to related safety [2], generate security threat [3, 4], increase security risk [5]. A significant increase in incidents in recent years has reminded that to overcome the increasing threat, in addition to technical solutions, necessary to provide

F. Saeed et al. (Eds.): IRICT 2019, AISC 1073, pp. 810–818, 2020.
https://doi.org/10.1007/978-3-030-33582-3_76

an information security policy [4]. However, during implementation user tend to ignore that policy because they are unaware and realize importance comply with information security policy. This opinion is in line with the research conducted by Abed to find out how the intention to comply with the policy or intention to refuse [6]. Conceptual attempt to encourage employees to comply with information security policies when organizational damage resulting from information violations becomes serious. Therefore, necessary to develop a substantial theoretical foundation that concerns about a social issue and human behavior [2]. Management as policymaker needs to encourage user confidence in the ISP to employees in the organization to ensure and achieve information security objectives, although the organization is conscious of the obstacles in information security related to unskilled employees of security technology because of a lack of security knowledge [7]. Many empirical studies have been conducted to find out and explain the human factors that influence compliance with information security policies [1–8].

However, the discussion about the importance of the value of information has less discussed explicitly human factors like the source of unpredictable and uncontrolled vulnerability in information security because of changes in human behavior. Research on information security behavior has the opportunity to combine people, technology, and organizations. Understanding individual behavior will enhance positive behavior while reducing harmful behavior. Understanding human behavior will be able to improve adherence to policies by knowing the motivation, modification, and prevention [21]. Therefore, necessary to understand how to change people in organizations becoming allied security, instead of just being a security risk. The policy of information significance can influence the level of awareness and behavior of users towards information security [8].

Level of awareness and security behavior problem occurs because it is often encountered Information security conflicts with the efficiency of information systems such as active sharing of critical information resources, standardizing business processes, and downloading applications or components to complete specific tasks. Consequently, challenging for the user to comply with information security policies. Employees prefer to complete their work and follow their colleagues to not complying with their policies [9]. There are some compliance regulatory from industrial and government such as the Federal Information Security Management Act (FISMA), U.S. Intelligence Community (IC), DoD Information Assurance Certification and Accreditation Process (DIACAP), Homeland Security/Presidential Directives and National Institute of Standards Technology (NIST) 800-XX series [10]. Unfortunately, this standard not always applicable to the university. Difference requirement and limited resources make the university can adopt an international standard to its security policy [9–11]. Universities have amount information technology resources, and access is given to the public and increase the risk for universities [12]. An academic institution with weak policies in information security creates threats for university data [13]. Universities with ineffective systems in information security create risks for university data.

The university is an organization with various activities and type of users. It becomes a challenge for universities to ensure organizational information, which saved, processed, and disseminated with information technology. Security incidents caused density, high dependence on the accuracy, integrity, and availability of technology-based information resources [27]. A well-known problem with information security is that it does not take into account user behavior and organization condition. The organization was more concerned about technical issues than the requirements of the organization or user habits that can have an impact on information security. Moreover, information security policy inefficiency because during implementation, not well-monitoring. Therefore, their need model to measure and evaluate the level of user comply with information security policy.

2 Literature Review

Information security policy is a document to ensure information security based on top management's vision related to key strategic business objectives within the scope of management goals and contains basic requirements. Information security policies contain rule, guidelines, and control to secure information asset [14]. Information security policies aim to provide management goals and support for information security [15]. Information security policy contained with Program Policy used to create an organization's IT security program, Issue-Specific Policies, and System-Specific Policies to address individual systems [16]. Security policies must continuously be reviewed and updated regularly by considering changes in circumstances, environment, changing needs in the business context, and identified risks [17]. Compliance with policies aims to ensure the application of organizational security standards [18]. Ross from the NIST lab conducted a study by performing a series of procedures to assess security controls and privacy controls used in federal information systems and organizations. Assessment procedures, carried out at various phases of the system development life cycle, are consistent with security and privacy controls using the NIST 800-53 standard [19]. Exactly the measuring instruments used can only be applied to organizations that have used these standards in the process of developing information security policies.

Sommestad added security awareness to developed a model of information security policy compliance. The results find other factors such as explanatory power and anticipated regret that have a significant effect While habit variables have a significant result to strengthen the model [20]. A different result was found by Moody [30] who combined eleven theories with building a unified model of information security policies compliance. His research found that habits can influence behavioral intentions to comply with information security policies. Information security behavior builds a relationship with people, technology, and organizations. Necessary for understanding individual behavior will enhance positive behavior while reducing harmful behavior. Understanding human behavior will be able to improve adherence to policies by knowing the motivation, modification, and prevention [21].

Vroom evaluated human behavior in safeguard organizational information and other valuable assets with audits and found that this method is not sufficient for testing human behavior on information security [22]. This interpretation contrasts with that of Kankanhalli who argue that Human factors, organizational factors can influence compliance with information security policies [23]. The significant impact of organizational factors on the effectiveness of information security management implementation [24]. Research on this subject has been mostly restricted to limited comparisons of theory about human behavior.

Lowry [21] develops a model for compliance with information security policies using organizational control theory. The results of empirical studies conducted by scenarios for professional workers show that there are reactions that arise when an organization applies threats and controls. Responses that appear related to anger and hatred towards the organization but are fully compliant [25]. Feasible regulation from organization mismatched with national culture and organization culture. Because according to Alshare [26], organizational culture is a significant predictor for determining crime against information security policies [26]. Likewise, the research conducted by Arage [37], states that national culture has an impact on information security compliance. Another factor within the organization is user involvement and leadership, this potential was seen by Amankwa [38] in his research on Establishing information security policy compliance culture in organizations using the theory of organizational behavior and organization culture. However, differences in results found by Sommestad [36] investigated the relationship between individual intentions in social groups rather than the workplace and the effects of this group by using the theory of planned behavior and information security culture.

The results found that culture does not have a significant impact on behavioral intention because the culture of information security is a concept that is used and interpreted in many different ways. [36]. The similarly discovered by Gerber [32] Management involvement is more needed to ensure user compliance. Participation from top management will influence user behavior intentions towards information security policies compared to organizational culture (2016).

3 Conceptual Framework

This research aims to construct a conceptual model presenting perspective or a way to understand practical phenomena. Therefore, a model was developed that showed the relationship between factors that influence compliance with information security policies. This model is the initial model used to prepare more complex models. The development of this model expected will support research problem, identifying relevant factors, and then providing a relationship to map the problem frame. Figure 1 will describe a conceptual model of information security compliance.

This conceptual model consists of ten variables composed of independent, dependent, and mediating variables. The dependent variable is information security policy compliance. Independent variables include organizational commitment, organization culture, reward, perceived cost, self-efficacy, attitude, moral belief, and habit. Moreover, behavioral intention as a mediating variable. This variable choosing based on literature review, conceptual survey, and interview.

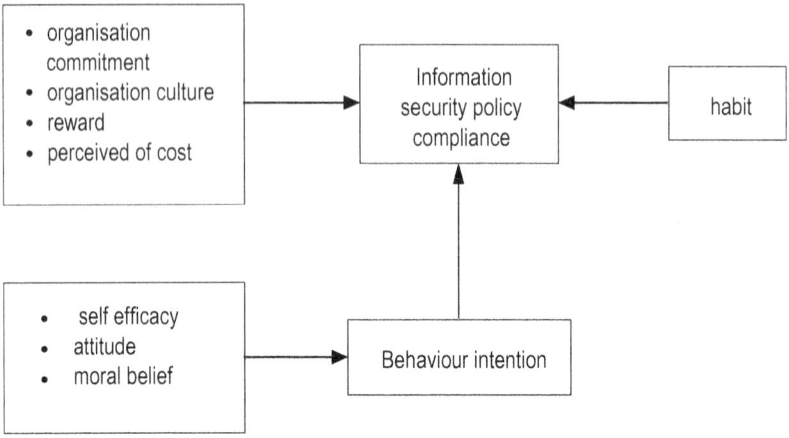

Fig. 1. Conceptual model of information security policy

The hypothesis for this study is as follows:

hypothesis 1. habit will influence a user to comply with information security policies
hypothesis 2. organizational commitment, organizational culture, reward, and perceived cost user comply with information security policies
hypothesis 3. behavioral intentions moderating self-efficacy, attitude, and moral belief affect users to comply with information security policy

Variable habits explain by interview can influence user behavior always to manage information security by an understanding of the pattern. According to the theory of interpersonal behavior, habit is behavior has become automatic due to continuous activity, and they conduct it attentively without instruction [29]. The assumption that habits are related to the complexity of ISS actions causes research to be conducted to examine the processes that lead employees to become non-compiler habits and determine how bad habits and noncompliance behavior can be changed [30]. The habit of transferring behavior can increase explanatory power became a reason to conduct further research about the relationship between habits and intentions [20]. Compliance with information security can be implemented entirely with the commitment of the organization. Commitments will improve the efforts and energy of individuals to support organizational policies. Safa study show a significant relationship between commitment and individual attitudes towards compliance with information security policies [31]. Organizational responsibility and commitment pretend an essential role in protecting information assets. Consequently, employees will realize that compliance with security policies is part of their work [32].

An organizational commitment will inspire employees in the organization. Empirical studies have confirmed a positive relationship between employee commitment to the organization [33]. An understanding of security policy provides a new lens for learning noncompliance behavior. Employees sometimes sacrifice compliance with information security policies to complete their duties [34]. The research conducted by

Sharma, which states that organizational commitment influences behavioral intentions to comply with the Security Policy Information and employee status has an impact on organizational commitment [35]. This study argues that organizational commitment can improve compliance directly without influencing with user intention or employee status. Cultural factors influence modeling human behavior. Compliance culture should be within the organization and play an essential role in increasing individual compliance and behavior [36]. This statement is supported by Arage [37], which states that national culture has more influence on compliance behavior [37]. However, different results shown by Gerber, management culture does not provide any improvement at all [32]. In a company with a stable culture, the appropriate policies and procedures, including sanctions and education programs regarding the policy, will be well implemented. Employee non-compliance with information security policies (ISP) can be overcome by maintaining an ISP compliance culture through organizational culture [38]. Further studies are needed to find out factors such as organizational culture to find out the involvement of organizational culture can influence compliance with security policies. Reward factors mentioned by interviewee will increase employee compliance with information security policies.

Although the different results shown by [30] found that punishment and reward/cost no significant impact on the intention to comply with the information security policy, this opinion supports by [4], which mention awards to comply with the information security policy is not significant. However, [32] indicates reward will support achieve performance goals and improve security compliance. This study attempts to explore the importance of rewards with information security policies compliance. Non-compliance with policies will pose a higher risk of increasing costs. The organization will incur higher costs when an incident occurs due to non-compliance of its employees. Kajtazi [39] examines cost factors as a mediating mechanism to explain intentional violations by people in information security policies.

The results of the study also show that perceptions of employee sunk costs predicted by the effects of settlement and physical mismatches [39]. The same thing was explained by Han that fulfilling psychological contracts could reduce the adverse impact of costs on ISP compliance intentions in the supervisor group [2]. Different perceptions are conveyed by Sommestad that Prediction of intention to adhere to information security policies is improved if compliance costs [20]. In this study, the perceived costs are the costs that arise due to the risk of non-compliance. The organization would become aware of the importance of employee compliance with information security policies if organizations spent a higher cost to secure their information. Self-Efficacy is a variable that is widely used by previous research to determine compliance with information security policies. Previous research has consistently stated that self-efficacy significantly influences employees' intention to comply with information security policies [37, 40, 41]. Self-Efficacy can increase employee awareness in handling security threats [28].

However, different results were found by Sikolia [42] who researched at a midwestern university by combining variables from Protection Motivation Theory (PMT), the Theory of Reasoned Action, and Cognitive Evaluation Theory. The results found that response-efficacy had more influence on behavioral intentions than self-efficacy. This study will further test the user's behavioral intentions using the compliance of information security policies using variable self-efficacy [42]. User attitudes and moral beliefs are also variables that are usually used by previous researchers to determine the user's intention to comply with information security policies. However, the empirical study conducted by Moody [30] the construct of moral beliefs does not significantly affect compliance. Moral cannot stand alone, so it needs to be united with self-concept, role, and influence, resulting in a new construct, namely the role value. Differences in moral definition affect by the culture of each organization. Therefore, in this study, it is still believed that moral constructs can affect user compliance.

4 Conclusion

This research has a contribution to theory with a proposed model that can be used by organizations to determine the level of compliance of their employees with information security policies. This model will develop using theories from a different field, combines several factors from several theories will provide additional knowledge in the field of information security behavior. This conceptual model uses organizational theory commitment, organizational culture, and cost with commitment organizational variables, culture organization, reward and perceived of cost. While for the theory of human behavior using planned behavior theory and additional habit variables.

The implementation model of information security policy compliance expected growth the effectiveness and efficiency of organizational performance. Moreover, changing user behavior will have a direct and indirect impact on an organization. Periodic assessment will be enhanced employee behavior and trigger organizational culture so that the organization can achieve its goal. This study will identify factors that influence compliance with information security policies. Formerly the variables found used to create models to evaluate employees the compliance of policy information security. The variables to be used are determined from the literature review and theories. Result of this study is a model that can be used by organizations to measure employee compliance with information security policies.

References

1. Bélanger, F., Collignon, S., Enget, K., Negangard, E.: Information & management determinants of early conformance with information security policies. Inf. Manag. **54**, 887–901 (2017)
2. Han, J.Y., Kim, Y.J., Kim, H.: An integrative model of information security policy compliance with psychological contract: examining a bilateral perspective. Comput. Secur. **66**, 52–65 (2017)
3. Pahnila, S., Siponen, M., Mahmood, A.: Which factors explain employees' adherence to information security policies? An empirical study. In: Pacis 2007 Proceedings, pp. 438–439 (2007)
4. Siponen, M., Adam Mahmood, M., Pahnila, S.: Employees' adherence to information security policies: an exploratory field study. Inf. Manag. **51**, 217–224 (2014)
5. Nasir, A., Arshah, R.A., Ab Hamid, M.R.: Information security policy compliance behavior based on comprehensive dimensions of information security culture. In: Proceedings of 2017

International Conference on Information System and Data Mining. - ICISDM 2017, pp. 56–60 (2017)
6. Abed, J., Dhillon, G., Ozkan, S.: Investigating continuous security compliance behavior : insights from information systems continuance model. In: Twenty-second Americas Conference on Information Systems, San Diego, pp. 1–10 (2016)
7. Humaidi, N., Balakrishnan, V.: Leadership styles and information security compliance behavior: the mediator effect of information security awareness. Int. J. Inf. Educ. Technol. **5**, 311–318 (2015)
8. Doherty, N.F., Tajuddin, S.T.: Towards a user-centric theory of value-driven information security compliance. Inf. Technol. People **31**, 348–367 (2018)
9. Hwang, I., Kim, D., Kim, T., Kim, S.: Why not comply with information security? An empirical approach for the causes of non-compliance. Online Inf. Rev. **41**, 2–18 (2017)
10. Andress, J., Winterfeld, S.: Cyber Warfare Techniques, Tactics and Tools for Security Practitioners, vol. 2. Elsevier Inc., Waltham (2014)
11. Gikas, C.: A general comparison of FISMA, HIPAA, ISO 27000 and PCI-DSS standards. Inf. Secur. J. Glob. Perspect. **19**, 132–141 (2010)
12. Katz, F.H.: The effect of a university information security survey on instruction methods in information security. In: Proceedings of 2nd Annual Conference on Information Security Curriculum Development, pp. 43–48 (2005)
13. Ayyagari, R., Tyks, J.: Disaster at a university: a case study in information security. J. Inf. Technol. Educ. Innov. Pract. **11**, 85–96 (2012)
14. BS ISO/IEC: ISO 27001 - Information Technology Security Techniques Information Security Management Systems, Requirements (2005)
15. Sommestad, T., Hallberg, J., Lundholm, K., Bengtsson, J.: Variables influencing information security policy compliance: a systematic review of quantitative studies. Inf. Manag. Comput. Secur. **22**, 42–75 (2014)
16. NIST: Glossary of Key Information Security Terms [NISTIR 7298 Rev 2] (2013)
17. Calder, A., Watkins, S.: It Governance an International Guide to Data Security and ISO 27001/ISO27002, vol. 6. Kopan Page, UK (2015)
18. Barry, L.: Information Security Policy Development for Compliance. CRC Press/Taylor & Francis Group, Boca Raton (2013)
19. Ross, R.S.: Assessing security and privacy controls in federal information systems and organizations: building effective assessment plans, pp. 1–487. NIST Special Publication (2014)
20. Sommestad, T., Karlzén, H., Hallberg, J.: The theory of planned behavior and information security policy compliance. J. Comput. Inf. Syst. **00**, 1–10 (2017)
21. Crossler, R.E., Johnston, A.C., Lowry, P.B., Hu, Q., Warkentin, M., Baskerville, R.: Future directions for behavioral information security research. Comput. Secur. **32**, 90–101 (2013)
22. Vroom, C., Von Solms, R.: Towards information security behavioural compliance. Comput. Secur. **23**, 191–198 (2004)
23. Kankanhalli, A., Teo, H.H., Tan, B.C.Y., Wei, K.K.: An integrative study of information systems security effectiveness. Int. J. Inf. Manag. **23**, 139–154 (2003)
24. Chang, S.E.: Organizational factors to the effectiveness of implementing information security management (2006)
25. Lowry, P.B., Posey, C., Bennett, R.B.J., Roberts, T.L.: Leveraging fairness and reactance theories to deter reactive computer abuse following enhanced organisational information security policies: an empirical study of the influence of counterfactual reasoning and organisational trust. Inf. Syst. J. **25**(3), 193–273 (2015)
26. Alshare, K.A., Lane, P.L., Lane, M.R.: Information security policy compliance: a higher education case study. Inf. Comput. Secur. **26**, 91–108 (2018)

27. Doherty, N.F., Anastasakis, L., Fulford, H.: The information security policy unpacked: a critical study of the content of university policies. Int. J. Inf. Manag. **29**, 449–457 (2009)
28. Hina, S., Dominic, D.D.: Information security policies: investigation of compliance in universities. In: 2016 3rd International Conference on Computer and Information Sciences. In: Proceedings, ICCOINS 2016, pp 564–569 (2016)
29. Bamberg, S., Schmidt, P.: Incentives, morality, or habit? Predicting students' car use for University routes with the models of Ajzen, Schwartz, and Triandis. Environ. Behav. **35**, 264–285 (2003)
30. Moody, G.D., Siponen, M., Pahnila, S.: Toward a unified model of information security policy compliance. MIS Q. **42**, 285–311 (2018)
31. Sohrabi Safa, N., Von Solms, R., Furnell, S.: Information security policy compliance model in organizations. Comput. Secur. **56**, 1–13 (2016)
32. Gerber, N., McDermott, R., Volkamer, M., Vogt, J.: Understanding information security compliance - why goal setting and rewards might be a bad idea. In: International Symposium on Information Assurance and Security, HAISA 2016, vol. 10, pp. 145–155 (2016)
33. Bulgurcu, B., Cavusoglu, H., Benbasat, I.: Information security policy compliance: an empirical study of rationality-based beliefs and information security awareness. MIS Q. **34**, 523–548 (2010)
34. Kajtazi, M., Cavusoglu, H., Benbasat, I., Haftor, D.: Escalation of commitment as an antecedent to noncompliance with information security policy. Inf. Comput. Secur. **26**, 171–193 (2018)
35. Sharma, S., Warkentin, M.: Do I really belong? Impact of employment status on information security policy compliance. Comput. Secur. (2018)
36. Sommestad, T.: Social groupings and information security obedience within organizations. In: International Federation for Information Processing, pp. 325–338 (2015)
37. Arage, T., Belanger, F., Beshah, T.: Influence of national culture on employees' compliance with information systems security (ISS) policies: towards ISS culture in Ethiopian companies. In: AMCIS 2015 Proceedings, pp. 1–7 (2015)
38. Amankwa, E., Loock, M., Kritzinger, E.: Establishing information security policy compliance culture in organizations. Inf. Comput. Secur. **26**, 420–436 (2018)
39. Kajtazi, M., Bulgurcu, B., Cavusoglu, H., Benbasat, I.: Assessing sunk cost effect on employees' intentions to violate information security policies in organizations. In: Proceedings of Annual Hawaii International Conference on System Sciences, pp. 3169–3177 (2014)
40. Sommestad, T., Karlzén, H., Hallberg, J.: The sufficiency of the theory of planned behavior for explaining information security policy compliance. Inf. Comput. Secur. **23**, 200–217 (2015)
41. Aurigemma, S., Mattson, T.: Privilege or procedure: evaluating the effect of employee status on intent to comply with socially interactive information security threats and controls. Comput. Secur. **66**, 218–234 (2017)
42. Sikolia, D., Twitchell, D., Sagers, G.: Employees' adherence to information security policies: a partial replication. In: Proceedings of the Americas Conference on Information Systems, pp. 1–9 (2016). https://doi.org/10.1109/ICMTMA.2009.433

A Framework for Preserving Location Privacy for Continuous Queries

Raed Saeed Al-Dhubhani[1(✉)], Jonathan Cazalas[2], Rashid Mehmood[3], Iyad Katib[3], and Faisal Saeed[4]

[1] University of Hafr Albatin, Hafar Al Batin, Kingdom of Saudi Arabia
raedsn@uhb.edu.sa
[2] Rollins College, Winter Park, USA
[3] King Abdul-Aziz University, Jeddah, Kingdom of Saudi Arabia
[4] Taibah University, Medina, Kingdom of Saudi Arabia

Abstract. The growth of the location-based services (LBSs) market in recent years was motivated by the widespread use of mobile devices equipped with positioning capability and Internet accessibility. To preserve the location privacy of LBS users, many mechanisms have been proposed to provide a partial disclosure by decreasing or blurring or the accuracy of the shared location. While these Location Privacy Preserving Mechanisms (LPPMs) have demonstrated effective performance with snapshot queries, this work shows that preserving location privacy for continuous queries should be addressed differently. In this paper, MOPROPLS framework is proposed with the aim to preserve location privacy in the specific case of continuous queries. As part of the proposed framework, a novel set of six requirements that any LPPM should meet in order to provide location privacy for continuous queries is proposed. In addition, a novel location privacy leakage metric and a novel two-phased probabilistic candidate selection algorithm are proposed. Comparing the performance of MOPROPLS framework with the geo-indistinguishability LPPM in terms of privacy (adversary estimation error) shows that the average of MOPROPLS framework improvement is 34%.

Keywords: MOPROPLS framework · Location privacy · Continuous queries · Location-based services and LBS

1 Introduction

The widespread use of mobile devices in recent years opened the doors to the growth of the location-based services (LBSs) market. LBSs provide customized services for users based on their current locations. A user sends the request to the LBS server either as a single request (snapshot query), or as a continuous query which will be evaluated continuously over a given interval. Directions and navigation applications, weather applications, venue finders, social games, and crowd-sensing applications are examples of LBSs [1].

Providing such context-aware services requires the users to share their private location data. Sharing a user's location data means revealing the user existence at

F. Saeed et al. (Eds.): IRICT 2019, AISC 1073, pp. 819–832, 2020.
https://doi.org/10.1007/978-3-030-33582-3_77

specific time and location to the service provider. Furthermore, using geographic maps and background knowledge to analyze the shared accurate locations of the user allows one to infer sensitive information about his/her life style, home and work addresses, and daily activities [2]. Even if it is assumed that service providers have no intention to misuse the collected locations data, their servers are still prone to the threat of malicious attacks to access the valuable data. According to [1], location privacy field has four open challenges: quantifying location privacy, the need for new protection mechanisms, the lack of large datasets, and the need to raise user awareness about over-sharing.

To preserve location privacy, many location privacy preserving mechanisms (LPPMs) have been developed which reduce or blur the user's location resolution prior to sending it to the service provider [1–3]. The caveat is that said blurring will, in fact, reduce query accuracy and may lower the utility level depending on the characteristics of the requested service. For example, sending a modified version of the user's location, ten kilometers farther from his/her accurate location, still provides the weather forecast service with an acceptable utility level, while using the same modified location version for a taxi booking service reduces its utility level significantly. While these mechanisms have demonstrated effective performance with snapshot queries, this work shows that preserving location privacy for continuous queries should be addressed differently.

In this paper, we aim to answer the following question: how to preserve location privacy for continuous queries? To accomplish that, we propose MOPROPLS framework. First, a novel set of requirements is proposed as a part of the framework, such that any LPPM should meet these requirements to provide location privacy for continuous queries. Second, a novel location privacy leakage metric is proposed to measure location privacy leakage of continuous queries. Third, a novel two-phased probabilistic candidate selection algorithm is proposed to preserve privacy for continuous queries, considering the correlation between the obfuscated locations. Comparing MOPROPLS framework with the geo-indistinguishability LPPM in terms of privacy (adversary estimation error) shows that the average of MOPROPLS framework improvement is 34%.

The rest of the paper is organized as follows. In Sect. 2, we present the related work. The proposed set of requirements are discussed in Sect. 3. In Sect. 4, we discuss the proposed framework. Section 5 presents the framework evaluation. Finally, Sect. 6 concludes the paper.

2 Related Work

To preserve the user's location privacy, many LPPMs have been proposed. These mechanisms reduce or blur the resolution of the user's location shared with the service provider [1–3]. Generally, LPPMs are classified as k-anonymity, obfuscation, mix zones, and cryptographic-based schemes.

K-anonymity is a concept which addresses the privacy issue in different domains. It depends on creating a set of k objects which are indistinguishable from each other. In the location privacy domain, k-anonymity is implemented by creating cloaking regions as a geographic area containing k users [4–8]. The accurate location in the user's query is then replaced by the cloaking region. Although generating effective cloaking regions for snapshots queries is feasible, reducing the cost of the continuous cloaking region extension process to resist potential attacks for continuous queries requires utilizing knowledge about the user movements which cannot be available in all cases [9, 10].

In addition to cloaking regions, position dummies are used to achieve the k-anonymity [11–14]. The user's request includes the accurate location of the user and k-1 faked locations (i.e. position dummies). In addition to the overhead resulted from using additional faked locations, the mechanisms of position dummies suffer from the potential threat of utilizing the geographic maps and background knowledge to identify the faked locations and infer user's accurate location.

Mix zone is another principle which is used to preserve location privacy [15–22]. In this principle, a set of zones are defined, such that the user suspends sending queries to the LBS provider when he/she is located inside them. Entering multiple users to a mix zone means suspending the process of sending the queries to the service provider by all users located inside that mix zone. Once the user leaves the mix zone, he/she is allowed to postpone sending the queries to the service provider. Entering the mix zone at the same time interval leads to mixing the identities of those users. The drawback of using mix zones is the extensive analysis of user's motion patterns and roads network characteristics required to construct effective mix zones [23]. In addition, blocking the queries inside the mix zone limits the service usage and has a negative impact on it.

Many proposed mechanisms use cryptographic tools to preserve location privacy. These mechanisms are proposed for specific types of LBSs. In [24], symmetric key cryptography is used to preserve location privacy for a nearby-friends finding service. In [25], BMobishare uses Bloom Filters to preserve the location privacy for the social location-enabled services. Private Information Retrieval is a cryptographic-based protocol, which is proposed to be used for querying Point of Interest (PoI) database without allowing the database server to identify the targeted data item [26, 27]. The drawback of cryptographic-based schemes is their limitation to support only particular kinds of LBS. Moreover, the high computation cost of these mechanisms limit their usage.

The aim of obfuscation-based LPPMs is to preserve location privacy by reducing the user's location accuracy before sharing it with the service provider [3, 28–34]. Using coordinate system transformations, caching, and generating obfuscation areas are examples of these mechanisms. In [35], a LPPM is proposed which takes into consideration the capabilities and knowledge of the adversary. The user profile is proposed to represent the adversary's knowledge as the probability distribution of the user to access the service from a definite location. A differential privacy based mechanism is proposed for location datasets in [36]. In [37], geo-indistinguishability is proposed to apply the concept of differential privacy for location privacy field. The proposed mechanism adds random noise to the user's location using Planar Laplacian distribution. In [38], an elastic distinguishability metric is introduced to handle the deviation in the density level of different areas and adjust the noise accordingly to provide the same level of privacy everywhere. For continuous queries, obfuscation-based mechanisms are vulnerable to the attack of correlation analysis. The noise-based obfuscation is also vulnerable to the attack of correlation analysis [39]. For area-based obfuscation, analyzing the intersection between the areas can allow an adversary to infer the accurate location of the user [30].

3 Proposed Set of Requirements

Although using the proposed LPPMs in Sect. 2 for snapshot queries is straightforward, applying these LPPMs for continuous queries is more complicated. In short, providing location privacy for a continuous query, which implies updating the service provider continuously about the user's location, introduces a higher level of risk for the user's location privacy, as compared to a snapshot query which implies sharing the user location for one time only. By using a LPPM, the user's reported location represents a modified version of his/her accurate location with reduced accuracy. To get a high location privacy level, the user can reduce the location accuracy significantly. Unfortunately, this is not applicable in all cases as a result of the potential loss of service utility. In reality, the user must choose a suitable reduction level, such that the query result is still useful when the service provider uses the reported location. Based on the service type and the reported location, the adversary can infer the potential area where the user's real location is located within. In the perfect situation, the location privacy is preserved by hiding the accurate location throughout the entire protection area. That means the adversary cannot use the prior knowledge to reduce or shrink this area by eliminating some of its sub-areas. Therefore, the adversary's ability to eliminate a sub-area from the protection area is considered as a privacy leakage. We call this type of leakage as intra-privacy leakage.

In inter-privacy leakage, the adversary reduces the user's protection area by analyzing the consequent protection areas produced by the user's reported locations. This can be achieved by analyzing the current protection area using the previous protection area to eliminate some sub-areas from the current protection area. The assumption is the adversary can estimate the user's traveled distance between the two consequent queries by measuring the elapsed time between the two queries and by estimating the user speed from the current region where the user is located within. Given the user's traveled distance, a sub-area is eliminated when it is unreachable from any of the sub-areas of the previous protection area. The sub-area is considered unreachable when the distance between this sub-area and all the sub-areas of the previous protection area exceeds the user's traveled distance. As a result, the inter-privacy leakage leads to shrinking the protection area and then reducing the location privacy level.

In this work, we introduce a novel set of six requirements, which we believe any LPPM should meet in order to provide location privacy for continuous queries. These requirements are discussed in the following sub-sections.

(I) Independency from Third-Party Servers. Some LPPMs depend on using a third-party server (a trusted server) to preserve the location privacy for mobile users. Although this can be acceptable for snapshot queries, which involve sending the user's accurate location only once to the third-party server, continuous queries require sending the user's accurate location to the third server frequently. If we assume that the third-party server is trusted, then receiving valuable data (the accurate locations of a large number of users) makes this server a target of attacks. In addition, there is no guarantee about the trust level of the third-party server. Therefore, we propose the independency from a third-party server as one of the requirements which should be met by any LPPM that targets the continuous queries, such that any query sent by the user's mobile device contain a protected version of the user's location.

(II) Ability to Provide Location Privacy based on Different Required Utility Levels. LPPMs should provide location privacy with the capability to customize it according to the required utility level of the LBS. Consequently, instead of providing one location privacy level for all LBSs, meeting this requirement ensures that each service provider gets only the required accuracy level of the user's location based on the service's utility requirements and hence reduces any unnecessary location privacy leakage. Although this requirement is essential, many LPPMs lack the ability to meet this requirement. Therefore, this requirement is important for LPPMs which address continuous queries, as the privacy leakage in continuous queries is more than the snapshot queries.

(III) Ability to Process User's Queries in All Cases. Some LPPMs are designed such that they require to block the user's queries in some situations to preserve the location privacy. This means the user must delay sending the query or simply drop the query in these situations. Although this blocking is essential in these LPPMs to preserve the location privacy, we think that any LPPM, which addresses the continuous queries, should be designed to avoid such situations. We think that blocking the user from using the LBS in some situations will affect the service usage negatively. Therefore, for continuous queries, such restriction on the service usage should be avoided.

(IV) Ability to Model and Proact All Sources of Location Privacy Leakage. Although many LPPMs are designed to adapt to one or both of the location privacy leakage sources, the adaptation is not systematic as a result of the lack of a proper privacy leakage metric. Essentially, many location privacy metrics are already proposed, but these metrics are proposed with the aim to evaluate and compare the performance of the different LPPMs. So, without a proper privacy leakage metric designed explicitly to measure the privacy leakage produced by reporting a specific location point, it will be difficult to provide an efficient LPPM for continuous queries. Accordingly, this requirement highlights the need to propose a location privacy leakage metric which has the ability to measure the privacy leakage of any potential reported location. Then, this metric must be integrated to the LPPM to adapt its operations to the measured privacy leakage in order to preserve the user's location privacy.

(V) The Output of the LPPM Should be a Geographic Point, Not an Area. Many LPPMs preserve the user's location privacy, where the output of the process is an area to be used instead of the accurate location of the user. Unfortunately, current LBS providers do not support this type of output, as they are developed to receive the user's location as a geographic point. So, to ensure that any LPPM can be used without any change to the LBS market, it is required that the LPPM output is a geographic point.

(VI) Independency from Any Intensive Computation. Mobile devices are well-known for their limitation in the computation and power capabilities. So, to propose an LPPM to run on such devices, intensive computations must be avoided. Many LPPM are proposed based on intensive computation procedures. Using public-key cryptosystems is an example of such intensive computation. Therefore, one of the requirements of any LPPM to preserve the location privacy of continuous queries which are expected to run for a long time in the user's mobile device is its independency from any intensive computation.

Table 1 shows a list of all LPPMs discussed in Sect. 2 and to what extent each LPPM satisfy the requirements proposed in this work. The table shows that the LPPMs discussed in Sect. 2 lack the support of all the proposed requirements. Accordingly, this emphasizes the need for a new LPPM which meets all requirements.

Table 1. The lack to meet the proposed requirements by current LPPMs.

LPPM		Requirements					
		I	II	III	IV	V	VI
k-anonymity using cloaking regions (Group1):	[4–6, 8]	×	×	✓	×	×	✓
k-anonymity using cloaking regions (Group2):	[7, 9, 10]	×	×	✓	✓	×	×
k-anonymity using position dummies (Group1):	[11–13]	✓	×	✓	×	✓	✓
k-anonymity using position dummies (Group2):	[14]	×	×	✓	×	✓	✓
Mix zones:	[15–23]	×	×	×	×	✓	✓
Cryptographic-based Schemes (Group1):	[24, 25]	×	×	✓	×	✓	✓
Cryptographic-based Schemes (Group2):	[26, 27]	✓	×	✓	×	×	×
Obfuscation (Group1):	[28, 29, 31, 35, 37, 38]	✓	✓	✓	×	✓	✓
Obfuscation (Group2):	[30, 33, 34]	✓	✓	✓	×	×	✓
Obfuscation (Group3):	[32]	✓	×	✓	×	✓	✓

4 Proposed Framework

The previous section shows that each LPPM has its own drawback. Geo-indistinguishability is an obfuscation-based mechanism which meets most of the proposed requirements. Despite its efficiency and simplicity, it lacks the capability to model and proact the potential intra-privacy and inter-privacy leakage.

As a result, we aim to propose an LPPM framework, namely MOPROPLS, which has the capability to model and proact the potential intra-privacy and inter-privacy leakage. MOPROPLS is based on the geo-indistinguishability notion. Thus, MOPROPLS fulfills all the requirements discussed in the previous section.

4.1 The Need for a New Privacy Leakage Metric

In this work, we aim to address the location privacy leakage of continuous queries. To achieve that, we propose a new privacy leakage metric, which captures the intra-privacy leakage and inter-privacy leakage of continuous queries. In Sub-Sect. 4.2, we discuss the preliminaries of the proposed metric, while Sub-Sect. 4.3 introduces our proposed privacy leakage metric.

4.2 Preliminaries

Let $\mathcal{X}$ be a set of user's possible locations, and x^t represents the user's location at time t, $x^t \in \mathcal{X}$. Let $c_{gU}(a)$ denotes the user's candidates generation function for an arbitrary location a, where $a \in \mathcal{X}$. $c_{gAdv}(a)$ represents the adversary's candidates generation function for an arbitrary location a, where $a \in \mathcal{X}$. $c_e(a)$ denotes the candidate evaluation function for an arbitrary location a, where $a \in \mathcal{X}$. $\mathcal{C}^t$ represents a set of all possible candidate locations for x^t, such that $\mathcal{C}^t = c_{gU}(x^t)$ and $c_i^t \in \mathcal{C}^t, i = 1, 2, \ldots, m$. z^t is defined as the user's reported location at time t, such that $z^t = c_s(\mathcal{C}^t)$, $z^t \in \mathcal{C}^t$, where c_s denotes the candidate selection function. d represents the distance function. l denotes the obfuscation level parameter. plm denotes the privacy leakage metric. plm_{intra} is defined as the intra-privacy leakage function, while plm_{inter} is defined as the inter-privacy leakage function. The user profile is defined by $prof_{exist}(a)$ as the user's existence probability at an arbitrary location a, and $prof_{move}(a, b)$ is defined as the user's movement probability from an arbitrary location a to an arbitrary location b. $\dot{\mathcal{C}}_i^t$ represents a set of all possible candidate locations for c_i^t, such that $\dot{\mathcal{C}}_i^t = c_{gAdv}(c_i^t)$ and $\dot{c}_{i,j}^t \in \dot{\mathcal{C}}_i^t, j = 1, 2, \ldots, n$. $\ddot{\mathcal{C}}^{t-1}$ represents a set of all possible candidate locations for z^{t-1}, such that $\ddot{\mathcal{C}}^{t-1} = c_{gAdv}(z^{t-1})$ and $\ddot{c}_k^{t-1} \in \ddot{\mathcal{C}}^{t-1}, k = 1, 2, \ldots, q$.

4.3 Proposed Privacy Leakage Metric

Given a user's location x^t, the privacy leakage metric of a user's candidate location c_i^t, where $c_i^t \in c_{gU}(x^t)$ is defined as:

$$plm(c_i^t) = \frac{1}{2}\left(plm_{intra}(c_i^t) + plm_{inter}(c_i^t)\right) \tag{1}$$

The intra-privacy leakage function, $plm_{intra}(c_i^t)$ is defined as:

$$plm_{intra}(c_i^t) = 1 - \frac{1}{n}\sum_{j=1}^{n} prof_{exist}\left(\dot{c}_{i,j}^t\right) \tag{2}$$

where $\dot{c}_{i,j}^t \in c_{gAdv}(c_i^t)$.

The inter-privacy leakage function, $plm_{inter}(c_i^t)$ is defined as:

$$plm_{inter}(c_i^t) = 1 - \frac{1}{n}\sum_{j=1}^{n} dig(\sum_{k=1}^{q}[1 : d\left(\dot{c}_{i,j}^t, \ddot{c}_k^{t-1}\right) \leq d(x_t, x_{t-1}) \bigvee 0 : d\left(\dot{c}_{i,j}^t, \ddot{c}_k^{t-1}\right) > d(x_t, x_{t-1})]) \tag{3}$$

where $\dot{c}_{i,j}^t \in c_{gAdv}(c_i^t)$, $\ddot{c}_k^{t-1} \in c_{gAdv}(z^{t-1})$ and $dig(v) = [1 : v > 0 \bigvee 0 : v = 0]$

4.4 MOPROPLS Framework

In this work, we propose a framework, namely MOdeling and PROacting to Privacy Leakage Sources (MOPROPLS). Our MOPROPLS framework has the capability to model and interact to all sources of privacy leakage using our proposed location privacy leakage metric. In addition, the framework includes a novel two-phased probabilistic candidate selection algorithm that takes into consideration the correlation between the obfuscated locations in order to preserve privacy for continuous queries. MOPROPLS consists of the following components: obfuscation candidates generation, obfuscation candidates evaluation, and two-phased probabilistic obfuscation candidate selection (Fig. 1).

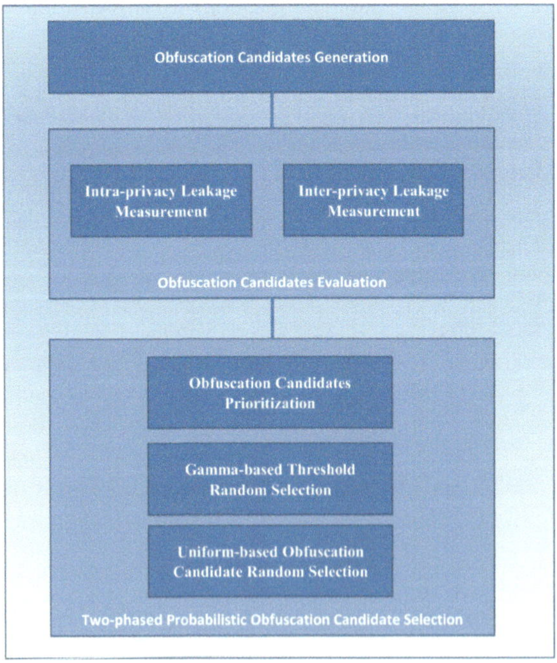

Fig. 1. MOPROPOLS framework

5 Evaluation and Results

5.1 Framework Implementation

We extended the LPM^2 tool [40] to implement our framework as a simulation tool. In addition to the MOPROPLS framework components, the tool has the experiments manager. It consists of dataset loading and parsing manager, user profile manager, and inference attacks manager.

5.2 Dataset

To evaluate the proposed framework, a real-world dataset is used to conduct the experiments. Epfl/mobility is a GPS trajectory dataset of more than 500 taxi cabs in San Francisco, USA [41]. The dataset represents the GPS trajectories of taxi cabs for more than thirty days. Twenty cabs were selected randomly from the dataset, where the trajectories of these users were used in the experiments.

5.3 Inference Attack

The prior knowledge of an adversary about an LBS user varies from no prior knowledge to full prior knowledge. In this paper, we focus on the extreme case, which is represented by an adversary who has full prior knowledge about the user. In this case, an adversary is assumed to have full knowledge about the user existence and transitions in the space, where the knowledge is formulated as a user profile. The user profile contains the probabilities of the user existence in any particular location x, and also the user's transition probability from any location x to any location y. The user profile is constructed from the user's real trajectories. Therefore, the aim of the inference attack is to estimate the user's real location based on the received obfuscated locations and the known user profile.

5.4 Experimental Setup

The experiments were conducted on an Intel i5 (3.10 GHz) machine with 16 GB RAM and Ubuntu 17.04 OS. The simulation area size is 50×88 km^2. Each conducted experiment was repeated ten times and the average of the ten measurements was taken to represent the experiment result. The results are discussed in the following section.

5.5 Experimental Scenarios and Results

To evaluate the performance of our framework, we compare the achieved location privacy level of MOPROPLS against the geo indistinguishability mechanism. Five obfuscation levels were used in the experiments. The strong adversary is used to run the inference attack, where the strong adversary has a full profile for each mobile user. The full profile represents the full knowledge about the user existence and transitions in the space.

Scenario 1. In scenario 1, the weak adversary attack and the strong adversary attack were used to run the inference attack against the geo-indistinguishability mechanism. The weak adversary attack uses partial prior knowledge which is the user's existence probability, while the strong adversary attack uses the full prior knowledge available, which is the user's existence probability and the user's movement probability. The objective of this experiment is to show the effect of using the user's movement probability in reducing the user's location privacy. The experiment shows that the strong adversary exploits the user's movement prior knowledge to reduce the location privacy level.

For obfuscation level 1, the average location privacy achieved against the weak adversary is approximately 0.469 with a 95% confidence interval between 0.45 and 0.487, while the average location privacy achieved against the strong adversary is approximately 0.297 with a 95% confidence interval between 0.277 and 0.316. Therefore, using the adversary to the full prior knowledge reduces the location privacy by approximately 37%.

For obfuscation level 2, the average location privacy achieved against the weak adversary is approximately 0.573 with a 95% confidence interval between 0.553 and 0.594, while the average location privacy achieved against the strong adversary is approximately 0.43 with a 95% confidence interval between 0.401 and 0.459. It is worth noting that the location privacy level is increased as a result of increasing the obfuscation level, which leads to generate more candidates. Using the adversary to the full prior knowledge reduces the location privacy by approximately 25%.

For obfuscation levels 3, 4, and 5, the average location privacy achieved against the weak adversary are approximately 0.618 with a 95% confidence interval between 0.598 and 0.639, 0.645 with a 95% confidence interval between 0.625 and 0.665, and 0.663 with a 95% confidence interval between 0.642 and 0.683 respectively. The average location privacy achieved against the strong adversary for obfuscation levels 3, 4 and 5 are approximately 0.51 with a 95% confidence interval between 0.483 and 0.538, 0.571 with a 95% confidence interval between 0.545 and 0.598, and 0.627 with a 95% confidence interval between 0.605 and 0.649. It is worth noting that the location privacy level is increased as a result of increasing the obfuscation level, which, again, leads to the generation of more candidates. Using the adversary to the full prior knowledge leads reduces the location privacy by approximately 17%, 11%, and 5% for obfuscation levels 3, 4, and 5, respectively.

Scenario 2. In scenario 2, the strong adversary attack was used to run the inference attack against the geo-indistinguishability and MOPROPLS frameworks. The strong adversary attack uses the full prior knowledge available, which is the user's existence probability and the user's movement probability. The objective of this experiment is to evaluate the performance of our proposed framework.

For obfuscation level 1, the experiment shows that MOPROPLS outperforms geo-indistinguishability significantly as a result to modeling and proacting the privacy leakage sources. The average location privacy achieved by the geo-indistinguishability is approximately 0.297 with a 95% confidence interval between 0.277 and 0.316, while the average location privacy achieved by MOPROPLS is approximately 0.577 with a 95% confidence interval between 0.56 and 0.593. Therefore, MOPROPLS increases the location privacy by approximately 94%. For obfuscation level 2, the experiment shows that MOPROPLS provides a better level of location privacy than geo-indistinguishability. The average location privacy achieved by geo-indistinguishability is approximately 0.43 with a 95% confidence interval between 0.401 and 0.459, while the average location privacy achieved by MOPROPLS is approximately 0.615 with a 95% confidence interval between 0.595 and 0.635. It is worth noting that the location privacy level is increased for both as a result of increasing the obfuscation level, which leads to generate more candidates. Therefore, MOPROPLS increases the location privacy level by approximately 43%. For obfuscation levels 3, 4, and 5, the experiment shows that MOPROPLS provides a better level of location privacy than geo-indistinguishability. Increasing the obfuscation level leads to improve location privacy for both; at the same time, the effect of exploiting the measured privacy leakage to improve the location privacy becomes less as a result of increasing the obfuscation level. The average location privacy achieved by geo-indistinguishability for Obfuscation Levels 3, 4 and 5 are approximately 0.51 with a 95% confidence interval between 0.483 and 0.538, 0.571 with a 95% confidence interval between 0.545 and 0.598, and 0.627 with a 95% confidence interval between 0.605 and 0.649, respectively. The average location privacy achieved by MOPROPLS for Obfuscation Levels 3, 4 and 5 are approximately 0.623 with a 95% confidence interval between 0.605 and 0.642, 0.629 with a 95% confidence interval between 0.61 and 0.648, and 0.643 with a 95% confidence interval between 0.624 and 0.662, respectively. It is worth noting that the location privacy level is increased as a result of increasing the obfuscation level, which leads to generate more candidates. Using MOPROPLS leads to increase the location privacy by approximately 22%, 10%, and 2% for Obfuscation Levels 3, 4, and 5, respectively. The average privacy improvement of MOPROPLS for all obfuscation levels is approximately 34% (Figs. 2 and 3).

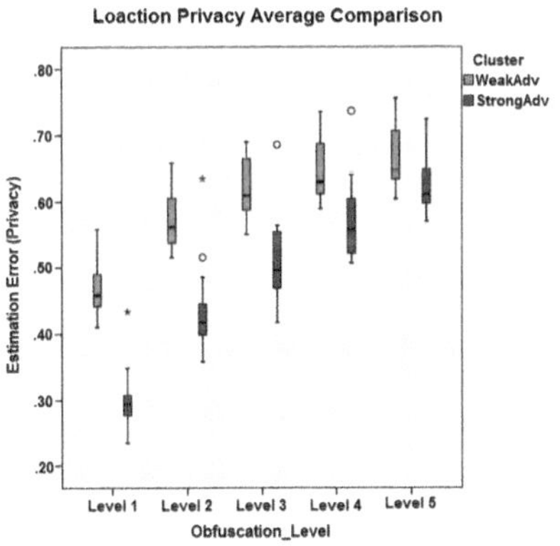

Fig. 2. Comparing the location privacy average using WeakAdv and StrongAdv (all obfuscation levels).

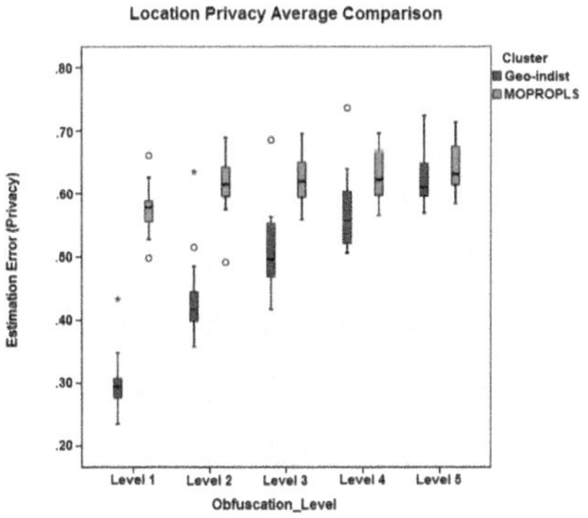

Fig. 3. Location privacy average comparison.

6 Conclusion

In this paper, we proposed the MOPROPLS framework, with the aim to preserve location privacy in the case of continuous queries. The performance for the framework was compared with geo-indistinguishability LPPM in terms of privacy (adversary estimation error), and the reported improvements average was 34%. For future work, we plan to extend our proposed framework to fit the requirements of preserving location privacy of smart city applications.

References

1. Primault, V., Boutet, A., Mokhtar, S.B., Brunie, L.: The long road to computational location privacy: a survey. IEEE Commun. Surv. Tutorials. 1 (2018). https://doi.org/10.1109/COMST.2018.2873950
2. Gupta, R., Rao, U.P.: An exploration to location based service and its privacy preserving techniques: a survey. Wirel. Pers. Commun. **96**, 1973–2007 (2017). https://doi.org/10.1007/s11277-017-4284-2
3. Talat, H., Nomani, T., Mohsin, M., Sattar, S.: A survey on location privacy techniques deployed in vehicular networks. In: 2019 16th International Bhurban Conference on Applied Sciences and Technology (IBCAST), pp. 604–613. IEEE (2019). https://doi.org/10.1109/IBCAST.2019.8667248
4. Gruteser, M., Grunwald, D.: Anonymous usage of location-based services through spatial and temporal cloaking. In: Proceedings of the 1st International Conference on Mobile Systems, Applications and Services - MobiSys 2003, pp. 31–42. ACM Press, New York (2003). https://doi.org/10.1145/1066116.1189037
5. Mokbel, M.F., Chow, C.-Y., Aref, W.G.: The new Casper: query processing for location services without compromising privacy (2006). https://dl.acm.org/citation.cfm?id=1164193
6. Gedik, B., Liu, L.: Protecting location privacy with personalized k-anonymity: architecture and algorithms. IEEE Trans. Mob. Comput. **7**, 1–18 (2008). https://doi.org/10.1109/TMC.2007.1062
7. Nguyen, N., Han, S., Shin, M.: URALP: unreachable region aware location privacy against maximum movement boundary attack. Int. J. Distrib. Sens. Netw. **11**, 246216 (2015). https://doi.org/10.1155/2015/246216
8. Lu, Q., Wang, C., Xiong, Y., Xia, H., Huang, W., Gong, X.: Personalized privacy-preserving trajectory data publishing. Chin. J. Electron. **26**, 285–291 (2017)
9. Ma, C., Zhou, C., Yang, S.: A voronoi-based location privacy-preserving method for continuous query in LBS. Int. J. Distrib. Sens. Netw. **11**, 326953 (2015). https://doi.org/10.1155/2015/326953
10. Wang, Y., Xia, Y., Hou, J., Gao, S., Nie, X., Wang, Q.: A fast privacy-preserving framework for continuous location-based queries in road networks. J. Netw. Comput. Appl. **53**, 57–73 (2015). https://doi.org/10.1016/J.JNCA.2015.01.004
11. Kido, H., Yanagisawa, Y., Satoh, T.: An anonymous communication technique using dummies for location-based services. In: Proceedings of International Conference on Pervasive Services, ICPS 2005, pp. 88–97. IEEE (2005). https://doi.org/10.1109/PERSER.2005.1506394

12. Shankar, P., Ganapathy, V., Iftode, L.: Privately querying location-based services with SybilQuery. In: Proceedings of the 11th international conference on Ubiquitous computing - Ubicomp 2009, p. 31. ACM Press, New York (2009). https://doi.org/10.1145/1620545.1620550
13. Gao, S., Ma, J., Shi, W., Zhan, G.: LTPPM: a location and trajectory privacy protection mechanism in participatory sensing. Wirel. Commun. Mob. Comput. **15**, 155–169 (2015). https://doi.org/10.1002/wcm.2324
14. Wei, W., Xu, F., Li, Q.: MobiShare: flexible privacy-preserving location sharing in mobile online social networks. In: 2012 Proceedings IEEE INFOCOM, pp. 2616–2620. IEEE (2012). https://doi.org/10.1109/INFCOM.2012.6195664
15. Beresford, A.R., Stajano, F.: Location privacy in pervasive computing. IEEE Pervasive Comput. **2**, 46–55 (2003). https://doi.org/10.1109/MPRV.2003.1186725
16. Palanisamy, B., Liu, L.: MobiMix: protecting location privacy with mix-zones over road networks. In: 2011 IEEE 27th International Conference on Data Engineering, pp. 494–505. IEEE (2011). https://doi.org/10.1109/ICDE.2011.5767898
17. Eckhoff, D., German, R., Sommer, C., Dressler, F., Gansen, T.: SlotSwap: strong and affordable location privacy in intelligent transportation systems. IEEE Commun. Mag. **49**, 126–133 (2011). https://doi.org/10.1109/MCOM.2011.6069719
18. Chen, Y.-S., Lo, T.-T., Lee, C.-H., Pang, A.-C.: Efficient pseudonym changing schemes for location privacy protection in VANETs. In: 2013 International Conference on Connected Vehicles and Expo (ICCVE), pp. 937–938. IEEE (2013). https://doi.org/10.1109/ICCVE.2013.6799933
19. Domenic, M.K., Wang, Y., Zhang, F., Memon, I., Gustav, Y.H.: Preserving users' privacy for continuous query services in road networks. In: 2013 6th International Conference on Information Management, Innovation Management and Industrial Engineering, pp. 352–355. IEEE (2013). https://doi.org/10.1109/ICIII.2013.6702947
20. Sun, Y., Zhang, B., Zhao, B., Su, X., Su, J.: Mix-zones optimal deployment for protecting location privacy in VANET. Peer-to-Peer Netw. Appl. **8**, 1108–1121 (2015). https://doi.org/10.1007/s12083-014-0269-z
21. Boualouache, A., Moussaoui, S.: Urban pseudonym changing strategy for location privacy in VANETs. Int. J. Ad Hoc Ubiquitous Comput. **24**, 49 (2017). https://doi.org/10.1504/IJAHUC.2017.080914
22. Ullah, I., Wahid, A., Shah, M.A., Waheed, A.: VBPC: Velocity based pseudonym changing strategy to protect location privacy of vehicles in VANET. In: 2017 International Conference on Communication Technologies (ComTech), pp. 132–137. IEEE (2017). https://doi.org/10.1109/COMTECH.2017.8065762
23. Palanisamy, B., Liu, L.: Attack-resilient mix-zones over road networks: architecture and algorithms. IEEE Trans. Mob. Comput. **14**, 495–508 (2015). https://doi.org/10.1109/TMC.2014.2321747
24. Mascetti, S., Freni, D., Bettini, C., Wang, X.S., Jajodia, S.: Privacy in geo-social networks: proximity notification with untrusted service providers and curious buddies. VLDB J. **20**, 541–566 (2011). https://doi.org/10.1007/s00778-010-0213-7
25. Shen, N., Yang, J., Yuan, K., Fu, C., Jia, C.: An efficient and privacy-preserving location sharing mechanism. Comput. Stand. Interfaces **44**, 102–109 (2016). https://doi.org/10.1016/J.CSI.2015.06.001
26. Ghinita, G., Kalnis, P., Khoshgozaran, A., Shahabi, C., Tan, K.-L.: Private queries in location based services: anonymizers are not necessary. In: Proceedings of the 2008 ACM SIGMOD International Conference on Management of Data - SIGMOD 2008, p. 121. ACM Press, New York (2008). https://doi.org/10.1145/1376616.1376631

27. Paulet, R., Kaosar, M.G., Yi, X., Bertino, E.: Privacy-preserving and content-protecting location based queries. IEEE Trans. Knowl. Data Eng. **26**, 1200–1210 (2014). https://doi.org/10.1109/TKDE.2013.87
28. Gutscher, A.: Coordinate transformation - a solution for the privacy problem of location based services? In: Proceedings 20th IEEE International Parallel & Distributed Processing Symposium, p. 7. IEEE (2006). https://doi.org/10.1109/IPDPS.2006.1639681
29. Yiu, M.L., Jensen, C.S., Møller, J., Lu, H.: Design and analysis of a ranking approach to private location-based services. ACM Trans. Database Syst. **36**, 1–42 (2011). https://doi.org/10.1145/1966385.1966388
30. Perazzo, P., Dini, G.: A uniformity-based approach to location privacy. Comput. Commun. **64**, 21–32 (2015). https://doi.org/10.1016/J.COMCOM.2015.02.014
31. Kachore, V.A., Lakshmi, J., Nandy, S.K.: Location obfuscation for location data privacy. In: 2015 IEEE World Congress on Services, pp. 213–220. IEEE (2015). https://doi.org/10.1109/SERVICES.2015.39
32. Zurbaran, M.A., Avila, K., Wightman, P., Fernandez, M.: Near-rand: noise-based location obfuscation based on random neighboring points. IEEE Lat. Am. Trans. **13**, 3661–3667 (2015). https://doi.org/10.1109/TLA.2015.7387946
33. Shahid, A.R., Jeukeng, L., Zeng, W., Pissinou, N., Iyengar, S.S., Sahni, S., Varela-Conover, M.: PPVC: privacy preserving voronoi cell for location-based services. In: 2017 International Conference on Computing, Networking and Communications (ICNC), pp. 351–355. IEEE (2017). https://doi.org/10.1109/ICCNC.2017.7876153
34. Schlegel, R., Chow, C.-Y., Huang, Q., Wong, D.S.: User-defined privacy grid system for continuous location-based services. IEEE Trans. Mob. Comput. **14**, 2158–2172 (2015). https://doi.org/10.1109/TMC.2015.2388488
35. Shokri, R., Theodorakopoulos, G., Troncoso, C., Hubaux, J.-P., Le Boudec, J.-Y.: Protecting location privacy: optimal strategy against localization attacks. In: Proceedings of the 2012 ACM Conference on Computer and Communications Security - CCS 2012, p. 617. ACM Press, New York (2012). https://doi.org/10.1145/2382196.2382261
36. Xiong, P., Zhu, T., Niu, W., Li, G.: A differentially private algorithm for location data release. Knowl. Inf. Syst. **47**, 647–669 (2016). https://doi.org/10.1007/s10115-015-0856-1
37. Andrés, M.E., Bordenabe, N.E., Chatzikokolakis, K., Palamidessi, C., Andrés, M.E., Bordenabe, N.E., Chatzikokolakis, K., Palamidessi, C.: Geo-indistinguishability: differential privacy for location-based systems. In: Proceedings of the 2013 ACM SIGSAC Conference on Computer & Communications Security - CCS 2013, pp. 901–914. ACM Press, New York (2013). https://doi.org/10.1145/2508859.2516735
38. Chatzikokolakis, K., Palamidessi, C., Stronati, M.: Constructing elastic distinguishability metrics for location privacy. Proc. Priv. Enhancing Technol. **2015**, 156–170 (2015). https://doi.org/10.1515/popets-2015-0023
39. Al-Dhubhani, R., Cazalas, J.M.: An adaptive geo-indistinguishability mechanism for continuous LBS queries. Wirel. Netw. 1–19 (2017). https://doi.org/10.1007/s11276-017-1534-x
40. Shokri, R., Theodorakopoulos, G., Le Boudec, J.-Y., Hubaux, J.-P.: Quantifying location privacy. In: 2011 IEEE Symposium on Security and Privacy, pp. 247–262. IEEE (2011). https://doi.org/10.1109/SP.2011.18
41. Piorkowski, M., Sarafijanovic-Djukic, N., Grossglauser, M.: A parsimonious model of mobile partitioned networks with clustering. In: 2009 First International Communication Systems and Networks and Workshops, pp. 1–10. IEEE (2009). https://doi.org/10.1109/COMSNETS.2009.4808865

Deliberate Exponential Chaotic Encryption Map

Aladdein M. S. Amro(✉)

Computer Engineering Department, Taibah University, Medina, Kingdom of Saudi Arabia
aamro@taibahu.edu.sa

Abstract. In the given work a new chaotic map is presented, referred to as deliberate exponential chaotic encryption map (DECM), the proposed map characteristics meets the recent demand of simple encryption algorithms with strong security. The methodology of iterative examination and correction was used to reach the optimal characteristics of the map. The chaotic behavior of the proposed map is examined, and its characteristics are proved experimentally, the achieved results shows advantages of the proposed map over the well-known chaotic maps. This gives the proposed map a potential to be implemented in real-time encryption algorithms for different kinds of information

Keywords: Chaotic map · S-unimodality · Chaotic behavior

1 Introduction

Different chaotic maps have been proposed recently in literature to meet the requirements of real-time data encryption [1–7]. The most important requirements are speed, computational complexity, high security, and computational power. The interest to chaotic based encryption is increasing for IoT devices because of their limited computational and power recourses. The chaotic based encryption is considered a suitable approach as it meets the mentioned requirements in addition to other properties like the initial condition's sensitivity and topological transitivity [4] achieving an encryption map that meets the mentioned requirements while avoiding the deficiencies in the existing maps is the purpose of this work, specially increasing the range of chaotic behavior was targeted, while the majority of well-known maps have small range of chaotic behavior which leads to weakening of the encryption algorithms [8].

This work proposes a novel chaotic map, named as deliberate exponential chaotic encryption map (DECM), the characteristics of the map are tested using specially for this purpose written MATLAB programs. The characteristics was tested repeatedly, and the map was corrected respectively until the optimal characteristics of chaotic behavior was achieved.

The paper consists of four sections: in Sect. 2 deliberate exponential chaotic-based map (DECM) is demonstrated while Sect. 3 inspects characteristics of DECM. Finally, conclusion and future work are presented in Sect. 4.

F. Saeed et al. (Eds.): IRICT 2019, AISC 1073, pp. 833–838, 2020.
https://doi.org/10.1007/978-3-030-33582-3_78

2 Deliberate Exponential Chaotic Map

Equation 1 shows the proposed Exponential Chaotic Map.

$$X_{n+1} = \begin{cases} \lambda x_n\left(\exp\left(\frac{\pi}{30} x_n\right)\right) & ,0 \leq x_n \leq 0.5 \\ \lambda(1 - x_n)\left(\exp\left(\frac{\pi}{30}(1 - x_n)\right)\right), & 0.5 < x_n \leq 1 \end{cases} \quad (1)$$

where $x_{n+1} \in [0, 1]$, both the initial condition x_0 and the control parameter λ can be used as part of the secret key for chaotic encryption.

3 DECM Characteristics Analysis

The aim of this section is to prove the characteristics of Deliberate Exponential Chaotic Map such as S-unimodality and chaotic behavior.

The iteration function of Exponential Chaotic Map is shown in Fig. 1. Figure 1 shows that the iteration function has a single maximum value, it starts from zero, gradually increases until it reaches its maximum value and then decrease back to zero. The iteration function shown in Fig. 1 proves the unimodality property of DECM at the selected value of λ. The Bifurcation diagram was built for DECM to find the range of control parameter λ for which DECM meets the unimodality property [13]. Figure 2 demonstrates the Bifurcation diagram of Exponential Chaotic Map. As Fig. 2 shows, DECM meets the unimodality property for $\lambda \in [1.7, 1.92]$.

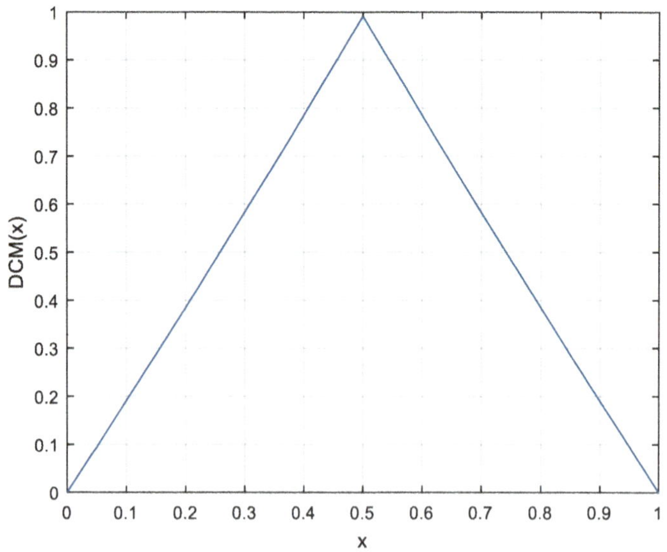

Fig. 1. The iteration function of DECM for $\lambda = 1.87$

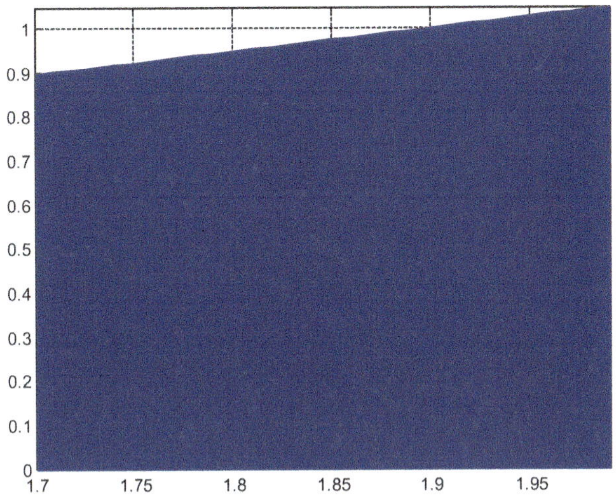

Fig. 2. The Bifurcation for DECM for $\lambda \in [1.7, 1.92]$.

The Schwarzian derivative was applied to examine the chaotic behavior of DECM using Eq. 2, [11]. Figure 3 demonstrates the results of the conducted computational experiment, all values of Schwarzian derivative for $\lambda = 1.37$ and are negative. these results show that the proposed map has robust chaotic behavior. In summary, the achieved computational results shows that the unimodality that the proposed DECM meets the S-unimodality characteristic at the selected value of secret key parameters.

$$S_{f(x)} \equiv \frac{f'''(x)}{f'(x)} - 1.5\left(\frac{f''(x)}{f'(x)}\right)^2 \tag{2}$$

Sensitivity to a small change of the initial condition was also tested.

Figure 4 compares two sequences generated using Eq. 1. The only difference between the two sequences is the initial condition. As the figure shows, the two sequences diverse after a few iterations which indicates the high sensitivity of Exponential Chaotic Map to small changes in the initial condition.

Another way to examine the chaotic behavior is the Lyapunov Exponent. It was calculated for the proposed map using Eq. 3, [10]. Equation 4 shows the derivative of Eq. 1. The achieved results are presented in Fig. 5 in the form of a graphed Lyapunov Exponent. The graph shows that the chaotic behavior is obtained when the Lyapunov Exponent is within the range of $[0, 0.82]$. where DECM provides chaotic behavior for $\lambda \in [1.0, 1.926]$.

The obtained results of Lyapunov exponent and Bifurcation diagram proves that Exponential Chaotic Map meets the S-unimodality, the achieved results confirm that the proposed chaotic map can be used to produce perfect random sequence for encryption proposes.

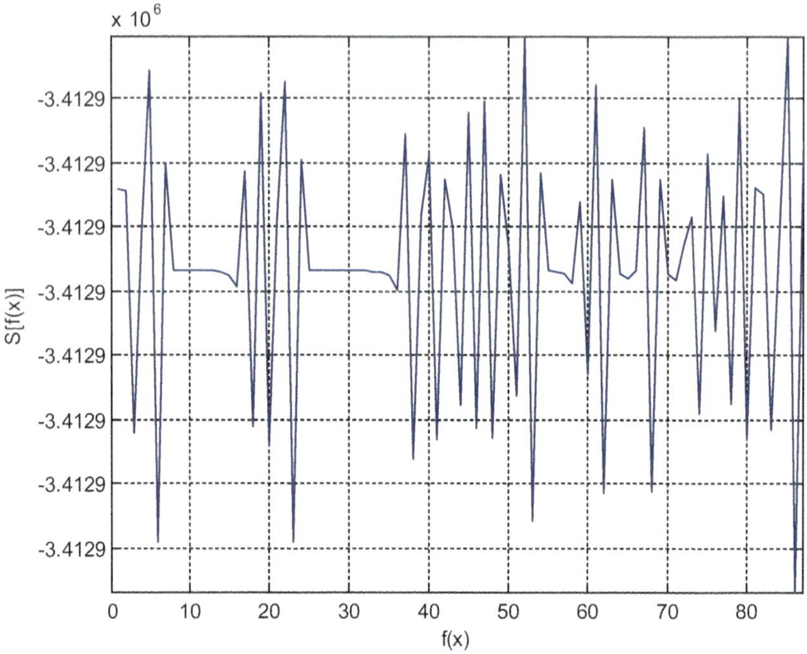

Fig. 3. The Schwarzian derivative of DECM for $\lambda = 1.37$

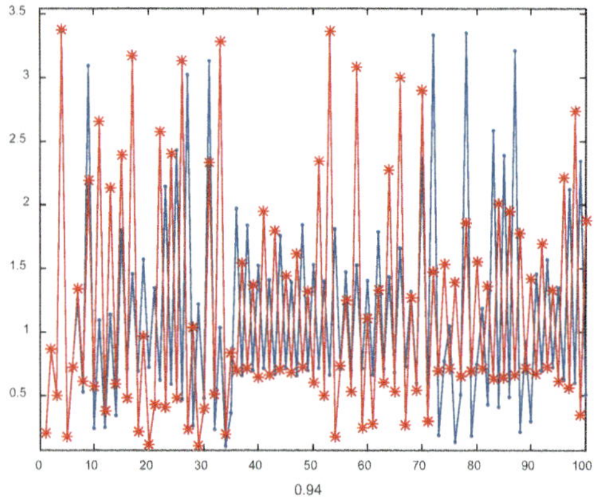

Fig. 4. Two sequences demonstrated are one for $(x_0, \lambda) = (0.94, 1.37)$, marked with dots·, and for $(x_0, \lambda) = (0.94001, 1.37)$, marked with $*$.

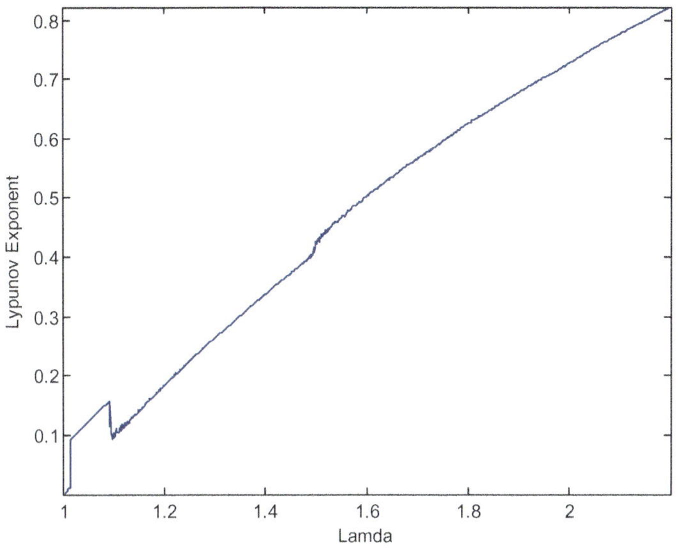

Fig. 5. The Lyapunov exponent of the DECM for $\lambda \in [1, 1.926]$.

Table 1. chaotic behavior comparison

Chaotic map	Tent map	Logistic map	Proposed exponential map (DECM)
Range of chaotic behavior	$\lambda \in [1.999, 2)$	$\lambda \in [3.96, 4]$	$\lambda \in [1.87, 1.92]$

In addition, DECM provides wide range of chaotic behavior when compared with well-known maps like Tent and Logistic Maps [12–14]. Table 1. Demonstrates the comparison of the ranges of chaotic behavior. This can be considered the main advantage of the proposed map.

$$\lambda_{LE}(x_0) = \lim_{N \to \infty} \frac{1}{N} \sum_{n=0}^{N-1} \ln|f'(x_n, \lambda)| \tag{3}$$

$$f'(x_n, \lambda) =$$

$$\begin{cases} \lambda\left(exp\left(\frac{\pi}{30}x_n\right)\right) + \frac{\pi}{30}\lambda x_n\left(exp\left(\frac{\pi}{30}x_n\right)\right) + \frac{\pi}{30}exp\left(\frac{\pi}{30}x_n\right), & 0 \le x_n < 0.5 \\ -\lambda exp\left(\frac{\pi}{30}(1 - x_n\right) - \frac{\pi}{30}\lambda(1 - x_n)exp\left(\frac{\pi}{30}(1 - x_n\right) - \frac{\pi}{30}exp\left(\frac{\pi}{30}(1 - x_n)\right), & 0.5 < x_n \le 1 \end{cases} \tag{4}$$

4 Conclusion and Future Work

A Deliberate Exponential Chaotic Map is developed. Comparing with well-known chaotic maps the proposed DECM avoids one of the most popular deficiencies of the well-known chaotic maps while it provides wide range chaotic behavior, the S-unimodality property of the DECM was proved, the high sensitivity to small changes in initial condition was demonstrated. In the future works the proposed map will be implemented for encryption of various types of information.

References

1. Peng, Y., Sun, K., Peng, D., Ai, W.: Dynamics of a higher dimensional fractional-order chaotic map. Phys. A **525**, 96–107 (2019)
2. Zhang, T., Yicong Zhou, C.L., Chen, P.: A new combined chaotic system for image encryption. In: 2012 International Conference on System Science and Engineering (ICSSE) (2012)
3. Kocarev, L.: Chaos-based cryptography a brief overview. In: Proceedings of the 5th WSEAS International Conference on Nonlinear Analysis, Non-Linear Systems and Chaos 2011, pp. 6–21 (2011)
4. Abandal, Y., Tiedeu, A.: Image encryption by chaos mixing. IET Image Process. **10**(10), 742–750 (2016)
5. Shyamsunder, S., Kaliyaperumal, G.: Image encryption and decryption using chaotic maps and modular arithmetic. Am. J. Sign. Process. **1**(1), 24–33 (2011)
6. Ye, R., Xi, Y., Ma, Y.: A chaotic image encryption scheme using swapping based confusion approach: a survey. In: First IEEE International Conference on Computer Communication and the Internet, pp. 374–377 (2016)
7. Hossain, B., Rahman, T., Rahman, S., Islam, S.: A new approach of image encryption using 3D chaotic map to enhance security of multimedia component. In: International Conference on Informatics, Electronics & Vision (ICIEV), pp. 1–6 (2014)
8. Hua, Z., Zhou, B., Zhou, Y.: Sine chaotification model for enhancing chaos and its hardware implementation. IEEE Trans. Ind. Electron. **66**(2), 1273–1284 (2019)
9. Kuo, C.L., Huang, L.C., Wang, S.J., Lin, J.S., Wang, C.C., Yan, J.J.: Image encryption based on fuzzy synchronization of chaos systems. In: IEEE 37th Annual Computer Software and Applications Conference (COMPSAC), pp. 153–154 (2013)
10. Glendinning, P.: Stability. Instability and Chaos. Cambridge University Press, Cambridge (1994)
11. Aguirregabiria, J.M.: Robust chaos with variable lyapunov exponent in smooth one-dimensional maps. Chaos, Solitons Fractals **42**(4), 2531–2539 (2009)
12. Dachselt, F., Schwarz, W.: Chaos and cryptography. IEEE Trans. Circ. Syst. I: Fundam. Theor. Appl. **48**, 1498–1501 (2001)
13. Shastry, M., Nagaraj, N., Vaidya, P.: The B-exponential map: a generalization of the logistic map, and its applications in generating pseudo-random numbers. Comput. Res. Repository, vol. 7. arXiv preprint cs/0607069 (2006)
14. Gao, H., Zhang, Y., Liang, S., Li, D.: A new chaotic algorithm for image encryption. Chaos, Solitons Fractals **29**, 393–399 (2006)

Using Hyperledger Fabric Blockchain to Maintain the Integrity of Digital Evidence in a Containerised Cloud Ecosystem

Kenny Awuson-David(✉), Tawfik Al-Hadhrami(✉), Olajide Funminiyi(✉), and Ahmad Lotfi(✉)

School of Science and Technology, Nottingham Trent University, Nottingham NG118 NS, UK
{kenny.awuson-david, Tawfik.al-hadhrami, funminiyi.olajide, ahmad.lotfi}@ntu.ac.uk

Abstract. As Cloud technology continues to evolve with a vast amount of data being transmitted daily, it has added another form of complexity in a forensic investigation. In order to mitigate the complexity of evidence acquisition in the Cloud, we need a method that can forensically maintain digital accounting and trustworthiness with data confidentiality and integrity in the Cloud environment. The current systems have severe security vulnerabilities as it does not provide data confidentiality, integrity and accountability of users and system logs. Due to the complexity in the Cloud ecosystem, acquired evidence can easily be contaminated and cannot be admissible in a court of law. The geolocation, multi-tenancy and political issues have added to the complexity in forensic investigations. Blockchain will facilitate trust between the Cloud Service Providers (CSP) and their customers by heightening the trustworthiness of data and its integrity both in transit and storage. In this paper, we are proposing an innovative permissioned Hyperledger Fabric Blockchain Containerised Cloud ecosystem that will secure and preserve the integrity of digital evidence both in transit or storage. Then, design and implement a Dockerised private Blockchain Cloud ecosystem architecture that will mitigate the challenges faced by forensic investigators in the Cloud ecosystem by preserving evidence integrity in a multi-tenancy and private Cloud environment.

Keywords: Blockchain · Cloud forensics · Digital forensic · Digital evidence · Docker container · Hyperledger Fabric

1 Introduction

Organisations today are deploying Cloud network due to its affordability and straightforwardness in implementing a system in such a way that has never been seen before, attracts cybercriminals [1]. According to the Cloud Security Alliance (CSA) [2] the impacts of cyber-attacks on the current Cloud ecosystem and how forensic investigations of those incidents are complex and challenging to investigate. Some of the evidence can span from multiple geolocations. This paper has implemented an architecture called Blockchain Cloud Forensics using the Hyperledger Fabric

F. Saeed et al. (Eds.): IRICT 2019, AISC 1073, pp. 839–848, 2020.
https://doi.org/10.1007/978-3-030-33582-3_79

Blockchain [3]. The focus of this paper is the novelty and contribution of using tamper-proof Blockchain evidence acquisition capability in a Cloud ecosystem. Furthermore, this service is a hybrid platform that ensures Cloud forensic evidence is secure, trustworthy and immutable. Due to the distribution ledger capability of Hyperledger Fabric, all participating organisations are able to view the secured transaction data during an incident. This will ensure the transparency and immutability of the chain of custody during the forensic acquisition process [4].

The importance of maintaining the integrity of acquired digital evidence in the Cloud ecosystem, both in transit or storage has been overlooked by previous studies. Although there are many automated analysis tools that forensic examiners can use for examination purposes, they are not sufficient as it is not a foolproof method for discriminating fake traffic produced by cybercriminal from legitimate traffic. Forensic examiners need to perform acquisitions in the cloud ecosystem, which identify the adversary and technique used to compromise the network. The remaining part of this paper is organised in sections: The first section takes a look at the current research and state of the art in the Blockchain Cloud ecosystem. The second section provides an overview of Hyperledger Fabric Blockchain components, the proposed architecture, implementation and evaluation of the framework. The third section provides an overview of Cloud computing and its services and deployment models, and an overview of digital forensics was introduced. Finally, the last section will include, the summary and conclusion of the paper.

2 Related Work

According to ENISA (2011), described "multi-tenant outsourced services typically cannot give access to the raw log data as it contains records of multiple users and thus would compromise the privacy of other customers [5] Consequently, structures identical with cloud ecosystem services such as multi-tenancy, data security, file encryption, and communications encryption also need to be addressed as part of a digital forensics investigation". Several applications require robust and tamper-proof evidence acquisition systems, e.g. Electronic voting or bank information systems [6]. A vital element of cloud computing is its "multi-tenancy" capability, which is described as a "shared pool of resources" in the NIST definition [7]. In the presence of the untrusted host, some threats could compromise data security. An adversary or insider can compromise a Cloud network using payload such as man-in-the-middle or distribution denial service attack (DDoS). However, the proposed mechanisms will be able to mitigate this attack by dropping such unknown packets due to the Blockchain consensus mechanism. The landscape of cloud computing not only presents doubt over where data is situated, but questions such as confidentiality and regulatory compliance are still a concern. For example, the European Data Protection Directive inferences on how a security breach should be handled. As a result, Cloud ecosystem, landscape, traditional digital forensic techniques may not be possible to capture evidence or other data, [8], According to [9, 10] who emphasised that the decentralised ledger (blockchain technology) provides auditable transaction logs, making legal disputes less likely and simpler to settle. In an increased cloud-oriented society, the ability to identify, obtain, preserve, and analyse potential digital evidence is valuable for business continuity and security. Whether reacting to a cyber intrusion, data breach, or in support of litigation, the ill-prepared establishment will find itself at a severe (and possibly expensive) disadvantage. (CSA 2013). Distributed consensus, data consistency, and immutability of processed transactions will solve the challenges of Cloud evidence admissibility. Furthermore, these features can make it nearly impossible to alter learning records on the network [16].

This paper described a framework based on a distributed Blockchain Cloud architecture, technology, which offers low-cost, secure with on-demand access to the utmost inexpensive computing IoT network. This was achieved by creating a distributed Cloud ecosystem, with cost efficiency and unlimited performance computing [11]. [12] discussed the importance and need for Blockchain and the capability of this technology to change the way we secure digital assets and present the current vulnerabilities in the Blockchain Cloud ecosystem. The paper goes on to propose a model block withholding (BWH) attack in a Blockchain Cloud ecosystem, considering distinct pool reward mechanisms. ProvChain mechanism, a framework that has the capability to collect, validate Cloud data origin, through embedding the source data into a Blockchain transaction. The ProvChain mechanism operates in three stages: Provenance data collection, Provenance data storage and Provenance data validation [13].

3 Background

Cloud Forensics consists of a hybrid forensic approach towards the generation of digital evidence [14]. It involves cross-domain analysis of virtual machines, physical machines and network and system interactions. Organisationally, it involves interactions among Cloud actors (i.e., Cloud service provider, Cloud consumer, Cloud broker, Cloud carrier and Cloud auditor) to facilitate both internal and external investigations. Legally, it often implies multi-jurisdictional and multi-tenant situations. Cloud forensics is regarded as being qualitatively different from traditional computer forensics because the Cloud encompasses multiple administrative domains, extensive data duplication, multi-tenancy and often operates across various jurisdictions [15]. Traditional approaches to evidence collection and recovery are not practical for Cloud as they rely on unrestricted access to the system and user data.

This paper will take a secure digital evidence-centric viewpoint and propose a new Cloud Hyperledger Fabric Blockchain Cloud mechanism. The proposed mechanism will acquire digital evidence in real-time streaming that can preserve accountability, transparency and trust in the cloud environment, designed on a Blockchain architecture. The proposed system will ensure the confidentiality, integrity and availability of evidence while in transit and storage.

4 Blockchain

In 2009, a whitepaper called Bitcoin: A Peer-to-Peer Electronic Cash System was published by Satoshi Nakamoto to resolve the current challenges faced in the monetary market. His main aim of releasing the whitepaper is to develop a technology that can allow electronic transaction from one party to another without going through financial institutions. One of the significant challenges addressed was the double method, which is used to avoid the double-spending (a unique problem with digital currency is the risk of reproducing the same amount, even after spending it) of Bitcoin. However, Bitcoin is a digital currency that has made it easier to solve the problem of double-spending, and technology that facilitate that is the Blockchain [16].

4.1 Hyberledger Fabric

Hyperledger Fabric is a permissioned Blockchain and was designed around some essential elements and use cases that were seen as vital for organisational users. However, at the centre of the design is the ledger, the ledger holds sets of transaction blocks. A transaction can be defined as the fundamental mechanisms that update the current state of the Blockchain. In turn, the transaction mechanism is facilitated by smart contract program codes installed on the Blockchain known as Chaincode [17]. It is vital to understand how the blocks and transactions are formed. In the formation, each block is arranged in a sequence and made up of established transactions. Then, the transactions are stored in a precise, controlled series. In contrast, it is different in permissionless Blockchain as the creation of the transaction and the sequence is given and not primarily done at the same time or on the same computer. This is because the ordering of transactions and the execution of transactions are separated. In Hyperledger Fabric, computers are being used to operate the blockchain can run in three different modes (node types); these are as follows:

Client: A client acts on behalf of users of the Blockchain and submits actions and events to the network as part of an application.

Peer: Peers process the incoming transactions for validating and handling updating state changes as a result of transactions and chaincode execution. Once they have executed a transaction, they broadcast the result to the network so that an orderer can handle the transaction.

Orderer: While peer nodes execute the transactions, the ordering service nodes decide the final order of events, and thus determine the final set of events that will be written on to the next block of the Blockchain. It is significant to understand that a single computer can support all three of these node types on a Fabric Blockchain, but this is not necessary.

4.2 The Proposed Hyberledger Fabric Cloud Ecosystem Framework

Hyperledger Composer makes Fabric deployment easier. For example, the proposed Hyperledger Cloud architecture is deployed using the participants, assets, transactions, and events model to describe our framework to customise the model to produce Hyperledger Fabric smart contracts, channel and ledgers. Then, the Hyperledger Composer model was used to generate a set of domain-specific APIs to access the transaction logs. Next, present a holistic view of the Blockchain Cloud network. The model will help the forensic investigator to acquire admissible evidence from a Cloud ecosystem. According to [18], implementing a Blockchain network in an organisation requires a balancing act. The Fabric Membership service integrates the transaction logs, Blockchain Cloud users, the peers, including the chaincode and also maintains the channel. The novelty of the paper is the transparent integration of the admissible evidence capability in a Cloud ecosystem. This model architecture will ensure that evidence is acquired in the Cloud ecosystem with minimum difficulty. Finally, the REST-Server and Playground will be used by Blockchain Cloud customers to access the Hyperledger Fabric Containers (Fig. 1).

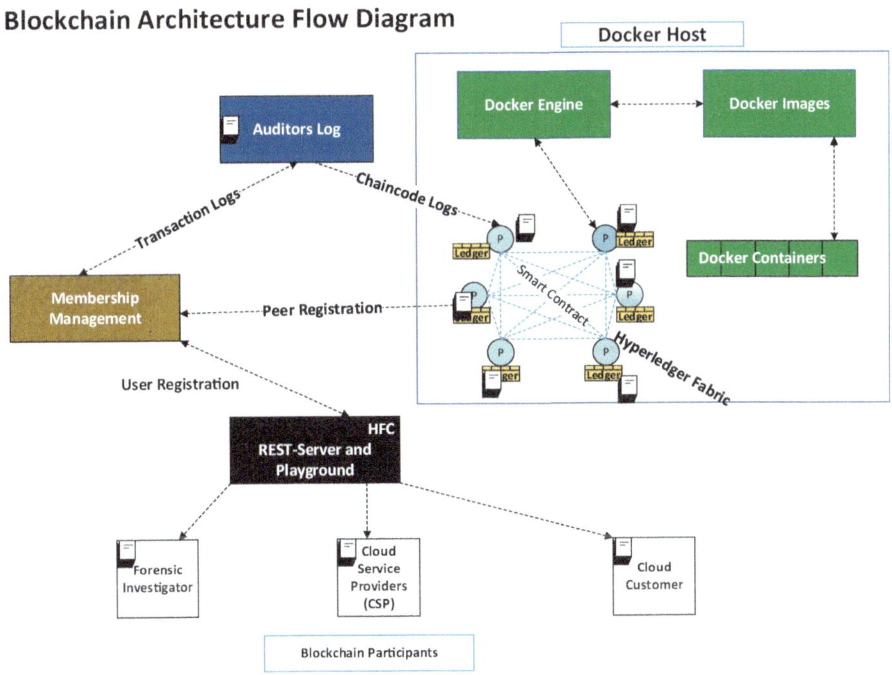

Fig. 1. The transaction model architecture of our blockchain cloud ecosystem

5 Docker and Hyperledger Fabric Stacks

The integration of the Docker containerisation system by big technology companies such as Microsoft, Apple, IBM, Google and Linux Red Hat has paved the way for a new form of network virtualisation. Research has shown that these companies have spent time and resource to make Docker containerisation more secure and resilient to today's dynamic cyber landscape. In March 2013, Docker virtualisation under the Apache 2.0 license was released as a dotCloudplatform. The Docker engine offers a RESTful interface to the outside world that can be accessed by any tool, such as the Docker CLI, Docker for Mac, and Docker for Windows or Kubernetes to name a few [19].

The Docker containerisation technology has a secure and resilient speed capability which compares favourably to the current VM technologies, most importantly, the area of bandwidth usage and speed. Figure 2 in the next section highlights some differences between container and virtual machine technology.

Docker container is a lightweight virtual machine and allows us to deploy applications in an original standard format such as building Hyperledger Fabric Blockchain. The proposed framework is integrated with Docker as the virtualisation backbone.

5.1 The Difference Between Docker Container and Virtual Machine

Figure 2, demonstrated a clear graphic view of the Docker Hyperledger Fabric integration in a physical laptop. The Docker host uses the Docker engine and images to facilitate the smooth running of our Hyperledger Fabric Cloud network. The Docker compose-e2e.yaml network configuration file is an important configuration file which is used to start up all the Docker containers. This file contains a critical program that uses Docker-Composer toolsets to enable level running of all our Hyperledger containers. The data itself depends on the statically configured file base/peer-based.yaml and base/docker-compose-base.yaml.

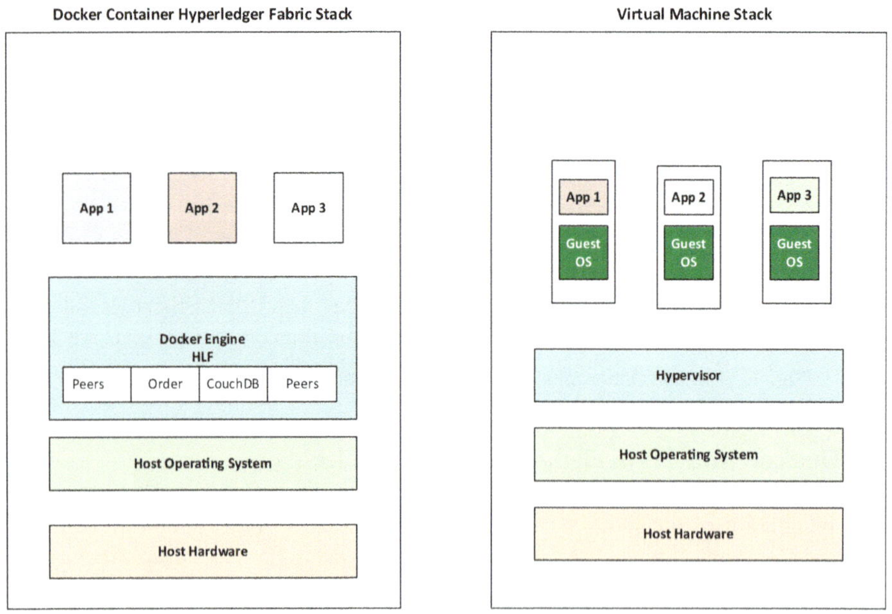

Fig. 2. Docker and the proposed hyperledger fabric integration

6 Cloud Computing

NIST SP 800-145 provides a one-sentence definition of cloud computing as "a model for allowing general, fit, on-demand network access to a shared pool of configurable computing resources". However, the NIST definition presents the supporting ideas of three cloud service models, five vital features, and four types of cloud deployments. NIST defines three service models, currently known as the SPI model, for SaaS, PaaS, and IaaS, or, respectively, software, platform, and infrastructure services.

The Cloud Security Alliance (CSA, 2009) then went on to define it as an evolving term that encompasses the development of many current technologies and methods for computing into something different. Cloud separates applications and information resources from the underlying infrastructure, and the mechanisms used to deliver them. Cloud enhances collaboration, agility, scaling, and availability, and provides the potential for cost reduction through optimised and efficient computing.

On-Demand Self-service: Consider the traditional IT ecosystem, where the economic cost and the deployment of devices has its complexity. Challenges such as troubleshooting hardware and software device is a daily challenge to network administrators. However, in a Cloud computing ecosystem, the Cloud customer is able to receive or handle extra computing resources in an automated fashion without having to call the Cloud customer service personnel for assistance. In practice, for many CSPs, the sign-up, payment, and deployment of applications are entirely on-demand and driven from their Website without the need for any human interaction, (CSA, 2016).

Blockchain has emerged as a suitable solution to solve this persistent issue by using its immutability and transparent capability to facilitate forensic evidence integrity and availability. As highlighted by [20] "Thanks to its immutability, transparency, and the consensus mechanism for avoiding double-spending, along with other clever designs such as blocks chained with the hashes of the previous blocks, the technology allows untrusting parties to trade with each other directly" Blockchain technology will disrupt current business models by making intermediary services obsolete.

7 Digital Forensics

Digital Forensics is a part of computer forensic and has emerged due to the escalation of electronic crime committed by the use of computers, networks and handheld devices. These devices are, in most cases, used as an instrument to facilitate a crime. The frameworks established a process that ensures the integrity of the evidence acquired and maintains the chain of custody at all levels during the acquisition processes [21, 22].

The DFRWS framework is a matrix with six classes: Identification, Presentation, Collection, Examination, Analysis, and Presentation [23]. According to Cory and Harlan (2012), the first Digital Forensics Research Workshop (DFRWS) in 2001, defined digital forensics as the use of systematically, resulting and established approaches towards the preservation, collection, validation, identification, analysis, interpretation, documentation and presentation of digital evidence acquired from digital sources for the drive of enabling or advancing the reconstruction of incident found to be criminal or helping to forestall unauthorised actions shown to be disruptive to planned operations.

According to [24], a number of forensic procedures and principals should always be applied when dealing with Digital evidence. These principals and procedures are: The integrity of the evidence should not be affected by the actions undertaken when securing and collecting evidence. Training should always be conducted for the personnel responsible for examining digital evidence.

Proper documentation should be done, preserved, and available for review of all the activities that relate to the seizure, examination, storage, and transfer of digital evidence. The [25] Best practices such as COBIT and ISO17799 do not consider the importance of putting in place the controls and procedures which would ensure successful investigations since they require that adequate controls are in place first. There are largely recognised standards, rules, and techniques that digital forensic examiner follow (ACPO, 2006; NIJ, 2004, 2008). [26] identifies and specifies four stages of a digital forensic investigation: "identification of digital evidence, preservation of digital evidence, analysis of digital evidence, and presentation of digital evidence."

In digital forensics cases involving data stored in the cloud, it becomes very challenging to extract and preserve data due to the difficulties of localising data. In addition to this, data in the Cloud are not necessarily stored physically on a single server but could be distributed onto many servers which are physically located in different countries and regions. This means that it can be challenging to collect complete data for digital forensic analysis. Even if complete data collection is possible, the vast amount of data stored in the Cloud requires a lot more computations.

8 Summary

Cloud computing offers customers on-demand shared resources in a virtual environment where human intervention is highly limited with the benefits like cost efficiency, scalability, agility, convenience and elasticity. However, this comes with risk, threats and challenges identified by a CSA cloud threats list. Therefore, these challenges and threats have been exploited by cybercriminals, which in turn added another layer of complexity in the cloud forensics investigation. With the involvement of Cloud computing as many organisations have adopted the technology, there is a need to define a forensic process technique suited for Cloud ecosystem that has the capability to persevered evidence privacy and integrity both on transits and storage.

The challenges faced by forensic investigators in acquiring and preserving evidence in the Cloud ecosystem have made it challenging to process Cloud artefacts in a sound forensic manner, free from contamination and being admissible in court. As of date, many researchers have focused on developing new tools in acquiring digital artefacts in the Cloud ecosystem with little research on the use of a new framework to acquire digital forensic evidence in Cloud ecosystem that is admissible in a court of law. This is the gap this paper seeks to fill, to achieve this, various Cloud frameworks were reviewed, and Blockchain was selected as an integrated technology in the Cloud ecosystem to maintain evidence integrity, privacy and trustworthiness.

9 Conclusions and Future Work

This paper first looked at related research that has been carried out in this area. Next, the paper defined Blockchain and how this is applied to a Cloud ecosystem to secure the integrity of evidence acquired in the Cloud. Then the paper went on to review Hyperledger Fabric and its components and the proposed framework. Next, the paper went on to examine the theoretical background and concepts of Cloud computing in line with National Institute of Standards and Technology (NIST) description. The computer and digital forensic were defined and highlighted the standards of forensics acquisition processes that are essential to be completed before carrying out digital forensics acquisition. Finally, In future works, we hope to develop a Design Science Methodology (DSM) that will incorporate our Blockchain Cloud forensics to enable the forensic investigators and Cloud stakeholders to acquire tamper-proof evidence.

Acknowledgements. We would also like to show our appreciation to all the anonymous reviewers for their valuable comments. We thank Nottingham Trent University for supporting this work and grateful for the resources provided by the University.

References

1. Farina, J., Scanlon, M., Le-Khac, N.A., Kechadi, M.T.: Overview of the forensic investigation of cloud services. In: 2015 10th International Conference on Availability, Reliability and Security, pp. 556–565. IEEE (2015)
2. Subashini, S., Kavitha, V.: A survey on security issues in service delivery models of cloud computing. J. Netw. Comput. Appl. **34**(1), 1–11 (2011)
3. Stanciu, A.: Blockchain based distributed control system for edge computing. In: 2017 21st International Conference on Control Systems and Computer Science (CSCS), pp. 667–671. IEEE (2017)
4. Abeyratne, S.A., Monfared, R.P.: Blockchain ready manufacturing supply chain using distributed ledger (2016)
5. Weissman, C.D., Bobrowski, S.: The design of the force. com multitenant internet application development platform. In: Proceedings of the 2009 ACM SIGMOD International Conference on Management of data, pp. 889–896. ACM (2009)
6. Cucurull, J., Puiggali, J.: Distributed immutabilization of secure logs. In: International Workshop on Security and Trust Management, pp. 122–137. Springer (2016)
7. AlJahdali, H., Albatli, A., Garraghan, P., Townend, P., Lau, L., Xu, J.: Multi-tenancy in cloud computing. In: 2014 IEEE 8th International Symposium on Service Oriented System Engineering, pp. 344–351. IEEE (2014)
8. Quick, D., Choo, K.K.R.: Big forensic data reduction: digital forensic images and electronic evidence. Cluster Comput. **19**(2), 723–740 (2016)
9. Ranjan, R., Buyya, R.: Decentralized overlay for federation of enterprise clouds. In: Handbook of Research on Scalable Computing Technologies, pp. 191–217. IGI Global (2010)
10. Yang, Z., Yang, K., Lei, L., Zheng, K., Leung, V.C.: Blockchain-based decentralized trust management in vehicular networks. IEEE Internet Things J. **6**(2), 1495–1505 (2018)
11. Sharma, P.K., Chen, M.Y., Park, J.H.: A software defined fog node based distributed blockchain cloud architecture for iot. IEEE Access **6**, 115–124 (2017)

12. Tosh, D.K., Shetty, S., Liang, X., Kamhoua, C.A., Kwiat, K.A., Njilla, L.: Security implications of blockchain cloud with analysis of block withholding attack. In: Proceedings of the 17th IEEE/ACM International Symposium on Cluster, Cloud and Grid Computing, pp. 458–467. IEEE Press (2017)
13. Liang, X., Shetty, S., Tosh, D., Kamhoua, C., Kwiat, K., Njilla, L.: Provchain: a blockchain-based data provenance architecture in cloud environment with enhanced privacy and availability. In: Proceedings of the 17th IEEE/ACM International Symposium on Cluster, Cloud and Grid Computing, pp. 468–477. IEEE Press (2017)
14. Ryan, M.D.: Cloud computing security: the scientific challenge, and a survey of solutions. J. Syst. Softw. **86**(9), 2263–2268 (2013)
15. Al Fahdi, M., Clarke, N.L., Furnell, S.M.: Challenges to digital forensics: a survey of researchers & practitioners attitudes and opinions. In: 2013 Information Security for South Africa, pp. 1–8. IEEE (2013)
16. Nakamoto, S., et al.: Bitcoin: a peer-to-peer electronic cash system (2008)
17. Androulaki, E., Barger, A., Bortnikov, V., Cachin, C., Christidis, K., De Caro, A., Enyeart, D., Ferris, C., Laventman, G., Manevich, Y., et al.: Hyperledger fabric: a distributed operating system for permissioned blockchains. In: Proceedings of the Thirteenth EuroSys Conference, vol. 30. ACM (2018)
18. Gaur, N., Desrosiers, L., Ramakrishna, V., Novotny, P., Baset, S.A., O'Dowd, A.: Hands-On Blockchain with Hyperledger: Building Decentralized Applications with Hyperledger Fabric and Composer. Packt Publishing Ltd (2018)
19. Schenker, D.G.N.: Deploy, scale, orchestrate, and manage containers with Docker and Kubernetes. PACKT (2018)
20. Xun (Brian) Wu, W.S.: A beginner's guide to developing enterprise-grade decentralized applications. PACKT (2018)
21. Lombardi, F., Di Pietro, R.: Secure virtualization for cloud computing. J. Netw. Comput. Appl. **34**(4), 1113–1122 (2011)
22. Ni, J., Zhang, K., Lin, X., Shen, X.S.: Securing fog computing for internet of things applications: challenges and solutions. IEEE Commun. Surv. Tutorials **20**(1), 601–628 (2017)
23. Garfinkel, S.L.: Digital forensics research: the next 10 years. Digit. Invest. **7**, S64–S73 (2010)
24. Casey, E.: Digital evidence and computer crime: Forensic science, computers, and the internet. Academic press (2011)
25. Grobler, C.P., Louwrens, C.: Digital forensic readiness as a component of information security best practice. In: IFIP International Information Security Conference, pp. 13–24. Springer (2007)
26. Taylor, M., Haggerty, J., Gresty, D., Hegarty, R.: Digital evidence in cloud computing systems. Comput. Law Secur. Rev. **26**(3), 304–308 (2010)

Phishing Email: Could We Get Rid of It? A Review on Solutions to Combat Phishing Emails

Ghassan Ahmed Ali(✉)

Department of Information Systems, College of Computer Sciences and Information Systems, Najran University, Najran, Kingdom of Saudi Arabia
gaabdulhabeb@nu.edu.sa, alhabeb@gmail.com

Abstract. From the inception of email in the last era until this hour, many stories about misleading victims in phishing emails are published. Phishing email has been considered to be one of the most common threats. Many researchers wonder why the phishing email still works. The problem lies in the strategies used by the attacker in the electronic trap and the lack of security awareness by the user at the same time. This paper presents stages and steps of phishing email and investigates the most tricking techniques used by the attacker to attract the user. The paper also motivates work on non-technical solutions and reviews the types of detection methods of phishing emails concentrating on methods related to message contents.

Keywords: Phishing emails · Detection methods · Cybercriminals

1 Introduction

Email has emerged in the last century and remains the most effective means of formal communication between individuals and businesses as well. However, email has become the most tools used for phishing attempts and malware propagation nowadays and in the future [1]. These attempts can be classified from unwanted email like spam email to more risky email like ransomware. Statistics show the increasing number of attacks using e-mail. According to the report of Kaspersky Lab, the statistics shows that in Q1 2019, the Anti-Phishing system stopped more than 111 million attempts of spam and phishing. Figure 1 below shows the amount of spam in mail traffic in Q1 2019. It can be seen the incremental percentage of spam is in March (56.33%) more than January and February of Q1 2019.

Reminder of this paper is organized as follows; Sect. 2 presents phishing process. The descriptions, strategies, and tools for phishing email is discussed in Sect. 3. Section 4 reviews previous works on phishing email tools related to attacker, victim, and email message, while Sect. 5 concludes the paper.

F. Saeed et al. (Eds.): IRICT 2019, AISC 1073, pp. 849–856, 2020.
https://doi.org/10.1007/978-3-030-33582-3_80

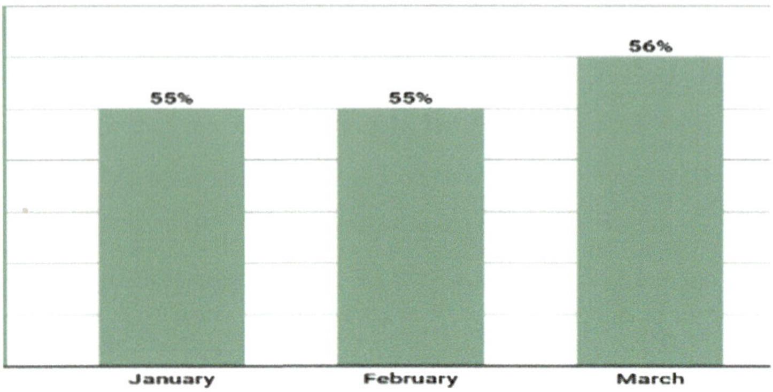

Fig. 1. Amount of spam in mail traffic, Q1 2019 [2].

2 Phishing Process

Generally, attacker sends illegitimate emails to be showed as authenticated emails originated from common companies that most people deal with such as banks, delivery posts, or famous companies. Mostly, emails request from a victim to reply, to click the included link, or to download attachment files. The first phase of the phishing email is to convince the victim to open and read the email by placing an attractive address for the sent email. If the phishing email is opened, it is considered by attacker as the most important step in the process of phishing email life cycle. Example of titles that may sent to the victim like a request of account verification, the victim name followed by he/she win prize. According to [3], opportunity of the phishing email will become more deceptive if the personal information of the user is mentioned. The second step is to redirect the victim to attack site. During that, the attacker try to exploit the vulnerabilities of the victims' system inorder to steal sensitive information, malware installed, or control of the victims' system. Figure 2 below shows steps of a typical phishing emails.

Fig. 2. Steps of phishing emails.

3 Phishing Email Stages

Attacker makes efforts to design attractive phishing email. Typically, require gathering information about the victim to mislead him/her to interact with the receiving email. Moreover, attacker skills are also required to bypass spam-filtered and to tailor the message content to convince the recipient. The more concentrating on message context leads to phishing success [3]. Mostly, a phisher customizes the contents of the message based on user profile that could be gathered from a social network or searching websites. Detail of phishing email stages is shown below in Fig. 3.

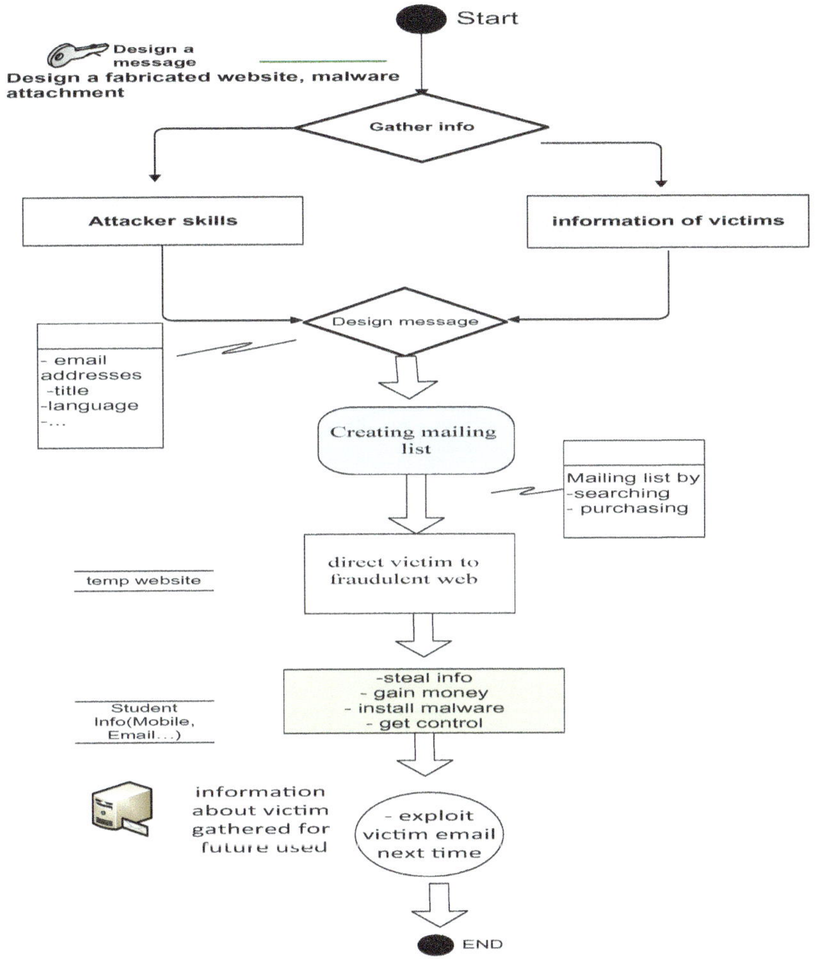

Fig. 3. Phishing emails stages

4 A Classification and Survey of Phishing Email

Typically, the three aspects of any phishing email are: sender or attacker, receiver or victim, and the message. This section discusses in details the descriptions, strategies, and tools for each aspect.

4.1 Attacker\Sender

Classification of email attacker is differ in literature based on different aspects. For example [4] constructed a model to characterize three types of attacks based on the action performed which are: email bomb, end malicious emails, crack password. Whereas [5] categorizes attacks based on effect on the network performance either to intercepts or disrupt data travelling. Furthermore, [6] investigated number of collected emails and concentrated on attacker behavior that specified on the email header.

Characteristically, email attackers can be categorized into three types: individuals, group, and software. The intention of individuals mostly is non-malicious whereas the goal of a group type is a malicious. The software attack uses extracted data from environment to expose coding or enter data. Figure 4 below shows the details of the attacks of email accounts based on the action performed.

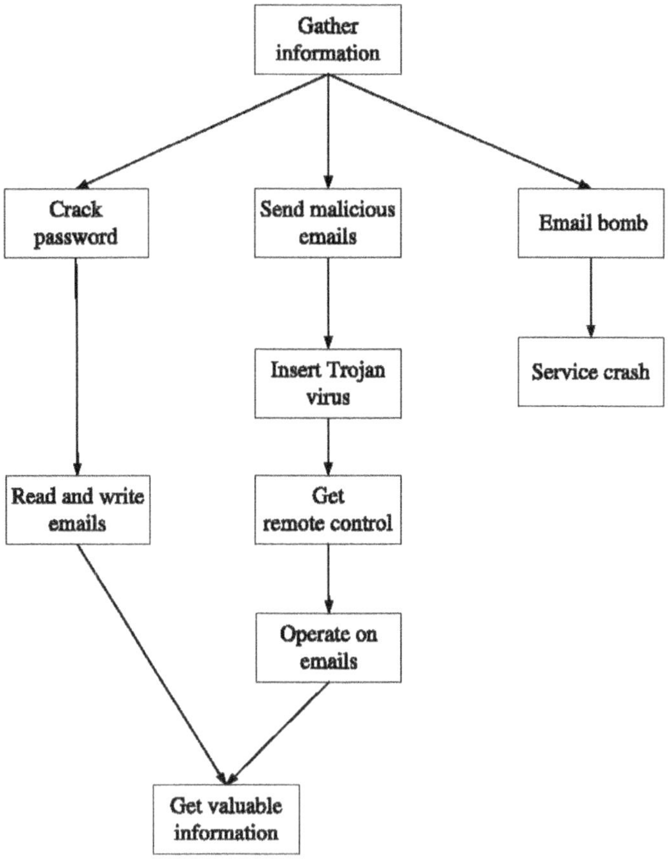

Fig. 4. Attacks of email accounts based on the action performed [4].

4.2 Victim\Receiver

Generally, any user may fall a victim of phishing attacks regardless of user background or level of education [7]. Sometimes expert user could be tricked and became a victim of phishing attack. According to [8], people with technical skills may also suspect to phishing. However, many studies such as [9, 10] concluded the importance of user awareness and education to detect phishing emails.

Victims can be categorized into individual and companies. It is stated by [11] that individuals and small businesses are the most targeted by phishing email.

Protection Users from Phishing Emails
In this subsection a protection tools based on user or receiver aspect are discussed. Numerous solutions are provided to prevent users for fall into phishing emails. Some protection focused on how to protect users using technology tools and others using education and training. Obviously, the integration of technical tools and other practices is the best solution. However, the non-technology is given more attention in this subsection since such technology solutions not provide full protection and need attention from users like update, send report, approve, etc. Many literatures conclude that the user remains "the weakest link in data protection". Therefore, user training and increasing level of security knowledge become highest priority for most IT department [12]. Table 1 below shows summary of previous works of non-technical solutions to combat phishing emails based on users aspect.

Table 1. Summary of previous works of non-technical solutions

Article	Users' education method	How the method works
Tally et al. (2004) [13]	Recommendations and Best Practices	A set of "best practices" for companies and clients to reduce effect of phishing attacks
Kumaraguru et al. (2007) [14]	Embedded training	Explain to users about phishing during regular use of email
Wen et al. (2019) [15]	Game playing	"What.Hack" game teaches and simulates phishing attacks to educate trainees
Werner & Courte (2010) [16]	Class based	Computer course teaches students how to detect a phishing email

4.3 Email Message

Commonly, phishers send authentic messages that seem like they initiated from respectable organizations like shopping companies, banks, and delivery services. The messages regularly request users to affirm data or to visit the attacker site by following the included link. Cybercriminals use many types of phishing messages to deceive victims such as: expiration date, billing issues, system infection or record alarm, government movement, and friend request. Mostly, phishing email messages frequently incorporate incorrect spellings and poor utilization of sentence structure [17].

Previous reviews on phishing have covered most phishing aspects. However, this section reviews phishing related to messages sent by email with no concentrating of other type of phishing such as website phishing, vishing, and SMiShing. Furthermore, this section focuses specifically on previous works of detection methods based on message content as summarized in Table 2.

Table 2. Summary of previous works of detection methods based on message content

Article	Detection method	How the method works	Features
Hamid & Abawajy (2014) [17]	Profiling phishing emails	Clustering forecasting to generate profile of phishing Emails	Efficient in data categorization, cluster and data sizes
Yearwood et al. (2012) [18]	Checking the embedded hyperlinks	Hyperlink and who is information using *BoosTexter* and *SVM* algorithms	Results show accuracy using hyperlink information
Basnet et al. (2008) [19]	Number of words used as features	Frequency of word (Six groups of keywords \number of words)	Classifying phishing emails using machine learning
Jain & Gupta (2019) [20]	Hyperlink analyzing with specific features	Classify features into twelve different class to train machine-learning algorithms	Independent and Multilanguage
Fang et al. (2019) [21]	THEMIS, used email header, body, character level, word level	Improved RCNN and attention mechanism	Detect phishing emails effectively
Park & Rayz (2018) [22]	Detector based on ontology	Ontological classifiers	Stability of ontological classifiers compared to lexical classifier
Chandrasekaran et al. (2006) [23]	Structural features	Using SVM as a classification tools	Efficiency of classification based on feature selection

Refer to the table above. Many studies have been contributed to detect phishing emails based on message contents. However, an attacker may violates user profile, modifies words or phrases to avoid detection. Further researches should be directed toward improving awareness of users to prevent phishing attacks.

5 Conclusion

Phishing is a critical challenging issue in past years and nowadays. From these phishing attacks, this paper focuses on phishing email because of its negative impacts on individuals and organizations. The present paper also provides a classification review based on the phishing email aspects: sender, receiver, and a message. Furthermore, a summary and comparison of the previous works of detection methods is provided with more focusing on methods related to message contents which are more proper to detect phishing with high rate accuracy. Steps and processes of phishing emails are also presented in this paper. The answering of "could we get rid of phishing email?" depends on the integration of countermeasures of the three aspects: attack prevention, messages detection, and users training. Additionally, an understanding of the attack processes and techniques is one of the important steps to combat phishing.

In the future, instructor based methods will be examined to deter phishing emails. Moreover, further classifications and solutions of phishing detection will be investigated.

References

1. Gupta, B.B., Tewari, A., Jain, A.K., Agrawal, D.P.: Fighting against phishing attacks: state of the art and future challenges. Neural Comput. Appl. **28**(12), 3629–3654 (2017)
2. KasperskyLab: Kaspersky lab: spam and phishing report 2019 (2019). https://securelist.com/spam-and-phishing-in-q1-2019. Accessed 10 June 2019
3. Chiluwa, I.: Congratulations, €1,000,000: analyzing the discourse structures of scam emails. In: The Palgrave Handbook of Deceptive Communication, pp. 897–912. Palgrave Macmillan, Cham (2019)
4. Wang, Y., Lin, C., Li, Q.L.: Performance analysis of email systems under three types of attacks. Perform. Eval. **67**(6), 485–499 (2010)
5. Pawar, M.V., Anuradha, J.: Network security and types of attacks in network. Proc. Comput. Sci. **48**, 503–506 (2015)
6. Hamid, I.R.A., Abawajy, J.: Hybrid feature selection for phishing email detection. In: International Conference on Algorithms and Architectures for Parallel Processing, pp. 266–275. Springer, Heidelberg, October 2011
7. Al-Hamar, M., Dawson, R., Al-Hamar, J.: The need for education on phishing: a survey comparison of the UK and Qatar. Campus-Wide Inf. Syst. **28**(5), 308–319 (2011)
8. Alsharnouby, M., Alaca, F., Chiasson, S.: Why phishing still works: user strategies for combating phishing attacks. Int. J. Hum.-Comput. Stud. **82**, 69–82 (2015)
9. Gupta, B.B., Arachchilage, N.A., Psannis, K.E.: Defending against phishing attacks: taxonomy of methods, current issues and future directions. Telecommun. Syst. **67**(2), 247–267 (2018)
10. Gupta, B.B., Tewari, A., Jain, A.K., Agrawal, D.P.: Fighting against phishing attacks: state of the art and future challenges. Neural Comput. Appl. **28**(12), 3629–3654 (2017)
11. Brewer, R.: Ransomware attacks: detection, prevention and cure. Netw. Secur. **2016**(9), 5–9 (2016)
12. Sasse, M.A., Brostoff, S., Weirich, D.: Transforming the 'weakest link'—a human/computer interaction approach to usable and effective security. BT Technol. J. **19**(3), 122–131 (2001)

13. Tally, G., Thomas, R., Van Vleck, T.: Anti-phishing: best practices for institutions and consumers. McAfee Research, March 2004
14. Kumaraguru, P., Rhee, Y., Acquisti, A., Cranor, L.F., Hong, J., Nunge, E.: Protecting people from phishing: the design and evaluation of an embedded training email system. In: Proceedings of the SIGCHI Conference on Human Factors in Computing Systems, pp. 905–914. ACM, April 2007
15. Wen, Z.A., Lin, Z., Chen, R., Andersen, E.: What. hack: engaging anti-phishing training through a role-playing phishing simulation game. In: Proceedings of the 2019 CHI Conference on Human Factors in Computing Systems, p. 108. ACM (2019)
16. Werner, L.A., Courte, J.: Analysis of an anti-phishing lab activity. Inf. Syst. Educ. J. **8**(11), n11 (2010)
17. Hamid, I.R.A., Abawajy, J.H.: An approach for profiling phishing activities. Comput. Secur. **45**, 27–41 (2014)
18. Yearwood, J., Mammadov, M., Webb, D.: Profiling phishing activity based on hyperlinks extracted from phishing emails. Soc. Netw. Anal. Min. **2**(1), 5–16 (2012)
19. Basnet, R., Mukkamala, S., Sung, A.H.: Detection of phishing attacks: a machine learning approach. In: Soft Computing Applications in Industry, pp. 373–383. Springer, Heidelberg (2008)
20. Jain, A.K., Gupta, B.B.: A machine learning based approach for phishing detection using hyperlinks information. J. Intell. Hum. Comput. **10**(5), 2015–2028 (2019)
21. Fang, Y., Zhang, C., Huang, C., Liu, L., Yang, Y.: Phishing email detection using improved RCNN model with multilevel vectors and attention mechanism. IEEE Access **7**, 56329–56340 (2019)
22. Park, G., Rayz, J.: Ontological detection of phishing emails. In: 2018 IEEE International Conference on Systems, Man, and Cybernetics (SMC), pp. 2858–2863. IEEE (2018)
23. Chandrasekaran, M., Narayanan, K., Upadhyaya, S.: Phishing email detection based on structural properties. In: NYS Cyber Security Conference, vol. 3 (2006)

Deauthentication and Disassociation Detection and Mitigation Scheme Using Artificial Neural Network

Abdallah Elhigazi Abdallah(✉), Shukor Abd Razak, and Fuad A. Ghalib

Information Assurance and Security Research, School of Computing, Faculty of Engineering, Universiti Teknologi Malaysia (UTM), Johor Bahru, Malaysia
abujwan78@gmail.com, shukorar@utm.my, fuadeng@gmail.com

Abstract. Wireless local area networks (WLAN) are increasingly deployed and widespread worldwide due to the convenience and the low cost that characterized it. However, due to the broadcasting and the shared nature of the wireless medium, WLANs are vulnerable to many kinds of attacks. Although there are many efforts to improve the security of a wireless network, some attacks are inevitable. Attackers can send fake de-authentication or disassociation frames to end the session a victim leading to a denial of service, stolen passwords, and leaks of sensitive information among many other cybercrimes. Effectively detecting such attacks is crucial in today's critical applications. However, the extant security standards are vulnerable to such an attack, and it is still an open research problem. In this paper, a scheme called D3MS is proposed to detect and mitigate de-authentication and disassociation attack effectively. The aim is to construct a model that can distinguish between benign and fake frames by recognizing the normal behavior of the wireless station before sending the authentication and de-authentication frames. The hypothesis is that the emulating the normal behavior of a benign station prior to the authentication and de-authentication attack is useless. The experimentation results showed the effectiveness of the proposed detection technique. The proposed scheme has improved the detection performance by 64.4% comparing to the related work.

Keywords: 802.11 · MAC, WLAN · Artificial Neural Network, DOS

1 Introduction

The convenience and low cost of 802.11 based wireless local area networks have led to widespread deployment worldwide. WLANs are widely operable in any kind of circumstances due to its availability, flexibility, and mobility (Misra et al. 2019). It also makes us more economical due to its easy configurability by which we can cover a larger span of transmission availability by limiting the additional wired networking infrastructure (Barroso et al. 2018). In WLAN, wireless nodes called clients stations (STA) are communicating through special wireless stations so-called access points

F. Saeed et al. (Eds.): IRICT 2019, AISC 1073, pp. 857–866, 2020.
https://doi.org/10.1007/978-3-030-33582-3_81

(AP). Access Point (AP) is considered a central point of communication for all STAs. STAs communicate with the APs in single hope transmission mode in a centralized architecture known as Basic Service Set (BSS). In a BSS network, wireless stations cannot communicate directly with each other. Instead, they first communicate with the AP, and then the AP will forward the frames to the destination. That is, every WLAN workstation (WS) communicates to any machine through an access point (AP).

According to the standard 802.11, WLAN stations (STAs and APs) exchange three types of frames, namely: data-frames, control-frames, and management frames, each of which has a different purpose. The data-frames are used to transmit the actual information produced from other layers while control-frames are used to coordinate access to the wireless medium and play a role in the delivery of data frames. Meanwhile, management frames allow STAs to establish communication with an AP and preserve connectivity with it. There are many types of management frames such as authentication, de-authentication, association request, association response, re-association request, re-association response, disassociation, probe request, probe response. Each type of these frames has its own purpose. For example, authentication request a client sends a request to the access point in order to be authenticated, AP in return sends authentication response to inform the STA whether the authentication process was succeeded or not.

Due to the broadcast nature of the wireless access and the fact that these frames are not protected (Bello et al. 2018) (Simic and Prodanovic 2007), attackers can exploit such vulnerability to send fake management frames to carry out a wide range of cyber-attacks that can disrupt any potential application of WLAN (Bicakci and Tavli 2009). An attacker can send de-authentication and disassociation frames to either the AP or the STA. In such a case, the attacker should forge the MAC address of the victim either the AP or the STA. Even now, skiable attackers can easily perform such attack with the availability of free attack tools such as aireplay-ng. The focus of this paper on detecting and mitigating the de-authentication and dis-association attacks.

Although numerous solutions have been proposed to protect management frames, these solutions are either simple that cannot effectively resolve such attack or affect the convenience and the flexibility of the WLAN. Due to the vast interest in WLAN technologies, these wireless networks have matured a lot since the ratification of the first 802.11 standard in 1997 (Bahl and Padmanabhan 2000; Ahlawat and Dulaney 2006). Since then, several amendments have been made to the base standard, out of which most have been to the physical (PHY) layer to increase the operating speeds and throughput of WLANs (Group 1997). However, one amendment-IEEE 802.11i was ratified in 2004 to address the threats of confidentiality, integrity, and access control in WLANs (Frankel et al. 2007; Hogan and Newton 2015). IEEE 802.11 has proposed many security extensions to secure wireless networks from malicious attacks that are possible due to the broadcast nature of wireless access. However, these extensions deal only with vulnerabilities related to unauthorized access and confidentiality issues (Arockiam and Vani 2013). There are many recent research works focused on security protocols and authentication mechanisms (Stubblefield et al. 2004; Ferrag et al. 2017). However, the IEEE 802.11 network is still vulnerable to de-authentication and dis-association attacks because these attacks take place before the security protocols are evoked (Lashkari et al. 2009). Hence the attack started by forging the MAC address of the clients, detection of the spoofed frames could be the suitable detection approach that can be used without modifying the legacy network client standards, disabling the open authentication system, or impose extra computational and communication overhead. However, existing detection approaches focusing on monitoring the consistency of the sequence number to detect the spoofed frames. Such an approach is vulnerable to the attackers who predict the next sequence number. It is worthy of mentioning that predicting the future sequence numbers is a trivial task even by non-skilled attackers.

In this paper, a scheme for detecting and mitigating fake de-authentication and disassociation frames was proposed. The concept is to train a model that can predict the class of the de-authentication and disassociation frames by driving more representative features for such attacks rather than the use of the sequence number. The aim is to learn the normal pattern of the behavior of the STAs prior to sending the de-authentication, and disassociation frames. The scheme consists of three main parts, features derivation, model training, and online operation. Firstly, the datasets contain frames of traffic flow of a realistic WLAN was collected from a real-world implementation of WLAN. Then, a list of the distinct features that can distinguish between normal and fake frames was driven using a heuristic approach. Next, the attackers' actions were simulated, and a number of fake-authentication, and disassociation attacks were injected so as to test the proposed detection model. Thus a labeled dataset that contains both fake and true frames were created in order to be used for training and validation of the proposed model. Then, a prediction model was trained using the artificial neural network (ANN) to predict the class (either false or genuine) of the de-authentication or disassociation frames, given features extracted from the previous behavior of the corresponding STA. Finally, a scheme for the online operation that describes the detection and mitigation of the fake de-authentication and disassociation frames was developed and tested. The experimentation results showed the effectiveness of the proposed detection technique. The proposed scheme has improved the detection performance by 64.4% comparing to the related work.

The rest of this paper is organized as follows. Section 2 highlights significant related work. In Sect. 3, the proposed solution was described. In Sect. 4, the experimental design is explained. Section 5 presents the results and discussion. Finally, the paper is concluded in Sect. 6.

2 Related Works

There are many solutions have been proposed to defend against de-authentication and disassociation DoS attack. These approaches can be categorized into six approaches: negotiation-based, network address based, interval-based, sequence number based, authentication bit, cryptographic based. Bellardo and Savage (2003) suggested a mechanism which delays the execution of de-authentication/disassociation requests to 5–10 s to see whether any data message is coming after that. The other methods suggest authentication of these two management frames, namely, de-authentication and disassociation (Bargh et al. 2004; Aslam et al. 2006). Usage of dynamic MAC address is proposed to prevent MAC spoofing DOS attacks (Bicakci and Uzunay 2008). Wang and Srinivasan (2010) proposed a 3-way handshake mechanism to prevent memory

DOS attacks by generating a nonce. Singh and Sharma (2015) presented a solution for Defending against DOS Attack on Wireless LAN; the proposed solution is an addition to the current 802.11-based protocol. To prevent the disassociation attack, we use the letter-envelop protocol to authenticate management frames in the association process. Samra and Abed (2010) proposed an algorithm to enhance the performance of the correlation of two Wireless Intrusion Detection Technique (WIDT) in the detection of MAC spoof DOS attacks. The two techniques are Received Signal Strength Detection Technique (RSSDT) and Round Trip Time Detection Technique (RTTDT). The detection results can be interrelated across the (WIDT) sensor this technique also does not provide a complete solution to prevent Da oS attacks in wireless infrastructure network.

Sequence number based methods (Guo and Chiueh 2005; Mar et al. 2010; Joshua Wright and Joshua 2003; Xia and Brustoloni 2004; Nandan et al. 2005; Ratnayake et al. 2014) have suggested different schemes for the detection of spoofing attacks based on the sequence number analysis. The sequence number is incremented by 1 in each frame. If the STA sends a frame with sequence number 'x' the successive frame is sent with the sequence number 'x + 1' and so on. If the next frame is sent with a sequence number greater than 'x + 1' it is quite likely that the frame is spoofed because the actual sequence must have been 'x + 1' under normal network circumstances. Though an intelligent attacker can predict the sequence number in advance to escape detection, sending a frame with sequence number 'x + 1' at the precise timing is often difficult if the number of frames to be sent are high. Singh and Sharma (2011) proposed The AP's security against DOS attack was not given priority due to computational limitations and enhanced flexibility requirement for adjusting existing authentication mechanisms. The proposed Detect and Reduce DOS Attack (DRDA) scheme not only detects the DOS attacks during the authentication phase but also reduces DOS impact at the AP and AS. The scheme also provides protection against flooding authentication/association and de-authentication/disassociation DOS.

From all the aforementioned approaches, detection the spoofed frames could be the suitable detection approach that can be used without modifying the legacy network client standards, disabling the open authentication system, or impose extra computational and communication overhead to the network due to the challenges and responses process. However, exiting research foxing on sequence number for detection the spoofed frames. Such an approach is vulnerable to the attackers who predict the next sequence number. It is worthy to note that predicting the next sequence number is a trivial task by even nonskilled attackers. This has motivated the authors of this paper to propose a solution for such an attack

3 Proposed Solution

As shown in Fig. 1, the proposed scheme consists of two main phases: offline training and online operation. In the offline operation phase, three tasks were performed as follows. The data was gathered, the distinction features were derived, and the model was trained. In the online operation, the online operation scheme has been tested. The attacker's data have been simulated according to the possible de-authentication and disassociation attack model. The attacker firstly monitors the traffic behavior. Then, the attack target is identified, which is the stations that are already associated with the access point. The attacker manipulates his own MAC address to impersonate the victim identity. Then, the attacker sends fake de-authentication/disassociation request to the access point. Access point examines the validity of the message using the model that is constructed in the following section.

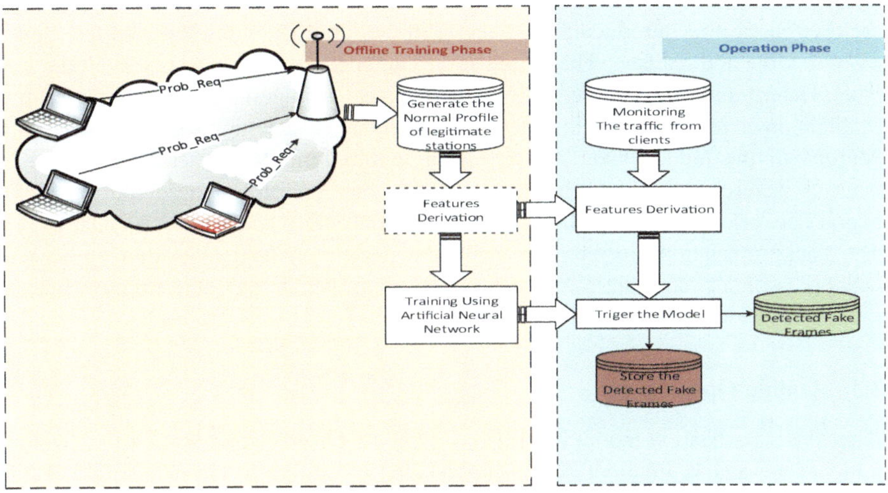

Fig. 1. The online and the offline phase of the scheme

3.1 Features Derivation

This section explains the features that have been extracted to represent normal DE-authentication and disassociation activities. These features are believed by the author that they have the ability to distinguish between the fake and genuine disassociation and de-authentication frames.

1. Time of Last Authentication/Association (TLA): TLA is the time span between last received authentication/association request-response and de-authentication/disassociation request.
2. Time to last data packet received (TLD): TLD can be defined as the time span between last received data packet and de-authentication/disassociation request.
3. Data Rate at De-authentication/Disassociation frames request (DR): DR is the average Station data rate measured within the last time window before receiving the de-authentication/disassociation request.

4. RSSI value of the De-authentication/Disassociation frames (RSSI): the signal strength at which the de-authentication/disassociation request frame was received.
5. Sequence Jumping Distance (SJD): the number span between the current and last sequence number.

3.2 Model Construction

The artificial neural network has been developed to achieve human-like performance. Three layers neural network has been used for the training, namely, input layer, hidden, and output layer. In each layer, different sets of neurons have been used. The input layer consists of 6 neurons, one for each feature and one bias neuron. The bias neurons are used in case of introducing zero vector in the input layer and to make the training more flexible and realistic. The output layer contains one neuron to represent the class label. Determining the number of the hidden layers is important to prevent overfitting problem in a neural network. Therefore, in order to determine the number of the neurons of the hidden layer N_h, a rule of thumb formula is used. Accordingly, 6 neurons were chosen for the hidden layer, and one neuron for the output layer. The neural network weights have been randomly initialized to small numbers near to zero. The backpropagation algorithm, namely the decent gradient algorithm updates the network weights and biases in the direction in which the performance function decreases most readily.

3.3 Online Operation

After the classification model is trained, the following procedure is used to detect and mitigate the attack. Figure 1 shows the detailed procedure of the detecting and mitigating the attack.

4 Experimental Design

There are two main procedure have been conducted. Firstly, normal traffic has been captured, and the normal profile is created. Secondly, the attack behavior is simulated and injected to the datasets samples. The detailed procedure is described in the following sub-sections.

4.1 Traffic Capturing and Normal Traffic Generation

The normal traffic behavior has been captured as follows. AP is a TP-LINK DG834GT wireless N router with MAC address F81A67DF22B2. The AP is configured with open access wireless LAN. Any wireless station can associate with the access point. Five user stations were used to communicate with the access points. One station was used to capture the wireless traffic. The capturing machine is MacBook Pro with OS X EI Caption v10.11.6 operating system. The wireless adapter in the machine was put into the monitoring mode so as to capture all frames types, including control and management. Monitoring mode can sniff the MAC layer traffic on the AP working channel. Unlike promiscuous mode, monitoring mode allows the adapter to sniff the traffic in the air without having to associate with the access point. Wire shark v2.6.5 was used for capturing the network traffic. Because only five stations are used for the data collection, there are two main replication procedures were conducted in order to generalize the finding based on this datasets. Firstly the dataset has been collected at a different time so as to increase the number of de-authentication and disassociation request in the data. Secondly, computer-generated data has been generated to simulate WLAN behavior during normal operation. Statistical techniques were used to model the normal station and access point behavior with regards to the captured traffic variables. Thus, using the client and access point model, one hundred workstations were simulated in the MATLAB environment. The simulation has been carried on utilizing the ground truth data collected from the real scenarios.

4.2 DE-Authentication/Disassociation Attack Simulation

The basic form of DE-authentication/Disassociation attack can be summarized as follows.

1. The attacker identifies the victims based on monitoring the network traffic for some time to identify that client which is associated with the access point.
2. The attacker manipulates the mac address of its own personating the identity of a genuine station which is associated with the access point.
3. The attacker sends false de-authentication/disassociation attack.
4. If the access point is not protected, it will respond by executing the request form the mac address associated with the request.

From the formation attack step, the attacker data is simulated by randomly sending fake de-authentication/disassociation messages. The model is tested by triggering the classification model upon receiving the de-authentication/disassociation messages. If the message is true, the driven feature vector will likely confirm with the pre-train model, and the label will high likely be assigned to normal. Otherwise, if the message is generated by an attacker, the driven message vector will deviate from the normal vector and the classification output will high likely biased to the abnormal class.

4.3 Performance Evaluation

To demonstrate the improvement gained by the proposed D3MS scheme, a comparison is made in terms of the overall accuracy, false-positive rate and false-negative rate and F-measure of the proposed scheme D3MS compared with the related scheme SeqNum Based Scheme (Ratnayake et al. 2014). The numerical results of the four evaluation metrics conducted on four scenarios are listed in Table 1. Figure 3 compares the performance of the proposed (Ratnayake et al. 2014) scheme SeqNum Based.

Table 1. Comparison between our proposed and SeqNum Based (Ratnayake et al. 2014)

Scheme	Scenario	Accuracy	FPR	FNR	F-score
D^3MS Scheme (the Proposed)	Scenario 1	88.87%	8.11%	14.43%	88.02%
	Scenario 2	86.96%	8.45%	17.81%	86.07%
	Scenario 3	87.26%	8.37%	18.38%	84.84%
	Scenario 4	89.30%	9.91%	11.17%	91.25%
Average		**87.97%**	**9.18%**	**15.09%**	**87.80%**
SeqNum Based (Baseline Scheme)	Scenario 1	59.91%	5.89%	78.79%	33.18%
	Scenario 2	52.78%	3.88%	91.16%	15.68%
	Scenario 3	48.80%	4.20%	83.31%	27.93%
	Scenario 4	58.86%	4.07%	84.36%	25.98%
Average		**55.64%**	**3.96%**	**95.47%**	**8.28%**

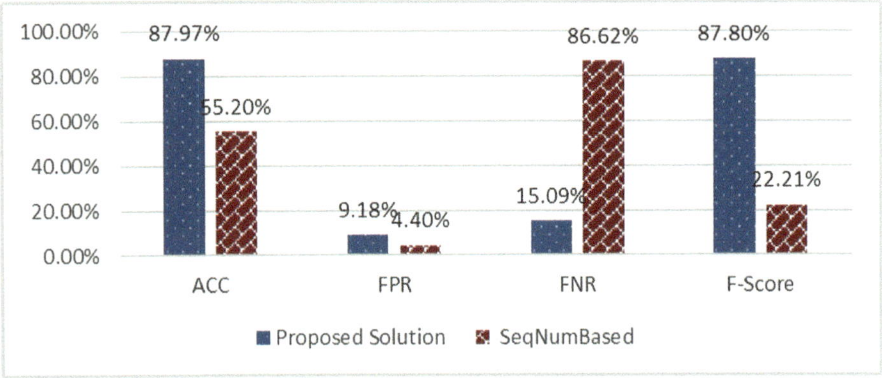

Fig. 2. Comparison between our proposed and SeqNum Based (Ratnayake et al. 2014)

As illustrated in Table 1 and Fig. 2, the proposed scheme (D3MS) achieved an average 88% accuracy, which is higher than of SeqNum Based, which is 56%. This is because SeqNum Based only considers the sequence number as the main factor that can distinguish the fake messages which can be predicted by the attacker. In contrast, the proposed scheme has been trained on more verbose features, which are well represented at the normal pattern of de-authentication/disassociation process. In terms of the false positive rate, the proposed scheme has achieved 9.18%, which is slightly higher than of that was achieved by the related work, namely the SeqNum based scheme. Although there is a 5.22% increment in the false positive rate, the detection accuracy has been improved by 32.33%. This is clear from the results in terms of the false-negative rate. The proposed scheme achieved 15.1% false-negative rate, which is lower than that of the related SeqNum Based scheme, which achieved 95.5%. The high false-negative rate of the SeqNum Based scheme is a clear indication of the ineffectiveness of the SeqNum based scheme. The results imply that the SeqNum Based scheme misclassify a percentage of 95.5% of fake messages. This leads to exposing the network to the attack. By comparing the false positive rate and false-negative rate of the SeqNum based scheme, it can be concluded that the SeqNum based scheme failed to a tradeoff between the false positive rate and false-negative rate. This interprets why the false positive rate of the SeqNum based scheme is lower than that of the proposed D3MS Scheme. The overall performance of both schemes has been evaluated in terms of F-score. The failure of the tradeoff between FPR and FNR lead to 22.2% overall performance of the SeqNum based scheme. Thus, an average of 64.4% of the overall performance improvement has been gained by applying the proposed model.

5 Conclusion

In this paper, we have proposed a MAC spoofing detection scheme that relies on Artificial Neural Networks (ANNs). ANNs provide the potential to identify and classify network behavior from limited, noisy, incomplete, and non-linear data sources. The high-speed processing of a large amount of information is another advantage of ANNs. The proposed method was validated through a prototype implementation. The validation results demonstrated that the proposed scheme achieves high detection rate for both legitimate and spoofed MAC addresses.

References

Ahlawat, R., Dulaney, K.: Magic quadrant for wireless LAN infrastructure 2006. Gartner Research (2006)

Arockiam, L., Vani, B.: Framework to detect and prevent medium access control layer denial of service attacks in WLAN. IRACST–Int. J. Comput. Netw. Wirel. Commun. (IJCNWC) **3**(2), 152–159 (2013)

Aslam, B., Islam, M.H., Khan, S.: Pseudo randomized sequence number based solution to 802.11 disassociation denial of service attack. In: 2006 Proceedings of the First Mobile Computing and Wireless Communication International Conference, MCWC 2006. IEEE (2006)

Bahl, P., Padmanabhan, V.N.: RADAR: An in-building RF-based user location and tracking system. In: Proceedings of Nineteenth Annual Joint Conference of the IEEE Computer and Communications Societies, INFOCOM 2000. IEEE (2000)

Bargh, M.S., Hulsebosch, R., Eertink, E., Prasad, A., Wang, H., Schoo, P.: Fast authentication methods for handovers between IEEE 802.11 wireless LANs. In: Proceedings of the 2nd ACM International Workshop on Wireless Mobile Applications and Services on WLAN Hotspots. ACM (2004)

Barroso, L.A., Hölzle, U., Ranganathan, P.: The datacenter as a computer: designing warehouse-scale machines. Synth. Lect. Comput. Archit. **13**(3), i–i189 (2018)

Bellardo, J., Savage, S.: 802.11 Denial-of-Service Attacks: Real Vulnerabilities and Practical Solutions. USENIX Security (2003)

Bello, L.L., Mariani, R., Mubeen, S., Saponara, S.: Recent advances and trends in on-board embedded and networked automotive systems. IEEE Trans. Ind. Inf. **15**(2), 1038–1051 (2018)

Bicakci, K., Tavli, B.: Denial-of-service attacks and countermeasures in IEEE 802.11 wireless networks. Comput. Stan. Interfaces **31**(5), 931–941 (2009)

Bicakci, K., Uzunay, Y.: Pushing the limits of address based authentication: how to avoid MAC address spoofing in wireless LANs. World Acad. Sci. Eng. Technol. **2**, 214–223 (2008)

Ferrag, M.A., Maglaras, L., Ahmim, A.: Privacy-preserving schemes for ad hoc social networks: a survey. IEEE Commun. Surv. Tutorials **19**(4), 3015–3045 (2017)

Frankel, S., Eydt, B., Owens, L., Scarfone, K.: Establishing wireless robust security networks: a guide to IEEE 802.11 i. National Institute of Standards and Technology (2007)

IW Group: Wireless LAN medium access control (MAC) and physical layer (PHY) specifications (1997)

Guo, F., Chiueh, T.-C.: Sequence number-based MAC address spoof detection. In: International Workshop on Recent Advances in Intrusion Detection. Springer (2005)

Hogan, M., Newton, E.: Supplemental information for the report on strategic US government engagement in international standardization to achieve US objectives for cybersecurity (2015)

Joshua Wright, G., Joshua, C.: Detecting wireless LAN MAC address spoofing. Cisco Certified Network Associate (2003)

Lashkari, A.H., Danesh, M.M.S., Samadi, B.: A survey on wireless security protocols (WEP, WPA and WPA2/802.11 i). In: 2nd IEEE International Conference on Computer Science and Information Technology, 2009, ICCSIT 2009. IEEE (2009)

Mar, J., Yeh, Y.-C., Hsiao, I.-F.: An ANFIS-IDS against deauthentication DOS attacks for a WLAN. In: 2010 International Symposium on Information Theory & Its Applications. IEEE (2010)

Misra, S., Sarkar, S., Chatterjee, S.: Sensors, Cloud, and Fog: The Enabling Technologies for the Internet of Things. CRC Press (2019)

Nandan, A., Das, S., Pau, G., Gerla, M., Sanadidi, M.: Co-operative downloading in vehicular ad-hoc wireless networks. In: Second Annual Conference on Wireless On-demand Network Systems and Services. IEEE (2005)

Ratnayake, D.N., Kazemian, H.B., Yusuf, S.A.: Identification of probe request attacks in WLANs using neural networks. Neural Comput. Appl. **25**(1), 1–14 (2014)

Samra, A.A., Abed, R.: Enhancement of passive mac spoofing detection techniques. Int. J. Adv. Comput. Sci. Appl. **1**(5) (2010)

Simic, D., Prodanovic, R.: A survey of wireless security. CIT J. Comput. Inf. Technol. **15**(3), 237–255 (2007)

Singh, R., Sharma, T.P.: Detecting and reducing the denial of service attacks in WLANs. In: 2011 World Congress on Information and Communication Technologies (WICT). IEEE (2011)

Singh, R., Sharma, T.P.: On the IEEE 802.11 i security: a denial-of-service perspective. Secur. Commun. Netw. **8**(7), 1378–1407 (2015)

Stubblefield, A., Ioannidis, J., Rubin, A.D.: A key recovery attack on the 802.11 b wired equivalent privacy protocol (WEP). ACM Trans. Inf. Syst. Secur. (TISSEC) **7**(2), 319–332 (2004)

Wang, L., Srinivasan, B.: Analysis and improvements over DoS attacks against IEEE 802.11 i standard. In: 2010 Second International Conference on Networks Security Wireless Communications and Trusted Computing (NSWCTC). IEEE (2010)

Xia, H., Brustoloni, J.: Detecting and blocking unauthorized access in Wi-Fi networks. In: International Conference on Research in Networking. Springer (2004)

Advances in Information Systems

Influence of Smart Interactive Advertising Based on Age and Gender: A Case Study from Sri Lanka

Wiraj Udara Wickramaarachchi[1,2(✉)], W. M. S. L. Weerasinghe[3], and R. M. K. T. Rathnayaka[3]

[1] School of Computer Science and Technology, Wuhan University of Technology, Wuhan, People's Republic of China
wirajudara@gmail.com

[2] Faculty of Applied Sciences, Rajarata University of Sri Lanka, Mihintale, Sri Lanka

[3] Faculty of Applied Sciences, Sabaragamuwa University of Sri Lanka, Belihuloya, Sri Lanka
sinethllllakshitha@gmail.com, kapila.tr@gmail.com

Abstract. Advertising media plays a major role in marketing. Sometimes, the consumers tend to get advertisements as a disturbance when they do not show what consumers need to know or buy at right time. Also, some advertisements cannot watch with the children together due to the reasons like nudity. The smart interactive advertising (SIA) concept is a solution to these issues. The study discusses the influence of SIA concept to Sri Lankan people. A prototype system of SIA to detect age and gender of the particular person and to shows advertisements according to detected age and gender has developed. A questionnaire was used to find the influence of SIA concept. The empirical findings suggest that the SIA concept is a good solution for the issues mentioned above. Specifically, it is influencing in better way to Sri Lankan People.

Keywords: Advertisements · Age estimation · Gender estimation · Image processing · Smart interactive advertising

1 Introduction

Advertising is audio and visual forms of promoting or selling products, services or ideas. A better explanation of usages of advertisements is given in [2]. Several advertisement classifications discussed in literature [1, 3] and [4]. The right type of content for the target audience makes maximum benefit of an advertisement.

The human face is a significant biometric trait of a person, which contains much valuable information about an individual, such as gender, age, and race. The age and gender estimation can be possibly applied in human-computer interaction and business analytics [6]. The image processing techniques are used in the estimation process so as they enhance image quality and identifying objects in images [5].

There is a considerable possibility to fail advertising campaigns because of low people interaction and not seen a specific advertisement by an expected set of

F. Saeed et al. (Eds.): IRICT 2019, AISC 1073, pp. 869–880, 2020.
https://doi.org/10.1007/978-3-030-33582-3_82

consumers. Moreover, there may be inappropriate advertisements which are not suitable to be seen by the children due to the way of Sri Lankan culture has grown up. In this case, the SIA concept could be used.

The purpose of the research study is to realize the influence of the SIA concept among Sri Lankan customers based on their age and gender. Moreover, understand the current status of the concept. Also, an initiative system that works as a SIA system has developed in order to get the responses from the general customers of different ages and genders.

2 Literature Review

A non-reflective similarity transformation is applied to normalize face images to improve the accuracy of facial component localization in the experiments of [7]. An enhanced Eigen-Face based face recognition method has introduced in [8]. Zhang, Wang, and Amin [9] have proposed a hierarchical approach for automatic age estimation and have provided an analysis of how aging influences individual facial components. An investigation of biologically inspired features (BIF) for human age estimation from faces has done in [10].

An age manifold learning method for mining face aging features and design a locally adjusted robust regressor for learning and forecast of human ages has introduced in [11]. A gender and age estimation method, which is very effective in real-time processing and robust against changes caused by light and facial expressions, has introduced in [12].

Huber, Lemmons, Zenoni and Hensgen [13] have explained about an enhanced content activator included in interactive advertisements. The effect of mobile advertising content on buying intention in a framework derived from past models and theories of technology acceptance has revealed in [14]. Park and Choi [15] have discussed an approach using secondary information tag for broadcasting interactive advertisements with detailed information.

A computerized method based on an embodiment of the invention for providing interactive advertisement is explained in [16]. The invention of [17] explained the way of enabling interactive advertisements to be loaded on mobile devices and the way to operate with such interactive advertisements on their mobile devices in an interactive style meanwhile in an off-line mode. Another innovation of [18] based on environmental conditions offers an interactive advertisement board.

Rafizadeh [19] has introduced an interactive system based on a business-to-business and business-to-consumer model for providing product information and rewards to consumers. A set of tools in [20] were used to support dynamically insertion of advertisements into a content stream distributed to viewers through an Internet Protocol (IP) network.

The article [21] have provided a computerized method of advertising an advertisement including data content or metadata linked with time in the future to a non-singularity of users connected via a computer network to a server. An advertisement swapping method has introduced in [22].

3 Methodology

3.1 The Survey

The first part was a survey to analyze the influence of SIA based on age and gender of Sri Lankan people. There were two parts in the questionnaire. The first part filled before interacting with the prototype and second part filled after interacting with the prototype. The main aspects were the attention of the person and relativeness of the advertisements according to Sri Lankan culture. The stat measures were obtained using Minitab by comparing responses of each user for the statements of the questionnaire. The sample population consisted of 100 users.

Design of the Questionnaire. There were two general demographic questions to collect the gender and age information. There were 14 statements included in part one, based on existing advertising media. Among them, five statements were to check the understanding of the nature of interaction of consumer with the advertisement media. Five statements were to check the understanding about the relativeness of the advertisements with themselves. The last four statements were to check the understanding of the consumers about the impact of advertisement media to Sri Lankan culture.

There were 12 questions or statements included in part two based on the prototype system. The first question was to write the user ID number which given by the prototype to each consumer. The next five statements were to check the attention of the consumers towards the advertisements they were shown. Moreover, the next five statements were to check their understanding of relativeness of shown advertisements with their selves. The last question was to write down the categories or names of the products or services which emphasized by the advertisements as he or she remembers. There were all kinds of advertisements included in the prototype. However, the statements of the questionnaire were designed to relate the part one of the questionnaire with part two.

The responses for the statements in both parts except first two general demographic questions and last question in part two, vary between 1 to 5. The response number of 1 to 5 describes the agreement or disagreement of the consumer with the given statements ranging from "strongly disagree" to "strongly agree" respectively (Five-point scales; 1-Strongly Agree, 2-Agree, 3-Neutral, 4-Disagree, 5-Strongly disagree). The responded numeric values were used in hypothesis tests in order to obtain stats.

3.2 The Prototype

The second part was a SIA prototype to get a response of consumers for the first part of the study. The prototype consists of three parts; digital camera, digital display, and control system. The real-time face images are captured using the camera, and selected advertisements are displayed on the digital display. Figure 1 shows the whole process of the proposed system.

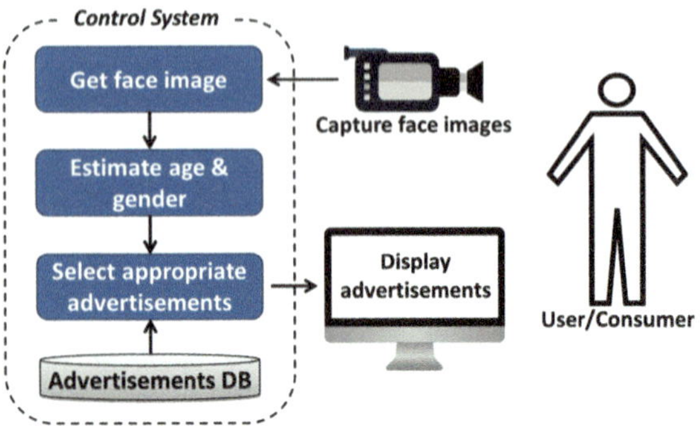

Fig. 1. Prototype design of the smart advertising system

The control system gets inputs of real-time face images and estimates the age group and the gender of the user(s). The estimated ages and genders for each user were stored in the database for the analysis.

The advertisement database (DB) consisted of 70 advertisements of different categories. The personal needs of people, age, education level, and the Sri Lankan culture were taken into consideration when deriving the advertisement categories.

After estimating age group and gender, the system chooses ten advertisements according to estimated age group and gender randomly from the advertisement DB. The advertisement categories based on age groups and genders are; $4 \leq$ age < 12 (male and female), $13 \leq$ age < 26 (female), $13 \leq$ age < 26 (male), $27 \leq$ age < 40 (female), $27 \leq$ age < 40 (male), $40 \leq$ age (male and female), and General advertisements.

Usually, 4 to 12 age group involves with primary education. Their needs are usually the same without caring the gender. The 13 to 26 age group involves with secondary and higher education. Gender-wise, their needs are different. Therefore, there were two separate categories for male and female of the 13 to 26 age group. The 27 to 40 age group consider their life goals. There were two separate advertisement categories for male and female of this age group since their needs are different gender-wise. The beyond 40 age group included both male and female since their needs seem to be the same.

The prototype has implemented using Python programing language with the support of OpenCV libraries. The generation of the advertisement database and the data retrieval process was done using SQL lite.

In real time, the system captures ten image frames to realize whether a person is standing or not in front of the camera. Then it estimates ages from detected faces and calculate the average age. According to detected age and gender, the viewer will be shown ten advertisements. The flow of the prototype system process is shown in Fig. 2.

There are ten advertisements for each age and gender groups and ten general advertisements. Five advertisements are selected randomly from the estimated age and gender group, and five advertisements are selected randomly from the general advertisement category. Altogether ten advertisements are shown to the viewer.

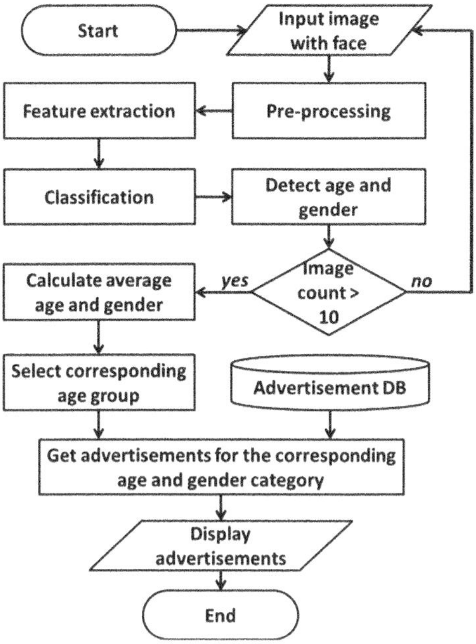

Fig. 2. The flow of process of the prototype system

The prototype system recognizes all identifiable faces on an image. For the easiness of getting responses of individuals separately and making analysis results clear, the single user was asked to use the prototype system at a time. They were given a user ID in order to make the analysis process convenient.

4 Results and Discussion

4.1 Accuracy of Gender and Age Group Estimation

There were two sets of values obtained from the experiment and the questionnaire in order to analyze the accuracy of age group estimation. The first set of values was obtained from the responses of the questionnaire and the second set of values was obtained from the estimated age groups of corresponding users by the prototype system. The user ID given by the system to each user was used to identify the correct user of each estimated age group with the corresponded response of the questionnaire.

The percentage of number of correctly estimated age groups over the total number of estimated age groups has been obtained as the accuracy of age group estimation. Therefore, the accuracy of age group estimation is 96%.

The same method was used in order to obtain the accuracy of gender estimation. The percentage of number of correctly estimated genders over the total number of estimated genders has been obtained as the accuracy of gender estimation. Thus, the accuracy of gender estimation is 100%.

4.2 Analysis of the Influence of SIA

The attractiveness, relativity, and effect to the culture are being discussed under this section. The hypotheses for attractiveness and relativity were tested using paired t-test. For the easiness of the comparison process, all statements of the questionnaire have renamed according to renaming method mentioned here; *Statement_Number* = *nXm*. Where, *n* - The part which contained the statement {part-one = 1, part-two = 2}. *X* - The aspect of the statement {attractiveness = A, relativity = R, effect to the culture = C}. *m* - The question number.

Attractiveness of SIA. The hypotheses were developed by assuming that, there were five statements in part one of the questionnaire were correlated with five statements in part two. The compared statements are shown in Table 1. And compared pairs of statements are; 1A1 and 2A1, 1A2 and 2A2, 1A3 and 2A3, 1A4 and 2A4, 1A5 and 2A5.

Table 1. The compared statements of the questionnaire to measure Attractiveness

#	Statement
1A1	You have absorbed any information about something from any type of advertisements
2A1	You have absorbed any information about something from the advertisements which were shown by the prototype system
1A2	Usually, you give attention to any kind of advertisements carefully
2A2	You gave more attention to the advertisements which were shown by the prototype system
1A3	You can keep the purposes of advertisements in your mind, usually
2A3	You can keep the purpose of the advertisements shown by the prototype system in your mind
1A4	You tend to buy the things that have shown by advertisements
2A4	You got any idea of buying the things that have shown by the prototype system
1A5	You give attention to advertisements that do not match with your age or gender
2A5	You gave careful attention to the advertisements shown by the prototype system because you got the advertisements according to your age and gender

The research hypothesis (H1) is $n2 - n1 < 0$. Where, the n1 and n2 are mean of value sets of part-one and part-two respectively. The common meaning of H1 of the research is; there is an influence of interactive smart advertising to people. The null hypothesis (H0) was the opposite condition ($n2 - n1 \geq 0$). The decisions have made using p values (see Table 2).

Table 2. The stats to measure attractiveness from the hypothesis test for statement pairs

Hypothesis (H1)	w value	p-value	Decision ($\alpha = 0.05$)
n2(2A1)-n1(1A1) < 0	137.772	0.6042	Denied
n2(2A2)-n1(1A2) < 0	13.253	0.041	Accepted
n2(2A3)-n1(1A3) < 0	12.346	0.039	Accepted
n2(2A4)-n1(1A4) < 0	11.77	0.0342	Accepted
n2(2A5)-n1(1A5) < 0	10.1	0.031	Accepted

The p value of the first hypothesis is 0.6042. This value is exceeding the significance level. So, the hypothesis (H1) is rejected. That means the users have not absorbed any information about something from the advertisements shown by the prototype system.

The p-value of 0.041 obtained for the second pair of 1A2 and 2A2 is less than α. Thus, the hypothesis is accepted. This depicts the users gave more attention to the advertisements which were shown by the prototype system. The p-value for the 1A3 and 2A3 is 0.039. Therefore, the hypothesis is accepted and it can be interpreted as the users could keep the purpose of the advertisements shown by the prototype system in their minds.

The hypothesis test for the fourth pair of 1A4 and 2A4 returned the p-value as 0.0342. Therefore, the hypothesis is accepted. The decision can be interpreted as the idea of buying the things shown by the prototype system has been spring out in their minds. The least p-value of 0.031 has been obtained for the fifth pair of 1A5 and 2A5. The value is an evidence to accept the hypothesis. Thus, it is proved that the users gave careful attention to the advertisements shown by the prototype system because they were viewed the advertisements according to their age and gender group.

Out of five hypotheses four hypotheses have been accepted in this section. It could be concluded that the attractiveness of SIA makes an influence than other media of advertising to the Sri Lankan people.

Relativity of SIA. The testing method of hypotheses is the same as the previous section and ten statements of the questionnaire associated with relativity aspect. The compared pairs of statements are 1R1 and 2R1, 1R2 and 2R2, 1R3 and 2R3, 1R4 and 2R4, 1R5 and 2R5. (see Table 3).

Table 3. The compared statements of the questionnaire to measure Relativity

#	Statement
1R1	You are seen the advertisements through regular advertising media which you are willing to buy
2R1	You were shown the advertisements through the prototype system which you are willing to buy
1R2	You are shown advertisements through regular advertising media according to your age
2R2	You were shown the advertisements through the prototype system according to your age
1R3	You are shown advertisements through regular advertising media according to your gender
2R3	You were shown the advertisements through the prototype system according to your gender
1R4	Usually, you give attention to advertisements shown through regular advertising media which do not match with your age or gender
2R4	You were not shown the advertisements through the prototype system which do not match with your age and gender
1R5	The advertisements shown through regular advertising media are advertisements what you are willing to know
2R5	All advertisements shown through the prototype system are advertisements what you are willing to know

The same research hypothesis (H1) used to find the relativity of SIA as $n2 - n1 < 0$ where n1 and n2 are mean of value sets of part-one and part-two respectively. The p values of each hypothesis tests were used to make the decision (see Table 4).

Table 4. The stats to measure relativity from the hypothesis test for statement pairs

Hypothesis (H1)	T value	p-value	Decision ($\alpha = 0.05$)
n2(2R1)-n1(1R1) < 0	13.344	0.041	Accepted
n2(2R2)-n1(1R2) < 0	10.686	0.034	Accepted
n2(2R3)-n1(1R3) < 0	12.903	0.0399	Accepted
n2(2R4)-n1(1R4) < 0	12.9825	0.0412	Accepted
n2(2R5)-n1(1R5) < 0	21.808	0.402	Denied

The first hypothesis for this section has compared the 1R1 and 2R1. The result suggests to accept the alternative hypothesis under the significance level of 0.05. This depicts the prototype system showed the advertisements which the users were willing to buy. This scenario makes a sense that consumers are affected with SIA instead of regular advertisement media since it shows only what they want to buy.

The hypothesis test for the second pair of statements returned the p-value as 0.034. Since this p-value is less than α, the hypothesis is accepted. This indicates the advertisements shown through the prototype system were matched with the ages of users, while regular advertisement media shows advertisements not related to users ages.

The p-value of 0.0399 has obtained for the third pair of statements named as 1R3 and 2R3. Moreover, the hypothesis is accepted. This proves the relativity of gender with the consumers would make more attention to the advertisements which are shown by the prototype system of smart interactive advertisements.

The fourth hypothesis is accepted under the 0.05 level of significance ($0.0412 < 0.05$). The accepted statement was about the relativity of age and gender group with the advertisements shown by the smart interactive advertisement media.

The p-value for the fifth hypothesis is 0.402. This value is out of the significance level. Hence, the fifth hypothesis is rejected. This explains that users have no problem to see any advertisements without relevance to them. This will make informative persons with more knowledge of product and services which they can buy. In this case, people do not want to look at the SIA media because it always gives information about what the people already know according to their age and gender group.

This concludes that the relativity of advertisements based on age, gender and willingness makes more effect on Sri Lankan people to ignore regular advertisement media whereas they make interest to watch advertisements on smart interactive advertisement media.

A highlighted fact is the consumers in Sri Lanka want to know about the product or service that they can buy because they could get a necessity of those products or services in the future. However, on the other hand, when they want to see only what they want to buy, it will be irritating to them to see what not related to them.

Influence to the Sri Lankan Culture. The traditional Sri Lankan culture tries to tie children and women in sensitive cultural matters. For example, commercial advertisements related to condoms are still not shown in regular advertising media in Sri Lanka. The culture still gives priority to children and women in many cases. It cares about morality and decorum in any activity. By considering this fact, the questionnaire was included four statements associated with Sri Lankan culture. They were based on commercial advertisements shown on regular advertisement media. The statements are shown in Table 5.

Table 5. The statements of the questionnaire to find the effect to the Sri Lankan culture

#	Statement
1C1	They do not show advertisements which cannot watch together with children
1C2	They do not show advertisements that make ladies' mind upset or inconvenient
1C3	They do show advertisements which can watch regardless of age or gender
1C4	They do not show unnecessary 18+ advertisements (e.g., sexual content)

The responses to the statements in this section have averaged and converted into percentages in order to understand results quickly. Pie charts of percentages for each statement are shown in Fig. 3.

The Fig. 3-(a) associates with adult advertisements that cannot watch with children together. There are 12% of strongly disagreements and 51% of disagreements (altogether 63%) depict the advertisement media show advertisements that cannot watch with children together. Only 14% of people (strongly agree 1% and agree 13%) agreed on the statement. That is the regular advertisement media do not show advertisements which cannot watch together with children.

According to the Fig. 3-(b), total 55% of people (strongly disagree 46% and disagree 9%) have disagreed to the statement 1C2. This depicts ladies might be viewed some advertisements through regular advertising media that make their mind upset or inconvenient. Only 21% of people (strongly agree 4% and agree 17%) seconded that the regular advertising media are not showing those kinds of advertisements.

The agreement and disagreement on 1C3 statement are likely to be equal. 30% of people (strongly agree 3% and agree 27%) are seconding that, the regular advertising media do show advertisement which can watch regardless of age or gender, while 37% of people (strongly disagree 4% and disagree 33%) disagree on the statement. Also, 33% of people are keeping neutral. According to these percentages, the conclusion can be made towards the disagreement of the statement. However, it shows the somewhat equal probability of being true or false of the statement.

45% of people (strongly disagree 10% and disagree 35%) have disagreed on the statement 1C4. This means the advertisements shown on regular advertising media include unfitting 18+ contents in their advertisements.

The average percentages for four statements have been calculated and shown in the pie chart in Fig. 4. The percentage values describe the agreement or disagreement of people for any statement among the above four statements.

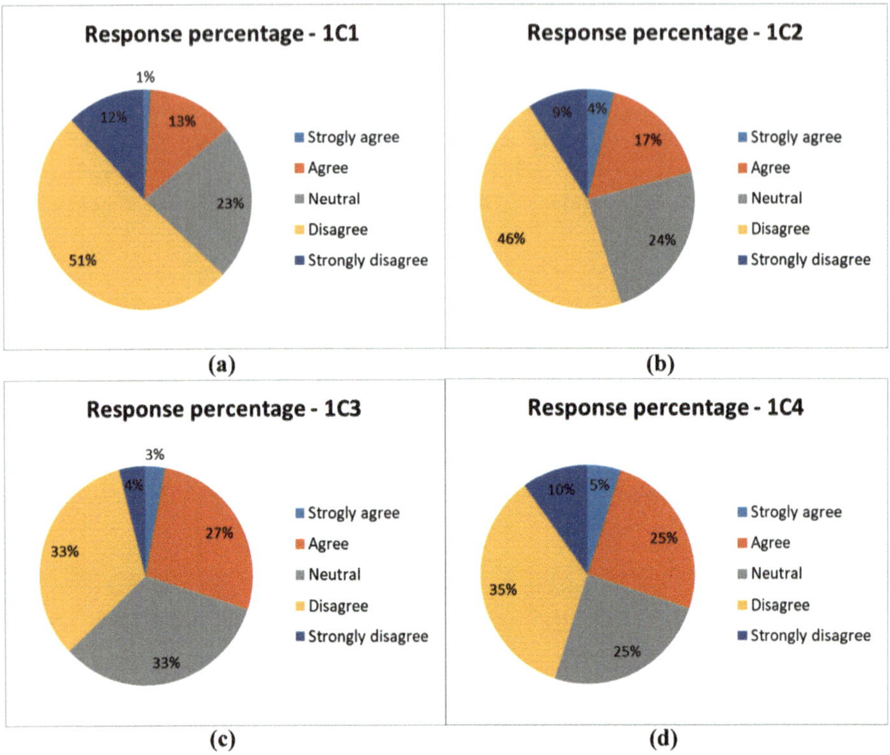

Fig. 3. Response percentages of statements related to Sri Lankan culture

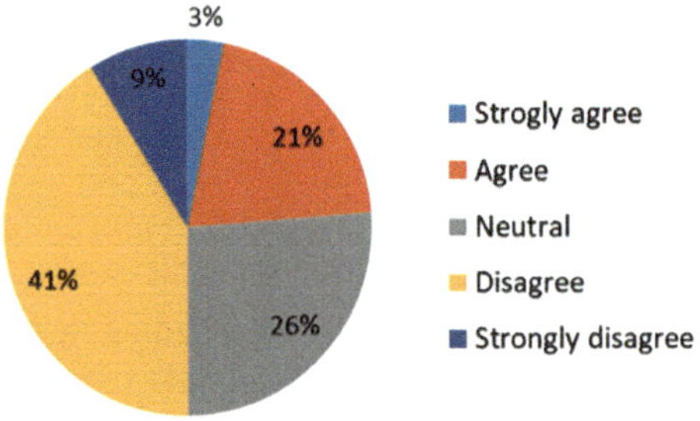

Fig. 4. Average percentages for four statements related to Sri Lankan culture

50% of disagreement (strongly disagree 9% and disagree 41%) seconding that the regular advertising media might be showing advertisements without considering the age groups and genders of people. Especially, the advertisements which make ladies' mind upset or inconvenient. Only 24% of people (strongly agree 3% and agree 21%) put their vote on the opposite side. It could be concluded that the regular advertisement media are affecting to people and culture critically.

On the other hand, the prototype system of SIA shows advertisements by classifying the ages and genders of the viewers. So, there is a very low possibility of showing inappropriate advertisements to the viewers by the prototype system. Therefore, the smart interactive advertisement could be used to reduce the bad influence of the advertisement to the Sri Lankan culture.

4.3 The Memorizing of Shown Advertisements

The final question of the questionnaire was to memorize the advertisements they were shown and write them on the answer sheet. Then the written advertisement names were compared with the shown advertisements to each person individually. The average percentage of remembered advertisements per person is 77.6%. The result shows that the number of remembered advertisements is higher than half a way out of 10 advertisements. This depicts the SIA makes people memorize the products or services other than regular advisement media do.

5 Conclusion

This novel study was to find the influence of SIA to Sri Lankan people. The prototype system for SIA estimates age groups with the accuracy rate of 96% while its gender estimation is 100% accurate. The influence has been evaluated in three aspects. They are attractiveness, relativity, and effect to the culture.

Out of 5 hypotheses based on the attractiveness of the advertisements, four hypotheses have been accepted. This proves the advertisements shown through smart interactive advertisement system are more attractive than the advertisements shown through regular advertisement media. In the relativity aspect of the advertisements, four hypotheses have been accepted out of 5 hypotheses. Thus, the relativity of the advertisements shown through smart interactive advertisement system is more relative to the subject than the advertisements shown through regular advertisement media.

50% of the sample set of people depicts the advertisements shown through regular advertisement media are affecting badly to the Sri Lankan culture. In this case, smart interactive advertisements might be useful to reduce the bad influence. Finally, the research study could be concluded as; smart interactive advertisements are influencing positively to the Sri Lankan people and the culture.

The topic is open for further studies. The methodology can be updated to provide the priority to children. Also the experiments can be executed to age classifications other than this study has been executed.

References

1. Jacobsen, T.: Customer Demographics - Age Demographics for Advertising (Updated 2018) —Marketing Artfully. Marketing Artfully (2018). https://marketingartfully.com/2012/12/13/customer-demographics-age-demographics-for-advertising/. Accessed 03 Mar 2018
2. Oubaiche, Z.: Advantages of Advertising: 12 Major Advantages of Advertising–Explained! (2015). www.linkedin.com, https://www.linkedin.com/pulse/advantages-advertising-12-major-explained-zahira-oubaiche/. Accessed 03 Mar 2018

3. Mooij, M.: Global Marketing and Advertising. Sage, Los Angeles (2013)
4. Curry, L.: What marketing content do different age groups like to consume? Digital Doughtnut (2016). https://www.digitaldoughnut.com/articles/2016/september/what-marketing-content-different-age-groups-like. Accessed 03 Mar 2018
5. Varshney, P.K., Arora, M.K.: Advanced Image Processing Techniques for Remotely Sensed Hyperspectral Data. Springer, Heidelberg (2004)
6. Yan, C., Lang, C., Wang, T., Du, X., Zhang, C.: Age estimation based on convolutional neural network. In: Pacific-RIM Conference on Advances in Multimedia Information Processing-PCM. Springer, New York, Inc. (2014)
7. Esme, B., Sankur, B.: Effects of aging over facial feature analysis and face recognition. Bogazici Un Electronics Eng Dept. (2010)
8. Heseltine, T., Pears, N., Austin, J.: Evaluation of image preprocessing techniques for eigenface-based face recognition. In: Second International Conference on Image and Graphics. International Society for Optics and Photonics, vol. 4875, pp. 677–686 (2002)
9. Zhang, Y.D., Wang, B., Amin, M.G.: Society of Photo-Optical Instrumentation Engineers (2008)
10. Guo, G., Mu, G., Fu, Y., Huang, T.S.: Human age estimation using bio-inspired features. In: IEEE Conference on Computer Vision & Pattern Recognition. IEEE (2009)
11. Guo, G., Fu, Y., Dyer, C.R., Huang, T.S.: Image-based human age estimation by manifold learning and locally adjusted robust regression. IEEE Trans. Image Process. Publ. IEEE Sig. Process. Soc. **17**(7), 1178–1188 (2008)
12. Guney, F.: Gender and Age Estimation Using Face Images. Bogazici University, Computer Engineering, Istanbul, Turkey, vol. 4 (2011)
13. Huber, T., Lemmons, T., Zenoni, I., Hansgen, D.: System and method for providing interactive advertisement. US Patent 9,865,017, 9 Jan 2018
14. Hashim, N.H., Kassim, N.M., George, R.A.: Revisiting the effect of mobile advertising content on buying intention. Gading J. Soc. Sci. **20**(01), 1–8 (2017)
15. Park, W., Choi, D.: Method of realizing interactive advertisement under digital braodcasting environment by extending program associated data-broadcasting to internet area. US Patent Application 11/575,975, 13 Mar 2008
16. Moshe, E.: Method and system for interactive advertisement. US Patent Application No. 10/700,476 (2005)
17. Aufricht, M., Weinstein, R., Broadwell, G.: Interactive advertisement mechanism on a mobile device. US Patent Application 09/864,293, 2 May 2002
18. Yen, D., Huang, T.H.: Interactive advertisement board. US Patent Application 11/429,986, 1 Feb 2007
19. Rafizadeh, K.: Interactive advertisement and reward system. US Patent Application 09/756,236, 10 Oct 2002
20. Casey, S.M., Selvadurai, G., Castro, F.: Dynamic interactive advertisement insertion into content stream delivered through ip network. US Patent Application 12/147,674 (2009)
21. Yerli, C.: Computing system method for advertising a future ad. US Patent Application 13/651,125, 18 Apr 2013
22. Miller, D.A.: Advertisement swapping using an aggregator for an interactive television system. US Patent 7,266,832, 4 Sept 2007

Eliciting Requirements for Designing Self-reflective Visualizations: A Healthcare Professional Perspective

Archanaa Visvalingam(✉), Jaspaljeet Singh Dhillon, Saraswathy Shamini Gunasekaran, and Alan Cheah Kah Hoe

Universiti Tenaga Nasional, Putrajaya, Malaysia
achavisva0196@gmail.com,
{jaspaljeet, sshamini, alancheah}@uniten.edu.my

Abstract. Self-care applications are commonly presented with visuals that empower users to reflect on their health data. Self-reflective visualizations play an important role in educating healthcare consumers to take proactive steps in managing their health independently. These visualizations are vital, and they should be carefully designed to meet the needs and requirements of users. We have developed a model to aid developers to design effective self-reflective visualizations. In this paper, we have explored the perspectives of healthcare professionals towards visuals offered in self-care applications. Semi-structured interviews were conducted with ten healthcare professionals which were then analyzed using a thematic analysis approach. In general, healthcare professionals were pleased with the visuals that are presented in existing self-care applications provided that they foster self-reflection amongst users. Several design factors were recommended to be included in self-care applications such as motivating and encouraging notifications, suitable colored graphs, emphasizing attitude change, reliable visuals, and respecting user privacy.

Keywords: Healthcare · Visualization · Self-care applications · Self-reflective · Design requirements

1 Introduction

The developments in technology have changed peoples' approach towards the healthcare industry. This includes the innovation of accessing healthcare information through healthcare applications. The incorporation of electronic communication and healthcare has conveyed a message that it is important to have knowledge about your own health progress at your fingertips when it comes to the decision-making process regarding your health.

Self-care applications are tools that enable healthcare consumers to manage their own health with, or at times without, the assistance of their healthcare providers. Healthcare providers find these applications to be empowering patients and their families to self-care and handle their own healthcare [1]. These self-care applications have the ability to reach many healthcare consumers and mostly feature graphical

F. Saeed et al. (Eds.): IRICT 2019, AISC 1073, pp. 881–893, 2020.
https://doi.org/10.1007/978-3-030-33582-3_83

representations of health-related data. The visuals are the center of attraction of a self-care application where users rely solely on them to comprehend their health progress.

Healthcare visualization in self-care applications can be categorized according to the type of platform on which it is delivered: (1) web-based health support applications, (2) mobile-based health support applications, (3) stand-alone health monitoring devices, and (4) wearable health monitoring devices [2]. Among all the categories of applications, it is apparent that visuals play a significant role in communicating the health progress of the users. Visuals in the web-based support applications are more complex than for the other categories since there is a bigger screen space to present the health information. The information can be made more accessible by adapting selected design principles from other categories of applications.

Healthcare consumers prefer to have their health information accessible at their fingertips anytime and anywhere therefore, self-care applications are becoming popular. For instance, *MyFitnessPal*, an application that helps people to reach their fitness goal by managing their food intake and keeping track of the number of calories they should consume after assessing their body mass index (BMI), has more than 200 million registered users that are using this app to inspire their own health and fitness journey [3]. *MyFitnessPal* has a rich database of food complemented with calorie counts that enable users to conveniently track their own food intake to live a healthier lifestyle. Consumers using this type of application will be conscious and eager to know about their health condition from time to time. They will be alert at all times to compare what they would like to consume to what and how much they are allowed to consume according to their body and health condition. Figure 1 shows a user's profile presenting his weekly calorie intake [4]. Moreover, *PatientsLikeMe* is a popular example that has more than 600,000 registered members. The users share their health data to track their progress and learn from the experiences of other patients who are experiencing similar health conditions [5]. Figure 2 shows two example features in the data entry and visualization system [6].

These applications have great potential in empowering consumers to manage, as well as prevent, life threatening diseases. Self-care applications present health information via various types of visuals. The data presented visually have a bigger impact on healthcare consumers than words alone that they may have difficulty in interpreting [7]. However, it is crucial that information presented is comprehensible and is interpreted correctly by the consumers. Misinterpretation of more complex health visuals leads to incorrect decision-making which can be harmful [8].

Visuals presented in self-care applications should ideally be self-reflective in ensuring that they are self-explanatory and effortlessly understandable. A consumer who is health literate will have the advantage of understanding the health information presented in self-care applications [9]. Otherwise, it can be difficult to gauge the information presented via the applications unless the visuals incorporated follow certain principles that would allow comprehensibility. Simple formatted graphs are generally preferred to more complex graphical information, irrespective of education level of the consumer [10].

Obtaining user requirements from target users of self-care applications is essential in providing high-quality care [11]. Often the requirements are focused towards the features that are to be included in an application, but less emphasis is given to the

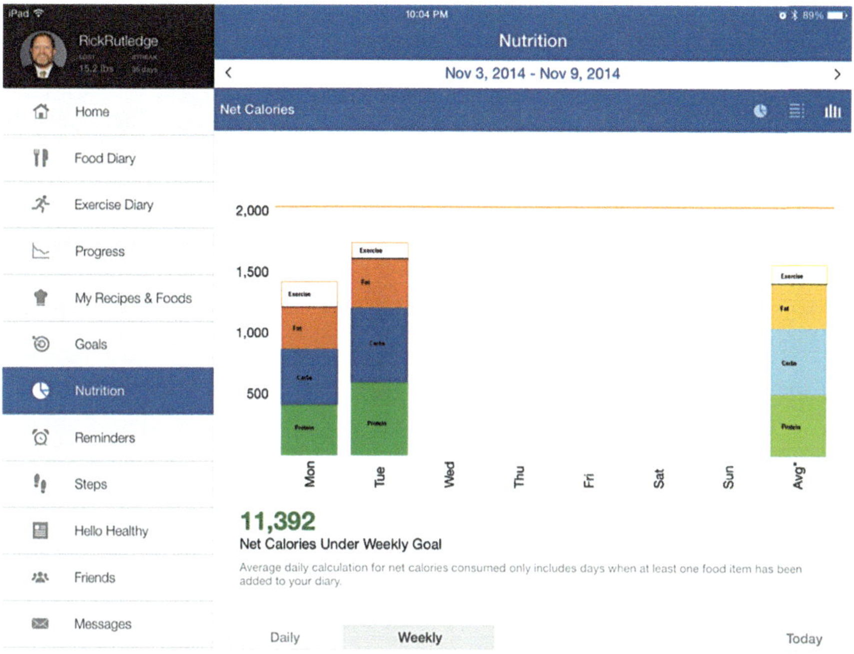

Fig. 1. Weekly calorie intake from a user's profile.

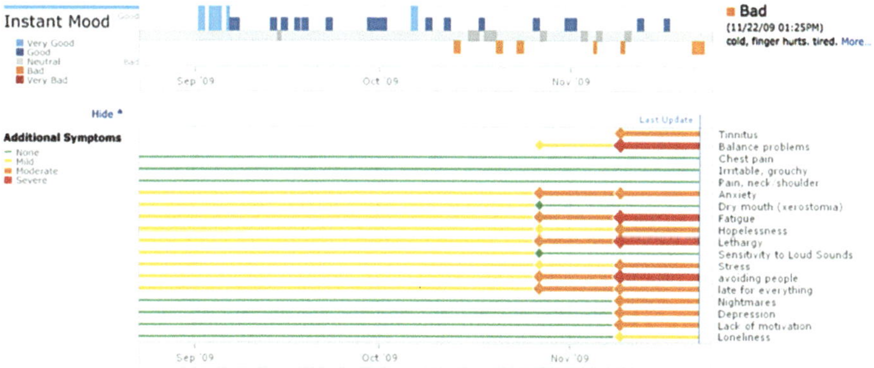

Fig. 2. Example features in the data entry and visualization system

design aspects of the visuals. We have conducted a Focus Group Discussion (FGD) study with a group of healthcare consumers in obtaining their needs and perception towards self-care visuals. Based on the responses from the FGD and literature survey, we developed a model consisting of important dimensions and related design principles that would aid application developers in designing self-reflective visuals [12].

Apart from the target users, we aim to incorporate the perspectives of healthcare professionals in the design process of the self-reflective visualization model. The goal is to validate the model in ensuring that all the necessary elements are included. We conducted semi-structured interviews with ten healthcare professionals. This paper offers and discusses the findings of the interviews and is organized as follows: Sect. 2 presents the details on the developed self-reflective visualization model. Section 3 describes the methodology that was used in steering the data collection session. Section 4 reports the findings from the interviews. Section 5 discusses the outcomes and implications of the semi-structured interviews. Finally, Sect. 6 summarizes the study and shapes the future work.

2 Self-reflective Visualization Model

The self-reflective visualization model, as illustrated in Fig. 3 is a conceptual model that presents a set of design principles to be considered in designing visuals in self-care applications [12]. This model was developed based on the analysis of existing design factors and user requirements.

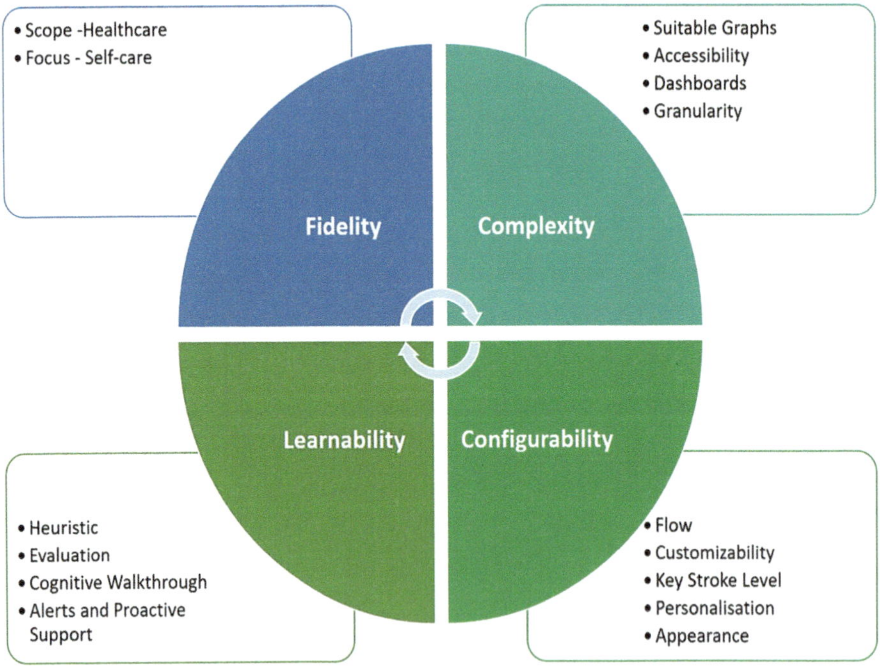

Fig. 3. Self-reflective visualization model with the dimensions and design principles for designing self-reflective visuals.

Ultimately, the aim of the model is to assist health application developers as a guideline in producing self-reflective health visuals in self-care applications. The model is split into four dimensions: *Fidelity*, *Complexity*, *Configurability* and *Learnability*. Each of these dimensions has a few design principles respectively.

The first dimension, *Fidelity,* is centering the degree of exactness of visuals which is the main purpose of the application. Thus, there are two principles namely *Scope towards Healthcare*, which ensures the visuals are primarily targeted to improve health and *Focuson Self-care*, which drives the application to focus on serving self-care as the ideal goal.

The second dimension, *Complexity,* reflects the details or the intricacy of the visuals that are presented within the application. Therefore, the principles are *Suitable Graphs* that makes sure the graphs displayed in the application are carefully formatted and *Accessibility* covers the choice of appropriate menus and toolbars, icons and screen & device compatibility of the application. Furthermore, *Dashboards* is a principle that ensures the application has a summary of all the elements that are used in the application in the home page and *Granularity* is the level of detailing of the data set, where the data visualized should be detailed yet kept simple and understandable.

The third dimension is *Configurability* that focuses on the arrangements of the elements within the application. Hence, the principles are *Flow* which keeps the steady movement within the elements in the app and *Customizability* ensures the application can be customized according to the users' taste and age group using the size and the color of the fonts used in the app. Additionally, *Key Stroke Level* maintains the minimal level of transition from one page to another, *Personalization* enables users' to change on a daily basis what they want to see on their feed and *Appearance* aims to present a simple and user-friendly overall impression of the application.

The last dimension, *Learnability,* targets the competence of the users to learn how to use the application. To enable quick learning, *Heuristic Evaluation* enables users to discover and learn some ways to use the application by themselves by depending only on the visuals. Also, *Cognitive Walkthrough* aids designers/developers to gauge understanding of potential users. In addition, the *Alerts and Proactive Support* feature helps users with prompted notifications and recommendations as necessary when using the application.

3 Methodology

A semi-structured interview method was selected for this study. This method gave us the flexibility to bring up new questions when prompted by the interviewee's responses [13]. This also enabled us to clarify the meaning of certain responses immediately. Open-ended questions were asked to motivate the interviewee's response. This technique allowed us to obtain the perspectives of the healthcare professionals effortlessly, since they could share their views freely [14, 15].

We conducted semi-structured interviews to explore healthcare professionals' understanding of the self-reflective visualization model. The semi-structured interview sessions were conducted among ten healthcare professionals from various health areas during the months of December 2018 and January 2019. The selected professionals are of various areas of expertise and represent different genders and ethnicities. They have knowledge of and have dealt with self-care applications and the visuals presented in them.

Each interview session lasted from 45 min to one hour. At the beginning of the interview session, a consent was taken from each professional. At the end of the session, a token of appreciation was given for their participation, time and effort. The interviews began with the health professionals presenting their credentials. This was followed by a pre-determined set of open-ended questions that were asked to gain their insights on the visuals of self-care applications.

Each discussion was audio-recorded, and later transcribed. We implemented the Thematic Analysis method to analyze the interview data [16]. Notes obtained from each session were compared with the audio-recordings. The transcripts from the interviews were read several times to highlight the themes, to identify the factors and assorted responses. The transcripts were coded, categorized and analyzed by identifying the recurring themes and assigning categorization codes. Each session was individually analyzed and was then compared with other sessions. The conclusions were extracted and verified individually.

4 Results

Ten healthcare professionals participated in the interview sessions. The characteristics of the professionals are displayed in Table 1. The outcome of the interview sessions is divided into six themes – motivation and encouragements, awareness and education, attractive and relatable graphs, limitation and reliability, attitude and family support, and privacy. These themes were developed from the responses of the healthcare professionals.

Table 1. Characteristics of ten healthcare professionals.

Category	Frequency	Percentage (%)
Affiliation		
Housemen	2	20
Medical officers	8	80
Gender		
Male	4	40
Female	6	60
Years of experience		
Less than 2 years	2	20
2–4 years	6	60
5–7 years	2	20

4.1 Motivations and Encouragements

All the healthcare doctors mentioned that is vital to monitor your health condition on a daily basis and there is no other way to do it easily than by using self-care applications. Self-care application visuals help healthcare consumers to further understand their health conditions. However, all the doctors felt that despite having these facilities in self-care applications, not many patients are exposed to or are using them. Therefore, they emphasized that it is necessary for the application to motivate patients to use it.

Three out of ten doctors suggested that it would be exciting for the patients to use the application if there were pop-ups visually motivating them to achieve better readings. Moreover, the visuals must be able to convey the message that they can do better to improve their readings. For instance, giving a user milestone targets to improve their vital readings by suggesting to they go for a 20-min average-paced walk on a daily basis. The patients should record their readings after the walk and the visuals must be able to compare their readings to show the differences before and after the walks. Once the patients reach a milestone, the application should probably give rewards to them in the form voucher redemptions, which they can use to shop for any medical supplies (e.g. walking sticks, crutches, milk formulated for diabetic patients, blood sugar level checking machine) through authorized online shopping applications. The vouchers would encourage them to put in more effort in taking care of their health and lead a healthier lifestyle.

Besides that, five of the interviewed doctors felt that the application should accurately compare the users' current vital reading with the average reading for a person of their age through suitable colored charts. This would keep the users aware of their readings and they would know they have to do something to improve them. One of the doctors said: *"If a user's vital readings are out of the normal range, there should be a pop-up message advising and alerting the user to visit the nearest clinician or healthcare provider. Their quick actions could save their life before it is too late"*. Another professional shared that they appreciate it if patients come in early knowing their condition is dangerous, so that they will be able to start immediately with possible treatments suggested by the doctors.

4.2 Education and Awareness

The awareness of the importance of self-care is low, as mentioned by two doctors. Most patients are less concerned about taking care of themselves without assistance from other people. Hence, one of the doctors interviewed said that self-care applications should educate users about their own health conditions in layman's terms to enable users to easily understand. He also added: *"Some would prefer to see and understand better with detailed simple visuals explaining about their health, rather than doctors telling them in complicated terms"*. This is because visuals give a clearer picture to patients, they are able to visualize their situation without getting confused. Seven out of ten doctors shared that notifications should be also sent to the patient's next of kin if their vital readings are out of the normal range, so that they too are aware, and they can take the patient to the nearest clinic or hospital to check on the patient.

Health literacy can increase the understanding of risks, symptoms and complications involved in one's health condition. All of the healthcare professionals agreed on this. Furthermore, one of them mentioned: *"When a patient or their family members know the risks and complications that they would have to face if one of the treatments that they tried did not work, it would be easier for us doctors to rectify this method and proceed with another option"*. He added that every patient is different, and every health condition is subjective to a patient's body condition and health capacity.

4.3 Attractive and Relatable Graphs

All the healthcare professionals mentioned that it would be better if the applications had bright and colored visuals. They advised that the colors will attract users and would inspire interest to use the application often. It will also help them to remember certain features and allow them to use these without memorizing how to maneuver them.

Another interviewee mentioned that even though the visuals displayed are colorful, they should not be unnecessarily colorful. She said: *"If the displays are too colorful, users of the older age group might feel out of place and childish to use the app. Therefore, the application should have a customizability feature, so that they are able to customize the colors that they want to see throughout the application. It can also be customized according to the users' age group"*.

Besides that, four of the doctors interviewed emphasized that it is not compulsory for the application itself to be colorful, but it is a must for the graphs presented to be as colorful as possible. They said that the different colors in the graphs can help a user to differentiate their vital readings easily without any confusion.

4.4 Limitation and Reliability

The healthcare professionals are well aware of the limitations in terms of the visuals that are available for themselves and healthcare consumers. Nearly all of them felt that it was acceptable and fair to limit the users with regard to the visuals that they see. The reason for this limitation is that certain visuals (e.g. scan reports, blood reports, x-ray charts, other test results) are specifically prepared for health professionals to see and they could be misinterpreted or misunderstood by non-medically trained consumers. One of the doctors shared: *"Even though patients are becoming health literate and curious to know what is happening in their body, knowing too many details as how we get to see will confuse them and make them anxious to understand their current health condition"*.

Another doctor mentioned that knowing too much about their health condition is dangerous. This happens when patients and their family members read up excessively about a particular disease or condition and they blindly come to conclusions. He added: *"Every symptom and medical condition is different; therefore, it is always necessary for patients to seek their healthcare provider's advice for a confirmed diagnosis. It is wrong to generalize a patient's symptoms"*.

Additionally, the currently available self-care applications and their visuals are based on US health guidelines. Malaysian based visuals and self-care applications will be helpful to enable Malaysian users to relate their health conditions and vital readings to how they are presented to them in hospitals when they go for their checkups. An interviewee said: *"If the scale range in a US-based app may show abnormal, according to the Malaysian-based scale reading it may show normal. Minute differences like this can cause the patient to get agitated and go into depression".*

4.5 Attitude and Family Support

Six out of ten interviewed healthcare professionals expressed that the patients' attitude towards their health and lifestyle is important. The visuals and the application itself can be very attractive and exciting, but if the users themselves have no interest in taking good care of their health then no one can do it for them. The change has to happen within themselves. A doctor mentioned that patients must realize that any treatment or remedy will not happen overnight. It takes gradual progress and also consistency in undergoing the treatment. He said: *"Patients come in for treatment with the expectation of getting 100% back to their normal-self overnight. It may look practical in movies, but in real-life, patience is virtue".*

Regardless of any health condition a patient is experiencing, they will feel better with their family members around them who play an important role in pushing them to try new possibilities to improve their health conditions. Another doctor expressed that some elderly patients may be old-fashioned and prefer to write down their medication intake in a book and are not interested in getting to know the new technology. However, their children can teach and support them in using a self-care application to note down their medication intake with ease. Elderly patients should also be taught how the visuals help to display their details.

4.6 Privacy

A huge amount of personal health data is generated or fed into self-care applications by the users, either manually or through sensors. It is important to ensure that these data are not shared with any other parties without receiving a consent from the patients. Unlike medical applications that are governed by hospitals, these applications should strictly adhere to privacy and security policies in order to gain the trust of the users so that they can make effective use of the application for the benefit of their health.

Three of the doctors expressed that the uptake of self-care applications would be greater if they collaborated with hospitals or health authorities. Not all patients are willing to share their health information with anyone, including family members. This could be due to the self-consciousness and embarrassment that they have to face due to their health condition. They may fear that others might avoid them or speak negatively about them. Although certain self-care applications encourage the involvement of family members to be part of the health improving process, this feature should be presented as an option to the users.

5 Discussion

These semi-structured interview responses emphasized that all the healthcare professionals are in favor of visuals presented in self-care applications, provided that they are interpreted well and educate users. They shared that it is important for the visuals to be accurate, informative, and comprehensive so that they benefit the patients that use the applications.

Besides having visuals that portray health trends, visual elements that motivate and encourage users are also important. The motivation could be as simple as alerting users with encouraging statements via notifications if there are changes in their health readings. They will feel like they have achieved something as they never have before and will be encouraged to do more to improve their health. The milestone system will also help users to focus on a target and strive to achieve it on a daily or weekly basis by keeping them on track when monitoring their health progress. The visuals should also motivate them when they fail to achieve their targets which will encourage them to try better the next time. For instance, if a user wanted to lose 10 kilograms (kgs) of weight in two months only managed to lose 4 kgs, the application should motivate him or her to not give up and to try other possible ways to achieve their body weight goal such as by reducing carbohydrate intake or trying other alternative exercise methods. Constant displays of prompts are known to motivate the health behavior of individuals who are obese to be active from time-to-time [17].

It was noted from the doctors that the majority of patients and their family members are not aware of the importance of using self-care applications to monitor health. Although some patients and their family members are aware of a few applications that are currently available, they are hesitant to use them. They mostly gave reasons such as the visuals presented in the application are ineffective and the terminology used is not comprehensible. Furthermore, they also feel inferior by using an application and worry that others will consider them to be not independent enough in looking after their health. Well-designed applications have great potential in improving health and giving users a sense of control over their health journey. Visuals presented in these applications are intended to educate patients in a fun and exciting manner about their health. Playful and appealing visuals make users more excited about using the applications [19].

Furthermore, health trends that are presented as graphs are best illustrated with suitable colors. Studies have shown that having variations in color when labelling graphs creates a bigger impact for users in understanding and differentiating the points in the graph [18]. It also reduces the inconvenience of understanding the trend changes, which can be confusing and time consuming. Users have mixed opinions about the choice of colors throughout the application. Some prefer muted colors and some like striking bright colors. The customizability of the colors will be handy for users who have different tastes about this.

Visuals in self-care applications do have some limits with regard to what is presented to the users. Users may desire to see the visuals that are offered to doctors in hospitals. There are many reasons for this potential limitation. Users should be able to accept the fact that even though the doctors are better qualified to understand the information and visuals that they see, ultimately, the health professionals will try their best to summarize and convey their diagnosis in a simple manner. This will avoid users misinterpreting the visuals and the details in them. The outcome of misinterpreting information could cause confusion or doubts when using the applications [20]. Hence, self-care applications should only present information that is relevant, sufficient and comprehensible to the users. These applications should be endorsed and approved by the authorities, such as the World Health Organisation (WHO) or the respective health ministry of a country. This would give patients trust in using the applications.

The change in a patient's life can only happen if they have the thought to do so within themselves. They should be proactive in taking care of their health by managing their life style with the support of self-care applications. They should be aware that there is support available in the form of technology that could be useful to them in meeting their health goals. Novel self-care applications encourage the involvement of a spouse or family members in helping to improve the patient's health condition. Family support is mostly the backbone in a patient's recovery from health conditions. Without family support, patients may feel demoralized and might go into depression assuming no one is interested about their health progress and they are no longer needed by anyone. It has been proven that family support has helped in improving medication adherence and mental status in patients [21].

Furthermore, applications developers should be sensitive towards the privacy needs of the self-care application users. The health information stored through these applications should solely be used for the patients unless consent is obtained from them to use it for other purposes. For instance, studies have shown that data privacy management is critical in creating trust towards Personal Informatics (PI) [22].

6 Conclusion and Future Work

The views and thoughts of healthcare professionals in designing visuals in self-care applications are undoubtedly necessary to collect appropriate and valid information. This is crucial as their ideas and suggestions can inspire changes in the visuals to improve the way a patient understands their health condition with ease.

Although our interview study is based on a small sample size, it has provided us with valuable understanding of ways to better convey health visuals to patients. These insights aid greatly to identify patients' visual preferences to be incorporated in a self-care application. The findings of these semi-structured interviews will be used to add value in terms of design principles to our existing model to better design visuals in self-care applications. The model will be validated and verified again. The refined principles will be transformed into a set questionnaire that will be given to healthcare consumers to investigate their perspectives on the principles that will be used when designing visuals in self-care applications. These outcomes will be used to finalize the model. The model can then be applied as a guideline for healthcare application developers to adapt and design self-reflective and comprehensible visualization in self-care applications for healthcare consumers.

Acknowledgement. This study is funded by Universiti Tenaga Nasional under the YCU Research Grant. We thank the healthcare professionals for their time and willingness to share their thoughts with us.

References

1. Silva, B.M.C., Rodrigues, J.J.P.C., De La, I., Díez, T., López-Coronado, M., Saleem, K.: Mobile-health: a review of current state in 2015. J. Biomed. Inform. **56**, 265–272 (2015)
2. Visvalingam, A., Dhillon, J.S., Gunasekaran, S.S., Hoe, A.C.K.: Self-reflective visualizations of patient-centered health information systems. In: Advances in Intelligent Systems and Computing, pp. 711–725. Springer, Cham (2018)
3. MyFitnessPal | MyFitnessPal.com. https://www.myfitnesspal.com/. Accessed 08 Feb 2019
4. Weekly Chart Details - App Suggestion for iPhone and iPad—MyFitnessPal.com. https://community.myfitnesspal.com/en/discussion/10016617/weekly-chart-details-app-suggestion-for-iphone-and-ipad. Accessed 08 Feb 2019
5. PatientsLikeMe | Conditions. https://www.patientslikeme.com/conditions. Accessed 21 Mar 2018
6. Brubaker, J.R., Lustig, C., Hayes, G.R.: PatientsLikeMe: empowerment and representation in a patient-centered social network. ResearchGate **15**, 715–722 (2010)
7. HealthCare Foundation, C.: Worth a Thousand Words: How to Display Health Data (2014)
8. Bantug, E.T., Coles, T., Smith, K.C., Snyder, C.F., Rouette, J., Brundage, M.D.: Graphical displays of patient-reported outcomes (PRO) for use in clinical practice: what makes a pro picture worth a thousand words? Patient Educ. Couns. **99**, 483–490 (2016)
9. Lee, S.-Y.D., Arozullah, A.M., Cho, Y.I.: Health literacy, social support, and health: a research agenda. Soc. Sci. Med. **58**, 1309–1321 (2004)
10. Brundage, M., Feldman-Stewart, D., Leis, A., Bezjak, A., Degner, L., Velji, K., Zetes-Zanatta, L., Tu, D., Ritvo, P., Pater, J.: Communicating quality of life information to cancer patients: a study of six presentation formats. J. Clin. Oncol. **23**, 6949–6956 (2005)
11. Åhlfeldt, R.M., Persson, A., Rexhepi, H., Wåhlander, K.: Supporting active patient and health care collaboration: a prototype for future health care information systems. Health Inform. J. **22**, 839–853 (2016)
12. Visvalingam, A., Dhillon, J.S., Gunasekaran, S.S., Hoe, A.C.K.: Designing self-reflective visualizations in patient-cantered systems. Int. J. Eng. Technol. **7**, 449–456 (2018)
13. Chauncey, W.: Semi-Structured Interviews. Interview Tech. UX Pract, pp. 23–41 (2014)
14. James, L.J., Wong, G., Craig, J.C., Ju, A., Williams, N., Lim, W.H., Cross, N., Tong, A.: Beliefs and attitudes to bowel cancer screening in patients with CKD: a semistructured interview study. Clin. J. Am. Soc. Nephrol. **12**, 568–576 (2017)
15. Ramirez-Rubio, O., Brooks, D.R., Amador, J.J., Kaufman, J.S., Weiner, D.E., Scammell, M. K.: Chronic kidney disease in Nicaragua: a qualitative analysis of semi-structured interviews with physicians and pharmacists. BMC Public Health **13**(1), 350 (2013)
16. Roberts, M., Pettigrew, S.: A thematic content analysis of children's food advertising. Int. J. Advert. **26**, 357–367 (2007)
17. Bond, D.S., Thomas, J.G., Raynor, H.A., Moon, J., Sieling, J., Trautvetter, J., Leblond, T., Wing, R.R.: B-MOBILE-a smartphone-based intervention to reduce sedentary time in overweight/obese individuals: a within-subjects experimental trial. PLoS ONE **9**, 8 (2014)

18. Arrue, M., Vigo, M., Abascal, J.: Including heterogeneous Web accessibility guidelines in the development process. Lecture Notes in Computer Science (including Subseries Lecture Notes in Artificial Intelligence and Lecture Notes Bioinformatics). LNCS, vol. 4940, pp. 620–637 (2008)
19. Tang, J., Abraham, C., Greaves, C.: Full Title: how can weight-loss app designers' best engage and support users? A qualitative investigation. Br. J. Health. Psychol. **20**(1), 151–171 (2015)
20. Wågbø, H.D.: The patient perspective : utilizing information visualization to present health records to patients, p. 133 (2014)
21. Tabasi, H.K., Madarshahian, F., Nikoo, M.K., Hassanabadi, M., Mahmoudirad, G.: Impact of family support improvement behaviors on anti diabetic medication adherence and cognition in type 2 diabetic patients. J. Diabetes Metab. Disord. **13**(1), 113 (2014)
22. Rapp, A., Cena, F.: Personal informatics for everyday life: how users without prior self-tracking experience engage with personal data. Int. J. Hum. Comput. Stud. **94**, 1–17 (2016)

A Conceptual Framework for Adopting Automation and Robotics Innovations in the Transformational Companies in the Kingdom of Saudi Arabia

Mohammed Aldossari(✉) and Abdullah Mohd Zin

Faculty of Information Science and Technology, Universiti Kebangsaan Malaysia, 43600 Bangi, Selangor, Malaysia
aldossari10102010@gmail.com, dftsm@ukm.edu.my

Abstract. Modern organizations are facing increasing labor costs and a lack of human resource, which has encouraged them to invest in robots, particularly as robots do not demand raises and can work 24/7. They are able to perform tasks that humans are unable to, like working in challenging conditions and achieving feats with accuracy. This paper aims to identify the factors influencing the behavioral intention to adopt robotics and automation among transformational companies in the kingdom of Saudi Arabia (KSA). This identification is to construct a conceptual framework for the proposed adoption of robotics and automation. The Technology Acceptance Model (TAM) and Technology, Organization, Environment (TOE) theory were the basis for the proposed conceptual framework. A literature review, including theory analysis, was used to identify factors and experts were consulted through interviews. The results show that the proposed conceptual framework was comprehensive and integral. All the prepositions were found to be important and supportive.

Keywords: Robotics · Automation · Adoption · Saudi Arabia · Transformational companies

1 Introduction

There is no doubt on the importance of manufacturing to the global economy and its significant contributions. In the last decade, the manufacturing sector lost almost 30% of its resources and investments, decreasing the quality of domestic goods, global share, and manufacturing capabilities. In answer to this, innovative methods have been proposed for the implementation of new technologies such as robotics and automation in the new manufacturing sector [1].

The current manufacturing industry is focused on enhancing competitiveness via the integration of ICT technologies to establish a direction for new growth. The fourth revolution in manufacturing, namely Smart Manufacturing, is also considered a new paradigm, and it refers to a collection of current technologies supporting effective and accurate engineering decision-making in real-time via the use of ICT technologies in combination with existing manufacturing technologies. Current organizations have

F. Saeed et al. (Eds.): IRICT 2019, AISC 1073, pp. 894–905, 2020.
https://doi.org/10.1007/978-3-030-33582-3_84

been urged to adopt such technologies to adapt to the ongoing changes brought on by IT [2].

Today's world is characterized by what would be science fiction a decade ago. In fact, there is increasing development of smart innovations, with scientist and engineer circles attempting to develop new and innovative devices, homes, factories, and cities. Regardless of ongoing progress in this field, these concepts are just visions of a future that needs significant efforts to be realized [3].

More importantly, the economy has been boosted by new robotics, although the potential benefits remain ambiguous. Therefore, it is crucial to determine the benefits and limitations of robotics in the early stages of implementation to establish additional employment for overall economic development and for limitations to be examined and minimized [4].

Specifically speaking, a robot refers to a system comprised of sensors, control systems, manipulators, power supplies, and software that works together to achieve a task [4]. Robotics is described as an independent innovation that comprises a sensor, actuator, microcomputer, and transceiver. 'Smart' is often utilized to describe an object fitted with additional features that introduce multi-platform communications and enhanced computational abilities. Such intelligent devices can be displayed through a crisscrossing network of other smart devices that are able to check system updates and decide on a course of action, which is commonly referred to as a smart network [3].

Industrial robot development and growth can be traced back to 1947, after which the development impetus was directed by technology developments. In modern times, industrial robots are widely applied, mostly as substitutes for human resources in various industries. Ongoing robotic technology growth and development has always posed challenges, such as the negative effect of robotics on employment. Because of cost reductions and production improvements, industrial nations have been urged to use robots to replace human laborer's in production lines [5]. Nevertheless, the extensive utilization of robotic systems in industry can be directed to ensure their successful adoption and to avoid adverse outcomes.

In the global robotics market, using computer-controlled robots to complete manual tasks is becoming more common. Robots call for innovative and creative ways to make use of technology to differentiate a company from its rivals and achieve sustainable advantage. The biggest obstacle to this is determining where and how to begin, and the development of such business model calls for knowledge to assist firms to move forward to a brighter and more dynamic future [6].

The focus of this paper is one the robotics and automation adoption in the transformative industry in Saudi Arabia, as a developing nation. Saudi Arabia is in urgent need to adopt the robotics and automation to help in the productivity of the transformational industrial sector [7]. It lays stress on the role of robotics in enhancing productivity, and on the framework importance in guiding effective and efficient technology adoption and use. On this basis, robotics and automation should be considered as crucial assets of transformative industry and its productivity, with the frameworks used as guidance. This paper is structured as follow; related works on robotics and automation adoption is presented after the introduction section. Thereafter, the methodology of the current study is accommodated in Sect. 3 while Sect. 4 is dedicated for the framework development. Section 6 is allocated for implication of the study and the last section is the conclusion of the paper.

2 Related Work on Robotics and Automation Adoption

Prior studies in the literature primarily focused on the advantages of technology adoption, while largely ignoring the determinants of adoption and the required technological tools [8]. Important promotion merits include high labor intensity, dependence on local skills and technology, entrepreneurship development, ingenuity and increasingly industrial networks, which are more accessible to developing nations [9]. Although the potential for robot adoption is high, current adoption is still lagging behind and to enhance it, it is required to that the determinants of adoption decisions are understood [10]. This study urgently calls for the identification of what factors influence behavioral intention to adopt and use robotics and automation in Saudi transformational companies.

On the basis of academic research on worldwide robotics, robotics may be the reason behind the 1/10th increase in productivity from 1993 to 2007 [11–14]. On the basis of the 2016 Economic Report of the President, the demand for robotics all over the world has almost doubled from 2010 to 2014 and this was accompanied by the number of robotic-oriented patents (CEA 2016), indicating the increased contribution of robots to productivity.

A majority of the developed frameworks for adoption and implementation process in developing nations were adapted from the experiences of Western countries [15–18]. Important lessons can be culled from the experiences of the West but the application of the frameworks developed therein are limited when juxtaposed to developing nations owing to different factors. In IS studies, the adoption framework covers factors assisting proper technology adoption [19–23].

Studies in the literature have accepted IT/IS models from other studies in their examination of technology acceptance. A majority of these studies were conducted in a specific environment. In this section, the studies dedicated to robotics and automation acceptance as well as relevant variables are presented and discussed.

A clear explanation for social robot acceptance and use calls for an understanding of the determinants for major acceptance variables. And in this regard, Conti, Cattani [24] and Conti, Di Nuovo [25] stated the way cultural backgrounds can affect the perception and intention of robot use as a system of instruments in the near future. The study sample comprised 37 Italian students and 37 U.K. students as future psychology professionals, who experienced the actual capabilities of a humanoid robot via a live demo. The study determined the UTAUT major factors in an attempt to determine cultural differences using a UTAUT questionnaire. The questionnaire was designed and validated for the examination of robot acceptance and use. The findings from the discriminant analysis indicated a high degree of separation between the two student groups, evidencing different approaches to using robotics in the two countries. The study factors included anxiety, attitude, facilitating conditions, intention to use, perceived adaptability, perceived enjoyment, perceived sociability, perceived usefulness, social influence, social presence, and trust.

In a study of the same caliber, De Graaf and Allouch [26] revealed that for the successful introduction of social robots, it is important to understand the underlying reasons upon which potential users accept them in their homes. An extensive literature review was conducted by the study to determine the variables that affect the acceptance of social robots, which were categorized into utilitarian variables, hedonic variables, user characteristics, social normative beliefs, and control beliefs. Sixty participants interacted with a social robot and the interaction experience was evaluated. Based on the findings, usefulness, adaptability, enjoyment, sociability, companionship, and perceived behavioral control are important to social robots acceptance. As such, this study is expected to contribute to the human-robot interaction literature with its determination of the variables that could facilitate the acceptance of social robots. This study is also expected to be a benchmark for developing an integral model that includes relevant social robot acceptance determinants.

In a review of the different political views regarding social robotics for the skill augmentation of children with autism and brain disabilities, Hashim and Yussof [27] analyzed article publications in peer-reviewed journals and conference proceedings. Three emerging themes were highlighted from the review and they are influence, acceptance, and adoption. The findings of their study revealed that the social skills of brain-impaired children may be impacted, but they are not present during the robot development and design process. Rather human social skills were appropriated to general robot capabilities. Thus, the necessity for social robots is aligned with societal changes, increased demographics, and demands for healthcare. More importantly, the conceptualization of brain-impaired children is that they are mentally and physically handicapped, helpless, and needing 24/7 care and robotic assistance. According to national status, the influence, acceptance, and adoption of social robotics are political and successful societal science calls for the re-examination and redefinition of the concept to leverage return on investments for robot manufacturing.

Studies dedicated to adopting robotics and automation are still few and far between and there is a dire need to examine technology to develop an appropriate framework that could assist in the proper use and enhancement of productivity among industrial firms.

ICT studies have highlighted different factors that significantly affect robotics and automation acceptance [4, 24, 26, 28–33], which include perceived ease of use, perceived usefulness, infrastructure, subjective norm, top management support, financial support, training, readiness, security, anxiety, efficacy, and reliability.

3 Methodology

In Fig. 1, the methodology phases that have been followed by this paper is shown. Three phases are included. The first phase is allocated to extract the factors that could influence the adoption of robotics and automation. Phase 2 is dedicated to develop framework constructions. It includes the factor classification and framework construction. There are several methods have been adopted in literature to determine the factors of technology adoption and implementation. Each approach is accompanied by strengths and weaknesses and a clarification of the techniques was made by [34]. In this paper, a methodology that suggested by Mukred, Yusof [35] as seen in Fig. 1.

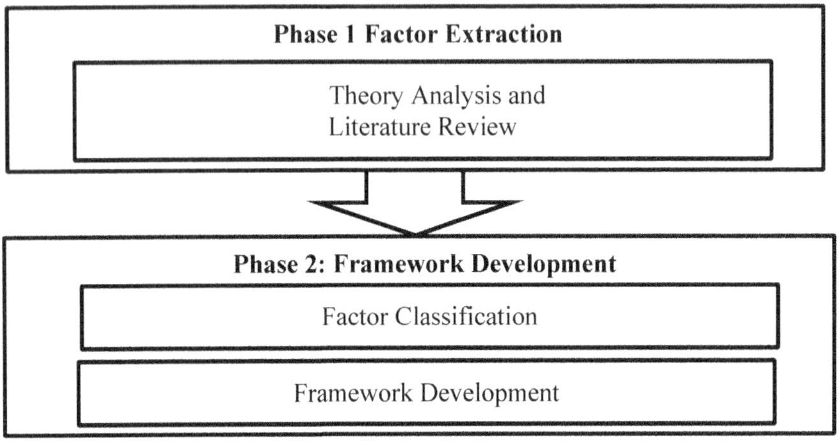

Fig. 1. Methodology adopted from [35]

In this paper, literature review method has been followed to determine the factors that influence the adoption of robotics and automation. The list of extracted factors was evaluated by group of 20 experts who are working in the same field. The experts ranked the factors and the final list of the extracted factors includes: perceived ease of use, perceived usefulness, infrastructure, subjective norms, top management support, financial support, training, readiness, efficacy, reliability, anxiety, and security.

4 Framework Development

This study adopted the Technology Acceptance Model 3 (TAM3) to examine intentions towards adopting automation and robotics based on specific technological, organizational, and trust variables. This study also integrated a taxonomic framework for automation and robotics to propose a framework that supports productivity among KSA industrial companies. The framework was established based on TOE theory. The framework was used to determine the influence of automation and robotics adoption on organizational productivity. This study is expected to contribute to the robotics adoption field of research, with the taxonomy development organizing required information. The study is also expected to help researchers as it determines and categorizes factors as well as relationships between factor categories.

Prior empirical studies have indicated several sub-important factors influence intentions towards adopting automation and robotics that are related to the trust dimension as a new dimension with technological and organizational dimensions. For the integration of the TOE framework with TAM3, the former's technology, organization, and trust dimensions were considered, which were all considered to contribute to successful robotics and automation technology adoption.

The trust dimension was included as a direct determinant of behavioral intentions towards adoption of technology as evidenced in prior studies [36, 37]. For example, in Alharbi [37] study, trust was established as a challenge towards technology adoption among users. Similarly, other scholars consider trust as a major aspect in e-commerce and e-government [38], while others [2, 3, 12, 39] evidenced that trust is a must for the adoption of robotics and automation technology.

The first framework level of factor dimensions (technology, organizational, and trust) are displayed in Fig. 2. It is evident that such factors represent exogenous variables directly influencing the adoption of robotics and automation technology (the endogenous variable). The second framework level (adoption stage) consists of intentions towards adopting robotics and automation technology, which represents the endogenous variable that directly influences organizational productivity. The third framework level (system use) consists of organizational productivity (dependent variable), which is also known as the outcome construct [40]. The third chapter furnishes the framework details and its factor classifications for framework development. Both TAM3 and TOE theory were adopted in this study to develop the proposed framework, with the latter used as a taxonomy for classification purposes. Figure 2 presents the study framework.

There is thus a dire need to examine intention to adopt technology factors in varying settings in terms of technology, organizations, and trust to analyze and determine productivity [33]. In the context of healthcare, [41] investigated the acceptance and adoption of care robots. In their study, personal behavioral intentions linked to the launching of robot applications in homecare were affected by a person's appreciation of the robots usefulness, the expectations of their colleagues and supervisors, and their perceptions of the learning required to use care robots. Specifically, the personnel cited the care robots value in reminding, guiding, and promoting older people's safety. It was confirmed by Heikkilä, Myyrä [33] that new technology adoption and investment aids that promoted investments were required in low-productivity areas where new technology investments still have the potential to maximize productivity, profitability, and competitiveness in the near future.

Past studies have evidenced that the application of automation technology and robotics is a growing trend in the livestock industry [42] and plays a key role in business productivity [12–14, 33]. Therefore, this study proposes that (Table 1):

Table 1. Preposition of the current study

No.	Preposition
P1	Technological factors have a positive influence on the adoption of robotics and automation
P1a	Perceived ease of use has a positive influence on the adoption of robotics and automation
P1b	Perceived usefulness has a positive influence on the adoption of robotics and automation
P1c	IT infrastructure has a positive influence on the adoption of robotics and automation
P1d	Subjective norm has a positive influence on the adoption of robotics and automation
P2	Organizational factors have a positive influence on the adoption of robotics and automation
P2a	Top management support has a positive influence on the adoption of robotics and automation
P2b	Financial support has a positive influence on the adoption of robotics and automation
P2c	Training has a positive influence on the adoption of robotics and automation
P2d	Readiness has a positive influence on the adoption of robotics and automation
P3	Trust factors have a positive influence on the adoption of robotics and automation
P3a	Efficacy has a positive influence on the adoption of robotics and automation
P3b	Reliability has a positive influence on the adoption of robotics and automation
P3c	Security has a positive influence on the adoption of robotics and automation
P3d	Anxiety has a positive influence on the adoption of robotics and automation
P4	Robotics and automation adoption has a positive impact on productivity

Despite the contribution of robots to productivity and management effectiveness, the mechanics behind how robots and automation contribute to firm-level productivity, the situations under which they complement or substitute labor, their effect on new firm formation, their contribution to effective and efficient management, and their shaping of regional economies still need in-depth examination. This is particularly evidenced by the scarcity of firm-level data when it comes to robotics and automation use [12], data which is necessary to answering the above questions and in assisting policymakers in laying down policies on robotics and the automation of economic and societal processes [43]. Studies should examine the relationship between robotics and automation adoption and industry productivity, and how adopting robotics will enhance Saudi transformational companies.

This paper is aimed to propose a framework for robotics and automation adoption in Saudi transformational companies to increase their productivity.

According to Cowan [44] study, today's world is characterized by quick changes in technology, which has transformed the world into a small village because of globalization, the communication revolution, and IT. Prior to automation, traditional management practices made it difficult for organizations to survive, much less thrive [45]. In the Arab world, specifically Saudi Arabia, western practices were adopted without unique considerations for the country's characteristics. In fact, the administrative and

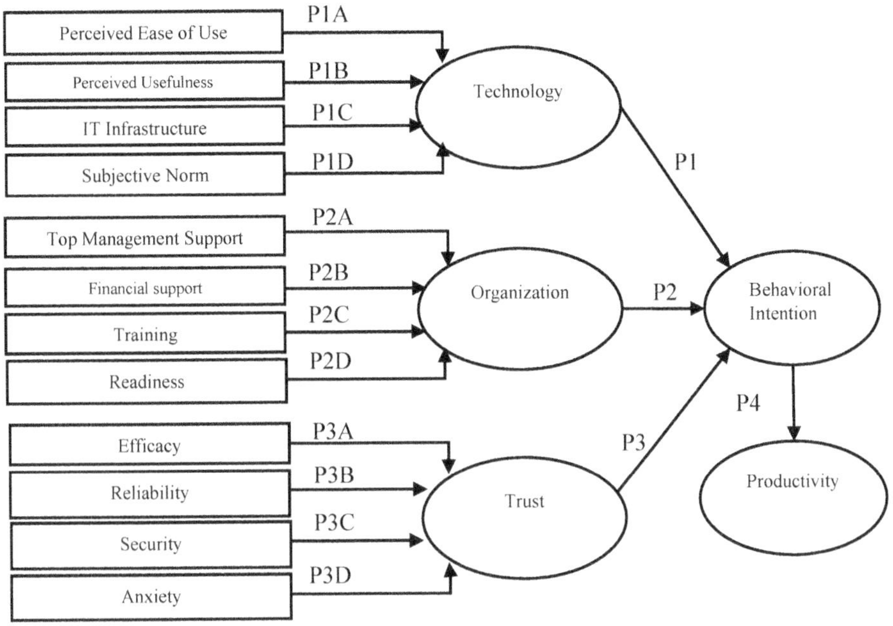

Fig. 2. The proposed framework of the study

managerial practices adopted in Saudi transformational companies lacked distinct frameworks for the adoption and use of robotics as well as innovative automation towards effective and efficient management and enhanced productivity.

5 Implication of the Study

This study is expected to practically contribute recommendations to the Saudi authorities when it comes to adopting robotics and automation technologies to increase the productivity of transformational industries. In addition to these practical contributions, this study aims to contribute to the literature as well.

The literature calls for the development of frameworks for adopting technology, particularly in the context of transformational companies. This research answers the call by adopting a distinct approach from existing studies to propose a new framework.

In the face of the recent dynamic technological developments in the field of robotics and automation, society is handed the potential to leverage robotics and automation to tackle several challenges. For the effective leveraging of such technologies, an in-depth understanding of the influence of robotics and automation on growth, productivity, labor, and equality has to be achieved. In this regard, there is a need for systematic data regarding robot and automation adoption and use, specifically at the business level, to provide insight into technologies role in the economy and society. The development and aggregation of such datasets, via census, surveys of public and private organizations, and internal data from individual firms, would provide the tools required by researchers and policymaker circles to conduct empirical investigations on the effect of robotics and automation as well as develop suitable responses to ongoing phenomenon.

The study meets the need for accurate data in this area to boost national competitiveness, especially for developing suitable policies. A minimal amount of information concerning robotics and automation could lead to an inability to prepare for technological advancements, which could eventually lead to missed opportunities and disastrous outcomes. For instance, decisions on taxing or subsidizing robots and automation depend on an understanding of whether or not technology should be used to substitute or complement labor. Such decisions can influence patterns of adoption. If such adoptions are made haphazardly without the right information, they could lower economic growth, lower employment levels, and lower wages. Additionally, data has to be properly used in response to technology adoption outcomes.

This study empirically tested the variables that influence the adoption of robotics and automation in transformational industries, which provide insights into increasing productivity. Although there is a lack in existing studies on automation and robotics in developing countries such as Saudi Arabia region, this study will provide the industrial sector with valuable information that can be used to guide the implementation of robotics and automation.

6 Conclusion

Without a doubt, robotics has a crucial role to play in industrial firms and in overall economic development. Robotics and automation hold the key to industrial company progress. For industry to be enhanced, new technologies have to be adopted to increase productivity. In effect, the adoption of new technologies calls for the examination of factors that facilitate their proper adoption and use. This study addressed major issues related to the factors that significantly influence robotics and automation adoption in transformative industries in Saudi Arabia. This study focused on the level to which technological factors in TAM and from the literature review (perceived ease of use, perceived usefulness, infrastructure, and subjective norms) were crucial to adoption. Aside from technological factors, organizational factors were culled from past models and the literature (top management support, financial support, training, and readiness) for their influence on robotics adoption in developing country public firms. Finally, trust factors were adopted from prior literature and they included efficacy, reliability, anxiety, and security are also found to be vital for successful adoption of robotics and automation in developing countries as KSA case. This paper highlighted the lack of studies dedicated to robotics and automation, particularly in their role of enhancing productivity and the developed framework for adoption success. The paper touched upon the limitations of prior studies when it comes to robotics. There is thus a need to develop and put forward a framework for proper robotic and automation adoption as the lack of such framework could lead to major challenges in successful system implementation.

References

1. Nagadi, K.: A framework to generate a smart manufacturing system configurations using agents and optimization (2016)
2. Kang, H.S., et al.: Smart manufacturing: past research, present findings, and future directions. Int. J. Precis. Eng. Manuf.-Green Technol. **3**(1), 111–128 (2016)
3. Radziwon, A., et al.: The smart factory: exploring adaptive and flexible manufacturing solutions. Procedia Eng. **69**, 1184–1190 (2014)
4. Qureshi, M.O., Syed, R.S.: The impact of robotics on employment and motivation of employees in the service sector, with special reference to health care. Saf. Health Work **5**(4), 198–202 (2014)
5. Basiri, A., Mousazadeh, R.: Industrial robotic systems & international human rights. J. Pol. L. **11**, 53 (2018)
6. Riben, M.: Laboratory automation and middleware. Surg. Pathol. Clin. **8**(2), 175–186 (2015)
7. Aldossari, M., Zin, A.M.: The use of automation and robotic innovations in the transformational companies: systematic literature review. J. Theor. Appl. Inf. Technol. **97** (13), 3661–3690 (2019)
8. Xu, X., Thong, J.Y., Tam, K.Y.: Winning back technology disadopters: testing a technology readoption model in the context of mobile internet services. J. Manag. Inf. Syst. **34**(1), 102–140 (2017)
9. Mettler, T., Sprenger, M., Winter, R.: Service robots in hospitals: new perspectives on niche evolution and technology affordances. Eur. J. Inf. Syst. **26**(5), 451–468 (2017)
10. Akmam Syed Zakaria, S., et al.: Contextual, structural and behavioural factors influencing the adoption of industrialised building systems: a review. Arch. Eng. Des. Manag. **14**(1–2), 3–26 (2018)
11. Graetz, G., Michaels, G.: Robots at work (2015)
12. Seamans, R., Raj, M.: AI, Labor, Productivity and the Need for Firm-Level Data. National Bureau of Economic Research (2018)
13. Robotics, M.C.f.A.I. Robots, Productivity, Competitiveness and Growth. 2018. http://www.mibizgroup.com/Technology/AI/artificial-intelligence
14. Robotics, I.F.O.: The impact of robots on productivity, employment and jobs in a positioning paper by the international federation of robotics (20170
15. AbuShanab, E., Pearson, J.M., Setterstrom, A.J.: Internet banking and customers' acceptance in Jordan: the unified model's perspective. Commun. Assoc. Inf. Syst. **26**(1), 23 (2010)
16. Mukred, M., et al.: The key factors in adopting an electronic records management system (ERMS) in the educational sector: A UTAUT-based framework. IEEE Access **7**, 35963–35980 (2019)
17. Mukred, M., Yusof, Z.M.: The performance of educational institutions through the electronic records management systems: factors influencing electronic records management system adoption. Int. J. Inf. Technol. Proj. Manag. (IJITPM) **9**(3), 34–51 (2018)
18. Mukred, M., et al.: A framework for electronic records management system adoption in the higher professional education: individual, technological and environmental factors. In: International Conference of Reliable Information and Communication Technology. Springer, Heidelberg (2018)
19. Mukred, M., Yusof, Z.M.: Electronic records management and its importance for decision making process in Yemeni higher professional education (hpe): a preliminary review. In: 1st International Conference of Recent Trends in Information and Communication Technologies (IRICT 2014), Johor Bahru, pp. 105–114 (2014)

20. Mukred, M., Yusof, Z.M.: The role of electronic records management (erm) for supporting decision making process in Yemeni higher professional education (hpe): a preliminary review. Jurnal Teknologi **73**(2), 117–122 (2015)
21. Mukred, M., Yusof, Z.M.: Factors influencing the adoption of electronic records management (ERM) for decision making process at higher professional education (HPE)'s institutions. In: 1st ICRIL-International Conference on Innovation in Science and Technology (IICIST 2015), Kuala Lumpur, pp. 399–403 (2015)
22. Mukred, M., et al.: Electronic records management system adoption readiness framework for higher professional education institutions in Yemen. Int. J. Adv. Sci. Eng. Inf. Technol. **6**(6), 804–811 (2016)
23. Mukred, M., Yusof, Z.M.: The DeLone–McLean information system success model for electronic records management system adoption in higher professional education institutions of Yemen. In: International Conference of Reliable Information and Communication Technology. Springer, Heidelberg (2017)
24. Conti, D., et al.: A cross-cultural study of acceptance and use of robotics by future psychology practitioners. In: 2015 24th IEEE International Symposium on Robot and Human Interactive Communication (RO-MAN). IEEE (2015)
25. Conti, D., et al.: Robots in education and care of children with developmental disabilities: a study on acceptance by experienced and future professionals. Int. J. Soc. Robot. **9**(1), 51–62 (2017)
26. De Graaf, M.M., Allouch, S.B.: Exploring influencing variables for the acceptance of social robots. Robot. Auton. Syst. **61**(12), 1476–1486 (2013)
27. Hashim, R., Yussof, H.: A review of the ambit of politics in social robotics. Procedia Comput. Sci. **105**, 316–321 (2017)
28. Kasemsap, K.: Robotics: theory and applications. In: Cybersecurity Breaches and Issues Surrounding Online Threat Protection, pp. 311–345 (2017)
29. de Boer, W., Åström, J.-M.: Robots of the future are coming, are you ready?: A study investigating consumers' acceptance of robotics (2017)
30. Pedersen, M.R., et al.: Robot skills for manufacturing: from concept to industrial deployment. Robot. Comput.-Integr. Manuf. **37**, 282–291 (2016)
31. Solbraa Bay, A.J.T.: Innovation adoption in robotics: consumer intentions to use autonomous vehicles (2016)
32. Royakkers, L., van Est, R.: A literature review on new robotics: automation from love to war. Int. J. Soc. Robot. **7**(5), 549–570 (2015)
33. Heikkilä, A.-M., Myyrä, S., Pietola, K.: Effects of economic factors on adoption of robotics and consequences of automation for productivity growth of dairy farms (2012)
34. Khandewal, V., Miller, J.: Information system study. opportunity management program (1992)
35. Mukred, M., et al.: Taxonomic framework for factors influencing ERMS adoption in organisations of higher professional education. J. Inf. Sci. **45**(2), 139–155 (2018)
36. van den Brule, R., et al.: Do robot performance and behavioral style affect human trust? Int. J. Soc. Robot. **6**(4), 519–531 (2014)
37. Alharbi, S.T.: Trust and acceptance of cloud computing: a revised UTAUT model. In: 2014 International Conference on Computational Science and Computational Intelligence (CSCI). IEEE (2014)
38. Lai, E.R.: Motivation: a literature review (2011)
39. Kohli, R.: Automation of academic libraries in ICT environment: issues and challenges. Asian J. Multidiscip. Stud. **2**(12) (2014)
40. Davis, F.D.: Perceived usefulness, perceived ease of use, and user acceptance of information technology. MIS Q., 319–340 (1989)

41. Rantanen, T., et al.: The adoption of care robots in home care—a survey on the attitudes of Finnish home care personnel. J. Clin. Nurs. **27**(9–10), 1846–1859 (2018)
42. Bhavya, Y.A.: Comprehensive study on the prevailing automation and infrastructure in urban dairy sector (2018)
43. Raj, M., Seamans, R.: AI, labor, productivity, and the need for firm-level data. In: Economics of Artificial Intelligence, University of Chicago Press (2018)
44. Cowan, R.S.: The "industrial revolution" in the home: household technology and social change in the twentieth century. In: The Routledge Companion to Modernity, Space and Gender, pp. 81–97. Routledge, Abingdon (2018)
45. Argyris, C.: Integrating the Individual and the Organization. Routledge, Abingdon (2017)

Smart Group Decision Making on Leadership Style Identification Using Bayes Theorem

Okfalisa[1(✉)], Frica A. Ambarwati[1], Fitri Insani[1], Toto Saktioto[2], and Angraini[3,4]

[1] Department of Informatics Engineering, UIN Sultan Syarif Kasim Riau, Pekanbaru, Indonesia
okfalisa@gmail.com
[2] Department of Physics, Universitas Riau, Pekanbaru, Indonesia
[3] Department of Information System, UIN Sultan Syarif Kasim Riau, Pekanbaru, Indonesia
[4] School of Computing, Faculty Engineering, Universiti Teknologi Malaysia, Skudai, Malaysia

Abstract. The un-synergistic criteria's of the candidate and the organization's vision and mission became the main problem. The subjectivity of certain group tended to bring the frustration hence it provoked the emergence of conflict interests. In order to cover such issues, this paper proposed the blending of an expert system using Bayes theorem and Group Decision Making in identifying the candidate's leadership style. The analysis was conducted in a series of activities through the development of knowledge base, inference engines, weighting, experts' alliance, decision tree diagram, intersection independent and mean sample calculation. As a result, eleven numbers of leadership styles were theoretically identified thus confirmed and weighted by experts. In addition, the correlation between the leadership style and organizational vision and mission was analyzed. The leaders' assessments were perceived by the candidate itself and supported team members. The calculation of estimated values proposed the percentage of candidate leadership style identification. This provided a smart group decision making in recommending organizational decision-makers towards the fit proper elected leader based on the situation. To automate the calculation, prototype assessment software was produced and tested. Black-box and User Acceptance Test found that this application was successfully applied in identifying the leadership style for future leaders.

Keywords: Bayes theorem · Multiple perspectives · Leader election · Leader assessment

1 Introduction

Leadership could be regarded as a process either individual or groups' activities in achieving the ultimate organization's goals through the creation of managing risks, administrative, regulation, creating harmony and maintains commitment within groups [1]. A leader in a nonprofit organization was responsible in generating innovative and effective management in the pursuit of success, motivating and triggering the

F. Saeed et al. (Eds.): IRICT 2019, AISC 1073, pp. 906–916, 2020.
https://doi.org/10.1007/978-3-030-33582-3_85

productivity of knowledge members thus the leadership styles, principles and practices towards the achievement of effective leadership became substantial [2, 3]. Moreover, the lacking fund in nonprofit organization enforced a leader to move towards increasing the diversity of funding sources to sustain the missions. A leader in a nonprofit organization played a significant role in inspiring, motivating, encouraging, and drivers of social changes [4]. Leadership style supported and coached the development of team builders through the close relationship between leaders and members, trusts and leader respects, wherewith impact on the production of decision making [5]. Vision and mission are considered as a strategic management process that plays a significant role in achieving the organizational goals thus impact on the performance of an organization. A leader takes a part in translating vision and mission into the day to day activities as well as key drivers on operational of members' work and team buildings [6, 7]. Therefore, the leadership style regarding the characteristics features carries weight on the creation of organizational culture towards the formulation and accomplishment of vision and mission. The emergence of conflicts in leaders' election was unavoidable as well as the political interference. Thus the objectivity of leadership style that meets to members' need, visions and mission are neglected. Effective leaders engage in the process of leadership assessment [8]. The contribution of the supported group as instead of self-assessment in gauging the leadership skills and competencies provides an unprejudiced decision on the leaders. Therefore, leadership competence and the potential need to be measured in recognizing their strength and identify areas toward the target for future growth. Formal procedures presume that collective decision making can be enhanced through the systematic approach in terms of group decision making. Group decision making aids decision-makers in facing fuzzy preferences for alternative and individual judgment towards the optimal solution [9].

In the computer science field, many techniques and algorithms have been applied in group decision making to ensure the election of leaders is efficacious and fairness. It is including Bayes Theorem, Decision tree, K-Mean Clustering, Support Vector Machine, Linear Regression, Learning Vector Quantization, K-Nearest Neighbors, Adaboost, and Random Forest [10], and [11]. Several comparative studies had been conducted to make an evidence of the best theory. Nevertheless, the applicability of the recommended techniques in supporting decision making is pointed out by the requirement of data analysis and proposed weights embedded to provide better results. Bayes theorem is one of the techniques that widely used and explored by researchers. The application of it in forecasting, pattern recognition, diagnostics, natural language processing, and classification problems increased the utilization of this method in many platforms. These above showed the capabilities of Bayes theorem in minimizing several errors in calculation and make them distorted from the real probabilistic results [12]. The Bayes rules in decision making are strong evidence could improve the quality and efficiency of decisions. Thus, the evolvement of it in Multi-Attribute Decision Making (MADM), as well as group decision-makers, became a challenge.

This paper tried to study the application of expert system using Bayes theorem in group decision making. Answering the research question on how a smart group decision-making system designed to solve the election problems. The smart leader assessment presented the identification of an appropriate leadership style based on the calculation of quantitative criteria's from multiple perspectives. As a limitation, a case study for the election of the Students Association Leader was conducted through the development of a smart system. Herein, the experts' justification, the lectures, and the colleagues' were asked their contribution in objectively assessing the candidate leaders. These multiple perspectives advanced the complexity of performing alternative priorities which were analyzed based on organizational vision and mission. As a result, the distinguishing of candidate leadership style was presented. This recommendation suggested the decision-makers in selecting the candidate and perceived the management capabilities of the candidate towards organizational success. Therefore, the conflict and political interest amongst candidates can be minimized and the elected leader is voted based on the quality of leadership management capabilities according to the organizational objectives.

2 Literature Review

2.1 Leadership Style

There are many different perspectives in viewing the characteristics of leaders. The shifted from traditional-based theories into situational theory impacts the determination of leadership skills and leader characteristics [13]. As reviews on leadership theories, the principles of styles are defined into eleven types including H1-Democratic, H2-Environmental oriented, H3-Human-oriented, H4-Transactional, H5-Service-oriented, H6-Charismatic, H7-Transformational, H8-Bureaucratic, H9-Task-oriented, H10-Autocratic, and H11-Laissez-faire [14] (See Table 1). The characteristics of the above style are described as sixty-seven criteria symbolized into E1–E67 [15]. Democratic (H1) is explained as the leaders who are including team members in the decision-making process, encouraging and motivating the creativity of team members, and highly engaging team members in the organizational day to day activity. Environmental oriented (H2) leaders support any kinds of environmentally action-oriented movements, the harmonious environment through ecologically friendly services, the application of environmental laws, and raises environmental awareness. Human-oriented (H3) is the opposite of task-oriented. This style involves supporting people in team members that require high participation of leaders and provide input on decision making thus a high level of communication performed. Transactional (H4) leaders start when the team members accept a job from leaders and return the payment for their effort and compliance. Service-oriented (H5) leaders care about the team member needs, constantly adjust, acclimate, and assert the strategies thus successfully accomplish for organizational survival and development. Charismatic (H6) is a leadership style that inspires and motivates the team members to move forward, pursuing the member's passion and commitment towards the goal achievement, high dependency on the leader, and the feelings of invincibility of leaders. Transformational (H7) style focuses on the performance and connection between leaders and team members, increasing of motivation, morality, and fulfilling their potential for goal achievement. Bureaucratic (H8) leaders ensure the procedures are followed precisely. Task-oriented style (H9) can be described as autocratic that leaders make decisions without consulting their team members. Task-driven leadership requires the leaders to a deep understanding of task productivity, roles, performance goals and deadlines, and motivated task to succeeding. Autocratic (H10) is kinds of style where leaders have full power over members, cut off the idea from members, provide quick decisions, act immediately upon decisions, provide incredibly efficient, and best used in a crises situation. Laissez-faire (H11) describes leaders who abdicate the responsibilities and complete freedom on members, provides the team members with resources and advice, and allows the job autonomy.

Table 1. Reviews of articles according to the proposed Leadership Style identification model

References	Method	Leadership style proposed	Case study
[13]	Styles and Principles of Educational Leadership	H10, H8, H6, H1, H11, H4	Management Education in Nigeria
[14]	Leadership theories and style	H4, H7	General Concept
[1]	Leadership theories and style	H10, H4, H8, H6, H7, H1, H11.	Management Education in India
[2]	Leadership Style affect group dynamics	H10, H1, H11,	Outdoor recreation students at Lakehead University
[4]	Authentic leadership development	H5, H6, H7, H3, H2, H9	University of Nebraska-Lincoln
[5]	The partial least squares method	H7, H6	Nonprofit organizations in Mexican Case

2.2 Bayes Theorem

Bayes theorem as one of the probabilistic models has been widely used in data analysis to evaluate decisions for decision making. One of the main is by determining the probability of belief from the numbers of decision-makers' inquiries and measuring the advantages of evidence values thus provides the numbers of the likelihood for the uncertain situation [16]. Bayes theorem is capable of combining new evidence and existing evidence as well as subjective probability and reviewing the previous one based on the information trace. This indicates that Bayes theorem is not only an ordinary calculus and measurement system. The uncertainty measured and probabilities are considered by Bayes theorem as shown in Eq. 1 where E_1, E_2, …, E_m are double evidence and H_1, H_2, …, H_m are a double hypothesis.

$$p(H_i|E_1E_2...E_m) = \frac{p(E_1E_2...E_m|H_i) \times p(H_i)}{\sum_{k=1}^{n} p(E_1E_2...E_m|H_k) \times p(H_k)} \tag{1}$$

In this paper, the combination of the experts perceives as well as multiple perspectives have been considered by combining the probability estimation in uncertainty situation. Thus, the analysis of probability assessments will be calculated from different sources into a single estimation using Bernoulli's probability formula [17]. This independently explains the context of the dichotomous perceptual decision by considering estimated on the experts' weights through the combination of ecological validity and subjective reliability. The probability estimation for multiple experts is determined in Eqs. 2 and 3.

$$\frac{pq}{[pq+(1-p)(1-q)]} \tag{2}$$

where:

p is the probability estimation of expert 1, and q as the estimation of expert 2.

In order to analyze the probability intersection between leadership style and vision and mission, Eq. 3 is applied.

$$\mathrm{P(A \cap B) = P(A) \cdot P(B)} \tag{3}$$

where:

A is defined as the probability estimation of vision and mission from the department, and B as the estimation from faculty.

The calculation of the mean sample is conducted to easily understand the central tendency from multiple observations. The following mean sample formula can be depicted in Eq. 4.

$$\overline{X} = \frac{X_1 + X_2 + X_3 + \ldots + X_n}{n} \tag{4}$$

where:

X_1, X_2, ... X_n is a random independent variable and n is a number of random variables.

3 Methodology

As a methodology, several activities are listed (see Fig. 1). Data collection was gotten hold of by reviewing the leadership theories, Bayes theorem and its application in decision making. To strengthen the theoretical foundation, interviews and observation were conducted with regards to students' affairs process in five departments of Faculty science and technology, UIN Suska Riau viz., Informatics Engineering (IF), Electrical Engineering (EE), Industrial Engineering (IE), Information System (IS) and Applied Mathematics (AM). Dean, deputy dean of students' affairs, head of the department, head of students association in faculty level, head of students association in department level, and team members of students association became target respondents in multiple viewing on the election process, the organizational needs, the objective, and goals. As an expert judgment, two experts from management leadership field were interviewed and knowledgeable transferred into knowledgebase development.

The analyses were performed through the development of knowledge base, inference engines, a combination of experts weighting, and decision tree diagrams. Herein, the calculation of probability values in leadership styles, criteria as indicators, vision and mission was carried out using Bayes theorem. Due to the complexity of multiple perspectives in group decision making, the estimation was figured out by applying the intersection independent formula in Eqs. 2 and 3. Intersection provided the dissections of probability values of leadership style over vision and mission in department and faculty level.

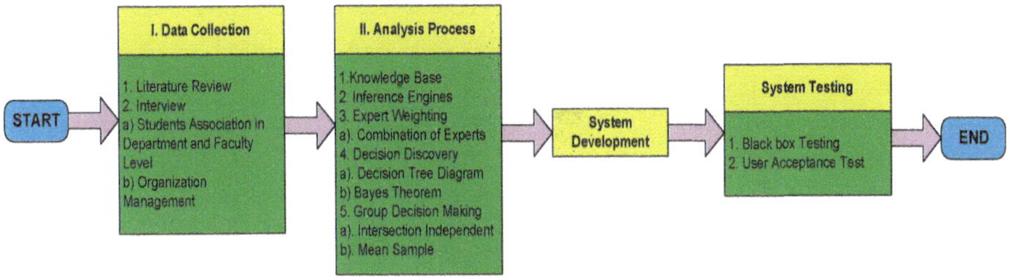

Fig. 1. Research procedure

Furthermore, the mean sample was then estimated based on the calculation of probability values from numbers of users involved in leaders' assessment. As a result, the appropriate leadership styles were identified and ranked. The recommendations were given as corrective action to aid decision-maker in making decisions in leader election.

To automate the analysis process, prototype software of leader election system was developed. The root process and calculation were worked out by the system. Twenty users from each department (2 candidates, 2 lectures, and sixteen colleagues) for totally one hundred respondents from the faculty were asked their leader assessment by answering the questionnaires. The questionnaire was made in "Yes" and "No" questions formatted based on the flow of forward chaining procedure and Bayes calculation. Black box testing and user acceptance test was performed in ensuring the capability of this system in measuring the leadership election and providing fit proper leader as well as the recommended leadership style.

4 Result and Analysis

4.1 Knowledge Base Development

The knowledge base was developed by analyzing the correlation between sixty-seven indicators (E1–E67) and eleven leadership style (H1–H11). Herein, H1 was described by fifteen criteria; H2 in ten indicators; H3 in 8 indicators; H4 in fourteen indicators; H5 in 8 indicators; H6 in 9 indicators; H7 in eleven indicators; H8 in 9 indicators; H9 was in 6 indicators; H10 was in fourteen indicators, and finally, H11 provided twelve indicators. In order to explain the knowledge base in forms of rule-based, the decision tree diagrams were generated (see Fig. 2).

Forward chaining procedure was followed as inference engines traced from up to down iteration. The example of the rule base as follows

The rule for H9: Task Orientation Style IF E1 AND E2 AND E3 AND E45 AND E46 AND E47.

This rule-based was intelligently generated in reliance on the users' answers during the leader assessment.

4.2 Expert Weighting

The analysis went forward into the leadership style and indicators' weighting by two experts. The probability values of indicators were estimated as well as subjective probability value against the hypothesis. It ranges from 0 to 1 which indicates the significant level of the indicator. Meanwhile, the probability values of experts weighing on leadership style (H1–H11) and 67 indicators (E1–E67) were distracted at Table 2 based on Eq. 1. Moreover, two experts were also given their probability values weighing on the correlation between organizational vision and mission of 5 departments (IF, EE, IE, IS, AM) and the leadership style.

To obtain a single subjective probability from multiple experts, Eqs. 2 and 3 are applied. Meanwhile, the probability value of experts' intersection between departments' vision and mission and leadership style was explained in Table 3. The mean score was provided as the calculation remark of intersection and Bayes probability values as referring to Eqs. 3 and 4.

Table 3 in mean score column identified the list of leadership style which accommodated the department's vision and mission. For example, Department Informatics Engineering (IF) provided the mean score ranging from the lowest in 1,084 to the highest 1, 44175. It meant that Human-oriented (H3) and Transformation (H7) styles became the most recommended of leadership style in this department. It was followed by Service-oriented (H5), Environmental-oriented (H2), Democratic (H1), Task-oriented (H9), Bureaucratic (H8), Laissez-faire (H11), Transactional (H7), Charismatic (H6), and Autocratic (H10) as the less one.

Referring to the input of leaders' assessment, the calculation of group decision making for 100 respondents was carried out. Equations 1, 2, 3 and 4 were initiated. The consideration of experts' weight enriched the multiple perspectives calculation in identifying the leadership style of the candidate. As a result, the decision-makers have suggested whether the candidate fitted into the flawless leaders as in the particular department.

4.3 Prototype System Leadership Style Identification

The architecture system development can be depicted (see Fig. 3). The System Interface was designed to accommodate the leader assessment form in forms of expert system-question and answer platform. The questionnaire was performed in accordance with the knowledge base rules and weights development. As a result, the list of the candidate leadership style was then ranked and identified its concordance with the expert recommendation. In order to test the functionality of the application, the Expert

Judgement, Black box and User Acceptance Test (UAT) were conducted. It found that the system application is running well and provided the analysis and result as the expert knowledge transferred.

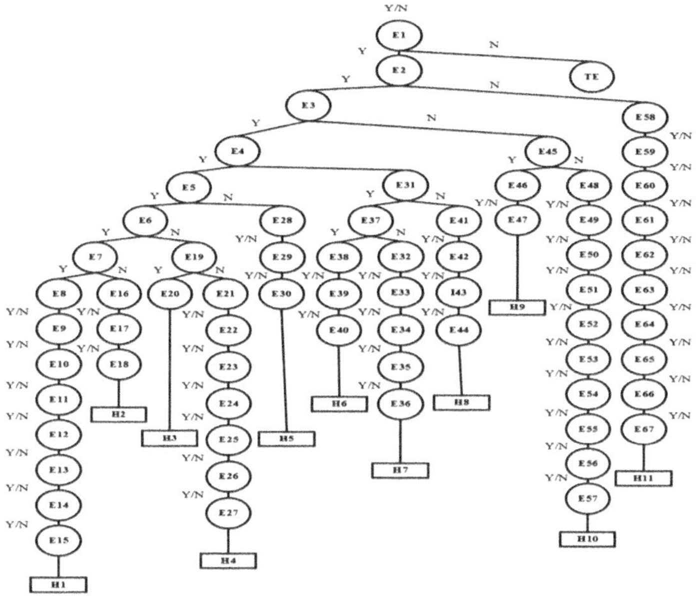

Fig. 2. Decision tree diagrams

Table 2. Probability hypothesis on indicators

$p(E_i \vert H_i)$	Probability	Hypothesis										
		i = 1	i = 2	i = 3	i = 4	i = 5	i = 6	i = 7	i = 8	i = 9	i = 10	i = 11
$p(E_1 \vert H_i)$	Expert One	0,2	0,2	0,2	0,2	0,2	0,2	0,2	0,2	0,2	0,2	0,2
	Expert Two	0,2	0,2	0,2	0,2	0,2	0,2	0,2	0,2	0,2	0,2	0,2
$p(E_2 \vert H_i)$	Expert One	0,6	0,5	0,5	0,2	0,5	0,2	0,6	0,2	0,4	0,5	0,3
	Expert Two	0,6	0,4	0,4	0,3	0,4	0,2	0,6	0,3	0,5	0,4	0,4
$p(E_3 \vert H_i)$	Expert One	0,8	0,4	0,4	0,3	0,4	0,3	0,6	0,4	0,4	0,5	0
	Expert Two	0,8	0,3	0,5	0,2	0,3	0,2	0,7	0,3	0,3	0,5	0
...	...	...	...	...	...	...	...	...	...	...	...	...
	...	...	...	...	...	...	...	...	...	...	...	...
$p(E_{47} \vert H_i)$	Expert One	0	0	0	0	0	0	0	0	0	0	0,8
	Expert Two	0	0	0	0	0	0	0	0	0	0	0,8

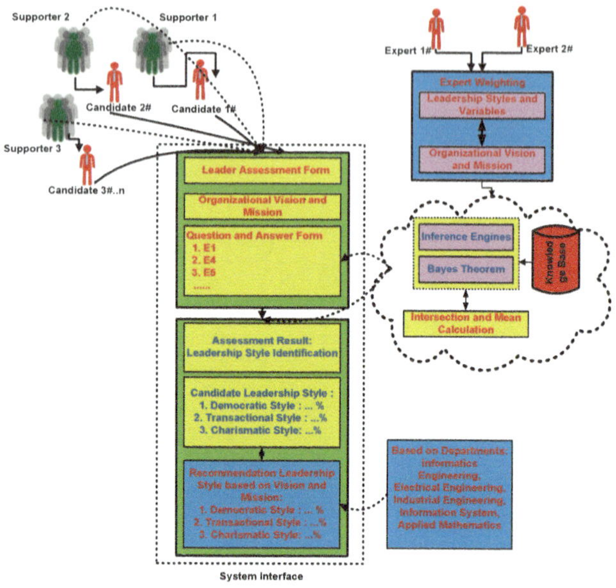

Fig. 3. Leadership style system architecture

Table 3. Probability intersection on hypothesis and mean score

Leadership Style	Department hypothesis					Probability mean score				
	IF	EE	IE	IS	AM	IF	EE	IE	IS	AM
H1	0,819	0,819	0,819	0,8463	0,8372	1,4095	1,4095	1,4095	1,42315	1,4186
H2	0,8366	0,799	0,846	0,846	0,846	1,4183	1,3995	1,423	1,423	1,423
H3	0,8835	0,9021	0,8835	0,9021	0,8928	1,44175	1,45105	1,44175	1,45105	1,4464
H4	0,6132	0,6132	0,6132	0,6132	0,584	1,3066	1,3066	1,3066	1,3066	1,292
H5	0,873	0,8633	0,873	0,8924	0,8924	1,4365	1,43165	1,4365	1,4462	1,4462
H6	0,5312	0,5184	0,5312	0,5056	0,48	1,2656	1,2595	1,2656	1,2528	1,24
H7	0,9935	0,9025	0,874	0,90025	0,8835	1,44175	1,45125	1,4365	1,45125	1,44175
H8	0,7304	0,7304	0,7304	0,7138	0,7138	1,3652	1,3652	1,3652	1,3569	1,3569
H9	0,748	0,748	0,7565	0,748	0,765	1,374	1,374	1,37825	1,374	1,3825
H10	0,168	0,165	0,165	0,18	0,195	1,084	1,084	1,0825	1,09	1,0975
H11	0,6688	0,6424	0,6688	0,6776	0,6424	1,335	1,3212	1,3344	1,3388	1,3212

5 Conclusion

Bayes theorem has been successfully applied in smartly identifying the leadership style in leader election. A group decision making, as well as multiple perspectives of the leader assessment process, turned over a complex decision making into unsophisticated adaptive problem-solving in the absence of respondents' vote. Two experts weighting towards the leadership styles and its relation to organizational vision and mission found the level significance of it based on the organizational situation. Thus, the assessment values became more immaculate not just in follow up respondents' vote as well as experts judgment confirmation.

Intersection independence and mean values calculation enriched the complexity of leadership style identification as an alternative priority. The recommendation of leadership style aids the decision-makers in making the right decision for leader election. Therefore, the quality of decisions was a guarantee, more objective, could minimize political intervention and ensuring the fit proper elected leader based on the situation. Moreover, the candidate leaders can measure his capabilities, skills, and knowledge-based upon organizational needs. To automate the calculation, the prototype system leadership style identification has been successfully developed and tested. It showed the significant tool in aiding the leader election process.

References

1. Monga, O.P.: Leadership theories and educational management: an insight. Biz Bytes: Biann. Manag. Technol. **6**, 46–55 (2015)
2. Val, C., Kemp, J.: Leadership styles. Pathways: Ontario J. Outdoor Educ. **24**(3), 28–31 (2012)
3. Karam, E.P., Hu, J., Davison, R.B., Juravich, M., Nahrgang, J.D., Humphrey, S.E., Scott DeRue, D.: Illuminating the 'Face' of justice: a meta-analytic examination of leadership and organizational justice. Manag. Stud. **56**(1), 134–171 (2019)
4. Avolio, B.J., Gardner, W.L.: Authentic leadership development: Getting to the root of positive forms of leadership. Leadersh. Q. **16**(3), 315–338 (2005)
5. Estrada, C.R., Carranza, M.T.G.: Leadership style in nonprofit organizations, the Mexican case. Int. J. Bus. Humanit. Technol. **6**(4) (2016)
6. Arar, K., Abu Nasra, M.: Leadership style, occupational perception and organizational citizenship behavior in the Arab education system in Israel. Emerald Insight: Educ. Adm. **57** (1), 85–100 (2019)
7. Taiwo, A., Agwu, E., Lawal, F.: Vision and mission in the organization: myth or heuristic device? Int. J. Bus. Manag. **4**(3), 127–134 (2016)
8. Cuseo, J.: Leadership self-assessment: assessing your leadership skills & leadership development (2015)
9. Zhang, G., Lu, J.: An integrated group decision-making method dealing with fuzzy preferences for alternatives and individual judgments for selection criteria. Group Decis. Negot. **12**, 501–515 (2003)
10. Reddy, S., Govindarajulu, P.: A Survey on data mining and machine learning techniques for internet voting and product/service Selection. Int. J. Comput. Sci. Netw. Secur. **17**(9), 76–84 (2017)
11. Okfalisa, Gazalba, I., Mustakim, M., Reza, N.G.I.: Comparative analysis of k-nearest neighbor and modified k-nearest neighbor algorithm for data classification. In: Information Systems and Electrical Engineering (ICITISEE), Yogyakarta, pp. 294–298 (2017)
12. Barbini, E., Manzi, P., Barbini, P.: Bayesian approach in medicine and health management. Intech (2013)
13. Amanchukwu, R.N., Stanley, G.J., Ololube, N.P.: A review of leadership theories, principles and styles and their relevance to educational management. Management **5**(1), 6–14 (2015)
14. Zakeer, A., Allah, N., Irfanullah, K.: Leadership theories and styles: a literature review. J. Res. Dev. Manag. **16**, 1–7 (2016)

15. Lorraine, D.C, Jealarr, J, Luis, L.M.L, Charmaine, A.M, Ria, A.S.: J. Adv. Comput. Netw. 2 (3) (2014)
16. Hummel, R.A., Landy, M.S.: A statistical viewpoint on the theory of evidence. IEEE Trans. Pattern Anal. Mach. Intell. **10**(2), 235–247 (1988)
17. Ordoobadi, S.M., Wang, S.: A multiple perspectives approach to supplier selection. Ind. Manag. Data Syst. **111**(4), 629–648 (2011)

Factors Influencing the Adoption of Social Media in Service Sector Small and Medium Enterprises (SMEs)

Alice Tabitha Ramachandran(✉), Norasnita Ahmad, Suraya Miskon, Noorminshah A. Iahad, and Nazmona Mat Ali

Azman Hashim International Business School, Universiti Teknologi Malaysia, 81310 Johor Bahru, Johor, Malaysia
alicetabitharam@gmail.com,
{norasnita, suraya, minshah, nazmona}@utm.my

Abstract. Social media has turned into an internet marvel in recent years. This fast technological evolution and intensity of competition oblige companies to constantly looking for different ways to set apart themselves and offer added values to their customers especially in small and medium sized enterprises (SMEs). The increasing popularity of social media has encouraged SMEs to start to consider this means as a new approach to communicate with clients and increase their brand reputation. In spite of the potential benefit and the developing numbers of adopters, previous studies show that only a few SMEs are concerned about social media adoption. Therefore, this forms a necessity to understand more on the factor influence in the decision making of the adoption in SMEs. Lack of literature review also creates another concern that require a thorough research. So the study will discuss the relationship between factors of influencing the adoption of social media in SMEs service sector. The data collected for this study was done by using online survey strategy. This study used 384 as a sample size with a population of total 809 126 Malaysia's SMEs in the service sector. From the result obtained, the three factors identified: ease of use, usefulness and trust have a significant relationship with adoption while usefulness is most contributed factor in the adoption of social media. This finding will be a valuable data in assisting new entrepreneurs in adopting appropriate social media for their business.

Keywords: Social media · Ease of use · Usefulness · Trust · SMEs · Service sector

1 Introduction

Social media has turned into an Internet marvel in recent years. In a report by Harvard Business Review, a comparable perspective was likewise communicated that the rise of Internet-based web-based life has begun another sort of discussion among consumers and organizations, challenging conventional ideas about brand management and marketing [1]. In the globalization era, organizations wind up in a domain of steady innovative transformation [2] which makes challenges in business environment. So the

F. Saeed et al. (Eds.): IRICT 2019, AISC 1073, pp. 917–925, 2020.
https://doi.org/10.1007/978-3-030-33582-3_86

business have to be join data and communications innovations in order to help their business exercises as the information innovation is considered as a basic instrument for economies improvement [3, 4] and a principal source of competitiveness.

With regard to social media adoption in their business settings, large organizations have the major focuses on the literature review [5, 6] and SMEs holds only little research on the social media adoption. In spite of the fact that having favorable circumstances for business purposes, it's stated in SME Corporation Malaysia that Malaysian SMEs are low in use of social media [8] just at 12.1%. Studies among SMEs on the adoption of social media is still in its early stages [7, 8]. Additionally, lack of studies on factor influencing in the adoption of social media in SMEs context especially in Malaysia initiate the motivation and enthusiasm in conducting this study. In this view, it is important for new entrepreneurs and new SMEs to have superior comprehension in social media adoption [9].

Therefore, this study aims to examine the influence of Ease of Use, Usefulness and Trust on the adoption of social media by SMEs in the service sector industry in Malaysia.

2 Literature Review

2.1 Social Media

In today's business world, online networking applications so-called social media permit organizations and customers to share information and discuss with one another on business marketing activities [8]. As an intelligent stage, web-based social networking empowers organizations to connect with clients customers and draw in new clients, build brand awareness, create more deals, and brand image. Most business people and promoters are going to internet based life since that is the place accomplices are to be found, including providers and clients. WhatsApp and Instagram taken over by Facebook have active user base of 1500 million and 1000 million respectively. The popular social media is Facebook with a billion dynamic users. As indicated by the digitalization Survey of SMEs in 2018 about 70.5% of the total respondent's used social media for their business, with Instagram, WhatsApp and Facebook being the most used social media principally for commitment or collaboration with customers [10].

2.2 Social Media Adoption

Utilization of web innovation has been turned into a typical process in working environment. The web empowered correspondence media, encourages associations to lead business whenever from anyplace. Various examinations researched the utilization of social media and discovered SMEs utilized social media for different purpose, for example, advertising, marketing, problem resolution, communications, customer services, innovations, driving cultural change, human resources, information technology and sales, advertising on the social network, and internet marketing. Dahnil et al. [12] distinguished five groups of external and internal elements which can influence social media adoption. The primary gathering of factors is identified with the end users themselves: preparing and learning on the web-based social networking condition and perceived usefulness. The second gathering is identified with authoritative assets: if the top administration allotted assets to social media advertising, for example, personnel, time and money. Third, is identified with innovative restrictions of the stages. For instance, hindrance in measuring business results. The last one is the business condition. In this gathering, competitors' conduct may impact just as nation foundation, similar to web broadband circulation.

2.3 Small and Medium Enterprises (SMEs)

SME is the perceived abbreviation for Small-and-Medium Enterprises. Various research have utilized the figure of staff as a proportion of enterprise size [13]. A few researches have connected both yearly turnover and number of staffs plays an indicator of the SMEs size. SMEs are the backbone of a country economy development. SME Corp. Malaysia, 2016 stated that 809 126 (89.2%) SMEs are under service sector [14].

2.4 Service Sector

The service sector segment comprises of soft parts of the economy, for example, education, retail, banking, social service, tourism and insurance. In a delicate part of work, individuals use time to convey learning resources, process engagement to make profitability, collaboration assets, sustainability, performance improvement potential and effectiveness. Service industry includes the arrangement of services to organizations as well as to the final consumers. Service may include transport, circulation and closeout of merchandise from maker to a shopper as may occur in wholesaling and retailing, or may include the arrangement of a service.

2.5 Technology Adoption Model

The Technology Acceptance Model or TAM [11] is a standout amongst the most compelling hypothesis to comprehend user acceptance of technology. TAM was proposed by Davis in 1986 [11] which clarify and foresee users' affirmation and appropriation or dismissal of new technologies. TAM is conceptually founded on Theory of Reasoned Action (TRA), which introduces two social convictions, perceived ease of use and perceived usefulness, as basic sources of attitudes towards behavioral expectations and utilization actual behavior (AB).

2.6 Hypotheses Development

Based on the previous research studies, the following hypotheses are proposed.

Hypothesis 1: Ease of use is significantly related to the adoption of social media in service sector SMEs. Previous research shown that ease of use is significantly related to the adoption of social media in service sector SMEs [9, 18–18]. Hence, the first hypothesis, H1 is formulated.

Hypothesis 2: Usefulness is significantly related to the adoption of social media in service sector SMEs.

Past studies revealed that usefulness is significantly related to the adoption of social media in service sector SMEs [2, 9, 15, 16, 18]. Therefore, second hypothesis, H2 is developed.

Hypothesis 3: Trust is significantly related to the adoption of social media in service sector SMEs.

Previous research pointed out that trust is significantly related to the adoption of social in service sector SMEs [2, 9, 19, 20]. Thus, third hypothesis, H3 is formulated.

3 Methodology

Questionnaire was distributed to the entrepreneurs, experience employees, experts, top management, and related workers who are adding to the selection of social media in service sector SMEs. Online internet survey method through Google Form is used to collect data. This research is conducted based on the context of Malaysian SMEs in service sector. Under SME corporation Malaysian there are a total of 809 126 SMEs in service sector. This research requires a total of 384 respondents as the sample size according to the method determined by Krejcie and Morgan [21]. Analysis of the data gathered was done by utilizing the Statistical Package for Social Science (SPSS), as it is powerful and proficient tool for statistical analysis in quantitative study.

4 Data Analysis

4.1 Descriptive Statistics

From the background questions, a general view on the respondents' behavior in social media is obtained. It is essential to get a better understanding on the utilization since identifies with trust on social media through user's encounters. The results of the study show that 28.4% of the respondents were regular users of WhatsApp while 12.5% constituted regular users of Facebook for business purpose. Regular users of Instagram 7.3% while regular users of other social media applications accounted for 1.3%. Regular users of both WhatsApp and Facebook 21.1% and the three main social media applications WhatsApp, Facebook and Instagram accounted for 15.1%. This indicates that WhatsApp is the dominant social media application among the respondents in SMEs. Among the respondents, 54 (14.1%) reported that they use social media 4–5 times per week, 31 (8.1%) use the medium for 2–3 times per week, 11 (2.9%) use once a week while 7 (1.8%) use it less than once a week.

4.2 Correlation Test

The correlation indicates the relationships between variables of the research study. Based on Table 1, all of the variable pairs have a moderate correlation which are between ± 0.41 to ± 0.70.

Table 1 Pearson's correlations

		Use of use	Usefulness	Trust
Ease of use	Pearson correlation	1	.705***	.549**
	Sig. (2-tailed)		.000	.000
	N	384	384	384
Usefulness	Pearson correlation	.705***	1	.673***
	Sig. (2-tailed)	.000		.000
	N	384	384	384
Trust	Pearson correlation	549**	.673**	1
	Sig. (2-tailed)	.000	.000	
	N	384	384	384

4.3 Multiple Regression Analysis

Ease of Use

Based on Table 3 and 4, the p-value is 0.000 which is less than 0.05. This outcome shows that hypothesis H1 is accepted. According to the Table 2 the r value is 0.844 while the r2 value is 0.712. It indicates that this factor has the very strong influence towards the adoption of social media (ß = 0.844). It implies that, 71% of the ease of use factor is influencing the adoption of social media.

Table 2 Model summary for ease of use

Model	R	R square	Adjusted R square	Std. error of the estimate
1	.844[a]	.712	.711	6.01117

a. Predictors: (Constant), Ease of Use
b. Dependent Variable: Adoption of Social Media

Table 3 ANOVA for ease of use

Model		Sum of squares	df	Mean square	F	Sig.
1	Regression	34069.199	1	34069.199	942.854	.000[a]
	Residual	13803.236	382	36.134		
	Total	47872.435	383			

a. Predictors: (Constant), Ease of Use
b. Dependent Variable: Adoption of Social Media

Table 4 Coefficients for ease of use

Model		Unstandardized coefficients		Standardized coefficients	t	Sig.
		B	Std. error	Beta		
1	(Constant)	8.652	1.719		5.032	.000
	Ease of use	2.435	.79	.844	30.706	.000

a. Dependent Variable: Ease of Use

Table 5 Model summary for usefulness

Model	R	R square	Adjusted R square	Std. error of the estimate
1	.904[a]	.817	.817	4.78876

a. Predictors: (Constant), Usefulness
b. Dependent Variable: Adoption of Social Media

Table 6 ANOVA for usefulness

Model		Sum of squares	df	Mean square	F	Sig.
1	Regression	39112.321	1	39112.321	1705.561	.000[a]
	Residual	8760.114	382	22.932		
	Total	47872.435	383			

a. Predictors: (Constant), Usefulness
b. Dependent Variable: Adoption of Social Media

Usefulness
Based on Tables 6 and 7, the p-value is 0.000 which is less than 0.05. This outcome shows that hypothesis H2 is accepted. According to the Table 5 the r value is 0.904 while the r2 value is 0.817. It indicates that this factor has the very strong influence towards the adoption of social media (ß = 0.904). It implies that, 82% of the usefulness factor is influencing the adoption of social media.

Table 7 Coefficients for usefulness

Model		Unstandardized coefficient		Standardized coefficient	t	Sig.
		B	Std. error	Beta		
1	(Constant)	9.138	1.270		7.196	.000
	Usefulness	2.434	.059	.904	41.298	.000

a. Dependent Variable: Adoption of Social Media

Table 8 Model summary for trust

Model	R	R square	Adjusted R square	Std. error of the estimate
1	.869[a]	.754	.754	5.54795

a. Predictors: (Constant), Trust
b. Dependent Variable: Adoption of Social Media

Table 9 ANOVA for trust

Model		Sum of square	df	Mean square	F	Sig.
1	Regression	36114.585	1	36114.585	1173.324	.000[a]
	Residual	11757.849	382	30.780		
	Total	47872.435	383			

a. Predictors: (Constant), Trust
b. Dependent Variable: Adoption of Social Media

Trust

Based on Tables 9 and 10, the p-value is 0.000 which is less than 0.05. This outcome shows that hypothesis H3 is supported. According to the Table 8 the r value is 0.869 while the r2 value is 0.754. It shows that this factor has the strong influence towards the adoption of social media (ß = 0.869). It means that, 76% of trust factor is influencing the adoption of social media.

Table 10 Coefficients for trust

Model		Unstandardized coefficients		Standardized coefficients	t	Sig.
		B	Std. error	Beta		
1	(Constant)	23.850	1.110		21.469	.000
	Trust	2.027	.059	.869	34.254	.000

a. Dependent Variable: Adoption of Social Media

5 Discussion of Finding and Conclusion

5.1 Hypothesis Test Result

For hypothesis 1 (H1), the result shows that the relationship between perceived ease of use and adoption of social media in service sector SMEs was significantly correlated as the p-value was 0.000, which is less than 0.05. Therefore, hypothesis H1 is accepted. The statement was in line with previous research findings [9, 18–18].

For hypothesis 2 (H2), study indicates that the relationship between usefulness and adoption of social media in service sector SMEs was significantly correlated as the p-value was 0.000, which is less than 0.05. Therefore, hypothesis H2 is accepted. The statement was in line with past research findings [2, 9, 15, 16, 18].

For hypothesis 3 (H3), result shows that the relationship between trust and adoption of social media in service sector SMEs was significantly correlated as the p-value was 0.000, which is less than 0.05. This result shows that hypothesis H3 is accepted, this means trust can be positively influenced adoption of social media in SMEs. The statement was in line with past research findings [2, 9, 19, 20].

5.2 Conclusion

According to the result generated, it indicates that all three factors positively influenced adoption of social media. Among the three independent variables, usefulness is the factor that influences most to the adoption of social media in service sector SMEs with the most astounding value of ß = 0.904 which near to one. The study also contributes to managerial scope which could benefit the industry and future studies in this research field. Besides that, the finding will also be a valuable data for Malaysian new entrepreneurs, especially in helping them to choose the most appropriate social media for their business.

Recommendation for the upcoming research should cover only single social media platform to investigate the acceptance or adoption of social media. By doing this, researcher can acquire a deeper understanding into the specific platform and which factors that influence the decision to adopt. The second recommendation is the study should cover the overall service sector as in this study its only cover service sector. By this, researcher can know which sector really need social media adoption and what are the importance factors for the influence. The finding of this study will add data to social media adoption in SMEs context.

References

1. Tran, N.: Adoption and Utilization of Social Media by Micro-enterprises in Jyväskylä (2015)
2. Al Rahbi.: Factors Influencing Social Media Adoption in Small and Medium Enterprises (SMEs) (2017)
3. Olawale, F., Garwe, D.: Obstacles to the growth of new SMEs in South Africa: a principal component analysis approach. Afr. J. Bus. Manag. **4**, 729–738 (2010)
4. Oliveira, T., Martins, M.F.: Literature review of information technology adoption models at firm level. Electron. J. Inf. Syst. Eval. **14**(1), 110 (2011)
5. Meske, C., Stieglitz, S.: Adoption and Use of Social Media in Small and Medium-Sized Enterprises (2013)
6. Zeiller, M., Schauer, B.: Adoption, motivation and success factors of social media for team collaboration in SMEs. In: Proceedings of the 11th International Conference on Knowledge Management and Knowledge Technologies (2011)
7. Verheyden, M., Goeman, K.: Does (company) size matter? Differences in social media usage for business purposes. Quant. Methods Inq. **8**(4), 3–15 (2013)
8. Razak, S.A., Azrin, N., Latip, B.: Factors that influence the usage of social media in marketing. J. Res. Bus. Manag. **4**(2), 1–7 (2016)
9. Ahamat, A., Ali, M.S.S., Hamid, N.: Factors influencing the adoption of social media in small and medium enterprises (SMEs). Int. E-J. Adv. Soc. Sci. **3**(8), 338–348 (2017)
10. HOME. In: Smecorp.gov.my. http://www.smecorp.gov.my/index.php/en/. Accessed 29 Jul 2019
11. Davis, F.D.: Perceived usefulness, perceived ease of use, and user acceptance of information technology. MIS Q. **13**(3), 319–339 (1989)
12. Dahnil, M.I., Marzuki, K.M., Langgat, J., Fabeil, N.F.: Factors influencing SMEs adoption of social media marketing. Procedia – Soc. Behav. Sci. **148**, 119–126 (2014)
13. Thong, J.Y.L., Yap, C.S.: CEO characteristics, organizational characteristics and information technology adoption in small businesses. Omega **23**(4), 429–442 (1995)
14. SME Statistics. In: Smecorp.gov.my. http://www.smecorp.gov.my/index.php/en/policies/2015-12-21-09-09-49/sme-statistics. Accessed 29 Jul 2019
15. Wirtz, B.W., Göttel, V.: Technology acceptance in social media: review, synthesis and directions for future empirical research. J. Electron. Commer. Res. **17**(2), 97 (2016)
16. Siamagka, N.T., Christodoulides, G., Michaelidou, N., Valvi, A.: Determinants of social media adoption by B2B organizations. Ind. Mark. Manag. **51**, 89–99 (2015)
17. Lorenzo-romero, C., Romero, C.L.: Determinants of use of social media tools in retailing sector (2014). https://doi.org/10.4067/s0718-18762014000100005
18. Sago, B.: Factors influencing social media adoption and frequency of use: an examination of Facebook, Twitter, Pinterest and Google+. Int. J. Bus. Commer. **3**(1), 1–14 (2013)

19. Weerasinghe, S., Chandanie, M., Hindagolla, B.: Technology acceptance model and social network sites (SNS): a selected review of literature. Glob. Knowl. Memory Commun. (2018). https://doi.org/10.1108/gkmc-09-2017-0079
20. Tripathi, S.: Factors influencing the technology acceptance of social media in India: a literature review and research agenda for future. J. Adv. Comput. Sci. Inf. Technol. (ACSIT) **1**(2), 43–47 (2014)
21. Krejcie, R.V., Morgan, D.W.: Activities **38**, 607–610 (1970)

Communication and Learning: Social Networking Platforms for Higher Education

Nani Amalina Zulkanain(✉), Suraya Miskon, Norris Syed Abdullah, Nazmona Mat Ali, and Norasnita Ahmad

Faculty of Computing, Universiti Teknologi Malaysia, 81310 Johor Bahru, Johor, Malaysia
amalinanan94@gmail.com,
{suraya,norris,nazmona,norasnita}@utm.my

Abstract. This study explored the use of social networks specifically for general communication and learning in higher education. The types of social networks used by most students in higher education were identified before examining the difficulties faced by these students in handling social networks. Quantitative and qualitative data were gathered through survey distribution to several faculties. The results indicate the use of WhatsApp as the preferred application by students. In general, the major difficulties faced by these students in using social networks were misunderstanding and a lack of internet connection. Most suggestions for future social network features were for the design of an application which offers ease of use for general communication and learning purpose which has a high storage capacity, links to an educational portal and comes with a good internet connection service. In summary, WhatsApp is currently a priority for communication among these students and is, therefore, a preferred platform for the development of effective learning applications.

Keywords: Social networks · Learning communities · Higher education

1 Introduction

A social networks can be defined as a medium to communicate with each other by posting messages through the combination of text, video, image, documents and music [1]. The word 'social' refers to the process of socialization with others and provides a platform to facilitate sharing processes [2]. Social networks such as Facebook, Twitter, WhatsApp, Instagram and YouTube are widely-used for users to connect and network with each other. Exposure to social networks has also facilitated the distribution of data to all social network communities [3]. Social networks provide an opportunity for people to communicate more actively without the need for physical meetings, promoting interactions with others without borders. If the updating of a status and instant messaging on social network platforms are perceived as social support, this also suggests improved relationships among others [4]. Social networks are cost-effective, user-friendly, flexible and facilitate collaboration [5–7]. For these reasons social networks have also been adopted for the purpose of education and have become an essential part

F. Saeed et al. (Eds.): IRICT 2019, AISC 1073, pp. 926–935, 2020.
https://doi.org/10.1007/978-3-030-33582-3_87

of distance learning, as a medium for delivering information and training [7, 8]. Of course, there are many functions of social networks such as: Instagram, which is mainly used to share images; YouTube, which is mainly used to share videos; and WhatsApp, which is mainly used to send text messages. Currently, most social networks, in fact, provide several functions, including the ability to send messages, videos and images through one application. This is an opportunity for current researcher to observe and study these function of social network in education specifically higher education. Since most media can be accessed using mobile phones which have optimized their design for this purpose, the demand for mobile applications (apps) in teaching and learning has, consequently, increased exponentially [9]. In 2016, WhatsApp is the most popular apps which 83.3% statistically count with people used in the world [10] that connect others with the main functions used as instant messages. However, it can be said that the potential of other social networks to become an effective aid to the learning of students today still needs to be fulfilled. A deep research of finding the best social network to compare their functions for best communication or learning need to be done. This study aims to discover the types of current social networks are used by students for general communication and learning; meaning specifically their discussions or queries related to course content. This study also examines the difficulties faced by them with a view to suggesting further improvements in adopting social networks for education.

2 Methods

The questions formulated for this study are related to research interest in finding specific descriptive data [11]. By referring to Loeb *et al.* [11], the questions created based on needs and opportunities in this research to discover answer from research objectives. Overall, the survey consisted of a questionnaire (for quantitative data) plus some open-ended questions (for qualitative data) where were related to the frequency of social network types used by students for communication (general communication) and learning purpose (discussion or asking questions which relate to their subjects). Descriptive analysis used to analyse questionnaires (quantitative data). Simple statistics using mean is very helpful in describing data. Then, the results are visualized using pie-chart to show the overall percentage. All data are categorized according to several themes using thematic analysis technique. For open-ended questions, all the comments and suggestions are categorized into themes and description as suggest by [12]. Two different methods of questionnaire distribution were used, based on the level of acceptance and cooperation from the respective faculties.

3 Registrar Officer

An email containing the link to the survey was given to the registrar. Then the registrar distributed the email to all of the lecturers of the respective faculty. The lecturer was then given the choice of distributing the questionnaire to their students through email or through any social network medium such as Facebook or WhatsApp.

4 Lecturers

The researcher met with the lecturers, asking for a phone number and sending them the survey link through the WhatsApp application. In addition, there was a lecturer who requested conducting the survey using paper.

5 Findings

The findings of this study were separated into two sections: the questionnaire and the open-ended questions.

6 The Questionnaire

This section was separated into two categories, the first being the type of social network which the students used for communication. The questions were: What is the main social network you use for communication with your classmates or lecturer?" and; "What is the main social network you use for learning purpose (discussion/asking question) with your classmates or lecturer?". The questionnaire consisted of demographic information requirements such as gender and faculty. Another two questions related to the choice of social network type used for communication or learning purpose. Overall, 94 responses were collected from various faculties. There were 49 females and 45 males involved. The open-ended questions also tended to have the same number of students responding, since both the questionnaire and the open-ended questions were in the same survey.

6.1 Types of Social Network for Communication

As stated above, the first questions were to ascertain the types of social networks commonly used for any kind of communication. The question stated, "What is the main social network you use for communication with your classmates or lecturer?" Figure 1 below shows the percentage for the types of social network used for communication. The result shows that the most popular social network, used by 61% of the students, was WhatsApp. The fact that the students who mostly used other platforms also used WhatsApp suggested the overall popularity of WhatsApp for communicating with others through contact lists. Surprisingly, this research also discovered the use of Telegram, which was widely used among students from faculty of Mechanical Engineering. The result shows that 10% of students used both WhatsApp and Telegram, indicating the emergence of Telegram as a popular platform for communication among students and lecturers. Next, 8% of students used both WhatsApp and Facebook only. Furthermore, 2% of student used WhatsApp and Instagram as a medium for interacting with their colleagues and lecturers. This suggests that Instagram, which is generally used to display photos or videos, can also be a platform for enhancing relationships among students and lecturers. Next, 4% of students used both WhatsApp and Twitter. Twitter is also a popular social network platform for updating status and messaging.

Another 2% of students used WhatsApp and WeChat applications. Students who used more than two types of social networks contributed to 1% of the results. Furthermore, the results indicate that 5 per cent of students used a combination of WhatsApp, Facebook and Telegram. In summary, WhatsApp is clearly the major social network used by students for communication. This result indicates the opportunity for WhatsApp to be utilized as a platform for learning since most students are already comfortable and familiar with its features.

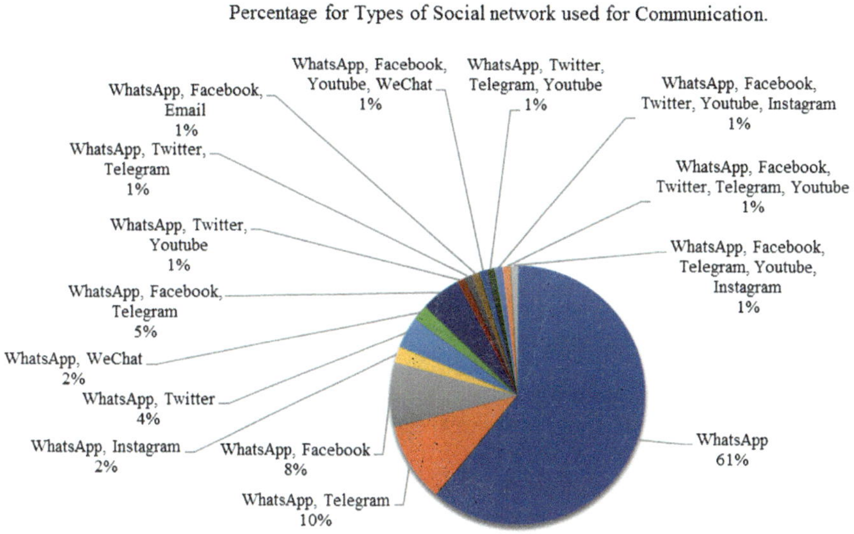

Fig. 1. Percentage for types of social network used for communication

6.2 Types of Social Network for Learning Purpose

'Learning' is interpret here to mean discussions or questions related with any learning topics; either with class members or lecturers. Figure 2 below, shows the percentages for the types of social network used specifically for learning purpose. The results show that 67% of students chose WhatsApp exclusively for learning purposes, illustrating the importance of WhatsApp as a medium of communication between students and lecturers for exchanging views and sharing information regarding lessons. 4% of students preferred both WhatsApp and Facebook, while 4% of students chose both WhatsApp and YouTube. These results indicate that Facebook and YouTube served as a supporting medium in addition to WhatsApp for learning purposes. In addition, 10% of students indicated use of both WhatsApp and Telegram, suggesting that Telegram can also be used as a platform for learning. Both WhatsApp and WeChat, and WhatsApp and E-Learning earn 1% respectively. In addition, 3% of student voted only for YouTube; showing that it was the exclusive platform for learning for several students, while one percent opted for face-to-face learning only. The rest of the results showed that students tended to use more than two types of social network, such as combinations of WhatsApp, Facebook, Twitter, YouTube, WeChat, Google, Gmail or Telegram, contributing to 1% of social network use. In line with the results for general communication, the combination of WhatsApp, Facebook and Telegram specifically achieved 3% for learning in this study. The analysis shows that WhatsApp certainly has potential as the preferred social network medium for learning purposes.

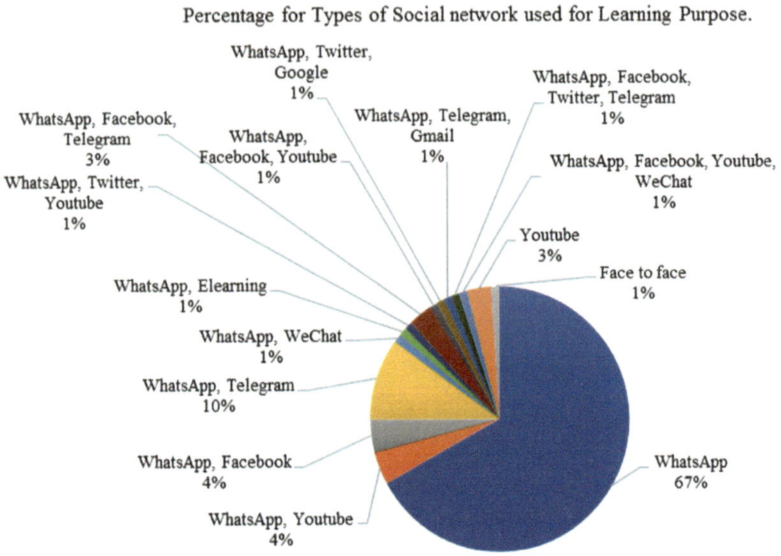

Fig. 2. Percentage for the different social networking platforms used for Learning Purpose

7 Open-Ended Questions

Results from the open-ended questions were analyzed and separated based on themes, since most answers were nearly identical. The questions suggested an understanding the problems faced by students with the purpose of developing a plan to enhance their learning via their use of a social network platform. This section was separated into three categories which are: difficulties in using social networks for communication; difficulties in using social networks for learning purposes and suggestions for the best social network features.

7.1 Difficulties in Using Social Network for Communication

The questions stated: "Do you find any difficulties in using social networks as a medium of communication? State your answer briefly". Table 1 below shows the themes derived from data analysis which includes several responses to the open-ended question. S stands for students and is followed by the queue number of the responses. For example, S1 refers to Student number one.

Table 1. Themes and responses for difficulties in using social networks for communication

Theme	Responses
No difficulties	Using social networks for study is much easier and time-saving (S5 and S31)
	Social networks are easy to use (S9 and S35)
	Social networks make communication faster (S11, S37, S38 and S69)
	When using social networks, one does not need to set an appointment date to communicate (S18)
	Messaging, voice notes, calls and video calls can be used to explain specific detail (S63)
	When using WhatsApp, only a little data is spent to communicate with friends (S68)
	There is the option to either write or to make a voice reply (S80)
No internet connection	Wi-Fi has been terminated and life has become more difficult (S6)
	Sometimes Wi-Fi in UTM is so bad that WhatsApp cannot be used to contact anyone (S27)
	Line coverage. The Telco company does not provide a reasonable price for data plans (S74)
	Not enough data (S86 and S87)
Lack of understanding	I would like to share something about an issue but others cannot fully understand it (S42)
	It is very hard for me to understand what my friends want to say because they often use short words (S70)
	Someone's intonation cannot be understood (S75)
Misunderstanding	Problems arise, leading to misunderstandings (S2)
	It is easy to be misunderstood (S28)
	Misinterpretation arises due to a person not conveying their message properly (S30)
	Information is misunderstood because of intonation of a text in Whatsapp and it difficult to know whether someone is angry or not (S40)
Lateness of reply	Some people do not always have their smartphone with them, so if I want an immediate answer, it is so difficult for me (S24)
	Delayed replies (S64)
	Lecturers normally reply late; possibly they are seldom online (S65)
	Lecturers seldom use their phones and rarely check their WhatsApp notifications (S82)

7.2 Difficulties in Using Social Networks for Learning

The questions stated: "Do you find any difficulties in using social networks as a medium for learning? State your answer briefly". Table 2 below shows the themes and responses for the difficulties in using social network for learning.

Table 2. Themes and responses for difficulties in using social networks for learning purpose

Theme	Responses
No difficulties	I think it is not difficult. We have instant contact with our teacher, so it is quite good (S5)
	Nowadays lecturers also prefer contacting him or her by WhatsApp (S21)
	People are more independent about looking for information rather than depending on lecturers (S24)
	Photos and voice notes can be used to explain something (S46)
	Even files or documents can be shared (S65 and S81)
Misunderstand	Sometimes the message does not come across as expected and it gets more confusing (S19)
	Knowledge looked up on social networks might be false (S31)
	Lecturers might not know exactly what we meant (S44)
	Miscommunication and different understanding of concepts may arise (S60)
	Social networks offer so much information that it is hard to decide whether the information is true or false (S70)
	The topic discussed is not understood because the explanation is not clear (S74)
No internet connection	When there is no internet connection (S6)
	If a person does not subscribe to the internet, how can the information be extracted and how the chat room even be entered? (S42)
	Not enough data and slow internet connection (S89)
Difficult to discuss	It is hard to ask questions and explain properly (S4)
	It is difficult when what is being conveyed needs a high degree of understanding (S8)
	Any hard question is difficult to discuss (S11)
	It difficult to explain compared to explaining face to face (S22)
	If something is not understood, it is hard to explain (S26)
	It is sometimes difficult to discuss things because it is difficult to explain things on social networks (S63)

7.3 Suggestions for Best Social Network Features

This section highlights the suggestions from students for enhancing the future potential of social networks. Table 3 below shows the themes and responses of suggestions for the best social network features.

Table 3. Themes and responses of suggestions for best social network features

Theme	Responses
Combine with an educational portal	Try to make an e-learning app but make it simpler. Such an app will be very useful for all students (S39)
	There is a need to have an education portal with a combination of features (S52)
A high storage application	Increase storage to make students more efficient in communicating with others (S37)
	Reduce the memory space needed for installation and usage (S40)
	Every social network must improve their app to make it easier for the user to send large files and high-quality images (S84)
Easy for communication and learning	Enable live communication so that communication can replicate face to face real-time communication (S11)
	Sharing learning materials (S16)
	Easy communication from anywhere at any time (S31)
	Reply with responsive answers (S43)
	Enable conference video calls (S61)
	Enable slide sharing so it can be viewed directly (S89)
Internet connection	Improve the Wi-Fi so that everyone can use it in more efficiently (S28)

8 Discussions

Based on the results, WhatsApp is a very widely-used application for both general communication and learning. Several of the students directly refer to WhatsApp to express social network benefits such as; "If we use WhatsApp, you only spend a little data to communicate with your friends" (S68) and "Nowadays lecturer also prefer contacting him or her via WhatsApp" (S21). The students highlight the use of WhatsApp in their daily learning sessions. In addition, they believe that social networks are easy to use and provide a fast response. This result highlights the function of social networks for sharing information, which is supported by results from the open-ended questions, which also state that this platform is preferred for sharing documents or materials easily. The function of sending messages through texting or by using voice audio was mentioned by two students; "We can either write or reply with voice" (S80) and "Because you can take a photo and use voice note to explain something" (S46), showing the variety of options for communicating with others. Other social networks were rated lower than WhatsApp. These findings show the benefits of social network that can be applied in education. In addition, recent study shows that social network lead to students' satisfaction, skills, enhance attention and experience [13]. In term of the difficulties faced by the students, lack of internet connection leads to difficulties in accessing information. Wi-Fi, or a good data plan is necessary to transfer and receive information. Furthermore, most learning session which are conducted using social networks should be fully prepared, with a good internet connection, in order to share materials and communicate effectively. This statement is supported by [14], that social networks can support collaboration in learning sessions but can also lead to misunderstandings since each has their own opinions and suggestions. In addition, it is hard to judge the accuracy of statements in social networks [15]. Based on results from the open-ended questions, for most students are not communicating face-to-face leads to difficulties in having discussions. Students do learn from collaborative work but feel uncertain about the right way to react in a group when social networking [16]. From the student suggestions in the open-ended section, many ask for an application which provides both an easy way to communicate and an easy way to learn. From all these difficulties, a recent study suggests for three points that need to be focused by institutions to overcome any difficulties which are, poor internet connection, rules and ethical code in implementing social network platform in education, and lastly is the instructor and students need to prepared with knowledge in using the social network as their teaching and learning tools [17]. As most social networks have almost identical functions and features, it is proposed that such an educational platform should apply a suitable pedagogy aligned with good learning concepts.

9 Conclusion

Based on results from the questionnaire showing that WhatsApp is the preferred platform and, there is a clearly a pressing need to test the capability of WhatsApp for learning purposes, stated on comments from the open-ended questions. Overall, the findings of this study certainly show that there are no critical difficulties in using social networks for both communication and learning purposes among the student population questioned. All suggestions for the best possible social network features show the suitability of WhatsApp, according to its many functions stated by participants. It is worth noting that several types of social network have been discovered in by this study to discover the current types of social networks used by students and examined the difficulties. However, WhatsApp ranks as the most widely-used social network by far for both general communication and for learning. As social networks were initially created for communication and sharing, this shows that this platform is indeed suitable for adaptation for learning purposes. Future study should focus with suitable pedagogical method that can be applied for social network platform.

Acknowledgement. The authors would like to thank the Ministry of Higher Education (MOHE) and the Universiti Teknologi Malaysia (UTM) for the Research University Grant Scheme (GUP) (vote number: I6H76) that had supported this research.

References

1. Brooker, P., Barnett, J., Cribbin, T.: Doing social media analytics. Big Data Soc. **3**, 1–12 (2016)
2. Burbules, N.C.: How we use and are used by social media in education. Educ. Theory **66**, 551–565 (2016). https://doi.org/10.1111/edth.12188

3. Tang, T., Cooper, R.: The most social games. Commun. Sport, 1–23 (2017). https://doi.org/10.1177/2167479516688438
4. Lu, W., Hampton, K.N.: Beyond the power of networks: differentiating network structure from social media affordances for perceived social support. New Media Soc. **19**, 861–879 (2017). https://doi.org/10.1177/1461444815621514
5. Mukherjee, D., Natrajan, N.S.: Comparative analysis of social media tool used in software projects deploying virtual teams. Vision **21**, 397–409 (2017). https://doi.org/10.1177/0972262917733180
6. Zaidieh, A.J.Y.: The use of social networking in education: challenges and opportunities. World Comput. Sci. Inf. Technol. J. **2**, 18–21 (2012)
7. De-Marcos, L., Domínguez, A., Saenz-de-Navarrete, J., Pagés, C.: An empirical study comparing gamification and social networking on e-learning. Comput. Educ. **75**, 82–91 (2014)
8. Sun, P.-C., Tsai, R.J., Finger, G., Chen, Y.-Y., Yeh, D.: What drives a successful e-Learning? An empirical investigation of the critical factors influencing learner satisfaction. Comput. Educ. **50**, 1183–1202 (2008)
9. So, S.: Mobile instant messaging support for teaching and learning in higher education. Internet High. Educ. **31**, 32–42 (2016)
10. Al-Emran, M., Elsherif, H.M., Shaalan, K.: Investigating attitudes towards the use of mobile learning in higher education. Comput. Hum. Behav. **56**, 93–102 (2016)
11. Loeb, S., Dynarski, S., McFarland, D., Morris, P., Reardon, S., Reber, S.: Descriptive Analysis in Education: A Guide for Researchers (2017)
12. Nassaji, H.: Qualitative and descriptive research: data type versus data analysis. Lang. Teach. Res. **19**, 129–132 (2015). https://doi.org/10.1177/1362168815572747
13. Stathopoulou, A., Siamagka, N., Christodoulides, G.: A multi-stakeholder view of social media as a supporting tool in higher education: an educator e student perspective. Eur. Manag. J. **37**, 421–431 (2019). https://doi.org/10.1016/j.emj.2019.01.008
14. Poellhuber, B., Roy, N., Anderson, T.: Distance students' readiness for social media and collaboration. Int. Rev. Res. Open Distance Learn. **12**, 102–125 (2011)
15. Ranginwala, S., Towbin, A.J.: Use of social media in radiology education. J. Am. Coll. Radiol. **15**, 190–200 (2017)
16. Neier, S., Zayer, L.T.: Students' perceptions and experiences of social media in higher education. J. Mark. Educ. **37**, 133–143 (2015). https://doi.org/10.1177/0273475315583748
17. Sobaih, A.E.E., Moustafa, M.A., Ghandforoush, P.: To use or not to use? Social media in higher education in developing countries. Comput. Hum. Behav. **58**, 296–305 (2016). https://doi.org/10.1016/j.chb.2016.01.002

The Role of Cloud Electronic Records Management System (ERMS) Technology in the Competency of Educational Institutions

Muaadh Mukred[1](✉), Zawiyah M. Yusof[1], Nor Azian Binti Md. Noor[2], Bakare Kazeem Kayode[2], and Ruqiah Al-Duais[2]

[1] Faculty of Information Science and Technology, Universiti Kebangsaan Malaysia, 43600 Bangi, Selangor, Malaysia
muaadh@ukm.edu.my
[2] Faculty of Education, Al-Madinah International University, Taman Desa Petaling, 57100 Kuala Lumpur, Malaysia

Abstract. Although the Electronic Records Management System (ERMS) is necessary for organizations to enhance their performance, many organizations are yet to embark on its implementation, while some others attempt at implementing it without model, thus leading to an ineffective and erroneous implementation. On the other side, researchers have already started applying cloud computing in ERMS implementations of education sector for its benefits. Therefore, this paper aims to propose a model for cloud ERMS implementation in an attempt to improve the Yemeni public sector educational institutions competency, using Technology Acceptance Model 3 (TAM3). The study used the quantitative approach through questionnaire that was used as instrument to gather data from 350 respondents working in the education sector of Yemen. Partial Least Square-Structural Equation Modelling (PLS-SEM) was used to predict the relationship between the constructs. The findings revealed that both dimensions of ERMS and Cloud are important in implementation of ERMS. The results also showed that ERMS has a significant relationship with the competency of the public sector educational institutions. This has arises implication whereby the public sector needs to pay attention to the factors found in the model in order to promote a successful implementation.

Keywords: Electronic Records Management System · Competency · Education · Yemen · Cloud computing · TAM model · Implementation

1 Introduction

Technology can be used to make information accessible to students and thus it can have a positive and influential effect on education. Technology can also be used to gather the information necessary to the decision making process [1, 2]. Educational institutions deal mostly with electronic records a form of strategic resource to enhance competency, improve and eventually make effective performance evaluation. There are many

F. Saeed et al. (Eds.): IRICT 2019, AISC 1073, pp. 936–946, 2020.
https://doi.org/10.1007/978-3-030-33582-3_88

institutions in around the world applying technologies which imply Electronic Records Management System (ERMS) is deemed important and should not be compromised [3].

According to Britland [4], the future of technology in education would be about anywhere access to resources for learning and collaboration. This suggested cloud computing as the future of technology in education. He further projected that "in the future, teaching and learning is going to be social". The students can use cloud services to aid independent learning, which enable them to study in their own way from anywhere [5]. The collaboration can be achieved using shared applications such as Google Apps and Office 365 which allow students and teachers to work on same documents from anywhere in the world. With cloud computing, teachers and students can be connected with each other within and outside their campuses, and classrooms can be everywhere since educational resources will be available around the clock.

Cloud computing addresses the problems above by providing services, resources, and IT infrastructures so that institutions can focus on teaching and research rather than on IT configuration and management [6, 7]. Cloud computing has the potential to reduce IT cost by virtualizing resources such as processing cycles and disk storage into an available and inexpensive operating expense [8].

Literature dedicated to cloud ERMS has been receiving increasing attention in the context of developed countries, with a focus on the practical solutions to records management and the way it can improve the performance and competency of organizations [9, 10]. In this regard, Adade, Quashigah [11] stated that public organizations should be related to government business in order to reinforce and maintain evidence-based governance and delivery of services and therefore, ERMS adoption is the answer.

In the context of developing nations, ERMS initiatives have been primarily short-term and unsustainable because of the barriers that exist [12]. These barriers can be exemplified by lack of models and several barriers in the form of new characteristics factors as reported by [13] as cloud.

Although there are various benefits, literature indicates that the implementation of ERMS is frequently followed by challenges owing to major changes required and sometimes end up in failure as evidenced in prior studies by many researchers [14]. In this regard new technology implementation (e.g., ERMS) entails monumental changes to be brought about in the routine of the organization.

Several authors evidenced that frameworks implementation can improve the process of ERMS implementation and lead to its implementation success [6, 9, 10]. Therefore, it is important to develop ERMS implementation model for organizations desirous of embarking on its implementation, adoption and use.

In this line of argument, no single model can be used by every organization as illustrated by Hasanain, Vallmuur [15] and this is attributed to the varying settings, environment and objectives. In relation to this, Venkatesh and Zhang [16] revealed that framework application in one specific culture may differ from those of other cultures, with some factors being more or less important in the former compared to the latter. Hence, the need arises to explore and examine the new factors that contribute to successful ERMS implementation [17]. As yet, no research has been conducted to developing a cloud ERMS implementation model, distinctly for the Yemeni public educational organizations.

The structure of the paper is as follows: Sect. 2 is allocated to elaborate on how ERMS elevates competency followed by framework development and hypothesis formulations in Sect. 3. The methodology of this research is described in Sect. 4, and the results and discussion are contained in Sect. 5. Lastly the conclusion is outlined in Sect. 6.

2 How could ERMS Elevates Competency?

If the initiatives of records and information management are improperly implemented, it will be difficult for the organization to achieve the requirements of adhering to the transparency and information obligations [3, 18–21]. Consequently, there will be risks when administrators and stakeholders are not provided access to the required information in a timely manner for the purposes of decision-making [9, 14, 22–24]. Such risks are the indirect result of the failed initiatives implementation that will lead to affect the competency.

Implementing cloud ERMS can assist public organization performance and enhance it, playing a significant role in planning and reliable information provision to reach informed and strategic decisions. In this case, when administrators and stakeholders are not provided with accurate and authentic information in a timely manner for their decision-making, consequences and risks will arise [22]. Because ERMS primarily enhances the performance of organizations and ensures their competencies, it is crucial to examine it in the context of Yemen as it is in dire need of performance enhancement.

Due to a growing competition in higher education environments, universities try to apply strategies and develop new instruments so to enhance the quality of teaching and research activities and provide the communities with relevant services and knowledge [23, 25]. Modern higher education institutions experience the need of effective decision support tools to accurately inform them, and assist in all managerial processes [26].

In recent years, interest in the competency of educational institutions, from the grade school to the university setting, has been amalgamated and articulated in various segments of society. Whether the focus of attention has been the efficiency of the institutions, the competency of the instructors, or the quality of the graduates, the renewed interest of the public in the educational institutions has put pressure on the administrators and teaching personnel to plan, organise, motivate and control for performance [27].

Competency is a characteristic of an employee that contributes to successful job performance and achievement of organizational results. These include knowledge, skill, ability, value, motivation, initiative and self-control. It comprises of several attributes such as cognitive (knowledge and skill), affective (attitude and value, behavioral and motivational (motive) and characteristic which enable an individual person to perform well. At organisational level, competency derives from a collective individual competency across the organization [28].

Records help organization to be seen as transparent, credible and accountable. This in turn helps organization to gain respect and trust from the clients. This could be materialized if the recorded information possessed is accurate and meet the requirement for decision making, relevant, ever ready available for access, updated and reliable [29].

As a whole, competency was found to be positively influenced by ERMS implementation, where the latter supports the former. ERMS facilitates quality, effective and accurate decisions that promote performance and competency. According to [30], ERMS usage in decision making has significantly developed, with the findings revealing that public educational sectors [11, 23, 31] perceive ERMS to play a crucial role in their decision making process that will lead to enhance competency.

3 Framework Development and Hypothesis Formulation

Studies dedicated to technology implementation and adoption of theories and models have been carried out mainly in developed nations. However, the models have rarely been examined and validated in developing nations. According to Venkatesh and Zhang [16], the models applications may differ from one environment to the next, with some of the variables taking a slightly less importance in an environment compared to others. Therefore, prior studies of this caliber have indicated mixed results in different contexts and environment [10]. It goes without saying that owing to the distinct environment in the Middle Eastern countries, like Yemen, which contrasts with those of the developed ones, it becomes a necessity to validate the models in the developing nations as the results obtained may be different.

The proposed framework of this study includes two dimensions that are: ERMS and Cloud characteristics in which many factors were included. The factors were extracted based on literature by contents as well as recommended by experts. The following section is dedicated for model construction and the hypothesis formulations of the study.

The reviewed literature shows that ERMS and Cloud characteristics dimension could determine ERMS implementation. The empirical studies in literature showed several sub-important factors that could impact the intention towards EMRS implementation relating to mentioned dimensions. Specific to this study, the ERMS dimension refers to a set of four factors, namely policy, roadmap, user involvement and technical support, while the cloud dimension refers to a set of three factors including security, reliability and integrity.

A good approach to explain the present state of ERMS adoption in the public firms is to identify the content and context of ERMS dimensions and to explain the shift from legacy shared-drive systems to ERMS [32] as well as explore the new environment of cloud computing factors to cope with modern development of the world. However, even though previous studies focused on several theories for explaining the influence of the usage or the adoption of technology such as TAM, DOI, TRA, UTAUT and TPB, this study concentrates on the TAM 3 as an integrated model to assess the behavioural intention towards ERMS implementation (Fig. 1).

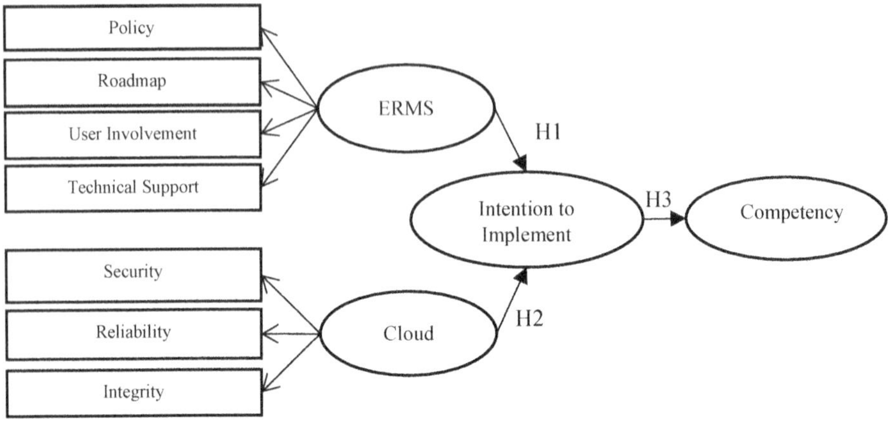

Fig. 1. Study framework with the hypothesis

This study proposes the following hypotheses for testing;

H1: ERMS characteristics factors positively affects intention to implement the system.
H2: Cloud characteristics factors positively affects intention to implement the ERM system.
H3: ERMS implementation positively affects competency of educational institutions.

4 Methodology

The aim of this paper is to explore and account for factors influencing the implementation of ERMS in the educational sector. The first step in the research method is to obtain a deeper insight into the domains of the main problems under examination. Therefore, the research carried out an extensive literature review of studies concerning ERMS in the public sector in general and also specifically in Yemen. This was taken proceeded to enumerate the barriers preventing the implementation of IT, and the acceptance and implementation of models in the public sector.

The study used a quantitative approach to collect data from 320 respondents from the educational institutions of Yemen through a questionnaire. The questionnaire items were adopted from previous researches as seen in Table 1. A probability sampling method was used in this study in consideration of the elements in the sampling frame in that to provide them with equal opportunities of selection. The generalization of analysis findings in this sampling method is deemed to be the most optimal [33]. The first phase in a quantitative data collection entails the questionnaire preparation and the survey obtained data is used to study the various relationships categories between the exogenous and endogenous variables [34].

Table 1. The source of the factors.

Dimension	Factor	No of items	Source
ERMS characteristics	Policy	5	[35, 36]
	Roadmap	5	[37]
	User involvement	5	[38]
	Technical Support	5	[39]
	Security	5	[40]
Cloud characteristics	Reliability	5	[41]
	Integrity	5	[42]

5 Results and Discussion

Quantitative data that has been collected was analyzed using Structure Equation Modeling (SEM) with Smart PLS 3.0. A common measure to establish convergent validity on the construct level is the average variance extracted (AVE). This criterion is defined as the grand mean value of the squared loadings of the indicators associated with the construct (i.e., the sum of the squared loadings divided by the number of indicators). Therefore, the AVE is equivalent to the communality of a construct. Using the same logic as that used with the individual indicators, an AVE value of 0.50 or higher indicates that, on average, the construct explains more than half of the variance of its indicators. Conversely, an AVE of less than 0.50 indicates that, on average, more error remains in the items than the variance explained by the construct [43].

The path coefficients were calculated as well as the path's significance in the model through the bootstrapping method. According to Hair Jr, Hult [43], internal consistency reliability as indicated by composite reliability (CR) value has to be 0.7, while convergent validity measured with the help of Average Variance Extracted (AVE) has to be at least 0.5, and Cronbach's alpha value has to exceed 0.7. Finally, factor loading has to obtain a value of at least 0.6.

[43] suggested assessing the structural model by looking at the beta (β), R^2 and the corresponding t-values via a bootstrapping procedure with a resample of 5,000. Moreover, they recommend reporting the effect sizes (f^2) as well as the predictive relevance (Q^2). As Sullivan and Feinn [44] argue that the p-value determine whether the effect exists but it does not reveal the size of the effect. Figure 2 shows the PLS results of both PLS algorithm and bootstring which were drawn on the version PLS 3.0.

In Table 2 above, the results showed sufficient values as established by [43]. More specifically, the values of factor loading varied from 0.729 to 0.895, while the values of Cronbach's alpha varied from 0.867 to 0.917. Moreover, the values of CR fell between 0.904 and 0.938, and AVE values fell between 0.654 and 0.751. Figure 2 presents the path model results.

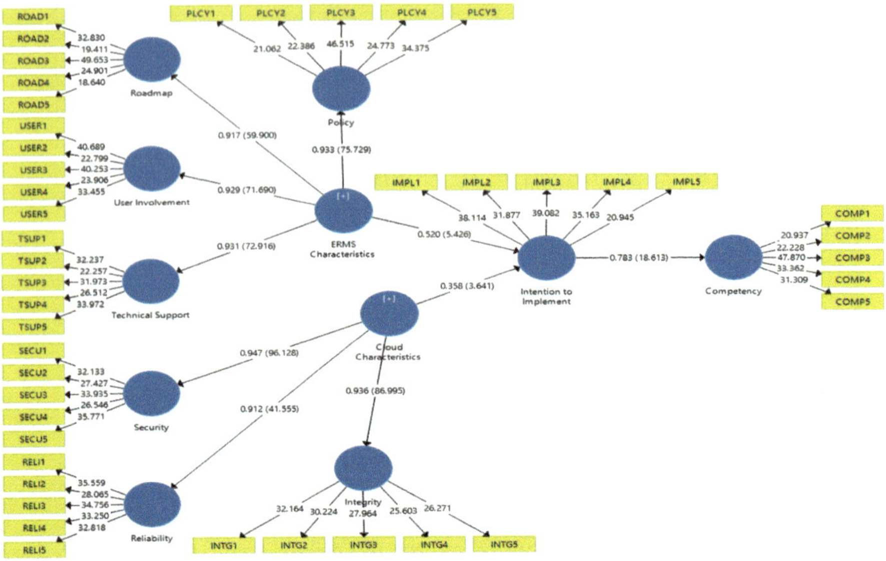

Fig. 2. Path coefficient model

Based on the results in Table 3 and Figure 2, the standardized path coefficient indicates the strengths of the relationship between exogenous and endogenous constructs, so the direct effects of ERMS characteristics on ERMS implementation is stronger than the direct effects of cloud characteristics variables.

For the improvement of education, implementation of new technologies is a must to improve the competency. Nevertheless, new technologies adoption calls for the examination of the factors that could contribute to it. Hence, in this paper, factors that are significant to ERMS adoption and implementation in the education sector of a developing country are identified. The specific identification of the factors influencing cloud ERMS implementation for the enhancing the organizations' competency involved the examination of the level of relationship among the implementation factors (ERMS characteristics factors and Cloud characteristics factors).

Thus, identifying the content and context dimensions of a record provides an approach to explaining the current state of cloud ERMS implementation within public sector organizations, as well as explaining the resistance to moving away from legacy shared-drive systems in favour of an ERMS.

Moreover, the paper reached to the conclusion that more studies are needed to examine the role of cloud ERMS in academic field. Hence, in this study, a framework is proposed to guide such implementation in order to tackle the challenge of system implementation. The literature review highlighted the relationship between successful proper cloud ERMS implementation and proper frameworks that guides it.

Table 2. Results of PLS analysis

Factor	Loading	Cronbach's alpha	Composite reliability	AVE
COMP1	0.780	0.893	0.921	0.701
COMP2	0.826			
COMP3	0.889			
COMP4	0.867			
COMP5	0.821			
IMPL1	0.881	0.917	0.938	0.751
IMPL2	0.875			
IMPL3	0.895			
IMPL4	0.868			
IMPL5	0.814			
INTG1	0.833	0.883	0.915	0.682
INTG2	0.831			
INTG3	0.841			
INTG4	0.813			
INTG5	0.810			
PLCY1	0.792	0.881	0.913	0.678
PLCY2	0.775			
PLCY3	0.870			
PLCY4	0.830			
PLCY5	0.847			
RELI1	0.845	0.905	0.929	0.725
RELI2	0.845			
RELI3	0.864			
RELI4	0.851			
RELI5	0.852			
ROAD1	0.820	0.867	0.904	0.654
ROAD2	0.798			
ROAD3	0.878			
ROAD4	0.812			
ROAD5	0.729			
SECU1	0.838	0.898	0.924	0.710
SECU2	0.835			
SECU3	0.866			
SECU4	0.817			
SECU5	0.856			
TSUP1	0.850	0.897	0.924	0.708
TSUP2	0.813			
TSUP3	0.859			
TSUP4	0.836			
TSUP5	0.849			
USER1	0.849	0.894	0.922	0.704
USER2	0.808			
USER3	0.866			
USER4	0.815			
USER5	0.855			

Table 3. Hypothesis testing results

Hypothesis		ß	t-value	Support	R^2
ERMS implementation					0.727
H1	ERMS Factors → intention to implement the system	0.520	5.439	Yes	
H2	Cloud Factors → intention to implement the system	0.358	3.632	Yes	
Organization's competency					0.614
H3	ERMS implementation → Competency	0.783	18.417	Yes	

6 Conclusion

Apparently, all organizations need to implement electronic records management systems with cloud computing technology to ascertain systematic, efficient, transparent and effective operations, functions and services. Cloud ERMS initiatives could also help organizations develop information system which becomes the basis in serving the public, enhance accountability and help nurturing the culture of building memory among the people other than to ensure the public awareness about their rights.

Records and information help to identify what information really needed by organizations and reduces the unnecessary duplication of information especially with cloud technology that allow to access such service anywhere at any time. Also, with cloud ERMS, it will identifies how long records are to be kept, what to be discarded, what to preserve and what precaution measures are taken to protect vital records without considering the fixed environment. This will hinder organization for destructing information unintentionally and reduce cost as well as keep them safe in case of disasters. In short, cloud ERMS is able to help organization to make planning for future needs, increase the utility value of available information, eliminate redundant information, ensure legal compliance, ensure information is accurate, relevant, ready available for access, updated, fulfill user requirement, reduce storage area improve efficiently and productivity, support decision making, help organization to recover from disaster, reduce operating costs, control the creation and growth of records, protect vital records, preserve corporate memory which finally improve operation and competitive edge.

Acknowledgment. We would like to thank Faculty of Information Science and Technology, Universiti Kebangsaan Malaysia by giving the authors an opportunity to conduct this research. This research is funded by Universiti Kebangsaan Malaysia under Research Grant Code: GUP-2017-046.

References

1. Laird, T.F.N., Kuh, G.D.: Student experiences with information technology and their relationship to other aspects of student engagement. Res. High. Educ. **46**(2), 211–233 (2005)
2. Turney, C., et al.: Using technology to direct learning in higher education: the way forward? Active Learn. High. Educ. **10**(1), 71–83 (2009)

3. Mukred, M., Yusof, Z.M.: The role of electronic records management (ERM) for supporting decision making process in Yemeni higher professional education (HPE): a preliminary review. Jurnal teknologi **73**(2), 117–122 (2015)
4. Britland, M.: What is the future of technology in education. The Guardian **19** (2013)
5. Benson, V., Morgan, S.: Student experience and ubiquitous learning in higher education: impact of wireless and cloud applications. Creative Educ. **4**(8a), 1–5 (2013)
6. Mohammed, F., Ibrahim, O., Ithnin, N.: Factors influencing cloud computing adoption for e-government implementation in developing countries: instrument development. J. Syst. Inf. Technol. **18**(3), 297–327 (2016)
7. Mohammed, F., et al.: Cloud computing fitness for e-government implementation: importance-performance analysis. IEEE Access **6**, 1236–1248 (2017)
8. Almazroi, A.A., Shen, H., Mohammed, F.: The impact of trust on the adoption of cloud computing services by university students. In: International Conference of Reliable Information and Communication Technology. Springer (2018)
9. Mukred, M., et al.: Taxonomic framework for factors influencing ERMS adoption in organisations of higher professional education. J. Inf. Sci. **45**(2), 139–155 (2018)
10. Mukred, M., et al.: The key factors in adopting an electronic records management system (ERMS) in the educational sector: a UTAUT-based framework. IEEE Access **7**, 35963–35980 (2019)
11. Adade, A., Quashigah, A.Y., Eshun, P.: Academic records management in Ghanaian basic schools: a study of basic schools in the Ashiedu Keteke sub-metro in the greater Accra region. Br. J. Educ. **6**(4), 33–49 (2018)
12. Ahmed, S.M., et al.: What are electronic information sharing and integration critical success factors that influence the participation in SMEs? J. Eng. Appl. Sci. **12**(16), 4234–4240 (2017)
13. Ross, L.E., et al.: Barriers and facilitators to primary care for people with mental health and/or substance use issues: a qualitative study. BMC Fam. Pract. **16**(1), 135 (2015)
14. Adu, K.K., Ngulube, P.: Key threats and challenges to the preservation of digital records of public institutions in Ghana. Inf. Commun. Soc. **20**(8), 1127–1145 (2017)
15. Hasanain, R., Vallmuur, K., Clark, M.: Implementation of electronic medical records in Saudi Arabia: a systematic review (2015)
16. Venkatesh, V., Zhang, X.: Unified theory of acceptance and use of technology: US vs. China. J. Glob. Inf. Technol. Manag. **13**(1), 5–27 (2010)
17. Al-Wazir, A.A., Zheng, Z.: E-government development in Yemen: assessment and solutions. J. Emerg. Trends Comput. Inf. Sci. **3**(4), 512–518 (2012)
18. de Mingo, A.C., Cerrillo-i-Martínez, A.: Improving records management to promote transparency and prevent corruption. Int. J. Inf. Manag. **38**(1), 256–261 (2018)
19. Mukred, M., Yusof, Z.M.: Electronic records management and its importance for decision making process in Yemeni higher professional education (HPE): a preliminary review. In: 1st International Conference of Recent Trends in Information and Communication Technologies (IRICT 2014), Johor Bahru, pp. 105–114 (2014)
20. Mukred, M., Yusof, Z.M.: Factors influencing the adoption of electronic records management (ERM) for decision making process at higher professional education (HPE)'s institutions. In: 1st ICRIL-International Conference on Innovation in Science and Technology (IICIST 2015), Kuala Lumpur, pp. 399–403 (2015)
21. Mukred, M., et al.: Electronic records management system adoption readiness framework for higher professional education institutions in Yemen. Int. J. Adv. Sci. Eng. Inf. Technol. **6**(6), 804–811 (2016)
22. Adu, K.K.: Framework for digital preservation of electronic government in Ghana (2015)

23. Mukred, M., Yusof, Z.M.: The DeLone–McLean information system success model for electronic records management system adoption in higher professional education institutions of Yemen. In: International Conference of Reliable Information and Communication Technology. Springer (2017)
24. Mukred, M., Yusof, Z.M.: The performance of educational institutions through the electronic records management systems: factors influencing electronic records management system adoption. Int. J. Inf. Technol. Proj. Manag. (IJITPM) **9**(3), 34–51 (2018)
25. Hase, S., Galt, J.: Records management myopia: a case study. Rec. Manag. J. **21**(1), 36–45 (2011)
26. Samsudin, A., Miah, S., McGrath, G.: An ontology-based record management systems approach for enhancing decision support (2014)
27. Kroll, M.: Depicting the organisational strategy of an educational institution from an overall perspective. Leadersh. Organ. Dev. J. **9**(1), 8–12 (1988)
28. Prion, S., et al.: Preceptor and self-evaluation competencies among new RN graduates. J. Contin. Educ. Nurs. **46**(7), 303–308 (2015)
29. Baron, J.R., Thurston, A., McLeod, J.: What lessons can be learned from the US archivist's digital mandate for 2019 and is there potential for applying them in lower resource countries? Rec. Manag. J. **26**(2), 206–217 (2016)
30. Luchenski, S., et al.: Survey of patient and public perceptions of electronic health records for healthcare, policy and research: Study protocol. BMC Med. Inform. Decis. Making **12**(1), 40 (2012)
31. Phiri, M.J.: Managing university records and documents in the world of governance, audit and risk: case studies from South Africa and Malawi. University of Glasgow (2016)
32. Lewellen, M.J.: The impact of the perceived value of records on the use of electronic recordkeeping systems (2015)
33. Creswell, J.W.: Research Design: Qualitative, Quantitative, and Mixed Methods Approaches. Sage Publications, Thousand Oaks (2013)
34. Kline, R.B.: Principles and Practice of Structural Equation Modeling. Guilford Publications, New York (2015)
35. Aziz, A.A., et al.: Establishing policy for the implementation of electronic document and records management system in public sector in Malaysia: the influencing factors. Adv. Sci. Lett. **23**(11), 10732–10736 (2017)
36. Salah, O.H., Yusof, Z.M., Mohamed, H.: A conceptual framework of CRM adoption among Palestinian SMEs. J. Theor. Appl. Inf. Technol. **97**(8), 2254–2267 (2018)
37. Barua, A., et al.: An empirical investigation of net-enabled business value. MIS Q. **28**(4), 585–620 (2004)
38. Leon, A.: The respective roles of user participation and user involvement in information system implementation success (1991)
39. Chen, H., Chiang, R.H., Storey, V.C.: Business intelligence and analytics: from big data to big impact. MIS Q. **36**, 1165–1188 (2012)
40. Da Veiga, A., Martins, N., Eloff, J.H.: Information security culture-validation of an assessment instrument. South. Afr. Bus. Rev. **11**(1), 147–166 (2007)
41. Li, H., Liu, Y., Suomi, R.: Measurement of e-service quality: an empirical study in online travel service (2009)
42. Zissis, D., Lekkas, D.: Addressing cloud computing security issues. Future Gener. Comput. Syst. **28**(3), 583–592 (2012)
43. Hair Jr., J.F., et al.: A primer on partial least squares structural equation modeling (PLS-SEM). Sage Publications, Thousand Oaks (2016)
44. Sullivan, G.M., Feinn, R.: Using effect size—or why the P value is not enough. J. Grad. Med. Educ. **4**(3), 279–282 (2012)

Computerized Decision Aid for First-Time Homebuyers

S. M. Sarif(✉), S. F. P. Mohamed, and M. S. Khalid

Universiti Utara Malaysia, Sintok, Kedah, Malaysia
{ctmahfuzah, shafinah, sukeri}@uum.edu.my

Abstract. Technology-mediated assistance in house purchasing decision is evidenced in many developed countries, however, in Malaysia less study was found although the decision challenges faced by first-time homebuyers are undeniable. This study attempts to embed technology assistance in the prominent consumer decision-making process model for the purpose of assisting first time homebuyers to make a house purchasing decision. This study employs mixed method approaches with Klang Valley, Malaysia as case study. 19 housing attributes under Locational, Neighborhood, Structural and Social Cultural group, are surveyed to validate the distinctive nature of the attributes to be part of the decision-making criteria. Factor analysis is performed on 320 data from the potential first-time homebuyers in Klang Valley. Two key factors are confirmed from the analysis; which are needs (i.e. Locational and Structural) and preferences (i.e. Neighborhood and Social Cultural). Both factors are later embedded in the proposed design model of computerized decision aid for homebuyers. The model is then evaluated through expert reviews and data was analysed qualitatively using thematic analysis. Four themes emerged from the analysis, which are Suggestion, Concern, Strength, and Limitation. Meaningful discovery on how would the future applications may have an impact is made through analysing themes with negative connotation like Concern and Limitation. The focused themes also reflect actual insights from the industry's key players, which are useful for improvement of the proposed design model and towards more effective computerized decision aid for first-time homebuyers.

Keywords: Computerized decision aid · First-time home buyer · Housing attributes · Factor analysis

1 Introduction

In general, house purchase is considered as high risk investment where chances of going bankruptcy are most likely [1]. Hence, it is imperative to understand the decision factors for house purchase to avoid such catastrophy. Housing attributes have been shown in many literatures to vary from intrinsic housing attributes such as interior living spaces [2], extrinsic attributes such as exterior design and exterior space [3] to neighborhood and locational attributes [4]. There is also relative importance of socio-cultural housing attributes in house purchase observed in settlement patterns and lifestyle [5]. This classification is in line with other scholarly views such as [6, 7]. In a more recent study, [8] investigates five factors that affect the consumer decision making

F. Saeed et al. (Eds.): IRICT 2019, AISC 1073, pp. 947–959, 2020.
https://doi.org/10.1007/978-3-030-33582-3_89

when it comes to purchasing a residential property; there are location factor, feature factor, financial factor, neighborhood factor and demographical factor. Previous studies implicate interesting debate about the relative importance of all these housing attributes in house buying preferences. However, very little is known about the relevant housing attributes [9, 22]. In a way, this study also intends to contribute to the literatures by developing an understanding on which housing attributes, contribute to house buying preferences among first-time home buyers. Ideally, with the knowledge about the housing attributes, providers of new housing would be better equipped to meet the demand of prospective home buyers and maximize construction efficiency [22].

One of the ways to assist home buyers in making decision is through technology-mediated decision aids. Technology-mediated assistance in house purchasing decision is evidenced [10] in many developed countries, however, in Malaysia less study was found although the decision challenges faced by first-time homebuyers are undeniable. The use of Internet, however, doesn't benefit homebuyers in terms of search time, flexibility and intuitive results. Even though it facilitates an intensive search and discovery of more properties, it also wastes more time and energy [8]. Reference [11] has verified this problem, arguing that due to each and every real estate company building its own website to advertise its products for the purpose of online buying and selling, customers are prone to get lost in searching among all those websites, and aside the time consuming of such processing, results can be very conflicting. In addition, it also noted that purchasing of a home is a multi-faceted decision, however search services in real estate websites are single-faceted (lack of multi-attribute search system), thus inhibiting prospective buyers when comparing properties; secondly, there is no semantic relationships between the various factors that constitutes intricately connected information about home assessments [11].

In view of the fact that home buyers need a better and practical assistance in house purchase, this study explores the potential of implementing a systematically designed technology-mediated decision aid for first-time homebuyers in Malaysia. In addition, this research aimed at examining the Consumer Decision Making Process (CDMP) model [17] and finding potential of embedding researcher's doctorate work, Computerized Personal Decision Aid (ComPDA) model [18], in attempt to coordinate the process of house purchasing decision for first-time buyers. CDMP model consists of the following components – input, information processing, decision process, and variables influencing the decision process. The decision process component consists of five following stages – need recognition, search, alternative evaluation, purchase, and outcomes. Meanwhile, ComPDA model conceptualises the decision-making process in three stages of activities: intelligence activity, design activity, and choice activity. The model agree with [23] that argues decision-making is a cognitive process that can be separated into simple, sequential steps. In ComPDA each stages are detailed out with sequential activities in decision-making - problem definition, information gathering, alternative generation, criteria identification, evaluation, and choice.

This paper is structured in such a way that it starts with some background study, followed by the process involved in constructing the design model, discussion of the evaluation results and ends with some concluding remarks.

2 The Material and Method

This study adapts the design science research methodology as the generally accepted methodology in information system. The methodology from [12] is adapted to achieve the proposed objectives. There are four main stages to accomplish the goal of the research, (i) awareness of problem; (ii) suggestion and development; (iii) evaluation; and, (iv) conclusion.

2.1 Awareness of Problem

The identification of this research problem came within the literature review activity. There are many attempts made in previous studies on how decision techniques can be used to assist homebuyers with their purchase decisions [13–15]. However, there is little evidence to examine first-time homebuyers' purchase decision in Malaysian context although the issues do exist. Therefore, this research intends to fill the gap that currently exists in the literature by exploring the potential of implementing a systematically designed computerized decision aid among first-time homebuyers in Malaysia, particularly in the area of Klang Valley.

Previous studies implicate interesting debate about the relative importance of all these housing attributes in house buying preferences. However, inadequate empirical study is found about the relevant housing attributes [9, 22]. This has led to a gap between consumers' expectations and developers' perceptions, which resulted in customer dissatisfaction [16].

2.2 Suggestion and Development

This study attempts to adapt the ComPDA model into CDMP model for the purpose of assisting first-time homebuyers to make purchase decision. This is justified by the fact that most people fail to bring the right information into their conscious awareness at the right time and their decision process seems affected by incomplete information and bounded rationality. Even more challenging when it is the first time of making a high risk purchase decision. The followings are stages in the development of the proposed solution:

Adaptation of CompDA into CDMP Model. Firstly, common stages are identified between the two models. Apparently, the common stages in both models are Problem/Need Recognition, Information Search, Evaluation of Alternatives and Purchase Decision. Likewise, in the ComPDA model, identical steps are categorised in three stages (i.e., Intelligence, Design, and Choice) – (i) problem definition, (ii) information gathering, (iii) alternative generation, (iv) criteria identification, (v) evaluation, and (vi) choice. While maintaining the main structure of the proposed model based on the three crucial stages in ComPDA (i.e. Intelligence, Design, and Choice), the additional variables in the CDMP model that influence the consumer purchase decision are also considered - Environmental Influences and Individual Differences. Thus, few modifications are made to the activities in the ComPDA model to fit the variables. Furthermore, the variables are unique information for each decision maker, which will make the process varies from one decision maker to another. Hence, a user profiling system is introduced in the decision process during the Intelligence stage. The Intelligence stage also concern with how the decision aid can help in understanding the problems by searching for conditions (i.e. threshold) that call for decisions. This initial concept of decision model is designed to ease the understanding in the decision problem identification and information gathering steps. Hence, the step of setting up the decision threshold is introduced to the stage. In the Design stage, the following steps remain - options identifications, criteria definition, and evaluation, however some updates and modifications are made to the criteria definition and evaluation activities, as further explained next.

In ComPDA, the evaluation activity comprises of two decision strategies known as non-compensatory (i.e., to do the filtering/short-listing process) and compensatory strategy (i.e., for trade off analysis of the remaining options), in which Take-the-Best (TTB) is used for the former and Pugh Matrix is used for the latter. In house purchase, the decision is considered as high risk since it involves a long-term commitment and huge monetary investment. Hence, Analytical Hierarchy Process (AHP) is embedded as the compensatory decision strategy to refine the process. The AHP uses a multi-level hierarchical structure of objectives, criteria, sub-criteria and alternatives. Also, AHP enforces the use of weighting factors for each criterion, in which the weights are used to compare the options. The strengths of AHP are mainly due to the detailed mathematical properties of the method and the ease to obtain required input. AHP is the most widely accepted method and is considered as the most effective MCDM technique [19].

In order to make a sensible decision, criteria or attributes of the options/alternatives under consideration need to be clearly identified. The presence of multiple criteria may intricate a decision process. In view of the matter mentioned above, a way to characterize the criteria used in each evaluation step (i.e., non-compensatory and compensatory) is needed. Through improvised TTB (in non-compensatory), all alternatives are evaluated as such if the criteria are recognized or identified then the alternatives will be included in the consideration set. This continues until all criteria are assessed. The evaluation in TTB shows that the type of assessment made to the criteria is more narrative. On the other hand, for the compensatory technique, it allows all alternatives to be evaluated through trade-off analysis of the criteria and its sub-criteria. Via the trade-off analysis the criteria help to determine rating values for all alternatives.

As stated earlier, clear identification of the criteria helps to make a sensible decision. Accordingly, from the discussion on narrative and rating criteria, it is necessary to clearly identify the criteria used in house purchasing decision. Hence, explained why it is important for this study to categorise the housing attributes/criteria into *needs* (i.e. narrative criteria) and *preferences* (i.e., rating criteria) (as discussed in the following section). Moreover, well-defined criteria not only help in structuring the decision process but also help to raise better awareness among decision makers about the alternatives under consideration.

Validation of the Decision Criteria. Needs and preferences are integral elements in the process of house purchasing decisions. This study makes a distinction between the needs and preferences of homebuyers as part of the decision-making criteria. A need is defined a homebuyer's desire for a property's specific benefit, whether that be functional or emotional. A preference is the desire for products or services that are not necessary, but which homebuyer's wish for.

A survey as published in [20] was conducted and scoped to Klang Valley area. The demographic statistics for first quarter of 2018 shows that Klang Valley has an estimated population of 8 million [21] consisting of a mixture of many different races and other demographics. The present study covered most of the majority race groups, various work sectors, and all income levels found in the population of Klang Valley. The only requirement was that respondents should have the financial means to buy a house in Klang Valley and legally eligible to do so. It was not known exactly what the total number of potential house buyers in Klang Valley and who they were. This meant that there was no sampling frame available for the potential house buyer's population in Klang Valley. Thus, Non-Probabilistic Convenient Sampling Method was adopted for this study. A total number of 400 sets of questionnaires were distributed and 388 questionnaires were collected from respondents in the area of Klang Valley (i.e., respond rate of 97%). However, from 388 questionnaires only 320 valid to be used for data analysis, which surpasses the sample size required.

Out of 320 respondents, 168 were male (52.5%) and the remaining 152 were females (47.5%), which correspond to the sex ratio of current population estimates in Malaysia [24]. In terms of age distribution, most of the respondents were in the age category of 26 to 40 (66.9%). Once again, this percentage seems to be consistent with the preliminary study. Malays were the overwhelming majority of the respondents at 267 (83.4%), followed by Chinese (8.1%), Indian (5.6%), Bumiputera (2.5%) and others (0.3%). The percentages are consistent with population distribution by ethnic group in Malaysia [21].

Factor Analysis was performed on four groups of housing attributes comprise seven items for Locational Attribute, four items for Neighborhood Attributes, six items for Structural Attributes, and two items for Social Cultural Attributes. Factor Analysis was employed to make sure that the number of items could be decreased to the number of factors (i.e. in this study if its fall under *Needs* or *Preferences*). The cut-off value for Kaiser–Meyer–Olkin (KMO) measure is 0.50 for Bartlett's test of sphericity to be significant. The minimum value of factor loading 0.4 on one factor should be achieved to consider it as significant. The KMO measure verified the sampling adequacy for the analysis, stand at 0.814, which falls into the range of great, so the sample size is adequate for Factor Analysis. Table 1 summarizes factor loadings for *Locational Attributes* (A1, A2, A3, A4, A5, A6, and A7), *Neighborhood Attributes* (B1, B2, B3, and B4), *Structural Attributes* (C1, C2, C3, C4, C5, and C6), and *Social Cultural Attributes* (D1 and D2) that were extracted from the Rotated Component Matrix. This study preset the first factor as Needs and second as Preferences. Based on Table 1, variables A1 to A7 and C1 to C6 loaded strongly on first factor (Needs) whereas variables B1 to B4 and D1 to D2 loaded strongly on second factor (Preferences). This survey shows that first-time homebuyers perceived Locational and Structural Attributes as Needs, where as Neighborhood and Social Cultural Attributes as Preferences.

Development of the Design Model. Having identified and validated the distinction between narrative and ratable criteria in the case of house purchase, this study proceed with the construction of the conceptual design model of computerized decision aid for

Table 1. Rotated Component Matrix a

	Component	
	1	2
A1	.530	
A2	.532	
A3	.596	
A4	.540	
A5	.669	
A6	.648	
A7	.586	
B1		.893
B2		.876
B3		.904
B4		.692
C1	.718	
C2	.721	
C3	.734	
C4	.755	
C5	.565	
C6	.552	
D1		.549
D2		.464

Extraction Method: Principal Component Analysis.
Rotation Method: Varimax with Kaiser Normalization.

house purchase. As described in earlier section, the principal of the design model follow the main framework of ComPDA, where it is outlined into three stages; Intelligence, Design and Choice. Figure 1 illustrates the conceptual design model of computerized decision aid for house purchase.

2.3 Evaluation

The proposed solution is evaluated both quantitatively and qualitatively through expert reviews. Expert reviews can generate valuable feedback on the semantic quality of the proposed model through the implementation of a working prototype. Meanwhile, experts who involved in this process were classified based on related academic background in real estate or related areas, have at least ten years industry background in real estate or related areas, and/or have been studying/researching in real estate or related areas for at least five years. Due to difficult-to-work-around schedules and workloads of the experts, only a 50% response rate (corresponding to 2 of 4 available key players in real estate area) was obtained, which is within expectations. The demographic profiles of the experts are as displayed in Table 2.

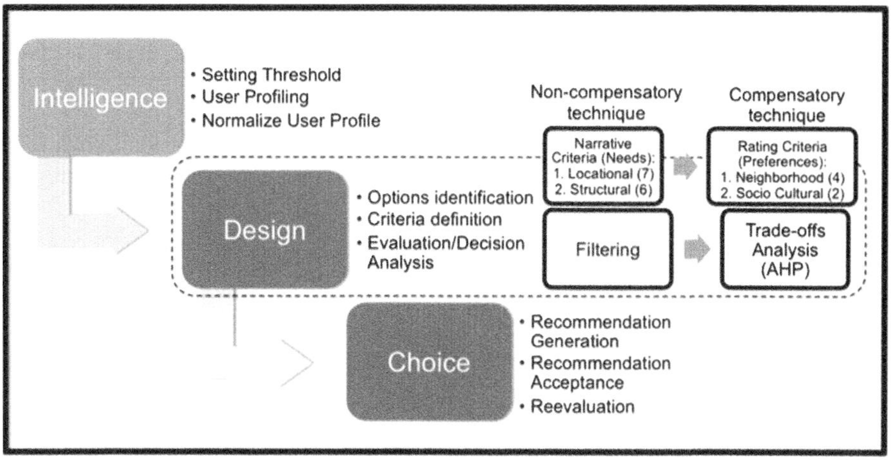

Fig. 1. The proposed design model of computerized decision aid for first-time home buyers

Table 2. Demographic profiles of experts

Gender	Age (Year)	Organisation	Position	Experience (Year)
E1: Male	43	Syarikat Perumahan Negara Berhad	Group Chief Marketing Officer	20
E2: Female	46	Pelaburan Hartanah Nasional Berhad	Manager	23

E1 = Expert 1, E2 = Expert 2

As shown in Table 2, the experts represent different position in the prestigious organization related to real estate area. Their years of experience in the area met with requirement for expert in this study. Furthermore, the age of experts involved also show the sufficient level of maturity in giving opinions and assessments.

The objective of the expert review was to validate the proposed solution and its components through simple survey (questionnaire) and discursive evaluation (semi-structured interview with open-ended questions). Before the expert review session begins, the background and objectives of the study were briefly explained to the experts. Then, the proposed model was introduced together with its working prototype. The session involved two-way interactions, where experts can ask questions and give opinions of the focused matter. Afterwards, the experts were required to respond to the simple survey (questionnaire). Next, during the interview, the experts were asked few semi-structured questions about the ability of the proposed solution, its foreseen benefits, anticipated obstacles, and ways for improvements.

Two types of data were obtained from the evaluation process; (1) survey data (quantitative), and (2) interview transcript (qualitative). Findings from the expert reviews are discussed in the next section.

3 Results and Discussion

For the survey data, descriptive analysis is used to explain the results. In general, both experts agree with inclusion of mostly all the components in the proposed model. However, there are conflicting feedbacks between Expert 1 (E1) and 2 (E2) about the first component in the proposed model (i.e., Intelligence). E2 disagrees with the component and stated a reason that there are too many questions to be answered for the user profiling. Researchers have clarified the issue, explaning that the user profile is only a one-off process during sign up, and it will not be asked every time user signs in to the system. Moreover, the user profile is important for a knowledge base recommender system to understand how the recommendation meets the user's needs. As for the last component (i.e., Choice), both experts have disagreement with the fact that the future developed recommender system will produce a single recommendation. Hence, they advice to provide more than one choices for the user to consider, which is doable in the system development and not going to effect the structure and flow of the conceptual design model.

For the interview part, face-to-face approach was opted because it took into account gestures, facial expressions and the interview environment and enables the underlying meaning of an interviewee's responses to be more deeply explored. All the interviews were done in groups; thus, it provides deeper and more engaging discussions among the researchers and industry players. This allows the flexibility to probe for details on the evaluated model and also to allow for focused and conversational two-way communication. Thematic analysis is used to classify and interpret the interview transcripts into codings to ensure they represent the whole content. Thematic analysis is a descriptive presentation of qualitative data. The purpose of categorising content into themes was also to present the different expressions of the data. Thematic analysis categorizes common thoughts provided in the data analysed. Emerging thematic ideas could be seen through this process. Initially, 32 codings were derived from the interviews transcriptions. The purpose of coding into a classified theme was to assist in interpreting the data and to look for patterns occurring from the data. These codings were then grouped into themes. The themes consisted of codings that were related and, thus, classified in the same category. Table 3 presents the classification of the codings and the formation of the themes. Codes belonging to the same theme were coded in the same row in the table. As a result, four themes were derived from the 32 codings (refer Table 3).

Based on the identified themes, apparently T2 and T3 have more negative connotations as compared to T4, where as T1 will be considered for future works. Thus, this paper extends the discussion for T2 and T3 to understand the effects on the performance of the future developed application.

3.1 T2: Concerns

This theme encapsulates each of the experts' concerns about the working prototype; a combination of concerns encompassing the features in the prototype, system performance, big data issue, and transparency of the system.

Table 3. The identified themes based on elicited codings

Theme	Elicited codings from the transcripts
T1: Suggestions	Pattern, lifestyle impression, include public data from iProperty, genY needs and preferences, separate group of house number for landed and stratified, more house features, PTPTN screening, indicative market value, banking criteria, real time data, hybrid application (web and mobile)
T2: Concerns	Set of questions for user profiling, criteria definitions not clear, competition with existing tool, time spent to use the system, sub-sale data, big data, user interface, AHP, transparency, competing with existing tool e.g. iProperty
T3: Limitations	Single recommendation, some criteria are too basic, no risk analysis
T4: Strengths	Interesting concept, paper-less technology, attraction to generate lead, user friendly system

In the beginning of the interview, when they were first introduced with the prototype, both experts felt that it could be tiresome for user to complete the entire user profiling questions since the number of questions are quite handful: *"How many questions are they? How long will this take?" [E1]* and *"Do I need to answer these all over again everytime I want to make a new search?" [E2]*. However, after explaining to them the purpose of user profiling and why we need to ask users all those questions, they started to grasp the concept. As mentioned in earlier section, user profile is important for a knowledge base recommender system to understand how the recommendation meets the user's needs. And the process is going to be a one-off process during the sign up.

One of the experts also commented on the vague differences between the need and preference criteria: *"For me the differences between needs and preferences are not that clear. Perhaps a better explanation could distinguish the differences between these two" [E1]*. In view of this issue, the differences between the needs and preferences criteria could be made clearer by some description or brief notes explaining the function of each in different stage of the decision making process.

As far as system performance is concerned, both experts are concerned about the time that needed to be spent in accomplishing the task: *"How long does it take to complete the process – until suggestion is given?" [E1]*. While the other is concerned on how the system can be a standout and compete with other existing tools: *"Ive seen similar systems in the market e.g. iProperties" [E2]*. To address the first concern on time spent to use the recommender system, the experts were then made to understand about the purpose of the user profiling questions and it wont effect the time spent to reuse the system for the second time onwards since its only a one-off process. In fact, a quick decision-making applies to any simple decision making with multiple options. This is especially important in control system environment. However to simplify a challenging decision making process, like making a high risk decision of purchasing a house, each specific parts of that process need to be presented at any one time on the screen. Instead of showing everything at once, it should be broke down. Reducing the number of perceived options on screen makes the interface more users' friendly. It is more likely that the user will accomplish the goal and not give up or get confused. If the user spends too little time, probably he/she has left without making a decision.

As far as the existing tool is concerned, iProperty is a search system where it probably uses a multi-agent based algorithm for the options to be retrieved. It does not involve a multi criteria decision-making technique that enable user to do trade off analysis of the options. Hence, it is not the same with what researchers are proposing in this study, where the core function is to assist user in making a decision through a trade off analysis.

As for the big data issue, both experts did raise their concern on the enormous amount of data that need to be registered in the database by the developer in order to keep the options active and relevant: *"How does this system works for the real estate developer to key in the updates?" [E1]* and *"I cant imagine how big the data will be like once the developers are given access to the database for data updates and housekeeping." [E2]*. This issue can be addressed with a concept called plug-in for big data management. Plug-in is a software component that adds a specific feature to an existing application/system. In fact this kind of plug-in applications are already available in the market.

Lastly, E1 raised some concerns regarding the user interface of the prototype, to him the user interface need to facilitate the user to comprehend the process better and at the same time make the system more transparent. Researchers agreed with the comment, and will take into consideration for future development of the full system.

3.2 T3: Limitations

This theme captured experts' dissatisfaction about the limitations in the model-driven prototype. The main issue raised by both experts is that the single recommendation produced by the model-driven prototype should be improved with more choices for the user to consider: *"So, basically this system can only produce one suggestion? Lets say, Ive made the search setting based of my preferred criteria, and then my dream house is no longer available in the market...I would probably want to have closest suggestions based on the criteria that I set earlier" [E1]* and *"I think, I would prefer a system that could give more than one suggestion" [E2]*. As discussed in earlier section, the suggestion is doable in the system development and it is not going to affect the structure and flow of the conceptual design model.

The experts also voiced out their dissatisfaction about the list of criteria available in the model-driven prototype, which is to them too basic. They would like to see more criteria that reflect the needs and preferences of the new generations: *"The criteria in this system are quiet general. Home buyers these days prefer to see lifestyle criteria in their dream house" [E1]* and *"I believe this system needs to consider the preferences of the new generations in housing attributes. I suggest, it would be helpful to include criteria like nearby facilities and amenities in the neighborhood" [E2]*.

In respect of the above, researchers do agree if the recommender system have richer data on the criteria it will have more impact on the recommendation. As already being elaborated in earlier section, 19 housing attributes (including those mentioned by the experts) have been categorised as needs and preferences. However, for prototyping purposes, researchers have to make do with the amount of data gained from the authorised sources, which apparently affects the performance of the prototype.

Some issue on the absence of risk analysis feature in the model-driven prototype also being highlighted by E1: *"Does this system also include the risk analysis part?"*. The expert was made to understand about the scope of the study, that risk analysis is beyond the scope. As far as first time homebuyers are concerned, to be able to make the first step of making decision is already a challenge for them, because of the information overload. Hence, by reducing the burden of processing too much information and one time will be the goal for this study. However, this can be considered in future enhancement of the model.

4 Conclusion

This paper explains about the process of developing and evaluating a design model, which is intended for future development of computerized decision aid for home-buyers. Part of the development process involves validating 19 housing attributes into categorical decision criteria through a survey among first time potential homebuyers in Klang Valley. The results of the survey revealed that housing attributes under Locational and Structural are perceived as needs where as attributes under Neighborhood and Socio-cultural are perceived as preferences. Understanding the differences between the criteria (i.e., needs and preferences) would assist homebuyer to make a more sensible decision. Moreover, well-defined criteria not only help in structuring the decision process but also help to raise better awareness among decision makers about the alternatives under consideration.

Another important process in the development of the design mode is the integration of ComPDA model [18] with CDMP model [17]. The integration is a matter of embedding technology assistance in consumer decision-making process. Upon completion of conceptual design model, it is then evaluated through expert reviews and data was analysed qualitatively using thematic analysis. Four themes emerged from the analysis, which are Suggestion, Concern, Strength, and Limitation. Meaningful discovery on how would the future applications may have an impact is made through analysing themes with negative connotation like Concern and Limitation. The focused themes also reflect actual insights from the industry's key players, which are useful for improvement of the proposed design model and towards more effective computerized decision aid for first-time homebuyers.

In general, findings from this study serve a twofold purpose, (i) the validation of the housing attributes provide better understanding to the property developers about the needs and preferences of first time homebuyers for better housing provisions, and (ii) the proposed design model could be used as guideline in development of applications like a recommendation system to assist the first time homebuyers with their purchase decision.

References

1. Letkiewicz, J.C., Stuart, J.H.: Homeownership among young Americans: a look at student loan debt and behavioral factors. J. Consum. Aff. **52**(1), 88–114 (2018)
2. Özsoy, A., Güçin, P.G.: Space use, dwelling layout and housing quality: an example of low-cost housing in İstanbul. In: Housing, Space and Quality of Life, pp. 17–28. Routledge (2017)
3. Bakar, A.A., Nurhayati, A.M., Mohamad, A.M., Rosniza, O., Aliyah Nur Zafirah, S.: Housing space quality towards quality of life: a case study of double storey terrace houses. In: MATEC Web of Conferences, vol. 66, pp. 00083. EDP Sciences (2016)
4. Tan, T.H.: Neighbourhood satisfaction: responses from residents of green townships in Malaysia. Int. J. Hous. Markets Anal. **9**(1), 137–155 (2016)
5. Shabani, M.M., Mazlan, M.T., Hoda, S., Honey, A., Fereshteh, M.: Relation of cultural and social attributes in dwelling, responding to privacy in Iranian traditional house. e-BANGI **8** (1) (2017)
6. Moghimi, V., Mahmud, M.J.: Priority of structural housing attribute preferences: identifying customer perception. Int. J. Hous. Markets Anal. **8**(1), 36–52 (2015)
7. Babawale, G.K., Koleoso, H.A., Otegbulu, C.A.: A hedonic model for apartment rentals in Ikeja area of Lagos metropolis. Mediterr. J. Soc. Sci. **3**(3), 109–120 (2012)
8. Hei, C.P., Omkar, D.: Buying a dream home-considerations of residential property consumers in Malaysia. Singaporean J. Bus. Econ. Manag. Stud. **51**(118), 1–17 (2017)
9. Zinas, B.Z., Mahmud, M.J.: Choice behaviour of housing attributes: theory and measurement. Asian J. Environ.-Behav. Stud. **2**(2), 23–37 (2017)
10. Yuan, X., Lee, J.H., Kim, S.J., Kim, Y.H.: Toward a user-oriented recommendation system for real estate websites. Inf. Syst. **38**(2), 231–243 (2013)
11. Alrawhani, E.M., Halizah, B., Zurina, S.: Real estate recommender system using case-based reasoning approach. J. Telecommun. Electron. Comput. Eng. **8**(2), 177–182 (2016)
12. Vaishnavi, V.K., William, K.: Design Science Research Methods and Patterns: Innovating Information and Communication Technology. CRC Press, Boca Raton (2015)
13. Sun, Z.H., Pan, L., Wang, Y.Y., Zhang, D.H.: The purchase house choice research based on the analytic hierarchy process. In: The 19th International Conference on Industrial Engineering and Engineering Management, pp. 897–902. Springer, Berlin (2013)
14. Wang, C.: Family house-purchase decision model based on analytic hierarchy process. In Applied Mechanics and Materials, vol. 423, pp. 2973–2976. Trans Tech Publications (2013)
15. Lin, C.T., Jun-Kun, L.: Fuzzy-GIS approach for applying the AHP multi-criteria decision-making model to evaluate real estate purchases. J. Test. Eval. **41**(6), 978–989 (2013)
16. Kam, K.J., Anthony, S.H.L., Karam, M.A., Tze, S.L.: Evaluating housing needs and preferences of generation Y in Malaysia. Plann. Pract. Res. **33**(2), 172–185 (2018)
17. Blackwell, R.D., Miniard, P.W., Engel, J.F.: Consumer Behavior. Thomson South-Western Publisher, Toronto (2006)
18. Sarif, S.M.: Conceptual design model of computerized personal-decision aid. Ph.D. Dissertation, UUM (2011)
19. Roy, S.: Comparative study of some MCDM techniques for e-commerce applications. Int. J. Adv. Res. Comput. Sci. **9**(1), 440–443 (2018)
20. Sarif, S.M., Mohamed, S.F.P, Khalid, M.S.: Assessing decision factors for house purchasing (needs vs. preferences) from case study of first-time potential homebuyers in Klang-Valley: towards technology-mediated individual decision aid. In: Proceedings of KMICE 2018, pp. 336–343 (2018)

21. Demographic Statistics Fourth Quarter 2018, Malaysia. Department of Statistics, Malaysia (2018)
22. Ismail, H., Shaari, S.M.: Housing decision: the choice between location, house and neighbourhood among malaysian generations. In: MATEC Web of Conferences, vol. 266, pp. 01026. EDP Sciences (2019)
23. Simon, H.A.: Theories of decision-making in economics and behavioral science. Am. Econ. Rev. **49**(3), 253–283 (1959)

Malaysian Health Centers' Intention to Use an SMS-Based Vaccination Reminder and Management System: A Conceptual Model

Kamal Karkonasasi(✉), Cheah Yu-N, Seyed Aliakbar Mousavi, and Ahmad Suhaimi Baharudin

School of Computer Sciences, Universiti Sains Malaysia, USM, 11800 George Town, Penang, Malaysia
asasi.kamal@gmail.com

Abstract. Most Malaysian health centers still apply a paper-based vaccination system to save a record of the vaccination milestones of infants, and to remind their parents to bring them to health centers for immunization. The current system causes massive workloads on nurses declining their efficiency in the workplace. Therefore, this study proposes an SMS-based vaccination reminder and management system called Virtual Health Connect (VHC) to assist the health centers by simplifying the management of infants' immunization records. This study aims to realize influential factors of intention to use VHC among Malaysian health centers by considering the technology acceptance model (TAM) as an underlying theory. The proposed model will be verified using data collected from self-directed questionnaires answered by nurses working in Malaysian government hospitals and government clinics. A research method based on the multi-analytical approach of Multiple Regression Analysis (MRA) and Artificial Neural Networks (ANN) will refine the results of this research. This study will contribute to the theoretical body of knowledge about influential factors of health centers' intention to use VHC in Malaysian government health centers by extending the TAM model by adding new variables. The practical contribution of this study will help software developers to develop VHC by considering essential factors that determine health centers to intend to use VHC.

Keywords: SMS-based vaccination reminder and management system · Technology acceptance model (TAM) · Artificial Neural Networks (ANNs) · Multiple Regression Analysis (MRA) · VHC

1 Introduction

Vaccination is one of the most productive and cost-effective public health interventions [1] to stay children healthy and to protect them from preventable diseases. However, the number of infants with delayed or missed vaccination milestones rises due to the growth of working parents [2, 3]. Therefore, the health centers' nurses are required to send several reminders to their parents to ensure on-time immunization of infants.

F. Saeed et al. (Eds.): IRICT 2019, AISC 1073, pp. 960–969, 2020.
https://doi.org/10.1007/978-3-030-33582-3_90

Currently, Malaysian health centers use a paper-based vaccination system to save track of the immunization and to remind parents by a telephone or a mail. The current system brings heavy workloads to the nurses. Besides, the system is not cost-effective due to imposing costs through phone calls or mails [3].

This study proposes VHC to replace the current paper-based system. VHC is a web-based system that assists nurses to record and save track of vaccination milestones automatically. Besides, it sends short message service (SMS) reminders to parents to take their infants to health centers to be vaccinated.

SMS has a remarkable and untouched capacity for illness management in low- and middle-income countries (LMICs) [1]. An SMS-based vaccination reminder is a beneficial and practical system to save time, labor, and cost of managing vaccination records [1, 3]. The effectiveness of SMS-based vaccination reminders has also been proved for different types of vaccinations around the world [1, 3–11]. The penetration rate for Malaysian cell phone subscriptions was over 144% in the year 2014 [12]. It represents that parents have at least one cell phone ready to receive SMS. Therefore, VHC might improve vaccination coverage between infants.

During the case study, nineteen managers of Malaysian health centers were enquired if they have already adopted an SMS-based vaccination reminder and management system. Many of them answered that they had not yet adopted it. This finding is supported by Ooi and Cheah [3], which represents that the adoption level of the system among Malaysian health centers is low.

Numerous studies about SMS-based reminder systems were carried out among industrialized countries [13]. Consequently, there are deficiencies in the literature to test the acceptance of these systems among health centers in LMIC [14, 15]. In response to the low-level adoption and insufficiencies in the literature, the objectives of this study are (1) to apprehend the significant factors that impact intention to use VHC among Malaysian health centers by considering the mediating role of attitude. (2) to contribute to the theoretical body of knowledge by extending the TAM model by adding new variables, i.e., accessibility, compatibility, and privacy concern. (3) to provide a practical contribution to assist software developers in the development of VHC by considering influential factors that impact health centers' intention to use VHC.

In the subsequent sections, a research model is proposed. In Sect. 3, each variable is defined, and hypotheses are stated. In Sect. 4, the proposed research methodology is discussed; finally, a conclusion is drawn in the last section.

2 The Proposed Research Model

Davis [27] proposed TAM as an analytical model to predict user acceptance to use computer technology. TAM is one of the most extensively used and acceptable models to study the behavioral and social intentions that affect the acceptance of technology [28]. It is the most capable model to predict the variables that influence the acceptance of technology among several contexts, including healthcare [26, 29, 30]. In the context of the present study, a TAM-based model is proposed to apprehend the factors that influence the intention to use VHC through the mediating role of Attitude. The model is illustrated in Fig. 1.

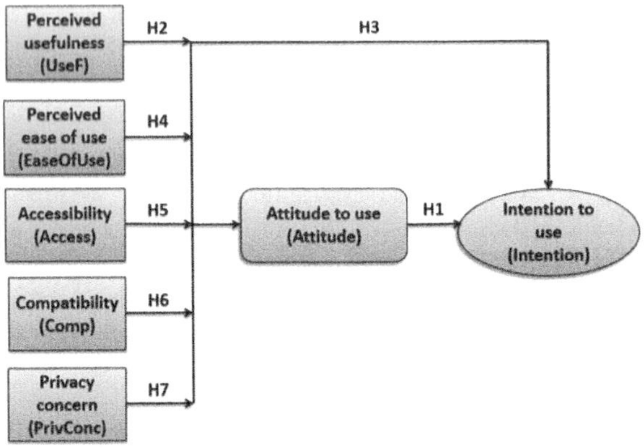

Fig. 1. The proposed research model (Note. H1 to H7 represent the hypotheses.)

3 Proposed Hypotheses

In the next sections, each factor is discussed in detail, and its influence is hypothesized.

3.1 The Relationship Between Attitude and Intention

Attitude towards using a system is defined as the individual's positive or negative feelings about working with the system, and intention to use a system is defined as the degree of the determination of the individual's aim to use the system [31]. Ajzen and Fishbein [41] indicated that attitude impacts behavioral intention. Besides, Davis [27] stated the significant and positive effect of Attitude on Intention. Kowitlawakul et al. [42] also indicated that attitude towards using the Electronic Health Records for Nursing Education (EHRNE) software program was a significant factor in the intention to use the EHRNE. Kijsanayotin et al. [43] also stated that attitude influenced the intention to use medical IT. Kim et al. [32] also indicated that attitude positively influenced the intention to use a mobile electronic medical record (MEMR). Hsiao and Chen [46] also stated that attitude towards using computerized clinical practice guidelines (CCPG) is a critical factor that impacts physicians' intention to use CCPG.

Therefore, it is claimed that if the end users of VHC have positive feelings to use the system, their degree of the determination to use the system is increased accordingly. Consequently, it is proposed that:

H1: Attitude will have a significant and positive effect on intention.

3.2 The Relationship Between Perceived Usefulness (UseF) and Attitude

Perceived usefulness (UseF) refers to the degree to an individual's faith that using a specific system will improve his or her job routine [31]. This variable repeatedly appears to be an essential predictor of Attitude [44]. For instance, Aldosari et al. [26] stated that UseF had a positive effect on nurses' acceptance toward using EMR. Elavarasen [24] also argued that UseF has a positive impact on attitude towards using e-Health Companion. Kim et al. [32] also reported that Performance Expectancy (PE) positively influenced the attitude towards using MEMR. PE has a similar concept to UseF.

Therefore, it is argued that if the end users of VHC believe that the system helps them to manage their daily routine, they have positive feelings to use it. Consequently, it is proposed that:

H2: UseF will have a significant and positive effect on Attitude.

3.3 The Relationship Between UseF and Intention

Earlier TAM-based models stated that UseF has a positive impact on Intention, both directly and indirectly [26, 45, 47]. Moreover, TAM and extended TAM models by Gagnon et al. [18, 25] stated UseF as a positive predictor of Intention to use Electronic Health Record (EHR). Elavarasen [24] also claimed that UseF has a positive impact on the intention to use e-Health Companion. Aldosari et al. [26] also stated that Usef had a positive effect on nurses' acceptance toward using EMR. Dünnebeil et al. [48] also indicated that UseF positively affected behavioral intention to use Electronic Health Services (EHS). Wahyuni [49] also pointed out that UseF positively affected intention to use e-health services. Moreover, Kuo et al. [40] reported that the Usef of mobile electronic medical record system (MEMRS) with nurses' work practices enhances nurses' intention to use the system. Aldosari [50] also proved that UseF of a picture archiving and communication system (PACS) positively affected the intention to use the system among staff. Dunnebeil et al. [48] also found out that UseF positively influenced the intention to use e-health applications. Esmaeilzadeh et al. [51] examined the intention to use a clinical decision support system (CDSS) and stated PE as an influential factor. Kijsanayotin et al. [43] also reported that PE impacted the intention to use medical IT. Hsiao and Chen [46] also stated that UseF of CCPG was a significant factor influencing physicians' intention to use CCPG. However, the integrated model and psychosocial model by Gagnon et al. [18] did not show any significant impact between two variables.

Therefore, it is claimed that if the end users of VHC sense that the system helps them to manage their daily routine, their degree of the determination to use the system are increased. Consequently, it is proposed that:

H3: UseF will have a significant and positive effect on Intention.

3.4 The Relationship Between Perceived Ease of Use (EaseOfUse) and Attitude

Davis et al. [31] define perceived ease of use (EaseOfUse) as a degree to an individual's faith that using a system would be free of struggle. Gagnon et al. [18, 25] stated that EaseOfUse has a significant and positive influence on intention to use EHR; however, the authors excluded attitude from the proposed model. Besides, Johnson et al. [17] stated that EaseOfUse is a significant and positive predictor of user satisfaction from using the CRS. Elavarasen [24] also suggested that EaseOfUse has a positive impact on attitude towards using e-Health Companion. Aldosari et al. [26] also reported that EaseOfUse has a positive influence on nurses' acceptance toward using EMR. Kim et al. [32] stated that Effort Expectancy (EE) influenced the attitude towards using a MEMR positively. EE has a similar concept to UseF.

In this study, it is claimed that when end-users of VHC assume that using the system is free of struggle, they have positive feelings to use it. Consequently, it is proposed that:

H4: EaseOfUse will have a significant and positive effect on Attitude.

3.5 The Relationship Between Accessibility of a System (Access) and Attitude

Accessibility of a system (Access) is defined as a capability of the system to be obtained, entered, affected, or comprehend [33]. Several studies represented that the perceived accessibility of information provided by a system is a more substantial factor than the quality of the information in that system for acceptance [34–36]. Elavarasen [24] suggested that accessibility of a system has a positive influence on intention toward using e-Health Companion.

In the current study, it is claimed that when VHC's end-users assume that the system is accessible, they have positive feelings to use it. Consequently, it is proposed that:

H5: Access will have a significant and positive effect on Attitude.

3.6 The Relationship Between Compatibility of a System (Comp) Comp and Attitude

Rogers [37] defines the compatibility of a system (Comp) as the extent an innovative system is sensed to be compatible with the current values, earlier experiences, and needs of possible adopters. Comp plays a vital role in making end-users feel positive about working with the system. Chen et al. [38] and Wu and Wang [39] stated that Comp is a more significant factor than UseF. The authors considered Comp as a significant driver in an online environment. Elavarasen [24] also claimed that Comp has a positive influence on attitude towards using e-Health Companion. Kuo et al. [40] also reported that the compatibility of MEMRS with nurses' work practices enhances nurses' willingness to use MEMRS. The lack of integration into healthcare practices lead to an electronic health system to fail [25].

In this study, it is proposed that when VHC's end-users think that the system is compatible with their current work practices, they have positive feelings to use it. Consequently, it is proposed that:

H6: Comp will have a significant and positive effect on Attitude.

3.7 The Relationship Between Privacy Concern (PrivConc) and Attitude

The sensed privacy and security anxieties of patients' personal information recorded on a health system by nurses act as a barrier to adoption of e-Health services [52]. It is crucial to protect the privacy of medical information of patients because a system that is vulnerable to hackers and unauthorized users' access dissatisfies its end users. The following hypothesis examines the negative effect of privacy concern about patients' personal information (PrivConc) on attitude towards using VHC:

H7: PrivConc will have a significant and negative effect on Attitude.

3.8 The Mediating Effect of Attitude

Attitude is included as a mediator in the proposed models by Davis [27], Venkatesh and Davis [45] and Venkatesh and Bala [47]. In this study, the mediating effect of attitude between the independent variables (UseF, EaseOfUse, Access, Comp, and PrivConc) and the dependent variable (Intention) is examined. Consequently, the following hypothesis is proposed:

H8: Attitude mediates the relationship between UseF and Intention.
H9: Attitude mediates the relationship between EaseOfUse and Intention.
H10: Attitude mediates the relationship between Access and Intention.
H11: Attitude mediates the relationship between Comp and Intention.
H12: Attitude mediates the relationship between PrivConc and Intention.

4 Proposed Research Methodology

In this study, an empirical study that is quantitative originally will apply; therefore, a self-directed questionnaire will be designed based on the variables to be measured. The items per each variable will be measured on a 5-point Likert scale ranging from "Strongly disagree" to "Strongly agree" using cross-sectional data. Moreover, some snapshots of VHC will be included in the questionnaire to explain how the system works. Therefore, the respondents may acquire enough knowledge about VHC to complete the survey.

The population of the study will cover all the government's hospitals and health clinics. As it is impractical to cover all of them in the sample, only 350 hospitals and health clinics will be selected based on a simple random sampling technique because many of them might be reluctant to participate in the survey. Simple random sampling technique gives every single item in the population the same probability of being chosen. Moreover, once an item in the population is chosen, it is impossible that the same item is reselected in the sample size [53]. In each health centers, a nurse will be asked to fill in the questionnaire, and the questionnaires will be delivered physically to them by a mail carrier.

In this study, two groups of multiple regression analysis will be tested. First, the relationships between independent variables (UseF, EaseOfUse, Access, Comp, and PrivConc) and Attitude will be examined. The sample size for this regression is calculated using G*Power analysis [54] at 80 percent with effect size 0.35 [55] because the minimum power required in social and behavioral science research is commonly 0.8 [56]. Since there are a total of five predictors, the minimum sample size is estimated at 43. In the second regression, the relationships between UseF and Attitude with the dependent variable (Intention) will be examined. The sample size is calculated by G*Power using the same criteria as mentioned above, but the number of predictors is considered as two. The suggested sample size is at 31. Forty-three is considered as the number of the needed sample size to fulfill the minimum requirement of sample size for both regressions mentioned above.

A research methodology, which is based on a mixed-method of MRA and ANNs, is proposed. In this respect, after examining the goodness of measures and analyzing the influence of independent variables on the related dependent variable using the MRA, ANNs will be applied. This method will enable further verification of the results provided by the MRA. Moreover, for testing the mediating role of Attitude on the relationship between UseF, EaseOfUse, Access, Comp, and PrivConc and Intention, the test for indirect effect suggested by Preacher and Hayes will be applied [57, 58]; therefore, this study will bootstrap the sampling distribution of the indirect effect [59].

5 Conclusion

An SMS-based vaccination reminder and management system called VHC is proposed to assist the health centers by making more straightforward the management of infants' immunization records in Malaysian health centers. A TAM-based model is proposed to study the influential factors of intention to use the system, and it is expected that the model will be applicable and will get predictive power in finding the significant factors that impact on intention to use the system.

References

1. Domek, G.J., Contreras-Roldan, I.L., O'Leary, S.T., Bull, S., Furniss, A., Kempe, A., Asturias, E.J.: SMS text message reminders to improve infant vaccination coverage in Guatemala: a pilot randomized controlled trial. Vaccine **34**(21), 2437–2443 (2016)
2. Kharbanda, E.O., Stockwell, M.S., Fox, H.W., Rickert, V.I.: Text4Health: a qualitative evaluation of parental readiness for text message immunization reminders. Am. J. Public Health **99**(12), 2176–2178 (2009)
3. Ooi, K.S., Cheah, Y.N.: Virtual health connect: a community-based immunisation scheduler and manager. In: Proceedings of the 4th International Conference on Computing and Informatics, pp. 441–446 (2011)

4. Kharbanda, E.O., Stockwell, M.S., Fox, H.W., Andres, R., Lara, M., Rickert, V.I.: Text message reminders to promote human papillomavirus vaccination. Vaccine **29**(14), 2537–2541 (2011)
5. Stockwell, M.S., Kharbanda, E.O., Martinez, R.A., Vargas, C.Y., Vawdrey, D.K., Camargo, S.: Effect of a text messaging intervention on influenza vaccination in an urban, low-income pediatric and adolescent population: a randomized controlled trial. J. Am. Med. Assoc. **207** (16), 1702–1708 (2012)
6. Wakadha, H., Chandir, S., Were, E.V., Rubin, A., Obor, D., Levine, O.S., Gibson, D.G., Odhiambo, F., Laserson, K.F., Feikin, D.R.: The feasibility of using mobile-phone based SMS reminders and conditional cash transfers to improve timely immunization in rural Kenya. Vaccine **31**(6), 987–993 (2013)
7. Kaewkungwal, J., Singhasivanon, P., Khamsiriwatchara, A., Sawang, S., Meankaew, P., Wechsart, A.: Application of smart phone in "Better Border Healthcare Program": a module for mother and child care. BMC Med. Inform. Decis. Mak. **10**(1), 69 (2010)
8. Vilella, A., Bayas, J.M., Diaz, M.T., Guinovart, C., Diez, C., Simó, D., Muñoz, A., Cerezo, J.: The role of mobile phones in improving vaccination rates in travelers. Prev. Med. **38**(4), 503–509 (2004)
9. Abbas, A.H., Yusof, Y.: Children vaccination reminder via SMS alert. In: International Conference on Research and Innovation in Information Systems, ICRIIS 2011, Kuala Lumpur, pp. 1–5 (2011)
10. Harvey, H., Reissland, N., Mason, J.: Parental reminder, recall and educational interventions to improve early childhood immunisation uptake: a systematic review and meta-analysis. Vaccine **33**(25), 2862–2880 (2015)
11. Jacobson, V.J.C., Szilagyi, P.: Patient reminder and recall systems to improve immunization rates. Cochrane Database Syst. Rev. **3**, 150–172 (2005)
12. MCMC: Hand phone users survey (2014). https://www.mcmc.gov.my/skmmgovmy/media/General/pdf/MCMC-Hand-Phone-User19112015.pdf. Accessed 13 Mar 2018
13. Cole-Lewis, H., Kershaw, T.: Text messaging as a tool for behavior change in disease prevention and management. Epidemiol. Rev. **32**(1), 56–69 (2010)
14. Gurman, T.A., Rubin, S.E., Roess, A.A.: Effectiveness of mHealth behavior change communication interventions in developing countries: a systematic review of the literature. J. Health Commun. **17**, 82–104 (2012)
15. Fjeldsoe, B.S., Marshall, A.L., Miller, Y.D.: Behavior change interventions delivered by mobile telephone short-message service. Am. J. Prev. Med. **36**(2), 165–173 (2009)
16. Tang, P.C., Clement, J.M.: Electronic health record systems. In: Biomedical informatics, pp. 447–475. Springer, New York (2006)
17. Johnson, M.P., Zheng, K., Padman, R.: Modeling the longitudinality of user acceptance of technology with an evidence-adaptive clinical decision support system. Decis. Support Syst. **57**, 444–453 (2014)
18. Gagnon, M.P., Ghandour, E.K., Talla, P.K., Simonyan, D., et al.: Electronic health record acceptance by physicians: testing an integrated theoretical model. J. Biomed. Inform. **48**, 17–27 (2014)
19. Triandis, H.C.: Values, attitudes and interpersonal behavior. In: Nebraska Symposium on Motivation Beliefs, Attitudes and Values, vol. 1, pp. 195–259. University of Nebraska Press (1980)
20. Orruno, E., Gagnon, M.P., Asua, J., Ben Abdeljelil, A.: Evaluation of teledermatology adoption by health-care professionals using a modified Technology Acceptance Model. J. Telemed. Telecare **16**(6), 303–307 (2011)

21. Godin, G., Bélanger-Gravel, A., Eccles, M., Grimshaw, J.: Healthcare professionals' intentions and behaviours: a systematic review of studies based on social cognitive theories. Implement. Sci. **3**, 36 (2008)
22. Gagnon, M.P., Orruño, E., Asua, J., Abdeljelil, A.B., Emparanza, J.: Using a modified technology acceptance model to evaluate healthcare professionals' adoption of a new telemonitoring system. Telemed. e-Health **18**(1), 54–59 (2012)
23. Gagnon, M.P., Sanchez, E., Pons, J.: From recommendation to action: psychosocial factors influencing physician intention to use Health Technology Assessment (HTA) recommendations. Implement. Sci. **1**(1), 8 (2006)
24. Elavarasen, M.: E-health companion: acceptance of e-health companion by potential users. Master thesis, The School of Computer Sciences, Universiti Sains Malaysia (2010)
25. Gagnon, M.P., Simonyan, D., Ghandour, E.K., Godin, G., Labrecque, M., Ouimet, M., Rousseau, M.: Factors influencing electronic health record adoption by physicians: a multilevel analysis. Int. J. Inf. Manag. **36**(3), 258–270 (2016)
26. Aldosari, B., Al-Mansour, S., Aldosari, H., Alanazi, A.: Assessment of factors influencing nurses acceptance of electronic medical record in a Saudi Arabia hospital. Inform. Med. Unlocked **10**, 82–88 (2018)
27. Davis, F.D.: Perceived usefulness, perceived ease of use, and user acceptance of information technology. MIS Q. **13**(3), 319–340 (1989)
28. Wu, J.H., Shen, W.S., Lin, L.M., Greenes, R.A., Bates, D.W.: Testing the technology acceptance model for evaluating healthcare professionals' intention to use an adverse event reporting system. Int. J. Qual. Health Care **20**(2), 123–129 (2008)
29. Chow, S.K., Chin, W.Y., Lee, H.Y., Leung, H.C., Tang, F.H.: Nurses' perceptions and attitudes towards computerisation in a private hospital. J. Clin. Nurs. **21**(11–12), 1685–1696 (2012)
30. Vathanophas, V., Pacharapha, T.: Information technology acceptance in healthcare service: the study of electronic medical record (EMR) in Thailand. In: Technology Management for Global Economic Growth (PICMET). 2010 Proceedings of PICMET 2010, pp. 1–5. IEEE (2010)
31. Davis, F.D., Bagozzi, R.P., Warshaw, P.R.: User acceptance of computer technology: a comparison of two theoretical models. Manag. Sci. **35**(8), 982–1003 (1989)
32. Kim, S., Lee, K.H., Hwang, H., Yoo, S.: Analysis of the factors influencing healthcare professionals' adoption of mobile electronic medical record (EMR) using the unified theory of acceptance and use of technology (UTAUT) in a tertiary hospital. BMC Med. Inform. Decis. Mak. **16**(1), 12 (2015)
33. The Concise Oxford Dictionary of Current English, 7th edn. Oxford University Press (1982)
34. O'Reilly, C.A.: Variations in decision makers' use of information sources: the impact of quality and accessibility of information. Acad. Manag. J. **25**(4), 756–771 (1982)
35. Culnan, M.J.: The dimensions of accessibility to online information: implications for implementing office information systems. ACM Trans. Off. Inf. Syst. **2**(2), 141–150 (1984)
36. Rice, R.E., Shook, D.E.: Access to, usage of, and outcomes from an electronic messaging system. ACM Trans. Off. Inf. Syst. **6**(3), 255–276 (1988)
37. Rogers, E.M.: Diffusion of Innovations, 4th edn. Free Press, New York (1995)
38. Chen, L., Gillenson, M.L., Sherrel, D.L.: Enticing online consumers: an extended technology acceptance perspective. Inf. Manag. **39**, 705–719 (2002)
39. Wu, J.H., Wang, S.C.: What drives mobile commerce? An empirical evaluation of the revised technology acceptance model. Inf. Manag. **42**, 719–729 (2005)
40. Kuo, K.M., Liu, C.F., Ma, C.C.: An investigation of the effect of nurses' technology readiness on the acceptance of mobile electronic medical record systems. BMC Med. Inform. Decis. Mak. **13**(1), 88 (2013)

41. Ajzen, I., Fishbein, M.: Understanding Attitudes and Predicting Social Behavior. Prentice Hall, Englewood Cliffs (1980)
42. Kowitlawakul, Y., Chan, S.W.C., Pulcini, J., Wang, W.: Factors influencing nursing students' acceptance of electronic health records for nursing education (EHRNE) software program. Nurse Educ. Today (2015). https://doi.org/10.1016/j.nedt.2014.05.010
43. Kijsanayotin, B., Pannarunothai, S., Speedie, S.M.: Factors influencing health information technology adoption in Thailand's community health centers: applying the UTAUT model. Int. J. Med. Inform. **78**(6), 404–416 (2009)
44. Venkatesh, V., Morris, M.G., Davis, G.B., Davis, F.D.: User acceptance of information technology: toward a unified view. MIS Q. **27**(3), 425–478 (2003)
45. Venkatesh, V., Davis, F.D.: A theoretical extension of the technology acceptance model: four longitudinal field studies. Manag. Sci. **46**(2), 186–204 (2000)
46. Hsiao, J.L., Chen, R.F.: Critical factors influencing physicians' intention to use computerized clinical practice guidelines: an integrative model of activity theory and the technology acceptance model. BMC Med. Inform. Decis. Mak. **16**(1), 3 (2015)
47. Venkatesh, V., Bala, H.: Technology acceptance model 3 and a research agenda on interventions. Decis. Sci. **39**(2), 273–315 (2008)
48. Dünnebeil, S., Sunyaev, A., Blohm, I., Leimeister, J.M., Krcmar, H.: Determinants of physicians' technology acceptance for e-health in ambulatory care. Int. J. Med. Inform. **81** (11), 746–760 (2012)
49. Wahyuni, R.: Explaining acceptance of e-health services: an extension of TAM and health belief model approach. In: 2017 5th International Conference on Cyber and IT Service Management (CITSM), pp. 1–7. IEEE (2017)
50. Aldosari, B.: User acceptance of a picture archiving and communication system (PACS) in a Saudi Arabian hospital radiology department. BMC Med. Inform. Decis. Mak. **12**(1), 44 (2012)
51. Esmaeilzadeh, P., Sambasivan, M., Kumar, N., Nezakati, H.: Adoption of clinical decision support systems in a developing country: antecedents and outcomes of physician's threat to perceived professional autonomy. Int. J. Med. Inform. **84**(8), 548–560 (2015)
52. Ponemon Institute LLC: Risks & Rewards of Online & Mobile Health Services: Consumer Attitudes Explored (2014). http://www.experian.com/assets/p/data-breach/experian-ehealth-study.pdf
53. Kumar, R.: Research Methodology. APH Publishing, New Delhi (2008)
54. Faul, F., Erdfelder, E., Buchner, A., Lang, A.G.: Statistical power analyses using G*Power 3.1: tests for correlation and regression analyses. Behav. Res. Methods **41**, 1149 1160 (2009)
55. Cohen, J.: Statistical Power Analysis for the Behavioural Sciences, 2nd edn. Psychology Press, New York (1998)
56. Rasoolimanesh, S.M., et al.: A revised framework of social exchange theory to investigate the factors influencing residents' perceptions. Tour. Manag. Perspect. **16**, 335–345 (2015)
57. Preacher, K.J., Hayes, A.F.: Asymptotic and resampling strategies for assessing and comparing indirect effects in multiple mediator models. Behav. Res. Methods **40**(3), 879–891 (2008)
58. Preacher, K.J., Hayes, A.F.: SPSS and SAS procedures for estimating indirect effects in simple mediation models. Behav. Res. Methods Instrum. Comput. **36**(4), 717–731 (2004)
59. Hair, J.F., Hult, G.T.M., Ringle, C.M., Sarstedt, M.: A primer on partial least squares structural equation modeling (PLS-SEM). Sage Publications, Thousand Oaks (2014). Incorporated

Current Knowledge Management Activities in a Manufacturing Company in Malaysia: A Case Study

Putri Norlyana Mustafa Kamal, Norlida Buniyamin(✉), and Azmi Osman

Faculty of Electrical Engineering, Universiti Teknologi MARA, Shah Alam, Malaysia
putrinorlyana@gmail.com, nbuniyamin@uitm.edu.my, azmiosman@hotmail.com

Abstract. The findings of a preliminary study conducted in a manufacturing company based in Selangor, Malaysia, are presented in this paper. There has been many instances where knowledge in an organization was lost, due to the knowledge not properly managed. One method of managing knowledge is by implementing a Knowledge Management System (KMS). The preliminary study was conducted in order to determine the current Knowledge Management (KM) activities in the company, and also to find out whether a KMS is needed in the organization. The study was conducted via formal and informal interviews, observations, and also sharing sessions. The study involved employees from various engineering departments in the company, as they are the target users for a KMS to be developed for the company. A survey is planned to be distributed among employees to further cement the findings in the near future. Based on the results obtained, it was found that KM is not yet a norm in the organization, but efforts to facilitate KM have been recently started. Hence, it can be concluded that a KMS is needed in the company, in order to maintain their intellectual assets.

Keywords: Knowledge Management · Preliminary study · Manufacturing company · Knowledge management capabilities

1 Introduction

At any given time, new knowledge is always being created. They are created through various ways, such as conducting research and experiments, sharing, or even by reading up on other people's works. In a manufacturing environment, new knowledge is created during the process of continuous improvements carried out in the company as the organization strive to improve product quality and manufacturing efficiency.

This knowledge then becomes embedded in organizations in the forms of documentations, the way employees work, organizational cultures, etc. [1–3]. These new knowledge, becomes an organization's intellectual assets if stored and re-used. Similar to physical assets, intellectual assets can contribute to a manufacturing organization if shared and re-used to improve processes that will eventually contribute to

F. Saeed et al. (Eds.): IRICT 2019, AISC 1073, pp. 970–979, 2020.
https://doi.org/10.1007/978-3-030-33582-3_91

manufacturing process efficiency and/or product quality. However, the contributions of intellectual assets are harder to measure compared to physical assets. Intellectual asset is the value of a firm's knowledge base and service capabilities [2, 4]. Intellectual assets has been said to be an important aspect in ensuring that a company maintains its competitive advantage [5, 6]. Realizing the importance of their intellectual assets, organizations has also been putting more effort into implementing KM, and KMS. A KMS is a system that supports KM. This means that it acts as a repository system to store the knowledge, or intellectual assets of an organization for other employees to access. Some examples of companies that has implemented KMS include [1]:

- McDonald's. They implemented KMS in order to ensure that their recipes are shared throughout all their branches across the world, and also to ensure best practices.
- Chevron. They reported tangible benefits through changing some processes, and also implementing best practices. This was made possible by implementing KMS.

There are companies that opt to buy off-the shelves KMS. Off-the-shelves KMS are KMS that are ready-made, and all that is left to do is implement the system in the organization. Some examples of off-the-shelves KMS are [7]: Evernote, Canva, and Instacart.

However, there may be cases where off-the-shelves KMS may be insufficient for specific applications, so researches have been conducted in order to improve existing KMS. The buzz on KMS started back in the 90s, where most of the efforts are on creating knowledge repositories. Back then, it was found that there were three main focus when creating these repositories [8]:

- External knowledge. For example: competitive intelligence.
- Structured internal knowledge. For example: reports, procedure, techniques.
- Informal internal knowledge. For example: tacit knowledge.

Current focus research areas in KMS includes:

- Extraction of tacit knowledge [9, 10]
- Frameworks for specific KMS applications. For example: medical, failure analysis [11, 12]
- KMS security [13, 14]

Due to the realization, that preserving and sharing knowledge is important within an organization, companies have strived to manage and preserve knowledge [15]. One of the most basic ways to preserve knowledge is by sharing one's knowledge to another, to ensure knowledge continuity.

The usual method of knowledge sharing between employees is by approaching the person who has the knowledge that is sought after, and then learning from the person. This exchange usually occurs between individuals, or between individuals within small groups. However, knowledge sharing between employees is a crucial process, but it is something that does not always happen in practice [16]. This may be due to the presence of barriers in knowledge sharing.

Examples of barriers to face to face knowledge sharing are [17–20]

- Jealousy
- Poor communication skills
- Working culture
- No self-confidence

When knowledge is not recorded where it may be preserved and spread for others to use, so in cases where an employee retires, his knowledge will be lost by the company. Hence, it is important for an organization to practice KM [21, 22].

1.1 Knowledge Management

Sharing, and preserving knowledge are usually the main purpose of KM. KM is an organization's concerted efforts in managing organizational knowledge [21, 23, 24]. KM process is a cycle of identifying, acquiring, disseminating and creating knowledge as Fig. 1 below.

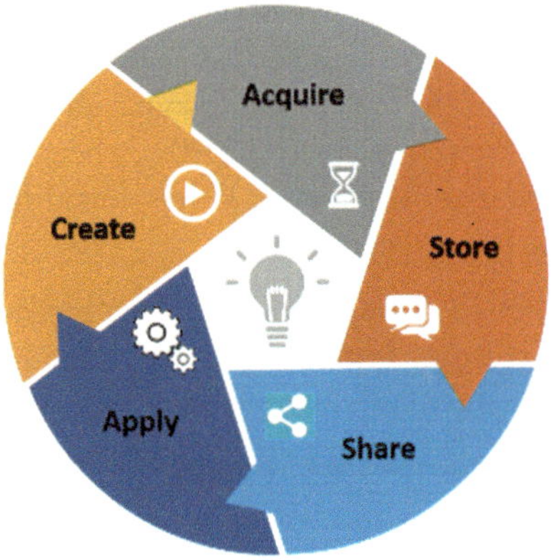

Fig. 1. KM process

As depicted in Fig. 1, KM process is a cyclic process. For example, when a new knowledge is acquired through conducting experiments, the knowledge can then be stored to be preserved. By storing the knowledge in a way so that it can be accessed by other employees, the knowledge can then be shared among employees, where the knowledge may then be applied where suitable. From doing so, new knowledge will be produced, i.e. the result of applying the knowledge. The cycle then keeps on repeating itself. The success of implementing a KMS in an organization depends on the KM practice in the organization, as KMS is a system that supports KM processes. In a study conducted by Al-Emran et al., it was found that knowledge sharing process is positively affecting the adoption of KMS in organizations [25].

Organizations that practice KM has reported results of practicing KM. Table 1 below shows the benefits of practicing KM in organizations [1].

Table 1. KM benefits

<table>
<tr><th colspan="6">PROCESS OUTCOMES</th></tr>
<tr><td colspan="3">Communication:
• Enhanced communication
• Faster communication
• More visible opinions of staff
• Increased staff participation</td><td colspan="3">Efficiency:
• Reduced problem solving time
• Shortening proposal times
• Faster results
• Faster delivery to market
• Greater overall efficiency</td></tr>
<tr><th colspan="6">ORGANIZATIONAL OUTCOMES</th></tr>
<tr><td colspan="2">Financial:
• Increased sales
• Decreased cost
• Higher profitability</td><td colspan="2">Marketing:
• Better service
• Customer focus
• Targeted marketing
• Proactive marketing</td><td colspan="2">General:
• Consistent proposals to multinational clients
• Improved project management
• Personnel reduction</td></tr>
</table>

From Table 1, it can be concluded that KM awareness is important in organizations, as the process not only helps preserve their intellectual assets, it also helps with the organizational growth by contributing new knowledge that, when applied, will be beneficial to the organization.

2 Case Study

The research conducted aims to produce a KMS framework for a manufacturing company located in Selangor, Malaysia. The case study company produces discrete products with an annual production capacity of 150,000 units, in two production working shifts. The number of employees in the company is approximately 12,000. The research conducted is mostly limited to eleven (11) engineering departments, such as the design, manufacturing, and testing departments.

This study is conducted to investigate the current status of KM in the organization and whether a bespoke KMS is required. As previously mentioned, KM is important in organizations if they want to preserve and re-use their knowledge. However, previous studies show that there are organizations that are still not fully aware, and have not adapted KM in their organizations.

2.1 Methodology

The method used for this study is data triangulation. This method uses multiple data sources in order to develop a comprehensive understanding of a situation, and also to ensure validity of the findings [26]. There are five types of data triangulation [27]:

- Same story, same meaning
- Same story, different interpretations
- Missing pieces
- Illuminating
- Unique information

Data triangulation is a method in which data from more than one source is compared, or cross-validated to determine validity of data. When data from one method of data collection is incomplete, the other methods used will be used as a backup in order to obtain a range of complete answers. Data triangulation is used in this research, where after conducting data collection by using different methods, the answers are then compared and cross-validated to obtain a conclusion.

The study was conducted to determine the organization's current sharing culture, and their awareness of the importance of KM. The objectives of conducting this study are as follows:

- To identify the current KM in the organization
- To identify the need for a KMS in the organization

The methods used to collect data during the preliminary study includes carrying out interviews; both formal and informal and observations of knowledge sharing. An example of observation is when the researcher observed activities in sharing sessions such as the daily morning meetings to update the team on events of the previous day. These methods are further explained below.

Method 1: Interviews
There are two types of interviews: formal and informal [28] that were used in this study as listed in Table 2.

Table 2. Formal and informal interviews

Type	Formal	Informal
Similarity	Conducted in order to achieve an objective	
Differences	Conducted in a formal, specific way	Conducted in a more laid back environment
Examples	Job interviews for roles that is highly technical	Interviews where skill sets are not clearly defined

Formal interviews were conducted on 15 employees, while informal interviews were conducted on another 30 employees. Some of the questions asked include how easy do they find to obtain documents or information that they do not possess, current knowledge sharing method, and also are they aware of the projects that other departments are conducting, or have conducted. Besides the questions that were already listed and prepared for the interviews, additional questions were also added based on how questions were answered by participants.

Method 2: Sharing Sessions
Sharing sessions were conducted, where a group of employees gathered and discussed problems they are currently facing, and also their general work flow. By conducting sharing sessions, knowledge was shared, new discoveries were made, and the researcher also obtained the employees' opinions on current KM in the organization.

Method 3: Observations
An observation of the employees' behavior was conducted during discussions, while they are in their comfort zone. The researcher was not directly involved in any conversations during observation activity. This method was supported in this paper by Baker et al. [29] as, in their opinion, researchers are most likely to get a more honest answers and findings when the employees are not being interviewed formally, as the employees may alter their answers to better suit their organization. Findings from this method were used to develop interview questions, should some clarifications were needed.

After obtaining findings from the study, conclusions, which will be discussed in the next section, were made. These conclusions were made by using the data triangulation method as mentioned earlier.

3 Findings

The results and findings of this study are divided into two parts corresponding to the objectives of the study, which is to find out the current KM in the organization, and also to find out whether a KMS is necessary for the organization. These findings are summarized based on results obtained from the various methods used for the study mentioned in the previous section.

3.1 Current KM in the Organization

Sharing sessions were conducted periodically by inviting speakers who are experts on a topic. These experts may belong to the organization, or there are a few sessions where experts from universities' research centers were invited as a speaker in the sharing sessions. There are also a few departments that have been conducting regular sharing sessions, where employees volunteer to share their knowledge with other in the departments. Usually, these sessions are held monthly.

It was also found that employees are often unaware of what other departments are doing, so there have been cases where the same experiment or research is conducted multiple times by different departments. Often, they discover that what they are doing has been done when they encounter a problem with the research and they start asking other departments for opinions.

Another finding is that when an expert retires from the company, his knowledge is not extracted for the company's future use. This is a loss to the company as they have lost their intellectual assets by not managing their organizational knowledge. To retain knowledge, the company have started compiling important information into handbooks, for new employees and vendors to refer to.

These findings are then analyzed to determine the organization's current KM capability level. According to [30], there are five KM capability levels, as listed in Table 3.

Based on Table 3, it was concluded that the organization's current KM capability level is in between levels 1 and 2. This is because, the culture here has started to encourage knowledge sharing, and the organization has started to recognize the value of knowledge assets. However, the knowledge assets have not yet been stored or tracked.

Table 3. KM capability levels

Capability level	General goals
Level 1: Possible	• Knowledge sharing is not discouraged • There is a general willingness to share • People who understand the value of sharing do it • Meaning of knowledge assets is understood
Level 2: Encouraged	• Culture encourages sharing of knowledge assets • Value of knowledge assets is recognized • Knowledge assets are stored/tracked in some fashion
Level 3: Enabled/practiced	• Sharing of knowledge assets is practiced • Systems/tools to enable KM activities exist • Rewards/incentives promote knowledge sharing
Level 4: Managed	• Employees expect to locate knowledge • Training is available • KM-related activities are part of work flow • Systems/tools for supporting KM activities are easy to use • KM capabilities and benefits are assessed • Leadership exhibits commitment to KM • Leadership provides KM strategy
Level 5: Continuous Improvement	• KM systems/tools are widely accepted, monitored/updated • KM processes are reviewed/improved • KM assessment generates realistic improvement

3.2 Existing Sharing Platforms

Another finding from the preliminary study is that the company has two existing repository system. The purpose of these two platforms is to make sharing documents and information between employees easier. Listed in Table 4 is a summary of the two platforms.

Based on the findings listed in Table 4, it was concluded that the organization does not have a system, or a platform, that manages and stores knowledge for future use. What they currently have are platforms that are used to store documentations.

It was also found that there have been a few other platforms that was later terminated in the organization due to lack of use. It was also mentioned that, most of the time, the way each department shares their documents differ. Some departments had

Table 4. Existing sharing platforms in the organization

Platform	Platform A	Platform B
Accessible by	All employees	Employees from engineering departments. However, not all engineers have full access
Content	• Employee details • Department details	• Bill of Material (BOM) • Product engineering design • Product specifications
Pros	All employees have access to this system	Is up to date
Cons	Is not properly maintained; some departments updated their documented years ago. List of employees is also not up to date	• Not all employees have access to this system • Some of the procedures of uploading documents are difficult, while some are repetitive

even discovered this fact during the sharing sessions conducted. There were also a few instances where some employees were unaware that they had access to a system, or sharing platform.

Files from sharing sessions that have been conducted, for example the PowerPoint presentation does not have a proper platform for it to be stored, causing the file to be stored individually, which has previously resulted in losing the file. In the long run, this will mean that although sharing sessions have been conducted in order to ensure knowledge has been passed on, the knowledge was still lost due to improper management.

4 Conclusion and Future Work

From the preliminary study, it can be concluded that KM awareness in the organization is still growing, and efforts are being done to improve KM awareness in the workplace. Some of their efforts include conducting knowledge sharing sessions by experts, where departments whose work is related to the topic discussed are invited, and compiling a handbook, which involves departments to provide important elements of their department for future references. It can also be concluded that there is currently no KMS implemented in the organization, which means that one of the steps for them to do in order to improve their KM capability is to employ a KMS in the organization.

In addition to the data that have been gathered, a survey will be distributed among the employees in the near future to further support the findings. This research will then attempt to produce a suitable framework for an industrial KMS.

One main limitation encountered while conducting the study is confidentiality concerns, which ended up being included as one of the findings. Due to the organization's lack of specification in access control, there were delays in obtaining some.

The study identified the problems of managing knowledge in the company, its current level of KM capability and what the company needs to do to better manage its knowledge assets.

References

1. Gray, P.: Knowledge management overview. McKinsey Q., 18 (2000)
2. Buniyamin, N., Barber, K.D.: The intranet: a platform for knowledge management system based on knowledge mapping. Int. J. Technol. Manag. **28**(7–8), 729–746 (2004)
3. Wong, T., French, L., Wickham, M.: Knowledge-management issues in China's hotel industry. Ann. Tour. Res. **57**, 257–259 (2016)
4. Li, Y.-Q., Liu, C.-H.S.: The role of problem identification and intellectual capital in the management of hotels' competitive advantage-an integrated framework. Int. J. Hosp. Manag. **75**, 160–170 (2018)
5. Wang, J.-F., Chen, M.-Y., Feng, L.-J., Yue, J.-J.: The construction of enterprise tacit knowledge sharing stimulation system oriented to employee individual. Procedia Eng. **174**, 289–300 (2017)
6. Sureephong, P., Chakpitak, N., Ouzrout, Y., Neubert, G., Bouras, A.: Knowledge management system architecture for the industry cluster. In: 2007 IEEE International Conference on Industrial Engineering and Engineering Management (2008)
7. Roosa, A.: 3 of the best knowledge management examples. Zendesk (2017). https://www.zendesk.com/blo/3-best-knowledge-management-examples/. Accessed 30 May 2019
8. Davenport, T.H., De Long, D.W., Beers, M.C.: Business building successful knowledge management projects, January 1997
9. Muthuveloo, R., Shanmugam, N., Teoh, A.P.: The impact of tacit knowledge management on organizational performance: evidence from Malaysia. Asia Pacific Manag. Rev. **22**(4), 192–201 (2017)
10. Ryan, S., O'Connor, R.V.: Acquiring and sharing tacit knowledge in software development teams: an empirical study. Inf. Softw. Technol. **55**(9), 1614–1624 (2013)
11. Wichawong, P., Chongstitvatana, P.: Knowledge management system for failure analysis in hard disk using case-based reasoning (2017)
12. Zhang, R., Chen, D., Shang, X.: A knowledge-constrained access control model for protecting patient privacy in hospital information systems. IEEE J. Biomed. Heal. Inform. **22**(3), 904–911 (2018)
13. Khan, M.F.F., Sakamura, K.: A secure and flexible e-Health access control system with provisions for emergency access overrides and delegation of access privileges. In: 2016 18th International Conference on Advanced Communication Technology (2016)
14. Malina, L., Dzurenda, P., Hajny, J., Martinasek, Z.: Secure and efficient two-factor zero-knowledge authentication solution for access control systems. Comput. Secur. **77**, 500–513 (2018)
15. Kianto, A., Ritala, P., Vanhala, M., Hussinki, H.: Reflections on the criteria for the sound measurement of intellectual capital: a knowledge-based perspective. Elsevier (2018)
16. Boer, N.: Knowledge sharing within organizations (2005)
17. De Long, D.W., Fahey, L.: Diagnosing cultural barriers to knowledge management. Acad. Manag. Perspect. **14**(4), 113–127 (2000)
18. Mezghani, E., Exposito, E., Drira, K.: A collaborative methodology for tacit knowledge management: application to scientific research. Futur. Gener. Comput. Syst. **54**, 450–455 (2016)

19. BenMoussa, C.: Barriers to Knowledge Management : A Theoretical Framework and a Review of Industrial Cases, pp. 901–912 (2009)
20. Riege, A.: Three-dozen knowledge-sharing barriers managers must consider. J. Knowl. Manag. **9**(3), 18–35 (2005)
21. Kumar, M.: Knowledge Management Practices in Indian software development companies: findings, vol. 10, no. 1, pp. 59–78 (2005)
22. Numprasertchai, S., Poovarawan, Y.: Enhancing university competitiveness through ICT based knowledge management system. In: ICMIT 2006 Proceedings - 2006 IEEE International Conference Management of Innovation and Technology, vol. 1, pp. 417–421 (2006)
23. Baxter, D., Gao, J.: Managing knowledge within the manufacturing enterprise : an overview, vol. 18, no. 2, pp. 183–209 (2009)
24. Bowman, B.J.: Building knowledge management. Inf. Syst. Manag. **19**(3), 32–40 (2006)
25. Al-emran, M., Mezhuyev, V., Kamaludin, A., Shaalan, K.: The impact of knowledge management processes on information systems: a systematic review. Int. J. Inf. Manag. **43** (July), 173–187 (2018)
26. Carter, N., Bryant-Lukosius, D., DiCenso, A., Blythe, J., Neville, A.J.: The use of triangulation in qualitative research. In: Oncology Nursing Forum (2014)
27. Sands, R.G., Roer-Strier, D.: Using data triangulation of mother and daughter interviews to enhance research about families. Qual. Soc. Work **5**(2), 237–260 (2006)
28. Gupta, P.: Formal interviews vs informal interviews: all you need to know! YOURSTORY (2017). https://yourstory.com/2017/02/formal-vs-informal-interviews. Accessed 29 May 2019
29. Baker, S.E., Edwards, R., Adler, P., Becker, H.S., Doucet, A.: How many qualitative interviews is enough ? Expert voices and early career reflections on sampling (2012)
30. Schwartz, D.G.: Measuring knowledge management capabilities. In: Encyclopedia of Knowledge Management, p. 608. Idea Group Inc. (2005)

Determinants of Users' Intention to Use IoT: A Conceptual Framework

Nura Muhammad Baba[1,2](✉) and Ahmad Suhaimi Baharudin[1]

[1] School of Computer Science, Universiti Sains Malaysia, USM, 11800 George Town, Malaysia
nabnurtrn@gmail.com, asuhaimi@usm.my
[2] Department of Office Technology and Management, School of Management Studies, Kano State Polytechnic, Kano, Nigeria

Abstract. The Internet of things (IoT) is realised as a potentially effective means of integrating multiple technologies to improve the quality of people's life and offering interesting and advantageous new services to individuals. However, it has emerged that consumers' acceptance of IoT is currently low despite its huge economic potentials and impacts, as well as high investment from the private and public sectors. Yet, few studies have investigated the perspectives of the users on IoT. Specifically, only a few empirical researches had examined the determinants of IoT service adoption from the user's perspective and research model were still not fully developed. Hence, there is a dearth of empirical research on IoT adoption in Malaysia. Therefore, this research aims to develop an integrative model of factors influencing users' acceptance of IoT. The research applied an integrated model from the combine theories of technology acceptance model (TAM), and the theory of perceived risk. The research hope to provide useful insight into the key driving factors with regard to understanding consumers' behavioural intention to use the IoT.

Keywords: IoT · TAM · Behavioural intention

1 Introduction

The advancement of ubiquitous computing technologies, such as wireless networks and mobile devices, has substantially increased the availability of digital information and services in our daily lives and changed how we access and use them. This has led to the rise of the number of people in utilizing the internet. Currently, there are about 3.4 billion internet users' among the 7.4 billion populace of the world [1]. With such an increase in the number of people gaining access to such a global information another big leap forward is in the making which is relatively related to the utilization of the Internet as a global rostrum for permitting communication, computation, dialogue, and coordination between machines and smart objects.

The growth potential of IoT is enormous, with 8.7 billion connected devices in 2012 to 50.1 billion in 2020 [2]. According to Cisco [3], billions of devices will be connected by 2030, with IoT taking half of the share. Having lots of things connected will change our life, and the IoT technologies have the potential to change the world,

F. Saeed et al. (Eds.): IRICT 2019, AISC 1073, pp. 980–990, 2020.
https://doi.org/10.1007/978-3-030-33582-3_92

just as the internet did [4]. Cisco [3] is backing their numbers with a bold statement, saying IoT will change everything, including ourselves. This is because IoT will help humanity take a huge leap in its ability to analysed and distribute data, which can be turned into information, knowledge and ultimately wisdom. The concept will blur the boundary between the physical and the digital world and create new relationships among people, things, and businesses. Such increases in the number of connected things, also, had led to the rise of IoT economic value to the world.

These opportunities will bring upon widespread impact to the entire marketplace across numerous sectors ranging from manufacturing and transportation to utilities and health care - fuelling gross domestic product (GDP), creating new job opportunities, and bolstering the global economy [5]. The worldwide IoT market spending will grow from $591.7 billion in 2014 to $1.3 trillion in 2019 with a compound annual growth rate of 17% [6]. Furthermore, International Data Corporation (IDC) report that the IoT market size in Asia Pacific will grow from USD 600.3 billion in 2013 to USD 2,602.6 billion in 2020, reporting CAGR of 23.3% [7].

Although, the rapidly increasing number of users has created demand for additional IoT services and resources, and new applications have recently been launched in a range of physical domains including shops, hospitals, airports, banks, restaurants, entertainment venues, supermarkets, bus stops, train stations, cafés, convenience stores, taxis, and campuses. Moreover, IoT technologies have been given wide attention, both in the field of research as well as media [5], and had the large number of applications in many fields [8]. However, these goals can only be achieved upon the acceptance of IoT services by the user. As such, Chong, Liu [9] assert that organizations benefits can only be achieved upon successful adoption and use of the IoT products by the individual. Furthermore, there's need to determine what factors drive the Malaysian users' to adopt IoT product because currently, the IoT services adoption is low [9].

Nevertheless, prior researches has provided a limited understanding of the key drivers in consumer acceptance of new IT technology such as IoT technology. Considering the significance of attracting and retaining IoT users, it is necessary to determine the factor affecting users' intention to adopt IoT product in Malaysia. Therefore, the paper aims to find the factors that affect users' intention to adopt IoT with a view to examines the impacts of innovation characteristics (perceived usefulness (PU), Perceived ease of use (PEOU), perceived cost, privacy perceived, and perceived security) and intention to adopt IoT. In doing so, the research will identifies the factors and propose a model on IoT adoption among Malaysian individuals.

2 Literature Review

2.1 Overview of IoT

The term internet of things (IoT) was first coined in 1999 by Kevin Ashton, a British technology pioneer who helped develop the concept [10] to describe a system in which objects in the physical world could be connected to the Internet by sensors [11] and referred to it as RFID tagged objects. Nowadays, the IoT has become a popular term for describing scenarios in which Internet connectivity and computing capability extend to a variety of objects, devices, and sensors [11], plus everyday items such as food, clothing, furniture, paper, landmarks, monuments, works of art, etc. [12]. Although there is no single contentious definition of IoT, however in 2014, the International Organization for Standardization (ISO) form a group of expert to come up with a common definition of IoT. They define IoT as "An infrastructure of interconnected objects, people, systems and information resources together with intelligent services to allow them to process information of the physical and the virtual world and react" [13].

In Malaysia, the economic potential of IoT based on the report of Gartner (2014) is anticipated to reach RM42.5 billion in 2025 from RM9.5 billion in 2020 [14]. Specifically, Apps, Services, and Analytics will contribute RM7.5 billion in 2020 and reach RM34.0 billion by 2025. The App and Service are the segment where the consumers come in and takes up 80% of the total share. Currently, the expected economic contribution to the Malaysian economy is worth RM1.78 billion [14].

2.2 Individual Adoption Theories

Individual adoption and use of information technologies is one of the most mature streams of information system (IS) research [15]. There have been several models that have been employed to predict behavioural intention to use a system and, consequently, system use [15, 16]. Using these theories, various IT innovations has been investigated, such as IoT technologies [e.g. 17], RFID technologies [e.g. 9], mobile commerce [e.g. 18], and smart watches [e.g. 19].

2.3 Empirical Studies Related IoT

The popularity and significance of IoT technologies to both individual and organization had made various scholars to call for the need to investigate the factors affecting IoT technology adoption. Currently, only a few studies have thus been conducted on user adoption, specifically from a Malaysian perspective, very few studies were conducted, and they are mostly conceptual studies. Therefore, it is vital to extend this streams of research by acquiring the empirical result of users' intention to use IoT products. Moreover, the success of new technologies depends on the user acceptance of the products [20]. Specifically, focusing on consumer view point is regarded as one of the pivotal success factors in adoption studies [21].

The author in [22] investigate the determinants of IoT adoption factors in logistics and supply chain management from the perspective of Taiwanese managers. A mixed approach was to determine the impending factors behind the lack of adoption. Results revealed that cost, benefits and external pressure were significant determinants. Just recently in the food retailing of supply chain, [23] investigated IoT adoption barriers among Indian retailer. Poor internet infrastructure and the lack of government policies emerges as the most significant deterrent to the IoT adoption. Hsu and Lin [24] investigate the factors contributing to IoT service adoption based on Value-based adoption model (VAM) in Taiwan. The author mention that the IoT companies' needs a better understanding of why individuals use their services because diffusion of innovation can be more efficiently achieved through user adoption. Similarly, the same author Hsu and Lin [17], examine the critical factors affecting consumer decisions to adopt IoT services from the perspective of network externalities and concern for information privacy. The authors emphasize that more studies are needed because only a few empirical researches had studied determinants of IoT service adoption from the user's perspective. Equally, Mani and Chouk [25] investigated why student resist IoT product. They added that the slow pace of consumer adoption of this innovation is a major disappointment for the IoT industry and thus understanding why consumers' resist the product, is a call for concern. Moreover, Gao and Bai [8] incorporated PBC, trust, SI, and PEJ to the original TAM model to explore user adoption of IoT products in China. The author asserts that academic enquiry on consumers behaviour had been generally neglected by earlier studies and therefore called for more studies on factors affecting users' intention to adopt IoT product.

Several scholars utilize a different approach to study and predict the adoption of IoT technology by combining many theoretical models. Kowatsch and Maass [26] integrate TAM with privacy calculus to study the critical privacy factors affecting IoT technology acceptance. They suggested that privacy concern is critical for using IoT services due to their characteristics and the way the data is collected which may pervade consumers' everyday life. Shin [27] proposes quality of experience model from the combine theories of TRA and TPB to explain how users developed quality experience in the use of IoT products. They argued that lack of studies currently exists on user experience and customer satisfaction from IoT services literature.

Liew, Ang [28] conducted a study in Malaysia from university students perspective to identify the influencing features of IoT technology acceptance by examining the consumer recognitions towards the technology. They infer that IoT technologies have the convenient for online ticketing services and thus allows users to effortlessly acquire the information they need. From the Indian context, Mital, Choudhary [29] empirically examined IoT adoption intention based on multiple theoretical perspectives. The author argued that only few IoT adoption studies were conducted from multiple theoretical approaches and were still limited to few applications. Also, Dong, Chang [30] explore the factors affecting consumers' IoT Intention to use and actual use based on TAM and experience theory. The author mentioned that past researchers had not investigated IoT studies extensively. Meanwhile, Balaji, Roy [31] investigate the consumers' intention to use IoT product in the retail store from the combine TAM model and task technology fit. The authors mention that it is perilous for scholars and retailers to comprehend the acceptance process of IoT technology in the retail industry. In the same vein, [32], consider how trust influences IoT adoption among USA farmers. The influence of trust, risk, and perceived value were investigated. their study shows the importance of applying IoT technology in improving the business profit of farmer. Similarly, the work of [33] integrated UTAUT 2 model with Social Influence, price, and IT knowledge to investigate Omanian entrepreneurs acceptance of IoT. The IT knowledge emerges as the most significant determinants of IoT acceptance. Another work in [34] also explore users attitudes towards IoT acceptance through the extended UTAUT model with social influence and hindering condition constructs to. Results obtained from 224 participants indicate that the identified factors had strong significant effect on their behavioural intention.

Several international recognised research firm conducted surveys to find the perception of consumers on IoT products. For example, in one of the biggest consumers' survey, the 2016 Accenture Digital Consumer Survey carried out a study from 28 different countries involving 28000 individual consumers' [35]. They investigated the purchase intent of five IoT products including smartwatch. Likewise, the same company Accenture [36] conducted another survey that aims to explore consumer adoption plan and likely barriers of IoT products, now and in the future. The survey suggest that IoT adoption rate can be increase through various promotion offers to consumers.

From the literature reviewed, this study is distinct from many of the researches in a number of ways. This study differs from other studies as it utilised an integrated model of TAM together with the theory of PR. By combining these two models to include the influences of security and PR, the model is designed to give a comprehensive perspective on adoption of IS studies. The model includes many of the antecedents used in the studies mentioned above in a more comprehensive perspective, making it possible to show the relative influences of antecedents related to user perceptions. Equally, the studies reviewed had found certain factors to be important in the adoption of IoT products, such as PR, however, to the authors' best knowledge, no past study has explored these construct in the same model. Hence, this study will serve to narrow the research gap found by examining the determinants of IoT acceptance using the integrated model of TAM with perceived risk theory.

3 Research Model

The research model propose in this study is constructed based on the literature reviewed from the previous chapter. The research model is depicted in Fig. 1.

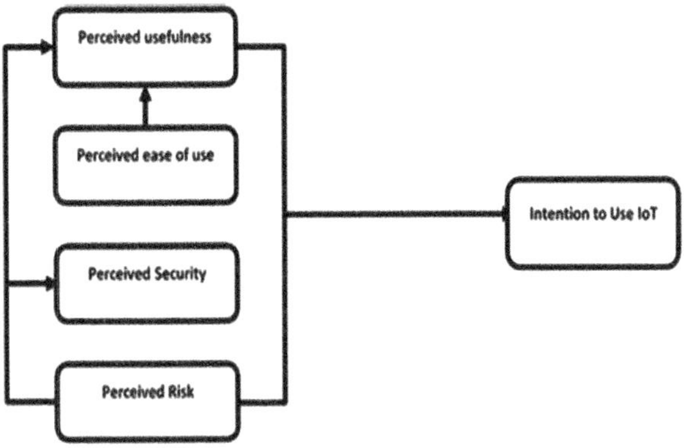

Fig. 1. Research model

3.1 Relationship Between Perceived Usefulness and Intention to Use IoT

According to the TAM theory, PU has been hypothesised to positively relate to user adoption behaviour. Many empirical IS studies had found the relationship between PU and intention to use [37–42]. Similarly, from IoT context, several studies also found the relationship between PU and intention to use [e.g. 8, 20, 43–47]. For instance, Liew, Ang [28], in their study of IoT product found to PU influence consumer acceptance.

Applied to IoT, PU relates to the benefits the consumer thinks he or she will take from the future use of a new product in terms of time saving, convenience, access to additional information [25] and improve efficiency and performance [43]. For example applied to IoT technology in railway section, maintenance staff can receive data from the transponders installed in trains through a mobile reader in their hand to decide whether or not they need maintenance, thereby improving the efficiency of maintenance tasks [8]. Hence, the higher the added value of the IoT, the more the consumer will be inclined to adopt it. Overall, if users feel that the use of IoT products or services, are not useful over their existing technology products, then the users might decide not to accept the product. Thus, it is expected that P*erceived usefulness may have positive effect on user's behavioural intention.*

3.2 Relationship Between Perceived Ease of Use and Intention to Use IoT

Perceived ease of use is one of the core variables of TAM [15, 48]. Several previous studies have state that PEOU is a significant determinant of behavioural intentions to use [8, 49, 50]. TAM stated that PEOU is positively related to intention to use [15, 50]. This relationship was also reported by previous IS studies. For instance, the study of [30, 51], PEOU was hypotheses to positively affect adoption intention. However, other scholars such as Di Pietro, Guglielmetti Mugion [52] in their studies on mobile ticketing have shown that PEOU has an indirect effect on behavioural intention through PU. The indirect effect stems from the situation where the easier a technology is to use, the more useful it is perceived to be, thus, the more positive one's attitude and intention toward using the technology [47, 50]. Therefore, building on the core TAM logic and the context above, it is expected that the association between ease of use and intention to use might also be significant.

3.3 Relationship Between Security and Intention to Use IoT

Security concern associated with a product or service has gained significance in consumer research [53] and has become one of the key issues in IS studies [51, 54]. Specifically, in the age of cybercrime, security is a major concern and barrier to any technology especially those that involve sharing and or using data or connected things [47].

Prior researchers in IS have found that security has a strong influence on whether the user will accept or not accept the technology product. For instance, the work of Khalilzadeh, Ozturk [45] and Oliveira, Thomas [55] found that security directly affects user intention. If the security doubt of a certain innovation is not addressed or solved, it will severely impact the user adoption intention towards the technology [47]. Security is relevant to introduce to our model because when one uses IoT, there is a perception of risk involved in transmitting sensitive information [56]. As an example, when a user uses IoT product such as NFC, there is a possibility that the credit card numbers might be stolen, thereby putting his bank account at risk. Potential adopters may perceived that their credit card information may be at risk and that they have no control over this [57]. As such, the users' intention to use technology product might be affected by their low level perception of that security. Such that users' with low tolerance of technological risks might defer the use of IoT if they have a low level of security perception.

3.4 Relationship Between Perceived Usefulness, Perceived Security with Perceived Risk

It is infer that as user's perceived risk or uncertainty in using IS services, they are likely to find that service more useful over their existing methods [59]. It is also anticipated that only those individuals that perceived using technology products as a low risk task, would have an inclination to recognised the system as being useful [60, 61]. Several studies had shown that the presence of PR in technology product would diminished the usefulness that user's see in that particular product [62–64]. For example, the study of Phonthanukitithaworn, Sellitto [59] on m-payment, revealed that PR has a subsidiary influence on PU. However, in their study on online shopping, Chang, Fu [65] did not found PU to negatively affect PR. Hence, the higher the levels of PU in using new innovative product like IoT, the lower the decrease in anxiety, worry or concern, which in turn will lowered the perceived risk of the innovation.

Equally, when the users of IoT believes that there is greater security in relation to their data regarding its integrity, confidentiality, authentication, and non-repudiation, then they will feel that their information is much safer [66, 67]. While users are required to secure their devices (i.e. smart phone) when using IoT with security measures like encryption software, antivirus, firewalls, the IoT firms can still reduce consumers security risk feelings with several security measures like intrusion-detection systems, firewall, Platform modules. Hence, when the consumer believes that the IoT firms have put adequate measures on securing the user data during transmission, their perception of risk will probably not increase. Recently, the study of D'Alessandro, Girardi [68] confirms that higher security reduces the risk of purchasing online. Therefore, in line with the work of D'Alessandro, Girardi [68], the study also infer that there might be a negative relationship between risk with usefulness and security.

3.5 Relationship Between Perceived Risk and Intention to Use IoT

The relationship between perceived risk and behavioural intentions is explained by the theory of planned behaviour [69]. The theory of planned behaviour postulated that consumers are more likely to buy a product or use a system if they perceived that the level of risk is low [63]. PR is proved to be the key antecedent of intention and mostly hypothesis to negatively influences behavioural intentions. Moreover, consumer perceptions of PR reduces consumer beliefs and intention of using IS [70, 71]. Phonthanukitithaworn, Sellitto [59] infer that in the DOI acceptance decision process, PR is argued to be a critical determinant especially in the pre-adoption stage of technology when people have no experience with it and are wary of the risks and consequences of its use.

The risk involved in the process of using IoT is a critical barrier to their adoption intention. Therefore, consumers will seek to use IoT that are more secured and useful to reduce their perceived risk, which will, in turn, increases the intention towards the use of the IoT. PR is encountered by customers consciously and unconsciously when they make purchase decisions [72, 73]. Since IoT involve some form of purchasing, it encompasses certain element of risk. Through risk reduction associated with their use, users will develop stronger adoption decision.

4 Conclusion

The study aim to find the factors that affects Malaysian individuals' intention to use IoT. Factors were identified through rigorous literature review which were link through the lens of TAM. The study will employed a quantitative survey to test the proposed study hypotheses through structural equation model technique. Recommendation will be given to the relevant stake holders based on the findings of the study.

Acknowledgement. This research is supported by Universiti Sains Malaysia through the USM Bridging Grant 2017 [A/C Number: 304.PKOMP.6316001].

References

1. Internet Live Stats: Internet Users (2017). http://www.internetlivestats.com/internet-users/malaysia/. Accessed 5 Mar 2017
2. Statista: Number of connected devices worldwide 2012–2020 (2017). https://www.statista.com/statistics/471264/iot-number-of-connected-devices-worldwide/. Accessed 6 Mar 2017
3. Cisco: Internet of Things: Connected Means Informed, p. 3. Cisco (2016)
4. Fantana, N., et al.: IoT applications–value creation for industry. In: Internet of Things: Converging Technologies for Smart Environments and Integrated Ecosystems, p. 153 (2013)
5. Lim, J.H.: Antedecents and outcome of Internet of Things adoption: a perspective of public listed companies on main market board of Bursa Malaysia. Universiti Sains Malaysia (2015)
6. Verizon: State of the Market: Internet of Things 2016. Verizon.com (2016)
7. Lund, D., et al.: Worldwide and regional Internet of Things (IoT) 2014–2020 forecast: a virtuous circle of proven value and demand. Technical Report, International Data Corporation (IDC) (2014)
8. Gao, L., Bai, X.: A unified perspective on the factors influencing consumer acceptance of Internet of Things technology. Asia Pac. J. Mark. Logist. **26**(2), 211–231 (2014)
9. Chong, A.Y.-L., et al.: Predicting RFID adoption in healthcare supply chain from the perspectives of users. Int. J. Prod. Econ. **159**, 66–75 (2015)
10. Gubbi, J., et al.: Internet of Things (IoT): a vision, architectural elements, and future directions. Future Gener. Comput. Syst. **29**(7), 1645–1660 (2013)
11. Rose, K., Eldridge, S., Chapin, L.: The Internet of Things: an overview, pp. 1–50. The Internet Society (ISOC) (2015)
12. Botta, A., et al.: Integration of cloud computing and Internet of Things: a survey. Future Gener. Comput. Syst. **56**, 684–700 (2016)
13. ISO: Internet of Things: Preliminary Report 2014, Switzerland, pp. 1–11 (2015)
14. Mimos: National Internet of Things Strategic Roadmap: A Summary, Malaysia, pp. 1–24 (2015)
15. Venkatesh, V., et al.: User acceptance of information technology: toward a unified view. MIS Q. **27**, 425–478 (2003)

16. Venkatesh, V., Davis, F.D., Morris, M.G.: Dead or alive? The development, trajectory and future of technology adoption research. J. Assoc. Inf. Syst. **8**(4), 267–286 (2007)
17. Hsu, C.-L., Lin, J.C.-C.: An empirical examination of consumer adoption of Internet of Things services: network externalities and concern for information privacy perspectives. Comput. Hum. Behav. **62**, 516–527 (2016)
18. Chong, A.Y.-L., Chan, F.T., Ooi, K.-B.: Predicting consumer decisions to adopt mobile commerce: cross country empirical examination between China and Malaysia. Decis. Support Syst. **53**(1), 34–43 (2012)
19. Kim, K.J., Shin, D.-H.: An acceptance model for smart watches: implications for the adoption of future wearable technology. Internet Res. **25**(4), 527–541 (2015)
20. Alolayan, B.: Do I really have to accept smart fridges? An empirical study (2014)
21. Dahlberg, T., et al.: Past, present and future of mobile payments research: a literature review. Electron. Commer. Res. Appl. **7**(2), 165–181 (2008)
22. Tu, M.: An exploratory study of Internet of Things (IoT) adoption intention in logistics and supply chain management: a mixed research approach. Int. J. Logist. Manag. **29**(1), 131–151 (2018)
23. Kamble, S.S., et al.: Modeling the Internet of Things adoption barriers in food retail supply chains. J. Retail. Consum. Serv. **48**, 154–168 (2019)
24. Hsu, C.-L., Lin, J.C.-C.: Exploring factors affecting the adoption of Internet of Things services. J. Comput. Inf. Syst. **58**(1), 49–57 (2018)
25. Mani, Z., Chouk, I.: Drivers of consumers' resistance to smart products. J. Mark. Manag., 1–22 (2016)
26. Kowatsch, T., Maass, W.: Critical privacy factors of internet of things services: an empirical investigation with domain experts. In: Knowledge and Technologies in Innovative Information Systems, pp. 200–211. Springer (2012)
27. Shin, D.H.: Conceptualizing and measuring quality of experience of the internet of things: exploring how quality is perceived by users. Inf. Manag. **54**(8), 998–1011 (2017)
28. Liew, C.S., et al.: Factors influencing consumer acceptance of Internet of Things technology. In: Handbook of Research on Leveraging Consumer Psychology for Effective Customer Engagement, p. 168. IGI Global (2016)
29. Mital, M., et al.: Adoption of Internet of Things in India: a test of competing models using a structured equation modeling approach. Technol. Forecast. Soc. Change (2017)
30. Dong, X., Chang, Y., Wang, Y., Yan, J.: Understanding usage of Internet of Things (IOT) systems in China: cognitive experience and affect experience as moderator. Inf. Technol. People **30**(1), 117–138 (2017)
31. Balaji, M., et al.: User acceptance of IoT applications in retail industry. In: Technology Adoption and Social Issues: Concepts, Methodologies, Tools, and Applications, pp. 1331–1352. IGI Global (2018)
32. Jayashankar, P., et al.: IoT adoption in agriculture: the role of trust, perceived value and risk. J. Bus. Ind. Mark. **33**(6), 804–821 (2018)
33. Abushakra, A., Nikbin, D.: Extending the UTAUT2 model to understand the entrepreneur acceptance and adopting Internet of Things (IoT). In: International Conference on Knowledge Management in Organizations. Springer (2019)
34. Lee, W., Shin, S.: An empirical study of consumer adoption of Internet of Things services. Int. J. Eng. Technol. Innov. **9**(1), 1 (2019)
35. The Acquity Group: Igniting Growth in Consumer Technology, United States (2016)
36. Accenture: The Internet of Things: The Future of Consumer Adoption, USA, pp. 1–12 (2014)
37. Arpaci, I.: Understanding and predicting students' intention to use mobile cloud storage services. Comput. Hum. Behav. **58**, 150–157 (2016)

38. Keong, W.E.Y.: The determinants of mobile shopping mall apps adoption intention in Malaysia: an empirical investigation. In: 2016 11th International Conference on Computer Science & Education (ICCSE). IEEE (2016)
39. Hew, T.-S., et al.: Predicting drivers of mobile entertainment adoption: a two-stage SEM-artificial-neural-network analysis. J. Comput. Inf. Syst. **56**(4), 352–370 (2016)
40. Chiyangwa, T.B., Alexander, P.T.: Rapidly co-evolving technology adoption and diffusion models. Telematics Inform. **33**(1), 56–76 (2016)
41. Ramayah, T., et al.: Explaining the adoption of Internet stock trading in Malaysia: comparing models. Asian J. Technol. Innov. **22**(1), 131–151 (2014)
42. Makki, A.M., Ozturk, A.B., Singh, D.: Role of risk, self-efficacy, and innovativeness on behavioral intentions for mobile payment systems in the restaurant industry. J. Foodserv. Bus. Res. **19**(5), 454–473 (2016)
43. Hsu, C.-L., Lin, J.C.-C.: Exploring factors affecting the adoption of Internet of Things services. J. Comput. Inf. Syst., 1–9 (2016)
44. Chuah, S.H.-W., et al.: Wearable technologies: the role of usefulness and visibility in smartwatch adoption. Comput. Hum. Behav. **65**, 276–284 (2016)
45. Khalilzadeh, J., Ozturk, A.B., Bilgihan, A.: Security-related factors in extended UTAUT model for NFC based mobile payment in the restaurant industry. Comput. Hum. Behav. **70**, 460–474 (2017)
46. Ooi, K.-B., Tan, G.W.-H.: Mobile technology acceptance model: an investigation using mobile users to explore smartphone credit card. Expert Syst. Appl. **59**, 33–46 (2016)
47. Zailani, S., et al.: Determinants of RFID adoption in Malaysia's healthcare industry: occupational level as a moderator. J. Med. Syst. **39**(1), 172 (2015)
48. Moore, G.C., Benbasat, I.: Development of an instrument to measure the perceptions of adopting an information technology innovation. Inf. Syst. Res. **2**(3), 192–222 (1991)
49. Davis, F.D.: Perceived usefulness, perceived ease of use, and user acceptance of information technology. MIS Q. **13**, 319–340 (1989)
50. Davis, F.D., Bagozzi, R.P., Warshaw, P.R.: User acceptance of computer technology: a comparison of two theoretical models. Manag. Sci. **35**(8), 982–1003 (1989)
51. Al-Momani, A.M., Mahmoud, M.A., Sharifuddin, M.: Modeling the adoption of internet of things services: a conceptual framework. IJAR **2**(5), 361–367 (2016)
52. Di Pietro, L., et al.: The integrated model on mobile payment acceptance (IMMPA): an empirical application to public transport. Transp. Res. Part C: Emerg. Technol. **56**, 463–479 (2015)
53. Schierz, P.G., Schilke, O., Wirtz, B.W.: Understanding consumer acceptance of mobile payment services: an empirical analysis. Electron. Commer. Res. Appl. **9**(3), 209–216 (2010)
54. Shin, S., Lee, W.-J.: The effects of technology readiness and technology acceptance on NFC mobile payment services in Korea. J. Appl. Bus. Res. **30**(6), 1615 (2014)
55. Oliveira, T., et al.: Mobile payment: understanding the determinants of customer adoption and intention to recommend the technology. Comput. Hum. Behav. **61**, 404–414 (2016)
56. Arpaci, I., Kilicer, K., Bardakci, S.: Effects of security and privacy concerns on educational use of cloud services. Comput. Hum. Behav. **45**, 93–98 (2015)
57. Salisbury, W.D., et al.: Perceived security and World Wide Web purchase intention. Ind. Manag. Data Syst. **101**(4), 165–177 (2001)
58. Chen, S.-C., Chen, H.-H., Chen, M.-F.: Determinants of satisfaction and continuance intention towards self-service technologies. Ind. Manag. Data Syst. **109**(9), 1248–1263 (2009)
59. Phonthanukitithaworn, C., Sellitto, C., Fong, M.W.: An investigation of mobile payment (m-payment) services in Thailand. Asia-Pac. J. Bus. Adm. **8**(1), 37–54 (2016)

60. Martins, C., Oliveira, T., Popovič, A.: Understanding the Internet banking adoption: a unified theory of acceptance and use of technology and perceived risk application. Int. J. Inf. Manag. **34**(1), 1–13 (2014)
61. Chan, S., Lu, M.: Understanding Internet Banking Adoption and Use Behavior: A Hong Kong Perspective (2004)
62. Cocosila, M., Archer, N.: Practitioner pre-adoption perceptions of Electronic Medical Record systems. Behav. Inf. Technol. **36**(8), 827–838 (2017)
63. Roy, S.K., et al.: Constituents and consequences of smart customer experience in retailing. Technol. Forecast. Soc. Change (2016)
64. Lim, N.: Consumers' perceived risk: sources versus consequences. Electron. Commer. Res. Appl. **2**(3), 216–228 (2003)
65. Chang, H.H., Fu, C.S., Jain, H.T.: Modifying UTAUT and innovation diffusion theory to reveal online shopping behavior: familiarity and perceived risk as mediators. Inf. Dev. **32**(5), 1757–1773 (2016)
66. Flavián, C., Guinalíu, M.: Consumer trust, perceived security and privacy policy: three basic elements of loyalty to a web site. Ind. Manag. Data Syst. **106**(5), 601–620 (2006)
67. Curry, P.: Consumer risk: the importance of privacy and security while connected to Wi-Fi hotspots: does location matter? In: AMCIS (2011)
68. D'Alessandro, S., Girardi, A., Tiangsoongnern, L.: Perceived risk and trust as antecedents of online purchasing behavior in the USA gemstone industry. Asia Pac. J. Mark. Logist. **24**(3), 433–460 (2012)
69. Pavlou, P.A., Fygenson, M.: Understanding and predicting electronic commerce adoption: an extension of the theory of planned behavior. MIS Q. **30**, 115–143 (2006)
70. Yang, Q., et al.: Exploring consumer perceived risk and trust for online payments: an empirical study in China's younger generation. Comput. Hum. Behav. **50**, 9–24 (2015)
71. Featherman, M.S., Pavlou, P.A.: Predicting e-services adoption: a perceived risk facets perspective. Int. J. Hum Comput Stud. **59**(4), 451–474 (2003)
72. Yu, J., et al.: User acceptance of media tablets: an empirical examination of perceived value. Telematics Inform. **34**(4), 206–223 (2017)
73. Shimp, T.A., Bearden, W.O.: Warranty and other extrinsic cue effects on consumers' risk perceptions. J. Consum. Res. **9**(1), 38–46 (1982)

Augmented Reality in Library Services: A Panacea to Achieving Education and Learning 4.0

Rifqah Okunlaya[1(✉)], Norris Syed Abdullah[2(✉)], and Rose Alinda Alias[2(✉)]

[1] School of Computing, Universiti Teknologi, Malaysia, Johor Bahru, Malaysia
okunlayarifqah@gmail.com

[2] Azman Hashim International Business School (AHIBS), Universiti Teknologi, Malaysia, Johor Bahru, Malaysia
{norris, alinda}@utm.my

Abstract. Augmented Reality (AR) concept is presently discussed as one of the foremost technologies of the next industrial revolution. AR has turned out to be a prevalent topic in educational research over the years as existing literature reflects the multiple benefits of its incorporation into educational setting however, the potential of achieving the Education 4.0 through the provision of AR based university library services is scarce in the literature. This paper conducts a content analysis of the literature to bridges the gap by discussing how AR usage for library services can serve as a remedy to achieve Education 4.0. We stand on the strength of service innovation model to point out numerous ways by which AR can be used to provide library services. This finding will assist university community as complimentary effort to achieving quality teaching, learning and research to foster the education and learning 4.0.

Keywords: Augmented Reality · Education 4.0 · Library services · 4IR

1 Introduction

University library is the heart of the university as it provides intellectual backing for quality teaching, learning and research [1]. The growth of digital technology has nevertheless fashioned disruptive innovation where novel prospects in communication and services has changed the way of learning [2]. Technological progressions are rapidly immersing and incorporating into the processes of education and electronic learning, which facilitated the growth of effective and innovative solutions [3]. This speedy expansion has steered the fourth Industrial Revolution (4IR) which altered the concept of work environs such that new training and learning concepts, future-ready curricula, new skills, as well as the tractability in education turns out to be essential. [4], mentions education as an Industry affected by Digital Transformation. This is supported by [5] who highlighted it as an attribute of the (4IR). [2] opines that if education is to be 4IR compliant, it has to be interactive and active in a way students can become part of higher-learning that calls for the need of a strong collaboration

F. Saeed et al. (Eds.): IRICT 2019, AISC 1073, pp. 991–998, 2020.
https://doi.org/10.1007/978-3-030-33582-3_93

between academic environment and industry in human resources formation. This type of education connected to 4IR has brought about the term Learning 4.0 and education 4.0 respectively [6]. Libraries as academic repository of the educational establishments has made better their services to attract new customers as well as keep existing users because they have the duty of gathering, processing, storing and making access to information to improve learning [7]. These current technological progressions has caused developments in University Library service delivery towards education 4.0 of which augmented reality is one of them. "Augmented Reality" (AR) has turn out to be a nascent technology which possesses potentials for its usage in education [8, 9]. It helps to enhance the actual world by means of computer-generated amplifications that unites flawlessly in the user's physical location in three-dimensional (3D) format [10].

AR is capable of being beneficial on several display place and gadgets such as tablets, computers, as well as smartphones [11]. Nonetheless, the ongoing digital revolution is no more limited to the application on the smart phone for daily requirements but now affects our professional life. In effect, the development of research in AR technology does not focus only on mobile applications and devices, but on an entire ecology built around smartphones, which includes applications for education and training [12]. Over the last years, AR was used in numerous fields [13]. Educational sector is one of the fields where AR is anticipated to make an impression in years to come in accompaniment of the current activities of education by providing supplementary information in a collaborating manner. Educational institutions will gain better accessibility to virtual technologies, which will enable teaching in virtual environments not envisage in physical classrooms. This simply indicates that the huge potentials of available virtual technologies will disrupt the borders of formal education [12].

Libraries as an essential part of universities facilitates the growth, exchange and curation of academic knowledge, information, and student's cultural interactions that goes on within community of the university. In these logic, libraries signifies a central power point of academic lifecycle hence assist a common drive of taking along individuals around common themes together through various channels including technology. University libraries can guide the entire campus into effective and efficient AR adoption. Despite the abundant prospects the university library has to support the future educational system, researchers continue to undermine libraries of universities for increased assessment and accountability [14]. Libraries are now becoming irrelevant, as university administrators been reducing the budgetary expenditures to libraries in the last few decades [15]. These factors signals an increasing burden for university libraries to demonstrate their continued relevance, especially on how emerging technology like AR in the library setting can support education. How AR supported library services can help to achieve Edu 4.0 is still missing in the literature thus this research fills this Gap by answering the question "How can the library use AR to provide services that can support education in order to prepare for 4IR.

The remainder of this paper is organized as follows: Sect. 2 gives a brief review of literature on various studies that have been done on AR especially in the context of Education and library services. Section 3, discusses the methodology adopted for this research. Following this, Sect. 4 discusses the findings of AR as an innovative technology that can support library services for education 4.0 while Sect. 5 is about conclusion, recommendation for future research.

2 Related Works

AR is significant in the provision of important tools required to progress the knowledge of interrelating by means of reality. Present literature reveals the numerous assistances of the incorporation of AR technologies in education. Research has been carried out to launch the capability, challenges and affordances of this technology for education [16]. AR is a prevalent technology that has turn out to be a given a significant emphasis in education and research lately. AR has the prospect to be an application that can be used to support pedagogical stance in education [17] because studies made us to understand how technology powered by AR can help student's learning [18]. In a current study, carried out by [19] on AR's usage in educational sceneries, they gave a report showing recent increase publications on AR concluding that AR technology has the tendency of supporting learning and teaching. In the same spirit, [20] reviewed 26 AR publications identifying possible of list things that impact educational effect of AR technology on activities of learning for students. His study settles that AR is supportive in promoting the motivation of students, increasing spatial capabilities, promoting collaboration in the midst of students and physical tasks performance improvement. It is however important to note that a gap of how AR can be used to enhance education 4.0. within the context of library services provision is yet to be filled.

University libraries are striving more actively in performing roles such as training and user education throughout the campus in order to deliver services to users at the point of need for academic fulfilment. Recently, faculty research is reinforced through devoted library services and embedded partnerships to manage access and preservation of research data [21]. It is obvious that information environs is continually changing hence academic libraries and librarians in the university settings are re-inventing themselves to fit into new teaching, learning and research environment in order to support the core goal and objectives of their parent institutions [22].

2.1 Theoretical Model

Four-Dimensional Model of Service Innovation

A service innovation is a service process or service product grounded on some technological or organized method. In services nevertheless, the novelty is not necessarily attached to the technology's newness itself but the creative ideas that is often embedded in the technologically oriented parts. In a nutshell, [23] defines service innovation as a "new or considerably patron's channel of interaction, changed service concept, service delivery system or technological concept that headed few or several redesigned service roles that are novel to the organisation". It is a service change that requires operationally new personnel, organizational or technological capabilities of the organization in the provision of its services.

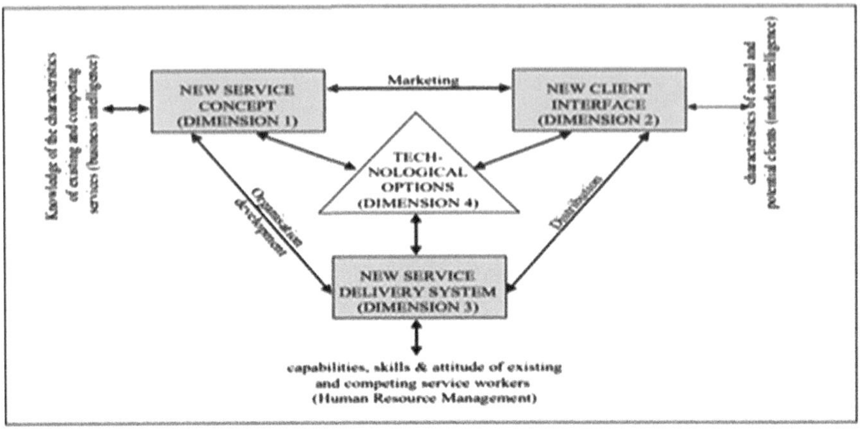

Fig. 1. den Hertog's model of service innovation model [24]

2.2 Service Innovation Research

For the actualization of the objective of the research, we adopt the "den Hertog's innovation model" [24] to explain how the AR as an emerging technology has brought about innovative ways of providing library services that can enhance education 4.0 in the university environment. [24] recognizes four "dimensions" of service innovation. "The Service Concept", "The Client Interface", "The Service Delivery System" and "Technological Options". The Service Concept refers to the concept of a service novel to its actual market a new service in weight, for instance; a "new value proposition". The Client Interface speaks of interface's innovation amid the customer and the provider of its service. Patrons are frequently entangled in service creation. "The Service Delivery System" likewise frequently connects to the link amid the client and its service provider, as delivery see to dealings through this interface. Technological possibilities look like most conversant process innovation. Novel information technology is particularly significant to services, as it permits for better productivity and efficiency in the predominant information-processing essentials that are in services sectors. In fact, the bulk of service innovations will closely include different blends of these four dimensions. For instance, innovation in Library service is a conceptual and repetitive activity of applying new methods, novel ideas or new technologies to progress and reconstruct present service. Four-Dimensional Model of Service Innovation, a conceptual model, can aid the integration and understanding of the concept of service innovation and expansively define service innovation model, which relates to service division and even other sectors, and can be used to analyse service innovation accomplishments of libraries.

3 Methodology

This research adopts a qualitative research approach to investigate and proposes solution to the gap identified in view of the usage of AR in the library to facilitate the education and learning 4.0. We conducted a literature review to collect relevant domain knowledge by identifying relevant publications on Augmented Reality. We made use of databases like Science Direct, Scopus, and Wiley online library, Web of science and AIS). Keywords like Augmented Reality, Education 4.0 and Library services were combined to look for articles using Boolean search. Boolean search permits users to make use of combination of keywords through operators such as "AND", "NOT" and "OR" to further produce results that are more relevant to those which may explain how AR can be used in the context of the library were selected to conduct this research. Adopting the content analysis of [25], a qualitative content analysis of articles collected from various databases was made. [25] explains content analysis to be a technique of research used in making valid and repeatable interpretations from texts based on the contexts of their use. Service innovation model of [24] is however used to explain the usability of AR in the context of the way it can be used in the library to achieve education 4.0.

4 Findings

In this section, we discuss how the qualitative analysis of the literature gave this research the opportunity to discover the following possible means of how AR compliant library services can facilitate education 4.0 using the Four Dimensional Model of Services Innovation (see Fig. 1) as a benchmark for evaluation. The findings goes thus:

The Service Concept

Many service innovations have been introduced to the field of library services by way of AR. Chen and Tsai [26] came up with a new "AR Library Instruction System" (ARLIS), which instructs some skills and ideas of how to use the library to students by means of a 3D technology that is interactive. ARLIS incorporates collaborative virtual 3D technology and the atmosphere of the physical library to produce a novel "context-aware" instruction mode of the library. Likewise, [27] used the technology of AR in proposition of a system where the location and information of library books used by users are shown. Both work of [26] and [27] evident that AR is able to facilitates the statistical analysis of library data concerning the provision and borrowing of electronic books in connection with relevant agencies.

The Client Interface

The library now provides a new interface of learning environment to students to fulfil their academic and leisure needs for accessing resources from the library. An example of the interface is "Flexible Display" which is a "paper- like" demonstration famous for getting the identical picture excellence either it is bent or folded. It is a technology replacing existing laptop display screens, PC monitors, and televisions because of the portability in screen size and capacity it signifies despite its reduction. "Transparent Display" which is a communal name for a display of technology that is totally translucent when switched off, and becomes incompletely transparent when it is switched on is another example. This technology is a combination of AR and touch screens which has numerous daily applications [28].

The Service Delivery System

AR is progressively becoming prevalent to unite users with "context-specific" content. AR has enormous possibilities with which libraries can occupy user community to achieve efficiency in education and learning. Computer-aided training for teaching rudimentary skills of the library on how students can access information for learning, research is an alluring method. It is opined by [25] that student's performance is enhanced in learning by incorporating AR in teaching library skills because it increasing the understanding of student's more than traditional instruction of the librarian.

Technology Options

Books tracking in the library is tailored towards the traditional static database systems by means desktop computing that only requires the 'Class Numbers' of the books instead of making available any additional direction to users for services and space alertness. It sometimes requires the librarian or library staff's helps to give meaning to the shelf numbers, which can end up to be time consuming. Inactive and fixed signs is what is adopted in many libraries. In replacement of this, "ShelvAR" of [29]; a shelf-reading AR system, which comprises of optically tracking books through book spines can be used. Tags on a row of books are read as ShelvAR uses augmentations in 2D to read spines of books by means of "check-marks" which helps to indicate to the user if the books are in right location or not. ShelvAR system calculates the most appropriate way to reorganize the books, and highlight those books that are need to be orderly positioned automatically.

5 Conclusion and Recommendation

AR as a technology is getting more attention in the setting of education and learning but little is said about it on how it can be used to provide library services in order to achieve education and learning 4.0. Standing on the strength of service innovation model, this study points out numerous ways by which AR can be used to provide library services to compliment the efforts of teaching, learning and research in the university community to foster the achievement of education and learning 4.0. Recommendation is hereby made that financial support be given to the library management to fully annex the usage of AR assisted library services. Future research should also look into how AR can be used to provide "Selective Information Dissemination" (SDI) for personalized library services.

References

1. Ifijeh, G.: Adoption of digital preservation methods for theses in Nigerian academic libraries: applications and implications. J. Acad. Librariansh. **40**(3–4), 399–404 (2014)
2. Ciolacu, M., Svasta, P.M., Berg, W., Popp, H.: Education 4.0 for tall thin engineer in a data driven society. In: 2017 IEEE 23rd International Symposium for Design and Technology in Electronic Packaging (SIITME), pp. 432–437 (2017)
3. Fernoagă, V., Stelea, G.A., Gavrilă, C., Sandu, F.: Intelligent education assistant powered by Chatbots. In: The International Scientific Conference eLearning and Software for Education, vol. 2, pp. 376–383. "Carol I" National Defence University (2018)
4. Ciolacu, M., Tehrani, A.F., Beer, R., Popp, H.: Education 4.0—fostering student's performance with machine learning methods. In: 2017 IEEE 23rd International Symposium for Design and Technology in Electronic Packaging (SIITME), pp. 438–443 (2017)
5. Baryshnikova, A., Taratukhin, V.: Digital transformation framework for smart factory. In: AMCIS (2017)
6. Klonn, M., Abke, J.: 'Learning 4.0': a conceptual discussion. In: 2018 IEEE International Conference on Teaching, Assessment, and Learning for Engineering (TALE), pp. 871–876 (2018)
7. Papic, A., Primorac, M.: Introducing e-CRM into academic libraries: exploration of needs and possibilities. In: 2014 37th International Convention on Information and Communication Technology, Electronics and Microelectronics, pp. 597–602 (2014)
8. Alkhattabi, M.: Augmented reality as e-learning tool in primary schools' education: barriers to teachers' adoption. Int. J. Emerg. Technol. Learn. (iJET) **12**(02), 91–100 (2017)
9. Chang, S.C., Hwang, G.J.: Impacts of an augmented reality-based flipped learning guiding approach on students' scientific project performance and perceptions. Comput. Educ. **125**, 226–239 (2018)
10. Cervera-Uribe, A.A.: [POSTER] The augmented library: an approach for improving users awareness in a campus library. In: 2017 IEEE International Symposium on Mixed and Augmented Reality (ISMAR-Adjunct), pp. 15–19 (2017)
11. Fombona Cadavieco, J., Vazquez-Cano, E., Del Valle Mejias, M.E.: Analysis of geolocation and augmented reality on mobile devices, social and educational proposals related to the environment and field trips. Profesorado-Revista de Curriculum y Formacion de Profesorado **22**(4), 197–222 (2018)
12. Martín-Gutiérrez, J., Mora, C.E., Añorbe-Díaz, B., González-Marrero, A.: Virtual technologies trends in education. EURASIA J. Math. Sci. Technol. Educ. **13**(2), 469–486 (2017)
13. Iftene, A., Trandabăț, D.: Enhancing the attractiveness of learning through augmented reality. Procedia Comput. Sci. **126**, 166–175 (2018)
14. Mylonas, G., Triantafyllis, C., Amaxilatis, D.: An augmented reality prototype for supporting IoT-based educational activities for energy-efficient school buildings. Electron. Notes Theor. Comput. Sci. **343**, 89–101 (2019)
15. Soria, K.M., Fransen, J., Nackerud, S.: Library use and undergraduate student outcomes: new evidence for students' retention and academic success. portal: Libr. Acad. **13**(2), 147–164 (2013)
16. Garzón, J., Acevedo, J.: A meta-analysis of the impact of augmented reality on students' learning effectiveness. Educ. Res. Rev. **27**, 244–260 (2019)
17. Garzón, J., Pavón, J., Baldiris, S.: Systematic review and meta-analysis of augmented reality in educational settings. Virtual Reality **23**, 1–13 (2019)

18. Ellen Frederick, D.: Libraries, data and the fourth industrial revolution (Data Deluge Column). Library Hi Tech News **33**(5), 9–12 (2016)
19. Akçayır, M., Akçayır, G.: Advantages and challenges associated with augmented reality for education: a systematic review of the literature. Educ. Res. Rev. **20**, 1–11 (2017)
20. Radu, I.: Augmented reality in education: a meta-review and cross-media analysis. Pers. Ubiquit. Comput. **18**(6), 1533–1543 (2014)
21. Kutay, S.A.: Advancing digital repository services for faculty primary research assets: an exploratory study. J. Acad. Librariansh. **40**(6), 642–649 (2014)
22. Jain, P., Akakandelwa, A.: Challenges of twenty-first century academic libraries in Africa. Afr. J. Libr. Arch. Inf. Sci. **26**(2), 147 (2016)
23. Van Ark, B., Broersma, L., den Hertog, P.: Services innovation, performance and policy: a review. Research Series No. 6, Directorate-General for Innovation, Ministry of Economic Affairs. Dialogic, The Hague (2003)
24. Hertog, P.D.: Knowledge-intensive business services as co-producers of innovation. Int. J. Innov. Manag. **4**(04), 491–528 (2000)
25. Krippendorff, K.: Content Analysis: An Introduction to Its Methodology. Sage Publications, Thousand Oaks (2018)
26. Chen, C.M., Tsai, Y.N.: Interactive augmented reality system for enhancing library instruction in elementary schools. Comput. Educ. **59**(2), 638–652 (2012)
27. Ha, S.H., Kim, D., Kim, J.: Mobile augmented reality interface (AR) design for user experience of library. In: KSDS Conference Proceeding 49 (2011)
28. Shatte, A., Holdsworth, J., Lee, I.: Mobile augmented reality based context-aware library management system. Expert Syst. Appl. **41**(5), 2174–2185 (2014)
29. Brinkman, B., Brinkman, S.: AR in the library: a pilot study of multi-target acquisition usability. In: 2013 IEEE International Symposium on Mixed and Augmented Reality (ISMAR), pp. 241–242 (2013)

A Systematic Review on Humanizing Factors for Online System

Lina Fatini Azmi[1(✉)] and Norasnita Ahmad[2]

[1] School of Computing, Faculty of Engineering, Universiti Teknologi Malaysia, 81310 Skudai, Johor, Malaysia
linaftni@gmail.com

[2] Azman Hashim International Business School, Universiti Teknologi Malaysia, 81310 Skudai, Johor, Malaysia
norasnita@utm.my

Abstract. Online systems have improved our lives in many ways compared to traditional method. Recently, humanizing online systems become an important issue and an interest of professionals and academician as it is believed can improve online systems become more user-centric. However, the systematic review to report the current state of humanizing online systems research is very limited. Thus, this paper aims to systematically review and contributes better understanding of humanizing for online systems research. To collect and review previous studies, we used systematic literature review method consisting three phases which are planning, conducting and reporting the reviews. A set of research questions for the review covers the meaning, problems and factors of humanizing online systems from various fields. As a result, we selected 33 papers that have relevant information to address the research questions been set. Based on the review, we also found that humanizing in online system is still undefined. Therefore, further study that investigate human behavior in online system is necessary.

Keywords: Humanizing · Online system · Human behavior

1 Introduction

1.1 Online System

Online systems have the primary tasks of interacting with customers, answering their questions and addressing their issues [1]. However, online systems were reported to be lack in humanness [1] in terms of communication, interaction, and response. Lack in humanness will lead to misunderstanding, dissatisfaction and uncontrollable customer's behavior. Hence, according to [2], there is a great opportunity in this research field to clearly define human and organization relationship boundaries.

Organizations able to enjoy countless advantages when customers perceive, connect and treat products and services favorably. These include loyalty and commitment, increased customer readiness to dismiss existing products and services, increased readiness to spread positive Word-of-Mouth (WOM) and a desire to pay a premium

F. Saeed et al. (Eds.): IRICT 2019, AISC 1073, pp. 999–1008, 2020.
https://doi.org/10.1007/978-3-030-33582-3_94

cost [2]. Humanizing online system will create a new approach to maximize user experience and organization profit as well as improving their business services.

Previous study shows that online system is an important chapter of study in technology that needs human-centric operations. However, report on the current state of humanizing online systems research is very limited. Therefore, this paper review systematically all previous studies related to humanizing aspects on online system, as well as formulate general definition of humanizing online systems based on humanizing factors identified. This contribution will help future researcher to understand the humanizing aspects in online systems.

The structure of this paper is according to the following arrangement. We start first section with brief introduction of online systems and followed by humanizing online systems. On the next section, there will be a discussion on methodology used in this study which is systematic literature review (SLR). Section 4 outlined the results of SLR by determining answers for each research question, followed by the discussion in Sect. 5. Lastly, Sect. 6 will conclude overall discussion of this paper.

2 Humanizing in Online Systems (SLR)

Humanizing refers to an act of making something more humane or civilized [3]. Although online system already makes our lives much better than traditional ways, unfortunately various obstacles identified in online system especially related to user of the online system. Humanizing online system can help to improve it to become more user-centric.

In online banking sector, technology could limit individuals, transmit a sense of incompetence, lead to waste of time and resources, raise customer worries about privacy and confidentiality and lead to the loss of true human interaction [4]. Therefore, in more personalized systems are needed to be implemented in order to provide the finest services and promote greater relationships. Another study found that humanize website help in improving the experience of question-and-answer, offering quicker data and making the experience more enjoyable to customers [5]. There are three ways to humanize online system suggested by [1]: first is the use of human figures as visual indications; second is by using names or identity associated with humans as identification signs; and third is to imitate human languages as conversational signals. There is a study that differentiates the function between technology and humanization of technology. Technology can identify and react intuitively to the individual requirements, while humanization of technology will reinforce service providers' capability to offer outstanding customer experience in the future [6].

3 Review Method

The systematic literature review (SLR) method for this study are based on the guideline of [7]. Table 1 shows the three phases of SLR. Each phase will be explained in the following subsection.

Table 1. Review method [7].

Phases	Steps
Planning the review	Step 1: Formulate the problem
	Step 2: Develop the review protocol
Conducting the review	Step 3: Search the literature
	Step 4: Screen the inclusion & exclusion
	Step 5: Assess quality
	Step 6: Extract data
	Step 7: Analyze and synthesis data
Reporting the review	Step 8: Report findings

3.1 Formulate the Problem

To fulfill the main goals for this study and formulate the problem by using root cause analysis, three research questions have been developed. Subsequently, these questions help the researchers to define humanizing online system, identify the problems in online system that need to humanize, and humanizing factors for online system. The three research questions are as follows:

RQ1: What are the meanings of humanizing online system in different area of studies?
RQ2: What are the problems in online systems that need to be humanize?
RQ3: What are the humanizing factors in context of online system that have human interactions?

3.2 Develop the Review Protocol

A protocol was developed to document the method of analysis and the criteria for inclusion. Review protocol refers to a flow that describing the plan of a proposed SLR process. Drafting the protocol allows the researcher to describe in detail the defined steps and methods to be implemented in the SLR. The method of evaluation this review protocol is by testing the protocol execution. The review was carried out by conducting a complete cycle of systematic reviews but using a smaller set of sources.

3.3 Search the Literature

Main source for the literature is electronic databases with high quality publications, and the techniques used were backward search and forward search [7]. Electronic databases used for these studies are ResearchGate, Scopus, IEEE Explore Digital Library, ScienceDirect, Taylor and Francis Online, Emerald Insight, Springer Link, and Google Scholar.

The keyword used to search the literature include "Humanizing", "Humanized", "Humanization", "Humanizing Online System", "Humanizing Online", "Human Online System Interaction", "Human Interaction Online System" and other combo were applied to gather as many related studies as possible such as "Humanizing" AND

"Online System", "Human" AND "Online System", "Humanized" AND "Online System", "Humanization" AND "Online System", "Human Interaction" AND "Online System". These keywords formulated based on the research question. Consequently, Mendeley Desktop was used to store the finding results of each database, as well as to avoid duplicate research studies and to store all citations.

3.4 Search for Inclusion and Exclusion

By having specified criteria for inclusion and exclusion in the protocol helps the researcher to focus only on relevant articles. Table 2 listed the inclusion and exclusion criteria for this study.

Table 2. Inclusion and exclusion criteria

Inclusion	Exclusion
Articles published between 2010–2019 as recent publication	Articles published did not within determined period
Journals, conference proceeding, book chapter, thesis	Websites, news, blogs, magazines
Full text available	Unavailable full text
Written in English	Written in other language
Articles related to humanizing, online system, human interaction	Article did not relate to the topic

3.5 Assess Quality

Quality assessment functions were used as a tool to refine the studies that available in full text. It is also the final step before providing the list of final studies for data extraction and synthesis step [7]. Table 3 shows the quality assessment questions used in this study in order to drive the exposition of the results and controlling the specialty of conclusions.

Table 3. Quality assessment questions

No	Questions
QA1	Are the research issues addressed in the paper straight linked to humanizing or human interaction with online system?
QA2	Does the studies mention about factors to humanized online system?

3.6 Extract Data

The data extraction process was carried out by cautiously screening each selected studies. Details for each study were recorded to ensure all selected papers are from high-quality studies. The purpose of this step is to outline extraction of data forms to record precisely the details acquired from the domain studies. The description of data extraction is shown in Table 4.

Table 4. Description of data extraction

Data	Description
Author(s) Name	Name of the authors
Title	Title of the paper
Source	The name of the journal or conference
Publication	Publication issues (volume, page number, issue number)
Publication Year	The publication year of the study
Database	The name of the database
Method	The methodologies applied in the studies
Citation count	The total number of citations of the studies
Issues/Topics	The issues or topic addressed in the studies (Domain area of study)
Theories	Theory used in the paper
Humanizing factors	Humanizing factors mentioned in the studies
Result/findings	Outcome of the studies
Limitations	Weakness of the studies
Future work	Recommendations

3.7 Analyze and Synthesis Data

Upon completion of the data extraction, we start organizing the list based on the selected studies. 63 studies were selected and were revised according to the inclusion and exclusion criteria. This process resulting in 17 studies were removed from the list, and 46 studies were remained. Next, the remaining 46 studies' references were checked thoroughly concerning the inclusion and exclusion criteria, and resulting in new additional of 8 related studies. This makes the total of 54 studies being selected. Afterwards, the assessment quality criteria were concerned where 21 studies were eliminated. Finally, only 33 studies were selected and created the base for the next step in the review.

4 SLR Result

4.1 Publication Sources

Figure 1 shows that most of the publications (between 2010–2019) were published in journals by majority distribution of 79%. This was followed by conference articles with 12%, and book chapter with 6%, while the remaining distribution 3% were graduate thesis.

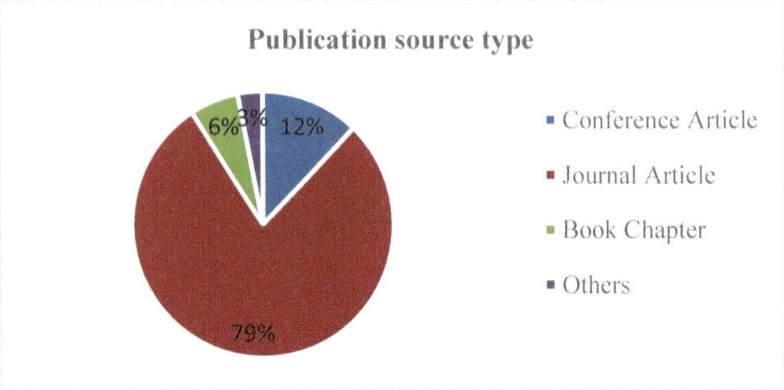

Fig. 1. Publication source type.

4.2 Citation Ranking

Figure 2 shows an overview of the selected studies' citation numbers. The citation numbers were obtained using Google Scholar, ScienceDirect, and ResearchGate. As shown in Fig. 2, three studies have no citation, while most of the selected studies were cited at least once. Nine studies were cited at least thirty-one times.

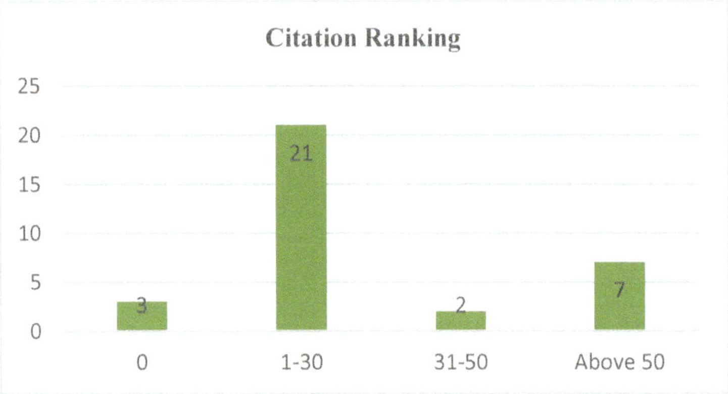

Fig. 2. Citation ranking.

4.3 Research Methodology Used for Humanizing Study

In reference to the methods of research, Fig. 3 shows that nine papers were not mentioned any methodology used in the study. While, qualitative and mixed mode method were used in four studies respectively. The most method used in this domain area is quantitative method with sixteen studies in related area. This concludes that quantitative method is the most suitable methodology to be used in this study.

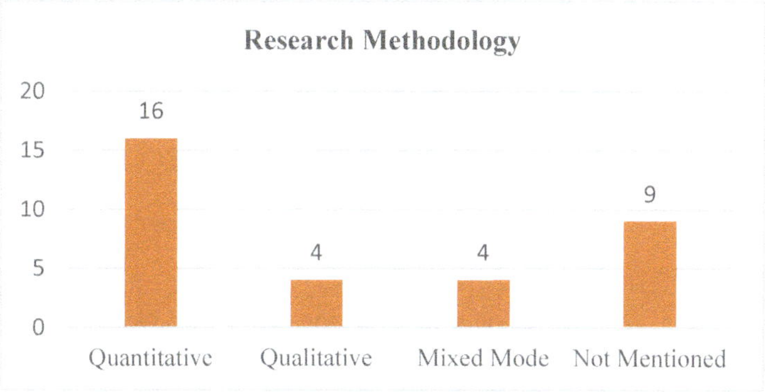

Fig. 3. Research methodology.

4.4 Results of Research Question

RQ1: What are the meanings of humanizing in different area of studies?
There are several studies in different field, where the main aim is to focus on human factors in the area. However, the definition of humanizing itself is very limited. Table 5 shows some meanings of humanizing in different area of studies including education, banking and shopping/retail.

Table 5. Meanings of humanizing

Sector	Meaning
Education	Refers to encourage by integrating voices of learners, engaging students in active knowledge building, promoting emotional connections and making choices available to the students [8]
Banking	Refers to more self-service of the customer as they can reduce the queuing at the bank and in the same time, effectively improves the service efficiency of the business [9]
Shopping/Retail	Describes that companies are trying to create better online experience for the customer while at the same time increasing the online productivity of the company [10]

RQ2: What are the problems in online systems that need to be humanize?
There are several problems found in different sectors. Some sectors have many problems to be addressed in terms of humanizing compared to the other sectors. Table 6 shows problems occur in education, banking and shopping/retail sector.

Table 6. Problems need to be humanized in online system

Sector	Problems need to be humanized
Education	Lack of humanity in the learning process [11], many student lack of online technology familiarity [12], lack of technical expertise among teachers, insufficient online learning orientation, lack of time for teachers to develop and design their online classes [13]
Banking	Marketing management lag, improving risk control, lack of professionals [9], lack of human interaction [14]
Shopping/Retail	High risk of electronic shopping, lack of human interaction, inherent risk to the Internet in terms of security, piracy, and illegal intrusion into the basics of customer data [15], lack of a physical environment [16]

RQ3: What are the humanizing factors in the context of online system that have human interactions?
Different areas of studies contribute various factors to humanize the online system. Thus, the researchers need to filter out the factors by selecting only relevant factors to online system based on their characteristics. Table 7 shows the humanizing factors identified in online systems for different sectors.

Table 7. Humanizing factors in online systems

Sector	Identified humanizing factor
Education	Interaction [1, 11–13, 17–19] Personalization [11, 17, 18, 20]
Banking	Service quality [4, 9] Personalization [14, 21]
Shopping/Retail	Personalization [5, 15, 22, 23] Web design [15, 22] Service quality [5, 15, 16] Interaction [3, 10, 15]

5 Discussion

Result shows that different sectors has different meaning of humanizing and different problems to be humanized. However, in terms of humanizing factors, some sectors share the same factors. Interaction is the factor that identified in education and shopping/retail sector, while service quality is identified as humanizing factor for banking and shopping/retail sector. There is very interesting when personalization exist in all sectors as the humanizing factor.

Based on these results, we can define humanizing online system as emotional interaction between users and online system through self-service system that create better user experience and improve online productivity and service efficiency of the organization. This definition clearly portrays humanizing factors in online systems namely: interaction that describe the interaction between user and organization through online system; personalization that describes the choices provided to the customer in the online system; service quality that describes self-service system which able to answer and responds customer requirement through online system; and lastly, the web design that describes the functionality of the online system and create a better customer experience.

6 Conclusion

This systematic literature review of humanizing online system was carried out for publication published between 2010 until 2019 from various sector. Although a thorough literature review about human interaction online system has been done, to date, there are no studies that directly focus in humanizing online system. Hence, the review is all about human interaction in various fields that relates to humanizing, with intention to produce customer-centered online system in the future. Out of all publications, only 33 studies were met the inclusion criteria for the review. This study also points out research issues for online system in various sectors, gaps, and some important guidance for future work in this area.

References

1. Go, E., ShyamSundar, S.: Humanizing Chatbots: the effects of visual, identity and conversational cues on humanness perceptions. Comput. Hum. Behav. **97**, 304–316 (2019)
2. Macinnis, D.J., Folkes, V.S.: Humanizing brands: when brands seem to be like me, part of me, and in a relationship with me. J. Consum. Psychol. **27**(3), 355–374 (2017)
3. McCool, L.B.: Humanizing advanced communication online writing instruction: developing social presence to communicate, collaborate, and connect. Iowa State University (2016)
4. Ahmad Al-Hawari, M.A.: Does customer sociability matter? Differences in e-quality, e-satisfaction, and e-loyalty between introvert and extravert online banking users. J. Serv. Mark. **28**(7), 538–546 (2014)
5. Etemad-Sajadi, R., Ghachem, L.: The impact of hedonic and utilitarian value of online avatars on e-service quality. Comput. Hum. Behav. **52**, 81–86 (2015)
6. Stanciulescu, G.C., Dumitrescu, F.: Optimizing the IT structures of tourism SMEs using modern applications and resources (cloud). Procedia Econ. Financ. **15**, 1769–1778 (2014)
7. Xiao, Y., Watson, M.: Guidance on conducting a systematic literature review. J. Plann. Educ. Res. **39**(1), 93–112 (2019)
8. AlQudah, A.A.: Models and frameworks for a successful virtual learning environment (VLE) implementation. Am. J. Softw. Eng. Appl. **3**(4), 33 (2014)
9. Yuan, Y.-L.: Analysis on the development of electronic banking business of commercial banks in China. In: International Conference on E-commerce and Contemporary Economic Development, pp. 153–156 (2018)
10. Etemad-Sajadi, R.: The impact of online real-time interactivity on patronage intention: the use of avatars. Comput. Hum. Behav. **61**, 227–232 (2016)
11. Khawla, S., Ayman, A.A.: High-tech human-touch for online courses: rules and principles for humanizing your online course. Eur. Sci. J. **11**(16), 114–133 (2015)
12. West, R.E.: A student's guide to strengthening an online community. TechTrends **54**(5), 69–75 (2010)

13. Keengwe, J., Kidd, T.T.: Towards best practices in online learning and teaching in higher education. J. Online Learn. Teach. **6**(2), 533–541 (2010)
14. Ho, J.S.Y., Chew, K.-W., Khan, N.: Humanizing websites: website personality for E-services. In: IEEE International Conference on Industrial Engineering and Engineering Management (IEEM), pp. 621–625 (2015)
15. Toufaily, E., Pons, F.: Impact of customers' assessment of website attributes on e-relationship in the securities brokerage industry: a multichannel perspective. J. Retail. Consum. Serv. **34**, 58–69 (2017)
16. Hsin Chang, H., Wang, H.: The moderating effect of customer perceived value on online shopping behaviour. Online Inf. Rev. **35**(3), 333–359 (2011)
17. Jowallah, R.: An investigation into the management of online teaching and learning spaces: a case study involving graduate research students. Int. Rev. Res. Open Distrib. Learn. **15**(4), 186–198 (2014)
18. Kop, R.: The challenges to connectivist learning on open online networks: learning experiences during a massive open online course. Int. Rev. Res. Open Distrib. Learn. **12**(3), 19–38 (2011)
19. Kilgore, W., Al-Freih, M., Bartoletti, R.: Design intent and iteration: the #HumanMOOC. In: European MOOC Stakeholder Summit, pp. 7–12 (2015)
20. Rhode, J., Richter, S., Miller, T.: Designing personalized online teaching professional development through self-assessment. TechTrends **61**(5), 444–451 (2017)
21. Ghalenooie, M.B., Sarvestani, H.K.: Evaluating human factors in customer relationship management case study: Private banks of shiraz city. Procedia Economics and Finance **36**, 363–373 (2016)
22. Valvi, A.C., West, D.C.: E-Loyalty is not all about trust, price also matters: extending expectation-confirmation theory in bookselling websites. J. Electron. Commer. Res. (JECR) **14**(1), 99–123 (2013)
23. Fang, Y.-H.: An app a day keeps a customer connected: explicating loyalty to brands and branded applications through the lens of affordance and service-dominant logic. Inf. Manag. **56**(3), 377–391 (2019)

Factors Influence the Intention to Use E-Portfolio in Saudi Technical and Vocational Training Corporation (TVTC) Sector: Pilot Review

Saeed Matar Alshahrani(✉), Hazura Mohamed, Muriati Mukhtar, and Umi Asma' Mokhtar

Faculty of Information Science and Technology, Universiti Kebangsaan Malaysia, 43600 Bangi, Selangor, Malaysia
smatar60@hotmail.com

Abstract. The major issue that institutions of technical and vocational training corporation (TVTC) are currently facing is inefficient management of information. In this context, the application of OBE entails the provision of several portfolios to students and they are used to generate daily information, with records stored as transaction evidence. Studies in literature on e-portfolio system framework in the education sector, and particularly the TVTC, are still few and far between and their extent failed to cover the determination of the factors that influence the system's adoption. Therefore, in this paper, the main significant factors might influence e-portfolio adoption among TVTC institutions were identified, on the basis of which, a framework was developed and proposed in order to facilitate technology adoption and enhance outcomes.

Keywords: e-Portfolio · Education productivity · Adoption · Technical and vocational training · Saudi Arabia

1 Introduction

In the present times technology developments have facilitated easy information access among students and this can positively affect their education process. In addition, technology can be employed to access information needed to enhance the education process Laird and Kuh [1, 2]. In fact, in the running of any educational institutions, education management constantly need to make informed decisions regarding issues in education and this is often facilitated via conjectural point of view and limited information. However, the issues to be tackled are significant to the field of education and thus, it becomes a necessity to carry out a critical evaluation of required information [3–7].

In the TVTC institutions context, information is considered as a significant resource that brings about and enhances the competency, performance evaluation and effective performance of the institutions [8]. Educational institutions all over the globe find the importance of e-portfolio system implementation and usage.

F. Saeed et al. (Eds.): IRICT 2019, AISC 1073, pp. 1009–1019, 2020.
https://doi.org/10.1007/978-3-030-33582-3_95

Additionally, the exponential continuous growth of information produced by electronic media along with the development of technological innovations is dynamically transforming the running of organizations, institutions and governments. This is evidenced by the technological developments that have already transformed records/archives storage process. Also, information and communication technologies (ICT) have led to other different ways that e-records can be generated via automated offices and digitalized instruments [9].

This is why, presently, institutions of higher education, such as TVTC largely depend on electronic tools to process and store data and the decision makers leverage new tools and strategies in order to convert data into information that can be invaluable for tackling and addressing institutional issues. Such decision makers employ technology in the form of software to test the institution's potential and to mitigate the possible risks [10, 11].

In the same line of argument, lack of relevant information will stump decision-makers and managers when it comes to planning and deciding, which indicates that information is the element that allows rational and informed decision making at the organizational level, and in turn, deliver high service quality that can improve the outcomes of the educational institutions and learning [12]. In relation to this, portfolios store all the required information for any kind of purposes by educational institutions. Ineffectively managed information will only lead to uninformed and inaccurate decisions. And for this reason, accurate decision-making can be brought about by information technology (IT), enabling the efficient and effective management of significant amounts of documents [13].

Past studies on the topic generally ignored technologies usage, like e-portfolio in the field of education [14–16], indicating that schools and educational institutions have been ignored when examining the relationship between computers, skills of communication, teaching experience and technologies that are effectively used in overcoming the challenges of tools integration. Moreover, past studies focused on the way schools determine how technology can be utilized for the improvement of the personal development of students and their educational outcomes through established strategic policies. This caliber of studies would facilitate the development of technologies that could meet the needs of educational institutions and their members and enable them to experience transformation and changes in technology. More importantly, the right technology and support can ensure that the user is using the technology effectively [3].

In the TVTC institutions context, information production is carried out periodically and this is why the available information can be the failure/success of the organization, highlighting the need to manage information and leveraging it to its optimal heights [17]. In e-portfolios information generated in TVTCs can be used as guidance by management in directing performance and provision of quality. Also, information should be effectively managed using a framework to ensure the feasibility of the whole process. Stated clearly, TVTC system needs robust framework for information management, and for decisions assessment, evaluation in order to establish educational relevancy and veracity [4, 5].

In information management system, Mohammed, Ibrahim [18], Mohammed, Alzahrani [19] and Almazroi, Shen [20] laid emphasis on new factors like cloud environment and big data. The system's low level of adoption could be detrimental to its implementation [21, 22], which means the factors that influence e-portfolios adoption have to be identified, especially due to the fact that works dedicated to the topic of e-portfolio effect on TVTC organizations performance and educational outcomes are still scarce. With the awareness and addressing of the issue, a framework can be developed to guide successful e-portfolio adoption among TVTCs.

2 The Importance of Electronic Portfolio for TVTC

The main advantage of e-portfolio is its provision of a scalable and comprehensive process for personal progress documentation towards the goals and objectives of the organization, evaluation of the business operations effectiveness, the projects oversight, learning support, professional development, public companies auditing and participation in assessments. A fully-developed e-portfolio offers several benefits as different users' types are addressing with different aims [23].

On the basis of the viewpoint of students, the academic e-portfolio development facilitates management of self-learning and learning autonomy while promoting evaluation with the guidance of the instructor. More importantly, e-portfolios enable self-learning regulation and learning participation among students. Generally, it is employed as a system of evaluation which is included in the process of teaching and learning [24].

Based on the above, an e-portfolio is neither an online course nor a content management system, but rather a personal student's tool that is user-centered. In this regard, users are generally the only actors involved but there are institutions, responsible for certifying the users' achievements that use the tool, along with tutors and instructors. However, all in all, it is the user/owner that has the last word when it comes to interaction levels enabling the two external actors in.

Furthermore, e-portfolios can be effectively utilized in all education levels, with the inclusion of young students (e.g., pre-schools), where parents out-of-class communication has led to enhanced process of learning [23].

In a related study, Yasin, Rahman [25] described learning portfolio as being characterized by flexibility, evidence-based process, which integrates reflection with documentation, and engages students in a continuous, reflective and collaborative learning analysis. Focus is built on selective results that improve and assesses learning. In another study, Barrett [26] argued that the learner's growth and development in a certain time can be assessed using a series of e-portfolio processes like reflection, setting of goals and conducting self-assessments. The author further explained students need to have the academic autonomy to include other works that presents their overall achievements and abilities in the best light.

Generally speaking all e-portfolios are digital storage, requiring file uploads or form-based templates, but there are four major characteristics, defining the tool's purpose that have to be taken into consideration. The four types can be present in a single program but some may dominate over others and they are assessment and evaluation, student learning, professional development and accreditation [27].

Aside from the above four major characteristics, the tools may also be categorized into four main types based on their software systems and they are; (1) E-portfolios

Management systems, which are designed and developed for institutional e-portfolio systems; (2) Learning Management Systems (LMS)/Virtual Classroom, which are capabilities providing e-portfolios or offering modules; (3) Integrated Systems, exemplified by Content Management System (CMS), providing the functions of e-portfolio and lastly; (4) other systems like the ERMS.

Nevertheless, in the case of vocational and education training, only a few countries have managed to implement and use the application to their e-learning education support initiative as evidenced by Altahawi, Sisk [28], Cameron [29], Fernández, Sanz [30], and Matsom, Duggan [31]. More specifically, e-portfolios are considered to be a different learning and assessment alternative that are attractive to the vocational instructor owing to their facilitation of skills assessment and performance assessment as opposed to merely recalling memorized facts [32]. Added to this, e-portfolios meet the needs of business and industry by linking classroom activities with actual life activities. But the successful achievement of expected outcomes hinges mainly on the aims, practices, and structures guiding the establishment of new learning and assessment form in vocational institutions.

In a related quantitative study, Turhan and Demirli [32] gathered the perceptions of vocational education teachers and students in light of the e-portfolio use in different countries, including the UK, Denmark, Romania, and Turkey using questionnaires. The authors found that teachers and students perceive e-portfolio process as significant in the vocational department as part of the process of learning, particularly to gather evidence, monitor progress in learning, assessment of knowledge and skills, and measuring the reflection of the students on his/her lessons.

3 Impact of the Research

Added to the present work's contribution to practice in providing government authorities of TVTC institutions in Saudi with the suggestions that could help in the application of e-portfolio plan that could enhance the educational outcomes, it also contributes to theory.

Literature still lacks e-portfolio adoption frameworks aimed at improving educational outcomes and this study contributes by adopting different approach from the extant technology adoption studies and education studies to develop an educational framework.

The study empirically examines the variables influencing e-portfolio adoption in TVTC context to provide deeper insight into the ability of the system to improve the outcomes of such institutions.

Because of the lack of studies of this caliber in developing nations, particularly in Saudi Arabia at the time the research is conducted, the study contributes enriching information the education field, which could guide e-portfolio implementation success.

4 Methodology

The first step in the methodology process is to determine the significant variables to the assessment of behavioral intention towards e-portfolio usage among Saudi TVTCs, after which, the factors are forwarded to experts in the field.

Figure 1 shows the study methodology comprising of four stages, which are; carrying out a thorough review of literature and identifying the factors, conducting experts' consultancy concerning the factors of the e-portfolio, and highlighting the significant factors, upon which the framework of the study are built and developed on.

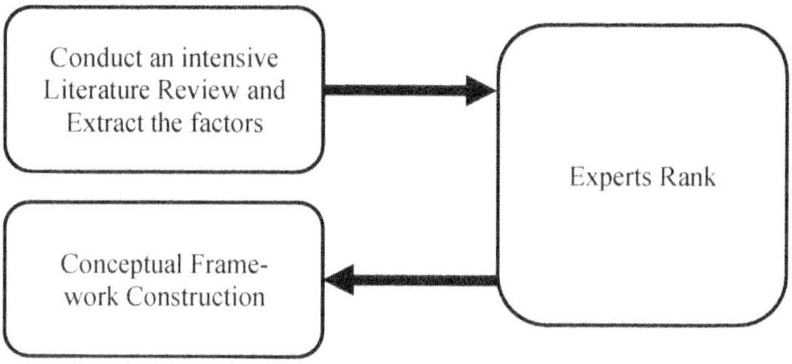

Fig. 1. Methodology of the study as adopted from [7]

4.1 Content Analysis Literature Review for Factor Extraction

This paper conducted an analysis of literature using the terms "e-portfolio adoption factors", "factors for technology adoption in education", and "e-portfolio and education productivity". Basically, the study conducted a review of the studies concerning e-portfolio to identify the factors mentioned by the authors. This was followed by the grouping and classification of factors into dimensions, after which, they were forwarded to the experts for review. There were 48 factors cited by the authors, but the study limited the selection of the top cited ones when it came to e-portfolio and technology adoption in the field of education (Table 1).

More specifically, the cited factors was confined to 22 different factors namely, perceived ease of use, perceived usefulness, trust, compatibility, privacy, system quality, service quality, information quality, top management support, social influence, facilitating conditions, financial support, training, government role, cloud computing ability, big data facility, change management, effective communication, IT infrastructure, interactivity, responsiveness, efficiency and effort expectancy.

4.2 Experts' Consultation and Factor Classification

Questionnaire copies were administered to the respondents regarding the e-portfolio in the education field and the list of factors affecting the use of e-portfolio were sent to the experts (Lecturers who use e-portfolio) for their comments and feedback. Eleven (11) factors were highlighted as the most significant when it comes to behavioural

Table 1. Cross reference factor exacted for content analysis literature review

Dimension	Factors	Total
Technology	Features used, trust, compatibility, security, information technology challenges, effort expectancy, personal normative, belief, self-identity, perceived consequences, system quality, perceived ease of use, perceived usefulness, service quality, information quality, reliability, technological readiness, privacy, efficiency, reliability, it infrastructure, interactivity, responsiveness, efficiency, effort expectancy	26
Organization	Human resources, motivation, resources available, outsourcing, social influence, facilitating conditions, training, perceived financial cost, information need, competition, top management support, standardization, change management, effective communication	14
Environment	Competitiveness pressure, clear vision and planning, laws and legislations, policy, government role, security concerns, cloud computing ability, big data facility	8
Total		48

intention towards using e-portfolio, and ultimately, actual use. The determination of the significance of factors followed the recommendation by [33] and [34] (Table 2).

Table 2. Factors as ranked by experts

No.	Underlying factors	Dimension	Reference
1	Perceived ease of uses	Technology	[35]
2	Perceived usefulness		[35]
3	System quality		[36, 37]
4	Service quality		[36, 38]
5	Information quality		[36, 37, 39]
6	Top management support	Organization	[38, 40]
7	Financial support		[38, 40]
8	Training		[40, 41]
9	Government role	Environment	[42]
10	Cloud computing ability		[18]
11	Big data facility		[43]

4.3 Framework Construction

To conduct any study, it is pertinent to examine and determine the underpinning theories/models to the topic under focus in order so that they may be used to guide the researcher in generating a premise of the relationships among the constructs when developing the framework. In case of e-portfolio adoption, the level of adoption can be increased if the adoption determinants are identified and examined. Therefore, several theories and models adopted or extended by prior studies in literature (e.g., [44–46]) were considered in this study to examine the technology adoption in institutions or end-users. The top theories reviewed were Technology Acceptance Model (TAM), Unified Theory of Acceptance and Use of Technology (UTAUT), Theory of Planned Behavior (TPB), Diffusion of Innovation (DOI) theory, and Technology-Organization-Environment (TOE) Framework.

This study proposes an e-portfolio framework by identifying five interrelated variables, which are, technological dimensions, organizational dimensions, environmental dimensions, e-portfolio adoption intention and outcome performance of educational institutions as per use as presented in Fig. 2. The variables are analyzed and clustered as technology adoption factors upon the framework platform.

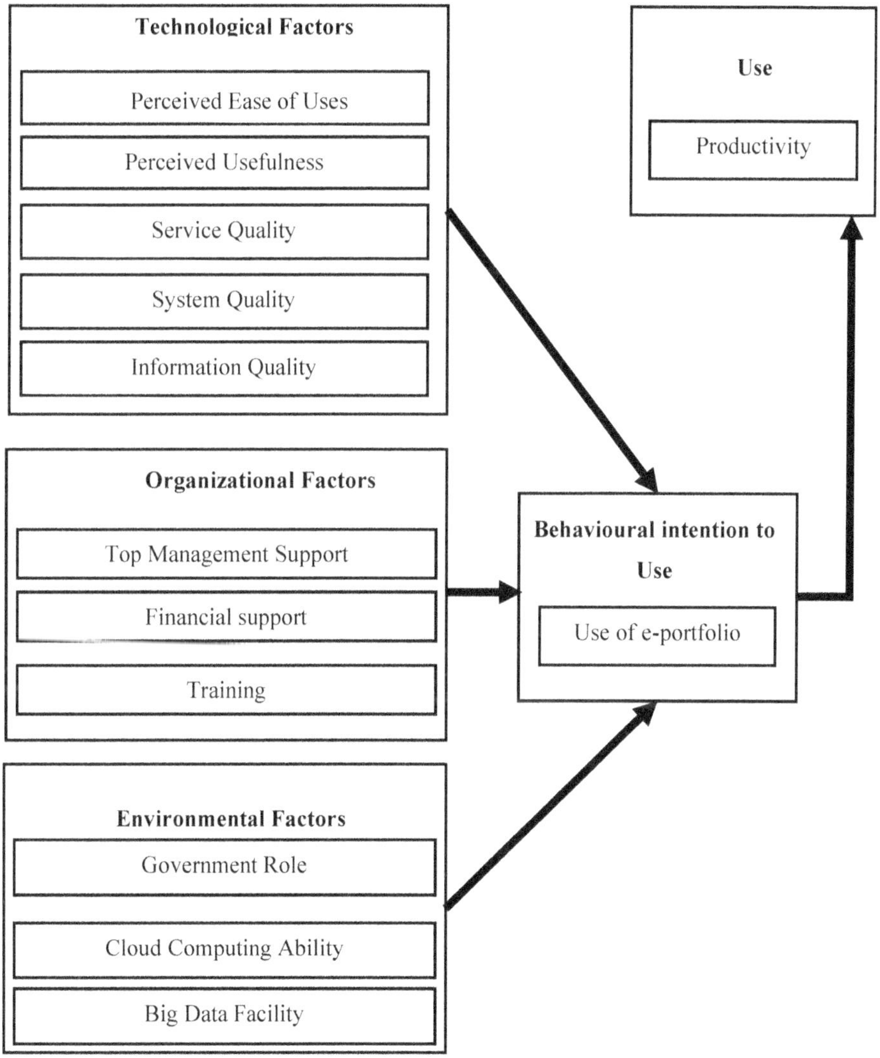

Fig. 2. The proposed model of the study

In relation to the above, Al-Gahtani [47] conducted an empirical study in the Saudi context that was dedicated to the e-learning acceptance and assimilation in an academic environment with the help of TAM3. The obtained findings supported a significant influence of image, perceived ease of use, job relevance and subjective norm on perceived usefulness, a significant influence of subjective norm on image, a significant influence of computer self-efficacy, external control perceptions, computer anxiety and perceived enjoyment on perceived ease of use, and finally, a significant influence of perceived usefulness, perceived ease of use and subjective norm on intention towards using e-learning system.

The review of the technology adoption theories and models were directed towards selecting the appropriate one among them that can achieve the study aims. It is also important to determine the top extensively used model in literature, specifically those that were proposed for the field of education such as TAM, TOE, UTAUT and DOI [44, 45].

On the basis of the TAM3 validation of the reviewed literature and its extensive use, appropriateness, validity and reliability in investigating technology adoption in various contexts (e.g., [35, 47–49]), this study adopted the model to examine the factors affecting the adoption of e-portfolio as suggested by Abdullah, Ward [46] in the context of TVTC in Saudi Arabia. The major features included in TAM3 are technological differences, characteristics of the organization and environmental settings and they are all considered to be determinants of adoption behavior towards e-portfolio in Saudi TVTC institutions. The model is suitable to be adopted as the underpinning theory based on the study aim and context.

5 Conclusion

This work highlighted the scarcity of studies on e-portfolio usage in literature and the important role it plays in lending support and enhancement to the organization/institution performance. It laid stress on the limitations of extant studies dedicated to examining e-portfolio usage, and to this end, it constructed and proposed a framework for such use. Owing to the e-portfolio importance to organizations/institutions, a robust model had to be proposed for its implementation success. Accordingly, the study conducted a thorough review of literature that supported a relationship between e-portfolio implementation and guidance using a study model.

Prior works were reviewed, particularly concerning the factors in e-portfolio, after which the factors were extracted and backed by experts. The factors were then classified into three dimensions namely, technological, organizational and environmental dimensions on the basis of their level of interaction. This work contributes to literature by determining the factors influencing behavioral intention towards e-portfolio usage and it contributes to practice by guiding the optimization of scarce resources of management on the key areas that would lend a hand to the process of successful implementation.

References

1. Laird, T.F.N., Kuh, G.D.: Student experiences with information technology and their relationship to other aspects of student engagement. Res. High. Educ. **46**(2), 211–233 (2005)
2. Turney, C., et al.: Using technology to direct learning in higher education: the way forward? Act. Learn. High Educ. **10**(1), 71–83 (2009)
3. Merigó Lindahl, J.M., López-Jurado, M.P., Gracia Ramos, M.C.: A decision making method for educational management based on distance measures. Revista de Métodos Cuantitativos para la Economía y la Empresa **8**, 29–49 (2009)
4. Mukred, M., Yusof, Z.M.: Electronic records management and its importance for decision making process in Yemeni higher professional education (HPE): a preliminary review. In: 1st International Conference of Recent Trends in Information and Communication Technologies (IRICT 2014), Johor Bahru, pp. 105–114 (2014)
5. Mukred, M., Yusof, Z.M.: The role of electronic records management (ERM) for supporting Decision making process in yemeni higher professional education (HPE): a preliminary review. Jurnal Teknologi **73**(2), 117–122 (2015)
6. Mukred, M., Yusof, Z.M.: Factors influencing the adoption of electronic records management (ERM) for decision making process at higher professional education (HPE)'s institutions. In: 1st ICRIL-International Conference on Innovation in Science and Technology (IICIST 2015), Kuala Lumpur, pp. 399–403 (2015)
7. Mukred, M., et al.: Electronic records management system adoption readiness framework for higher professional education institutions in Yemen. Int. J. Adv. Sci. Eng. Inf. Technol. **6**(6), 804–811 (2016)
8. Mukred, M., Yusof, Z.M.: The performance of educational institutions through the electronic records management systems: factors influencing electronic records management system adoption. Int. J. Inf. Technol. Proj. Manag. (IJITPM) **9**(3), 34–51 (2018)
9. Eze Asogwa, B.: The readiness of universities in managing electronic records: a study of three federal universities in Nigeria. Electron. Libr. **31**(6), 792–807 (2013)
10. Şuşnea, E.: Improving decision making process in universities: a conceptual model of intelligent decision support system. Procedia-Soc. Behav. Sci. **76**, 795–800 (2013)
11. Samsudin, A., Miah, S., McGrath, G.: An ontology-based record management systems approach for enhancing decision support (2014)
12. Atulomah, B.C.: Perceived records management practice and decision making among university administrators in Nigeria (2011)
13. Eusoff, R., Yusof, Z.M.: Development of records management system for matriculation colleges in Malaysia. Asia-Pac. J. Inf. Technol. Multimedia **11**(1) (2011)
14. Gröger, S., Decker, J., Schumann, M.: Do universities get the hang of working efficiently?–a survey of the influencing factors on the adoption of electronic document and workflow management in German-speaking countries (2014)
15. Tzeng, W.-S., et al.: Do ePortfolios contribute to learners' reflective thinking activities?: a preliminary study of nursing staff users. J. Med. Syst. **39**(9), 1–10 (2015)
16. Alharthi, E.F.J.: Teacher evaluation in the Kingdom of Saudi Arabia's (KSA) schools-moving forward. University of Southampton (2017)
17. Hase, S., Galt, J.: Records management myopia: a case study. Rec. Manag. J. **21**(1), 36–45 (2011)
18. Mohammed, F., Ibrahim, O., Ithnin, N.: Factors influencing cloud computing adoption for e-government implementation in developing countries: instrument development. J. Syst. Inf. Technol. **18**(3), 297–327 (2016)

19. Mohammed, F., et al.: Cloud computing fitness for E-Government implementation: importance-performance Analysis. IEEE access **6**, 1236–1248 (2017)
20. Almazroi, A.A., Shen, H., Mohammed, F.: The impact of trust on the adoption of cloud computing services by university students. In: International Conference of Reliable Information and Communication Technology. Springer (2018)
21. Moseti, I.: Digital preservation and institutional repositories: case study of universities in Kenya. J. S. Afr. Soc. Arch. **49**, 137–154 (2016)
22. Mosweu, O., Bwalya, K.J., Mutshewa, A.: A probe into the factors for adoption and usage of electronic document and records management systems in the Botswana context. Inf. Dev. **33** (1), 97–110 (2017)
23. Buzzetto-More, N.: Understanding e-portfolios and their applications. The E-Portfolio paradigm: Informing, educating, assessing and Managing with E-Portfolios. Santa Rosa/CA, pp. 1–17 (2010)
24. Amaya, P., et al.: Educational e-portfolios: uses and tools. Procedia-Soc. Behav. Sci. **93**, 1169–1173 (2013)
25. Yasin, R.M., Rahman, S., Ahmad, A.R.: Framework for reflective learning using portfolios in pre-service teacher training. Procedia-Soc. Behav. Sci. **46**, 3837–3841 (2012)
26. Barrett, B.: Using e-portfolios to evaluate intellectual capital of online learners. In: Proceedings of the 2nd European Conference on Intellectual Capital: ISCTE (2010)
27. Cheal, C.: Implications of open source E-Portfolios. The E-Portfolio Paradigm: Informing, Educating, Assessing, and Managing with E-Portfolios, p. 35 (2010)
28. Altahawi, F., et al.: Student perspectives on assessment: experience in a competency-based portfolio system. Med. Teach. **34**(3), 221–225 (2012)
29. Cameron, R.: Recognising workplace learning: the emerging practices of e-RPL and e-PR. J. Workplace Learn. **24**(2), 85–104 (2012)
30. Fernández, L.M.S., Sanz, C.C., Pérez, M.A.N.: The presence of ICT in vocational guidance: a training proposal. Problems of Education in the 21st Century, vol. 44 (2012)
31. Matsom, H., et al.: E-PORTFOLIO development and implementation in Malaysian technical and vocational education training (TVET): a mixed methods analysis of Stakeholders' and students' perceptions. Int. J. Arts Sci. **8**(1), 243 (2015)
32. Turhan, M., Demirli, C.: The study on electronic portfolios in vocational education: the views of teachers and students in United Kingdom, Denmark, Romania and Turkey. Sci. Res. Essays **5**(11), 1376–1383 (2010)
33. Hawking, P., Sellitto, C.: Business intelligence (BI) critical success factors. In: 21st Australian Conference on Information Systems (2010)
34. Ahmad, M.M., Cuenca, R.P.: Critical success factors for ERP implementation in SMEs. Robot. Comput.-Integr. Manuf. **29**(3), 104–111 (2013)
35. Venkatesh, V., Bala, H.: Technology acceptance model 3 and a research agenda on interventions. Decis. Sci. **39**(2), 273–315 (2008)
36. Delone, W.H., McLean, E.R.: The DeLone and McLean model of information systems success: a ten-year update. J. Manag. Inf. Syst. **19**(4), 9–30 (2003)
37. Mukred, M., Yusof, Z.M.: The DeLone–McLean information system success model for electronic records management system adoption in higher professional education institutions of Yemen. In: International Conference of Reliable Information and Communication Technology. Springer (2017)
38. Mukred, M., et al.: Taxonomic framework for factors influencing ERMS adoption in organisations of higher professional education. J. Inf. Sci. **45**(2), 139–155 (2018)

39. Mukred, M., et al.: A framework for electronic records management system adoption in the higher professional education: individual, technological and environmental factors. In: International Conference of Reliable Information and Communication Technology. Springer (2018)
40. Mohammed Aldossari, A.M.Z.: The use of automation and robotic innovations in the transformational companies: systematic literature review J. Theor. Appl. Inf. Technol. **97** (13), 3661–3690 (2019)
41. Alsharari, J.: The perceived training needs of female head teachers compared with the training needs of male head teachers in the government public schools in Saudi Arabia, Durham University (2010)
42. Mukred, M., et al.: The key factors in adopting an electronic records management system (ERMS) in the educational sector: a UTAUT-based framework. IEEE Access **7**, 35963–35980 (2019)
43. Chen, H., Chiang, R.H., Storey, V.C.: Business intelligence and analytics: from big data to big impact. MIS Q. **36**, 1165–1188 (2012)
44. Al-Jabri, I.M.: The perceptions of adopters and non-adopters of cloud computing: application of technology-organization-environment framework (2014)
45. Alharbi, S.T.: Users' acceptance of cloud computing in Saudi Arabia: an extension of technology acceptance model. Int. J. Cloud Appl. Comput. (IJCAC) **2**(2), 1–11 (2012)
46. Abdullah, F., Ward, R., Ahmed, E.: Investigating the influence of the most commonly used external variables of TAM on students' Perceived Ease of Use (PEOU) and Perceived Usefulness (PU) of e-portfolios. Comput. Hum. Behav. **63**, 75–90 (2016)
47. Al-Gahtani, S.S.: Empirical investigation of e-learning acceptance and assimilation: a structural equation model. Appl. Comput. Inf. **12**(1), 27–50 (2016)
48. Agudo-Peregrina, Á.F., Hernández-García, Á., Pascual-Miguel, F.J.: Behavioral intention, use behavior and the acceptance of electronic learning systems: differences between higher education and lifelong learning. Comput. Hum. Behav. **34**, 301–314 (2014)
49. Faqih, K.M., Jaradat, M.-I.R.M.: Assessing the moderating effect of gender differences and individualism-collectivism at individual-level on the adoption of mobile commerce technology: TAM3 perspective. J. Retail. Consum. Serv. **22**, 37–52 (2015)

Design and Development of Knowledge Management System in the Small and Medium-Scale Enterprises Base on Mobile Apps (SMEs at Indonesia)

Junita Juwita Siregar[1(✉)] and R. A. Aryanti Wardaya Puspokusumo[2]

[1] Computer Science Department, School of Computer Science, Bina Nusantara University, Jakarta 11480, Indonesia
juwita_siregar@binus.ac.id

[2] International Marketing Program, Management Department, BINUS Business School, Undergraduate Program, Bina Nusantara University, Jakarta 11480, Indonesia
rpuspokusumo@binus.edu

Abstract. Indonesia as a country has the potential to build a creative industry in the Asia region and the world market. However, technology and human resources in Indonesia and creative economy development are still constrained by the limited quantity and quality of business-owner skills in the creative sector and also operations and business management. This impacts the development of the creative industry which is currently sluggish. The aim of this research is to analyze the problems in building knowledge management technology based on a mobile app for SMEs. Such applications are aimed at being able to capture, regulate, store, and spread knowledge with practitioners and disseminate existing SMEs knowledge. The SEM method is applied in inferential analysis to process data obtained through surveys and questionnaires with a total of 204 samples conducted in the city of JABODETABEK. Research and development and software development methodologies were used to develop a knowledge management system for SMEs. Based on the results of an investigation of benefits (perceive usefulness), obtained from the use of knowledge management technology in terms of uploading knowledge in the form of videos, discussion forums, and articles on SMEs will influence the perceived behavioral intention to use (ITU) an app application for knowledge management, thus giving rise to increased usage of knowledge management technology, thus giving rise to a positive attitude towards usage (ATU) of knowledge management technology.

Keywords: Knowledge management · SMEs · Mobile application · SEM

1 Introduction

The economic development of creative sector enterprises in Indonesia is a strategic goal from various aspects of human resources, cultural diversity, and the large domestic market [16]. According to data from the Ministry of Trade, in 2009 the number of SMEs reached 48.9 million business units. Meanwhile, persons in the workforce

F. Saeed et al. (Eds.): IRICT 2019, AISC 1073, pp. 1020–1030, 2020.
https://doi.org/10.1007/978-3-030-33582-3_96

reached 91.8 million people [4]. As such, SMEs as an aggregate are one of the largest contributors to GDP in Indonesia. From 2007 to 2012, there has been an increase in the number of SMEs contributing to GDP in the amounts of Rp. 2,107,868.10 to Rp. 4,869,568.10 billion [16]. While there has been robust growth, SMEs in Indonesia are still weak and undermanaged in several aspects, with limited access to financial institutions, especially banks [16]. SMEs development is limited by several factors, including mastery of technology and entrepreneurial capabilities. Limited business facilities and infrastructure further make it difficult to access information related to the advancement of science and technology. The uneven quality, quantity, suitability, and distribution of formal and non-formal educational institutions has caused the quality of human resources to be far behind those of countries with mature creative economies [16]. Going forward, SMEs must be able to act innovatively and think creatively. At present, the most important asset of such an industry is knowledge. Nonaka and Takeuchi [5] showed that new knowledge arises because of a continuous interaction between explicit and tacit knowledge. Continuous interaction includes four stages that form the SECI[5] model: socialization (tacit to explicit), externalization (tacit to explicit), combination (explicit to expliciti), and internalization (explicit to tacit).

However, most SMEs in Indonesia do not implement knowledge management systems that are commonly applied in the modern business world. Accordingly, a knowledge management system (KMS) [13] is needed to facilitate the capture, storage, search, transfer and reuse of information or documentation in the field of small and medium enterprises (SMEs).

Indonesian SMEs need to increase their competitive advantage through the implementation of knowledge management. Such efforts to implement a knowledge management system (KMS) [2] can be supported by web or mobile application technology.

2 Literature Review

2.1 Creative Industry

There are several definitions of the creative industry put forward by various experts. The Ministry of Commerce of the Republic of Indonesia defines the creative industry as enterprises that utilize creativity, skills and talents of individuals to provide welfare and employment by generating and exploiting an individual's creativity and creative power [4].

2.2 Knowledge

Knowledge is closely related to doing and implies information and understanding. The products of experience that encompass norms used to evaluate new facts from the environment form the knowledge of each individual. Knowledge [12] is an ability to utilize information that enables action and decisions or information with direction. Knowledge can be thought of as in two forms: tacit and explicit [10]. Tacit knowledge [5] refers to knowledge that is based on personal experience and activities and tends to be personal, featuring insight, intuition, and expertise which is difficult to transfer into a formal form. This knowledge is difficult to express and formalize and hence is difficult to distribute.

Explicit knowledge [10] refers to knowledge that is in the form of data, images, audio and video, manuals, or computer programs that can be formally shared, easily collected and systematically accessed. This is often referred to as 'knowing that', or declarative knowledge.

2.3 Knowledge Management

Knowledge management (KM) can be defined as a process within an organization that coordinates activities to retrieve, use, share, create and store individual and group knowledge and expertise [12]. The knowledge management process has several stages, namely: knowledge discovery, knowledge capture, knowledge sharing, and knowledge application [13].

2.4 Knowledge Management System

A knowledge management system (KMS) utilizes information-technology based systems to support and enhance the organizational processes of knowledge creation, storage, transfer, and application [2]. A KMS facilitates a cycle that captures the process of gathering information up and eventual leveraging and sharing of such information [13] (Fig. 1).

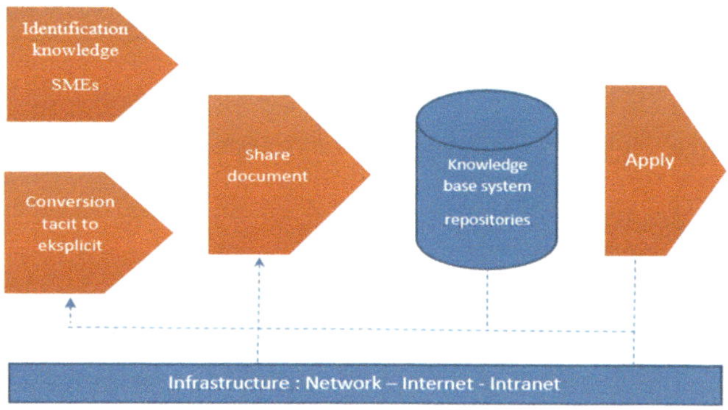

Fig. 1. Knowledge management role [10] of SMEs

2.5 SECI Model

According to Nonaka [5], new knowledge arises because of the continuous interaction between explicit (actualized) and tacit (hidden) knowledge. This continuous interaction includes four stages: socialization (tacit to tacit), externalization (tacit to explicit) [14], combination (explicit to tacit), and internalization (explicit to tacit). These interactions form the socialization, externalization, combination, internalization (SECI) [5] model.

2.6 Knowledge Management Technologies

Knowledge management technology [2] is used to primarily facilitate communication, collaboration, and content management for better knowledge capture, sharing, dissemination, and application [12, 17]. Knowledge management technologies are information technologies that can be used to facilitate knowledge management [10]. KM technologies constitute a key component of KM systems [2] (Fig. 2).

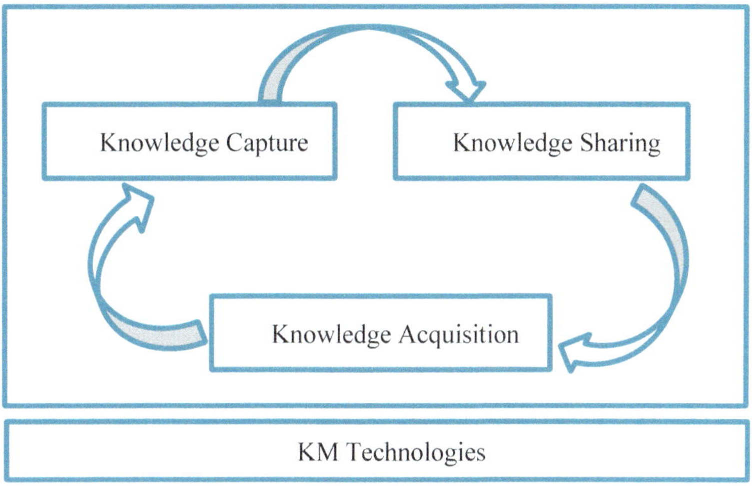

Fig. 2. Technology components of KM.

3 Research Methodology

The current research pursued research and development (R&D) with an explanatory research approach.

3.1 Data Collection Method

Survey questionnaires were distributed and collected from creative-industry enterprises and businesses and were sent directly to participants who responded through an online survey phone and by direct Whatsapp to provide a total of 204 samples. The city of JABODETABEK in Indonesia were chosen as the research. location. The SEM [1] method in inferential analysis was used to process the data thru a random sampling technique.

3.2 Design Methodology

A software engineering development methodology was adopted to create a knowledge management app thru an object-oriented analysis design with a waterfall [8] meothodology.

3.3 Problem Analysis

Analysis looking at the problem with knowledge management for SMEs used a fishbone [8] methodology. At this stage, the analysis of user needs in terms of technology in knowledge management systems showed variable benefits perceived usefulness (PU) [11], perceived behavioral intention to use (ITU), attitude towards usage (ATU), perceived ease of use (PEOU) [11] and Mobile Usage (MU). There was a relationship between these various components. We also looked at the relationship between culture in terms of a reliance on knowledge sharing, and the problem of resource expertise and limited infrastructure (Fig. 3).

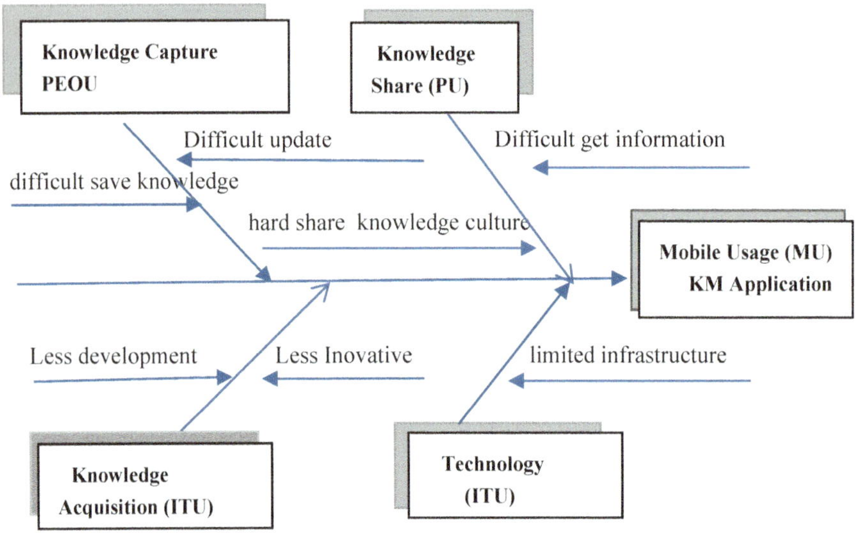

Fig. 3. Fishbone methodology of the SMEs

3.4 Hypotheses Research

According to the problem analysis, this study tested the following hypotheses:

H1: PEOU as an exogenous variable is expected PU variable.
H2: PU as an endogenous variable is expected ATU variable.
H3: PU as an endogenous variable, is expected ITU variable.
H4: ATU as an endogenous variable, is expected MU variable.
H5: ITU as an endogenous variable, is expected MU variable.

3.5 Previews Research

This section discusses some of the issues related to knowledge management practice.

Kurniawati [3] *et al.* analyzed the implementation of knowledge management in SMEs in Indonesia by using indicators of knowledge creation, knowledge acquisition, storage, identification and retrieval knowledge.

Nousala [9] stated that to actively, simultaneously and continuous support sustainable development for SMEs organizations it is necessary to identify the value of the structure of SMEs elements of the tacit knowledge network.

Abdullah [15] *et al.*, validated concepts and models for a mobile computing system named mKaasS, that can be used to facilitate knowledge sharing of services and to utilize an organization's knowledge services for best learning practices, in particular related to environment.

While previous research has begun looking at knowledge sharing for SMEs organizations, there is limited work detailing management processes, in particular for capturing knowledge, sharing processes and the process of applying knowledge. More specifically, our work aims at developing android-based knowledge management application technology that provides discussion forums, article sharing functions and video sharing to support the process management.

4 Result and Discussion

4.1 Data Analysis

This research used AMOS software to test and analyze a development which was done by applying the technology acceptance model (TAM) [11] in design and development of a mobile app for SME knowledge management [13]. The AMOS software was used for analysis of statistical inferentials of the relationship among PEOU, ITU, ATU, PU [11] and Mobile Usage variables (MU).

Data was collected and interpreted using AMOS software, as shown in (Fig. 4), with the following results: a chi-square [1] value of sig ($p = 0.009$) means that the test model is fit. The RMSEA [7] value obtained = 0.049, which is < 0.08 showing that this model is also fit. The obtained PNFI [7] value = 0.724 also produces a fit model. The overall SEM [6] model is fit namely suitability between the model and the data.

4.2 Design Analysis

The design of a SME knowledge management application system adopted an OOAD [8] approach which includes analysis and design of a system with an object approach, utilizing object-oriented analysis (OOA) [8]. System design was done with a UML [8] tool system to provide a Business Use Case, Sequence Diagram, Activity Diagram and Class Diagram. This stage also produced a document called Software Requirements (Fig. 5).

4.3 Design User Interface

In this stage the program design was translated into code using the java programming language.

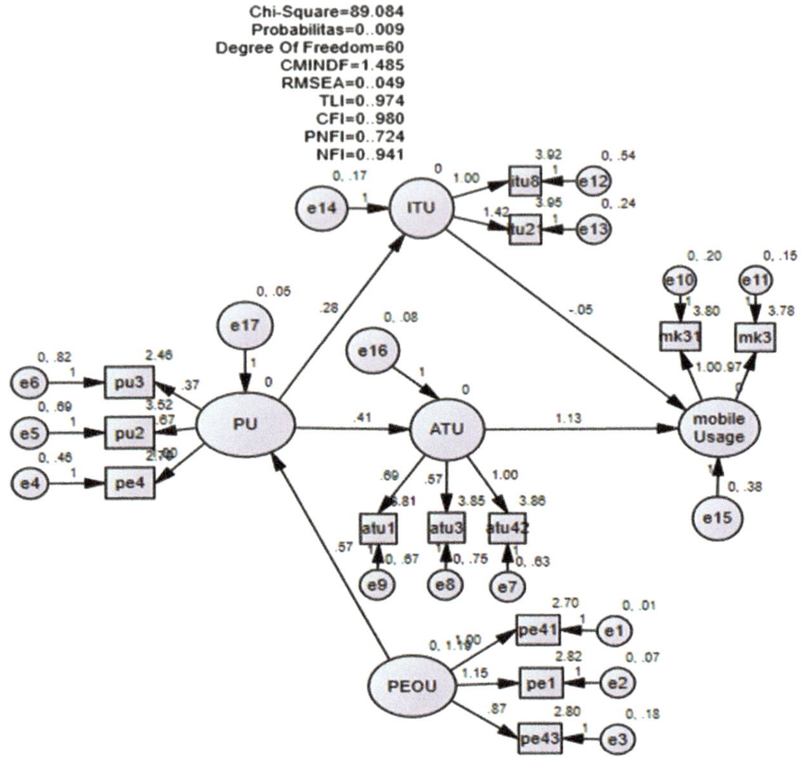

Fig. 4. TAM application knowledge management for SME model. Reference: Siregar, J.J., ICEAST 2018

4.3.1 Main Page

On this page, users who have become SMEs members can use any menu function available in the SMEs web knowledge management platform and interact with fellow members (Fig. 6).

4.3.2 Video Tutorial Page

On this page, SMEs members can manage the process of capturing, sharing and applying knowledge, where knowledge is reflected in the form of experience, expertise and experience, which can be communicated through uploaded video tutorials (Fig. 7).

4.3.3 Forum Page

On this page, the knowledge creation process known as socialization, externalization, combination and internalization (SECI) is possible with fellow SMEs, namely:

1. Socialization [5] which is a process of sharing and creating tacit knowledge through direct interaction and experience. Externalization [5, 14] which is receiving and sharing knowledge explicitly that an individual has with others. Combination [5], which is a process that manipulates explicit knowledge possessed by individuals by

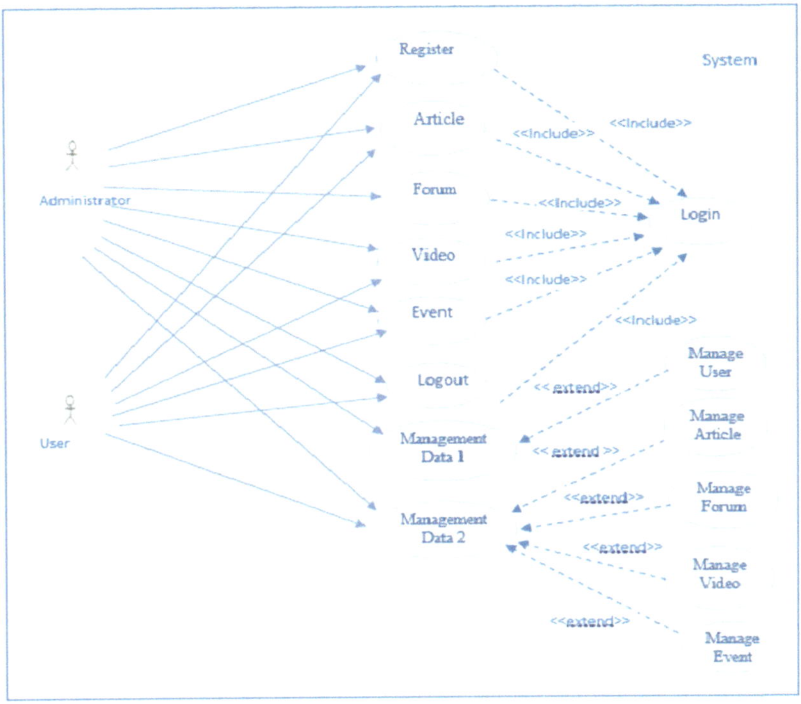

Fig. 5. Use case diagram for SME knowledge management app

Fig. 6. Main page user interface for SME knowledge management app

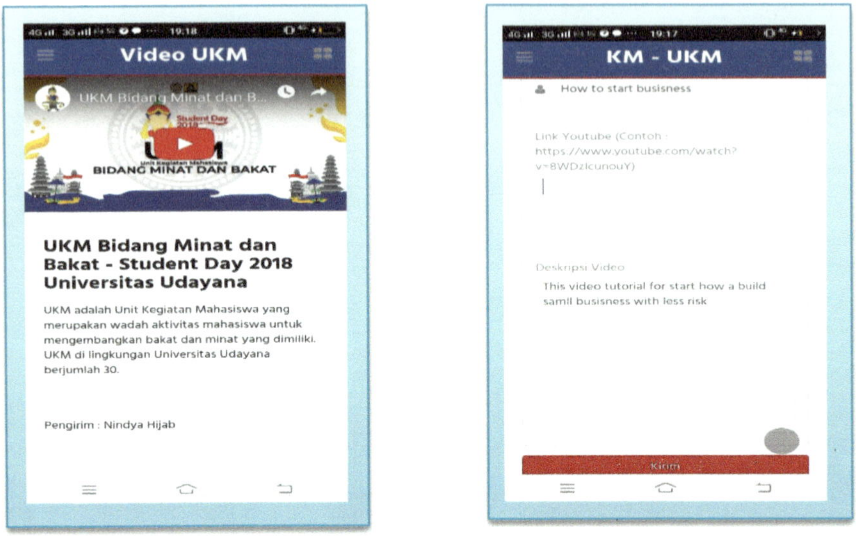

Fig. 7. Video tutorial page for SME knowledge management app

sorting, adding or combining some explicit knowledge, into new explicit knowledge. Internalization [5] of explicit knowledge is then learned, understood and developed in accordance with the capabilities of each individual. The discussion forum can facilitate the process of transferring tacit and explicit knowledge (Fig. 8).

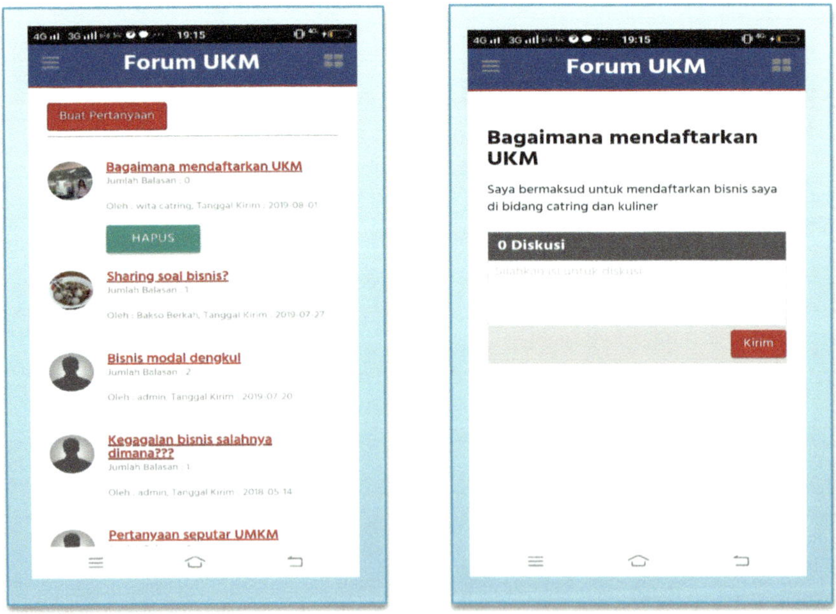

Fig. 8. Forum page for SME knowledge management app

5 Conclusion

The availability of a mobile application for knowledge management based on a community of SMEs can support the knowledge capture process and retention of knowledge for access by all SMEs practitioners, thus creating a knowledge culture [18] of sharing among SMEs practitioners to explicitly share all the tacit knowledge possessed by members.

Benefits (perceived usefulness) obtained from the use of knowledge management technology in terms of uploading knowledge in the form of videos, discussion forums, articles on SMEs will influence the perceived behavioral intention to use (ITU) for an app application for knowledge management, thus giving rise to a positive attitude towards usage (ATU) of knowledge management technology.

Research Limitations and Future Research
This research does not cover all SMEs in Indonesia. It will be important for future research to investigate organizational knowledge management readiness among other SMEs. Further research needs to expand the scope to all SMEs in Indonesia and evaluate the implementation of knowledge management in SMEs.

Acknowledgement. This research was fully funded by the Directorate General of Strengthening for Research and Development, Ministry of Research, Technology, and Higher Education, Republic of Indonesia, as a part of *Penelitian Terapan Unggulan Perguruan Tinggi* Research Grant to Bina Nusantara University titled "Analisis dan Penerapan Knowledge Management pada sektor Industri Kreatif dengan pendekatan TAM" with contract number 23/AKM/MONOPNT/2019 and contract date: March, 27th 2019.

References

1. Byrne, B.M.: Structural Equation Modeling with AMOS Basic Concepts, Applications, and Programming, 2nd edn. Routledge Taylor & Francis Group, New York (2010)
2. Becerra-Fernandez, I., Sabherwal, R.: Knowledge Management Systems and Processes. M. E. Sharpe, London (2010)
3. Kurniawati, A., Samadhi, T.M.A., Wiratmadja, I.I.: Indicators of knowledge management cycle in Indonesian small and medium enterprises. In: International Conference on Management of Innovation and Technologies (2016)
4. Departemen Perdagangan R.I.: Rencana Pengembangan Industri Kreatif Indonesia 2015 (2008)
5. Nonaka, I., Takeuchi, H.: The Knowledge-Creating Company: How Japanese Companies Create the Dynamics of Innovation. Oxford University Press, Oxford (2000)
6. Hoyle, R.H.: Hand Book Structure Equation Modeling. The Guilford Press, New York (2012)
7. Siregar, J.J, Aryanti Wardaya Puspokusumo, R.A.: The analysis of technology acceptance model in implementing knowledge management for small medium sized enterprises (SMEs) in a creative industry base on mobile application. In: International Conference on Engineering, Applied Science and Technology (2018)
8. Pressman, R.S.: Software Engineering a Practitioner's Approach, 7th edn. McGraw-Hill, New York (2010)

9. Nousala, S.: The value of sustainable knowledge transfer methods for SMEs, utilizing socio-technical networks and complex systems. Syst. Cybern. Inf. **8**(6) (2010). ISSN 1690-4524
10. Davenport, T.H., Prusak, L.: Working Knowledge-How Organizations Manage What They Know. Harvard Business School Press, Boston (2000). Revised vol. 2000, no. August
11. Park, S.Y.: An analysis of the technology acceptance model in understanding university students. Behavioral intention to use e-Learning. Educ. Technol. Soc. **12**, 150–162 (2009)
12. Cheng, E.C.K.: Knowledge strategies for enhancing school learning capacity. J. Educ. Manag. **26**(6), 557–592 (2012)
13. Becerra-Fernandez, I., Rajiv, S.: Knowledge Management System and Process. M.E. Sharpe, London (2010)
14. Sameon, G., Garfield, M., Mashesha, M.: Towards enabling tacit knowledge externalization using mobile phone. The Case of Participatory Agricultural Innovation in Eth, Iopia. IEEE (2015)
15. Abdullah, R., Yousef, A., Alsharaei, M.: A mobile knowledge as a service (mKaaS) model of knowledge management system in facilitating knowledge sharing of cloud education community environment. In: Third International Conference on Information Retrieval and Knowledge Management (2016)
16. BEKRAF: Rencana Strategis Ekonomi Kreatif Indonesia 2015–2025 (2017). https://www.bekraf.go.id/berita/page/17/rencana-strategis-badan-ekonomi-kreatif-2015-2019
17. Calvo, N., Rodeiro-Pazos, D., Rodríguez-Gulías, M.J., Fernández-López, S.: What knowledge management approach do entrepreneurial universities need? J. Inf. Syst. **85**, 21–29 (2019)
18. Seong, B.H., Lee, H., Kim, T.: A study on knowledge management system for knowledge competitiveness with one stop knowledge service. In: 1st IEEE International Conference on Knowledge Innovation and Invention (2018)

A Review on the Methods to Evaluate Crowd Contributions in Crowdsourcing Applications

Hazleen Aris(✉) and Aqilah Azizan

Universiti Tenaga Nasional, 43000 Kajang, Selangor, Malaysia
hazleen@uniten.edu.my

Abstract. Due to the nature of crowdsourcing that **openly** accepts contributions from the crowd, the need to evaluate the contributions is obvious to ensure their reliability. A number of evaluation methods are used in the existing crowdsourcing applications to evaluate these contributions. This study aims to identify and document these methods. To do this, 50 crowdsourcing applications obtained from an extensive literature and online search were reviewed. Analysis performed on the applications found that depending on the types of crowdsourcing applications, whether simple, complex or creative, three different methods are being used. These are expert judgement, rating and feedback. While expert judgement is mostly used in complex and creative crowdsourcing initiatives, rating is widely used in simple ones. This paper is the only reference known so far that documents the current state of the evaluation methods in the existing crowdsourcing applications. It would be useful in determining the way forward for research in the area, such as designing a new evaluation method. It also justifies the need for automated evaluation method for crowdsourced contributions.

Keywords: Evaluation method · Crowdsourcing evaluation · Survey · Systematic review · Grounded theory

1 Introduction

Crowdsourcing was first coined in 2006 to define a concept where wisdom of the crowd is leveraged in solving problems and executing tasks [1]. Crowdsourcing seeks solutions to problems and completion to tasks by outsourcing them to the open crowd. Human intelligence and expertise are openly and under certain circumstances and extent, globally sought in crowdsourcing, regardless of the geographical boundaries that separate them. The ability of crowdsourcing in harnessing the power of crowd to complete tasks that were traditionally performed by salaried employees has been proven in a number of successful crowdsourcing applications. Threadless, Waze and iStockPhoto are amongst the manifestations of this success. Quality of the solutions offered or tasks performed by the crowd (referred to as **contributions** from this point onwards) is an important success factor in any crowdsourcing initiatives, without which the initiatives can fail horribly. Therefore, assessing or evaluating the contributions from the crowd is necessary to ensure the quality of the contributions with respect to

F. Saeed et al. (Eds.): IRICT 2019, AISC 1073, pp. 1031–1041, 2020.
https://doi.org/10.1007/978-3-030-33582-3_97

the fulfillment of the defined criteria. In [2], a system is regarded as a crowdsourcing system if it addresses the following four fundamental challenges.

- How to recruit and retain users?
- What contributions can users make?
- How to combine user contributions to solve the target problem?
- How to **evaluate** users and their contributions?

Thus, it can be seen that evaluation is regarded as one of the fundamental issues in crowdsourcing, which further emphasises the importance of evaluating crowd contributions. Furthermore, due to the open nature of crowdsourcing, there is no guarantee that the contributions obtained from the crowd are always acceptable, even for the simplest tasks. Thus, on top of evaluating against the defined criteria or requirements, contributions from the crowd also need to be evaluated with respect to their reliability, i.e. the extent to which the contributions are trustworthy.

However, the extent to which the evaluation is being performed in the existing crowdsourcing applications, such as the evaluation methods used, is thus far not properly studied and documented. Due to the importance of evaluation in ensuring the survivability of crowdsourcing initiatives, information on the current state of its implementation is needed, so that means to improve the current methods, if any, can be made. The objective of this study is therefore to identify and document the methods used by the existing crowdsourcing applications in evaluating the contributions from the crowd. Listed below are the research questions addressed in this study.

1. How many crowdsourcing applications are available to date?
2. How many of these applications evaluate the contributions from the crowd?
3. What are the methods used in evaluating the contributions?

To answer the first research question, a systematic literature search was performed to obtain a list of crowdsourcing applications to date. The final list of crowdsourcing applications was carefully populated to ensure that only applications that fulfill the criteria for crowdsourcing applications were included.

Results obtained show that most of the existing crowdsourcing applications included in the study evaluate the contributions received from the crowd. Three types of evaluations methods are found; expert judgement, rating and feedback. These methods are used depending on the types of the crowdsourcing initiative, either complex, creative or simple. While complex and creative crowdsourcing initiatives mostly use expert judgement, rating is widely used in simple crowdsourcing initiatives. It is also found that most of these evaluation methods are manually performed by humans, which suggests research opportunities on the development of automated evaluation of crowdsourcing contributions, particularly for simple crowdsourcing. The rest of this paper is organised as follows. Section 2 describes the method used in the study. Results obtained are presented in Sect. 3 and discussed in Sect. 4. Finally, Sect. 5 concludes the paper.

2 Method

In order to answer the research questions, a list of existing crowdsourcing applications needed to be identified. For this purpose, our approach was to look for the applications from the secondary sources that had already gathered them, because most of them had already performed the necessary steps to ensure the inclusion of as many crowdsourcing applications as possible in their work. Exhaustive search was therefore performed, covering the academic (scholarly) databases as well as online search engines. These included conference papers, journals, dissertations as well as personal web pages. Based on our preliminary literature review on the topic, it was found that at least four terms were interchangeably used in the literature to mean crowdsourcing applications. Therefore, the following four terms were used in performing the search to avoid from missing the needed samples.

- crowdsourcing initiative
- crowdsourcing system
- crowdsourcing project
- crowdsourcing application

The lists of crowdsourcing applications from various sources found were then compiled into one raw list of crowdsourcing applications. From this point onwards, the preferred reporting items for systematic reviews and meta-analyses (PRISMA) [3] guideline was followed. From the compiled list, duplicate applications were removed. Then, the presence of each application was verified by going to the primary source of the application where further information about the application can be obtained. This is important because in the final selection process, elaborate description of each application is needed. The final selection process is to compare each application, based on their respective description, against the eight crowdsourcing characteristics in [4] as listed below.

1. clearly defined crowd,
2. task with clear goal,
3. clear recompense received by the crowd,
4. clearly identified crowdsourcer,
5. clearly defined compensation to be received by the crowdsourcer,
6. online assigned process of participative type,
7. uses an open call of variable extent and
8. uses the Internet

The eight crowdsourcing characteristics were derived by studying crowdsourcing definitions taken from the literature. These characteristics help in differentiating between what-is and what is-not crowdsourcing. This step is important to be certain that the applications included in our study are indeed crowdsourcing applications and that they were not mistakenly labelled as crowdsourcing applications. Due to its fast growing popularity ever since its formal introduction in 2006, there is the tendency to label any applications with crowd involvement as crowdsourcing applications, although some may not fulfil the characteristics of crowdsourcing. This has subsequently led to some confusion on what a crowdsourcing application actually is or should be. Therefore, comparing against the eight characteristics in making the selection helped in ensuring the validity of the analysis performed. Non-crowdsourcing applications that were mistakenly labelled as crowdsourcing applications, as far as the eight characteristics are concerned, were excluded. Furthermore, the selection was also made based on the availability of the descriptions of the applications in English. Once the list of the crowdsourcing applications was finalised, the analysis began by looking at each application one at a time, focusing on the methods used in evaluating the contributions from the crowd, if any, i.e. answer to the second and third research questions.

3 Results and Analysis

The raw search results comprised a total of 103 applications (records). From these, 39 records were obtained from the literature (n_1) and the remaining 64 from the online search (n_2) as shown in Fig. 1. After duplicates removal, 90 unique records (n) were obtained. These 90 records were further filtered based on the eight characteristics of crowdsourcing listed earlier and the availability of the descriptions in English. At the end of the selection process, only 50 crowdsourcing applications made it to the analysis stage. The remaining 40 applications (m) were excluded as shown in Fig. 1.

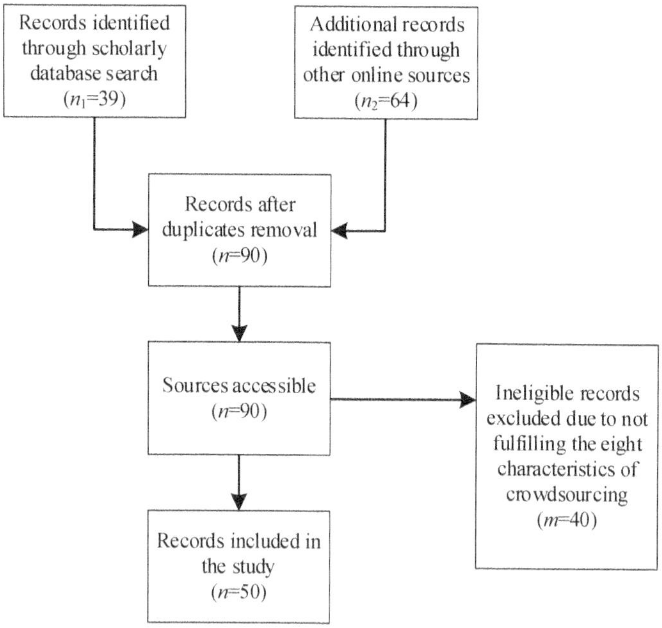

Fig. 1. Selection process using PRISMA guideline

In interpreting the third characteristic of the crowdsourcing application, which is the 'recompense received by the crowd', there is a need to clarify our approach. Recompense in this context means the rewards that the crowd would receive in return for their contributions. The rewards can be divided into monetary rewards and non-monetary rewards. The monetary rewards are either a sum of money given at once as a winning reward or the revenue share from a crowdsourcing project that implemented ideas or solutions given by the crowd. The non-monetary rewards include prizes, recognition by companies, self-recognition and gamification elements such as points, ranking and rating [5]. The rewards are also seen categorised into extrinsic (seen) and intrinsic (unseen, but felt) [6, 7]. Thus, seen monetary and non-monetary rewards are extrinsic rewards. In deciding upon this characteristic, only crowdsourcing applications that offer extrinsic rewards were included in the list. Applications that imply intrinsic reward such as recognition and self-satisfaction were excluded due to the subjective nature of the recompense.

Another clarification is concerning the sixth characteristic, which requires the crowdsourcing applications to be of participative nature. Participatory means that users are actively engaged in giving the contributions, as opposed to the opportunistic crowdsourcing where contributions are automated with little or no user involvement [8]. Based on this characteristic, opportunistic crowdsourcing applications were further excluded from the analysis. The remaining characteristics are straightforward. The 50 crowdsourcing applications that managed to proceed to the analysis stage are shown in Table 1, labelled C1 to C50 respectively.

Table 1. The short listed crowdsourcing applications

Application		Type	Evaluation method
C1	Innocentive	Complex	Experts
C2	Ideaken	Creative	Experts
C3	Nesta UK	Creative	Experts
C4	Humanitarian Innovation	Complex	Experts
C5	Unilever (via yet2.com)	Creative	Experts
C6	Innoget	Creative	Experts
C7	GE Open Innovation Ecomagination	Complex	Experts
C8	Innovation Community (via Hyvecrowd)	Creative	Experts & feedback
C9	Kickstarter	Simple	Not evaluated
C10	Quirky	Creative	Experts
C11	Cambrian House	Creative	Rating
C12	99 Design	Creative	Experts, rating & feedback
C13	Mobile 3D Modeler	Simple	Rating
C14	Greenpeace (via Jovoto)	Creative	Experts, rating & feedback
C15	Kopernik	Complex	Experts
C16	Dell Social Innovation Challenge	Complex	Experts
C17	G-Win	Complex	Experts
C18	Postcode Lottery Green Challenge	Complex	Experts

(continued)

Table 1. (*continued*)

Application		Type	Evaluation method
C19	Open Ideo	Complex	Experts
C20	Idea Connection	Complex	Experts
C21	Samsung	Creative	Experts
C22	Challenge.gov	Complex	Experts
C23	Idea Bounty	Complex	Experts
C24	Innovation Exchange	Complex	Experts
C25	eYeka	Complex	Experts
C26	The Cairo Transport App Challenge	Complex	Rating
C27	Crowdspring	Complex	Feedback
C28	Lego	Creative	Experts & rating
C29	MIT Climate Colab	Complex	Experts, rating & feedback
C30	Idea Storm	Creative	Experts, rating & feedback
C31	Ideas Brewery	Creative	Experts
C32	Lays	Creative	Experts
C33	Kraft	Creative	Experts
C34	McDonalds Burger Builder	Creative	Rating
C35	CrowdCrafting	Complex	Experts
C36	Scientific American (via Innocentive)	Complex	Experts
C37	UbiAsk	Simple	Experts
C38	Blablacar	Simple	Rating & feedback
C39	Librorium	Simple	Rating
C40	Unbabel	Simple	Rating
C41	Stereopublic	Simple	Rating
C42	CityExplorer	Simple	Rating
C43	Go Kamino	Simple	Rating
C44	Sit or Squat	Simple	Rating
C45	Waze	Simple	Rating
C46	Placemeter	Simple	Not evaluated
C47	Jana.com	Simple	Rating
C48	Gigwalk.com	Simple	Experts
C49	AirBnB	Simple	Rating
C50	Arcbazar	Creative	Rating

Elaborate description of each evaluation method identified will be discussed in the next section. In order to obtain a more meaningful insight to the information on the evaluation methods used in existing crowdsourcing applications, they were further categorised into simple task, complex task and creative task as shown in Table 1. The categorisation is based on the definitions given in [4] and practices defined in [9] and [10], which is also seen used in [11]. Simple crowdsourcing tasks, also known as distributed human intelligence, are coordinated series of simple tasks that need to be accomplished by humans. These tasks cannot be performed by computers, despite being simple, such as determining the cleanliness of a washroom (C44). Due to their simplicity too, these tasks can be performed by virtually any crowd. Complex crowdsourcing tasks are tasks in the form of more complicated problems to be solved and wisdom of the crowd is needed to come up with the solutions. The types of solution sought usually require the crowd to possess certain skills and knowledge and not just any crowd can participate. Finally, in creative crowdsourcing tasks, the crowdsourcer is looking for creative contributions from the crowd in the forms of ideas or designs.

4 Discussion

From the analysis performed on the 50 selected applications, it can be seen that the majority (96%) of the crowdsourcing initiatives do evaluate the contributions given by the crowd. Different methods were used in evaluating the contributions from the crowd and at least three types of method were extracted from the analysis. These are expert judgement, rating and feedback. These evaluation methods are used either independently or in combination with one another.

4.1 Expert Judgement

In this type of evaluation, experts who evaluate the contributions are specialists in the respective fields. They consist of any of the following groups of expert; the applications' team reviewer, specialist reviewer, global support team, sponsors, challenge creators, brand owners and researchers or scientists related to the projects or problems. They will evaluate the contributions based on pre-defined criteria. There could be more than one experts involved in the evaluation, especially for large scale projects. For example, Open Ideo (C19) is a crowdsourcing application that allows people from all corners of the world to collaboratively tackle some of the toughest global issues. The solutions to the global issues delivered by the crowd will be evaluated by experts from the Open Ideo team and the sponsors. However, under certain circumstances, when a crowdsourcer is seeking solution for a small project, he can be the only evaluator for the project. For example, CrowdCrafting (C35) is a crowdsourcing application where volunteers contribute to the scientific projects developed by the citizens, professionals or institutions who need help to solve problems, analyse data or complete challenging tasks. The crowdsourcer who is requesting for the solutions will evaluate the contributions given by the crowd.

The advantage of having subjective expert judgement is that the results of the evaluation are certainly more reliable compared to the other evaluation methods, as the experts are deemed knowledgeable, skillful and experienced in their respective fields. Therefore, solutions to the issues discussed are more dependable and accurate. However, the disadvantage is that the experts might be biased in evaluating the contributions depending on their viewpoints and stands.

4.2 Rating

Rating is an instrument to determine ones' preferences based on designated scales. Rating scales can be of two types [12]. The first type is done by assigning a specific descriptor for each value in the rating scale such as the 5-point Likert scale. The second type is done by specifying only the end points of the rating scale, such as allowing people to rate based on a scale of one to ten, where one is for the lowest preference and ten is for the highest preference. Rating can be done by experts and/or peers.

Experts. In this case, rating is done by the experts in the related fields depending on the requirements of the contributions. Rating scale used for each crowdsourcing application varies. For example, Arcbazar (C50) is a crowdsourcing application that uses this evaluation method because the contributions are related to designs, such as options for remodelling, interior designs, landscape designs, home designs and commercial design projects. Expert voters in Arcbazar are the designers who are listed among the top 100 of the overall designer's chart in that application. These expert voters carry heavier weight than the peer (non-expert) voters.

Peers. In peer rating, the evaluation is done by the users of the crowdsourcing applications based on a given scale. One of the crowdsourcing applications, which allows peers to rate is Blablacar (C38), a carpooling service where drivers offer a ride for passengers who are going to the same destinations as them. In this application, the drivers and passengers are allowed to rate each other regarding their experience in carpooling together. The rating scale used in Blablacar is a 5-point Likert scale of outstanding, excellent, good, poor and very disappointing [13].

The advantage of implementing rating method is that the crowd will be more confident in making the decisions to trust the contributions by looking at the given ratings because they are standardised based on the scales provided in the crowdsourcing applications. Users will be able to know the reliability of the contributions from the ratings shown. On the contrary, the disadvantage is that the crowd may rate the other crowd based on different opinions and this limits the reliability of the ratings. There are possibilities where these voters might not rate accurately as they are either not serious in rating the contributions or not familiar with them [14]. Lastly, there may be malicious users who give high ratings to low quality data and vice-versa [15].

4.3 Feedback

Feedback is defined as the information provided by an agent, e.g. teacher, peer, book, parent and self, regarding aspects of one's performance or understanding [16]. It is an open remark given by the crowd to provide their opinions regarding the contributions submitted by the other crowd. It is essential for identifying the strengths and providing comments on areas for improvement and development [17]. However, it may or may not affect the final evaluation results. There are several crowdsourcing applications that take feedbacks into account in making decisions to ensure that the contributions are reliable, while some feedbacks in crowdsourcing applications do not affect the evaluation results. An example of a crowdsourcing application that implemented feedbacks is Crowdspring (C27), where the project owners post project details to receive feedbacks and revisions from the crowd. They will collaborate with the chosen project inputs. The crowd may leave comments or feedbacks to the project owners and it is up to the project owners to consider the feedbacks received. The advantage of requesting feedbacks from the users is that the issues discussed by the users will be from various aspects and this will expand the perspective of each issue. On the other hand, the disadvantage is that the public might not be well versed on the discussed issues and the opinions given may be of little relevance. Besides, users may not express themselves precisely [14].

Table 2 summarises the evaluation methods implemented in the crowdsourcing applications under review. As can be seen from the table, expert judgement (experts) is the most widely used evaluation method. Out of the 50 crowdsourcing applications reviewed, 25 of them are evaluated by experts. Experts evaluate submitted contributions in a subjective manner and the most relevant contributions will be chosen. A further 13 crowdsourcing applications use rating as the evaluation method. Those who rate the applications are either other users of the applications (peers) or the experts in the field. While expert evaluation is mostly used in complex and creative crowdsourcing applications, rating is used mostly in simple crowdsourcing applications. There is only one crowdsourcing application that is found to evaluate crowd contribution using the feedback method.

Table 2. The types of evaluation method

Evaluation mechanism	Number of applications		
	Simple	Complex	Creative
Experts	2	15	9
Rating	10	1	3
Feedback	0	1	0
Experts & rating	0	0	1
Experts & feedback	0	0	1
Rating & feedback	1	0	0
Experts, rating and feedback	0	1	3
Not evaluated	2	0	0
Total	15	18	17

It can also be seen that a number of crowdsourcing applications use more than one evaluation method. There is one crowdsourcing application in this study that evaluates crowd contributions using rating and feedback methods. Users can rate the crowdsourcing applications along with giving opinions regarding their experiences of using the crowdsourcing application. Another crowdsourcing application that has two evaluation methods allows subjective evaluation by experts and rating by users. The crowd will first rate for the best ideas, and the ideas will also be evaluated by the experts based on some specific criteria. There is also a combination of experts and feedback evaluation methods used in one of the crowdsourcing applications analysed in this study. The experts will subjectively evaluate the contribution and the crowd may provide feedback by sharing their opinions. The experts evaluation will affect the end results of the contributions. However, it is up to the experts to consider the feedback from the crowd in their evaluation. Four crowdsourcing applications implemented three evaluation methods, which allow subjective evaluation by experts, rating by the users or experts, and comments or feedbacks by the users. Experts will evaluate the data based on certain criteria and rating may or may not affect the evaluation results. The same goes to feedback. It may or may not affect the final results. Rating and comments or feedbacks will assist experts in evaluating the data, but the experts may not solely rely on them.

From the 50 crowdsourcing applications reviewed, there are two crowdsourcing applications that do not evaluate crowd contributions. In such a crowdsourcing application, the contributions from the crowd will be accepted or made available to the users or the public instantly without any filter. Thus, reliability of such crowdsourcing applications is questionable. It can also be seen that the evaluation given to the contributions may not directly influence the final decision, or may not affect the decision at all. This is especially the case when more than one evaluation methods are implemented in the crowdsourcing application. For example, in the Lego challenge (C28), two evaluation methods are used; expert judgement and rating. Other users can freely rate the design and the experts may or may not consider the rating given in judging the contributions and making the decision. Thus, the evaluation methods were found used in two different ways; direct and indirect. Direct use of the evaluation method is when the evaluation affects the final decision on the contributions. Indirect use of the evaluation method is when the evaluation of the contributions does not affect the selection process. Finally, from the above discussion, it can also be seen that most of the evaluation methods implemented in the existing crowdsourcing initiatives are performed manually, i.e. requiring users intervention for the evaluation to be done. Manual evaluation has several disadvantages. One of the disadvantages is that the evaluators are prone to making human errors. It is also time consuming as it depends on the availability of the evaluators. Besides that, manual evaluation involving human is also subject to bias. These disadvantages can be overcome by having an automated evaluation method to evaluate the contributions. Using automated evaluation, human involvement can be kept to the minimum or can be removed altogether.

5 Conclusion

In this paper, the results of a review performed on the evaluation methods used in the systematically selected list of crowdsourcing applications are presented, analysed and discussed. The majority of these applications use subjective evaluation by the experts in evaluating the contributions from the crowd. A significant portion uses rating, either by experts or by peers. The presence of evaluations in most of the applications under review emphasises the potential risks associated with obtaining inaccurate and false contributions from the crowd in such an open environment like crowdsourcing. Different evaluation methods are also found used depending on the types of the crowdsourcing. Findings from this study serves as a reference for research work on the evaluation methods in crowdsourcing. Future work includes the design of a fully automated evaluation method to address the identified gaps, i.e. dependent on manual evaluation by humans.

Acknowledgement. Information presented in this paper forms part of the research work funded by Universiti Tenaga Nasional entitled Formulation of a Trust Building Mechanism for Trustworthy Non-profit Mobile Crowdsourcing Initiatives Using Beta Reputation System (J510050668).

References

1. Howe, J.: Crowdsourcing: a definition (2015). http://www.crowdsourcing.com
2. Doan, A., Ramakrishnan, R., Halevy, A.Y.: Crowdsourcing systems on the world-wide web. Commun. ACM **54**, 86–96 (2011)
3. Moher, D., Liberati, A., Tetzlaff, J., Altman, D.G., The PRISMA Group: Preferred reporting items for systematic reviews and meta-analyses: the PRISMA statement. Ann. Intern. Med. **151**, 264–269 (2009)
4. Estellés-Arolas, E., González-Ladrón-de-Guevara, F.: Towards an integrated crowdsourcing definition. J. Inf. Sci. **38**, 189–200 (2012)
5. Borromeo, R.M., Toyama, M.: An investigation of unpaid crowdsourcing. Hum.-Centric Comput. Inf. Sci. **6**, 11 (2016)
6. Rogstadius, J., Kostakos, V., Smus, B., Laredo, J., Vukovic, M.: An assessment of intrinsic and extrinsic motivation on task performance in crowdsourcing markets. In: Fifth International AAAI Conference on Weblogs and Social Media, pp. 321–328. Association of Advanced Artificial Intelligence (2011)
7. Zhao, Y.C., Zhu, Q.: Effects of extrinsic and intrinsic motivation on participation in crowdsourcing contest: a perspective of self-determination theory. Online Inf. Rev. **38**, 896–917 (2014)
8. Lane, N.D., Miluzzo, E., Lu, H., Peebles, D., Choudhury, T., Campbell, A.T., College, D.: ADHOC and sensor networks a survey of mobile phone sensing. IEEE Commun. Mag. **48**, 140–150 (2010)
9. Brabham, D.C.: Moving the crowd at threadless. Inf. Commun. Soc. **13**, 1122–1145 (2010)
10. Smeets, B.C.W.: Crowdsourcing: the process of innovation (2011)
11. Kleemann, F., Voß, G.G., Rieder, K.: Un(der)paid innovators: the commercial utilization of consumer work through crowdsourcing. Sci. Technol. Innov. Stud. **4**, 5–26 (2008)
12. Brown, T.C., Daniel, T.C.: Scaling of ratings: concepts and methods. USDA For. Serv. Res. Pap. RM-293, pp. 1–24 (1990)
13. Müller, N.: BlaBlaCar – business model and empirical analysis of usage patterns (2018)
14. Pan, W., Xiang, E.W., Yang, Q.: Transfer learning in collaborative filtering with uncertain ratings transfer by integrative factorization. In: Proceedings of the Twenty-Sixth AAAI Conference on Artificial Intelligence, pp. 662–668, Toronto (2011)
15. Wang, G., Xie, S., Liu, B., Yu, P.S.: Review graph based online store review spammer detection. In: 2011 IEEE 11th International Conference on Data Mining, Vancouver (2011)
16. Hattie, J., Timperley, H.: The power of feedback. Rev. Educ. Res. **77**(1), 81–112 (2007). https://doi.org/10.3102/003465430298487
17. Leibold, N., Schwarz, L.M.: The art of giving online feedback. J. Eff. Teach. **15**, 34–46 (2015)

Exploring Process of Information Systems and Information Technology for Enterprise Agility

Olatorera Williams, Funminiyi Olajide, Tawfik Al-Hadhrami, and Ahmad Lotfi

School of Science and Technology, Nottingham Trent University, Nottingham NG11 8NS, UK
olatorera.williaims@my.ntu.ac.uk,
{funnminiyi.olajide, tawfik.al-hadhrami, ahmad.lotfi}@ntu.ac.uk

Abstract. System Engineering is the ability to establish several questions in system development, for efficient delivery of business system operations. Automation of business processes has been found common today with resolution of common issues of enterprise agility with the use of AI/ML. Robotic process automation (RPA) was reviewed as capable to handle high-volume, repeatable tasks required by humans to perform. Several tasks of business queries, calculations, maintenance of records, process and business transactions continues to experience changes in IS/IT environment. In today's digital world, organisations are under pressure, to utilize systems to ensure continuous adaptability of business processes for agility. This paper provided insights to systemic approach of IS/IT, by exploring optimum systems configuration of performances and hybridization of Information Systems Technology for enterprise agility. Enterprise for business organisation respond to changing environment and with the aim of business adaptation to various changes are caused by the emergent of IS/IT systems. Social media adoption are amongst the tools identified for business agility and for IS/IT systems capabilities. This paper explores the process and reviews emergent of IS/IT systems integration as potential impact for enterprise agility.

Keywords: IS/IT · Enterprise · Systems · Process · Organisation · Emergent · Integration · Business · Agility

1 Introduction

Alignment of operational systems approach helps to improve organizational performance and thereby, provided policies and procedures that are mostly necessary for incident responses, depending on organisational readiness for improvement and delivery of business processes. A paper on enterprise paradigm shifts, due to social media and robotic technologies were from process of dealing with mass production via traditional market competitions [1]. These are related to cyber-physical techniques and as relevant to incidence solutions, while maintaining agility complex to systems

F. Saeed et al. (Eds.): IRICT 2019, AISC 1073, pp. 1042–1051, 2020.
https://doi.org/10.1007/978-3-030-33582-3_98

development. The effect knowledge is traceable to social media marketing on brand loyalty for customers [2]. However, there have been considerable amount of efforts in research development, with issues remained unsolved for systems performance and operational delivery. Hence, cyber-crimes related activities on IT/IS applications environment are crucial for enterprise to adapt to changing business environment. Whilst, increasing productivity and customers' satisfaction thereby reducing the level of complexity of the technological systems errors. These challenges slow down digital position in the business market world.

Enterprises have kept on raising their practice standards by development and this paper explores exploitation of operational systems capabilities and designed models for enterprise agility. It is essential that systems engineering in business organisational projects flows well with agile technology. Past research paper of [3], responding to operational requirements for efficient products delivery and as presented in related examples of [4]. In the past decades, business operations are determined based on the use of IT technology, hence played vital role in IS enterprise agility. A research of [5], described people, technology and process as a major technology enablement tool for determinant of systems value. It is obvious, that through IT-enabled enterprise agility argument research idea of [6]. Of recent years, process happened by intelligence systems of robotic technology, running in higher capacity of AI and ML algorithms. It is evidence that a research work of [7] and [8] presented that agility has the ability for an organisation to sense and respond to changes caused by turbulent technological environment. Therefore, IT/IS, are motivated development with progress growth for solutions, based on IT integration and capabilities. This is for performance efficiency while, several studies pointed out the importance of IT systems in enabling agile enterprise. These are vital understanding, for exploring a dynamic relationship amongst IS/IT agility, in process performances, based on operational delivery impacts.

Exploration research adopted several approaches to learning by assessment of qualitative and quantitative approaches and thus, focusing on the understanding of dynamic correlations between enterprise and open system platform. It can be said therefore, that the impact of IS/IT agility enhances operational delivery and thus, reduces unforeseen hindrances that can are caused by internal or external influences. This paper review IS/IT with impact of enterprise agility to maintain and increase business operational performances in a business organisation.

2 Related Work

As researcher reviews, agility was used and defined in several disciplines [9] and as defined such that agility is a measure of ability to recognize, act and benefit from changing business environments. This paper further described agility from an IT perspective as the ability to pivot and change directions in response to market pressure. This is obviously, requires distinct patterns of Information Technology (IT) capabilities with specific positioning in enterprise level within its business process. As reported by paper of [10], agility is important in the contemporary business environment which is characterized by an intense global competition with short product life cycle. These are relevant to an increased technological innovation and time sensitive customer demands of [6]. A research by [11] described ideas on various characteristics of agility. The ideas are characterized on agility with distinctive features in collaborative efforts, for cooperative solutions by strategic leaders. The incremental and adaptable facts are iterative with constant testing of emergency and feedback incorporation, whereby, leanness, modularity and people orientation matters in reflection and introspection relationship to small teams, time-boxing and transparency of technological usage in business. Obviously, using enterprise concept as a qualifier for agility can at some stage lead to some constraint on the focus of how an organization possesses the attributes of agility [12].

Digital transformation solution is followed on today's digital marketing, adaptiveness and responsiveness to changes, with continuous feedback. A research work of [13] reflected on agility priority as attributes, to responsiveness, competency, flexibility, quickness, cost-effectiveness, integration and technology. Table 1 below presents past reviews of enterprise agility model and measures specific to organisation timely delivery of products in reference to speed, success, indicator, simplicity, weight and social media access.

Table 1. Enterprise agility model and measures.

Past and current work	A sample research work of [12]	A sample research work of [13]
Enterprise agility model and measures	Enterprise handle change, uncertainty and unpredictability within the business environment and make an appropriate response	Agility assessment model in reflection of customers' needs with an indicator of project simplicity and weight
Speed	X	X
Success		X
Indicator		X
Simplicity		X
Weight		X
Social media		X

3 Review of Enterprise Agility Enabler

System development and planning by evaluation of the business process for enterprise agility are enabler solution for systems engineering. Table 2 presented sample research work on enablers of enterprise agility in accordance to ranks of hierarchy and agility attributes by enablers determinants which are relevant to the research idea of this paper for reviews.

Table 2. Enterprise agility enablers.

Research study	Enterprise agility enablers/determinants	Description
Urgent and emergency care: an academic application system case study [14]	Systems engineering	A systemic solution in reference to process and in simplicity of the use of Big Data and as useful for potential block chain ideas for digital transformation in emergent technologies
Towards usability model [15]	Systems engineering	Process and practice adoption are key elements in modern software process improvement initiatives, many of them fail
Evaluation of agility in supply chains [16]	Systems engineering of supply chain	Process requirements for integrating IT/IS organizational units as data flows to improve product competitiveness and achieve success in responding to customer's demands
IT-enabled operational agility: an interdependencies perspective [17]	Systems engineering	This research work focused on clear dynamic network of organisation and systems linkages to different business processes
Obstacles in moving to agile software development methods; at a glance [18]	Systems engineering	A research study of agility Assessment Model useful to measure agility in certain degree of tools and development using Agile Software solutions
Enterprise agility [19]	Information technology	IS/IT enhances sensing and responding capabilities of firms to respond to rapidly changing environmental conditions

As described in Table 2 above, the impact of emergent IS/IT Systems are related to functional and non-functional requirement indicated by [20], a research work investigated on agile enablers. This is described as vital role, in facilitating features for business to enhance its performances. IS/IT integration may include supply chain management, customer management systems, for enterprise resource planning in business organisations with concurrent engineering, cloud computing integration, project management, DevOps, CRMS, SCMS, that include systems/hardware engineering for team building and knowledge management. As referenced to a paper of [13], the research idea by [21] classified on hierarchy with priority for enablers of agility using a fuzzy multiple criteria decision approach with seven determinants business focus.

3.1 Digital Impacts of IS/IT for Enterprise Agility

Researcher explores idea of digital options and as classified as set of IT enabled capabilities to digitized enterprise work processes with knowledge systems. The idea of the review associated propositions with added understanding about critical linkages between IS/IT investments and organisation performances based on the use of digital technology. A paper of [22] created a multi-theoretical approach towards understanding, the strategic role of IS/IT investments and capabilities in organisation process performances. Various research works reviewed based on three significant organizational capabilities (agility, digital options, and entrepreneurial alertness), based on strategic processes. As indicated by [3] enabling role of IS/IT are key components of enterprise agility. Therefore, organisation intentions to adapt to changes are important and thus, specified business processes for system development and enabling environment capabilities for efficient solutions development and implementation of IS/IT-enabled services of technology ventures. IS/IT capabilities enhances knowledge process capabilities thereby, maintaining technology-driven environments for social virtual cycle. On reviews, the research idea supported strategic agility of organisation performance that was presented by [23], and as related to recent advances in service computing for enterprise agility. Computing services are the combination of service-oriented systems, service-oriented architecture, service science, management and systems engineering.

As digital transformation enhances, this is classified as another step to enterprise agility, meeting changing customer needs based on new advances of social media in business technology. Further research work of [24], proposed two distinctive roles for good understanding of how IT competency shaped up organisation performance and alignment for IS/IT enterprise agility.

3.2 New Trends and Emergent of IS/IT Systems

Digital technology cases indicated businesses need to adapt to current technological trends to improve their business agility. Emergent system development described how IS/IT rapid growth has a significant degree of coherence and as persisting over time, with strong impact on the socio-economic environment [25]. Therefore, intelligent or autonomous capabilities are required through the simulation of human intelligence, computing deployment architectures, data consumption, with user-friendly technology interfaces of key elements of social networks for business agility based on business processes.

This research paper reviewed emergent of IT technologies to include big data platforms and analytics Internet of things (IoT) and Industrial Internet of Things (IIoT), as key concept to this presented by [26], where block chain technology is the key component of business agility.

As this paper builds on a paper of [26], and as related to sample Fig. 1, where data analysis of many IS/IT solution providers, promises that emergent technologies can enhance enterprise agility, with sound capabilities performances, based on social media. Therefore, the new trend of artificial intelligence of research work of [27] are reviewed and as presented on business operational solutions for machine learning,

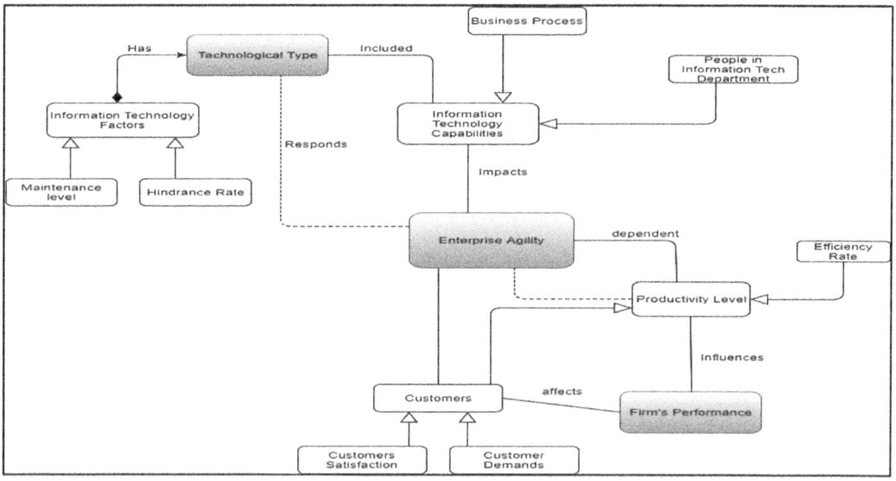

Fig. 1. Conceptual model of the research design for emergent IS/IT systems

which are focused on emergent of social media integration which was earlier discussed by [28]. These research work are characterized by intelligent informative systems and hence, focused on the societal environment of emergent IS/IT systems. As reviewed, this key area of research is seen as a promising research work for enterprise agility and with strategic operational procedures requirements for enabling systems for business processes. To claim that the emergent of IS/IT trends has a strong impact on enterprise agility, orientation of various business products and technology interfaces must be well understood by the business leaders, including social media technology enabler, which are considered as responding to both external and internal change agents. Obviously, data analysis and cognitive aspect of systems computing and organisation readiness to business operations are determined and as referenced to the research work of [29] and [30], which argued consistently, that business enterprise are moving towards a new wave of digital transformation of information systems based on product development and robotics solutions including the use of automated AL/ML algorithms. Questions are raised as there are research gaps in the literature review of how related benefits are determined to businesses and as quantified.

This exploration research review presents how business processes are quantified with values of technology-type for enterprise agility. The conceptual model in Fig. 1, is a visual representation of how business agility is impacted by technology and how enterprise agility can positively effect changes to performance alignment of IS/IT. Solution alignment for business enterprise with effect to customers demand and supply are both directly and indirectly technology factors related and as created with component of technology maintenance for enterprise agility [31]. This research idea reviewed enterprise agility influences for technology solution of business performances, leading to increased productivity in conceptual design of Fig. 1 and thus presented some models intended for the future research work and as represented in Fig. 2.

4 Research Intended Model and Process

As research paper reviewed on the need to build a detailed level of systems, using the stock and flow model described in Fig. 1, there is need for improvements and systems development with counts of systems failures in relations to input process with impacts. Exploration approach of use cases is determined based on systems interdependencies and relationships within the study. As indicated in Sect. 3 above, the research review are relevant to necessary analysis based on solving problems. There are formulated metrics considered below for discussions being the key focus of research work, for further investigations relevant to systems implementation of Fig. 2 below.

(a) What effect of Big Data/Cloud Computing technology for enterprise agility?
(b) How emergent IS/IT systems creates impacts for enterprise agility needs.
(c) What criteria of emergent IT/IS systems are in high priority to be considered for business organisations while selecting on choices of technology for business operational environment.
(d) How can business organisation operational technology choice help to maximize influences on positivity of enterprise agility, whilst an alternate technology parallel to mitigation plan for in case of IT/IS failures.
(e) What are the existing deployment topology design and to investigate architectural studies on key impact of enterprise agility including mixture topologies to actualise benefits for overall agility?
(f) Build on previous studies of theoretical or empirical (qualitative or quantitative) research from a static point of view and to improve on illustrated conceptual model of Fig. 1 with series of tests of experiments.
(g) Knowledge application approach of the conceptual model in Fig. 1 to be reviewed based on new design idea as in (f) above.
(h) Involved simulation of data and analysed impacts caused by technologies on enterprise agility.
(i) In clear term, of new design models, series of test simulations will be created and run parallel to related hypothetical views of enterprise agility.
(j) Effect of cloud computing as referenced to existing technologies of Amazon and Walt Mart SCMS and to explore on the value chain improvement of enterprise agility.

Measuring impact of IT/IS capabilities are key to emergent technology including deployment of architectural topology. System dynamics modelling of Fig. 1 would offer an insight to different emergent technologies for enterprise agility with good understanding of the distant impact of enterprise performance and its existence responds to customer needs. This exploration research reviews intention focus on key needs for digital transformation and as applicable to business organisation for enterprise agility which are represented in Fig. 2. The simulation model of the research may offer a reasonable level of flexibility, and hence, organizations can update impact by evaluation of technology performances on business products and thus, manipulated process constraints to the model under different emergent circumstances. Further approach considered would offer significant results in understanding the relation between IT/IS capabilities and enterprise agility. The applicability by classification and taxonomy effect of changes for enterprise agility can be evaluated upon certain gaps, gathering data for analysis of quantitative and qualitative assessment of operational and strategical data fact of technology-type, resulting to service performance based on emergent IT/IS systems to increasing enterprise agility.

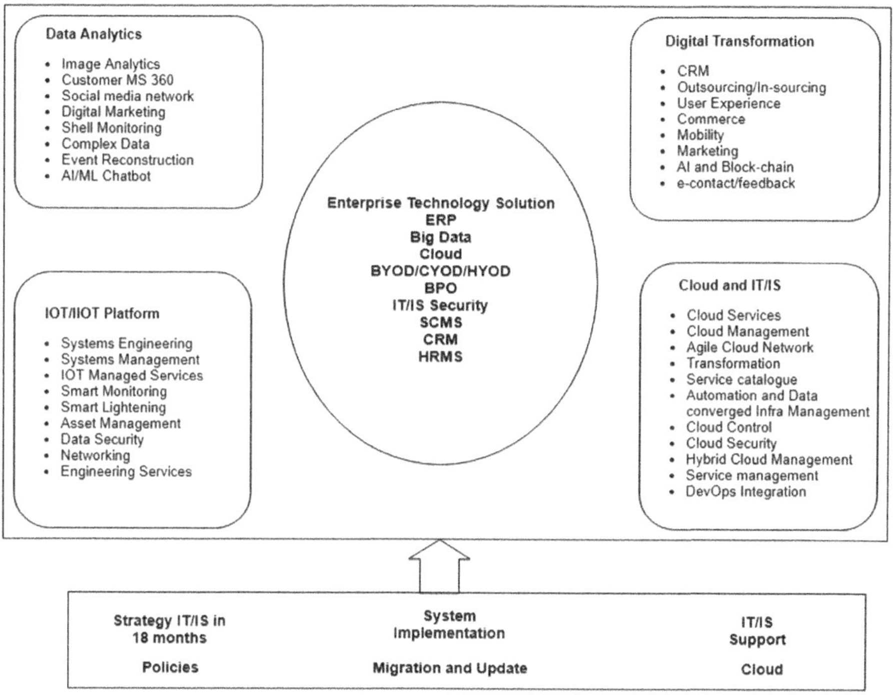

Fig. 2. Innovative digital transformation solutions for enterprise agility

5 Conclusion

Simulation of the model needs to run with real time data from both small and medium organizations. In turn, this may not support claims of adoption of enterprise agility if the IS/IT digital technologies are not in good process of operational systems. Therefore, relevant theories of improved processes and based on hypothetical views of design model adopted may be expanded to include other technological types of emergent technologies such as Internet of Things (IoT) and Industrial Internet of Things (IIoT), referenced to critical natural infrastructure. This aspect of the research considered in the simulation model can improve business processes of relevant technology for positive impact of enterprise agility.

6 Future Work

Possible research work streamlines data gathering to generate automated privileges of system user in enterprise systems with technological enablers of IS/IT framework to be built on data classification based on Fig. 2 and this will provide measures for IS/IT impacts of enterprise agility, for performance enabler of organizational data which can be considered for digital marketing and technology-driven solutions.

Subsequent paragraphs, however, are indented.

References

1. Necula, S., Popescul, D.: Determining the discovery and delivery skills for the Romanian employees – an useful method for organizational decision-making. In: 19th International Business Information Management Association Conference (Innovation Vision 2020, 12–13 November 2012, Barcelona, Spain (2012)
2. Erdomuú, I.E., Çiçek, M.: The impact of social media marketing on brand loyalty. Procedia - Soc. Behav. Sci. **58**, 1353–1360 (2012)
3. Overby, E., Bharadwaj, A., Sambamurthy, V.: Enterprise agility and the enabling role of information technology. Eur. J. Inf. Syst. **15**, 120–131 (2006)
4. Vidru, L.: Utilizarea social media de catre company (2012). http://marketingportal.manager.ro/articole/ultima-ora-74/utilizarea-social-media-de-catre-companii-unde-care-cat-4433. Accessed 08 May 2019
5. Wang (Zack), Z., Pan, S.L., Ouyang, T.H., Chou, T.: Achieving IT-enabled enterprise in China: an IT organizational identity perspective. IEEE Trans. Eng. Manag. **61**(10), 225–400 (2014)
6. Wu, X., Fang, G., Wu, Z.: The dynamic IT capabilities and firm agility: a resource-based perspective. In: International Technology and Innovation Conference (2006)
7. Karvonen, T., Sharp, H., Barroca, L.: Enterprise agility: why is transformation so hard? In: LNBI Processing, vol. 314, pp. 131–145 (2018)
8. Nwokeji, J.C., Aqlan, F., Clark, T., Barn, B., Kulkarni, V.: A modelling technique for enterprise agility. In: Proceedings of the 51st Hawaii International Conference on System Sciences, vol. 9, pp. 4631–4640 (2018). https://doi.org/10.24251/hicss.2018.584
9. Zhu, P.: An agile IT: how to define IT agility (2011). http://futureofcio.blogspot.co.uk/2013/10/how-to-define-it-agility.html. Accessed 25 Apr 2019
10. Varhol, P.: To agility and beyond: the history—and legacy—of agile development. TechBeacon (2017). https://techbeacon.com/agility-beyond-history%E2%80%94-legacy%E2%80%94-agile-development. Accessed 22 Apr 2019
11. Anagnostou, A., Taylor, S.J.E.: Towards a methodology for building large-scale distributed hybrid agent-based and discreet-event simulations: the case of emergency medical services. In: Proceedings of the Operational Research Society Simulation Workshop (2014)
12. Srinivasan, J.: Exploring sources of enterprise agility in software organizations. Mälardalen University Press-School of Innovation Design and Engineering, Mälardalen University Press (2009)
13. Avazpour, R., Ebrahimi, E., Fathi, M.R.: Prioritizing agility enablers based on agility attributes using fuzzy prioritization method and similarity based approach. Int. J. Econ. Manag. Soc. Sci. **3**(1), 143–153 (2014)

14. McCoy, D.W., Plummer, D.C.: Defining, cultivating and measuring enterprise agility. Gartner Research (2006)
15. Tseng, Y., Lin, C.: Enhancing enterprise agility by deploying agile drivers, capabilities and providers. J. Inf. Sci. **181**(1), 3693–3708 (2011)
16. da Silva, D.A., et. al.: Urgent and emergency care: an academic application system case study. In: 6th International Conference on Information Technology-New Generations (ITNG 2019) (2019)
17. Bottani, E.: A fuzzy QFD approach to achieve agility. Int. J. Prod. Econ. **119**, 380–391 (2009)
18. Fontdevila, D., Genero, M., Oliveros, A.: Towards a usability model for software development process and practice. In: International Conference on Product-Focused Software Process Improvement (2017)
19. Patel, B.S., Samuel, C., Sharma, S.K.: Evaluation of agility in supply chains: a case study of an Indian manufacturing organization. J. Manuf. Technol. Manag. **28**(2), 212–231 (2017)
20. Tan, F.T.C., et al.: IT-enabled operational agility: an interdependencies perspective. J. Inf. Manag. Syst. **54**(3), 292–303 (2017)
21. Gandomani, T.J., et al.: Obstacles in moving to agile software development methods-at a glance. J. Comput. Sci. **9**(5), 620–625 (2017)
22. Huang, C., Nof, S.Y.: Enterprise agility: a view from the PRISM lab. Int. J. Agile Manag. Syst. **1**(1), 51–60 (1999)
23. Sambamurthy, V., Bharadwaj, A., Grover, V.: Shaping agility through digital options: reconceptualising the role of information technology in contemporary firms. MIS Q. **27**(2), 237–263 (2003)
24. Chakravarty, A., Grewal, R., Sambamurthy, V.: Information technology competencies, organizational agility and firm performance: enabling and facilitating roles. Inf. Syst. Res. **24** (4), 976–997 (2013)
25. Rotolo, D., Hicks, D., Martin, B.R.: What is an emerging technology? Res. Policy **44**(10), 1827–1843 (2015)
26. Lakhrouit, J., Benhaddi, M., Baina, K.: Enterprise architecture approach for agility evaluation. In: IEEE International Conference on Cloud Technologies and Applications (2015)
27. Information Age: How the Internet of Things is impacting enterprise networks (2017). http://www.information-age.com/internet-things-impacting-enterprise-networks-123463122/. Accessed 18 Mar 2019
28. Varhol, P.: TechBeacon. To agility and beyond: the history—and legacy—of agile development (2017). https://techbeacon.com/agility-beyond-history%E2%80%94-legacy%E2%80%94-agile-development. Accessed 11 Apr 2019
29. Dunlop-Hinkler, D., Parente, R., Marion, T.J., Friar, J.H.: The role of technology agility on business processes and organizational agilities. In: IEEE International.-Technology Management Conference (ITMC) (2011)
30. Hindustantimes: Artificial Intelligence to dramatically impact businesses: TCS (2017). http://www.hindustantimes.com/business-news/artificial-intelligence-to-dramatically-impact-businesses-tcs/story-1ALbDKKR8lCIaD2sLtDZNM.htm. Accessed 19 June 2019
31. Tan, F.T.C., Wang, W., Tan, B.: Towards management innovation for IT enabled operational agility: a dependencies perspective. In: PACIS 2014 Proceedings Paper 345 (2014)

Cloud Computing Services Adoption by University Students: Pilot Study Results

Abdulwahab Ali Almazroi[1(✉)], Haifeng Shen[2], Fathey Mohammed[3], and Nabil Hasan Al-Kumaim[4]

[1] Faculty of Computing and Information Technology, University of Jeddah, Khulais, Saudi Arabia
aalmazroi@uj.edu.sa

[2] Faculty of Law and Business, Australian Catholic University, Brisbane, Australia

[3] School of Computing, Universiti Utara Malaysia, Bukit Kayu Hitam, Malaysia

[4] Faculty of Technology Management and Technopreneurship, Universiti Teknikal Malaysia, Melaka, Malaysia

Abstract. Cloud computing applications provide computing resources and services on demand via the Internet. Implementation and adoption of cloud services can lead to improved performance and reduced cost related to the computing services. In developing countries, more specific in the high education sector, cloud computing services adoption is low. This paper presents the pilot study which is conducted to validate the instrument developed for examining the factors affecting the adoption of cloud computing services by university students. A small sample of university students in Saudi Arabia was targeted to collect the data for validating the questionnaire which was developed by adapting TAM3 and related studies constructs' items. The results indicate that all the measures used in the questionnaire are reliable.

Keywords: Cloud services · University students · Technology acceptance model · Adoption

1 Introduction

Cloud computing is an advanced technology that transformed different areas of businesses including, commerce, healthcare and education. In the context of education, cloud computing has been widely used. It provides students with scalable on demand and affordable computing resources to store, manipulate and share their data on the cloud [1]. Cloud-based services can offer students cost savings and access to scalable computing power [2–4]. The computing power provided by the cloud offers the learner greater participation and control of the learning process. 'Microsoft live@edu' giving students access to email, Skydrive and online office package are among the most cloud services adopted by students. These services provide students free access to the most recent Microsoft items from anywhere. Google App Education (GAE) is another cloud-based service for students. It includes components such as Talk, Docs, and Mail which offer rich advantages upgrading the experience of learning and online student joint

F. Saeed et al. (Eds.): IRICT 2019, AISC 1073, pp. 1052–1060, 2020.
https://doi.org/10.1007/978-3-030-33582-3_99

efforts. Google also provides Google Apps for Education Suite including Google Docs, classroom management system, email service, shared storage (Google Drive), site creation, hosting, and tools for collaboration. Although cloud computing applications in the education context may still at an early stage of development, several advantages for the learning process are identified by the existing research. González-Martínez, Bote-Lorenzo, Gómez-Sánchez, and Cano-Parra, [5] explored the benefits of cloud computing services for the learning process in terms of flexible learning environment, online availability, computing-intensive support for learning; mobile learning, scalability of learning applications; and cost savings.

Higher education institutions in Saudi Arabia adopt cloud computing services to transform their learning and teaching activities to provide a more competitive and robust environment [6–9]. However, existing literature is mainly focusing on the adoption of cloud computing from organizations perspective (IT personnel and decision makers) considering security concerns, implementation, skills, etc. and there is a lack of studies address the adoption and use of the cloud computing services by students [6, 10, 14]. Ibrahim, Salleh, and Misra, [15] conducted a systematic literature review and indicated that there is a lack of empirical studies that identify factors affecting the adoption of cloud computing by students. Therefore, to deliver the maximum benefit of cloud computing applications in the education context, more research is required to identify the factors influencing students' adoption of cloud services. This research aims at presenting the pilot study results as a process of validating the questionnaire used in [16] to identify and examine the factors affecting adoption of cloud computing services by students.

The rest of the paper is organized as follows. Section 2 presents a review of studies on cloud computing adoption. Section 3 justifies the selection of Technology Acceptance Model 3 (TAM3). Section 4 presents the research methodology. Section 5 presents the pilot study results. Finally, Sect. 6 concludes the paper.

2 Related Work

In the literature, factors influencing cloud computing services adoption have been investigated from two different perspectives; organizations and individuals [17, 18]. From the organizations view of point, previous studies identified and examined implementation related factors such as infrastructure, security, top management support, innovative, content design, technology readiness, etc. [19–22]. On the other hands, investigating cloud services from individuals perspective considered different groups including teachers [23–25], staff [25–28], IT professionals [12, 14, 19, 20] and students [27–33]. Table 1 summarizes the related studies indicating the country, the subject, the context and the theory applied. It can be concluded that there is a lack of studies investigating the factors influencing cloud services adoption by students of Saudi universities. Identifying the factors influencing the adoption can promote cloud computing services and applications. Previous studies examined these factors by applying different theories and models. As Table 1 show, the most frequent model applied in the context of the cloud services adoption by individuals is TAM. Table 2 represents the factors identified in the previous studies.

Table 1. Summary of the previous studies on cloud computing services adoption

Author(s)	Country	Subject	Context	Theory used
(Ratten 2016)	Australia	Managers	Organization	Social Cognitive Theory (SCT)
(Arpaci 2016)	Turkey	Students	Education	TAM
(Hew & Kadir 2016)	Malaysia	Teachers	Education	Self determination theory, and channel expansion theory
(Alhammadi et al. 2015)	Saudi Arabia	Computing professionals	Organization	DOI and TOE
(Park et al. 2015)	China and Korea	College students and employees	End-user	-
(Ratten 2015)	USA & Turkey	Students	End-user	SCT and TAM
(A. Tashtandi & I.M. Al-Jabri 2015)	Saudi Arabia	IT decision makers	Education	TOE
(Abu-Shanab & Qasem 2014)	-	Individuals	End-user	-
(Burda & Teuteberg 2014)	German	Students and staff	End-user	TAM
(Al-Jabri 2014)	Saudi Arabia	IT managers. IT consultants, and IT professionals	Organization	TOE
(Alotaibi 2014)	Saudi Arabia	IT professionals and end users	End-user	TAM
(Nguyen et al. 2014)	Vietnam	Users and potential users of cloud based e-Leaming systems	Education	UTAUT2
(Trenz et al. 2013)	German	Students	End-user	Principal-Agency Theory
(Cao et al. 2013)	China	College students, faculty members, and other people	End-user	UTAUT
(Alharbl 2012)	Saudi Arabia	Employees of an IT organizations	Organization	TAM
(Li & Chang 2012)	Taiwan	Students	End-user	TAM and TPB
(Behrend et al. 2011)	USA	Student	Education	TAM3

Table 2. Factors influencing cloud computing services adoption by individuals

Factors examined	(Arpaci 2016)	(Park et al. 2015)	(Men 2015)	(Burda & Teuteberg 2014)	(Trenz et al. 2013)	(Cao et al. 2013)	(Li & Chang 2012)	(Behrend et al. 2011)
Perceived ease of use	✓		✓	✓			✓	✓
Perceived ubiquity	✓							
Perceived security	✓		✓		✓		✓	
Perceived privacy	✓		✓		✓		✓	
Perceived usefulness	✓			✓			✓	✓
Trust	✓	✓		✓	✓			
Subjective norm	✓					✓	✓	
Attitude	✓						✓	
Risk		✓	✓	✓		✓	✓	
Self-efficacy			✓					
Perceived behavioural control							✓	
Awareness			✓					
Performance expectancy			✓			✓		
Reputation				✓			✓	
Costs					✓	✓		
Availability				✓	✓			✓
Uncertainty					✓			
Satisfaction				✓	✓			
Facilitating conditions								
Effort expectancy								
Personal innovativeness								✓
Technology anxiety								✓

3 Research Model

The common applied models in studying technology adoption from the perspective of individuals and more specific in cloud computing services adoption literature are TAM, UTAUT, and DOI [20, 22, 24]. However, these models still need further investigation to improve their ability of predicting outcomes, identifying beliefs related to specific behaviour, and considering factors such as personal control and social factors [34–36]. TAM has been adapted in the literature and two extensions were proposed and validated; TAM2 by Venkatesh and Davis [37] and TAM3 by Venkatesh and Bala [38].

TAM3 adapts the original TAM by identifying and classifying the influencing factors into four groups social, technology characteristics, individual differences, and facilitating conditions. This can provide valuable insights into how a new technology can be adopted Al-Gahtani [39]. TAM3 has been evaluated and validated across different contexts including health informatics, m-commerce, e-Learning [39–42] as well as cloud services adoption by students Behrend, Wiebe [33]. Potentials of TAM3, its comprehensiveness, suitability and reliability makes it the suitable model for examining factors affecting students' use of cloud computing in Saudi universities. Technology characteristics, social influence, facilitating conditions and individual differences (included in TAM3) were proposed to understand the factors influencing cloud computing applications adoption behaviour by university students.

TAM3 was adapted by incorporating constructs related to the context. In the context of cloud computing adoption, various studies showed that trust influences the adoption process [43, 44]. In addition, objective usability construct which was used to measure the novice-to-expert ratio of effort can be ignored since, in the current study, Google Docs is used as a cloud service application, which does not support the novice-to-expert ratio of effort to measure objective usability. Furthermore, voluntariness construct is omitted because of voluntariness nature of cloud services use by students. Finally, experience as a moderator is adapted to be Internet experience reflecting the technology under investigation (cloud computing applications) as it is Internet-based applications and determining the effect of Internet experience may help explain the behavioural intention better.

4 Research Methodology

To evaluate the feasibility of studying the influencing factors, a small scale preliminary study is conducted. A survey questionnaire is used to collect the data for the purpose of this pilot study. The questionnaire was developed by adapting the constructs scales of the previous related studies [38, 45, 52]. The questionnaire was verified by 9 experts and modified based on their feedback.

Undergraduate students of a Saudi university were the respondents of the questionnaire. 31 students were selected randomly to fill up the questionnaire based on a practical session. The practical session aimed at introducing and demonstrating a cloud computing application (Google Docs). The students were asked to perform simple tasks on Google Docs to be familiar with the cloud services, so they can answer the questionnaire efficiently. Out of the 31 distributed questionnaires, 29 were usable for reliability test. The measurement reliability was evaluated by assessing the internal consistency of each construct measurement scale. Cronbach's alpha coefficients and factor loading were used to assess the internal consistency of the construct measures.

5 Pilot Study Results

Table 3 represents the results of reliability assessment using SPSS version 19. The results (see Table 3) shows that Cronbach's alpha for all factors measurements are above 0.7 which indicate the reliability of the measurement based on the criteria of Sekaran [53] and Hair, Black [54]. In addition, the results of factor loadings for all items show the value above 0.5 which indicate that all the items are reliable to measure the corresponding construct based on the criteria of Hair, Black [54].

Table 3. Cronbach's alpha and factor loadings results

Construct	No. of items	Factor loading	Cronbach's Alpha (α)
Perceived usefulness	6	0.694, 0.666, 0.649, 0.699, 0.568, 0.617	0.849
Perceived ease of use	6	0.718, 0.550, 0.747, 0.652, 0.759, 0.677	0.861
Behavioural intention	3	0.714, 0.788, 0.728	0.862
Self-efficacy	5	0.768, 0.746, 0.673, 0.608, 0.678	0.865
Playfulness	4	0.525, 0.631, 0.592, 0.783	0.799
Perceptions of external control	5	0.501, 0.675, 0.636, 0.668, 0.530	0.802
Anxiety	4	0.758, 0.789, 0.782, 0.766	0.896
Perceived enjoyment	3	0.722, 0.668, 0.647	0.822
Subjective norm	3	0.765, 0.775, 0.755	0.875
Image	3	0.749, 0.677, 0.695	0.839
Job relevance	3	0.737, 0.771, 0.718	0.862
Output quality	3	0.720, 0.750, 0.665	0.843
Result demonstrability	4	0.693, 0.765, 0.747, 0.758	0.875
Trust	5	0.674, 0.726, 0.737, 0.654, 0.698	0.871

6 Conclusion

Identification of the factors affecting university students' adoption of cloud services is critical for promoting higher education institutions in Saudi Arabia to implement and support cloud services. Understanding the factors that influence the adoption can help in reducing the rejection rate. This study validated the instrument used to examine the impact of PU, PEOU, output quality, job relevance, result demonstrability, self-efficacy, anxiety, perceptions of external control, playfulness, perceived enjoyment, subjective norm, image, and trust on university students' adoption of cloud applications. The related measurements were adapted from the literature related to TAM3 by considering the context of the study. A pilot test was conducted to examine the reliability of the questionnaire. The results showed that all the measurement scales were reliable. Thus, the questionnaire is reliable and can be used to collect data in a large scale. This study can be considered one of the few studies that examined the reliability of a questionnaire for an extended TAM3 model in the context of cloud computing adoption by students. Certainly, this may add to the body of knowledge on cloud services adoption in developing countries especially in Saudi Arabia in the context of higher education institutions.

References

1. Sultan, N.: Cloud computing for education: a new dawn? Int. J. Inf. Manag. **30**(2), 109–116 (2010)
2. Buyya, R., Broberg, J., Goscinski, A.M.: Cloud Computing: Principles and Paradigms, vol. 87. Wiley, Hoboken (2010)
3. Armbrust, M., et al.: Above the Clouds: a Berkeley View of Cloud Computing 2009. University of California at Berkeley, Berkeley (2016)
4. Motta, G., Sfondrini, N., Sacco, D.: Cloud computing: an architectural and technological overview. In: 2012 International Joint Conference on Service Sciences. IEEE (2012)
5. González-Martínez, J.A., et al.: Cloud computing and education: a state-of-the-art survey. Comput. Educ. **80**, 132–151 (2015)
6. Tashkandi, A.N., Al-Jabri, I.M.: Cloud computing adoption by higher education institutions in Saudi Arabia: an exploratory study. Cluster Comput. **18**(4), 1527–1537 (2015)
7. AlCattan, R.F.: Integration of cloud computing and web 2.0 collaboration technologies in E-learning. Int. J. Comput. Trends Technol. **12**(1), 46–55 (2014)
8. Ministry of Communications and Information Technology: $29 billion, the size of Saudi Arabia's investments in cloud (2014). http://mcit.gov.sa/En/InformationTechnology/Pages/ReportsandStatistics/Tech-Reports23112014_356.aspx. Accessed 5 Feb 2015
9. Alsanea, M., Wainwright, D.: Identifying the determinants of cloud computing adoption in a government sector – a case study of Saudi organisation. Int. J. Bus. Manag. Stud. **6**(2), 29–43 (2014)
10. Sabi, H.M., et al.: Conceptualizing a model for adoption of cloud computing in education. Int. J. Inf. Manag. **36**(2), 183–191 (2016)
11. Irshad, M.B.M., Johar, M.G.M.: A study of undergraduate use of cloud computing applications: special reference to Google Docs. Eur. J. Comput. Sci. Inf. Technol. **3**(4), 22–32 (2015)
12. Tashkandi, A., Al-Jabri, I.M.: Cloud computing adoption by higher education institutions in Saudi Arabia. In: 2015 International Conference on Cloud Computing, Riyadh, Saudi Arabia (2015)
13. Hashim, H.S., Hassan, Z.B., Hashim, A.S.: Factors influence the adoption of cloud computing: A comprehensive review. Int. J. Educ. Res. **3**(7), 295–306 (2015)
14. Alotaibi, M.B.: Exploring users' attitudes and intentions toward the adoption of cloud computing in Saudi Arabia: an empirical investigation. J. Comput. Sci. **10**(11), 2315–2329 (2014)

15. Ibrahim, M.S., Salleh, N., Misra, S.: Empirical studies of cloud computing in education: a systematic literature review, In: The 15th International Conference on Computational Science and Its Applications, pp. 725–737. Springer, Alberta, Canada (2015)
16. Almazroi, A.A., Shen, H., Mohammed, F.: The impact of trust on the adoption of cloud computing services by university students. In: International Conference of Reliable Information and Communication Technology. Springer (2018)
17. AlAjmi, Q., et al.: A conceptual model of e-learning based on cloud computing adoption in higher education institutions. In: 2017 International Conference on Electrical and Computing Technologies and Applications (ICECTA). IEEE (2017)
18. Al-Sharafi, M.A., Arshah, R.A., Abu-Shanab, E.A.: Factors influencing the continuous use of cloud computing services in organization level. In: Proceedings of the International Conference on Advances in Image Processing. ACM (2017)
19. Alhammadi, A., Stanier, C., Eardley, A.: The determinants of cloud computing adoption in Saudi Arabia. In: Second International Conference on Computer Science & Engineering. AIRCC Publishing Corporation, Duba, United Arab Emirates (2015)
20. Al-Jabri, I.M.: The perceptions of adopters and non-adopters of cloud computing: application of technology-organization-environment framework. In: 14th International Conference on Electronic Business, Taipei, Taiwan (2014.)
21. Opala, O.J., Rahman, S.M.: An exploratory analysis of the influence of information security on the adoption of cloud computing. In: 2013 8th International Conference on System of Systems Engineering. IEEE, Wailea-Makena, Hawaii, USA (2013)
22. Alharbi, S.T.: Users' acceptance of cloud computing in Saudi Arabia: an extension of technology acceptance model. Int. J. Cloud Appl. Comput. **2**(2), 1–11 (2012)
23. Hew, T.-S., Kadir, S.L.S.A.: Predicting the acceptance of cloud-based virtual learning environment: the roles of self determination and channel expansion theory. Telemat. Inf. **33** (4), 990–1013 (2016)
24. Cao, Y., Bi, X., Wang, L.: A study on user adoption of cloud storage service in China: a revised unified theory of acceptance and use of technology model. In: 2013 International Conference on Information Science and Cloud Computing Companion. IEEE, Guangzhou, China (2013)
25. Nguyen, T.D., et al.: Acceptance and use of E-learning based on cloud computing: the role of consumer innovativeness. In: International Conference on Computational Science and Its Applications. Springer, Guimarães, Portugal (2014)
26. Ratten, V.: Continuance use intention of cloud computing: innovativeness and creativity perspectives. J. Bus. Res. **69**(5), 1737–1740 (2016)
27. Park, S.-T., et al.: Factors affecting the continuous use of cloud service: focused on security risks. Cluster Comput. **19**(1), 485–495 (2015)
28. Burda, D., Teuteberg, F.: The role of trust and risk perceptions in cloud archiving—results from an empirical study. J. High Technol. Manag. Res. **25**(2), 172–187 (2014)
29. Arpaci, I.: Understanding and predicting students' intention to use mobile cloud storage services. Comput. Hum. Behav. **58**, 150–157 (2016)
30. Ratten, V.: Factors influencing consumer purchase intention of cloud computing in the United States and Turkey: the role of performance expectancy, ethical awareness and consumer innovation. EuroMed J. Bus. **10**(1), 80–97 (2015)
31. Trenz, M., Huntgeburth, J.C., Veit, D.: The role of uncertainty in cloud computing continuance: antecedents, mitigators, and consequences. In: 21st European Conference on Information Systems, Utrecht, Netherlands (2013)
32. Li, Y., Chang, K.-C.: A study on user acceptance of cloud computing: a multi-theoretical perspective. In: Eighteenth Americas Conference on Information Systems. Association for Information Systems AIS Electronic Library (AISeL), Washington, USA (2012)

33. Behrend, T.S., et al.: Cloud computing adoption and usage in community colleges. Behav. Inf. Technol. **30**(2), 231–240 (2011)
34. Clarke, R.: A primer in diffusion of innovations theory (1999). http://www.rogerclarke.com/SOS/InnDiff.html. Accessed 15 Mar 2016
35. Liu, J.: E-learning in english classroom: Investigating factors impacting on ESL (english as second language) college students' acceptance and use of the modular object-oriented dynamic learning environment (Moodle). Iowa State University - Ames, Iowa, USA (2013)
36. Rui, G.: Information systems innovation adoption among organizations - a match-based framework and empirical studies, in Information System. National University of Singapore – Singapore (2007)
37. Venkatesh, V., Davis, F.D.: A theoretical extension of the technology acceptance model: four longitudinal field studies. Manag. Sci. **46**(2), 186–204 (2000)
38. Venkatesh, V., Bala, H.: Technology acceptance model 3 and a research agenda on interventions. Dec. Sci. **39**(2), 273–315 (2008)
39. Al-Gahtani, S.S.: Empirical investigation of E-learning acceptance and assimilation: a structural equation model. Appl. Comput. Inf. **12**(1), 27–50 (2014)
40. Chang, S.J., Im, E.-O.: A path analysis of internet health information seeking behaviors among older adults. Geriatric Nurs. **35**(2), 137–141 (2014)
41. Faqih, K.M., Jaradat, M.-I.R.M.: Assessing the moderating effect of gender differences and individualism-collectivism at individual-level on the adoption of mobile commerce technology: TAM3 perspective. J. Retail. Consum. Serv. **22**, 37–52 (2015)
42. Agudo-Peregrina, Á.F., Hernández-García, Á., Pascual-Miguel, F.J.: Behavioral intention, use behavior and the acceptance of electronic learning systems: differences between higher education and lifelong learning. Comput. Hum. Behav. **34**, 301–314 (2014)
43. Van der Schyff, K., Krauss, K.E.: Higher education cloud computing in South Africa: towards understanding trust and adoption issues. S. Afr. Comput. J. **55**, 40–55 (2014)
44. Alharbi, S.T.: Trust and acceptance of cloud computing: a revised UTAUT model. In: 2014 International Conference on Computational Science and Computational Intelligence. IEEE, Las Vegas, Nevada, USA (2014)
45. Davis, F.D.: Perceived usefulness, perceived ease of use, and user acceptance of information technology. MIS Q. 319–340 (1989)
46. Compeau, D.R., Higgins, C.A.: Computer self-efficacy: development of a measure and initial test. MIS Q. 189–211 (1995)
47. Mathieson, K.: *Predicting user intentions: Comparing the technology acceptance model with the theory of planned behavior*. Information systems research **2**(3), 173–191 (1991)
48. Taylor, S., Todd, P.A.: Understanding information technology usage: a test of competing models. Inf. Syst. Res. **6**(2), 144–176 (1995)
49. Heinssen, R.K., Glass, C.R., Knight, L.A.: Assessing computer anxiety: development and validation of the computer anxiety rating scale. Comput. Hum. Behav. **3**(1), 49–59 (1987)
50. Davis, F.D., Bagozzi, R.P., Warshaw, P.R.: Extrinsic and intrinsic motivation to use computers in the workplace. J. Appl. Soc. Psychol. **22**(14), 1111–1132 (1992)
51. Moore, G.C., Benbasat, I.: Development of an instrument to measure the perceptions of adopting an information technology innovation. Inf. Syst. Res. **2**(3), 192–222 (1991)
52. Pavlou, P.A.: Consumer acceptance of electronic commerce: Integrating trust and risk with the technology acceptance model. Int. J. Electron. Commer. **7**(3), 101–134 (2003)
53. Sekaran, U.: Research Methods for Business: A Skill Building Approach, 4th edn. Wiley, Hoboken (2003)
54. Hair, J.F., et al.: Multivariate Data Analysis. Pearson, London (2010)

Development of a Theoretical Framework for Customer Loyalty in Australia

Hassan Shakil Bhatti(✉), Ahmad Abareshi, and Siddhi Pittayachawan

RMIT University, Melbourne, VIC 3000, Australia
Hassan.shakilbhatti@rmit.edu.au

Abstract. Companies increasingly recognise customer loyalty as one of the most important measures for the business of an organisation, and informed knowledge of factors underpinning customer loyalty may help managers build customer retention strategies. Telecommunication ombudsman reports demonstrate that there are service quality and customer retention issues in the mobile telecommunication sector in Australia which can distress end customers and businesses. This study classifies and discovers key elements impacting customer loyalty in mobile telecommunication services in Australia. Moreover, antecedents such as behavioural intention which leads to customer intention to stay loyal with the services are measured through the Unified Theory of Acceptance and Use of Technology (UTAUT2), Marketing Mix Theory and Expectation Confirmation Theory (ECT). Furthermore, investigators have studied the habit, hedonic motivation, customer satisfaction, customer experience, marketing mix factors relationship by empirical testing. It has been determined that very little research is done on customer loyalty in mobile telecommunication services in Australia. Illustration upon theories of the marketing mix, ECT and UTAUT2, this study aims to determine what factors affect customer loyalty (CL) in mobile telecommunication services in Australia. This research outline delivers an extension to the current UTAUT2 model, and this model also offers a strategy for sustaining customer loyalty in mobile service businesses. This theoretical framework can be used in any Telecommunication and IT business and it can help in increasing and retaining the customer base.

Keywords: UTAUT2 · Customer loyalty · Mobile service · Australia

1 Introduction

There is a substantial reputation for customer loyalty in the service industries [1, 2]. Customer loyalty has been observed as the forte of the relationship between a person's relative attitude and repeats patronage. Customers are experiencing digital presence through mobile and wearable internet technologies by allowing them to network with emerging and new technologies [3]. Nowadays customers are more connected with these technologies, which can deliver services, relevant communication through geo-locations, banking information, health services and behaviour monitoring. The mobile and telecommunication technology has been advanced with the passage of time and service providers are doing network upgrades. There is a furious competition among service providers for more competitiveness and customers' acquirement. In Australia,

F. Saeed et al. (Eds.): IRICT 2019, AISC 1073, pp. 1061–1075, 2020.
https://doi.org/10.1007/978-3-030-33582-3_100

there is an independent body Telecommunications Industry Ombudsman (TIO), which focuses on resolving customer complaints effectively. In the latest TIO report TIO [4], there are 158,016 mobile phone and internet complaints received by an ombudsman. Moreover, as per Telecommunication industry reports TIO [5], customers are providing grievances statements about their mobile network service, network quality, legal contract documents, customer service and mobile billing issues such as over charge and un clear hidden charges. The network service quality has a significant positive influence on customer satisfaction and customer loyalty [6]. The independent variable for this research customer satisfaction is gauged in the industry by a tool called Net Promoter Score (NPS) in the service industry. The TIO intelligence documents suggest that NPS which is used to measure customer satisfaction for many service providers has been dropped intensely down, because of it there is a considerable increase in customer service complaints. This study assesses the antecedents which can affect customer loyalty in the mobile telecommunication sector. The theoretical underpinning and research framework for this study is grounded on the unified theory of acceptance and use of technology (UTAUT 2), marketing mix theory (MT) and expectation confirmation theory (ECT). The acknowledged factors for this study through the extensive literature review such as customer experience, customer satisfaction, effort expectancy, performance expectancy, facilitating conditions, social influence, price value, habit and hedonic motivation are resultant from previous studies [7–12]. Now-a-day, business models are rapidly adopting the strategy of encompassing new technologies with behavioural marketing and analytics reaching new stages of sophistication [3]. The research tool which is an online survey has been established for this study with 64 items (Appendix 1). The online research study is conducted with 364 valid participants via Qualtrics tool (www.qualtrics.com) for collecting the data. The research analysis for this study has been completed in smart PLS version 3.0 using partial least square path modelling technique (PLS-PM). This study will give future directions to businesses and researchers. This study will help in the formation of a theoretical model which can be implemented in service industries.

2 Literature Review

Economic growth boosted mobile telecommunication users and encouraged service companies to launch new services one by another in recent years. Changes in technology and the arrival of new compatible handsets in the market were essential to drive mobile companies to introduce innovative and competitive services. Customers nowadays value more on service quality, experience and satisfaction. Thus, this study explored the impacts of key factors in customer repurchase behaviour from Australian mobile service industry. Customer satisfaction, customer intention and customer experience have long attracted the researchers [13] due to the prime impact on the customer post-purchase behaviour such as repurchase behaviour. In previous studies, customer repurchase behaviour is not investigated with marketing mix factor, customer experience, habit and hedonic motivation. According to the consumer behaviour theories [14–17], satisfaction and retention are investigated but other factors such as the potential commercial and the marketing mix factors, product price customer experience, habit, hedonic motivation, effort expectancy, performance expectancy, facilitating conditions and social influence are less investigated. However, Venkatesh, Thong and Xu [11] empirically tested in UTAUT 2 model the relationship of user behaviour and behavioural intention but issues such as customer satisfaction and customer experience are not investigated in this framework.

Park and Kim [18] have investigated online shopping attributes for repurchasing services. Furthermore, it is discussed that customers tend to engage in relational behaviours to acquire a more cognitive approach in their decisions for repurchasing a service. Moreover, it helps in reducing the perceived risks associated with future adoptions [19]. Additionally, it is further argued that the availability of information is crucial for making a purchasing decision [20]. Therefore, not only the availability of information for products and service is important, but it also provides ease and personalization in service provisioning which also helps in retaining the customers. Francken [15] argued that customer buying decision for a product or service discreetly depends upon three things such as category (generic choice), product/service class (model choice), and in the end brand (specific choices). So, marketing mix factors such as product or service orientation and facilitating conditions including, availability of information are considered as very important factors.

In different studies, relationships among corporate image, alternative attractiveness, pricing, social influence, trust, perceived value and service quality which are strongly related to behavioural factors such as customer loyalty, continuance usage and relationship commitment have been observed in past literature [21–26], but not as much of attention has been given to the customer repurchase behaviour in the mobile telecommunication service industry. There is no experiential study to date has been conducted which examined these emerging factors in a single framework of the study, as all the above-mentioned studies investigated the relationship with customer loyalty in a different framework. An insufficient statistical examination has been done to classify factors impacting on customer loyalty. Venkatesh, Thong and Xu [11] UTAUT 2 model only focuses on user behaviour but not on customer loyalty. This study will aid to mature a complete model which will emphasis on technical service quality and customer behavioural matters which can help businesses progress customer loyalty for repurchasing a mobile service. Consequently, this study is intended to report the research issues in the literature by leading empirical research using Unified Theory of Acceptance and Use of Technology (UTAUT2) on customer behavioural intention to stay loyal with the mobile service provider. These findings from this study will also observe retention with the help of linked factors in the background of telecommunication mobile services in Australia.

3 Research Questions and Objectives

This study venture aims to discover the issues that affect customer loyalty in the telecommunication sector of Australia. The chief objective of this study is to develop a theoretical framework to examine the factors impacting customer loyalty in the mobile telecommunication service business. In order to achieve this objective, numerous queries have been expressed, as follows:

1. What are the variables that effect customer loyalty in the Australian telecommunication segment?
2. To what degree do these variables impact customer loyalty?
3. What is the impact of age, gender and experience as a moderating factor between independent variables and dependent variable in the extended Unified Theory of Acceptance and Use of Technology model?

These research questions are achieved with the help of some specific aims and objectives as mentioned below.

Aims and objectives

- This study will discover the variables which can impact customer loyalty.
- This research study agenda will be used by telecommunication business managers and business components in articulating their business plans and strategies.
- This study model will help the companies in refining their client base.
- This learning framework will suggestively contribute to the body of knowledge by statistically testing a new model for customer repurchase behaviour.

4 Theoretical Framework

The information system (IS) academics and marketing researchers have used an unlike model for challenging customer behaviour and implementation of new technologies. There are many established models which have explained the areas and extents of psychology, information systems, business and sociology. Previous studies have used a numerous range of theories and models to scrutinize the adoption, usage and behaviour intention of customers towards cyber security, broadband, IT, banking, mobile telecommunication services and technologies. The research framework for these studies includes models including the Theory of Reasoned Action (TRA) [27]; the Diffusion of Innovation Theory [28]; the Theory of Planned Behaviour (TBP) [29]; the Technology Acceptance Model (TAM) [30]; the Motivational Model [31]; the Model of PC Utilisation[32]; the Social Cognitive Theory (SCT) [33]; the Technology Organisation Environment (TOE) [34]; and the UTAUT [10, 11]. Adaption of TAM has been used to explain technology acceptance in the field of telecommunication and IT [35]. There is a fairly valuable feature of TAM, which is the effective extension of essential constructs of the model to hold additional exterior variables that can be applicable to the background of different studies [36, 37].

Grounded on numerous aspect, evaluation and combination of different theories, Venkatesh, Morris, Davis and Davis [10] proposed the Unified Theory of Acceptance and Use of Technology (UTAUT), which has been systematically and rigorously used in many areas of IS/IT acceptance and adoption. UTAUT model explained an acceptable amount of variance in behavioural intention and usage. Additionally, the model conceptualised some associations that may not be appropriate to all situations and circumstances, omitted some associations that may be hypothetically vital, and also omitted some variables that be significant for updating IS/IT acceptance and use [38].

Furthermore, the UTAUT model is reviewed with some supplementary variables i.e. Habit, Hedonic Motivation and Price Value etc. [11].

Besides this it is important to explain the theory and classifying the forthcoming guidelines, the researcher for this study has endeavoured to critically review and boost the revised original UTAUT model with an extension. In this theoretical model, the researcher has claimed that this model is not applicable in all areas and contexts and modification for the current UTAUT model is necessary. The foremost dependent variable behavioural intention is modified to customer loyalty with the accumulation of two new variables such as customer satisfaction and customer experience. In the UTAUT model, individual features such as attitudinal loyalty, customer experience and customer satisfaction not conceptualised in the original UTAUT model should be introduced. In this study, customer loyalty has been inspected with reviewed extended UTAUT model with an amalgamation of IS and marketing concept. The marketing notions such as product price value and customer loyalty are formed by the use of marketing mix theory. Likewise, the part of customer satisfaction is introduced in the original revised UTAUT model with the help of customer expectation theory. This extended revisited model by the researcher is empirically verified in telecommunication setting using Smart PLS method.

Consequently, the theoretical research outline for this study is built on the perceptions of particular marketing-mix variables from 7Ps such as product, place, promotion, physical evidence, process and price [39] along with the SERVQUAL model which will measure service quality for customer experience, Unified Theory of Technology Acceptance (UTAUT 2) and Expectation Confirmation Theory (ECT). The following segment will confer the detail of these theories which comprises of the theories of IS/IT acceptance, marketing and customer satisfaction with consideration to the UTAUT 2 model and formalises the revised research theoretical model.

4.1 Marketing-Mix Theory

The marketing mix perception is one of the essential notions of marketing theory. McCarthy [40] enlightened the knowledge of chief marketing mix 4P's (product, price, promotion and place). Likewise, different alterations to 4P's marketing mix framework have been planned in several studies [41–43]. McCarthy [40, p. 35] further articulates the marketing mix as a mixture of all the factors at the marketing manager's command to content the target market. McCarthy and Perreault [44] have cleared the marketing mix as the manageable variable that an organization can organize to satisfy its target market. This classification of this phenomenon is more expanded by Kotler and Armstrong [45] as the set of governable marketing variables that the company balances to produce the answer it needs in the target market. Manufacturing marketers have demanded that the marketing mix makes it exclusive due to its structures and hence it becomes diverse from consumer marketing [46]. Bitner, Booms and Tetreault [47] have adapted the outdated 4P's framework to 7P's marketing mix. Hereafter, this alteration comprises the service marketing obligation including physical evidence and process. Furthermore, their framework contains the product, price, place, promotion, participants, physical evidence and process. Consequently, marketing mix elements such as price, product, promotion, physical evidence, process, and place will be used to control customer retention in the telecommunication sector. The unified theory of acceptance and use of technology along with the support of the marketing mix fundamentals in the revised model (UTAUT 2) Venkatesh, Thong and Xu [11] explains the theoretical framework for this study.

4.2 The Unified Theory of Acceptance and Use of Technology

The Unified Theory of Acceptance and Use of Technology (UTAUT) is a creation of the study directed by [10]. Venkatesh, Morris, Davis and Davis [10] studied and compared the eight dominant models that have been used to determine technology acceptance behaviour. These models are Theory of Reasoned Action (TRA) [48], the Technology Acceptance Model (TAM) [49], the Motivational Model [31], the Theory of Planned Behaviour (TPB) [29], the combined Technology Acceptance Model and Theory of Planned Behaviour [29, 50], the Model of PC Utilization [32], the Diffusion of Innovation Theory [51, 52] and the Social Cognitive Theory [33, 53]. Consequently, after combining and synthesising all these models, a new model was made which was drawn as the Unified Theory of Technology Acceptance and Use of Technology (UTAUT) [10].

4.3 Planned Model

The extended model for this research has been created on the grounds of the UTAUT 2 model, marketing mix theory, expectation confirmation theory and previous studies [10]. Cronin, Brady and Hult [54] reflected four models in their research which display that the service value has a significant influence on behavioural intention. Correspondingly, customer satisfaction is assertively associated with behavioural intention. Additionally, Ryu, Han and Kim [55] discover the relationship between perceived value and behavioural intention. Also, customer satisfaction and Image are completely linked to behavioural intention. In this research, variables such as customer experience, customer satisfaction and selected marketing mix factors facilitating conditions, product price value [11] will be tested along with UTAUT2 model which include current seven factors such as performance expectancy, effort expectancy, social influence, facilitating condition, price value, hedonic motivation and habit. Facilitating conditions will emphasis on telecommunication service marketing matters. There is an accumulation of moderating factor customer experience. Moreover, in the preceding studies [56, 57], established a customer experience model which include variables that affect the customer experience. The phenomenon of the behavioural, attitudinal loyalty or customer loyalty in light of customer experience is not very well examined in current studies [57–60]. According to the past research, there is no sole framework which is being teste or used to test UTAUT with customer loyalty and the factors affecting customer retention with customer experience. Accordingly, in this study age, gender and length experience will act as a moderating factor. A moderating variable is an intermingling term which is said to indicate the relationship between the dependent and independent variable is unexpectedly weak, inconsistent, unreliable and no relationship at all. Hence, the moderating variable is familiarized to support the relationship [61].

The theoretical model is also founded on the philosophies of the SERVQUAL model, which will help to evaluate the customer's experience. Past studies discovered that SERVQUAL helps to regulate the relationship between service quality and customer satisfaction. Service quality has been explained as a result derived from the assessment of customer expectation with customer performance [62, 63]. Additionally, it was debated that service quality is not directly linked to customer satisfaction, but it is a form of attitude which leads to satisfaction [62–65]. Besides this, there are studies which show that there are ten determinants in the beginning to quantify service quality which were later condensed to five specific determinants [66]. Furthermore, these five dimensions are tangibility, reliability, responsiveness, assurance and empathy. The SERVQUAL model helps to govern the difference between the customer's expectation and perception [66]. Later, this model proposes to measure overall customer satisfaction and its influence on behavioural intention and customer loyalty.

These variables perform an important part in the pre-& post–purchase decision making [67–72]. Consequently, this study model has combined these variables along with the other new variables as discussed earlier such as customer experience, customer satisfaction, designated marketing mix factors such as product price value, and the UTAUT 2 model factors in order to assess customer loyalty.

The completeness of the research model along with the validity and reliability of the UTAUT model has stimulated the current author of this study to implement and validate it in the context of mobile telecommunication services retention in Australia. The research model is revised to collaborate with the context of this study in telecommunication outlook. The UTAUT 2 model also accomplish the obligation of marketing mix factors such as product, price, plan, place, process and promotion for this study. This research model also provisions the impact of social influence through social networking using marketing and advertising apparatuses for the promotion of product and services. Consequently, the revised model will also deliberate the influence of moderators that are thought would influence the three direct elements: gender, age and the length of experience.

In past studies, Madan and Yadav [73] prominence on the importance of the marketing literature has been recognized with some mistakes and discrepancies in the adoption of IS/IT patterns worldwide, and these inconsistencies and complications are hard to explain with a single model or theory. Moreover, with this drawback, the current researcher has adopted the framework of current models with the incorporation of new two variable i.e. customer satisfaction and customer experience. Hence, this model also studies the moderating effect of age, gender and experience on the planned relationships.

The following mathematical formulas explain the research model for this study.

Mathematical Model and Conceptual Framework

In mathematical terms, the model would be:

$$\mathrm{B} \approx \mathrm{BI} = [\sum(\boldsymbol{CL})\boldsymbol{W1} + \sum(\boldsymbol{CR})\boldsymbol{W2}] \tag{1}$$

Where B = behaviour: BI = behavioural intention; CL = customer loyalty associated with attitudinal loyalty: CR = customer retention associated with repetitive behaviour, W1 and W2 are the regression weights for the customer loyalty and the customer retention.

For this model, the analysis will be done in SMART PLS and this tool uses the regression algorithm. Therefore, the mathematical regression equation for this current study would be:

$$CL=\beta0 +\beta1(FC) +\beta2(EE)+ \beta2(EE)+ \beta3(PE)+ \beta4(SI)+ \beta5(PV)+ \beta6(HM)+ \beta7(HT)+ \beta8(CS)+ \beta9\ (CE)+ ȩ \tag{2}$$

Where ȩ is the error, FC is the facilitating conditions, EE is effort expectancy, PE is the performance expectancy, SI is the social influence, PV is the price value, HM is the hedonic motivation, HT is the habit, CS is the customer satisfaction and CE is the customer experience. β0 is the slope of Y-intercept and β1-9 is the population of slope constant.

The conceptual model for this study is illustrated in Fig. 2. Figure 1 is emphasing on the behaviour intention formation, as BI is the behavioural intention, CR is customer retention and CL is customer loyalty.

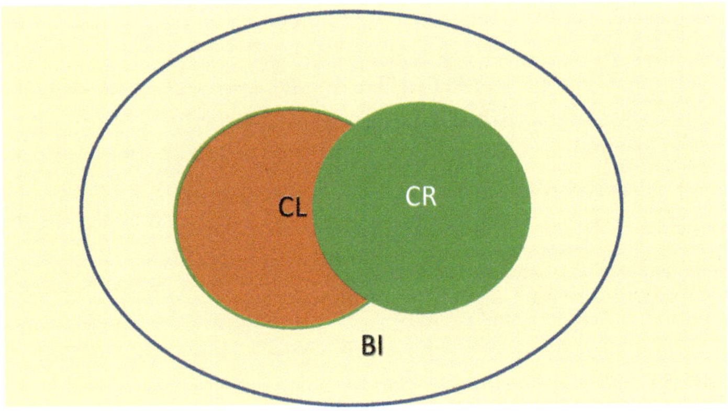

Fig. 1. Conceptual diagram for behavioural intention

5 Discussion and Conclusion

In the next phase of this research, this model will be empirically tested, and all the hypotheses will be tested through the statistical procedure. This study provides empirical validation of consumer behaviours in the mobile telecommunication services repurchase and usage context. In this study, a theoretical underpinning including the unified theory of technology acceptance and use of technology (UTAUT), marketing mix theory and expectation theory of satisfaction/dissatisfaction are adapted to explain the consumer response to mobile telecommunication services. Furthermore, there are several factors identified in this study, which are affecting customers' loyalty. This

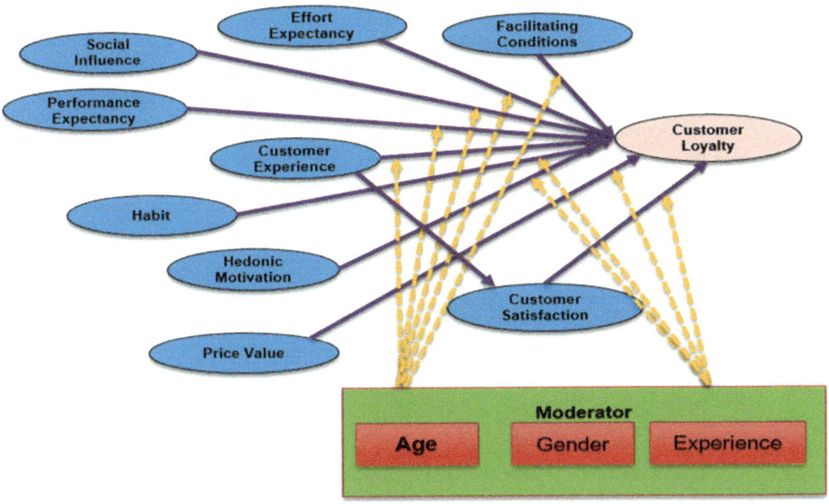

Fig. 2. Conceptual Model for the theoretical framework for customer loyalty study

research has provided a complete framework for customer loyalty along with the variables which effect customer behaviour. Furthermore, this study and research framework will give a new course of action to upcoming researchers in the field of mobile telecommunication business studies and industries can also apply this model in their trades.

Appendix 1: Research Instrument

Item code	Survey questions
EE1	It is easy for me to become skilful at using mobile services such as voice and data
EE2	My interaction with mobile services such as voice and data are clear and understandable
EE3	Mobile services such as voice and data services
EE4	Mobile online billing & payment service
EE5	Overall, mobile services
EE6	Mobile internet services
EE7	The terms and conditions of mobile services
PE1	Using mobile services has saved time during my routine tasks
PE2	The mobile value-added services are useful in my daily routine
PE3	Using mobile services increases my productivity in daily life
PE4	The use of mobile services will increase my chances to improve my overall performance in daily life
PE5	The mobile services enable me to accomplish tasks more quickly

(continued)

(*continued*)

Item code	Survey questions
PE6	Mobile services have helped me generate more innovative ideas
CL1	I will repurchase the offerings from my service provider
CL2	I will recommend the mobile products and services of my service provider to others
CL3	My provider is the primary service provider I consider when I want mobile communication products and services
CL4	I consider my service provider to be my first choice when I need mobile communication products and services
CL5	If the price is reduced, I will be glad to use the mobile services of another service provider
SI1	My friends can influence me in choosing the mobile data services of their service provider
SI2	My family has influenced when choosing the mobile service provider
SI3	Whenever I feel difficulties in investigating mobile products or services, I contact my friends to take their opinion
SI4	My colleagues have influenced me to choose the mobile service such as voice, data and bundle mobile broadband plans
SI5	Social media friends have affected my decision to stay with my service provider
SI6	My family and friends can influence me in choosing the data services of their service provider
FC1	My service provider has provided resources such as network coverage, necessary to use my mobile service products
FC2	My service provider facilitates the necessary technical customer support for the use of their mobile phone services
FC3	I cannot get help from my service provider when I have difficulties using mobile services
FC4	Whenever I have encountered difficulty with the use of mobile service, I can use the information from a service provider to solve the problems
FC5	Whenever I connected my mobile data service from my location, network service is available
FC6	Whenever I use my mobile data service from my location, network service is available
FC7	Mobile broadband service connectivity from my home location to service providers' network is
HM1	New data technologies such as (3G/4G/LTE) in mobile service can increase my interest to use the mobile service
HM2	Using mobile services can bring some recreation. -Never: Always
HM3	Using the mobile services of my service provider is: -Very boring: Very interesting
HM4	My mobile service providers' offers are: -Very Unappealing: Very appealing
HM5	Using mobile services of my service provider, including voice and data is-Boring: Interesting

(*continued*)

(*continued*)

Item code	Survey questions
HM6	Using the mobile services of my service provider, including voice and data is-Tiring: Effortless
HM7	The mobile value-added services of a service provider are -Enjoyable: Unenjoyable
HT1	The use of mobile services has become routine for me. -Always: Never
HT2	I am addicted to using mobile services such as voice and data-Always: Never
HT3	I must use mobile services including mobile data services. -Never: Always
HT4	I would not find hard to use the mobile service. -Always: Never
HT5	Using mobile services has become natural to me. -Always: Never
HT6	Using mobile service would require minimum effort to use it. -Never: Always
CS1	The mobile services such as voice, data and broadband services that you use
CS2	The performance of mobile services
CS3	The speed of mobile phone data services offered by the service provider
CS4	The accurate billing information related to usage by the service provider
CS5	Overall, I am satisfied with my mobile service provider
CS6	I received adequate required information for the service that I needed
CE1	The employees of my service provider as per my experience are willing to help, whenever I need any assistance
CE2	My experience with mobile services of my service provider is-Very bad: Very good
CE3	I will rate my experience with my previous service provider plan as-Very bad: Very good
CE4	My information privacy protection experience as a customer is-Very less protected: Highly protected
CE5	My customer care experience with my service provider, such as (receiving outage notification mess…-Stressful: Enjoyable
CE6	The mobile value-added service -Less than my expectation: Higher than my expectation
CE7	The quality of voice calls such as (voice clarity, signal strength)-Less than my expectation: Higher than my expectation
CE8	The resolution of service complaints-Less than my expectation: Higher than my expectation
CE9	The reliability of data services-Less than my expectation: Higher than my expectation
PV1	My mobile services are a good value for the money
PV2	The service provider provides a variety of price plan
PV3	The mobile plan offered by my service provider is reasonably priced
PV4	At the current price, my mobile service provider provides good value for a data bundle plan (mobile phone, broadband and home broadband)
PV5	The service provider provides the flexibility of changing the price plan, i.e. plan upgrade and downgrade

References

1. Caruana, A.: Service loyalty. Eur. J. Mark. **36**(7/8), 811–828 (2002)
2. Rai, A.K., Srivastava, M.: Customer loyalty attributes: a perspective. NMIMS Manag. Rev. **22**, 49–76 (2012)
3. Doorey, A.M., Wilcox, G.B., Eastin, M.S.: Consumer privacy and the new mobile commerce. The Dark Side of Social Media: A Consumer Psychology Perspective (2017)
4. TIO: TIO_AnnualReport_2017 (2017)
5. TIO: Telecommunications Industry Ombudsman 2014. Annual Report (2014)
6. Bhatti, H., Abareshi, A., Pittayachawan, S.: An empirical examination of customer retention in mobile telecommunication services in Australia. In: Proceedings of the 13th International Joint Conference on e-Business and Telecommunications (ICETE 2016), vol. 2, pp. 72–77 (2016)
7. Brown, S.A., Venkatesh, V.: Model of adoption of technology in households: a baseline model test and extension incorporating household life cycle. MIS Q. **29**, 399–426 (2005)
8. Escobar-Rodríguez, T., Carvajal-Trujillo, E.: Online purchasing tickets for low cost carriers: an application of the unified theory of acceptance and use of technology (UTAUT) model. Tour. Manag. **43**, 70–88 (2014)
9. Escobar-Rodríguez, T., Carvajal-Trujillo, E.: Online drivers of consumer purchase of website airline tickets. J. Air Transp. Manag. **32**, 58–64 (2013)
10. Venkatesh, V., Morris, M.G., Davis, G.B., Davis, F.D.: User acceptance of information technology: toward a unified view. MIS Q. **27**, 425–478 (2003)
11. Venkatesh, V., Thong, J.Y., Xu, X.: Consumer acceptance and use of information technology: extending the unified theory of acceptance and use of technology. MIS Q. **36**(1), 157–178 (2012)
12. Wirtz, J., Brah, S., Yanamandram, V., White, L.: Switching barriers in business-to-business services: a qualitative study. Int. J. Serv. Ind. Manag. **17**(2), 158–192 (2006)
13. Kuo, T.-P., Yang, C.-Y.: The study of consumer intention, service quality cognition and repurchase behavior of consumers in exotic restaurants–mexican restaurant. J. Inf. Optim. Sci. **32**(3), 551–560 (2011)
14. Howard, J.A., Sheth, J.N.: The Theory of Buyer Behavior. Wiley, New York (1969)
15. Francken, D.A.: Postpurchase consumer evaluations, complaint actions and repurchase behavior. J. Econ. Psychol. **4**(3), 273–290 (1983)
16. Paulssen, M., Birk, M.M.: Satisfaction and repurchase behavior in a business-to-business setting: Investigating the moderating effect of manufacturer, company and demographic characteristics. Ind. Mark. Manag. **36**(7), 983–997 (2007)
17. van der Heijden, H., Verhagen, T., Creemers, M. (eds.): Predicting Online Purchase Behavior: Replications and Tests of Competing Models, p. 10. IEEE (2001)
18. Park, C.-H., Kim, Y.-G.: Identifying key factors affecting consumer purchase behavior in an online shopping context. Int. J. Retail Distrib. Manag. **31**(1), 16–29 (2003)
19. Sheth, J.N., Parvatiyar, A.: The evolution of relationship marketing. Int. Bus. Rev. **4**(4), 397–418 (1995)
20. Wolfinbarger, M., Gilly, M.C.: Shopping online for freedom, control, and fun. Calif. Manag. Rev. **43**(2), 34–55 (2001)
21. Kim, M.-K., Park, M.-C., Jeong, D.-H.: The effects of customer satisfaction and switching barrier on customer loyalty in Korean mobile telecommunication services. Telecommun. Policy **28**(2), 145–159 (2004)
22. Liu, C.-T., Guo, Y.M., Lee, C.-H.: The effects of relationship quality and switching barriers on customer loyalty. Int. J. Inf. Manag. **31**(1), 71–79 (2011)

23. Zhou, T., Lu, Y.: Examining postadoption usage of mobile services from a dual perspective of enablers and inhibitors. Int. J. Hum.-Comput. Interact. **27**(12), 1177–1191 (2011)
24. Wu, Y.-L., Tao, Y.-H., Li, C.-P., Wang, S.-Y., Chiu, C.-Y.: User-switching behavior in social network sites: a model perspective with drill-down analyses. Comput. Hum. Behav. **33**, 92–103 (2014)
25. Lee, Y.-K., Ahn, W.-K., Kim, K.: A study on the moderating role of alternative attractiveness in the relationship between relational benefits and customer loyalty. Int. J. Hosp. Tour. Adm. **9**(1), 52–70 (2008)
26. Sharma, N., Patterson, P.G.: Switching costs, alternative attractiveness and experience as moderators of relationship commitment in professional, consumer services. Int. J. Serv. Ind. Manag. **11**(5), 470–490 (2000)
27. Warshaw, P.R., Davis, F.D.: Disentangling behavioral intention and behavioral expectation. J. Exp. Soc. Psychol. **21**(3), 213–228 (1985)
28. Roger, E.: Diffusion of Innovations. The Free Press, New York (1995)
29. Ajzen, I.: The theory of planned behavior. Organ. Behav. Hum. Decis. Process. **50**(2), 179–211 (1991)
30. Davis, F.D.: Perceived usefulness, perceived ease of use, and user acceptance of information technology. MIS Q. **13**, 319–340 (1989)
31. Davis, F.D., Bagozzi, R.P., Warshaw, P.R.: Extrinsic and intrinsic motivation to use computers in the workplace. J. Appl. Soc. Psychol. **22**(14), 1111–1132 (1992)
32. Thompson, R.L., Higgins, C.A., Howell, J.M.: Personal computing: toward a conceptual model of utilization. MIS Q. **15**, 125–143 (1991)
33. Compeau, D.R., Higgins, C.A.: Computer self-efficacy: development of a measure and initial test. MIS Q. **19**, 189–211 (1995)
34. Tornatzky, L.G., Fleischer, M., Chakrabarti, A.K.: Processes of Technological Innovation. Lexington Books, Lanham (1990)
35. Jang, H.Y., Noh, M.J.: Customer acceptance of IPTV service quality. Int. J. Inf. Manage. **31** (6), 582–592 (2011)
36. Chow, M., Herold, D.K., Choo, T.-M., Chan, K.: Extending the technology acceptance model to explore the intention to use Second Life for enhancing healthcare education. Comput. Educ. **59**(4), 1136–1144 (2012)
37. Ghazizadeh, M., Lee, J.D., Boyle, L.N.: Extending the technology acceptance model to assess automation. Cogn. Technol. Work **14**(1), 39–49 (2012)
38. Dwivedi, Y.K., Rana, N.P., Jeyaraj, A., Clement, M., Williams, M.D.: Re-examining the unified theory of acceptance and use of technology (UTAUT): towards a revised theoretical model. Inf. Syst. Front. **21**(3), 719–734 (2019)
39. Deng, Z., Lu, Y., Wei, K.K., Zhang, J.: Understanding customer satisfaction and loyalty: an empirical study of mobile instant messages in China. Int. J. Inf. Manag. **30**(4), 289–300 (2010)
40. McCarthy, E.J.: Basic marketing: a managerial approach. Richard D. Irwin, Inc., Homewood (1979). 1960
41. Barich, H., Kotler, P.: A framework for marketing image management. Sloan Manag. Rev. **32**(2), 94 (1991)
42. Kotler: Marketing Management (2003)
43. Anderson, E.W., Fornell, C., Lehmann, D.R.: Customer satisfaction, market share, and profitability: findings from Sweden. J. Mark. **58**, 53–66 (1994)
44. McCarthy, E.J., Perreault, W.D.: Basic Marketing. Richard D. Irwin', Inc., Homewood (1984). (Warwick: HO 5000 M2)
45. Kotler, P., Armstrong, G. (eds.): Principles of Marketing, 5th edn. Prentice Hall, Englewood Cliffs (1991)

46. Rafiq, M., Ahmed, P.K.: Using the 7Ps as a generic marketing mix: an exploratory survey of UK and European marketing academics. Mark. Intell. Plan. **13**(9), 4–15 (1995)
47. Bitner, M.J., Booms, B.H., Tetreault, M.S.: The service encounter: diagnosing favorable and unfavorable incidents. J. Mark. **54**, 71–84 (1990)
48. Fishbein, M., Ajzen, I.: Belief, Attitude, Intention and Behavior: An Introduction to Theory and Research. Addison-Wesley, Reading (1975)
49. Venkatesh, V., Davis, F.D.: A theoretical extension of the technology acceptance model: four longitudinal field studies. Manag. Sci. **46**(2), 186–204 (2000)
50. Taylor, S., Todd, P.A.: Understanding information technology usage: a test of competing models. Inf. Syst. Res. **6**(2), 144–176 (1995)
51. Rogers, E.M. (ed.): Diffusion of Innovations. Free Press, New York (2003/2008)
52. Moore, G.C., Benbasat, I.: Development of an instrument to measure the perceptions of adopting an information technology innovation. Inf. Syst. Res. **2**(3), 192–222 (1991)
53. Bandura, A.: Perceived self-efficacy in the exercise of control over AIDS infection. Eval. Program Plan. **13**(1), 9–17 (1990)
54. Cronin, J.J., Brady, M.K., Hult, G.T.M.: Assessing the effects of quality, value, and customer satisfaction on consumer behavioral intentions in service environments. J. Retail. **76**(2), 193–218 (2000)
55. Ryu, K., Han, H., Kim, T.-H.: The relationships among overall quick-casual restaurant image, perceived value, customer satisfaction, and behavioral intentions. Int. J. Hosp. Manag. **27**(3), 459–469 (2008)
56. Anaman, M.: Toward a model of customer experience. Brunel University, School of Information Systems, Computing and Mathematics (2010)
57. Maklan, S., Klaus, P.: Customer experience: are we measuring the right things? Int. J. Mark. Res. **53**(6), 771–792 (2011)
58. Islam, M.B., Rima, A.R.: Factors affecting customer experience in telecommunication services and its importance on brand equity: a study on telecommunication companies in Bangladesh. J. Contemp. Res. Bus. **5**(8), 254–262 (2013)
59. Meyer, C., Schwager, A.: Understanding customer experience. Harv. Bus. Rev. **85**(2), 116 (2007)
60. Richardson, A.: Understanding customer experience (2010). Accessed 26 Sep 2010
61. Uma, S., Roger, B.: Research Methods for Business: A Skill Building Approach. Wiley, New York (2003)
62. Cronin Jr., J.J., Taylor, S.A.: Measuring service quality: a reexamination and extension. J. Mark. **56**(3), 55–68 (1992)
63. Bolton, R.N., Drew, J.H.: A longitudinal analysis of the impact of service changes on customer attitudes. J. Mark. **55**(1), 1–9 (1991)
64. Marshall, G.W., Shepherd, C.D.: Selling and sales management in action: service quality and the sales force: a tool for competitive advantage. J. Pers. Sell. Sales Manag. **19**(3), 73–82 (1999)
65. Parasuraman, A., Zeithaml, V.A., Berry, L.L.: SERVQUAL - a multiple-item scale for measuring consumer perceptions of service quality. J. Retail. **64**(1), 12–40 (1988)
66. Parasuraman, A., Zeithaml, V.A., Berry, L.L.: A conceptual model of service quality and its implications for further research. J. Mark. **49**, 41–50 (1985)
67. Ajzen, I.: Theory of planned behavior. Handb. Theor. Soc. Psychol. **1**, 438 (2011)
68. De Canniere, M.H., De Pelsmacker, P., Geuens, M.: Relationship quality and the theory of planned behavior models of behavioral intentions and purchase behavior. J. Bus. Res. **62**(1), 82–92 (2009)

69. Han, H., Kim, Y.: An investigation of green hotel customers' decision formation: developing an extended model of the theory of planned behavior. Int. J. Hosp. Manag. **29**(4), 659–668 (2010)
70. Hung, S.-Y., Chang, C.-M.: User acceptance of WAP services: test of competing theories. Comput. Stand. Interfaces **27**(4), 359–370 (2005)
71. Ya-Yueh, S., Kwoting, F.: The use of a decomposed theory of planned behavior to study Internet banking in Taiwan. Internet Res. **14**(3), 213–223 (2004)
72. Luarn, P., Lin, H.-H.: Toward an understanding of the behavioral intention to use mobile banking. Comput. Hum. Behav. **21**(6), 873–891 (2005)
73. Madan, K., Yadav, R.: Behavioural intention to adopt mobile wallet: a developing country perspective. J. Indian Bus. Res. **8**(3), 227–244 (2016)

Exploring the Software Quality Criteria and Sustainable Development Targets: A Case Study of Digital Library in Malaysian Higher Learning Institution

Masrina A. Salleh[1(✉)], Mahadi Bahari[2(✉)], and Waidah Ismail[3]

[1] School of Computing, Faculty of Engineering, Universiti Teknologi Malaysia, Skudai, Malaysia
nikakmalsalleh@yahoo.com

[2] Department of Information Systems, Azman Hashim International Business School, Universiti Teknologi Malaysia, Skudai, Malaysia
mahadi@utm.my

[3] Centre of Holistic Intelligence Institut Sains Islam, Universiti Sains Islam Malaysia, Nilai, Negeri Sembilan, Malaysia

Abstract. Being sustainable within social, environmental, and the economy is still a major issue to almost every company globally. Although on-going actions have been implemented via information and communication technology (ICT), studies and development efforts in seeking solutions are in scarcity, particularly for the present use of software to perform its intended task. The lack of technological innovation and development motivate us to explore the present use of software with the focus on software quality criteria (SQC) in attaining sustainable development targets (SDTs). The understanding of SQC is significant for the purpose of meeting the needs and operating environment of software that are continually changed in the future. This paper presents a semi-structured interview involving eight digital library's stakeholders in one of Malaysian higher learning institution. By applying thematic analysis, the study finds six digital library software quality criteria (DL-SQC) and seven digital library sustainable development targets (DL-SDTs) that were heavily stressed by more than half of participants. The DL-SQC are usability, maintainability, functionality, portability, efficiency, and reliability. While the DL-SDTs are equal access to information/knowledge, energy efficiency, persistent updates the information/knowledge, collaboration & partnership, funding, green technology innovation, and information/knowledge sharing. The study contributes as a guide to researchers who are motivated to position their further studies into this topic and for a practitioner; it is to summarize the SQC and their relevant SDTs that may be applied to make decisions regarding the next level of software development and maintenance process towards being sustainable.

Keywords: Software quality criteria · Sustainable development goals · Information service sustainable development targets · Thematic analysis

F. Saeed et al. (Eds.): IRICT 2019, AISC 1073, pp. 1076–1086, 2020.
https://doi.org/10.1007/978-3-030-33582-3_101

1 Introduction

The world leaders have come to realize the importance of being sustainable within social, environmental, and economic but it is still a major challenge that remains up to the present time [1–3]. On-going actions have been implemented to address relevant critical issues via information and communication technology (ICT) as a tool toward being sustainable [4]. In the context of ICT as a tool for sustainability, ICT by itself is never a solution, but it must be properly combined with the goals of being sustainable [4–6]. The sustainable development agenda by the United Nation [3] consists of 17 sustainable development goals (SDGs) with its 169 sustainable development targets (SDTs) are targeted to be achieved by the year 2030. These SDGs and its SDTs are complex and with left eleven years to achieve it, requires greater participation from all nations across the globe to play their role in addressing the uprising issue. Although UN2030 Agenda claimed that ICT is a key enabler for being sustainable, studies and development efforts in seeking solutions are in scarcity [4, 5] within the ICT area, particularly for the present use of software to perform its intended task [5, 6]. Within the software area, understanding the software quality criteria (SQC) is important as a fundamental knowledge that can affect the quality of software in meeting its intended task [5, 6] as well as it's evolving of the needs and operating environment [8]. As revealed by Garcia-Mireles [9], performance efficiency has been the main SQC focused on prior studies and the need for more investigations on other SQC should be carried out to comprehend the extent to which the other SQC might be relevant to sustainability development [9].

Thus, this study investigated the SQC and its relevant SDTs in the information service field which is in line with the UN2030 Agenda [3], that reckoned the access to information as a target under SDG 16: "to promote peaceful and inclusive societies for sustainable development, to provide access to justice for all and to build effective, accountable, and inclusive institution at all levels", with Target 16.10: "Ensure public access to information and protect fundamental freedoms, in accordance with national legislation and international agreements", wherein universal literacy has been acknowledged in the vision for UN2030 Agenda [3]. Comprehending that being sustainable is everyone's role together with this area has strong motivations for technological innovation and development, which are needed to support all nations in achieving the stipulated 17 top global goals within another eleven years, hence the motivation to perform this study is by continuing the previous years initiated by Garcia-Mireles [9]. The primary aim of this study is to contribute to the body of knowledge in the area of SQC and SDTs by providing the list of SQC and its relevant SDTs. With that, this study may function as a guide to researchers who are motivated to further probe into this topic. While for a practitioner, the summarize list of the SQC and its relevant SDTs may be applied to make decisions regarding the next level of software development and maintenance process in accelerating movement towards being sustainable. The remainder of this study is structured as follows: Sect. 2 presents the brief overview of background related studies; Sect. 3 explains the research method used to conduct this study; Sect. 4 reveals the research outcomes, and finally, Sect. 5 provides the discussion, conclusion and proposes future work.

2 Literature Review

SQC is an indicator of software in meeting user needs in a number of areas, including usability, sustainability, and maintainability of the software [10]. SQC differ from one software program to another, depending on its intended purposes [8]. Among the SQC discussed in the prior studies are: *reliability* that refers to the ability of the software to provide functionalities based on specific needs and its capability of failure-free operation for a specific period in a specific environment [9]; *maintainability* is the ease with which modification can be made to the software to enhance software functionalities [10]; *portability* is the ease with which the software can be transferred from one platform to another without requiring major rework [10]; *performance* is how fast the software can generate outputs [9]; *accuracy* is the degree of how the system is able to capture and display data accurately [11] and many more. The SQC is vital to characterize the usefulness of the software in a given environment as well as in characterizing software for the purpose of software quality control, software quality assurance, and software process improvement [12]. In the context of this study, comprehending the SQC offers the basis to determine if the running software is consistent with the institution overall desired SDTs. Software failure or errors due to lack of criteria in meeting user needs have caused more than an inconvenience to end-users. Indeed, disruption of the original design and structure, which makes it difficult for further changes to take place, will be minimized after determining the SQC [13]. Software with appropriate criteria, if ideally applied, can be a more important and strategic competitive advantage in today's economic scenario [14]. Hence, it is significant to understand the SQC due to its evolving needs and operating environment [15].

In the context of information service for 17 SDGs, SDGs 1, 2, 3, 8 and 9 focus more on economic sustainable with information service used to expend productive economic opportunities. Among the relevant SDTs were the technology innovation [14–16], increased food security [17, 18], connectivity in health ecosystem [18], transforming the business operation [19] as well as maintaining resilient infrastructure [19]. While SDGs 4, 5, 10, 11, 16 and 17 were impacting social and political sustainability via information service by offering the stability, security, and openness of our social and political institutions. Among the relevant SDTs were the technology innovation in information communication technology to ensure the information production, distribution and access equality throughout the globe are attained [20–22]. While the information service used to assist on waste and water management [23], intelligent transport system [24], smart building [25] and environment [25] are also highlighted goals for the betterment of community healthy living. SDGs 6, 7, 12, 13, 14 and 15 focus more on the importance of environmentally sustainable. Among the relevant SDTs were smart information service applications [26] on water, e-waste carbon emission, energy consumption that can promote and protect biodiversity. All these SDTs need to periodically reassess to align with social, political, cultural and economic need changes.

As digital library was one of the entities in the information service field, supported by Witt [27] and the Business & Sustainable Development Commission [28] asserted that more researches are in need in identifying digital library related to software in attaining its SDTs as align with the UN2030 Agenda [29], increased the significance of this study. A semi-structured interview method to collect rich data and insightful evidence about how people feel, act and perceive a situation [30] was employed to generate the digital library software quality criteria (DL-SQC) and its sustainable development targets (DL-SDTs). Two groups of the profession from information technology and managerial were involved. The high level of their responsibility is the key to the success of the current system in attaining sustainable development targets. This leads the researchers to identify all possible factors that need to be considered as important in balancing the user needs and the global agenda in accelerating movement towards a green world.

3 Methodology

Identify the criteria which are relevant to the study can be done by a survey or group problem-solving technique [30]. In this study, a single case study at one of the digital library from a higher institution in Malaysia was conducted due to it is a good way to describe case for an easier understanding [31]; it is meeting the study's digital library definition which is defined as an institution that made accessible the online information through the collection of diversifying information to serve their respective community of users; and as Malaysian government integrating the SDGs into its Eleventh Malaysian Plan with information/knowledge anchoring the people growth [32]. The digital library in this case study had been transformed from a traditional library into a computing environment since the year 1986. Most of its operations and services are handled by the Computerized Library Integrated System. It offers information searching through a wide range of international and local databases that can be accessed online in the fields of science, technology, and the social sciences. In 1998, this digital library was awarded MS ISO 9002 Quality Award as recognition to its initiative in delivering quality services. The services provided including consultation, information and content collection searching via OPAC, database usage, reference material, borrowing and returning. Figure 1 depicts the flow of identifying SQC and SDTs relevant to the real environment for this study.

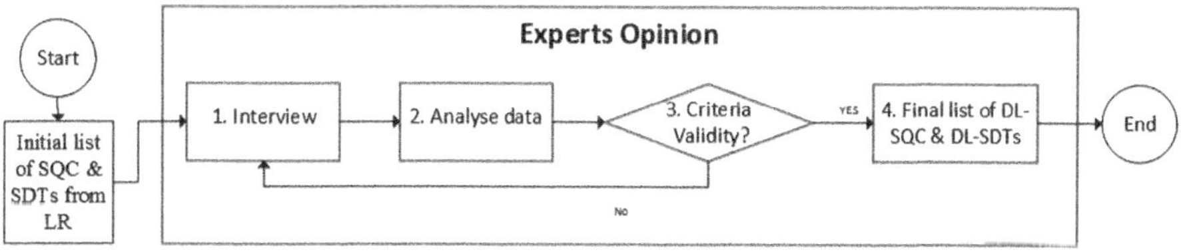

Fig. 1. The operational process for generating experts' opinion.

Step 1: *Interview* - We employed one-to-one semi-structured interview with an expert in the field in order to understand the existing knowledge of the digital

library's SQC and its SDTs. The goal of this session is to generate the maximum amount of discussion and opinions within a given period of time. Each of the conversations in the interview session was recorded with a digital recorder.
Step 2: *Analyse Data* - the transcribing of the interview process was conducted using NVIVO to ensure all important matters discussed are not left behind. The interview data were then analyzed by using thematic analysis [33] for exploring the perspective of different research participants. In this case, the participants involved were three managerial personnel and five information technology personnel, thus can produce similarities and differences, and create unanticipated insights. Here, all the interviews' data were systematized and structured according to clusters. In doing so, the interview transcripts were read in search of incidents and facts. The incidents were then compared with each other to discover or redevelop the clusters.
Step 3: *Criteria Validity* - The validation process with the same experts on the SQC and its SDTs were followed to check whether it meeting their needs or not [30].
Step 4: *Final DL-SQC and DL-SDTs* – Only the DL-SQC and DL-SDTs with more than half participants discussed it, were compiled as a finding of the study [30].

4 Results

There are six DL-SQC and seven DL-SDTs emerged from the data analysis.

4.1 Usability

The result from the case study shows that seven (87.5%) of the participants agreed on the importance of the software be able to accomplish the desired task; user-friendly; easy to learn; navigation is simple; consistent user interface standards and easy to new or infrequent users to use the software. As the participants stated that, *"We focus on the user-friendliness of our digital library system to ensure our users will always utilize our system"* and *"We keep updating on user new needs through the user satisfaction survey and we want all users will experience simple browsing with easy steps to remember."*

4.2 Maintainability

The result from the case study shows that six (75%) of the participants agreed on the importance of the software be able to support changes; easy to add code to existing system; easy to upgrade to add new features and new technologies related to new business needs; should be cost-effective and easy to correct defects or errors on the effected system components or separate methods. As the participants stated that *"Our digital library system should easy to maintain in terms of a simple function can be added and modify, the current modules are capable to new technology with minor updating configuration so that any changes or modification can be done internally"* and "*Our IT group normally work together with the outside IT groups to ensure our digital library system can provide the best service to our users.*"

4.3 Functionality

The result from the case study shows that all eight (100%) participants agreed on the importance of the software in terms of the degree of software performance to its intended purpose. As the participants stated that *"The correctness of the functions in our digital library system is crucial to ensure users will get what they search for; the suitability of the functions will help users to easily understand the flow of our digital library system"* and *"We need to continuously improve our digital library functions accuracy"*.

4.4 Portability

The result from the case study shows that five (62.5%) of the participants agreed on the importance of the software in terms of cost, technical or behavioral issues related to porting. As the participants stated that *"We are in the process of saving our budget wherever applicable. Thus our digital library system should be at very optimum in terms of being portable specifically in costing and technical issues"* and *"Our system still need to focus on how to make transferring of software easier"*.

4.5 Efficiency

The result from the case study shows that seven (87.5%) of the participants agreed on the importance of the software be able to complete any task given at specific time required such as efficient in utilizing processor capacity, disk space and memory so that it can be used in real-time applications. As the participants stated that *"Any task assign should be done in an expected period"* and *"Frequently monitor on the processor capacity, storage, and memory speed to ensure the system running efficiently"*.

4.6 Reliability

The result from the case study shows that all eight (100%) participants agreed on the importance of the software be able to operate correctly; consistently produce correct results; able to operate and sustain under different working environment; able to access the external elements such as databases, systems and network connections and able to automatically recovery to reduce the mean time between failure. As the participants stated that *"The correctness of the results always being among the user main concern"* and *"To sustain in this industry, we need to ensure our system always capable in adapting different environment needs"*.

4.7 Equal Access to Information/Knowledge

The result from the case study shows that seven (87.5%) of the participants agreed on the importance of every user have equal access to the resources regardless of the gender and demographic. As the participants stated that *"Every user need to allow an equal opportunity in term of information collection access accordingly to their category registered, for example, every member registered as a student either they choose in-campus or off-campus study, should eligible to access any information collection specifically meant for them regardless of the gender and demographic. We keep monitoring this aspect as being equal is important for our users"*. The participants also agreed that by providing equal access to the information resources will increase users' satisfaction as the participants stated that "*From our latest survey, all participated users either male or female are satisfied with the access to our information collections which can be accessed twenty-four hours in seven days a week*" and "*Having equal access is always being our main priority targets*".

4.8 Energy Efficiency

The result from the case study shows that five (62.5%) of the participants agreed on the importance of energy efficiency practiced. As the participants stated that *"We need to minimize our basic consumption such as water and electricity so that the budget allocation can be efficiently distributed to other expenses as well", "Now we set our computing peripherals with either sleeping or standby mode as one of our initiatives in reducing electricity consumption waste. We also produce a duty roster on ensuring the building and other electronic equipment facilities are properly switch off* "and *"We strongly emphasize through various activities to ensure the awareness among our staffs and user for being responsible on the energy consumption."*

4.9 Persistent Update the Information/Knowledge

The result from the case study shows that seven (87.5%) of the participants agreed on the importance of persistently update the content collection. As the participants stated that *"Despite a very tight budget, we will always find a way to renew the journals subscription and online database to ensure our users be updated and diversify of information collection sustain*" and "*Monitoring and updating our content collections is one of our major services to our users"*.

4.10 Collaboration and Partnership

The result from the case study shows that seven (87.5%) of the participants agreed on the importance of collaboration and partnership approach to leverage their technology, content collection, staff training and other. As the participants stated that "*Partnership and collaboration with other companies in any community programs that relate to awareness on sustainability is always our target"*, "*We do collaborate with other digital libraries to leverage on the information collection*" and *"Being having a tight budget, we look forward to leveraging the digital library technology with other institutions"*.

4.11 Funding

The result from the case study shows that five (62.5%) of the participants agreed on the funding become one of their SDTs. As the participants stated that "*Being in non-profit business, we need to focus on the money resources to ensure our services are sustained*" and "*We also come out with activities that will generate some money for us while keep searching for the funders*".

4.12 Green Technology Innovation

The result from the case study shows that seven (87.5%) of the participants agreed on the importance of employing green technology innovation. As the participants stated that "*Since 2010, only computing peripherals with an inverter will be purchased*", "*Not only we focus on physical computing devices but also on the response rate performance such as of the searching, navigation, browsing and downloading the information collection*" and "*Our aims are not only meeting changing user information needs but also balancing with being environmental concerns*".

4.13 Information/Knowledge Sharing

The result from the case study shows that six (75%) of the participants agreed on the importance of sharing their information collection and services. As the participants stated that "*Sharing the information resources with other institution really help us improve in our service in the context of diversifying the information collection*" and "*The concept of share the information stored should be prolonging for the benefit of users as well as for us as the information service provider*".

5 Discussions and Conclusion

This study helps to identify SQC and SDTs concerning digital library from a qualitative perspective. The study is novel in that it provides some of the participants' views on the digital library system with a focused on SQC in attaining its SDTs aspect. Figure 2 depicts the case study data analysis that was group into its respective clusters with the percentage of participants discussed it. The DL-SQC includes usability (87.5%), maintainability (75%), functionality (100%), portability (62.5%), efficiency (87.5%) and reliability (100%) were placed in the center of all SDTs indicate that those six SQC are the most frequently mentioned by participants in attaining the respective SDTs. In terms of 17 SDGs by United Nation, we can conclude that this digital library focus on the SDTs related to SDG4, SDG5, SDG7, SDG8, SDG9, SDG10, and SDG17. SDG4 is to ensure inclusive and equitable quality education and promotes life-long learning opportunities for all [3]. In this regard, the equal access of the information/knowledge becomes a target as participants agreed that equal access on the information/knowledge resources regardless of the technology, format, and methods of delivery can help in life-long education enhancement to all people.

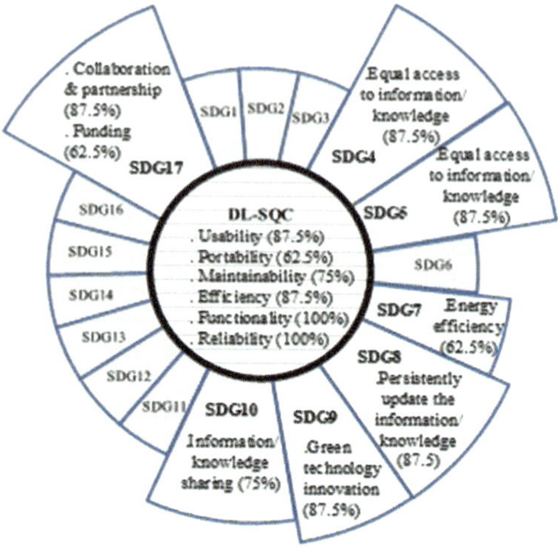

Fig. 2. Clustering of each DL-SQC and DL-SDTs and percentage of participants discussed it.

SDG5 is to achieve gender equality and empowerment, especially women, girls, and those with disability [3]. In this regard, again the equal access to the information/knowledge resources become a target as participants agreed that all library users are entitled to seek, receive and impart information effectively regardless of their skills, education, gender, age, race or physical limitations. Participants also aware the people with disabilities are often neglected part of society and some of them face personnel challenges like the economic inequity, illiteracy, cultural isolation, discrimination in education and the broad range of societal activities. Thus, the library institution plays a catalytic role in lives by facilitating their full participation in their society. SDG7 is to improve energy efficiency, lower energy consumption, and reduce emissions [3]. In this regard, energy efficiency becomes a target as participants agreed that energy consumption and minimize gas emissions help in attaining green environments. SDG8 is to promote inclusive, sustainable economic growth and productive employment for all for decent work [3]. In this regard, the persistently update the information/knowledge become a target as participants agreed that they are responsible for preparing their users to enhance self-development and performance via providing the updated quality information. SDG9 is to build and maintain a resilient infrastructure to promote inclusive and sustainable industrialization [3]. In this regard, adopting the green technology innovation became a target as participants agreed that by minimizing the environmental impact as well as transformation and development towards environmentally friendly and less carbon-intensive will help in creating a healthy environment. SDG10 is to reduce inequalities within and between nations [3]. In this regard, the information/knowledge sharing is a target as participants agreed that enabling information/knowledge sharing will empower and reduce information divide among users. SDG17 is to strengthen the means of implementation and partnership globally for sustainable development [3]. In this regard, the collaboration, partnership, and funding are the target as participants agreed that by leveraging the information collection and technologies, partnership in any sustainability relevance programs, as well as search for funders, will help them sustain in this non-profit information service field.

All these findings can lead a digital library towards better recognition of the current software employed in which subsequently can result in the growth of information services in the higher learning institution [21]. Since the study methods employed greatly depend on the expert opinion, the findings may be only applicable to this particular digital library at this period. Therefore, there is a need to periodically reassess all the findings to align with the future need changes. The future research shall be considered in this topic for the purpose of sustaining the knowledge sharing. It will help increase awareness in the field among researchers and practitioner especially with the continuous technology advancement for digital library [27].

Acknowledgement. The author wants to appreciate the Editor and anonymous referees for their constructive comments and criticism. This work was supported by the International Grant USIM/INT-NEWTON/FST/IHRAM/053000/41616 under Newton-Ungku Omar Fund.

References

1. Koukios, E., Monteleone, M., Carrondo, M.J.T., Charalambous, A., Girio, F., Hernández, E. L., Zabaniotou, A.: Targeting sustainable bioeconomy: a new development strategy for Southern European countries. J. Clean. Prod. **172**, 3931–3941 (2018)
2. Xue, L., Weng, L., Yu, H.: Addressing policy challenges in implementing sustainable development goals through an adaptive governance approach: a view from transitional China. Sustain. Dev. **26**(2), 150–158 (2018)
3. Holden, E., Linnerud, K., Banister, D.: Sustainable development: our common future revisited. Glob. Environ. Change **26**, 130–139 (2015)
4. Nica, E.: ICT innovation, internet sustainability, and economic development. J. Self-Gov. Manag. Econ. **3**(3), 24–29 (2015)
5. Lago, P., Koçak, S.A., Crnkovic, I., Penzenstadler, B.: Framing sustainability as a property of software quality. Commun. ACM **58**(10), 70–78 (2015)
6. Buchalcevova, A.: Analysis of the management of business informatics framework from the green ICT viewpoint. Int. J. Inf. Technol. Manag. **15**(1), 41–58 (2016)
7. Al-Sarayreh, K. T., Meridji, K., Issa, G., Hadi, M. A. A. W.: Procedural model of software quality improvements for sustainable applications process of renewable energy systems (Development & enhancement). In: 2nd International Conference on the IEEE, pp. 1–6 (2017)
8. Jnr, B.A., Majid, M.A., Romli, A.: Application of intelligent agents and case-based reasoning techniques for green software development. TTEM **12**, 30 (2017)
9. García-Mireles, G.A., Moraga, M.A., Garcia, F., Calero, C., Piattini, M.: Interactions between environmental sustainability goals and software product quality: a mapping study. Inf. Softw. Technol. **95**, 108–129 (2018)
10. Jung, H.W., Kim, S.G., Chung, C.S.: Measuring software product quality: a survey of ISO/IEC 9126. IEEE Softw. **21**(5), 88–92 (2004)
11. Kern, E., et al.: Sustainable software products: Towards assessment criteria for resource and energy efficiency. Future Gener. Comput. Syst. **86**, 199–210 (2018)

12. Salleh, M.A., Bahari, M., Zakaria, N.H.: An overview of software functionality service: a systematic literature review. Procedia Comput. Sci. **124**, 337–344 (2017)
13. Vijayasarathy, L.R., Butler, C.W.: Choice of software development methodologies: do organizational, project, and team characteristics matter? IEEE Softw. **33**(5), 86–94 (2016)
14. Paris, D.L., Bahari, M.: Exploring implementation factors influencing business-to-customer (B2C) e-Commerce. Adv. Sci. Lett. **21**(5), 1455–1459 (2015)
15. Khare, S.R., Shrivastava, R.: Literature survey on agile information systems development. PIET **2**(8) (2015)
16. Bi, T., Yang, X., Ren, M.: The design and implementation of smart campus system. JCP **12** (6), 527–533 (2017)
17. Leng, K., et al.: Research on agricultural products supply chain inspection system based on internet of things. Cluster Comput. 1–9 (2018)
18. Huang, J. et al.: Cloud services platform based on big data analytics and its application in livestock management and marketing. In: Proceedings of Science (2015)
19. Ivezic, N., Srinivasan, V.: On architecting and composing engineering information services to enable smart manufacturing. J. Comput. Inf. Sci. Eng. **16**(3), 31002 (2016)
20. Li, D., et al.: Software and applications of spatial data mining. Wiley Interdisc. Rev.: Data Min. Knowl. Discov. **6**(3), 84–114 (2016)
21. Chowdhury, G.G.: How to improve the sustainability of digital libraries and information services? J. Assoc. Inf. Sci. Technol. **67**(10), 2379–2391 (2016)
22. Li, C., Liu, Y.C., Yan, X.: Optimization-based resource allocation for software as a service application in cloud computing. J. Sched. **20**(1), 103–113 (2017)
23. Dai, J., Xu, D.: Detection of underground pipeline based on Golay waveform design. In: AIP Conference Proceedings, p. 020151(2017)
24. Hasegawa, T.: Toward the mobility-oriented heterogeneous transport system based on new ICT environments-understanding from a viewpoint of the systems innovation theory. IATSS Res. **42**, 40–48 (2017)
25. Dong, N., et al.: Research on information security system of smart city based on information security needs. J. Phys.: Conf. Ser. **1069**, 012040 (2018)
26. Xia, K., Gao, L., Wang, L., Li, W., Chao, K.M.: A semantic information services framework for sustainable WEEE management toward cloud-based remanufacturing. J. Manuf. Sci. Eng. **137**(6), 061011 (2015)
27. Witt, S.W., Rudasill, L.M.: World sustainable development web archive: preserving and disseminating knowledge for sustainable growth (2015)
28. Corporate-citizenship. http://www.corporate-citizenship.com. Accessed 16 May 2019
29. Kear, R.L.: Libraries, development, and implementation of the UN2030 agenda. Int. Inf. Libr. Rev. **50**(1), 60–62 (2018)
30. Fu, C., Yang, J.B., Yang, S.L.: A group evidential reasoning approach based on expert reliability. Eur. J. Oper. Res. **246**(3), 886–893 (2015)
31. Gustafsson, J.: Single case studies vs. multiple case studies: a comparative study. Academy of Business, Engineering, and Science, Halmstad University Sweden, pp. 1–15 (2017)
32. Malaysia Sustainable Development Goals Voluntary National Review (2017). http://www.epu.gov.my
33. Braun, V., Clarke, V.: Using thematic analysis in psychology. Qual. Res. Psychol. **3**, 77–101 (2006)

Motivations of Teaching in Massive Open Online Course: Review of the Literature

Muhammad Aliif Ahmad(✉), Ab Razak Che Hussin, and Ahmad Fadhil Yusof

Universiti Teknologi Malaysia, 81310 Johor Bahru, Malaysia
maliif.ahmad@gmail.com, {abrazak,ahmadfadhil}@utm.my

Abstract. The emergence of Massive Open Online Course (MOOC) had changed the landscape of Higher Education in the world. To date, studies in MOOC have focused largely on student perspectives. However, instructor in MOOC plays a critical role in facilitating the courses and should also had a proper attention. One of the key aspects regarding instructors is their experiences in teaching MOOC, especially their motivations. This is because motivations can positively influence them in teaching MOOC and it is important to recognize it in order to encourage and leverage the instructor's motivation. This study aims to shed lights on this issue by investigating the literature on the motivations that affects instructor in teaching MOOC. Systematic Literature Review (SLR) is conducted on the recent literature from 2015 to 2018. 39 instructor's perspectives studies in MOOC had been discovered where 10 studies were selected as primary studies. The results provide the current characteristics of studies and motivations of instructor in teaching MOOC. There were 5 intrinsic and 17 extrinsic motivations that had been identified. Based from the results, we categorize them to 3 main themes of motivations; enjoyment in helping students, professional development and reputation enhancement.

Keywords: Disruptive innovation · SLR · MOOC · Instructor perspectives · Motivations

1 Introduction

The rapid growth of Massive Open Online Course (MOOC) in recent years has attracted substantial interest within academia. Despite being a nascent field, it had grown rapidly from 50 peer-reviewed studies in 2013 to 1500 in 2015 [1]. Currently, many studies in MOOC focuses on understanding student's perspectives such as behavior patterns [2] and enhancing teaching practices to suit students' needs [3, 4]. There is a lack of consideration given to instructors who play a crucial role in facilitating MOOC courses. One important aspects of instructors are their experiences in teaching MOOC, especially their motivations. The newness of MOOC paradigm makes it harder for instructors to understand what is the best practices for them [3, 5]. By understanding their motivations in teaching MOOC, institution can help design appropriate support for instructors. Furthermore, little effort had been done to review and synthesize MOOC instructor's motivation systematically. This process is important

F. Saeed et al. (Eds.): IRICT 2019, AISC 1073, pp. 1087–1097, 2020.
https://doi.org/10.1007/978-3-030-33582-3_102

in order to understand and convey a clear view of their motivations especially for academicians and practitioners. Consequently, this study address this issue using Systematic Literature Review (SLR) to investigate the motivation of instructors in teaching MOOC. It will systematically gather, analyses and synthesizes current studies in literature and provides a comprehensive list of instructor's motivations. To achieve this study's main objective, we propose two key research questions. Research questions for this study are stated below:

RQ1: What are the characteristics of the primary studies? What are the method used and who are the participants in the study?
RQ2: What are the motivations of instructor in teaching MOOC? What are the main theme of motivations that can be identified?

Generally, this study has two main contribution. First, from analysis results of 10 out of 39 studies, this review delivers readers with broad understanding of motivations of instructors in teaching MOOC. Secondly, for practitioners, this review provides them up to date with the current state of studies that emphasis on the motivations of MOOC instructors. The remainder of this study is as follows: Sects. 2 and 3 will discuss on the problem background and the review method used in this study. Section 4 reveals the research question results. Finally, Sect. 5 presents the discussion and conclusion.

2 Motivations for Instructors in MOOC

Numerous topics throughout MOOC had been discussed extensively such as topics regarding student's perspectives [6]. However, there are still a paucity of studies covering MOOC from instructor's perspectives based on a few SLR that had been conducted recently [7, 8]. MOOC's instructor interest and drive in teaching and interacting with students is essential for student's need for competency. It increases positive feeling the students have with the MOOC and assist them to stay involved [9]. Instructors play a major role in MOOC, therefore it is important to addressed this gap in the literature, especially their experiences. One of the most important experiences are their motivations in teaching. The instructor's motivations are closely related towards the challenges of teaching MOOC. If the value that the instructor gets from teaching MOOC are not worth the cost that they need to endure, instructor may reevaluate their decision to teach in MOOC [10]. So, it is important to understand their motivations in order to prevent this scenario from happening. Several researchers have investigated the motivations of instructors in teaching MOOC but it is done as secondary consideration to primary focus on the student's perspectives. Some of the identified motivations include intrigue [11], personal rewards [12], and altruism [13]. But, these types of studies have some limitations. First, the student-focused outcomes present little clarification of the justification behind the instructor's motivation in teaching MOOC. This is because the outcomes were usually based on the overall experiences and as secondary consideration behind student's perspectives. Second, this type of studies also does not provide a comprehensive list of motivations that affect MOOC instructors which is vital towards keeping them to stay engaged in teaching duties. We aim to address this limitation by using SLR because teaching MOOC can be completely challenging for instructors. It is essential to recognize their motivations to provide appropriate support and alleviate their difficulties.

3 Systematic Literature Review

To answer the two research questions above, this study uses SLR method [14]. The method is used because it addresses specific research questions and reduces the review bias by being methodical and explicit about how the review is done. Besides that, this method is used to analyze evidence of specific method advantages, summarize evidence on a treatment or technology and offer deep understanding of new phenomenon [15]. It is suitable with the aim of our review. Generally, this study follows review stages by [16]. Figure 1 below shows and explained the three main stages.

3.1 Stage 1: Initial Search

The initial search stage comprises of developing research protocol by establishing a set of clearly defined research questions (as in the introduction section), developing search string (as in Fig. 1), finding the bibliographic databases to be used: Web of Science (WOS), Scopus, ScienceDirect, Emerald, SAGE and AIS Electronic Library (AISeL). Next, we conduct searches in six bibliographic databases to acquire the initial studies. There is no inclusion and exclusion criteria being applied on the results at this stage.

3.2 Stage 2: Literature Selection

The literature selection stage comprised of examining the studies titles, abstracts, and keywords found in terms of the inclusion criteria: written in English, publish within selected period (2015–2018), publish in the selected databases, full paper availability, MOOC domain studies, and studies associated to instructor perspectives in MOOC. Next, we applied the exclusion criteria: rejecting non-English studies, studies outside chosen time, duplicated and uncompleted studies and studies not related to instructor perspectives in MOOC. Table 1 below shows the results of this stage.

3.3 Stage 3: Data Extraction and Synthesis

The data extraction and synthesis stage comprised of examining the studies full content in terms of inclusion criteria: relevant studies contain motivation of instructors in teaching MOOC; and then acquiring them from the databases as our primary studies. Table 1 shows the results of this stage. Next, we designed data extraction form to record all the information precisely. This process was executed by reading each study carefully, and extracting related data using Microsoft Excel. Research framework proposed by [17] was adopted but the selected items were personalized towards this review objective and research questions. The subsequent columns were considered for data extraction: the study ID, authors, title, source, year of publication, participant's types and methodology used.

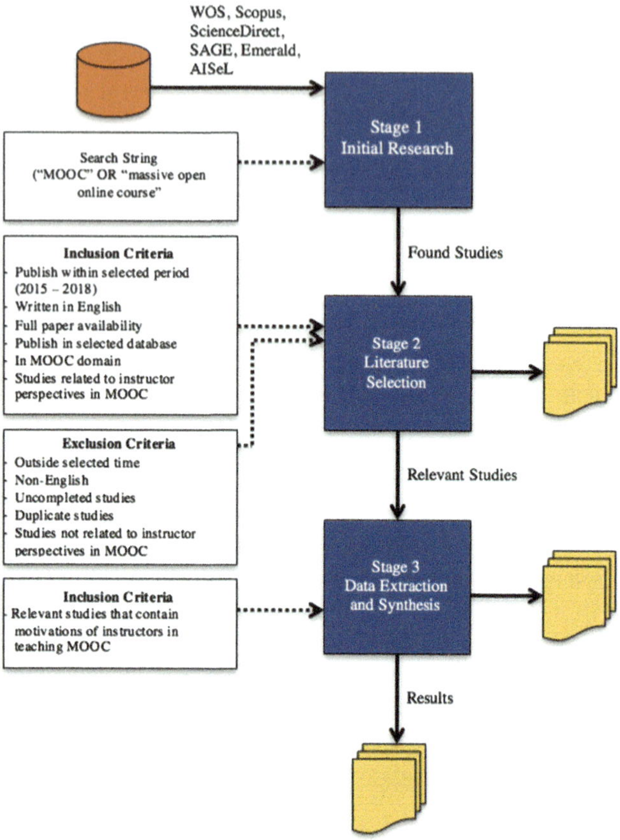

Fig. 1. Flow of research methodology.

Ratio of Studies that Comprise of Motivation. Figure 2 below shows that there are 10 studies from 39 instructor perspectives studies that discuss the motivation of MOOC instructor in teaching MOOC. This represents 26% of the total studies that emphasis on MOOC instructor perspectives. It clearly displays that there was average attention given towards understanding the motivations of the instructor. The percentage should and need to be higher because the novelty of MOOC currently makes it a difficult and complex situation for instructors. They will face various challenges and it is imperative to identify and nurture their motivation to provide suitable support and encouragement for them to stay engage in teaching MOOC.

Table 1. Selected instructor perspective studies & primary studies (motivation)

Bibliographic database	Instructor perspective studies				
	2015	2016	2017	2018	2015 - 2018
WOS	3	5	3	3	14
Scopus	4	3	6	6	19
ScienceDirect	1	0	0	0	1
Emerald	0	0	2	0	2
SAGE	1	0	2	0	3
AISeL	0	0	0	0	0
TOTAL	9	8	13	9	39
Bibliographic database	Primary studies (Motivations)				
	2015	2016	2017	2018	2015 - 2018
WOS	2	2	0	1	5
Scopus	1	0	2	1	4
ScienceDirect	0	0	0	0	0
Emerald	0	0	0	0	0
SAGE	0	0	1	0	1
AISeL	0	0	0	0	0
TOTAL	3	2	3	2	10

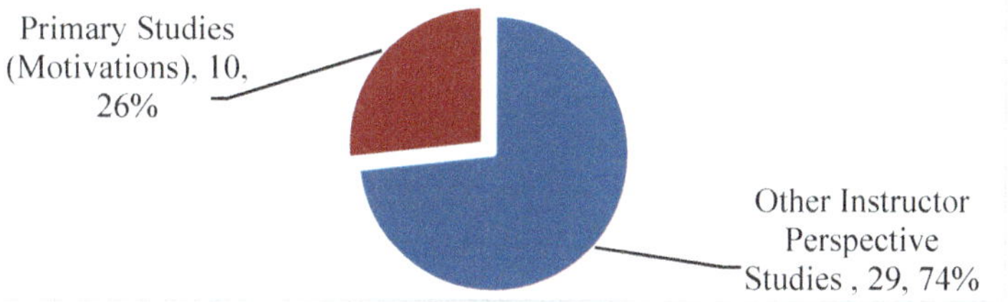

Fig. 2. Ratio of primary studies (motivations) with instructor perspectives studies.

Publication Sources Overview. Figure 3 shows that majority of studies in this review were published in indexed journal such as WOS and Scopus. This adds to the increase significance of this review by ensuring the high quality of the primary studies that were selected. Most of the publication sources were journals with (8) studies and followed by (2) conference. Instructor perspectives article were included for comparison. It displays similar trend as journals had (28) studies followed by (9) conference, (1) workshops and (1) magazines studies.

Temporal View of Publication. The selected period for this review is 2015 to 2018. Figure 3 shows the distribution of the studies over the years. As seen in the figure, the publications of studies comprising instructor's motivation in teaching MOOC have remain stagnant from 2015 to 2018. The highest number of publication was recorded in 2015 and 2017 with (3) studies and followed by 2016 and 2018 with (2) studies. This shows that there is a need for more studies to explore instructor's motivation because MOOC will become increasingly popular among institutions and more instructors will participate in teaching MOOC. Figure 3 also comes with publication view of instructor perspectives article for comparison. Highest number of publication is in 2017 with (13) studies, trailed by 2018 and 2015 with (9) and 2016 with (8) studies.

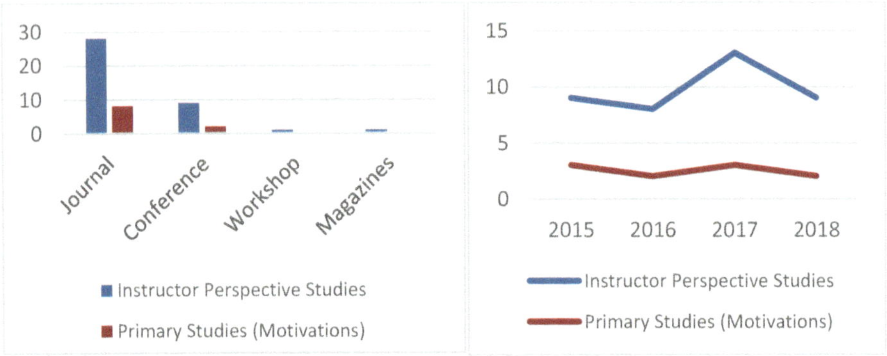

Fig. 3. Publication sources overview and temporal view publication.

4 Analysis and Results Discussion

4.1 RQ1: What Are the Characteristics of the Primary Studies? What Are the Methods Used and Who Are the Participants in the Study?

Characteristics of the primary studies are shown in Table 2. It comprises of the authors, sources and type of studies. Next, we identified the methodology used for obtaining the motivations and the type of participants used in these studies. 2 from 10 studies used mixed method to extract motivation from participants, 1 use quantitative while 7 of them use qualitative method. 9 of 10 studies target participants were instructors that had experience in designing or teaching MOOC while study by [20] had MOOC mentors (assistants or students) as their target participants. Study by [12] has exclusive conditions where instructors need to had experiences teaching MOOC, traditional classroom and regular online teaching format. This is done to provide MOOC specifics artifacts and not blurring it with other method of teaching.

Table 2. Characteristics of the primary studies

Author/Sources/Type	Methodology	Participants types
[18]/WOS/Journal	Qualitative – 20 participants in 4 focus group	All participants working as MOOC mentors
[10]/WOS/Conference	Qualitative - semi-structured and face-to-face interviews	Participants from large public university in United States. Two inclusion criteria (1) Teaching MOOC at least 2 sessions (2) Experience in traditional teaching. Ensure motivations, challenges exclusive for MOOCs
[19]/WOS/Journal	Mixed - online survey, open ended question, interviews	Participants listed as teacher on Coursera and edX's. Email sent and 162 completed open ended survey. 5 participants in interviews
[20]/WOS/Journal	Qualitative - semi-structured interview	8 U of T MOOC instructors participate in the study
[21]/Scopus/Journal	Mixed - tool by MOOC experts	203 facilitator volunteers responded to the tool
[22]/Scopus/Conference	Qualitative - World Café method (i1), interviews (i2)	All MOOCs staff invited to i1. 34 attended In i2, staff directly involved in MOOC attended The questions same as in the first stage in i1
[23]/Scopus/Journal	Qualitative - interviews	Interviews conducted with 19 instructors. Participants were self-selected and high probability they more motivated to teach
[24]/SAGE/Journal	Qualitative - semi-structured interviews in 90 min	Details of interviewed participant: Michael, instructor, nonprofit owner, journalist, educator
[25]/WOS/Journal	Qualitative - online, offline interview	8 professors who had experiences to create and teach MOOC
[26]/Scopus/Journal	Quantitative - online survey	45 full-time university teachers that thought at least 1 MOOC

4.2 RQ2: What Are the Motivations of Instructor in Teaching MOOC? What Are the Main Theme of Motivations that Can Be Identified?

Table 3 shows the motivations of instructors from 10 identified studies. We found 31 motivations with 5 of them can be considered as intrinsic motivation and the rest as extrinsic motivation. Based from analyzing these two type of motivations and the description of each motivation, we categorized them into 3 main motivation themes as shown in Table 4. The first theme is instructor's enjoyment in helping students learn. This theme consists of 5 motivations. The second theme is based on the professional development gained by instructors. This theme consists of 13 motivations. The third theme is based on the enhancement of the instructor's reputation and it consists of 4 motivations. These 3 main themes play a huge role in increasing instructor's motivation especially professional development theme. This theme of motivation should be a treated with importance from institutions that want to improve the quality of support for their instructors in teaching MOOC. Besides that, 9 motivations found in study by [21] and [26] was not categorized into respected theme because it lacks description that was needed in order to decide a suitable theme for them.

Table 3. Motivations of instructors in teaching MOOC.

No	Authors	Intrinsic motivations
1	[10]	Global impact on students
2	[19]	Reach enormous numbers of students
3	[22]	Social mission, free education
4	[24]	Passion to educate
5	[25]	Altruism
No	**Authors**	**Extrinsic motivations**
1	[18]	Confidence
2		Skills
3		Knowledge
4	[10]	Professional growth
5		Research opportunities
6		Enhanced name recognition
7	[19]	Produce more successful material
8	[20]	Contribute to open educational resources
9		Academic relevance
10		Recognition received
11	[21]	To learn about open education resources
12		To update knowledge (participants with prior knowledge)
13		To improve teaching practices
14		To learn new things
15		Professional reasons
16	[23]	Professional development
17		Flexibility in class time
18	[24]	Career enhancement
19	[25]	Research exposure
20		Experiment with new technology
21		Public recognition
22	[26]	Pedagogical experiment and innovation
23		Reaching different audience
24		Making area of knowledge visible
25		Research
26		Publicizing other courses I teach

Table 4. Main themes of motivations.

No	Intrinsic/Extrinsic	Motivations	Main themes of motivations
1	Intrinsic motivations	Global impact on students	Enjoyment in helping students
2		Reach enormous numbers of students	
3		Social mission, free education	
4		Passion to educate	
5		Altruism	
1	Extrinsic motivations	Confidence	Professional development
2		Skills	
3		Knowledge	
4		Professional growth	
5		Produce more successful material	
6		Academic relevance	
7		Professional development	
8		Flexibility in class time	
9		Career enhancement	
10		Research opportunities	
11		Research exposure	
12		Experiment with new technology	
13		Pedagogical experiment and innovation	
14		Enhanced name recognition	Reputation enhancement
15		Contribute to open educational resources	
16		Recognition received	
17		Public recognition	

5 Conclusion

This study provides an overview of the motivations of MOOC instructors in teaching MOOC. We consider this effort could be valuable and impactful for both academicians and practitioners. This study can help them to appropriately support instructors by understanding their motivations and reduce their difficulty in teaching MOOC. The motivations themes found improve comprehensions of academicians towards instructors teaching in MOOC and provide opportunity for further exploration. Practitioners will have a clearer picture on how to implement certain policies and requirement to attract instructors to teach MOOC for their organizations. The conclusion above shows the importance of this study as instructors are facing a lot of challenges because of MOOCs novelty and by promoting their motivations, it can reduce their effort in teaching and lead to a better quality MOOC.

Acknowledgement. This authors would like to extend gratitude towards MyBrain15 scholarship from the Ministry of Higher Education (MOHE) and Universiti Teknologi Malaysia (UTM).

References

1. Breslow, L.: MOOC research: some of what we know and avenues for the future. In: De Corte, E., Engwall, L., Teichler, U. (eds.) From Books to MOOCs? Emerging Models of Learning and Teaching in Higher Education, pp. 57–67. Portland Press, London (2016)
2. Clow, D.: MOOCs and the funnel of participation. In: Proceedings of the Third International Conference on Learning Analytics and Knowledge, pp. 185–189 (2013)
3. Zheng, S., Rosson, M.B., Shih, P.C., Carroll, J.M.: Understanding student motivation, behaviors and perceptions in MOOCs. In: The Proceedings of the 2015 ACM Conference on Computer Supported Cooperative Work (2015)
4. Kizilcec, R.F., Schneider, E.: Motivation as a lens to understand online learners: toward data-driven design with the OLEI scale. ACM Trans. Comput.-Hum. Interact. (TOCHI) **22** (2), 6 (2015)
5. Zheng, S., Rosson, M.B., Shih, P.C., Carroll, J.M.: Designing massive open online courses as interactive places for collaborative learning. In: The Proceedings of the 2015 ACM Conference on Learning at Scale (2015)
6. Kizilcec, R.F., Piech, C., Schneider, E.: Deconstructing disengagement: analyzing learner subpopulations in massive open online courses. In: Proceedings of the Third International Conference on Learning Analytics and Knowledge, pp. 170–179 (2013)
7. Liyanagunawardena, T.R., Adams, A.A., Williams, S.A.: MOOCs: a systematic study of the published literature 2008–2012. Int. Rev. Res. Open Distrib. Learn. **14**(3), 202–227 (2013)
8. Veletsianos, G., Shepherdson, P.: A systematic analysis and synthesis of the empirical MOOC literature published in 2013–2015. Int. Rev. Res. Open Distrib. Learn. **17**(2) (2016)
9. Hew, K.F.: Promoting engagement in online courses: what strategies can we learn from three highly rated MOOCS. Br. J. Educ. Technol. **47**(2), 320–341 (2016)
10. Zheng, S., Wisniewski, P., Rosson, M.B., Carroll, J.M.: Ask the instructors: motivations and challenges of teaching massive open online courses. In: Proceedings of the 19th ACM Conference on Computer-Supported Cooperative Work & Social Computing, pp. 206–221. ACM, February 2016
11. Roth, M.S.: My modern experience teaching a MOOC. Chron. High. Educ. **59**(34), B18–B21 (2013)
12. Hew, K.F., Sum, C.W.: Students' and instructors' use of massive open online courses (MOOCs): motivations and challenges. Educ. Res. Rev. **12**, 45–58 (2014)
13. Kolowich, S.: The professors who make the MOOCs. Chron. High. Educ. **59**(28), A20–A23 (2013)
14. Hanafizadeh, P., Keating, B.W., Khedmatgozar, H.R.: A systematic review of Internet banking adoption. Telemat. Inf. **31**(3), 492–510 (2014)
15. Kitchenham, B., Charters, S.: Guidelines for performing systematic literature reviews in software engineering, Keele University and University of Durham, Vol. 2, pp. 1051 (2007)
16. Sanchez-Gordon, S., Luján-Mora, S.: Research challenges in accessible MOOCs: a systematic literature review 2008–2016. Univ. Access Inf. Soc. 1–15 (2017)
17. Liang, T.-P., Turban, E.: Introduction to the special issue social commerce: a research framework for social commerce. Int. J. Electron. Commer. **16**(2), 5–14 (2011)

18. Urrutia, M.L., Fielding, S., White, S.: Professional development through MOOCs in higher education institutions: challenges and opportunities for Ph.D. students working as mentors. J. Interact. Media Educ. **1**, 1–11 (2016)
19. Evans, S., Myrick, J.G.: How MOOC instructors view the pedagogy and purposes of massive open online courses. Distance Educ. **36**(3), 295–311 (2015)
20. Najafi, H., Rolheiser, C., Harrison, L., Håklev, S.: University of Toronto instructors' experiences with developing MOOCs. Int. Rev. Res. Open Distrib. Learn. **16**(3) (2015)
21. Mercado-Varela, M.A., Beltran, J., Perez, M.V., Vazquez, N.R., Ramirez-Montoya, M.S.: Connectivity of learning in MOOCs: facilitators' experiences in team teaching. Turkish Online J. Distance Educ. **18**(1), 143–156 (2017)
22. Urrutia, M.L., Cobos, R., Dickens, K.: Internal perspectives of MOOCs in universities. In: Proceedings of EMOOCs 2017 (2017)
23. Griffiths, R., Mulhern, C., Spies, R., Chingos, M.: Adopting MOOCS on campus: a collaborative effort to test MOOCS on campuses of the university system of Maryland. Online Learn. **19**(2), n2 (2015)
24. Loizzo, J., Watson, S.L., Watson, W.R.: Examining instructor and learner experiences and attitude change in a journalism for social change massive open online course: a mixed-methods case study. J. Mass Commun. Educ. **73**, 392–409 (2017)
25. Blackmon, S.: MOOC makers: professors' experiences with developing and delivering MOOCs. Int. Rev. Res. Open Distrib. Learn. **19**(4) (2018)
26. Gil-Jaurena, I., Domínguez, D.: Teachers' roles in light of massive open online courses (MOOCs): evolution and challenges in higher distance education. Int. Rev. Educ. **64**, 197 (2018)

Green Information Technology Adoption Antecedence: A Conceptual Framework

Hussein Mohammed Esmail Abu Al-Rejal[1(✉)], Zulkifli Mohamed Udin[1], Mohamad Ghozali Hassan[1], Kamal Imran Mohd Sharif[1], Waleed Mugahed Al-Rahmi[2], and Nabil Hasan Al-kumaim[3]

[1] School of Technology Management and Logistic, College of Business, Universiti Utara Malaysia, 06010 Sintok, Kedah, Malaysia
abualrejal@uum.edu.my,
{zulkifli,ghozali,kamalimran}@uum.edu.my

[2] Faculty of Social Sciences and Humanities, School of Education, Universiti Teknologi Malaysia, 81310 Johor Bahru, Johor, Malaysia
waleed.alrahmi@yahoo.com

[3] Faculty of Technology Management and Technopreneurship, Universiti Teknikal Malaysia Melaka (UTeM) Melaka, Hang Tuah Jaya, 76100 Melaka, Malaysia
nabil@utem.edu.my

Abstract. Green Information Technology (Green IT) has been adopted in developed countries, as a strategic initiative consideration for developing sustainable business practices, through the balancing economic and environmental performance of an organization. However, there is still a low adoption in developing countries like Malaysia, and limited empirical research on the Green IT adoption, despite it being a necessity even though numerous government incentives were provided to adopt the green IT in Malaysia. The aims of this research are to develop a conceptual framework to determine the relationship between learning capability, innovation capability, emotional capability, and government incentives as drivers toward green IT Adoption. In order to develop the green IT adoption's antecedents' relationships in Malaysia, this research will rely on some technology adoption theories such as Technology-organization-environment (TOE), Resource-based theory (RBT), Institutional theory (IT). Based on previous literature, there is a relationship among green IT adoption with organization capabilities, and government incentives, but in a scattered manner. Furthermore, this study has provided integrated framework for researchers and business homes, it showed that internal and external factors play major role in green IT adoption within organizations, So, they can start their green initiatives to cope with requirements in a new industrial revolution 4.0. Also, economic and environment sustainability should be considered as consequences factors in future researches.

Keywords: Green IT adoption · Organizational capabilities · Government incentives · Malaysia

F. Saeed et al. (Eds.): IRICT 2019, AISC 1073, pp. 1098–1108, 2020.
https://doi.org/10.1007/978-3-030-33582-3_103

1 Introduction

Green IT is becoming a strategic consideration in the development of sustainable business practices [1]. It also aims at giving direction to implement programs and activities related to green technology [2]. In developing countries, Green IT development still at an early stage [3], Even Malaysian renewable energy development still in the beginning phase [4]. Add to that, little common understanding for green IT concept [5], lack of awareness, and concerning environmental sustainability in Malaysia and it has never been the main concern of IT departments in most companies [6]. Companies have to obtain insight on how to be green with avoiding spending resources for Green IT initiatives without being sure of the future expected returns [7]. Therefore, it is required to have a clear understanding of Green IT adoption and how companies can obtain the value of Green initiatives through improving their capabilities and utilizing incentives.

Learning capability is the firm's capacity and processes that will provide guidelines for organizational action [8], enhance performance [9], and influences the deployment of IS as it has a positive impact on the green practices adoption intention [10]. Innovation capability is defined as an idea, practice, object or a unit of adoption that is perceived as a novelty to an individual or organization [11]. Complex innovation calls for superior resources and skills for adoption and it reduce the inclination of new technology adoption [12]. A key barrier to IT adoption comes from the complexity of technology itself [13].

Organizational emotional capability is firm competency [14] and inherent concept of individual-level [15]. It linked to organizations and not just individuals [16]. Viewed this way, an individual's decision to join the green movement hinges on both motivation and capability. In the investigation of the relationship between environmental attitudes and behavior, emotional appeal showed greater influence compared to both logical reasoning or factual description of the adverse impact of environmental pollution upon convincing people to act green [17].

The majority of regulations and legislation concerning green initiatives presently enforced are not binding [18]. The government provide technical and financial support such as, tax-cut incentives and the industrial complexes infrastructure development that are environmentally friendly. Most importantly, the government is expected to play a strategic role in urging SME suppliers to pay more attention to green initiatives [19]. These support and regulations have a key role in triggering environmental innovations in the context of SMEs [20]. Through regulations, the government may encourage or discourage innovation adoption [21]. The government may also provide financial incentives, pilot projects and tax breaks to trigger technological innovation [22]. Therefore, this study aims to propose framework of Green IT adoption antecedents in Malaysian companies certified with EMS ISO 14001 and assist them to utilize their capabilities and utilize government incentives to achieve a high level of Green IT adoption. This should help to avoid Green IT adoption failure and achieve better organizational sustainability. The research also aimed at promoting a better understanding of the benefits of Green IT that can contribute toward enhancing organizational sustainability.

2 Related Work on Green IT Adoption

Prior studies in the literature that focused on green IT adoption have investigated some factors that support green IT adoption and end with improvement in organization performance. Amon these factors are motivations [23], institutional pressures [24], employees' intention, Social norms [25], organizational and technology factors [26], and internal attitudes and external pressures [27]. However, some factors were investigated separately and more clear insight about green IT initiatives drivers still needed. Therefore, current study will consider organizational capabilities such as learning, innovation, emotional, and government incentives as important factors that related to green IT adoption success.

Learning capability enables organizations in educating their workforce with the skills and knowledge required for cleaner production and better manufacturing practices due to the increase of climate change and the introduction of a carbon emissions trading scheme [28]. Furthermore, it regarded as a competency that plays a key factor in influencing information system deployment [29]. It impacts new technology adoption and improved business performance [30]. Moreover, the absorptive capacity is revealed to be a significant element in IT management practice and IT use understanding, and the improvement of the ability to implement new IT in an effective manner [31].

In the same line, innovation capability increased cognitive effort from the side of the potential adopter, reducing the inclination of new technology adoption [12], helps organizations to modify present technology in order to enhance and upgrade function [32]. Also, emotional capability plays an important role in ICT implementation [33]. In literature, organizational behavior and management stresses that emotions linked to organizations and individuals [16]. Moreover, firm competency can be considered as tacit and ingrained in the group of individuals' social system and their relationships, networks, and experiences implied through the processes of interaction [34]. Also, when employees are forcibly mandated to display a confined group of prescribed emotions, literature evidenced that it would likely lead to emotional stress, resulting in emotional burnout [35], and most commonly revealed job stressors in organizations are getting into arguments, fight, and conflicts [36], which culminate in anxiety and decreasing the innovativeness level [37].

Furthermore, studies have stressed on the role of government to encourage or discourage innovation adoption, it plays key role in triggering environmental innovation [21]. Also, regulatory forces are important in creating adherence to green initiatives [19]. Therefore, governments support and inforce firms to adopt Green IT through issue laws to caps in greenhouse gas emissions, and mandating the restrictions of e-waste disposal [38], and encourage green practices [39].

Some studies that concern Green IT have highlighted different factors that related to Green IT adoption [23], institutional pressures [24], employees' intention, Social norms [25], organizational and technology factors [26], and internal attitudes and external pressures [27], which include learning, innovation, and emotional capabilities and government incentives.

3 Methodology

The methodology of this paper has three phases. The first phase is assigned to extract factors that could influence Green IT adoption. Second Phase devote to develop integrated framework, in this phase factors have classified followed by framework construction. In literature, there are some methods have been adopted to examine adoption and implementation of technology, they are varying in their strength and weaknesses. This paper followed methodology suggested by Deng [23]. Also, to ensure selection for most related factors intensive literature has been followed, a group of 12 experts in same area were evaluated and ranked the factors and final list of extracted factors are learning capability, innovation capability, emotional capability, and government incentives. To measure factors, survey was adopted from previous studies. Green IT Adoption are adopted from Molla [40], learning capability from Akgün [41], innovation capability from Wang [32], emotional capability from Akgün [41, 42], and government incentives from Lin [43]. For data collection, the suitable person environmental management representative and IT managers [44].

4 Framework Development

There are numerous previous studies that have investigated the Green IT Adoption setting and employed one of the technology acceptance models or some mixture of two theories [45]. The Green technology adoption is to some extent, similar to other technologies' adoption [46]. Therefore, this study used an integrated model. Technology-organization-environment (TOE) as it used to investigation of Green IT [47]. Resource-based theory (RBT) used in analysis of the Green IT-grid technology relationship [48], this study used these theories in investigating the role of learning, innovation, emotional capabilities in adopting green IT.

Moreover, institutional theory (IT) is concern with the study of the influence of external social pressure on an organization's behavior and policy-making [38]. Several studies have taken an institutional theory to investigate managerial attitudes towards green initiatives [49], and external forces in the adoption of technology [50]. This theory is used in investigating the role of government incentives in adopting green IT. Based on theories discussed above and literature review, this study proposes integrated framework to test the relationship between antecedent factors and green IT adoption. In the following hypotheses dimensions are developed to test the relationships among proposed antecedents and green IT adoption.

4.1 The Relationship Between Organizational Learning Capabilities and Green IT Adoption

Some previous studies focus on learning capabilities that facilitate the diffusion of IT [51]. In the context of adopting green practices, learning capability have positive impacts and influencing the deployment of IS [52]. Regardless of IT's pervasiveness in current workplaces, evidence point to the failure to optimize organizational effectiveness owing to poor employee's new technologies acceptance [53]. Learning availability and significant technical expertise are important for firms employing IT as their core competencies [54]. Also, support from top management is important to the IT diffusion within organizations to achieve environmental progress [55]. In literature of learning capability human resource quality and organizational encouragement showed significant impact to adopt green practices [48]. This leads to the develop hypothesis H1: There is a relationship between organizational learning capability and Green IT adoption.

4.2 The Relationship Between Innovation Capability and Green IT Adoption

Earlier studies showed that an innovation calls for superior resources and skills for adoption and it calls for increased cognitive effort from the side of the potential adopter, reducing the inclination of new technology adoption [12]. Furthermore, improvement in organizational innovation comes from the firm's encouragement of innovation and reinforcement of innovative resources, and management skills are revealed to lend a hand to organizational innovation [56]. Technology innovation adoption is significantly influenced by the quality of human resources, leadership behavior, communication and informal linkages among employees, and the internal resources and has a positive impact upon the green practices adoption intention [57]. This eventually leads to the static situation of organizational innovation at the initial stage without the help of dedicated champions [58]. This leads to develop hypothesis *H2: There is a relationship between organizational innovation capability and Green IT adoption.*

4.3 The Relationship Between Emotional Management Capability and Green IT Adoption

In the literature, emotions may be linked to organizations and not just individuals [35]. As firm competency, it can be considered as tacit and ingrained in the group of individuals' social system and their relationships, networks, and experiences implied through the processes of interaction [34]. Literature evidenced that, arguments, fight, and conflicts [36], which culminate in anxiety, depression, emotional frustration [38] resulting in emotional burnout [35] and decreasing the innovativeness level [37]. Emotional appeal factual description of the adverse impact of environmental pollution upon convincing people to act green [24]. In ICT studies, adopting new technology make employees afraid from losing their jobs as they feel may lack the skills to cope. Hence, emotion plays an important role in ICT implementation [33]. This lead to develop hypothesis H3: There is a relationship between organizational emotion capability and Green IT adoption.

4.4 The Relationship Between Government Incentives and Green IT Adoption Relationship

In literature, it showed that government could provide various direct technical and financial support to indirect encouragement through tax-cut incentives and development of infrastructure for environmentally friendly units [59]. To adopt the green initiative, government involvement and support are evident through various ways including funding, taxation policy, import duty, business training in the quest to promote SMEs

[60]. In the same line, governments should convince firms to adopt Green IT, lay down rules and laws allowing and prohibiting some practices like caps in greenhouse gas emissions, institutionalizing trading emissions, and mandating the restrictions of e-waste disposal [38]. Government incentives are considered as one of the motivator factors for Malaysian manufacturing companies to adopt an Environmental Management System [61]. Owing to the issue, a significant number of individuals are still pessimistic concerning government regulations or incentives driving their Green IT strategy [62], and question arises whether government-driven initiatives are helping the many SMEs in moving towards greener enterprises [63]. Hence, it is expected that governmental support influences Green IT adoption. This leads to develop hypothesis H4: There is a relationship between Government incentives and Green IT Adoption.

The proposed framework of this study was developed based on literature review and above discussion of the TOE theory, RBT, and institutional theory. It tests the relationship between Learning, innovation, emotional capabilities, and government incentives, should work in harmony in order to improve green IT adoption. The following Fig. 1. shows the relationships between all proposed factors.

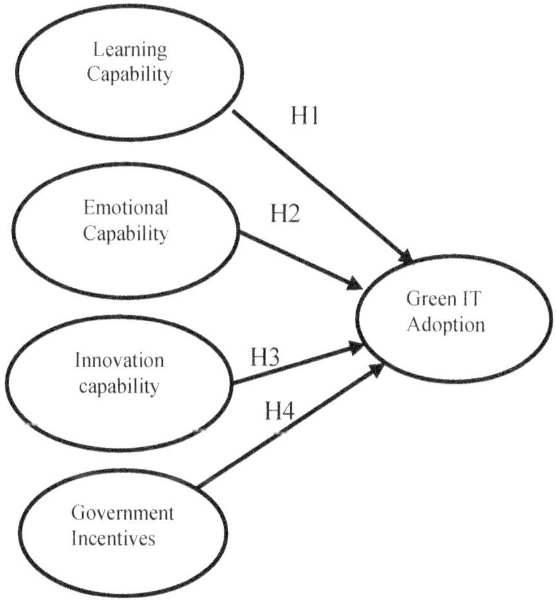

Fig. 1. The proposed framework of the study

5 Implication of the Study

This paper presents the value of learning capability, emotional capability, innovation capability, and government incentive for adopting green IT in Malaysia. The innovation and adoption of green IT offerings proficiently for Malaysian organizations especially for organizations services [64]. However, the adoption in Malaysian organization adoption of green technology [65]. Therefore, the investigation shows that there is a big need for support of government incentives to improve and overcome this issue into manipulating and increase the Green IT adoption in Malaysia [66]. To encapsulate, the consequences of the exploration of the determining influences for green technology revolution designated that organizational learning, innovation capability, emotional capability, and government incentives as special drivers in the direction of green IT Adoption. Song and Yu [67] highlights that for improving the capabilities of their adoption of green IT in an organization is very essential. Though, green technology adoption can resolve many complex problems related to the process and development of the organization [68]. So, the use of green IT adoption results concluded that there is more need adopt in Malaysian companies.

This result shows that for the adoption of advanced green technology the need for governments support is necessary factors that can proceed with these adoption issues. The relation of organizations and their drivers have also investigated the positive development for the future. The theories that are also mentioned in this study focus on the green technology adoption and its driver's perspective for the future success of organization especially in Malaysia.

6 Conclusion and Further Research

For that, this research is providing a very good contribution to both academic and practical purposes. By providing a framework for drivers of Green IT adoption, the future researchers can have a clearer picture while they pursue research in the same field. Moreover, it can motivate companies and introduce the expected benefit and how they can be obtained from such adoption. In conclusion, this study originated that green technology revolution adoption in organizational capabilities perspective in Malaysia is positively related to the complementary economic and environmental performance of an organization. These discoveries of the study found that actual green applies innovation is critical and less moderated in aligning the organization's necessities according to this new era of technologies. It has been stated that green adoption is a necessary part for resolving the organizational issue nowadays and remain a substantial basis of environmental impact and also for business. Therefore, green technology improvement in the organization must deliver quick development, as well as the growth of innovative systems to preserve ordinary environmental and economic development resources. This will decrease the undesirable effect on the organization for new adoption. This study discloses the results that the organization determinant factors aimed at green technology adoption in the organization, and thus it is able to advance the knowledge of organizational learning, innovation capability, emotional capability, and government incentives as drivers toward GIT Adoption (GITA). Comprehend the prominence and value of green practices and the relationship with these organizational factors, this accepting is crucial to escalation the indulgent of environmental issues and their economic significance. In addition, it will improve the competitive influence of the organization for both internationally and provincially. This study helps to avoid Green IT adoption failure and recommend further study on how to achieve better organizational sustainability. The future research should promote for better understanding of the benefits of Green IT that can contribute toward enhancing organizational sustainability. The author will conduct empirical study to investigate the outcomes of green IT adoption.

References

1. Zaman, B., Sedera, D.: Green information technology as administrative innovation-organizational factors for successful implementation: literature review research in progress. In: Australasian Conference on Information Systems (2015)
2. Anthony Jnr., B., Abdul Majid, M., Romli, A.: Green information technology adoption towards a sustainability policy agenda for government-based institutions. J. Sci. Technol. Policy Manag. (2018). https://doi.org/10.1108/JSTPM-11-2017-0056
3. Foo, K.Y.: A vision on the opportunities, policies and coping strategies for the energy security and green energy development in Malaysia. Renew. Sustain. Energy Rev. **51**, 1477–1498 (2015)
4. Bekhet, H.A., Othman, N.S.: The role of renewable energy to validate dynamic interaction between CO_2 emissions and GDP toward sustainable development in Malaysia. Energy Econ. **72**, 47–61 (2018)
5. Angappa, G., Subramanian, N., Rahman, S.: Green supply chain collaboration and incentives: current trends and future directions. Transp. Res. Part E: Logist. Transp. Rev. **74**, 1–10 (2015)
6. Ghazilla, R.A.R., Sakundarini, N., Taha, Z., Abdul-Rashid, S.H., Yusoff, S.: Design for environment and design for disassembly practices in Malaysia: a practitioner's perspectives. J. Clean. Prod. **108**, 331–342 (2015)
7. Abdullah, M., Zailani, S., Iranmanesh, M., Jayaraman, K.: Barriers to green innovation initiatives among manufacturers: the Malaysian case. Rev. Manag. Sci. **10**(4), 683–709 (2016)
8. Ortiz, B., Donate, M.J., Guadamillas, F.: Inter-organizational social capital as an antecedent of a firm's knowledge identification capability and external knowledge acquisition. J. Knowl. Manag. **22**(6), 1332–1357 (2018)
9. Yeniyurt, S., Wu, F., Kim, D., Cavusgil, S.T.: Information technology resources, innovativeness, and supply chain capabilities as drivers of business performance: a retrospective and future research directions. Ind. Mark. Manag. **79**, 46–52 (2019)
10. Spyropoulou, S., Katsikeas, C.S., Skarmeas, D.: Strategic goal accomplishment in export ventures: the role of capabilities, knowledge, and environment. J. Acad. Mark. Sci. **46**(1), 109–129 (2017)
11. Panopoulos, A., Theodoridis, P., Poulis, A.: Revisiting innovation adoption theory through electronic public relations. Inf. Technol. People **31**(1), 21–40 (2018)
12. Dyck, B., Silvestre, B.S.: A novel NGO approach to facilitate the adoption of sustainable innovations in low-income countries: lessons from small-scale farms in Nicaragua. Organ. Stud. **40**(3), 443–461 (2019)
13. Garrone, P., Grilli, L., Groppi, A., Marzano, R.: Barriers and drivers in the adoption of advanced wastewater treatment technologies: a comparative analysis of Italian utilities. J. Clean. Prod. **171**, S69–S78 (2018)
14. Russell Li, S.: The relationship between organizational emotional capability and employee innovation behavior of S&T enterprise. In: Proceedings of the 3rd International Seminar on Education Innovation and Economic Management (SEIEM 2018) (2019)

15. Russell, Z.A., Steffensen, D.S., Ellen, B.P., Zhang, L., Bishoff, J.D., Ferris, G.R.: High performance work practice implementation and employee impressions of line manager leadership. Hum. Resour. Manag. Rev. **28**(3), 258–270 (2018)
16. Troth, A.C., Lawrence, S.A., Jordan, P.J., Ashkanasy, N.M.: Interpersonal emotion regulation in the workplace: a conceptual and operational review and future research agenda. Int. J. Manag. Rev. **20**(2), 523–543 (2018)
17. Hafezalkotob, A., Zamani, S.: A multi-product green supply chain under government supervision with price and demand uncertainty. J. Ind. Eng. Int. **15**(1), 193–206 (2019)
18. McIntyre, O.: Transnational environmental regulation and the normativisation of global environmental governance standards. J. Property Plann. Environ. Law **10**(2), 92–112 (2018)
19. Jugend, D., Jabbour, C.J.C., Alves Scaliza, J.A., Rocha, R.S., Junior, J.A.G., Latan, H., Salgado, M.H.: Relationships among open innovation, innovative performance, government support and firm size: comparing Brazilian firms embracing different levels of radicalism in innovation. Technovation **74–75**, 54–65 (2018)
20. Kivimaa, P., Kern, F.: Creative destruction or mere niche support? Innovation policy mixes for sustainability transitions. Res. Policy **45**(1), 205–217 (2016)
21. Braunerhjelm, P., Henrekson, M.: An innovation policy framework: bridging the gap between industrial dynamics and growth. In: Essays in Public Sector Entrepreneurship, pp. 95–130. Springer, Cham (2016)
22. Zhao, Z.-Y., Chen, Y.-L., Chang, R.-D.: How to stimulate renewable energy power generation effectively? – China's incentive approaches and lessons. Renew. Energy **92**, 147–156 (2016)
23. Deng, Q., Ji, S.: Organizational green IT adoption: concept and evidence. Sustainability **7** (12), 16737–16755 (2015)
24. Ainin, S., Naqshbandi, M.M., Dezdar, S.: Impact of adoption of Green IT practices on organizational performance. Qual. Quant. **50**(5), 1929–1948 (2016)
25. Aziati, H., Ping, C.S., Ahmad, M.F., Abdullah, N.H., Ismail, S.N.Z.: Assessing the factors of green computing adoption among manufacturing employees: an analysis of the electrical and electronic sector. Adv. Sci. Lett. **24**(6), 4594–4602 (2018)
26. Ahmed, A.I.: Understanding the factors affecting the adoption of green computing in the Gulf Universities. Int. J. Adv. Comput. Sci. Appl. **9**(3), 304–311 (2018)
27. Baggia, A., Maletič, M., Žnidaršič, A., Brezavšček, A.: Drivers and outcomes of green IS adoption in small and medium-sized enterprises. Sustainability **11**(6), 1575, 1–19 (2019)
28. Hassan, M.G., Nordin, N., Ashari, H.: Sustainable manufacturing practices implementation in Malaysia industries. Jurnal Teknologi **77**(4), 49–56 (2015)
29. Radulovich, L., Javalgi, R.(Raj)G., Scherer, R.F.: Intangible resources influencing the international performance of professional service SMEs in an emerging market. Int. Mark. Rev. **35**(1), 113–135 (2018)
30. Ahmad, S.Z., Ahmad, N., Abu Bakar, A.R.: Reflections of entrepreneurs of small and medium-sized enterprises concerning the adoption of social media and its impact on performance outcomes: evidence from the UAE. Telemat. Inf. **35**(1), 6–17 (2018)
31. Papa, A., Dezi, L., Gregori, G.L., Mueller, J., Miglietta, N.: Improving innovation performance through knowledge acquisition: the moderating role of employee retention and human resource management practices. J. Knowl. Manag. (2018). https://doi.org/10.1108/JKM-09-2017-0391
32. Wang, Y.-B., Lin, K.-Y., Chang, L., Hung, J.C.: A diffusion of innovations approach to investigate the RFID adoption in Taiwan logistics industry. J. Comput. **6**(3), 441–448 (2011)
33. Huang, K.-T., Cotten, S.R., Rikard, R.V.: Access is not enough: the impact of emotional costs and self-efficacy on the changes in African-American students' ICT use patterns. Inf. Commun. Soc. **20**(4), 637–650 (2017)

34. Rutten, W., Blaas-Franken, J., Martin, H.: The impact of (low) trust on knowledge sharing. J. Knowl. Manag. **20**(2), 199–214 (2016)
35. Ghadri, J.R., Sarcon, A., Diekmann, J., Bataiosu, D.R., Cammann, V.L., Jurisic, S., Napp, L. C., Jaguszewski, M., Scherff, F., Brugger, P., Jäncke, L., Seifert, B., Bax, J.J., Ruschitzka, F., Lüscher, T.F., Templin, C., Schwyzer, M., Franke, J., Katus, H.A., Burgdorf, C., Schunkert, H., Thiele, H., Bauersachs, J., Tschöpe, C., Rajan, L., Michels, G., Pfister, R., Ukena, C., Böhm, M., Erbel, R., Cuneo, A., Kuck, K.-H., Jacobshagen, C., Hasenfuß, G., Karakas, M., Koenig, W., Rottbauer, W., Said, S.M., Braun-Dullaeus, R.C., Cuculi, F., Banning, A., Fischer, T.A., Vasankari, T., Juhani Airaksinen, K.E., Fijalkowski, M., Rynkiewicz, A., Opolski, G., Dworakowski, R., MacCarthy, P., Kaiser, C., Osswald, S., Galiuto, L., Crea, F., Dichtl, W., Franz, W.M., Empen, K., Felix, S.B., Delmas, C., Lairez, O., Erne, P., Prasad, A.: Happy heart syndrome: role of positive emotional stress in Takotsubo syndrome. Eur. Heart J. **37**(37), 2823–2829 (2016)
36. Bliese, P.D., Edwards, J.R., Sonnentag, S.: Stress and well-being at work: a century of empirical trends reflecting theoretical and societal influences. J. Appl. Psychol. **102**(3), 389–402 (2017)
37. Honkaniemi, L., Lehtonen, M.H., Hasu, M.: Well-being and innovativeness: motivational trigger points for mutual enhancement. Eur. J. Train. Dev. **39**(5), 393–408 (2015)
38. Gareau, B.J., Lucier, C.A.: Neoliberal restructuring of the world polity: the weakening of the Montreal Protocol and Basel Convention in historical perspective. Environ. Sociol. **4**(3), 325–342 (2018)
39. Han, H., Hyun, S.S.: What influences water conservation and towel reuse practices of hotel guests? Tour. Manag. **64**, 87–97 (2018)
40. Molla, A.: Organizational motivations for Green IT: exploring Green IT matrix and motivation models. In: PACIS 2009 Proceedings, p. 13 (2009)
41. Akgün, A.E., Keskin, H., Byrne, J.C., Aren, S.: Emotional and learning capability and their impact on product innovativeness and firm performance. Technovation **27**(9), 501–513 (2007)
42. Akgün, A.E., Keskin, H., Byrne, J.: Organizational emotional capability, product and process innovation, and firm performance: an empirical analysis. J. Eng. Technol. Manag. **26** (3), 103–130 (2009)
43. Lin, C.-Y., Ho, Y.-H.: An empirical study on logistics service providers' intention to adopt green innovations. J. Technol. Manag. Innov. **3**(1), 17–26 (2008)
44. Darko, A., Chan, A.P.C., Owusu, E.K.: What are the green technologies for sustainable housing development? An empirical study in Ghana. Bus. Strategy Dev. **1**(2), 140–153 (2018)
45. Davis, R., Campbell, R., Hildon, Z., Hobbs, L., Michie, S.: Theories of behaviour and behaviour change across the social and behavioural sciences: a scoping review. Health Psychol. Rev. **9**(3), 323–344 (2015)
46. Rezvani, Z., Jansson, J., Bodin, J.: Advances in consumer electric vehicle adoption research: a review and research agenda. Transp. Res. Part D: Transp. Environ. **34**, 122–136 (2015)
47. Bose, R., Luo, X. (Robert): Green IT adoption: a process management approach. Int. J. Acc. Inf. Manag. **20**(1), 63–77 (2012)
48. Versteeg, T., Baumann, M.J., Weil, M., Moniz, A.B.: Exploring emerging battery technology for grid-connected energy storage with constructive technology assessment. Technol. Forecast. Soc. Change **115**, 99–110 (2017)
49. Erdogan, B., Bauer, T.N., Taylor, S.: Management commitment to the ecological environment and employees: implications for employee attitudes and citizenship behaviors. Hum. Relat. **68**(11), 1669–1691 (2015)
50. Grover, V., Chiang, R.H.L., Liang, T.-P., Zhang, D.: Creating strategic business value from big data analytics: a research framework. J. Manag. Inf. Syst. **35**(2), 388–423 (2018)

51. Santoro, G., Vrontis, D., Thrassou, A., Dezi, L.: The Internet of Things: building a knowledge management system for open innovation and knowledge management capacity. Technol. Forecast. Soc. Change **136**, 347–354 (2018)
52. Govindan, K., Muduli, K., Devika, K., Barve, A.: Investigation of the influential strength of factors on adoption of green supply chain management practices: an Indian mining scenario. Resour. Conserv. Recycl. **107**, 185–194 (2016)
53. Tweedie, D., Wild, D., Rhodes, C., Martinov-Bennie, N.: How does performance management affect workers? Beyond human resource management and its critique. Int. J. Manag. Rev. **21**(1), 76–96 (2019)
54. Cavusgil, S.T., Knight, G.: The born global firm: an entrepreneurial and capabilities perspective on early and rapid internationalization. J. Int. Bus. Stud. **46**(1), 3–16 (2015)
55. Liu, W., Atuahene-Gima, K.: Enhancing product innovation performance in a dysfunctional competitive environment: the roles of competitive strategies and market-based assets. Ind. Mark. Manag. **73**, 7–20 (2018)
56. Hannah, D., Parent, M., Pitt, L., Berthon, P.: Secrets and knowledge management strategy: the role of secrecy appropriation mechanisms in realizing value from firm innovations. J. Knowl. Manag. **23**(2), 297–312 (2019)
57. Maduku, D.K., Mpinganjira, M., Duh, H.: Understanding mobile marketing adoption intention by South African SMEs: a multi-perspective framework. Int. J. Inf. Manag. **36**(5), 711–723 (2016)
58. Koster, M., Vos, B., Schroeder, R.: Management innovation driving sustainable supply management. Bus. Res. Q. **20**, 240–257 (2017)
59. Prasad, M., da Silva, M.B.M., Nickow, A.: Approaches to corruption: a synthesis of the scholarship. Stud. Comp. Int. Dev. **54**(1), 96–132 (2019)
60. Tu, Z., Hu, T., Shen, R.: Evaluating public participation impact on environmental protection and ecological efficiency in China: evidence from PITI disclosure. China Econ. Rev. **55**, 111–123 (2019)
61. Salim, H.K., Padfield, R., Lee, C.T., Syayuti, K., Papargyropoulou, E., Tham, M.H.: An investigation of the drivers, barriers, and incentives for environmental management systems in the Malaysian food and beverage industry. Clean Technol. Environ. Policy **20**(3), 529–538 (2018)
62. York, J.G., Vedula, S., Lenox, M.J.: It's not easy building green: the impact of public policy, private actors, and regional logics on voluntary standards adoption. Acad. Manag. J. **61**(4), 1492–1523 (2018)
63. Ponte, S.: Green capital accumulation: business and sustainability management in a world of global value chains. New Polit. Econ. 1–13 (2019)
64. Zailani, S., Iranmanesh, M., Nikbin, D., Jumadi, H.B.: Determinants and environmental outcome of green technology innovation adoption in the transportation industry in Malaysia. Asian J. Technol. Innov. **22**(2), 286–301 (2014)
65. Ibrahim, I., Sundram, V.P.K., Omar, E.N., Yusoff, N., Amer, A.: The determinant factors of green practices adoption for logistics companies in Malaysia. A case study of PKT logistics group Sdn. Bhd. J. Emerg. Econ. Islamic Res. **7**(1), 14–23 (2019)
66. Ojo, A.O., Raman, M., Downe, A.G.: Toward green computing practices: a Malaysian study of green belief and attitude among Information Technology professionals. J. Clean. Prod. **224**, 246–255 (2019)
67. Song, W., Yu, H.: Green innovation strategy and green innovation: the roles of green creativity and green organizational identity. Corp. Soc. Responsib. Environ. Manag. **25**(2), 135–150 (2018)
68. Yin, S., Li, B.: Transferring green building technologies from academic research institutes to building enterprises in the development of urban green building: a stochastic differential game approach. Sustain. Cities Soc. **39**, 631–638 (2018)

Software Engineering

Search Space Reduction Approach for Self-adaptive Web Service Discovery in Dynamic Mobile Environment

Salisu Garba(✉), Radziah Mohamad, and Nor Azizah Saadon

School of Computing, Faculty of Engineering,
Universiti Teknologi Malaysia UTM, 81310 Skudai, Johor, Malaysia
Salisu.garba@graduate.utm.my,
{radziahm,azizahsaadon}@utm.my

Abstract. The proliferation of functionally similar Mobile Web Service (MWS) result in huge search space, the discovery of MWS on such large space increases the response time and probability of discovering irrelevant MWS irrespective of the matchmaking algorithm. The existing research on MWS discovery mostly focused on applying coarse-grained search space reduction that fails to deal with cold-start and data sparsity challenges at the expense of large computing resources. The proposed search space reduction is achieved by subsuming k-means in the modified negative selection algorithm (M-NSA) to place the service in an appropriate category so that the matching is only performed on the MWS in the target category. The experimental results show significant improvement in terms of accuracy of the categorization which can improve the MWS discovery in in a dynamic mobile environment (DME).

Keywords: Mobile Web Service · Discovery · Search space reduction · Categorization algorithm

1 Introduction

The fundamental notion of Service Oriented Architecture (SOA) is to confront interoperability and heterogeneity challenges through Web Service. It employs loosely coupled services, standard interface and protocols to deliver seamless cross-platform integration. Web Services are self-contained modular applications that are defined, published and accessed across the Internet using standard protocols [1]. Web services can be as simple as Microsoft.Net passport or Google MAP API. Mobile Web Services (MWS) are deployed in a dynamic mobile environment (DME) where devices may play the role of consumer, broker, or provider. The mobile era facilitates the delivery of services to the right person, at the right time to perform critical operations regardless of location.

Web services clustering is a process of distributing web services into categories such that the similarity between web services within one category is high and the similarity between dissimilar categories is low. Complications caused by the proliferation of functionally similar web services and the semantic sparsity issue because of the

F. Saeed et al. (Eds.): IRICT 2019, AISC 1073, pp. 1111–1121, 2020.
https://doi.org/10.1007/978-3-030-33582-3_104

short messages used in describing web service [2] makes it difficult to achieve an accurate categorization. Although the results of applying NSA tend to show substantial improvement in low computational time, high precision when compared to other methods [3, 4].

In this research, the natural language processing tool kit (NLTK) is used to extract and integrate MWS goals and tags. After that, the modified negative selection algorithm that subsumes the k-means is used to categorized services with similar functionalities. The rest of this paper is organized as follows: The related works are discussed in Sect. 2; The proposed algorithm is discussed in Sect. 3; The empirical evaluation and results obtained are presented in Sect. 4; Lastly, the conclusions are presented Sect. 5.

2 Related Work

Because the functionality of the web service is always the primary target of a request and the changes associated with non-functional properties makes it appropriate to use functionality as the basis for search space reduction as a phase preceding the discovery of more relevant web services [5]. So far, many solutions are proposed to handle different challenges caused by the persistent influx of web service with similar functionality.

In an effort to reduce the search space, [6] proposed Mashup service clustering by using mashup service APIs' description and tags to build Latent Dirichlet allocation (LDA) topic model to be used for computing the mashup service similarity, after which K-Means and Agnes algorithm are employed to cluster the mashup services based on domains. The semantic-based solution to service clustering proposed by [7], in which natural language processing is used to mine service goals from RESTful service description document, followed by extending the service goals semantically, after which K-Means algorithm is applied to functionally-cluster the web services.

Classic clustering algorithms such as k-means [8], agglomerative [9], density-based spatial clustering of applications with noise (DBSCAN) [10], are used to cluster web services, however, the accuracy of k-means, the speed of DBSCAN and the complexity of agglomerative left more to be desired. Besides, meta-heuristics clustering algorithms such as Cat Swarm Optimization [11], are special forms of approximate algorithms that are adopted to solve numerous classification and optimization problems but are sometimes unpredictable when it comes to performance evaluation of results. Although, the rapidly growing literature on negative selection algorithm (NSA) indicates its applicability in data partitioning, categorization, clustering, and classification [12–15] due to its powerful information processing capability, unlike other meta-heuristic algorithms.

The computational efficiency, effectiveness for intended application domain and dataset plays an important role, as the best clustering algorithm may not always produce the best results. [12] points out that the quality of the feature extraction technique outweighs the performance of the clustering algorithm in real-world application since it portrays the fundamentals of the dataset. [16] argued that it is significant to recognize the difference between two constructs of the clustering algorithm (distance and similarity) and misappropriation of the measure can lead to loss of semantic information due to heterogeneity of service publishers, besides, the balance between optimal solution and acceptable execution time is assessed only to a very limited extent. Therefore, the time taken and the accuracy of the categorization prior do discovery can be enhanced to achieve less computation time and more accurate solution.

3 The Proposed Search Space Reduction Approach

The proposed search space reduction algorithm for self-adaptive web service discovery in a dynamic mobile environment is designed to deal with the proliferation of functionally similar web services by minimizing the search space, minimizing the computation time, reducing the workload of the matchmaking algorithm, and paving the way for efficient and more accurate MWS discovery. As illustrated in Fig. 1, the proposed search space reduction algorithm comprises of three stages (pre-processing, initial detector generation, search space reduction by categorization). The initial detector generation is based on a modified k-means and a modified negative selection algorithm (M-NSA) to perform the categorization thereby reducing the search space. The three stages are explained in Sects. 3.1, 3.2 and 3.3 respectively.

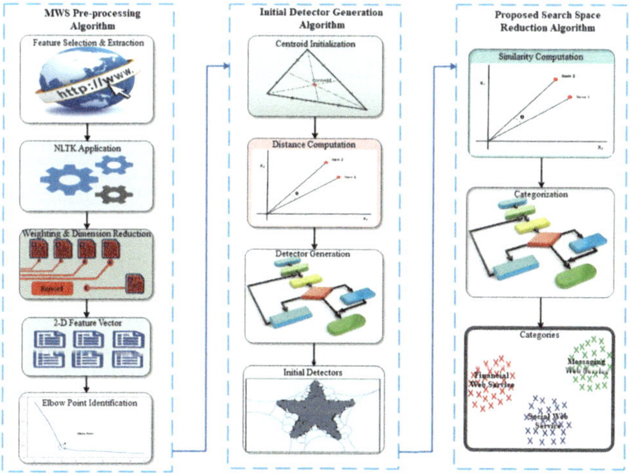

Fig. 1. The proposed search space reduction methodology.

3.1 Pre-processing Stage

The challenges in MWS or RESTful web service discovery is compounded by the lack of WSDL-like document to provide a standard definition/description of the RESTful web service. However, API information directories such as ProgrammableWeb (PW) provide relevant information on different categories of RESTful APIs, therefore, a web crawler tool is employed in acquiring all the necessary information such as APIs' title, tags, categories, description, etc. Because most of the information is in form of natural language, it undergoes through redaction process to remove APIs that are not described, categorized, or no longer available in addition to filtering out duplicates, filters such as stop-words are also applied to ensure undistorted feature space. Subsequently, NLTK lemmatizer is applied as it is more semantic-aware than NLTK stemmer to ensure the efficient production of the new vector.

MWS (Goals and Tags) are considered the most important features since the service functionality is the basis for search space reduction. Service Goals (functionalities that are provided by a service). [7] defined service goal as a "relation between a verb or a verb phrase, which represents an executable action, and a noun or a noun phrase, which represents an operational objective in a service". The tags provide additional information, especially the categorical meta-data of the RESTful web service.

Before categorization, similarity measures are used to determine the degree of closeness or separation of the target services or features within a service [17], therefore, TF-IDF vectorizer is used to convert the service goals and tags to vector space using Eq. (1).

$$tf - idf_{gt,S,CS} = tf_{gt,d} * idf_{gt,CS} \quad (1)$$

Where tf is calculated using Eq. (2), idf is calculated using Eq. (3) in which gt denotes the goals and tags; S denotes the description of each web service; CS denotes the collection of the web service description.

$$tf_{gt,S} = \frac{Goals\ \&\ tags(gt)\ frequency\ in\ document(S)}{Summation\ of\ words\ in\ document(S)} \quad (2)$$

$$idf_{gt,CS} = \log_2\left(\frac{Summation\ of\ documents(CS)}{Documents\ with\ term(t)}\right) \quad (3)$$

Due to the large dimension created, a non-linear dimensionality reduction technique known as multidimensional scaling (MDS) is used to place each web service data point in 2-dimensional space so as to preserve the distance between web service, tackle outliers and to have a better visualization of the similarities between the web service. Equation (6) shows how the MDS is calculated, where M is the matrix of the distance between services, S_{ij} is the distance between the coordinates of the ith and j^{th} service.

$$M = [S_{ij}] = \sqrt{(x_i - x_i)^2 + (y_i - y_i)^2} \quad (4)$$

Since a modified k-means is used to generate the initial detectors, Elbow method is used to identify the optimal number of categories (elbow point). The preprocessed web service data in the 2-D feature vector is run using k-means and within-category sum of squared errors (WCSSE) is plotted. The x-axis shows within-category sums of squares (WCSSE) which is used to determine the variability within each category while the y-axis represents the number of categories, the elbow point is considered to be the appropriate number of categories for the given dataset. The pre-processing algorithm is shown in Fig. 2.

Algorithm 1: Pre-processing
Input: Crawled WS Data
Output: 2-dimensional Feature Vector and Number of Categories Z_j
1 Extract goals from WS description and tags parameters
2 Convert the goals and tags into token using NLTK tokenization
3 Remove stop words from the goal and tags corpus
4 Apply lemmatization on the corpus using NLTK lemmatizer.
5 Compute the TI-IDF weight using Equation 1
6 Reduce the dimension of the service corpus using multidimensional scaling (MDS) and place each web service data point in 2-dimensional space
7 Plot the service corpus data using Elbow method;
8 Identify the Elbow Point (No. of Categories)

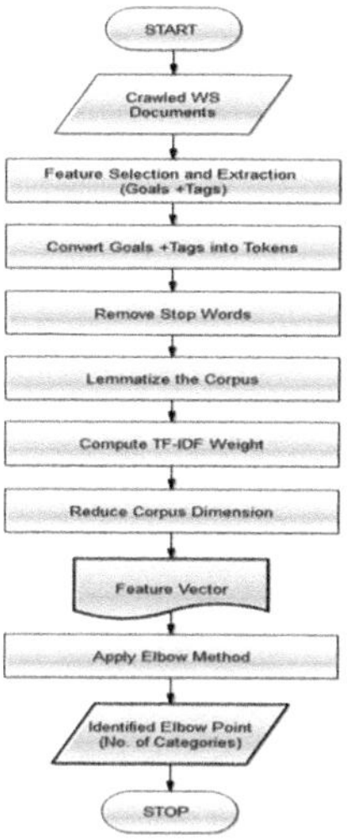

Fig. 2. Pre-processing algorithm

3.2 Initial Detector Generation Stage

Detectors are data generated to differentiate between self and non-self data. The size of the detectors (constant or variable) is very important as it can affect the performance of M-NSA, variable-sized detectors are used in this case because there is no need to set the number of detectors in advance, the large area of non-self space can be covered by a limited number of variable-sized detectors, likewise, smaller detectors can cover the gaps, thus reducing the categorization error unlike constant-size detectors.

Since the number of service goals and tags extracted from RESTful web service descriptions is in form of text, cosine similarity is used to calculate the distance between two services as shown in Eq. (5) in which S1 and S2 are two service vectors under consideration; $S1_i$ and $S2_i$ are components of vector S1 and S2 respectively; and n is the number of components available. Thus, the result ranges between -1 and 1, where -1 is perfectly dissimilar and 1 is perfectly similar.

$$Similarity(S1, S2) = \frac{S1 * S2}{||S1|| * ||S2||} = \frac{\sum_{i=1}^{n} S1_i * S2_i}{\sqrt{\sum_{i=1}^{n} S1_i^2} * \sqrt{\sum_{i=1}^{n} S2_i^2}} \quad (5)$$

The conventional k-means categorize the input data X into K (calculated using Elbow Method to determine the appropriate number of categories based on the dataset) distinct sets (categories Z) by randomly choosing a centres C in each category Z, followed by assigning object (web service data) x_i into a Category Z_j based on the proximity between centres C of Category Z_j and the object (web service data) x_i. The centres C of each category Z is recomputing after every assignment of the new object until no change is observed. The conventional k-means is modified to perform a comparison between a new web service data x_i and a threshold value τ (the boundary and outside the category), if the web service data is less than the threshold value, then the data will be discarded as it does not satisfy the condition of being a detector, otherwise, it will be added into the initial set of detectors (D) as shown in Fig. 3.

Algorithm 2: Initial Detector Generation

Input: Training Data X & Number of Categories z_j (Based on Elbow Point)

Output: Initial set of detectors (D)

Phase 1:

1 Randomly choose K variable in X as initial centres C, $1 \le i \le K$

2 For each object x_i in X find the nearest centre C using COSINE Distance and assign object x_i to the Category z_j

3 Recompute the centres based on the assignment of objects

4 Repeat 2 and 3 if the centre changes; Else Go to Phase 2.

Phase 2:

5 Check the distance between x_i and threshold τ in Category z_j

6 If x_i < threshold τ then remove the new x_i;

7 Else Add x_i to D;

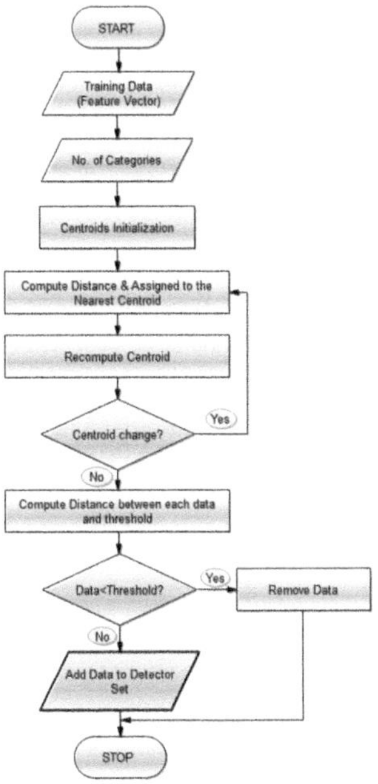

Fig. 3. Initial detectors generation algorithm

3.3 The Search Space Reduction Stage

WS represents the web service to be categorized, X represents the input variable. The maximum number of detectors, N_{max} and the threshold value, τ set at 0.5. The output D represent the set of detectors that satisfy the categorization condition and Z represent the categories of functionally similar web services.

The modified NSA (M-NSA) is also made up of two modules based on the "Censoring phase" and "Monitoring phase". The censoring process involves random generation of detector candidates *d* in the entire feature space (S). It involves data features conversion where the features will be represented by numerical values and mapped into the N-dimensional hyperspace (X = 16). After the conversion, the numerical values will then go through the complementary process and become the detectors (d). The initial set of detectors generated in algorithm 1 are used to kickstart the process. The detectors are created based on how many MWS that needs to be classified. During this process, self-data (WS data) are compared with the detector candidates to evaluate the similarity values.

In the observation process, matching rule is determined by comparing the affinity value obtain in censoring process $f_{aff}(d, x)$ with a threshold value. If the distance is less than the threshold value, then the input data is added to the category, otherwise, it is added to the set of detectors (D) for that category as shown in Fig. 4. The Censoring phase and Monitoring phase will be iterated until all web services are categorized.

Algorithm 3: Search Space Reduction

Input:

WS to be classified and Input Variable $X = (x_1, x_2, \ldots x_n)$ *where* $\forall x \in S$

Initial set of detectors, N_{max} *and the threshold* r

Output:

Set of detectors $D = (d_1, d_2, \ldots d_n)$ *that satisfy the categorization condition*

Categories of functionally similar WS $Z = (z_1, z_2, \ldots z_n)$

Start

Phase1: Censoring phase

Use the x_i *generated by Algorithm 1 as initial candidate detectors* d;

Select web service data x, *where* $x \in S$;

Calculate the similarity of x *with every detector* d *in* $D, \forall d \in D$ *using Equation (5)*

Phase2: Monitoring phase

Check the Distance $f_{aff}(d, x)$;

If $f_{aff}(d, x) < r$ *then Add the new* x *to the Category Z;*

Else Add x *to* D;

End if

Finish

Repeat phase 1 and 2 until $|D| = N_{max}$ *or D cover all WS*

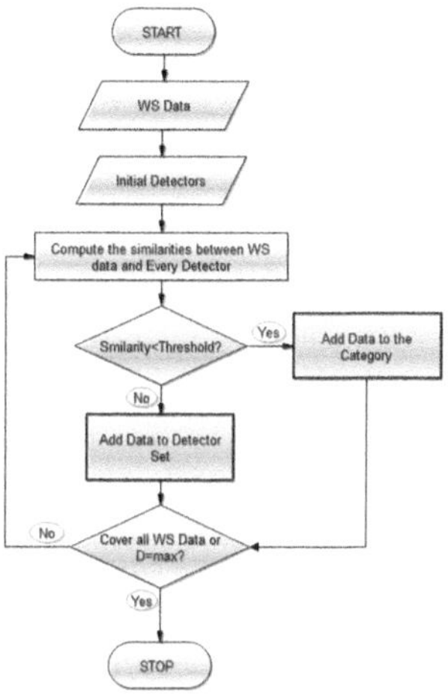

Fig. 4. The search space reduction algorithm

4 The Empirical Evaluation

To evaluate the proposed MWS search space reduction approach, web service repositories such as seekda.com, woogle.com, xmethods.net, webservicelist.com are considered for RESTful web service data acquisition, however, these repositories are no longer operational. Alternatively, ProgrammableWeb.com provide relevant information on different categories of RESTful APIs, therefore, a web crawler tool is employed in acquiring all the necessary information such as APIs' title, tags, categories, description, etc. as used in [6, 18]. In view of the fact that most of the information is unstructured and is in the form of natural language, it undergoes through redaction process to remove APIs that are not described, categorized, or longer available in addition to filtering out duplicates after which 4,854 from related categories are selected for the experiment (3,236 for training, 1618 for actual censoring).

$$Accuracy(\%) = \frac{Services\ obtained\ in\ category\ of\ WS}{Actual\ services\ in\ category\ of\ WS} * 100\% \tag{6}$$

The proposed algorithms discussed above was applied to the datasets, the accuracy of the categories is calculated using Eq. (6). Table 1 presents the accuracy obtained using the proposed approach. The domain to which a Web service belongs to is taken into account for calculating the accuracy. Domains are "Email", "Financial", "Games",

"Mapping", "Messaging", "Photos", "Social", and "Video". The reason for selecting these domains because of the high level of relatedness between them.

Table 1. Categorization accuracy.

Category	Actual categories	Obtained categories	Accuracy (%)
Email	187	174	93%
Financial	140	124	89%
Games	166	136	82%
Mapping	185	156	84%
Messaging	343	331	97%
Photos	168	160	95%
Social	261	223	85%
Video	168	161	96%

Figure 5 shows the marginal difference between the actual categories and the categories obtained using the proposed algorithm in which the x-axis shows the number of categories while the y-axis represents the number of services. Figure 6 depicts the visualization of the data points after the dimension reduction. Similarly, Fig. 7 shows the plot of the categories in 2-dimensional feature vectors (Table 2).

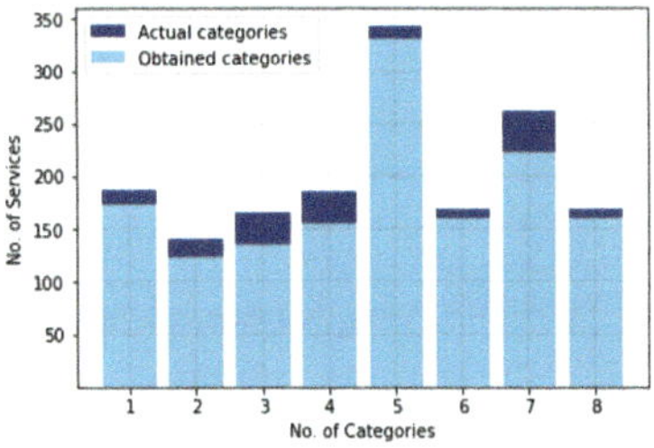

Fig. 5. The actual categories and the categories obtained

The results obtained approach shows significant improvement in terms of accuracy compared with similar approaches. This will improve the MWS discovery in DME by bootstrapping the discovery process to avoid a cold star, and also deal with the proliferation of functionally similar web services. Thought web service description is different from other datasets, it's observed that the quality of the categorization is directly proportional to the quality of the web service descriptions.

Since different approaches choose the dataset from almost 400 different domains for different reasons such as domain popularity, a number of services per domain, due to the objective constraint of the study or to avoid the complexity of intersections among the domains. The dataset from eight (8) closely related domains as shown in Table 1 is used for the experiment in this work and the compared approaches to ensure a fine-grained solution. Amongst the difference and the reason why the proposed approach

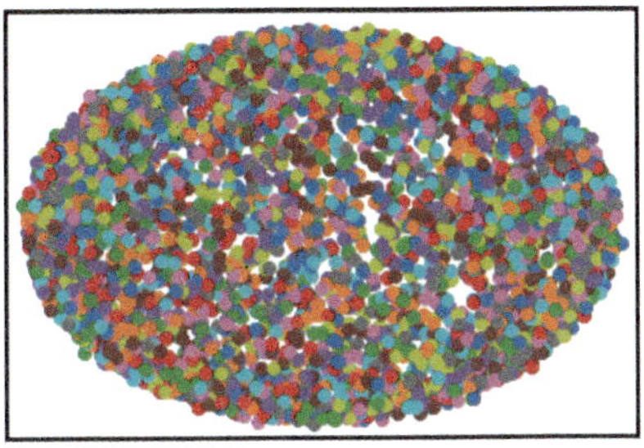

Fig. 6. The visualization of search space

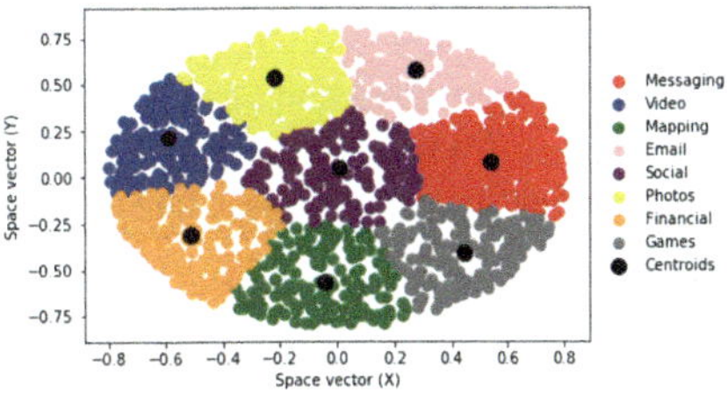

Fig. 7. The categories in 2-dimensional feature vectors

performed better than the existing approaches is the use of dimension reduction and incorporation of tags as the textual description of MWS contained very limited terms. Moreover, many existing studies either choose a single classical algorithm such as k-means which has a number of cluster issue, empty cluster issue, or a meta-heuristic algorithm and similarity measure that are incompatible with the data.

Table 2. The comparison of the result with similar approaches.

Approaches	Algorithm	Average accuracy (%)
[11]	K-means & CSO	74%
[6]	K-means & Agnes	71%
[18]	K-means++	86%
Proposed Approach	**M-NSA**	**90.1%**

5 Conclusion and Future Work

This paper proposed search space reduction approach for self-adaptive web service discovery in a dynamic mobile environment to deal with the proliferation of functionally similar web services by minimizing the search space, minimizing the computation time, reducing the workload of the matchmaking algorithm for efficient and more accurate MWS discovery. The algorithm was applied to eight categories of the programmableWeb dataset, the result shows significant improvement in terms of accuracy compared to related works that employed other approaches. In the future, more dataset can be tested and the feature reduction technique can be fine-tuned to enhance the accuracy. Other methods such as deep neural network can be explored for extraction of service goals and tags. Moreover, integration between the proposed algorithm and MWS discovery algorithm to measure its impact on the discovery process will be done in the future.

Acknowledgements. We would like to thank the Ministry of Education (MOE) Malaysia for sponsoring the research through the Fundamental Research Grant Scheme (FRGS) with vote number 5F080 and Universiti Teknologi Malaysia for providing the facilities and supporting the research. In addition, we would like to extend our gratitude to the lab members in the EReTSEL Lab, School of Computing, Faculty of Engineering, Universiti Teknologi Malaysia for their invaluable ideas and support throughout this study.

References

1. Alonso, G., Casati, F., Kuno, H., Machiraju, V.: Web services. In: Web Services: Concepts, Architectures and Applications, pp. 123–149. Springer, Heidelberg (2004)
2. Tian, G., Liu, P., Peng, Y., Sun, C.: Tagging augmented neural topic model for semantic sparse web service discovery. Concurr. Comput. **30**(16), 1–10 (2018)
3. Hayyolalam, V., Pourhaji Kazem, A.A.: A systematic literature review on QoS-aware service composition and selection in cloud environment. J. Netw. Comput. Appl. **110**, 52–74 (2018)
4. Chifu, V.R., Pop, C.B., Salomie, I., Chifu, E.S.: Hybrid honey bees mating optimization algorithm for identifying the near-optimal solution in web service composition. Comput. Inf. **36**(5), 1143–1172 (2017)
5. Sellami, M., Bouchaala, O., Gaaloul, W., Tata, S.: Communities of web service registries: construction and management. J. Syst. Softw. **86**(3), 835–853 (2013)
6. Cao, B., Frank Liu, X., Liu, J., Tang, M.: Domain-aware Mashup service clustering based on LDA topic model from multiple data sources. Inf. Softw. Technol. **90**, 40–54 (2017)
7. Jiang, B., Ye, L., Wang, J., Wang, Y.: A semantic-based approach to service clustering from service documents. In: 2017 IEEE International Conference on Services Computing (SCC), pp. 265–272 (2017)
8. Shi, M., Liu, J., Cao, B., Wen, Y., Zhang, X.: A prior knowledge based approach to improving accuracy of web services clustering. In: 2018 IEEE International Conference on Services Computing (SCC), pp. 1–8 (2018)
9. Rupasingha, R.A.H.M., Paik, I., Kumara, B.T.G.S.: Improving web service clustering through a novel ontology generation method by domain specificity. In: Proceedings - 2017 IEEE 24th International Conference on Web Services, ICWS 2017, pp. 744–751 (2017)
10. Chen, F., Li, M., Wu, H., Xie, L.: Web service discovery among large service pools utilising semantic similarity and clustering. Enterp. Inf. Syst. **11**(3), 452–469 (2017)
11. Kotekar, S., Kamath, S.S.: Enhancing web service discovery using meta-heuristic CSO and PCA based clustering. Prog. Intell. Comput. Tech. Theory Pract. Appl. **519**, 393–403 (2018)
12. Nanda, S.J., Panda, G.: A survey on nature inspired metaheuristic algorithms for partitional clustering. Swarm Evol. Comput. **16**, 1–18 (2014)
13. Dasgupta, D., Yu, S., Nino, F.: Recent advances in artificial immune systems: models and applications. Appl. Soft Comput. J. **11**(2), 1574–1587 (2011)
14. Bereta, M., Burczyński, T.: Immune K-means and negative selection algorithms for data analysis. Inf. Sci. (Ny) **179**(10), 1407–1425 (2009)
15. Li, Z., Tan, H.-Z.: A combinational clustering method based on artificial immune system and support vector machine. In: International Conference on Knowledge-Based and Intelligent Information and Engineering Systems, pp. 153–162 (2006)
16. Xu, D., Tian, Y.: A comprehensive survey of clustering algorithms. Ann. Data Sci. **2**(2), 165–193 (2015)

17. Ahmed, M.S., Khan, L.: SISC: a text classification approach using semi supervised subspace clustering. In: ICDM Workshop 2009 - IEEE International Conference on Data Mining, pp. 1–6 (2009)
18. Shi, M., Liu, J., Cao, B., Wen, Y., Zhang, X.: A prior knowledge based approach to improving accuracy of web services clustering. In: 2018 IEEE International Conference on Services Computing, pp. 1–8 (2018)

Modeling Reliable Intelligent Blockchain with Architectural Pattern

Sin-Ban Ho[1], Nur Azyyati Ahmad[1(✉)], Ian Chai[1], Chuie-Hong Tan[2], and Swee-Ling Chean[3]

[1] Faculty of Computing and Informatics, Multimedia University, 63100 Cyberjaya, Selangor, Malaysia
{sbho,azyyati.ahmad,ianchai}@mmu.edu.my
[2] Faculty of Management, Multimedia University, 63100 Cyberjaya, Selangor, Malaysia
chtan@mmu.edu.my
[3] Lee Kong Chian Faculty Engineering and Science, Universiti Tunku Abdul Rahman, 43000 Kajang, Selangor, Malaysia
karencheean@gmail.com

Abstract. The emerging of blockchain technology has been associated with the revolution of World Wide Web that going to transform from Web 2.0 into Web 3.0. The blockchain network reduces the dependency of centralized and turning into decentralized network in the Web 3.0 era. In the decentralized network, the threat of data breaching can be eliminated where none of party has the authority to control over user digital badges in education. In this case, all the data will be distributed over the whole network. Furthermore, the weak vulnerability of Internet of Things (IoT) toward Advanced Persistent Threat (APT) has stimulated the development of blockchain technology as a solution. Blockchain technology has released a computer program as smart contract with the generation of token to authorize the flow of user into IoT network for strengthen the security and privacy of IoT network. Last but not least, the challenge to develop an effective blockchain network is the most challenging task since it requires deep understanding of blockchain technology.

Keywords: Blockchain · Component and connector · Architectural pattern

1 Introduction

The business network is often inefficient due to each participant in the network store the record or a ledger of all transactions between all the parties that the business interacts with. This process has cost lots of money due to duplication of effort and intermediaries adding costs for their service.

Therefore, a solution has been developed to solve this issue which is called blockchain. Blockchain is an emerging technology which provides a shared ledger technology that allows any participant in the blockchain network to see the whole system of a record or ledger. In conjunction with the development of blockchain technology, the central trusted authority system has been eliminated and replaced via a

F. Saeed et al. (Eds.): IRICT 2019, AISC 1073, pp. 1122–1131, 2020.
https://doi.org/10.1007/978-3-030-33582-3_105

decentralized system on data sharing across a large network [1]. With the emerging of blockchain technology, the development of decentralized application or known as DApps will rise. Therefore, a particular plugin system has been invented to facilitate the communication between the current traditional browser and DApps. As an example, MetaMask [2] is one of the popular blockchain browser plugin. In this case, the plugin designed has transformed the authentication of user personal data into more secure handling via encryption.

The first generation of the establishment of blockchain technology has been vastly related to the electronic coin named Bitcoin. Bitcoin has been started proposed by a pseudonymous software developer, Satoshi Nakamoto in the year of 2008 [3]. The mechanism behind Bitcoin was to produce a means of exchange without relying on any central authority, especially the bank. Bitcoin cannot be printed physically like money notes or coin but will be created and stored in the computer system that being referred as cryptocurrencies [3].

The motivations and challenges of this paper are:

Firstly, designing applications in blockchain requires deep understanding and development experience to use the blockchain environments [4]. For instance, one may overlook vulnerable and attackable code directly. Secondly, modeling a set of architectural patterns for the reliable and intelligent blockchain application is challenging because it delivers the module pattern, and the Component & Connector (C&C) pattern to facilitate the design of blockchain applications for improving the scalability and privacy of blockchain applications. One can group architectural patterns by the significant elements type, which they indicate module patterns with layers. Other consideration include components and connectors (C&C) patterns. The integration of modules, C&C and non-software elements finally represent the allocation patterns. C&C patterns are the most widely used patterns as compared to module and allocation patterns. C&C views use particular type elements of component and connector, which follow specific guidelines on how to combine these elements. This paper is different from the previous work because it presents C&C patterns that consider system runtime aspects, with association of the C&C style with an intelligent model, which outlines how flow of data and control throughout the systems modeled in such view.

The contents of this paper are organized as follows. Section 2 reviews the background on the reliable blockchain layer architecture. Section 3 explains the C&C methodology of the education domain case study. The methodology is validated against security and reliability aspects in Sect. 4. Finally, Sect. 5 gives conclusions and recommendations of this research.

2 Background and Reliable Blockchain Layer Architecture

2.1 Building a Software Application with Blockchain Technology

With the emerging of blockchain technology, many software developers as well as researchers are finding ways to develop software application using blockchain technology. The usage of blockchain application is not limited to the Internet of Things (IoT), digital currency, management of security, supply chain, and taxation. For instance, applying blockchain technology to develop a system to manage personal data, the users must be able to control and possess their own data in a decentralized platform [5]. At the same time, ProvChain demonstrates the capability of data analytics using cloud provenance to enhance the privacy of data [6].

2.2 Challenges of Blockchain Application Development

This section presents the current challenges in developing a blockchain application.

- Data Privacy: A particular blockchain participant may have certain sensitive data in blockchain application. Nevertheless, the blockchain made available those information accessible to the entire participants. In this case, one does not distinguish whether the blockchain user is private or public, with no privileged user. This paper proposes a controlled access with different privileges of users in a seminar management case study highlighted in Sect. 3.1.
- Permission Control: By default, the blockchain participants can access and request the entire smart contracts available in the blockchain participants. The unauthorized users might accidentally trigger permission-less function, which forms a blockchain application vulnerability. For instance, the Parity multi-sig wallet [7] used a permission-less function from a smart contract library that caused about 500K Ether freezing. As such, this control is solved with the peer to peer C&C view in Sect. 3.2.
- Key Management: Digital signatures contribute to authentication on the blockchain. Nevertheless, a compromised or lost private key is not recoverable in the current blockchain mechanism. As such, one permanently lose control an account if the key is lost, and results in losing smart contracts referred to it. The call-and-return style highlighted in Sect. 3.3 ensures the results of smart contracts are verified with feedback.
- Scalability: Since blockchain consists all the transactions full history of all blockchain network participants, the blockchain size would keep growing. Furthermore, the network control in limiting the blockchain blocks size [8] poses a challenge to store huge data amounts. Thus, it can be impossible to deploy large smart contracts for a transaction. For the novelty of the research, the C&C view of storage consensus mechanism to cater scalability is highlighted in Sect. 4.

2.3 Layer Pattern of a Reliable Blockchain

The architectural pattern refers to discovering a reusable solution to a recurring problem for the context of designing a software. Bartoletti and Pompianu [9] gathered hundreds of smart contracts and analyzed them empirically in order to segregate them into multiple categories, i.e. randomness, time constraint, token, math, termination, authorization, poll, oracle, and fork check. The module pattern is outlined with its five different layers, which help one to achieve a reliable blockchain (see Fig. 1).

The first application layer consists of DApps (Decentralized Applications), application hosting, DApp browser, programming languages, and user interfaces [10]. This application layer acts as the user interface that combines business logic and customer interactions. DApp is a general-purpose blockchain that allows individual users to interact with a distributed ledger using smart contracts and web interface frameworks.

The second layer, i.e. Services and Optional Components provides the services to enable application operations with a views to connecting with other technologies and platform, which covers all the important tools that function in creating and running the DApp layer. It usually covers data feeds, off-chain computing, oracles, wallets, digital assets, multi organizations signatures, smart contracts, state channels, and governance or DAOs (Decentralized Autonomous Organizations). DAO consists contracts set for the Ethereum blockchain participants, which do not own a physical address. They are also not in formal management roles. Ethereum is a second-generation permissionless blockchain that transforms the way blockchain has perceived in a wider world as a cryptocurrency. Unlike Bitcoin, Ethereum is both a cryptocurrency and programmable blockchain for building a decentralized application and facilitate a transaction in a trustless network. Ethereum uses a token called Ether.

Layer 1 Application (DApps, hosting, programming languages, DApp browsers)
Layer 2 Services & Optional Components (data feeds, wallets, smart contract, DAOs, Oracles)
Layer 3 Protocol (consensus algorithm, permissioned & permissionless, side chains, EVM)
Layer 4 Network (RPLx, roll your own, block delivery networks, peer-to-peer)
Layer 5 Infrastructure (BaaS, mining, virtualization, nodes, storage, tokens)

Fig. 1. The layers pattern of a reliable blockchain.

Smart contract would help to get rid of the middleman and no need have to deal with trust issue and will be able to exchange valuables asset such as money, shares or property in conflict-free solution. The DAO theory removed the directors' power, which delegate it to the owners directly. DAO prevents the fund managers and directors from wasting and misdirecting investor funds.

The third protocol layer decides the methods of consensus and network participations. This layer covers the different consensus algorithms, network participation requirements such as permissioned and permissionless, side chains, and virtual machines such as EVM (Ethereum Virtual Machine). Blockchains apply consensus algorithms to ensure that the nodes come to an agreement.

The fourth network layer acts as a transportation medium, and interface for the Peer-to-Peer network and decides how data is packetized, addressed, transmitted, routed, and received. This layer consists of RPLx specification (where Ethereum uses Kademlia protocol), roll your own, and block delivery networks.

The fifth infrastructure layer provides in-house infrastructure or Blockchain as a Service (BaaS) to control the nodes. This service helps users to build efficient blockchain application. Amazon, IBM, and Microsoft Azure [11] service platforms are some most common BaaS examples used for direct and fast deployment. The infrastructure layer includes mining, virtualization, nodes, tokens, and storage.

3 Component and Connector Methodology

Producers and consumers interact blockchain services or data through the possibility of client-server, peer-to-peer (P2P), and call-and-return styles. A software developer needs to be aware of using the appropriate style. C&C view analysis is crucial to ensure consistency of a request in a harmonious flow direction. Table 1 lists the difference between client server and P2P architecture [12]. The prominent definition of P2P is every node is either the server or the client depending on whether the node is providing or requesting service. In contrast, client server architecture has a dedicated server that provides services and store data for the clients to use.

Table 1. Characteristics of P2P and Client-Server architecture.

Aspect	Peer-to-Peer (P2P)	Client-Server
Basic	The P2P node each acts as client and server, with no difference of client and server	Dedicated clients and specific server serving the client
Focus	Connectivity	Sharing the information
Service	Each node is able to request and provide services	Only client requests service, while server provides services
Server	The services distributed across the P2P system prevents the bottlenecked problem of a server	Bottleneck can happen in a server when several client request for its servers simultaneously
Stability	P2P is unstable as the peers numbers increase tremendously	Client-Server is more stable and scalable
Data	Each peer owns data	The centralized server store the data
Expense	Cheaper relatively without the need of servers	More expensive relatively to implement

Service Oriented Architecture (SOA) pattern is described as a form of the proposed solution. Services are provided and consumed through distributed components collection. To form a correct data flow and to ensure that the data flow is not unidirectional, the architect uses diagrams to determine the data flow. With proper use of component diagrams, the architect can distinguish between the styles.

3.1 Client Server C&C View

The education domain is used as a case study for a client server view for the SOA pattern. The clients in Fig. 2 are Student Administration, Seminar Management, Security, Persistence and University DB. The servers are Facilities, Student, Seminar and Schedule. The Seminar Management client has access to the Student, Seminar and Schedule server as that client only need to view the data stored in the servers. At the same time, the servers have access to the security and persistence clients. Figure 2 aims to show the control access of the seminar management with request from the student to reserve their proposed seminar venue and schedule. Subsequently, the student administration department would verify the validity of the request with blockchain data feeds service against the schedule assets. Upon the confirmation of the digital asset block, the particular venue would be confirmed and persistently stored into the university Oracle database using Java Database Connectivity (JDBC).

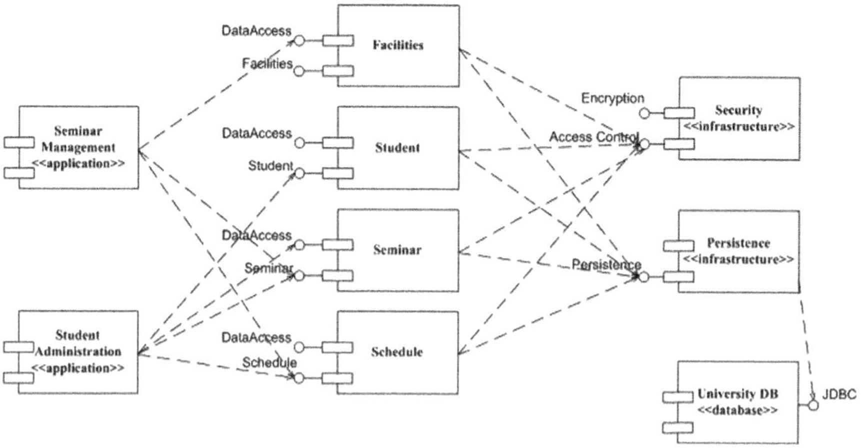

Fig. 2. Client-server in SOA pattern.

3.2 Peer to Peer C&C View

Figure 3 shows all the components (peers) have their own roles forming interconnectivity between them all. For example, in the Generic Servant sub-system, one component is acting as a router while in the Remote Access sub-system, one component dedicated itself to finding other peers.

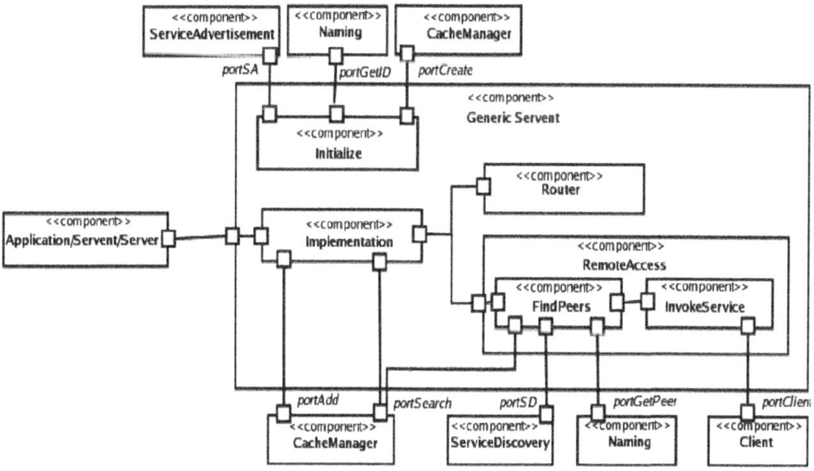

Fig. 3. Peer to peer in SOA pattern.

3.3 Call and Return Style

Figure 4 shows the illustration on how call-and-return style works. The call-and-return style is characterized through putting a main entity as the system. This main entity is able to call finer sub-entities to carry out an action. The main entity receives the action outcome. To download the block file, a participant invokes a DApp browser or DApp. For the browser, a plug-in for the file type can be retrieved so that the file type would be able to be handled, before the download process.

(a) Before verification (b) After verification

Fig. 4. Simulation to prove the case study validity. (a) A main entity making a call for verification. (b) Clients return verification approval back to the main entity.

The implementation of call-and-return style in blockchain is when a transaction is performed, the server has to broadcast the transaction, which is also known as "block" to all the clients in the network for them to verify and approve the transaction. After all the clients approve and verify that the transaction is valid, they will return the data back to the server, which is the main entity.

4 Discussion

A C&C view is shown to emphasize the blockchain system security and reliability aspects. With this, this section discusses the kinds of properties that one might associate with the components and the connectors. The process of creating a block is initiated by a transaction party, where thousands of nodes (computers) would verify this block. Next, the verified block is integrated into the chain to form blockchain. This block is actually stored across the network, decentralized rather than centralized. It is not possible to forge a single record since this involves forging thousand instances of the entire chain. Block uses peer to peer architecture instead of server-client architecture and that is one of the features that makes it different and very well secured.

The heterogeneous node types in the network enable one of them becoming full nodes, which are responsible for holding and distributing copies of the entire blockchain ledger and be able to validate transactions throughout the genesis block as a return. These nodes then propagate the blockchain to all other nodes on the network to ensure the most trusted blockchain is maintained. The more nodes, the more decentralized the network, and the harder it is to hack, another type of nodes are super nodes that generally operate around the clock to help connect other full nodes to each other and spread the blockchain across the entire network. They serve as information or redistribution relays to ensure everyone has the right copy of the blockchain, Light nodes on the other hand perform a similar function to full nodes, but rather than holding an entire copy of the blockchain, they only contain a portion of it. Light nodes only download the block header of previous transactions, to confirm the validity of the blockchain, and to pass this information on to other nodes, the block header is a summary of a given block, it includes information about the previous block to which it is hashed.

Blockchain consists recorded transactions linked list, which associated with one another through hash pointers. Each block has its previous block hash value. As such, this makes the cryptographic hash function as one of blockchain fundamentals. Bitcoin and other cryptocurrencies uses SHA-256. For example, when a hacker intends to attack the seventh block in a chain, he would change the data by altering the data in eighth block, and thus making a change in ninth block too, and so on. So, basically, the hacker has to change all following blocks data, as their hash would also different. Therefore, it is not possible to change any blockchain data recorded.

Hashing the whole block, which contains thousands of transactions would be impractical, because of that blockchain uses Merkle tree. To validate the original transaction data, the system runs a block of transactions by generating a hash through an algorithm. It hashes each transaction, linking those transactions and hashing together, instead of running the hash function on an entire transactions block at once. Next, the Merkle root hash is located in a block header with one hash of the previous block discussed above. The key components to allow blockchain technology to work include nonce, timestamp, current difficulty target, block version number, Merkle hashes and trees. This provides irrefutability, security, integrity, reliability, and consensus protocols, which contributes towards the intelligent aspects of blockchain design.

Peer-to-peer made sharing easier. Everyone can join the network without any requirements. To prevent the intruders to tamper the block. Therefore, in order to establish mutual trust, consensus mechanism in blockchain is used [13]. From point of view of C&C, blockchain consensus mechanism can use publish-subscribe pattern as shown in Fig. 5.

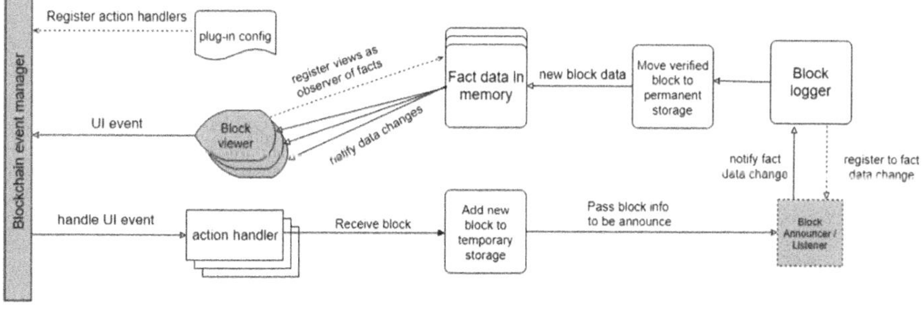

Fig. 5. C&C view of consensus mechanism of blockchain.

One new block will be created for every transaction made. The block will be added to temporary storage and the block info will be passed to the block announcer to announce to every member that is in the P2P network. After the P2P client receive the info from the announcer, they will check whether if the block they received matches the one announced by the block announcer. If it is the same, then the client will agree on the info they received and move on to the next step, else disagree and the block will be removed from temporary storage. All the data change or modification will be logged by block logger. Then the block will be move to permanent storage.

For the observer who wants to view the block, they will have to register themselves as an observer and the system will notify them the changes made to the block every time they view the block. This measurement step is to enhance the security of the block and the timestamp is referred to check on who is the last person view the block before anything happens to the block. The data logged from the viewer also will be store on the permanent storage for future references.

Multiple layer of encryption to the block is required such as RSA key or Triple DES to protect the safety and privacy of public. Although it might sacrifice the overall usage experience due to the decryption time, but better safe than sorry. Other than that, although blockchain uses P2P network as their medium to share the network, user identification should be implement in order for them to access the network of the blockchain. The system will have to trace and log the IP address, MAC address of the devices of the user when they try to use log in to blockchain system.

5 Conclusion

In this research work, we discover and understand how blockchain actually could change the way we store files in the future. Although the popularity of the blockchain only started to rise in recent few years but we can be actually moving one-step forward to the future.

In the aspect of reliable intelligent informatics, choosing the right architectural patterns are crucial for recent blockchain and education research [14–16] because there is no right or wrong, it is just good or bad pattern to become a solution for a particular problem.

Blockchain uses P2P network, and it allows everyone to joins. Although the creation might look good for now, however, there is still a lot of work to do for the design to become perfect. Developers need to take the pros and cons of usage into consideration such as security risks, privacy, reliability and intelligence before everyone started to use it widely.

References

1. Tschorsch, F., Scheuermann, B.: Bitcoin and beyond: a technical survey on decentralized digital currencies. IEEE Commun. Surv. Tutor. **18**(3), 2084–2123 (2016)
2. MetaMask Explained (2017). https://cryptocurrencyfacts.com/metamask-explained. Accessed 28 May 2019

3. Nakamoto, S.: Bitcoin: a peer-to-peer electronic cash system (2008). https://bitcoin.org
4. Siegel, C.: Understanding the DAO attack (2016). https://www.coindesk.com/understanding-dao-hack-journalists. Accessed 28 May 2019
5. Zyskind, G., Nathan, O., Pentland, A.: Decentralizing privacy: using blockchain to protect personal data. In: Proceedings of 2015 IEEE Security and Privacy Workshops (SPW 2015), San Jose, USA, 21–22 May 2015. https://doi.org/10.1109/spw.2015.27
6. Liang, X., Shetty, S., Tosh, D., Kamhoua, C., Kwiat, K., Njilla, L.: ProvChain: a blockchain-based data provenance architecture in cloud environment with enhanced privacy and availability. In: Proceedings of 17th IEEE/ACM International Symposium on Cluster, Cloud and Grid Computing (CCGRID), Madrid, Spain, 14 May 2017, pp. 468–477 (2017). https://doi.org/10.1109/ccgrid.2017.8
7. Anderson, L., Holz, R., Ponomarev, A., Rimba, P., Weber, I.: New kids on the block: an analysis of modern blockchains. The Computing Research Repository (CoRR), abs/1606.06530 (2016)
8. Weber, I., Gramoli, V., Staples, M., Ponomarev, A., Holz, R., Tran, A., Rimba, P.: On availability for blockchain-based systems. In: Proceedings of 2017 IEEE International Symposium on Reliable Distributed Systems (SRDS 2017), Hong Kong, China, 26–29 September 2017, pp. 64–73 (2017). https://doi.org/10.1109/srds.2017.15
9. Bartoletti, M., Pompianu, L.: An empirical analysis of smart contracts: platforms, applications, and design patterns. In: Financial Cryptography Workshops, pp. 494–509, ArXiv e-prints. Springer, Cham, March 2017. https://doi.org/10.1007/978-3-319-70278-0_31
10. Web 3.0 will be powered by blockchain technology stack (2019). https://hackernoon.com/web-3-0-will-be-powered-by-blockchain-technology-stack-626ce3f828c7. Accessed 28 May 2019
11. Blockchain as a BaaS (2019). https://azure.microsoft.com/en-us/solutions/blockchain. Accessed 28 May 2019
12. Different between client server and Peer to Peer network (2019). https://www.techrepublic.com/article/understanding-the-differences-between-client-server-and-peer-to-peer-networks. Accessed 28 May 2019
13. Li, X., Jiang, P., Chen, T., Luo, X., Wen, Q.: A survey on the security of blockchain systems. Future Gener. Comput. Syst. **13** (2017). https://doi.org/10.1016/j.future.2017.08.020
14. Tee, W.J., Murugesan, R.K.: A theoretical framework to build trust and prevent fake news in social media using blockchain. In: Proceedings of 3rd International Conference of Reliable Information and Communication Technology 2018 (IRICT 2018). Advances in Intelligent Systems and Computing (AISC), vol. 843, pp. 955–962 (2019). https://doi.org/10.1007/978-3-319-99007-1_88
15. Mikroyannidis, A., Domingue, J., Bachler, M., Quick, K.: Smart blockchain badges for data science education. In: Proceedings of IEEE Frontiers in Education Conference (FIE 2018), San Jose, USA, 3–6 October 2018, pp. 1–5 (2018). https://doi.org/10.1109/fie.2018.8659012
16. Borrás-Gené, O.: Use of digital badges for training in digital skills with higher education. In: Proceedings of 2018 International Symposium on Computers in Education (SIIE 2018), Jerez, Spain, pp. 19–21, September 2018. https://doi.org/10.1109/siie.2018.8586734

The Organisational Factors of Software Process Improvement in Small Software Industry: Comparative Study

Shuib Basri[1], Malek Ahmad Almomani[1(✉)], Abdullahi Abubakar Imam[1], Murugan Thangiah[1], Abdul Rehman Gilal[2], and Abdullateef Oluwagbemiga Balogun[1,3]

[1] Department of Computer and Information Sciences, Universiti Teknologi PETRONAS, 31620 Seri Iskandar, Perak, Malaysia
{shuib_basri,malek.ahmad_G02668}@utp.edu.my, aiabubakar3@gmail.com, tm_gun@hotmail.com, bharlow058@gmail.com

[2] Sukkur IBA University, Sukkur, Pakistan
a-rehman@iba-suk.edu.pk

[3] Department of Computer Science, University of Ilorin, Ilorin, Nigeria

Abstract. Small and Medium Enterprises (SMEs) are a great contribution to the international economy and have also been considered an important component in today's world business. Thus, in order to be more competitive, it is necessary for these companies to deliver their products with high-quality. However, despite their importance, small software companies still face myriad challenges and barriers in producing high-quality products. The objective of this study is to identify the organizational factors that have a positive impact to enable Software Process Improvement (SPI) effort in the small software industry. A Systematic Literature Review (SLR) was conducted to achieve the main objective of this study. The findings from this study provide a roadmap to guide future research in order to enable SPI effort in the small software development industry. We believe that findings from this study will give interesting insights to encourage researchers in using compromise technique to analyze future empirical studies based on a specific region to validate the suitability of identified factors in the specific country.

Keywords: Small and Medium Enterprises (SMEs) · Software Process Improvement (SPI) · Systematic Literature Review (SLR) · Critical Success Factors (CSF) · Organizational Factors (OF)

1 Introduction

Nowadays, software development companies are prioritizing the release of top quality software products with suitable cost, resources, high productivity, and within the right scope. In order to develop a top quality software product, the software development process must be of a high standard and well organized [1, 2]. Hence, continuous improvement in the software development processes will ensure an optimal software

F. Saeed et al. (Eds.): IRICT 2019, AISC 1073, pp. 1132–1143, 2020.
https://doi.org/10.1007/978-3-030-33582-3_106

quality process. Consequently, there are several standards introduced to support organizations in steering their software development processes to achieve top quality software products. These international models are: (1) "the Software Process Improvement and Capability determination (SPICE)" [3], (2) "Bootstrap" [4], (3) "International Organization for Standardization ISO 9000" [5], (4) "Six Sigma" [6] and (5) "the Capability Maturity Model [7] and advanced version the Capability Maturity Model Integration (CMMI)" [8]. However, existing literature, including [9–11] emphasize that these models are not appropriate to be implemented in Small and Medium-sized software development Enterprises (SMEs) although they constitute most of the companies in the marketplace [12–14].

SMEs are a great contribution to international economic and it considered as an important component in worldwide business [15]. United Nations industrial development organization [16], reported that the majority of companies worldwide fall under SMEs category, with representing more than 90% of total existing enterprises. Altenburg and Eckhardt [16] stressed the important role of small enterprises in creating employment opportunities and for providing workout events alongside relevant services for deprived employees. However, despite their contribution to economic growth and employment opportunities, SMEs still face myriad challenges and barriers. For example, lack of awareness, time-consuming, lack of resources, and lack of support. Baharorn et al. [17] emphasized that the failure of implementing good software development practices could be a major factor contributing to quality problems in software products. Almomani et al. [18] also indicated that factors such as time-consuming, lack of resources, and lack of support are preventing SME software development companies from implementing SPI in their activities. Accordingly, studies conducted by Almomani et al. [18, 19] stressed that the utilization level of SPI activities in SME software development companies is below par.

Therefore, it is imperative to analyze relevant studies to identify organizational factors of SPI in SMEs. Understanding of organizational factors will help the SPI manager to gain a deeper understanding of process improvement. Hence, the main objective of this study is to identify organizational factors that have a positive impact to enable SPI in the SMEs software development industry.

This paper is structured to cover the follows sections: Background and motivation of the study will be presented in Sect. 2. Section 3 presents the research methodology of dataset selection. The result of the study will be presented in Sect. 4. Section 5 addresses the result discussion and research findings of the study. Finally, Sect. 6 presents the conclusion as well as direction for future work.

2 Motivation

Effective factors outline aspects which are essential to assuring the successful implementation of SPI amongst SMEs software development industry. It is defined by Bullen and Rockart [20], as a mechanism to identify information needed in specific areas and where more attention is needed. Knowledge of effective factors helps to develop process improvement successfully to achieve organizations' objectives [20]. In fact, effective factors influence SPI from several dimensions [21]. Therefore, it is very important to address them appropriately in order to enable SPI effort in SMEs software development industry. From the last decade, there have been several studies conducted to investigate and analyze these effective factors. For example, organizational size [22–27], environmental conditions [18, 25, 26], social capital and team dynamics [28, 29], or geographical locations [30–32]. However, with the use of standard methods and research findings from these studies, much focus was not placed on organizational perspective as an effective factor.

Organizational perspective is defined to include all those factors impacting organizational politics, structure, and culture [33]. The organizational factors have a direct impact on SPI and its sustainable growth. Hence, El-Emam et al. [34] pointed out that the organizational dimension is a factor that characterizes organization undergoing SPI and the characteristics of the initiatives itself. However, despite their importance, SPI, organizational perspective, and small software industry are rarely found together in recent literature studies.

Over the last decades, several scientific reviews have been conducted in software process improvement field. For example, Staples and Niazi [35] conducted a review to investigate the organization and their related motivations in adopting CMM SPI model. They identified 43 primary studies related to CMM-based SPI model. They found that business benefits are the primary motivation for companies. Also, they mentioned that "organizations reported customer reasons infrequently and employee reasons very rarely".

Another review performed by Lavallée and Robillard [36], to identify the impact of SPI on software developers, 26 primary studies were found related to the context of review objectives. The lists of positive and negative SPI impact on software developers were the main findings. The positive impact such as an increase in morale, communication among the team, better requirements and documentation through a decrease in the number of problems occurring during the development process and the negative impact such as an increase in overhead through the production of documents that have no perceived value to developers. Almomani et al. [11], conducted a systematic literature review in 2014, to identify SPI efforts in SME software firms. Their results indicated that most of process improvement initiatives are carried out in America and Europe. Furthermore, there are few studies done in Asia context.

However, these aforementioned reviews do not clearly consider organizational factors that positive impact on SPI implementation and they are mostly conducted in large and medium-size companies. Furthermore, none of them observed clearly SPI efforts amongst SMEs software development industry. Therefore, the objective of this review is to identify and analyze existing literature to fill this gap.

Our systematic review differs from the previous systematic review from several perspectives. Firstly, the focus of this review to gain an in-depth understanding of the organizational factors that have a positive impact to enable SPI in SMEs software development industry. Secondly, previous systematic reviews investigated generic impacts factors while our review is solely focusing on the specific domain, therefore, it will be more specific than the reviews. Thirdly, our systematic review focuses on SMEs software development industry, while other reviews do not indicate the size of the organization, owing to the fact, large and small companies differ from a different perspective, therefore, their results cannot generalize to all companies' sizes.

3 Research Methodology

Systematic Literature Review (SLR) is defined by Kitchenham et al. [37] as "A systematic literature review is a means of identifying, evaluating and interpreting all available research relevant to a particular research question, or topic area, or phenomenon of interest". The guideline template and systematic review protocol proposed by Kitchenham et al. [37] were adopted in this review. The Kitchenhand's protocol consists of three major stages as explained in the following subsections.

3.1 Planning for Review

During this stage, we developed a review protocol, defined research questions, identified and determined search strategy for data sources and search terms to retrieve relevant studies. Study selection criteria and filter were performed, relevant studies were retrieved based on paper title and abstract. Final selection of studies was performed for precise filtering based on information in the whole of research papers. Also, it important to note that the first activity 'identify the need for review' of this stage has been discussed in the background and motivation section.

Formulate Research Questions: Developing significant research questions is considered an important part of any successful systematic literature review. In order to address the objective of this study, we have formulated the following research questions:

- (Q1) What organizational factors do small industry need to address in order to enable SPI effort in their activities?
- (Q2) Do the identified factors differ from developed to developing countries?

Development of Review Protocol

Sources of data: The search sources included two main categories to determine related work: publishers' sites (Online Search Engines) and index engines (Online Databases). The sources were validated by experts in the respective software engineering areas. Several data sources have been used to determine related work such (Google Scholar, Research Gate, IEEE Xplore, Science Direct, ACM Digital library, ISI Web of Science, SCOPUS, Springer Link and The IEEE Computer Society Digital Library).

Search Strategy: The Search Strategy for this SLR consists of four main steps. **Step (1)** Derive keywords for the research questions by using Petticrew's and Roberts' method (Population, Intervention, Comparison, and Context - PICOC). **Step (2)** Identify terms derived from synonyms to perform searches in digital libraries. **Step (3)** Group synonyms and related words with the identifier "OR". **Step (4)** Group each set of terms with the identifier "AND".

After adopting Steps 3 and 4, the following results were determined for the search string ("Software Improvement OR Software Process Improvement OR Software Process Enhancement OR Software Process OR Improvement OR SPI") AND ("Small and Medium OR SMEs OR Very Small entities OR VSEs, Micro") AND ("Software Company OR Firms OR Enterprise OR Company OR Setting OR Businesses") AND

("Element OR Component OR Constituent OR Item OR Feature OR Aspect OR Characteristic OR Influence OR Circumstance OR Determinant OR Parameters OR Indicator OR Practices OR Impact OR Catalyst") AND ("Organizational OR Structural OR Managerial").

Inclusion Criteria: Studies consisting of SPI effort that focuses on organizational perspective, Studies must be based on small software development industry, Studies should include a list of factors that effect to implement SPI effort, Studies those are largely related to the research content.

Exclusion Criteria: Studies are not written in English language, Studies are limited and short or the full text not available, Studies which are not relevant and related to the objective of this study, Studies out of research scope.

Quality Criteria: Using quality assessment during the data extraction process is important, to evaluate the quality of research studies and to determine the elaborate inferences strength. Therefore, quality assessment checklists were proposed to assess the quality of primary studies according to existing studies in software engineering field [37, 38]. The checklists were adopted by using questions from Kitchenham et al. [37] to reflect the review context questions and are accordingly adjusted: (1) Is the objective of the paper sufficiently explained? (2) Is the idea of paper clearly presented? (3) Is the problem of paper clearly stated? (4) Is the methodology used suitably to address the stated research questions? (5) Are the data collection methods adequately defined? (6) Are the findings of the research relevant to answer the research questions? (7) Is the knowledge claimed validated?

3.2 Conducting the Review

This stage involved four main activities which are the selection of primary studies after adopted exclusion and inclusion criteria, assessment of the study quality is performed through developed quantitative checklists, data extraction objective to accurately the information and final process in execution is data synthesis where the data collected and summarized the results of the primary studies.

Primary Sources: After applying the procedure to gather initial research studies, a total of 8002 articles was found. Removal of duplicate studies and applying previous inclusion and exclusion criteria, 1085 studies were selected as being relevant studies. After reading full-text papers of relevant studies, only 81 studies are related to research questions. The selected studies were published over the last decade. Table 1 presents the number of selected articles found in the literature.

Quality Assessment: The quality assessment checklists are rated according to Kitchenham et al. [37], as follows: "Yes", "No" or "Partially" scale where a score of one was allocated for "Yes", half for "Partially" and zero for "No". Based on the quality assessment checklists, the score was 84% and this percentage score indicated stronger evidence to the relevant context.

Data Synthesis: A list of Organizational factors from the 81 primary studies was formed. A primary reviewer has performed data synthesis on useful information to provide a foundation for phase 3 (report review). The syntheses of data are performed to identify the organizational factors that have a positive impact on implementing SPI efforts in the SMEs software development industry, and a list of 11 factors was identified, as tabulated in Table 2.

Table 1. List of primary studies

DB. name	Found studies	Not. related	Primary studies	Selected studies
IEEE explore	2362	2154	208	19
ACM library	1238	1044	194	6
ScienceDirect	343	191	152	9
Wiley	360	244	116	9
Springer link	245	149	96	19
Emerald insight	241	234	7	1
Other	3213	2901	312	18
Total	**8002**	**6917**	**1085**	**81**

4 A Review Report

In this systematic review, research questions were developed to gain insight into the current state of organizational factors that have a positive effect to enable SPI in the small software development industry. In order to collect relevant articles that afforded the best answers for the research questions, a total of 81 articles were Selected and read in depth. The following subsections present the results based on our formulated questions:

RQ1: What organizational factors do small industry need to address in order to enable SPI effort in their activities?

In order to answer this research question, Table 2 shows the list of organizational factors identified through the SLR along with the frequency and percentage of the total number of selected studies.

Business Orientation: From the results, 'Business Orientation' is identified as the most critical factor in the SLR study, i.e. (78%, 63 articles). This indicates that understanding projects goals amongst every member of a software organization has a positive impact on implementing SPI effort in SMEs software development industry. Due to the importance of this factor, process improvement objectives should be associated directly with business orientations [P1]. Therefore, considering business orientation will provide the best way to enable the SPI strategy in small companies [P15].

Management Commitment: The results of this research study also indicated that 'Management Commitment' is the second most important factor (74.1%, 60 articles). According to Guerrero and Eterovic [P64], commitment from management helps throughout the project by providing resources and helping to obtain buy-in from developers. Several researchers emphasize that organization's SPI efforts have a strong influence on management commitment [P3] [P4] [P36] [P50]. Therefore, senior management commitment is required to ensure that the improvement of the process is coordinated as a strategic project.

Table 2. List of identified factors

Factor	Studies	Total	%
Business orientation	[P1]; [P2]; [P3]; [P4]; [P5]; [P6]; [P7]; [P8]; [P9]; [P10]; [P11]; [P12]; [P13]; [P14]; [P15]; [P16]; [P17]; [P18]; [P19]; [P20]; [P21]; [P22]; [P23]; [P24]; [P25]; [P26]; [P27]; [P28]; [P29]; [P30]; [P31]; [P32]; [P33]; [P34]; [P35]; [P36]; [P37]; [P38]; [P39]; [P40]; [P41]; [P42]; [P43]; [P44]; [P45]; [P46]; [P47]; [P48]; [P49]; [P50]; [P51]; [P52]; [P53]; [P54]; [P55]; [P56]; [P57]; [P58]; [P59]; [P60]; [P61]; [P62]; [P63];	63	78.0
Management commitment	[P1]; [P3]; [P4]; [P5]; [P6]; [P7]; [P9]; [P10]; [P11]; [P12]; [P13]; [P14]; [P16]; [P17]; [P18]; [P19]; [P22]; [P23]; [P24]; [P25]; [P26]; [P27]; [P28]; [P29]; [P30]; [P31]; [P32]; [P35]; [P36]; [P37]; [P38]; [P39]; [P40]; [P41]; [P42]; [P43]; [P44]; [P45]; [P46]; [P47]; [P48]; [P49]; [P50]; [P51]; [P54]; [P56]; [P62]; [P63]; [P64]; [P65]; [P66]; [P67]; [P68]; [P69]; [P70]; [P71]; [P72];	60	74.1
Organizational culture	[P1]; [P4]; [P9]; [P13]; [P14]; [P15]; [P16]; [P20]; [P22]; [P25]; [P27]; [P28]; [P32]; [P35]; [P42]; [P47]; [P48]; [P49]; [P51]; [P64]; [P69]; [P76]; [P77]; [P78];	24	30
Organizational politics	[P4]; [P6]; [P16]; [P18]; [P24]; [P29]; [P30]; [P31]; [P42]; [P48]; [P49];	11	14
Organization structure	[P1]; [P6]; [P10]; [P13]; [P42]; [P48];	6	7.4
Resource allocation	[P4]; [P6]; [P10]; [P11]; [P13]; [P16]; [P18]; [P22]; [P24]; [P26]; [P28]; [P29]; [P31]; [P32]; [P35]; [P36]; [P37]; [P41]; [P42]; [P48]; [P50]; [P52]; [P53]; [P54]; [P69]; [P70]; [P79];	27	33.3
SPI benefit	[P6]; [P9]; [P13]; [P15]; [P16]; [P19]; [P22]; [P28]; [P31]; [P35]; [P36]; [P38]; [P42]; [P44]; [P46]; [P47]; [P48]; [P49]; [P50]; [P52]; [P57];	21	26
Roles and responsibility	[P5]; [P6]; [P11]; [P18]; [P22]; [P23]; [P25]; [P26]; [P28]; [P30]; [P31]; [P32]; [P41]; [P43]; [P46]; [P48]; [P51]; [P54]; [P55]; [P61]; [P77]; [P80];	22	27
Political stability	[P5]; [P42]; [P48]; [P49]; [P69]; [P78];	6	7.4
Effective management	[P4]; [P18]; [P17]; [P19]; [P22]; [P31]; [P37]; [P40]; [P44]; [P51]; [P53]; [P63]; [P76]; [P78]; [P81];	15	18.5
Organization vision	[P10]; [P15]; [P22]; [P31]; [P38]; [P44]; [P47]; [P48]; [P49]; [P60]; [P64];	11	14

Resource Allocation: 'Resource Allocation' was identified as effective factors, representing 33.3% of the selected studies. The researchers argued the importance of this factor [P22] [P24] [P28]. Therefore, in order to increase an understanding of resource allocation for the different roles, the minimum overhead resources should be allocated based on experience [P42] [P70].

Organizational Culture: our findings also showed that 'Organizational Culture' is an important factor (30%, 24 articles). According to GarzáS et al. [P22], one of the main reasons for small companies to tackle the improvement effort is to establish a culture of quality. Muñoz et al. [P15] argued that companies should analyze organizational culture properly in order to establish SPI efforts successfully. Furthermore, Albuquerque et al. [P16], conducted a study to understand the factors that can influence the success of SPI effort in small companies. The results indicated that factors of organizational culture could directly influence SPI implementation.

Roles and Responsibility: Almost a quarter of the selected articles in this systematic review point to the importance of 'Roles and Responsibility' factor for the success of the improvement process. Garcia et al. [P23], conducted a study with the aim to support small software development industry to adopt iterative approach by using SPI web tool. The results emphasize that responsibility was one of the factors that strongly contributed to support the web tool to achieve the main goal. Similarly, assigning responsibility and authority play important roles in increasing the capacity level of the process [P18] [P22]. Other research studies, including [P6] and [P26], emphasizes that clearly defined roles and their responsibilities in SPI efforts will lead SMEs companies to successfully implement the improvement strategy.

SPI Benefit: This factor was cited in 26% of the total articles as a critical factor in the organizational domain to enable SPI effort in the small software development industry. The awareness of SPI benefit amongst organization members will be a guarantee to implement efficient SPI strategy [P42]. According to Sulayman et al. [P49], management in small software companies needed to convince clients about SPI benefits in order to include the SPI related project cost in the overall project cost. Therefore, in order to convince small companies about the expected benefits in the business, it is necessary to recognize the importance of minimizing the costs of improvement to make SPI benefits visible in shortest time frame [P19] [P52].

Effective Management: 18.5% of the studies cited 'Effective Management' as an essential factor to deploy the process improvement over the small software development industry. According to Pino et al. [P44], "the process improvement is not an isolated activity, but is closely related to other activities of the software process management". Therefore, having strong configuration management play a critical role in the SPI effort within small companies [P22].

Organizational Politics and Vision: The findings of this study also indicated that both 'Organizational Politics and Organization Vision' were cited by (14%, 11 articles) of the total selected papers. Other factors that were cited by articles in the review were Political stability and Organization Structure and they had the least significant percentage (7.4%, 6 articles) compared with other organizational factors.

RQ2: Do the identified factors differ from developed to developing countries?

In order to answer research question 2, Table 3 demonstrates the list of organizational factors identified based on various developed and developing countries. All identified articles through the SLR were classified into three clusters based on the condition of the country's economy where the selected study was performed, namely, developed, developing and semi-industrialized. In total, 81 research articles were included in this review process, with 44 of the research conducted in developed countries, while 10 articles were in developing countries and 27 of the articles were conducted in semi-industrialized countries. From the results, of the eleven factors, two factors were cited by more than 50% of the groups. The 'Business Orientation' factor was the most cited factor, with 84.1% percentage of articles in developed countries group. The second highest factor was 'Management Commitment' factor and it has 72.7% of total articles in this group.

The main objective of this question is to investigate if there is any difference between developed, developing and semi-industrialized countries in order to consider the identified factors in the particular country. To identify if there exists a significant difference across the groups, the Chi-square test linear-by-linear association was deployed. From the statistical test results, there was a significant difference in 'organization structure' factor only.

Table 3. List of identified factors across country-wise

Factor	Developed (n = 44)	Developing (n = 10)	Semi industrialized (n = 27)	Chi-square test (linear by linear association) $df = 1$; $\alpha = 0.05$	
				X^2	ρ
F1	37 (84.1%)	8 (80%)	18 (66.7%)	2.829	.093
F2	32 (72.7%)	9 (90%)	19 (70.4%)	.013	.910
F3	10 (22.7%)	4 (40%)	10 (37%)	1.783	.182
F4	5 (11.4%)	1 (10%)	5 (18.5%)	.665	.415
F5	**0 (0%)**	**3 (30%)**	**3 (11%)**	**3.874**	**.049**
F6	11 (25.0%)	4 (40%)	12 (44%)	2.929	.087
F7	9 (20.5%)	5 (50%)	7 (25.9%)	.442	.506
F8	13 (29.6%)	2 (20%)	7 (25.9%)	.142	.707
F9	3 (6.8%)	0 (0%)	3 (11%)	.339	.561
F10	6 (13.6%)	2 (20%)	7 (25.7%)	1.670	.196
F11	4 (9.1%)	3 (30%)	4 (14.8%)	.665	.415

5 Discussion

In this SLR study, we have identified SPI factors that are faced by SMEs software development industry in the context of organizational perspective. Extracted factors help SMEs companies to implement SPI strategy successfully. Therefore, for better understanding, we have classified and analyzed the identified factors into three clusters based on the condition of the country's economy where the selected study was performed. These clusters help to represent the identified factors with respect to a particular country. Therefore, small software companies in those particular countries should consider identifying factors with a positive impact in the specific country.

To broaden the relevance of the identified factors, definite criteria were entrenched to deduce the level at which factors were recognized as influential to the implementation of an SPI strategy with respect to the professional characteristics of the practitioners. The level of importance of each factor was considered in the assessment of such a factor. That is if a factor is cited more (greater than 50% of the identified articles), such factor will be recognized as important. Several research studies in same domain have used the same criterion [22, 39–41].

Therefore, to answer research question one, by using these criteria we have identified two organizational factors to be critical for SMEs software development industry. These factors are: 'Business Orientation' and 'Management Commitment'. Similarly, to address research question two, the results showed that out of the eleven factors, 'Organization Structure' had a significant difference across developed, developing and semi-industrialized countries when compared country-wise.

The construct validity of this study is based on SLR study and its data extraction procedures were assessed by expert researchers. Furthermore, to arrive at the acceptable assessment of the investigated effective factors, this study adopted the results of the literature studies as highlighted in the research methodology section (See Sect. 3). The majority of our primary research studies were selected from recognized scientific research sources from authors with extensive experience in SPI.

6 Conclusion and Future Direction

This study was conducted to identify the effective factors that have a positive impact to enable SPI efforts in SMEs software development industry. The identified factors in this review were investigated based on organizational context perspective. Consequently, systematic review technique as mentioned by Kitchenham et al. [37] was adopted. In order to answer our research questions 'What organizational factors do small industry need to address in order to enable SPI effort in their activities? And Do the identified factors differ from developed to developing countries?'. From a total of 1085 studies, 81 articles were found to have a significant contribution to the research context. List of eleven factors has identified as mentioned in Table 2. From the results, 'Business Orientation' and 'Management Commitment' factors were the most critical organizational factors that have a direct effect on the successful implementation of SPI in SMEs software development industry. The Chi-square test was used to identify if there is any significant difference in term of developed, developing and semi-industrialized countries. 'Organization Structure' factor has a significant difference amongst the three groups when compared country-wise and it is important to note that this factor ('Organization Structure') did not appear in the developed countries where the reviewed studies were performed.

In conclusion, the finding of this research will help SMEs software companies to understand and respond positively to organizational factors related to SPI efforts. In addition, the results and findings from this study will give interesting insights to encourage researchers in using compromise technique to analysis future empirical studies based on a specific region to validate the suitability of these factors in a specific country. For the next step, our future work is to investigate other possible perspectives; human, process, project management, knowledge management, external and internal support. The ultimate future goal to develop Malaysian SME Software Process Improvement model (MSME-SPI). This research paper contributes only one part of the model. Identifying these factors will positively assist in developing the MSME-SPI model based on the specific context, especially for Malaysian SMEs software development industry.

References

1. Deming, W.E.: Out of the Crisis. MIT Press, Cambridge (2000)
2. Juran, J., Godfrey, A.B.: Quality Handbook. Republished McGraw-Hill (1999)
3. Dorling, A.: SPICE: software process improvement and capability determination. Softw. Qual. J. **2**, 209–224 (1993)
4. Bicego, A., Khurana, M., Kuvaja, P.: Bootstrap 3.0—software process assessment methodology. In: Software Quality Management VI, pp. 26–37. Springer, London (1998)
5. Murphy, C.N., Yates, J.: The International Organization for Standardization (ISO): Global Governance Through Voluntary Consensus. Routledge (2009)
6. Pande, P.S., Neuman, R.P., Cavanagh, R.R.: The Six-Sigma Way. McGraw-Hill (2000)
7. Paulk, M.: Capability maturity model for software. Encycl. Softw. Eng. (1993)
8. CMMI Product Team: Capability Maturity Model® Integration (CMMI), Version 1.1–Continuous Representation (2002)
9. Sánchez-Gordón, M.-L., O'Connor, R.V.: Understanding the gap between software process practices and actual practice in very small companies. Softw. Qual. J. **24**, 549–570 (2016)
10. Staples, M., Niazi, M., Jeffery, R., Abrahams, A., Byatt, P., Murphy, R.: An exploratory study of why organizations do not adopt CMMI. J. Syst. Softw. **80**, 883–895 (2007)
11. Almomani, M.A.T., Basri, S., Mahamad, S., Bajeh, A.O.: Software process improvement initiatives in small and medium firms: a systematic review. In: 2014 3rd International Conference on Advanced Computer Science Applications and Technologies (ACSAT), pp. 162–167 (2014)
12. Ireland, E.: Software Industry Statistics (1991–2005). Accessed Jan 2007
13. Department of Statistics, Malaysia (DOSM): Media Conference Census Report on SMEs, May 2011. http://www.smecorp.gov.my/index.php/en/policies/2015-12-21-09-09-49/sme-statistics
14. Airaksinen, A., Luomaranta, H., Alajääskö, P., Roodhuijzen, A.: Dependent and independent SMEs and large enterprises, 5 April 2015. http://ec.europa.eu/eurostat
15. Smallbone, D., Wyer, P.: Growth and development in the small firm. In: Enterprise and Small Business, pp. 100–126 (2000)
16. Altenburg, T., Eckhardt, U.: Productivity enhancement and equitable development: challenges for SME development. UNIDO, Vienna (2006)
17. Baharorn, F., Deraman, A., Hamdan, A.R.: A survey on the current practices of software development process in Malaysia. J. ICT **4**, 57–76 (2006)
18. Almomani, M.A., Basri, S., Mahmood, A.K.B., Bajeh, A.O.: Software development practices and problems in Malaysian small and medium software enterprises: a pilot study. In: 2015 5th International Conference on IT Convergence and Security (ICITCS), pp. 1–5 (2015)
19. Nasir, M., Ahmad, R., Hassan, N.H.: Issues in the implementation of software process improvement project in Malaysia. WSEAS Trans. Inf. Sci. Appl. **5**, 1031–1043 (2008)
20. Bullen, C.V., Rockart, J.F.: A primer on critical success factors (1981)
21. Clarke, P., O'Connor, R.V.: The situational factors that affect the software development process: towards a comprehensive reference framework. Inf. Softw. Technol. **54**, 433–447 (2012)

22. Rainer, A., Hall, T.: Key success factors for implementing software process improvement: a maturity-based analysis. J. Syst. Softw. **62**, 71–84 (2002)
23. Hall, T., Rainer, A., Baddoo, N.: Implementing software process improvement: an empirical study. Softw. Process Improv. Pract. **7**, 3–15 (2002)
24. Niazi, M., Ali Babar, M., Katugampola, N.M.: Demotivators of software process improvement: an empirical investigation. Softw. Process Improv. Pract. **13**, 249–264 (2008)
25. Dybå, T.: Factors of software process improvement success in small and large organizations: an empirical study in the scandinavian context. In: ACM SIGSOFT Software Engineering Notes, pp. 148–157 (2003)
26. Dybå, T.: An empirical investigation of the key factors for success in software process improvement. IEEE Trans. Softw. Eng. **31**, 410–424 (2005)
27. Sulayman, M., Mendes, E., Urquhart, C., Riaz, M., Tempero, E.: Towards a theoretical framework of SPI success factors for small and medium web companies. Inf. Softw. Technol. **56**, 807–820 (2014)
28. Yilmaz, M., O'Connor, R.V.: Social capital as a determinant factor of software development productivity: a empirical study using structural equation modelling. Int. J. Hum. Cap. Inf. Technol. Prof. **13**(2), 40–62 (2012)
29. O'Connor, R.V., Basri, S.: The effect of team dynamics on software development process improvement. Int. J. Hum. Cap. Inf. Technol. Prof. **13**(3), 13–26 (2012)
30. Montoni, M., Rocha, A.R.: A methodology for identifying critical success factors that influence software process improvement initiatives: an application in the Brazilian software industry. In: European Conference on Software Process Improvement, pp. 175–186 (2007)
31. Coleman, G., O'Connor, R.: Using grounded theory to understand software process improvement: a study of Irish software product companies. Inf. Softw. Technol. **49**, 654–667 (2007)
32. Nasir, M.H., Ahmad, R., Hassan, N.H.: Resistance factors in the implementation of software process improvement project in Malaysia. J. Comput. Sci. **4**, 211 (2008)
33. Kaltio, T., Kinnula, A.: Deploying the defined SW process. Softw. Process Improv. Pract. **5**, 65–83 (2000)
34. El-Emam, K., Goldenson, D., McCurley, J., Herbsleb, J.: Modelling the likelihood of software process improvement: an exploratory study. Empir. Softw. Eng. **6**, 207–229 (2001)
35. Staples, M., Niazi, M.: Systematic review of organizational motivations for adopting CMM-based SPI. Inf. Softw. Technol. **50**, 605–620 (2008)
36. Lavallée, M., Robillard, P.N.: The impacts of software process improvement on developers: a systematic review. In: Proceedings of the 34th International Conference on Software Engineering, pp. 113–122 (2012)
37. Kitchenham, B., Brereton, O.P., Budgen, D., Turner, M., Bailey, J., Linkman, S.: Systematic literature reviews in software engineering – a systematic literature review. Inf. Softw. Technol. **51**, 7–15 (2009)
38. Sommerville, I.: Software Engineering. Addison-Wesley, New York (2010)
39. Niazi, M., Babar, M.A.: Identifying high perceived value practices of CMMI level 2: an empirical study. Inf. Softw. Technol. **51**, 1231–1243 (2009)
40. Almomani, M.A., Basri, S., Mahmood, A.K.B., Baashar, Y.M.: An empirical analysis of software practices in Malaysian small and medium enterprises. In: 2016 3rd International Conference on Computer and Information Sciences (ICCOINS), pp. 442–447 (2016)
41. Almomani, M.A., Basri, S., Kamil, A.: Factors influencing the implementation of software process improvement in small and medium enterprises: an empirical study. Adv. Sci. Lett. **24**, 1716–1722 (2018)

Missing Data Imputation Techniques for Software Effort Estimation: A Study of Recent Issues and Challenges

Ayman Jalal Hassan Almutlaq(✉) and Dayang N. A. Jawawi(✉)

School of Computing, Faculty of Engineering,
Universiti Teknologi Malaysia (UTM), Skudai, Johor, Malaysia
aymanjmatlak@yahoo.com, dayang@utm.my

Abstract. Software effort estimation is one the critical aspects of software engineering. It revolves around predicting the required efforts needed to complete a software task. However, any estimation technique or model relies on an input data in which it defines and predicts future values. Missing data and values within such data is a common occurrence in the software development industry and thus it leads to inaccurate predictions or misleading results. Thus, Missing Data is an important aspect of effort estimation models that is required to be addressed. However, Missing Data is not without its gaps and issues. This review aims at elaborating the recent issues and gaps that exist within the missing data and software effort estimation field. This may allow future researchers to get a better grasp and understanding of the inner workings of Missing Data and the methods through which these challenges can be addressed.

Keywords: Missing Data · Software Effort Estimation · Imputation methods · Literature review

1 Introduction

Software Engineering (SE) is a large research field that is concerned with the several aspects of developing and engineering a software. However, on an industrial level, one of the most important challenges is software cost estimation [1–3]. In such scenarios, projects that have already been developed provide their data sets so that the building of estimation models can ultimately enable better decision making as well as proper research allocation.

Effort estimation in software development is defined as a process in which the amount of effort needed to complete a software related task is predicted by using previously collected information at the earlier stages of the development [4]. There are several software effort estimation models that have resulted from the research. These models have been also tested in the field. Initially, most of the proposed estimation models had a regression based analysis concept. However, recently, the techniques proposed are more projected towards Machine Learning, Search-based approaches among others [5]. Machine Learning, in particular, has proved to be one of the most popular effort estimation techniques. This is due to the fact that it has been utilized in

F. Saeed et al. (Eds.): IRICT 2019, AISC 1073, pp. 1144–1158, 2020.
https://doi.org/10.1007/978-3-030-33582-3_107

various applications such as the use of the Bayesian Network or Support Vector Regression [6, 7].

One of the main issues faced in software effort estimation models is the challenge of missing data. This is considered to be a relatively common occurrence in the software development process. In most cases, there are missing attributes in project tables. The absence of these tables can mislead results and reduce the accuracy of the prediction models. Thus, to deal with Effort Estimation, Missing Data has to be addressed as one of its core concerns [8].

Past data is used as one of the main attributes to construct the project missing or incomplete data [9]. The issue of missing data or the incompleteness of data is a common industrial issue as data sets can often have missing components or values. This can be caused due to a variety of reasons namely combined projects or different data sorting requirements that ultimately leads to gaps within the dataset and the system. Each of the Data collection techniques has its own set of outputs which change based on the needs of the system [10]. This can lead to different collection techniques as there are useful and non-useful information that can be used for predication models. However, these techniques might also have internal gaps. This problem is now one of the most highlighted challenges in software engineering and several methods were proposed to address it [11].

Thus, it is not uncommon to turn to Machine Learning as a prediction model since, in most cases, such techniques can be used for addressing the issue of missing data via a variety of different techniques and processes. Machine Learning (ML) is a popular technique which is often used in Software Engineering for the purpose of predicting future projections [12]. It has been used to handle various types of SE challenges such as cost prediction [1] and effort estimation [13].

It has been shown that Machine Learning as a technique is far more suitable as a prediction model than statistical analysis techniques are. This is mainly because in a Supervised Learning (SL) environment, the decisions make more sense from a human perspective than it does from a straightforward component analysis used in statistical methods.

Understanding the problems of software effort estimation and specifically the missing data allows for a deeper understanding of the field. It also helps in categorizing and highlighting of top issues and challenges of missing data. This would enable future researchers to get a better understanding of missing data and effort estimation field. Moreover, it also helps to propose techniques that can more effective in terms of addressing the different facets and aspects of missing data.

In order to provide a more structured flow for the review, three research questions (RQ) were designed in order to specify the scope of the research, as well as to gain a better understanding of software effort estimation and its relation to missing data. The research questions are as following:

(RQ1) What are the gaps and issues in the most recently published research on Missing Data Imputation for software effort estimation?

(RQ2) What are the top concerns and challenges in the field of Software Effort Estimation?

(RQ3) How can the challenges and the issues present in Missing Data Imputation for software effort estimation be categorized and grouped?

In this paper we perform a Literature review because of the limited published recent papers in this area (missing data imputation methods in software effort estimation) in the last four years (2015–2019), we found about 15 recent paper tackling this problem. The aim is to get an understanding of the inner workings of missing data imputation techniques and effort estimation. Thus, to perform this task, the research questions were used as a guideline that would lead to an understanding of these challenges and issues related to effort estimation and missing data better.

This paper is divided in the following way. The first section elaborates on the related works and the existing reviews on missing data imputation effort estimation and other literatures that are similar in concept. This section is needed, as it would explain what sets this review apart from all the other existing ones that have been done on the topic. The second section aims at answering the first research question by listing and critically analyzing the existing studies and their limitations. The fourth section discusses upon the challenges and issues faced in software effort estimation and missing data imputation. This section addresses both second and third research questions. In the fifth and final section, the conclusion is written with a review of the research questions and the method by which they were answered. It also illustrates what is expected to be yield from their discovery. The study closes by reflecting on the possible scenarios in which the review can help upcoming researches propose new methods.

2 Related Work

There are several existing reviews and mapping studies in both software effort estimation and missing data. A systematic literature review on ensemble missing data imputation techniques for effort estimation was published in 2012 [14]. That work conducted a deep literature review based on Kitchenham's guideline on performing a systematic literature review in software engineering [15]. The review proposed several research questions in which the main goal revolved around ensemble effort estimation models and their role in software effort estimation. Although the review was systematic and deep, it is outdated because it only covered the period from 2000 to 2012 and reviewed only 8 primary studies.

Another study focused on performing an empirical study on software cost estimation using machine learning [1]. That study was published in 2015. The first part of the paper introduced an in-depth analysis of data prepressing techniques for machine learning based models and their role in software effort estimation. The content of the review which consisted of about 28 primary studies mainly included LD (listwise deletion) and MI (Mean Imputation). Even though, their review covered the research studies from 2005 to 2014, the several new techniques and models proposed after that made the study outdated. They also took missing data treatment as part of the data preprocessing as a whale. Unlike these review, the current paper mainly focuses on and addresses missing data imputation.

However, the most relevant paper published was a review in 2019. It focused on an analysis and review on missing value imputation [16]. That study provided a decade study (2006–2017) on missing value imputation and its techniques. It was published in order to identify its issues and the direction in which the field is taking. However, they included missing data imputation of many domains (Medical, Image, software, Financial, questionnaires), they did not pay much attention on software effort estimation. Our scope in this study is to focus on missing data imputation for software effort estimation only.

Therefore, unlike previous surveys, this survey reviews the most recent techniques in missing data imputation for effort estimation. In particular, 15 papers published over the years from 2015 to 2019 were reviewed and analyzed. In addition, this paper synthesizes the strength and weakness points of the most recent existing techniques and elaborates the challenges in adopting these techniques in software effort estimation. The aim is to fill the gap that was found in other reviews and to conduct a deep review that highlights the issues, strength and weakness of the techniques, and challenges in the most recent published papers in the field of missing data imputation for software effort estimation.

3 Results

3.1 Review of Missing Data Imputation Techniques

In this section, the results of the study review the various types of recent techniques that have been conducted in the past four years in an attempt to answer RQ 1. The reviewed papers are listed in Table 1. The first column of the table lists the related references and study. The second column presents the effort estimation models that are applied in the experiment study. The third column highlights Missing Data Treatment (MDT) which is either k-NN, Mean Imputation (MI) or Listwise Deletion (LD). The fourth column is the missing data mechanism applied in the study which can be classified as MCAR (Missing Completely at Random), MAR (Missing at Random), and NIM (Not Ignorable Missing. Finally, he fifth column highlight the proposed technique in the study. Various types of methods and techniques are used in order to process data and perform missing data. The findings identify the current problem or overview of the technique. However, each of these techniques is further elaborated and explained in details in terms of their aspects.

Table 1. Data imputation techniques used by ML methods for software effort estimation

Reference	Effort technique	MDT	Mechanism	Method
[1]	ANN CBR SVR CART BBN	LD MI	N/A	Listwise Deletion (LD) Mean Imputation

(continued)

Table 1. (*continued*)

Reference	Effort technique	MDT	Mechanism	Method
[17]	LR BR SVR M5 RT CBR	MI CMI MINI kNNI BMI	MAR	BREM (Bayesian Regression and Expectation Maximization)
[18]	–	MI k-NNI	MCAR MAR NIM	DICkNNI Dynamic incomplete case KNNI
[19]	LPR NPR	MI k-NNI	MCAR	KNN and parametric regression
[20]	CART OLS RBNF MARS RoR	MEI RI MI MLI HI	N/A	LRSRI (low-rank recovery and semi supervised regression imputation)
[21]	KNN, Discriminant analysis, and Naï Bayes	k-NNI	MCAR MAR NIM	GRA-KNN Grey Relation Analysis
[22]	EBA (Classical, Fuzzy)	Toleration, Deletion, KNN Imputation	MCAR MAR NIM	EBA (Classical, Fuzzy) (KNN)
[23]	Fuzzy EBA	Toleration, Deletion, KNN Imputation	MCAR MAR NIM	EBA/Fuzzy (KNN)
[24]	Regression analysis	Multiple Imputation	N/A	Multiple Imputation Strategy/regression analysis
[25]	Discriminant analysis, K NN, Naive Bayes and SVM	k-NNI	MCAR MAR NIM	Cross-validation based *K* NN imputation (CVB *k* NNI)
[26]	EBA	LD (listwise deletion), MDI (missing data imputation)	N/A	KNN Mean/mode imputation (MEI)
[27]	EBA (classical)	Toleration, Deletion, KNN imputation, SVR imputation	MCAR MAR NIM	EBA/SVR
[28]	EBA (Classical and Fuzzy)	k-NNI SVRI	MCAR MAR NIM	EBA/SVR
[29]	Multiple regression	Multiple Imputation	N/A	MR/MI Multiple Regression/Multiple Imputation
[30]	Least squares regression	Multinomial logistic regression, listwise deletion, mean imputation, expectation maximization and regression imputation	N/A	*Multinomial Logistic* Regression (MLR)

In order to go over each of the studies, the techniques have been chronologically ordered in Table 1. This would provide a better context as to the growth of the techniques over the past four years (2015–2019).

The experiment requires to be expanded further using different datasets and the machine learning methods which can be more advanced in nature. Moreover, it requires the use of other unbiased measurements to validate and generalized the results in these studies [1, 17, 19, 20]. Further experiment investigations using various missing data imputation techniques could improve the performance of BREM (Bayesian Regression and Expectation Maximization) algorithm in [17]. The experiment in [19] is limited to one data set and alternative linear regression model could be investigated. Experimenting the proposed method in situations with higher missing rate need to be validated in [20].

Higher variety is required when setting the parameters that can improve the results drastically in [18, 21, 25]. The extension of the proposed method in [18] to handle various missing data patterns is to increasing the capacity of dynamic incomplete-case *k*NNI. Alternative imputation techniques are needed in order to validate the performance better [21]. CVBKNNI can be further improved by testing the results with more varied datasets in different software engineering domains [25].

Alternative imputation techniques are required in order to further validate the results and also extend the empirical experiment to handle mixed data (numerical and categorical) [22, 23, 26–28]. Extending the experiment to different variant of Estimation by Analogy (EBA) in [27] was for the purpose of validating the findings. [26] Handled mixed data in their experiment and recommended to pay attention to the characteristics of the data set for data prepressing combination.

[24, 29] Used multiple imputation missing data technique to handle the missing data on the ISBSG dataset. [24] concluded that multiple regression analysis results did not find evidence that ES and ET have any impact on the EI (Effort Implement) or EP (Effort Plan) parameters. In contrast, they appeared to have an impact on the EB (Effort Build) parameter. [29] concluded that the simulated results require replication of the simulation on the same dataset (ISBSG) in order to confirm the results.

Finally, [30] concluded that the proposed method is required to be tested with other prediction models like EBA in order to get a better understanding and idea of how effective the technique is. The proposed technique can be further adopted in other software engineering fields like defect and quality prediction.

As we can see from Table 1, the EBA (Estimation by Analogy) is the common effort estimation model among this literature [22, 23, 26–28], and KNN is the common missing data imputation technique adopted in software effort estimation [1, 18, 19, 21–23, 25, 26]. The distribution of the concretization type over the years is shown in Fig. 1.

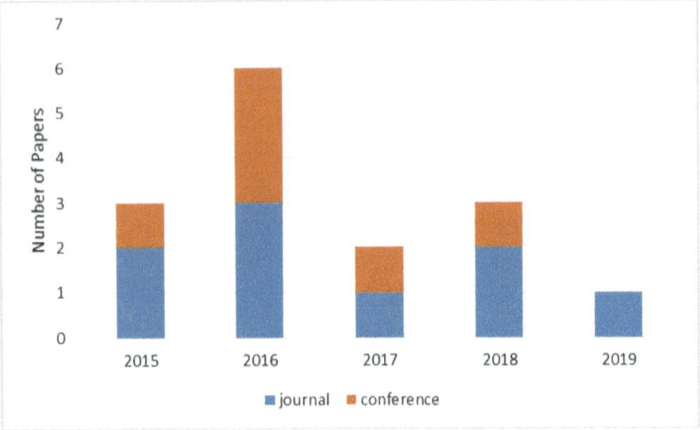

Fig. 1. Primary study year wise trend

3.2 Strengths and Weaknesses of Existing Techniques

In this section, strengths and weaknesses of the existing techniques elaborated earlier are listed and explained to answer the second part of RQ 1 as illustrated in Table 2.

Table 2. Strengths and weaknesses of existing missing data techniques

Techniques	References	Strengths and weaknesses
Listwise Deletion *Mean Imputation*	[1]	(−) Listwise Deletion may delete some data which are relevant to the prediction and consequently effect of the results (−) Mean Imputation can alter the variance of the data (+) Easily to implement
BREM Algorithm (Bayesian Regression and Expectation Maximization)	BREM [17]	(−) Semi-supervised learning techniques, use of the labeled and unlabeled data one after another in learning process (−) Will attenuate the variance of the data (+) Outperforms Linear Regression (LR), Bayesian Regression (BR), Support Vector Regression (SVR) and M5 Regression Tree in software effort Prediction.
KNN Algorithm	DICkNNI [18] KNN and Parametric regression [19] GRA KNN [21] EBA+KNN [22, 23] CVB KNN [25] KNN [26]	(−) Not efficient for large datasets (−) It imputes values based on the neighbors which may or may not be the related projects for donor values (−) Depend on parameter setting (+) Simple to implement (+) Can be applying to different type of variables

(*continued*)

Table 2. (*continued*)

Techniques	References	Strengths and weaknesses
SVR *Support Vector Regression*	SVR+EBE [27, 28]	(+) More accurate than KNN (−) Avoids the most relevant projects as a donor for imputing the missing values
LRSRI *Low-rank recovery and semi supervised regression imputation*	LRSRI [20]	(−) When the missing rate is high is a challenging (−) Depend on the correlation between cost drivers (+) Take attention to the characteristics of the data set
Multiple imputation (MI)	MI [24, 29]	(+) Proved unbiased estimates (+) Uses all available data, preserving sample size and statistical power (−) Suffers from large influences of different data missing mechanisms (−) Outlier Problem
Multinomial Logistic Regression (MLR)	MLR [30]	(−) Statistical Model depend on the linear relationships between attributes (+) Can handle categorical data imputation

In total, there are seven categories of techniques listed. The first one is Listwise deletion and Mean imputation. It is easy to implement but it deles more relevant data and changes the variance of the data threatening the precision of the results.

The second listed technique is BREM (Bayesian Regression and Expectation Maximization) Algorithm. It works based on machine learning and probability based algorithms. The third is KNN Algorithm, which is relatively effective and widely used, suffers from inefficiency when it comes to larger datasets. The fourth technique is SVR which is more accurate than KNN, however it has a habit of avoiding the most relevant donors for imputation of the missing values. The fifth is LRSRI (low-rank recovery and semi-supervised regression imputation) and it has subpar performance when the missing rate is high. Multiple Imputation or (MI) is also a good technique, however it is known to suffer from large influences of different data missing mechanisms. The MLR or Multinomial Logistic Regression is the last listed technique which can handle categorical imputation, however it is ultimately a statistical Model that depends on the linear relationships among attributes. It also has all the drawbacks that a statistics based model has.

Figure 2 shows the distribution of the existing missing data imputation techniques in these works. It is highlighted that KNN has the highest majority of concern among others.

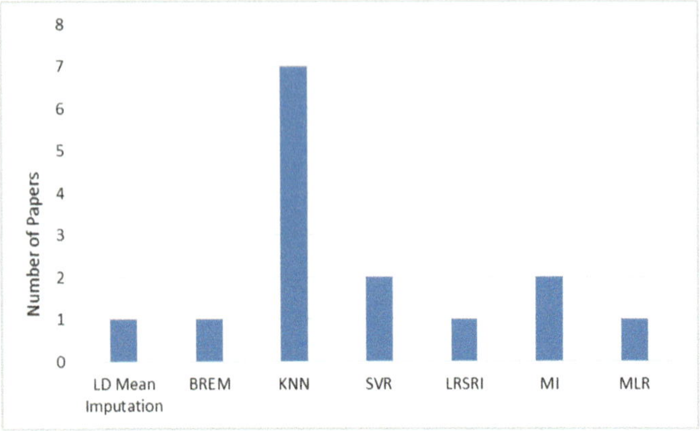

Fig. 2. Number of solution techniques with respect to existing studies

4 Discussion

4.1 Software Effort Estimation Issues

In this section we are attempting to answer (RQ2). There are several issues that exist in software effort estimation preventing it from obtaining relatively accurate results. The first issue is data availability which is considered crucial. It refers to verifying the degree of results validity. Many models are largely dependent on this availability in order to make proper adjustments to the data that available.

Another difficulty in the estimation process is the fact that the software environment itself is considered to be largely unstable. Moreover, the continuous changing environment poses challenges for the existing models to predict the upcoming issues and challenges that exist within the system. Thus, in many scenarios, there may not be enough time to fix or correct the results of the model according, as change is the only constant in software development. This is true as most software systems are rapidly growing and improving day by day. It also becomes a monumental task for it to be able to measure its degree of effort over an extended period of time.

Change in software development also leads to other issues. One of these issues is uncertainty that leads to new requirements being proposed every day due to the changing of the scope and the size of the software. Thus, this uncertainty can affect the final results of estimation in a negative way. These results are not often directly factored into the estimation models [4].

Another problem is data sensitivity which refers to the type of data used in each company that might be different from other companies. Since most companies do not share the same type of data, researchers should take into account that the type of data can change and thus precautions are needed to be in place to deal with such change [31].

Overall, software effort estimation deals with a lot of problems that are abstract in nature and are considered to be macro (large scale). These problems often stem from other sources that are part of software engineering as a whole and they cannot be addressed directly. However, these problems should be considered for any proposed software estimation model so that they can perform optimally over time.

4.2 Missing Data Imputation Challenges

In this section, we attempt to answer (RQ 3). Missing Data/Value Imputation is plagued by mainly two category of technical challenges. These challenges are mainly related to how these groups of techniques operate and how they can be measured in a way for this procedure to be properly outlined. Table 3 lists all the related challenges. These challenges mainly stem from two sources. The first is the list of reviewed papers that are included in this study. The second is the review papers and related articles that include the past knowledge that exists on missing data. It is important to have both sources available since the first source might give off the impression that which of these challenges are more of the focus, while the second would help in identifying challenges that might not be of concern anymore. Thus, it is needed to have both groups in order to be able to identify the existing issues.

Table 3. Missing data/value imputation challenges

Category	Challenge	References
Performance challenges	Missing data accuracy	[18, 21, 25, 26]
	Classifier accuracy	[1, 17, 19–30]
	Missing data and classifier accuracy	[21, 25, 26]
	Time efficiency	[25]
Dataset challenges	Numerical data imputation	[18–24, 28, 29]
	Categorical data imputation	[30]
	Mixed data imputation	[1, 17, 25–27]
	Dataset characteristics and size variety	[1, 17, 22, 24, 25, 28–30]
	MD mechanism variety	[18, 21–23, 25, 27, 28]

These challenges are mainly divided into two categories: Performance oriented challenges and Dataset oriented challenges. Performance oriented challenges refer to challenges and issues that exist within the techniques themselves on a performance level. While the dataset challenges revolve around the role of the dataset and its effect on the missing data imputation techniques.

The first performance challenge is accuracy; this is an important aspect of the missing data imputation as it measures whether or not the technique is accurate. Missing data accuracy evaluate the difference between the actual values in the collected data set and the predicted values by the classifier. It is measured by using the percentage of correct predictions (PCP) for when the values are discrete, and mean absolute percentage error (MAPE) or root-mean-square error (RMSE) for when the values are continuous. Accuracy is the corner stone of MDI techniques since the usefulness of any technique depends on how accurate it is in terms of its correct predictions.

The second performance challenge is also related to accuracy, however this specific form of accuracy is related to the effort estimation and the classifier accuracy which is measured via the Standard Accuracy (SA) or Magnitude Relative Error (MRE). This form of accuracy measurement has been used throughout all the effort estimation techniques because it provides an unbiased metric for measuring the degree of effort estimation.

It is recommended to measure both accuracy metrics in the prediction model to give a deep insightful assessment. Figure 3 shows the distribution of measuring accuracy among the selected studies.

The third Performance challenge is the time efficiency. This challenge is based on how quick the technique is in terms of its computational time. This is particularly important when it comes to Machine Learning techniques as most of them go over several iterations. If a technique takes an absurdly long amount of time, then it is not considered acceptable even if the result were to be relatively accurate. The goal has always been to be able to balance the time consumption with the accuracy of the predictions to provide a more well-rounded technique. This challenge is rarely confronted by researchers as seen in the literature.

The second group of challenges are the dataset related challenges. This is an important issue as a missing value/data imputation technique operates solely on the basis of its input data, which in this case are the datasets used.

The first issue in dataset challenges is a lack of variety in the type of data imputation features from the dataset. This issue is relatively an extension of the first issue, as different domains, favor specific types of data. These types are the categorical, numerical and mixed data features. Each has a different performance. While a technique might excel in numerical datasets, others might have a higher accuracy in categorical datasets. Thus, this is an issue that can be addressed through the inclusivity of these various types of data features in the original data set. Imputation of both types of data (Mixed data) will be recommended to improve the quality of the prediction model. Figure 4 present data type imputation challenge in the selected studies.

The second challenge in the dataset category is related to its relation to the size and the characteristics of the dataset. This is an important factor affecting the prediction models, as the size of the available data can dictate how the technique can learn from (in case of machine learning techniques), Based on that learning, it can correctly or accurately complete the missing values. Most of the researches focus on a specific sample size and do not test different sizes in order to ensure that the technique can operate optimally at all levels. The size and the heterogeneity characteristics of the ISBSG data gain attention by most of the researchers in the field. Thus, making this an important issue that is required to be addressed as well.

The third and final challenge is related to the lack of variety of different Missing data mechanisms. These refer to the types of missing mechanism that can cause incompleteness in a dataset. The different types are Missing Completely at Random (MCAR), missing at Random (MAR), and not missing at Random (NMAR). Each of these types dictates the method in which the technique can operate and predict the values. Which means the technique would operate based on known or unknown values. Many of the techniques mentioned in this review adamantly point out that they require testing with more missingness types and that their testing was done solely with a singular type and not the others. This ultimately refers to the capacity of the technique and to the types it can handle. In most cases, it is recommended for all the other types to be used and measured.

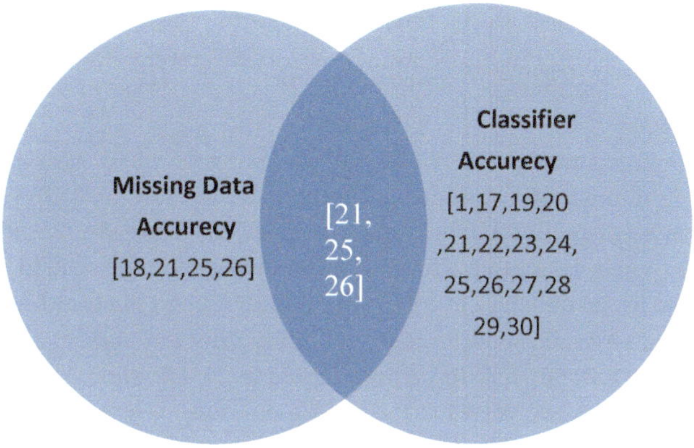

Fig. 3. Measuring accuracy challenge

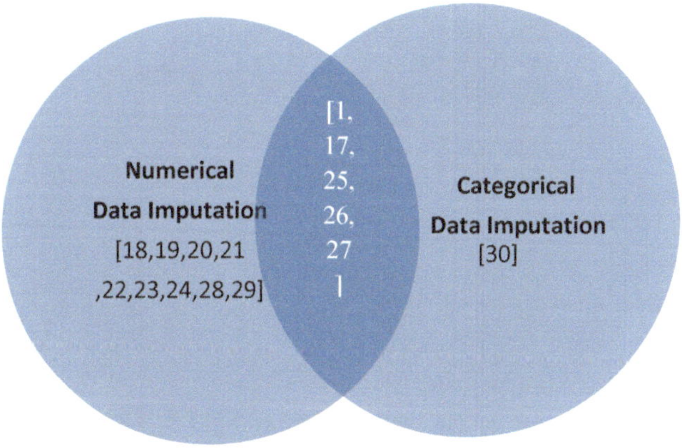

Fig. 4. Data type imputation challenge

5 Conclusion

This review covers three research questions that aim to identify the roots of software effort estimation as well as missing data imputation issues and challenges. There are three research questions addressed in this review. The first research question highlights most important issues related to missing data imputation in the most recent published papers such as Effort technique used, MDT, missing mechanism, and the proposed method. Furthermore, we elaborate the strengths and weaknesses of each proposed method. As observed, there is a micro problems challenging the researchers as high missing rate, dealing with different types of missing mechanisms, and technical challenges in the proposed methods need to be addressed to get more accurate effort estimation.

In second research question it should be noted that while Software Effort Estimation issues focused on the macro issues, the missing value imputation gaps focus on micro issues. This has led to the discovery and proposal of new techniques that would ultimately lead to identification of attributes. This is expected to solve the problem of data identification.

Finally, the third research question involves listing the challenges and issues related to missing data imputation techniques. These challenges are extracted from the recent techniques that were reviewed. Each of these provides a better understanding into the missing data as a whole. Both performance oriented and data oriented challenges need to be addressed by researchers to gain improvement in estimation accuracy.

Further research is required to address highlighting the various technique oriented challenges. If a technique were to be highlighted using the aforementioned challenges, then the technique's weaknesses are highlighted. This is expected to allow for the proposal of new techniques and solutions that would avoid or address those issues. Furthermore, considerations to characteristics and nature of the data before applying imputation techniques must be taken as a guideline to improve the accuracy of the effort estimation model.

Acknowledgments. The authors fully acknowledge Universiti Teknologi Malaysia for UTM-TDR Grant Vot No. 06G23, and Ministry of Higher Education (MOHE) for FRGS Grant Vot No. 5F117, which have made this research endeavor possible.

References

1. Huang, J., Li, Y.-F., Xie, M.: An empirical analysis of data preprocessing for machine learning-based software cost estimation. Inf. Softw. Technol. **67**, 108–127 (2015)
2. Dizaji, Z.A., Gharehchopogh, F.S.: A hybrid of ant colony optimization and chaos optimization algorithms approach for software cost estimation. Indian J. Sci. Technol. **8**(2), 128 (2015)
3. Papatheocharous, E., et al.: An investigation of effort distribution among development phases: a four-stage progressive software cost estimation model. J. Softw. Evol. Process **29** (10), e1881 (2017)
4. Trendowicz, A., Jeffery, R.: Software project effort estimation. In: Foundations and Best Practice Guidelines for Success, Constructive Cost Model – COCOMO 2014, pp. 277–293 (2014)
5. Wen, J., et al.: Systematic literature review of machine learning based software development effort estimation models. Inf. Softw. Technol. **54**(1), 41–59 (2012)
6. Song, L., Minku, L.L., Yao, X.: Software effort interval prediction via Bayesian inference and synthetic bootstrap resampling. ACM Trans. Softw. Eng. Methodol. (TOSEM) **28**(1), 5 (2019)
7. Azzeh, M., Nassif, A.B., Banitaan, S.: Comparative analysis of soft computing techniques for predicting software effort based use case points. IET Softw. **12**(1), 19–29 (2017)

8. Twala, B., Cartwright, M.: Ensemble missing data techniques for software effort prediction. Intell. Data Anal. **14**(3), 299–331 (2010)
9. Stephens, M., Scheet, P.: Accounting for decay of linkage disequilibrium in haplotype inference and missing-data imputation. Am. J. Hum. Genet. **76**(3), 449–462 (2005)
10. Lethbridge, T.C., Sim, S.E., Singer, J.: Studying software engineers: data collection techniques for software field studies. Empir. Softw. Eng. **10**(3), 311–341 (2005)
11. Mockus, A.: Missing data in software engineering. In: Guide to Advanced Empirical Software Engineering, pp. 185–200. Springer (2008)
12. Srinivasan, K., Fisher, D.: Machine learning approaches to estimating software development effort. IEEE Trans. Softw. Eng. **21**(2), 126–137 (1995)
13. Sarro, F., Petrozziello, A., Harman, M.: Multi-objective software effort estimation. In: 2016 IEEE/ACM 38th International Conference on Software Engineering (ICSE). IEEE (2016)
14. Andrew, B., Selamat, A.: Systematic literature review of missing data imputation techniques for effort prediction. In: International Conference on Information and Knowledge Management, Singapore (2012)
15. Keele, S.: Guidelines for performing systematic literature reviews in software engineering. In: Version 2.3, EBSE Technical report (2007)
16. Lin, W.-C., Tsai, C.-F.: Missing value imputation: a review and analysis of the literature (2006–2017). Artif. Intell. Rev. 1–23 (2019)
17. Zhang, W., Yang, Y., Wang, Q.: Using Bayesian regression and EM algorithm with missing handling for software effort prediction. Inf. Softw. Technol. **58**, 58–70 (2015)
18. Huang, J., et al.: An empirical study of dynamic incomplete-case nearest neighbor imputation in software quality data. In: 2015 IEEE International Conference on Software Quality, Reliability and Security. IEEE (2015)
19. Soltanveis, F., Alizadeh, S.H.: Using parametric regression and KNN algorithm with missing handling for software effort prediction. In: 2016 Artificial Intelligence and Robotics (IRANOPEN). IEEE (2016)
20. Jing, X.-Y., et al.: Missing data imputation based on low-rank recovery and semi-supervised regression for software effort estimation. In: Proceedings of the 38th International Conference on Software Engineering. ACM (2016)
21. Huang, J., Sun, H.: Grey relational analysis based k nearest neighbor missing data imputation for software quality datasets. In: 2016 IEEE International Conference on Software Quality, Reliability and Security (QRS). IEEE (2016)
22. Idri, A., Abnane, I., Abran, A.: Missing data techniques in analogy-based software development effort estimation. J. Syst. Softw. **117**, 595–611 (2016)
23. Abnane, I., Idri, A.: Evaluating fuzzy analogy on incomplete software projects data. In: 2016 IEEE Symposium Series on Computational Intelligence (SSCI). IEEE (2016)
24. Bala, A., Abran, A.: Use of the multiple imputation strategy to deal with missing data in the ISBSG repository. J. Inf. Technol. Softw. Eng. **6**, 171 (2016)
25. Huang, J., et al.: Cross-validation based K nearest neighbor imputation for software quality datasets: an empirical study. J. Syst. Softw. **132**, 226–252 (2017)
26. Huang, J., et al.: An empirical analysis of three-stage data-preprocessing for analogy-based software effort estimation on the ISBSG data. In: 2017 IEEE International Conference on Software Quality, Reliability and Security (QRS). IEEE (2017)
27. Abnane, I., Idri, A.: Improved analogy-based effort estimation with incomplete mixed data. In: 2018 Federated Conference on Computer Science and Information Systems (FedCSIS). IEEE (2018)
28. Idri, A., Abnane, I., Abran, A.: Support vector regression-based imputation in analogy-based software development effort estimation. J. Softw. Evol. Process **30**(12), e2114 (2018)

29. Bala, A., Abran, A.: Impact analysis of multiple imputation on effort estimation models with the ISBSG repository of software projects. Softw. Meas. News **23**(1), 17–34 (2018)
30. Chatzipetrou, P.: Software cost estimation: a state-of-the-art statistical and visualization approach for missing data. Int. J. Serv. Sci. Manag. Eng. Technol. (IJSSMET) **10**(3), 14–31 (2019)
31. Padhy, N., Singh, R., Satapathy, S.C.: Software reusability metrics estimation: algorithms, models and optimization techniques. Comput. Electr. Eng. **69**, 653–668 (2018)

CarbonFree – A Multi-platform Application for Low Carbon Education

Han Xin Hui[1], Noraini Ibrahim[1,2](✉), and Fatin Aliah Phang[2]

[1] School of Computing, Faculty of Engineering, Universiti Teknologi Malaysia, Skudai, Malaysia
xhhan1227@gmail.com

[2] Centre for Engineering Education, Universiti Teknologi Malaysia, Skudai, Malaysia
{noraini_ib, p-fatin}@utm.my

Abstract. Malaysia had experienced a warming trend for the past decades. The main contributor of this problem is carbon emissions which caused climate change. Despite the magnitude of the problem, the level of Low Carbon Society (LCS) awareness among Malaysia citizens is unsatisfied. Due to this situation, CarbonFree, a multi-platform application for low carbon education is proposed to curb this severe issue. CarbonFree is inspired by the program of Iskandar Malaysia Eco-Life Challenge (IMELC). Therefore, users can complete the program workbook through CarbonFree. Users are also able to retrieve LCS knowledge from the application. From the perspective of development of CarbonFree, the methodology used is Rational Unified Process (RUP), which including inception, elaboration, construction and transition stages. Furthermore, design and testing of CarbonFree is explained in Software Requirement Specification (SRS), Software Design Document (SDD) and Software Testing Document (STD) accordingly. Based on the user Acceptance Testing that conducted during testing phase, most of the participants agreed that CarbonFree is able to raise LCS awareness among students, teachers and their families as well as to promote knowledge and practice of LCS by providing information to students, teachers and their families which mean that the main purpose of this project is achieved.

Keywords: Global warming · Reduce carbon emission · Low Carbon Society (LCS) · Learning application · Multi-platform

1 Introduction

For the past decades, the world had experienced a warming trend. The main contributor of this problem is carbon emissions which caused climate change [1]. Therefore, carbon reduction is part of the nation effort to address the threat of the carbon discharge on human health and environment. Majority of the human activities which release carbon emission have caused the increase of temperature globally [2]. Although Malaysia's economy is developing rapidly, public concern and awareness are not

F. Saeed et al. (Eds.): IRICT 2019, AISC 1073, pp. 1159–1169, 2020.
https://doi.org/10.1007/978-3-030-33582-3_108

evolving in parallel [3]. Hence, it is essential to develop an informative application system to reach public's attention towards LCS [5].

Johor government continues committed in its effort in reducing carbon emission by 50% in 2025 as compared to 2005 levels through the implementation of Iskandar Malaysia Eco-Life Challenge (IMELC) [4]. Still, there is no tendency that can show that the carbon emission has been decreased in these years and this indicates lack of the effectiveness in these programs.

IMELC was introduced to school students and teachers by using a workbook that include knowledge and information concerning environment issues and activities that can build LCS awareness [6]. The awareness of students and teachers were measured before and after the program. However, there are several problems in the existing system as it was conducted manually. For example, printed workbook which lead to carbon waste, time consuming for workbook distribution and submission, failed to monitor students' progress on workbook completion and others human errors. There is a needed for implementation of automated system to reduce human workload as well as human error.

Therefore, CarbonFree, a multi-platform application for low carbon education is proposed to solve the problems mentioned above. This project will show that with the assistant of CarbonFree, it is more effective and reachable as users can learn the knowledge and the practice on LCS anytime, anywhere. When public have access adequate information of LCS, they are able to pay more attention to this issue and hence raise public awareness [7].

2 Literature Review

2.1 Comparison Between Existing System

The comparison between similar existing system which is Carbon Footprint, CarbonSins and Equivalent Carbon Content (ECC) based on the functionality is conducted. These three applications are created for the purpose of increase awareness on LCS.

A table of summarization of comparison between existing systems which are Carbon Footprint, CarbonSins as well as ECC and CarbonFree are created and shown as below.

From the comparison above which shown in Table 1, it can be concluded that all three applications had covered different functionalities. CarbonSins had covered the majority of the functionalities while ECC covered the least functionalities among the systems. However, based on the proposed functionalities, CarbonFree had covered all the functionalities that listed out at the table.

Table 1. Comparison among existing system

Features	Carbon Footprint	CarbonSins	ECC
Carbon footprint calculation method			
Calculate carbon footprint referring to particular items	Yes	No	No
Calculate carbon footprint by key in numbering data	No	Yes	Yes
Reward system			
Ranking reward	Yes	No	No
Pointing reward	No	Yes	No
History			
Carbon footprint in detail	Yes	No	No
Carbon footprint in total	No	Yes	No
Tips on knowledge of LCS			
Tips on knowledge of LCS	No	Yes	No
Mobile operating system			
iOS	No	No	Yes
Android	Yes	Yes	No

3 Project Methodology

3.1 Requirement Phase

The system requirements are elicited through a requirement elicitation workshop involving school teachers, UTM project managers, Iskandar Regional Development Authority (IRDA) officers and application developers. Figure 1 below illustrates the current process flows of IMELC As-Is system.

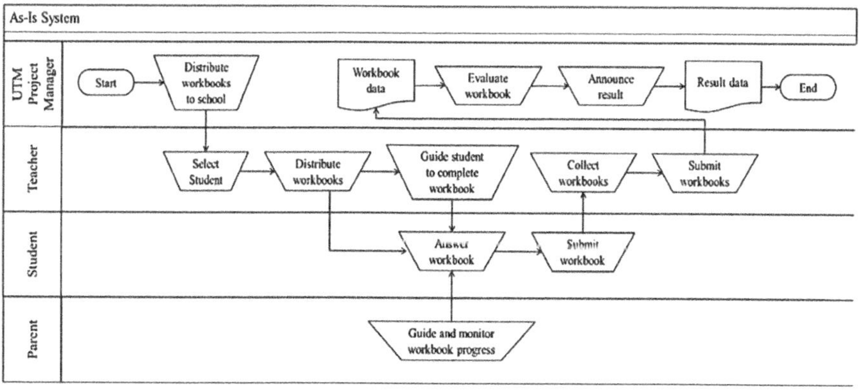

Fig. 1. Current process flows of IMELC

The selected teachers will then select a group of five of students to form a team to be participated in IMELC [8]. After this, teachers distribute the workbooks to each student. Moreover, teachers provide guidelines to students to complete the workbooks.

Students can discuss among themselves to complete the rest of the workbooks. Meanwhile, parents are able to guide and monitor the students' workbook progress as well. After done the workbook, students submit them to teachers. Teachers collect the workbooks completed and then submit to UTM project manager.

UTM project manager evaluates the workbook submitted. Finally, UTM project manager will announces the final result to related teachers and students.

Analysis Phase. A use case diagram is created to represent the interactions between the users and the actions performed by the developed system.

Figure 2 display the use case diagram of the developed system. The relationship between CarbonFree and users are shown as well. There are 3 main end users, including Students, Teachers and UTM Project Manager will interact directly with CarbonFree. Besides, there are total of 12 use cases are included.

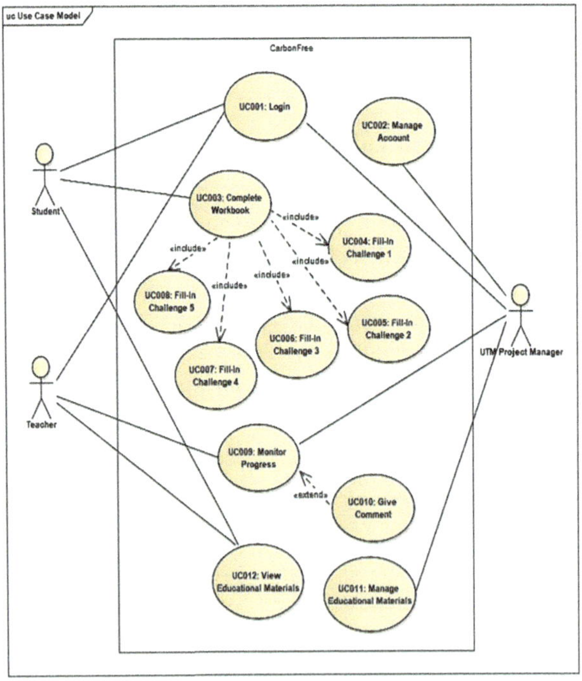

Fig. 2. Use case diagram for CarbonFree

The detail descriptions of use cases are listed as below:

1. Login - This use case allows student, teacher, and UTM project manager to login the application by using the given account credentials.

2. Manage Account - This use case allows UTM project manager to create, edit and delete user accounts for student and teacher as they are not allowed to create their own account.
3. Complete Workbook - This use case allows student to complete the workbook of IMELC.
4. Fill-In Challenge 1 - This use case allows student to fill-in the challenge 1 which is electric appliances usage module.
5. Fill-In Challenge 2 - This use case allows student to fill-in the challenge 2 which is electric and water usage module.
6. Fill-In Challenge 3 - This use case allows student to fill-in the challenge 3 which is vehicle mileage usage module.
7. Fill-In Challenge 4 - This use case allows student to fill-in the challenge 4 which is recycle module.
8. Fill-In Challenge 5 - This use case allows student to fill-in the challenge 5 which is renewable energy module.
9. Monitor Progress - This use case allows teacher and UTM project manager to monitor the workbook progress of students.
10. Give Comment - This use case allows teacher and UTM project manager to give comment based on the workbook progress of students.
11. Manage Educational Materials - This use case allows UTM project manager to manage educational materials by adding, editing and deleting materials.
12. View Educational Materials - This use case allows student and teacher to view the educational materials which related to LCS.

Design Phase. Model-View-Controller (MVC) is implemented as the architecture design of CarbonFree. The components of CarbonFree are separated into three main interconnected logical components: model, view and controller as displayed in Fig. 3.

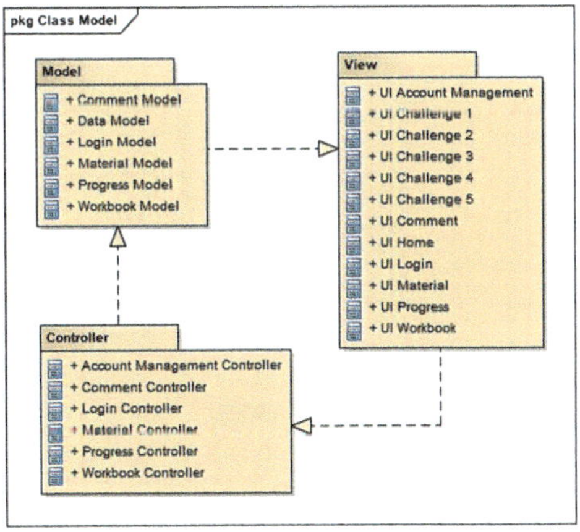

Fig. 3. Architecture of CarbonFree

The components of CarbonFree are separated into three main interconnected logical components: model, view and controller. Model serves the purpose of managing the data of the application while view is responsible to the presentation of the application. For controller, it handles the user input and performs interactions towards objects of model.

4 Implementation

CarbonFree is developed into cross-platform mobile application. Without the use of cross-platform tools it would be necessary to produce the same application twice to reach both marketplaces [9]. The development is done using Ionic framework which takes HTML, CSS and JavaSript as development programming languages. It enables developer to build the mobile application on multiple platforms by providing open source framework and developer-friendly tools. CarbonFree is able to be run on iOS, android and web platform. Besides, tool that used for the develop system is Window Visual Code respectively which is a lightweight but powerful source code editor.

Besides, there are some gamification elements are implemented in CarbonFree, such as badges, drag and drop, levelling up and progress display. Based on a study which carried out on 2 group of students with different education score, it is proven that game-based education is able to help students on learning. The improvement rate of the students in the low-score group was clearly greater than that of the high-score group, thereby indicating that the game-based instruction increased concentration [10].

Figures 4, 5 and 6 display the user interfaces of manage account, containing create, view and delete account. It allows users with the role of UTM project manager to create, read, update and delete user accounts.

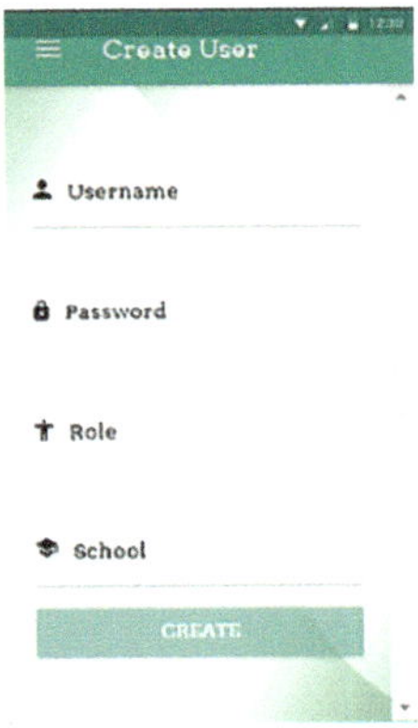

Fig. 4. User interface of create account

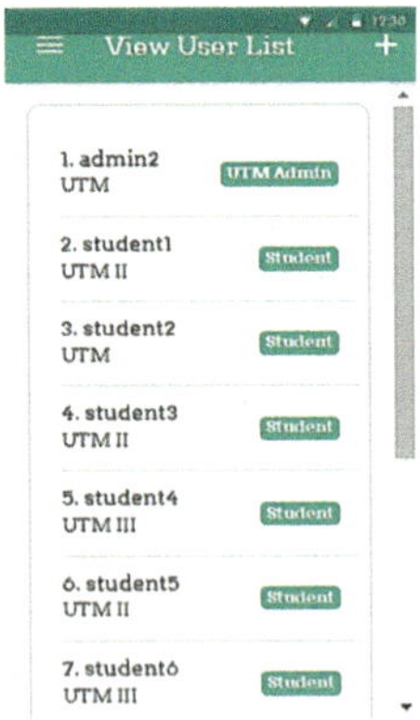

Fig. 5. User interface of view

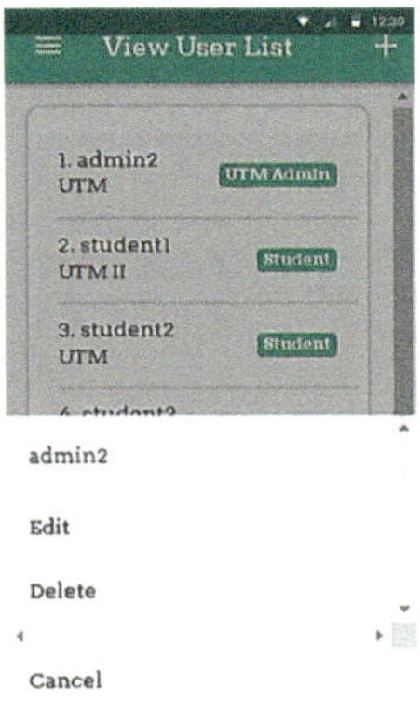

Fig. 6. User interface of delete account

Figures 7, 8, 9, 10 and 11 display the user interfaces of workbook assessments namely Challenge 1, Challenge 2, Challenge 3, Challenge 4 and Challenge 5. It allows users with the role of student to complete the workbook of IMELC.

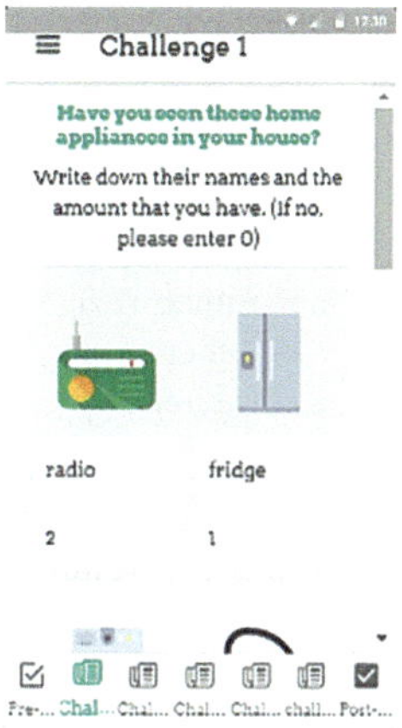

Fig. 7. User interface of challenge 1

Fig. 8. User interface of challenge 2

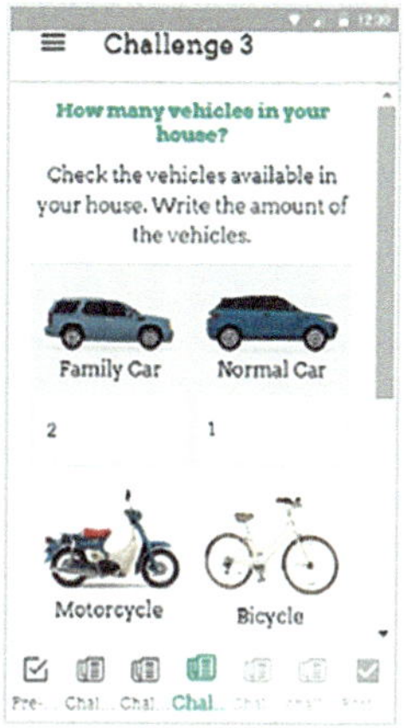

Fig. 9. User interface of challenge 3

Fig. 10. User interface of challenge 4

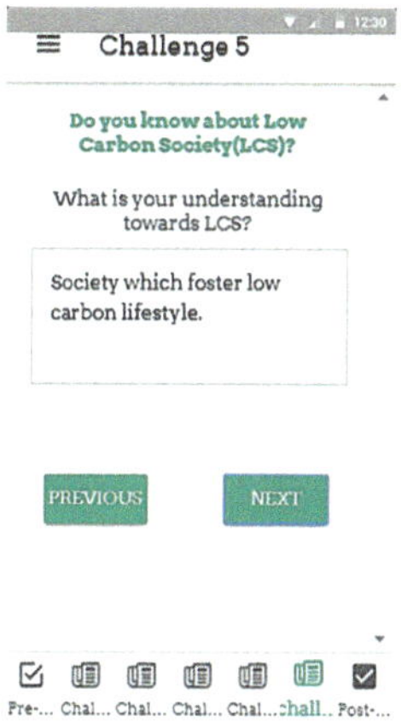

Fig. 11. User interface of challenge 5

Table 2. List of tasks for user with different roles

Roles	Admin	Teacher	Student
Tasks	• View user list • Add a new user account • Edit the role and school of the user account created • Delete the account created • View material list • Add a new educational material • Delete any educational material • View progress list • View progress details of any student • Give the comment to student selected	• View user list • View material list • View progress list • Select a student to view the progress details • Give the comment to the student selected	• View material list • Fill-in challenge 1 and submit • Fill-in challenge 2 and submit • Fill-in challenge 3 and submit • Fill-in challenge 4 and submit • Fill-in challenge 5 and submit

5 Testing and Results

5.1 User Acceptance Testing (UAT)

User acceptance testing (UAT) were conducted to identify whether the requirements of the system are achieved or not. It is focusing the usability of CarbonFree. Table 2 below illustrates the three different lists of tasks are assigned to different roles of users, namely admin, teacher and student.

A questionnaire was given to each of 6 participants which including 1 admin, 1 teacher and 4 students during UAT to gather the feedbacks and comments on the usability and functionalities of CarbonFree. The questionnaire contains five-choice scale to measure the degree of agreement on usability of each use case as well as some open-ended questions are created in order to obtained the opinion from participants.

The following are questions used in UAT.

1. How do you rate the functionality of the use cases (features) in terms of performance?
2. How do you rate the usability of the use cases (features) below?
3. How do you rate the interface design of the use cases (features) below?
4. Did you understand the concept of the system delivered? If no, please state the reason.
5. Did the system satisfy your requirements for carbon awareness learning? If no, please state the reason.
6. Do you think the system have improved the current manual workbook program? If no, please state the reason.
7. Do you think the gamification elements applied in the system is an enhancement features for this system? If no, please state the reason.
8. Did you encounter any difficulty while using this system? If yes, please state the reason.
9. Kindly recommend any improvement(s) for this system (if applicable).

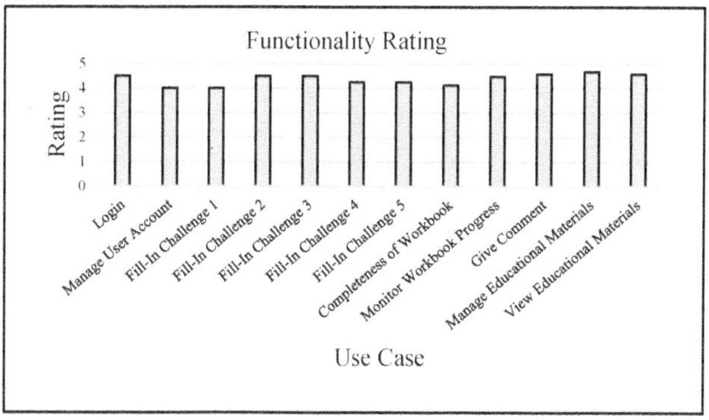

Fig. 12. Functionality rating

Results. The data collected from questionnaires including rating and comment from participants on the satisfactory from the aspect of functionality, usability and user interface of CarbonFree. All participants show positive feedback to the result of UAT. The degree of satisfactory from the all aspects are above average which are 4 and above. The results of data are illustrated using appropriate charts.

From Fig. 12, CarbonFree has an above average degree of satisfactory of functionality on each use case. The lowest degree of satisfactory of functionality are use cases of Manage User Account and Fill-In Challenge 1. This is most probably because of creating user account manually will increase the workload to UTM Admin, while for Fill-In Challenge 1, it can be due to lack of interaction of user interface of Challenge 1. Besides the highest degree of satisfactory of functionality is use case of Manage Educational Materials. This may due to UTM Admin is able to display educational materials with certain format according to the materials' file type once done uploaded.

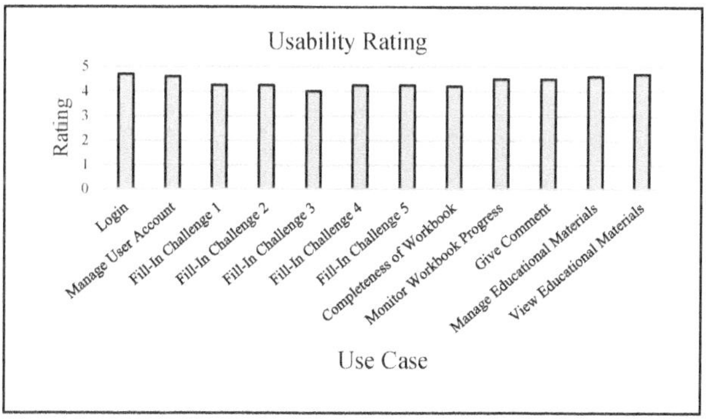

Fig. 13. Usability rating

From Fig. 13, CarbonFree has an above average degree of satisfactory of usability on each use case. The lowest degree of satisfactory of usability is use case of Fill-In Challenge 3. This is most probably due to the complexity of Challenge 3 itself. On the other hand, the highest degree of satisfactory of functionality are use cases of Login and

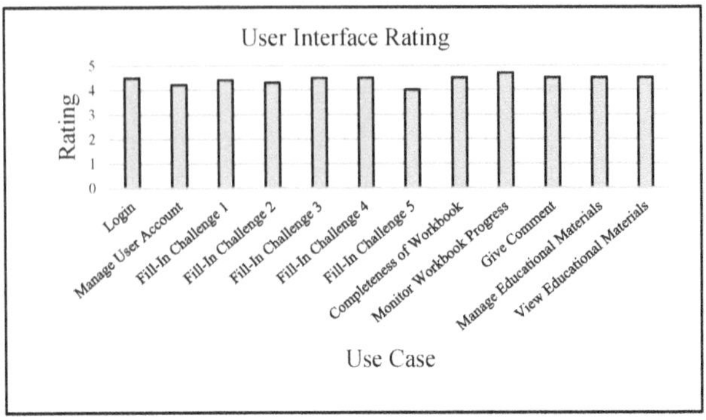

Fig. 14. User interface rating

View Educational Materials. This may due to the simplicity of these two functions which allow users to use them easily.

From Fig. 14, CarbonFree has an above average degree of satisfactory of user interface on each use case. The lowest degree of satisfactory of usability is use case of Fill-In Challenge 5. This is most probably due to the question structures of Challenge 5 itself which containing only open-ended questions. On the other hand, the highest degree of satisfactory of functionality is use case of Monitor Workbook Progress. This may due to the use of elements like progress bar for this function to display the progress that have been done by students graphically.

Participants involved pointed out several feedbacks and suggestions on the functionality of CarbonFree which listed as below.

1. The details of progress of student should be displayed after selecting the progress of particular students.
2. There are few pictures used in CarbonFree are confusing.
3. The workload of admin on creating users' account can be reduces by importing the Excel file which consists of user's details to perform the function of creating accounts.
4. Users should be able to manage their profile by uploading their profile pictures.
5. The performance of each student should be able to evaluate in all challenges.

The implementations have been done to perfect CarbonFree. The details of progress of student will be displayed when users select the particular student's progress. There are some pictures are replaced to reduce the confusing of users. However, due to the time constraint, the rest of the implementations that have not developed yet will be recorded into recommendation for future development of CarbonFree.

6 Conclusion

In Malaysia, the trend of global warming is likely to worsen. However, most of the Malaysians are still unaware and unconcerned about this dangerous trend. The sustainable actions against global warming should be implemented by each individual in

Malaysia, especially from schools [11]. There are many simple daily habits which can result in big changes. Therefore, CarbonFree, a multi-platform application for low carbon education is developed. CarbonFree enables our future leaders, mostly are students and their families to foster LCS habits with effective and error-free computerized method. In the hope of that, CarbonFree is effective and reachable for users to learn the knowledge and practice on LCS anytime, anywhere.

Acknowledgements. This study is partially funded by the Ministry of Education Malaysia's Research University Grant (RUG) and High Impact Grant (HIG) of Universiti Teknologi Malaysia (UTM) under Cost Centre No. R.J130000.7728.4J237 & Q.J130000.2451.04G70. In particular, the authors wish to thank especially the involving school teachers, UTM project managers, Iskandar Regional Development Authority (IRDA) officers for their involvement during requirements elicitation activities and survey for the development of CarbonFree application.

References

1. Lee, C.T., Klemeš, J.J., Hashim, H., Ho, C.S.: Mobilising the potential towards low-carbon emissions society in Asia. Clean Technol. Environ. Policy **18**(8), 2337–2345 (2016)
2. Phang, F.A., Wong, W.Y., Ho, C.S., Musa, A.N.: Achieving low carbon society through primary school ecolife challenge in Iskandar Malaysia. Chem. Eng. Trans. **56**, 415–420 (2017)
3. Agamuthu, P., Fauziah, S.H.: Challenges and issues in moving towards sustainable land filling in a transitory country-Malaysia. Waste Manag. Res. **29**(1), 13–19 (2011)
4. Kili, K.A.: Johor targets to reduce carbon emission by 50% in 2025. The star, 3 November 2016. https://www.thestar.com.my/metro/community/2016/11/03/on-a-green-mission-johor-targets-to-reduce-carbon-emission-by-50-in-2025/
5. Shimada, K., Tanaka, Y., Gomi, K., Matsuoka, Y.: Developing a long-term local society design methodology towards a low-carbon economy: an application to Shiga Prefecture in Japan. Energy Policy **35**(9), 4688–4703 (2007)
6. Phang, F.A., Wong, W.Y., Ho, C.S., Fujino, J., Musa, A.N., Suda, M.: Iskandar Malaysia ecolife challenge: low carbon education for teachers and students. Clean Technol. Environ. Policy **18**(8), 2525–2532 (2016)
7. Takahashi, K., Sato, M., Hijioka, Y.: Trends and issues of climate change education in Japan. In: Implementing Climate Change Adaptation in Cities and Communities, pp. 303–320. Springer, Cham (2016)
8. Phang, F.A., Wong, W.Y., Tee, X.R., Musa, A.N.: Fostering higher order thinking skills among primary students through Iskandar Malaysia ecolife challenge. Man India **97**(17), 1–14 (2017)
9. Gonsalves, M.: Evaluating the mobile development frameworks apache cordova and flutter and their impact on the development process and application characteristics (2019)
10. Chen, M.H.M., Tsai, S.T., Chang, C.C.: Effects of game-based instruction on the results of primary school children taking a natural science course. Educ. Sci. **9**(2), 79 (2019)
11. Phang, F.A., Pusppanathan, J., Musa, N.A., Baharom, M.M., Rahman, N.F.A., Arsat, M., Nawi, N.D.: Framework to develop low carbon society among school community. Int. J. Recent Technol. Eng. **7**(6S5), 773–776 (2019)

Load Balancing Approach of Protection in Datacenters: A Narrative Review

Legenda Prameswono Pratama[1(✉)], Safaa Najah Saud[2(✉)], Risma Ekawati[1(✉)], and Mauludi Manfaluthy[1(✉)]

[1] Department of Electrical Engineering, Institut Teknologi dan Kesehatan Jakarta, DKI, Jakarta 17411, Indonesia
{legenda, risma, mauludi}@itkj.ac.id

[2] Faculty of Information Sciences and Engineering, Management and Science University, 40100 Shah Alam, Selangor, Malaysia
safaa_najah@msu.edu.my

Abstract. The stability of load balancing in the routers is a major problem for traffic flow survivability. Arrangements usually utilize an Equal Cost Multi Path (ECMP) mechanism, which basically emphasizes the load balancing network by equally splitting flows to the accessible concise paths. Empirical research conducted by previous researchers has provided new proof of techniques and methods solutions for more efficient load balancing schemes. However, a survey that examines the practical recommendation has rarely been considered. The surveys on which previous papers have focused give an experimental confirmation for load balancing in a datacenter with regards to descriptive investigation to order to suggest a scheme of best practices. The investigation of load balancing in every telecommunication layer technology is the premise for this descriptive survey. The outline arrangement of the advance level path parallelism in a datacenter and redundancy of transmission to accomplish low Flow Completion Times (FCTs) are the goals of the researcher. More research is expected to assess the effect of packet distribution against FCT. The terms usually used are load balancing, survivability, ECMP, network congestion, link criticality, and link failure protection. Load balancing or ECMP mechanism that are not focused on alternate congestion scheme of the network were exempted from the review.

Keywords: Load balancing · Datacenter · ECMP · Link failure protection · Network congestion

1 Introduction

The load balancing concept is utilized on routing multipath by dividing traffic between a source path and destination path becoming a multiple path schematic [1]. The default mechanism in multipath routing standard is Equal Cost Multi Path (ECMP). The ECMP is the protocol utilized in OSPF and IS-IS [2]. ECMP standard strategy implies that when a router has more than one various track paths to go to a destination, it uniformly balances the traffic between the link connections by looking for the shortest path [3]. Traffic flows between the different sources and destinations are

F. Saeed et al. (Eds.): IRICT 2019, AISC 1073, pp. 1170–1183, 2020.
https://doi.org/10.1007/978-3-030-33582-3_109

created progressively where total flows are closely related to size (rate) and duration [4]. Optimization on multiple path routing enhances link utilization and mitigates congestion [5]. Apart from the setting of computation packet ECMP, all schemes of routing select a path for bundles of packets in the balancing switch and destination pair. These are known as congestion control schemes. Furthermore, all connections have the same possibility to reduce or raise the flow completion time (FCT) in datacenter topology. It is noteworthy that it is necessary to balance protection of the Datacenter as if the scheme does not ensure complete security of each single connection the whole backup scheme will be in vain [6].

The ECMP cannot fulfill imperative execution necessities in datacenters. The hash collision impacts cause serious problems of packet congestion, causing decreased throughput in elephant flows [7] and tail latency for mouse flows [8]. The previous surveys of review papers have aim to exploring the plan of load balancing in datacenters. The datacenter depends on a variety of paths with service being measurable, flexible, and cost effective [4]. Therefore, there are many papers on planning by calculation and algorithms of traffic flow that utilize the activity of traffic burst and static unbalance [9] or designing new load balancing solutions (e.g., TCP-Path [2], Expeditus [10], DiffFlow [11], etc.). The review papers that discuss the theoretical performance of ECMP recommendation are offer limited opportunities for learning [12, 13].

This paper limits research from 2015 to 2019 and there usage additionally further research. Because capability of load balancing techniques has been studied in existing researches, this examination varies from previous surveys by aiming to explore the work of load balancing in different scheme approaches on data center networks. Accordingly, this paper's purpose is to provide a narrative result, explain the implementation which has previously existed, and describe the recommendations for every technique. The load balancing application potential that has been designed is useful for reducing congestion and packet loss. Joining packet loss and aims to become the research benchmarks for developing future techniques.

2 Methods: Selection Data

2.1 Information Criteria

The related article based on search to load balancing, survivability, ECMP, network congestion, and link failure protection on Science Direct, IEEE and SIGCOMM was explained. This paper research written in English and published from 2015 to 2019. This time frame and related content have been determined subjectively, the reason being the necessity to analyze current research and implementation of load balancing to reduce network congestion that utilizes different approaches. The advance searching of all databases generated 65 articles in April 2019.

The first selection filtering from abstract and title were conducted manually. The appropriate paper was chosen based on empirical study. Several criteria which were set this review papers required it to (a) express load balancing technique and mechanism which aimed to provide a solution to congestion, packet losing, traffic burst and unbalance, (b) to evaluate a new concept or technique to give a new result, and (c) to have the potential to be developed.

2.2 Data Analysis

This review presents a qualitative comparison approach to metadata analysis. Based on the basis of qualitative comparison, the author has studied the papers by systematically reading then selecting based on concept, algorithm, topologies, and so on. A particular analytic of results has been obtained as an illustration of the throughput system. Finally, all of the selection papers were divided into groups according to the review conduct implementation.

3 Survey Results

This section is divided into five topics according to the following criteria: issue and goal, method and algorithm, evaluation of performance, topological, and result/recommendation of load balancing approach in datacenter networks.

3.1 Selected Papers

The authors selected a total of 15 papers that met the criteria already discussed and the content of which was justified for reviews. The previous papers represented results related to the possibility of significant reduction of congestion and recommended good load balancing in datacenters.

The authors arranged the 15 papers into selective sub-groups based on the aspect associated with the purpose of the review that the paper reflected: Issue and Goal study on recent load balancing mechanism. Method or routing algorithm in a datacenter is studied. Evaluation of performance load balancing approach. Datacenter topologies of load balancing approach and result/recommendations of load balancing. These groups are represented in tables with the diversity of subcategory in each field of discussion.

3.2 Issue and Goal of Load Balancing Approach in Datacenter Networks

A number of research paper which were reviewed emphasized the main challenge that has been faced by researchers, whose objective is to provide an ideal solution for each problem. Normally, minimizing the link congestion in the network or minimizing the sum of link costs as a result of link congestion. But each research takes a different approach to answer each challenge. In the first result in Table 1, the authors present 15 selected papers based on the issue and goal of the paper.

Based on the research on load balancing mechanism in DCN which is the basis of the review of paper, it can be grouped into two main issues; the ECMP weakness in handling topology asymmetry and adaptive load balancing at the network layer.

It can be seen that the main problem of congestion is error of the estimating infrastructure, no routing multi-rooted paths, and absence of multi-link failure protection. This becomes an indicator that load balancing planning and implementation are

Table 1. Issue and goal of load balancing mechanism in data center networks

Authors	Issue	Goal
Shi et al. [1]	Erroneousness identification and inadequate density response illustrate improper blockage asymmetry status	The adaptive mechanism in load balancing runs at each far-end destination which can adjust the imbalance of network load
Alvarez-Horcajo et al. [2]	The advantage of loops or multipath topologies to balance traffic load	To design information multilayer in the source and destination about MAC and TCP ports
Liu et al. [3]	Tail latency in poor production networks (10x worse) based on short TCP flows which cause stunted mouse flow resulting in long flow queues and increased FCT	To reduce performance degradation that occurs randomly by replicating the flow in the application layer
Dong et al. [6]	Load balancing in DCN requires high costs to be applied to the characteristics of traffic flow	To propose as a low-cost approach but there are rack-local flows or many-to-one short flows
Jiang et al. [9]	Chaotic static sequences of MPTCP flow and router sub-flow traffic loads that do not match the total capacity on the network	To combine different satellite networks and improve user mobility performance
Wang et al. [10]	The satisfaction of performance and Hash collisions cause congestion	To monitor local congestion on uplinks to the upper tier switches
Carpio et al. [11]	The decreasing throughput and path latency.	To achieve the fast FCT and high throughput desired
Luiz et al. [14]	The elephant flows in Data Center Networks are responsible for a bottleneck in the network, causing influence on load path routings that are sensitive to delays	To reroute the long flows in DCNs with contrasting the collective utilization of EFM and ECMP
Oljira et al. [15]	The bottleneck in the cloud datacenter causes the increasing latency, low network optimization of protocols, and decreased throughput	To transport multipath to the cloud datacenter based on the MP-TCP sub-flows uses ECN signal that aims to break down the queue of congestion, lower latency and maximize network useable
Khabbaz et al. [16]	Motivated by the weakness of D-MQPM and increasing latency due to congestion in the queue on the switch	To trim the delay in time in the Flow Mice and increase performance on the datacenter base on the QoS metric
Gao et al. [17]	Other only discuss the flow size, regardless of the existing network bandwidth, which makes the flow performance lower and makes the system overhead	To schedule the SRTF distribution base on the need for the package of each flow at senders based on the surviving time

(*continued*)

Table 1. (*continued*)

Authors	Issue	Goal
Li et al. [18]	The fault of calculation schemes that have a harmful impact on real schemes in the datacenter and all are not flexible in both the topology and cost changes	To provide the best scheme in resolving of traffic balancing from virtual switches
Gao et al. [19]	Disadvantages of a single queue scenario in maintaining congestion control performance in maintaining low latency and high throughput	To control congestion and packets encapsulation that carry information on congestion levels for the TCP sender
Ghorbani et al. [20]	The micro load balancing can achieve performance without reduce the link failures	To prove switch level stability and guarantees of admissible traffic
Abdelmoniem et al. [21]	The traditional congestion control such as TCP make to lead several congestion in traffic in DCN	To calculate a balancing share in each flow that traversed and sent back of information to the source

the main issues in DCN [16, 22]. Research development is continuously seeking the latest technique to give the best solution to this unfinished problem.

The development of complex data center devices can trigger the researchers to undertake continuous ongoing research. For example, ECMP mechanism existing development W-ECMP mechanism estimating to change network topology [23], or focusing controlling traffic load balance [24], and to reduce on path latencies when unbalancing happens [11]. The comparison of differential mechanism, network topology, controlling, and minimizing unbalance as studied in previous work is an ongoing issue and one that should be observed more deeply.

3.3 Method and Routing Algorithm of Load Balancing Approach in Datacenter Networks

Other variables are method and algorithm which have been applied with load balancing in VM simulation. Table 2 illustrates methods of load balancing applicable and routing algorithm.

Based on Table 2, it can be seen that load balancing of the Routing Algorithm can be divided into 5 focus area: reduce congestion, re-routing network, flow schematic, reverse path, and adaptation schedule. Based on previous work focusing on method, each paper has provided a good contribution to load balancing results, but the application is still empirical research on the subject matter in the datacenter.

Table 2. Method and routing algorithm control of load balancing applicable

Authors	Load balancing method	Routing algorithm
Shi et al. [1]	ABL	- Directing detected bottlenecks based on latency in the load path diverted to spacious load lines
Alvarez-Horcajo et al. [2]	TCP-Path	- Establishing in TCP connection over the fastest available path
Liu et al. [3]	RepNet	- Breaking down the congested pathway in the network into several paths and applying simple flow replication on less dense paths
Dong et al. [6]	DCMPTCP	- Reverting the schematic ToR Switch communication from server to client - To enforcement at MPTCP carrier - Congestion is divided into several sub-flows with balanced loads
Jiang et al. [9]	scMPTCP	- Widening the range of control on SDN to the destination host. - Routing selection of sub-flow using mptcp-lba and route adjustments of sub-flow using mptcp-lba-Adjust when traffic congestion occurs
Wang et al. [10]	Expeditus	- Performing local congestion monitoring for all uplinks to upstream switches
Carpio et al. [11]	DiffFlow	- Forwarding short flows in network system - Sending advertises of MAC and IP addresses in all aggregation
Luiz et al. [14]	EFM	- Rearranging the largest elephant flow (Biggest label) - Changing the route of the smallest label
Oljira et al. [15]	MDTCP	- Developing sub-flow based on ECN signals on congestion before queue overflow
Khabbaz et al. [16]	DAFS	- Shorter delivery time packages are placed afore of the package priority which has a long time - When entering the expiration time, the priority of packet delivery is increased, the package with a short time limit will be prioritized first
Gao et al. [17]	D-SRTF	- Classifying streams marking each package into several categories - Setting different priorities based on priority (category ID) on the package
Li et al. [18]	VMS	- A TCP stream is propagated to several different forwarding paths - Deployment selects adaptive paths based on the size of virtual windows from different paths
Gao et al. [19]	DemePro	- Setting the lowest limit for each total queue length and each time the congestion increases, a jam signal is issued - Marking the head pack in the densest queue whose length exceeds the most weighted part
Ghorbani et al. [20]	DRILL	- Schedule acts without adaptation among switches or controllers
Abdelmoniem et al. [21]	RWNDQ	- Rewriting of ACK fair share along the reverse path - Reflecting the bottleneck of link fair share

3.4 Evaluation of Performance Load Balancing Approach in Datacenter Networks

The performance of each load balancing approach can be defined as an evaluation to propose an algorithm through implementations the testbed implementation in their topologies. The value parameter of FCT (Flow Completion Time) provides an illustration of throughput on load balancing for long flows backup workloads and data network utilization [17]. The short value of FCT is poor, because it means the networks have high latency and low throughput. Table 3 shows the result of all load balancing approaches in studies conducted by previous researchers.

Table 3. Result performance of load balancing approach

Authors		Flow Scheduling			Throughput	FCT
		Elephants	Rabbit	Mice		
Shi et al.	[1]	✓		✓	–	7% to 40% (Decreases)
Alvarez-Horcajo et al.	[2]	✓	✓	✓	5% to 25% (Increases)	5% to 55% (Decreases)
Liu et al.	[3]	✓		✓	Decreases	15% to 69.9% (Decreases)
Dong et al.	[6]	✓		✓	–	6%–15% (Decreases)
Jiang et al.	[9]			✓	Increases	–
Wang et al.	[10]	✓		✓	23% - 42% (Increases)	14%–30% (Decreases)
Carpio et al.	[11]	✓		✓	13% (Decreases)	7% (Decreases)
Luiz et al.	[14]	✓			Increases	16.7% to 54.4% (Decreases)
Oljira et al.	[15]			✓	15%-50% (Increases)	0.3 ms–0.8 ms (Decreases)
Khabbaz et al.	[16]			✓	7% - 29% (Increases)	52% (Decreases)
Gao et al.	[17]			✓	–	8%–12% (Decreases)
Li et al.	[18]	✓		✓	Increases	–
Gao et al.	[19]	✓		✓	9.74/10 GBPS (Increase)	16%–34% (Decreases)
Ghorbani et al.	[20]	✓		✓	20% (Increases)	–
Abdelmoniem et al.	[21]	✓		✓	–	10 ms–33 ms (Decrease)

Table 3 shows the research on the performance of load balancing approach, which is the basis of the review of this paper, with the three main parameters affecting performance grouped: flow scheduling, throughput, and FCT. Based on the studies that have been conducted, it has been found that the purpose of each research in generating load balancing techniques and mechanisms will be based on the results of the throughput value and flow completion time level. In each research the algorithm used will always be more effective than using the multipath routing ECMP mechanism.

Regardless of load balancing scheme of flow measured, there are three flow scheduling elements used on the datacenter, namely elephant flow, rabbit flow, and mice/mouse flow. Elephant flow and mice flow are commonly used in research. The load balancing handling in flow scheduling in existing papers always demonstrate a good performance and high bandwidth availability that are focused on high throughput in elephant flows and low-latency in mice flow [10].

3.5 Topologies of Load Balancing Approach in Datacenter Networks

There are several topologies that should be built to conduct experiments through the load balancing approach. Based on the existing topology, the proper use of the architecture will have an impact on reducing the computational paradigm that is more environmentally friendly. Various datacenter topologies that have been proposed focus on designing of scalability, cost, latency, and extensibility [25]. Table 4 shows the datacenter topology of the load balancing approach. From the data above, it can be recognized that this load balancing approach conducted by the researchers includes each topology concept: Fat-Tree, Clos, Dcell, and Mininet etc.

Table 4. Datacenter Topology of Load Balancing Approach

Authors	Tree-based			Recursive	Hybrid	
	FatTree	Clos (spine-leaf)	Dumbbell	DCell	Lattice-like	Mininet
Shi et al. [1]		✓				
Alvarez-Horcajo et al. [2]	✓	✓				
Liu et al. [3]		✓				✓
Dong et al. [6]		✓				
Jiang et al. [9]					✓	✓
Wang et al. [10]	✓	✓				
Carpio et al. [11]	✓			✓		
Luiz et al. [14]	✓					✓
Oljira et al. [15]	✓					
Khabbaz et al. [16]		✓				
Gao et al. [17]		✓				
Li et al. [18]	✓					
Gao et al. [19]		✓				
Ghorbani et al. [20]	✓	✓				
Abdelmoniem et al. [21]			✓			

From all topology concepts that continue to develop, basically there are only three types of topology: star, hub and spokes (H&S), and ladder [26]. Star topologies are the concept that all nodes are connected in a small hops the other center hub nodes [27]. Hub and Spokes topologies are an expansion concept of star topology that forms a connection between hubs by making direct interface nodes that impact on traffic load which tend to concentrate at hub nodes. Ladder topologies are a concept of freeways, where there is no hub node and combination topology exists between the hub nodes of the loop. All topology networks used have high-radix switches for various communication paths between pairs (primary and secondary) to servers where evenly balanced bandwidth is required [28].

In experiments conducted by previous researchers, Fat-Tree and Clos became topologies that were used more often. The superiority of Fat-Tree is due to it being a highly structured topology which consists of a collection of edge, aggregation, and core switches and offers equal length path between the edge switch [4] whereas Clos is comprised of spine and leaf layers. Servers are connected to leaf switches and each leaf is connected to all spines and direct leaf-to-leaf and spine-to-spine connections [29]. There are at least 6 differences between data center topologies based on Network on Chip (NoC) that have been developed over the years, differences in the number of high radix routers, link bandwidths, routing algorithms, flow control, routing delay, and link latency [25].

3.6 Result and Recommendations of Load Balancing Approach in Datacenter Networks

The researchers from the 15 selected papers achieved results and recommendations from the research that they conducted. Below is a table for future research based on a load balancing approach.

The results in Table 5 illustrate that it can be recognized that there are positive results in the improved throughput and decrease in FCT value in all research conducted. There is also evidence that load balancing approaches to using the correct topology, mechanism, and algorithm can affect the value used to resolve a focus major problem in research (e.g. congestion, high latency, delay, or performance metrics).

Alternatively, multipathing systemic in load balancing different paths is also investigated. First, solution classified in stochastic load balancing, migration and centralized scheduling [10]. Second, replication is an intuitive approach to combat unpredictable performance degradations [30]. Lastly, hash based load balancing to distribute the load over multiple paths to efficiently spread the load [31, 32].

Table 5. Result and recommendations of load balancing approach

Authors	Result	Recommendation/Gap
Shi et al. [1]	ALB give reasonable execution compared with plans that need particular proceed or complex measurement settings	Not specific
Alvarez-Horcajo et al. [2]	Throughput enhancement in all flow scheme (Elephant, Rabbit and Mice) is obtained using the TCP-Path algorithm	The incretion in simulations uses TCP-Path algorithm compared to ECMP in datacenters that have the conformable link speed (for example in 10 GB or more)
Liu et al. [3]	Low latency transport layer applications allow replication of the flow in the real testbed and Mininet experimental by encouraging RepSYN to minimize performance deterioration in mice flows	The simple concept designed in degrading FCT without modification in TCP or Switches and executed using protocols at the transport layer
Dong et al. [6]	DCMPTCP shows achievement of ~65-771X and ~10-15X better than MPTCP in terms of FCT for each rack-local and cross-shelf scheme	The utilization of DSCP bits in the IP header which makes DSCP and DCMPTCP perfect with one another and builds the work layer with multiple layers topological
Jiang et al. [9]	The scMPTCP scheme has been validated with evidenced by a much higher total system throughput than the disjoint path and shortest path	Route selection algorithms based on performance metrics need to be further improved to consider DASH (Dynamic Adaptive Streaming over HTTP) on the network
Wang et al. [10]	Expeditus approach fix implementation network like short flow and long flow to enabling congestion aware load balancing whose presents on the scalability and effectiveness of the design	Not specific
Carpio et al. [11]	DiffFlow solution surpass of RPS and ECMP, although the overall throughput that is accomplished at only proportional value against RPS	The improvement in current DCN are weak to applying new modification of network transformation in protocols
Luiz et al. [14]	EFM indicates re-routing solution demonstrated to be better when contrasted to routing by ECMP	Not specific

(continued)

Table 5. (*continued*)

Authors	Result	Recommendation/Gap
Oljira et al. [15]	MDTCP shows better degradation flow completion time oppose to DCTCP or NDP and has a higher traffic throughput value	MDTCP assumes that the results are equivalent to MPTCP such as BCube topology and Jellyfish topology because asymmetric can control path-by-path congestion in the next work
Khabbaz et al. [16]	DAFS decreases potential deadline which nonconformities and cuts down on the completion period and increases throughput	Not specific
Gao et al. [17]	D-SRTF works better than the PIAS Scheme in terms of comparison of FCT in WEB search and performs better than pFabric in terms of data mining workload	Identification in flow classification and compatibility based on the sequence of TCP/IP, so that a less effective workload threshold can be minimized by providing a fixed threshold
Li et al. [18]	VMS fix the better balance in traffic density in a virtual datacenter, and VMS performance is equivalent to MPTCP in most experiments but there is a VMS overhead with a reasonable tolerance	Not specific
Gao et al. [19]	DemePro has preferable execution over MQ-ECN which is at present the DCN plot for the multiple-service multiple-queue situation, and DemePro can likewise well ensure reasonableness	Not specific
Ghorbani et al. [20]	DRILL, out-performs the state-of-the-art load balancing schemes in Clos networks, particularly under heavy load	The investigation of micro load balancing in other (especially asymmetric) topologies
Abdelmoniem et al. [21]	The simulations showed that RWNDQ can achieve its goals efficiently while outperforming the most prominent alternative approaches	The dominant factor affecting the RWNDQ mechanism on the link utilization bottleneck must closely due discretion during incase

4 Discussion

This review paper is focused on the previous research and empirical evidence for a load balancing approach in the context of narrative analysis to explain, describe, and compare from previous research in this field. There are actually seven main parts that must be prepared before conducting research, namely problem, purpose, method, algorithm, topologies, flow scheduling, and parameter FCT or throughput.

It has been stated repeatedly that the major problem of the load balancing approach is congestion and error of the estimating infrastructure, no routing multi-rooted paths, and absence of multi-link failure protection. From Table 1, it can be seen that the ECMP weakness in handling topology asymmetry and adaptive load balancing at network layer are two main problems that can be recognized and from this kind of problem, there are some load balancing methods and routing algorithms that further can be utilize to provide a solution. Some examples of load balancing activities that have been recognized are reducing congestion, re-routing network, flow schematic, reverse path, and adaptation schedule.

Table 2 gives an overview of the active mechanism in setting priority goals based on each technique used in each of the problems to be resolved for example transmission control, multipathing, package priority, and scheduling. The focus of this paper is to obtain a higher value or equivalent to existing techniques and to establish a large impact on the scheme changes in increasing traffic throughput.

The evaluation performance in this narrative review of load balancing approach has been fully discussed in Table 3, as a protection in all aspects that aims to reduce network congestion. All methods show better resource utilization of increasing throughput and reduction FCT than many other backup protection schemes. From all research, we did not get details of reducing the backup cost of load balancing in a datacenter network and this is a possible area to be investigated, improved, or enhanced in future research.

Also, the concept of the data center always evolves over time. High radix routers, link bandwidths, routing algorithms, flow control, routing delay, and link latency are the common elements that can be implemented and become the basis of future research to challenge the weakness of previous topology concepts.

5 Conclusion

This review paper shows that latterly from 2015 to 2019 the development of a load balancing approach in datacenters has a positive influence on evolution of ECMP mechanism. All research fields focus on generating decongestion of network using various combinations for solving the problem; purpose, method, algorithm, topologies, flow scheduling, and parameter FCT or throughput. In this review, we try to explain and split them into several discussion sub topics then categorize them for investigation.

References

1. Shi, Q., Wang, F., Feng, D., Xie, W.: ALB: adaptive load balancing based on accurate congestion feedback for asymmetric topologies. In: 2018 IEEE/ACM 26th International Symposium on Quality of Service (IWQoS), pp. 1–6 (2018)
2. Alvarez-Horcajo, J., Lopez-Pajares, D., Arco, J.M., Carral, J.A., Martinez-Yelmo, I.: TCP-path: improving load balance by network exploration. In: Proceedings 2017 IEEE 6th International Conference Cloud Networking, CloudNet 2017, pp. 0–5 (2017)

3. Liu, S., Xu, H., Liu, L., Bai, W., Chen, K., Cai, Z.: RepNet: cutting latency with flow replication in data center networks. IEEE Trans. Serv. Comput. **1374**(c), 1–14 (2018)
4. Shafiee, M., Ghaderi, J.: A simple congestion-aware algorithm for load balancing in datacenter networks. IEEE/ACM Trans. Netw. **25**(6), 3670–3682 (2017)
5. Maheswaran, M., Asirvatham, D.: Real-time business analytical model using big data strategies for telecommunication industry, vol. 863 (2019)
6. Dong, E., Fu, X., Xu, M., Yang, Y.: DCMPTCP: host-based load balancing for datacenters. In: Proceedings - International Conference on Distributed Computing Systems 2018, pp. 622–633, July 2018
7. Shakeri, A., et al.: Modeling the effect of wavelength selective switch latency on optical flow switching performance. J. Opt. Commun. Netw. **10**(12), 924 (2018)
8. De Pellegrini, F., Maggi, L., Massaro, A., Saucez, D., Leguay, J., Altman, E.: Blind, adaptive and robust flow segmentation in datacenters. In: Proceedings - IEEE INFOCOM, 2018, pp. 10–18, April 2018
9. Jiang, Z., Wu, Q., Li, H., Wu, J.: ScMPTCP: SDN cooperated multipath transfer for satellite network with load awareness. IEEE Access **6**(c), 19823–19832 (2018)
10. Wang, P., Xu, H., Niu, Z., Han, D., Xiong, Y.: Expeditus: congestion-aware load balancing in Clos data center networks. IEEE/ACM Trans. Netw. **25**(5), 3175–3188 (2017)
11. Carpio, F., Engelmann, A., Jukan, A.: DiffFlow: differentiating short and long flows for load balancing in data center networks. In: 2016 IEEE Global Communications Conference GLOBECOM 2016 - Proceedings (2016)
12. Alsalem, M.A., et al.: A review of the automated detection and classification of acute leukaemia: coherent taxonomy, datasets, validation and performance measurements, motivation, open challenges and recommendations. Comput. Methods Programs Biomed. **158**, 93–112 (2018)
13. Jamal, A., Kumar, D., Helmi, R.A.A., Fong, S.L.: Portable TOR router with Raspberry Pi, pp. 533–537 (2019)
14. Rocha, A.L.B., Verdi, F.L.: EFM: improving DCNs throughput using the transmission rates of elephant flows. In: Proceedings - IEEE Symposium on Computers and Communications, June 2018, no. 1, pp. 154–157 (2018)
15. Oljira, D.B., Grinnemo, K.J., Brunstrom, A., Taheri, J.: MDTCP: towards a practical multipath transport protocol for telco cloud datacenters. In: Proceedings of the 2018 9th International Conference on the Network of the Future, NOF 2018, pp. 9–16 (2018)
16. Khabbaz, M., Shaban, K., Assi, C.: Delay-aware flow scheduling in low latency enterprise datacenter networks: modeling and performance analysis. IEEE Trans. Commun. **65**(5), 2078–2090 (2017)
17. Gao, C., Lee, V., Li, K.: D-SRTF: distributed shortest remaining time first scheduling for data center networks. IEEE Trans. Cloud Comput. 1 (2018)
18. Li, Z., Bi, J., Zhang, Y., Dogar, A.B., Qin, C.: VMS: traffic balancing based on virtual switches in datacenter networks. In: Proceedings - International Conference on Network Protocols, ICNP, October 2017
19. Gao, C., Lee, V.C.S., Li, K.: DemePro: DEcouple packet marking from enqueuing for multiple services with PROactive congestion control. IEEE Trans. Cloud Comput. 1 (2017)
20. Ghorbani, S., Godfrey, B., Ganjali, Y., Firoozshahian, A.: Micro load balancing in data centers with DRILL. In: Proceedings 14th ACM Workshop Hot Topics Networks - HotNets-XIV, pp. 1–7 (2015)
21. Abdelmoniem, A.M., Bensaou, B.: Reconciling mice and elephants in data center networks. In: 2015 IEEE 4th International Conference Cloud Networking, CloudNet 2015, no. Dc, pp. 119–124 (2015)

22. Tahaei, H., Bin Salleh, R., Ab Razak, M.F., Ko, K., Anuar, N.B.: Cost effective network flow measurement for software defined networks: a distributed controller scenario. IEEE Access **6**(1), 5182–5198 (2018)
23. Ye, J.L., Chen, C., Huang Chu, Y.: A weighted ECMP load balancing scheme for data centers using P4 switches. In: Proceedings of the 2018 IEEE 7th International Conference on Cloud Networking, CloudNet 2018, pp. 1–4 (2018)
24. Chrysos, N., et al.: Large switches or blocking multi-stage networks? An evaluation of routing strategies for datacenter fabrics. Comput. Networks **91**, 316–328 (2015)
25. Lebiednik, B., Mangal, A., Tiwari, N.: A survey and evaluation of data center network topologies, pp. 1–12 (2016)
26. Haeri, S., Trajkovic, L.: Virtual network embeddings in data center networks. In: Proceedings - IEEE International Symposium on Circuits and Systems, 2016, pp. 874–877, July 2016
27. Bakopoulos, P., et al.: NEPHELE: an end-to-end scalable and dynamically reconfigurable optical architecture for application-aware SDN cloud data centers. IEEE Commun. Mag. **56** (2), 178–188 (2018)
28. Maksić, N.: Two-phase load balancing for data center networks using OpenFlow. In: 2017 25th Telecommunications Forum, TELFOR 2017 - Proceedings, 2018, Janua 2017, pp. 1–4 (2017)
29. Dhaliwal, H.K., Lung, C.H.: Load balancing using ECMP in multi-stage Clos topology in a datacenter. In: DSC 2018 – 2018 IEEE Conference on Dependable and Secure Computing, pp. 1–7 (2019)
30. Xie, J., Guo, D., Zhu, X., Ren, B., Chen, H.: Minimal fault-tolerant coverage of controllers in IaaS datacenters. IEEE Trans. Serv. Comput. **1374**(1), 1–14 (2017)
31. Yang, G., Jiang, Y., Li, Q., Jia, X., Xu, M.: Cross-layer self-similar Coflow scheduling for machine learning clusters. In: Proceedings - International Conference on Computer Communications and Networks, ICCCN, 2018, July 2018
32. Yajid, M.S.A.: Factors influencing customers' satisfaction on Malaysian telecommunication service providers, vol. 14, pp. 155–178 (2016)

Author Index

F. Saeed et al. (Eds.): IRICT 2019, AISC 1073, pp. 1185–1188, 2020.
https://doi.org/10.1007/978-3-030-33582-3

GPSR Compliance

The European Union's (EU) General Product Safety Regulation (GPSR) is a set of rules that requires consumer products to be safe and our obligations to ensure this.

If you have any concerns about our products, you can contact us on ProductSafety@springernature.com

In case Publisher is established outside the EU, the EU authorized representative is:

Springer Nature Customer Service Center GmbH
Europaplatz 3
69115 Heidelberg, Germany

Batch number: 10755090

Printed by Printforce, the Netherlands